The Coelomycetes (1980)

Corrigenda

Contents - Blastostromatineae p.128
p.18, l. 9 - couplet 4, Blastostromatineae p.128
p.31, l. 7 - *asperulina* not *tomispora*
p.56, l. 26 - (Pollack) not (Pollack
p.61, l. 33 - 15 not 16
p.116, l. 9 - 3-4-euseptate not 4-euseptate
p.128, l. 9 - p.129 not p.125
p. 130, last line - 22 not 12
p. 134, last line - 50 51 not 5051
p. 151, l. 3 - *Pyracantha* not *Pyrenacantha*
p. 177, l. 26 - doliiform not dolliform
p. 178, l. 20 - 1888 not 1881
p. 180, l. 1 - 1888 not 1881
p. 208, l. 16 delete 21
p. 240, l. 19 - *Ost. Bot. Z.* **66**: 106 (1916) not *Annls Mycol.* **22**: 210 (1924)
p. 240, l. 29 - *ibid.*, 107 (1916) not *ibid.*, 201 (1924)
p. 348, l. 3 - (Fig. 207A) not (Fig. 207A,C)
p. 348, l. 13 - (Fig. 207B,C) not (Fig. 207B)
p. 349, legend to read Fig. 207. A, *Neohendersonia*.....conidioma; B,C, *N. congoensis*...
p. 387, legend - *leveillei* not *levellei*
p.391, l. 35 - 1923 not 1909
p. 391, l. 36 - *philippinensis* not *philippines*
p. 438, l. 45 - hysteriiform not hysteriform
p. 441, l. 46 - β-conidia not 3-conidia
p.471, l. 6 - *caricis* not *australis*
p. 504, legend - *Discosporina deplanata* not *Discosporium deplanatum*
p. 530, l. 39 - (Penz.) Penz. & Sacc., *Atti R. Ist. Ven. Sci. Lett. Art.* Ser. 6, **2**(5): 670 (1884) not (Penz.) Sacc., *Fung. Agrum.* **2**: 6 (1882)
p. 537, l. 34 - (H. Syd.) not (Syd.)
p. 662, middle line under *Alnus* - *Disculina vulgaris* not *Discula vulgaris*
p. 667, middle line under *Malus* - *Phlyctema vagabunda* not *Phylctema vagabunda*
p. 684, second column - LEPTOTHYRIUM, 167, 212 not **LEPTOTHYRIUM, 167, 212**
p. 685, first column - *philippinensis* not *philippines*

The Coelomycetes

Fungi Imperfecti with Pycnidia
Acervuli and Stromata

by
BRIAN C. SUTTON
Principal Mycologist
Commonwealth Mycological Institute

First published 1980

CABI *Publishing*

First published 1 June 1980 by the Commonwealth Mycological Institute under the authority of the Executive Council, Commonwealth Agricultural Bureaux, Farnham Royal, Slough, England

CABI Publishing is a division of CAB International

CABI Publishing
CAB International
Wallingford
Oxon OX10 8DE
UK

Tel: +44 (0)1491 832111
Fax: +44 (0)1491 833508
Email: cabi@cabi.org
Web site: www.cabi-publishing.org

CABI Publishing
875 Massachusetts Avenue
7th Floor
Cambridge, MA 02139
USA

Tel: +1 617 395 4056
Fax: +1 617 354 6875
Email: cabi-nao@cabi.org

ISBN 0 85198 446 0

First published 1980
Reprinted 1998
Transferred to print on demand 2004

Printed and bound in the UK by Antony Rowe Limited, Eastbourne.

CONTENTS

PREFACE

It is probably only an author who can fully appreciate the deficiencies of his own work. No statement could be more apt to describe my own feelings about 'The Coelomycetes'. My research on taxonomy and identification over a period of several years has demonstrated two basic problems in the group. One concerns the proper nomenclatural typification of genera, without which it is extremely difficult to apply names with any degree of confidence. The other is the lack of any single illustrated modern account of the group where genera and species can be directly compared and contrasted.

I made an attempt to solve the difficulties of typification and acceptance of published coelomycete names (Sutton, 1977a), and, although it was possible to reject 720 names of the 1336 then proposed for coelomycetes, there remained 226 names for which little or no data are available. Consequently the acceptance of 393 generic names and the synonymies that have been verified must be to some extent conditional upon an assessment of the residual genera that have not yet been re-examined.

This present volume on the coelomycetes is the logical extension of that preliminary work on typification, synonymy and acceptance, inasmuch as descriptions, illustrations and keys are provided for most of the accepted genera.

As will be readily apparent from the text and particularly the treatment of species within genera, there still remains an enormous amount of revisionary work to be done. I have been reluctantly forced to state on several occasions that no modern revisionary account of species is available for a particular genus. As a consequence, I have been unable to give much detail about species in certain genera because the only existing accounts are either inaccurate or simply lack the necessary information required for separating taxa according to present-day standards. The bias of 'The Coelomycetes' therefore is towards the genus rather than the species.

'The Coelomycetes' is a summation of modern thought and practice on this group, and since there are now more active mycologists involved in coelomycete research than there were when I started, the book should only be visualized as a template, from which a more satisfactory, comprehensive and detailed treatment of individual species may evolve.

ACKNOWLEDGEMENTS

These studies have required the examination of many species not represented by collections in herb. IMI and so it is with a deep sense of gratitude that I acknowledge permission to examine material in the keeping of the Directors and Curators of the following herbaria (abbreviated according to *Index Herbariorum* Part I. The Herbaria of the world (comp. P.K. Holmgren & W. Keuken: ed. F.A. Stafleu), 1974): — ABD, AMH, B, BERN, BISH, BM (now incorporated in K), BP, BPI, BR, BUCA, C, CBS, CFB, CUP, DAOM, DAVFP, E, FH, G, GRO, H, HBG, HCIO, ILLS, K, L, LE, LEP, LISE, LISU, LPS, M, MA, MPU, NY, NYS, PAD, PC, PM, PR, PRE, S, SIENA, STE-VB, STR, TENN, TO, UC, UPS, URM, VER, W, WINF(M) (now incorporated in CFB), WIS, WSP, ZT, Vouaux Herbarium, J. Kohlmeyer Herbarium, Victoria Plant Research Institute, Burnley, Australia, the Tea Research Institute, Talawakelle, Sri Lanka, and Plantenziektenkundige Dienst, Wageningen, The Netherlands.

Although this volume could not have been published without the cooperation of curators of the above herbaria, it would eventually have been published, even without the help of Mrs Gill Butterfill and Miss Nazira Bhimani. However, their unstinted assistance respectively in providing technical and typing services has considerably shortened the period of gestation, and it is in no small measure to them that I have been able to accomplish my aim within a reasonable period of time.

During the final stages of preparation I have had much useful discussion on principles of taxonomy and identification, and problems of interpretation and definition with Dr Barbara J. Dyko (N.S.F. Post-Doctoral Fellow, 1977-1978 at CMI) to whom I extend due thanks.

I am also pleased to acknowledge the permission of the following to reproduce various figures that I have published previously:— Dr H.J. Hudson (Fig. 2, *Trans. Br. mycol. Soc.* **47**: 201, 1964); The British Mycological Society (fig. 2, *T.B.M.S.* **47**: 500, 1964); fig. 1, *T.B.M.S.* **48**: 613, 1965; parts of figs. 1-3, *T.B.M.S.* **55**: 37-44, 1970; fig. 1, *T.B.M.S.* **57**: 542, 1971; parts of figs. 2-4, *T.B.M.S.* **59**: 285-294, 1972; figs. 1, 2, *T.B.M.S.* **59**: 296-297, 1972; figs 1, 2, *T.B.M.S.* **66**: 300-301, 1976; fig. 1A, B, *T.B.M.S.* **67**: 245, 1976); The National Research Council of Canada (figs. 1, 2, *Can. J. Bot.* **44**: 1506-1507, 1966; figs. 1-7, *C.J.B.* **46**: 181-196, 1968; figs. 2-7, *C.J.B.* **45**: 1249-1263, 1967; fig. 2, *C.J.B.* **47**: 2098, 1969; figs. 1, 5, *C.J.B.* **47**: 2085 and 2092, 1969; fig. 2, *C.J.B.* **55**: 329, 1977).

INTRODUCTION

In a general purpose classification of the fungi, Ainsworth (1966) proposed the separation of asexual (or imperfect) fungi into the subdivision Deuteromycotina. This was subdivided into three classes, the Hyphomycetes, Coelomycetes and Agonomycetes (Mycelia sterilia). Such organisms are the asexual reproducing states (anamorphs) of ascomycetes and basidiomycetes, although in the majority of cases the sexual states (teleomorphs) have become permanently separated, temporarily disconnected and now either do not exist or never existed in the first place. Only a comparatively small number of the Deuteromycotina have been correlated with their sexual states, and so far, for Coelomycetes, these have been referred to the Ascomycotina without exception. The possibility that some Coelomycetes may yet be correlated with the Basidiomycotina cannot be excluded, for recently at least two newly described coelomycetes have been found with clamp connexions (Nag Raj, 1978a; Dyko & Sutton, 1979a).

The name 'Coelomycetes' was first introduced by Grove (1919) when he sought a comprehensive term to apply to *Phyllosticta* auct., *Phomopsis* and *Phloeospora*, three genera which showed considerable diversity in their fructifications (conidiomata). This initial concept was later expanded (Grove, 1935, 1937) when he used 'Coelomycetes' for those imperfect fungi (Deuteromycotina) in which conidia are formed within a cavity lined by either fungal tissue, host tissue, or a combination of both. In this sense it is now used to distinguish such fungi from the Hyphomycetes where conidia are formed from aggregated or separated modified hyphae (conidiogenous cells) borne on the exterior face of substrates and not enclosed by additional fungal or host tissue.

It has become increasingly apparent that rather than continuing as a formal taxonomic category at the class level, the name Coelomycetes should be relegated to use as a popular collective term for the description of pycnidial, acervular, stromatic and pycnothyrial Deuteromycotina. This is a reversion to the original concept of Grove. Precedents for such use are found, for example, in ascomycetes, basidiomycetes and phycomycetes where nothing more than a very general indication of the major fungal category to which a taxon belongs is indicated by the use of such a name.

The reasons for the rejection of Coelomycetes, and indeed Hyphomycetes also, as formal taxonomic ranks can be justified by two arguments. Firstly, both Mason (1937) and Hughes (1953a) have clearly demonstrated the anomalies arising from the undue emphasis placed on conidiomatal structure in taxonomy of the Deuteromycotina. When behaviour in culture is also considered (Sutton, 1973a) it becomes obvious that there are neither real generic distinctions between acervular and sporodochial fungi, nor between many moniliaceous, dematiaceous and synnematous fungi. Mason (1937) interpreted all such conidiomata as different forms of a basic fungal stroma. Individual extremes may be recognisable and to some extent rationalized but there are many intermediates and the majority merge into one another. A separation at any point along this continuum is bound to meet serious problems in practical application. The second factor is that conidiogenesis is now universally established as a fundamental taxonomic character in the Deuteromycotina, and although there are still certain basic problems to be solved, it indicates relationships between taxa far more satisfactorily than conidiomatal structure ever did. Conidiogenesis has frequently been used as a basis for classificatory schemes in the Hyphomycetes (Hughes, 1953a; Tubaki, 1958; Subramanian, 1962; Barron, 1968; Dong, 1972) and since a similar, though not quite as extensive, range of conidial ontogeny is now known to occur in coelomycetes (Sutton, 1971a, 1973a) it is logical to use a classificatory scheme for the Deuteromycotina as a whole, with conidiogenesis accorded prime importance in distinguishing major taxonomic categories.

Coelomycetes are known mainly from tropical and temperate regions, less so from the antarctic and arctic, and are capable of growing, reproducing and surviving in a wide variety of ecological situations. They are commonly found and recovered from cultivated and uncultivated soil of different types (Domsch & Gams, 1970; Gilman, 1956), leaf litter and other organic debris from both natural (Hudson, 1968) and manufactured sources (biodeteriorative and biodegradative species), from saline- and fresh-water (Hughes, 1975), on other fungi and lichens (Hansford, 1946; Seeler, 1943; Keissler, 1930), and as parasites of vertebrates (Anon., 1977; Punithalingam & English, 1975; Punithalingam, 1976a; Borelli, 1976) and insects (Madelin, 1968; Petch, 1921). Though largely producing slimy conidia that are not forcibly released but dispersed by water in various ways, they do constitute a small proportion of the air spora (Gregory, 1973). Even more diverse are the activities of coelomycetes in relation to plant substrates. The capacity of species or strains within species to function somewhat variably along the spectrum from saprophytism to parasitism is shown by their ubiquity in disease syndromes. Members of the group are consistently isolated from diseased conditions together with other organisms, and without adequate experimental proof it is difficult to ascertain if they are functioning as primary pathogens, secondary inhabitants or fulfil a more complex role as one of several different factors leading to the diseased condition.

Some certainly colonize roots and cause disease (Garrett, 1970) but the majority are benign as constituents of the rhizosphere or as latent infections in root systems. There are many species pathogenic to economically important agricultural and forestry crops (Peace, 1962), ornamental plants (Wager, 1970), and weeds. Other members are also capable of attacking the majority of plants that are of little economic significance. As such they may function as reserves or alternative sources of inoculum. Such diseases may be leaf, stem and root lesions, blights and anthracnoses, cankers and galls. In addition to the diseases of vegetative tissues, there are coelomycetes responsible for flower blights, anther hypertrophy, rot and blemish of growing and harvested fruit (Cook, 1975), and reduction of viability of seed, and seed transmission of disease (Neergaard, 1978). Many are quite capable of existing as latent infections in fruit and leaves (Simmonds, 1941; Carroll, Müller & Sutton, 1977) for considerable periods. A large group of genera in the coelomycetes is also responsible for many dieback and deteriorative conditions in shrubs and trees (Peace, 1962). Here the fungi vary in pathogenicity inasmuch as the ability to enter host tissues and cause disease is often dependent on existing or coincident physiological stress or mechanical damage to the host.

In reviewing the taxonomy of the coelomycetes (Sutton, 1973a), I pointed out that the group is in a transitional state as new information (particularly concerning conidiogenesis) is assimilated into the existing framework. The point in time has now been reached when there are more mycologists actively interested in using modern concepts to revise the group and many more workers, such as pathologists, ecologists and those interested in biodeterioration, require accurate names for the organisms with which they are dealing. To provide a basis on which further revisionary work may be accomplished must be one of the main purposes of the book. The other purpose is to enable coelomycetes to be identified at least to generic level and thence to species, with guides to literature for those genera not covered in depth. Much of the historical information relating to the recognition of the coelomycetes as an entity, and the development of classificatory systems in the Deuteromycotina, has already been covered (Sutton, 1973a) and it would be superfluous to repeat it here. I have decided however, to provide an adequate background to the classificatory system adopted in this book. In addition a section on methods and techniques of examination is included because of the small conidiomata, minute conidia and conidiogeneous cells or conidiophores that are relatively difficult to find and hard to interpret. Characters of diagnostic value are described, illustrated and discussed, and as an aid to the identification of taxa and use of the descriptive account, a glossary of terms is also provided.

COLLECTION OF MATERIAL

It is a relief to receive a coelomycete specimen and by a single character name it to species immediately. Unfortunately this is the exception rather than the rule. The less one recognizes in an unknown organism the more one must do to obtain the rest of the information required for identification. Because most coelomycetes have simple brown conidiomata rarely exceeding 300 µ in diameter and conidia frequently less than 15 µ in length it follows that identification is accomplished at the microscopic rather than macroscopic level.

The basic equipment required for finding and collecting coelomycetes in the field is as follows. More general hints on collecting are to be found in Savile (1962) and Hawksworth (1974).

Collecting container (haversack, hold-all, carrier bag, satchel)
A compartmentalized type is recommended so that the various smaller pieces of necessary equipment may be kept separately from each other and from the specimens collected.

Lens (× 10)
Since few coelomycetes can be identified with any accuracy in the field a higher magnification is not required.

Small sheath-knife or pen-knife
Useful for trimming specimens from wood, bark or twigs. A small axe is also recommended for sampling bulkier specimens, or for simply acquiring large samples of wood or bark which are studied in more detail in the laboratory.

Pruning shears or secateurs
Possibly the most valuable piece of equipment because many coelomycetes inhabit small branches, twigs and leaves.

Envelopes (15 × 9 cm)
Since most collections of microfungi are maintained in herbaria in packets of this sort of size or smaller, it is convenient in the field to trim stem and twig material accordingly. Ample material for examination can be contained in such envelopes, although much larger samples have to be kept separate. On few occasions it is required to protect coelomycetes from damage in the field, as for example myxomycetes or discomycetes. They are robust and stand comparatively rough handling so it is not usually necessary to carry tins or boxes.

Pencils
An important part of any collecting procedure is the annotation of material in the field. It is wise to keep several pencils available in the collection container in case of loss.

Elastoplast
Not obligatory but definitely of value in case of accidents when using a sheath knife or pen-knife.

Coelomycetes grow on stems, branches, twigs, leaves, bark, floral structures and fruit. These are either in a living, moribund or decayed condition. Evidence of coelomycetes on living leaves, petioles and stems is usually by the presence of necrotic lesions, whereas on twigs and branches they may be associated with distinct cankers or lesions. In searching for coelomycetes on these substrates one looks for symptoms initially then confirms that conidiomata are present by using a hand lens. Such symptoms are absent for example on leaf litter or decaying or moribund bark tissue and here one must look directly for the individual conidiomata. In both living and dead plant material the conidiomata of different genera may be superficial, semi-immersed, or immersed, and consequently members in the last two categories are difficult to find. Sometimes the only clue that conidiomata are present may be the appearance of papillate or rostrate ostioles protruding slightly above the substrate. In some cases conidial masses emerge through concealed ostioles and either spread over the surface of the substrate, or, depending on the humidity, form tendrils. Both conidial masses and tendrils vary from black, through the spectrum of pigmentation, to white, and often the colour is characteristic of the species.

It is recommended in collecting ascomycetes to cut the ascomata horizontally or vertically to ascertain that the specimen is in good condition and not effete. A similar procedure is also useful in coelomycete collecting, although of less wide application, for the majority of coelomycetes have much smaller conidiomata and it is difficult to manipulate such material in the field.

As the specimen is collected, the data should be recorded on the envelope used to transport the material back to the laboratory (Savile, 1962; Hawksworth, 1974). It is essential that the identity of the substrate should be accurately known, for the identification of many species is still regrettably on a host basis. Should there be any doubt as to the identity of the host, then adequate material such as leaves or flowers must be taken to facilitate naming of the substrate at a later date. Unless material is examined immediately it should be dried thoroughly to prevent the growth of moulds and unwanted saprophytes. For leaf and flower material the procedure is precisely the same as that used for the preparation of phanerogamic samples. Pressing and drying are fully described by Savile (1962). For bark, stem and twig samples the material should be either simply air-dried or dried with gentle heat. If material seems to be immature it is worthwhile placing it in a damp chamber for a few days to induce conidial production, after which the sample can be dried for incorporation in the herbarium.

EXAMINATION OF MATERIAL

The comparatively small size of conidia, conidiophores and conidiomata of coelomycetes means that, except in a few cases, it is necessary to examine them by microscopy before there is any real hope of making an identification. Since in many species and genera the conidia and conidiophores lack any pigmentation, it is desirable to stain these structures to make them visible for examination. At what level of microscopy one conducts the examination is clearly dependent on the facilities available.

The aim of preparation, staining, mounting and examination is to obtain all the information required to identify the organism, and this can be done at different levels depending on the expertise of the identifier.

Preparation

All techniques require the use of a low power dissecting microscope (× 20- × 25) where a conidioma may be selected for examination. If it is superficial then there are few problems in removing it from the substrate with a sharp scalpel moistened with either water or a drop of the mountant to be used. However, if it is immersed or semi-immersed it is then necessary to extricate it from the substrate. The least amount of host tissue as possible should be taken with the conidioma because the fungus is the organism one is interested in at this stage rather than its relationship to the substrate. The more host material on the slide preparation the thicker it will be and therefore the less informative. It is very useful to work on a white background from this stage on because darker backgrounds make it difficult to manipulate such small amounts of tissue. Once the selected conidioma has been removed from the substrate it is transferred to a clean microscope slide, in the centre of which is placed a small drop of mountant.

Whole preparation. This is the least valuable of preparations although one may observe details of the ostioles, setae and conidiomatal wall structure etc. in correctly prepared mounts. Generally, however, such preparations are thick and rarely yield any more information than can be obtained by the following procedure.

Squash preparation. Once the conidioma has been mounted, cleared and examined, the preparation should be tapped vigorously to break the conidioma and then reheated. In this way the conidia should be released and if one is diligent some conidiophores or conidiogenous cells will be evident with developing conidia attached. Another technique is to chop up the conidioma before clearing and sealing.

Hand sections. It is insufficient just to know that there are conidiogenous cells and/or conidiophores present in the conidioma. Their orientation is also an important criterion in identification, so at this stage a rough longitudinal (not transverse) section can be made with a scalpel or a razor blade.

Microtome sections. Although useful information about orientation of conidiogenous structures can be obtained with hand sections, the latter are no substitute for microtome sections, 5-10 µ in thickness. These are used for determining the different types comprising the conidiomatal walls, the conidiogenous layers, the ostiolar regions and the relationship of the conidioma to the substrate on or in which it occurs. At CMI we have used a freezing microtome operated either by carbon dioxide or the 'Pelcool' cooled water system, both with consistently good results. If a conidioma is spherical then a median vertical section in any plane, but passing through the ostiole, is satisfactory. If the conidioma is elongated, hysteriform or eustromatic with locules orientated in various directions, it may be necessary to take vertical sections across the width and along the length to obtain the required information. To see the relationship of the conidiomata and vegetative mycelium to the substrate it is necessary to dissect a piece of the supporting or surrounding host tissue together with the conidioma.

Stains and mountants

A distinction should be made between those preparations that are required for temporary use and those intended for permanent deposit in an herbarium for future use.

Temporary preparations. Water or KOH may be used without any added stains, and as such are quite useful for the observation of conidial appendages and mucilaginous sheaths. However I have found 3% erythrosin in 10% NH_4OH solution to be far more satisfactory. It stains the cytoplasm and gives good differentiation between cells walls, cytoplasm and the mountant. It is particularly useful for studies of conidiogenesis where water or KOH are not nearly as satisfactory.

It also replaces negative staining with dyes and indian ink as a means of distinguishing appendages of various types. Its great drawback is that it cannot be made permanent, although sealing will preserve the preparation for 1-3 days. Shoemaker (1964c) recommends the use of 1% aqueous azur A (CAz-4, Coleman & Bell) which can be made semi-permanent by adding a drop of saturated solution of azur A in glycerol to replace the water and then sealing.

Permanent preparations. Microscopic preparations can be made permanent by sealing coverslips with either two applications of commercial nail varnish or a single application of a proprietary fluid sealant such as Glyceel (Gurrs Ltd). It is important to clear preparations thoroughly by gentle heat before sealing, otherwise air bubbles quickly appear.

With pigmented coelomycetes there is no need to stain tissues and the preparations can be mounted directly in lactophenol (Ainsworth, 1971). However, the majority of coelomycetes require staining to demonstrate guttules and septa in conidiophores and conidia and differences in structure of conidiomatal wall types. The standard stain and mountant is cotton blue (concentration variable according to requirements) in lactophenol. When using a blue stain however I have usually resorted to trypan blue in lactophenol. This immediately stains darker than cotton blue but over a period of time seems to result in greater shrinkage of cytoplasm even than one gets with cotton blue. Neither stain is satisfactory over extended periods of time. I have preferred to use lactofuchsin (Carmichael, 1955) for routine permanent preparations for more than 5 years now and the results over this period are consistently better than with the blue stains in lactophenol. By mixing equal parts of lactophenol with Gurr's water mounting medium a permanent preparation can be made. However, within 1-2 years such preparations have developed a sufficiently high proportion of globules to negate their value.

Necol preparations. This technique is of particular value in the study of coelomycetes forming superficial colonies on leaf and stem surfaces. It is fully described by Ellis (1950) and basically consists of using a cellulose acetate (necol) preparation to remove the colonies from the surface of the substrate. The necol is dissolved with acetone and the remaining fungus is stained, mounted and sealed by one of the techniques already enumerated.

CULTURES

It is obligatory in genera such as *Phoma* and *Colletotrichum* to grow the organisms in artificial culture for purposes of identification. This approach is likely to be of increasing relevance in the future as more coelomycetes are isolated into pure culture and their behaviour and relationships documented.

Isolation. Booth (1971) has summarized methods of isolation, culture and stimulation of sporulation in *Fusarium.* Those principles are applicable to coelomycetes, for with *Fusarium,* they are amongst the commonest of organisms isolated from soil and plant material. However, in comparison with *Fusarium,* coelomycetes are quite slow-growing so may be overtaken in general isolations by *Penicillium, Aspergillus, Fusarium,* Mucorales etc. By serially diluting on a minimal medium such as tap water agar there is a reasonable chance of isolating individual colonies before they are covered by adjacent species.

Stimulation of sporulation. All cultures are submitted to 12 h near ultra violet (black light) irradiation/12 h dark (Leach, 1962) in plastic rather than glass petri dishes at 25°C. This stimulates sporulation whilst avoiding the germicidal wavelengths. Full details are found in Booth (1971). In addition to this standard technique which stimulates sporulation better than any other general procedure so far used it has been found advantageous to incorporate sterilized natural substrata such as wheat straw, macerated leaves and lupin stems into the basic agar. Many species sporulate better on the included material than the agar itself.

Media. The media employed in the growth of coelomycetes are varied, and it is fair comment that standardized cultural conditions for different coelomycete genera have not been explored fully. This is simply because the taxonomies of very few genera have been investigated in this manner. Dorenbosch (1970) in discussing *Phoma-* like fungi from soi' recommends the use of oatmeal agar because it stimulates production of pycnidial conidiomata and conidia, and malt agar because it stimulates mycelial growth, formation of chlamydospores and crystal formation (both recipes according to Ainsworth, 1971). Growth on these agars has been used by Boerema & Höweler (1967), Boerema, Dorenbosch & van Kesteren (1965, 1968, 1971, 1973, 1977), Boerema & Dorenbosch (1973) and Boerema (1976) to distinguish several species within *Phoma.* Sutton (1965a, 1968a) used potato dextrose agar, potato carrot agar and oatmeal agar (Booth, 1971) to distinguish graminicolous *Colletotrichum* species, and Simmonds (1965) used potato dextrose agar in separating fruit-rotting *Colletotrichum* species in Queensland, Australia. van der Aa (1973) recommends cornmeal agar, cherry decoct agar, and oatmeal agar, all with sterile lupin stems, for the growth of *Phyllosticta* sensu stricto (*Phyllostictina*). Punithalingam (1970, 1974a) who has been mainly concerned with the correlation of pycnidial fungi with their ascomatal states has used potato dextrose agar and oat agar.

Slide cultures. This technique is particularly applicable to the study of appressoria in *Colletotrichum* (Sutton, 1962, 1968a), although it is used more extensively in hyphomycetes for direct observation of growth in microculture. It is fully described in Sutton (1968a) and Hawksworth (1974) and consists of inoculating a small cube of agar at the sides and allowing the mycelium to grow out over the glass surface. Appressoria are formed on the glass and these can be mounted for examination.

SUPRAGENERIC TAXA IN DEUTEROMYCETES

The ideal suprageneric classificatory scheme for deuteromycetes (imperfect fungi, anamorphs, anamorphoses) would be one based on the major recognized categories of perfect fungi (teleomorphs, teleomorphoses). In the majority of cases, however, the teleomorphs are not known, so we are forced to use artificial systems for separating taxa above the level of genus.

Many of the earlier classificatory schemes for anamorphs were comprehensive in that they dealt with all types of imperfect state. These were mostly based on differences in conidiomata, conidial morphology, and spatial relationships of the conidiomata to the substrates on which they occurred. A review of these morphological systems of classification was given by Sutton (1973a). Since the work of Hughes (1953a), an alternative approach has achieved acceptance, in which conidiogenesis has played a prime role in separating taxa in ranks at generic level and above. These schemes have been suggested for the separation of hyphomycetes, but apart from that proposed by Höhnel (1923a) none have so far taken account of the coelomycetes. Again the ontogenetic systems of classification, both for hyphomycetes and coelomycetes were reviewed by Sutton (1973a).

It is evident from the work published since 1953 that the basis for any comprehensive classification of anamorphs must be conidial ontogeny. The breakdown of developmental types suggested by Ellis (1971) following Kananaskis I (Kendrick, 1971) would be a logical basis for a scheme and indeed such a system was suggested in 1973 by Sutton (in press).

The primary dichotomy (Fig. 1) of the subdivision Deuteromycotina is based on the two fundamental origins of conidia, which are blastic and thallic. This gives two classes, Blastodeuteromycetes and Thallodeuteromycetes. Each class is further divided according to the origin of the conidium wall. Holothallic and enterothallic origins give the subclasses Holothallomycetidae and Enterothallomycetidae, while holoblastic and enteroblastic origins give the Holoblastomycetidae and the Enteroblastomycetidae. Three of these subclasses contain a single order, Thallales, Enterothallales and Blastales, but the Enteroblastomycetidae contains two orders which reflect the differences between phialidic and tretic ontogeny. These are the Phialidales and the Tretales. So far the scheme is a simple transformation of the Kananaskis I recommendations into an hierarchical system for segregating suprageneric taxa. It is at the subordinal level that concepts concerning the conidiomata are introduced. This is a complex issue, but reduced to its fundamentals is as follows.

In the Deuteromycotina certain types of conidiomata are readily recognised—the separate conidiophore or conidiogenous cell, the pycnidium and the pycnothyrium. Others, such as synnemata, sporodochia, acervuli and stromata of different types are not readily separable into distinct groups, particularly when behaviour in culture is considered. Following the precedents suggested by Saccardo (1880a), Potebnia (1910) and Höhnel (1911a, 1923a), four types of conidiomata have been recognised in the Deuteromycotina—pycnidia, pycnothyria, separate conidiophores and stromata. It has been necessary to use a broad concept for the stroma which includes acervuli, sporodochia, synnemata, loculate stromata and cupulate conidiomata. The plethora of different loculate and cupulate conidiomatal types requires further analysis. It is possible that additional research will result in more recognizable conidiomatal types but since most work at the moment is concerned with amassing primary data, it would be premature to erect more categories at subordinal rank based on different conidiomata. Thus the Thallales, Blastales and Phialidales are each divided into four suborders based on these types of conidiomata. The Tretales contains a single suborder, the Tretohyphineae because it is only in hyphomycetes with separate conidiophores that the tretic mode of conidiogenesis is known. Within the various suborders many additional criteria can be used in defining the family names available for use in the Deuteromycotina (Cooke & Hawksworth, 1970).

This scheme is experimental, and, despite clear anomalies which will inevitably arise in any system, it has proved workable over several years. It has three clear advantages compared with previous schemes. If a genus is assigned to the Blastopycnidiineae, the Phialostromatineae or the Thallohyphineae, then by these terms one will understand precisely the type of conidial ontogeny and conidiomatal structure it is supposed to have. Another advantage is that it does away with the unnatural distinctions perpetuated by previous systems. Although still an artificial system, it does group together those organisms by diagnostic criteria that at the moment are considered to be the best available. Thirdly, it is comprehensive for all the deuteromycotina, not just the hyphomycetes or just the coelomycetes, and as such is a major advancement over previous suprageneric schemes.

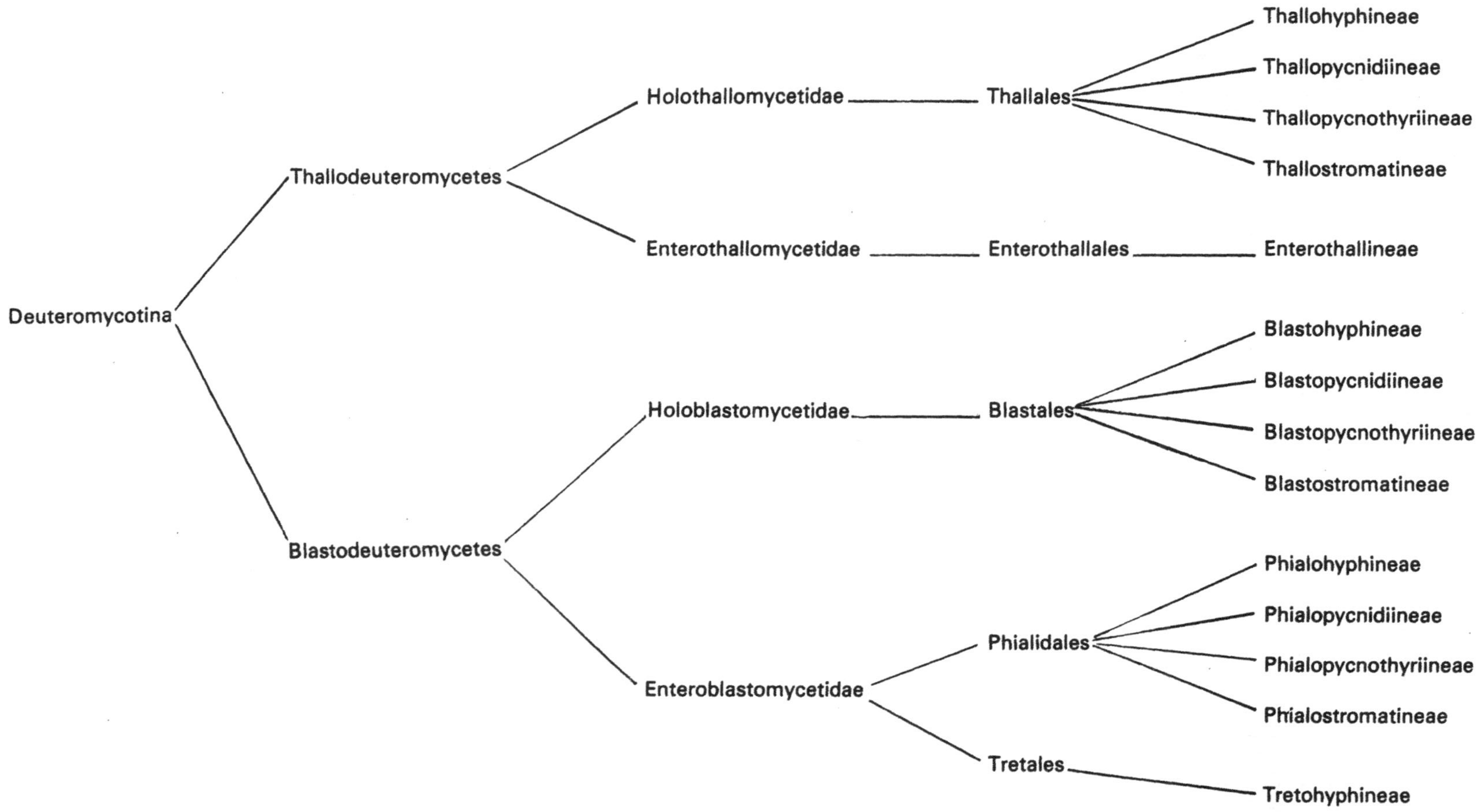

Fig. 1. Suprageneric classification of coelomycetes.

KEY TO SUBORDERS

1. Conidiogenesis thallic 2
1. Conidiogenesis blastic 3
   2. Conidiomata pycnidial .... Thallopycnidiineae p. 24
   2. Conidiomata stromatic .... Thallostromatineae p. 26
3. Conidiogenesis holoblastic 4
3. Conidiogenesis enteroblastic 5
   4. Conidiomata pycnidial .... Blastopycnidiineae p. 41
   4. Conidiomata stromatic .... Blastostromatineae p. 129
5. Conidiomata pycnidial .... Phialopycnidiineae p. 358
5. Conidiomata stromatic .... Phialostromatineae p. 438

SUBORDINAL KEYS AND DIAGNOSTIC CRITERIA

The daunting tasks of an author preparing, and the reader using, a single key to the 370 genera or so covered in this book have been avoided by two solutions. Based on suprageneric classification outlined on p. 17 a dichotomous key to suborders of the coelomycetes is given (p. 18). Such an initial breakdown reduces the groups of genera keyed out in any one section to manageable proportions. Whilst the information used in these keys has been obtained by direct observation of voucher or type material, it has not been possible to extend such treatment fully to the majority of fungi formerly included in the Pycnothyriales (superficial coelomycetes with scutate conidiomata). In particular, there is virtually no published information on conidiogenesis in the group. Genera formerly placed in the Pycnothyriales will, in future, be accommodated in the Thallopycnothyriineae, Blastopycnothyriineae and Phialopycnothyriineae when their modes of conidiogenesis have been more fully worked out. At present however, they are omitted from this treatment of the coelomycetes.

The pycnidial, acervular and stromatic genera which form the main body of this book are keyed out within the suborders by two different methods. The traditional dichotomous keys reflect a reasonable compromise between the need to identify an organism to genus and the taxonomist's attempt to group closely related taxa together. An alternative approach, the synoptic key (Korf, 1972), serves one purpose, that is to lead the user to the correct taxon as quickly as possible. In discussing genera of Pezizales in this context, Korf (1972) suggested that thirty taxa is a limit for use, beyond which the synoptic key becomes cumbersome. He dealt with 93 genera and in doing so had to provide an analytical breakdown at a preliminary stage to identify unusual characters. In coelomycetes there is a paucity of characters for identification, and this fact, coupled with the large number of taxa dealt with in this book, means that a complete synoptic key to all genera would be so unwieldy that it would defeat the prime objective, which is speedy identification. Consequently synoptic keys are provided to each suborder as alternatives to the dichotomous keys.

The data used in both types of key are precisely the same but they are organized quite differently. They basically comprise criteria of the mycelia, conidiomata, conidiogenous cells, conidiophores, conidia and paraphyses, and each one of these types shows a varying range of additional differential criteria.

A review of the criteria used in classification of the coelomycetes has been presented by Sutton (1973a). This included discussions on fructifications (conidiomata), conidiogenous cells and conidium development, conidial morphology, and behaviour in artificial culture. Little change in the actual scope of diagnostic criteria has emerged since that time, although modifications in terminology as applied to fructifications (conidiomata), and the increasing complexity in concepts of conidiogenesis are two areas where more recent information should be mentioned.

Conidiomata

For several years a need has been evident for a comprehensive noun that can be applied to the fruit-bodies in which, or on which conidiogenous cells and conidia develop in the Deuteromycotina. This requirement was met by Kananaskis II (Kendrick, 1979) where the term conidioma (pl. conidiomata) was introduced. Its status is precisely analogous to the terms basidioma and ascoma. Thus now, instead of referring to pycnothyria, pycnidia, sporodochia and acervuli, we may use pycnothyrial, pycnidial, sporodochial, and acervular conidiomata. Based on such examples an argument against the introduction of apparently superfluous terms could be advanced. However, the value of this term becomes clear when one wishes to refer to the fructifications of such widely different form-genera as *Dinemasporium*, *Ceuthospora* and *Dilophospora* etc., where conventional nouns cannot be applied. In such cases conidiomata can be qualified by adjectives such as cupulate, pseudostromatic and eustromatic and thereby fulfil a useful function.

Another need which has recently been realized is the recognition of different tissue types in conidiomata and the necessity of more precision in their terminology. Within conidiomata there is often distinct differentiation of tissue types. These have recently proved exceptionally useful in comparing and contrasting genera, even though some tend to intergrade. The use of the terminology described by Korf (1973) and attributed to Starbäck (1895) and Korf (1952) in discomycete taxonomy has been comparatively easy to apply during the preparation of this work on coelomycetes. Its adoption has led to some degree of uniformity in preparation of descriptions which results in easier comparison between taxa.

Conidiogenesis

The publication of '*Dematiaceous Hyphomycetes*' by Ellis (1971) put to practical application the recommendations of Kananaskis I (Kendrick, 1971) concerning conidiogenesis and the preferred terminology. Although the concepts advanced by these two authors are readily applicable to the taxonomy of hyphomycetes, it is somewhat more difficult in the coelomycetes, for conidiogenous structures are much smaller, and consequently more difficult to interpret. The number of electron microscopy studies of conidiogenesis in coelomycetes is small and time-lapse photomicrography, so readily applied in the study of hyphomycetes, has not been used at all in the coelomycetes. Consequently certain basic problems still need to be solved.

The full range of developmental types in coelomycetes, as outlined by Sutton (1971a, 1973a) and Morgan-Jones (1971a, b, c) has not been enlarged, inasmuch as neither the tretic nor the basauxic type have been confirmed in the group. The multiplicity of other enteroblastic, holoblastic and thallic types is now well-documented but there remains a single difficult problem which has both theoretical and practical implications. This is the distinction, if any, to be made between percurrently proliferating phialides and annellides. It is a difficulty peculiar to the coelomycetes for it does not appear to be quite such a problem in the larger hyphomycetes. Percurrently proliferating phialides (as in *Chloridium* and *Catenularia*), where more than one conidium is clearly formed from successively higher fixed conidiogenous loci, and annellides (as in *Stigmina* and *Scopulariopsis*) are comparatively easy to distinguish in the hyphomycetes, and difficulties in interpretation can largely be solved by the use of time lapse photomicrography techniques. No such techniques have been devised for coelomycetes yet because conidiogenesis usually takes place within a closed conidioma, consequently it is not yet possible to observe and directly measure exact sequences in conidiogenesis with the precision that has proved to be so useful in hyphomycetes.

Morgan-Jones, Nag Raj & Kendrick (1972a) state that the crux of the problem concerning proliferating phialides and annellides concerns the transverse layer or layers separating the conidium from the conidiogenous cell and whether any layers or parts thereof are left on the conidiogenous cell when the conidium secedes. If a wall or a layer remains, then development is intepreted as annellidic. Conversely if no wall or layer is left enclosing the protoplast remaining in the conidiogenous cell after conidial secession then development is classed as phialidic. If the latter occurs and is combined with conidial formation at progressively higher levels by percurrent growth then development is suggested as being percurrently phialidic. These are the criteria suggested for use in distinguishing between the two types. It is acknowledged however that the precise nature of the periclinal wall and its contribution to or lack of involvement in collarette formation may be of more significance.

The arguments advanced by Morgan-Jones, Nag Raj & Kendrick (1972a) substantiating the presence of percurrently proliferating phialides in coelomycetes are plausible, yet more questions are really raised than answers given.

The evidence presented for a circumscissile break in the periclinal wall at or just below the level of the septum separating the conidium from the conidiogenous cell and for the retraction of the naked protoplast back into the conidiogenous cell is not convincing. There are cases however where retraction is known to occur (e.g. in *Gloeocoryneum cinereum*) but unfortunately here it is not combined with percurrent growth. The qualifying terms 'seems to' and 'apparently' do nothing to strengthen the arguments. The studies by Brewer & Boerema (1965), Sutton & Sandhu (1969a), Griffiths & Campbell (1972) and Boerema & Bollen (1975) show very clearly the presence of wall material or a layer surrounding the young conidial primordium following secession of the preceding conidium, though its precise structure has not been investigated fully.

Furthermore, it is by no means clear whether a single conidium or a plurality of conidia is formed from successively higher conidiogenous loci in the coelomycetes. Thus if the presence or absence of a wall layer remaining over the protoplast in the conidiogenous cell is in doubt, and the number of conidia formed at any single conidiogenous locus is not precisely known, then it is rather difficult to be precise about the type of development one is dealing with.

The variation in points of secession of the conidia in relation to the periclinal wall of the conidiogenous cell (Sutton & Sandhu, 1969a) shows that regular percurrent proliferation as opposed to secession of conidia at the same level or variable levels cannot itself be diagnostic of a particular type of conidiogenesis, and although annellations and collarettes are useful for identification they may not be quite so useful in the taxonomy of the group.

The degree to which collarettes, ridges or annellations are flared, shows that there is considerable diversity both within individual species and between genera. Morgan-Jones, Nag Raj & Kendrick (1972a) suggest that the double septum at the conidial base in an annellide precludes the presence of a flared collarette. However this is certainly not the case in *Stigmina robusta* (Sutton, 1973b), a dematiaceous hyphomycete where the conidiogenous cells are cupulate and percurrent with a widely flared collarette. To what extent similar exceptions occur in the coelomycetes is conjectural, although it is by no means impossible for a proportion of the genera with flared conidiogenous cells to develop like *S. robusta*. Although the apparent absence of a residual layer round the protoplast and the presence of flared collarettes may be an indication of phialidic development, the converse may also apply, in that there are quite possibly species and genera at present classed as phialidic which may eventually prove to be annellidic because the collarettes, ridges or annellations may not be flared and there may be wall material left after conidial secession!

It seems abundantly clearly from these points that arguments for and against particular interpretations are hindered by a serious lack of data. Much of the evidence is conflicting and it is fairly simple to find exceptions to generalizations.

Until techniques are evolved for the solution of these developmental problems little advance will be made. Detailed transmission electron microscopy of a wide range of genera and species seems the most obvious requirement at the moment.

For the practical purposes of identification it is almost impossible to make a distinction between annellides and proliferating phialides as Morgan-Jones, Nag Raj & Kendrick (1972a) advocate. The point at which circumscissile rupture occurs, the absence of a 'half septum' or wall layer and the retraction of the protoplast are ultrastructural criteria and as such are difficult to discern with optical microscopy. It is only the flared collarettes that can be used at this level, but if one takes the degree of flare as an indication of phialidic as opposed to annellidic development, then it is extremely difficult to make an absolute judgement. It is acknowledged that there are different types of phialide and probably (but by no means certainly) different types of percurrently proliferating conidiogenous cell. However, I propose to refer in the keys for identification to all percurrently proliferating conidiogenous cells as annellidic, unless it is clear from the presence of a channel, collarette and periclinal thickening that phialidic development is operating. In either case the genera are included in the dichotomous keys for both annellidic and phialidic groups.

CONSTRUCTION AND USE OF SYNOPTIC KEYS

Construction

In constructing a synoptic key the characters used in distinguishing genera are grouped firstly according to whether they pertain to mycelium, conidioma, conidiogenous cell, conidiophore, conidium or paraphysis. Then, within each of these groups, the characters are listed, so that, for instance, all criteria of the conidium such as pigmentation, shape, septation, size, wall thickness, guttulation, ornamentation etc. are included in that character group.

Each genus which is to be keyed out, is provided with a key number, and for each positive attribute that the genus displays, its key number is inserted next to that particular character. Thus the members of the Thallostromatineae with hyaline conidia are genera with key numbers 1 2 3 4 5, brown conidia are genera 6 7 8 9, and versicoloured conidia genus 7.

Use

In using a synoptic key it is theoretically possible to enter at any point, but clearly it is quicker to enter at a point where there are only a few numbers between which one needs to further discriminate. These numbers usually apply to unusual characters and of course the genera characterized by unusual or odd criteria are quicker and easier to identify. For genera which can be separated only by a combination of characters there is almost no substitute to the time-consuming procedure of starting at the beginning of the synoptic key and working through it until the genus is identified. It is of considerable help, however, and greatly speeds up the process of identification, if distinctive characters are used to reduce the number of initial possibilities. Thus, if the conidia are distoseptate, or have appendages, or if the pycnidia are setose, or if the conidiomata are eustromatic with a particularly odd locular configuration, or if the conidiomatal tissue type is textura prismatica or textura oblita, then such features are found only in a comparatively small number of genera, and in entering the key at such stages, the mass of genera lacking such characters can be eliminated from consideration.

For example, in attempting to key out *Pyrenochaeta* which by its simple pycnidia and phialidic conidiogenesis is known to belong in the Phialopycnidiineae, one has the option of starting with mycelial characters and working slowly through the key or selecting a distinctive character and trying a short-cut.

Using mycelial characters, the first criterion is relationship to the substrate—superficial or immersed. Pycnidia in *Pyrenochaeta* are immersed so there are 42 possible genera to be considered:—

6	10	11	12	13	14	15	18	19	20	21	22	23	24	25
26	28	29	30	31	32	33	34	35	36	37	38	39	40	41
43	44	45	46	47	48	49	50	51	52	53	54			

The second group of characters concerns the conidiomata, and again the first criterion is immersed, semi-immersed or superficial. Pycnidia in *Pyrenochaeta* are immersed, so a few genera which have immersed mycelium and pycnidia that are superficial or semi-immersed may be eliminated:—

(6) 10 11 12 13 14 15 (18) 19 20 21 22 (23) (24) 25
26 28 29 30 (31) (32) 33 34 (35) 36 37 38 (39) (40) (41)
43 44 45 46 (47) 48 49 50 51 52 (53) 54

The tissue type in *Pyrenochaeta* is textura angularis and only genus 33 is not included here so it is eliminated:—

10 11 12 13 14 15 19 20 21 22 25 26 28 29 30
(33) 34 36 37 38 43 44 45 46 48 49 50 51 52 54

The ostiole in *Pyrenochaeta* may be non-papillate or papillate but several genera have other types of dehiscence and are eliminated:—

(10) (11) (12) 13 14 15 19 20 21 22 (25) (26) 28 29 30
34 36 37 38 43 44 45 (46) 48 49 50 51 52 54

Pycnidial shape in *Pyrenochaeta* is globose, but since all the remaining genera also fall in this category, none can be eliminated here:—

13 14 15 19 20 21 22 28 29 30 34 36 37 38 43
44 45 48 49 50 51 52 54

The pycnidial walls are comparatively thin in *Pyrenochaeta*, rarely exceeding 3 cells in width, and many genera with thick-walled pycnidia may be eliminated at this stage:—

13 14 (15) (19) 20 (21) (22) (28) (29) (30) 34 (36) (37) 38 43
44 (45) (48) 49 (50) (51) (52) (54)

All the remaining genera, like *Pyrenochaeta*, have brown pigmented pycnidia so none is removed:—

13 14 20 34 38 43 44 49

The pycnidia of *Pyrenochaeta* are setose. Only one genus, number 20 has this character and it proves to be *Pyrenochaeta*.

(13) (14) 20 (34) (43) (44) (49)

The correct generic name has been arrived at by 7 steps in this synoptic key.

The alternative is to select key characters and start with those, irrespective of where they are situated in the key. Two such characters for *Pyrenochaeta* are the setose pycnidia and the filiform, septate, branched conidiophores. Thus, commencing with characters of the setae, *Pyrenochaeta* certainly possesses them:—

1 2 3 4 5 20 33 47 52 54

They are pigmented, but genera 2 and 52 are eliminated for they have hyaline setae:—

1 ② 3 4 5 20 33 47 (52) 54

Setae are unbranched in *Pyrenochaeta*, but in genus 3 they are branched so it is eliminated:—

1 ③ 4 5 20 33 47 54

Turning now to characters of the conidiophores, in *Pyrenochaeta* they are filiform and branched, but none of the other genera with pycnidial brown unbranched setae have this type of conidiophore, so genus 20 (*Pyrenochaeta*) is left as the identified genus.

① ④ ⑤ 20 ㉝ (47) (54)

The correct generic name has been arrived at by 4 steps, each of which is of much smaller magnitude than the 7 steps used in starting from the beginning of the synoptic key.

THALLOPYCNIDIINEAE subord. nov.

Cellulae conidiogenae holothallicae; conidiomata pycnidialia.
Conidiogenesis holothallic, conidiomata pycnidial.

This suborder contains a single genus, *Vouauxiella* (1).

1. VOUAUXIELLA Petrak & Sydow, *Beih. Rep. spec. nov. regni veg.* **42**: 482 (1927).
Alysia Cavalcante & Silva, *Publ. Inst. Mic. Univ. Recife* 647: 32 (1972).
Original spp.: *V. verrucosa* (Vouaux) Petrak & Sydow, *V. lichenicola* (Linds.) Petrak & Sydow.
Lectotype sp.: *V. verrucosa* (Vouaux) Petrak & Sydow (syn. *Torula verrucosa* Vouaux), fide Clements & Shear (1931).
Illustration: Morgan-Jones (1971c).

Mycelium immersed, branched, septate, brown. *Conidiomata* pycnidial, separate, globose to flask-shaped, dark brown, unilocular, immersed; wall thick, composed of thick-walled, dark brown textura angularis to intricata. *Ostiole* absent, dehiscence by irregular breakdown of upper tissue. *Conidiophores* short, branched and septate irregularly at the base, pale brown, verrucose, formed from the inner cells of the pycnidial wall. *Conidiogenous cells* holothallic, integrated or more frequently discrete, determinate, hyaline to brown, verrucose, consisting of basal thicker-walled cells which become flared at the inception of conidiogenesis (? similar to the conidiophore mother cell in *Arthrinium*). *Conidia* arthric, formed by disarticulation of the conidial chain, produced in simple unbranched chains with the oldest conidium at the apex, brown or dark brown, verrucose or smooth, cylindrical or long cuneiform, both ends truncate except for the apical conidium which is obtuse at the apex, aseptate.

Morgan-Jones (1971c) has described conidiogenesis in *V. lichenicola* and interpreted the species as producing meristem thalloconidia. These data are confirmed, but the presence of 'conidiophore mother cells' or their equivalent in *V. verrucosa* and *V. lichenicola* suggests that conidiogenesis is more complex than is found in genera like *Phacidiella*. The absence of such cells in *V. uniseptata* and the 1-septate conidia suggest that this species may be better referred to a separate genus.

V. VERRUCOSA (Vouaux) Petrak & Sydow, *ibid.*: 483 (1927). (Fig. 2B)
Torula verrucosa Vouaux, *Bull. trimest. Soc. mycol. Fr.* **54**: 697 (1907).

Conidiomata up to 150 µ diam. Conidia 5-9 × 4.5-5 µ, coarsely verrucose, dark brown.

On apothecia of *Lecanora chlarotera*, *L. subfusca*, *L. laevis*, G.B., France, Pakistan (IMI 189067, holotype ex VOUAUX, 186236, 199857c, 186861b).

V. LICHENICOLA (Linds.) Petrak & Sydow, *ibid.*: 484 (1927). (Fig. 2A, D)
Torula lichenicola Linds., *Trans. R. Soc. Edinb.* **25**: 515 and 530 (1868-1869).
Sirothecium lichenicola (Linds.) Keissl., *Öst. Bot. Z.* **60**: 61 (1910).
Sirothecium lichenicola (Linds.) Keissl. var. *bisporum* Keissl., *Centralbl. Bakt.* **27**: 210 (1910).

Conidiomata 75-90 µ. Conidia 6.5-9 × 3-3.5 µ, finely verrucose, pale to medium brown.

On apothecia of *Lecanora chlarotera*, U.K. (IMI 188759, 185887, 186861a, 178325, 224494, holotype *T. lichenicola* ex E).

V. PITHOSPORA (Cavalcante & Silva) comb. nov.
Alysia pithospora Cavalcante & Silva, *Publ. Inst. Mic. Univ. Recife* 647: 33 (1972).

Conidiomata up to 60 µ diam. Conidia 3-3.5 × 2.5 µ, smooth.

On indet. lichen thallus on Palmae, Brazil.

V. UNISEPTATA D. Hawksw., *Notes R. Bot. Gdn. Edinb.* **36**: 185 (1978). (Fig. 2C)

Conidiomata 40-70 µ diam. Conidia 1 septate, smooth to slightly verrucose, 12-15 × 5-7 µ.

On thallus of *Parmelia laevigata*, U.K. (IMI 212668, holotype, 228667).

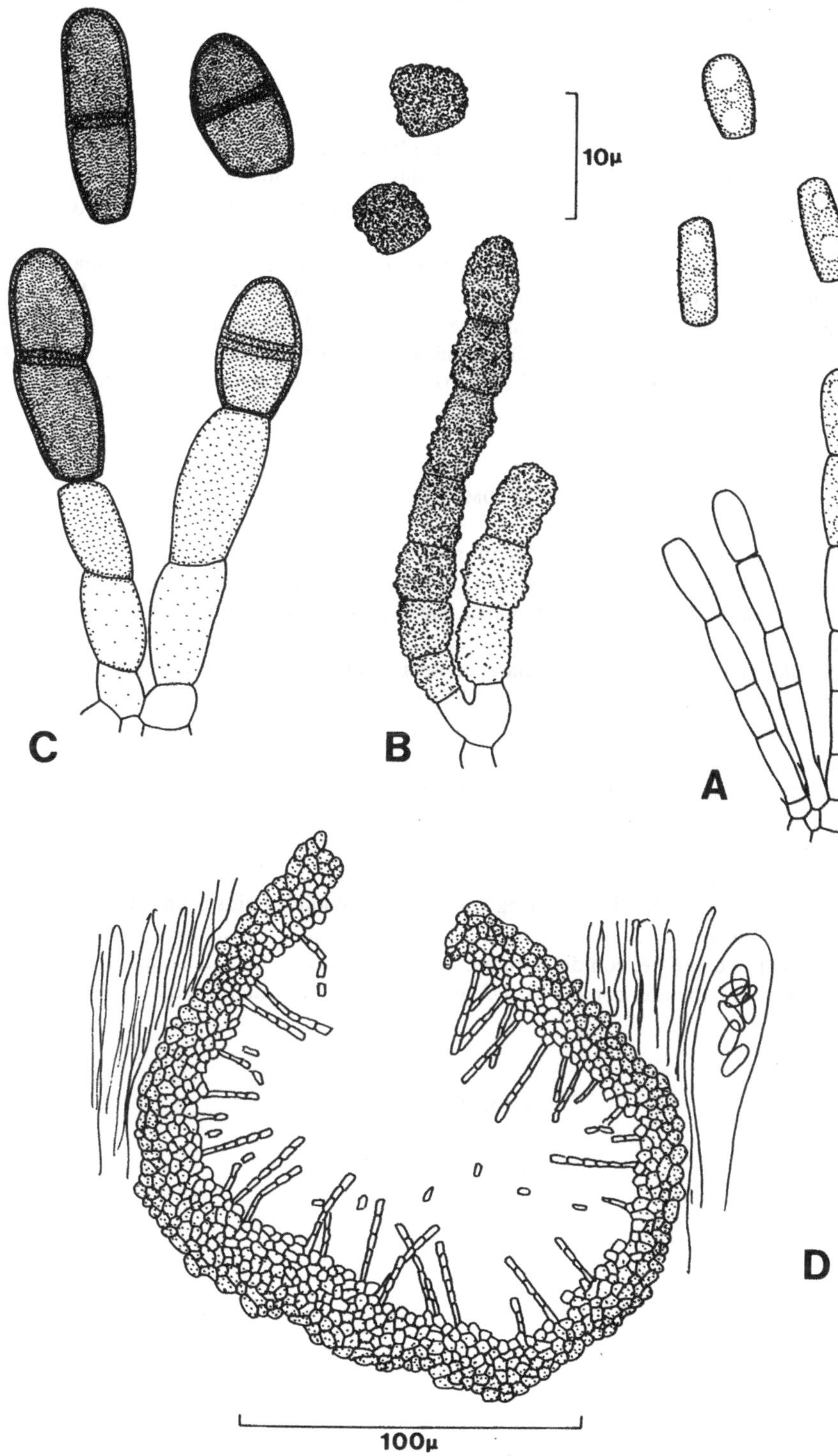

Fig. 2.A, *Vouauxiella lichenicola*, conidiophores and conidia; B, *V. verrucosa*, conidiophore and conidia; C, *V. uniseptata*, conidiophores and conidia; D, *V. lichenicola*, vertical section of conidioma.

THALLOSTROMATINEAE subord. nov.

Cellulae conidiogenae holothallicae; conidiomata stromatica.
Conidiogenesis holothallic; conidiomata stromatic.

This suborder contains nine genera, of which one (*Staninwardia*) is acervular, one is locular and stromatic (*Barnettella*) and the rest are simple stromatic or cupulate stromatic. It is heterogeneous in conidial morphology as well as conidiomatal type. One genus, *Barnettella*, is anomalous in that the large spherical muriform conidia are formed in chains and each is separated by several disjunctor cells. This is the only example in the coelomycetes, and indeed, in the deuteromycetes as a whole, *Dactuliophora* Leakey (1964) is the only other genus where propagules (or conidia) secede by the simultaneous fracture of several cells.

Dothioropsis is difficult to separate from *Neozythia* and is distinguished by the more regular shape of the conidia and the fact that only the larger conidia become 1 septate. The distinction between *Phacidiella* and *Sirozythiella* is also marginal in that conidia are either aseptate or septate respectively. There is also a greater tendency for the conidiophores to branch irregularly at the base in *Sirozythiella*. It is quite possible that *Sirozythiella*, the later name, may be relegated to synonymy, should sufficient material be available in the future to enable a reassessment to be made.

DICHOTOMOUS KEY

1. Conidia separated by groups of hyaline disjunctor cells, muriform .... *Barnettella* (9)
1. Disjunctor or separating cells between conidia absent 2
   2. Conidiomata acervular; conidia 0-1 septate, verruculose, enclosed in a mucilaginous sheath .... *Staninwardia* (8)
   2. Conidiomata eustromatic 3
3. Conidiomata subglobose, flattened or discoid to cupulate; conidia oblong, pale brown, aseptate .... *Trullula* (6)
3. Conidiomata discoid to cupulate; conidia variable, simply transversely septate, or muriform, or versicoloured and muriform with median projections .... *Phragmotrichum* (7)
3. Conidia hyaline 4
   4. Attenuated, cellular, unbranched appendages formed laterally on 1 septate conidia .... *Acarosporium* (5)
   4. Conidia lacking appendages 5
5. Conidiomata simple, globose to subglobose, very pale brown; conidial chains branched irregularly; conidia 0-1 septate, irregular in shape .... *Neozythia* (1)
5. Conidiomata separate or confluent, longitudinal or irregular, brown; conidial chains branched; conidia usually aseptate, ellipsoid-oblong, larger ones 1-septate .... *Dothioropsis* (2)
5. Conidiomata finally collabent to cupulate; conidia aseptate .... *Phacidiella* (3)
5. Conidiomata finally collabent to cupulate; conidia (0) 1-2 septate. .... *Sirozythiella* (4)

SYNOPTIC KEY

Conidiomatal characters

Stromatal type					
a. acervular	8				
b. cupulate	3	4	5	6	7
c. multilocular	2	9			
d. unilocular	1				
Position in relation to substrate					
a. superficial	9				
b. immersed	1	2	3	7	8
c. initially immersed, finally superficial	4	5	6		

Conidiophore characters

Type of branching							
a. irregular	1	2	4	6			
b. restricted to the base	3	7	8				
c. dichotomous or subdichotomous	5						
d. none	9						
Pigmentation							
a. hyaline	1	2	3	4	5	6	7
b. brown	8	9					
Disjunctor cells between conidia	9						

Conidial characters

Pigmentation								
a. hyaline	1	2	3	4	5			
b. brown	6	7	8	9				
c. versicoloured	7							
Septation								
a. aseptate	1	2	3	6				
b. 1-septate	1	2	5	8				
c. 2-multiseptate	4	7						
d. with longitudinal septa	7	9						
Shape								
a. ovoid to ellipsoid	1	2	7	8				
b. fusiform	3	4	7					
c. cylindrical	3	5	6					
d. globose	9							
Ornamentation								
a. smooth	1	2	3	4	5	6	7	9
b. verruculose	8							
Symmetry								
a. asymmetrical	1	5	7					
b. symmetrical	2	3	4	6	7	8	9	
Appendages								
a. cellular	5							
b. mucilaginous sheath	8							
Guttulation	6							

1. NEOZYTHIA Petrak, nom. nov., *Sydowia* **11**: 351 (1958).
Sclerozythia Petrak, *Sydowia* **9**: 539 (1955), non Petch, 1937.
Sp. typ.: *N. handelii* (Bub.) Petrak (syn. *Roumegueriella handelii* Bub., *N. nectrioidea* (Petrak) Petrak).

Mycelium immersed, branched, septate, hyaline. *Conidiomata* eustromatic, separate or aggregated, immersed, subepidermal, very pale brown, globose to subglobose, thin-walled; wall of hyaline, thin-walled textura angularis. *Ostiole* absent, dehiscence by breakdown of upper wall. *Conidiophores* irregularly branched and septate, hyaline, smooth, formed from the inner cells of the conidiomatal wall. *Conidiogenous cells* meristematic, thallic, integrated, fragmenting, hyaline. *Conidia* hyaline, 0-1 septate, thin-walled, smooth, eguttulate, irregular in shape, cylindrical, ellipsoid, continuous or constricted at the septa.

In proposing *Neozythia* as a nomen novum, Petrak (1958a) took up his own name *Sclerozythia nectrioidea* for the type species. Later (Petrak, 1959a) he uncovered an earlier name for it, *Roumegueriella handelii* Bub., and therefore made the necessary combination. So far the genus contains a single species.

N. HANDELII (Bub.) Petrak, *Sydowia* **13**: 8 (1959). (Fig. 3)
Roumegueriella handelii Bub., *Ann. Naturk. Mus. Wien* **28**: 214 (1914).
Neozythia nectrioidea (Petrak) Petrak, *Sydowia* **11**: 351 (1958).
Sclerozythia nectrioidea Petrak, *Sydowia* **9**: 539 (1955).

Conidiomata up to 150 µ diam. Conidia 5.5-10.5 × 3-4.5 µ.

On dead stems of *Astragalus* sp., Kurdistan (IMI 94447).

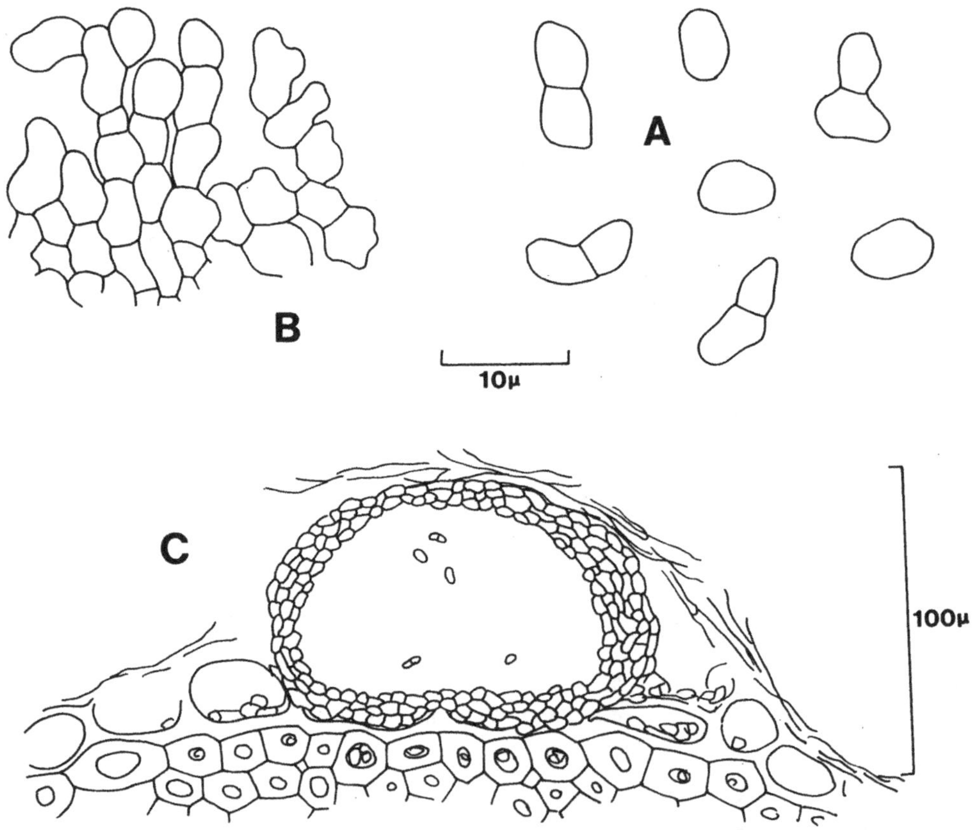

Fig. 3. *Neozythia handelii*, A, conidia; B, conidiophores; C, vertical section of unopened conidioma.

2. DOTHIOROPSIS Riedl, *Phyton (Austria)* **16**: 222 (1974).
Sp. typ.: *D. corni* Riedl.
Illustration: Riedl (1974).

Mycelium immersed, branched, septate, hyaline. *Conidiomata* eustromatic, immersed, subcuticular and peridermal, separate or confluent, brown, ± circular, longitudinal or irregular; structure diffuse, basal wall of pale brown, ± horizontally-orientated textura prismatica, lateral and upper walls of brown, variously sized textura angularis. Dehiscence by irregular rupture of the upper wall. *Conidiophores* hyaline, repeatedly branched throughout and loosely compacted, septate, formed from the basal and lateral walls. *Conidiogenous cells* holothallic, integrated, hyaline, smooth. *Conidia* arthric, formed by disarticulation of the conidiogenous cells comprising the conidiophores, produced in compacted irregularly branched chains (? acropetal or basipetal), hyaline, smooth, thin-walled, ellipsoid to oblong, often the upper layers of the conidiomata produce larger 1-septate arthroconidia.

The genus is monotypic.

D. CORNI Riedl, *ibid.*: 222 (1974).

Conidiomata 180-200 µ diam., 70-75 µ high. Conidia 6-8 × 3-4 µ (aseptate), 10-13 × 5-6 µ (1-septate).

On twigs of *Cornus sanguinea*, Austria.

3. PHACIDIELLA Karst., *Hedwigia* **23**: 85 (1884).
Ramulariospora Bub., *Ann. Naturk. Mus. Wien* **28**: 216 (1914).
Desmopatella Höhn., *Mitt. bot. Inst. tech. Hochsch. Wien* **1**: 76 (1924).
Schizothyrellina Petrak, nom. nud.
Sp. typ.: *P. salicina* Karst.

Mycelium immersed, branched, septate, hyaline. *Conidiomata* eustromatic, immersed, peridermal, separate, rarely aggregated, dark brown, at first globose and closed, later opening and splitting irregularly to become collabent to cupulate; basal wall of pale brown, thin-walled textura angularis, lateral walls of medium brown textura porrecta approximately 5 cells thick. *Conidiophores* hyaline, branched at the base, less so above, septate, formed mainly from the base but also from the lateral wall. *Conidiogenous cells* holothallic, integrated, hyaline, smooth, cylindrical. *Conidia* arthric, formed by disarticulation of the conidiogenous cells from the conidiophore, produced in long, branched chains with youngest conidia at the base, hyaline, straight, smooth, both ends truncate except apical conidia which are subacute at the apices, long doliiform, aseptate.

Desmopatella is clearly congeneric with *Phacidiella* because both are based on the same species. Although *Ramulariospora* was described with 1-septate conidia, the type only shows aseptate conidia, consequently this places it into synonymy with *Phacidiella*. *Sirozythiella* Höhn. is only maintained as distinct because its conidia are 1-septate.

KEY

1. Conidia 12-17 × 2.5 µ .... *P. asperulina*
1. Conidia 5-8 × 1.5-2 µ 2
    2. Conidiophores up to 105 µ long; conidia long doliiform .... *R. salicina*
    2. Conidiophores up to 60 µ long; conidia cylindrical .... *R. tomispora*

P. SALICINA Karst., *ibid.*: 85 (1884). (Fig. 4C, D)
Desmopatella salicis Höhn., *Mitt. bot. Inst. tech. Hochsch. Wien* **1**: 76 (1924).
Schizothyrellina moravica Petrak, *Flor. Boh. Morav. exs.* II/1. No. 1486 (nom. nud.).

Teleomorphosis: *Pyrenopeziza salicis* (Feltg.) Nannf., *Nova Act. Reg. Soc. Scient. Upsal.*, ser. IV, **8**: 160 (1932).

Conidiomata up to 300 µ diam. Conidiophores up to 105 × 2 µ. Conidia 5-8.5 × 2 µ.

On twigs of *Salix viminalis, S. caprea*, Finland, Austria (IMI 200036 ex H, holotype *P. salicina*, 31056).

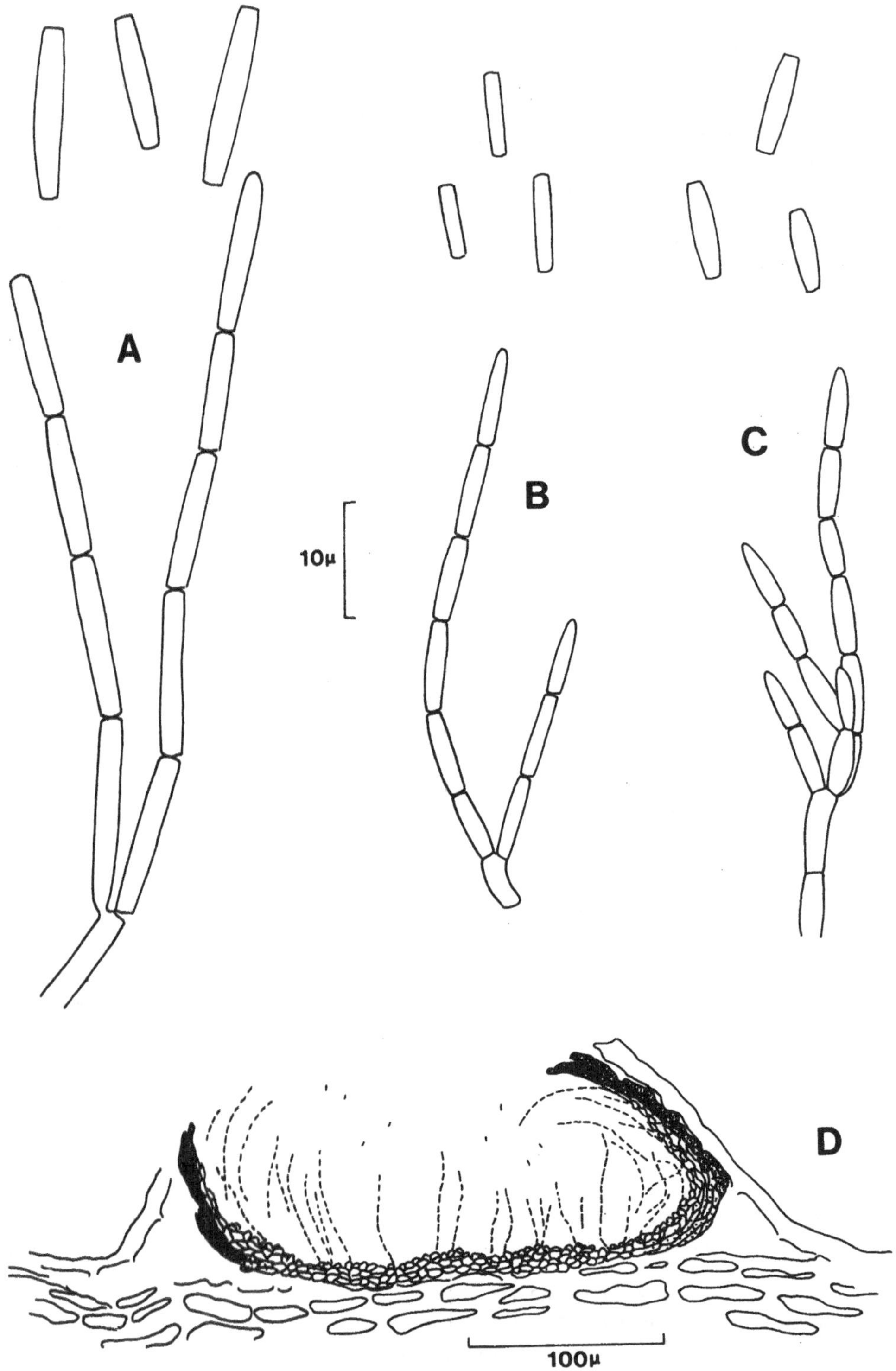

Fig. 4. A, *Phacidiella asperulina*, conidiophores and conidia; B, *P. tomispora*, conidiophores and conidia; C, *P. salicina*, conidiophores and conidia; D, *P. salicina*, vertical section of conidioma.

P. **TOMISPORA** (Berl. & Bres.) comb. nov. (Fig. 4B)
Rhabdospora tomispora Berl. & Bres., *XIV Ann. Soc. Alp. Trid.* 76 (1889).
Desmopatella tomispora (Berk. & Bres.) Petrak, *Annls mycol.* **29**: 355 (1931).

Conidiomata up to 400 µ. Conidiophores up to 60 × 1.5 µ. Conidia 5-8 × 1.5-2 µ.

On dead stems of *Artemisia vulgaris*, Sweden, Latvia (IMI 30079, 33480).

P. **ASPERULINA** (Bub.) comb. nov. (Fig. 4A)
Ramulariospora tomispora Bub., *Ann. Naturk. Mus. Wien* **28**: 216 (1914).

Conidiomata up to 200 µ diam. Conidiophores up to 80 × 2.5 µ. Conidia 12-17 × 2.5 µ.

On dead stems of *Asperula asperne*, Kurdistan (IMI 194251, holotype ex BPI).

4. SIROZYTHIELLA Höhn., *Sber. Akad. Wiss. Wien* **118**: 1532 (1909).
Sp. typ.: *S. sydowiana* (Sacc.) Höhn. (syn. *Schizothyrella sydowiana* Sacc.).

Mycelium immersed, branched, septate, hyaline to pale brown. *Conidiomata* eustromatic, initially immersed, epidermal to subepidermal, separate, dark brown, at first globose and closed, later opening and splitting irregularly to become collabent to cupulate; basal wall of pale brown textura intricata extending into the lateral walls of similar tissue, lined at the base and at the sides by smaller celled textura angularis. *Conidiophores* hyaline, branched at the base and repeatedly above, septate, formed mainly at the base and also from the lateral walls. *Conidiogenous cells* hollothallic, integrated, hyaline, smooth, cylindrical. *Conidia* arthric, formed by disarticulation of the conidiogenous cells from the conidiophore, produced in long, branched chains with the youngest conidia at the base, hyaline, straight or slightly curved, smooth, both ends truncate except apical conidia which are subacute at the apex, long doliiform, 0-2 euseptate.

Sirozythiella is monotypic and can only be distinguished from *Phacidiella* Karst. by the 1-2 septate conidia. *Siropatella* Höhn. which could have predated *Sirozythiella* appears likely to have been based on discordant elements and should be rejected.

S. **SYDOWIANA** (Sacc.) Höhn., *ibid.*: 1532 (1909). (Fig. 5)
Schizothyrella sydowiana Sacc., *Annls mycol.* **3**: 233 (1905).

Conidiomata up to 400 µ diam. Conidiophores up to 90 µ long × 1.5-3.5 µ wide. Conidia 9.5-19.5 × 2.5-3.5 µ.

On leaves and culms of *Phragmites communis*, W. Germany (IMI 205840, holotype ex PAD).

5. ACAROSPORIUM Bub. & Vleugel apud Bub., *Ber. dtsch. Bot. Ges.* **39**: 385 (1911).
Ciliosira Syd., *Annls mycol.* **40**: 212 (1942).
Chaetalysis Peyron., *Bull. trimest. Soc. mycol. Fr.* **38**: 141 (1922).
Sp. typ.: *A. sympodiale* Bub. & Vleugel.
Illustrations: Bubák (1911); White & Whetzel (1938).

Mycelium immersed, branched, septate, hyaline to pale brown. *Conidiomata* initially subepidermal, becoming erumpent, separate, rarely confluent, dark brown to black, globose to collabent or irregularly collapsed and finally cupulate, basal wall flat, thin, of thin-walled, hyaline to pale brown textura angularis, lateral and upper wall of dark brown pseudoparenchyma on the outside, becoming thinner-walled and paler towards the inside. *Ostiole* absent, dehiscence by irregular to stellate splitting of the upper wall and recurvature. *Conidiophores* hyaline, septate, dichotomously to sub-dichotomously branched, smooth, thin-walled, restricted to the basal wall. *Conidiogenous cells* holothallic, integrated, determinate, hyaline, smooth. *Conidia* arthric, formed by disarticulation of the conidial chain, produced in complex, repeatedly branched chains with the oldest conidium at the base, hyaline, medianly 1 euseptate, smooth, cylindrical, both ends truncate but eventually obtuse, with 0-3 lateral, unbranched, ± straight, attenuated, cellular appendages.

From the published accounts of *A. sympodiale* by Bubák (1911) and White & Whetzel (1938), *Chaetalysis myrioblephara* by Peyronel (1922) and *Ciliosira hederae* by Sydow (1942) it seems clear that these binomials refer to a single species. In spite of the fact that Höhnel (1920a) introduced *A. austriacum*, measurements from the type indicate that this taxon is not distinct.

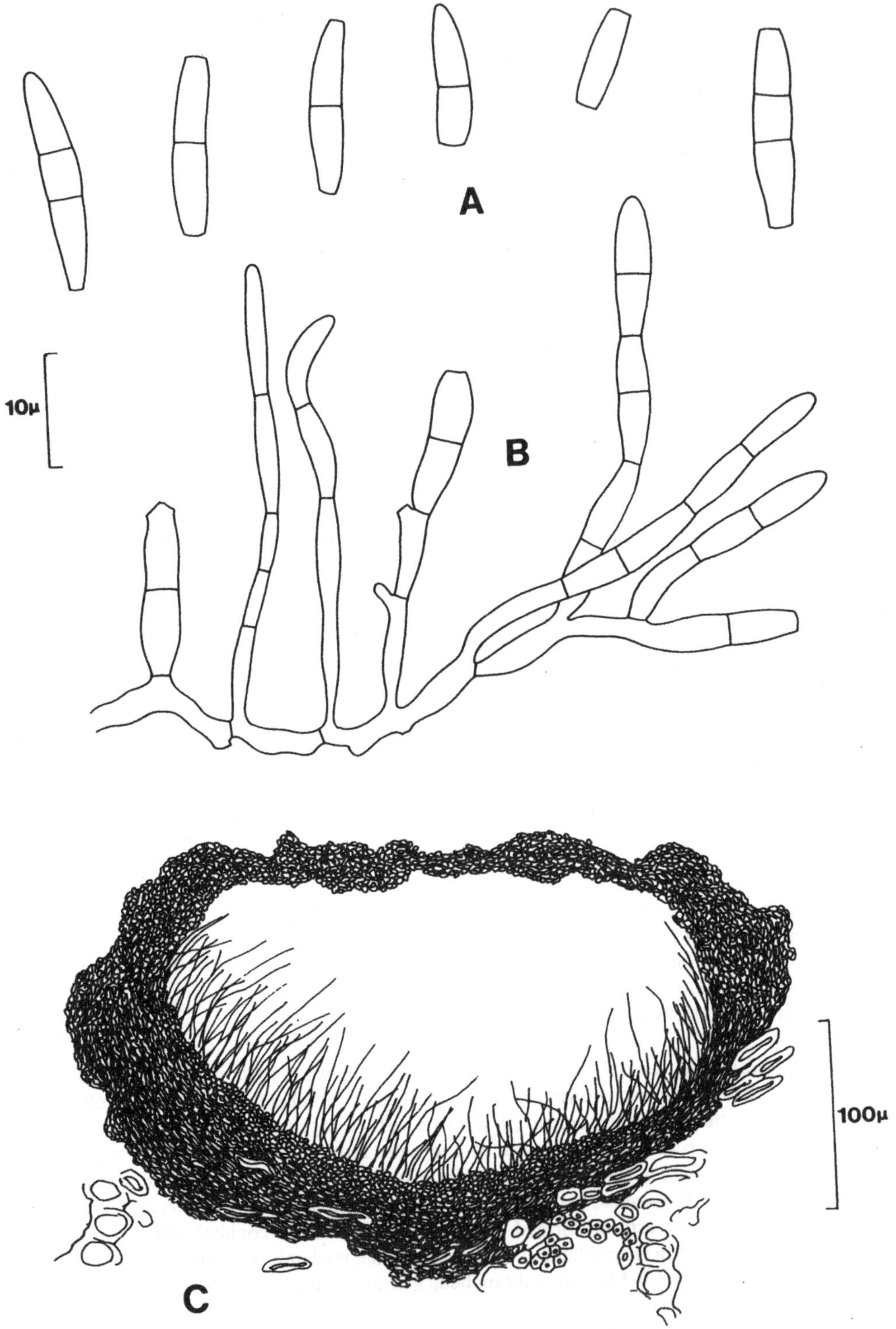

Fig. 5. *Sirozythiella sydowiana*, A, conidia; B, conidiophores; C, unopened conidioma.

White & Whetzel (1938) and Whetzel & White (1940) correlated two *Acarosporium* species with the newly introduced Helotiaceous genus *Pycnopeziza*. Certain nomenclatural errors necessitate the following amendments.

KEY

Conidia with appendages .... *A. sympodiale*
Conidia lacking appendages .... *A. quisquiliaris*

A. SYMPODIALE Bub. & Vleugel, *ibid.*: 385 (1911). (Fig. 6)
Acarosporium austriacum Höhn., *Hedwigia* **62**: 83 (1920).
Chaetalysis myrioblephara Peyron., *Bull. trimest. Soc. mycol. Fr.* **38**: 141 (1922).
Ciliosira hederae Syd., *Annls mycol.* **40**: 212 (1942).
Acarosporium hederae (Syd.) Petrak, *Sydowia* **14**: 350 (1960).

Fig. 6. *Acarosporium sympodiale*, conidiophores with attached conidia (holotype of *A. austriacum*).

Teleomorphosis: **Pycnopeziza sympodialis** White & Whetzel ex Sutton
P. sympodialis (Bub. & Vleugel) White & Whetzel, *Mycologia* **30**: 190 (1938), non rite publ.

Apothecia 3-5 mm diam., stipite .5-1.5 mm long. × 1 mm diam. Paraphyses 2-3 septatae, 75-85 × 2-5 µ. Asci 75-85 × 6.5-7.5 µ. Ascosporae 7-9 × 3-3.5 µ.
In inflorescentiis *Aceris rubri*, Malloryville Bog, Ithaca, N.Y., U.S.A., CUP 25859, holotypus.

Conidiomata 400-1000 µ diam. Conidia 10-30 × 2-3 µ, appendages 1-4, 20-29 µ long.

On buds and staminate inflorescences of *Acer rubrum*, catkins of *Alnus rugosa*, buds of *Populus candicans*, *P. tremuloides*, leaves of *Betula odorata*, fruits of *Cornus mas*, wood of *Salix caprea*, runners of *Hedera helix*, Italy, Sweden, Austria, Germany, N.Y., W.Va, U.S.A., Man., Que., Ont., Canada (Holotype *A. austriacum* ex FH).

A. QUISQUILIARIS White & Whetzel ex Sutton
A. quisquiliaris White & Whetzel, *Mycologia* **30**: 192 (1938), non rite publ.

Teleomorphosis: *Pycnopeziza pachyderma* (Rehm) Whetzel & White, *Mycologia* **32**: 617 (1940).

Conidiomata 400-1000 µ diam., solitaria, nigra, globosa, collabentes, stellatim dehiscentes. Conidia laevia, hyalina, cylindrica, 1 septata, 15-22 × 2.5-3 µ, 16-20 in catenis 2-3 ramosis dichotomis formata; sine appendices.
In gemmis *Aceris rubri*, CUP 25824, holotypus.

Conidiomata 400-1000 µ diam., solitary, black, globose, collapsed, dehiscing stellately. Conidia smooth, hyaline, cylindrical, 1 septate, 15-22 × 2.5-3 µ, formed in chains of 16-20, branched 2-3 times; lacking appendages.

On buds of *Acer rubrum* and decaying leaves, Iowa, N.Y., W.Va, U.S.A.

6. TRULLULA Ces., *Bot. Z.* **10**: 287 (1852).
Cesatia Rabenh., *Rabenh.-Klotzsch, Herb. viv. mycol. Cent. XV.* Nr. 1443 (1850), non Endl., 1838 (= *Trachymene*).
Trullula Ces. subgen. *Cesatia* (Rabenh.) Sacc., *Syll. fung.* **3**: 732 (1884).
Hormodochium (Sacc.) Höhn., *Sber. Akad. Wiss. Wien* **120**: 465 (1911).
Hormodochium (Sacc.) Trotter in Sacc., *Syll. fung.* **25**: 987 (1931).
Epidochium Fr. subgen. *Hormodochium* Sacc., *Syll. fung.* **4**: 749 (1886).
Hormodochis Clements, *Genera of Fungi*: 163 and 176 (1909).
Original spp.: *T. oreoselini* Ces., *T. leguminum* Ces. a) *spartii* Ces., *T. leguminum* Ces. b) *robiniae* Ces.
Lectotype sp.: *T. oreoselini* Ces.
Illustration: Pirozynski & Morgan-Jones (1968).

Mycelium immersed, branched, septate, hyaline to pale brown. *Conidiomata* eustromatic, scattered or gregarious, at first immersed, later erumpent, sugblobose or flattened and discoid, dark brown to black, thick-walled, eventually becoming cupulate; wall at the base of pale brown to subhyaline, thin-walled textura angularis, at the sides and above of dark brown, thick-walled textura angularis to intricata. *Ostiole* absent, dehiscence by irregular rupture of the overlying wall. *Conidiophores* cylindrical, hyaline, branched, septate near the base and above, formed from the inner cells of the conidiomatal wall. *Conidiogenous cells* holothallic, integrated, hyaline, smooth, cylindrical. *Conidia* arthric, formed by disarticulation of the conidiogenous cells from the conidiophores, produced in long, branched chains with the youngest conidia at the base, regularly oblong, truncate at each end, pale brown, smooth, often guttulate, aseptate.

Although a full treatment of the genus was provided by Höhnel (1915a) and several taxa referred to other genera, neither he, nor Clements & Shear (1931) chose the correct lectotype species. *Trullula olivascens* (Sacc.) Sacc. was not one of the two original species assigned to the genus so it cannot serve as a lectotype. Sutton (1977a) suggested *T. oreoselini* Ces. as the lectotype species since it was the first listed in the protologue, but the species has not been re-examined, so the characteristics of the genus are still largely unknown. The only modern account of species is by Pirozynski & Morgan-Jones (1968) and it is from this description and illustration that the generic criteria are taken. If, however, *T. oreoselini* eventually proves to be different generically from *T. olivascens* then an alternative generic name would have to be found for the latter species.

So far, more than 35 species have been described in the genus and bearing in mind the aforementioned problems over identity of the lectotype species, it seems clear that a detailed revision of the genus is required.

T. MELANOCHLORA (Desm.) Höhn., *Sber. Akad. Wiss. Wien* **124**: 97 (1915).
Epidochium melanochlorum Desm., *Ann. Sci. nat.*, ser. 3, **16**: 327 (1848).
Hormodochium melanochlorum (Desm.) Höhn., *Sber. Akad. Wiss. Wien* **120**: 465 (1911).
Hormococcus olivascens Sacc., *Michelia* **1**: 94 (1877).
Trullula olivascens (Sacc.) Sacc., *Syll. fung.* **3**: 731 (1884).
Sirozythia olivacea Höhn., *Sber. Akad. Wiss. Wien* **119**: 653 (1910).
Hormodochium olivaceum (Höhn.) Höhn., *Sber. Akad. Wiss. Wien* **120**: 466 (1911).

Conidiomata 230-300 µ wide × 175-200 µ high. Conidiophores 18-47 × 1.5-2 µ. Conidia 3-6.5 × 1.5-2.5 µ.

Described and illustrated by Pirozynski & Morgan-Jones (1968) under the name *T. olivascens* (Sacc.) Sacc. On cones of *Abies*, Italy (IMI 124616 ex K).

7. PHRAGMOTRICHUM Kze in Kze & Schm., *Mykol. Hefte* **2**: 84 (1823).
Alysisporium Peyronel, *Bull. trimest. Soc. mycol. Fr.* **28**: 140 (1922).
Seiridiella Karst., *Meddn. Soc. Fauna Flora fenn.* **18**: 67 (1892).
Taeniophora Karst., *Meddn. Soc. Fauna Flora fenn.* **13**: 163 (1886).
Sp. typ.: *P. chailletii* Kunze.

Mycelium immersed, brown, branched, septate. *Conidiomata* eustromatic, immersed to semi-immersed, separate, sometimes gregarious, dark brown to black, at first closed, later opening to become disciform to cupulate; upper wall composed of thick-walled, dark brown textura angularis, periclinal and basal wall composed of similar but thin-walled tissue. *Ostiole* absent, dehiscence by irregular breakdown of the upper wall. *Conidiophores* hyaline, branched only at the base, septate, smooth, cylindrical, formed from the base of the conidioma. *Conidiogenous cells* holothallic, probably meristematic, integrated, hyaline, smooth, cylindrical, producing a chain of conidia the oldest of which is at the apex. *Conidia* brown, concolourous or versicolourous, 1- many transverse and longitudinal eusepta, some species lacking longitudinal septa, constricted or continuous, apices and bases truncate except for the apical conidium which is obtuse at the apex, thin or thick-walled, smooth, eguttulate, straight or curved, cymbiform, fusiform or ellipsoidal, sometimes muriform with median projections.

Sutton & Pirozynski (1965) redescribed the genus with three accepted species and Sutton & Sandhu (1969) added a fourth. Morelet (1968a) introduced *P. complanatum* (Nees ex Fr.) Morelet (syn. *Didymosporium complanatum* Nees ex Fr.) and *P. fresenii* Morelet nom. nov. (syn. *Didymosporium complanatum* Fresenius, non Nees ex Fr.). The species to which these names refer are only tenuously related to *Phragmotrichum*. *D. complanatum* Nees ex Fr., illustrated by Nees & Henry (1837) has 1-septate hyaline conidia and their development does not appear to be thallic. Until the ontogeny is precisely established no redisposition of the species can be made. *D. complanatum* Fresenius was described as related to *Coniothecium* Cda and from the illustration would seem best referred to *Trimmatostroma* Cda since it shows the irregularity in conidial shape and development that characterise this genus.

KEY

1. Conidia lacking longitudinal septa 2
1. Conidia with longitudinal septa 3
  2. Conidia irregular in shape, dark reddish brown, 18-36 × 9-11.5 µ, (0-) 3 septate .... *P. pini*
  2. Conidia ellipsoid, pale brown, 17-32 × 5.5-8 µ, (2-) 3 (-7) septate .... *P. rivoclarinum*
3. Conidia ellipsoid, pale brown, 16.5-34 × 6.5-12 µ, (2-) 3-5 (-8) septate .... *P. platanoidis*
3. Conidia versicoloured 6- pointed, muriform, 24-35 × 14.5-20 µ .... *P. chailletii*

P. CHAILLETII Kze in Kze & Schm., *Mykol. Hefte* **2**: 84 (1823). (Fig. 7C, D)
Gymnosporangium chailletii (Kze) Sprengel, *Syst. Veg.* Ed. 16, **4**: 562 (1827).

Conidiomata 350-550 µ diam. Conidia 24-35 × 14.5-20 µ, muriform, apical and basal cells hyaline, median cells longitudinally divided, brown.

On cones scales of *Picea excelsa*, U.S.S.R., Switzerland, Romania (IMI 22666, 24783, 44725).

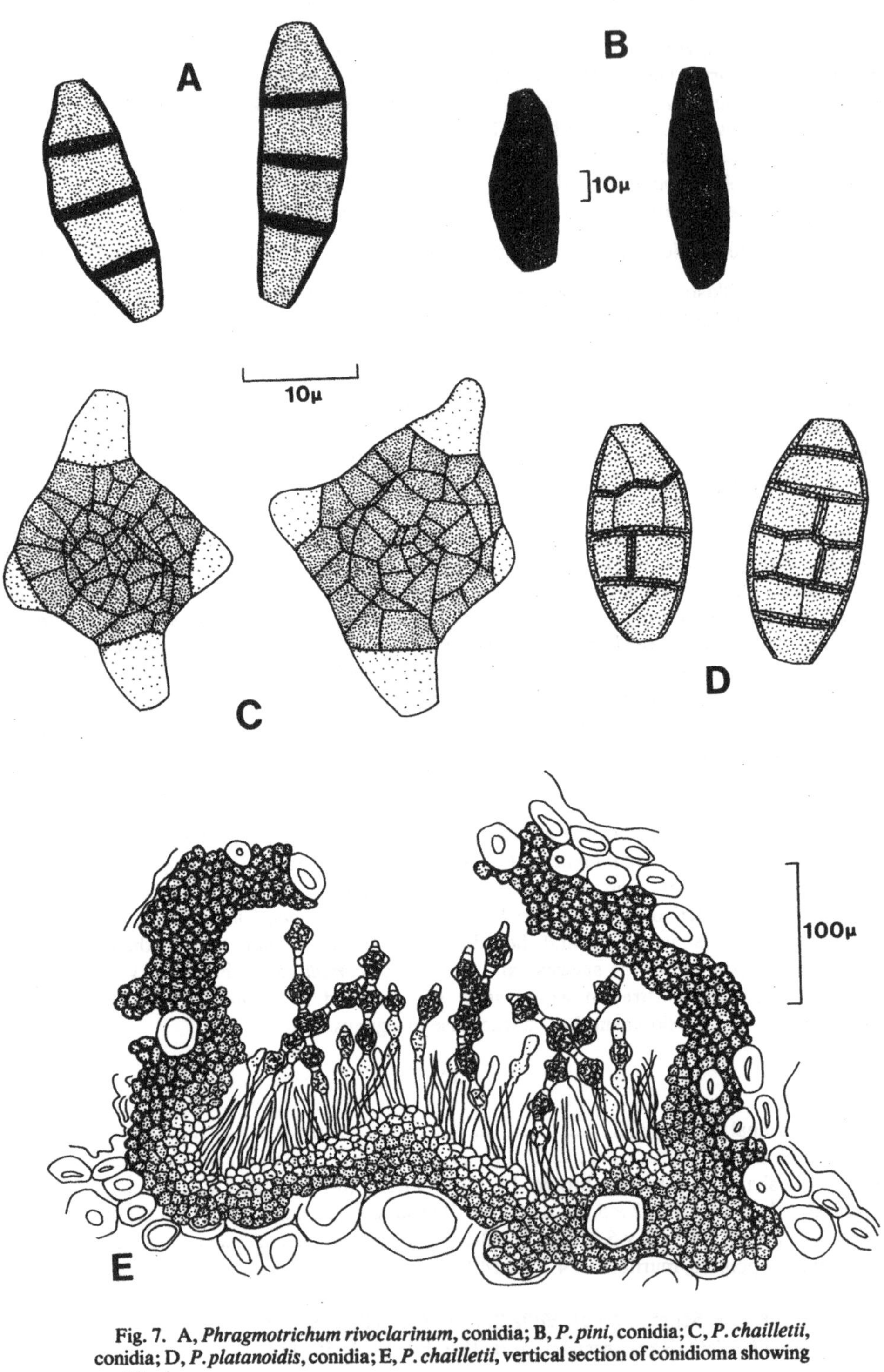

Fig. 7. A, ***Phragmotrichum rivoclarinum***, conidia; B, ***P. pini***, conidia; C, ***P. chailletii***, conidia; D, ***P. platanoidis***, conidia; E, ***P. chailletii***, vertical section of conidioma showing conidiophores.

P. PINI (W.B. Cooke) Sutton & Sandhu, *Trans. Br. mycol. Soc.* **52**: 67 (1969). (Fig. 7B)
Trimmatostroma pini W.B. Cooke apud Cooke & Shaw, *Mycologia* **44**: 808 (1952).

Conidiomata 300-500 µ diam. Conidia 18-36 × 9-11.5 µ, 0-3 septate, dark reddish brown.

The species is fully illustrated in Sutton & Sandhu (1969).
On twigs of *Pinus monticola*, Wash., U.S.A. (WINF(M) 3409, holotype).

P. PLATANOIDIS Otth, *Mitt. naturf. Ges. Bern* 723: 111 (1870). (Fig. 7D)
Phragmotrichum spiraeae Vestergr., *Ofv. Vet. Akad. Forh., Stockh.* **10**: 46 (1897).
Phragmotrichum flageoletianum Sacc., *Annls mycol.* **6**: 560 (1908).

Conidiomata up to 300 µ diam. Conidia 16.5-34 × 6.5-12 µ, with several longitudinal eusepta and 3-5 (2-8) transverse septa.

On twigs of *Acer platanoides*, *Spiraea*, *Carpinus betulus* and *Sambucus nigra*, Switzerland, Sweden, France, Germany (IMI 108390, holotype *P. platanoidis* ex BERN, 108388, holotype *P. spiraeae* ex UPS, 108389, holotype *P. flageoletianum* ex PAD, 106836 ex B).

P. RIVOCLARINUM (Peyronel) Sutton & Pirozynski, *Trans. Br. mycol. Soc.* **49**: 522 (1966). (Fig. 7A)
Alysisporium rivoclarinum Peyron., *Bull. trimest. Soc. mycol. Fr.* **28**: 140 (1922).
Taeniophora acerina Karst., *Meddn. Soc. Fauna Flora fenn.* **13**: 163 (1886), non *Phragmotrichum acerinum* Fr., *Summa Veg. Scand.* 474 (1846).
Phragmotrichum karstenii Sutton & Pirozynski, *Trans. Br. mycol. Soc.* **48**: 359 (1965).
Septotrullula bacilligera Höhn. var. *cambrica* Grove & Rhodes apud Grove, *J. Bot. Lond.* **70**: 33 (1932).
Seiridiella ramealis Karst., *Meddn. Soc. Fauna Flora Fenn.* **18**: 67 (1892).

Conidiomata up to 300 µ diam. Conidia (2-) 3 (-7) septate, 17-32 × 5.5-8 µ.

On twigs of *Acer platanoides*, *Alnus glutinosa*, *Salix capraea* and catkins of *Alnus*, Finland, G.B., Italy (IMI 104795, holotype *T. acerina* ex H, 45140, holotype *S. bacilligera* var. *cambrica* ex K, 104976, holotype *A. rivoclarinum* ex TO, 73430a, 54096).

8. STANINWARDIA Sutton, *Trans. Br. mycol. Soc.* **57**: 540 (1971).
Sp. typ.: *S. breviuscula* Sutton.

Mycelium immersed, pale brown, branched, septate. *Conidiomata* acervular, epidermal, separate, black, formed of pale brown, smooth-walled textura angularis. *Dehiscence* irregular. *Conidiophores* hyaline to pale brown, branched only near the base, septate, smooth or verruculose, consisting of individual conidiogenous cells enclosed in a mucilaginous sheath, meristematic at the base, formed from the upper pseudoparenchyma. *Conidiogenous cells* holothallic, arthric, integrated, indeterminate, ± cylindrical, hyaline or pale brown, verruculose, the oldest at the apex, youngest at the base. *Conidia* formed in simple unbranched chains of up to 3 conidia, pale brown, thick-walled, verruculose, doliiform to clavate, medianly 1-euseptate and constricted, truncate except for the apical conidium which is rounded at the apex, with a mucilaginous sheath.

The genus is monotypic.

S. BREVIUSCULA Sutton, *ibid.*: 540 (1971). (Fig. 8)

Conidiomata 50-290 µ diam. Conidia 7-10.5 × 3.5-5 µ; sheath 2 µ thick.

Causing chocolate leaf spot of *Eucalyptus* spp., Mauritius, Sri Lanka, India (IMI 139473 (holotype), 173404, 174792, 174820).

9. BARNETTELLA Dev Rao & Raghuveer Rao, *Mycopath. Mycol. appl.* **22**: 56 (1964).
Vinculum Roy, Dwivedi & Khanna, *Trans. Br. mycol. Soc.* **48**: 113 (1965).
Sp. typ.: *B. speciosa* Dev Rao & Raghuveer Rao.

Mycelium immersed or superficial, branched, septate, pale brown to hyaline. *Conidiomata* stromatic, superficial, large, circular to effuse, brown, consisting of a basal region of thin-walled, pale brown textura angularis, and an upper

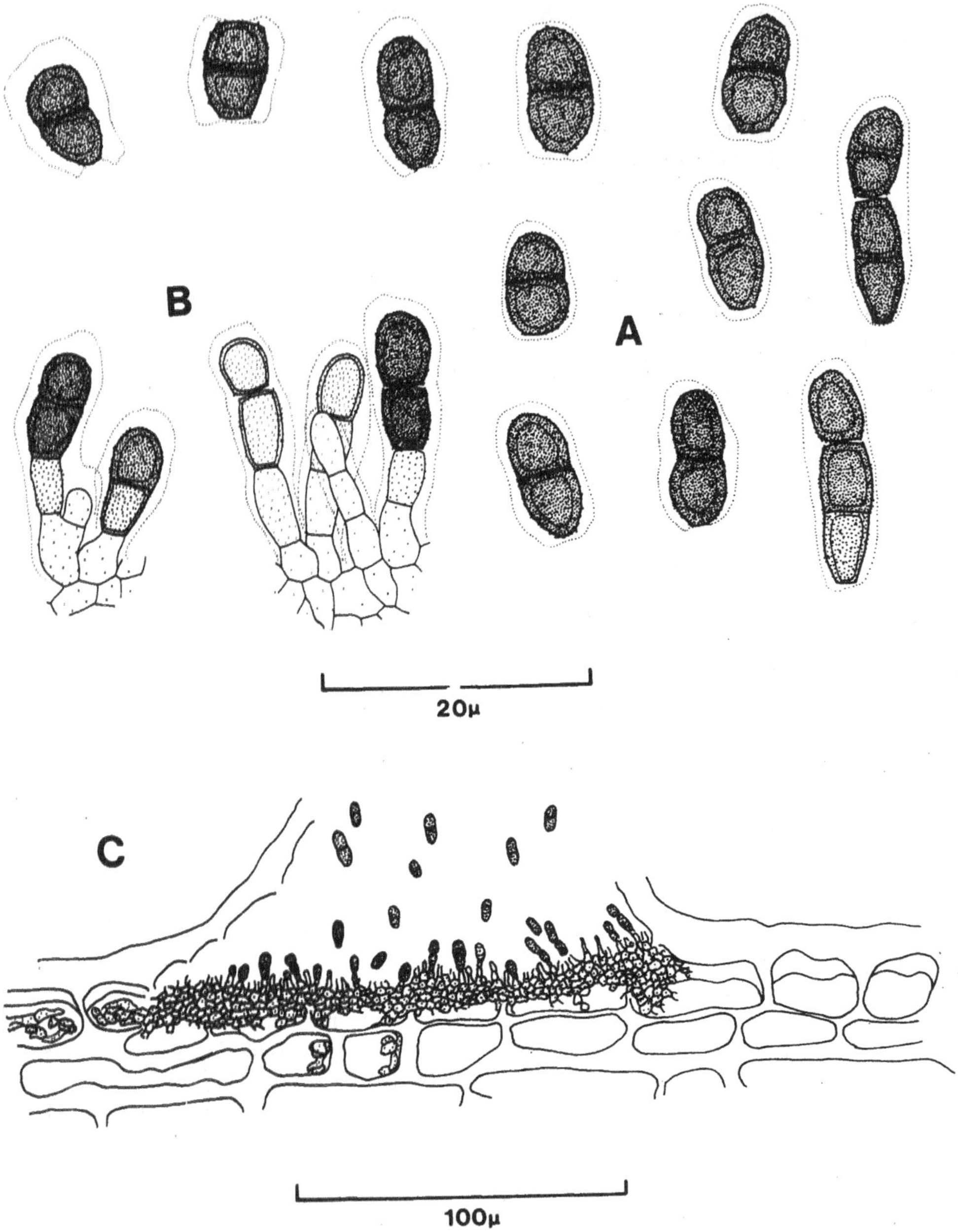

Fig. 8. *Staninwardia breviuscula*, A, conidia; B, conidiophores; C, vertical section of conidioma.

region in which several pseudolocular areas are separated by darker brown, thicker-walled, vertically elongated pseudoparenchyma; pseudolocular tissue of isodiametric thin-walled pseudoparenchyma. Dehiscence by breakdown of upper tissues. *Conidiophores* macronematous or semi-macronematous, meristematic, unbranched, straight, pale brown, smooth, formed from the upper cells of the pseudolocular tissue. *Conidiogenous cells* holothallic, integrated, intercalary, fragmenting. *Conidia* formed in long, unbranched, basipetal chains, schizogenous, dry, dark brown, muriform, smooth, connected by several disjunctor cells to the preceding and subsequent conidia.

The structure of the conidiomata and method of conidial development are most unusual. Roy, Dwivedi & Khanna (1965) placed *Vinculum* in the Tuberculariaceae and Dev Rao & Raghuveer Rao (1964) assigned *Barnettella* to the

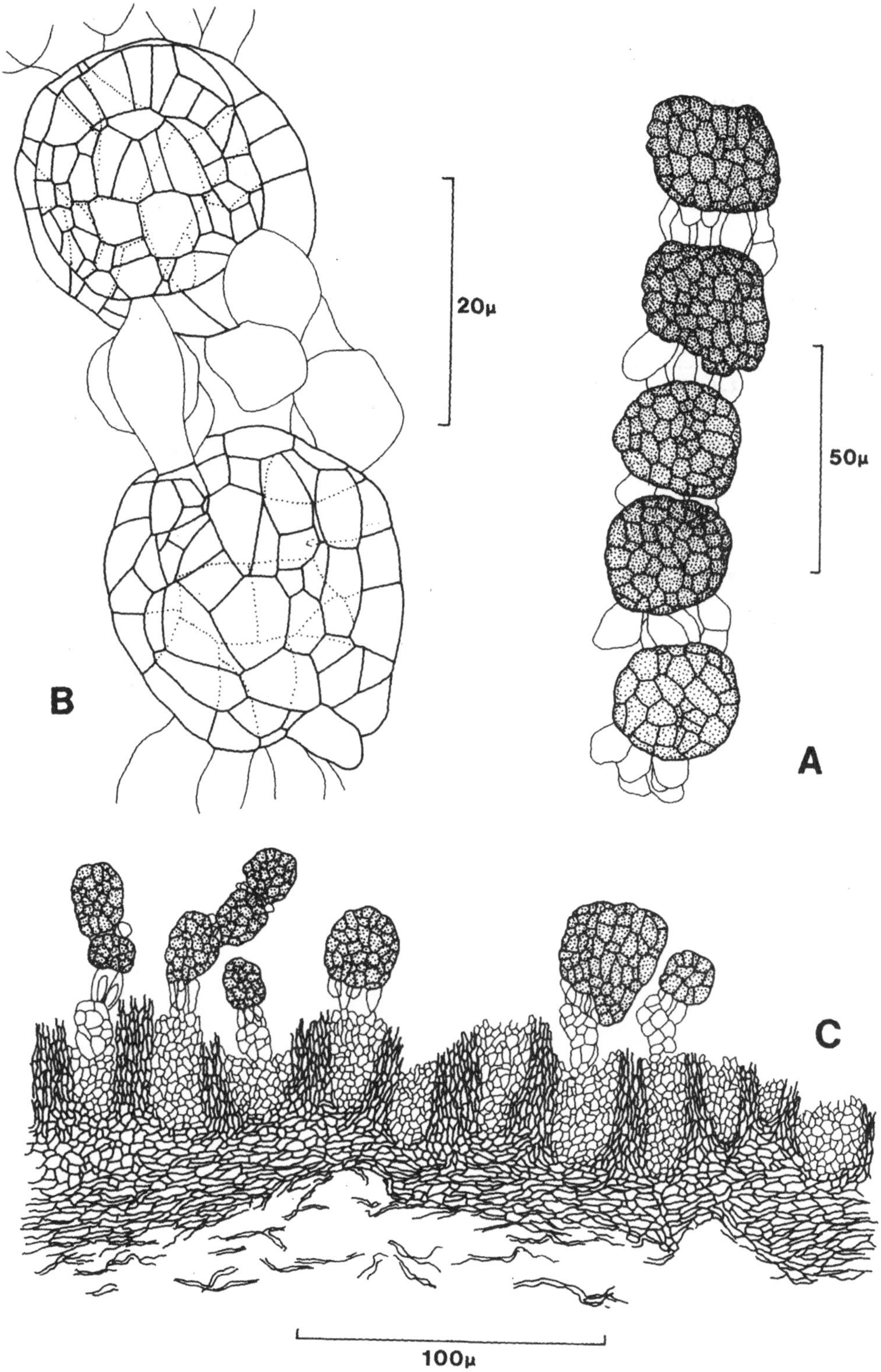

Fig. 9. *Barnettella indica*, A, B, conidial chains; C, vertical section of conidioma from culture.

Melanconiaceae. The conidioma is however not typically acervular, for it is superficial, neither is it really strictly sporodochial for there are certainly pseudolocular cavities in it. Although it is here considered to be a stroma, there is no specialized mode of dehiscence and conidia are never covered by overlying tissue. The genus probably represents an intermediate between the Tuberculariaceae and the stromatic coelomycetes. Conidial development is best interpreted as similar to that in *Coniosporium* Lk ex Fr. and *Trimmatostroma* Cda but differing inasmuch as the connective or disjunctor between conidia is multicellular. As the conidial chain primordium becomes differentiated, certain cells are destined to become conidia whilst others do not divide or develop pigment but become the connectives between the conidia. In immature stages the connective and conidial cells cannot be distinguished.

Dev Rao & Vasant Rao (1973) provided a review of *Barnettella* and accepted 3 species separated as follows.

KEY

Conidia ± cylindrical, l/b ratio 3:1 .... *B. speciosa*
Conidia ± globose, l/b ratio 1:1 .... *B. indica*
Conidia ± flattened, l/b ratio 3-4:1 .... *B. narasimhanii*

B. INDICA (Roy et al.) Raghuveer Rao, *Mycopath. Mycol. appl.* **45**: 268 (1971). (Fig. 9)
Vinculum indicum Roy et al., *Trans. Br. mycol. Soc.* **48**: 114 (1965).

Conidiomata up to 700 μ diam.; pseudolocules 20-25 μ wide. Conidia 24-29 × 27-28 μ; disjunctors 10-15 μ long.

From Gramineae, *Zea mays*, India (IMI 103904, holotype *V. indicum*, 183839).

BLASTOPYCNIDIINEAE subord. nov.

Cellulae conidiogenae holoblasticae; conidiomata pycnidialia.
Conidiogenesis holoblastic; conidiomata pycnidial.

This suborder contains a large number of genera and it is convenient to break them down into three groups based on the inability or ability of the conidiogenous cells to produce more than one conidium. The simplest type of conidiogenous cell forms a single conidium. Some produce a plurality of conidia by percurrent proliferation and as such are annellides, whilst others develop several conidiogenous loci either synchronously or sympodially.

It is difficult to recommend a formal taxonomic rank for the three groups since they should be placed between suborder and family. Although the *International Code of Botanical Nomenclature* allows supplementary ranks to be intercalated between those more formally recognised, there are no recommendations concerning the suffixes applicable to supplementary ranks. In Zoological nomenclature (Jeffrey, 1977) the superfamily has a recommended suffix of '-oidea' but unfortunately in Botanical nomenclature the suffix for subfamily is '-oideae', and to employ the former for superfamilies may cause unnecessary confusion. Consequently the three types of conidiogenesis in this suborder are keyed as follows into informal groups.

KEY TO GROUPS WITHIN THE BLASTOPYCNIDIINEAE

Conidiogenous cells holoblastic, monoblastic .... p. 41
Conidiogenous cells holoblastic, annellidic .... p. 95
Conidiogenous cells holoblastic, sympodial or synchronous .... p. 88

BLASTOPYCNIDIINEAE, MONOBLASTIC

In this group there are certain well-defined associations of genera. The most clear-cut is the '*Kellermania – Piptarthron – Alpakesa*' complex where the division between genera could arbitrarily be drawn on the basis of either conidial septation or number and presence of appendages. A simpler solution would be to include the taxa at present disposed in three seperate genera in a single genus. The earliest name would be *Kellermania*.

Genera with fairly large hyaline septate conidia such as *Apiocarpella*, *Dearnessia* and *Aristastoma* are again closely related, distinctions being made by presence or absence of setae on the pycnidia and size and shape of conidia. Were species to be grown in culture the presence of setae may not be a sufficiently stable character to maintain *Aristastoma* separate.

Several genera are characterized by conidia with mucilaginous appendages derived from the splitting or eversion of a sheath initially completely enclosing the developing conidium. *Tiarosporella*, *Tiarospora* and *Neottiosporina* are only separated by the septation of conidia and form a very natural group of mainly monocotyledonous-inhabiting species.

DICHOTOMOUS KEY

1. Conidia with a mucilaginous appendage at the apex and/or base 2
1. Conidia with cellular appendage(s) at the apex and/or base 3
1. Conidia lacking appendages 11
   2. Conidia hyaline, aseptate, thin-walled; appendages apical, infundibuliform or tentacular .... *Tiarosporella* (12)
   2. Conidia hyaline, medianly 1 septate, thick-walled; appendages apical and basal, cap-like, convoluted, rarely tentacular .... *Tiarospora* (14)
   2. Conidia hyaline, 1(2-3) septate, thin-walled; appendages apical, only occasionally basal, infundibular to cap-like .... *Neottiosporina* (15)
3. Pycnidia setose, thin-walled; conidia hyaline, 4 septate, apical cell with 3-4 cellular, 1-2 septate, unbranched appendages .... *Chaetospora* (20)
3. Pycnidia not setose 4
   4. Conidial appendages at the apex and the base or just above the base 5
   4. Conidial appendages restricted to the apex 6

5. Conidia hyaline, aseptate, fusiform, with 1(-2) apical, branched or unbranched appendage(s); paraphyses with irregularly lobed apices .... *Mycotribulus* (16)
5. Conidia hyaline, medianly 2 septate, apical and basal cells modified as long filiform appendages; conidiophores filiform, branched, septate, loci terminal and lateral below septa .... *Scopaphoma* (18)
5. Conidia hyaline, aseptate, cylindrical, with 2 unbranched subapical appendages, one suprabasal appendage, and one basal appendage formerly attached to the conidiogenous cell .... *Monodia* (17)
   6. Conidia aseptate 7
   6. Conidia septate 8
7. Conidiogenous cells lageniform; conidia hyaline, with 3-7 fine extracellular setulae at the apex .... *Giulia* (13)
7. Conidiogenous cells doliiform; conidia hyaline, with several apical filiform appendages; pycnidia very thick-walled .... *Alpakesa* (10) (see also 10)
   8. Apical appendage single 9
   8. Apical appendages several 10
9. Apical appendage simple, rarely branched; conidia hyaline, 1-2 septate; pycnidia very thick-walled .... *Kellermania* (9)
9. Apical appendage simply a cellular extension of the apical cell; conidia brown, occasionally versicoloured, multiseptate, sometimes with longitudinal septa .... *Scolecosporiella* (32)
9. Apical appendage irregularly branched; conidia hyaline, 2 septate .... *Japonia* (19)
   10. Conidia brown, multiseptate; appendages corniform, dichotomous or distinctly branched along one side .... *Labridella* (23)
   10. Conidia hyaline, with several filiform appendages; conidiogenous cells doliiform; pycnidia very thick-walled .... *Alpakesa* (10) (see also 7)
   10. Conidia pale brown; apical appendages 2-4, arising very close to the point of origin; conidiophores filiform .... *Hyalotiella* (21)
   10. Conidia pale brown; apical appendages 2-4 which each form 2-4 branches shortly above the point of origin; conidiogenous cells lageniform to short cylindrical .... *Hyalotiopsis* (22)
11. Conidia aseptate, hyaline; conidiophores repeatedly branched, septate; conidiomata pale yellow, globose at the base, with a long neck .... *Hyalopycnis* (11)
11. Conidia aseptate, brown, with a transverse, hyaline band, apex hyaline, mammillate; conidiogenous cells doliiform; conidiomata superficial .... *Capnodiastrum* (24)
11. Conidia hyaline, septate 12
11. Conidia brown, septate 17
   12. Conidia 1 septate 13
   12. Conidia multiseptate 14
13. Pycnidia superficial, colonies setose; conidia cylindrical, medianly 1 septate .... *Chaetodiplodina* (1)
13. Pycnidia immersed, glabrous; conidia clavate, septum towards the base .... *Apiocarpella* (2)
   14. Pycnidia setose .... *Aristastoma* (4)
   14. Pycnidia with an elevated clypeus .... *Cytostagonospora* (7)
   14. Pycnidia glabrous 15
15. Conidia with 2-3 primary branches and 2 secondary branches, each branch 1-3 septate .... *Dendroseptoria* (5)
15. Pycnidia thin-walled; conidia 2-8 septate .... *Dearnessia* (3)
15. Pycnidia thick-walled 16
   16. Conidiogenous cells ampulliform to lageniform; conidia 3-4 septate, filiform .... *Jahniella* (6)
   16. Conidiogenous cells doliiform; conidia 0-4 septate, often constricted .... *Piptarthron* (8)
17. Conidia with 3 transverse eusepta only .... *Sclerostagonospora* (31)
17. Conidia 1 septate 18
17. Conidia with longitudinal septa 19
   18. Septum median, conidia smooth; conidiophores filiform, septate, branched; pycnidia thick-walled .... *Diplodia* (25)
   18. Septum dividing a smaller basal cell; conidiogenous cells filiform, swollen at the base; pycnidia thin-walled .... *Kamatella* (26)
19. Conidia distoseptate .... *Camarosporellum* (30)
19. Conidia euseptate 20
   20. Conidia vertically flattened, reddish brown .... *Nummospora* (27)
   20. Conidia spherical .... *Stigmella* (29)
   20. Conidia with swollen hyaline apical cell, rest up to 7 septate with longitudinal septa .... *Benjaminia* (28)

SYNOPTIC KEY

Mycelial characters

a. superficial 1 17 19 20 23 24
b. immersed 2 3 4 5 6 7 8 9 10 11 12 13 14 15 16 18 21 22 25 26 27 28 29 30 31 32
c. setose 1

Conidiomatal characters

Position in relation to substrate
a. superficial 1 19 23 24
b. immersed 2 3 4 5 6 7 8 9 10 12 13 14 15 16 17 18 20 21 22 25 26 27 28 29 30 31 32
c. semi-immersed 11 17 18 23

Dehiscence
a. ostiolate, nonpapillate 1 2 3 4 5 6 7 8 9 10 11 12 13 14 15 17 18 20 21 25 26 27 28 29 30 31 32
b. ostiolate, papillate 6 11 12 13 14 15 17 18 21 25 26 27 30 31 32
c. ostiolate, depressed 7
d. linear-longitudinal 23
e. stellate 24

Wall thickness
a. thin-walled (1-3 cells wide) 1 2 3 4 5 11 15 17 20 24 26 27 29 30 31 32
b. thick-walled (5 cells wide or more) 6 7 8 9 10 12 13 14 16 18 19 21 22 23 25 28

Pigmentation
a. dark brown 1 2 3 4 5 6 7 8 9 10 12 13 14 15 16 17 18 19 20 21 22 23 24 25 26 27 28 29 30 31 32
b. very pale brown (hypocreaceous) 11

Other characters
a. setose 4 20
b. clypeate 7 13
c. rostrate 11 12 14 15 17

Conidiogenous cell characters

Insertion
a. discrete 1 2 3 4 5 6 7 8 9 10 12 13 14 15 20 22 23 24 26 27 28 29 30 31 32
b. integrated 11 16 17 18 19 21 25

Shape

a. ampulliform-doliiform 1 2 3 4 8 9 10 14 15 24 27 28 29 30 31 32

b. lageniform 3 6 7 12 13 20 26 27

c. cylindrical 5 22 23

Conidial characters

Pigmentation

a. hyaline 1 2 3 4 5 6 7 8 9 10 11 12 13 14 15 16 17 18 19 20

b. brown 14 21 22 23 24 25 26 27 28 29 30 31 32

c. versicoloured 21 22 23 28

d. median hyaline band 24

Septation

a. distoseptate 30

b. euseptate 1 2 3 4 5 6 9 14 15 18 19 20 21 22 23 24 25 26 27 28 29 31 32

c. aseptate 7 8 10 11 12 13 16 17

d. medianly 1 septate 1 4 9 10 14 18 25

e. unequally 1 septate 2 15 24 26

f. 2-multiseptate 3 4 5 6 7 8 9 10 15 19 20 21 22 23 31 32

g. muriform 4 27 28 29 30 32

Shape

a. oval-ellipsoid 4 11 12 14 25 29 30

b. cylindrical 1 3 4 8 9 10 13 17 18 20 22 28 31

c. fusiform 4 8 9 10 15 16 19 20 21 22 23 24 32

d. pyriform 2 12 15 26

e. filiform 6 7

f. branched 5

g. flattened 27

Ornamentation

a. verruculose 28 31

b. smooth 1 2 3 4 5 6 7 8 9 10 11 12 13 14 15 16 17 18 19 20 21 22 23 24 25 26 27 29 30 32

Appendages

a. apical 9 10 12 13 14 15 16 17 18 19 20 21 22 23 32

b. basal 14 16 17 18

c. single 9 12 14 15 16 17 18 19 23 32

d. several 10 13 16 17 20 21 22

e. cellular 9 10 13 16 17 18 19 20 21 22 23 32

f. simple 9 10 13 16 17 18 20 21 32

g. branched 9 19 22 23

h. mucilaginous 12 14 15

i. infundibular 12 15

j. tentacular 12

Guttulation

a. guttulate 2 4 5 8 9 18 26

Paraphysis characters

Paraphyses present 16

1. CHAETODIPLODINA Speg., *An. Mus. Nac. Buenos Aires* **20**: 368 (1910).
Sp. typ.: *C. graminicola* Speg.

Mycelium superficial, branched, septate, brown, with erect, straight or curved, tapered, septate, smooth, brown setae. *Conidiomata* pycnidial, superficial, separate, brown, globose, unilocular, thin-walled; wall 1-2 cells thick, of brown thin-walled textura angularis. *Ostiole* central, circular. *Conidiophores* absent. *Conidiogenous cells* holoblastic, discrete, doliiform, thin-walled, hyaline, lacking any proliferations, formed from the inner cells of the pycnidial wall. *Conidia* hyaline, medianly 1 euseptate, thin-walled, smooth, eguttulate, ± straight, obtuse at the apex, slightly tapered to a truncate base.

Petrak & Sydow (1935) maintained *Chaetodiplodina* as a good genus. A second species was added by Batista & Vital (1957) but has not been re-examined.

C. GRAMINICOLA Speg., *ibid.*: 368 (1910). (Fig. 10)

Conidiomata up to 100 µ diam.; setae up to 180 µ long × 10 µ wide at the base. Conidiogenous cells 5-6 × 6.5-8.5 µ wide. Conidia 30-34 × 4.5-5.5 µ.

On leaves of *Panicum laxum*, Argentina (IMI 196761 ex LPS, holotype).

2. APIOCARPELLA H. & P. Syd. nom. nov., *Annls mycol.* **17**: 43 (1919).
Apiosporella Speg., *An. Mus. Nac. Buenos Aires* **20**: 364 (1910), non Speg., 1912, nec Höhn., 1909.
Ascochyta Lib. subgen. *Apiocarpella* (H. & P. Syd.) Melnik, *Mikol. i Fitopatol.* **5**: 21 (1971).
Sp. typ.: *A. macrospora* (Speg.) H. & P. Syd. (syn. *Apiosporella macrospora* Speg.).

Mycelium immersed, pale brown to hyaline, branched, septate. *Conidiomata* pycnidial, separate, globose, brown, immersed, unilocular, thin-walled, of textura prismatica. *Ostiole* central, circular. *Conidiophores* absent. *Conidiogenous cells* holoblastic, determinate, discrete, doliiform to ampulliform, hyaline, smooth, formed from the inner cells of the pycnidial wall. *Conidia* hyaline, 1 euseptate, separating two unequal cells, continuous, apex obtuse, tapered to a truncate base, thin-walled, smooth, clavate, guttulate.

Petrak (1952b) regarded *A. macrospora* as a *Diplodina* or a *Stagonospora* species, but Melnik (1971) preferred to include it as a subgeneric taxon of *Ascochyta* Lib. As interpreted by Sprague (1950) the genus is heterogeneous because species with enteroblastic phialidic conidiogenesis are included with *A. macrospora* which is holoblastic. Species to be removed are *A. agropyri* Sprague and *A. minor* Sprague. Later Melnik (1976) accepted the generic name for 8 species including *A. macrospora* and extended the range to include species on Ranunculaceae and Leguminosae as well as Gramineae.

A. MACROSPORA (Speg.) H. & P. Syd., *ibid.*: 43 (1919). (Fig. 11)
Apiosporella macrospora Speg., *An. Mus. Nac. Buenos Aires* **20**: 364 (1910).
Ascochyta macrospora (Speg.) Melnik, *Mikol. i Fitopatol.* **5**: 21 (1971).
Apiocarpella bromi Sprague, *Northw. Sci.* **29**: 109 (1955).

Conidiomata up to 150 µ diam. Conidiogenous cells 4.5-9 × 4.5-7 µ. Conidia 27-30.5 × 5.5-8.5 µ.

On leaves of *Hordeum jubatum* and *Bromus anomalus*, Argentina, Utah, U.S.A. (IMI 109471, holotype *A. macrospora* ex LPS, 110460, isotype *A. bromi* ex WSP).

3. DEARNESSIA Bub., *Hedwigia* **58**: 25 (1917).
Sp. typ.: *D. apocyni* Bub.

Mycelium immersed, pale brown to hyaline, branched, septate. *Conidiomata* pycnidial, amphigenous, separate,

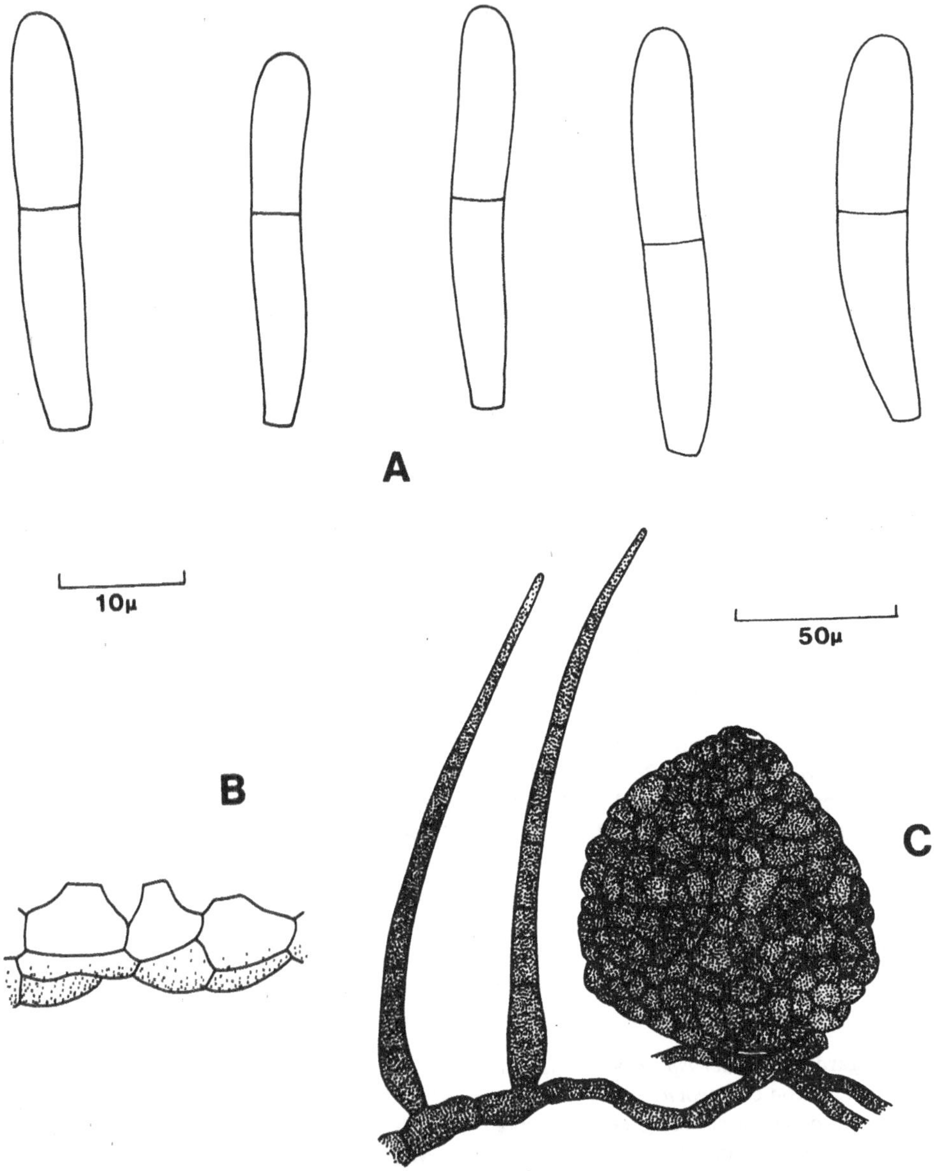

Fig. 10. *Chaetodiplodia graminicola*, A, conidia; B, conidiogenous cells; C, conidioma.

globose, brown, immersed, unilocular, thin-walled, of textura angularis. *Ostiole* central, circular, papillate. *Conidiophores* absent. *Conidiogenous cells* holoblastic, determinate, discrete, doliiform to ampulliform, hyaline, smooth, formed from the inner cells of the pycnidial wall. *Conidia* hyaline, 2- numerous transverse eusepta, continuous, not constricted, base truncate, apex obtuse, thin-walled, smooth, often guttulate, cylindrical or irregular.

This genus is closely related to *Aristastoma* Tehon and only differs in having simple pycnidial conidiomata which lack any setae around the ostioles. It is monotypic.

D. APOCYNI Bub., *ibid.*: 25 (1917). (Fig. 12)

Conidiomata up to 100 µ diam. Conidiogenous cells 7.5-10 × 5-8 µ. Conidia up to 52 µ long × 6-6.5 µ wide, up to 8 septate.

Causing leaf lesions on *Apocynum androsaemifolium*, Canada (IMI 21555, 25900, 25901, 159696).

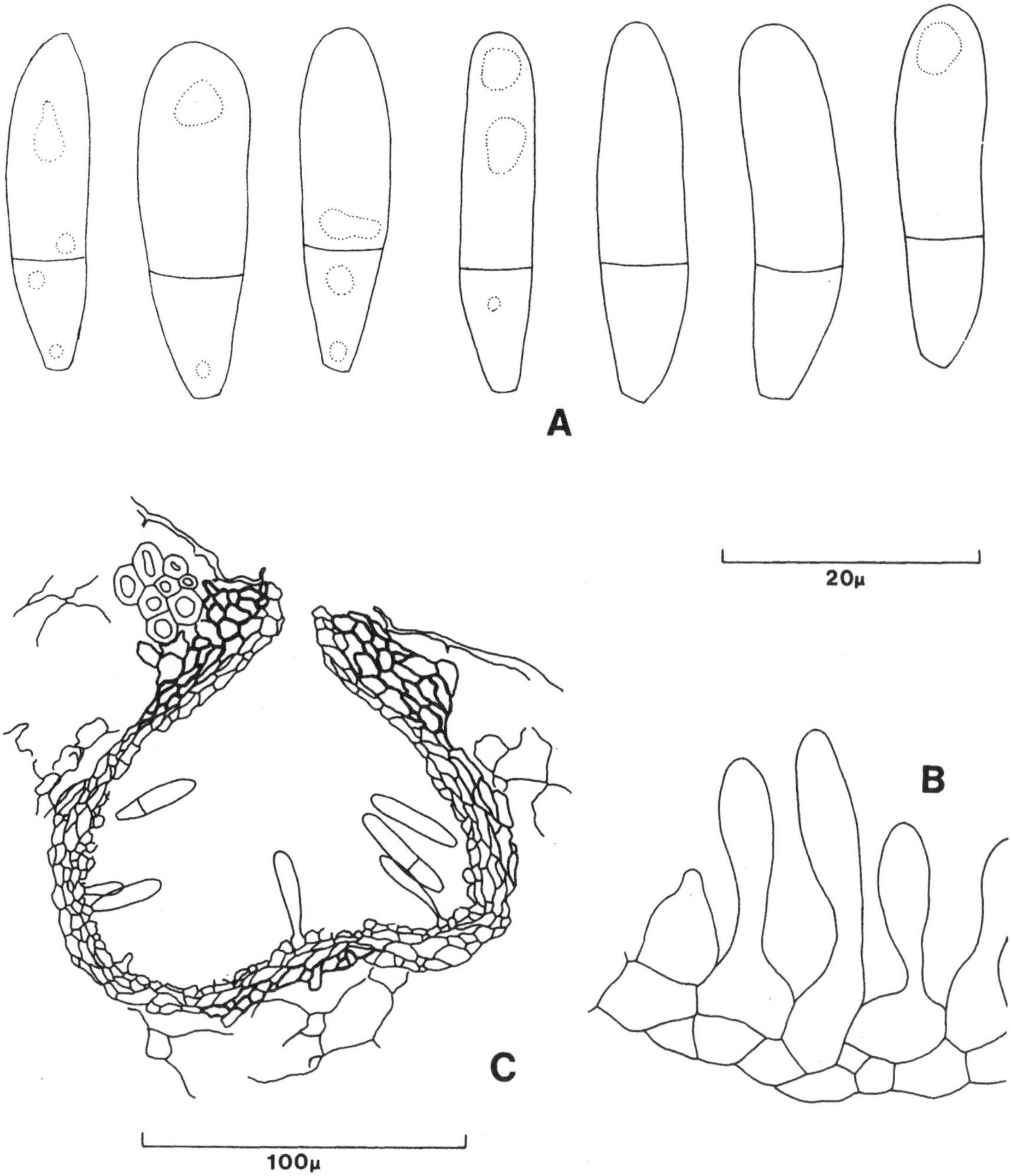

Fig. 11. *Apiocarpella macrospora*, A, conidia; B, young conidia and conidiogenous cells; C, vertical section of conidioma.

4. ARISTASTOMA Tehon, *Mycologia* **25**: 248 (1933).

Sp. typ.: *A. oeconomicum* (Ell. & Tracy) Tehon (syn. *Amerosporium oeconomicum* Ell. & Tracy, *Aristastoma concentricum* Tehon).

Illustrations: Sutton (1964a).

Mycelium immersed, pale brown to hyaline, branched, septate. *Conidiomata* pycnidial, amphigenous, separate, globose, brown, immersed, unilocular, thin-walled, of textura angularis. *Ostiole* central, circular, papillate. *Setae* ostiolar,

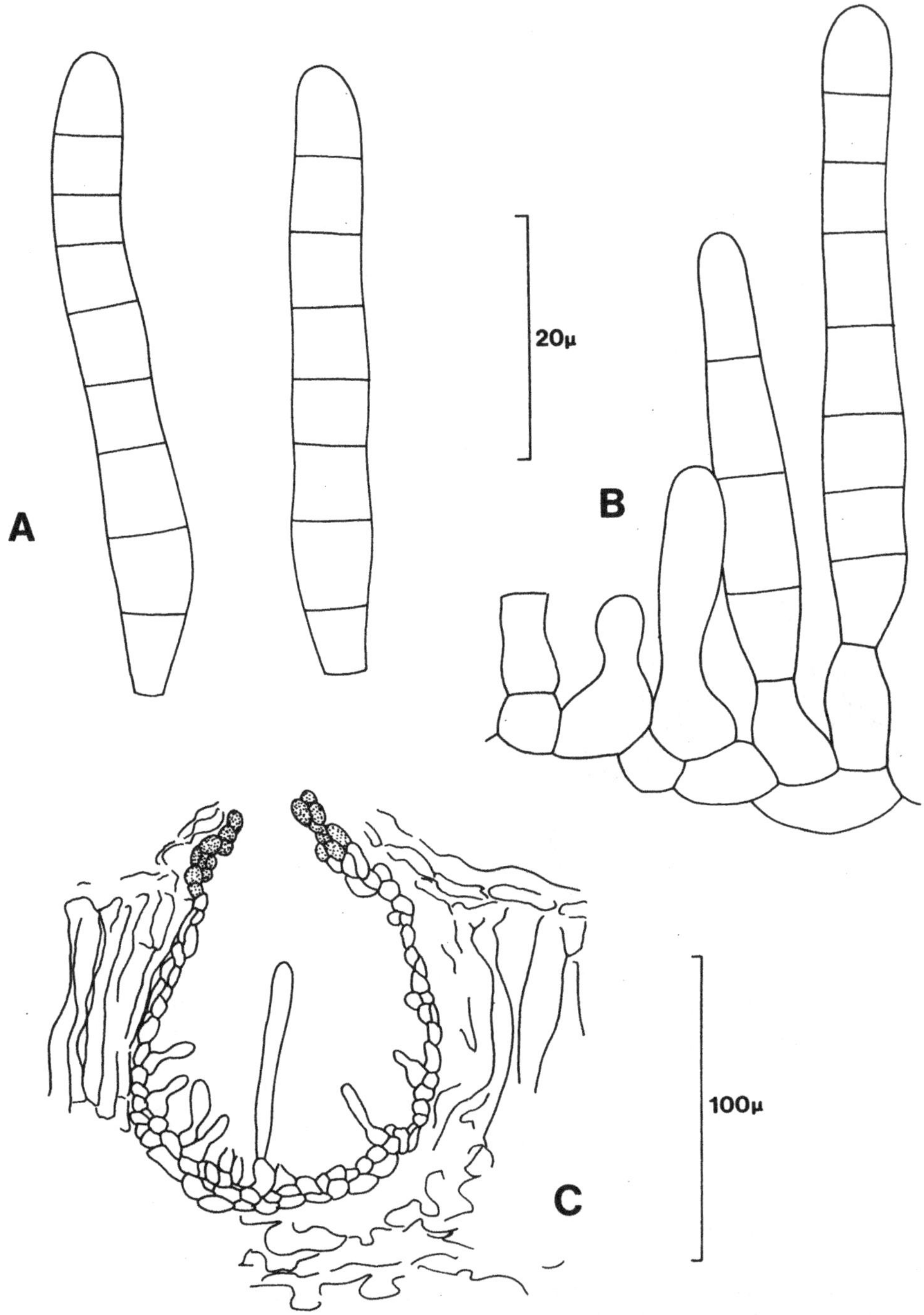

Fig. 12. *Dearnessia apocyni*, A, conidia; B, developing conidia and conidiogenous cells; C, vertical section of conidioma.

± straight, unbranched, tapered to the apex, dark brown, smooth, thin-walled, septate. *Conidiophores* absent. *Conidiogenous cells* holoblastic, determinate, discrete, doliiform to ampulliform, hyaline, smooth, formed from the inner cells of the pycnidial wall. *Conidia* hyaline, 1- numerous transverse eusepta, rarely with 1-2 longitudinal eusepta, continuous or constricted, often tapered to the apex, base truncate, thin-walled, smooth, ±guttulate, occasionally branched, cylindrical to irregular.

Sutton (1964a) distinguished 5 species.

KEY

1. Conidia up to 5 septate 2
1. Conidia more than 5 septate 3
   2. Pycnidia 200 µ diam. Conidia 3 septate, navicular, guttulate .... *A. guttulosum*
   2. Conidia 2-4 septate, clavate, rarely navicular, rarely guttulate .... *A. oeconomicum*
   2. Pycnidia 100 µ diam. Conidia 1 septate, clavate .... *A. minimum*
3. Conidia 7-11 septate, fusiform-sigmoid, 48-70 µ long .... *A. jaculum*
3. Conidia 4-6 septate, clavate, 30-45 µ long .... *A. camarographioides*

A. OECONOMICUM (Ell. & Tracy) Tehon apud Lefebvre & Stevenson, *Mycologia* **37**: 42 (1945). (Fig. 13C)

Conidiomata 110-200 µ diam.; ostiole 15-20 µ diam.; setae 6-8, 0-4 septate, 80-160 × 5-10 µ. Conidiogenous cells 5.5-6 × 4-7 µ. Conidia 2-4 septate, navicular-clavate, tapered to the apex, 19-32 × 3.5-5.5 µ.

Causing lesions on leaves of *Vigna sinensis*, *Dolichos arvensis*, U.S.A. (IMI 87017 (holotype), 40060).

A. JACULUM Sutton, *Mycol. Pap.* 97: 5 (1964). (Fig. 13A, F, G)

Conidiomata 100-200 µ diam.; ostiole 25-35 µ diam.; setae 8-12, 2-5 septate, 120-180 × 5-10 µ. Conidiogenous cells 2.5-4 × 3 µ. Conidia 7-11 septate, occasionally 1 longitudinal, straight, curved or sigmoid, abruptly tapered at apex, 48-70 × 6.5-8.5 µ.

Causing lesions on leaves of *Mucuna pruriens*, *M. pruriens* var. *utilis*, Malawi (IMI 86585 (holotype), 52908, 59539).

A. CAMAROGRAPHIOIDES Sutton, *Mycol. Pap.* 97: 7 (1964). (Fig. 13B)

Conidiomata 110-215 µ diam.; ostiole 20-30 µ diam.; setae 2-5, 0-4 septate, 85-120 × 5-8 µ. Conidiogenous cells 2-4 × 3-4 µ. Conidia 4-6 septate, often 1-2 longitudinal, straight or curved, clavate, tapered to the base, 33-45 × 7.5-9 µ.

Causing lesions on leaves of *Pueraria phaseoloides*, *P. javanica*, New Guinea, Solomon Isl. (IMI 81527 (holotype), 81420, 112152, 180648).

A. GUTTULOSUM Sutton, *Mycol. Pap.* 97: 10 (1964). (Fig. 13D)

Conidiomata 200 µ diam.; ostiole 20-30 µ diam.; setae 4-9, 0-4 septate, 100-150 × 5-10 µ. Conidiogenous cells 4.5-6.5 × 5-7 µ. Conidia 3 septate, straight or curved, tapered to the apex, guttulate, 16-29 × 2.5-3.5 µ.

Causing lesions on leaves of *Vigna sinensis*, *Phaseolus lunatus*, Nigeria (IMI 75957 (holotype), 79038*b*, 199121, 199123, 211679).

A. MINIMUM Sutton, *Mycol. Pap.* 97: 8 (1964). (Fig. 13E)

Conidiomata 50-110 µ diam., ostiole 10-15 µ diam.; setae 4-21, 0-2 septate, 50-100 × 4-7 µ. Conidiogenous cells 4-6 × 5-7 µ. Conidia 1 septate (rarely 0 or 2 septate), cylindrical, obtuse at apex, 14.5-22 × 4-5.5 µ.

Causing lesions on leaves of *Setaria homonyma*, S. Rhodesia (IMI 98932, holotype).

5. DENDROSEPTORIA Bausa Alcalde, *An. Jard. bot. Madrid* **7**: 489 (1948).
Sp. typ.: *D. arrhenatherae* Bausa Alcalde.
Illustration: Bausa Alcalde (1948).

Mycelium immersed, branched, septate, hyaline to pale brown. *Conidiomata* pycnidial, immersed, separate, globose, pale brown, unilocular, thick-walled, at the base and sides of textura intricata, in the ostiolar region of textura angularis. *Ostiole* single, central, circular. *Conidiophores* absent. *Conidiogenous cells* holoblastic, discrete, determinate, short cylindrical, hyaline, smooth, formed from the inner cells of the pycnidial wall. *Conidia* hyaline, smooth, thin-walled, minutely guttulate, consisting of a basal cell from which 2-3 divergent, primary branches arise; primary branches in turn may form up to 2 secondary branches; branches 1-3 euseptate, straight, apical cells tapered and acute.

This description has been based on the excellent account and figures provided by the original author.

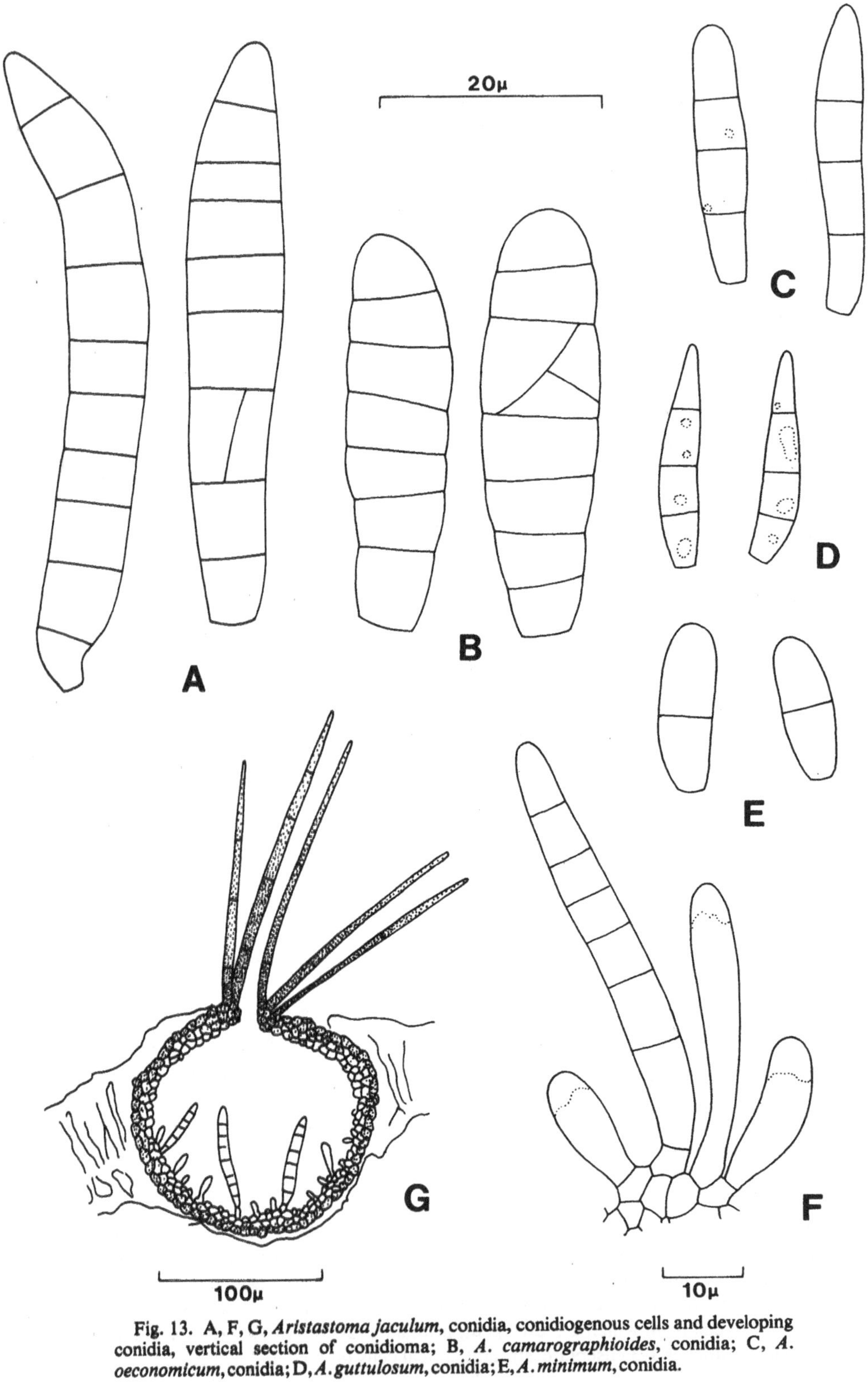

Fig. 13. A, F, G, *Aristastoma jaculum*, conidia, conidiogenous cells and developing conidia, vertical section of conidioma; B, *A. camarographioides*, conidia; C, *A. oeconomicum*, conidia; D, *A. guttulosum*, conidia; E, *A. minimum*, conidia.

D. ARRHENATHERI Bausa Alcalde, *ibid.*: 489 (1948).

Conidiomata 135-200 µ diam. Conidia up to 80 µ long × 1.5-2 µ.

On dry leaves of *Arrhenatherum elatium*, Spain.

6. JAHNIELLA Petrak, *Annls mycol.* **18**: 123 (1920).
Sp. typ.: *J. bohemica* Petrak.

Mycelium immersed, branched, septate, brown. *Conidiomata* pycnidial, subepidermal, immersed, separate, globose, papillate, dark brown, thick-walled, sclerenchymatic; wall consisting of an outer layer 1 cell thick of dark brown, thick-walled textura angularis, a middle layer 8 cells thick, of hyaline to pale brown, thick-walled cells, and an inner layer of thin-walled, hyaline irregular cells. *Ostiole* single, circular, papillate, with a distinct channel. *Conidiophores* absent. *Conidiogenous cells* holoblastic, determinate, discrete, hyaline, ampulliform, lining the wall of the pycnidium. *Conidia* straight or slightly curved, hyaline, thin-walled, smooth, 3-4 euseptate, eguttulate, truncate at the base, slightly tapered to the apex.

Jahniella shows a similar relationship to *Septoria* as does *Plenodomus* Pr. to *Phoma* Sacc. inasmuch as it is the sclerenchymatous counterpart. Boerema & Bollen (1975) reduced *Plenodomus* to synonymy with *Phoma*, and the logical step would be to make *Jahniella* a synonym of *Septoria*. However, the types of *Jahniella* and *Septoria* have not been studied in culture, so for the moment they are maintained separately on account of their differences in pycnidial structure.

Apart from the type, two further species have been added, *J. campanulae-cervicariae* (Vestergr.) Petrak and *J. hordei* Petrak, but neither have been confirmed as congeneric by type studies.

J. BOHEMICA Petrak, *ibid.*: 123 (1920). (Fig. 14)

Conidiomata up to 400 µ diam. Ostiole 45 µ diam. Conidiogenous cells 4-9 × 2-4.5 µ. Conidia 33-40 × 2 µ.

On dead stems of *Scrophularia nodosa*, Czechoslovakia (IMI 204129 ex BPI, Holotype).

7. CYTOSTAGONOSPORA Bub., *Annls mycol.* **14**: 150 (1916).
Cytostaganis Clements & Shear, *Genera of fungi*: 367 (1931).
Sp. typ.: *C. photiniicola* Bub.

Mycelium immersed, dark brown, branched, septate. *Conidiomata* pycnidial, amphigenous, separate, globose, dark brown to black, immersed, unilocular, thick-walled, clypeate; walls of dark brown, thick-walled, textura angularis to textura globulosa, becoming hyaline towards the conidiogenous region, extended in the upper part to become a circular clypeus of similar thickness to the wall. *Ostiole* central, circular, papillate to shortly rostrate, depressed, situated down within the clypeus. *Conidiophores* absent. *Conidiogenous cells* holoblastic, determinate, discrete, lageniform, hyaline, smooth, formed from the inner cells of the pycnidial wall. *Conidia* hyaline, 0-2 euseptate, not constricted, base truncate, apex obtuse, thin-walled, eguttulate, smooth, filiform, often curved.

Apart from the type, two other species have been added to the genus. *C. cryptica* was described from *Arbutus unedonis* in Italy by Curzi & Barbiani (1927) and from the excellent account and illustration quite probably is a good species. Pycnidia were described as 110-130 µ diam. with a depressed ostiole and clypear tissue, and conidia as 2-3 septate and 45-60 × 3 µ. Whether the species is distinct from *C. photiniicola* is debatable. *C. traversiana* described from leaves of *Hedera helix* in Portugal by Dias & Camara (1955) is either *Ceuthospora lauri* (Grev.) Grev. or possibly an aggregated *Coleophoma empetri* (Rostr.) Petrak. Conidiomata are stromatic with more than one pycnidial cavity each measuring 155-220 × 125-180 µ, with cylindrical conidia measuring 10-13.5 × 2.5-2.6 µ.

C. PHOTINIICOLA Bub., *ibid.*: 150 (1916). (Fig. 15)
Cytostaganis photiniicola (Bub.) Clements & Shear, *Genera of fungi*: 367 (1931).

Conidiomata 150-190 µ diam. Conidiogenous cells 3.5-8 × 2-3.5 µ. Conidia 53-65 × 3 µ.

On dead leaves of *Photinia serrulata*, Austria (IMI 194179, holotype).

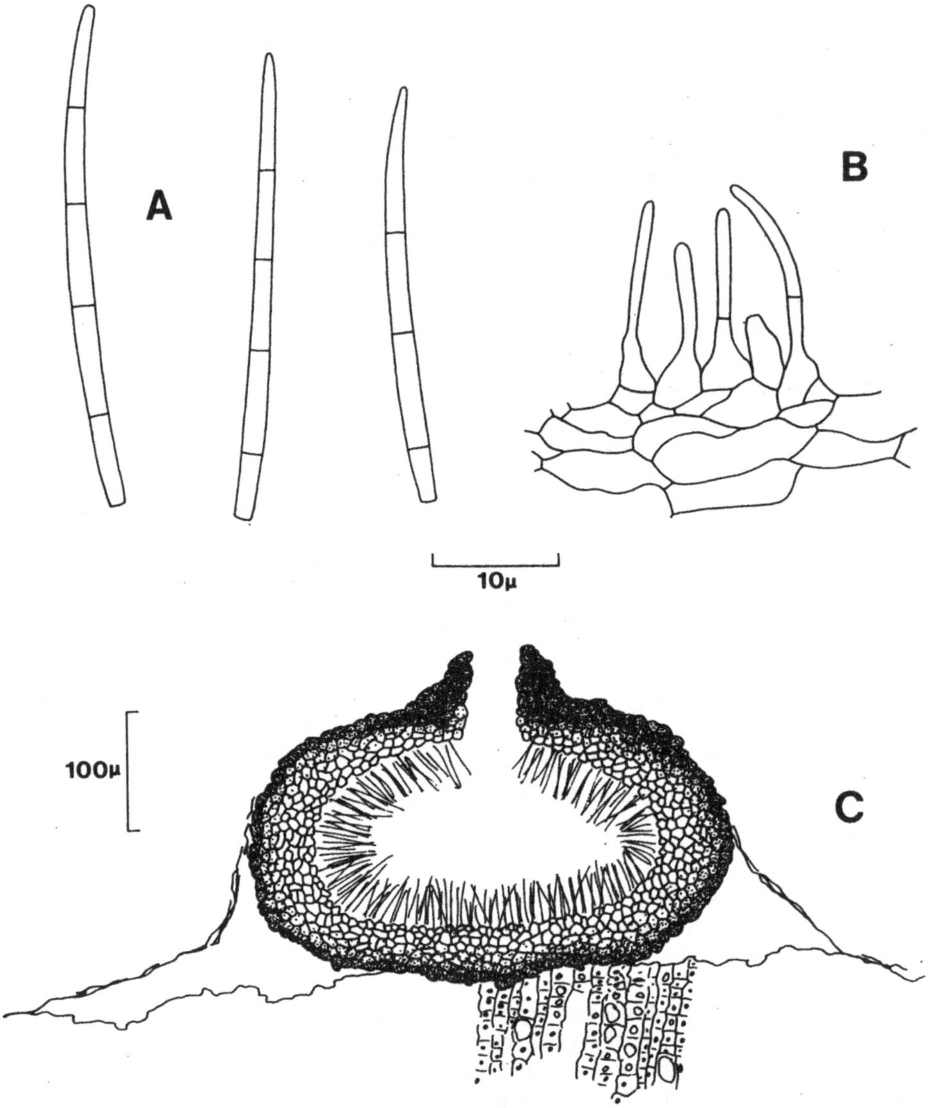

Fig. 14. *Jahniella bohemica*, A, conidia; B, conidiogenous cells and developing conidia; C, vertical section of conidioma.

8. PIPTARTHRON Mont. ex Höhn., *Hedwigia* **60**: 203 (1918).
Piptarthron Mont. nom. provis., *Syll. Pl. Crypt.* 277 (1861).
Septoplaca Petrak, *Sydowia* **17**: 271 (1964).
Sp. typ.: *P. macrosporum* (Dur. & Mont.) Höhn. (syn. *Septoria macrospora* Dur. & Mont.).
Illustrations: Morgan-Jones, Nag Raj & Kendrick (1972c).

Mycelium immersed, dark brown, branched, septate. *Conidiomata* pycnidial, amphigenous, separate or aggregated, globose, dark brown, immersed, unilocular or multilocular, thick-walled; wall composed of ± loosely aggregated, dark brown, thick-walled textura angularis. *Ostiole* central, papillate or depressed, circular, wall tissue often elongated. *Conidiophores* absent. *Conidiogenous cells* holoblastic, determinate, discrete, doliiform, hyaline, smooth, formed from the inner cells of the pycnidial wall. *Conidia* hyaline, 0-4 euseptate, often constricted, base truncate with a distinct marginal frill, thick-walled, smooth, often guttulate, cylindrical, fusiform, straight or curved.

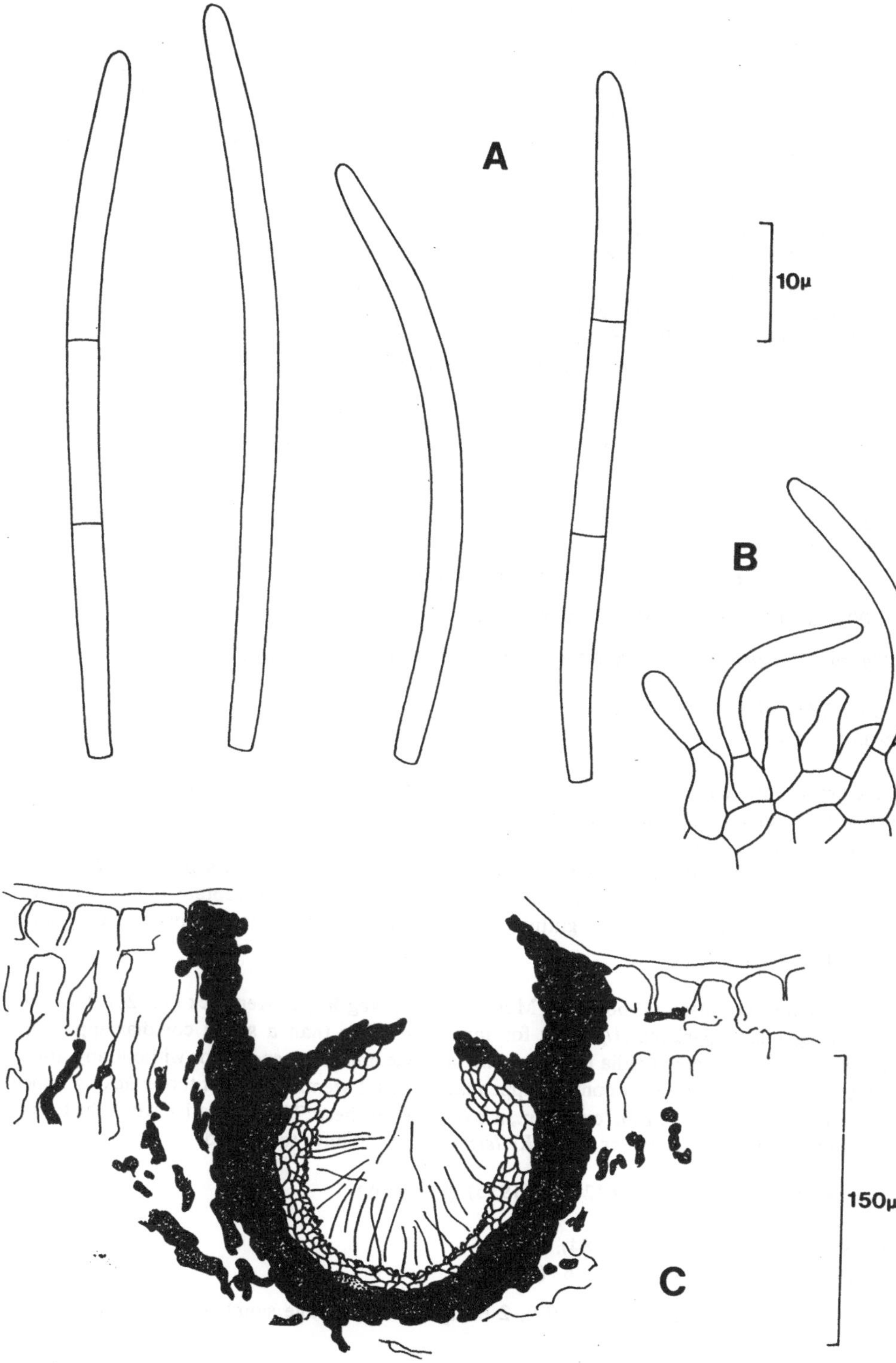

Fig. 15. *Cytostagonospora photiniicola*, A, conidia; B, conidiogenous cells and developing conidia; C, vertical section of conidioma.

The generic and specific names were validated by Höhnel (1918a). Morgan-Jones, Nag Raj & Kendrick (1972c) included two new species in a revised account of *Kellermania* Ell. & Ev. Neither of these had appendaged conidia. *K. attenuata* moreover was aseptate, but *K. multiseptata* as the name implies, was 3-4 septate. The latter species is but a later name for *P. macrosporum*, the type of *Piptarthron*. If the circumscription of *Kellermania* suggested by Morgan-Jones et al. (1972c) is accepted, then *Kellermania* must be re-instated as the earliest generic name. However I prefer a more restricted delimitation of *Kellermania* in accordance with my earlier treatment (Sutton, 1968b) where only two species with appendaged conidia are maintained. The generic synonymy involving *Septoplaca* has been fully explained by Sutton (1977a).

P. MACROSPORUM (Dur. & Mont.) Höhn., *ibid.*: 203 (1918).
Septoria macrospora Dur. & Mont., *Expl. scient. Alg.* **1**: 589 (1846).
Stagonospora macrospora (Dur. & Mont.) Sacc., *Syll. fung.* **3**: 450 (1884).
Hendersonia piptarthra Sacc., *Michelia* **2**: 111 (1880).
Hendersonia montagnei Cke, *Nuovo G. Bot. Ital.* **10**: 19 (1878).
Kellermania multiseptata Morgan-Jones, Nag Raj & Kendrick, *Can. J. Bot.* **50**: 1644 (1972).

Conidiomata 550-600 μ diam. Conidia 50-68 × 6-7.5 μ, 2-4 euseptate.

On leaves of *Yucca macrocarpa*, *Y. brevifolia*, *Agave americana*, *A.* sp., Crete, Italy, Texas and Ariz., U.S.A. (IMI 202435, 202434 ex K).

P. LIMBATUM (Petrak) comb. nov.
Septoplaca limbata Petrak, *Sydowia* **17**: 271 (1964).
Kellermania attenuata Morgan-Jones, Nag Raj & Kendrick, *Can. J. Bot.* **50**: 1643 (1972).

Conidiomata 500-600 μ diam. Conidia 70-85 × 3-5 μ, aseptate.

On leaves of *Yucca filifera*, *Y. angustifolia*, *Y. brevifolia*, *Y.* spp., Nev., Colo, Fla, U.S.A., Mexico.

9. KELLERMANIA Ell. & Ev., *J. Mycol.* 1: 153 (1885).
Sp. typ.: *K. yuccigena* Ell. & Ev.

Mycelium immersed, dark brown, branched, septate. *Conidiomata* pycnidial, amphigenous, separate or aggregated, globose, dark brown, immersed, unilocular or multilocular, thick-walled; wall composed of ± loosely aggregated, dark brown, thick-walled textura angularis. *Ostiole* central, papillate-depressed, circular, wall tissue elongated. *Conidiophores* absent. *Conidiogenous cells* holoblastic, determinate, discrete or sometimes integrated on short 1-septate conidiophores, doliiform, hyaline, smooth, formed from the inner cells of pycnidial wall. *Conidia* hyaline, 1-2 euseptate, often constricted, base truncate with a marginal frill, thick-walled, smooth, ± guttulate, cylindrical, fusiform or long clavate, with 1 apical extracellular simple or branched appendage.

Previous accounts have been given by Sutton (1968b) and Morgan-Jones, Nag Raj & Kendrick (1972c). The latter authors accepted 4 species in the genus, separating *Alpakesa* for taxa with more than a single conidial appendage. *Kellermania* was maintained for taxa with septate conidia either bearing a single apical appendage or without any appendage. If generic taxa are to be separated on the basis of conidial appendages then it seems clear that those lacking appendages should be segregated. The name *Piptarthron* Mont. ex Höhn., is available for these. The alternative would be to refer all species in this complex to a broadly expanded genus for which *Kellermania* would have to be used.

K. ANOMALA (Cke) Höhn., *Sber. Akad. Wiss. Wien* **124**: 84 (1915). (Fig. 16A)
Discella anomala Cke, *Grevillea* **7**: 11 (1878).
Discula anomala (Cke) Sacc., *Syll. fung.* **3**: 677 (1884).
Kellermania major Dearn. & Barth., *Mycologia* **16**: 163 (1924).

Conidiomata 250-300 μ diam. Conidia 55-70 × 8-10 μ, 2 euseptate; appendage simple or branched, 12-25 μ long.

On leaves of *Yucca* spp., Calif., Fla, Ariz., N.M., Texas, U.S.A. (WINF(M) 3419).

K. YUCCIGENA Ell. & Ev., *J. Mycol.* **1**: 154 (1885). (Fig. 16B, C, D)

Conidiomata 250-300 μ diam. Conidia 50-61 × 9-10 μ, 1 euseptate; appendage 22-32 μ long.

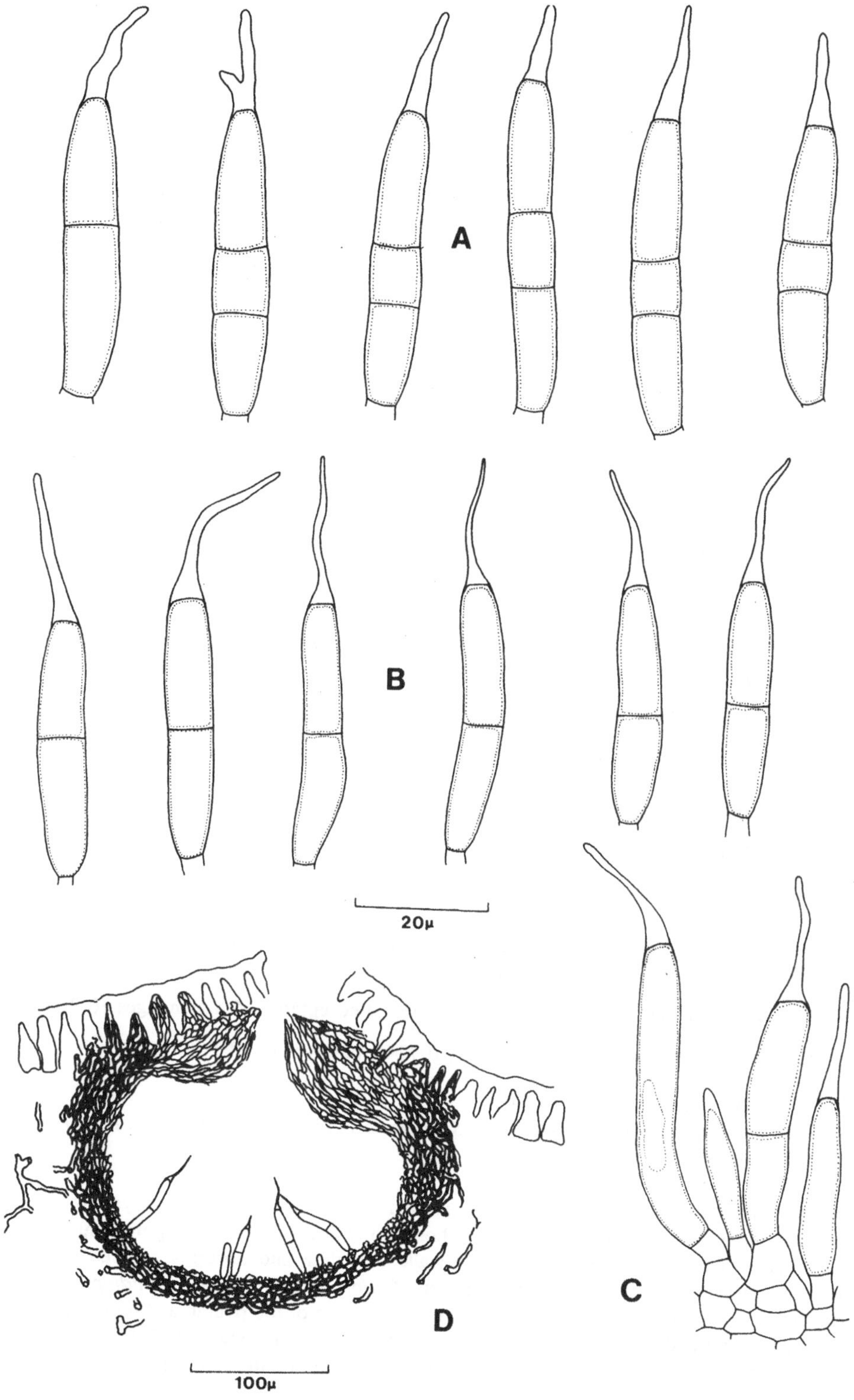

Fig. 16. A, *Kellermania anomala*, conidia; B, C, D, *K. yuccigena*, conidia, conidiogenous cells and developing conidia, vertical section of conidioma.

On leaves of *Yucca* spp., Calif., N.M., Nev., N. Dak., Wyo, U.S.A. (WINF(M) 4226, IMI 97350, 97349, 21864, 97060, 202436 ex K).

10. ALPAKESA Subram. & Ramakr., *J. Ind. Bot. Soc.* **33**: 204 (1954).
Sp. typ.: *A. yuccifolia* (Hall) Subram. & Ramakr. (syn. *Neottiospora yuccifolia* Hall).
Illustrations: Morgan-Jones, Nag Raj & Kendrick (1972b).

Mycelium immersed, dark brown, branched, septate. *Conidiomata* pycnidial, amphigenous, separate, globose, dark-brown, immersed, unilocular, thick-walled, composed of loosely aggregated, dark brown, thick-walled textura angularis. *Ostiole* single, circular, papillate or depressed. *Conidiophores* absent. *Conidiogenous cells* holoblastic, determinate, discrete, doliiform, hyaline, smooth, formed from the inner cells of the pycnidial wall. *Conidia* hyaline, 0-3 euseptate, occasionally constricted, base truncate with a distinct marginal frill, thick-walled, smooth, often guttulate, cylindrical, straight, with several apical, filiform, unbranched appendages.

The genus, its type species, and two additional species were described by Morgan-Jones, Nag Raj & Kendrick (1972b). It was maintained as distinct from *Kellermania* from which it differs in the number of conidial appendages and perhaps their origin.

KEY

Conidia aseptate .... *A. yuccifolia*
Conidia 1 septate .... *A. uniseptata*
Conidia 3 septate .... *A. nolinae*

A. YUCCIFOLIA (Hall) Subram. & Ramakr., *ibid.*: 205 (1954). (Fig. 17A, B, E, F)
Neottiospora yuccifolia Hall, *Phytopath.* **5**: 57 (1915).

Conidiomata 160-300 µ diam. Conidia 23-50 × 6.5-13 µ, aseptate; 2-5 appendages 16-43 µ long.

On leaves of *Yucca filamentosa*, *Y. gloriosa*, *Y.* sp., Wash., Kans., Ga, Ala, U.S.A. (IMI 177975).

A. UNISEPTATA Morgan-Jones, Nag Raj & Kendrick, *Can. J. Bot.* **50**: 879 (1972). (Fig. 17D)

Conidiomata 140-280 µ diam. Conidia 17-40 × 9-11 µ, 1 euseptate; 3-5 appendages 26-39 µ long.

On leaves of *Yucca rupicola*, Tex., U.S.A.

A. NOLINAE (Pollack Morgan-Jones, Nag Raj & Kendrick, *Can. J. Bot.* **50**: 879 (1972). (Fig. 17C)
Bartalinia nolinae Pollack, *Mycologia* **39**: 620 (1947).

Conidiomata 250-400 µ diam. Conidia 36-50 × 6.5-10 µ, 3-euseptate; 3-5 appendages 10-25 µ long.

On leaves of *Nolina microcarpa*, Ariz., U.S.A. (IMI 107246, holotype ex BP1).

11. HYALOPYCNIS Höhn., *Hedwigia* **60**: 152 (1918).
Sp. typ.: *H. blepharistoma* (Berk.) Seeler (syn. *Sphaeronaema blepharistoma* Berk., *H. vitrea* (Cda) Höhn.).
Illustration: Seeler (1943).

Mycelium immersed, branched, septate, hyaline. *Conidiomata* pycnidial, separate, pale yellow, translucent, immersed to semi-immersed, unilocular, thin-walled, swollen and globose at the base but with a long neck surmounted by outwardly curving, parallel, scarcely septate hyphae that are thin-walled, hyaline and abruptly tapered to obtuse apices; walls composed of very pale brown to yellowish textura angularis. *Ostiole* single, circular. *Conidiophores* hyaline, repeatedly branched, septate, smooth, with acrogenous conidia, formed at the base of the pycnidia. *Conidiogenous cells* holoblastic, determinate, integrated, cylindrical, hyaline, smooth, producing a single apical conidium. *Conidia* hyaline, aseptate, oval to obovoid, straight, apex obtuse, tapered to a flattened base, thin-walled, eguttulate, smooth.

Seeler (1943) redescribed the genus and elaborated upon the synonymy. Most herbarium material is in poor condition and details of conidial ontogeny could not be verified from material in K. The genus is monotypic.

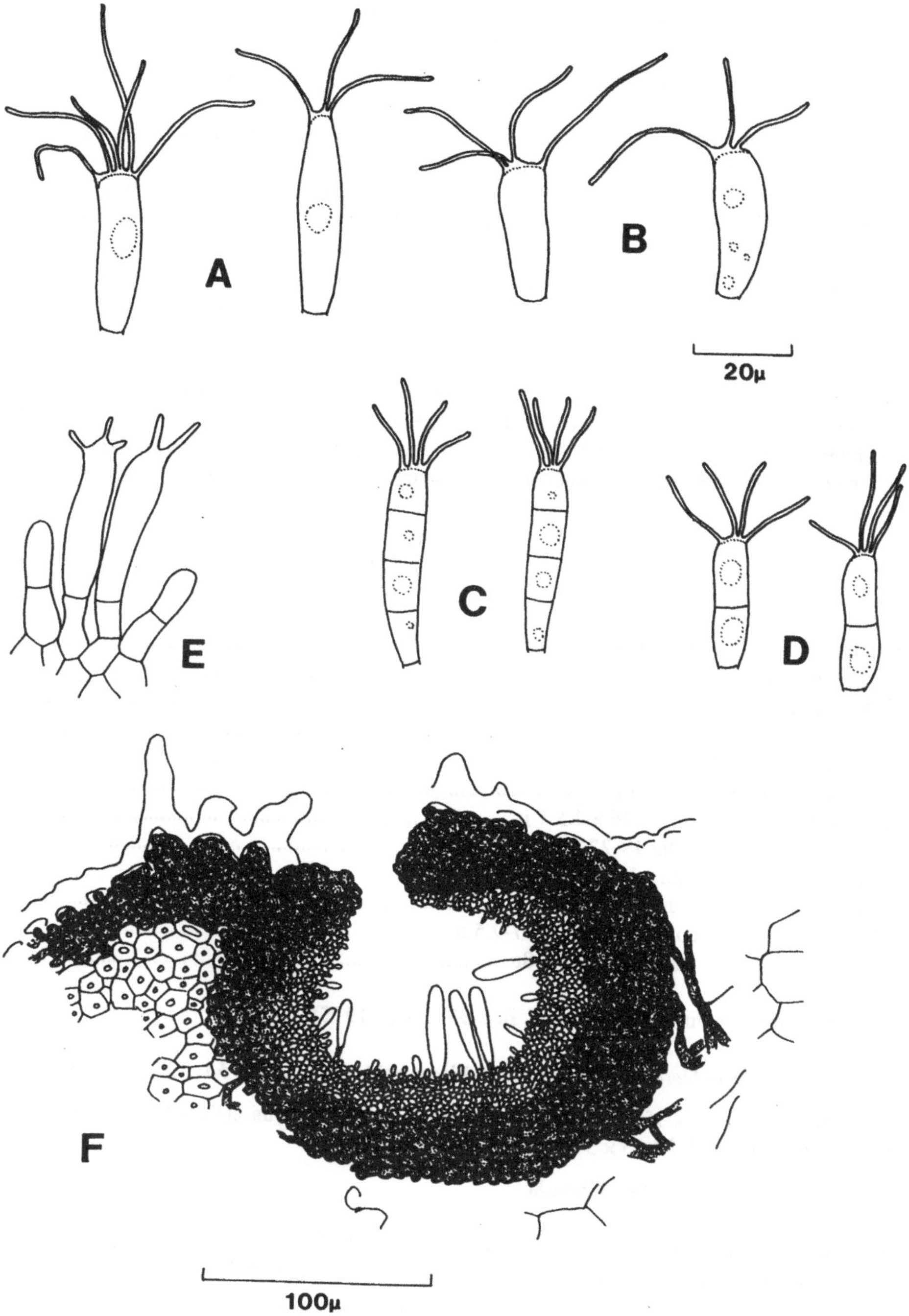

Fig. 17. A, B, E, F, *Alpakesa yuccifolia*, conidia, conidiogenous cells and developing conidia, vertical section of conidioma; C, *A. nolinae*, conidia; D, *A. uniseptata*, conidia.

H. BLEPHARISTOMA (Berk.) Seeler, *Farlowia* 1: 124 (1943).
Synonymy in Seeler (1943).

Conidiomata 100-200 µ diam. at the base, neck almost as long. Conidia 9-13 × 3.5-5.5 µ.
On *Agaricus adustus*, U.K., but recorded by Seeler (1943) from *Urtica, Daucus* and various agaric species from N. Hamp., Mass., N.Y., U.S.A., Austria, Czechoslovakia.

12. TIAROSPORELLA Höhn., *Ber. dtsch. Bot. Ges.* **37**: 159 (1919).
Sakireeta Subram. & Ramakr., *J. Ind. bot. Soc.* **36**: 83 (1957).
Sp. typ.: *T. paludosa* (Sacc. & Fiori) Höhn. (syn. *Neottiospora paludosa* Sacc. & Fiori).

Mycelium immersed, pale brown to hyaline, branched, septate. *Conidiomata* pycnidial, separate or linearly aggregated, globose, dark brown, immersed, unilocular, thick-walled; wall composed of an outer layer of slightly elongated, dark brown, thick-walled textura angularis and an inner layer of hyaline thin-walled tissue. *Ostiole* central, circular, papillate. *Conidiophores* absent. *Conidiogenous cells* holoblastic, determinate, discrete, lageniform, hyaline, smooth, with acrogenous conidia, formed from the inner cells of the pycnidial wall. *Conidia* hyaline, septate, base truncate, apex flattened, thin-walled, smooth, eguttulate, cylindrical to fusiform, straight or slightly curved, with an apical gelatinous appendage originating by the splitting from the base of an envelope enclosing young conidia, finally dividing into 2-4 unbranched tentacles.

Recent revisionary accounts of the genus have been made by Pirozynski & Shoemaker (1971), Nag Raj (1973a) and Sutton & Marasas (1976). The latter authors accepted four species and one variety. Whitney, Reid & Pirozynski (1975) described three coniferous species and van der Aa (1977) added *T. phaseoli* (Maubl.) van der Aa. Although these four species have not been examined, data on the coniferous taxa have been incorporated in the following key.

KEY

1. Conidia averaging more than 10 µ in width .... *T. tritici*
1. Conidia averaging less than 10 µ in width 2
  2. Conidia clavate to cylindrical 3
  2. Conidia ellipsoid 5
3. Conidia not exceeding 45 µ in length 4
3. Conidia consistently more than 45 µ long .... *T. pseudotsugae*
  4. Appendages tentacular; conidia 24-46 × 4.5-7 µ .... *T. paludosa*
  4. Appendages infundibular; conidia 19-27 × 4.5-6 µ .... *T. madreeya*
  4. Appendages infundibular to tentacular; conidia 23-40 × 4-6 µ .... *T. parca*
  4. Appendages infundibular; conidia 29-45 × 5-8 µ .... *T. abietis*
5. Appendages 15-19 (17) µ wide; conidia 21.5-31 × 7-9.5 µ .... *T. graminis*
5. Appendages 6-17 (12) µ wide; conidia 21-28 × 5-8µ .... *T. graminis* var. *karoo*

T. ABIETIS Whitney, Reid & Pirozynski, *Can. J. Bot.* **53**: 3055 (1975).

? Teleomorphosis: *Darkera abietis* Whitney, Reid & Pirozynski, *Can. J. Bot.* **53**: 3052 (1975).

Conidiomata 550 µ long × 350 µ wide × 225 µ deep. Conidiogenous cells 5.5-10 × 3-7 µ. Conidia cylindrical, straight or slightly curved, 29-45 × 5-8 (-9) µ.

On needles of *Abies balsamea*, Ont., Canada.

T. GRAMINIS (Pirozynski & Shoemaker) Nag Raj, *Can. J. Bot.* **51**: 2469 (1973). (Fig. 18B)
Tiarosporella paludosa (Sacc. & Fiori) Höhn. var. *graminis* Pirozynski & Shoemaker, *Can. J. Bot.* **49**: 538 (1971).

Conidiomata up to 150 µ diam. Conidiogenous cells 12-15 × 1.5-2.5 µ. Conidia ellipsoid-fusiform, straight, 20-29.5 × 7-9 µ; appendage 15-19 (17) µ wide.

On *Calamovilfa longifolia, Triticum aestivum*, Wisc., U.S.A., S. Africa (IMI 193727, holotype, 186784, 186785).

T. GRAMINIS (Pirozynski & Shoemaker) Nag Raj var. KAROO Sutton & Marasas, *Trans. Br. mycol. Soc.* **67**: 73 (1976). (Fig. 18C)

Conidiomata rostrate. Conidiogenous cells 12-18 × 1.5-2.5 µ. Conidia ellipsoid-fusiform, straight, 21-28 × 5-8 µ, appendage 6-17 (12) µ wide.

From *Tribulus terrestris, Eriocephalus* sp., *Nestlera* sp., South Africa (IMI 184559, holotype, 186782, 186783).

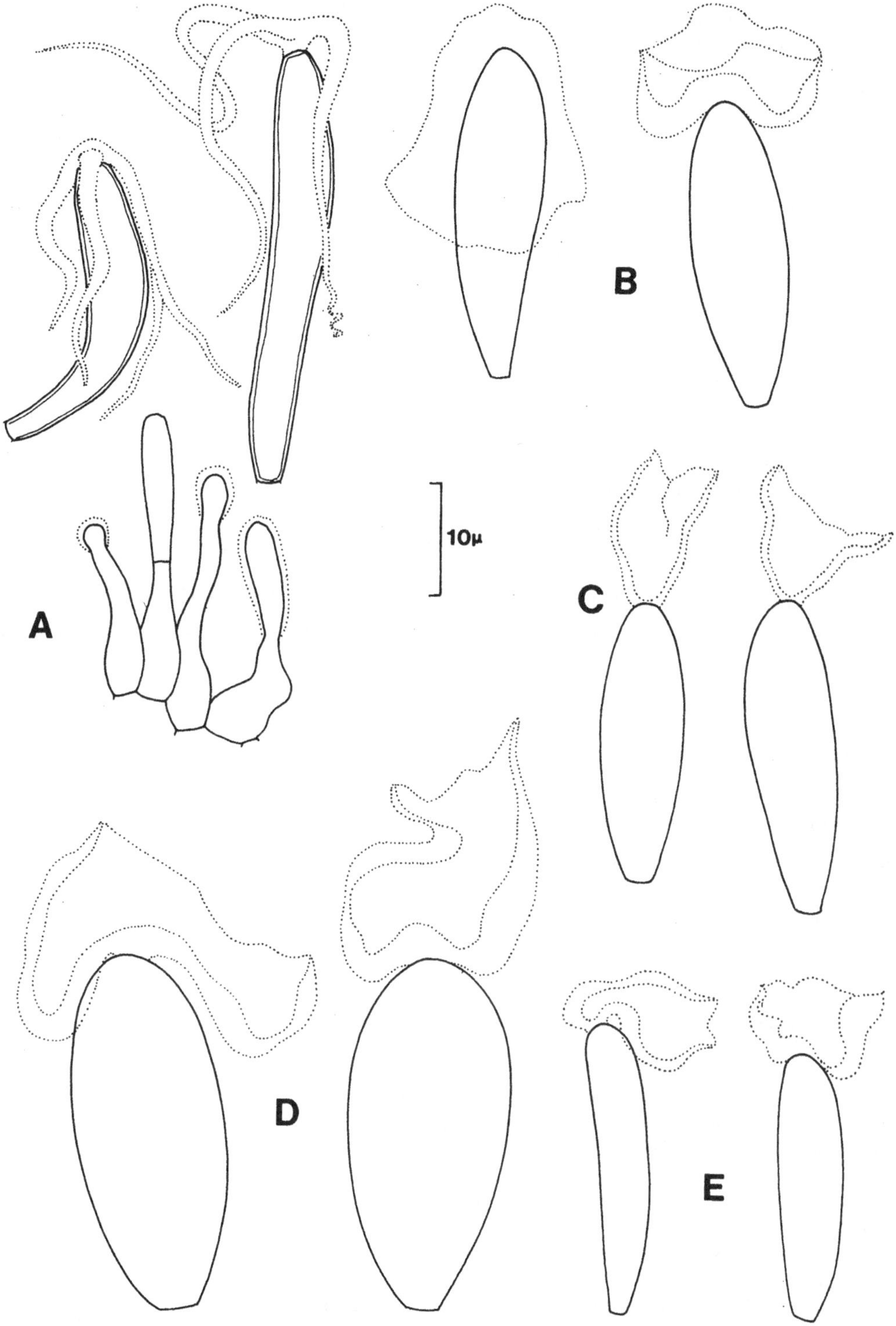

Fig. 18. A, *Tiarosporella paludosa*, conidia and conidiogenous cells; B, *T. graminis*, conidia; C, *T. graminis* var. *karoo*, conidia; D, *T. tritici*, conidia; E, *T. madreeya*, conidia.

T. MADREEYA (Subram. & Ramakr.) Nag Raj, *Can. J. Bot.* **51**: 2470 (1973). (Fig. 18E)
Sakireeta madreeya Subram. & Ramakr., *J. Ind. Bot. Soc.* **36**: 84 (1957).

Conidiomata up to 200 µ diam. Conidiogenous cells 8-10 × 2.5-4 µ. Conidia cylindrical, straight or slightly curved, 19-27 × 4.5-6 µ.

On dead culms of *Aristida setacea*, India (IMI 196129, holotype).

T. PALUDOSA (Sacc. & Fiori) Höhn., *Ber. dtsch. Bot. Ges.* **37**: 159 (1919). (Fig. 18A)
Neottiospora paludosa Sacc. & Fiori, *Hedwigia* **38**: 137 (1899).
Neottiospora schizochlamys Ferd. & Winge, *Svensk Bot. Tidskr.* **28**: 255 (1908).
Tiarosporella schizochlamys (Ferd. & Winge) Höhn., *Ber. dtsch. Bot. Ges.* **37**: 59 (1919).
Neottiospora arenaria Syd., *Annls mycol.* **10**: 448 (1912).

Conidiomata 200-250 µ diam. Conidiogenous cells 6-9 × 1.5-3 µ. Conidia cylindrical, straight or slightly curved, 24-46 × 4.5-7 µ, appendages tentacular.

On *Scirpus caespitosus*, U.K. (IMI 92315).

T. PARCA (Berk. & Br.) Whitney, Reid & Pirozynski, *Can. J. Bot.* **53**: 3055 (1975).
Sphaeropsis parca Berk. & Br., *Ann. Mag. Nat. Hist.*, ser. 2, **5**: 375 (1850).
Phoma parca (Berk. & Br.) Sacc., *Syll. fung.* **3**: 100 (1884).
Macrophoma parca (Berk. & Br.) Berl. & Vogl., *Syll. fung.* add. I-VI: 313 (1886).

? Teleomorphosis: *Darkera parca* Whitney, Reid & Pirozynski, *Can. J. Bot.* **53**: 3053 (1975).

Conidiomata 500 × 350 × 120-200 µ deep. Conidiogenous cells 4-12 × 3-5 µ. Conidia straight, cylindrical, 23-40 × 4-6 µ.

On needles of *Picea abies*, *P. engelmannii*, *P. glauca*, U.K., Czechoslovakia, Alta, Canada.

T. PSEUDOTSUGAE Whitney, Reid & Pirozynski, *Can. J. Bot.* **53**: 3057 (1975).

Conidiomata up to 600 µ diam. × 300 µ wide × 150-250 µ deep. Conidiogenous cells 3-9 × 3-5 µ. Conidia cylindrical, straight, often with a swollen apex, 42-65 × 4-5 µ.

On needles of *Pseudotsuga menziesii*, Alta, Canada.

T. TRITICI Sutton & Marasas, *Trans. Br. mycol. Soc.* **67**: 74 (1976). (Fig. 18D)

Conidiomata up to 200 µ diam. Conidiogenous cells 9-14 × 4-5 µ. Conidia oval to fusiform, straight, 29-38 × 12-17 µ.

From *Triticum aestivum*, South Africa (IMI 186786, holotype).

13. GIULIA Tassi, *Bull. Lab. Ort. bot., Siena* **6**: 93 (1904).
Nematospora Tassi, *Bull. Lab. Ort. bot., Siena* **6**: 92 (1904).
Sp. typ.: *G. tenuis* (Sacc.) P.A. & D. Sacc. (syn. *Leptostroma tenue* Sacc.).
Illustration: Pirozynski & Shoemaker (1971).

Mycelium immersed, branched, septate, brown. *Conidiomata* pycnidial, separate or gregarious, dark brown, immersed, subperidermal, unilocular, globose, with a distinct lateral neck, the apex of which is ostiolate; thick-walled, composed of an outer region of dark brown, irregular, thick-walled textura angularis and an inner layer of hyaline, thin-walled-cells; neck composed entirely of dark brown cells, flattened at the apex. *Ostiole* circular, ± papillate, almost clypeate. *Conidiophores* absent. *Conidiogenous cells* holoblastic, determinate, discrete, lageniform, hyaline, smooth, formed from the inner pycnidal wall cells. *Conidia* hyaline, aseptate, base truncate, apex obtuse, straight or curved, cylindrical, thin-walled, smooth, eguttulate, with 3-7 fine, unbranched, apical, extracellular setulae arising from a common papilla.

A redescription from the type was provided by Pirozynski & Shoemaker (1971) who removed the genus from the Leptostromataceae to the Sphaeropsidales.

G. TENUIS (Sacc.) P.A. & D. Sacc., *Syll. fung.* **18**: 435 (1906).
Leptostroma tenue Sacc., *Michelia* **2**: 282 (1882).
Leptostromella tenuis (Sacc.) Sacc., *Syll. fung.* **3**: 660 (1884).
Nematospora tenuis (Sacc.) Tassi, *Bull. Lab. Ort. Bot. Siena* **1**: 92 (1904).

Conidiomata 500 µ wide × 200 µ deep. Conidiogenous cells 4-7 × 3-4 µ. Conidia 17-23 × 3.5-4.5 µ; setulae up to 20 µ long.

On dead stems of *Lepidium graminifolium*, Italy.

14. TIAROSPORA Sacc. & March., *Rev. mycol.* **7**: 148 (1885).
Sp. typ.: *T. perforans* (Rob.) Höhn. (syn. *Sphaeria perforans* Rob., *T. westendorpii* Sacc. & March.).

Mycelium immersed, pale brown to hyaline, branched, septate. *Conidiomata* pycnidial, separate, gregarious, globose, with a short neck, subepidermal, immersed, unilocular, brown, thick-walled; wall composed of an outer layer of thick-walled, dark brown, somewhat flattened textura angularis, and an inner layer of thinner, paler tissue. *Ostiole* central, circular, papillate. *Conidiophores* absent. *Conidiogenous cells* holoblastic, sympodial, indeterminate, discrete, doliiform, cylindrical or lageniform, hyaline, smooth, with 1-3 unthickened proliferations, formed from the inner cells of the pycnidial wall. *Conidia* hyaline, later very pale brown, 1 median euseptum, continuous, obtuse to flattened at apex and base, thick-walled, smooth, guttulate, broadly ellipsoid, with a short tentacular, irregular or cap-like gelatinous appendage at each end. Origin of the appendages is at present undetermined.

Apart from the type species only one other has been described in the genus, *T. deschampsiae* Kruschewa (1959). Pirozynski & Shoemaker (1971) gave an illustrated account of *T. perforans* and related genera with appendaged conidia, but no-one has re-examined *T. deschampsiae*. It was described with pycnidia of similar size but smaller, narrower conidia measuring 12-20 × 3-5 µ.

T. PERFORANS (Rob.) Höhn., *Hedwigia* **60**: 141 (1918). (Fig. 19)
Sphaeria perforans Rob., *Ann. Sci. nat.* **19**: 357 (1843).
Sphaerella perforans (Rob.) Sacc., *Syll. fung.* **1**: 538 (1882).
Ascochyta perforans (Rob.) Sacc., *Syll. fung.* **3**: 406 (1884).
Diplodina ammophilae Trail, *Scott. Nat.* **8**: 76 (1885).
Darluca ammophilae Sacc., Bomm. & Rouss. in Sacc., *Syll. fung.* **10**: 311 (1892).
Tiarospora westendorpii Sacc. & March., *Rev. Mycol.* **7**: 148 (1885).
Ascochyta psammae Oud., *Ned. Kruidk. Arch.* ser. 3, **2**: 263 (1903).

Conidiomata 300 × 200 µ. Conidiogenous cells 9-22 × 5-8 µ. Conidia 20-27 × 10-15 µ.

On culms of *Ammophila arenaria*, *A. arundinacea*, U.K., Australia (IMI 50141, 82813, 95735*a*, 122101*b*, 115445, 213047, 214369, 214370).

16. NEOTTIOSPORINA Subram., *Proc. nat. Inst. Sci. India* **27B**: 238 (1961).
Sadasivanella Agnihothrudu, *Proc. Ind. Acad. Sci. sect. B*, **60**: 81 (1964).
Sp. typ.: *N. apoda* (Speg.) Subram. (syn. *Cryptostictis apoda* Speg.).

Mycelium immersed, pale brown to hyaline, branched, septate. *Conidiomata* pycnidial, separate, globose, brown, immersed, unilocular, thin-walled; wall of textura angularis. *Ostiole* central, circular, papillate. *Conidiophores* absent. *Conidiogenous cells* holoblastic, determinate, discrete, doliiform, hyaline, smooth, formed from the inner cells of the pycnidial wall. *Conidia* hyaline, 2-3 euseptate, continuous or constricted, apex obtuse, sometimes papillate, base truncate, thin-walled, smooth, cylindrical, navicular to obovoid, at first enclosed or partially enclosed in a gelatinous sheath that may become everted as an apical and basal appendage.

Sutton & Alcorn (1974) redescribed the genus and illustrated six species, and Sutton & Marasas (1976) added a seventh.

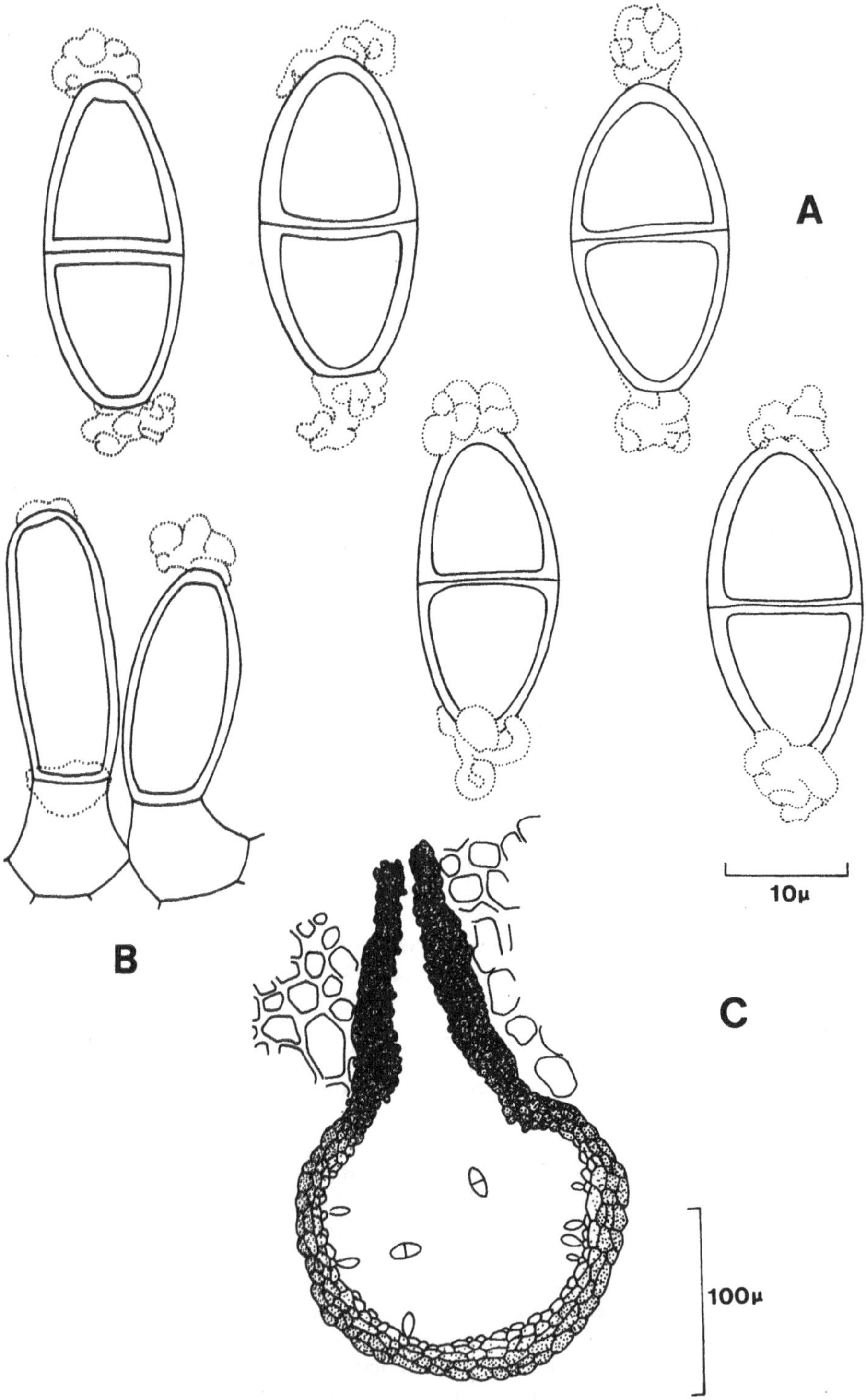

Fig. 19. *Tiarospora perforans*, A, conidia; B, conidiogenous cells and developing conidia; C, vertical section of conidioma.

KEY

1. Conidia 1 septate .... *N. masonii*
1. Conidia 2 septate 3
1. Conidia 3 septate 2
   2. Conidia 14-24 × 3-5 µ .... *N. apoda*
   2. Conidia 26.5-32.5 × 7.5-8.5 µ .... *N. australiensis*
   2. Conidia 45-51 × 10-13 µ .... *N. clavata*
3. Median conidial cell larger than end cells 4
3. Median conidial cell smaller or the same size as end cells 5
   4. Appendage excentric, conical when everted; conidia 21-28 × 8-10 µ .... *N. asymmetrica*
   4. Appendage centric, not often everted; conidia 36-41.5 × 9.5-10.5 µ .... *N. phragmiticola*
5. Appendage 6-13 µ diam.; conidia 6.5-8 µ wide .... *N. paspali*
5. Appendage 16-24 µ diam.; conidia 8-10.5 µ wide .... *N. infundibularis*

N. APODA (Speg.) Subram., *ibid.*: 239 (1961).
Cryptostictis apoda Speg., *An. Mus. Nac. B. Aires* **20**: 374 (1910).

Conidiomata 125-145 µ diam. Conidia 3 septate, 2 median cells shorter than the end cells, 14-24 × 3-5 µ; apical appendage infundibuliform, 4-10 µ diam.; basal appendage 4 µ long sometimes present.

On *Achyrocline saturejoides*, Argentina (LPS 21809, holotype).

N. ASYMMETRICA Sutton & Alcorn, *Austr. J. Bot.* **22**: 523 (1974). (Fig. 21B)

Conidiomata up to 325 µ diam. Conidia 2 septate, median cells longer than the end cells, 21-28 × 8-10 µ; apical appendage conical or displaced to one side and lateral to excentric.

From leaf lesions on *Themeda australis*, Qd., Australia (IMI 176626, holotype, 207695).

N. AUSTRALIENSIS Sutton & Alcorn, *Austr. J. Bot.* **22**: 526 (1974). (Fig. 20B)

Conidiomata 310-390 µ diam. Conidia 3 septate, cells of equal length, 26.5-32.5 × 7.5-8.5 µ; apical appendage infundibuliform, 6-8 µ diam.

From leaf lesions on *Phragmites australis*, Qd., Australia (IMI 176624, holotype).

N. CLAVATA sp. nov. (Fig. 21C)

Coloniae in MA et PDA tarde crescentes. Mycelium superficiale modice abundans, griseum vel album et cremeum, compactum, appressum. Conidiomata superficialia vel semi-immersa sparsa, solitaria, atro brunnea vel nigra, globosa vel subglobosa, 250-350 µ diam.; parietes ex textura angulari, medio brunnea, tenui compositi. Ostiolum singulum, centrale, circulare, 20-40 µ diam. Cellulae conidiogenae holoblasticae, simplices, determinatae, discretae, doliiformes, ex cellulis interioribus parietis pycnidii oriundae, 6.5-9.5 × 8-9 µ. Conidia acrogena, solitaria, hyalina, cylindrica vel clavata, basim truncata, apicem obtusa, semper 2 septata, deinde 3-septata, cellulis basim versus gradatim parvioribus, 45-51 × 10-13 µ; ad apices cum appendice gelatinosa eversa, infundibuliformi, usque ad 9 µ diam.

Ex *Phragmitis australis*, Ormiston, Qd., Australia, 24 Jan. 1977, J.L. Alcorn, 7705b, IMI 211607, holotypus.

Colonies on MA and PDA slow-growing, superficial, moderately abundant, dull grey with irregular white or cream patches, compact, often appressed to the agar. Conidiomata superficial to semi-immersed, comparatively sparse, solitary, dark brown to black, globose to subglobose, 250-350 µ diam.; walls composed of thin, medium brown textura angularis. Ostiole central, circular, 20-40 µ diam. Conidiogenous cells holoblastic, simple, determinate, discrete, doliiform, formed from the inner cells of the pycnidial wall, 6.5-9.5 × 8-9 µ. Conidia acrogenous, solitary, hyaline, cylindrical to clavate, base truncate, apex obtuse, often 2-septate, but finally 3-septate, the cells progressively smaller towards the base, 45-51 × 10-13 µ; apical appendage everted, gelatinous, infundibuliform to irregular, up to 9 µ diam.

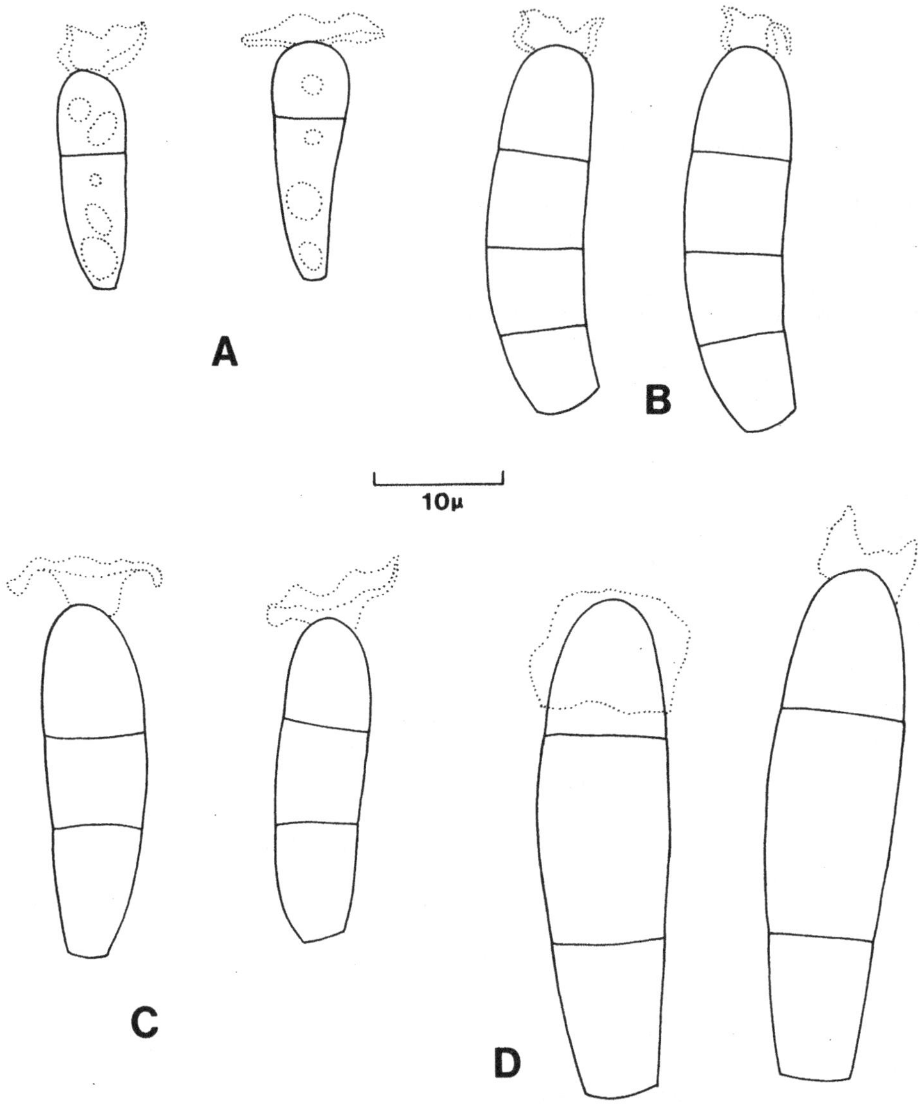

Fig. 20. A, *Neottiosporina masonii*, conidia; B, *N. australiensis*, conidia; C, *N. paspali*, conidia; D, *N. phragmiticola*, conidia.

The species is distinguished from those already described by the very large conidia bearing a comparatively small apical appendage. Most *Neottiosporina* species are consistent in septation, but *N. clavata* may be 2 or 3 septate.

N. INFUNDIBULARIS Sutton & Alcorn, *Austr. J. Bot.* **22**: 521 (1974). (Fig. 21A)

Conidiomata 110-240 μ diam. Conidia 2 septate, median cells smaller than the end cells, apex and base sometimes papillate, 26-33 × 8-10.5 μ; apical appendage infundibuliform, 16-24 μ diam.; basal appendage 2-4 μ diam.

From leaf lesions on *Paspalum distichum*, Qd., Australia (IMI 173925, holotype).

N. MASONII Sutton apud Sutton & Marasas, *Trans. Br. mycol. Soc.* **67**: 69 (1976). (Fig. 20A)

Conidiomata up to 300 μ diam. Conidia 1 septate, apical cell smaller than the basal cell, 15.5-19 × 4.5-6 μ; apical appendage infundibuliform, 6.5-11.5 μ diam.

From *Pinus caribaea*, Tanzania (IMI 118368, holotype).

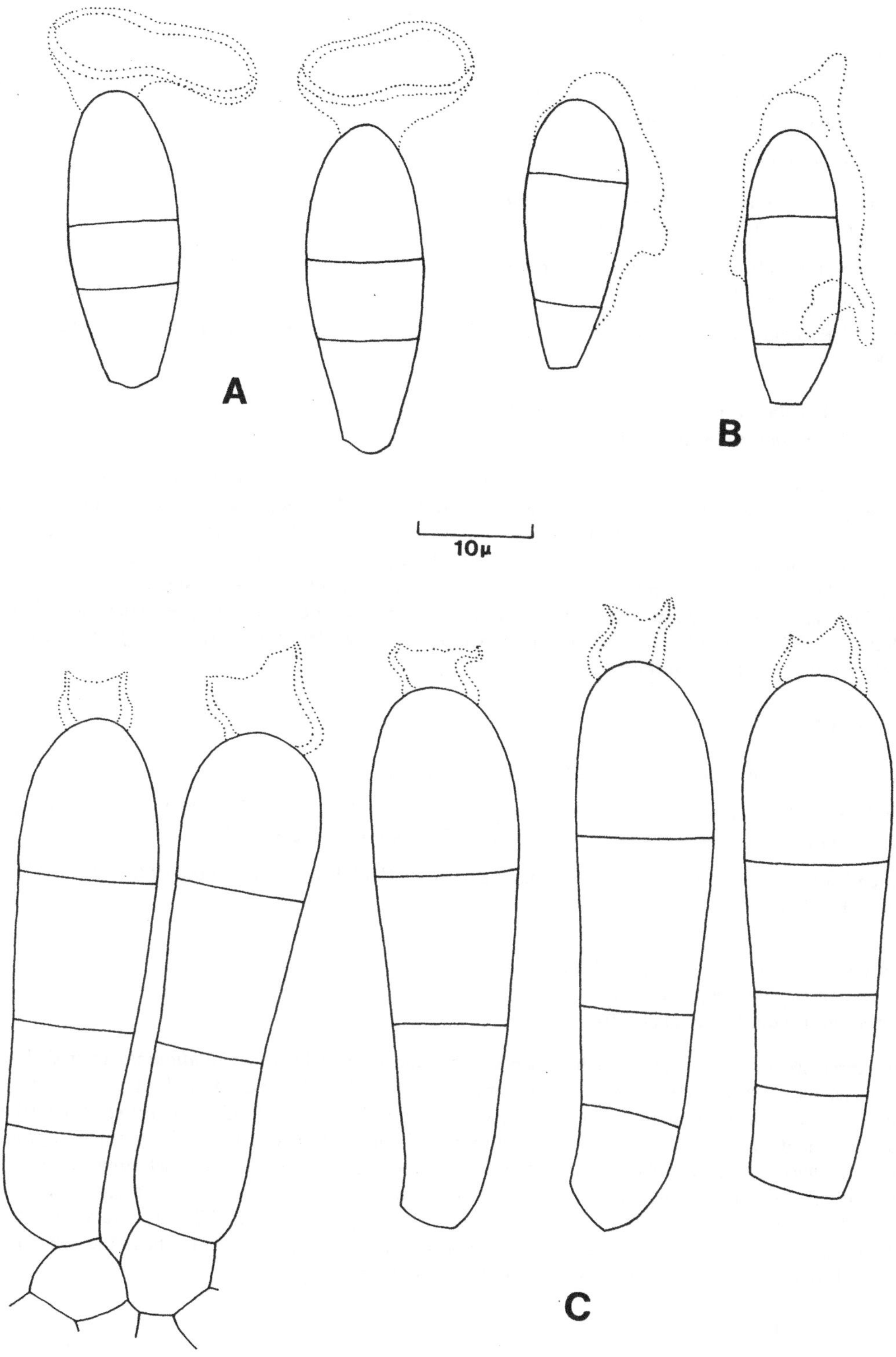

Fig. 21. A, *Neottiosporina infundibularis*, conidia; B, *N. asymmetrica*, conidia; C, *N. clavata*, conidia.

N. PASPALI (Atk.) Sutton & Alcorn, *Austr. J. Bot.* **22**: 519 (1974). (Fig. 20C)
Stagonospora paspali Atk., *Bull. Cornell Univ.* **3**: 33 (1897).

Conidiomata up to 115 µ diam. Conidia 2 septate, cells of equal length or the median one smaller, 26-33 × 6.5-8 µ; apical appendage infundibuliform, 6.5-13 µ diam.

On leaves of *Paspalum laeve*, Ala., U.S.A. (IMI 175641, holotype ex CUP).

N. PHRAGMITICOLA (Hansf.) Sutton & Alcorn, *Austr. J. Bot.* **22**: 525 (1974). (Fig. 20D)
Stagonospora phragmiticola Hansf., *Proc. Linn. Soc. Lond.* **157**: 38 (1945).

Conidiomata 80-110 µ diam. Conidia 2 septate, median cell longer than the end cells, 36-41 × 9.5-10.5 µ; apical appendage infundibuliform, up to 12 µ diam.

On leaves of *Phragmites mauritianus*, *P. communis*, Uganda, Sudan, Ethiopia (IMI 4743, holotype, 34250, 34253, 40125, 88571).

16. MYCOTRIBULUS Nag Raj & Kendrick, *Can. J. Bot.* **48**: 2219 (1970).
Sp. typ.: *M. mirabilis* Nag Raj & Kendrick

Mycelium immersed, pale brown to brown, branched, septate. *Conidiomata* pycnidial, amphigenous, separate or confluent, globose, dark brown, immersed, unilocular, thick-walled; wall of two layers, outer of brown, thick-walled, irregular cells, inner of hyaline, thin-walled, textura angularis. *Ostiole* absent, dehiscence by irregular rupture. *Conidiophores* hyaline, branched and septate at the base, smooth, with acrogenous conidia, formed from the inner cells of the pycnidial wall. *Conidiogenous cells* holoblastic, integrated, determinate, cylindrical, hyaline, smooth. *Conidia* hyaline, aseptate, navicular, base truncate, apex acute with a single simple or branched cellular appendage, smooth, thin-walled, guttulate, with 2-4 basal unbranched cellular appendages inserted laterally. *Paraphyses* hyaline, septate, smooth, simple or branched, deeply lobed and irregular at the apices.

The genus is monotypic.

M. MIRABILIS Nag Raj & Kendrick, *ibid.*: 2219 (1970). (Fig. 22)

Conidiomata 200-220 × 230-275 µ diam. Conidiophores 7.5-16.5 × 1-2 µ. Conidia 9-15 × 2-4.5 µ; apical appendage 7.5-11 µ long, basal appendages 7.5-10 µ long. Paraphyses 33-55 × 1-1.5 µ.

On rotten leaves of *Eucalyptus grandis*, *E.* sp., Brazil, India (IMI 128041, holotype, 192514*a*, 194561*e*, 186979*c*, 186978*l*, 186977*p*, 196484*i*, 196481*m*, 196482*b*).

17. MONODIA Breton & Faurel, *Rev. Mycol.* **35**: 23 (1970).
Sp. typ.: *M. elegans* Breton & Faurel
Illustration: Breton & Faurel (1970).

Mycelium immersed to superficial, branched, septate, hyaline to pale brown. *Conidiomata* pycnidial, separate or gregarious, globose, dark brown, immersed or semi-immersed, unilocular, thin-walled; wall of pale brown, thin-walled textura angularis. *Ostiole* single, circular, distinctly papillate or with a short rostrum. *Conidiophores* branched at the base, hyaline, septate, ± cylindrical, smooth, formed from the inner cells of the pycnidial wall, with acropleurogenous conidia. *Conidiogenous cells* illustrated as holoblastic but quite possibly enteroblastic and phialidic, determinate, integrated, cylindrical, hyaline, smooth, conidia formed immediately below transverse septa. *Conidia* hyaline, aseptate, cylindrical, apex obtuse, thin-walled, smooth, eguttulate, formed as in *Strasseria* at the apex of a basal filiform, cellular appendage; with 2 divergent, unbranched, subapical cellular appendages, 1 similar appendage inserted laterally above the conidial base, and a single basal appendage at the base of which the conidium becomes abstricted.

The genus is monotypic.

M. ELEGANS Breton & Faurel, *ibid.*: 24 (1970).

Conidiomata up to 300 µ diam., shortly rostrate. Conidiophores up to 70 µ long. Conidia 18-25 × 5-8 µ; appendages 35-40 µ long.

On dung of *Procavia rufipes*, Chad Republic.

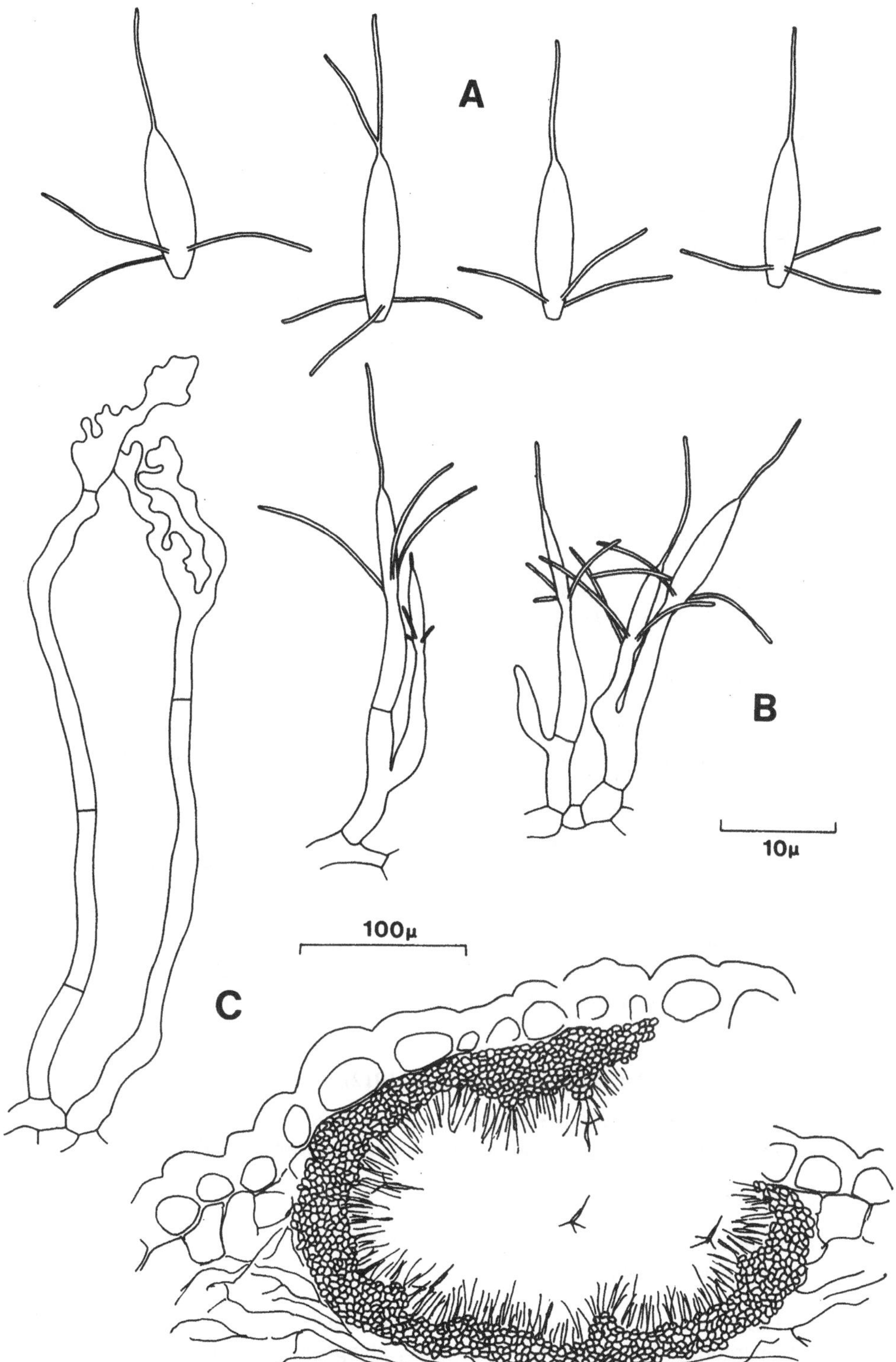

Fig. 22. *Mycotribulus mirabilis*, A, conidia; B, conidiophores and developing conidia; C, paraphyses; D, vertical section of conidioma.

18. SCOPAPHOMA Dearn. & House, *Bull. N.Y. St. Mus.* 266: 83 (1925).
Sp. typ.: *S. corioli* Dearn. & House

Mycelium immersed, branched, septate, hyaline. *Conidiomata* pycnidial, aggregated but not confluent, globose, black; walls of even thickness throughout, composed of an outer layer 1-2 cells thick, of dark brown, thick-walled textura angularis, a middle layer of much larger, thin-walled, paler tissue, and an inner layer of elongated, hyaline, thin-walled cells. *Ostiole* central, circular, solitary, almost papillate. *Conidiophores* 1-2 septate, tapered to the apices, branched at the base and just below transverse septa, hyaline, formed from the cells lining the conidiomatal wall. *Condiogenous cells* holoblastic, integrated, determinate, hyaline, forming conidia apically or laterally just below transverse septa. *Conidia* hyaline, smooth, thin-walled, consisting of two, cylindrical, guttulate, central cells with cytoplasmic contents, and a median septum; the basal cell is thinner and devoid of contents; the apical cell is long conic, the apex drawn out into an unbranched filiform appendage.

The genus is monotypic.

S. CORIOLI Dearn. & House, *ibid.*: 83 (1925). (Fig. 23)

Conidiomata up to 600 μ high × 450 μ wide. Conidiophores up to 30 μ long × 1.5-3.5 μ. Conidia: median cells 17-19.5 × 2-2.5 μ; basal cell 9.5-11.5 × 1 μ; apical cell (appendage) 12-45 μ long.

On upper surface of fructification of *Coriolus versicolor*, N.Y., U.S.A. (IMI 197180, holotype ex NYS).

19. JAPONIA Höhn., *Sber. Akad. Wiss., Wien* **118**: 879 (1909).
Yoshinagamyces Hara, *Bot. Mag., Tokyo* **26**: 143 (1912).
Sp. typ.: *J. quercus* Höhn.
Illustration: Morgan-Jones & Kendrick (1972).

Mycelium partly immersed, partly superficial, pale brown to hyaline, branched, septate. *Conidiomata* pycnidial, epiphyllous, separate, globose to discoid, dark brown, superficial, unilocular, thick-walled, outer layer of dark brown, thick-walled textura angularis. *Ostiole* absent, dehiscence by irregular rupture. *Conidiophores* hyaline, unbranched, septate, smooth, with acrogenous conidia, formed from the base and sides of the pycnidial wall. *Conidiogenous cells* holoblastic, determinate, integrated, cylindrical, hyaline, smooth. *Conidia* hyaline, 2 euseptate, median cell longer and guttulate, ± continuous, navicular, thin-walled, smooth, base truncate, apex extended into a single cellular irregularly branched appendage.

Japonia was figured and described without comment by Morgan-Jones & Kendrick (1972). Hara (1954) included *J. quercus* and *Y. quercus* in synonymy with *Monoloculia* Hara and although previous references to the name are cited, none have been traced. Whether this is a name for a perfect state or an earlier name for this conidial fungus cannot be ascertained in the absence of a description. The genus contains a single species.

J. QUERCUS Höhn., *ibid.*: 879-880 (1909).
Yoshinagamyces quercus Hara, *Bot. Mag., Tokyo* **26**: 143 (1912).

Conidiomata up to 550 μ diam. × 200 μ high. Conidiophores 50-65 × 3 μ. Conidia 44-56 × 8.5-9.5 μ; appendage branches up to 28 μ long (much longer, up to 100 μ in original description of *Y. quercus*).

On leaves of *Quercus glauca*, Japan.

20. CHAETOSPORA Faurel & Schotter, *Rev. Mycol.* **30**: 149 (1965).
Sp. typ.: *C. quezelii* Faurel & Schotter

Mycelium immersed to superficial, branched, septate, hyaline to pale brown. *Conidiomata* pycnidial, innate then erumpent, subglobose to pyriform, black, separate, thin-walled, composed of textura angularis. *Setae* ostiolar, rigid, straight or flexuous, acuminate, dark brown. *Ostiole* large, circular, central. *Conidiophores* absent. *Conidiogenous cells* holoblastic, discrete, determinate, lageniform, smooth, pale brown, formed from the inner cells of the pycnidial wall. *Conidia* hyaline, smooth, eguttulate, cylindrical to fusiform, 4 euseptate, constricted at the septa, median cells unequal, the lower being longer, apical cell modified with 3-4, cellular, 1-2 septate, filiform, straight or curved, unbranched appendages.

The genus is monotypic.

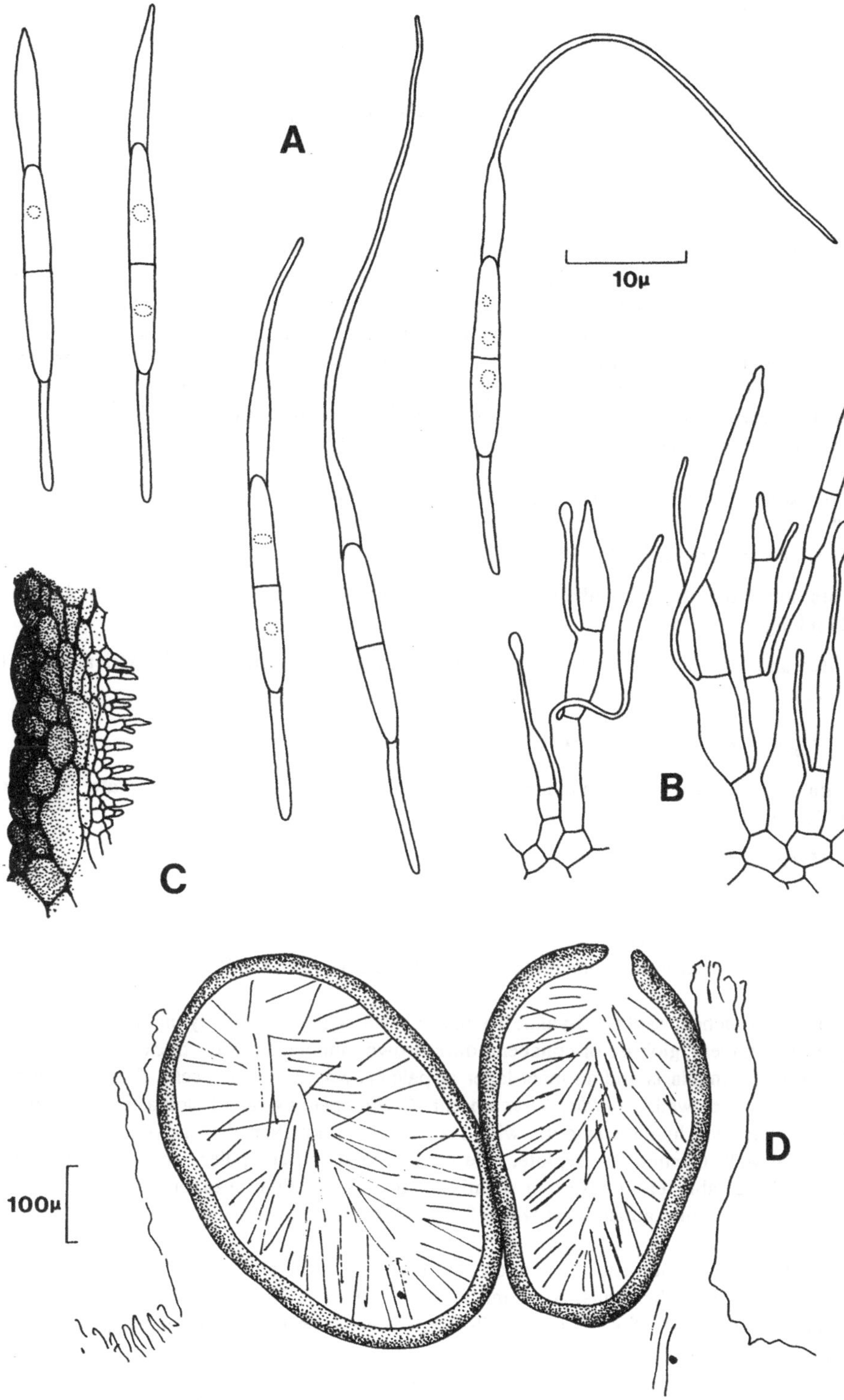

Fig. 23. *Scopaphoma corioli*, A, conidia; B, conidiophores and developing conidia; C, detail of conidiomatal wall; D, vertical section of conidiomata.

C. QUEZELII Faurel & Schotter, *ibid.*: 150 (1965). (Fig. 24)

Conidiomata 125-175 µ diam.; ostiole 25-50 µ diam.; setae 50 × 5-7 µ. Conidia 45-60 × 6-8 µ; appendages 60-80 µ long.

On dung of *Massoutiera harterti*, Central Sahara (IMI 199625, holotype).

21. HYALOTIELLA Papendorf, *Trans. Br. mycol. Soc.* **50**: 69 (1967).
Sp. typ.: *H. transvalensis* Papendorf

Mycelium immersed, branched, septate, hyaline to pale brown. *Conidiomata* eustromatic, immersed, subepidermal, separate or aggregated, globose, lobed or irregular, dark brown, mostly unilocular but sometimes divided; walls thick, composed of an outer layer of dark brown, thick-walled textura angularis and an inner layer of paler thin-walled tissue. *Ostiole* present or absent, papillate, circular, central. *Conidiophores* filiform, septate, branched only at the base, hyaline, formed all round the pycnidial wall. *Conidiogenous cells* holoblastic, occasionally sympodial, indeterminate, discrete or integrated, cylindrical, hyaline, smooth. *Conidia* 3- euseptate, pale brown except the apical cell which is hyaline with a single cellular appendage with 2-4 branches near the point of origin, smooth, eguttulate, cylindrical or slightly curved, 2 median cells much longer than the end cell.

The genus was redescribed by Nag Raj (1975) who accepted two species.

H. TRANSVALENSIS Papendorf, *ibid.*: 69 (1967). (Fig. 25)

Conidiomata 200-800 µ diam. Conidiophores 8-20 × 1.5-4.5 µ. Conidia 3-septate, 19-26 × 2.5-3 µ; apical appendage 12-27 µ long.

From leaf litter and leaves of *Acacia auriculaeformis*, *Eucalyptus citriodora*, S. Africa, India (IMI 161774, 137470, 210125).

H. ORIENTALIS Nag Raj, *Can. J. Bot.* **53**: 1619 (1975).

Conidiomata 170-270 µ diam. Conidiophores up to 20 µ long. Conidia 3(-4) septate, 24-29 × 2-4 µ; apical appendage 12-22 µ long.

On leaves of *Mimusops hexandra*, India.

22. HYALOTIOPSIS Punithalingam, *Mycol. Pap.* 119: 12 (1969).
Sp. typ.: *H. subramanianii* Punithalingam (as '(Agnihothrudu) comb. nov.').

Mycelium immersed, branched, septate, hyaline to pale brown. *Conidiomata* pycnidial, immersed, subepidermal, separate or aggregated to confluent, globose or collabent, dark brown, unilocular; walls thick, composed of an outer layer of brown, thick-walled textura angularis and an inner layer of pale brown, thin-walled cells. *Ostiole* absent, dehiscence by irregular breakdown of the upper wall. *Conidiophores* absent. *Conidiogenous cells* holoblastic, occasionally sympodial, indeterminate, discrete, cylindrical, hyaline, smooth, with 1-2 irregular unthickened conidiogenous loci, formed from the inner cells of the pycnidial wall. *Conidia* 3-euseptate, pale brown, each end cell paler, smooth, eguttulate, cylindrical or slightly curved, size of cells variable (end ones often slightly longer than median ones, or, penultimate apical cell smaller than the rest), apical cell with 2 cellular appendages which form 2-4 branches a short distance from the point of origin.

Thaung (1975) has clarified the nomenclatural anomalies concerning the publication of the name of the type species and Nag Raj (1975) has redescribed the genus. A second species *H. borassi* Thaung (1975) was added. It differs from *H. subramanianii* by smaller conidia (14-18 × 4-5 µ) with 2-4 separately inserted apical appendages. It is doubtfully congeneric with *H. subramanianii*.

H. SUBRAMANIANII Punithalingam (as '(Agnihothrudu) comb. nov.'), *ibid.*: 14 (1969). (Fig. 26)
Hyalotiopsis subramanianii Agnihothrudu & P. Luke, *Proc. Ind. Acad. Sci.* **71**: 48 (1970).

Teleomorphosis: *Lepteutypa indica* (Punithalingam) Arx, *The genera of fungi sporulating in pure culture*: 118 (1970).

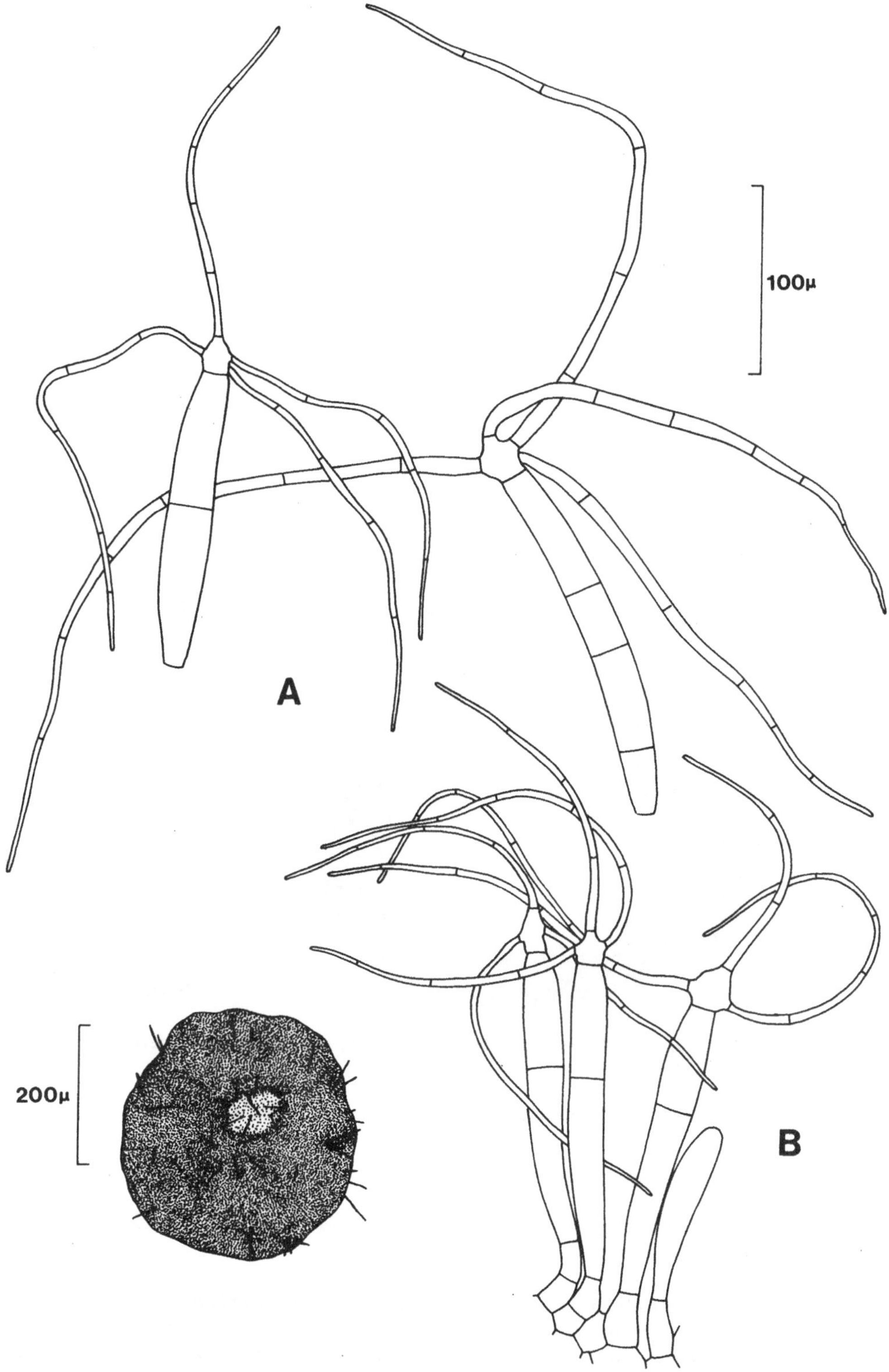

Fig. 24. *Chaetospora quezelii*, A, conidia; B, conidiogenous cells and developing conidia; C, surface view of a conidioma.

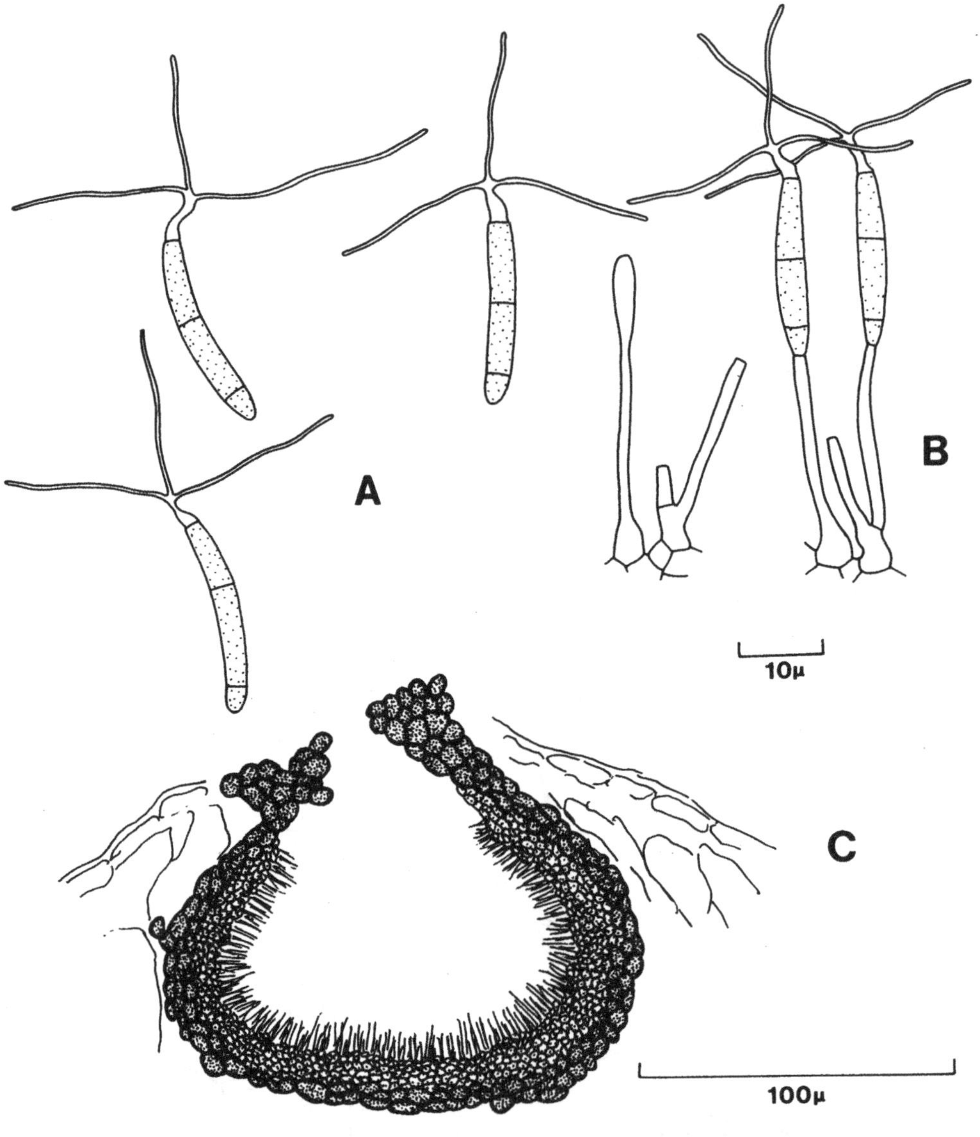

Fig. 25. *Hyalotiella transvalensis*, A, conidia; B, conidiophores and developing conidia; C, vertical section of a conidioma.

Conidiomata up to 250 µ diam. Conidiogenous cells 6-8 × 3-4 µ. Conidia 22-40 × 4.5-5.5 µ; appendages 30-50 µ long.

On and isolated from *Tamarindus indica*, India (IMI 136542, 142374).

23. LABRIDELLA Brenckle, *Fungi Dakotenses* 663 (Oct. 1929), and *Mycologia* **22**: 161 (1930).
Sp. typ.: *L. cornu-cervae* Brenckle

Mycelium partly superficial, mostly immersed, brown, septate, branched. *Conidiomata* pycnidial, superficial to semi-immersed, separate, globose, dark brown, unilocular, thick-walled; walls formed of dark brown, thick-walled textura intricata. *Ostiole* absent, dehiscence by irregular to longitudinal fissure. *Conidiophores* absent. *Conidiogenous cells* holoblastic, discrete, determinate, cylindrical, hyaline, smooth, with acrogenous conidia, formed from the inner cells of the pycnidial wall. *Conidia* fusiform, straight or slightly curved, 5 euseptate, ± constricted, 5 lower cells medium brown, thick-walled, smooth, basal cell truncate, apical cell hyaline, thin-walled, smooth, prolonged into a tapered, cellular appendage that has 1-5 lateral branches usually on one side, main and lateral branches may be dichotomously branched.

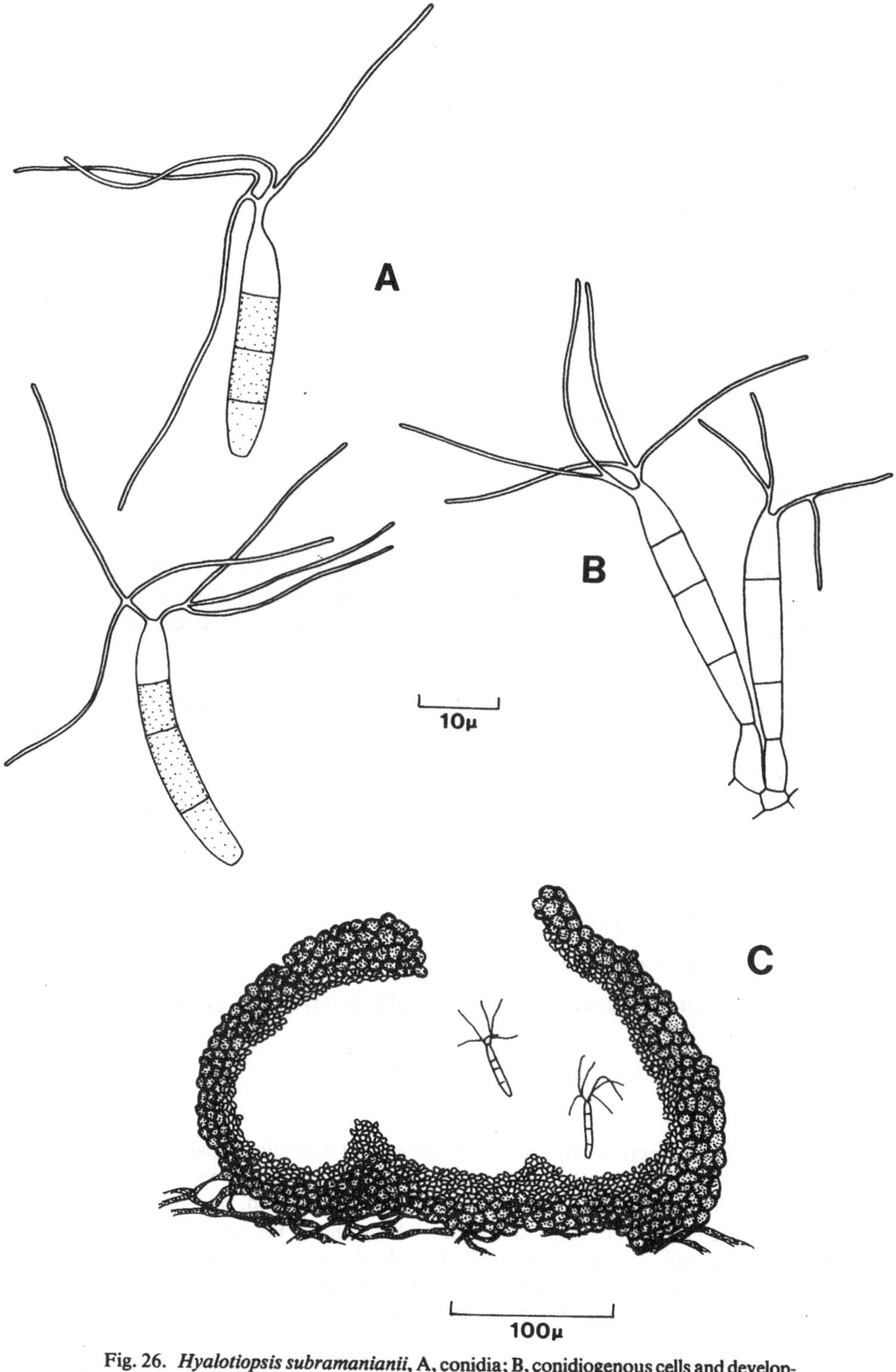

Fig. 26. *Hyalotiopsis subramanianii*, A, conidia; B, conidiogenous cells and developing conidia; C, vertical section of a conidioma.

This monotypic genus was redescribed and illustrated by Sutton (1969a) and the correlation with *Griphosphaerioma* made by Shoemaker (1963).

L. CORNU-CERVAE Brenckle, *Fungi Dakotenses* 663 (Oct. 1929), and *Mycologia* **22**: 161 (1930). (Fig. 27)
Pestalotia cornu-cervae (Brenckle) Guba, *A monograph of Monochaetia and Pestalotia*: 251 (1961).
Pestalotia pezizoides de Not. forma *longiseta* Dearn. in Bisby et al., *Fungi of Manitoba*: 135 (1929).

Teleomorphosis: *Griphosphaerioma kansensis* (Ell. & Ev.) Shoemaker, *Can. J. Bot.* **41**: 1419 (1963).

Conidiomata 600 × 200 µ. Conidiogenous cells 5-15 × 2 µ. Conidia 8-10 µ; appendages 10-35 µ long.

On twigs of *Symphoricarpos occidentalis*, S. Dak., N. Dak., U.S.A., Man., Canada (IMI 26628, 13975, 26637, 10883, 92706 ex DAOM, 92707 ex DAOM, 92708 ex DAOM, 10882, isotype).

24. CAPNODIASTRUM Speg., *An. Soc. Cient. Arg.* **22**: 145 (1886).
Original spp.: *C. guaraniticum* Speg., *C. paraguayense* Speg.
Lectotype sp.: *C. guaraniticum* Speg., fide Arnaud (1918), Clements & Shear (1931).
Illustrations: Arnaud (1918).

Mycelium superficial, dark brown, branched, septate, with alternate, lateral, irregular, often recurved hyphopodia. *Conidiomata* pycnidial, superficial, separate, globose, dark brown, unilocular, thin-walled; walls 1 cell thick, of thin-walled textura angularis or textura globulosa. *Ostiole* absent, dehiscence by irregular to stellate rupture of the upper wall. *Conidiophores* absent. *Conidiogenous cells* holoblastic, determinate, discrete, doliiform, hyaline, smooth, formed from the inner cells of the pycnidial wall. *Conidia* pale brown, 1 euseptate, base truncate, slightly protuberant, apex obtuse to conic, obovoid, thick-walled, with a central guttule and a median transverse hyaline band; upper cell small, pale brown, lower cell much larger.

Arnaud (1918) referred to *C. guaraniticum* as the 'pseudotype' and regarded both original species as probably identical. He scarcely differentiated *Capnodiastrum* from *Asterostomella*, considering *Capnodiastrum* merely as an *Asterostomella* without the radial Leptostromataceous conidioma. In conidiogenesis and conidial morphology the two genera are identical.

Additional references to *Capnodiastrum* are provided by Farr (1973) and it would be superfluous to repeat the superb illustrations by Arnaud (1918). So far 12 species have been described in the genus.

C. GUARANITICUM Speg., *ibid.*: 145 (1886). (Fig. 28)
C. paraguayense Speg., *ibid.*: 146 (1886).

Conidiomata 50-60 µ diam. Conidia 16-18 × 9-11 µ.

On leaves of *Celtis boliviensis*, Paraguay (IMI 196608 ex LPS, holotype of *C. guaraniticum*).

25. DIPLODIA Fr. apud Mont., *Ann. Sci. nat.*, ser. 2, **1**: 302 (1834).
Sp. typ.: *D. mutila* Fr.

Mycelium immersed, branched, septate, dark brown. *Conidiomata* pycnidial, separate or aggregated, globose, dark brown to black, immersed, unilocular, thick-walled; wall of an outer layer of dark brown, thick-walled textura angularis, a median layer of dark brown thin-walled cells, and an inner layer of thin-walled, hyaline cells. *Ostiole* single, circular, central, papillate. *Conidiophores* hyaline, branched and septate above and at the base, smooth, cylindrical, formed from the inner cells of the pycnidial wall. *Conidiogenous cells* holoblastic, integrated or discrete, determinate, cylindrical, hyaline, smooth, forming a single apical conidium. *Conidia* at first hyaline, with a central guttule, thick-walled, aseptate, smooth, later becoming dark brown and medianly 1 euseptate, apex obtuse, base truncate.

The history of *Diplodia mutila* has been reviewed by Stevens (1933) and basically I agree with his interpretation but differ in some minor points. Fries (1823) described *Sphaeria mutila* from *Populus* but gave no details of spores, only of the fructifications and stromata. The two exsiccati distributed under this name as *Scler. Suec.* 164 and 385 were cited later by Fries (1849) when he included *D. mutila* Fr. amongst the several species referred to *Diplodia* Fr. Meanwhile, Montagne sent Fries a fungus on *Populus* which he identified as *S. mutila*. The record was listed by Montagne (1834) under *S. mutila*

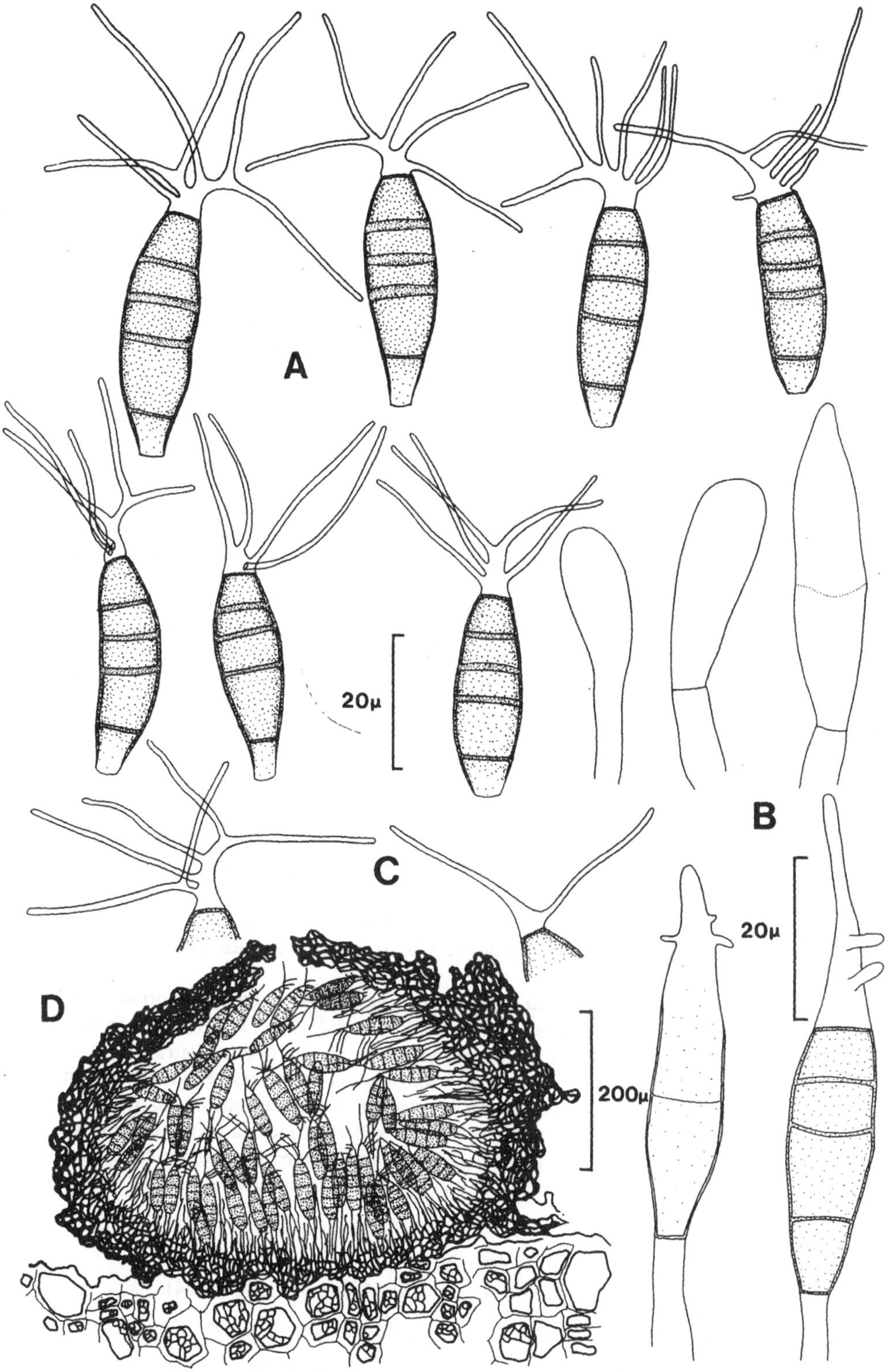

Fig. 27. *Labridella cornu-cervae*, A, conidia; B, developing conidia; C, branched conidial appendages; D, vertical section of a conidioma.

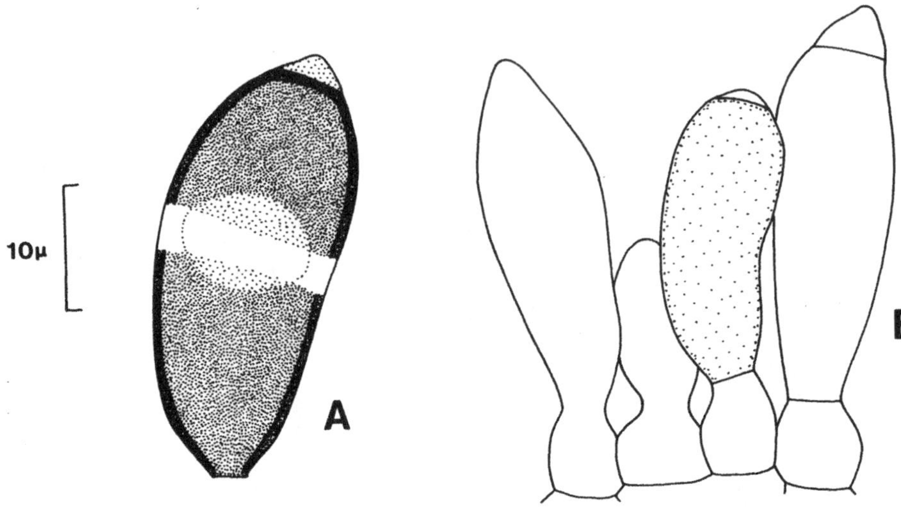

Fig. 28 *Capnodiastrum guaraniticum*, A, conidium; B, conidiogenous cells with developing conidia.

Fr. with the note that this species would become the type of a new genus, *Diplodia*, characterised by Fries. There followed a generic description and citation of Montagne's specimen.

Stevens (1933) examined material of *Scler. Suec.* 164 and 385 from BPI and found no spores. Similar material in K showed 385 to be quite effete. It is difficult to say whether the material is pycnidial or ascomycetous. One sample of 164 was an ascomycete of the *Botryosphaeria* type whilst another looked very much like an effete *Tympanis.* Thus no fertile pycnidial fungus was found by Stevens (1933) and from examination of additional material I confirm this. On the other hand Stevens (1933) examined several slides from Montagne's exsiccatus in STR and found pycnidia with a mixture of hyaline and 1-septate brown conidia, mostly 25-27 × 10-12 µ. A specimen in K labelled '*Diplodia mutila* Fr. in litt. ad cort. *Populi nigr.* (Ardenne) herb Montagne' has been studied and agrees in all respects with the account by Stevens (1933). It may well be a piece of Montagne's original material.

Stevens (1933) adopted the name *Diplodia mutila* (Fries) Mont. for the pycnidial fungus with 1- septate conidia. However, this conclusion denies the fact that Fries' material is not pycnidial. A lectotype should be chosen for Fries' *Sphaeria* name and *S. mutila* should be adopted for that ascomycetous element that it represents. This clearly has a bearing on the nomenclature of *Botryosphaeria.*

The name of the pycnidial fungus dates from Montagne (1834) and it is typified by Montagne's material. The correct citation is therefore *Diplodia mutila* Fr. apud Montagne (1834) and this species, being the only one mentioned in this publication is the type.

Clements & Shear (1931) adopted *Diplodia* dating from Fries (1849) as most authorities have done. The citation of the type species as '*D. mutica* F. & M.' is obviously an error. Petrak & Sydow (1927) placed *Sphaeria malorum* Berk. and *Sphaeropsis malorum* Berk. in synonymy with *Botryodiplodia malorum* (Berk.) Petrak & Syd., but these names are regarded as synonyms of *D. mutila* by Stevens (1933) who quoted an extensive synonymy. Shoemaker (1964b) has clarified the identity of certain *Botryosphaeria* species; in which *B. stevensii* Shoem. was regarded as the ascomycetous state of *D. mutila.* Zambettakis (1954) in a reassessment of generic concepts in the didymosporous Sphaeropsidales erroneously adopted *D. conigena* Desm. as the type for *Diplodia* and referred *D. mutila* to the genus *Metadiplodia* Syd.

The genus is in need of revision. Zambettakis (1954) accepted 24 species and so far about 900 names have been proposed for taxa at species level and below.

D. MUTILA Fr. apud Mont., *ibid.*: 302 (1834). (Fig. 29)
Teleomorphosis: *Botryosphaeria stevensii* Shoemaker, *Can. J. Bot.* **42**: 1299 (1964).

Conidiomata up to 600 µ diam. Conidiophores up to 45 µ long × 3-5 µ wide. Conidia (mature) 27-31 × 12-13.5 µ.

On *Acacia arabica*, *A. farnesiana*, *A. modesta*, *A.* sp., *Fraxinus excelsior*, *F.* sp., *Lycopersicon esculentum*, *Malus pumila*, *M. sylvestris*, *Pinus caribaea*, *P. patula*, *Piper betle*, *Populus nigra*, *Prunus armeniaca*, *P. persica*, *Pyrus communis*, *P.* sp., *Spilanthes urens*, *Solanum tuberosum*, *Syringa vulgaris*, *Thuja compacta*, *Ulmus* sp., *Vitis* sp.,

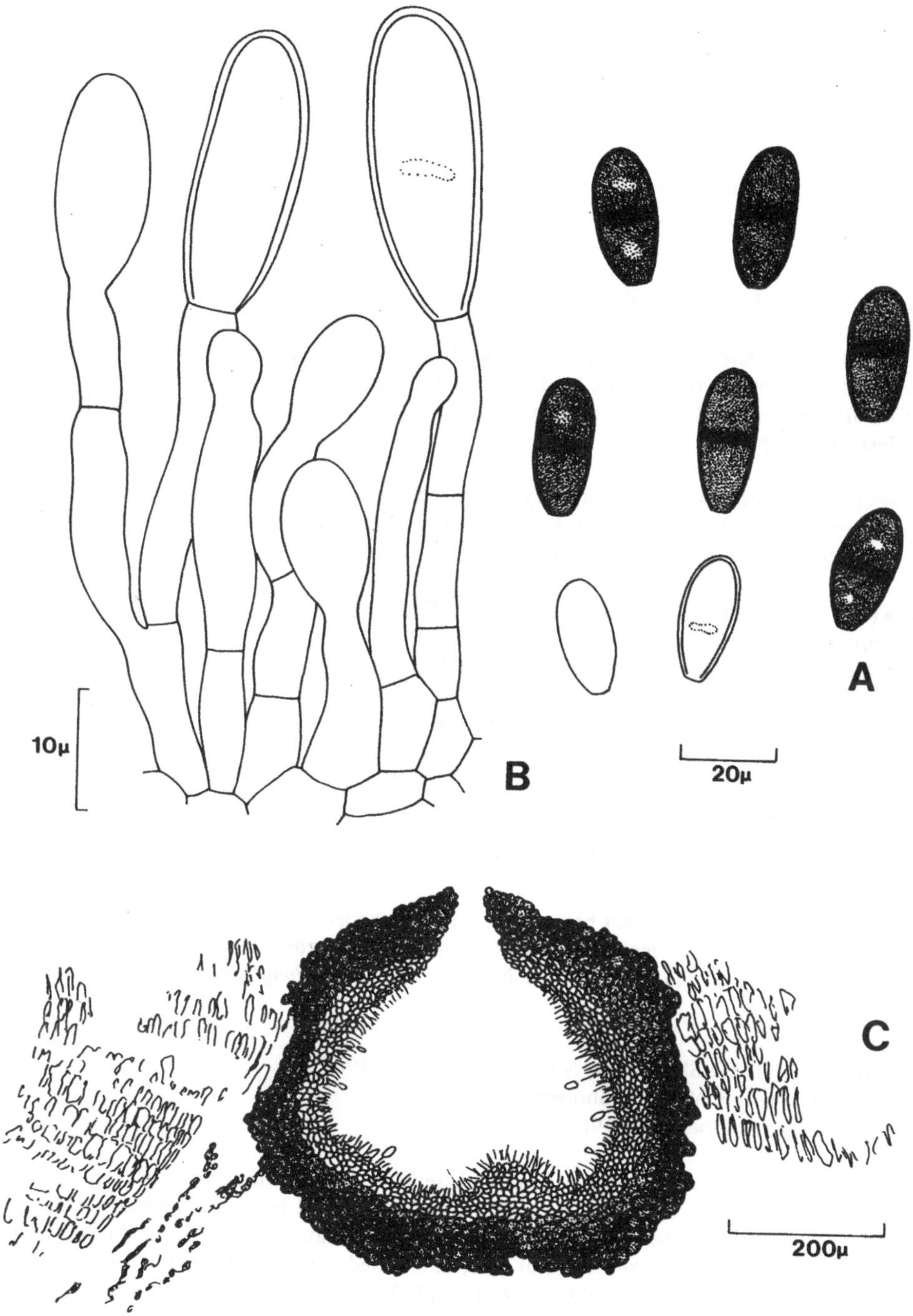

Fig. 29. *Diplodia mutila*, A, conidia; B. conidiophores with developing conidia; C, vertical section of a conidioma.

G.B., France, Eire, Cyprus, U.S.S.R., New Zealand, Sarawak, Australia, Kenya, India, Pakistan, Cuba (IMI 140246, 83958, 85821, 93004, 138614, 133495, 31012, 110089b, 105573, 6402-8, 369, 14504, 69466, 71040b, 57101, 68039a, 69064, 124598, 33448, 16316, 144071a, 155567b, 155571, 134860, 152159, 199755-7, 129977, 130302a, 115484, 70292, 180998, 136965, 124531, 130227, 12937, 139150, 1386, 191526, 181015, 200274, isotype of *D. mutila* ex K).

26. KAMATELLA Anahosur, *Bull. Torrey bot. Cl.* **96**: 207 (1969).
Sp. typ.: *K. apiospora* (Cke & Mass.) Melnik & Sutton (syn. *Ascochyta apiospora* Cke & Mass., *K. indica* Anahosur).

Mycelium immersed, pale brown to hyaline, branched, septate. *Conidiomata* pycnidial, epiphyllous, separate, globose to irregular, brown, immersed, unilocular but convoluted, thin-walled; walls composed of textura angularis, thicker and darker around the ostiole. *Ostiole* circular, papillate, central or excentric. *Conidiophores* absent. *Conidiogenous cells* holoblastic, determinate, discrete, swollen at the base, filiform above, hyaline, smooth, formed from the inner cells of the pycnidial wall. *Conidia* 1 euseptate, lower cell smaller, conic, but truncate at the base, upper cell larger, globose, with a guttule, smooth, dark brown, thick-walled.

The type species often occurs mixed with *Mycohypallage congesta* (Berk. & Br.) Sutton and *Deshpandiella jambolana* (Ramakr. et al.) Kamat & Ullasa apud Ullasa & Rao (1973) and it seems not inconceivable that two, if not all three, are different stages in the life cycle of a single species.

K. apiospora (Cke & Mass.) Melnik & Sutton comb. nov. (Fig. 30)
Ascochyta apiospora Cke & Mass., *Grevillea* **15**: 98 (1886).
Kamatella variispora (Died.) Sutton, *Mycol. Pap.* 141: 105 (1977).
Diplodia variispora Died., *Annls mycol.* **14**: 196 (1916).
Muelleromyces variisporus (Died.) Ullasa, *Bull. Torrey bot. Cl.* **98**: 3 (1971).
Kamatella indica Anahosur, *Bull. Torrey bot. Cl.* **96**: 207 (1969).
Diplodia longipedicellata T.S. & K. Ramakr., *Proc. Indian Acad. Sci.* sect. B, **32**: 78 (1950).
Kamatia longipedicellata (T.S. & K. Ramakr.) Ullasa, *Bull. Torrey bot. Cl.* **98**: 3 (1971).

Conidiomata 200-250 µ diam. Conidiogenous cells 14-21 × 3.5-5 µ wide at the base, tapering to 1 µ. Conidia 14-15 × 12-12.5 µ.

The nomenclature of this species has been discussed by Seshadri & Shaw (1978).
Often associated with lesions on leaves of *Eugenia jambolana, E. grandis* and *E. cordatum*, India, S. Africa (IMI 145503, holotype of *Kamatella indica*, 101866, holotype of *Diplodia longipedicellata*, 45164b).

27. NUMMOSPORA E. Müller & Shoemaker, *Nova Hedwigia* **7**: 1 (1964).
Sp. typ.: *N. hexamera* E. Müller & Shoemaker
Illustration: Müller & Shoemaker (1964).

Mycelium immersed, pale brown to hyaline, branched, septate. *Conidiomata* pycnidial, separate, globose, medium brown, immersed, unilocular, thin-walled; wall of pale brown, thin-walled textura angularis. *Ostiole* central, circular, papillate. *Conidiophores* absent. *Conidiogenous cells* holoblastic, determinate, discrete, ampulliform to lageniform, hyaline, smooth, formed from the inner cells of the pycnidial wall, and mixed with hyphae that ramify through the developing conidial mass. *Conidia* vertically flattened, reddish brown, with 3 transverse eusepta and 2 longitudinal eusepta in the middle cells, not constricted, circular in shape, wall thick, smooth, guttulate.

The genus is monotypic.

N. HEXAMERA E. Müller & Shoemaker, *ibid.*: 1 (1964).

Conidiomata 200-320 µ diam. Ostiole 15-30 µ diam. Conidia 9-11 µ diam. × 3-4 µ wide.

On dead leaves of *Carex inflata*, Switzerland (DAOM 91070, holotype).

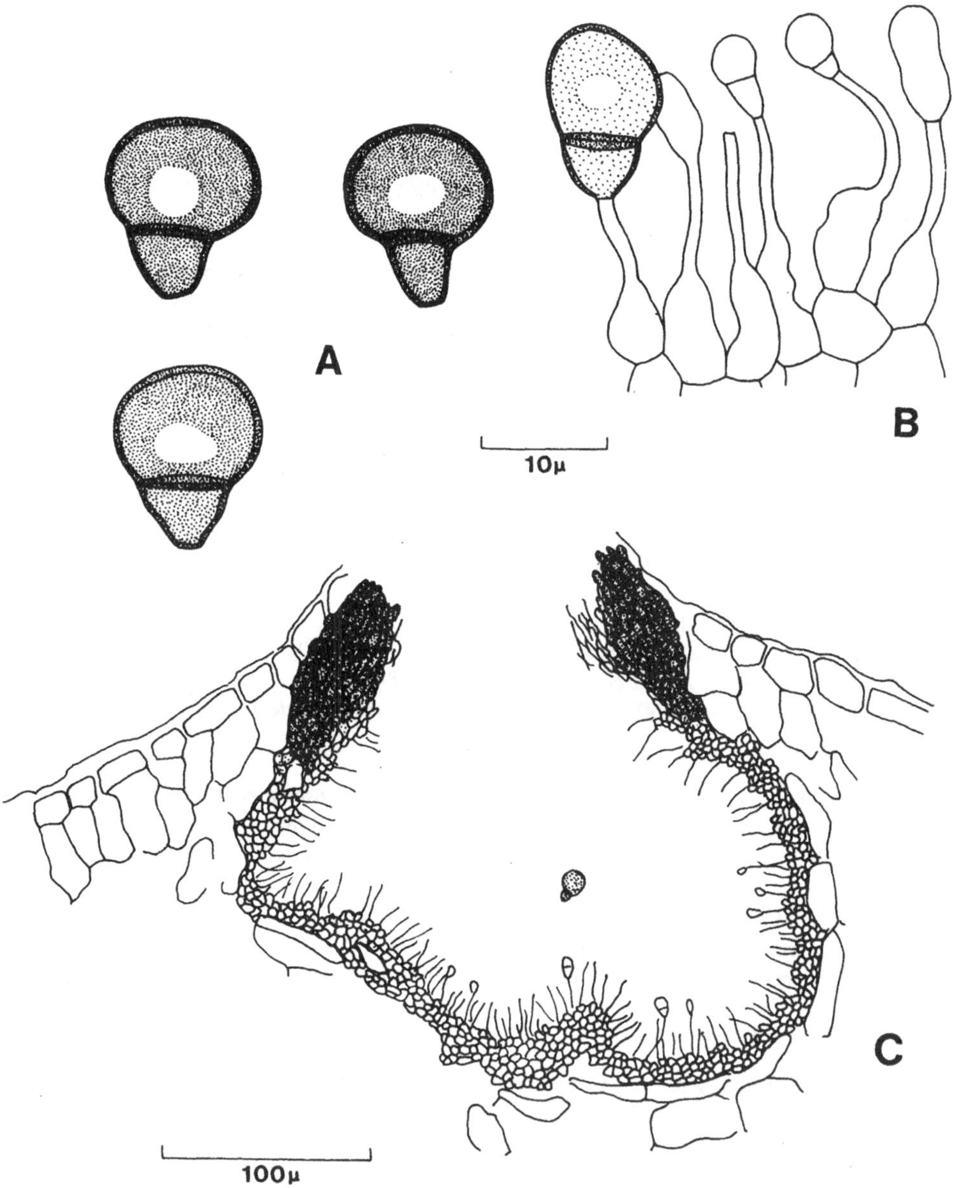

Fig. 30. *Kamatella apiospora*, A, conidia; B, conidiogenous cells and developing conidia; C, vertical section of a conidioma.

28. BENJAMINIA Ahmad, *Biologia, Lahore* **13**: 21 (1967).
Sp. typ.: *B. aristata* Ahmad
Illustration: Ahmad (1967).

Mycelium immersed, pale brown to hyaline, branched, septate. *Conidiomata* pycnidial, separate, globose, brown, immersed, unilocular, thick-walled, of 3-5 layers of polygonal, brown textura angularis. *Ostiole* central, circular. *Conidiogenous cells* holoblastic, determinate, discrete, papillate, hyaline, smooth, formed from the inner cells of the pycnidial wall. *Conidia* dark brown except for the obtuse slightly swollen apical hyaline cell, 7 transverse eusepta, several longitudinal eusepta, ± constricted, tapered to the truncate base, verrucose, straight or slightly curved.

The genus is monotypic, and I have seen no material.

B. ARISTATA Ahmad, *ibid.*: 21 (1967).

Conidiomata up to 140 µ diam. Conidia 30 × 9-10 µ.

On graminicolous culms and *Apluda aristata*, Pakistan.

29. STIGMELLA Lév. in de Demidoff, *Voyage dans la Russie meridionale* **2**: 111-112 (1842).
Sp. typ.: *S. effigurata* (Schw.) Hughes (syn. *Coryneum effiguratum* Schw., *Stigmella dryina* Lév.).
Illustrations: Hughes (1952), Sutton (1975a).

Mycelium immersed, branched, septate, hyaline to pale brown. *Conidiomata* pycnidial, hypophyllous, separate, globose, pale brown, immersed, unilocular, thin-walled; wall composed of pale brown textura angularis. *Ostiole* central, circular, not papillate. *Conidiophores* absent. *Conidiogenous cells* holoblastic, determinate, discrete, doliiform to ampulliform, hyaline, smooth, formed from the inner cells of the pycnidial wall. *Conidia* brown, muriform, up to 6 transverse eusepta and some longitudinal septa, constricted, apex and base ± obtuse, thick-walled, smooth, clavate to ellipsoid, variable in shape.

Hughes (1952; 1958) has contributed a modern description and illustration of this genus.

S. EFFIGURATA (Schw.) Hughes, *Can. J. Bot.* **36**: 814 (1958).
Coryneum effiguratum Schw., *Trans. Am. phil. Soc.* **2**: 4 (1832).
Dicoccum dryophyllum Cda, *Icon. fung.* **1**: 6 (1837).
Stilbospora dryophylla (Cda) Cda, *Icon. fung.* **5**: 4 (1842).
Stigmella dryophylla (Cda) Lindau, *Rabenh. Krypt. Fl.* 2, **9**: 191 (1910).
Sporidesmium foliicola Desm., *Ann. Sci. nat.*, ser. 2, **14**: 8 (1840).
Stigmella dryina Lév. in de Demidoff, *Voyage dans la Russie meridionale* **2**: 111-112 (1842).

Conidiomata up to 100 µ diam. Conidia 12-20 µ × 7-10 µ.

Causing lesions on leaves of *Quercus macropoda*, *Q. pubescens*, *Q. pseudorubra*, *Q. pedunculata*, *Q. robur*, *Q.* sp., Italy, France, Albania, Switzerland, Austria, N. America (IMI 74986, isotype *C. effiguratum*, 45470, 45465, 45466, 45467, 45462, 48844, 45463, 45469, 45464, 45471, 45468, all ex K, 30671, 22295, 10414, 16472).

30. CAMAROSPORELLUM Tassi, *Bull. Lab. Ort. Bot. Siena* **5**: 62 (1902).
Original spp.: *C. nervisequum* (Tassi) Tassi, *C. eucalypti* (Wint.) Tassi
Lectotype sp.: *C. nervisequum* (Tassi) Tassi (syn. *Camarosporium nervisequum* Tassi), fide Sutton & Pollack (1974).
Illustration: Sutton & Pollack (1974).

Mycelium immersed, branched, septate, pale brown to hyaline. *Conidiomata* pycnidial, separate, immersed, globose, black, unilocular, of thin-walled textura angularis. *Ostiole* single, circular, ± papillate. *Conidiophores* absent. *Conidiogenous cells* holoblastic, determinate, discrete, hyaline, smooth, ampulliform, formed directly from the inner cells of the pycnidial wall. *Conidia* muriform, distoseptate, lumina reduced, smooth, brown, ellipsoid, base often truncate.

Sutton & Pollack (1974) lectotypified the generic name and added a second species, *C. cercocarpi*. This differed from *C. nervisequum* in having septate conidiophores and annellidic conidiogenous cells and at that time the generic limits of *Camarosporellum* were enlarged to include such diversity. The species could be separated from *C. nervisequum* by the only occasionally longitudinally septate conidia and the annellidic conidiogenous cells that sometimes arise on short 1-septate, sparsely branched conidiophores. Until the generic limits of this group are more satisfactorily known *C. cercocarpi* is retained tentatively in *Camarosporellum*. The only other *Camarosporellum* species, *C. eucalypti*, was referred to *Dichomera* Cke by Sutton (1975a).

C. NERVISEQUUM (Tassi) Tassi, *ibid.*: 62 (1902). (Fig. 31)
Camarosporium nervisequum Tassi, *Bull. Lab. Ort. Bot. Siena* **3**: 19 (1900).

Conidiomata 110-130 µ diam. Conidiogenous cells 3-5 × 4-6 µ. Conidia 9-12 × 5.5-7.5 µ with 3 transverse distosepta and several longitudinal or oblique distosepta.

On leaves of *Magnolia pumila*, Italy (IMI 164048 ex Siena).

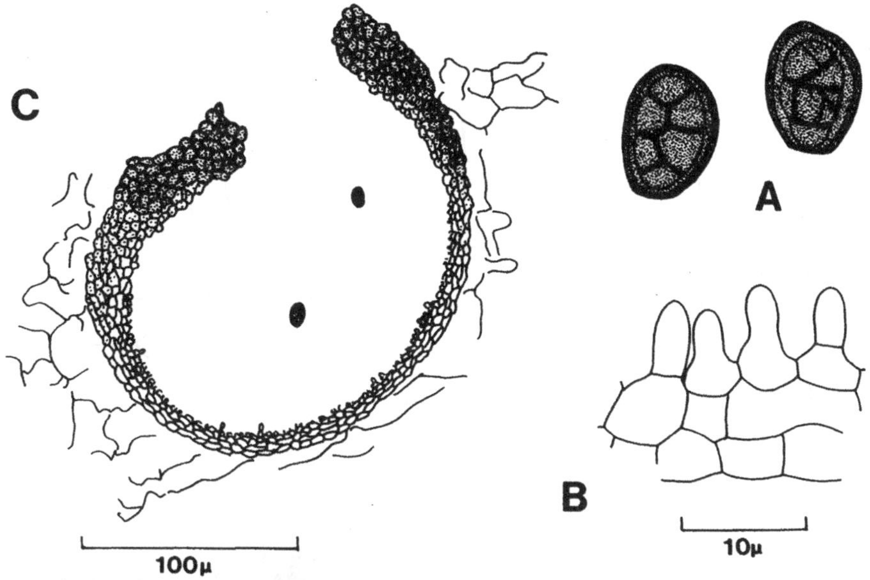

Fig. 31. *Camarosporellum nervisequum*, A, conidia; B, conidiogenous cells with developing conidia; C, vertical section of a conidioma.

31. SCLEROSTAGONOSPORA Höhn., *Hedwigia* **59**: 252 (1917).
Sp. typ.: *S. heraclei* (Sacc.) Höhn. (syn. *Hendersonia heraclei* Sacc.).

Mycelium immersed, branched, hyaline, septate. *Conidiomata* pycnidial, immersed, separate, dark brown to black, globose, unilocular; walls thin, composed of thick-walled, dark brown textura angularis with a hyaline lining 1 cell thick bearing conidiogenous cells. *Ostiole* single, circular, central, papillate. *Conidiophores* absent. *Conidiogenous cells* holoblastic, determinate, discrete, hyaline, smooth, ampulliform to irregular, formed from the inner cells of the pycnidial wall. *Conidia* pale brown, thin-walled, minutely verruculose, 3 euseptate, sometimes slightly constricted at the septa, base truncate, apex obtuse, cylindrical or tapered slightly to the base.

Petrak (1925) regarded *Sclerostagonospora* as no different from *Stagonospora* Sacc., a conclusion accepted by Clements & Shear (1931). It clearly differs, however, in pigmentation of the conidia and in future may well be the recipient of taxa that have to be removed from the nomen rejiciendum *Hendersonia*.

S. HERACLEI (Sacc.) Höhn., *ibid.*: 252 (1917). (Fig. 32)
Hendersonia heraclei Sacc., *Michelia* **1**: 213 (1878).
Stagonospora pulchra Bub. & Krieger, *Annls mycol.* **10**: 51 (1912).

Conidiomata up to 180 µ diam. Conidiogenous cells 5-8 × 7-10 µ. Conidia 24-27 × 5-6.5 µ.

On dead stems of *Heracleum sphondylium*, France (IMI 200664 ex K).

32. SCOLECOSPORIELLA Petrak, *Annls mycol.* **19**: 30 (1921).
Urohendersoniella Petrak, *Sydowia* **9**: 513 (1955).
Brencklea Petrak, *Annls mycol.* **21**: 326 (1923).
Sp. typ.: *S. typhae* (Oud.) Petrak (syn. *Hendersonia typhae* Oud.).

Mycelium immersed, pale brown to hyaline, branched, septate. *Conidiomata* pycnidial, separate, globose, brown, immersed, unilocular, thin-walled, composed of pale brown, thin-walled textura angularis. *Ostiole* central, circular, papillate. *Conidiophores* absent. *Conidiogenous cells* holoblastic, determinate, discrete, doliiform to ampulliform, hyaline, smooth, formed from the inner cells of the pycnidial wall. *Conidia* pale brown, 3- numerous transverse eusepta, sometimes with longitudinal septa, continuous or constricted, apical cell tapered to a short or long filiform unbranched, cellular appendage, base truncate, thin-walled, smooth, fusiform.

Sutton (1968b) redescribed the genus and a fourth species was added by Sutton & Alcorn (1974).

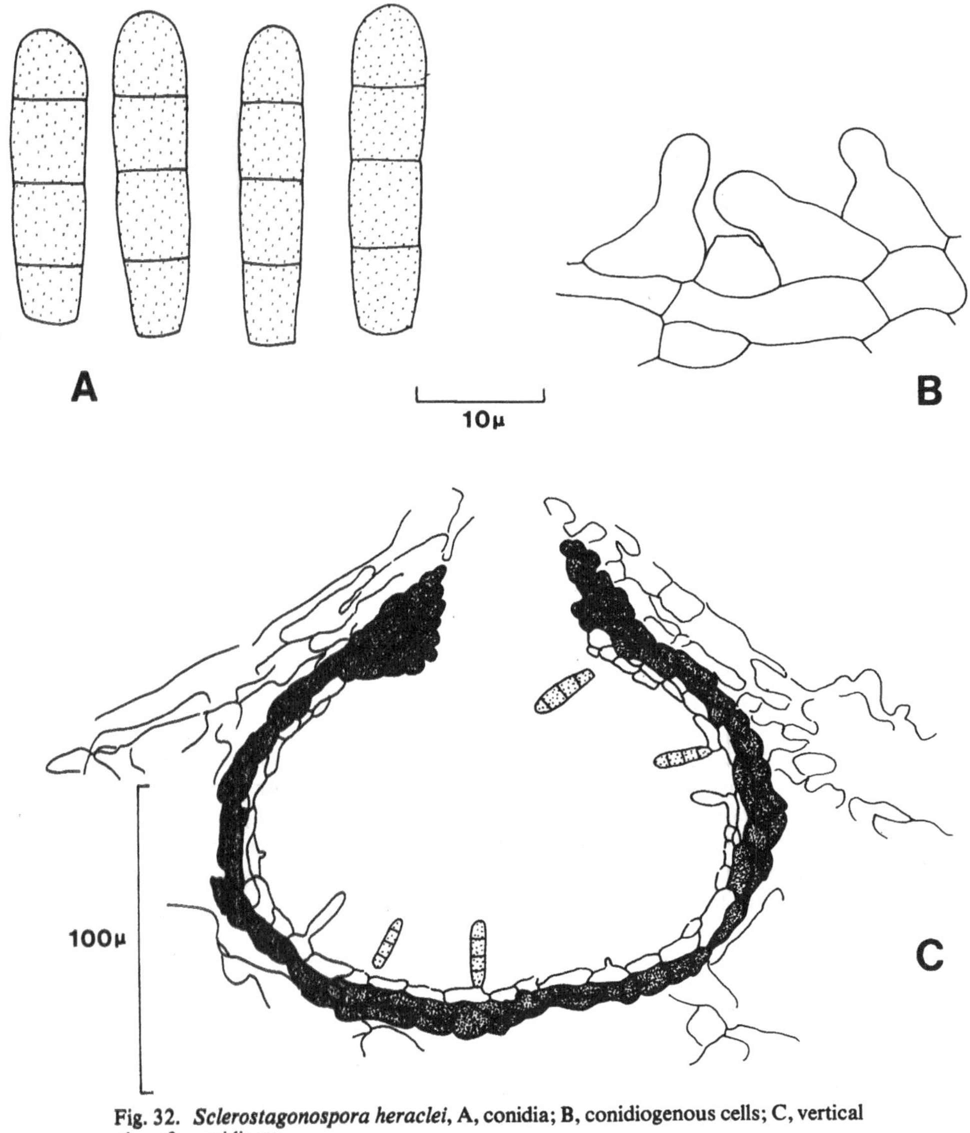

Fig. 32. *Sclerostagonospora heraclei*, A, conidia; B, conidiogenous cells; C, vertical section of a conidioma.

KEY

1. Conidia with longitudinal septa .... *S. kranzii*
1. Conidia transversely septate only 2
   2. Conidia constantly 3 septate .... *S. sisyrinchii*
   2. Conidia constantly 7 septate .... *S. mastigospora*
   2. Conidia with 0-7 septa, variable 3
3. Conidia 4-7 septate .... *S. typhae*
3. Conidia 0-3 septate .... *S. spraguei*

S. KRANZII Sutton, *Can. J. Bot.* **46**: 191 (1968). (Fig. 33)

Conidiomata 95-100 µ diam. Ostiole 20-25 µ diam. Conidia 12-15 transversely septate, 0-13 longitudinally septate, 48-70 × 12-14 µ; apical appendage 18-20 µ long.

On dead leaves of *Andropogon tectorum*, Guinea (IMI 103919, holotype).

S. SISYRINCHII (Ell. & Ev.) Sutton, *Can. J. Bot.* **46**: 191 (1968). (Fig. 34)
Kellermania sisyrinchii Ell. & Ev., *J. Mycol.* **2**: 111 (1886).
Brencklea sisyrinchii (Ell. & Ev.) Petrak, *Annls mycol.* **21**: 326 (1923).

Conidiomata 120-150 µ diam. Ostiole up to 70 µ diam. Conidia 3 septate, 18-22 × 4-5 µ; apical appendage 7-20 µ long.

On dead leaves and culms of *Sisyrinchium angustifolium*, *S. bellum*, U.S.A. (IMI 108534, lectotype, 21863).

S. SPRAGUEI Sutton & Alcorn, *Austr. J. Bot.* **22**: 528 (1974).

Conidiomata up to 200 µ diam. Ostiole 15 µ diam. Conidia 0-3 septate (variable), 14-34 × 4.5-7 µ (including apical appendage).

Causing lesions on *Paspalum dilatatum*, *P. floridanum* var. *glabratum*, U.S.A. (IMI 176768, holotype, 176770, 176771, 176772 ex WIS).

S. TYPHAE (Oud.) Petrak, *Annls mycol.* **19**: 30 (1921). (Fig. 35)
Hendersonia typhae Oud., *Arch. néerl. Sci.* **8**: 361 (1873).
Scolicosporium typhae (Oud.) Höhn., *Sber. Akad. Wiss. Wien* **118**: 405 (1909).

Teleomorphosis: *Leptosphaeria typharum* (Desm.) Karst., *Bidr. Finl. Nat. Folk*: 100 (1873).

Conidiomata 100-120 µ diam. Ostiole 20 µ diam. Conidia 48-68 × 6.5-8.5 µ (including apical appendage).

On *Typha angustifolia*, *T. latifolia*, Germany, U.K. (IMI 108536, holotype, 108537, 88870b).

S. mastigospora (Petrak) comb. nov. (Fig. 36)
Urohendersoniella mastigospora Petrak, *Sydowia* **9**: 513 (1955).

Conidiomata up to 170 µ diam. Conidia 42-59.5 × 6.5-7.5 µ (including apical appendage), often apical and basal cell paler.

On leaf lesions of *Dianella revoluta*, S. Australia (IMI 197517 ex K).

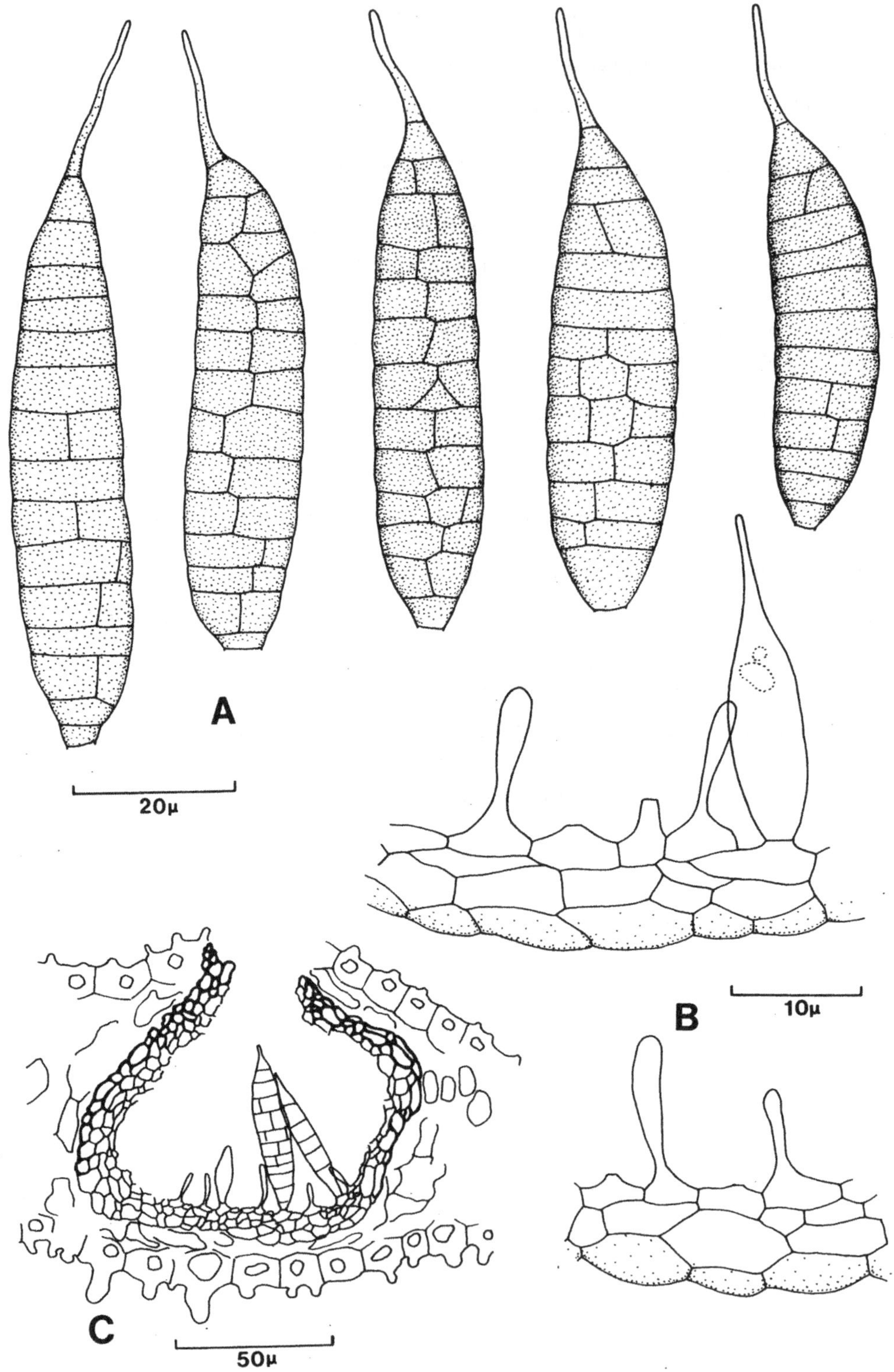

Fig. 33. *Scolecosporiella kranzii*, A, conidia; B, conidiogenous cells and developing conidia; C, vertical section of a conidioma.

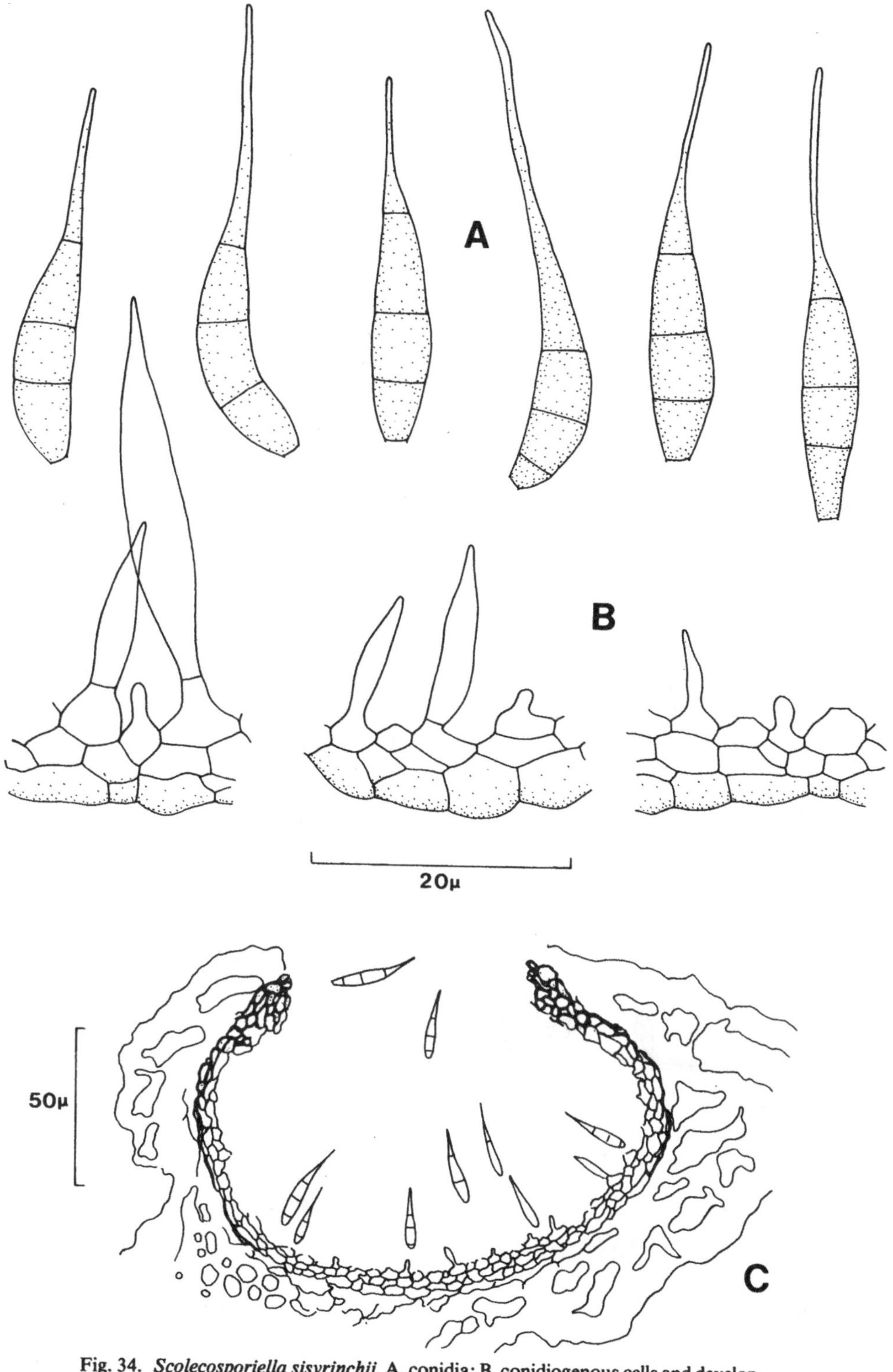

Fig. 34. *Scolecosporiella sisyrinchii*, A, conidia; B, conidiogenous cells and developing conidia; C, vertical section of a conidioma.

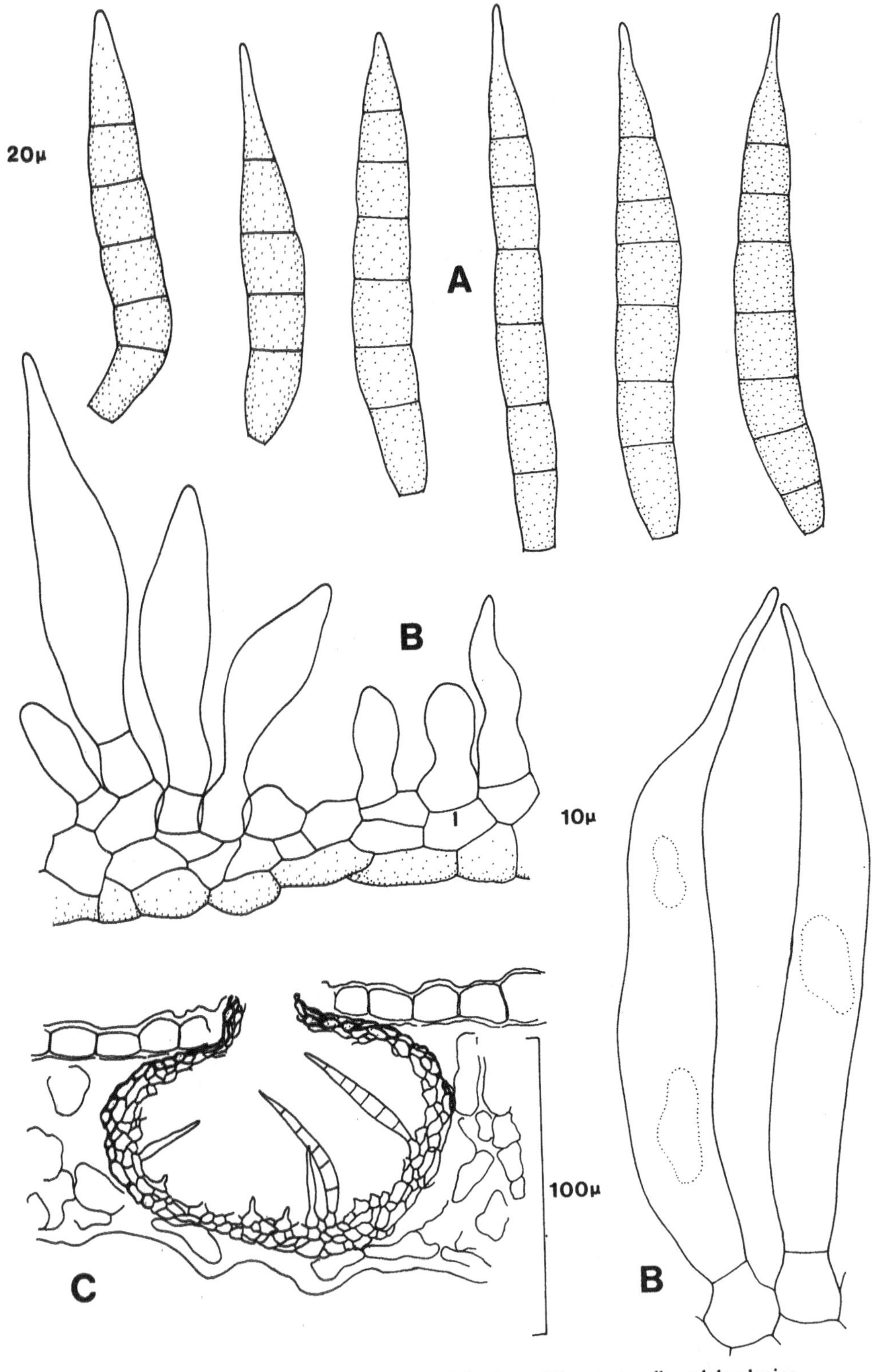

Fig. 35. *Scolecosporiella typhae*, A, conidia; B, conidiogenous cells and developing conidia; C, vertical section of a conidioma.

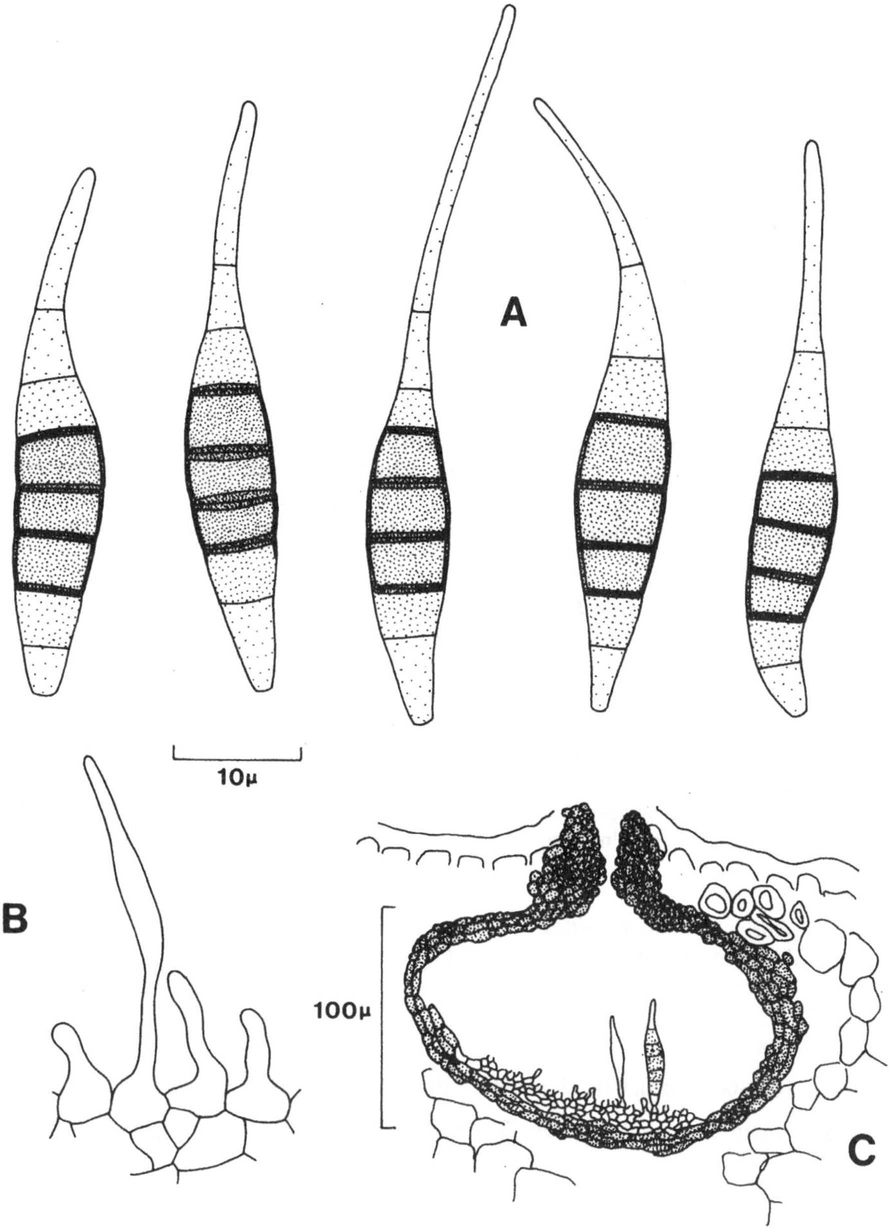

Fig. 36. *Scolecosporiella mastigospora*, A, conidia; B, conidiogenous cells and developing conidia; C, vertical section of a conidioma.

BLASTOPYCNIDIINEAE, SYMPODIAL

The sympodial or synchronous type of holoblastic conidial ontogeny is quite rare in coelomycetes and is represented in pycnidial fungi by only five genera. These are diverse in conidial morphology and clearly do not represent a natural group. The generic taxa are so few and of such a distinctive nature that it is felt unnecessary to prepare a synoptic key for their identification.

DICHOTOMOUS KEY

1. Conidia septate 2
1. Conidia aseptate 3
    2. Pycnidia simple, thin-walled; conidia hyaline, multi-septate, filiform, separate .... *Septoria* (4)
    2. Pycnidia clypeate; conidia often remaining attached to the conidiogenous cell and then appearing tufted, when separate they are filiform, septate .... *Eriospora* (5)
3. Pycnidia simple, thin-walled; conidia unicellular, hyaline, small .... *Alveophoma* (1)
3. Pycnidia hyperparasitic on other fungi or lichens; conidia with apex extended into 2 short divergent appendages, base with an excentric, lateral, short appendage .... *Cornutispora* (2)
3. Conidia with an apical or subapical cellular appendage which branches near the point of insertion .... *Pestalozziella* (3)

1. ALVEOPHOMA Bausá Alcalde, *Ann. Inst. bot. A.J. Cavanilles* **10**: 247 (1952).
Sp. typ.: *A. caballeroi* Bausá Alcalde
Illustrations: Bausá Alcalde (1952); Sutton (1964b).

Mycelium immersed, pale brown to hyaline, branched, septate. *Conidiomata* pycnidial, separate or closely aggregated, globose, brown, immersed, unilocular, thin-walled, of textura angularis. *Ostioles* 1-3, circular, ± papillate. *Conidiophores* hyaline, branched only at the base, septate, cylindrical, smooth, formed from the inner cells of the pycnidial wall. *Conidiogenous cells* holoblastic, sympodial, indeterminate, integrated or discrete, cylindrical, hyaline, smooth, with 1-2 sympodial proliferations. *Conidia* hyaline, aseptate, base truncate, apex obtuse, thin-walled, often guttulate, clavate, ellipsoid or irregular.

Redescription and illustration was provided by Sutton (1964b). The only other species placed in *Alveophoma* is *A. arundinis* Ahmad (1971). However, this species has very pale brown, smooth conidia which are formed from phialides in pycnidia. It is therefore referred to *Microsphaeropsis* Höhn.

A. CABALLEROI Bausá Alcalde, *ibid.*: 247 (1952). (Fig. 37)

Conidiomata 100-300 µ diam. Conidiophores 5-10 × 1-2 µ. Conidia 3-4.5 × 1.5 µ.

On dead twigs of *Fontanesia phyllyraeoides*, Spain (IMI 103845*a*, holotype ex MA).

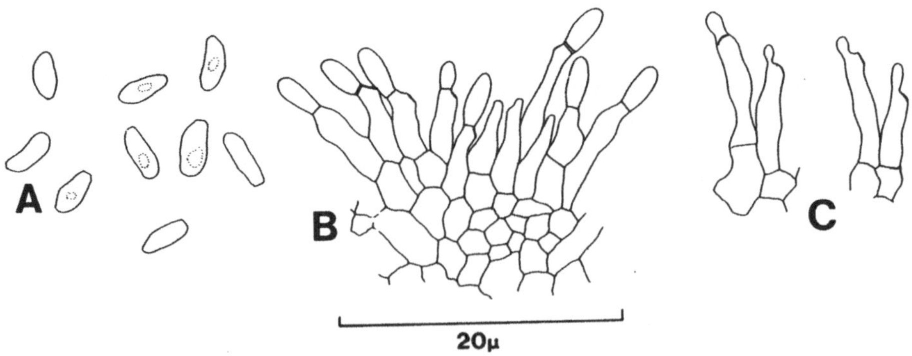

Fig. 37. *Alveophoma caballeroi*, A, conidia; B, conidiogenous cells and developing conidia.

2. CORNUTISPORA Pirozynski, *Mycologia* **65**: 763 (1973).
Sp. typ.: *C. limaciformis* Pirozynski

Mycelium immersed, branched, septate, hyaline. *Conidiomata* pycnidial, globose, pale amber or honey-coloured and translucent, immersed to superficial, unilocular, thick-walled; wall of thin-walled, irregular, indistinct textura angularis, often with collapsed cells. Dehiscence by disintegration of upper tissue. *Conidiophores* hyaline, branched irregularly, septate, smooth, with acropleurogenous conidia, formed from the inner cells of the pycnidial wall. *Conidiogenous cells* polyblastic, indeterminate, integrated, cylindrical, expanding towards the apices, hyaline, smooth, with 1-4 apical unthickened sympodial or synchronous denticles. *Conidia* hyaline, aseptate, straight, cylindrical or long cuneiform, thin-walled, smooth, eguttulate, with 2 apical divergent cellular unbranched appendages and a single basal, exogenous cellular unbranched appendage (as in a *Fusarium* foot cell).

A second species was added by Hawksworth & Sutton apud Hawksworth (1976) which is distinguished by the much smaller conidia.

KEY

Conidia 12-22 µ long, appendages 7-11 µ long .... *C. limaciformis*
Conidia 6-11.5 µ long, appendages 2.5-6 µ long .... *C. lichenicola*

C. LIMACIFORMIS Pirozynski, *ibid.*: 763 (1973). (Fig. 38A)

Conidiomata 80-120 µ diam. Conidiophores 6-9 × 1.5 µ. Conidia 12-22 × 2.5-3 µ; apical appendages 7-14 × 1 µ; basal appendage 5-7 µ long.

Hyperparasitic in apothecia of *Therrya fuckelii*, B.C., Canada (IMI 144358, holotype).

C. LICHENICOLA Hawksworth & Sutton apud Hawksworth, *Trans. Br. mycol. Soc.* **67**: 51 (1976). (Fig. 38B)

Conidiomata up to 110 µ diam. Conidiophores up to 30 × 1.5-2 µ. Conidia 6-11.5 × 1.5-2 µ; apical appendages 2.5-6 µ; basal appendage 2.5-3 µ long.

Hyperparasitic in thalli of *Parmelia sulcata*, U.K. (IMI 186831, holotype).

3. PESTALOZZIELLA Sacc. & Ell. apud Sacc., *Michelia* **2**: 575 (1882).
Sp. typ.: *P. subsessilis* Sacc. & Ell.
Illustrations: Nag Raj & Kendrick (1972a).

Mycelium immersed, hyaline to pale brown, branched, septate. *Conidiomata* pycnidial, separate or aggregated, globose, medium brown, immersed, unilocular, epiphyllous, subepidermal, thin-walled; wall composed of thin-walled, brown textura angularis. *Ostiole* single, circular, wide. *Conidiophores* absent. *Conidiogenous cells* holoblastic, sympodial, indeterminate, discrete, or very rarely on 1-septate conidiophores, ampulliform, hyaline, smooth, with 1- many, unthickened conidiogenous loci, formed from the inner cells of the pycnidial wall. *Conidia* hyaline to very pale brown, aseptate, base abruptly tapered, truncate, apex obtuse, thin-walled, smooth, eguttulate, straight or slightly curved, ± cylindrical, with a subapical, dichotomously branched, cellular appendage.

In spite of the recent treatment by Nag Raj & Kendrick (1972a) this genus is obviously heterogeneous. Four species were accepted, of which 3 were pycnidial and one acervular. In addition some appendages are dichotomous, others branching from a single point, and some conidiogenous cells are sympodial while others are solitary and holoblastic. Appendages are apical or subapical and some are separated by a distinct septum. The above generic description is based on *P. subsessilis*.

P. SUBSESSILIS Sacc. & Ellis apud Sacc., *ibid.*: 575 (1882). (Fig. 39A, B)
Synonymy in Nag Raj & Kendrick (1972a).

Conidiomata 85-220 µ diam. Conidiogenous cells 5-11 µ × 1.5-6.5 µ. Conidia 14.5-22.5 × 4-8 µ; appendages 10-30.5 µ long.

On fading leaves of *Geranium* sp. (IMI 33577).

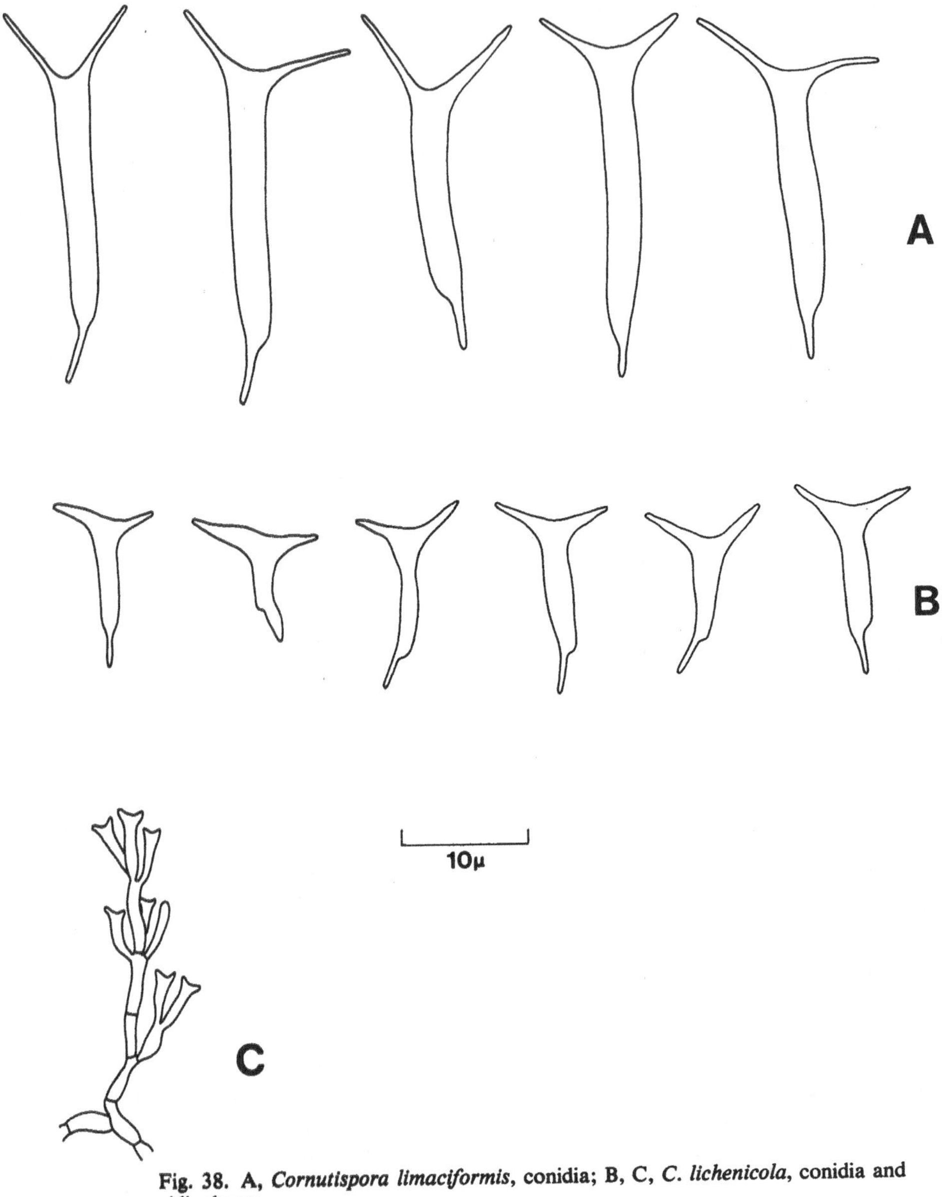

Fig. 38. A, *Cornutispora limaciformis*, conidia; B, C, *C. lichenicola*, conidia and conidiophores.

P. ARTOCARPI Nag Raj & Kendrick, *Can. J. Bot.* **50**: 609 (1972). (Fig. 39C, D)
Conidiomata 300-400 μ diam. Conidiogenous cells 11.5-17 × 2-2.5 μ. Conidia cylindrical, straight, 20-22.5 × 4.5-5.5 μ; appendages up to 24 μ long.

Conidiophores in this species were described as apical and holoblastic, but I have found the conidiogenous cell apices to be very irregular. This is indicative of irregular sympodial or polyblastic conidiogenesis rather than simple holoblastic. The appendages are inserted apically rather than subapically as in *P. subsessilis*.

On *Cynodon dactylon*, Azores (IMI 193803).

4. SEPTORIA Sacc. *Syll. fung.* **3**: 474 (1884), nom. cons.
Septaria Fr., *Syst. Myc.* **1**: XL (1821), nom. rej.
Septoria Fr., *Syst. Orbis Veg.*: 119 (1825), nom. rej.
? *Spilosphaeria* Rabenh., *Bot. Z.* **15**: 173 (1857).
Sp. typ.: *S. cytisi* Desm.

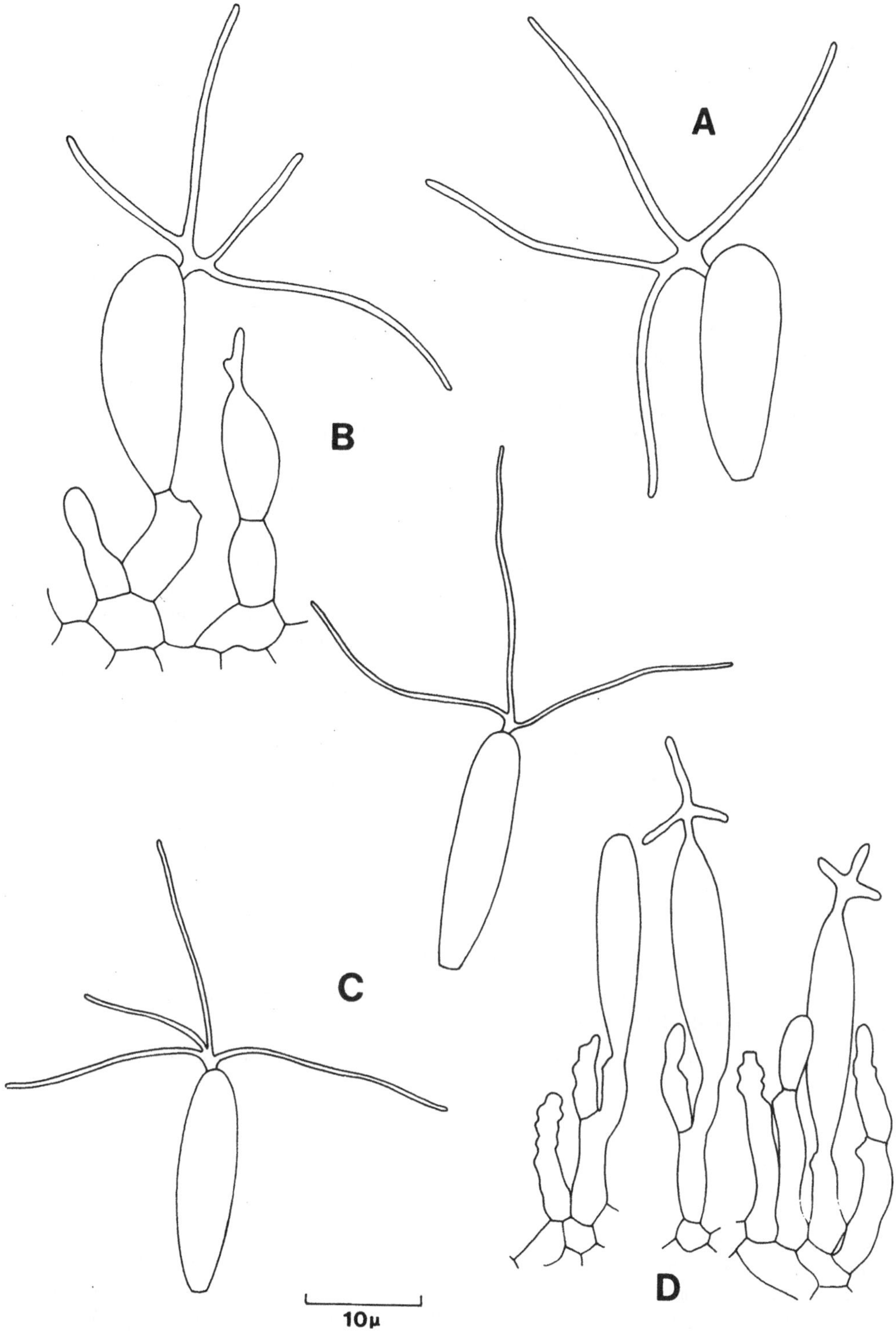

Fig. 39. *Pestalozziella subsessilis*, A, conidium; B, conidiogenous cells and developing conidia; *P. artocarpi*, C, conidia; D, conidiogenous cells and developing conidia.

Mycelium immersed, branched, septate, pale brown. *Conidiomata* pycnidial, immersed, separate or aggregated, but not usually confluent, globose, papillate or not, brown, thin-walled; wall of pale brown textura angularis, often with a smaller-celled inner layer, somewhat darker and more thick-walled around the ostiole. *Ostiole* single, circular, central, sometimes papillate. *Conidiophores* absent. *Conidiogenous cells* holoblastic, determinate or indeterminate with a limited number of sympodial proliferations and then each locus with a broad, flat, unthickened scar, discrete, hyaline, smooth, ampulliform, doliiform or lageniform to short cylindrical. *Conidia* hyaline, multiseptate, filiform, smooth, continuous or constricted at the septa.

The typification of *Septoria* and the arguments surrounding proposals for conservation are found in Wakefield (1940), Rogers (1949) and Donk (1964).

The genus is extremely large and contains more than 2000 described taxa. So far there has been no effort to evolve a practical system of identification and taxonomy for the genus, although individual species and groups of species from certain geographic areas have been studied in detail. None of these, however, take into account relationships based on conidiogenesis so the genus is heterogeneous and still requires considerable revision.

In the *Phloeospora – Cylindrosporium – Septogloeum* complex it has been found that a range of species produce conidia from simple, sympodial or annellidic holoblastic conidiogenous cells. Although such genera all have acervular conidiomata there is no reason to disbelieve the probability that a similar range in conidiogenesis is present in *Septoria*. Indeed, the pathogenic species described in detail by Sutton & Waterston (1966), Punithalingam (1967), Punithalingam & Waterston (1970), Holliday & Punithalingam (1970) and Punithalingam & Holliday (1972) may be placed in one of three groups based on conidiogenesis. Species congeneric with the type, *S. cytisi*, in having holoblastic sympodial conidiogenesis included *S. chrysanthemella* Sacc., *S. obesa* Syd., *S. passerinii* Sacc. and *S. helianthi* Ell. & Kell. Species with phialides are *S. apiicola* Speg. and *S. tritici* Rob. Species with simple holoblastic development and no sign of sympodial or percurrent proliferation are *S. adanensis* Petrak, *S. leucanthemi* Sacc. & Speg., *S. socia* Pass., *S. lactucae* Pass., *S. glycines* Hemmi and *S. lycopersici* Speg. It is clear that members of the latter two groups could form the basis of segregates from *Septoria* established primarily on conidiogenesis, but until many more species are revised it would be premature to suggest any nomenclatural changes.

Apart from the classic floras (Diedicke, 1914a; Allescher, 1900; Grove, 1935) there are several useful compilations on *Septoria* apart from the species included in Saccardo's *Sylloge* and Oudemans' *Enumeratio*. Sprague (1950) deals with Graminicolous species and Mäkelä (1977) fully illustrates the species occurring in Finland. Teterevnikova-Babayan (1974, 1976) has revised the species on Rosaceae and Salicaceae in the U.S.S.R. respectively. Byzova et al. (1970) dealt with many species from Kazakhstan, U.S.S.R., and Radulescu, Negru & Docea (1973) enumerated and described more than 300 species from Roumania. Jørstad (1965, 1967) has described the species of *Septoria* and related genera on Dicotyledons and Gramineae respectively in Norway, but unfortunately this excellent work is not illustrated.

In the same way that a comprehensive account of *Ascochyta* and *Phomopsis* etc. cannot be presented at the present time the type species of *Septoria* is only described below.

S. CYTISI Desm., *Ann. Sci. nat.*, ser. 3, **8**: 24 (1847). (Fig. 40)

Conidiomata 150-175 μ diam. Conidiogenous cells 9-15.5 × 3-5.5 μ. Conidia tapered gradually towards the obtuse apex, base truncate, 4-6 septate, 58-93 × 4-5.5 μ.

Causing lesions on leaves of *Cytisus laburnum*, *C. nigricans*, Austria, Hungary, Italy, Roumania (IMI 33367, 11193, 202214, 20217).

5. ERIOSPORA Berk. & Br., *Ann. Mag. nat. Hist.*, ser. 2, **5**: 455 (1850).
Sp. typ.: *E. leucostoma* Berk. & Br.

Mycelium immersed, branched, septate, extensively ramifying, hyaline to pale brown. *Conidiomata* pycnidial, immersed, separate, dark brown to black, globose, unilocular, but sometimes convoluted, thick-walled, composed of an outer region of very loose textura intricata, hyaline at the base, but pale to dark brown and thicker-walled in the upper region and particularly so near the ostiole, almost forming a clypeus in this region, and an inner region of hyaline, loose, reticulately branched textura intricata of minute diameter. *Ostiole* single, circular, central, papillate, appearing white in contrast to the pycnidial wall. *Conidiophores* repeatedly branched to form an apparently reticulate layer lining the whole pycnidial wall, hyaline, septate. *Conidiogenous cells* holoblastic, sympodial, integrated, indeterminate, comprising the ultimate branches of the conidiogenous region, with several (6 or more) small, unthickened, conidiogenous loci towards the apices. *Conidia* often persistent, hyaline, filiform, straight or curved, eguttulate, up to 8 euseptate.

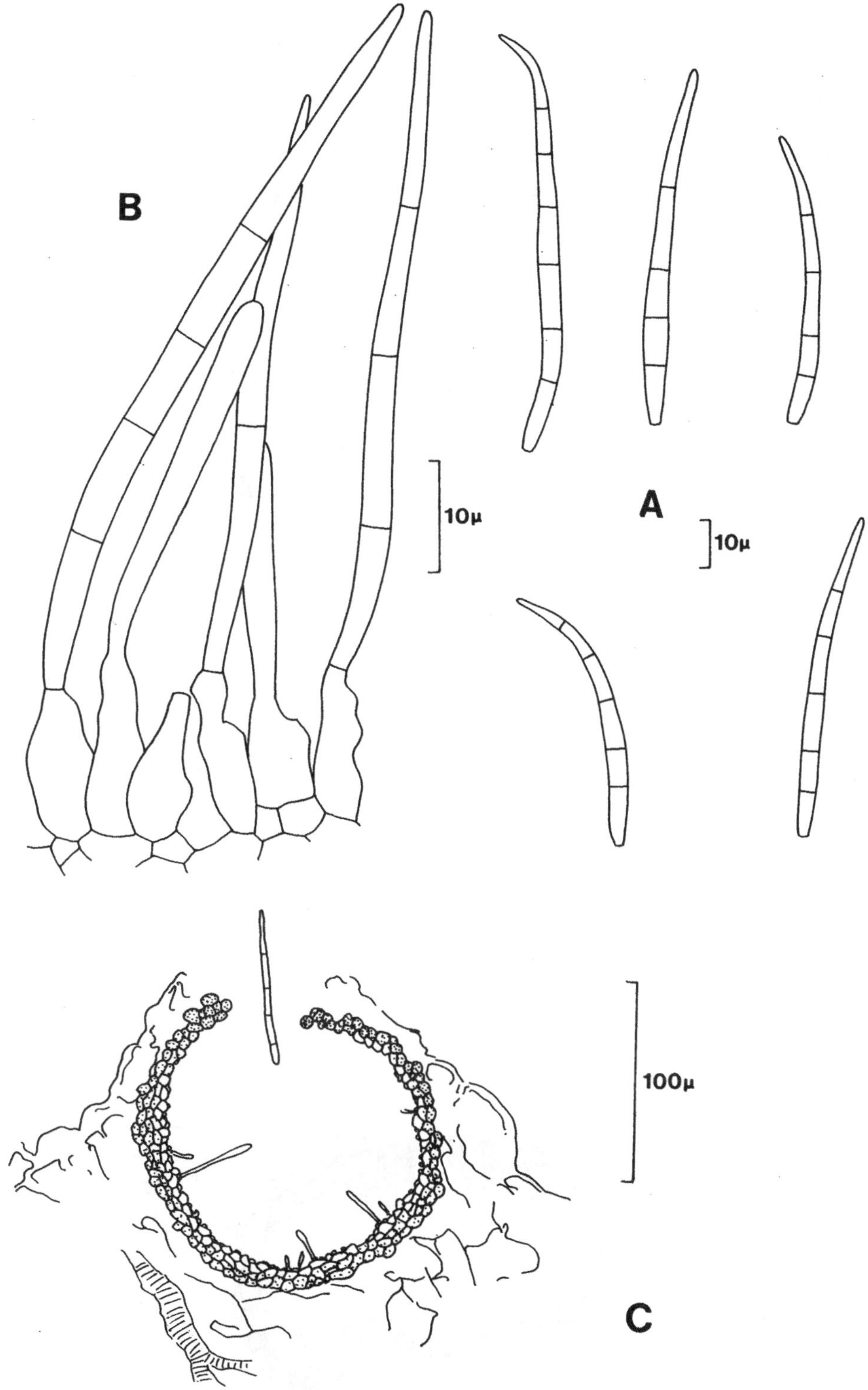

Fig. 40. *Septoria cytisi*, A, conidia; B, conidiogenous cells and developing conidia; C, vertical section of a conidioma.

Additional accounts of this monotypic genus have been given by Grove (1937), Petch (1943), and Petrak (1947b). Although conidia are referred to as stellate by Clements & Shear (1931), they are in fact filiform, remaining attached to the conidiogenous cell for a long time. Often preparations show individual conidiogenous cells with several attached conidia and this feature led Fuckel (1870) to describe spores as a 'short handled cat-o' nine tails'.

E. LEUCOSTOMA Berk. & Br., *ibid.*: 455 (1850). (Fig. 41)

Conidiomata up to 600 µ diam.; ostiole 50 µ diam. Conidiogenous cells 7-10 × 1.5 µ. Conidia 70-100 × 1 µ.

On dead leaves of *Carex paniculata*, U.K. (IMI 34837b).

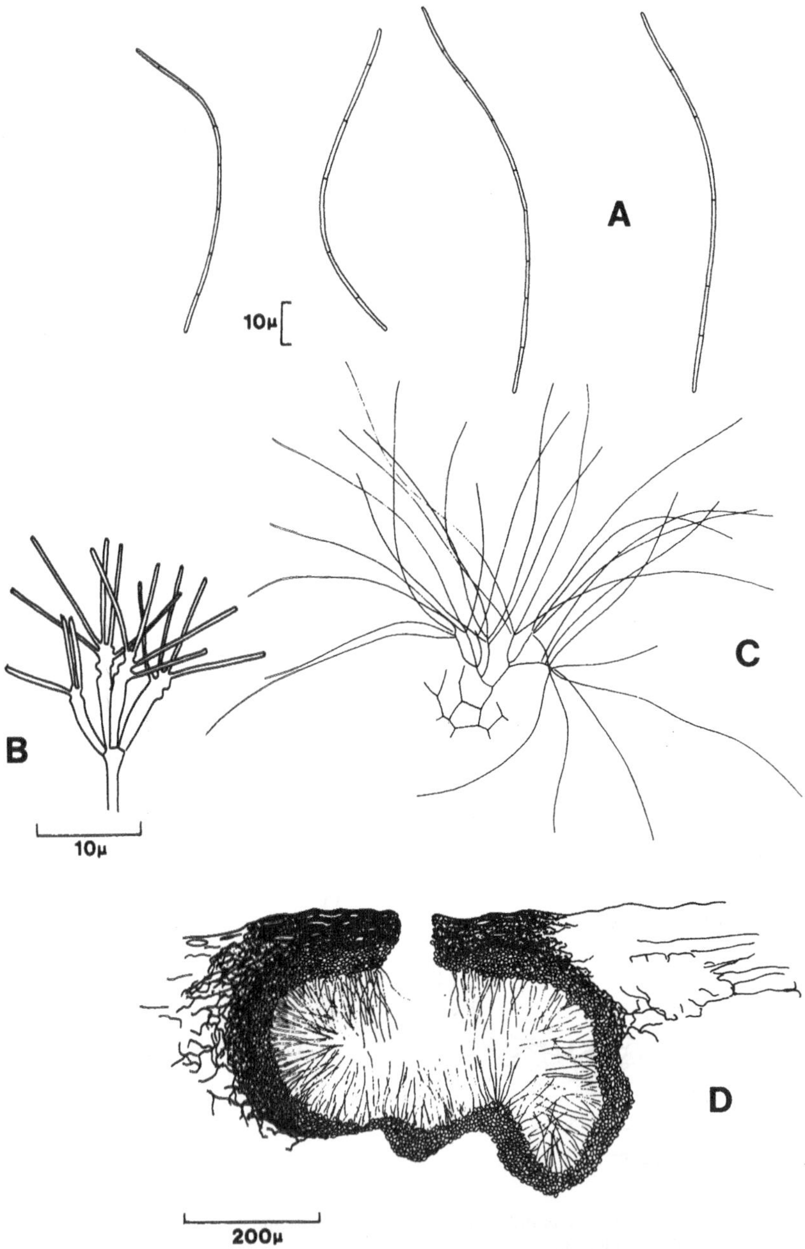

Fig. 41. *Eriospora leucostoma*, A, separate conidia; B, details of conidiogenous cells; C, dichotomous branching of conidiophores; D, vertical section of a conidioma.

BLASTOPYCNIDIINEAE, ANNELLIDIC

This is a heterogeneous group in which the genera are largely unrelated. The exceptions are *Bartalinia* and *Doliomyces*, and *Sphaeropsis* and *Coniothyrium* which belong together. The latter two genera with aseptate conidia are simple forms of *Camarosporium* which has muriform conidia. *Lichenoconium* is also close to this group but differs in having pigmented cylindrical annellides.

DICHOTOMOUS KEY

1. Conidia hyaline 2
1. Conidia pigmented 5
 2. Conidia aseptate 3
 2. Conidia with 1 or more septa 4
3. Conidiogenous cells ampulliform with several annellations; conidia falcate and fusiform .... *Pseudoseptoria* (1)
3. Conidiogenous cells lageniform, rarely percurrent, with marked periclinal thickening .... *Trematophoma* (see p. 371)
3. Conidiophores septate and branched sparingly; conidia long and gradually attenuated towards the apex . *Groveolopsis* (2)
3. Conidiogenous cells with widely-spaced, flared annellations; conidia clavate to ellipsoid, with the remains of the conidiogenous cell remaining attached .... *Idiocercus* (3)
 4. Pycnidia superficial, formed in an anastomosing mycelium; conidiogenous cells ampulliform with several annellations; conidia medianly 1 septate, fusiform .... *Dialaceniopsis* (4)
 4. Pycnidia immersed; conidiogenous cells ampulliform to doliiform with several annellations; conidia medianly 1 septate, with 1(2-) apical cellular branched appendages; paraphyses present .... *Vasudevella* (5)
 4. Pycnidia immersed; conidia multiseptate, often guttulate .... *Stagonospora* (6)
5. Conidia either with appendages or enclosed in a mucilaginous sheath 6
5. Conidia lacking either appendages or sheaths 9
 6. Conidia distoseptate, enclosed in a mucilaginous sheath .... *Macrodiplodiopsis* (7)
 6. Conidia euseptate, with an apical mucilaginous appendage .... *Urohendersonia* (8)
 6. Conidia with cellular appendages 7
7. Conidia concolourous, euseptate but with reduced lumina; individual appendages inserted apically, subapically or suprabasally .... *Seimatosporiopsis* (9)
7. Conidia versicolourous 8
 8. Conidia brown; basal appendage absent; apical appendage single, unbranched, inclined laterally .... *Bleptosporium* (10)
 8. Appendages several, inserted apically .... *Bartalinia* (11)
 8. Appendages several, inserted subapically to laterally .... *Doliomyces* (12)
9. Conidia aseptate 10
9. Conidia 0-1 septate 11
9. Conidia muriform, thick-walled, with oblique and longitudinal septa .... *Camarosporium* (17)
 10. Conidia deltoid; conidiogenous cells with marked periclinal thickening .... *Readeriella* (see p. 427)
 10. Conidia large, thick-walled, with ornamentation on the inner face of the wall .... *Sphaeropsis* (14)
 10. Conidia variable in size, often developing a median septum; conidiogenous cells short, otherwise difficult to distinguish from *Lichenoconium* .... *Coniothyrium* (15)
 10. Conidia small, verruculose, ornamentation on the outer face of the wall; conidiogenous cells cylindrical; lichenicolous .... *Lichenoconium* (13)
11. Conidiogenous cells cylindrical, with ragged verruculose annellation(s); conidia pale brown with irregularly verruculose ornamentation on the outside of the wall .... *Schwarzmannia* (16)
11. Conidiogenous cells doliiform to short cylindrical, annellations smooth; conidia brown 0-1 septate, with verruculose ornamentation on the outside of the wall .... *Coniothyrium* (15)

SYNOPTIC KEY

Mycelial characters

Position in relation to substrate

a. superficial	4	10													
b. immersed	1	2	3	5	6	7	8	9	11	12	13	14	15	16	17

Conidiomatal characters

Position in relation to substrate	
a. superficial	4 10
b. immersed	1 2 3 5 6 7 8 9 11 12 13 14 15 16 17
c. semi-immersed	13
Dehiscence	
a. ostiolate, non papillate	1 4 6
b. ostiolate, papillate	2 3 5 6 7 8 9 10 11 14 15 17
c. irregular breakdown of upper wall	12 13 16
Wall thickness	
a. thin-walled (1-3 cells wide)	1 2 3 4 15 16
b. thick-walled (5 cells wide or more)	5 6 7 8 9 10 11 12 13 14 15 17
Other characters	
a. Setose	4

Conidiogenous cell characters

Insertion	
a. discrete	1 3 4 5 6 7 8 9 11 12 13 14 15 16 17
b. integrated	2 3 10 12
Shape	
a. ampulliform-doliiform	1 4 5 6 8 9 11 14 15 17
b. lageniform	9 11 12 14 15 17
c. cylindrical	2 3 7 10 11 12 13 16 17
Pigmentation	
a. hyaline	1 2 3 4 5 6 7 8 9 10 11 12 13 14 15 17
b. brown (at least near apices)	8 13 15 16
Ornamentation	
a. smooth	1 2 3 4 5 6 7 8 9 10 11 12 13 14 15 17
b. verruculose	13 16

Conidial characters

Pigmentation	
a. hyaline	1 2 3 4 5 6
b. brown	7 8 9 13 14 15 16 17
c. versicoloured	10 11 12

Septation														
a. distoseptate	7													
b. euseptate	4	5	6	8	9	10	11	12	15	16	17			
c. aseptate	1	2	3	13	14	15	16							
d. 1-septate	4	5	15	16										
e. 2- to multiseptate	6	7	8	9	10	11	12							
f. muriform	7	17												
Shape														
a. oval-ellipsoid	3	6	7	9	13	14	15	16	17					
b. cylindrical	6	8	9	10	11	12	14							
c. fusiform	1	4	5	6	8	11	12	17						
d. falcate	1													
e. pyriform	8	9	10	13	15	16	17							
f. obpyriform	2													
g. globose	13	15	16											
Ornamentation														
a. verrucose	8	13	14 (on inner face)	15	16	17								
b. smooth	1	2	3	4	5	6	7	8	9	10	11	12	13	17
Appendages														
a. apical	5	8	9	10	11									
b. subapical	12													
c. basal	3	9	11	12										
d. single	3	5	8	10	11	12								
e. several	5	9	11	12										
f. cellular	3	5	9	10	11	12								
g. simple	3	8	9	10	11	12								
h. branched	5													
i. mucilaginous	7	8												
j. tentacular	8													
k. completely enclosing conidium	7													
Wall thickness														
a. thin	1	2	4	5	6	8	11	12	13	16				
b. thick	3	7	9	10	13	14	15	17						
Guttulation	2	3	5	6	8									

Paraphysis characters

Paraphyses present	5

1. PSEUDOSEPTORIA Speg., *An. Mus. Nac. Buenos Aires* **20**: 388 (1910).
Lunospora Frandsen, *Medd. Vet. Hojsk. plantepat. Afd., Kbh.* 26: 70 (1943).
Sp. typ.: *P. donacis* (Pass.) Sutton (syn. *Septoria donacis* Pass., *P. donacicola* Speg.).

Mycelium immersed, branched, septate, pale brown. *Conidiomata* pycnidial or linearly aggregated, immersed, brown, globose, unilocular, thin-walled; walls of thin-walled, pale brown textura angularis. *Ostiole* distinct, central, circular. *Conidiophores* absent. *Conidiogenous cells* holoblastic, simple or sympodial, discrete, determinate or indeterminate, hyaline, smooth, ampulliform with a prominent cylindrical annellate papilla from which several conidia are formed in succession, produced from the inner cells of the pycnidial wall. *Conidia* falcate, fusiform, acute at each end, hyaline, aseptate, ± guttulate, smooth, thin-walled.

Hitherto *Pseudoseptoria* has been regarded as synonymous with *Selenophoma* Maire by Sprague & Johnson (1950), Petrak (1951a) and Luc (1953). The name of the type species has been placed in synonymy with *Selenophoma donacis* by Sprague & Johnson (1950) and Petrak (1951a), but kept distinct as *Selenophoma donacicola* (Speg.) Luc by Luc (1953). That these graminicolous species with simple pycnidia and falcate conidia are true *Selenophoma* species is a misconception. The pycnidial structure is different, and more importantly, the method of conidiogenesis is quite distinct. Since *Neopatella* Sacc. is a distinct phialidic genus and *Falcispora* Bub. & Serebr. is congeneric with *Selenophoma*, the graminicolous species must be referred to *Pseudoseptoria*, of which *Lunospora* is a later synonym.

Additional data on host range and disease syndromes are found in Sprague (1950) and Punithalingam & Waller (1969).

P. DONACIS (Pass.) Sutton, *Mycol. Pap.* 141: 169 (1977). (Fig. 42D, E)
Septoria donacis Pass., *Atti Soc. Critt. ital.* **2**: 44 (1879).
Selenophoma donacis (Pass.) Sprague & Johnson, *Mycologia* **32**: 415 (1940).
Septoria oxyspora Penz. & Sacc., *Syll. fung.* **3**: 565 (1884).
Pseudoseptoria donacicola Speg., *An. Mus. Nac. Buenos Aires* **20**: 388 (1910).
Lunospora oxyspora (Penz. & Sacc.) Frandsen, *Medd. Vet. Hojsk. plantepat. Afd., Kbh.* 26: 71 (1943).
Septoria curva Karst., *Medd. Soc. Fauna Flora fenn.* **14**: 103 (1887).
Rhabdospora curva (Karst.) Allesch. in Rabenh., *Krypt. Fl.* **1**(6): 886 (1901).
Selenophoma curva (Karst.) Petrak, *Annls mycol.* **39**: 238 (1940).
Lunospora curva (Karst.) Frandsen, *Medd. Vet. Hojsk. plantepat. Afd., Kbh.* 26: 71 (1943).
Septoria nebulosa Rostr., *Medd. Groenl. Kjbh.* **3**: 575 (1888).
Lunospora baldingerae Frandsen, *Medd. Vet. Hojsk. plantepat. Afd. Kbh.* 26: 75 (1943).

Conidia 20-23 × 2-2.5 µ.

Causing leaf lesions on *Bromus inermis, Agropyron repens, Aeluropus lagopoides, Arundo donax, Hordeum bulbosum, Triticum* sp., U.K., Cyprus, Ethiopia, Calif., U.S.A., Pakistan, Hungary (IMI 11763, 188970, 137038, 154546, 154545, 88365, 20102, 166082).

P. stomaticola (Bauml.) comb. nov. (*Fig.* 42B, F)
Phyllosticta stomaticola Bauml., *Öst. Bot. Z.* 289 (1889).
Selenophoma donacis (Pass.) Sprague & Johnson var. *stomaticola* (Bauml.) Sprague & Johnson, *Mycologia* **37**: 639 (1945).
Septoria culmifida Lind, *Annls mycol.* **5**: 276 (1907).
Septoria oxyspora Penz. & Sacc. var. *culmorum* Grove, *J. Bot.* **54**: 192 (1916).
Septoria oxyspora Penz. & Sacc. var. *penniseti* Trotter, *N. giorn. bot. ital.* **23**: 26 (1916).
Septoria suboxyspora Lobik, *Morbi Pl. Leningr.* **17**: 178 (1928).
Septoria lunata Grove, *British Stem- and Leaf-fungi* **1**: 458 (1935).
Lunospora avenae Frandsen, *Medd. Vet. Hojsk. plantepat. Afd. Kbh.* 26: 74 (1943).
Lunospora penniseti (Trotter) Frandsen, *ibid.*: 72 (1943).
Lunospora culmifida (Lind) Frandsen, *ibid*: 72 (1943).
Lunospora culmorum (Grove) Frandsen, *ibid.*: 72 (1943).
Lunospora suboxyspora (Lobik) Frandsen, *ibid.*: 72 (1943).
Lunospora lunata Frandsen, *ibid.*: 72 (1943).

Conidia 16-18.5 × 2 µ.

Causing leaf lesions on *Agropyron spicatum, Dactylis glomerata, Phleum pratense, Triticum sp.*, Gramineae indet., U.K., Eire, Wyo, U.S.A. (IMI 96964, 147124, 147123, 119506, 101340).

P. bromigena (Sacc.) comb. nov. (Fig. 42C)
Septoria bromigena Sacc., *Annls mycol.* **13**: 123 (1915).
Selenophoma bromigena (Sacc.) Sprague & Johnson, *Mycologia* **32**: 415 (1940).
Lunospora bromigena (Sacc.) Frandsen, *Medd. Vet. Hojsk. plantepat. Afd. Kbh.* 26: 72 (1943).

Conidia 20-23 × 2-2.5 µ.

Causing leaf lesions on *Bromus inermis, B.* sp., N. Dak., U.S.A. Sask., Man., Canada (IMI 20091, isotype of *S. bromigena*, 20092, 29943, 20094, 20093, 94045).

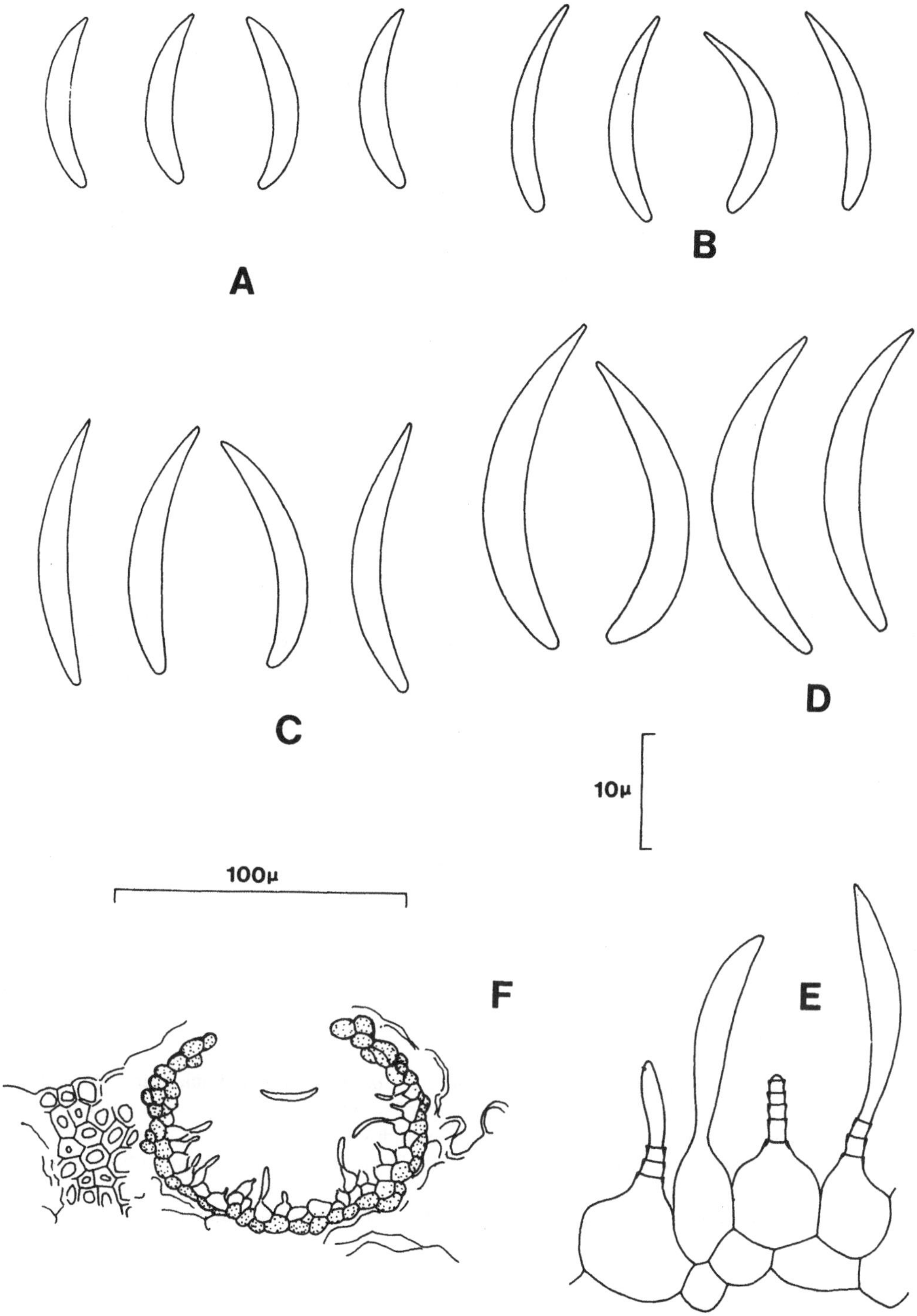

Fig. 42. A, *Pseudoseptoria everhartii*, conidia; B, F, *P. stomaticola*, conidia and vertical section of conidioma; C, *P. bromigena*, conidia; D, E, *P. donacis*, conidia and conidiogenous cells.

P. everhartii (Sacc. & Syd.) comb. nov. (Fig. 42A)
Septoria everhartii Sacc. & Syd., *Syll. fung.* **16**: 973 (1902).
Selenophoma everhartii (Sacc. & Syd.) Sprague & Johnson, *Mycologia* **37**: 639 (1945).
Septoria calamagrostis Ell. & Ev., *Bull. Torrey bot. Cl.* **25**: 511 (1898).

Conidia 14-16 × 2 µ.

Causing leaf lesions on *Hordeum vulgare*, from *H.* sp., Eire, Libya (IMI 104186, 82646).

An additional species accepted by Sprague (1950) but not represented in Herb. IMI is **Pseudoseptoria obtusa** (Sprague & Johnson) comb. nov. (syn. *Selenophoma obtusa* Sprague & Johnson, *Mycologia* **37**: 638, 1945). Originally described from *Sitanion hystrix* in Calif., U.S.A., the species was subsequently recorded by Sprague (1950) from *Agropyron, Elymus, Poa* and *Stipa* species in various states of the U.S.A.

2. GROVEOLOPSIS Boedijn nom. nov., *Sydowia* **5**: 225 (1951).
Septoriopsis Höhn., *Sber. Akad. Wiss. Wien* **129**: 168 (1920), non Fragoso & Paul apud Fragoso (1915), nec Stevens & Dalbey (1919).
Sp. typ.: *G. pandani* (Höhn.) Boedijn (syn. *Septoriopsis pandani* Höhn.).

Mycelium immersed, branched, septate, pale brown. *Conidiomata* pycnidial, brown, globose, immersed, unilocular; wall thin, composed of pale brown, elongated, thin-walled textura angularis. *Ostiole* single, circular, central, papillate. *Conidiophores* hyaline, septate and branched at the base, short, formed from the inner cells of the pycnidial wall. *Conidiogenous cells* holoblastic, annellidic, discrete or integrated, hyaline, smooth, cylindrical, with 1-2 prominent annellations. *Conidia* hyaline, aseptate, thin-walled, smooth, occasionally guttulate, base abruptly tapered and truncate, tapered gradually towards the apex.

The genus is monotypic.

G. PANDANI (Höhn.) Boedijn, *ibid.*: 225 (1951). (Fig. 43)
Septoriopsis pandani Höhn., *Sber. Akad. Wiss. Wien* **129**: 168 (1920).
Rhabdospora pandani (Höhn.) Petrak, *Sydowia* **5**: 351 (1951).

Conidiomata up to 150 µ diam. Conidiophores 8-16 × 3-4 µ. Conidia 43-57 × 5-6 µ, tapering to 2-2.5 µ.

On dead leaves of *Pandanus*, Java (IMI 202940 ex FH, holotype).

3. IDIOCERCUS Sutton, *Can. J. Bot.* **45**: 1255 (1967).
Sp. typ.: *I. pirozynskii* Sutton

Mycelium immersed, hyaline to pale brown, branched, septate. *Conidiomata* pycnidial, immersed, separate, globose, brown, dark brown at the ostiole, unilocular, thin-walled, formed of brown, thin-walled textura angularis. *Ostiole* single, circular, papillate, punctiform. *Conidiophores* hyaline, smooth, cylindrical, 1-2 septate, branched at the base and above, formed from the inner cells of the pycnidial wall. *Conidiogenous cells* holoblastic, percurrent with 1-4 flared annellations, discrete or integrated, terminal or lateral, indeterminate, cylindrical, hyaline, smooth. *Conidia* hyaline, guttulate, apex obtuse, base truncate with the remains of the conidiogenous cell attached, aseptate, smooth, cylindrical, long clavate to ellipsoid.

The genus only contains two species.

I. PIROZYNSKII Sutton, *ibid.*: 1256 (1967). (Figs. 44, 45)

Conidiomata 240-265 µ diam. Conidiophores 15-21.5 × 2.5-5 µ. Conidia 31.5-46 × 8.5-12 µ; basal frill 1-5 µ long.

Causing lesions on *Harungana madagascariensis*, Tanzania, Sierra Leone (IMI 107154a, holotype, 51782, 106156).

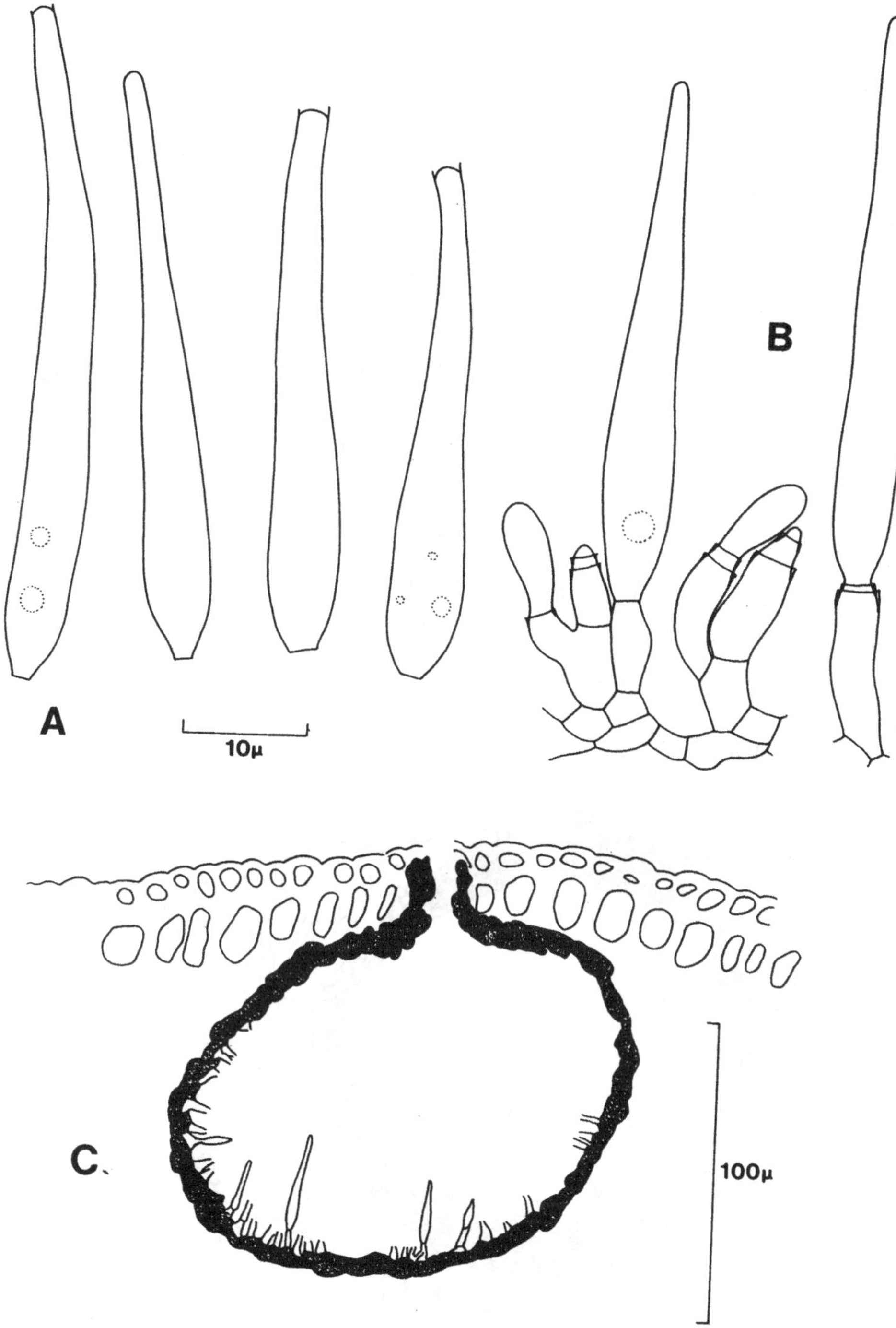

Fig. 43. *Groveolopsis pandani*, A, conidia; B, conidiophores and developing conidia; C, vertical section of a conidioma.

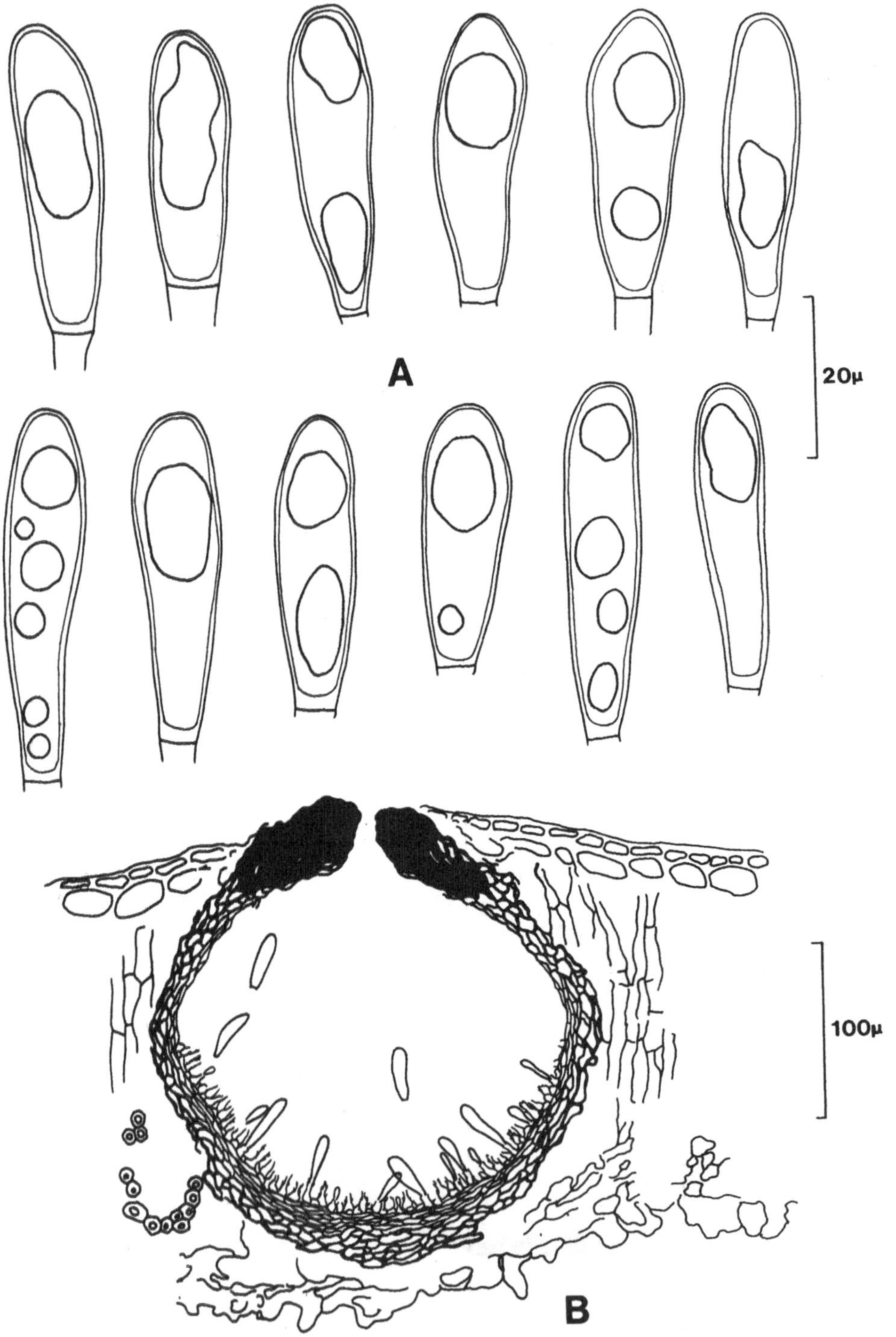

Fig. 44. *Idiocercus pirozynskii*, A, conidia; B, vertical section of a conidioma.

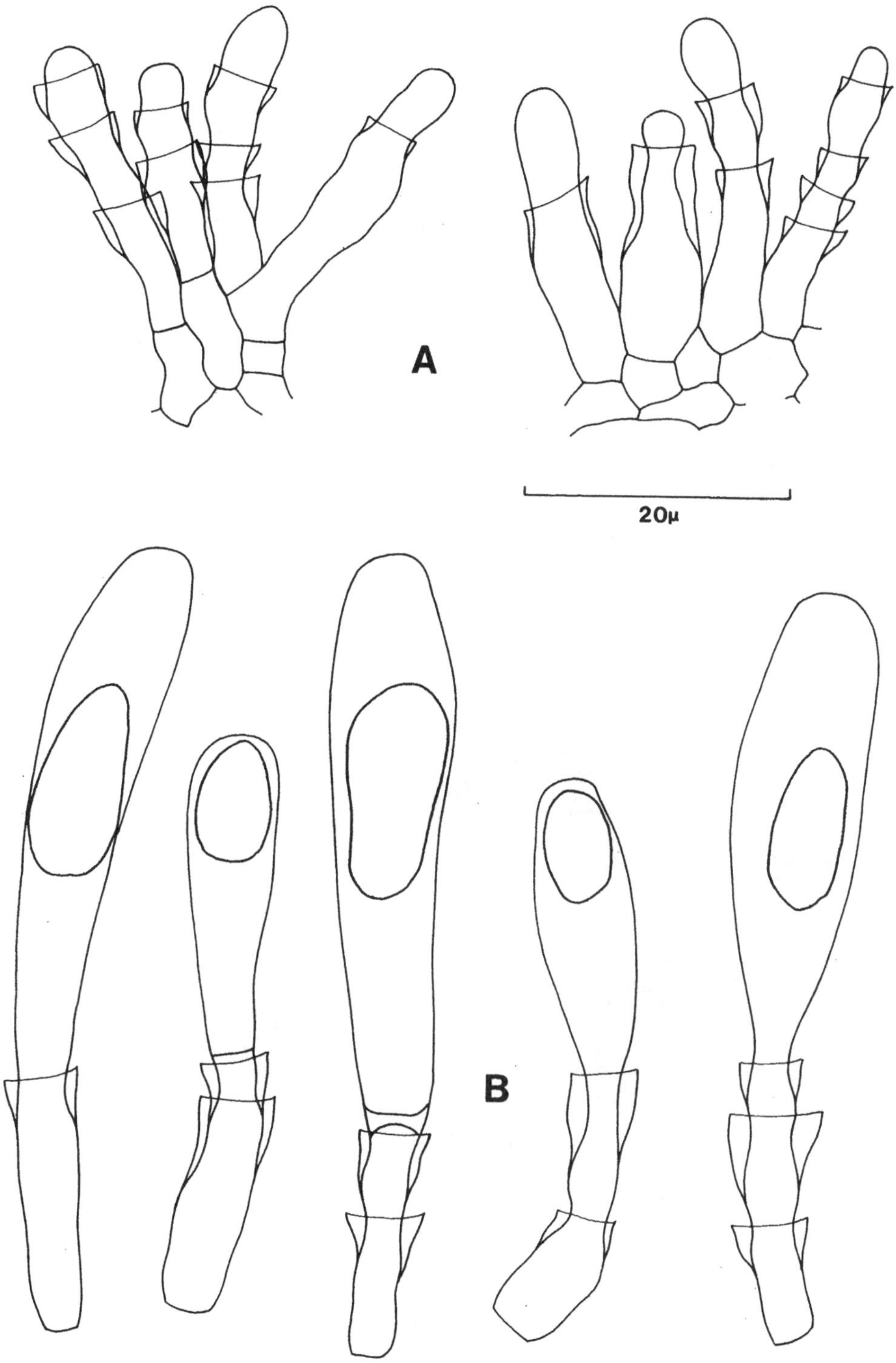

Fig. 45. *Idiocercus pirozynskii*, A, B, young and older conidiogenous cells.

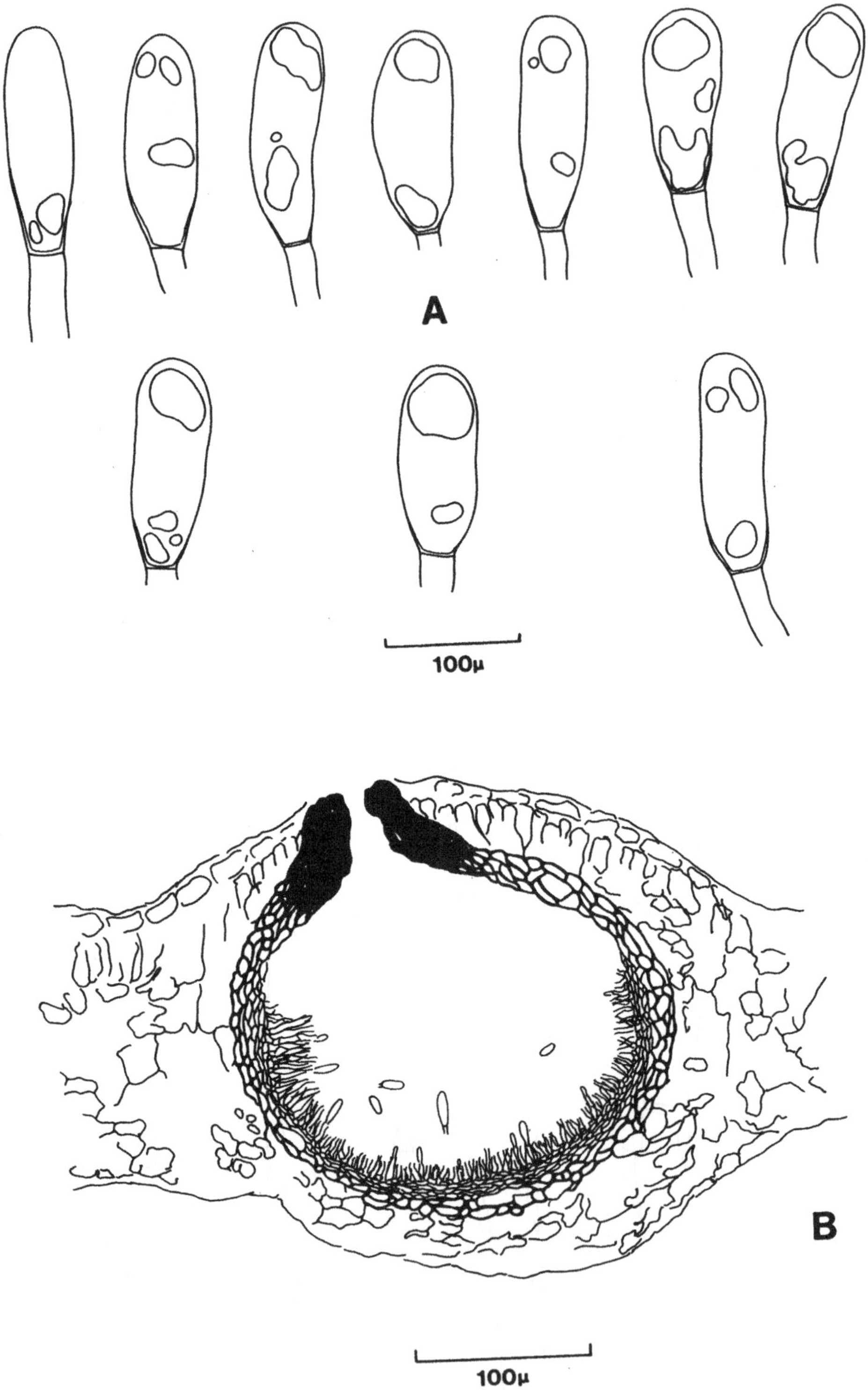

Fig. 46. *Idiocercus macarangae*, A, conidia; B, vertical section of a conidioma.

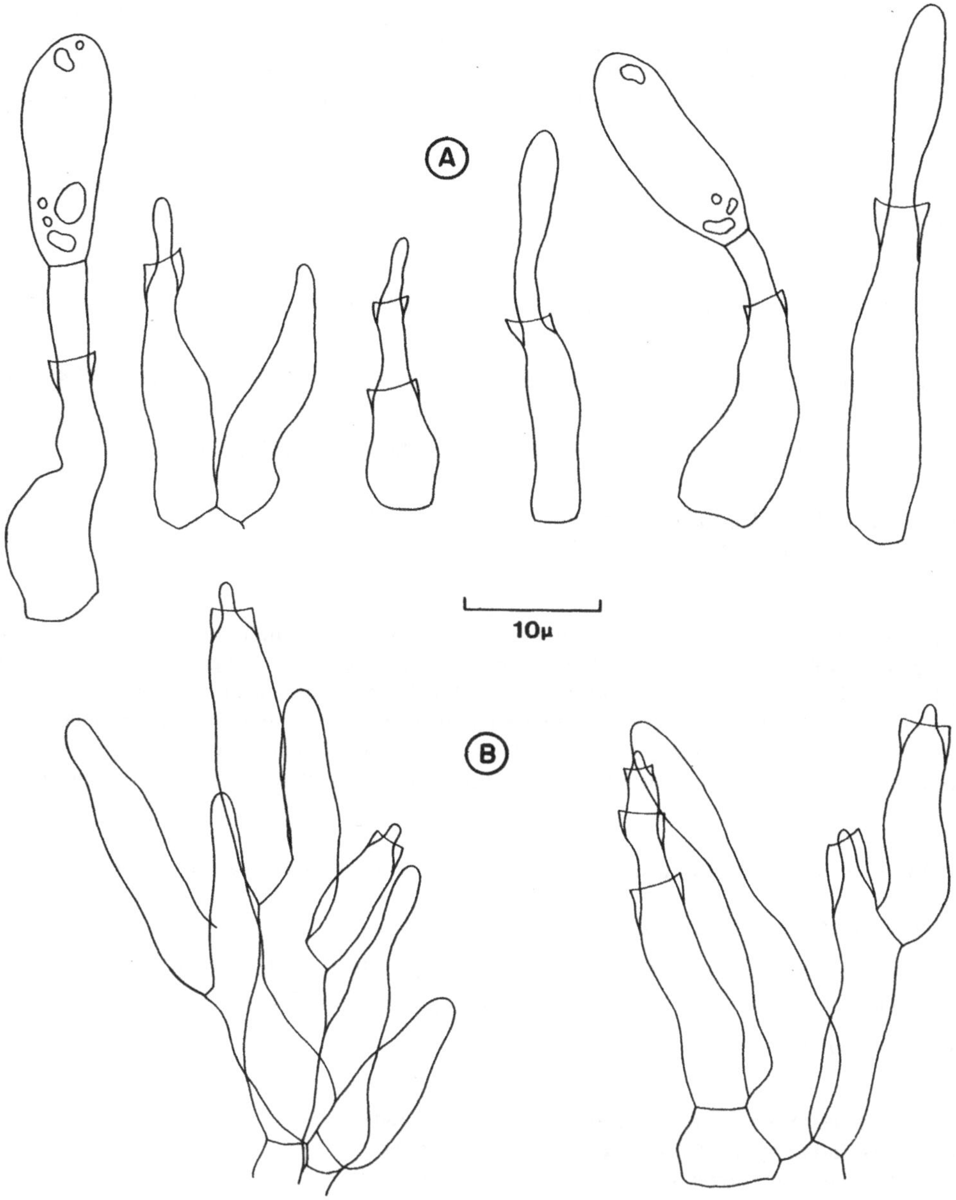

Fig. 47. *Idiocercus macarangae*, conidiophores, conidiogenous cells and developing conidia.

I. MACARANGAE (T.S. Ramakr.) Sutton, *ibid.*: 1257 (1967). (Figs. 46, 47)

Conidiomata 230-275 μ diam. Conidiophores 15-40 × 3-5 μ. Conidia 15-18.5 × 5-6 μ; basal frill 2-7.5 μ long.

Causing lesions on *Macaranga* sp., Sierra Leone (IMI 41330).

4. DIALACENIOPSIS Batista, *An. Soc. biol. Pern.* **16**: 141 (1959).
Sp. typ.: *D. landolphiae* Batista

Mycelium superficial, branched, septate, brown, smooth, anastomosing, with opposite or alternate, 1-3 septate, tortuose to sinuate hyphopodia. *Conidiomata* pycnidial, superficial, brown, globose, collapsing to collabent at maturity, unilocular, thin-walled, wall of dark brown textura angularis. *Setae* radiating to reflexed, straight or flexuous, dark brown, smooth, septate, paler near the apices, often adherent to the superficial mycelium. *Ostiole* single, ± circular. *Conidiophores* absent. *Conidiogenous cells* holoblastic, annellidic, pale brown, hyaline above, smooth, discrete or occasionally with a single supporting cell, doliiform to ampulliform, indeterminate, with 1-4 apical annellations, formed from the inner cells of the pycnidial wall. *Conidia* hyaline, smooth, medianly 1 euseptate, contents distinctly granular, base truncate, apex obtuse, fusiform, eguttulate.

This generic name was introduced to accommodate the pycnidial state of *Dialacenium landolphiae* Hansf. which was fully described by Hansford (1945). The genus is monotypic.

D. LANDOLPHIAE Batista, *ibid.*: 143 (1962). (Fig. 48)
Teleomorphosis: *Rhytidenglerula landolphiae* (Hansf.) Arx apud Müller & Arx, *Beitr. Krypt. Fl. Schw.* **11**: 159 (1962).

Conidiomata up to 100 µ diam. Setae up to 150 µ long × 3-4 µ wide. Conidiogenous cells 7-10 × 4-7 µ (2 µ wide at the apices). Conidia 25-27 × 5-7 µ.

Hypophyllous on *Landolphia ugandensis*, Uganda (IMI 47886, holotype, 4753g, 21)

5. VASUDEVELLA Chona, Munjal & Bajaj, *Indian Phytopath.* **9**: 188 (1957).
Sp. typ.: *V. sporoboli* Chona, Munjal & Bajaj
Illustration: Nag Raj (1973b).

Mycelium immersed, pale brown, branched, septate. *Conidiomata* pycnidial, amphigenous, separate, globose, brown, immersed, unilocular, thick-walled; wall composed of dark brown, thick-walled textura angularis. *Ostiole* papillate, central or excentric, circular. *Conidiophores* absent. *Conidiogenous cells* holoblastic, annellidic, indeterminate, discrete, doliiform to ampulliform, hyaline, smooth, up to 5 percurrent proliferations, formed from inner cells of pycnidial wall. *Conidia* hyaline, 1 euseptate, ± constricted, navicular, occasionally slightly curved, base truncate with a marginal frill, apex obtuse, thin-walled, smooth, guttulate, with 1(-2) apical cellular branched appendages. *Paraphyses* hyaline, filiform, unbranched, smooth, aseptate.

Nag Raj (1973b) redescribed and illustrated *V. sporoboli*, the type and only species so far described in the genus.

V. SPOROBOLI Chona, Munjal & Bajaj, *ibid.*: 188 (1957).

Conidiomata up to 800 µ long × 130-180 µ wide × 110-125 µ high; ostiole 15-20 µ diam. Conidiogenous cells 5.5-8 × 3-3.5 µ. Conidia 16-25 × 2.5-3.5 µ; apical appendage branches 15-25 µ long. Paraphyses 15-30 × 1 µ.

On leaves of *Sporobolus* sp., India.

6. STAGONOSPORA (Sacc.) Sacc., nom. cons., *Syll. fung.* **3**: 445 (1884).
Hendersonia Berk., nom. rej., *Ann. Mag. Nat. Hist.* **6**: 430 (1841).
Hendersonia Berk. subgen. *Stagonospora* Sacc., *Michelia* **2**: 8 (1880).
Gymnosphaera Tassi, *Bull. Lab. Ort. bot. Siena* **5**: 78 (1902).
Stagonospora (Sacc.) Sacc. subgen. *Gymnosphaera* (Tassi) Sacc. & D. Sacc., *Syll. fung.* **18**: 361 (1906).
Stagonosporella Tassi, *Bull. Lab. Ort. bot. Siena* **5**: 50 (1902).
? *Stagonosporina* Tassi, *Bull. Lab. Ort. bot. Siena* **5**: 51 (1902).
? *Tetradia* T. Johnson, *Sci. Proc. R. Soc. Dublin* **10**: 157 (1904).
? *Diedickella* Petrak, *Annls mycol.* **20**: 305 (1922).
Original spp.: 55 species listed.
Lectotype sp.: *S. paludosa* (Sacc. & Speg.) Sacc. (syn. *Hendersonia paludosa* Sacc. & Speg.), fide *International Code of Botanical Nomenclature* (1952).
Illustrations: Cunnell (1956, 1957b, 1961).

Mycelium immersed, branched, septate, pale brown. *Conidiomata* pycnidial, separate, often interveinal and linearly arranged, immersed, globose, black, unilocular, thick-walled; walls on the outside of dark brown, thick-walled textura angularis and on the inside of hyaline, thin-walled smaller cells. *Ostiole* single, circular, central, occasionally papillate. *Conidiophores* absent. *Conidiogenous cells* holoblastic, occasionally annellidic, with a single proliferation, discrete, indeter-

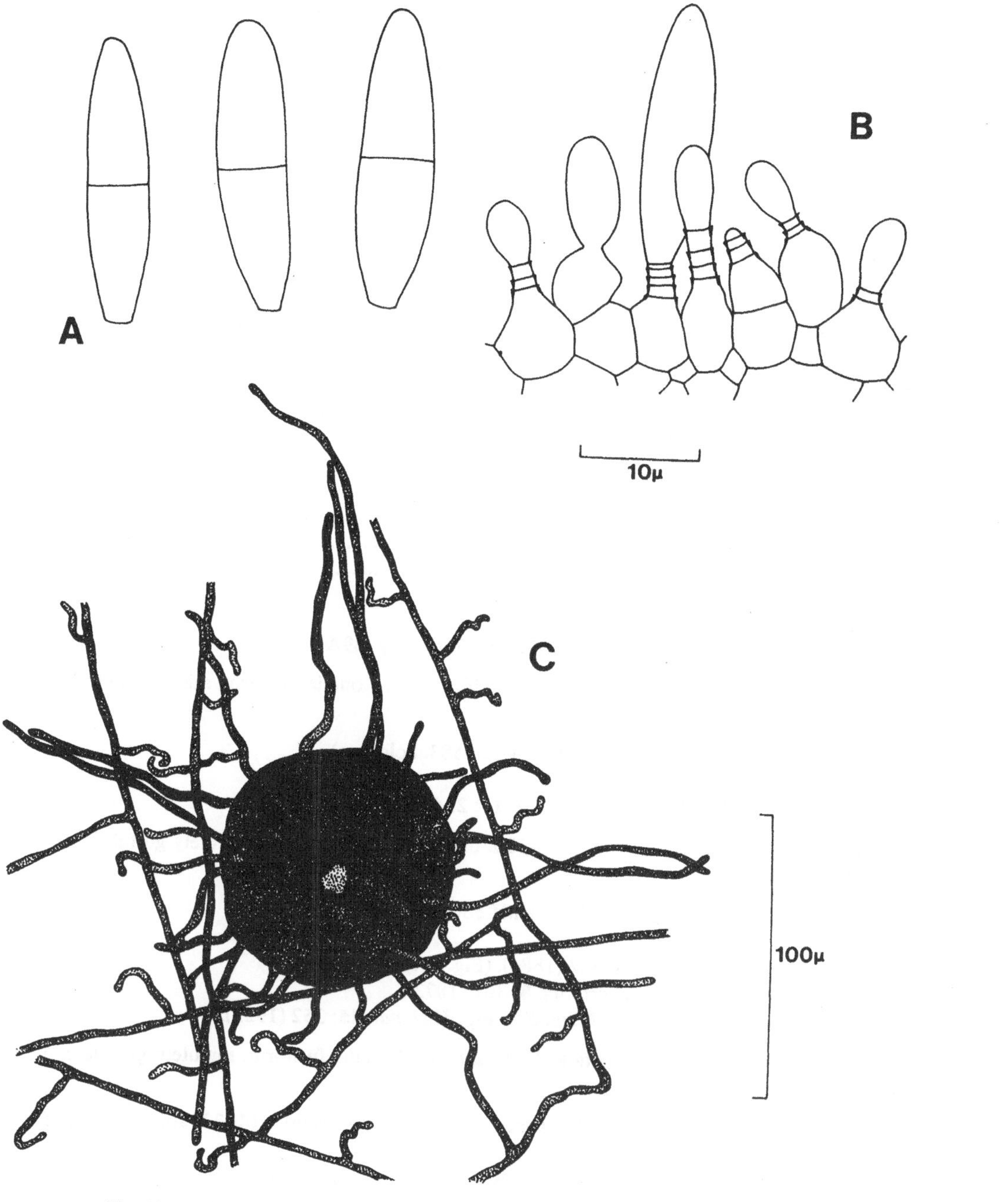

Fig. 48. *Dialaceniopsis landolphiae*, A, conidia; B, conidiogenous cells and developing conidia; C, surface view of conidioma and superficial mycelium.

minate, doliiform, hyaline, formed from the inner cells of the pycnidial wall. *Conidia* hyaline, smooth, with several transverse eusepta, continuous or constricted, cylindrical or fusiform, straight or slightly curved, often with many minute guttules.

The generic synonymy has been discussed by Sutton (1977a) under the individual generic names.

With about 350 species described in the genus there is evidently considerable opportunity for revisionary studies. Little has been done in this area, apart from the work of Cunnell (1956, 1957b, 1961) and Castellani & Germano (1977). Cunnell directed his attention to species occurring on *Carex* and *Phragmites* and illustrated and described 9 species from these hosts. A very broad concept of the genus was made by Castellani & Germano who accepted 77 *Stagonospora* species from Gramineae. Their results show several nomenclatural shortcomings and the illustrations are particularly poor. Nevertheless their account is the only comprehensive treatment of this genus on Gramineae and is useful for the excellent bibliography. Moreover 37 *Stagonospora* names are reduced to synonymy.

KEY

1. Pycnidia up to 1250 µ diam. × 120-400 µ high 2
1. Pycnidia rarely exceeding 500 µ diam. 3
   2. Conidia 25-70 × 2.5-5 µ, 3(-5) septate .... *S. macropycnidia*
   2. Conidia 52-84 × 8.5-14 µ, (3) 4-6 septate .... *S. elegans*
   2. Conidia 48-78 × 8-11.5 µ, 3(-4) septate .... *S. cylindrica*
3. Conidia mostly <30 µ long 4
3. Conidia >30 µ long 5
   4. Conidia cylindrical, 18-32 × 4-6 µ, 2-3 (-4) septate .... *S. vitensis*
   4. Conidia fusiform, 14-27 × 4-6 µ, 2 (3-4) septate .... *S. caricinella*
5. Conidia 5-8 µ wide .... *S. caricis*
5. Conidia >8 µ wide 6
   6. Conidia 41-67 × 8-11 µ (mean 53.5 × 9.5 µ) .... *S. paludosa*
   6. Conidia 58-84 × 10-14 µ (mean 74.5 × 12.6 µ) .... *S. gigaspora*
   6. Conidia 68-109 × 14-18 µ (mean 87.5 × 16 µ) .... *S. anglica*

S. ANGLICA Cunnell, *Trans. Br. mycol. Soc.* **39**: 40 (1956). (Fig. 49A)

Conidiomata 310-500 × 170-340 µ; ostiole 20-30 µ diam. Conidia fusiform, minutely guttulate, 5-8 septate, constricted, 68-109 × 14-18 µ.

On dead leaves of *Carex acutiformis*, U.K. (IMI 59583, holotype).

S. CARICINELLA Brun., *Bull. Soc. bot. Fr.*, ser. 2, **15**: 224 (1893). (Fig. 49H)

Conidiomata 90-120 µ diam.; ostiole 15-20 µ diam. Conidia fusiform, minutely guttulate, 14-27 × 4-6 µ, 2(3-4) septate, constricted.

On dead leaves of *Carex acutiformis*, U.K. (IMI 58658).

S. CARICIS (Oud.) Sacc., *Syll. fung.* **3**: 452 (1884). (Fig. 49I)
Hendersonia caricis Oud., *Arch. neerl. Sci.* **8**: 361 (1873).
? *Stagonospora asturica* Unamuno, *Mem. Soc. esp. Hist. nat.* **15**: 352 (1929).

Conidiomata 110-220 µ diam.; ostiole 10-15 µ diam. Conidia fusiform, minutely guttulate, 5 (-7) septate, constricted, 25-45 × 4-8 µ.

On leaves of *Carex sylvatica*, *C. pseudocyperus*, *C.* sp., U.K., Germany (IMI 22920b, 35333, 59582).

S. CYLINDRICA Cunnell, *Trans. Br. mycol. Soc.* **40**: 451 (1957). (Fig. 49B)

Conidiomata 590-1020 × 200-400 µ; ostiole 35-70 µ. Conidia cylindrical, guttulate, continuous, 48-78 × 8-11.5 µ, 3(-4) septate.

On dead stems of *Phragmites communis*, U.K. (IMI 63875, holotype).

S. ELEGANS (Berk.) Sacc. & Trav., *Syll. fung.* **20**: 878 (1911). (Fig. 49D)
Hendersonia elegans Berk., *Ann. Nat. Hist.* **6**: 430 (1841).

Conidiomata 600-1260 × 190-340 µ often aggregated and complex; ostiole 15-70 µ diam. Conidia fusiform, minutely guttulate, constricted, (3) 4-6 septate, 52-84 × 8.5-14 µ.

On stems of *Phragmites communis*, U.K. (IMI 63876, 186877).

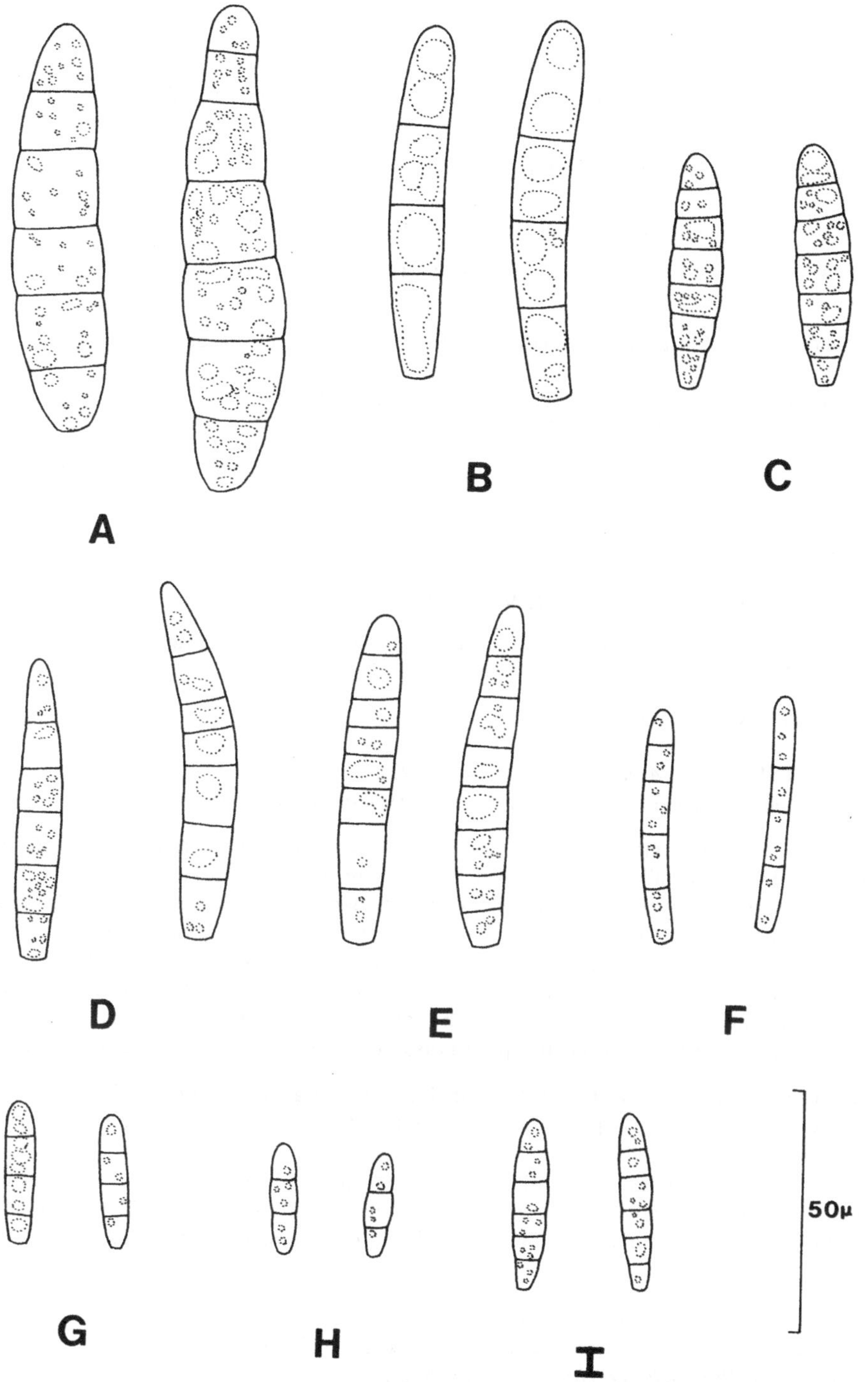

Fig. 49. A, *Stagonospora anglica*, conidia; B, *S. cylindrica*, conidia; C, *S. paludosa*, conidia; D, *S. elegans*, conidia; E, *S. gigaspora*, conidia; F, *S. macropycnidia*, conidia; G, *S. vitensis*, conidia; H, *S. caricinella*, conidia; I, *S. caricis*, conidia.

S. GIGASPORA (Niessl) Sacc., *Syll. fung.* **3**: 453 (1884). (Fig. 49E)
Hendersonia gigaspora Niessl, *Hedwigia* **22**: 181 (1883).
Stagonospora paludosa (Sacc. & Speg.) Sacc. var. *caricis-pendulae* Unam., *Bol. Soc. esp. Hist. nat.* **29**: 401 (1929).

Conidiomata 150-240 µ diam., ostioles 20-25 µ diam. Conidia fusiform, guttules large, constricted, 6-9 septate, 58-84 × 10-14 µ.

On leaves of *Carex riparia*, U.K. (IMI 13695).

S. MACROPYCNIDIA Cunnell, *Trans. Br. mycol. Soc.* **44**: 87 (1961). (Fig. 49F)

Conidiomata 410-1020 × 120-380 µ. Conidia cylindrical, continuous minutely guttulate, 3(-5) septate, 25-70 × 2.5-5 µ.

On *Carex pseudocyparus*, *C. riparia*, U.K. (IMI 79901, holotype, 28537, 34283).

S. PALUDOSA (Sacc. & Speg.) Sacc., *Syll. fung.* **3**: 453 (1884). (Fig. 49C)
Hendersonia paludosa Sacc. & Speg., *Michelia* **1**: 353 (1879).

Conidiomata 100-200 µ diam.; ostioles 20-30 µ diam. Conidia fusiform, minutely guttulate, constricted, 6-8 septate, 41-67 × 8-11 µ.

On leaves of *Carex pseudocyparus*, *C. riparia*, U.K. (IMI 58655, 58657).

S. VITENSIS Unam., *Bol. Soc. esp. Hist. nat.* **29**: 401 (1929). (Fig. 49G)

Conidiomata 150-350 µ diam.; ostioles 30-40 µ diam. Conidia cylindrical, minutely guttulate, continuous, 2-3(-4) septate, 18-32 × 4-6 µ.

On *Carex disticha*, U.K. (IMI 58656).

7. MACRODIPLODIOPSIS Petrak, *Annls mycol.* **20**: 343 (1922).
Sp. typ.: *M. desmazieresii* (Mont.) Petrak (syn. *Hendersonia desmazieresii* Mont.).

Mycelium immersed, branched, septate, brown. *Conidiomata* separate or gregarious, immersed, peridermal to subperidermal, globose to collabent, papillate, dark brown to black, unilocular, thick-walled; walls of outer, thick-walled textura porrecta except at the base where they are of textura angularis, becoming progressively pale and more hyaline towards the conidiogenous region. *Ostiole* single, circular, papillate. *Conidiophores* absent. *Conidiogenous cells* holoblastic, annellidic, discrete, indeterminate, cylindrical, hyaline, smooth, thick-walled, with 1-2 percurrent proliferations, formed from the inner cells of the pycnidial wall. *Conidia* ellipsoid to obovoid, or clavate, 3 distoseptate, occasionally with a longitudinal septum, lumina very much reduced and often surrounded by dark brown wall deposits, continuous, pale brown, thick-walled, base truncate, apex obtuse, surrounded by a large gelatinous sheath.

This genus was redescribed and illustrated by Morgan-Jones, Nag Raj & Kendrick (1972d) who interpreted conidial development as phialidic with percurrent proliferations, and the distosepta and reduced lumina as endoconidia. No mention was made of the gelatinous sheath.

Wall structure of conidia is complex. The lumina are surrounded by a layer with dark brown granular wall deposits, and both are situated within a common body of paler brown less dense granular material. This is surrounded by a definite wall lacking pigmentation or granular material. This is the basic conidium. It is produced in and retains an irregular gelatinous sheath.

M. DESMAZIERESII (Mont.) Petrak, *ibid.*: 343 (1922). (Fig. 50)
Hendersonia desmazieresii Mont., *Ann. Sci. nat.* 3 ser., **12**: 310 (1848).
Stegonsporium platani Preuss, *Fungi Hoyersw.* 336 (1853).
Hendersonia platani Pk, *25th Rep. St. Mus., N.Y. 1871*: 86 (1872).

Conidiomata 500-1000 µ diam. Conidiogenous cells 14-24 µ × 4.5-8 µ. Conidia 40-46 × 16-20.5 µ (excluding sheath).

On dead branches of *Platanus orientalis*, *P. occidentalis*, Austria, Roumania, Ont., Canada and Calif., U.S.A. (IMI 30495, 21822, 21821, 154395).

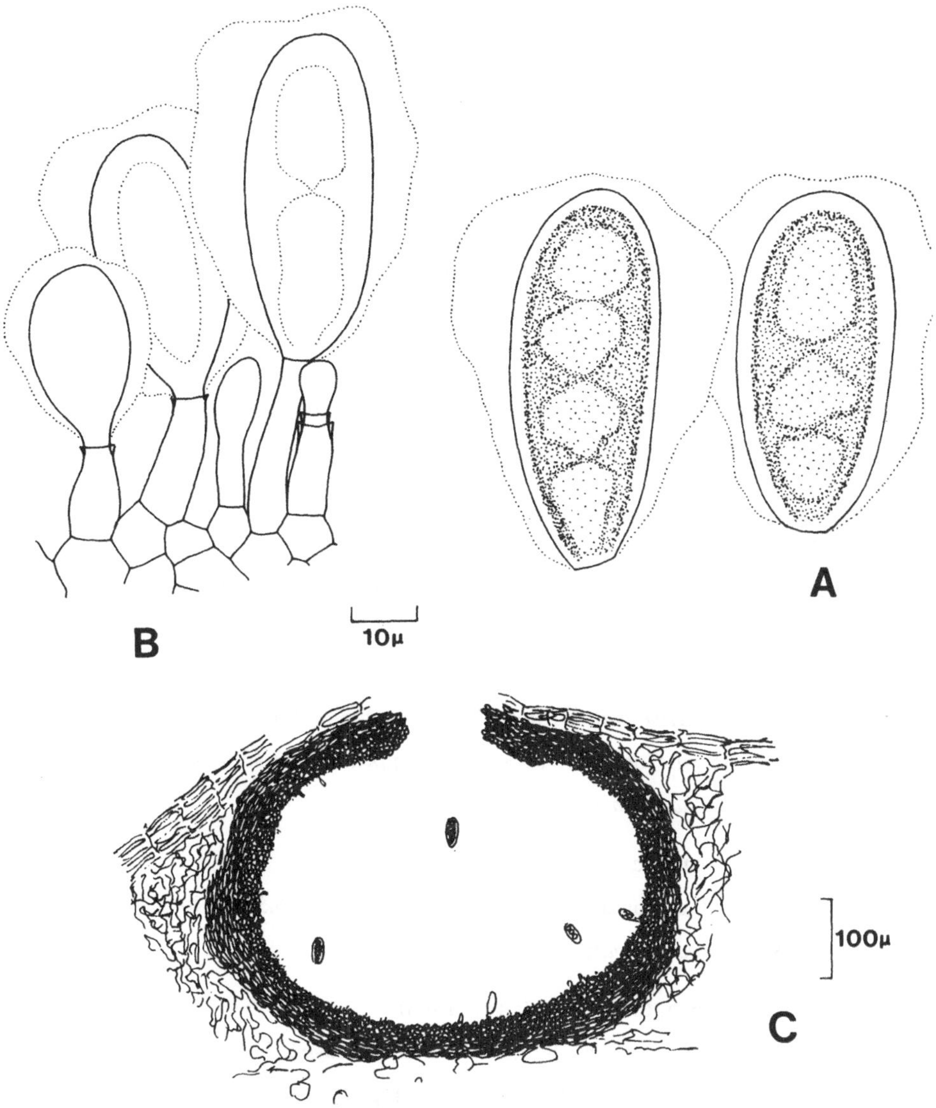

Fig. 50. *Macrodiplodiopsis desmazieresii,* A, conidia; B, conidiogenous cells and developing conidia; C, vertical section of a conidioma.

8. UROHENDERSONIA Speg., *An. Mus. Nac. Hist. Nat. Buenos Aires* **3**, ser. 1-84 (1902).
Uroconis Clements, *Genera of Fungi*: 126 (1909).
Sp. typ.: *U. platensis* Speg.
Illustrations: Nag Raj & Kendrick (1971).

Mycelium immersed, pale brown to hyaline, branched, septate. *Conidiomata* pycnidial, amphigenous, separate, gregarious or confluent, globose, brown, immersed, subepidermal, unilocular, thick-walled; wall composed of brown, thick-walled textura angularis to intricata. *Ostiole* circular, single, ± papillate. *Conidiophores* absent. *Conidiogenous cells* holoblastic, determinate or annellidic and indeterminate, discrete, ampulliform to doliiform, pale brown to hyaline, smooth, with 0-2 percurrent proliferations, formed from the inner cells of the pycnidial wall. *Conidia* pale brown, 3 euseptate, continuous or constricted, cylindrical, clavate or fusiform, apex obtuse, base truncate often with a marginal frill, thin-walled, smooth or verruculose, occasionally guttulate, with an extracellular, gelatinous, unbranched, single, apical appendage.

The genus was monographed by Nag Raj & Kendrick (1971) and amendment made by Sutton, Ghaffar & Abbas (1972).

KEY

1. Conidia smooth 2
1. Conidia verruculose .... *U. stipae*
   2. Appendage >10 µ long 3
   2. Appendage <10 µ long .... *U. pongamiae*
3. Appendage consistently recurved back towards conidial base .... *U. platensis*
3. Not so 4
   4. Conidia 6.5-10 µ wide .... *U. indica*
   4. Conidia 5-6.5 µ wide .... *U. subterranea*

U. INDICA Syd., *Annls mycol.* **14**: 209 (1916).

Conidiomata 90-215 × 100-230 µ. Conidia 20-31 × 6.5-10 µ; appendages 7-32 µ long.

Causing leaf lesions on *Erythrina crista-galli*, Burma (IMI 170562).

U. SUBTERRANEA Panwar & Bilgrami, *Labdev. J. Sci. Techn.* **7B**: 240 (1969).
U. mysorensis Nag Raj & Kendrick, *Can. J. Bot.* **49**: 1856 (1971).

Conidiomata 90-180 × 90-225 µ. Conidia 17-35 × 5-6.5 µ; appendages 29-54 µ long.

On stems of *Striga lutea*, indet. leaf, branches of *Peganum harmala*, and from rhizosphere of *Dactyloctenium aegyptium*, India(IMI 133177, holotype *U. subterranea*).

U. PLATENSIS Speg., *An. Mus. Nac. Hist. Nat. Buenos Aires* **3**, ser. 1, 84 (1902).

Conidiomata 60-170 × 50-150 µ. Conidia 19-30 × 6-8.5 µ; appendage 23-45 µ long.

On rotten leaves of *Manihot carthagenensis*, Argentina.

U. STIPAE Greene, *Trans. Wisc. Acad. Sci. Arts Lett.* **42**: 77 (1953).

Conidiomata 70-130 × 60-120 µ. Conidia 23-32 × 7.5-9 µ; appendages 6.5-47 µ long.

Causing leaf lesions on *Sporobolus elongatus*, Australia (IMI 212663).

U. PONGAMIAE Nag Raj & Ponnappa, *Curr. Sci.* **37**: 417 (1968).

Conidiomata 60-160 × 90-190 µ. Conidia 13-25.5 × 5-7 µ; appendage 2.5-10.5 µ long.

On leaf lesions on *Pongamia pinnata*, India (IMI 130733a, holotype *U. pongamiae*).

9. SEIMATOSPORIOPSIS Sutton, Ghaffar & Abbas, *Trans. Br. mycol. Soc.* **59**: 295 (1972).
Sp. typ.: *S. salvadorae* Sutton, Ghaffar & Abbas

Mycelium immersed, brown, branched, septate. *Conidiomata* pycnidial, gregarious, globose, dark brown, immersed, subperidermal, unilocular, thick-walled; wall composed of an outer layer of dark brown thick-walled textura angularis and an inner layer of thin-walled tissue. *Ostiole* single, circular, papillate. *Conidiophores* absent. *Conidiogenous cells* holoblastic, annellidic, indeterminate, discrete, lageniform to doliiform, hyaline, smooth, with 1-3 percurrent proliferations, formed from the inner cells of the pycnidial wall. *Conidia* dark brown, 1-4 euseptate with reduced lumina, continuous, cylindrical or slightly curved, apex obtuse, base truncate, thick-walled, smooth, with 1 apical or subapical and 1-2 lateral basal cellular, filiform, occasionally branched appendages.

The genus is monotypic.

S. SALVADORAE Sutton, Ghaffar & Abbas, *ibid.*: 295 (1972). (Fig. 51)

Conidiomata 177-279 µ diam. Conidia mostly 3 septate, 12.5-21.5 × 6-8 µ; appendages 6-37 µ long.

On stems of *Salvadora oleioides*, *S.* sp., Pakistan (IMI 148620a, holotype, 178114).

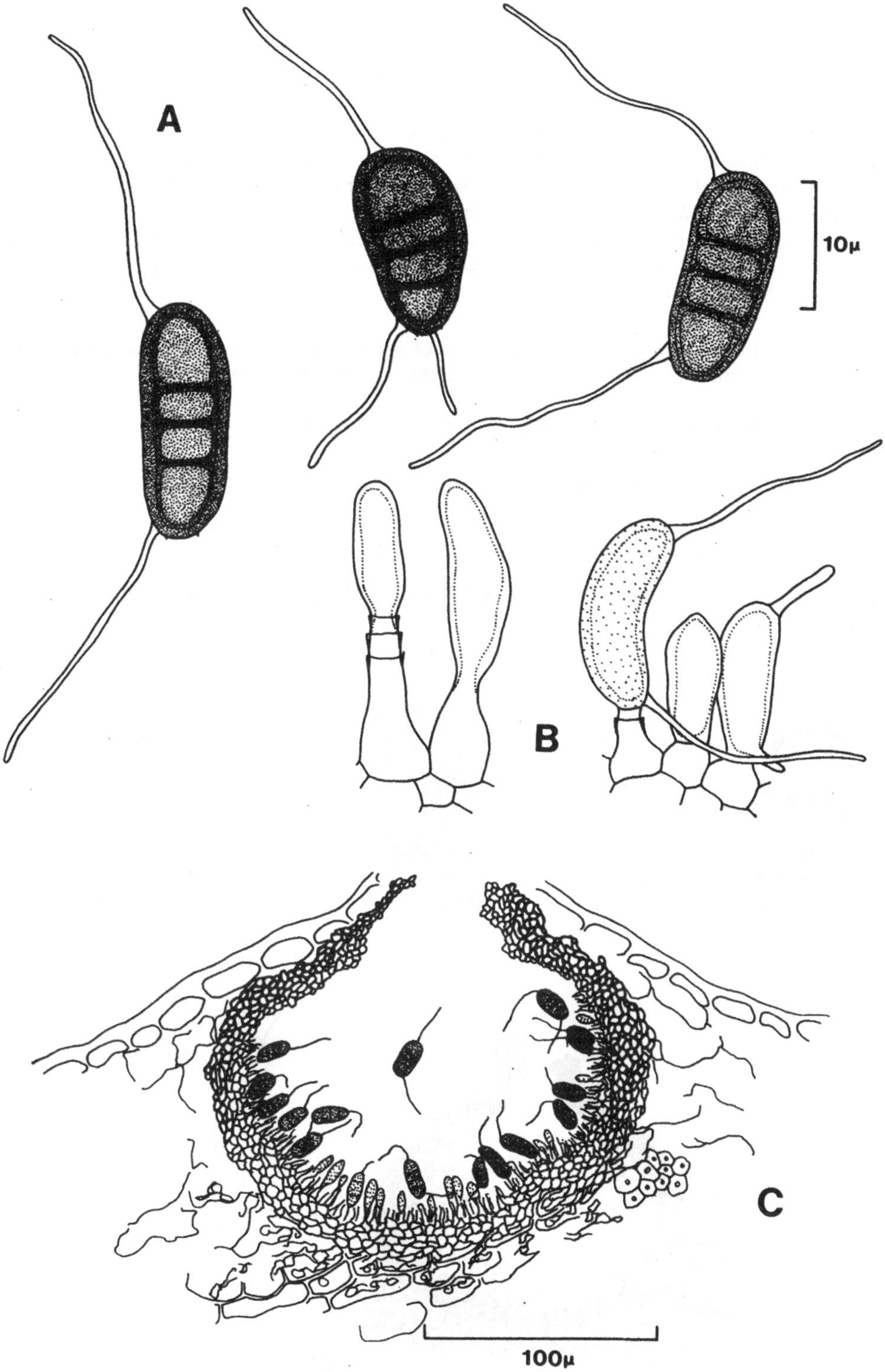

Fig. 51. *Seimatosporiopsis salvadorae*, A, conidia; B, conidiogenous cells and developing conidia; C, vertical section of a conidioma.

10. BLEPTOSPORIUM Steyaert, *Darwiniana, B. Aires* **12**: 171 (1961).
Monochaetina Subram., *Proc. nat. Inst. Sci. India, B, Biol. Sci.* **27**: 243 (1961).
Sp. typ.: *B. pleurochaetum* (Speg.) Sutton (syn. *Cryptostictis pleurochaeta* Speg., *B. montteae* Steyaert).
Illustrations: Sutton (1963); Steyaert (1961).

Mycelium superficial and immersed, brown, septate, branched. *Conidiomata* pycnidial, superficial, separate, gregarious or confluent, globose, dark brown, unilocular, thick-walled, formed of dark brown, thick-walled textura angularis. *Ostiole* circular, central, papillate. *Conidiophores* hyaline, branched at the base, smooth, septate, formed from the inner cells of the pycnidial wall. *Conidiogenous cells* holoblastic, annellidic, integrated, indeterminate, cylindrical, hyaline, smooth, with 1-3 percurrent proliferations. *Conidia* cylindrical to clavate, 4-euseptate, ± constricted, 3 median cells dark brown, thick-walled, each end cell hyaline, thin-walled, smooth, base truncate, apex broadly obtuse with a single, cellular, unbranched appendage inserted laterally.

Sutton (1963) reviewed the nomenclatural entanglements of *Bleptosporium* and *Monochaetina* Subram. and Nag Raj (1977a) described the perfect state as *Amphisphaeria argentinensis*. The genus is monotypic.

B. PLEUROCHAETUM (Speg.) Sutton, *Mycol. Pap.* 88: 7 (1963). (Fig. 52)

Teleomorphosis: *Amphisphaeria argentinensis* Nag Raj, *Can. J. Bot.* **55**: 757 (1977).

Conidiomata 100-150 µ diam. Conidiophores 25 × 3-5 µ. Conidiogenous cells 10-15 × 3-5 µ. Conidia 26-34 × 9.5-13 µ; median cells 20-25 µ long; appendage 25-45 µ long.

On rotten twigs of *Monttea aphylla*, Argentina (IMI 87351 (holotype *Cryptostictis pleurochaeta*), 93992 (holotype of *B. montteae*)).

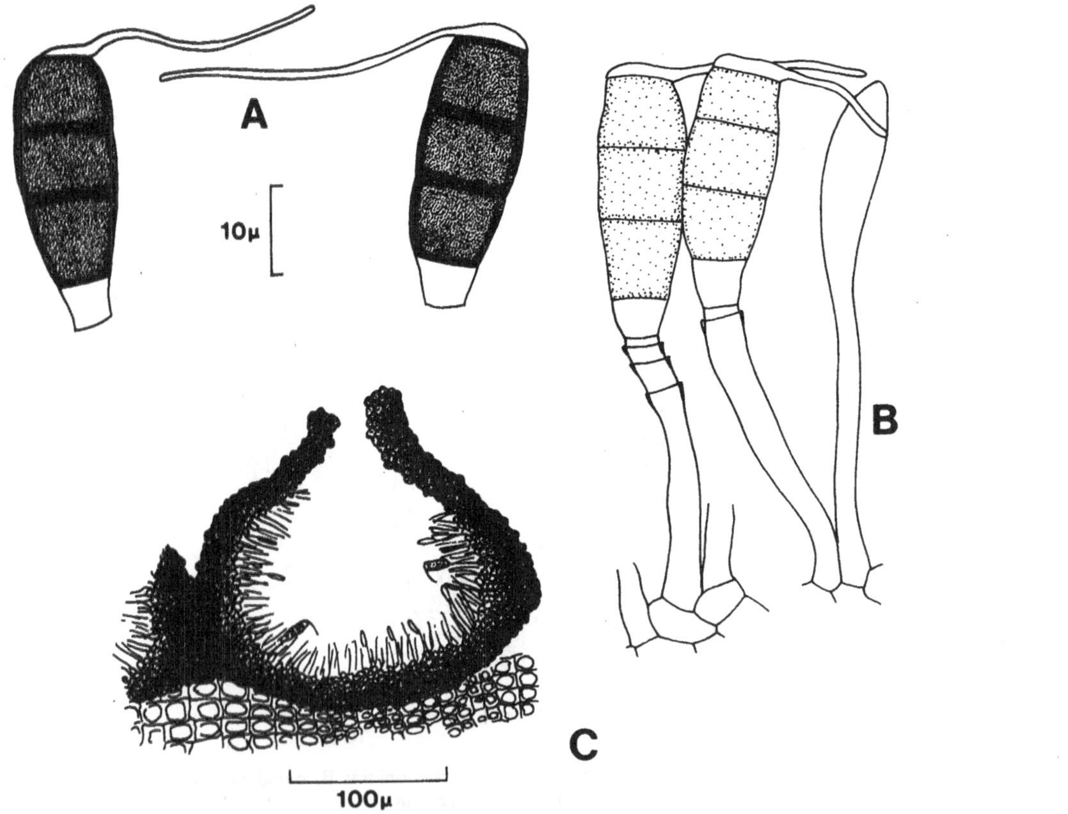

Fig. 52. *Bleptosporium pleurochaetum*, A, conidia; B, conidiogenous cells and developing conidia; C, vertical section of a conidioma.

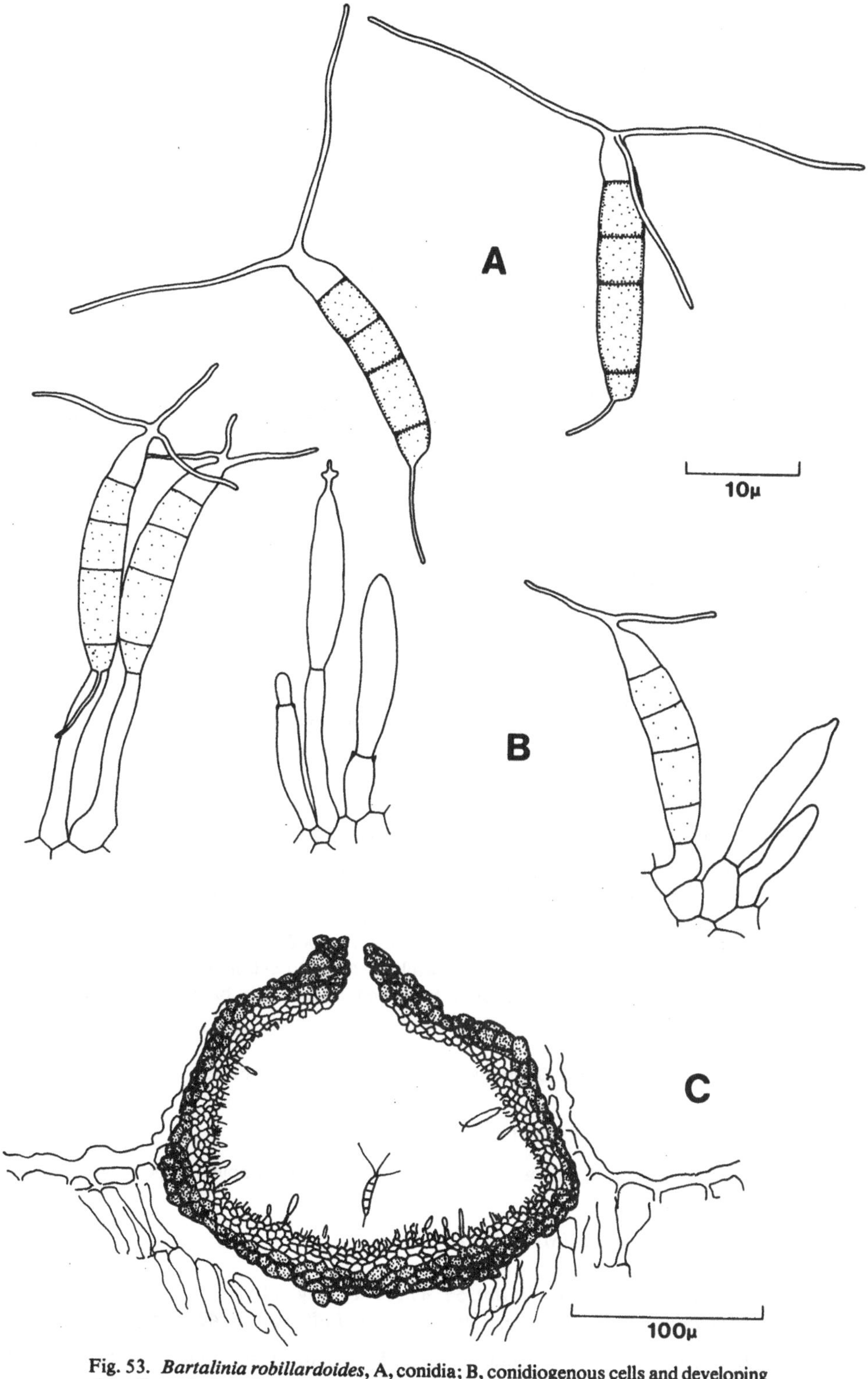

Fig. 53. *Bartalinia robillardoides*, A, conidia; B, conidiogenous cells and developing conidia; C, vertical section of a conidioma.

11. BARTALINIA Tassi, *Bull. Lab. Ort. bot. Siena* **3**: 3 (1900).
Hyalotia Guba, Monograph of *Monochaetia* and *Pestalotia*: 292 (1961).
Sp. typ.: *B. robillardoides* Tassi
Illustration: Morgan-Jones, Nag Raj & Kendrick (1972b).

Mycelium immersed, branched, septate, pale brown. *Conidiomata* pycnidial, separate, globose, dark brown, immersed, subepidermal, unilocular, thick-walled, composed of pale brown, thin-walled textura angularis, darker and thicker at the ostiole and the periphery of the pycnidia. *Ostiole* single, circular, slightly papillate. *Conidiophores* absent. *Conidiogenous cells* holoblastic, annellidic, indeterminate, discrete, cylindrical to lageniform, hyaline, smooth, with 1-2 percurrent proliferations, formed from the inner cells of the pycnidial wall. *Conidia* hyaline to pale brown, 4-euseptate, the penultimate basal cell longer than the rest, continuous, cylindrical to fusiform, straight or slightly bent, base truncate, apex conic, thin-walled but the median cells slightly thicker than the end cells, eguttulate, with 2-3 apical cellular unbranched appendages and a single exogenous cellular unbranched appendage.

A modern description of this genus has been given by Morgan-Jones, Nag Raj & Kendrick (1972b), and although it contains 9 species, most have not been revised.

B. ROBILLARDOIDES Tassi, *ibid.*: 5 (1900). (Fig. 53)

Conidiomata 200-300 µ diam. Conidiogenous cells 6-8.5 × 3-4.5 µ. Conidia 19-25 × 4-4.5 µ; apical appendages 10-21 × 1-1.5 µ; basal appendage 6.5 µ long.

From and on 41 different hosts, Italy, Burma, India, Zambia, Pakistan, Hong Kong, Egypt, Venezuela, Nigeria, Guyana, Australia, 54 collections including IMI 93089, holotype.

12. DOLIOMYCES Steyaert, *Darwiniana, B. Aires* **12**: 169 (1961).
Bartaliniopsis S.S. Singh, *Proc. natn. Acad. Sci. India* sect. B, **42**: 395 (1974).
Sp. typ.: *D. senegalensis* (Speg.) Steyaert (syn. *Pestalotia senegalensis* Speg.).
Illustration: Nag Raj & Kendrick (1972b).

Mycelium immersed, pale brown to hyaline, branched, septate. *Conidiomata* pycnidial, amphigenous, separate, globose, dark brown, immersed, unilocular; thick-walled, outer layer of dark brown thick-walled textura angularis becoming thinner and hyaline towards the inside. *Ostiole* absent, dehiscence by irregular rupture. *Conidiophores* hyaline, branched and septate at the base, smooth, formed from the inner cells of the pycnidial wall. *Conidiogenous cells* holoblastic, annellidic, integrated, indeterminate, cylindrical, hyaline, smooth. *Conidia* cylindrical to navicular, unequally 3-5 euseptate, constricted, median cells dark brown, thick-walled, each end cell hyaline to very pale brown, thin-walled, smooth, apex obtuse with 1 laterally inserted, cellular, branched appendage, base truncate with 1 lateral, cellular, unbranched appendage.

Nag Raj & Kendrick (1972b) redescribed *Doliomyces* and added a second species.

D. SENEGALENSIS (Speg.) Steyaert, *ibid.*: 170 (1961). (Fig. 54B, C)

Conidiomata 100-150 µ diam. Conidia 4 septate, 11.5-18 × 3.5-6 µ; apical appendage branches 11-18 µ long; basal appendage 4.5-6 µ long.

From *Elaeodendron glaucum*, India (IMI 110644).

D. MYSORENSIS Nag Raj & Kendrick, *Can. J. Bot.* **50**: 47 (1972). (Fig. 54A)
Bartaliniopsis saksenaensis S.S. Singh, *Proc. natn. Acad. Sci. India* sect. B, **42**: 395 (1974).

Conidiomata 345-600 × 150-460 µ. Conidia 18.5-21.5 × 4-5 µ; apical appendage branches 16.5-28 µ; basal appendage 5.5-9.5 µ long.

On leaves of *Eucalyptus* sp., and from soil, India (IMI 113705 (holotype of *B. saksenaensis*), 183930b).

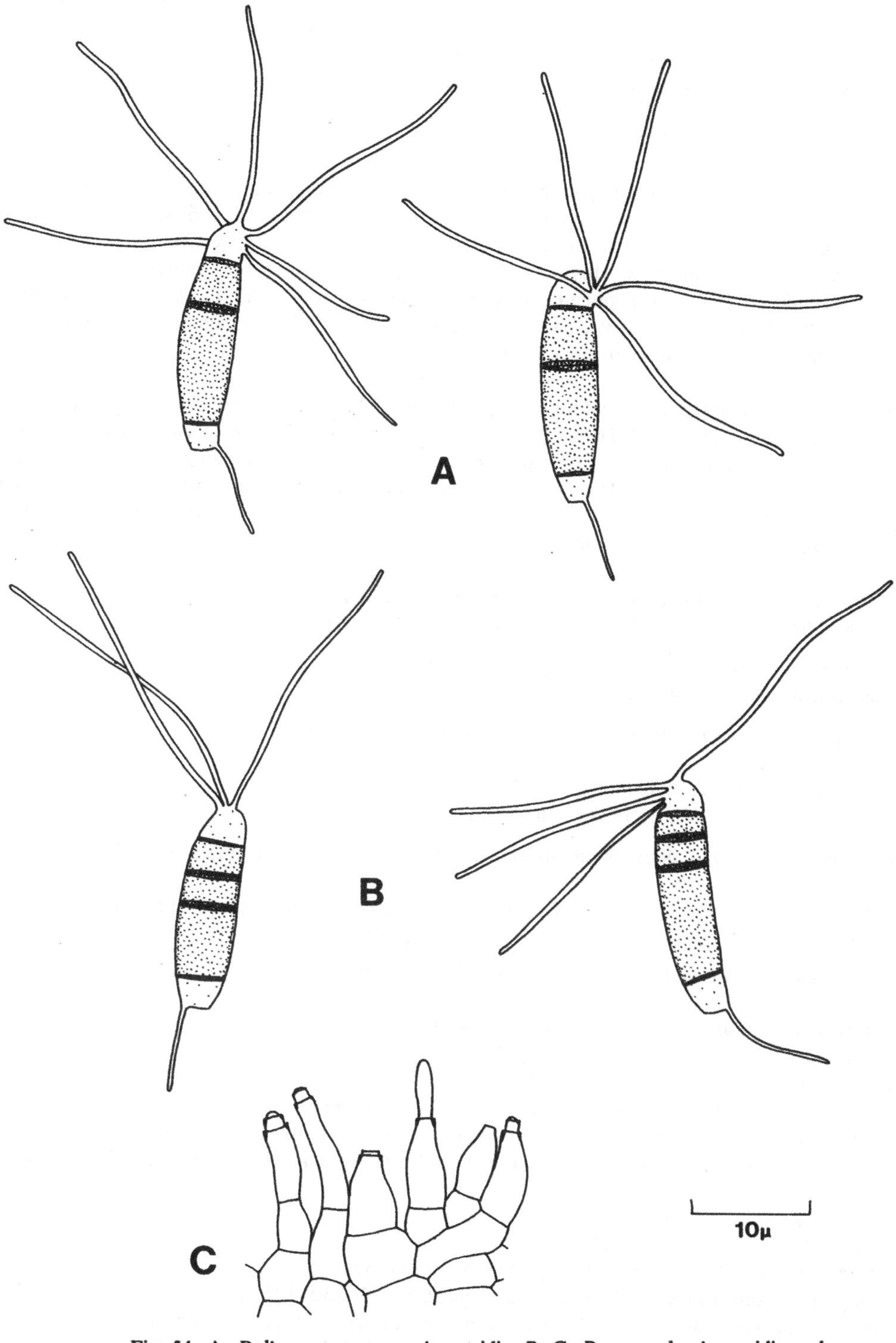

Fig. 54. A, *Doliomyces mysorensis*, conidia; B, C, *D. senegalensis*, conidia and conidiophores.

13. LICHENOCONIUM Petrak & Syd., *Beih. Rep. spec. nov. regni veg.* **42**: 432 (1927).

Original spp.: *L. lichenicola* (Karst.) Petrak & Syd., *L. cladoniae* (Ell. & Ev.) Petrak & Syd., *L. lecanoracearum* (Vouaux) Petrak & Syd., *L. pyxidiatae* (Oud.) Petrak & Syd., *L. jaapii* (Died.) Petrak & Syd., *L. resinae* (Sacc. & Berl.) Petrak & Syd,, *L. boreale* (Karst.) Petrak & Syd.

Lectotype sp.: *L. lichenicola* (Karst.) Petrak & Syd. (syn. *Coniothyrium lichenicola* Karst.), fide Clements & Shear (1931).

Illustration: Hawksworth (1977).

Mycelium immersed, branched, septate, pale brown. *Conidiomata* pycnidial, immersed to semi-immersed, separate, occasionally gregarious, dark brown to black, globose, unilocular, thick-walled; wall consisting of medium brown, thick-walled textura angularis, somewhat darker in the upper region. *Ostiole* absent, dehiscence by breakdown of upper wall, finally appearing cupulate. *Conidiophores* absent. *Conidiogenous cells* holoblastic, annellidic, indeterminate, discrete, cylindrical or long lageniform, hyaline or pigmented towards the apices, smooth, with 0-3 apical annellations formed from the inner pycnidial wall cells. *Conidia* aseptate, base truncate, often with a frill, apex obtuse, thick-walled, various shades of brown, verrucose, eguttulate, globose, obovate, obpyriform.

Originally the genus contained lichenicolous and non-lichenicolous species, but of the latter, *L. resinae* was later referred to *Epithyrium* (Sacc.) Saccardo & Trotter (1913) and has been selected as lectotype species for that genus, and *L. boreale* with smooth conidia does not at the moment appear to be congeneric although there appears to be no other satisfactory available generic name for it. This leaves *Lichenoconium* as a genus of solely lichenicolous hyperparasites. Hawksworth (1977) revised the genus and accepted 10 species, including *L. boreale* which occurs on decorticated wood of *Pinus*. The species were fully keyed out and amply illustrated.

L. LICHENICOLA (Karst.) Petrak & Syd., *ibid.*: 432 (1927). (Fig. 55)
Coniothyrium lichenicola Karst., *Meddel. Soc. fauna flora fenn.* **14**: 104 (1887).

Conidiomata up to 150 µ diam. Conidiogenous cells 8-15 × 3-4.5 µ, narrowing to 2.5 µ. Conidia 5.5-7 × 3.5-4 µ, verrucose, brown to dark brown.

On ascocarps of *Physcia aipolia*, Finland (IMI 186890, holotype ex H).

L. BOREALE (Karst.) Petrak & Syd., *ibid.*: 436 (1927). (Fig. 56)
Levieuxia borealis Karst., *Hedwigia* **26**: 126 (1887).

Conidiomata 150-250 µ diam. Conidiogenous cells 7-9 × 1.5-2 µ. Conidia 3.5-4.5 × 2-2.5 µ, smooth pale brown.

Although this species does not strictly belong in *Lichenoconium* it cannot be retained in *Levieuxia*, for, as Sutton (1977) indicated, this generic name is based on sterile stromata.

On decorticated *Pinus* wood, Finland (IMI 198492, holotype ex H).

14. SPHAEROPSIS Sacc., nom. cons., *Michelia* **2**: 105 (1880).

Sphaeropsis Lév. in Demidoff, *Voy. Russ.* **2**: 112 (1842), nom. rej.
? *Gyratilium* Preuss, *Linnaea* **26**: 722 (1855).
Macroplodia Westd., *Bull. Acad. Sci. Belg.* **2**: 562 (1857), nom. rej.
Botryosphaerostroma Petrak & Syd., *Beih. Rep. spec. nov. regni veg.* **42**: 126 (1927).
Macrophoma (Sacc.) Berl. & Vogl., *Atti Soc. Venet. – Trent. Sci. Nat.* **10**: 4 (1886), and Sacc., *Syll. fung. addit.* **1-4**: 306 (1886).
Phoma Westd. subgen. *Macrophoma* Sacc., *Syll. fung.* **3**: 65 (1884).
Catosphaeropsis Tehon, *Mycologia* **31**: 542 (1939).
Phomatosphaeropsis Ribaldi, *Ann. Sper. agr. N.S.* **7**: 847 (1953).
Granulodiplodia Zambettakis, *Bull. trimest. Soc. mycol. Fr.* **70**: 330 (1955).
Granulodiplodia Zambettakis ex Morelet, *Bull. Soc. Sci. nat. Archéol. Toulon Var* **203**: 12 (1973).

Sp. typ.: *S. visci* (Fr.) Sacc. (syn. *Sphaeria atrovirens* Alb. & Schw. ex Fr. α visci Fr.).

Illustrations: Punithalingam & Waterston (1970); Ribaldi (1953).

Mycelium immersed or superficial, dark brown, branched, septate. *Conidiomata* pycnidial, immersed, often appearing superficial, separate or aggregated, globose, dark brown, unilocular, thick-walled; wall formed of dark brown, thick-walled,

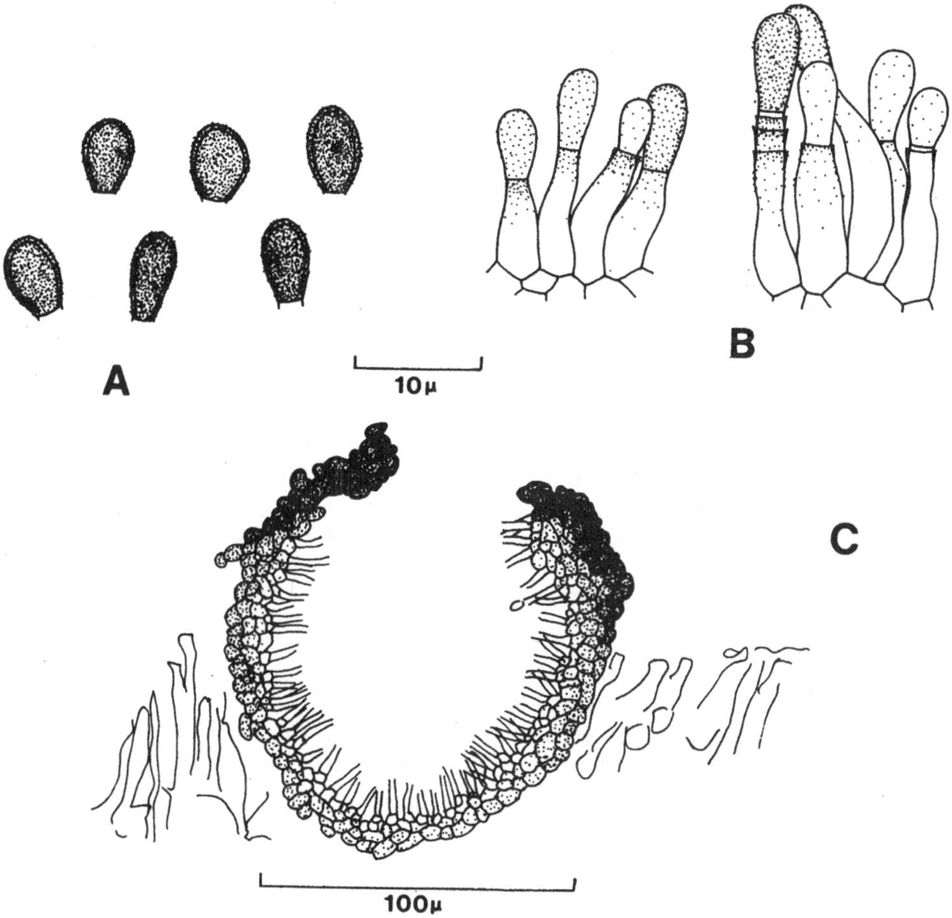

Fig. 55. *Lichenoconium lichenicola*, A, conidia; B, conidiogenous cells and developing conidia; C, vertical section of a conidioma.

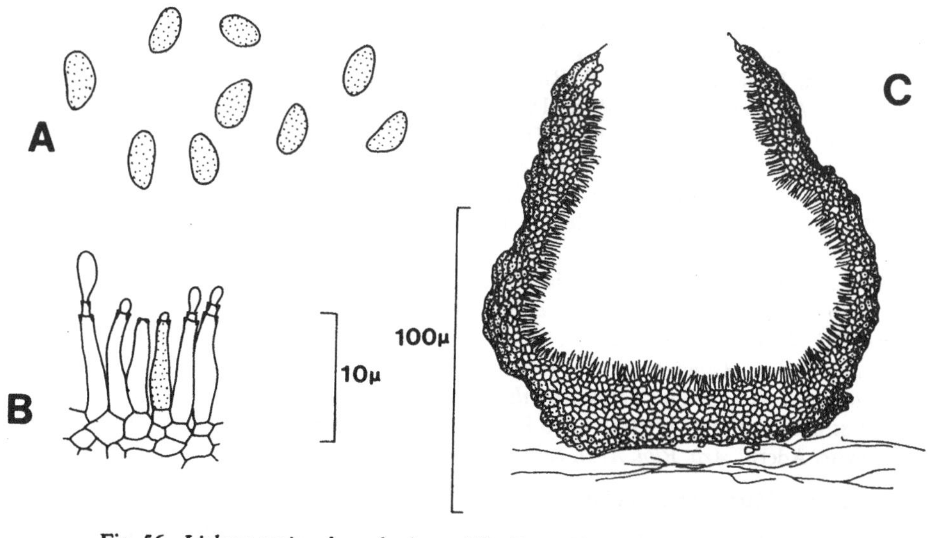

Fig. 56. *Lichenoconium boreale*, A, conidia; B, conidiogenous cells and developing conidia; C, vertical section of a conidioma.

textura angularis, darker round the ostiole. *Ostiole* central, circular, single, papillate. *Conidiophores* absent. *Conidiogenous cells* holoblastic, discrete, determinate or indeterminate with a single percurrent proliferation, swollen at the base, long lageniform, hyaline, smooth, with acrogenous conidia and only occasionally percurrent growth, formed from the cells lining the inner wall of the pycnidium. *Conidia* oblong to clavate, straight, aseptate (but developing eusepta prior to germination), thick-walled, ornamented on the inner face of the wall, apex obtuse, tapered to a truncate base.

Sutton (1977) considered *Sphaeropsis* and *Macrophoma* to represent quite different genera, but examination of the two type species shows close agreement in conidiomatal structure, conidiogenesis and conidial morphology. The fact that in *M. sapinea* the conidia tend to develop a single septum later in development or just prior to germination is not considered sufficient to separate this species into a genus distinct from *S. visci*. Consequently *Macrophoma* is here regarded as a later synonym of the conserved name *Sphaeropsis* Sacc.

The genera *Macrophoma* and *Sphaeropsis* have been the subject of detailed revision by Petrak & Sydow (1927), and in fact *Sphaeropsis* was made a later synonym of *Haplosporella* (*Aplosporella*). However the conidiomata in *Aplosporella* are eustromatic not pycnidial, and the conidial ornamentation is extracellular rather than on the inside of the wall as in *Sphaeropsis*. For these reasons the synonymy is rejected. Many more than 500 taxa have been described in *Sphaeropsis* and several of these have been referred to other genera by Petrak & Sydow (1927).

Macrophoma has been used for more than 600 *Phoma*-like species with relatively large conidia, and despite the work of Petrak & Sydow (1927) later authors have indiscriminately used the name in a Saccardoan sense. Many genera are available to accommodate the heterogeneous species that have been, and still are, described in *Macrophoma*, as Petrak & Sydow (1927) have ably shown.

Punithalingam & Waterston (1970), under the name *Diplodia pinea*, and Petrak (1961) have illustrated *S. sapinea* and provided an extensive bibliography.

Sphaeropsis sapinea (Fr.) Dyko & Sutton comb. nov. (Fig. 57A)
Sphaeria sapinea Fr., *Systema mycologicum* **2**: 491 (1823).
Synonymy listed in Punithalingam & Waterston (1970) under *Diplodia pinea* (Desm.) Kickx, Petrak & Sydow (1927) under *Macrophoma pinea* (Desm.) Petrak & Sydow, and Petrak (1961) under *M. sapinea* (Fr.) Petrak.

Conidiomata up to 250 µ diam. Conidiogenous cells 15-20 µ long. Conidia often yellowish, ultimately dark brown, 30-45 × 10-16 µ.

Petrak & Sydow (1927) arbitrarily took the first listed species of *Macrophoma* as the lectotype, *M. macrosperma* (Karst.) Berl. & Vogl., but showed that *Sphaeria pinea* Desm. provided an earlier epithet and consequently proposed the binomial *M. pinea* (Desm.) Petrak & Syd. Subsequently Petrak (1961) showed that *M. pinea* is a later homonym of *M. pinea* Pass. (≡ *Dothiorella pinea* (Pass.) Petrak & Syd.), consequently he replaced the name of the lectotype species by *M. sapinea* (Fr.) Petrak.

Occurring on bark, shoots, needles and cones of *Abies excelsa*, *A. procera*, *Cupressus lusitanica*, *Larix*, *Pinus austriaca*, *P. coulteri*, *P. caribaea*, *P. caribaea* var. *hondurensis*, *P. halepensis*, *P. elliottii*, *P. patula*, *P. radiata*, *P. douglasiana*, *P. strobus*, *P. sylvestris*, *P.* sp., *Pseudotsuga menziesii*, U.K., Australia, Austria, Eire, New Zealand, Selangor, Malaysia, India, Swaziland, S. Rhodesia, Tanzania, Mauritius, Uganda, Malawi, Kenya, S. Africa, Brazil, Calif., U.S.A., Canada (IMI 6688, 184547b, 127164, 89587, 6687, 142194, 164169b, 6857, 89615a, 114778, 114777, 127153, 6859, 6689, 172716a, 143522, 127151, 132977a, 146755a, 6858, 6860, 38803, 53097, 77886, 150095, 162499, 165511, 111949, 127324, 127326-9, 156992, 172141, 181955, 121050a, 60521, 58773, 69023-6, 69028, 38761, 57822, 79267c, 89063, 110448, 114387, 6686, 17512, 60171, 94421, 178242, 49364, 76786, 76965, 69467, 178384, 184299a, 28090, 197536c, 186707, 6064, 127325).

S. VISCI (Fr.) Sacc., *Michelia* **2**: 105 (1880). (Fig. 57B, C, D)
Sphaeria atrovirens Alb. & Schw. ex Fr. *α visci* Fr., *Syst. Myc.* **2**: 501 (1823).
Sphaeria visci DC., *Fl. franç.* **6**: 146 (1815).
Ceuthospora visci Sollm., *Hedwigia* **2**: 187 (1863).
Botryosphaerostroma visci (Sollm.) Petrak, *Annls mycol.* **21**: 120 (1923).
Botryosphaerostroma visci (DC.) Petrak apud Petrak & Sydow, *Beih. Rep. spec. nov. regni veg.* **42**: 127 (1927).
Macrophoma visci Aderh., *Arb. Biol. Anst. Land – u. Forstw.* **4**: 462 (1905).

Conidiomata up to 350 µ diam. Conidiogenous cells 4-8 × 4-6 µ. Conidia 40-54 × 20-22 µ.

On leaves and twigs of *Viscum album*, Egypt, Roumania, Czechoslovakia, Austria (IMI 11770, 11771, 61584, 31009, 6110).

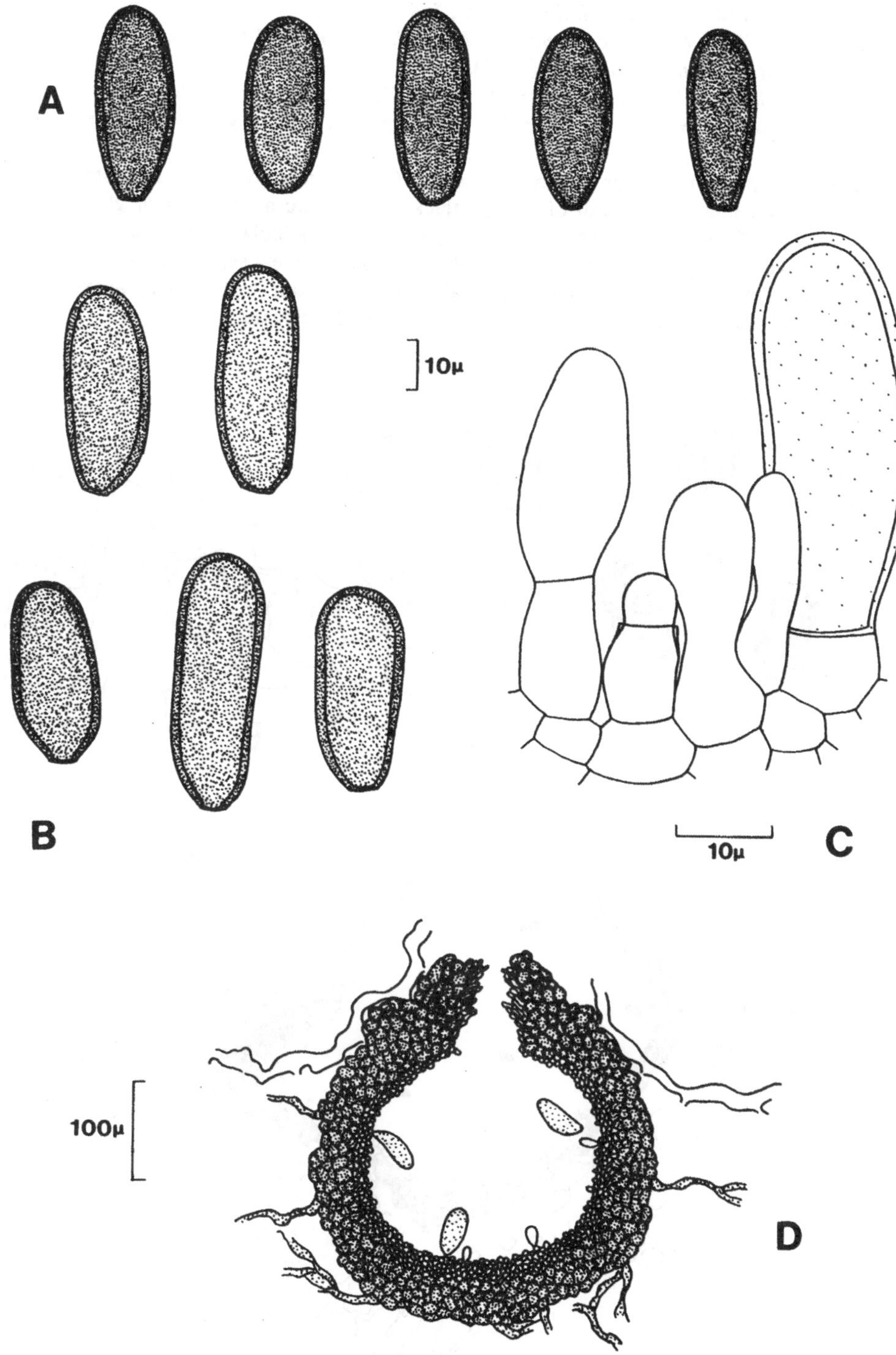

Fig. 57. A, *Sphaeropsis sapinea*, conidia; B, C, D, *S. visci*, conidia, conidiogenous cells, vertical section of a conidioma.

15. CONIOTHYRIUM Cda, nom. cons., *Icones fungorum* **4**: 38 (1840).

Clisosporium Fr., nom. rej., *Syst. myc.* **1**: XLVII (1821).

? *Monoplodia* Westd., *Bull. Acad. R. sci. Belg.*, 2 ser., **7**: 91 (1859).

? *Asteropsis* G. Frag., *Trab. Mus. Nac. Cienc. Nat. Madrid*, ser. bot. **12**: 50 (1917).

? *Coniothyrinula* Petrak, *Annls mycol.* **21**: 2 (1923).

Original spp.: *C. pini* Cda, *C. palmarum* Cda, *C. subtile* Cda, *C. glomeratum* Cda

Lectotype sp.: *C. palmarum* Cda, *Int. Code Bot. Nom.* App. III. Nomina generica conservanda et rejicienda.

Mycelium immersed, septate, brown or hyaline, branched. *Conidiomata* pycnidial, separate, globose, dark or pale brown, immersed, unilocular, thin-walled; wall of brown, thick-walled textura angularis or globulosa. *Ostiole* circular, central, sometimes papillate. *Conidiophores* absent. *Conidiogenous cells* holoblastic, annellidic, indeterminate, discrete, doliiform to cylindrical, hyaline or pale brown, smooth, with 1-4 close or widely separated annellations, formed from the inner cells of the pycnidial wall. *Conidia* brown, thick-walled, 0-1 euseptate, verruculose, apex obtuse, base truncate (sometimes with a marginal frill), cylindrical, spherical, elliptical or broadly clavate.

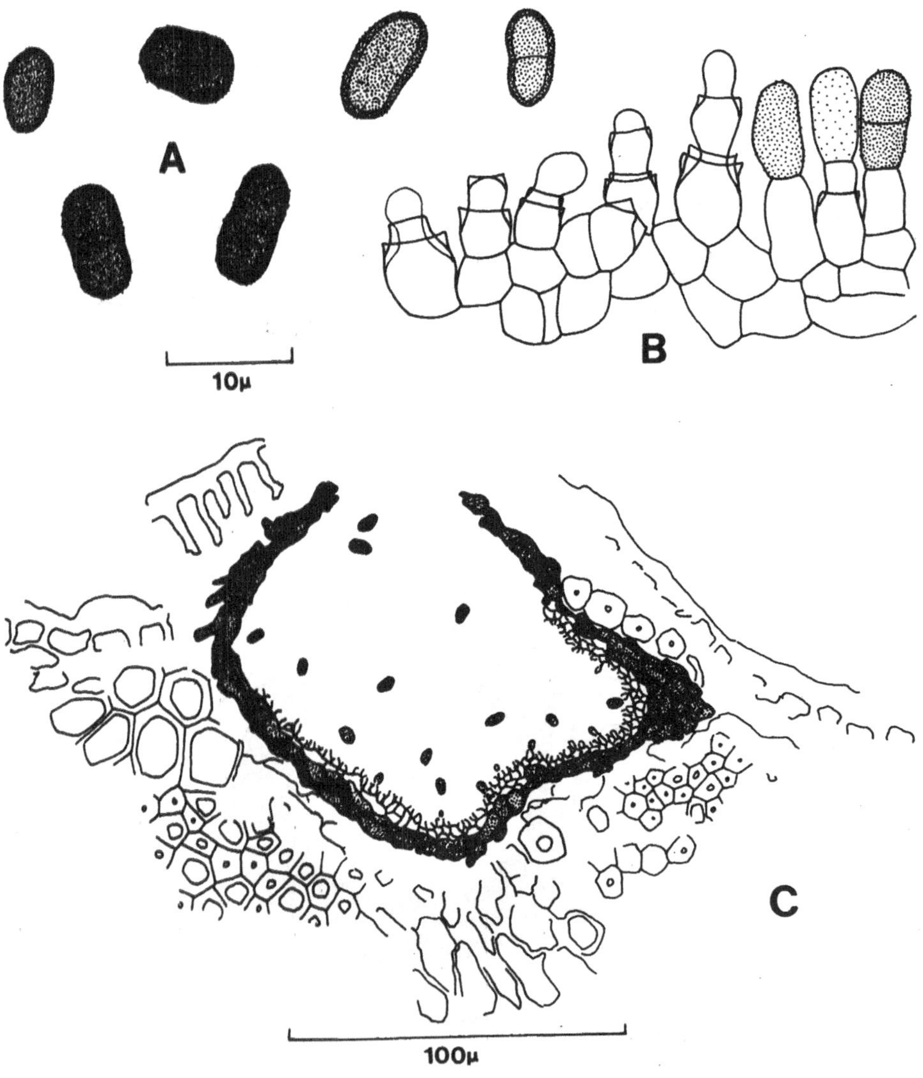

Fig. 58. *Coniothyrium palmarum*, A, conidia; B, conidiogenous cells; C, vertical section of a conidioma.

Sutton (1971b) discussed the nomenclatural status of *C. palmarum* and suggested that the majority of species described in *Coniothyrium* were not congeneric with the type species. It was proposed to use *Microsphaeropsis* Höhn. for those species with phialides and retain the species with annellides in *Coniothyrium*.

Morgan-Jones (1975a) and Sutton (1974) dealt with several species but since the genus contains in excess of 800 described taxa it is clear that nothing short of a major revision aimed at determining which species should be retained in *Coniothyrium* and which should be moved to *Microsphaeropsis* is required. Biga, Ciferri & Bestagno (1959) keyed out a large number of species by host substrate and conidial size but unfortunately this work is largely an uncritical compilation based on published data rather than examination of original material.

C. PALMARUM Cda, *Icones Fungorum* **4**: 38 (1840). (Fig. 58)

Conidiomata up to 150 µ diam. Conidiogenous cells 5-12 × 3-5 µ. Conidia 0-1 septate, 6-8.5 × 4-5 µ.

On leaves of *Phoenix dactylifera*, Palmae indet., India, Cyprus (IMI 16478, 21325).

C. EUCALYPTICOLA Sutton, *Mycol. Pap.* 123: 34 (1971). (Fig. 59)

Conidiomata up to 30 µ diam. Conidiogenous cells 2.5-7 × 4-6 µ. Conidia aseptate, with a distinct basal frill 8.5-10 × 6-7.5 µ.

On leaves of *Eucalyptus leptophylla*, Australia (IMI 71347, holotype).

C. AHMADII Sutton, *Nova Hedwigia* **25**: 163 (1974).
Coniothyrium eucalypti Ahmad, *Biologia, Lahore* **17**: 69 (1971), non Thümen, *Instituto Coimbra* **27**: 44 (1880), nec G. Fragoso, *Bol. Soc. Brot.*, ser. 2, **2**: 59 (1924).

Conidia thick-walled, medium brown, elliptical, 6-7 × 3.5-4.5 µ.

On branches of *Eucalyptus*, Pakistan (IMI 92059, holotype).

16. SCHWARZMANNIA Pissareva, *Bot. Mater.* **5** 72 (1968).
Sp. typ.: *S. goebeliae* Pissareva

Mycelium immersed, branched, septate, pale brown. *Conidiomata* pycnidial, separate or aggregated, immersed, dark brown, subepidermal, globose, unilocular, thin-walled; wall of dark brown, thick-walled textura angularis. *Ostiole* absent, dehisence by irregular breakdown of the upper wall, opening widely. *Conidiophores* absent. *Conidiogenous cells* holoblastic, annellidic, indeterminate, discrete, cylindrical, pale brown, verruculose, with a single ragged percurrent proliferation, formed from the inner cells of the pycnidial wall. *Conidia* globose to pyriform, very pale brown, thin-walled, verruculose, often medianly 1 euseptate, not constricted, base ± truncate.

Two species were originally described in the genus, *S. goebeliae* Pissareva and *S. ammodendri* Kuznetzova. Examination of the isotypes shows them to be identical, consequently *S. ammodendri* is reduced to synonymy.

S. GOEBELIAE Pissareva, *ibid.*: 74 (1968). (Fig. 60)
S. ammodendri Kuznetzova apud Pissareva, *ibid.*: 76 (1968).

Conidiomata 200-250 µ. Conidiogenous cells 8-20 × 2-3.5 µ. Conidia 10-15 × 8.5-10 µ.

On living leaves of *Goebelia alopecuroides, Ammodendron argenteum*, Kazakhstan, U.S.S.R. (IMI 199149, isotype of *S. goebeliae*, IMI 199148, isotype of *S. ammodendri*).

17. CAMAROSPORIUM Schulz., *Verh. Zool.-Bot. Ges. Wien* **20**: 641 (1870).
? *Hyalothyridium* Tassi, *Bull. Lab. Ort. Bot. Siena* **3**: 91 (1900).
Camarosporulum Tassi, *Bull. Lab. Ort. Bot. Siena* **5**: 63 (1902).
? *Hyalothyris* Clements, nom. nov., *Genera of Fungi*: 127 (1909).
? *Pseudodichomera* Höhn., *Hedwigia* **60**: 186 (1918).
Sp. typ.: *C. quaternatum* (Hazsl.) Schulz. (syn. *Clinterium quaternatum* Hazsl.).

Mycelium immersed, brown, branched, septate. *Conidiomata* pycnidial, separate, immersed, subperidermal, globose, dark brown to black, unilocular, occasionally papillate; walls thick, dark brown, of textura angularis, becoming paler and

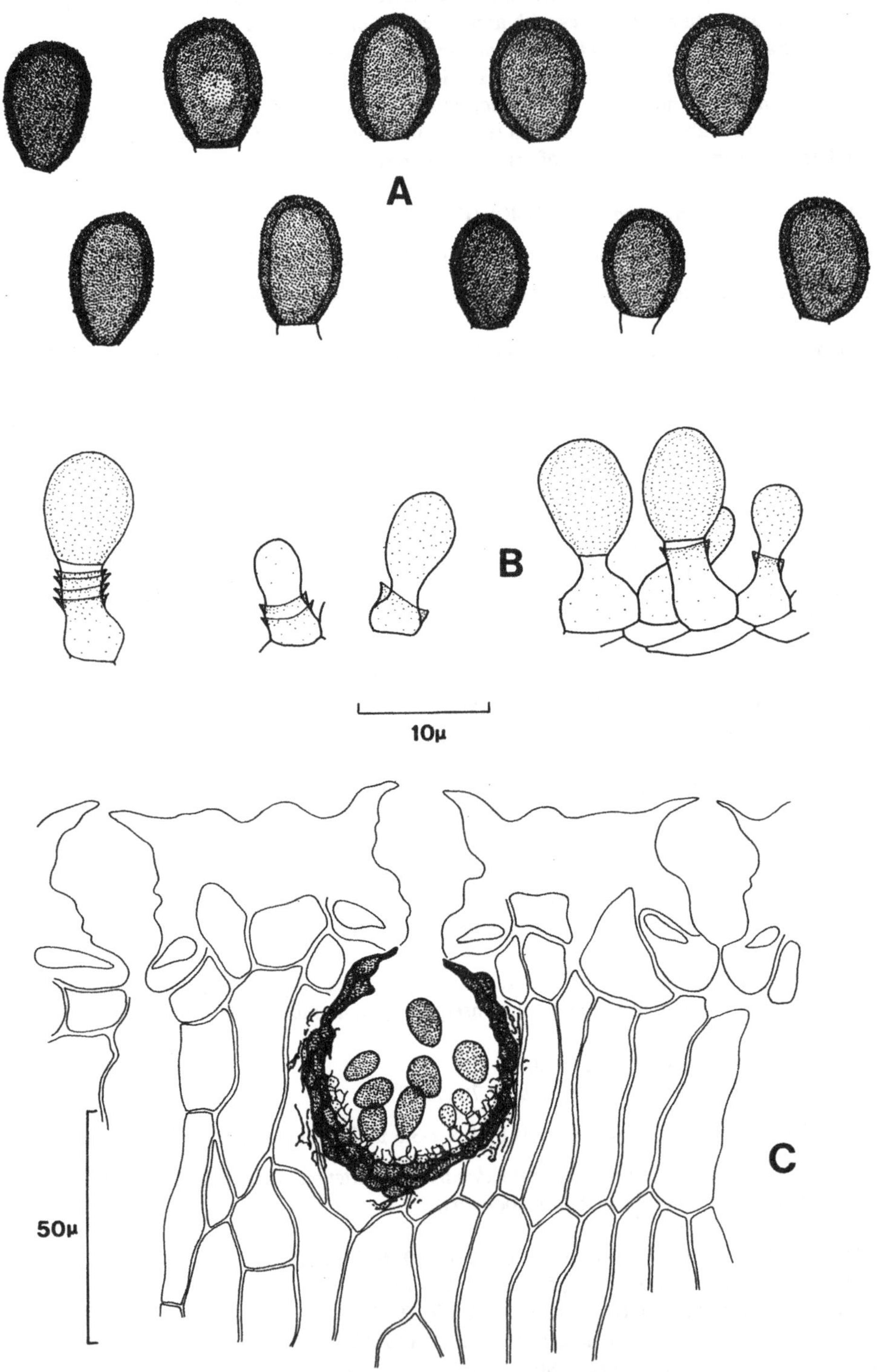

Fig. 59. *Coniothyrium eucalypticola*, A, conidia; B, conidiogenous cells, C, vertical section of a conidioma.

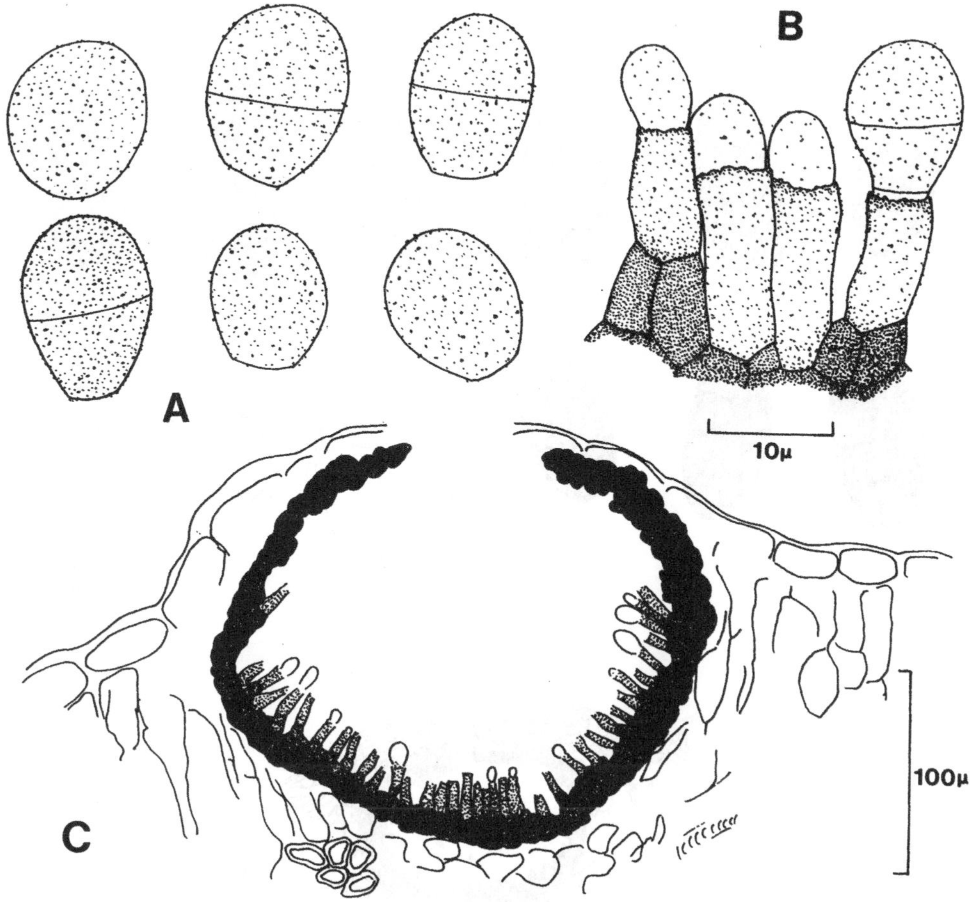

Fig. 60. *Schwarzmannia goebeliae*, A, conidia; B, conidiogenous cells; C, vertical section of a conidioma.

thin-walled in the conidiogenous region. *Ostiole* single, circular, central, sometimes papillate. *Conidiophores* absent. *Conidiogenous cells* holoblastic, annellidic, discrete, indeterminate, doliiform, lageniform or cylindrical, smooth, hyaline, formed from the inner cells of the pycnidial wall. *Conidia* brown, muriform, with transverse, longitudinal and oblique eusepta, smooth or verruculose, base truncate, apex obtuse, continuous or constricted at the septa, variable in shape.

The generic synonymy has been discussed in Sutton (1977a), and the type species described and illustrated by Sutton & Pollack (1974). Even as interpreted by Sutton & Pollack (1974), Sutton (1975a) and Sutton (1977a) the genus is heterogeneous. Not only is there the conflicting status of annellidic conidiogenesis in *C. quaternatum* as opposed to what could be interpreted as proliferating phialidic conidiogenesis in *C. propinquum* (Sacc.) Sacc., but also the inclusion of species with smooth and verruculose conidia in the same genus. Until the generic limits are more clearly defined, such anomalies will persist.

The genus contains about 400 names, many of which refer to species that have not been re-examined since their introduction. Many taxa have been proposed on the basis of their occurrence on hosts other than those known already to support *Camarosporium* species, so the genus is in urgent need of revision, discounting any approach based on host specificity.

C. QUATERNATUM (Hazsl.) Schulz., *Verh. Zool.-Bot. Ges. Wien* **20**: 641 (1870). (Fig. 61D, E)
Clinterium quaternatum Hazsl., *ibid.*: **15**: 451 (1865).

Conidiomata 450 μ diam. Ostiole papillate, 50-60 μ diam. Conidiogenous cells 12.5-22.5 × 3-5 μ. Conidia very variable, fusiform, ovate or clavate, 3-4 primary septa, several secondary longitudinal or oblique septa, smooth, 15.5-26 × 6.5-10.5 μ.

On twigs of *Lycium halimifolium*, *L. barbarum*, Germany, Austria, Roumania (IMI 30306, 31020, 50077).

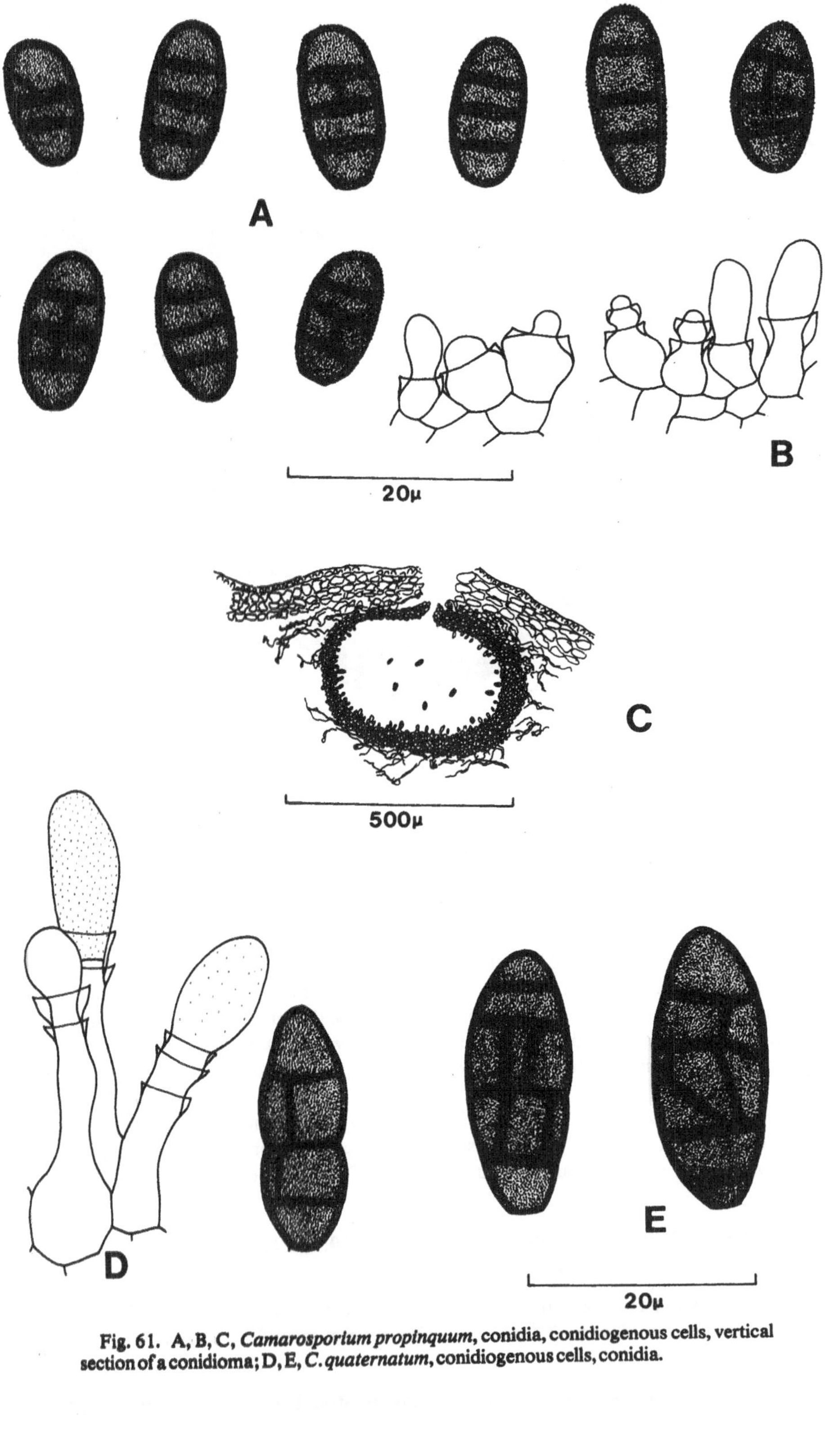

Fig. 61. A, B, C, *Camarosporium propinquum*, conidia, conidiogenous cells, vertical section of a conidioma; D, E, *C. quaternatum*, conidiogenous cells, conidia.

C. PROPINQUUM (Sacc.) Sacc., *Syll. fung.* **3**: 464 (1884). (Fig. 61A, B, C)
Hendersonia propinqua Sacc., *Michelia* **1**: 516 (1879).
Coryneum populi Oud., *Nederl. Kruidk. Archf.*, ser. 3, **1**: 510 (1898).
Camarosporium populi Oud., *Verh. k. Akad. Wet.*, ser. 2, **11**: 541 (1904).

Conidiomata 450 µ diam. Ostiole 20 µ diam. Conidiogenous cells 3.5-7.5 × 4-6.5 µ. Conidia ellipsoidal to oval, 2-3 primary transverse septa, 1-2 longitudinal septa, verruculose, 11.5-17 × 6.5-7.5 µ.

On branches of *Populus canadensis*, *P.* sp., *Corylus*, *Salix*, Netherlands, Austria, U.K. (IMI 176914, holotype *Coryneum populi*, 180381, 180382, 180380).

BLASTOSTROMATINEAE subord. nov.

Cellulae conidiogenae holoblasticae; conidiomata stromatica
Conidiogenesis holoblastic; conidiomata stromatic

This suborder contains a large number of genera and it is therefore convenient to break them down into three groups based on the ability or inability of the conidiogenous cells to produce more than one conidium. As in the Blastopycnidiineae some conidiogenous cells form a single conidium, while others form a plurality of conidia by percurrent proliferation (annellides) and a further group develops several conidiogenous loci either synchronously or sympodially.

KEY TO GROUPS WITHIN THE BLASTOSTROMATINEAE

Conidiogenous cells holoblastic, monoblastic .... p. 125
Conidiogenous cells holoblastic, annellidic........ p. 251
Conidiogenous cells holoblastic, sympodial or synchronous........ p. 205

BLASTOSTROMATINEAE, MONOBLASTIC

Again, this group within the suborder is far from homogeneous, the genera varying considerably, especially in different types of stromata. The most easily recognisable group of related genera is that containing *Myxocyclus*, *Endobotryella*, *Cheirospora* and *Endobotrya*, all of which have acervular to cupulate conidiomata, long filiform conidiophores and dictyosporous conidia often enclosed in a mucilaginous sheath. *Myxocyclus* is separated by its distoseptate verruculose conidia but otherwise shows a remarkable resemblance to the other genera in this group. Another fairly easily definable group contains *Harknessia*, *Endomelanconium*, *Cymbothyrium* and *Lasmenia*. Differences in structure of the conidiomata are important in separating these genera, as is the conidial morphology. Only in *Harknessia* are the conidiogenous cells contributory towards the conidia inasmuch as by fracture of the usually long conidiogenous cell the conidia are furnished with a basal appendage. The species with very short appendages, like *H. eucalypti*, *H. gunnerae* and *H. insueta* clearly are closer to the other genera in the group where conidia completely lack any basal appendage at all. There is another group containing genera with stellate or branched conidia. *Asterosporium* is distinct owing to the distoseptate conidia, but *Prosthemium* and *Psammina* are closely related, and *Digitosporium* seems little more than an irregularly branched *Sirothecium*.

The rest of the genera in this suborder are seemingly unrelated apart from their stromatic conidiomata and simple holoblastic mode of conidiogenesis.

DICHOTOMOUS KEY

(numbers in parentheses are species numbers)

1. Conidia branched 2
1. Conidia with short projections from the apex, complete with cellular contents 6
1. Conidia with appendages (setulae, cellular appendages, mucilaginous caps) 7
1. Conidia not so 10
   2. Conidia stellate 3
   2. Conidia cheiroid or digitate 4
   2. Conidia not so 5
3. Conidiomata acervular; conidia distoseptate, with 4 arms .... *Asterosporium* (1)
3. Conidiomata acervular; conidiophores long, filiform; conidia brown, of 10-14 radiating arms, each of which is 2-3 euseptate .... *Prosthemium* (2)
3. Conidiomata acervular; conidiophores short; conidia hyaline or very pale brown, of up to 30 radiating arms, each of which is up to 6 euseptate .... *Psammina* (3)
   4. Conidiomata eustromatic, unilocular; conidia brown, cheiroid, with 3 or more upright branches .... *Sirothecium* (4)
   4. Conidiomata eustromatic, irregularly unilocular; conidia brown, irregularly digitate .... *Digitosporium* (5)
5. Conidiomata eustromatic, unilocular, linear; conidia hyaline, consisting of a basal cell from which 3 lateral and one vertical arms are formed .... *Tetranacrium* (6)
5. Conidiomata eustromatic, cupulate, setose; conidia brown, consisting of 4 arms, the two upper tapered, divergent and 2 septate, the two lower ones parallel, one longer, 3-4 septate, the other shorter, 2-3 septate by which the conidium is attached to the conidiogenous cell .... *Ypsilonia* (8)
5. Conidiomata superficial, globose, unilocular; conidia hyaline, consisting of 6-14 arms each up to 15 septate, formed by close dichotomous branching at the base .... *Geastrumia* (7)
   6. Conidiomata acervular; conidia aseptate, with reduced lumina, with 3-4 conical projections .... *Asteroconium* (9)
   6. Conidiomata acervular; conidia 2 septate, with 2-3 short apical projections .... *Thrinacospora* (10)
7. Conidiomata eustromatic, applanate, circular, multiloculate; conidia with a subapical and suprabasal setula, 3-4 septate, pale brown .... *Discosia* (11)
7. Conidia with cellular appendages 8
7. Conidia with a mucilaginous cap or filiform mucilaginous appendages 9
   8. Conidiomata eustromatic, unilocular to multilocular, globose; appendage single, but branched 1-5 times near point of insertion; conidia 1 septate, brown .... *Mycohypallage* (12)
   8. Conidiomata acervular; conidia hyaline, with a large upper and lower cell and 2-3 smaller, median cells; appendages single on apical and median cells .... *Entomosporium* (13)
   8. Conidiomata acervular; conidia hyaline, aseptate, with tufts of cellular (in one sp. with setulose) appendages at each end .... *Chaetospermum* (14)

9. Conidiomata, eustromatic, multilocular, locules linearly arranged; conidia pale brown, multiseptate, smooth or minutely verruculose, with a mucilaginous cap at the apex, occasionally at the base .... *Septoriella* (15)
9. Conidiomata eustromatic, convoluted, irregularly multilocular; conidia versicoloured, 3 septate, septa black-banded; appendage as a mucilaginous cap .... *Toxosporiopsis* (16)
9. Conidiomata eustromatic, columnar to infundibuliform; conidia dark brown, 1 septate, septum near the apex; appendages as subapical and suprabasal mucilaginous tufts .... *Stevensonula* (17)
   10. Conidia aseptate, hyaline 11
   10. Conidia septate or muriform, hyaline 15
   10. Conidia aseptate, brown 19
   10. Conidia septate or muriform, brown 23
11. Conidiomata acervular 12
11. Conidiomata eustromatic, locular 13
   12. Conidiophores branched and septate, loci immediately below transverse septa; conidia thin-walled .... *Catenophora* (18)
   12. Conidiogenous cells filiform; conidia thick-walled .... *Leptodermella* (19)
13. Conidiomata eustromatic, immersed, irregularly multilocular, with a large cap of textura oblita; conidiophores branched and septate; conidia ellipsoid, thick-walled .... *Rabenhorstia* (21)
13. Conidiomata eustromatic, only the base immersed, multilocular, the locules peripheral; conidiogenous cells short, simple; conidia globose to ellipsoid, very thick-walled; phialidic microconidial state present .... *Cytosphaera* (20)
13. Conidia thin-walled 14
   14. Conidiomata eustromatic, immersed, then appearing superficial, linear, bi-, tri- or quadrilabiate, unilocular to convoluted; conidiophores branched and septate at the base; conidia ovoid .... *Polymorphum* (22)
   14. Conidiomata eustromatic, immersed, multilocular; conidiophores short, cylindrical; conidia fusiform .... *Fusicoccum* (23)
   14. Conidiomata pseudostromatic, irregularly multiloculate, effuse, black; conidiophores cylindrical; conidia ellipsoid .... *Hysterodiscula* (24)
   14. Conidiomata with an upper wall of dark brown hyphae converging towards the centre; conidiogenous cells ampulliform; conidia falcate, aseptate or unevenly 1 septate .... *Kabatia* (25)
15. Conidia 1 septate 16
15. Conidia muriform 17
15. Conidia multiseptate, filiform 18
   16. Conidiomata superficial, eustromatic, multilocular, with peripheral locules; conidiogenous cells ampulliform to lageniform; conidia fusiform, medianly 1 septate .... *Microperella* (27)
   16. Conidiomata semi-immersed to superficial, uniloculate, conidial mass reddish; conidiophores verruculose, pale brown; conidia ellipsoid, medianly 1 septate .... *Cystotricha* (26)
17. Conidiomata acervular; conidiophores filiform; conidia with 4-6 transverse and 1-3 longitudinal septa .... *Hyalodictyum* (28)
17. Conidiomata acervular; conidiophores branched and ramified amongst conidia; conidia thick-walled, muriform, most septa oblique .... *Thyrsidina* (29)
   18. Conidiomata acervular to eustromatic and then multilocular; conidiogenous cells ampulliform ...*Dothistroma* (30)
   18. Conidiomata eustromatic, applanate, linear; conidiophores branched and septate; conidia apical and lateral on loci immediately below transverse septa .... *Pleurothyrium* (31)
   18. Conidiomata eustromatic, unilocular to multilocular; conidiophores branched at the base, septate above, enclosed in a sheath; conidia apical and lateral on loci immediately below transverse septa .... *Placothyrium* (32)
19. Conidiomata acervular 20
19. Conidiomata eustromatic 21
   20. Conidia brown, short cylindrical, verruculose, formed from brown ampulliform conidiogenous cells .... *Phaeolabrella* (33)
   20. Conidia brown, smooth, scyphoid, with a transverse germination slit in the upper part, formed from filiform brown conidiophores .... *Scyphospora* (34)
21. Conidia often with the persistent conidiogenous cell as a basal appendage, smooth .... *Harknessia* (35)
21. Conidia verrucose, ornamentation as angular, reticulate ridges; paraphyses present; conidiomata eustromatic, multilocular but with a single ostiole .... *Aplosporella* (40)
21. Conidia pale brown with a median hyaline band .... *Perizomella* (39)
21. Conidia not so 12

22. Conidiomata eustromatic, irregularly multilocular; conidia dark brown, broadly fusiform to ventricose, with a vertical germination slit .... *Endomelanconium* (36)
22. Conidiomata eustromatic, unilocular to convoluted; conidia dark brown, cymbiform .... *Cymbothyrium* (37)
22. Conidiomata eustromatic, unilocular, lifting a flap of tissue at dehiscence; conidia pale brown but thick-walled, ellipsoid .... *Lasmenia* (38)
23. Conidia 1 septate 24
23. Conidia multiseptate 25
23. Conidia muriform 26
24. Conidiomata eustromatic, uni- or multilocular; conidia dark brown, with longitudinal striations .... *Lasiodiplodia* (41)
24. Conidiomata pseudostromatic, convoluted and irregularly multilocular, apex rostrate; conidia smooth, dark brown .... *Pellionella* (42)
25. Conidiomata acervular. Conidia fusiform, pale brown, end cells hyaline or pale, with several eusepta .... *Scolicosporium* (44)
25. Conidiomata pulvinate to bullate. Conidia pale brown, 1 euseptate nearer the obtuse apex, tapered to a truncate base... .... *Oncospora* (43)
25. Conidiomata infundibuliform, 1 cell thick. Conidia nidulant, at first strongly recurved, then curved or straight, distoseptate .... *Shawiella* (45)
26. Conidia enclosed in a gelatinous sheath 27
26. Conidia not enclosed in a gelatinous sheath 29
27. Conidia distoseptate, verruculose .... *Myxocyclus* (48)
27. Conidia euseptate 28
28. Conidia irregular, of up to 6 central cells, each with several short acropetal branches forming a mass of thick-walled brown cells comprising the conidium .... *Cheirospora* (50)
28 Conidia ellipsoid to cylindrical, of 3-4 central cells, from which are formed columns of several lateral cells .... *Endobotrya* (51)
29. Conidia euseptate, brown, variable in shape, smooth. Conidiomata eustromatic, multilocular, immersed to superficial.. .... *Dichomera* (46)
29. Conidia distoseptate, brown, cylindrical, clavate, obclavate or irregular. Conidiomata acervular .. *Camarosporula* (47)
29. Conidia cylindrical, of up to 4 central cells, each with several short lateral cells forming a mass of thick-walled spherical cells comprising the conidium. Conidiomata eustromatic, finally cupulate .... *Endobotryella* (49)

SYNOPTIC KEY

Mycelial characters

Position in relation to substrate

a. superficial	7	8	17	45											
b. immersed	1	2	3	4	5	6	9	10	11	12	13	14	15	16	17
	18	19	20	21	22	23	24	25	26	27	28	29	30	31	32
	33	34	35	36	37	38	39	40	41	42	43	44	45	46	47
	48	49	50	51											

Conidiomatal characters

Position in relation to substrate

a. superficial	4	6	7	8	17	20	26	27	45						
b. immersed	1	2	3	5	9	10	11	12	13	14	15	16	18	19	20
	21	22	23	24	25	28	29	30	31	32	33	34	35	36	37
	38	39	40	41	42	43	44	46	47	48	49	50	51		
c. semi-immersed	4	5	26	40	43	46									

Dehiscence	
a. ostiolate, non-papillate	20 35 40 46
b. ostiolate, papillate	11 12 15 46
c. ostiolate, rostrate	42
d. linear-longitudinal	4 6 22 24 26
e. stellate	22
f. irregular rupture	1 2 3 4 5 7 8 9 10 13 14 16 17 18 19 21 23 25 27 28 29 30 31 32 33 34 36 37 39 41 43 44 45 47 48 49 50 51
g. lifting a flap of host tissue	38
Tissue type	
a. textura angularis	1 2 4 5 6 7 8 9 10 11 12 13 14 15 16 17 18 19 20 21 22 23 24 25 26 27 28 30 31 32 33 34 35 36 37 38 39 40 41 42 43 44 45 46 47 48 51
b. textura globulosa	7 47
c. textura intricata	3 8 21 26 39 42 49 50
d. textura oblita	14 21 29
e. textura porrecta	11 49
f. textura prismatica	22 27 34
Type of stroma	
a. acervular	1 2 3 9 10 13 14 18 19 28 29 30 33 34 44 47 48 49 50 51
b. cupulate	1 7 8 14 26 43 48 49
c. simple eustromatic	4 5 6 7 12 21 35 37 38
d. complex eustromatic	16 23 36 39
e. complex eustromatic, pulvinate	20 22 27 30 40 41 46
f. tubular	17 45
g. flattened, pycnidial-like	11
h. linear	15 31 32
i. flattened, upper wall of radiating hyphae	25
j. pseudostromatic	24 62
Locular configuration	
a. unilocular	1 2 3 4 7 8 10 12 13 14 17 18 19 25 26 28 29 30 32 33 34 35 37 38 43 44 45 47 48 49 50 51
b. multilocular	11 12 15 20 23 27 30 31 32 35 40 41 46
c. multilocular, convoluted	5 6 12 16 21 22 24 36 39 42
Other criteria	
a. dark brown pigmentation	4 5 6 7 8 11 12 15 16 17 20 21 22 23 24 25 26 27 30 31 32 35 36 37 38 39 40 41 42 43 45 46 48 49
b. forming tar spots	24
c. setose	8
d. furfuraceous	35

e. clypeate	37
f. rostrate	42
g. gelatinous cap over dehiscent area	21

Conidiogenous cell characters

Insertion	
a. discrete	4 6 9 10 11 12 15 19 25 27 30 33 35 36 37 38 39 40 41 42 44 45 46 47 51
b. integrated	1 2 3 5 7 8 13 14 16 17 18 20 21 22 23 24 26 28 29 31 32 34 38 43 44 48 49 50
Shape	
a. ampulliform-doliiform	4 6 15 25 26 27 30 33 38 45 46 47
b. lageniform	11 27 36 38 43 44
c. cylindrical	1 2 3 4 5 6 7 8 9 10 12 13 14 16 17 18 19 20 21 22 23 24 28 29 31 32 34 35 36 37 38 39 40 41 42 43 44 46 47 48 49 50 51
Ornamentation	
a. enclosed in mucilage	16 32 48
b. verrucose	33 34 48
c. smooth	1 2 3 4 5 6 7 8 9 10 11 12 13 14 15 16 17 18 19 20 21 22 23 24 25 26 27 28 29 30 31 32 35 36 37 38 39 40 41 42 43 44 45 46 47 49 50 51

Conidiophore characters

Shape	
a. irregularly branched and ramified	3 26 29
b. long cylindrical to filiform, unbranched	1 2 14 17 20 24 28 34 43 48 49 50
c. filiform, branched	5 7 8 13 16 18 20 21 22 23 31 32 38 44 46

Conidial characters

Pigmentation	
a. hyaline	6 7 9 10 13 14 18 19 20 21 22 23 24 25 26 27 28 29 30 31 32
b. brown	1 2 3 4 5 8 11 12 15 17 33 34 35 36 37 38 39 40 41 42 43 45 46 47 48 49 50 51
c. versicoloured	16 44
Septation	
a. distoseptate	1 45 47 48
b. euseptate	2 3 4 5 6 7 8 10 11 12 13 15 16 17 25 26 28 29 30 31 32 44 46 49 50 51
c. aseptate	9 14 18 19 20 21 22 23 24 25 33 34 35 36 37 38 39 40 43

d. medianly 1 septate	12 26 27 32 41 42 43
e. unequally 1 septate	17 25
f. 2—multiseptate	1 2 3 4 5 6 7 8 10 11 15 16 30 31 32 44 45
g. muriform	13 28 29 46 47 48 49 50 51
h. septa black-banded	16
i. aseptate but with a hyaline band	39
Shape	
a. oval-ellipsoid	12 14 18 20 21 22 24 25 26 28 35 36 38 39 40 41 42 46 47 48 49 51
b. cylindrical	10 11 15 33 43 44 45 46 47 49
c. fusiform	16 23 25 27 35 36 43 44 45 46 47
d. falcate	25 43 44
e. pyriform	12 18 24 29 35 36 38 46 48
f. filiform	15 30 31 32
g. branched	1 2 3 4 5 6 7 8
h. stellate	1 2 3 6
i. palmate-cheiroid	4
j. digitate	5
k. dichotomously branched at the base	7
l. H-shaped	8
m. globose	50
n. corniform	9 10
o. reniform	14 17 35 37
p. turbinate	34
q. two main cells and 2-3 equatorial cells	13
Ornamentation	
a. verrucose (random)	15 32 40 48
b. longitudinally striate	41
c. horizontal or longitudinal germination slit	34 35 36
d. smooth	1 2 3 4 5 6 7 8 9 10 11 12 13 14 15 16 17 18 19 20 21 22 23 24 25 26 27 28 29 30 31 32 34 35 36 37 38 39 42 43 44 45 46 47 49 50 51
Appendages	
a. apical	11 12 13 14 15 16 17 35
b. basal	11 14 15 17 35
c. lateral	13 14
d. single	11 12 13 14 15 16 35
e. several at one point	14 17
f. cellular	12 13 14 35
g. simple	11 13 14 15 16 17 35
h. branched	12
i. mucilaginous	15 16 17
j. tentacular to conic	15 16 17
k. setulose	11
l. completely enclosing conidium	35 48 5051

Wall thickness

a. thin-walled	2	3	4	5	6	7	8	10	11	12	13	14	15	17	18
	22	23	24	25	26	27	30	31	32	39	43	44	46	49	50
	51														
b. thick-walled	1	9	14	19	20	21	28	29	33	34	35	36	37	38	40
	41	42	45	46	47	48									

Guttulation

a. eguttulate	1	2	3	4	5	6	7	8	9	10	11	12	13	14	15
	16	18	19	24	25	26	27	28	29	30	31	32	33	34	37
	39	40	41	42	45	46	47	48	49	50	51				
b. guttulate	17	20	21	22	23	35	36	38	43	44					

Paraphysis characters

a. present	16	17	40	41
b. enclosed in a mucilaginous sheath	16			
c. simple	16	40	41	
d. branched	17			

1. ASTEROSPORIUM Kze ex Wallr., *Flora Crypt. Germanica* **2**: 179 (1833).

Asterosporium Kze, *Flora* **1**: 225 (1819).

Sp. typ.: *A. asterospermum* (Pers. ex Gray) Hughes (syn. *Stilbospora asterosperma* Pers. ex Gray, *Asterosporium hoffmannii* Kze ex Wallr.).

Mycelium immersed, branched, septate, pale brown. *Conidiomata* acervular, subperidermal, separate or confluent, formed of brown, thin-walled textura angularis, sometimes extending laterally to appear cupulate. Dehiscence irregular. *Conidiophores* cylindrical, septate, only branched at the base, hyaline to pale brown, formed from the upper pseudoparenchyma. *Conidiogenous cells* holoblastic, integrated, determinate, cylindrical, hyaline to pale brown, smooth with no percurrent proliferation. *Conidia* terminal, smooth, transversely distoseptate, lumina reduced, brown, consisting of 4 arms at 90° to each other, connected to the conidiogenous cell at the point where they meet.

KEY

Conidia 45-53 μ between widest points .... *A. asterospermum*
Conidia 30-44 μ between widest points .... *A. betulinum*

A. ASTEROSPERMUM (Pers. ex Gray) Hughes, *Can. J. Bot.* **36**: 738 (1958). (Fig. 62, 64B)
synonymy as in Hughes (1958).

Conidiomata up to 1000 μ diam. Conidia 45-53 μ between widest points, arms up to 18 μ wide at the base.

Material of *A. strobilorum* Roum. & Fautr., an isolate from *Abies alba*, could not be distinguished from *A. asterospermum.*

On *Fagus grandifolia, F. pedunculata, F. sylvatica* and *F.* sp., G.B., France, Germany, Hungary, Ont., Canada (26 colln. in IMI).

A. BETULINUM Pk, *33rd Rept. St. Mus. Nat. Hist. 1879*: 26 (1880). (Figs. 63, 64A)

Conidiomata often cupulate, up to 135 μ diam. Conidia 30-44 μ between widest points, arms up to 10 μ wide at the base.

On twigs of *Betula lutea*, N.S., N.B., Canada (IMI 32284, 32285).

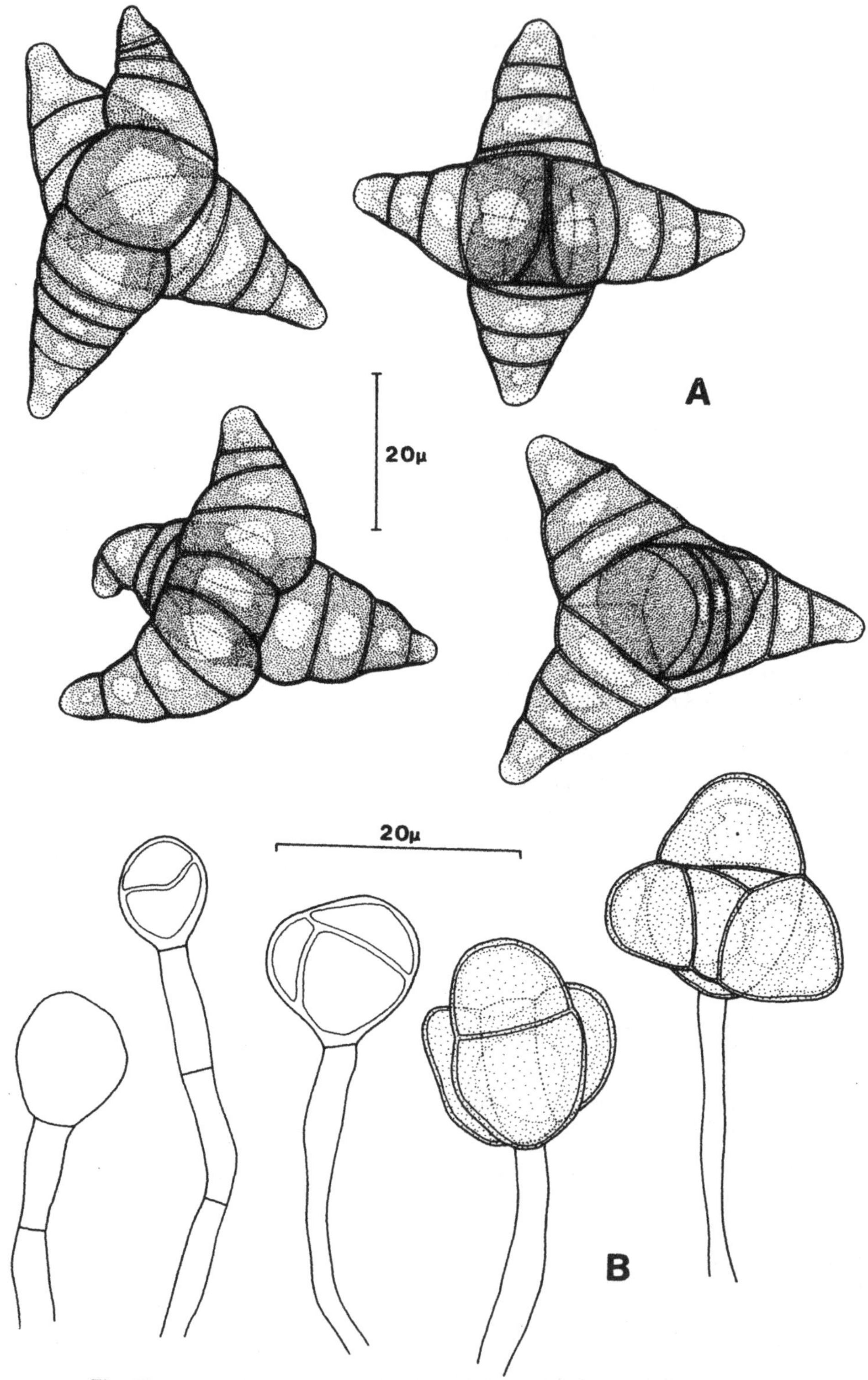

Fig. 62. *Asterosporium asterospermum*, A, conidia; B, conidiogenous cells and developing conidia.

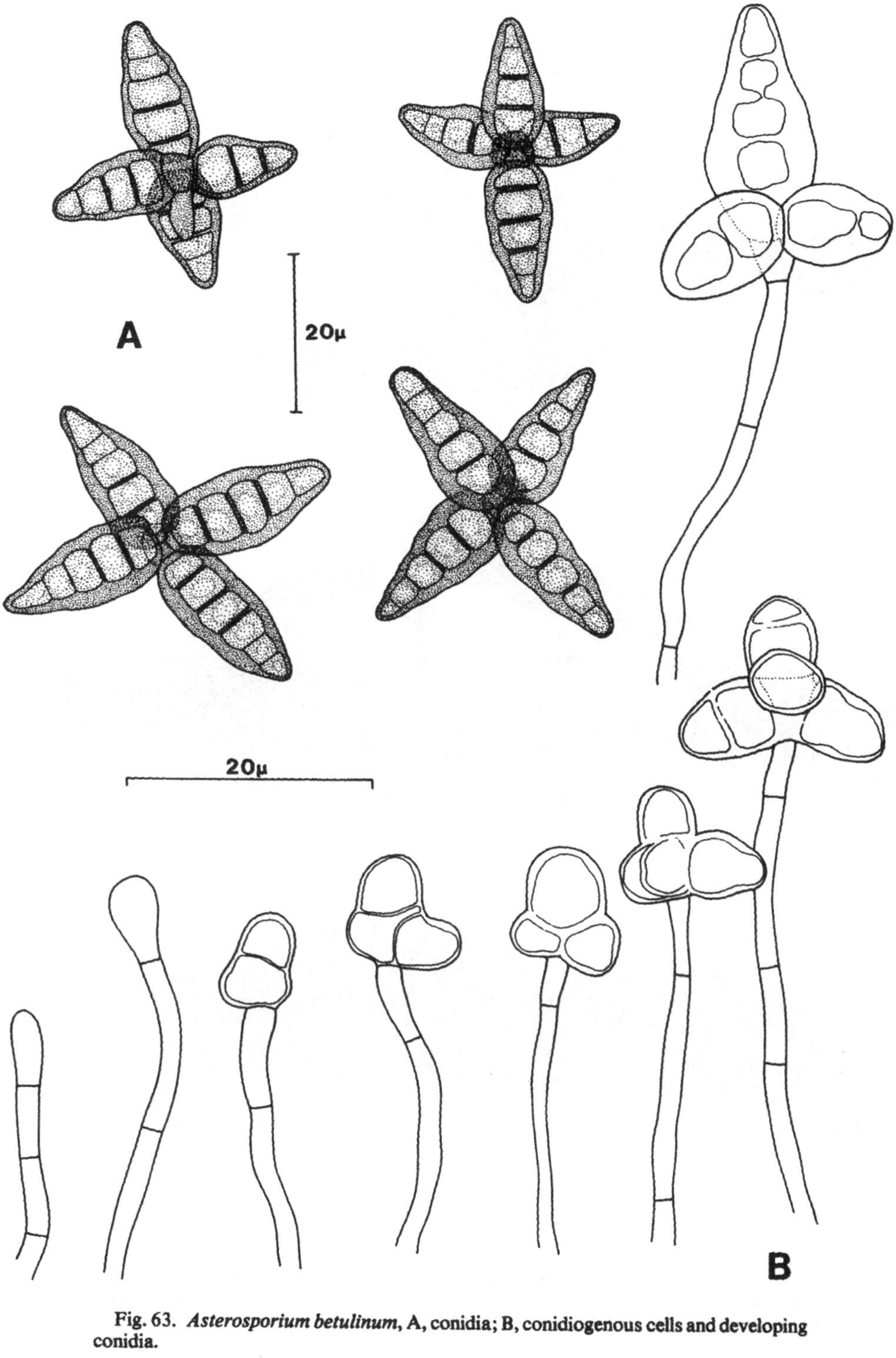

Fig. 63. *Asterosporium betulinum*, A, conidia; B, conidiogenous cells and developing conidia.

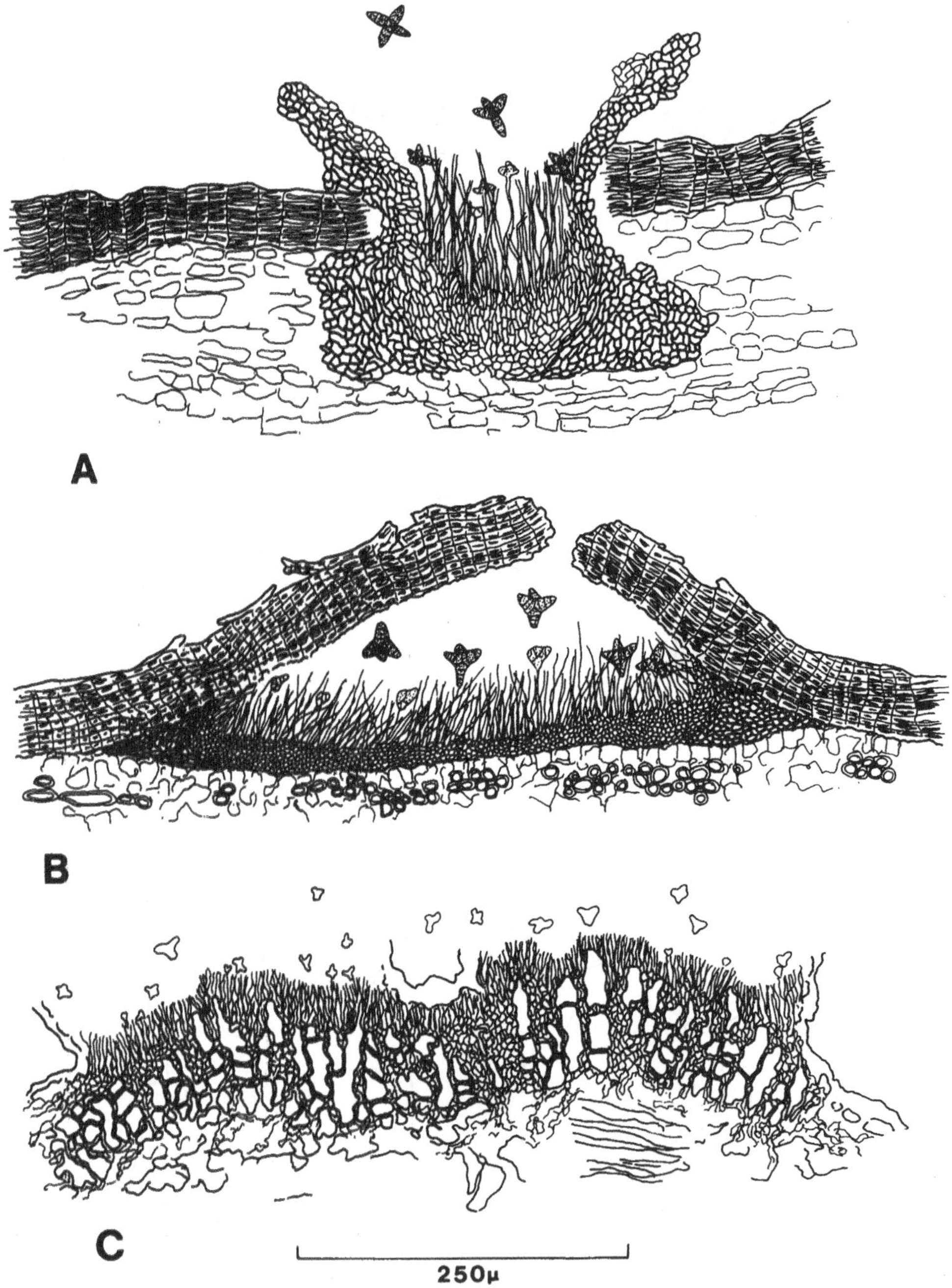

Fig. 64. A, *Asterosporium betulinum*, vertical section of a conidiaoma; B, *A. asterospermum*, vertical section of a conidioma; C, *Asteroconium saccardoi*, vertical section of a conidioma.

2. PROSTHEMIUM Kze ex Schlecht., *Syn. Pl. Crypt.* **2**: 101 (1824).
Prosthemium Kze apud Kze & Schmidt, *Myk. Hefte* **1**: 17 (1817).
Sp. typ.: *P. betulinum* Kze ex Schlecht.

Mycelium immersed, branched, septate, pale brown, *Conidiomata* acervular to eustromatic, subperidermal, circular, discrete, only occasionally confluent, dark brown, thick-walled; basal wall of hyaline to pale brown, thin-walled textura angularis, upper wall when present, of similar pale to medium brown, thin-walled tissue. *Ostiole* absent, dehiscence by irregular rupture of the overlying tissues. *Conidiophores* hyaline, unbranched, septate, filiform, confined to the basal wall. *Conidiogenous cells* holoblastic, integrated, determinate, hyaline, smooth, terminal, cylindrical, forming a single conidium at the apex. *Conidia* smooth, brown, consisting of 10-14, radiating, 2-3 transversely euseptate arms connected to a central cell.

So far five taxa have been described in *Prosthemium* and none have so far been revised. *P. formosum* Lib., which is an unpublished name, formed the basis of *Prosthemiella formosa* Sacc. & Malbr., type species of *Prosthemiella* Sacc. The only information on this genus is from the original description.

P. STELLARE Riess, *Bot. Zeit.* 130 (1853). (Fig. 65)

Teleomorphosis: *Massaria holoschista* (Berk. & Br.) Tul., *Sel. Fung. Carpol.* **2**: 222 (1863).

Conidiomata up to 500 μ diam. Conidiophores up to 60 μ long × 2.5-3 μ wide. Conidia up to 60 μ diam., each arm 2-3 septate, 25-35 μ long × 7-8 μ wide.

Grove (1937) erroneously interpreted the individual arms of the spores as separate conidia. This species is distinguished from the type, *P. betulinum*, which has 3-5 septate arms measuring 40-55 × 13-16 μ, altogether a larger-spored species.

On twigs of *Alnus* sp., U.K. (IMI 110034, 105287a).

2. PSAMMINA Rouss. & Sacc., *Bull. Soc. Bot. Belg.* **29**: 295 (1901).
Sp. typ.: *P. bommerae* Rouss. & Sacc.

Mycelium immersed, branched, septate, hyaline. *Conidiomata* acervular, separate, epidermal, ± circular, pale brown, composed of hyaline to pale brown, thin-walled textura intricata. *Conidiophores* irregularly branched, septate, hyaline, ramifying over the acervular tissue. *Conidiogenous cells* holoblastic, determinate, integrated, formed as small lateral branches on the ramified conidiophores, cylindrical to doliiform, hyaline. *Conidia* hyaline or very pale brown, smooth, consisting of many (more than 30), separate, up to 6 septate arms radiating from a central complex of short branched cells.

Apart from the type, there is only one other species described in the genus, *P. mariae-theresiae* Dias & Teixeira (1963), and this seems acceptably included from the description and illustration. Whether it is distinct from *P. bommerae* is questionable, but the conidial branches were described as 27.5-45 × 2.5-3 μ which is appreciably longer than in the type species.

P. BOMMERAE Rouss. & Sacc., *ibid.*: 295 (1901). (Fig. 66)

Conidiomata up to 120 μ diam. Conidiogenous cells 2-3 × 1.5-3 μ. Conidia up to 45 μ diam.; individual arms 3-6 septate, 12-25 × 2-2.5 μ.

On leaves of *Ammophila arenaria, Juncus effusus*, U.K. (IMI 89624a, 35328, 214334, 223474, 44094, 89610b).

4. SIROTHECIUM Karst., *Medd. Soc. Fauna Flora Fenn.* **14**: 105 (1887).
Sp. typ.: *S. saepiarium* Karst.
Illustration: Sutton (1978a)

Mycelium immersed, branched, septate, pale brown to hyaline. *Conidiomata* eustromatic, separate, rarely confluent, semi-immersed to superficial, flattened, oval to elongated, unilocular, thick-walled; walls composed of brown, thick-walled textura angularis at the base and sides, becoming darker and more sclerotioid in the upper wall which is incurved over the conidiogenous region. *Ostiole* absent, dehiscence a more or less circular to longitudinal rupture. *Conidiophores* absent.

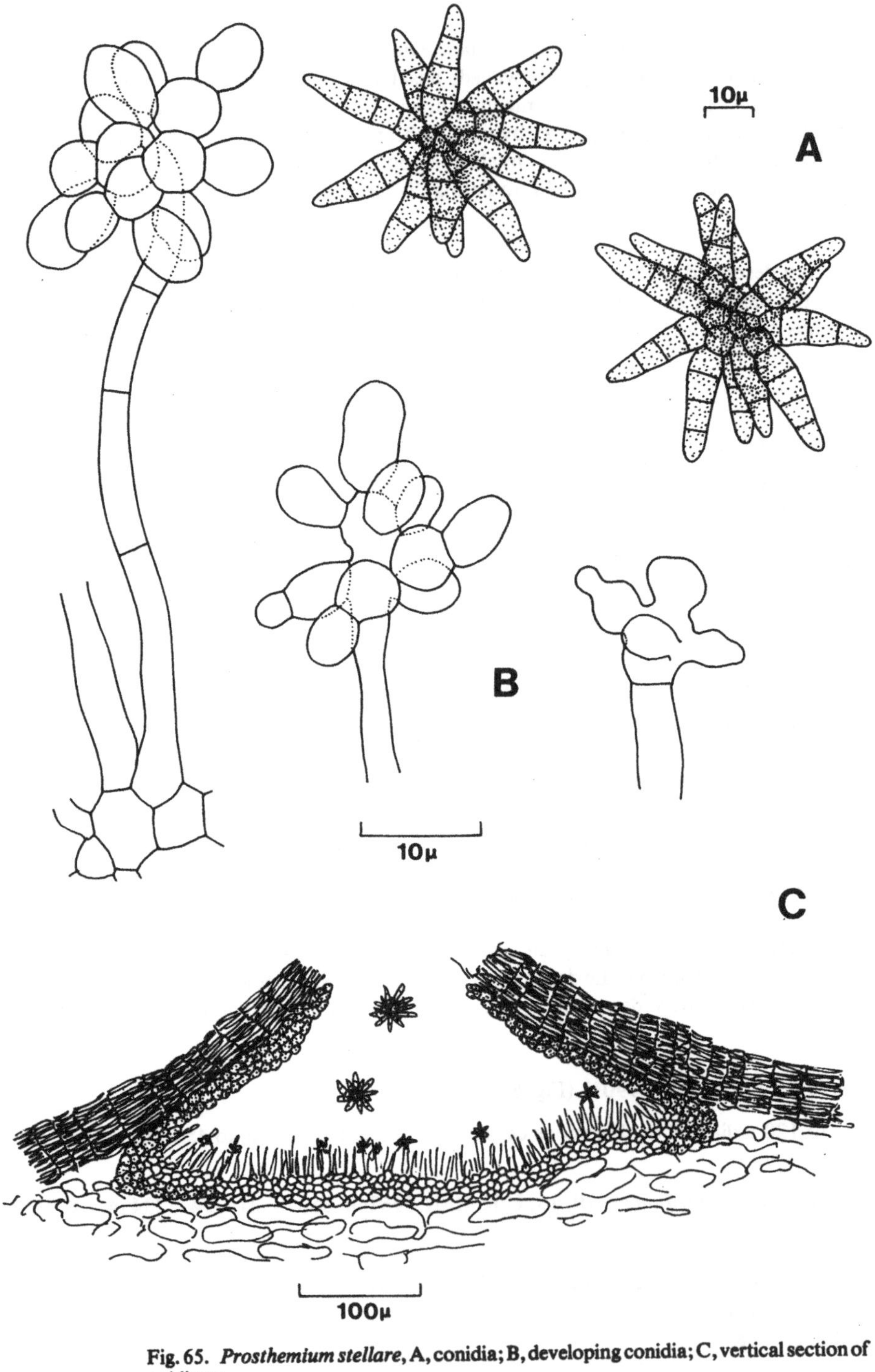

Fig. 65. *Prosthemium stellare*, A, conidia; B, developing conidia; C, vertical section of a conidioma.

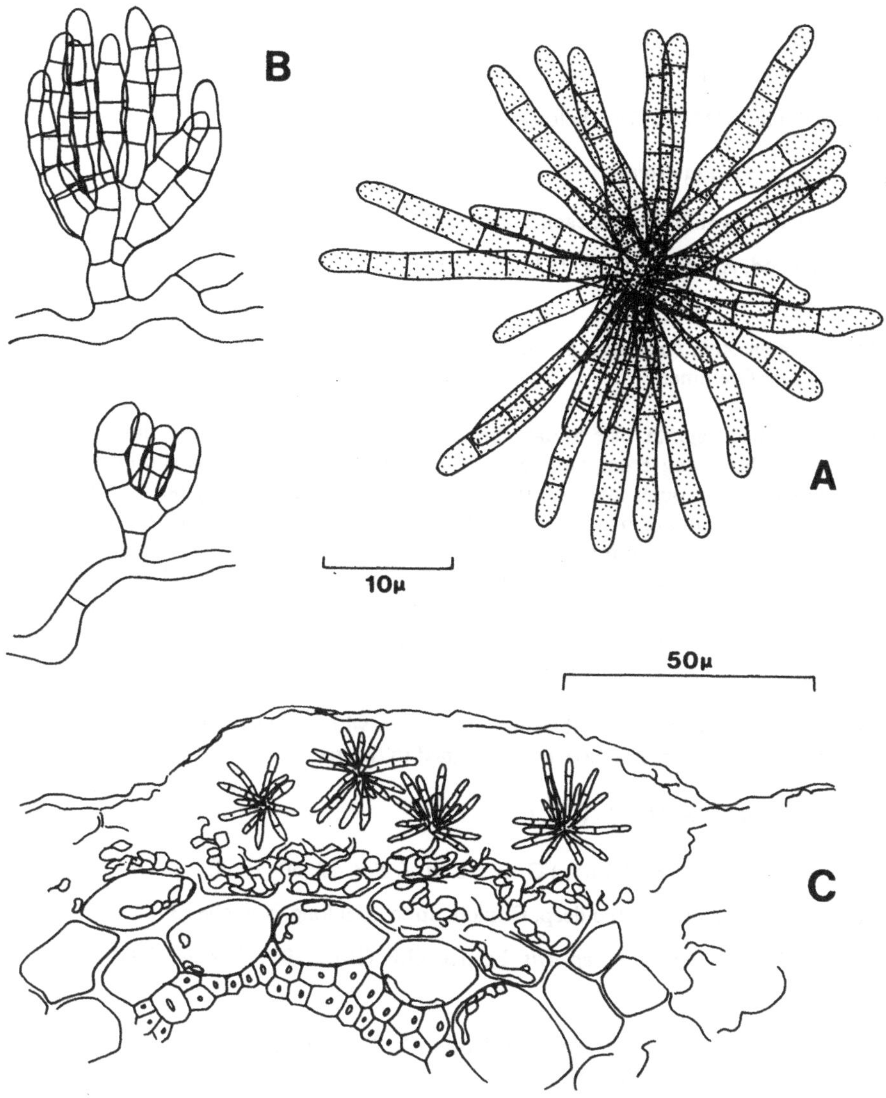

Fig. 66. *Psammina bommerae*, A, conidium; B, developing conidia; C, vertical section of a conidioma.

Conidiogenous cells holoblastic, occasionally annellidic, discrete, indeterminate, hyaline to pale brown, smooth, with 1 percurrent proliferation, formed from the basal cells of the conidioma. *Conidia* pale brown, smooth, consisting of a basal cell from which 3 or more septate upright branches arise, cheiroid, slightly constricted at the septa, apices obtuse.

Petrak & Sydow (1925, 1927) showed that the original description of catenate brown conidia was quite inaccurate, and subsequently Morgan-Jones, Nag Raj & Kendrick (1972e) described and illustrated the type species from the holotype specimen with cheiroid conidia. Most species placed in *Sirothecium* have catenate conidia and should be removed to other genera. Sutton (1978a) accepted only two species.

KEY

1. Conidia 30-33 × 12-19 μ .... *S. saepiarium*
1. Conidia 12-17 × 10-13 μ .... *S. minor*

S. SAEPIARIUM Karst., *ibid.*: 105 (1887).

Conidia 30-33 × 12-19 µ, with 3-5 arms, each 3-5 septate.

On old *Pinus* wood, Finland (IMI 209763, holotype ex H).

S. MINOR Sutton, *Kew Bull.* **33**: 343 (1978).

Conidia 12-17 × 10-13 µ, with 3-7 arms, each 2-3 septate.

On old *Populus alba* wood, Azores (IMI 197191, isotype).

5. DIGITOSPORIUM Gremmen, *Acta bot. neerl.* 2: 233 (1953).

Sp. typ.: *D. piniphilum* Gremmen

Mycelium immersed, branched, septate, pale to dark brown. *Conidiomata* eustromatic, immersed to semi-immersed, subepidermal to subperidermal, becoming erumpent, separate or confluent, medium brown, unilocular, multilocular or irregular, thick-walled, composed of an outer layer of irregular, dark brown textura angularis restricted to the upper and lateral walls and an inner layer, extending over the base, of pale brown, thin-walled tissue. Dehiscence by irregular breakdown of upper wall. *Conidiophores* pale brown, irregularly branched, septate, smooth, formed from the base of the stromata. *Conidiogenous cells* holoblastic, determinate, discrete or integrated, cylindrical or doliiform, pale brown to hyaline, smooth. *Conidia* pale brown, smooth, with several eusepta, digitate, with several branches arising frequently but not exclusively immediately below septa at different levels, branches of unequal length, 1-5 septate, the apical cell long conic.

Van Vloten & Gremmen (1953) correlated this imperfect state with *Crumenula sororia* Karst. and provided an account of the pathology. No additional species have been assigned to the genus.

D. PINIPHILUM Gremmen, *ibid.*: 233 (1953). (Fig. 67)

Teleomorphosis: *Crumenulopsis sororia* (Karst.) Grove, *Can. J. Bot.* **47**: 50 (1969).

Conidiomata up to 620 µ diam. × 250 µ deep. Conidia up to 60 µ long, branches 3-4 µ wide.

On bark and cones of *Pinus laricis*, *P. halepensis*, France, G.B. (IMI 91018, 86854, 199465).

6. TETRANACRIUM Hudson & Sutton, *Trans. Br. mycol. Soc.* 47: 202 (1964).

Sp. typ.: *T. gramineum* Hudson & Sutton
Illustration: Hudson & Sutton (1964).

Mycelium immersed, branched, septate, dark brown. *Conidiomata* eustromatic, subcuticular, becoming superficial, sessile, separate or gregarious, hysteriform, carbonous, blackish brown, unilocular; walls of dark brown, thick-walled textura angularis in the upper and lateral walls, and pale brown to hyaline, thin-walled tissue at the base and lining the upper and lateral walls. Dehiscence by a longitudinal fissure. *Conidiophores* absent. *Conidiogenous cells* holoblastic, discrete, determinate, doliiform to cylindrical, hyaline, smooth, with an acrogenous conidium formed from the basal and lateral pseudoparenchyma. *Conidia* tetraradiate, consisting of 1 vertical arm and 3 ± equidistant horizontal arms, arms arising from a central globose cell which is truncate at the base, hyaline, smooth; each arm up to 4 euseptate, tapered towards the apex, not constricted except at a point of attachment to the central cells.

Apart from the type species the only other taxon described in *Tetranacrium* is *T. malpighiacearum* Bezerra & Poroca (1969) from Brazil, but this has not yet been reexamined.

T. GRAMINEUM Hudson & Sutton, *ibid.*: 202 (1964). (Fig. 68)

Conidiomata 310-500 × 125-200 µ. Conidiogenous cells 4-5 × 3 µ. Conidial arms 15-36 × 2-3.5 µ.

On dead leaves of *Saccharum officinarum*, dead culms of *Sorghum vulgare* and *Cocos nucifera*, Jamaica, India, N. Borneo, Scotland (IMI 83001, 100035, 103686c, 189481).

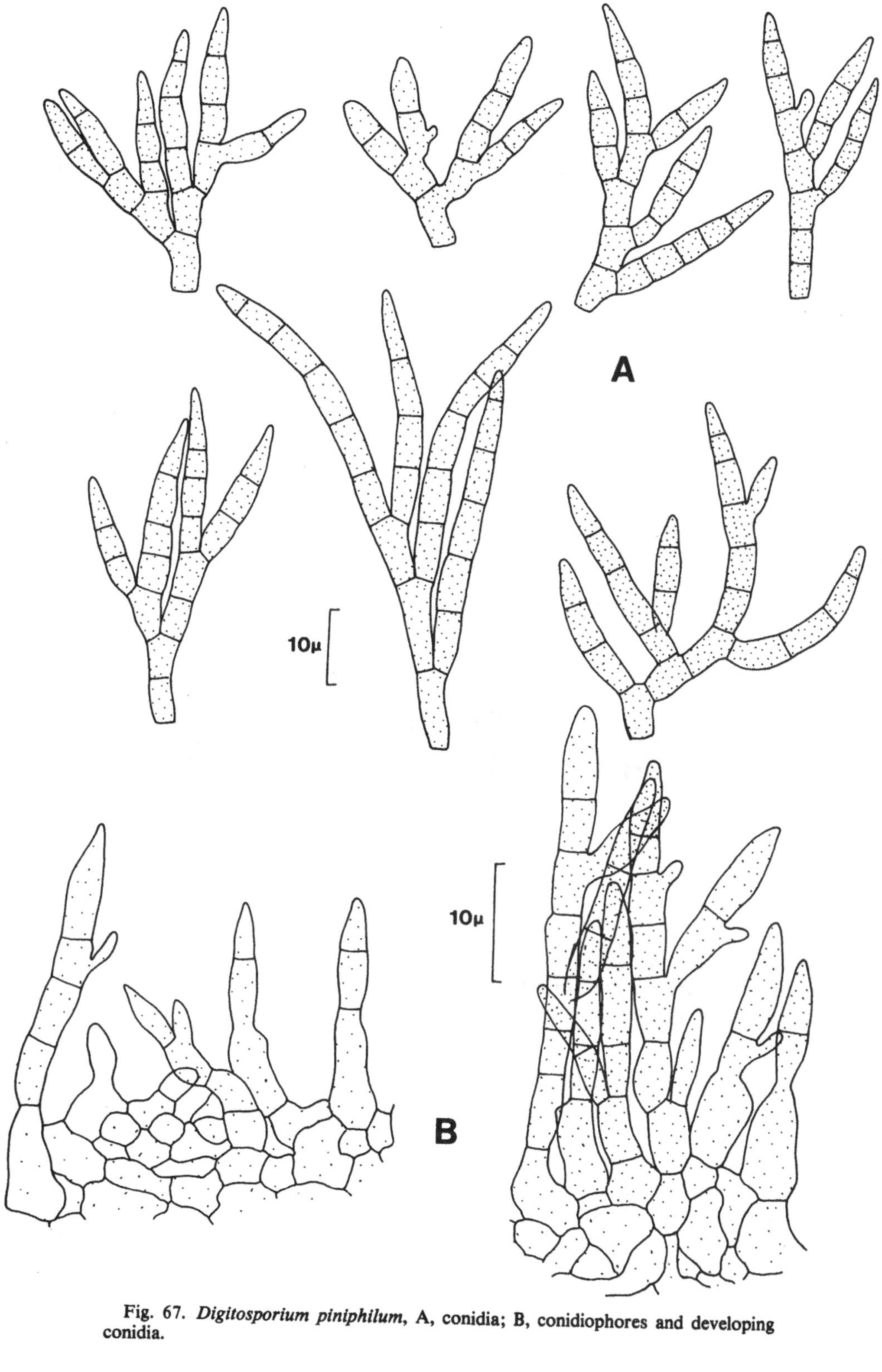

Fig. 67. *Digitosporium piniphilum*, A, conidia; B, conidiophores and developing conidia.

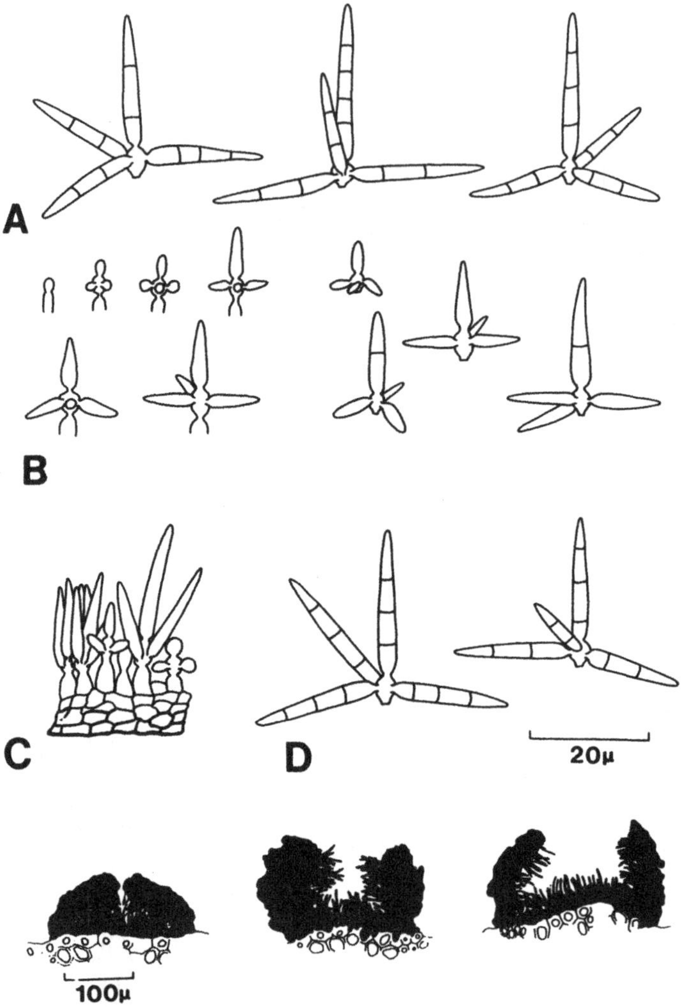

Fig. 68. *Tetranacrium gramineum*, A, conidia; B, developing conidia; C, conidiogenous cells; D, vertical sections of conidiomata.

7. GEASTRUMIA Batista, *Saccardoa* **1**: 71 (1960).
? *Benekea* Batista & Bezerra, *Publ. Inst. Mic. Univ. Recife* 299: 5 (1960).
Sp. typ.: *G. polystigmatis* Batista & Farr
Illustration: Pirozynski (1971).

Mycelium superficial, composed of brown to pale brown, branched, septate hyphae. *Conidiomata* superficial, amphigenous, separate, dark brown to black, sessile, at first globose and closed, later opening by irregular dehiscence of the overlying wall to become cupulate, situated directly on the effuse subicle; basal wall multicellular, pseudoparenchymatous, of pale brown, thin-walled textura angularis, lateral and upper walls 1 cell thick, composed of rounded, thick-walled, dark brown textura globulosa. *Conidiophores* hyaline, branched and septate mainly at the base, smooth, tapered towards the apex, formed from the basal region only. *Conidiogenous cells* holoblastic, determinate, integrated or discrete, cylindrical or tapered, hyaline, smooth, with acrogenous conidia. *Conidia* hyaline, thin-walled, smooth, consisting of 6-14, fusiform, straight or slightly curved arms that are each up to 15-euseptate and produced by ± dichotomous branching of a basal cell that remains pedicellate.

The genus was redescribed by Pirozynski (1971) and the possible synonymy of *Benekea* briefly discussed by Sutton (1977a).

G. POLYSTIGMATIS Batista & Farr apud Batista, *ibid.*: 72 (1960).

Conidiomata 50-100 µ diam. Conidiophores 5-20 × 1.5-2 µ.

On leaves of *Andira jamaicensis, Costus afer* and *Hymenocardia acida*, Tanzania (IMI 105978a, 105979b).

8. YPSILONIA Lév., *Ann. Sci. nat.*, ser. 5, **3**: 284 (1846).
Acanthothecium Speg., *Boln. Acad. Nac. Cienc. Cordoba* **11**: 228 (1889).
Psalidosperma Syd., *Annls mycol.* **12**: 571 (1914).
Sp. typ.: *T. cuspidata* Lév.
Illustrations: Nag Raj (1977b).

Mycelium superficial, dark brown, septate, radiating often in adpressed strands from the centre. *Conidiomata* superficial, amphigenous, separate, rarely aggregated, dark brown, sessile, at first globose and closed, later opening to become cupulate, situated on a flat, effuse stroma composed of dark brown, thick-walled textura angularis; periclinal wall composed of textura angularis towards the inside and textura intricata eventually extending into setae on the outside. *Ostiole* absent. *Setae* restricted to periclinal wall, incurved over the conidiogenous region, later becoming divergent, acuminate, apex obtuse or acute, septate, thick-walled, dark brown. *Conidiophores* pale brown, branched at the base, septate, smooth, cylindrical, formed mainly from the base. *Conidiogenous cells* holoblastic, integrated or discrete, pale brown, smooth, cylindrical. *Conidia* formed singly from the apex of each conidiogenous cell, smooth, pale brown, consisting of 4 arms, the two upper ones tapered, divergent and 2-septate, the two lower ones parallel, one of which is longer, tapered and 3-4 septate, the other 2-3 septate, shorter, cylindrical with the truncate base indicating the point of secession from the conidiogenous cells.

Nag Raj (1977b) revised *Acanthotheciella* Höhn. and its conidial states, *Ypsilonia*. Four species were accepted in *Ypsilonia*, *Y. mirabilis* (Speg.) Nag Raj, *Y. tropicalis* Nag Raj (= *Psalidosperma mirabile* Syd.), *Y. cuspidata* Lév., and the *Ypsilonia* state of *A. barbata* (Pat.) Höhn. Unfortunately revision of the *Acanthotheciella* species did not take account of *Setella xyphopaga* Viégas (1944) and *Gilmania pulchra* Batista & Cavalcanti apud Batista et al. (1962), so there is still a need for a reassessment of nomenclature following the excellent taxonomic account by Nag Raj (1977b).

YPSILONIA CUSPIDATA Lév., *Ann. Sci. nat.*, ser. 5, **3**: 284 (1846).

Colonies 3-5 mm diam.; conidiomata up to 265 µ diam.; setae reaching 300 µ in length. Conidiophores up to 50 µ long. Conidia 85 µ long overall, short basal arm 20-25 µ long, long basal arm up to 40 µ long, upper arms divergent, variable in length, each arm up to 3.5 µ wide.

On leaves of *Mitrephora* sp., Philippines (IMI 22777).

9. ASTEROCONIUM H. & P. Syd., *Annls mycol.* **1**: 36 (1903).
Sp. typ.: *A. saccardoi* H. & P. Syd.

Immersed mycelium branched, septate, hyaline. *Conidiomata* acervular to subepidermal, separate or confluent, formed of hyaline to pale brown, thin-walled textura angularis. Dehiscence irregular. *Conidiophores* absent. *Conidiogenous cells* holoblastic, monoblastic or sympodial, discrete, indeterminate, cylindrical, flexuous, hyaline, thick-walled, smooth, sometimes with 1-2 sympodial or synchronously formed conidiogenous loci, formed from the upper pseudoparenchyma. *Conidia* hyaline, very thick-walled with much reduced lumina, smooth, base truncate or uneven, tapered markedly towards the base, expanding in the upper part of conidia and forming 3 stout, obtuse, rounded or tapered, short projections at approximately the same level.

A second species, *A. nothopegiae* Ramakrishnan, Srinivasan & Sundaram (1953) seems to be based on insect galls (IMI 153719, holotype).

A. SACCARDOI H. & P. Syd., *ibid.*: 36 (1903). (Figs. 64C, 69)

Conidiomata up to 400 µ diam. Conidiogenous cells up to 75 µ long × 1.5-4 µ wide, often bifurcate, each branch producing a conidium. Conidia 12-18 µ wide between the apices of 2 projections, 2-3 µ wide at the base, 13-15 µ long.

On leaves of *Litsea glaucescens*, Mexico (IMI 153220, holotype).

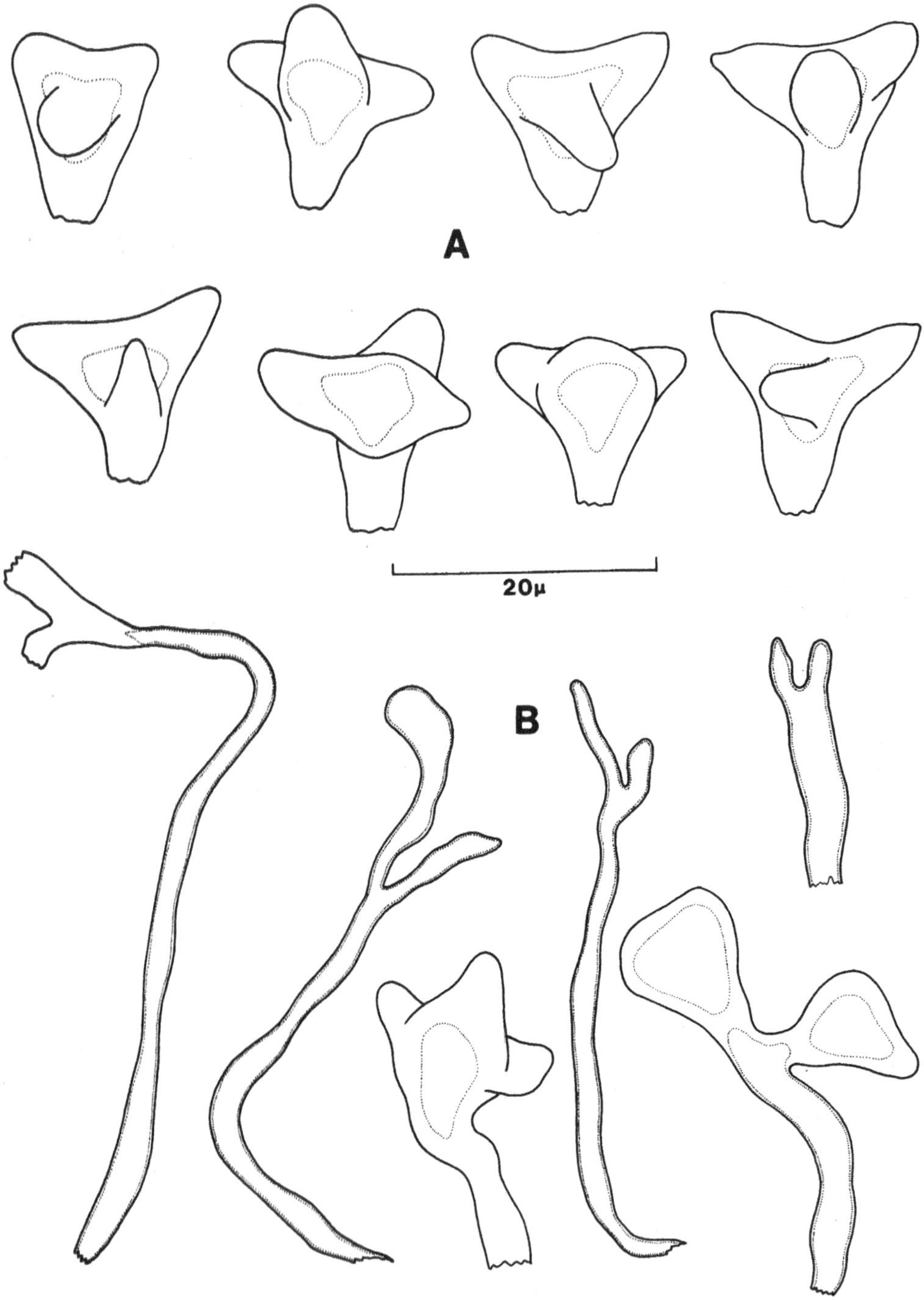

Fig. 69. *Asteroconium saccardoi*, A, conidia; B, developing conidia.

10. THRINACOSPORA Petrak, *Sydowia* **2**: 49 (1948).
Sp. typ.: *T. insignis* Petrak

Mycelium immersed, branched, septate, hyaline to pale brown. *Conidiomata* acervular, epidermal to subepidermal, separate, circular, formed of pale brown, thin-walled textura angularis. Dehiscence irregular. *Conidiophores* absent. *Conidiogenous cells* holoblastic, discrete, determinate, forming a single apical conidium, hyaline, smooth, filiform, swollen at the base, formed from the upper cells of the pseudoparenchyma. *Conidia* cuneiform to cylindrical, straight, 2 euseptate, smooth, hyaline, eguttulate, apical cells with 2-3 short projections.

The genus is monotypic.

T. INSIGNIS Petrak, *ibid.*: 49 (1948). (Fig. 70)

Conidiomata up to 210 µ diam. Conidiogenous cells up to 20 µ long × 1.5 µ wide. Conidia 15.5-20 × 1.5-2 µ wide at the base; distance between projections 4-5.5 µ.

On dead leaves of *Coriaria thymifolia*, Ecuador (IMI 202259, holotype ex ZT).

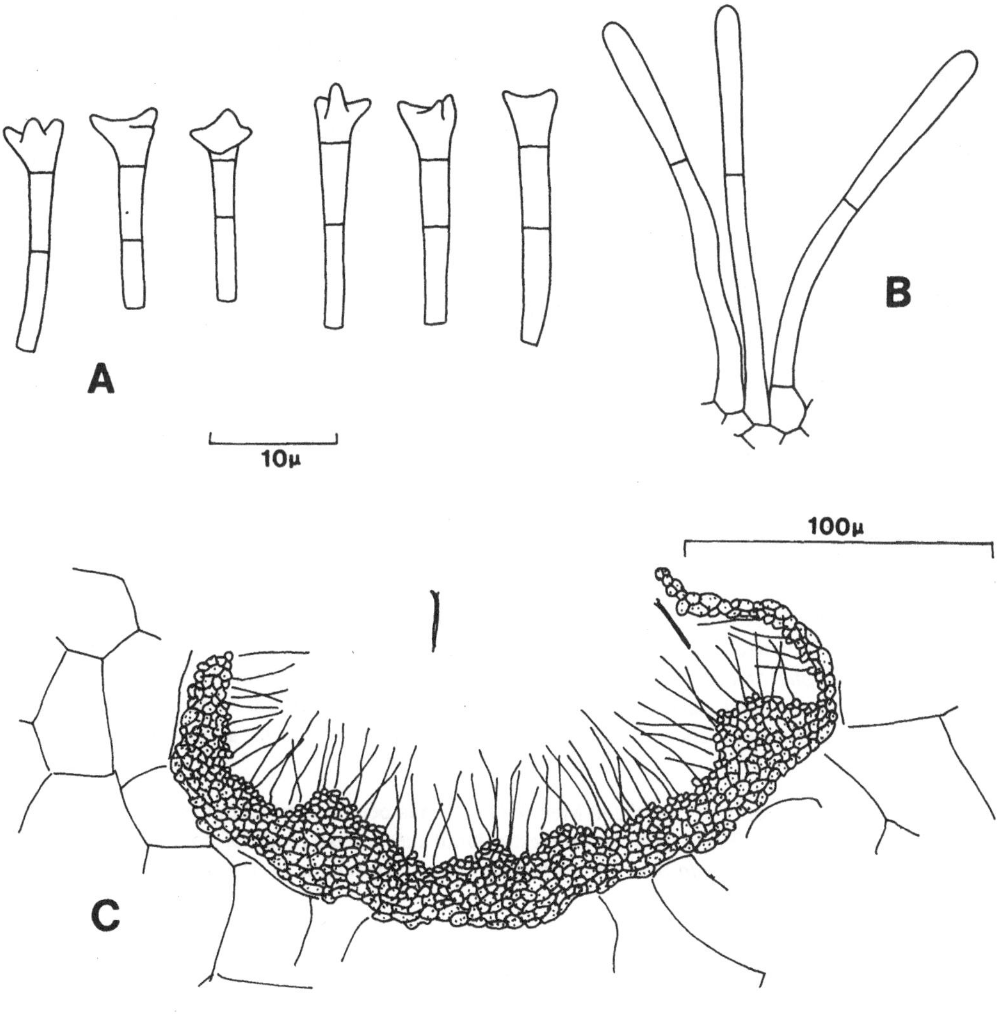

Fig. 70. *Thrinacospora insignis*, A, conidia; B, conidiogenous cells and developing conidia; C, vertical section of a conidioma.

11. DISCOSIA Lib., *Pl. Crypt. Ard.* exs. 346 (1837).
Cryptostictella Grove, *J. Bot., Lond.* **50**: 52 (1912).
? *Amphiciliella* Höhn., *Ber. dtsch. Bot. Ges.* **37**: 160 (1919).
? *Leptina* Batista & Peres apud Batista et al., *Saccardoa* **1**: 25 (1960).
Original spp.: *D. strobilina* Lib., *D. fagina* Lib.
Lectotype sp.: *D. strobilina* Lib., fide Subramanian & K.R.C. Reddy (1974).

Mycelium immersed, branched, septate, pale brown. *Conidiomata* eustromatic, amphigenous, separate or aggregated and confluent, flattened, black, epidermal, unilocular or multilocular, locules separated by textura porrecta; wall very thick at the base, of pale brown, thin-walled textura angularis, upper wall thinner, but of dark brown, thick-walled textura angularis in which lumina are occluded. *Ostioles* 1-many to each conidioma, circular, papillate. *Conidiophores* absent. *Conidiogenous cells* holoblastic, each forming a single conidium, discrete, determinate, conical, hyaline to pale brown, smooth, restricted to the basal conidiomatal wall. *Conidia* straight or slightly curved, dorsiventral, 3-4 euseptate, smooth, tapered to a truncate base, apex obtuse, hyaline to pale brown, cells of varying length; apical and basal cells each with a subapical and suprabasal unbranched, filiform setula formed on the concave side of the conidium.

More than 60 taxa have been described in this genus, the most commonly reported species being *D. artocreas* (Tode ex Fr.) Fr. but the status of this species is in doubt. It is evident from the excellent work of Subramanian & Chandra Reddy (1974) that a number of distinct taxa occur but without more extensive type studies the application of specific names is difficult. They recognised four sections within the genus and five described species (*D. strobilina* Lib., *D. vagans* de Not., *D. julia* Speg., *D. aquatica* Fautr. and *D. grammita* Berk. & Curt.) plus three other species (*D.* tax. sp. I, II and III) to which they could not confidently apply epithets. Taxa are distinguished by conidial septation and size, and the varying proportional lengths of the conidial cells. The type of *D. artocreas* was not examined by Subramanian & Chandra Reddy (1974) but Morgan-Jones (1964) did see Fries's material of the species and his account of *D. artocreas* and *D. strobilina* indicate two different taxa. Unfortunately the illustrations of *D. strobilina* by Subramanian & Chandra Reddy (1974) and by Morgan-Jones (1964) are not in agreement so the status of *D. artocreas* is still open to question.

D. STROBILINA Lib., *Pl. Crypt. Ard.* exs. 346 (1837). (Fig. 71)
D. theae Cavara, *Mat. Myc. Comb. Toulouse* 120 (1889).
D. virginiana Thüm., *Mycotheca Univ.* exs. 1985 (1882).
D. himalayensis Died., *Annls mycol.* **14**: 218 (1916).

Conidiomata up to 450 µ diam. Conidiogenous cells 6-8 × 4 µ tapering to 1.5 µ at the apices. Conidia 21-25 × 3-3.5 µ; setulae 10-12 µ long.

On dead cones of *Abies* sp., France (IMI 107250, isotype *D. strobilina* ex BR).

12. MYCOHYPALLAGE Sutton, *Mycol. Pap.* 88: 4 (1963).
Sp. typ.: *M. congesta* (Berk. & Br.) Sutton (syn. *Pestalotia congesta* Berk. & Br.).
Illustration: Sutton (1963).

Mycelium immersed, pale brown to hyaline, branched, septate. *Conidiomata* eustromatic, amphigenous, separate, globose to irregular, brown, immersed, unilocular or multilocular, or convoluted, thin-walled. *Ostiole* circular, papillate, central or excentric. *Conidiophores* absent. *Conidiogenous cells* holoblastic, determinate, discrete, cylindrical, filiform, hyaline, smooth, formed from the inner cells of the conidiomatal wall. *Conidia* medium brown, lighter around the middle, 1 euseptate, constricted, base truncate, apex obtuse, clavate, thick-walled, smooth, ± guttulate; apical appendage 1, branched 1-5 times, cellular.

Apart from the type species, only *M. northeae* Melnik (1970) has been described in the genus. The pale conidia with obtuse ends are unlikely to be congeneric with *M. congesta*. Until the precise mode of conidiogenesis is determined the reassignment of *M. northeae* to *Robillarda* Sacc. or *Pseudorobillarda* Morelet is premature, if indeed the species belongs there.

M. CONGESTA (Berk. & Br.) Sutton, *ibid.*: 5 (1963).
Pestalotia congesta Berk. & Br., *J. linn. Soc.* (Bot.), **14**: 89 (1875).
Neobarclaya congesta (Berk. & Br.) Petch, *Ann. R. bot. Gdn, Peradeniya* **9**: 165 (1924).
Neobarclaya natalensis Syd., *Hedwigia* **38**: 134 (1899).
Pestalotia evansii P. Henn. in Engler's *Bot. Jahrb.* **41**: 273 (1908).

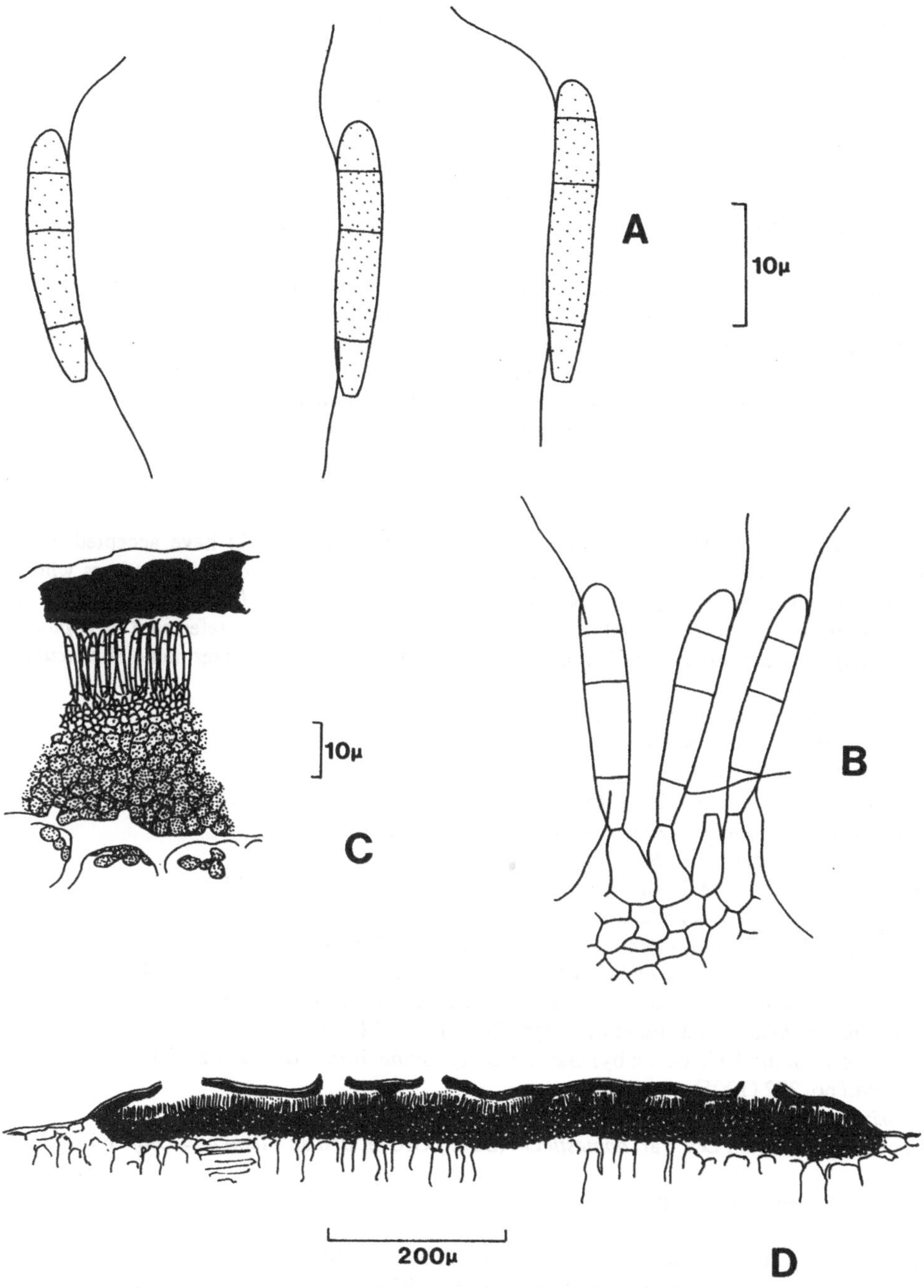

Fig. 71. *Discosia strobilina*, A, conidia; B, conidiogenous cells and developing conidia; C, detail of conidiomatal walls; D, vertical section of a conidioma.

Teleomorphosis: *Plagiostigma jambolana* (Ramakr. et al.) Ullasa, *Sydowia* **24**: 162 (1971).

Conidiomata 200-250 μ diam. Conidiogenous cells 10-120 × 2-3 μ. Conidia 23-29 × 7.5-9 μ; branches of the apical appendage 12-30 μ long.

Causing lesions on leaves of *Syzygium jambolanum*, *S. cordatum*, *S. guineense* var. *macrocarpum*, *S.* sp., Sri Lanka, India, Uganda, S. Africa, Zambia (IMI 84983, holotype *P. congesta* ex K, 84984, 200257, 4503g, 45164c, 22411, holotype *N. natalensis*, 163089a, 92824, 159080, 137712, 174769, 174787, 89987a, 110746, 163366).

13. ENTOMOSPORIUM Lév., *Bull. Soc. bot. Fr.* **3**: 31 (1856).
Morthiera Fckl, *Jb. nassau. Ver. Naturk.* **23-24**: 382 (1870).
Sp. typ.: *E. mespili* (DC. ex Duby) Sacc. (syn. *Xyloma mespili* DC. ex Duby, *E. maculatum* Lév.).

Mycelium immersed, branched, septate, pale brown to hyaline. *Condiomata* acervular, subcuticular, separate or aggregated and confluent, ± circular, composed of hyaline to pale brown textura angularis. Dehiscence irregular. *Conidiophores* septate, branched at the base, hyaline, cylindrical, straight or flexuous, smooth, formed from the upper acervular cells. *Conidiogenous cells* holoblastic, determinate, integrated or discrete, terminal, hyaline, smooth, cylindrical, with a single apical conidium. *Conidia* hyaline, smooth, thin-walled, eguttulate, consisting of a larger basal and upper cell and 2 or more smaller lateral cells arising from the upper region of the lower cell; apical cell globose, apex obtuse, basal cell ± short cylindrical with the base truncate, lateral cells ± globose; apical and lateral cells are each furnished with a single, cellular, unbranched, flexuous appendage.

Jørstad (1945), Stowell & Backus (1966) and Boerema & Verhoeven (1972) have accepted a single species in *Entomosporium* in contrast to Grove (1937), Vassiljevsky & Karakulin (1950) and earlier authors who separated such fungi on different Rosaceous hosts into distinct species or varieties. *Fabraea maculata* Atk. is generally accepted as the perfect state of this composite species, but Jørstad (1945) and Nannfeldt (1932) have referred it to *Diplocarpon* Wolf. An account of the pathology has been provided by Sivanesan & Gibson (1976) under the later name *D. maculatum*.

E. MESPILI (DC. ex Duby) Sacc., *Michelia* **2**: 115 (1880). (Fig. 72)
Xyloma mespili DC. ex Duby, *Bot. Gall.* ed. 2, 2 (1830).
Xyloma mespili DC. in DC. & Lamarck, *Fl. fr.* **5**(6): 158 (1805).
Morthiera mespili (DC. ex Duby) Fckl, *Jb. nassau. Ver. Naturk.* **23-24**: 382 (1870).
Entomosporium maculatum Lév., *Bull. Soc. bot. Fr.* **3**: 31 (1856).
Entomosporium brachiatum Lév., *Bull. Soc. bot. Fr.* **3**: 31 (1856).
Morthiera thuemenii Cke in de Thümen, *Mycotheca univ.* cent. 9, No. 895 (1877).
Entomosporium thuemenii (Cke) Sacc., *Syll. fung.* **3**: 657 (1884).
Morthiera mespili (DC. ex Duby) Fckl var. *cydoniae* Cke & Ell., *Grevillea* **6**: 84 (1878).
Entomosporium maculatum Lév. ♉ *cydoniae* (Cke & Ell.) Sacc., *Syll. fung.* **3**: 657 (1884).
Entomosporium maculatum Lév. var. *cydoniae* (Cke & Ell.) Grove, *British Stem- and Leaf-fungi* **2**: 191 (1937).
Entomosporium cydoniae (Cke & Ell.) Sacc., *Syll. fung.* **15**: 135 (1901).
Entomosporium mespili (DC. ex Duby) Sacc. var. *cydoniae* Briosi & Cavara, *I Funghi parasiti delle piante coltivate od utili* no. 372 (1905).
Entomosporium maculatum Lév. *β domesticum* Sacc., *Syll fung.* **3**: 657 (1884).
Entomosporium maculatum Lév. var. *domesticum* (Sacc.) Grove, *British Stem- and Leaf- Fungi* **2**: 191 (1937).

Teleomorphosis: **Diplocarpon mespili** (Sorauer) comb. nov.

Stigmatea mespili Sorauer, *Handb. der Pflanzenkr.* ed. 2, 374 (1878).
Entomopeziza mespili (Sorauer) Höhn., *Hedwigia* **62**: 51 (1920).
Entomopeziza sorauerі Kleb., *Vortr. Gesamtgeb. Bot.* **1**: 33 (1918).
Diplocarpon sorauerі (Kleb.) Nannf., *Nov. Acta Reg. Soc. scient. Ups.* ser. 4, **8**(2): 173 (1932).
Fabraea maculata Atk., *Science* **30**: 452 (1909).
Diplocarpon maculatum (Atk.) Jørst., *Meld. St. plpatol. Inst.* **1**: 163 (1945).

Conidiomata up to 200 μ diam. Conidiophores up to 15 × 3-4 μ. Conidia 15-22.5 μ long × 5.5-9.5 μ; lateral cells 4-7 μ diam.; appendages 7-15.5 μ long.

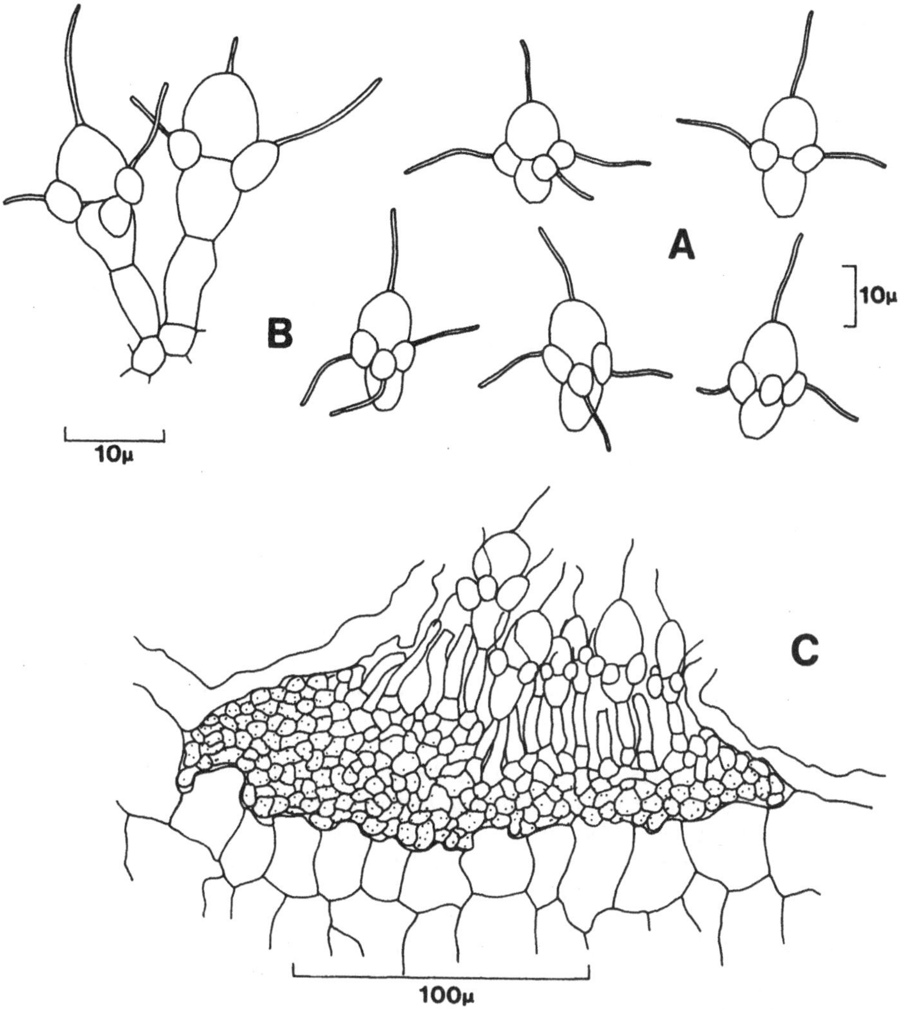

Fig. 72. *Entomosporium maculatum*, A, conidia; B, conidiogenous cells and developing conidia; C, vertical section of a conidioma.

Causing leaf and fruit lesions on *Amelanchier oreophila*, *A. alnifolia*, *Aronia melanocarpa*, *Cydonia vulgaris*, *C. oblonga*, *Crataegus*, *Eriobotrya japonica*, *Heteromeles arbutifolia*, *Malus pumila*, *M.* sp., *Mespilus germanica*, *Pyrenacantha*, *Pyrus communis*, *P. pashia*, *Sorbus tianschanica*, G.B., France, Germany, Roumania, Austria, Italy, U.S.S.R., Colo., N. Dak., S. Carol. and Fla, U.S.A., Ont., Canada, Malawi, Kenya, Cyprus, S. Rhodesia, India, Australia (IMI 29702, 85231, 29700, 97283, 29701, 75979, 11195, 21657, 21656, 135630, 159718, 159717, 168037, 178382, 180119, 172779, 78555, 94665, 95939, 50656, 18735, 41452, 160971, 110484, 95937, 81035, 154368, 76275, 95938, 21655, 95944, 41457, 76276, 146717a, 147584, 155441).

14. CHAETOSPERMUM Sacc., *Syll. fung.* **10**: 706 (1892).

Ciliospora Zimmerman, *Zbl. Bakt.* **8**: 217 (1902).

Mastigonema Speg., *Bol. Acad. Nac. Cienc. Cordoba* **29**: 177 (1926).

Entomopatella Petrak, *Annls mycol.* **25**: 215 (1927).

Chaetospermella Naumov, *Mater. Mycol. Phytopath. Leningr.* **8**: 142 (1929).

Sp. typ.: *C. chaetosporum* (Pat.) A.L. Smith & Ramsbottom (syn. *Tubercularia chaetospora* Pat., *C. tubercularioides* Sacc., nom. nov.).

Mycelium immersed, branched, septate, pale brown to hyaline. *Conidiomata* eustromatic, epidermal to subepidermal or peridermal, initially immersed but finally appearing superficial, separate or aggregated and confluent, initially closed, later rupturing irregularly to expose an orange-yellow conidial mass, often unilocular but occasionally convoluted or divided irregularly; wall at the base of hyaline or very pale brown thin-walled textura angularis, the cells becoming elongated at the periphery and extending into heavily gelatinized textura oblita. Dehiscence irregular. *Conidiophores* hyaline, branched irregularly and only sparingly septate, smooth, formed from the inner cells of the conidiomatal wall. *Conidiogenous cells* holoblastic, discrete or integrated on conidiophores, determinate, cylindrical or slightly irregular, smooth, hyaline. *Conidia* hyaline, smooth, aseptate, thin-walled, eguttulate, with several cellular, unbranched, filiform appendages inserted subapically and suprabasally.

The several generic synonyms have been discussed briefly by Sutton (1977a) but really need to be confirmed by type studies. The only developmental study on the genus has been by Fonseka (1960) but the conidiomata are so variable that species have been placed in the Sphaeropsidales, Melanconiales and Tuberculariales. In describing *C. camelliae* from India, Agnihothrudu (1962) gave a key to 5 species. It seems possible, however to split collections into three taxa, mainly based on conidial size, and although the following synonymies are suggested they are only tentative. Of the three taxa, *C. gelatinosum*, seems slightly different from the other two, inasmuch as the conidia are much thicker-walled and the appendages are very thin. Perhaps they are not cellular in origin, unlike *C. chaetosporum* and *C. carneum* where they are clearly cellular. This may necessitate the removal of *C. gelatinosum* to a distinct genus in the future, but for the moment it is maintained in *Chaetospermum*.

KEY

1. Conidia >35 µ long 2
1. Conidia <26 µ long .... *C. carneum*
    2. Conidia 6-10 µ wide .... *C. chaetosporum*
    2. Conidia 16-21 µ wide .... *C. gelatinosum*

C. CHAETOSPORUM (Pat.) Smith & Ramsbottom, *Trans. Br. mycol. Soc.* **4**: 328 (1914). (Fig. 73C, D, E)
Chaetospermum chaetosporum (Pat.) Höhn., *Mitt. bot. Inst. tech. Hochsch. Wien* **1**: 87 (1924).
Tubercularia chaetospora Pat., *Bull. Soc. mycol. Fr.* **4**: 39 (1888).
Chaetospermum tubercularioides Sacc., nom. nov., *Syll. fung.* **10**: 706 (1892).
Dilophospora albida Massee & Crossl., *Naturalist* 3 (1904).
Ciliospora albida (Massee & Crossl.) Grove, *British Stem- and Leaf-Fungi* **2**: 117 (1937).
Pestalozziella ambigua Höhn., *Sber. Akad. Wiss. Wien* **116**: 145 (1907).
Chaetospermum ambiguum (Höhn.) Höhn., *Mitt. bot. Inst. tech. Hochsch. Wien* **1**: 88 (1924).
Entomopatella mirabilis Petrak, *Annls mycol.* **25**: 215 (1927).

Conidiomata 400-2000 µ diam. Conidiophores 10-30 × 2 µ. Conidia 26-45 × 8-15 µ; appendages 3-10, up to 45 µ long.

On stems, twigs and leaves of *Alnus glutinosa, Cupressus macrocarpa, Dahlia, Elaeis guineensis, Epilobium hirsutum, Glyceria maxima, Prunus serotina, Theobroma cacao, Triticum, Valeriana officinalis*, indet. substrate, India, Kenya, Tanzania, N.Y., U.S.A., New Hebrides, U.K. (IMI 104674, 74743, 14125, 101571, 20955, 1391, 20954, 81768, 105789d, 95833, 40097, 76793).

C. CARNEUM Tassi, *Bull. Lab. Ort. Bot., Siena* **3**: 130 (1900). (Fig. 73B)
Ciliospora gelatinosa Zimm., *Zentr. f. Bakt. Abt.* 2, **8**: 217 (1902).
Chaetospermum elasticae Koorders, *Bot. Untersuch. Java* 244 (1907).
Chaetospermum elasticae Koorders var. *artocarpi* Nag Raj, *Proc. Ind. Acad. Sci.*, B, **59**: 50 (1964).
Chaetospermella populina Naumov, *Mater. Mycol. Phytopath. Leningr.* **8**: 142 (1929).
Chaetospermum camelliae Agnihothrudu, *Mycopath. Mycol. appl.* **16**: 115 (1962).

Conidiomata up to 200 µ diam. Conidiophores up to 12 × 1-2 µ. Conidia 16-26 × 3-4(-6) µ; appendages 5-9, up to 14-18 × 1 µ.

On leaves, twigs and stems of *Artocarpus integrifolia, Archontophoenix alexandrae, Coffea robusta, Cordia myxa, Hevea, Olea cuspidata, Oryza sativa, Thea sinensis*, indet. substrates, India, Pakistan, Malaysia, Bermuda, Brazil, Venezuela (IMI 76076, 99087, 75840, 209035, 193391, 160969, 193383, 86739, 1392, 99088 holotype of *C. elasticae* var. *artocarpi*).

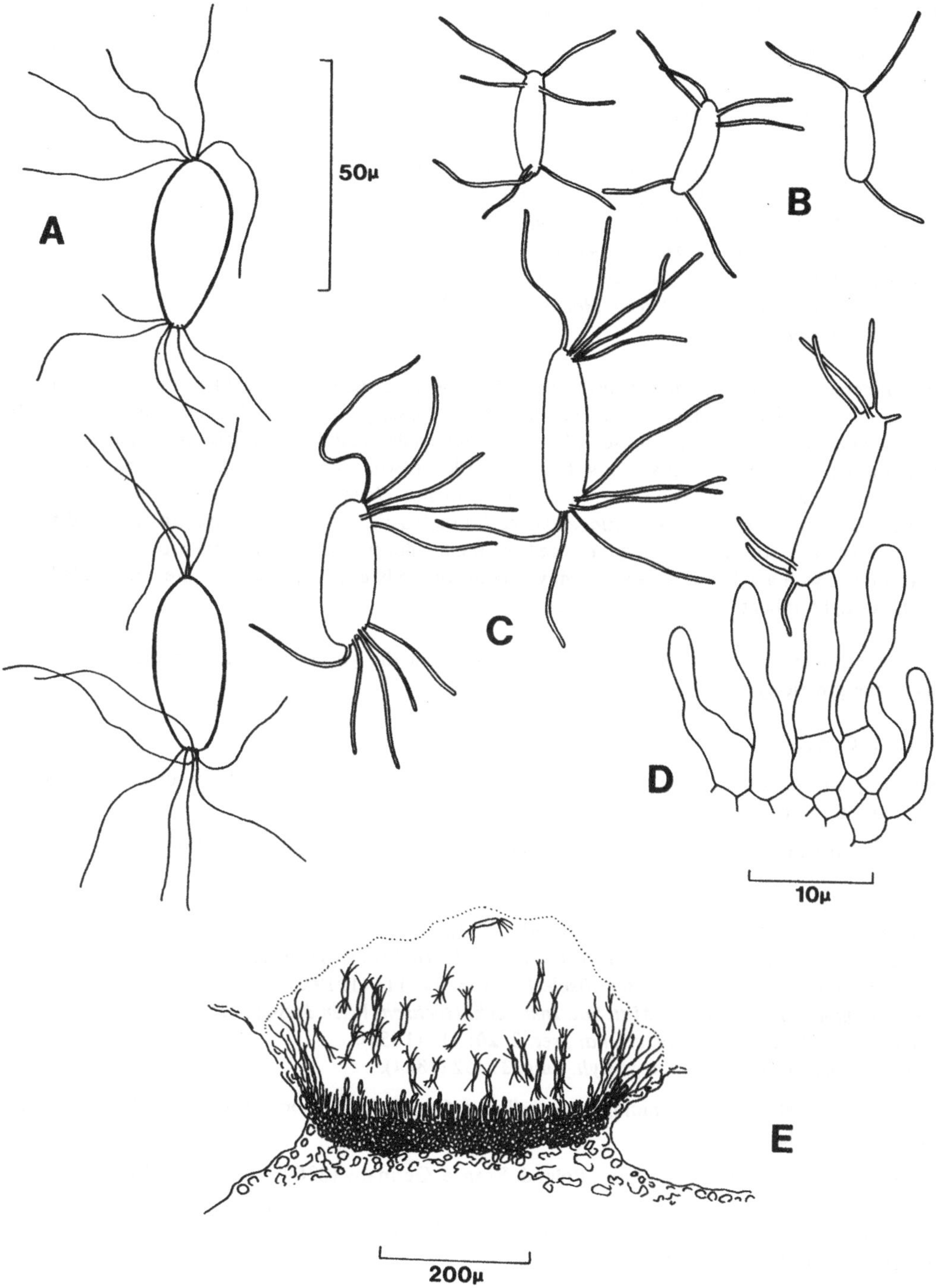

Fig. 73. A, *Chaetospermum gelatinosum*, conidia; B, *C. carneum*, conidia; C, D, E, *C. chaetosporum*, conidia, conidiophores and developing conidia, vertical section of a conidioma.

C. GELATINOSUM Petch, *Ann. R. bot. Gdn, Peradeniya* **6**: 255 (1917). (Fig. 73A)
Dinemasporium gelatinosum Berk. & Br., *J. Linn. Soc., Lond.* **14**: 88 (1873).
Chaetospermum gelatinosum (Berk. & Br.) Petch, *Ann. R. bot. Gdn, Peradeniya* **9**: 174 (1924).

Conidiomata 400-600 µ diam. Conidiophores up to 30 × 3 µ. Conidia 36-44 × 16-21 µ; appendages 6-12, up to 80 µ long.

On dead twigs, Sri Lanka (IMI 94747, holotype of *D. gelatinosum*, ex K).

15. SEPTORIELLA Oud., *Ned. Kruidk. Arch.*, 2 ser., **5**: 504 (1889).
Septosporiella Oud. in Saccardo, *Syll. fung.* **10**: 403 (1892).
? *Hysteridium* Karst., *Acta Soc. Fauna Flora fenn.* **27**: 10 (1905).
Linochorella Syd., *Annls mycol.* **10**: 43 (1912).
Naemostroma Höhn., *Ber. dtsch. Bot. Ges.* **37**: 114 (1919).
Nemostroma Höhn., *Mitt. bot. Inst. tech. Hochsch. Wien* **3**: 83 (1925).
Sp. typ.: *S. phragmitis* Oud.

Mycelium immersed, hyaline to pale brown, branched, septate. *Conidiomata* like solitary pycnidia or loosely aggregated into linear eustromata; when pycnidial, separate, globose, papillate, dark brown, immersed, subepidermal, unilocular or convoluted, thick-walled, composed of brown thick-walled textura angularis somewhat thinner and paler at the base; stromata linear between viens, consisting of dark brown, thick-walled textura angularis in the epidermis and hypodermis and looser tissue connecting the pycnidial locules. *Ostioles* single to each locule or pycnidium, central, circular, papillate. *Conidiophores* absent. *Conidiogenous cells* holoblastic, determinate, discrete, ± cylindrical but swollen at the base, hyaline, smooth, formed from the inner cells of the locular walls. *Conidia* pale brown, up to 8 euseptate, not constricted, apex obtuse with a gelatinous cap that may become partially everted, base truncate, thin-walled, smooth, eguttulate, cylindrical, straight or bent somewhat irregularly.

More than ten species have been described in *Septoriella* but the genus has not been extensively revised. Sutton (1977a) discussed the generic synonymy.

S. PHRAGMITIS Oud., *ibid.*: 504 (1889). (Fig. 74)

Locules up to 320 µ diam. Conidiogenous cells 3-6 × 3-4 µ. Conidia up to 85 µ long × 4 µ wide, 5-7 septate; gelatinous appendage up to 9 µ long.

On culms of *Phragmites communis*, Austria, Netherlands (IMI 192310, holotype, 12464).

S. junci (Desm.) comb. nov. (Fig. 75)
Septoria junci Desm., *Ann. sci. nat.*, ser. 3, **20**: 85 (1853).
Rhabdospora junci (Desm.) Allesch. in Rabenh. *Krypt.-Flora* **1**(6): 910 (1901).
Naemostroma junci (Desm.) Höhn., *Ber. dtsch. Bot. Ges.* **37**: 114 (1919).
Nemostroma junci (Desm.) Höhn., *Mitt. bot. Inst. tech. Hochsch. Wien* **3**: 84 (1925).
Hendersonia insidiosa Desm., *Ann. sci. nat.*, ser. 3, **20**: 213 (1853).
Stagonospora insidiosa (Desm.) Sacc., *Syll. fung.* **3**: 452 (1884).

Locules up to 400 µ diam. Conidiogenous cells 3.5-6.5 × 3-5.5 µ. Conidia 48-69 × 2.5-3 µ, 1-5 septate; small gelatinous appendage at each end.

On stems of *Juncus effusus*, *J. maritimus*, Poland, France, Germany (22698, 201971, 201970).

16. TOXOSPORIOPSIS Sutton & Sellar, *Can. J. Bot.* **44**: 1505 (1966).
Sp. typ.: *T. capitata* Sutton & Sellar

Mycelium immersed, branched, septate, hyaline to pale brown. *Conidiomata* eustromatic, immersed, subperidermal, separate or confluent, irregular in shape, hyaline to pale brown, irregularly locular; wall composed of small-celled, hyaline to pale brown, thin-walled textura angularis. *Ostiole* absent, dehiscence irregular. *Conidiophores* hyaline, branched at the base, septate, smooth, formed from the pseudoparenchyma. *Conidiogenous cells* holoblastic, determinate, integrated or discrete, cylindrical, hyaline, smooth. *Conidia* 3-euseptate, septa dark brown, median one black and thick, continuous or slightly constricted, base truncate, apex conic to papillate, thick-walled, smooth, broadly fusiform, apical and basal cells

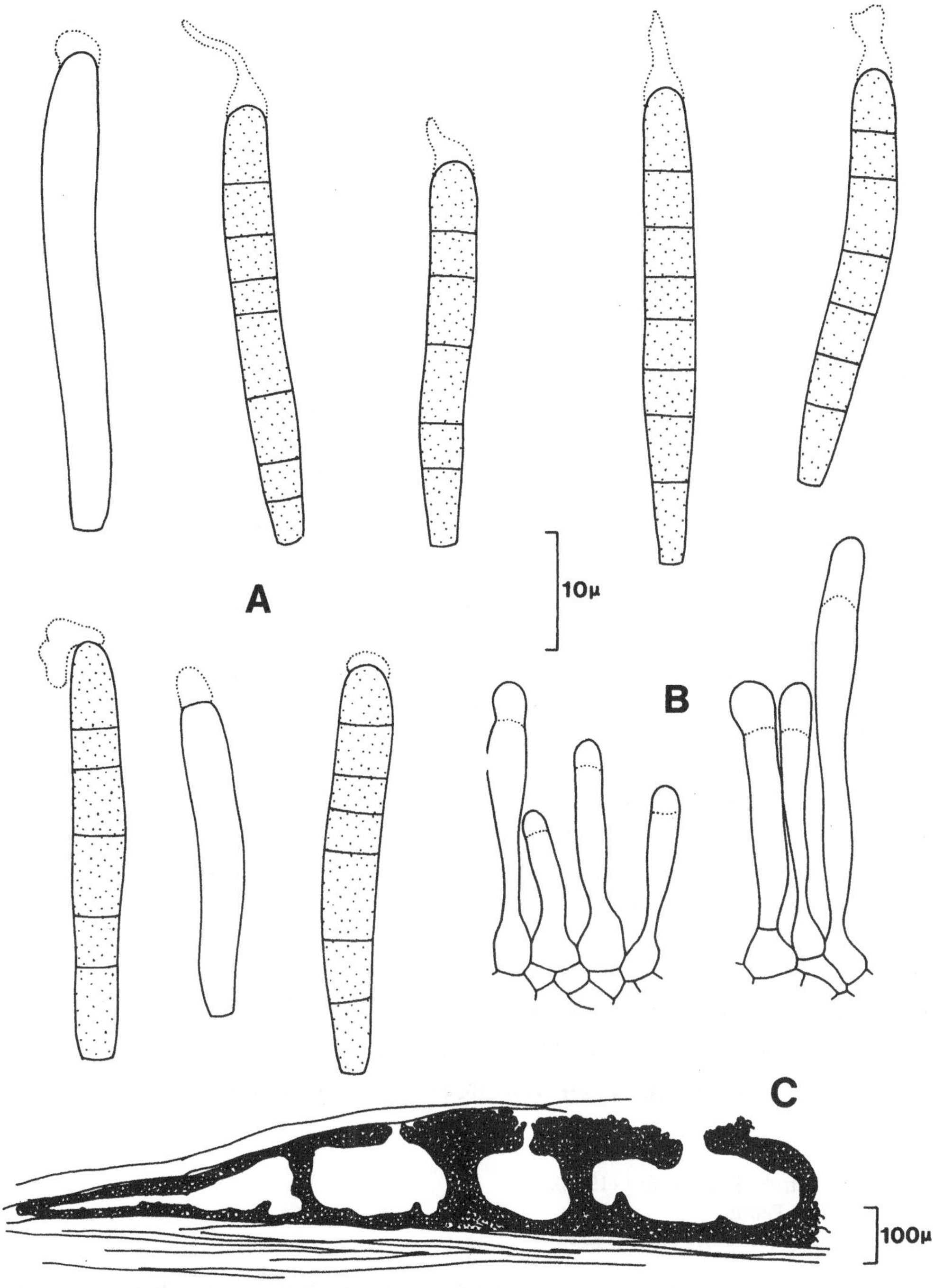

Fig. 74. *Septoriella phragmitis*, A, conidia; B, conidiogenous cells and developing conidia; C, vertical section of a conidioma.

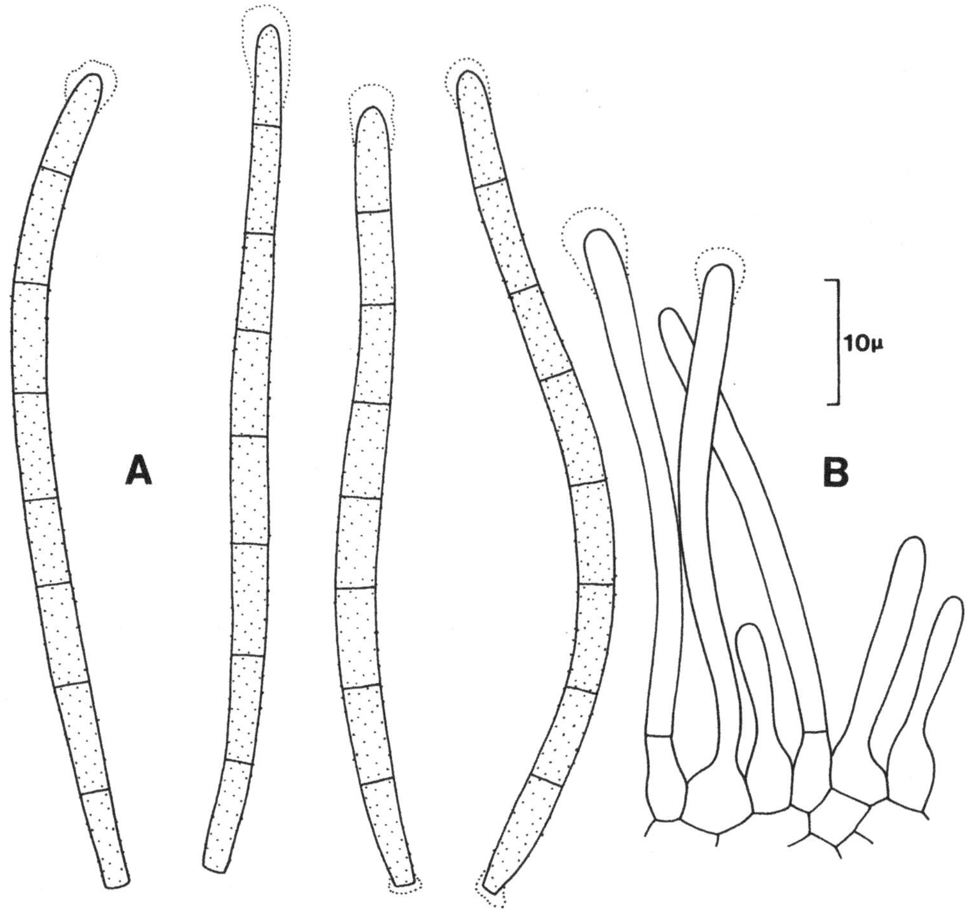

Fig. 75. *Septoriella junci*, A, conidia; B, conidiogenous cells and developing conidia.

thinner-walled and paler. *Conidia* develop in a gelatinous sheath that finally reduces to an apical, attenuated appendage. *Paraphyses* hyaline, aseptate, flexuous, enclosed in a gelatinous sheath.

The genus is monotypic.

T. CAPITATA Sutton & Sellar, *ibid.*: 1507 (1966). (Figs. 76, 77)

Conidiophores 3-10 × 2-3 μ. Conidia 29.5-37.5 × 11.5-15 μ. Paraphyses 15-75 μ.

On branches of *Ceiba pentandra*, Sierra Leone (IMI 103428b, holotype, 103429).

17. STEVENSONULA Petrak, *Sydowia* **6**: 1 (1952).
Sp. typ.: *S. ciliata* Petrak
Illustration: Sutton (1975b); Petrak (1952a).

Mycelium immersed or superficial, dark brown, branched, septate. *Conidiomata* columnar, ± infundibuliform, superficial, separate, dark brown, sessile; base broad, cylindrical, solid, of dark brown textura angularis, upper periclinal wall 1 cell thick with divergent, olivaceous, flexuous, septate hyphae producing a shallow cupulate wall. *Conidiophores* long, septate, hyaline, cylindrical, inflated at the apices, formed from the upper cells of the broad base. *Conidiogenous cells* holoblastic, integrated, determinate, hyaline, smooth. *Conidia* dark brown, thick-walled, 1 euseptate towards the apex, straight, slightly curved or reniform, smooth, truncate at the base, larger cell guttulate, with several unbranched

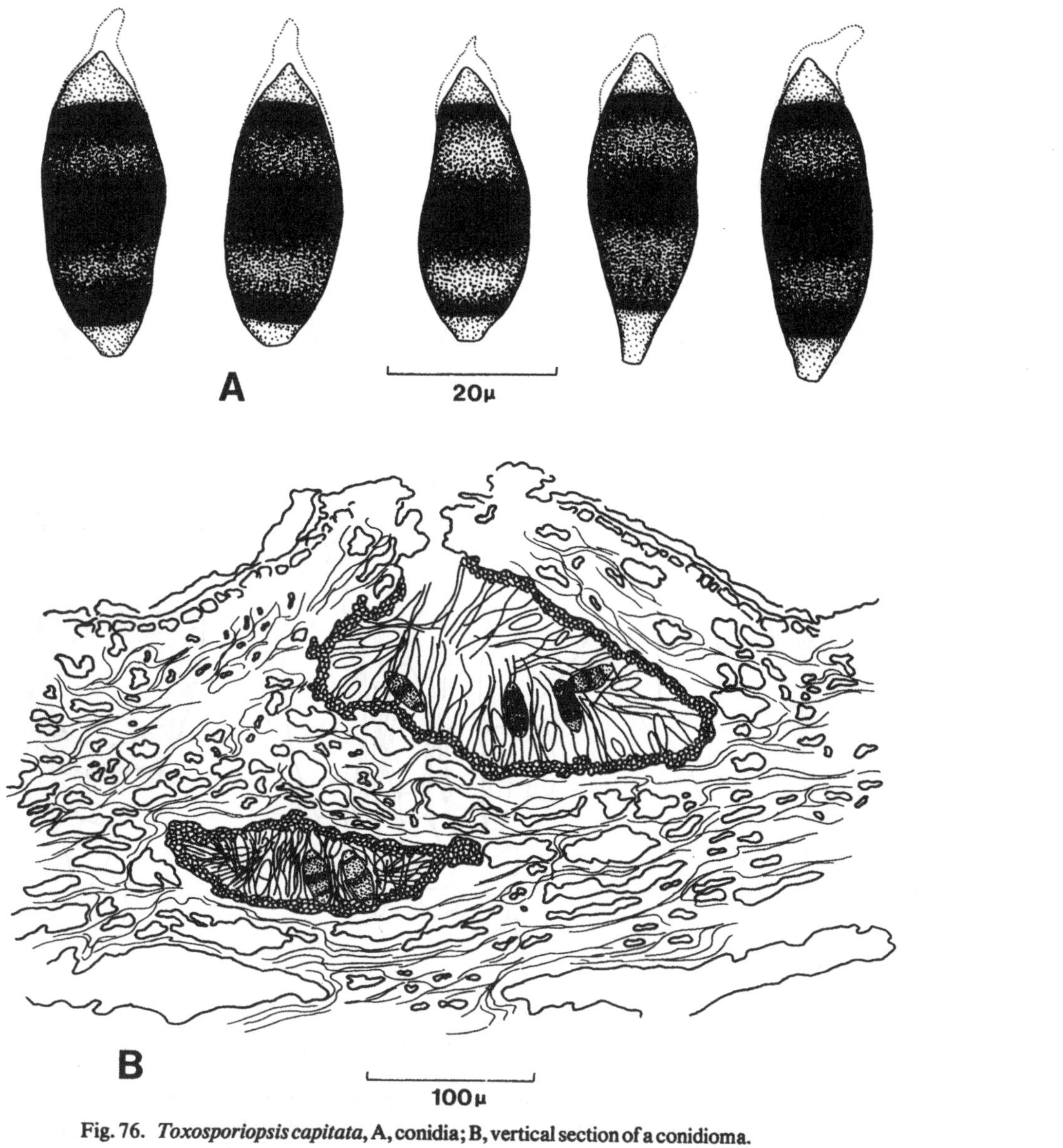

Fig. 76. *Toxosporiopsis capitata*, A, conidia; B, vertical section of a conidioma.

extracellular mucilaginous appendages inserted subapically and laterally near the base. *Paraphyses* hyaline, branched, septate, filiform.

Sutton (1975b) redescribed the genus and single species.

S. CILIATA Petrak, *ibid.*: 2 (1952).

Conidiomata 225 μ high × 310 μ wide. Conidiogenous cells up to 90 μ long × 2 μ wide (6-8 μ wide apices). Conidia 38-45 × 14-17 μ; appendages up to 42 μ long. Paraphyses up to 150 μ long × 2 μ wide.

On dead stems of *Baccharis halimifolia* and Gramineae, New Caledonia (IMI 102785a).

18. CATENOPHORA Luttrell, *Mycologia* **32**: 535 (1940).
Sp. typ.: *C. pruni* Luttrell

Immersed mycelium pale brown to hyaline, branched, septate. *Conidiomata* acervular, peridermal, separate, rarely confluent, formed of pale brown, thin-walled textura angularis. Dehiscence irregular. *Conidiophores* hyaline, branched at

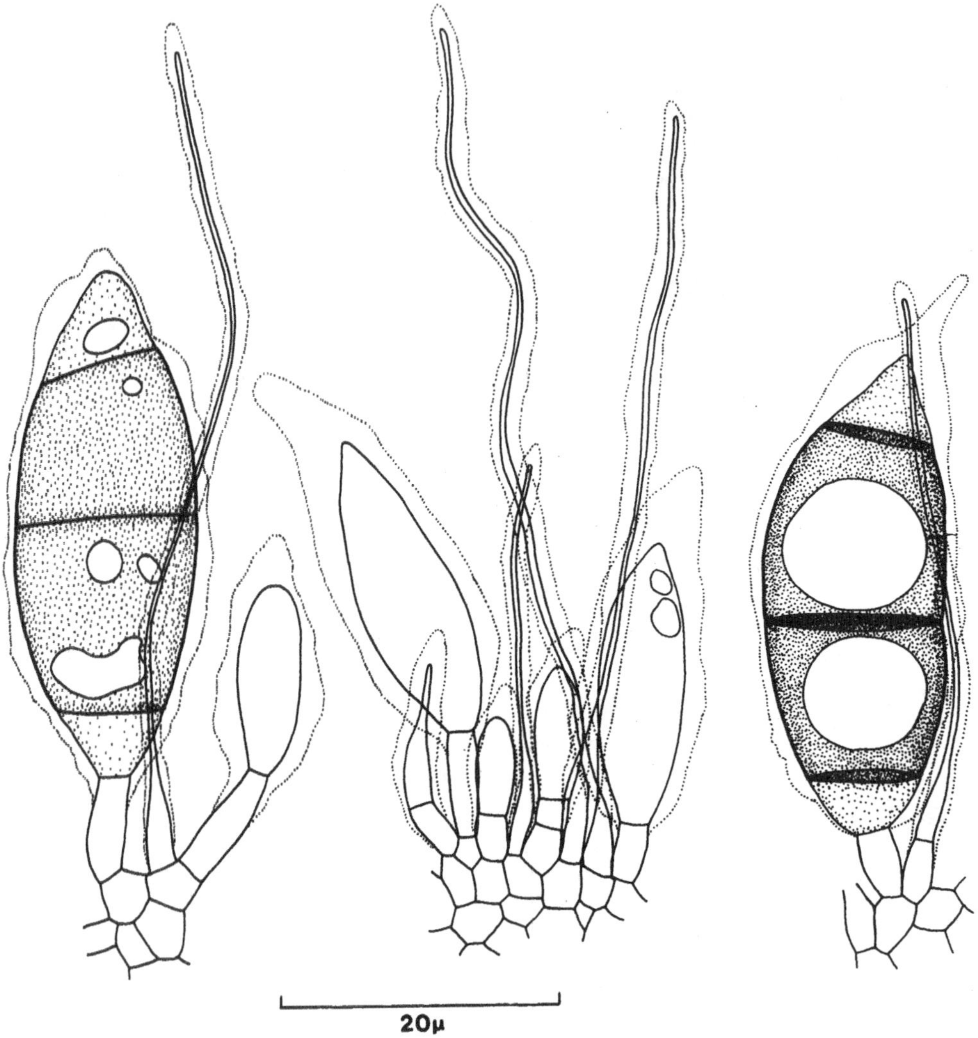

Fig. 77. *Toxosporiopsis capitata*, developing conidia, conidiophores and paraphyses.

the base, septate, separating small cells, smooth, with acropleurogenous conidia, formed from the upper pseudoparenchyma. *Conidiogenous cells* holoblastic, determinate, integrated, cylindrical to doliiform, hyaline, smooth, producing single conidia from a short terminal or lateral peg formed immediately below a septum. *Conidia* hyaline to subhyaline, aseptate, apex obtuse, base truncate, thin-walled, smooth, eguttulate, elliptical.

A second species, *C. yuccae*, was added by Nag Raj (1977a), with larger conidiomata, 170-420 μ wide × 150-620 μ high, and larger conidia, 6.5-13 × 4-7 μ. The species was recorded on *Yucca* spp. from Colo and Ariz., U.S.A.

C. PRUNI Luttrell, *ibid.*: 536 (1940). (Fig. 78)

Conidiomata 80-100 μ diam. Conidiophores up to 50 μ long. Conidia 6-7 × 3-3.5 μ.

On twigs of *Prunus serrulata*, N. Carol., U.S.A. (IMI 150996, holotype).

19. LEPTODERMELLA Höhn., *Z.f. Garungsph.* **5**: 212 (1914).
Sp. typ.: *L. incarnata* (Bres.) Höhn. (syn. *Zythia incarnata* Bres.).

Mycelium immersed, branched, septate, hyaline. *Conidiomata* acervular, separate, epidermal to subepidermal, circular, pale orange, formed of hyaline, thin-walled pseudoparenchyma. Dehiscence irregular. *Conidiophores* absent.

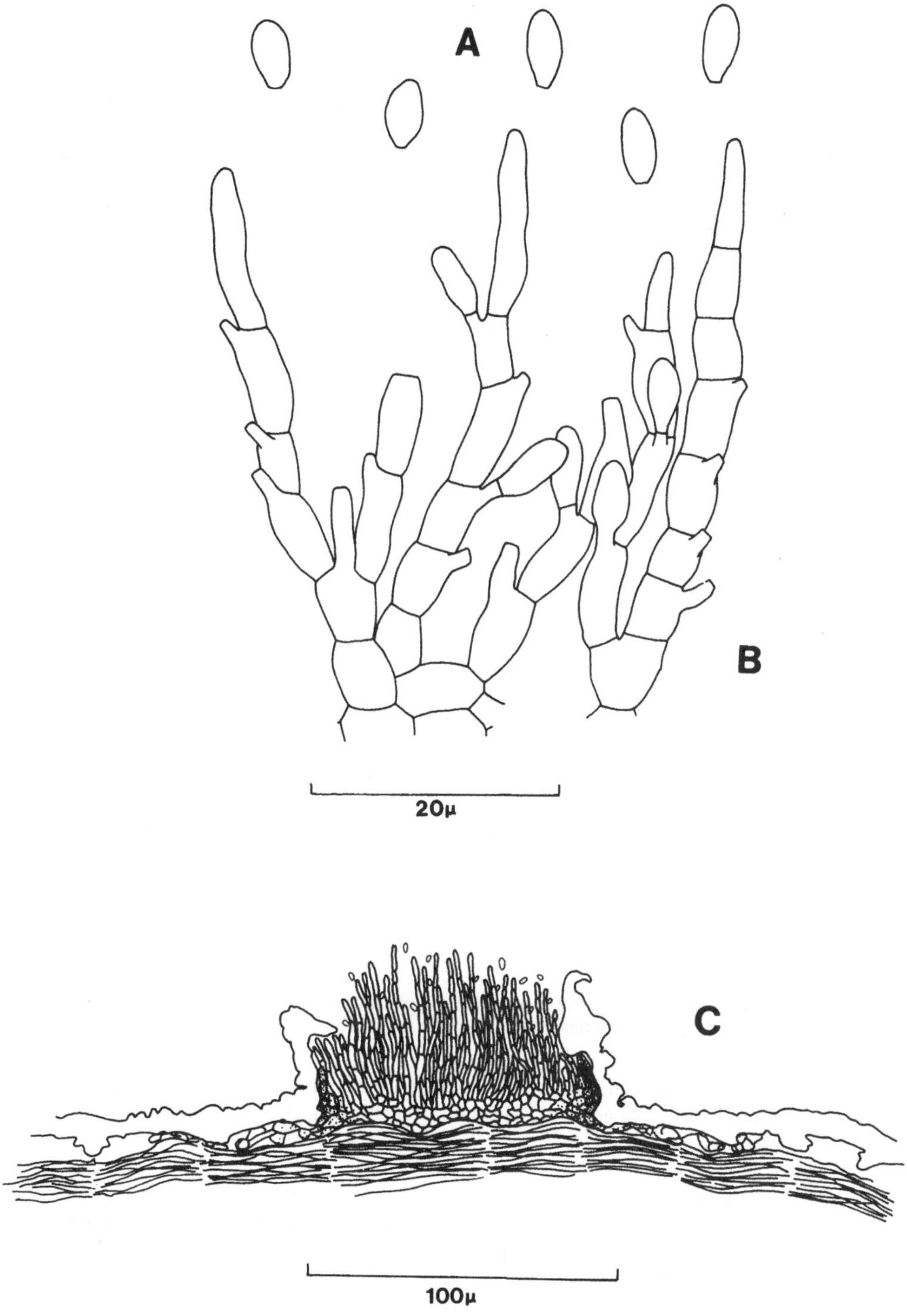

Fig. 78. *Catenophora pruni*, A, conidia; B, conidiophores; C, vertical section of a conidioma.

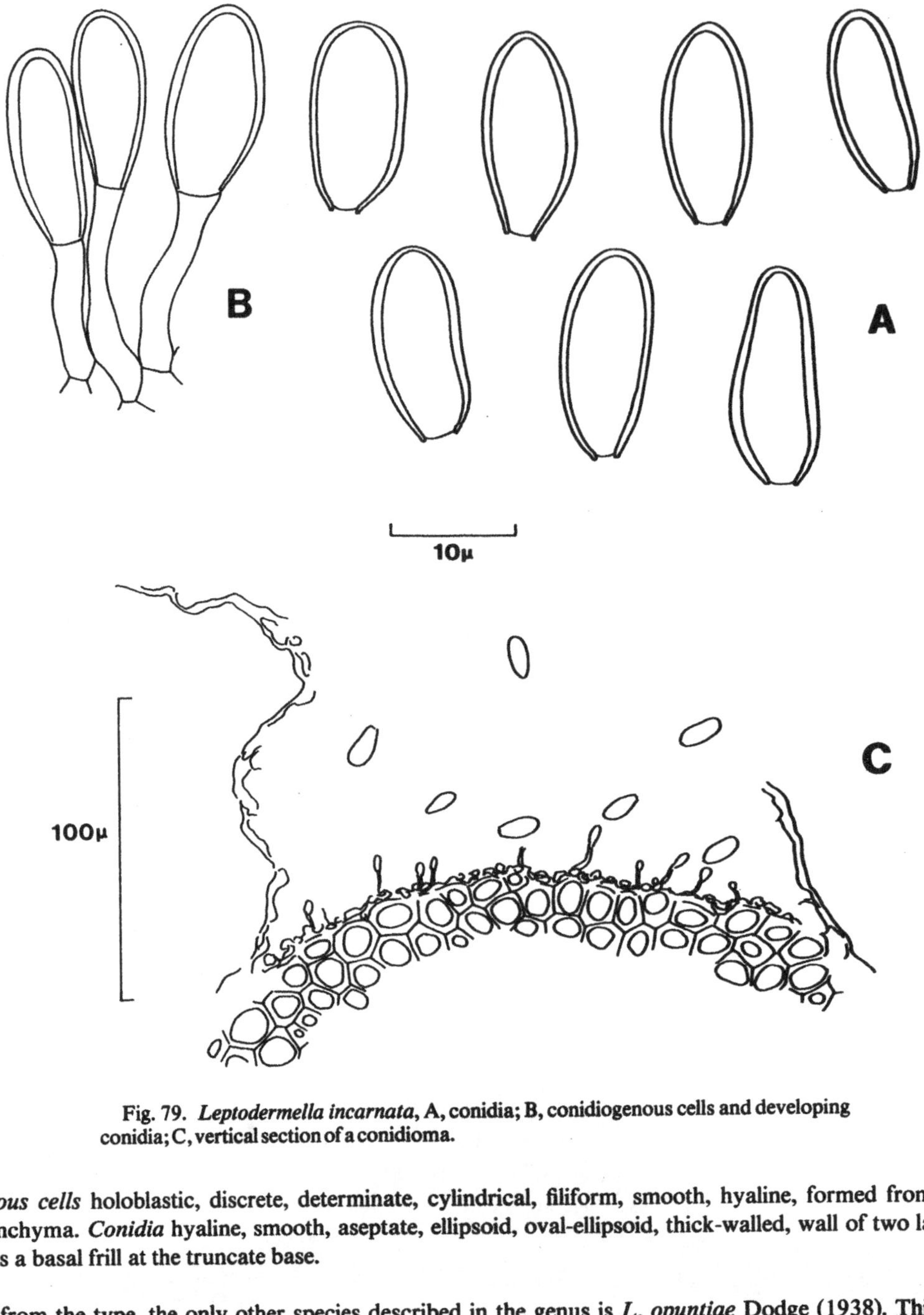

Fig. 79. *Leptodermella incarnata*, A, conidia; B, conidiogenous cells and developing conidia; C, vertical section of a conidioma.

Conidiogenous cells holoblastic, discrete, determinate, cylindrical, filiform, smooth, hyaline, formed from the acervular pseudoparenchyma. *Conidia* hyaline, smooth, aseptate, ellipsoid, oval-ellipsoid, thick-walled, wall of two layers, the outer projecting as a basal frill at the truncate base.

Apart from the type, the only other species described in the genus is *L. opuntiae* Dodge (1938). This was reported with pinkish fleshy pycnidia 100-170 µ diam., filiform conidiophores and aseptate conidia measuring 3-4.5 × 1-1.5 µ. The conidial size and conidiomatal structure alone preclude the retention of this fungus in *Leptodermella*, but until the type is seen no decision on its relationships can be made.

L. INCARNATA (Bres.) Höhn., *ibid.*: 212 (1914). (Fig. 79)
Zythia incarnata Bres., *Hedwigia* **39**: 352 (1900).

Conidiomata up to 190 µ diam. Conidiogenous cells 11-18 × 2-2.5 µ. Conidia 15-18 × 6-8 µ.

On dead stems of *Matricaria suaveolens*, Latvia, U.S.S.R. (IMI 30063).

20. CYTOSPHAERA Died., *Annls mycol.* **14**: 205 (1916).
Sp. typ.: *C. mangiferae* Died.

Mycelium immersed, branched, septate, dark brown. *Conidiomata* eustromatic, separate, occasionally confluent, irregularly pulvinate, dark brown to black, initially subperidermal but strongly erumpent, finally appearing superficial, multilocular; locules simple, globose, peripheral in the upper part of the stromata; tissue of thick-walled dark brown textura angularis, darker and more sclerotioid at the base and periphery. *Ostioles* single to each locule, circular, not papillate. *Conidiophores* sometimes present, 1-2 septate, hyaline, smooth, sparingly branched, formed from inner cells of the locular walls. *Conidiogenous cells* 1) macroconidial, holoblastic, determinate, integrated or more frequently discrete, straight, cylindrical, hyaline, smooth, 2) microconidial, enteroblastic, phialidic, determinate, discrete, straight, cylindrical, hyaline, smooth, channel minute, collarette visible, periclinal thickening considerable. *Macroconidia* hyaline, aseptate, with very thick walls, ± guttulate, smooth base truncate. *Microconidia* hyaline, aseptate, thin-walled, guttulate, ellipsoid to cylindrical.

Petrak & Sydow (1927) placed this generic name in synonymy with *Aplosporella* Speg., but *C. mangiferae* can be distinguished from species of that genus by the hyaline, thick-walled conidia and the stromata of different structure. The genus is monotypic.

C. MANGIFERAE Died., *ibid.*: 205 (1916). (Fig 80)
Aplosporella mangiferae (Died.) Petrak & Syd., *Rep. spec. nov. regni veg.* **42**: 102 (1927).

Conidiomata up to 1250 µ diam. Conidiophores 10-18 × 2-3 µ. Macroconidia 15.5-25.5 × 10-12 µ. Microconidia 2.5-3 × 1 µ.

On *Mangifera indica*, *Aquilaria agallocha*, Guinea, Pakistan, Australia (IMI 95644, 152939, 195540).

21. RABENHORSTIA Fr., *Summa Veg. Scand.*: 410 (1849).
Naemosphaera (Sacc.) Sacc., *Syll. fung.* **10**: 259 (1892).
Original spp.: *R. clandestina* (Fr.) Fr., *R. rudis* (Fr.) Fr., *R. personata* (Fr.) Fr., *R. tiliae* (Fr.) Fr., *R. adpressa* Fr.
Lectotype sp.: *R. tiliae* (Fr.) Fr. (syn. *Sphaeria tiliae* Fr.), fide Clements & Shear (1931).

Mycelium immersed, branched, septate, hyaline to pale brown. *Conidiomata* eustromatic, immersed, subperidermal, dark blackish brown, applanate but markedly erumpent in the centre, separate, unilocular, multilocular or convoluted, thick-walled; wall several cells thick, of dark brown, thick-walled textura intricata above and at the sides, at the base of pale brown textura angularis, upper wall consisting entirely of textura oblita and projecting as a cap over the dehiscent area. *Ostiole* absent, dehiscence irregular. *Conidiophores* branched extensively at the base, less so above, hyaline, septate, smooth, often developing in mucilage, formed at the base and sides of the conidiomatal wall. *Conidiogenous cells* holoblastic, determinate, integrated, filiform, straight or curved, hyaline, smooth, each forming a single conidium at the conidiophore apex or from a short locus immediately below a transverse septum. *Conidia* hyaline, smooth, aseptate, thick-walled, contents granular, often with a central guttule, straight, ellipsoid, base occasionally truncate, produced in mucilage but without any mucilaginous envelope or appendage at maturity.

At least 17 species have been described in *Rabenhorstia*, and considering the unusual nature of the stromatic conidiomata, it would be surprising if most of these were congeneric with the lectotype species. Some have already been removed to genera like *Aplosporella*, but the genus is in need of revision.

R. TILIAE (Fr.) Fr., *ibid.*: 410 (1849). (Fig. 81)
Sphaeria tiliae Pers. ex Fr., *Syst. Myc.* **2**: 485 (1823).
Sphaeria tiliae Pers., *Syn. Meth. fung.* 84 (1801).
Spilobolus tiliae (Pers. ex Fr.) Lk, *Hand. Erkenn. Nutzb. & Hanfig. Vorkom. Gewachse* **3**: 380 (1833).
Oudemans (1922) lists binomials involving '*tiliae*' in *Torsellia*, *Stigmatosphaeria* and *Haplosporium* as synonyms for this species. The citations have not been verified.

Teleomorphosis: *Hercospora tiliae* (Pers. ex Fr.) Tul., *Sel. Fung. Carp.* **2**: 154 (1863).

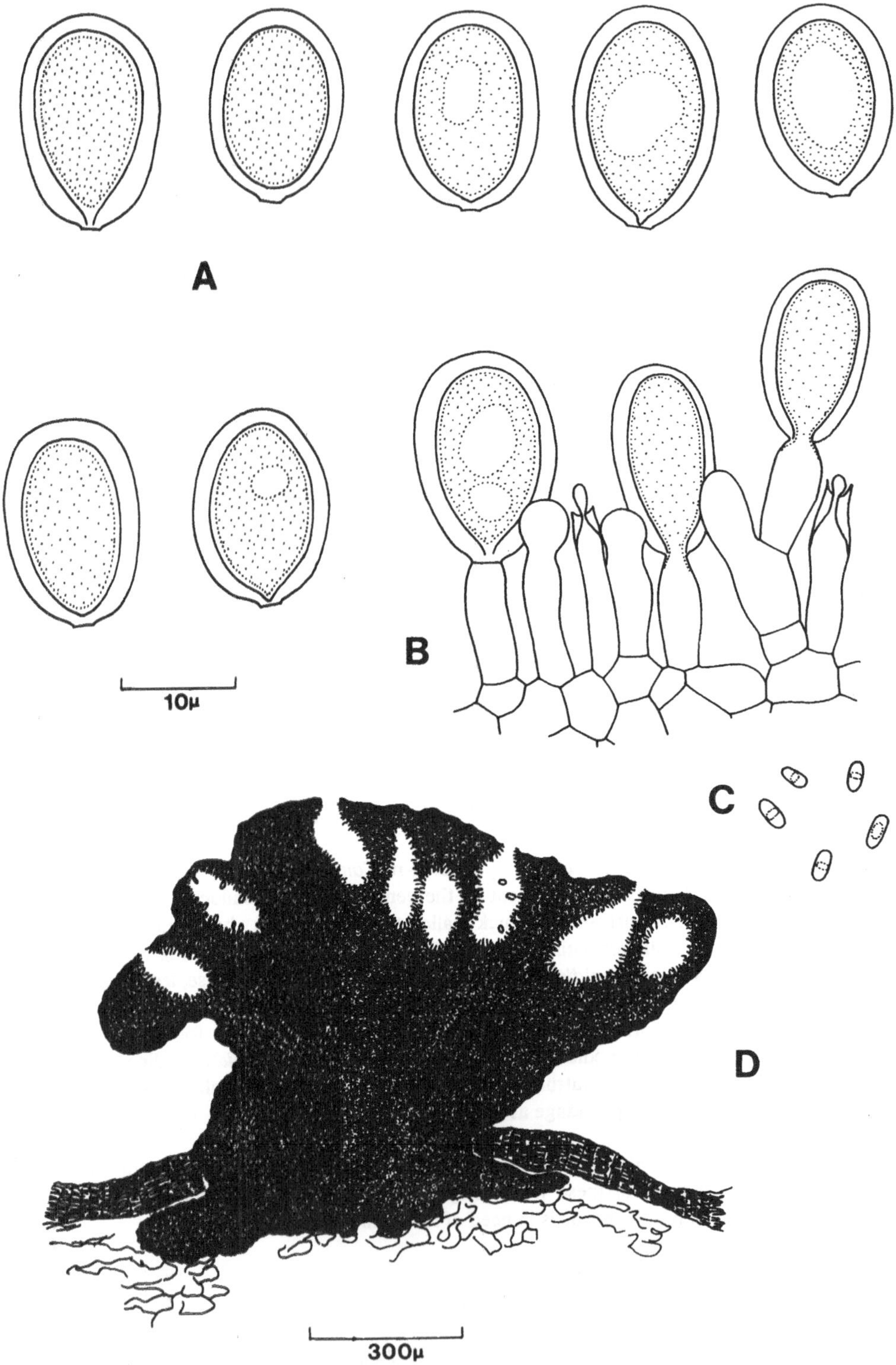

Fig. 80. *Cytosphaera mangiferae*, A, conidia; B, conidiogenous cells, conidiophores, with developing conidia and microconidia; C, microconidia; D, vertical section of a conidioma.

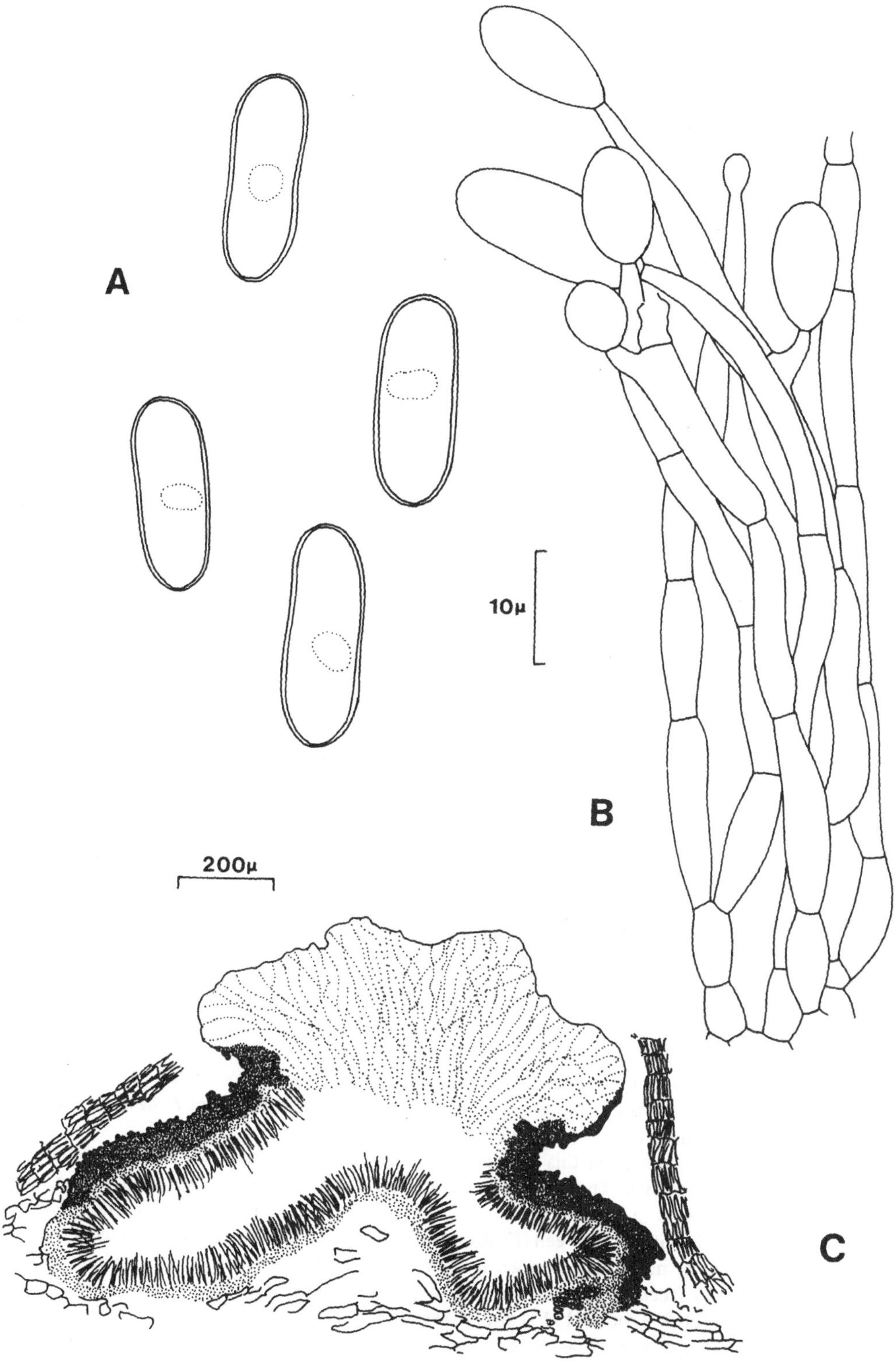

Fig. 81. *Rabenhorstia tiliae*, A, conidia; B, conidiophores and developing conidia; C, vertical section of a conidioma.

Conidiomata up to 1250 μ diam. at the base; mucilaginous cap up to 80 μ diam. Conidiophores up to 90 μ long × 3-4 μ wide. Conidia 18-20 × 6-7.5 μ.

On dead branches of *Tilia platyphylla, T. vulgaris, T. europaea, T.* sp., Austria, Norway, U.K. (IMI 5990, 6484, 3068, 16695, 31599b, 40961b, 193129, 127050, 24795, 21918, 16696, 12686, 57157, 94432).

22. POLYMORPHUM Chev., *J. Phys. Chem. Hist. Nat.* **94**: 32 (1822).
Heterographa Feé, Essai sur les Cryptogames XXX (1824).
Dichaena Fr., *Syst. Orb. Veg.* **1**: 108 (1825).
Psilospora Rabenh., *Hedwigia* **1**: 107 (1856)
Psilosporina Died., *Annls mycol.* **11**: 534 (1913).
Original spp.: *P. quercinum* Chev., *P. fagineum* Chev.
Lectotype.: *P. rugosum* (Fr.) Hawksworth & Punithalingam (syn. *Dichaena rugosa* Fr., *P. quercinum* Chev.).
Illustration: Hawksworth & Punithalingam (1973a).

Mycelium superficial, dark brown to black, forming a subicle of angular pseudoparenchyma, or immersed, hyaline to pale brown, branched and septate. *Conidiomata* eustromatic, immersed, erumpent and finally appearing superficial, separate or densely gregarious, dark brown to black, sessile, navicular, hysterioid and bilabiate to oval and disciform and trilabiate or quadrilabiate, unilocular but often convoluted, dehiscing by a longitudinal fissure; wall composed of an outer layer of dark brown thick-walled textura angularis, a median layer of paler textura prismatica and an inner layer of pale brown textura angularis that lines the lateral walls and forms the base of the conidioma. *Conidiophores* hyaline, branched only at the base, septate, smooth, ± cylindrical, possibly basauxic, formed from the inner layer of the wall. *Conidiogenous cells* holoblastic, determinate, integrated, rarely discrete, cylindrical, hyaline, smooth, with acrogenous conidia. *Conidia* hyaline, aseptate, ovoid, base truncate, often with part of the conidiogenous cell remaining attached, thin-walled, smooth, ± guttulate.

The taxonomic and nomenclatural history of this genus has been fully described by Hawksworth & Punithalingam (1973a), and Butin (1977) described the perfect state.

P. RUGOSUM (Fr.) D. Hawskw. & Punithalingam, *Trans. Br. mycol. Soc.* **60**: 503 (1973).
Synonymy in Hawksworth & Punithalingam (1973).

Teleomorphosis: *Ascodichaena rugosa* Butin, *Trans. Br. mycol. Soc.* **69**: 249 (1977).

Conidiomata 0.5-1 (-1.5) mm long × 0.2-0.4 mm wide. Conidiophores 15-25 × 3-4 μ. Conidia 15-25 × 10-15 μ.

On trunks of *Fagus grandifolia, F. sylvatica, F.* sp., *Quercus ilex, Q. petraea, Q.* sp., represented by 36 collections in herb. IMI from Ont., Canada, U.K., Germany, France, Pakistan.

23. FUSICOCCUM Cda in Sturm, *Deutschlands Flora* **2**: 111 (1829).
Sp. typ.: *F. aesculi* Cda

The status of *Fusicoccum* and its type species has been the source of confusion in the past, and this can be mainly attributed to the fact that no-one reports having re-examined Corda's original material of *F. aesculi.* So far more than 200 taxa have been ascribed to the genus and very few have been revised.

Saccardo (1880b) described conidia in this species as 23-30 × 5 μ and conidiophores as 20-30 × .5-2 μ, and later (Saccardo, 1886) illustrated the species. These data however were from material collected in Padua, and although he mentioned Czechoslovakian material in another account (Saccardo, 1884) it is by no means certain that he saw Corda's specimen. Höhnel (1916a) distinguished between *Fusicoccum* Cda based on *F. aesculi* Cda which he interpreted as Melanconiaceous with aseptate, spindle-shaped, large, hyaline conidia, and *Fusicoccum* Sacc. which should be dropped. Precisely what was meant by *Fusicoccum* Sacc., is difficult to know because Saccardo (1880b, 1884, 1895), Saccardo & Sydow (1899, 1902) and Saccardo & Trotter (1913) always referred to the genus as *Fusicoccum* Cda.

Petrak (1921) collected a fungus on twigs of *Aesculus hippocastanum* which he interpreted as the conidial state of *Diaporthe aesculi* (Fckl) Höhn. This ascomycete was later referred to by Arx (1957a, 1963, 1970) as *Cryptodiaporthe aesculi* (Fckl) Petrak. The conidia developed sometimes three faint septa and measure 14-32 × 3-5 μ. This conidial state was thought to have been first published as *Myxosporium* ? *aberrans* Sacc. and later as *Septomyxa aesculi* Saccardo (1884) who described conidia as 1 septate and 15-16 × 3-4 μ. Since Petrak doubted the existence of two separate fungi on

Aesculus, the one with aseptate conidia called *F. aesculi* Cda by Höhnel (1916a) and the other with septate conidia collected by Petrak and named *S. aesculi* by Saccardo (1884), he proposed uniting the two species under the name *Septomyxa aesculi* (Cda) Petrak. In doing so he used *F. aesculi* as the basionym for the transfer into *Septomyxa*, thereby creating a later homonym. Thus not only is *S. aesculi* (Cda) Petrak illegitimate but if Petrak's conclusions are correct, *Fusicoccum* should be taken up as the earlier available generic name. Later, Petrak (1922) removed the fungus called *F. aesculi* by Saccardo (1880b, 1884, 1886) to *Dothiorella*, citing the species as the conidial state of *Botryosphaeria berengeriana* de Not. The correct citation for the fungus is *D. aesculi* Petrak since Saccardo's identification was regarded as a misapplication.

It is evident that the conidial state of *Cryptodiaporthe aesculi* is a 1-septate species now referable to *Diplodina* Westd., of which *Discella* and *Septomyxa* are synonyms. The use of *Fusicoccum* in this context is erroneous, for in *F. aesculi* conidia are not septate.

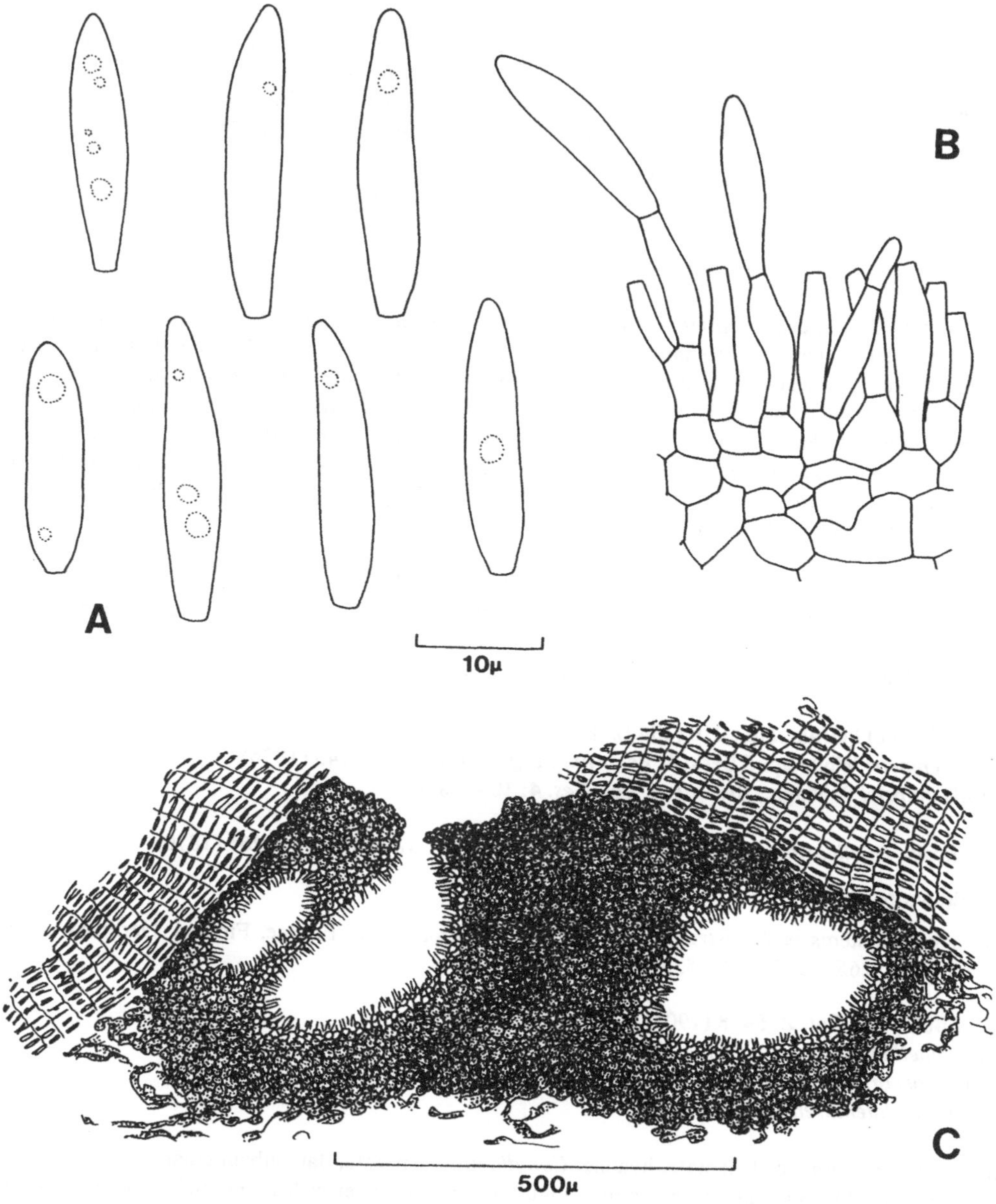

Fig. 82. *Fusicoccum aesculi*, A conidia; B, conidiophores and developing conidia; C, vertical section of a conidioma.

The fungus described as *F. aesculi* by Saccardo (1880b) and later by Höhnel (1916a) is the conidial state of *Botryosphaeria berengeriana*, a synonym of *B. dothidea* (Moug. ex Fr.) Ces. & de Not., according to Arx & Müller (1954). Until Corda's material is examined it is not possible to say if the determination is correct. However, there is sufficient similarity in the illustrations by Corda (1829) and Saccardo (1886) to suggest that *Fusicoccum* is in contention for the conidial state name of ascomycetes such as *B. dothidea* and *B. ribis* Grossenb. & Dugg.

To facilitate future developments in this problem the following illustrated account of Saccardo's specimen of *F. aesculi* is given.

F. AESCULI Cda apud Sturm, *ibid.*: 111 (1829). (Fig. 82)

Mycelium immersed, branched, septate, pale brown. *Conidiomata* eustromatic, up to 1000 µ diam., immersed, subperidermal, separate, dark brown to black, multilocular (locules up to 250 µ diam.); walls of locules composed of large-celled, brown, rather irregular textura angularis becoming looser at the periphery, and smaller, more hyaline towards the conidiogenous region. *Ostioles* indistinct, dehiscence by rupture of the overlying host tissues. *Conidiophores* hyaline, sparingly septate and branched at the base, cylindrical, formed from the inner locular wall cells, 14-24 × 2-3 µ. *Conidiogenous cells* holoblastic, determinate, discrete or integrated, hyaline, smooth, cylindrical, producing a single apical conidium. *Conidia* hyaline, thin-walled, aseptate, fusiform, straight, irregularly guttulate, apex obtuse, base distinctly truncate, 18-25 × 4-4.5 µ.

On branches of *Aesculus*, Italy (IMI 201771 ex PAD).

24. HYSTERODISCULA Petrak, *Bot. Arch.* **43**: 210 (1942).
Sp. typ.: *H. empetri* (White) Petrak (syn. *Rhytisma empetri* White).

Mycelium immersed, branched, septate, brown. *Conidiomata* pseudostromatic, black, effuse, often confluent, extensive, often encircling stems, irregular in shape but frequently linear, irregularly multilocular; basal and lateral walls of hyaline to pale brown, thin-walled textura angularis, upper wall thicker, composed of dark brown, thick-walled pseudoparenchyma that extends beyond the locules in the epidermis and subepidermis to cover the substrate. *Ostiole* absent, dehiscence by longitudinal fissure. *Conidiophores* hyaline, 1-2 septate, cylindrical, straight, branched sparingly at the base, with acrogenous conidia, formed all round the locular cavity. *Conidiogenous cells* holoblastic, determinate, integrated or discrete, cylindrical, hyaline, smooth. *Conidia* hyaline, aseptate, smooth, thin-walled, straight, apex obtuse, base truncate, tapered towards the base, cylindrical to clavate.

The type species has been described by Wilson (1921) and Eriksson (1974). Apart from the difference in conidiogenesis, there is considerable resemblance between *Hysterodiscula* and *Apomelasmia* Grove. If conidial development in *H. empetri* is ultimately proved to be phialidic then I can see no justification for keeping the two genera separate, but on the available material no conidia were found attached to conidiophores.

H. EMPETRI (White) Petrak, *ibid.*: 210 (1942). (Fig. 83)
Rhytisma empetri White apud Berk. & Br., *Ann. Mag. Nat. Hist.* **17**: 145 (1876).
Melasmia empetri P. Magn., *Ber. dtsch. Bot. Ges.* **4**: 104 (1886).

Teleomorphosis: *Duplicaria empetri* (Fr.) Fckl, *Jb. nassau. Ver. Naturk.* **23-24**: 265 (1870).

Conidiophores 20-32.5 × 3-4 µ. Conidia 12-16.5 × 6-7 µ.

On tar spots on stems of *Empetrum nigrum*, *E. hermaphroditicum*, France, Finland, Germany, Latvia, Sweden (IMI 46099, 175265, 22354, 22355, 33170, 17239).

25. KABATIA Bub., *Öst. Bot. Z.* **54**: 8 (1904).
Colletotrichella Höhn., *Sber. Akad. Wiss. Wien* **125**: 73 (1916).
Pseudogloeosporium Jaczewski, *Fungi Imperfecti* **2**: 136 (1917).
Sp. typ.: *K. latemarensis* Bub.

Mycelium immersed, branched, septate, hyaline. *Conidiomata* separate, flat, subcuticular, unilocular, consisting of a basal wall of large-celled, hyaline, thin-walled textura angularis, and an upper wall 1 cell thick of dark brown, thick-walled cells converging towards the centre and distinctly hyphal. *Ostiole* absent, dehiscence by radial splitting between the hyphae of the upper wall from the centre to the periphery. *Conidiophores* absent. *Conidiogenous cells* holoblastic, determinate, discrete, doliiform to cylindrical, hyaline, smooth, with no proliferation, restricted to the basal wall. *Conidia* hyaline, thin-walled, smooth, rarely guttulate, aseptate or with 1-2 septa near the apices, base truncate, apex obtuse, pointed or rostrate, clavate, fusiform, cylindrical or falcate.

Conners (1959) revised species of *Kabatia* on *Lonicera* and placed the septate species in *Kabatia* and the aseptate species in *Leptothyrium* Kunze. However, *Leptothyrium* has superficial pycnothyria in which conidia are enteroblastic and phialidic, and it lacks the radiate hyphal structure in the upper wall that is so characteristic of *Kabatia.* Reusser (1964) correlated some species with *Guignardia* ascomatal states.

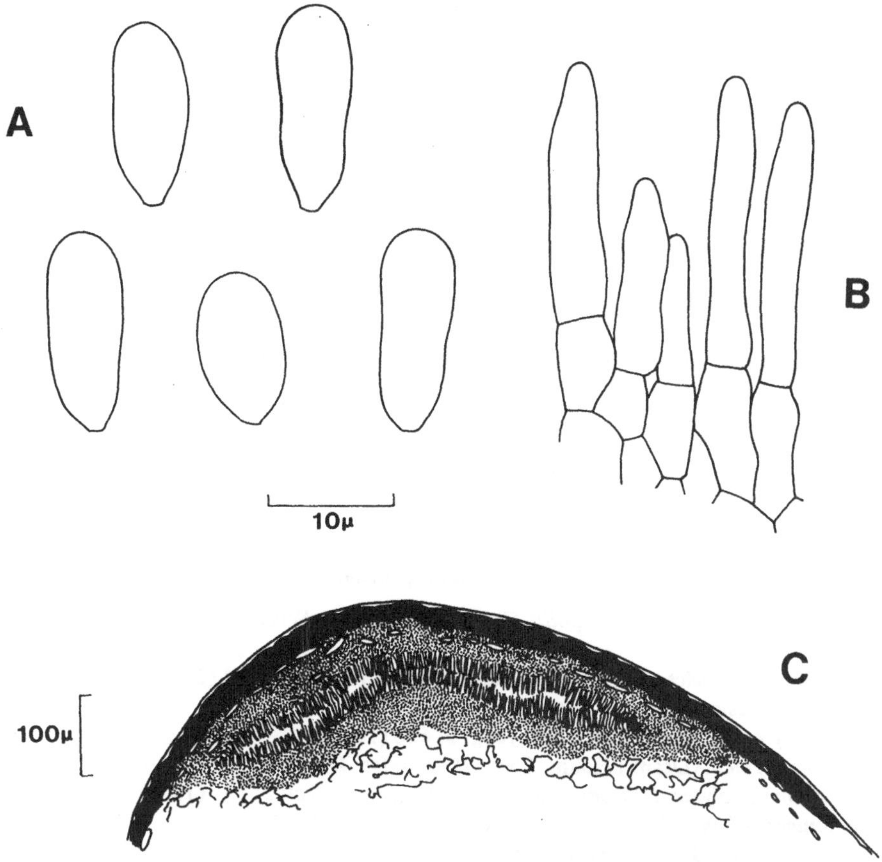

Fig. 83. *Hysterodiscula empetri*, A, conidia; B, conidiophores; C, vertical section of a conidioma.

I have found the separation of taxa proposed by Conners (1959) to be effective in identifying species and his system is followed, with the modifications of nomenclature that place the species formerly included in *Leptothyrium* and *Colletotrichella* Höhn. now in *Kabatia*. Krug (1977) has given a brief review of the genus and added *K. valpellinensis* (Trav.) Krug on *Salix reticulata.* Other taxa included in the genus but so far unexamined are *K. silenes* Lobik, *K. fragariae* Solheim and *K. cucubali* Bukhalo.

K. LONICERAE (Harkn.) Höhn., *Ber. dtsch. Bot. Ges.* **37**: 106 (1919).
Marssonia lonicerae Harkn., *Calif. Acad. Sci. Bull.* **1**: 31 (1884).
Gloeosporium lonicerae (Harkn.) Ell. & Ev., *J. Mycol.* **1**: 116 (1885).

Conidia 1 septate, 27-31.5 × 5.5-8.5 μ; apical cell 7-14.5 μ long; basal cell 13-20.5 μ long.

Causing lesions on leaves of *Lonicera conjugialis*, Calif., U.S.A.

K. LONICERAE (Harkn.) Höhn. var. AMERICANA (Ell. & Ev.) Conners, *Can. J. Bot.* **37**: 424 (1959). (Fig. 84E)
Leptothyrium periclymeni (Desm.) Sacc. var. *americanum* Ell. & Ev., *J. Mycol.* **8**: 116 (1890).

Conidia 1-septate, 20-27.5 × 7-9 μ, strongly curved; apical cell 7.5-11 μ long; basal cell 15-18 μ long. Placed by Reusser (1964) in synonymy with *K. latemarensis* Bub.

Causing leaf lesions on *Lonicera canadensis*, Canada (IMI 125755, 122990).

K. LATEMARENSIS Bub., *Öst. Bot. Z.* **54**: 8 (1904). (Fig. 84D)
K. lonicerae (Harkn.) Höhn. var. *latemarensis* (Bub.) Conners, *Can. J. Bot.* **37**: 425 (1959).

Teleomorphosis: *Guignardia latemarensis* Müller, *Sydowia* **7**: 329 (1953).

Conidia 1-septate, 32-46 × 7.5-9 μ, straight or strongly curved; apical cell 15-21 μ long; basal cell 22-28 μ long.

Causing leaf lesions on *Lonicera coerulea*, Germany, Switzerland (IMI 17252, 163035).

K. LONICERAE (Harkn.) Höhn. var. INVOLUCRATAE Conners, *Can. J. Bot.* **37**: 425 (1959). (Fig. 84F, G)

Conidia 1-septate, falcate, 40-50 × 8-12 μ; apical cell 11-16.5 μ long; basal cell 28-38 μ long.

Causing leaf lesions on *Lonicera involucrata*, Mont., U.S.A. (IMI 169925, 169926, 92766).

K. MIRABILIS Bub., *Öst. Bot. Z.* **55**: 241 (1905).

Teleomorphosis: *Guignardia mirabilis* Müller, *Phytopath. Z.* **34**: 412 (1959).

Conidia 1(-2) septate, often strongly curved, 21.5-45.5 × 6.5-13 μ; apical cell 11.5-21.5 μ long; basal cell 17-27 μ long.

Causing leaf lesions on *Lonicera* spp., Italy, Switzerland, Czechoslovakia.

K. MIRABILIS Bub. var. OBLONGIFOLIAE Conners, *Can. J. Bot.* **37**: 427 (1959).

Conidia 1-septate, 19-33 × 7-10 μ; apical cell 3-6 μ long; basal cell 16-27 μ long.

Causing leaf lesions on *Lonicera oblongifolia*, Ont., Canada, N.Y., Wisc., U.S.A.

K. PERICLYMENI (Desm.) Morelet, *Bull. Soc. Sci. nat. Archéol. Toulon Var* **31** (212): 9 (1975). (Fig. 84B, H, I)
Labrella periclymeni Desm., *Ann. Sci. nat.*, 3 ser., **10**: 358 (1848).
Leptothyrium periclymeni (Desm.) Sacc., *Syll. fung.* **3**: 626 (1884).
Colletotrichella periclymeni (Desm.) Höhn., *Sber. Akad. Wiss. Wien* **125**: 100 (1916).
Pseudogloeosporium periclymeni (Desm.) Jaczewski, *Fungi Imperfecti* **2**: 137 (1917).
Gloeosporium lonicerae J.W. Ell., *Trans. Br. mycol. Soc.* **4**: 295 (1913).
Leptothyrium pictum Berk. & Br., *Ann. Nat. Hist.* 1450 (1875).
Phyllosticta vossii Thüm., *Verh. Zool.-Bot. Ges. Wien* **28**: 48 (1878).

Conidia aseptate, straight or curved, 18-28 × 7-8 μ.

Causing leaf lesions on *Lonicera utahensis*, *L. caprifolium*, Hungary, Mont., U.S.A. (IMI 11416, 97171).

K. periclymeni (Desm.) Morelet var. **xylostei** (Pass.) comb. nov. (Fig. 84C)
Depazea xylostei Pass., in Rabenh., *Fungi europaei* 2220 (1877).
Labrella xylostei Fautr., *Rev. mycol.* **17**: 168 (1895).
Colletotrichella xylostei (Fautr.) Höhn., *Ber. dtsch. Bot. Ges.* **37**: 106 (1919).
Leptothyrium periclymeni (Desm.) Sacc. var. *xylostei* (Pass.) Conners, *Can. J. Bot.* **37**: 423 (1959).

Teleomorphosis: *Guignardia xylostei* Reusser, *Phytopath. Z.* **51**: 223 (1964).

Conidia aseptate, curved, 17-35 × 5.5-13 μ.

Causing leaf lesions on *Lonicera xylosteum*, *L. tatarica*, and *L.* sp., U.S.S.R., Austria, Germany, Roumania, Italy, Sweden (IMI 16248, 12358, 150620, 22066, 22065, 22067, 22068, 30147).

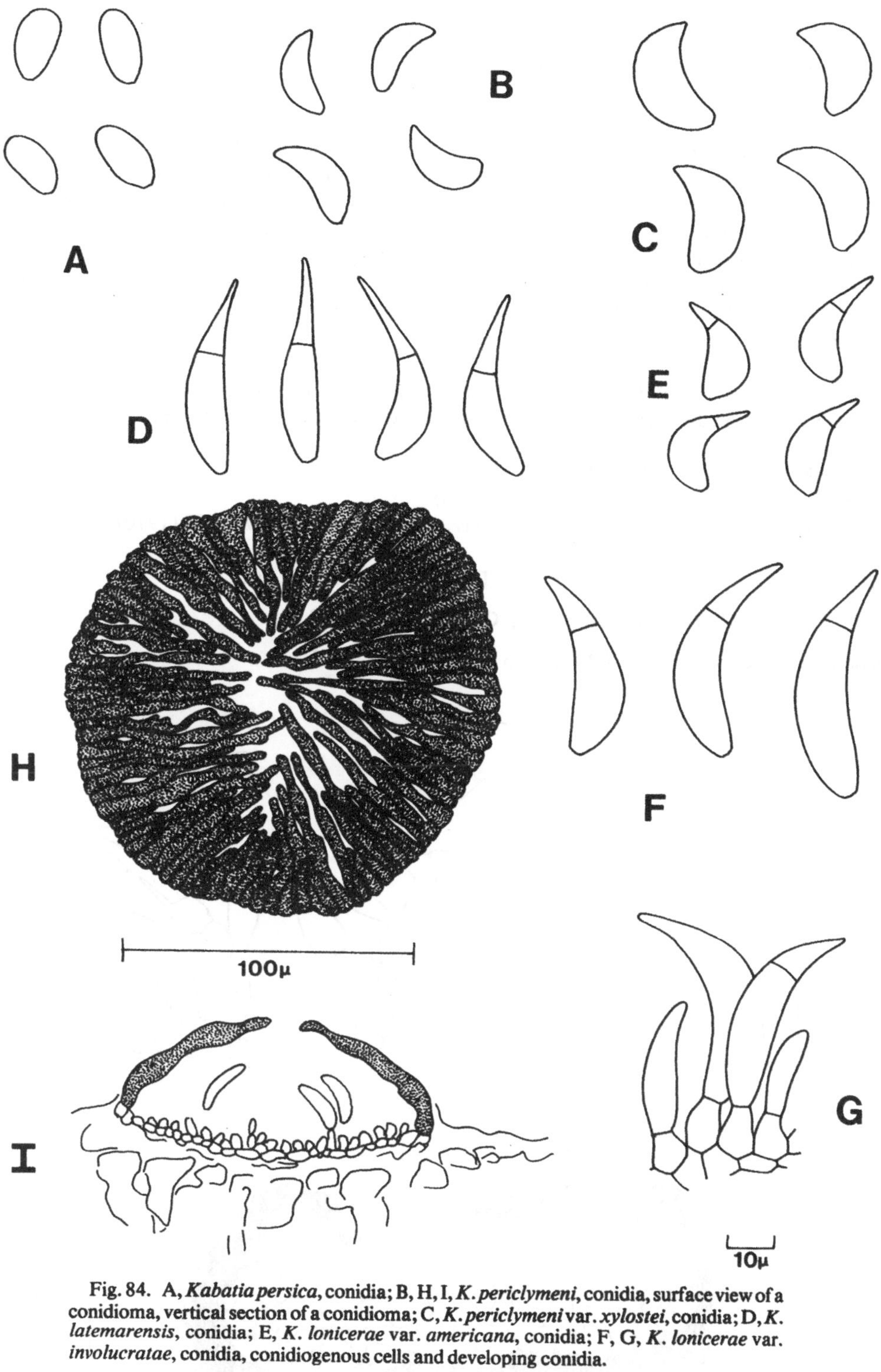

Fig. 84. A, *Kabatia persica*, conidia; B, H, I, *K. periclymeni*, conidia, surface view of a conidioma, vertical section of a conidioma; C, *K. periclymeni* var. *xylostei*, conidia; D, *K. latemarensis*, conidia; E, *K. lonicerae* var. *americana*, conidia; F, G, *K. lonicerae* var. *involucratae*, conidia, conidiogenous cells and developing conidia.

K. persica (Petrak) comb. nov. (Fig. 84A)
Colletotrichella persica Petrak, *Ann. Naturh. Mus., Wien* **50**: 483 (1939).

Conidia aseptate, straight, 14-17 × 6-8 µ.

Causing leaf lesions on *Lonicera* sp., Iran (IMI 31038, isotype).

26. CYSTOTRICHA Berk. & Br., *Ann. Mag. nat. Hist.*, 2 ser., **5**: 457 (1850).
Pseudopatella Sacc., *Syll. fung.* **3**: 688 (1884).
Sp. typ.: *C. striola* Berk. & Br.

Mycelium immersed, occasionally becoming superficial, septate, branched, pale brown. *Conidiomata* separate, rarely confluent, semi-immersed to superficial, flattened, oval to elongated, walls dark brown, exposed conidial mass reddish; walls thick, of loose or compact textura intricata or angularis over the conidiogenous region. *Ostiole* absent, dehiscence by a longitudinal fissure which widens. *Conidiophores* pale brown, irregularly branched, obscurely septate, rough-walled, produced from the inner walls of the conidioma and ramifying singly or in groups to form a network in the lumen. *Conidiogenous cells* holoblastic, integrated, determinate, lateral or intercalary, pale brown, rough-walled, producing a single conidium. *Conidia* hyaline, medianly 1 euseptate, often slightly constricted, smooth, eguttulate, cylindrical to ellipsoid, apex obtuse, base obtuse to truncate.

Cystotricha was accepted as a distinct genus by Diedicke (1912), Petch (1943) and Petrak (1963a), although it was Höhnel (1910a) who elaborated the accepted synonymy for the species. It remains monotypic.

C. STRIOLA Berk. & Br., *ibid.*: 457 (1850). (Fig. 85)
Pseudopatella tulasnei Sacc., *Syll. fung.* **3**: 688 (1884).
Cystotricha compressa Höhn. (as '(P.) Höhn.'), *Sber. Akad. Wiss. Wien* **119**: 632 (1910).

Teleomorphosis: *Durella compressa* (Pers. ex Fr.) Tul., *Sel. fung. carp.* **3**: 163 (1865).

Conidiomata up to 280 µ wide × 600 µ long. Conidia 9-11.5 × 3-3.5 µ.

On decorticated indet. wood, U.K. (IMI 200151, lectotype *C. striola*).

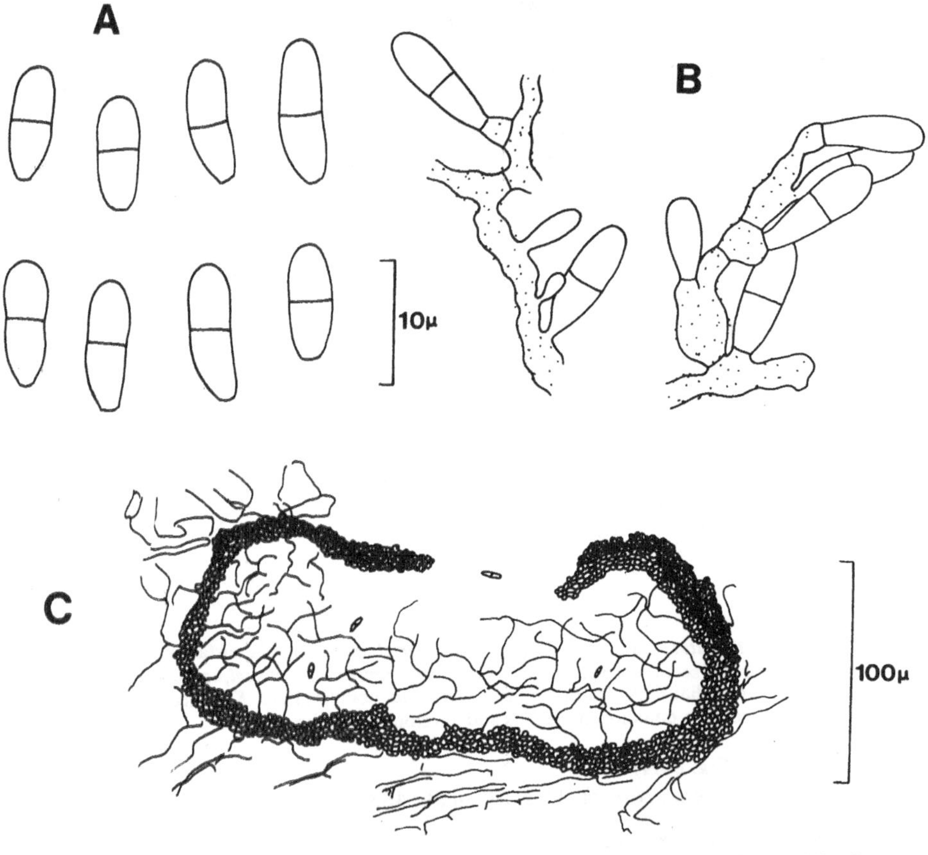

Fig. 85. *Cystotricha striola*, A, conidia; B, conidiophores and developing conidia; C, vertical section of a conidioma.

27. MICROPERELLA Höhn., *Sber. Akad. Wiss. Wien* **118**: 67 (1909).
Sp. typ.: *M. quercus* Höhn.

Mycelium immersed, branched, septate, pale brown. *Conidiomata* eustromatic, separate, only rarely confluent, irregularly pulvinate to hemispherical, often irregularly lobed and elevated at the periphery and depressed towards the centre, dark brown to black, superficial, attached to the substrate only in the centre, multilocular; locules simple, globose to irregular but undivided, situated in the upper part of the stroma; tissue mostly of thick-walled medium to pale brown textura prismatica or textura angularis with distinctly vertical parallel arrangement beneath the locules, dark brown and more thick-walled at the periphery. *Ostioles* absent, dehiscence by breakdown of tissue overlying locules. *Conidiophores* absent. *Conidiogenous cells* holoblastic, determinate, discrete, hyaline, ampulliform to lageniform, smooth, each with a single apical conidium, formed from the cells lining the locules. *Conidia* hyaline, fusiform, medianly 1 euseptate, slightly constricted at the septum, smooth, thin-walled, eguttulate, attenuated towards the truncate base and obtuse apex.

The genus is monotypic.

M. QUERCUS Höhn., *ibid.*: 67 (1909). (Fig. 86)

Conidiomata up to 1150 µ diam. × 450 µ high. Conidiogenous cells 9-15 × 5-7.5 µ. Conidia 26.5-31.5 × 5-7.5 µ.

On leaves of *Quercus glauca*, Japan (Holotype ex FH).

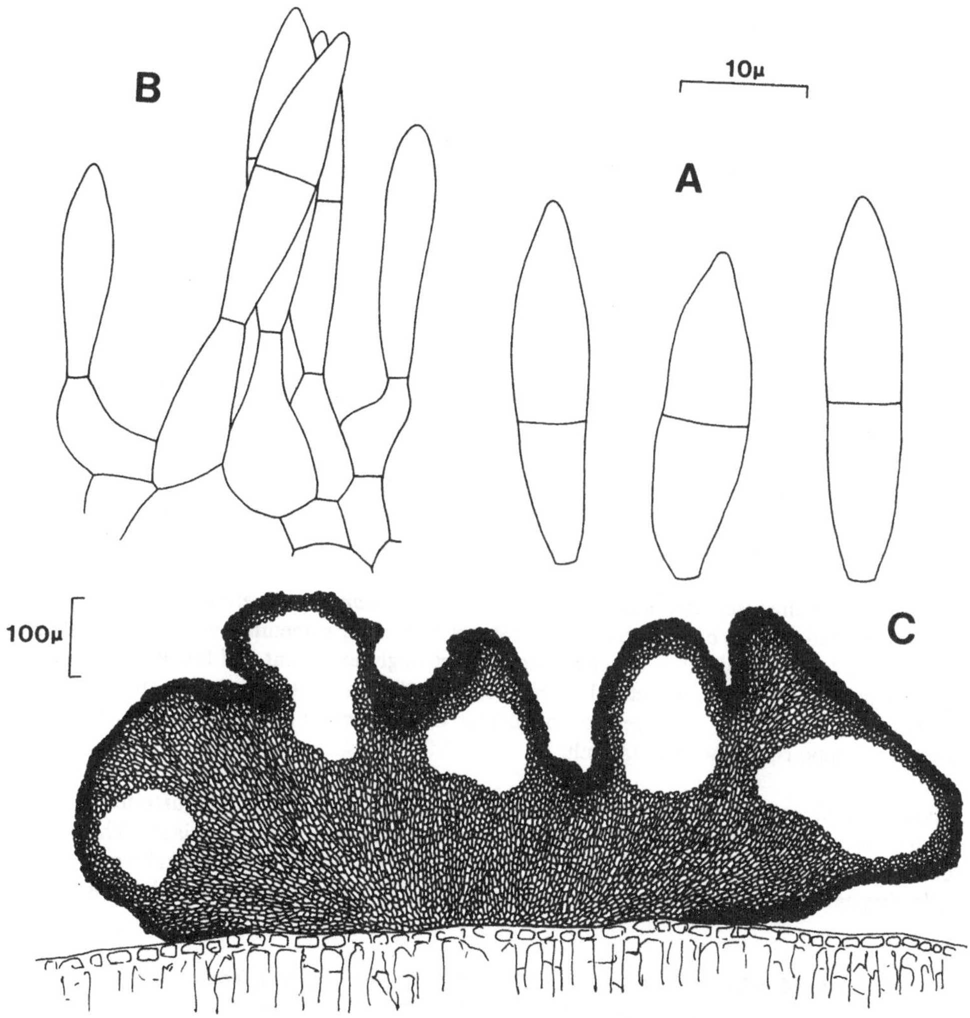

Fig. 86. *Microperella quercus*, A, conidia; B, conidiogenous cells and developing conidia; C, vertical section of a conidioma.

28. HYALODICTYUM Woronichin, *Bull. Mus. Caucase* **10**: 30 (1916).
Sp. typ.: *H. colchicum* Woronichin
Illustration: Vassiljevsky & Karakulin (1950).

Mycelium immersed, branched, septate, hyaline to pale brown. *Conidiomata* acervular, epidermal to subepidermal, separate, formed of hyaline to pale brown textura angularis. Dehiscence irregular. *Conidiophores* filiform, hyaline, septate, formed from the upper parenchyma. *Conidiogenous cells* presumably holoblastic, determinate, integrated or discrete, cylindrical, hyaline, with acrogenous conidia. *Conidia* hyaline, with 4-6 transverse and 1-3 longitudinal septa (? distoseptate and lumina reduced), straight or slightly irregular, both ends obtuse, elliptical, ovate to irregular, smooth.

Apart from the original account, later reproduced by Vassiljevsky & Karakulin (1950), this monotypic genus has not been either collected or redescribed.

H. COLCHICUM Woronichin, *ibid.*: 30 (1916).

Conidiomata 60-150 µ diam. Conidia 22-30 × 10-15 µ.

Causing lesions on leaves of *Quercus hartwissiana*.

29. THYRSIDINA Höhn., *Annls mycol.* **3**: 337 (1905).
Sp. typ.: *T. carneominiata* Höhn.

Mycelium immersed, branched, septate, hyaline, thick-walled. *Conidiomata* acervular, subperidermal, separate, irregular in shape, formed of hyaline, septate, thick-walled textura oblita. Dehiscence irregular. *Conidiophores* hyaline, branched irregularly, septate, thick-walled, gelatinous, variable in length, growing amongst the conidial mass, formed from the upper prosenchymatic cells. *Conidiogenous cells* holoblastic, indeterminate, integrated, cylindrical, hyaline, smooth, thick-walled, forming apical and/or lateral conidia. *Conidia* hyaline, smooth, euseptate, eguttulate, thick-walled, muriform and irregular in shape, septa longitudinal, transverse and oblique.

The genus is monotypic.

T. CARNEOMINIATA Höhn., *ibid.*: 337 (1905). (Fig. 87)

Conidiomata up to 800 µ diam. × 500 µ deep. Conidiophores up to 250 µ long × 4 µ wide. Conidia 15-19 × 12-15 µ.

On dead branches of *Acer ? pseudoplatanus*, Austria (IMI 202985 ex FH).

30. DOTHISTROMA Hulbary, *Bull. Ill. nat. Hist. Surv.* **21**: 235 (1941).
Sp. typ.: *D. septospora* (Dorog.) Morelet (syn. *Cytosporina septospora* Dorog., *D. pini* Hulbary).
Illustrations: Thyr & Shaw (1964); Ivory (1967); Punithalingam & Gibson (1973).

Mycelium immersed, branched, septate, pale brown to hyaline. *Conidiomata* sometimes acervular, initially subepidermal later erumpent, composed of pale brown, thin-walled textura angularis, sometimes eustromatic, multilocular and of darker brown, thick-walled tissue. Dehiscence irregular, stromata strongly erumpent and finally pulvinate. *Conidiophores* absent. *Conidiogenous cells* holoblastic, discrete, determinate, ampulliform, hyaline, smooth, non-proliferating, formed from the upper cells of stroma or from inner cells of the locular walls. *Conidia* acrogenous, hyaline, straight or curved, filiform, 1-5 euseptate, continuous, thin-walled, smooth.

The single species, which causes red band needle blight has been fully described and illustrated by Punithalingam & Gibson (1973) under the name *Scirrhia pini* Funk & Parker. Three varieties are recognised. Whether this distinction between subspecific taxa as advocated by Ivory (1967) will survive the scrutiny of additional world-wide collections is debatable. The following new combinations place the nomenclature on a rational basis although it may need to be revised in the future.

D. SEPTOSPORA (Dorog.) Morelet var. SEPTOSPORA, *Bull. Soc. Sci. nat. Archéol. Toulon Var* 177: 9 (1968).
Cytosporina septospora Dorog., *Bull. trimest. Soc. Mycol. Fr.* **27**: 106 (1911).
Septoriella septospora (Dorog.) Sacc. apud Trotter, *Syll. fung.* **25**: 480 (1931).
Dothistroma pini Hulbary, *Bull. Ill. nat. Hist. Surv.* **21**: 235 (1941).

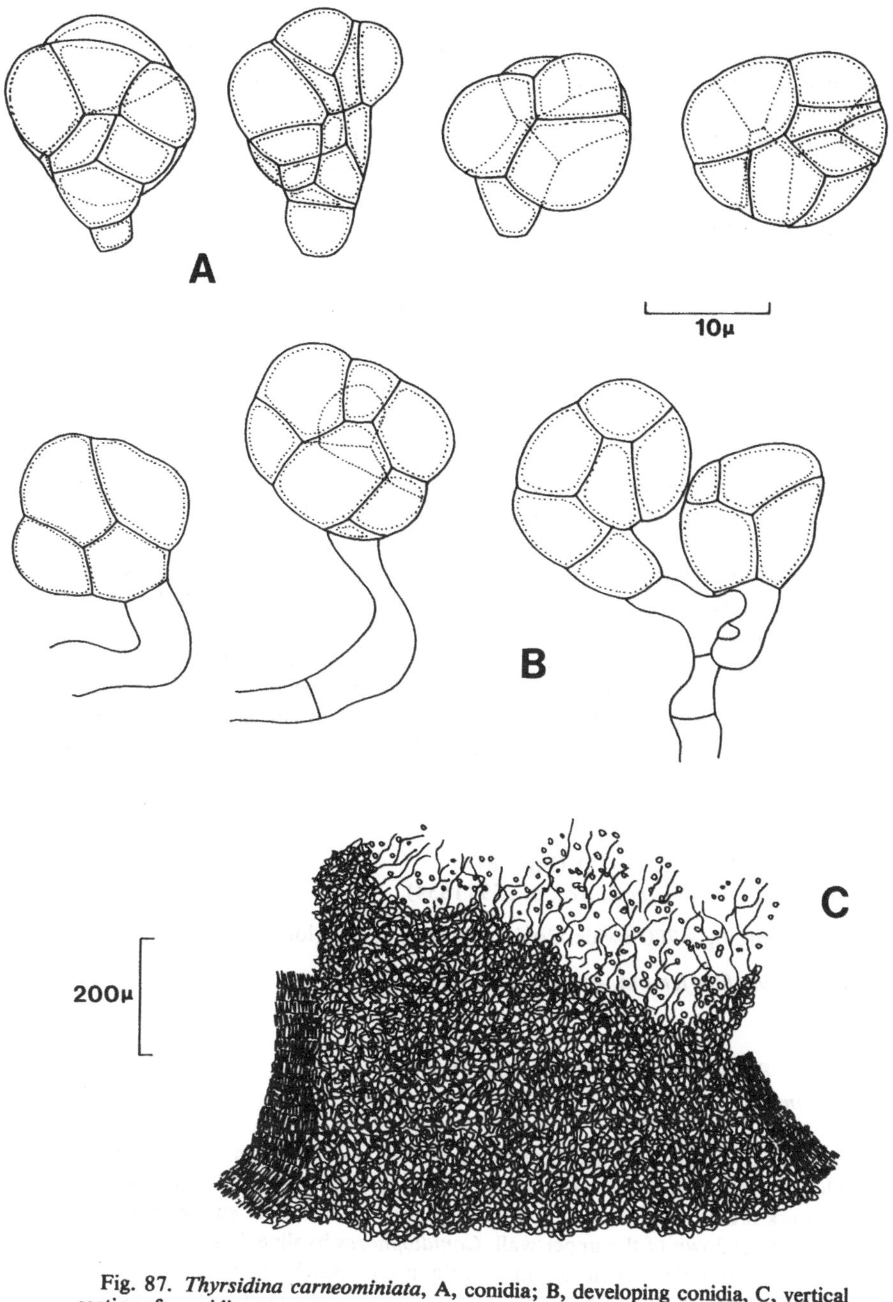

Fig. 87. *Thyrsidina carneominiata*, A, conidia; B, developing conidia, C, vertical section of a conidioma.

Teleomorphosis: *Scirrhia pini* Funk & Parker, *Can. J. Bot.* **44**: 1171 (1966).

Conidia 12.5-32.5 (22) × 2.5-4 (3) μ.

On needles of *Pinus* spp., world-wide in distribution.

D. septospora var. **lineare** (Thyr & Shaw) comb. nov.
D. pini var. *lineare* Thyr & Shaw, *Mycologia* **56**: 107 (1964).

Conidia 20-67.5 (37.5) × 2-3 (2.5) μ.

On needles of *Pinus ponderosa*, *P. nigra*, *P. caribaea*, *P. clausa*, *P. contorta*, *P. hondurensis*, *P. nigra* var. *austriaca*, *P. khasya*, *P. pithyusa*, *P. pinaster*, *P. radiata*, *P. bungeana*, *P. roxburghii*, *P. sylvestris*, *P. wallichiana*, *P.* sp., Ill., Calif., Wash., U.S.A., B.C., Canada, Brunei, India, New Zealand, Australia, Brazil, Chile, Tanzania, Kenya, S. Rhodesia, Malawi, U.S.S.R., France, Austria, G.B., Yugoslavia (IMI 91341, holotype *A. marginatum* ex PAD, 154072, 148402, 91062, 120997, 95451a, 124346, 110078, 94427, 117564, 187703, 103849, 90163, 87295b, 86986a, 87378c, 89678b, 89679a, 91124a, 91806a, 92163, 92161a, 52059, 94935a, 94934, 95012a, 84872, 99961a, 102321, 142142, 142204, 120997, 123290, 156990, 156991, 156988, 149163, 159264, 149265, 159266, 184162, 144847, 124961).

D. septospora var. **keniense** (Ivory) comb. nov.
D. pini var. *keniense* Ivory, *Trans. Br. mycol. Soc.* **50**: 294 (1967).

Conidia 15-47.5 (29) × 1.5-4 (2.5) µ.

On needles of *Pinus radiata*, Kenya (IMI 116919, holotype).

31. PLEUROTHYRIUM Bub., *Ber. dtsch. Bot. Ges.* **34**: 322 (1916).
Sp. typ.: *P. longissimum* (Lib.) Bub. (syn. *Leptostroma longissimum* Lib.).

Mycelium immersed, hyaline to pale brown, branched, septate. *Conidiomata* eustromatic, separate, linear, sometimes confluent, applanate, dark brown, subepidermal, multilocular, thick-walled; upper wall of dark brown, thick-walled textura angularis concentrated in the epidermis, lower wall of similar but hyaline, thin-walled tissue; connective tissue between locules of pale brown irregular pseudo-parenchyma. *Ostiole* absent, dehiscence by breakdown of the upper wall. *Conidiophores* hyaline, branched at the base, septate, smooth, irregular or sharply bent to one side, formed from the base of the locular cavities. *Conidiogenous cells* holoblastic, integrated, terminal or lateral immediately below transverse septa, smooth, hyaline. *Conidia* hyaline, up to 8 euseptate, continuous, filiform, curved irregularly, truncate at the base.

The genus is monotypic.

P. LONGISSIMUM (Lib.) Bub., *ibid.*: 322 (1916). (Fig. 88)
Leptostroma longissimum Lib., *Pl. Crypt. Ard.* **3**: 259 (1834).

Conidiomata up to .3 cm long × 700 µ wide. Conidiophores up to 15 µ long × 2-3 µ wide. Conidia 45-85 × 1.5-2 µ.

On dead stems of *Athyrium filix-femina*, Austria (IMI 194254 ex BPI).

32. PLACOTHYRIUM Bub., *Ber. dtsch. Bot. Ges.* **35**: 302 (1916).
Sp. typ.: *P. athyrium* Bub.

Mycelium immersed, hyaline to pale brown, branched, septate. *Conidiomata* eustromatic, separate, linear, often aggregated and then confluent, applanate, dark brown, epidermal to subepidermal, unilocular or occasionally multilocular, thick-walled; upper wall of dark brown, thick-walled textura angularis, basal wall of similar but hyaline, thin-walled tissue. *Ostiole* absent, dehiscence by breakdown of the upper wall. *Conidiophores* hyaline, branched at the base and above, up to 5 septate, smooth, straight or irregular, enclosed in a gelatinous sheath, formed all around the locular cavity. *Conidiogenous cells* holoblastic, integrated, terminal or lateral immediately below transverse septa, smooth, hyaline, tapered abruptly to an apiculate conidiogenous locus. *Conidia* hyaline, 0-2 euseptate, continuous, filiform, straight or curved, tapered at each end.

Bubák (1916) made no mention of the gelatinous covering to the conidiophores. After conidial formation the sheath appears with a V-shaped depression at the apex of the conidiophore. Conidia lack any gelatinous appendage and no conidiogenous cells have been observed developing sympodially or percurrently. The genus is monotypic.

P. ATHYRIUM Bub., *ibid.*: 302 (1916). (Fig. 89)

Conidiomata 250-500 µ long × 150 µ wide. Conidiophores up to 25 µ long × 2.5-3 µ wide. Conidia 29-34 × 1 µ.

On dead stems of *Athyrium filix-femina*, Germany (IMI 194252 ex BPI).

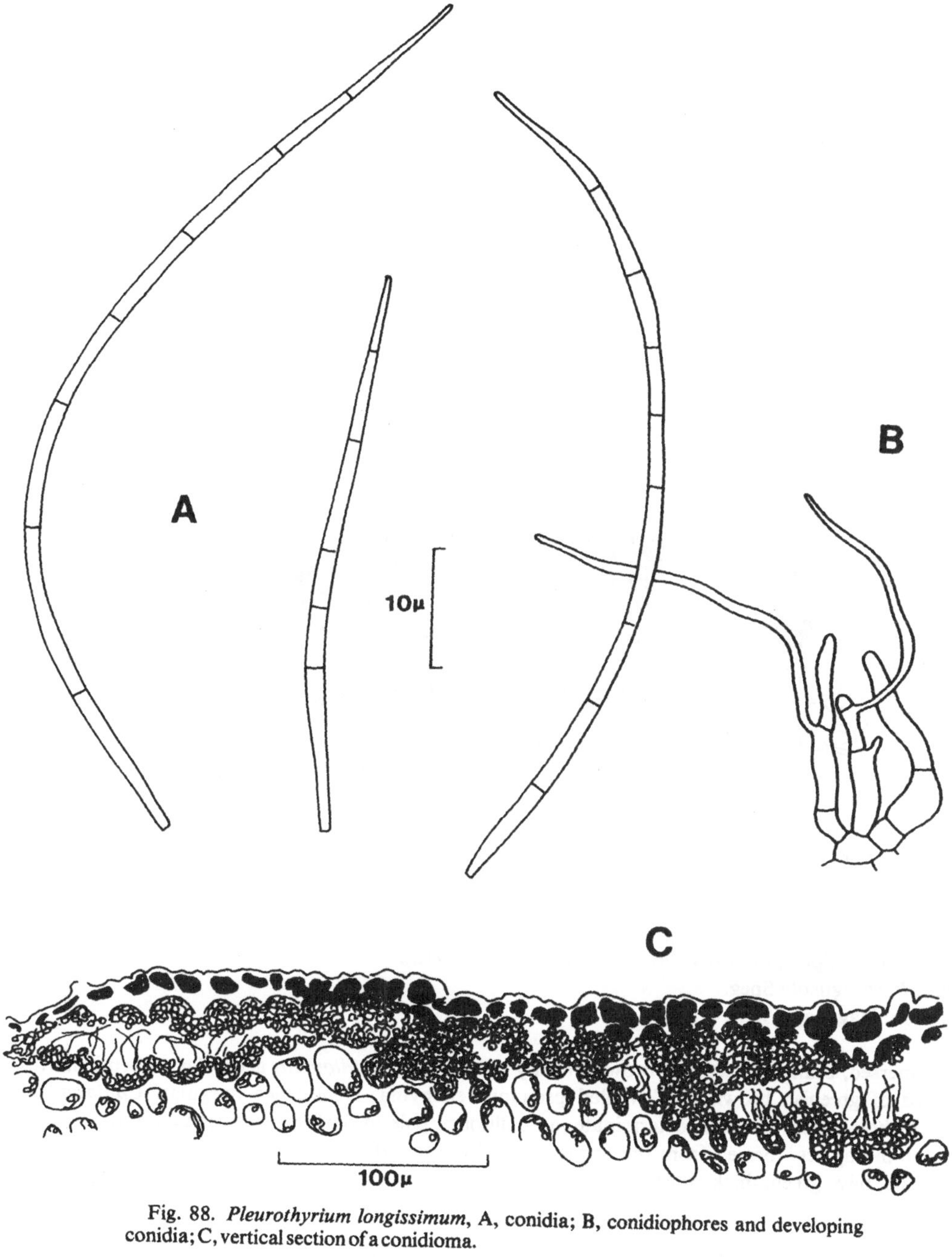

Fig. 88. *Pleurothyrium longissimum*, A, conidia; B, conidiophores and developing conidia; C, vertical section of a conidioma.

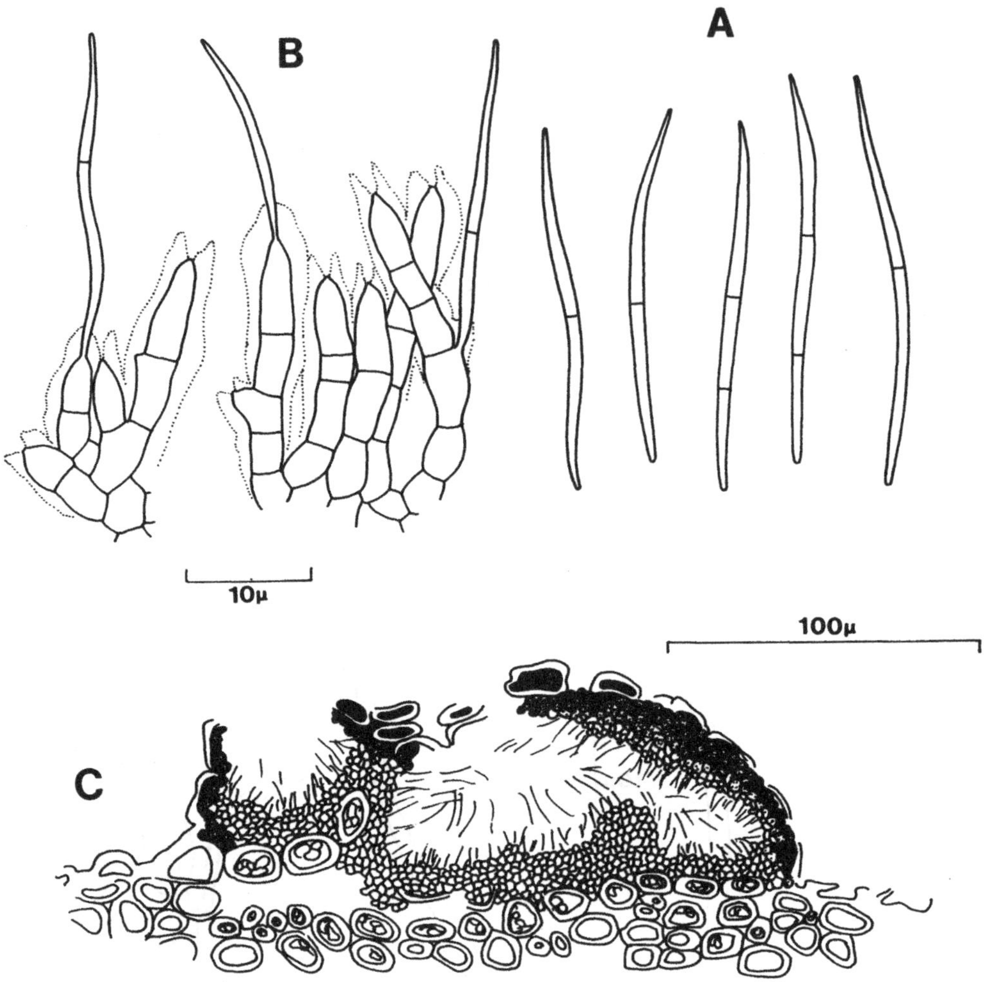

Fig. 89. *Placothyrium athyrium*, A, conidia; B, conidiophores and developing conidia; C, vertical section of a conidioma.

33. PHAEOLABRELLA Speg., *An. Mus. Nac. B. Aires* **23**: 117 (1912).
Sp. typ.: *P. eryngiicola* Speg.
Illustration: Nag Raj (1974).

Mycelium immersed, branched, septate, brown, mostly epidermal. *Conidiomata* acervular, subcuticular, separate or aggregated, formed of greenish brown, thick-walled pseudoparenchyma. Dehiscence irregular. *Conidiophores* absent. *Conidiogenous cells* holoblastic, discrete, determinate, ampulliform, pale to dark brown, verrucose, paler, thinner-walled and smoother near the apices, formed from the upper cells of the pseudoparenchyma. *Conidia* cylindrical, straight or slightly bent, aseptate, olive green, thick-walled, verrucose, apex obtuse or flattened; base ± truncate, thinner-walled, paler and smoother.

Petrak & Sydow (1935) maintained this as a distinct genus. It is monotypic.

P. ERYNGIICOLA Speg., *ibid.*: 17 (1912).

Conidiomata up to 300 μ diam. Conidiogenous cells 4.5-6 × 3-6 μ. Conidia 16.5-29 × 3-4.5 μ.

On dead stems of *Eryngium pandanifolium*, Argentina.

34. SCYPHOSPORA Kantschaveli, *Bol. Rast.* 17: 87 (1928).
Sp. typ.: *S. phyllostachydis* Kantschaveli
Illustration: Nag Raj (1974).

Mycelium immersed, branched, septate, pale to dark brown. *Conidiomata* acervular, subepidermal, dark brown, separate or gregarious, pulvinate, composed of subhyaline to pale brown textura angularis which is darker in the conidiogenous region and develops into textura prismatica. Dehiscence irregular. *Conidiophores* septate, rarely branched, consisting of an ampulliform pale basal cell and an upper cylindrical brown verrucose cell, formed from ± parallel cells in the upper pseudoparenchyma. *Conidiogenous cells* holoblastic, determinate, integrated, cylindrical, thick-walled, brown. *Conidia* acrogenous, solitary, brown, smooth, thick-walled, base truncate, apex broadly obtuse, aseptate, turbinate, with a periclinal slit situated near the apex.

Hino & Katumoto (1961) briefly redescribed and illustrated the species from Japan and Nag Raj (1974) provided more accurate details of conidiogenesis. The genus is monotypic.

S. PHYLLOSTACHYDIS Kantschaveli, *ibid.*: 87 (1928).

Conidiomata .8-1.2 mm long × 400-600 µ wide × 200-250 µ deep. Conidiogenous cells up to 35 µ long × 3-4 µ wide. Conidia 16.5-23.5 × 13-18.5 µ.

On stems of *Phyllostachys nigra* var. *henosis, P.* sp. and *Shibataea kumasaca,* Japan, U.S.S.R. (IMI 241647).

35. HARKNESSIA Cke apud Cke & Harkn., *Grevillea* **9**: 85 (1881).
Mastigonetron Kleb., *Myc. Centralbl.* **4**: 17 (1914).
Caudosporella Höhn., *Sber. Akad. Wiss. Wien* **123**: 135 (1914).
Sp. typ.: *H. eucalypti* Cke

Mycelium immersed, branched, septate, hyaline to pale brown. *Conidiomata* eustromatic, immersed, globose, unilocular, more rarely convoluted and multilocular, pale brown; walls composed of thin-walled, pale brown to hyaline, small-celled textura angularis. *Ostiole* central, circular, wide, unspecialised, furfuraceous. *Conidiogenous cells* holoblastic, discrete, very rarely integrated on 1-2 septate supporting hyphae (conidiophores), determinate, unbranched, filiform, swollen at the base, formed from the wall lining the locule, often carried away by conidia at secession. *Conidia* acrogenous, globose, subglobose, lenticular, gibbose, oblique-napiform, oval, dolliform or reniform, brown, smooth, sometimes longitudinally striate, guttulate, with a cellular unbranched basal appendage resulting from the persistent conidiogenous cell, some species with an apical, subapical or lateral hyaline appendage.

Sutton (1971b) monographed the genus, accepting 16 species. Subsequently *H. fumaginea* Sutton & Alcorn apud Sutton (1975b), *H. renispora* Swart (1972), *H. gharsei* Golatkar (1976) and *H. araucariae* Sutton & Hodges (1977) have been added. *H. gharsei* has not yet been verified as an accepted species.

KEY

1. Conidia with a filiform apical appendage 2
1. Conidia with a small indistinct apical appendage 3
1. Apical appendage absent 4
 2. Conidial body 14.5-17.5 × 7.5-9 µ .... *H. caudata*
 2. Conidial body 19-24 × 10-12.5 µ .... *H. americana*
3. Conidia lenticular, basal appendage 19-29 µ long .... *H. deightonii*
3. Conidia oblique napiform to gibbose, dark brown, basal appendage very short .... *H. insueta*
3. Conidia ± gibbose, pale brown, basal appendage 2 µ long .... *H. gunnerae*
 4. Conidia less than 15 µ long 5
 4. Conidia more than 15 µ long 8
5. Conidia globose 6
5. Conidia gibbose 7
 6. Conidia 14-16 × 12-13.5 µ .... *H. globosa*
 6. Conidia 9.5-11 × 8-9 µ .... *H. hawaiiensis*

7. Conidia 12.5-15 × 8.5-9.5 µ .... *H. fumaginea*
7. Conidia 11-13 × 6.5-7.5 µ .... *H. shearii*
   8. Conidia formed in a persistent mucilaginous sheath .... *H. podocarpi*
   8. Mucilaginous sheath evanescent, absent at maturity .... *H. eucalypti*
   8. Mucilaginous sheath absent 9
9. Basal appendage more than 40 µ long .... *H. uromycoides*
9. Basal appendages 15-40 µ long 10
9. Basal appendage usually less than 15 µ long 11
   10. Conidia bluntly apiculate, 19.5-22 × 9-10 µ .... *H. thujina*
   10. Conidia bluntly apiculate, 25-31 × 9.5-13 µ .... *H. ventricosa*
   10. Conidia oval .... *H. rhoina*
11. Conidia reniform .... *H. renispora*
11. Conidia broadly oval, 20-25 × 11.5-12 µ .... *H. arctostaphyli*
11. Conidia oval, 17.5-21.5 × 8-9 µ .... *H. fuegiana*
11. Conidia oval, 18.5-23 × 9-10.5 µ, ± bluntly apiculate, microconidia present .... *H. antarctica*
11. Conidia globose, 16-19 × 14.5-20 µ .... *H. araucariae*

H. AMERICANA (Mont.) Sutton, *Mycol. Pap.* 123: 4 (1971). (Fig. 90A)

Conidia 19-24 × 10-12.5 µ; apical appendage 16-27 µ long; basal appendage 24-40 µ long.

On twigs of *Liquidambar* spp., S. Carol., Fla, N.J., U.S.A., Chile (IMI 151335, 146766, 150958, 150957 ex K, NY, PC).

H. ANTARCTICA Speg., *Boln. Acad. nac. Cienc. Cordoba* **11**: 291 (1881). (Fig. 90D)

Conidia 18.5-23 × 9-10.5 µ; basal appendage 5-15 µ long. Microconidia 2.5-5 × 2 µ.

On dead leaves of *Fagus betuloides*, Argentina (IMI 11818, holotype ex LPS).

H. ARAUCARIAE Sutton & Hodges, *Mycologia* **69**: 831 (1977). (Figs. 90B, 94C)

Conidia globose, 16-19 × 14.5-20 µ; basal appendage 10-17 µ long.

On dead leaves of *Araucaria angustifolia*, Brazil (IMI 192520, holotype, 212205).

H. ARCTOSTAPHYLI Cke & Harkn., *Grevillea* **13**: 111 (1885). (Fig. 90C)

Conidia 20-23 × 11.5-12 µ; basal appendage 7-18 µ long.

On dead leaves of *Arctostaphylos pumila*, *A. myrtifolia*, *A. viscida*, *A.* sp., *Arbutus menziesii*, Calif., U.S.A. (IMI 146767, 146768, 146769, 146770, 146771 ex K, 142228, 142229).

H. CAUDATA Ell. & Ev., *J. Mycol.* **1**: 92 (1885). (Fig. 90E)

Conidia 14.5-17.5 × 7.5-9 µ; basal appendage 17-25 µ long; apical appendage 24-39 µ long.

On leaves, twigs and petioles of *Quercus alba*, *Q. coccinea*, S. Carol., U.S.A. (IMI 146763, 146764, 146765 ex K).

H. DEIGHTONII Sutton, *Mycol. Pap.* 123: 11 (1971). (Fig. 91A)

Conidia 9.5-11 × 6-9 µ; apical appendage 1-2 µ long; basal appendage 19-29 µ long.

On dead leaves of *Klainedoxa gabonensis*, Sierra Leone (IMI 21817, 21818).

H. EUCALYPTI Cke apud Cke & Harkn., *Grevillea* **9**: 85 (1881). (Fig. 91B)

Conidia 18-25 × 11-14.5 µ; basal appendage 2-12 µ long.

On dead leaves of *Eucalyptus globulus*, *E.* sp., S. Carol., U.S.A., New Zealand, Australia (IMI 146779, 146778, 146782, 146774a ex K, 190685c, 189775a).

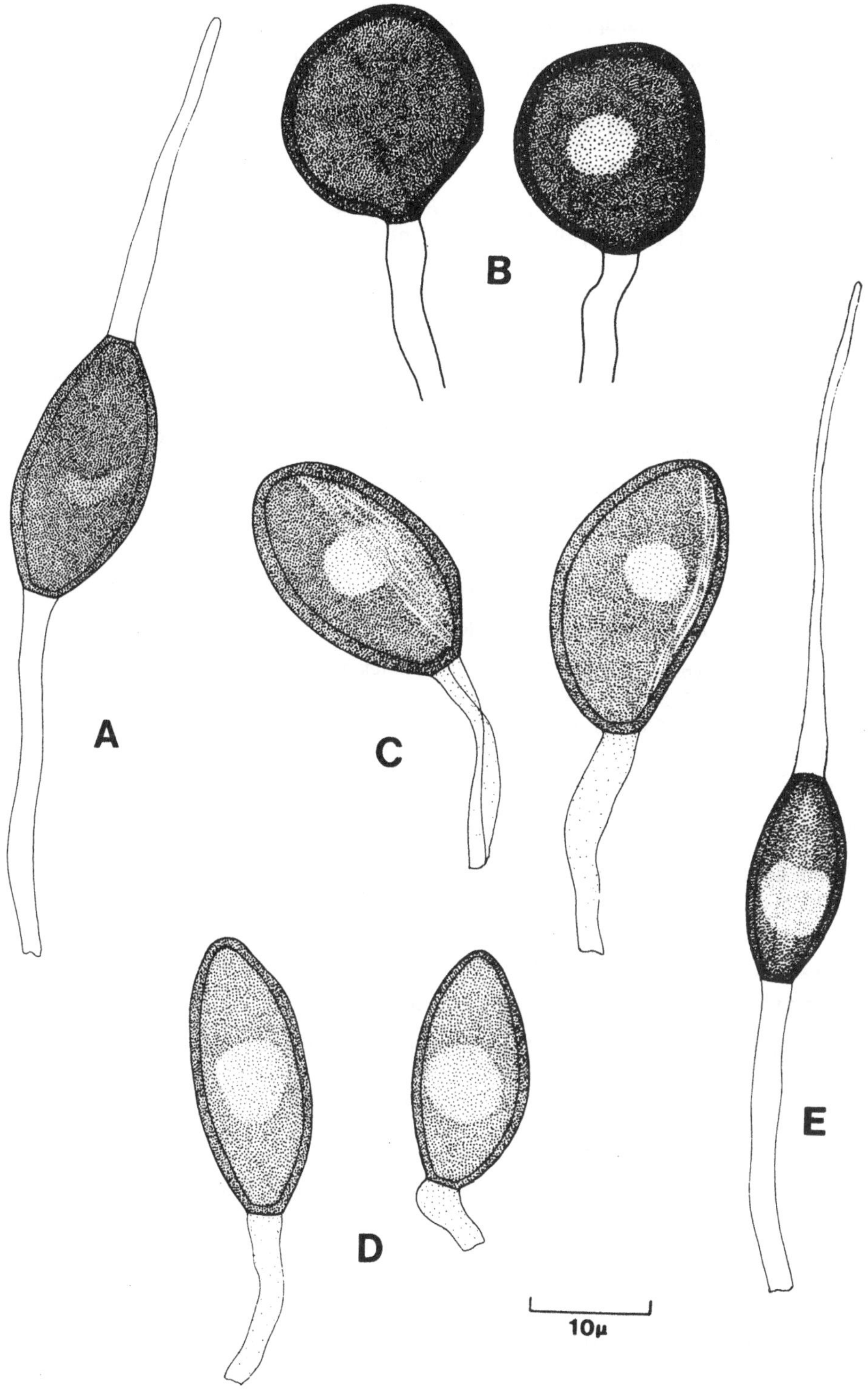

Fig. 90. A, *Harknessia americana*, conidia; B, *H. araucariae*, conidia; C, *H. arctostaphyli*, conidia; D, *H. antarctica*, conidia; E, *H. caudata*, conidia.

H. FUEGIANA Speg., *Boln. Acad. nac. Cienc. Cordoba* **11**: 291 (1881). (Fig. 91E)

Conidia 17.5-21.5 × 8-9 µ; basal appendage 8-17 µ long.

On leaf lesions of *Maytenus magellanica*, Argentina (IMI 148591 ex LPS).

H. FUMAGINEA Sutton & Alcorn apud Sutton, *Nova Hedwigia* **26**: 14 (1975).

Conidia 12.5-15 × 8.5-9.5 µ; basal appendage 2-7.5 µ long.

On dead leaves of *Eucalyptus grandis*, *E. pilularis*, *E. saligna*, *E.* sp., Brazil, Australia (IMI 196488g, 167044, 197557b, 184647d, 190855f, 196481i, 196584b, 196482b).

H. GLOBOSA Sutton, *Mycol. Pap.* 123: 15 (1971). (Fig. 91D)

Conidia 14-16 × 12-13 µ; basal appendage 2-7 µ long.

On dead leaves of *Eucalyptus globulus*, *E. robusta*, New Zealand, Brazil (IMI 21815, 192513b).

H. GUNNERAE Stev. & Young, *Bull. Bernice P. Bishop Mus.* **19**: 136 (1925). (Fig. 91C)

Conidia 11-13.5 × 6-7 µ; apical appendage 1 µ long; basal appendage 2 µ long.

On dead leaf of *Gunnera petaloidea*, Hawaii, U.S.A. (IMI 148756 ex BISH).

H. HAWAIIENSIS Stev. & Young, *Bull. Bernice P. Bishop Mus.* **19**: 136 (1925). (Fig. 91F)

Conidia 9.5-11 × 8-9 µ; basal appendage 2-7 µ long.

On dead leaves of *Eucalyptus robusta*, *E. grandis*, *E. paniculata*, *E. tereticornis*, *E.* spp., Zambia, Brazil, Fla and Hawaii, U.S.A. (IMI 148757a, 93924, 173093a, 181532d, 186979i, 202857b, 202856a, 202855a, 186981e, 202853a, 202854, 202851, 186878o, 186976n).

H. INSUETA Sutton, *Mycol. Pap.* 123: 20 (1971). (Fig. 92A)

Conidia 10-11.5 × 8-9 µ; apical appendage 1.5 µ long.

On dead leaf lesions of *Eucalyptus robusta*, *E. grandis*, *E. camaldunensis*, *E.* spp., Brazil, Mauritius, U.S.A. (IMI 192516b, 174680d, 22697, 192512b, 184647b, 196481c, 202857f).

H. PODOCARPI Lindquist & Sutton apud Sutton, *Mycol. Pap.* 123: 20 (1971). (Fig. 92B)

Conidia 17-21 × 10.5-13 µ; basal appendage 7-12.5 µ long.

On dead leaves of *Podocarpus parlatorei*, Argentina (IMI 148596).

H. RENISPORA Swart, *Trans. Br. mycol. Soc.* **59**: 309 (1972). (Fig. 92C)

Conidia 14-18 × 10-12 µ; basal appendage 2-6 µ.

On dead leaves of *Melaleuca lanceolata*, indet., Australia (IMI 163449, 168732).

H. RHOINA Ell. & Ev., *J. Mycol.* **10**: 168 (1904). (Fig. 92E)

Conidia 19.5-21.5 × 9-10.5 µ; basal appendage 27-38 µ long.

On dead leaves of *Rhus integrifolia*, Calif., U.S.A. (IMI 148814 ex NY).

H. SHEARII Petrak, *Sydowia* **7**: 404 (1953). (Fig. 92F)

Conidia 11-13 × 6-7.5 µ; basal appendage 7-13 µ long.

On dead leaves of *Metrosideros* sp., Hawaii, U.S.A. (IMI 150959).

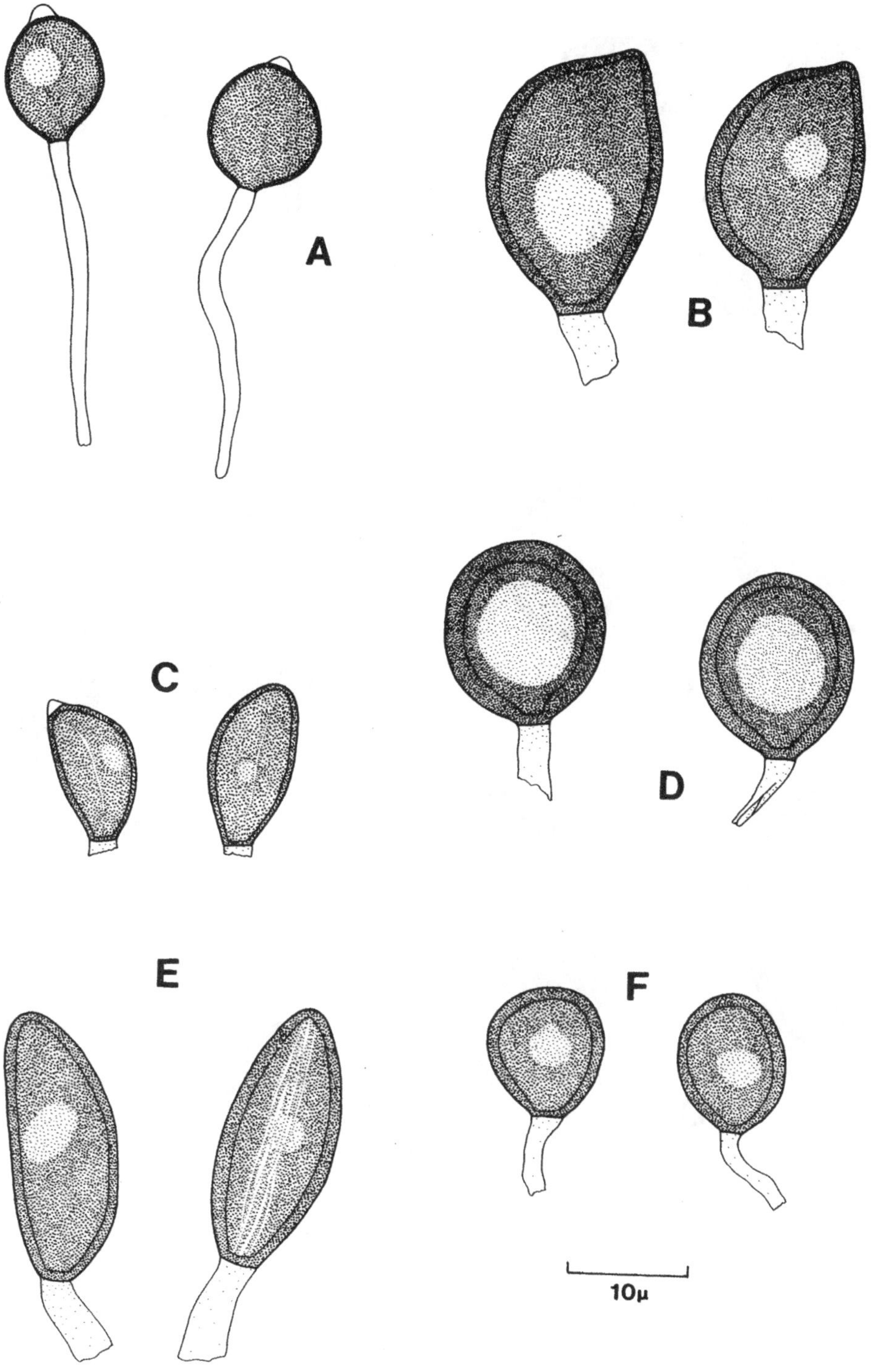

Fig. 91. A, *Harknessia deightonii*, conidia; B, *H. eucalypti*, conidia; C, *H. gunnerae*, conidia; D, *H. globosa*, conidia; E, *H. fuegiana*, conidia; F, *H. hawaiiensis*, conidia.

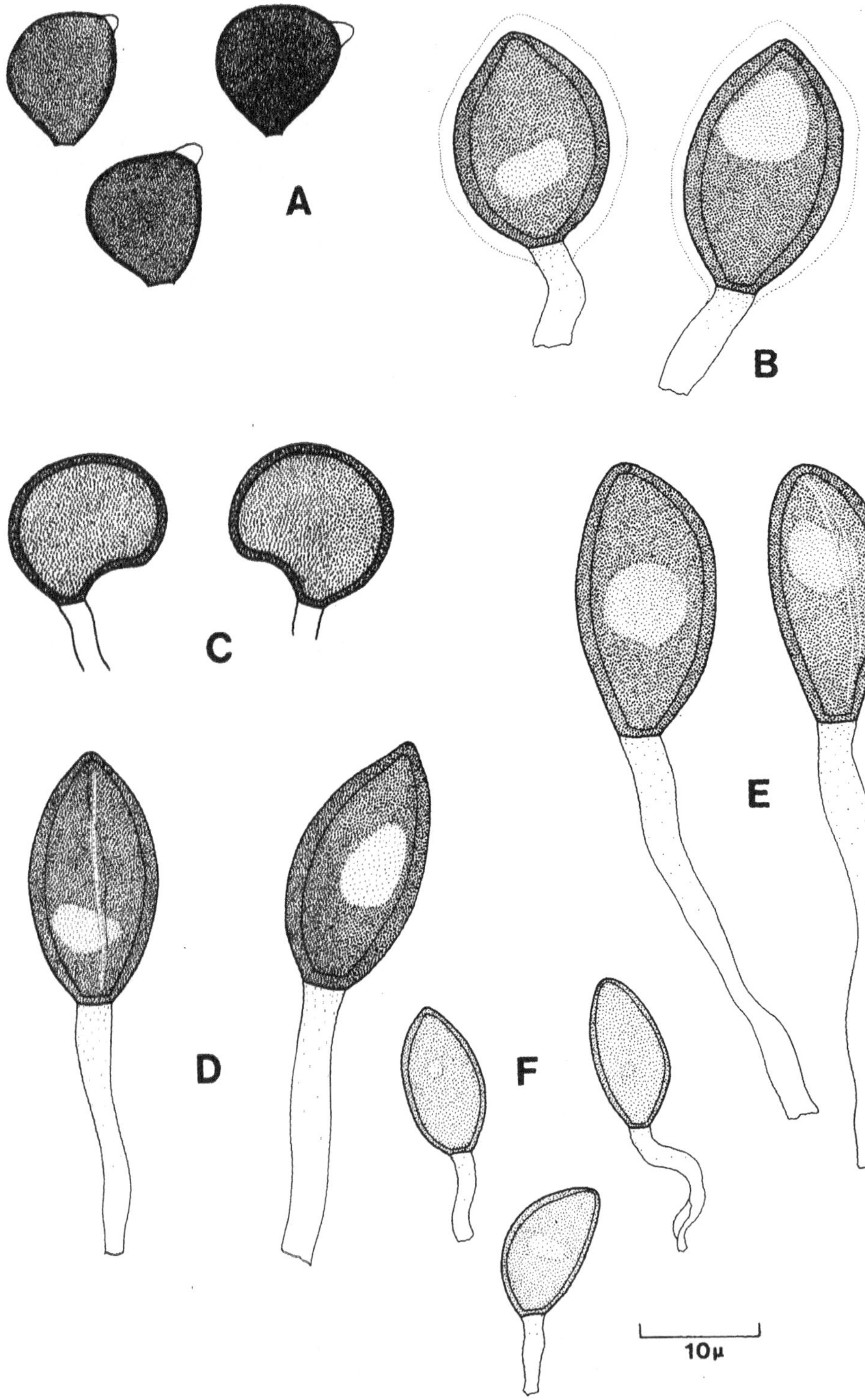

Fig. 92.1 A, *Harknessia insueta*, conidia; B, *H. podocarpi*, conidia; C, *H. renispora*, conidia; D, *H. thujina*, conidia; E, *H. rhoina*, conidia; F, *H. shearii*, conidia.

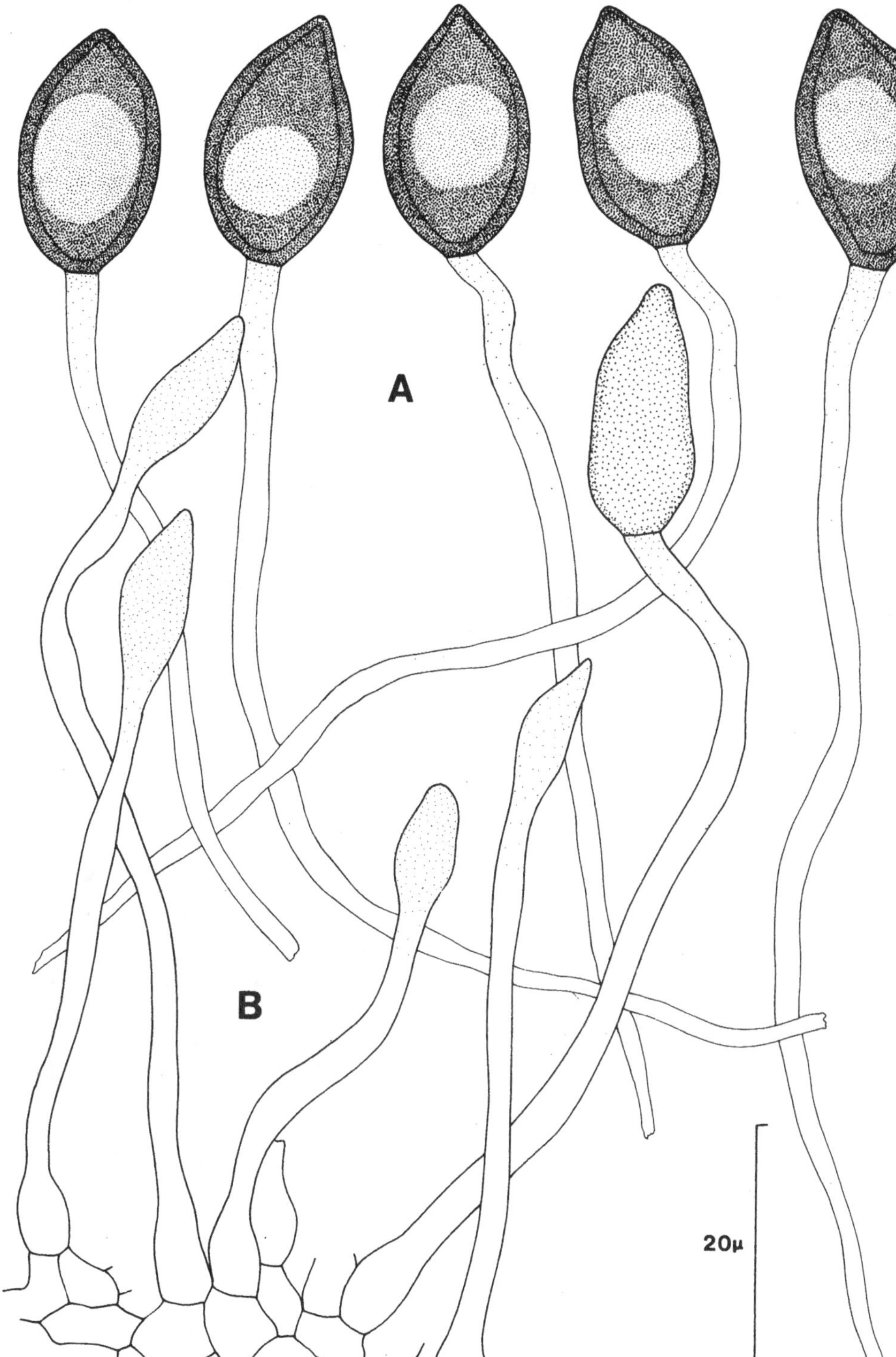

Fig. 93. *Harknessia uromycoides*, A, conidia; B, conidiogenous cells and developing conidia.

H. THUJINA Ell. & Ev., *Proc. Acad. nat. Sci. Philad. 1893*: 459 (1893). (Fig. 92D)

Conidia 19.5-22 × 9-10 µ; basal appendage 20-31 µ long.

On bark of *Thuja occidentalis*, *Cupressus thujoides*, N.J., U.S.A., Ont., Canada (IMI 146762, 21816 ex K).

H. UROMYCOIDES (Speg.) Speg., *An. Soc. Cient. Argent.* **13**: 21 (1882). (Fig. 93)

Conidia 18-26 × 9-13 µ; basal appendage 30-90 µ long.

On dead parts of *Eucalyptus* spp., Argentina, Portugal, Australia, U.S.A., Peru, Bolivia, S. Africa, represented by 22 collections in herb. IMI.

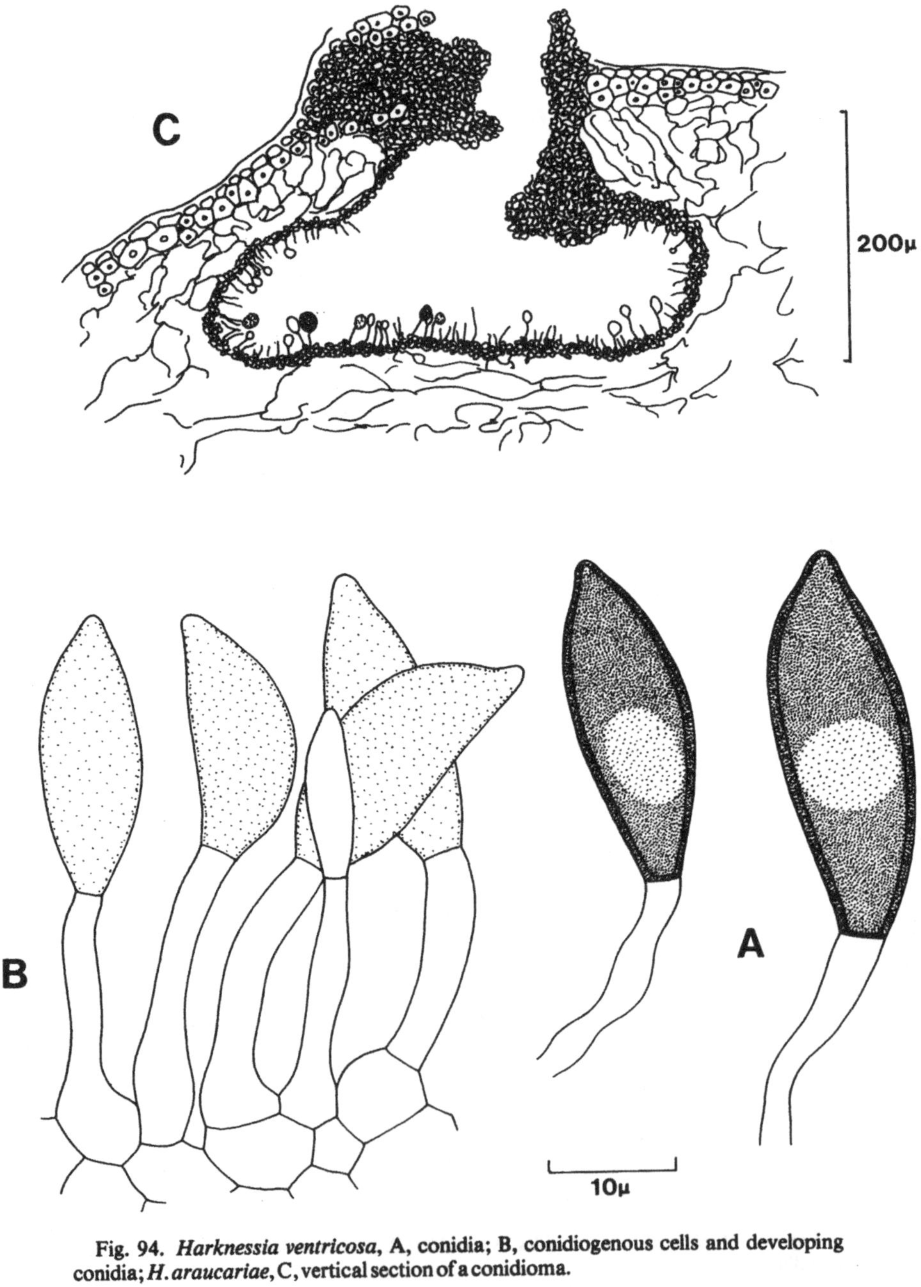

Fig. 94. *Harknessia ventricosa*, A, conidia; B, conidiogenous cells and developing conidia; *H. araucariae*, C, vertical section of a conidioma.

H. ventricosa Sutton & Hodges n. sp. (Fig. 94A, B)

Fructificationes abundantes, separatae, discretae, immersae, subepidermales, globosae vel subglobosae, uniloculares, erumpentes et punctatae, usque ad 300 µ diam, omnes massae nigrae conidiorum semper circumcinctae margine distincto furfuraceo; parietes basales et laterales stromatum 5-7 cellulis crassis, ex cellulis subhyalinis vel pallide brunneis pseudoparenchymaticis compositi, qui sunt hyalinis versus partem cellulorum conidiogenorum; ostioli distincti absentes, sed pseudoparenchyma pallide brunneum texturis incohaerentibus marginem furfuraceum facit. Cellulae conidiogenae holoblasticae, determinatae, discretae ex cellulis interioribus parietis stromatibus formatae, lageniformes vel cylindricae, ad bases 5-7 µ crassae et apices versus 2-3.5 µ crassae, hyalinae, laeves, non ramosae, 20-25 µ longae. Conidia solitaria, aseptata, ventricosa, atrobrunnea, apicem versus apiculata, cum guttula centrali globosa, laevia, 25-31 × 9.5-13.5 µ, appendice hyalina non ramosa basali ex cellulis conidiogenis persistentibus oriunda, 10-19 µ longa. In foliis emortius *Eucalypti globuli*, Hawaii, U.S.A., C.S. Hodges Jr, 27 Aug. 1976, IMI 209222i, holotypus.

This species is chiefly distinguished from similar taxa by the long ventricose conidia. They most closely resemble *H. thujina* but are longer and wider and with a shorter appendage. Of the species known on *Eucalyptus* it is intermediate between *H. uromycoides* which has a much longer appendage and *H. eucalypti* where the appendage is very short or missing entirely. The conidia in *H. ventricosa* are longer than in either *H. uromycoides* or *H. eucalypti*.

36. ENDOMELANCONIUM Petrak, *Annls mycol.* **38**: 206 (1940).
Sp. typ.: *E. pini* (Cda) Petrak (syn. *Melanconium pini* Cda).

Mycelium immersed, branched, septate, hyaline to pale brown. *Conidiomata* eustromatic, immersed, peridermal to subperidermal, separate, irregularly multilocular; walls composed of small-celled, pale brown, thin-walled textura angularis, becoming hyaline towards the conidiogenous region. Dehiscence irregular. *Conidiophores* absent. *Conidiogenous cells* holoblastic, determinate, discrete, cylindrical, tapered markedly or gradually towards the apices, hyaline, smooth, thin-walled, formed from the walls of the locules. *Conidia* aseptate, pyriform to limoniform, dark brown, thick-walled, smooth, base often protruding and papillate, often with a central guttule and a longitudinal striation.

The genus is monotypic.

E. PINI (Cda) Petrak, *ibid.*: 206 (1940). (Fig. 95)
Melanconium pini Cda, *Icon. fungorum* **1**: 3 (1837).

Conidiomata up to 1000 µ diam. Conidiogenous cells 7-16 × 3-7 µ, tapering to 1.5 µ wide. Conidia 11.5-13.5 × 6.5-7.5 µ.

On dead trunks and branches of *Abies pectinata, A. alba, A. excelsa*, Switzerland, Austria, Germany, Czechoslovakia (IMI 200665, 200659 ex K, 12068).

37. CYMBOTHYRIUM Petrak, *Sydowia* **1**: 148 (1947).
Sp. typ.: *C. sudans* Petrak

Mycelium immersed, branched, septate, hyaline to pale brown. *Conidiomata* eustromatic, immersed, globose to applanate, unilocular, more rarely convoluted, pale brown; walls composed of thin-walled, pale brown to hyaline, small-celled textura angularis; clypeus of small-celled, dark brown, rather loose pseudoparenchyma. *Ostiole* absent, dehiscence by irregular rupture of overlying host tissues. *Conidiophores* absent. *Conidiogenous cells* holoblastic, discrete, determinate, filiform, tapered gradually towards the apices, hyaline, smooth, formed from the inner cells of the locular wall. *Conidia* acrogenous, cymbiform, thick-walled, dark brown, smooth, aseptate, initially truncate at the base.

Cymbothyrium is distinguished from *Harknessia* only by the clypeate stromata and non persistent conidiogenous cells. It is monotypic.

C. SUDANS Petrak, *ibid.*: 148 (1947). (Fig. 96)

Stromata 250-500 µ diam. Conidiogenous cells 14-19 × 3-4 µ. Conidia 12.5-14 × 7-9 µ.

On dead stems of *Smilax aspera*, France (IMI 201888, holotype ex ZT).

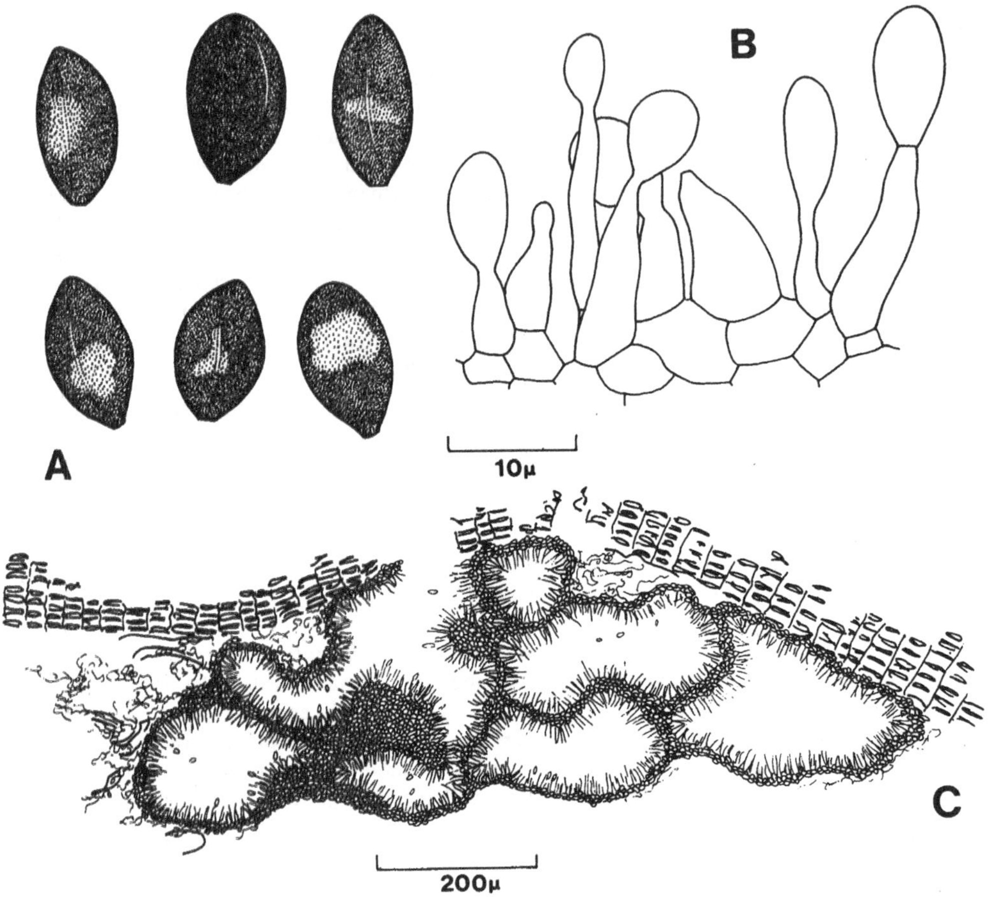

Fig. 95. *Endomelanconium pini*, A, conidia; B, conidiogenous cells and developing conidia; C, vertical section of a conidioma.

38. LASMENIA Speg., *An. Soc. Cient. Arg.* **22**: 152 (1886).
Original spp.: *L. balansae* Speg., *L. guaranitica* Speg.
Lectotype sp.: *L. balansae* Speg., fide Höhnel (1910).

Mycelium immersed, branched, septate, hyaline. *Conidiomata* eustromatic, irregular in outline, separate or coalescing, lifting a flap of epidermis and hypodermis at maturity; wall of hyaline, thin-walled textura angularis, except at the periphery where a line 3-4 cells thick of dark brown cells distinguishes the conidioma from healthy tissue. Dehiscence by elevation of a flap of host tissue hinged at one side. *Conidiophores* hyaline, sparingly branched, septate, cylindrical, formed from upper and lower acervular tissue which is somewhat irregular in outline. *Conidiogenous cells* holoblastic, determinate, integrated or discrete, hyaline, smooth, cylindrical. *Conidia* pale brown but thick-walled, smooth, aseptate, truncate and protuberant at the base, apex obtuse, ellipsoid, often with an irregular central guttule.

Höhnel (1910b) selected *L. balansae* as the lectotype species and Petrak & Sydow (1927) reaffirmed this by removing the second species, *L. guaranitica*, to *Lasmeniella* Petrak & Syd. Tassi (1904) placed *L. balansae* in *Auerswaldia* Sacc., Pleosporales, but Höhnel (1910b) considered Tassi to have mixed up the imperfect and perfect states of the fungus.

More than 10 taxa have been described in *Lasmenia*, but apart from the revision by Petrak & Sydow (1927) where four species were removed to *Lasmeniella*, the genus has not been extensively re-examined.

L. BALANSAE Speg., *ibid.*: 152 (1886). (Fig. 97)

Conidiomata up to 350 µ diam. Conidiophores 7-19 × 3.5-5 µ. Conidia 16-19 × 8.5-11 µ.

On leaves of *Eugenia* sp., Paraguay (IMI 200651, holotype ex K).

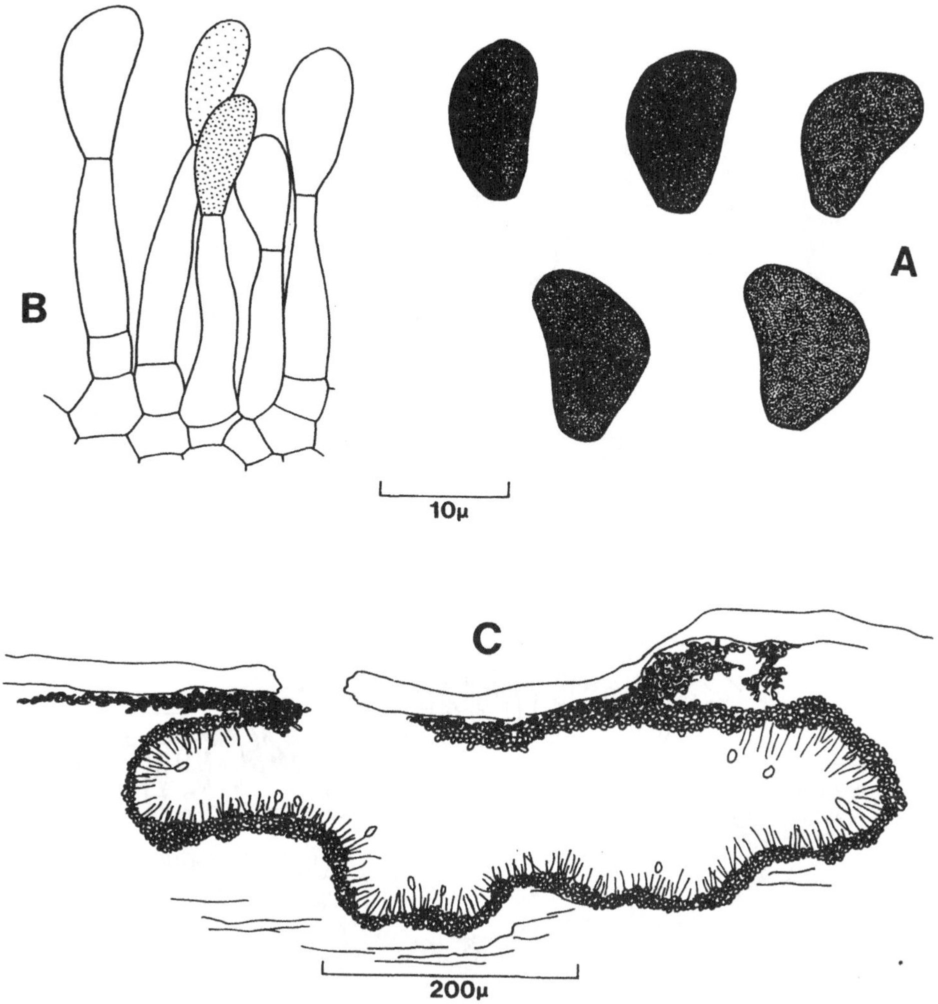

Fig. 96. *Cymbothyrium sudans*, A, conidia; B, conidiogenous cells and developing conidia; C, vertical section of a conidioma.

39. PERIZOMELLA H. Syd., *Annls mycol.* **25**: 106 (1927).
Sp. typ.: *P. inquinans* H. Syd.

Mycelium immersed, branched, septate, hyaline to pale brown. *Conidiomata* eustromatic, immersed in leaf tissue amongst *Phyllachora* stromatic tissue, separate or confluent, unilocular or multilocular and convoluted, thin-walled; wall composed of hyaline, thin-walled textura angularis to textura intricata. *Ostiole* absent, dehiscence by irregular rupture of the overlying host and *Phyllachora* tissue. *Conidiophores* hyaline, cylindrical, septate only at the base, formed from the inner cells of the locular walls. *Conidiogenous cells* holoblastic, integrated or discrete, determinate, cylindrical, hyaline, smooth, producing a single acrogenous conidium, often with the cytoplasmic contents retracted after conidium formation. *Conidia* pale brown but with a hyaline equatorial band, smooth, ellipsoid, eguttulate, aseptate, base truncate, apex flattened.

The genus is monotypic and the type species was reported to be a hyperparasite on *Phyllachora*. While no fertile ascocarps were found on the material examined, there were certainly locules enclosed by dark brown stromatic tissue in the epiphyllous region of the leaves. The conidiomata of *P. inquinans* were consistently situated between such stromata and dark brown fungus tissue concentrated in the hypodermis. However it could not be determined whether the hypoder-

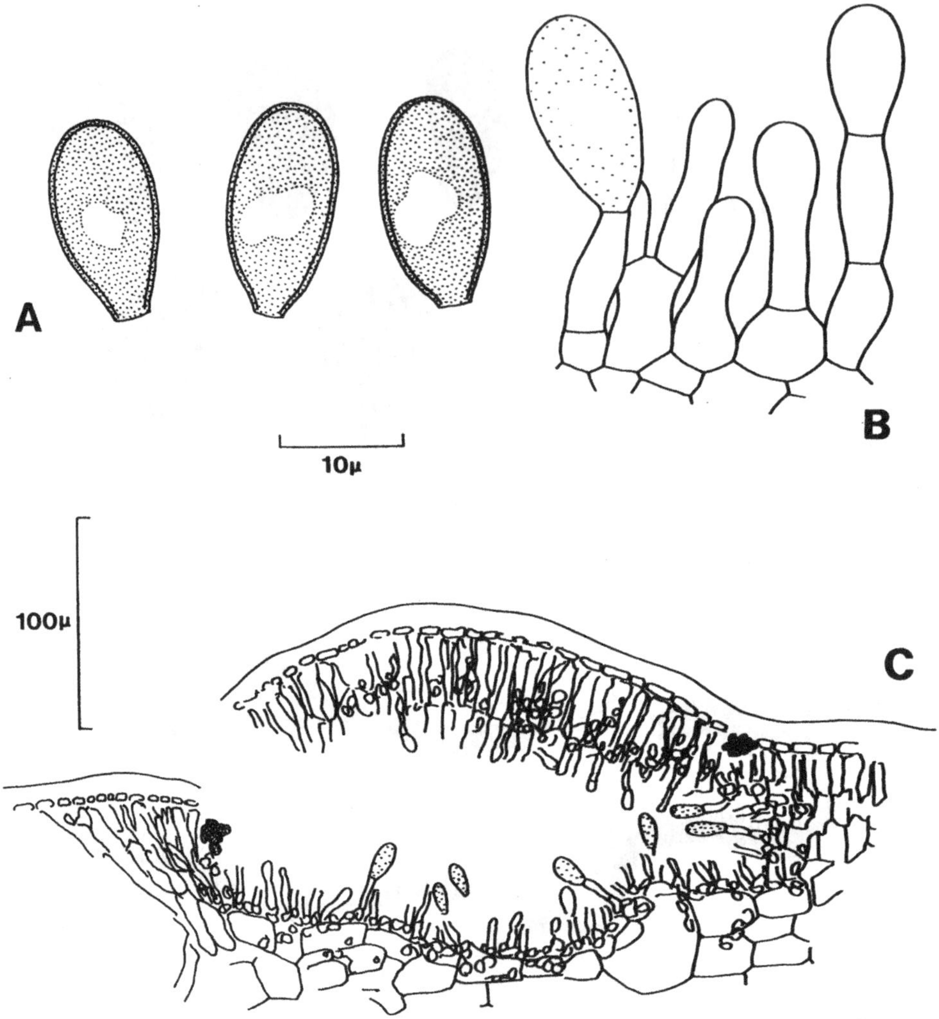

Fig. 97. *Lasmenia balansae*, A, conidia; B, conidiophores and developing conidia; C, vertical section of a conidioma.

mal tissue was structurally connected with *Perizomella* conidiomata or the sterile epiphyllous stromata. Consequently the generic description of this taxon is restricted to that tissue known to bear conidia.

P. INQUINANS H. Syd., *ibid.*: 106 (1927). (Fig. 98)

Conidiomatal locules 100-150 µ diam. Conidiogenous cells 8-13.5 × 3-4 µ. Conidia 7-8 × 3.5-4 µ.

In ? *Phyllachora* sp. on *Phoebe costaricana*, Costa Rica (IMI 22428, 22429).

40. APLOSPORELLA Speg., *An. Soc. Cienc. Arg.* **10**: 157 (1880).
Podosporium Bonord., *Handb. der Allgem. Myk.* 227 (1851).
Podosporium Sacc. & Schulz., *Rev. mycol.* **6**: 72 (1884).
Pleosphaeropsis Died., *Annls mycol.* **14**: 203 (1916).
Epicyta Petrak, *Annls mycol.* **24**: 413 (1926).
Lectotype sp.: *A. chlorostoma* Speg., fide Höhnel (1918a).

Mycelium immersed, branched, septate, pale brown. *Conidiomata* eustromatic, semi-immersed, erumpent and finally superficial, separate, often aggregated, pulvinate, rarely applanate, carbonous, dark brown, multilocular, locules divided by

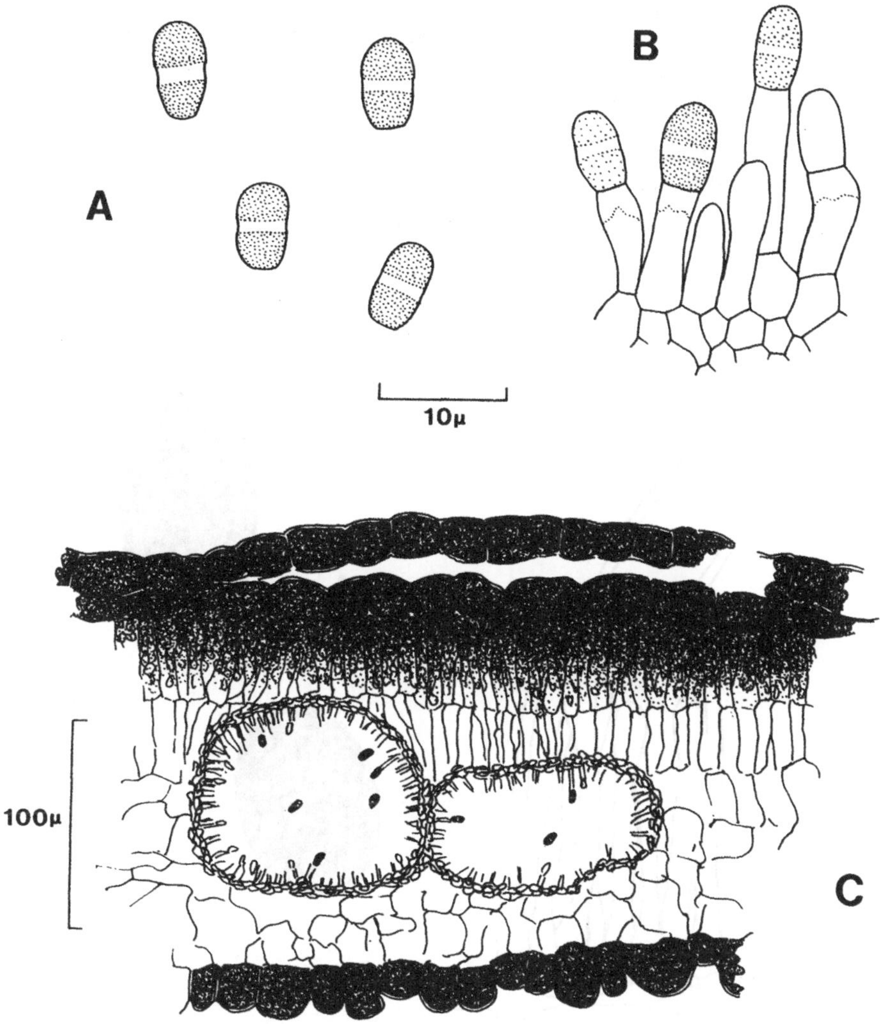

Fig. 98. *Perizomella inquinans*, A, conidia; B, conidiogenous cells and developing conidia; C, vertical section of conidiomata.

pale brown elongated textura prismatica, but uniting in the ostiolar region; basal, lateral and upper walls of brown, thick-walled textura angularis. *Ostiole* single, circular, central. *Conidiophores* absent. *Conidiogenous cells* holoblastic, determinate, discrete, hyaline, filiform, smooth, formed from the walls of the locules. *Conidia* dark brown, thick-walled, aseptate, ellipsoid, occasionally guttulate, verrucose, the ornamentation as angularly reticulate ridges. *Paraphyses* filiform, hyaline, aseptate, occasionally swollen at the apices.

The lectotypification of *Podosporium* Bonord. (see Sutton, 1977a) provides an earlier name for *Aplosporella* (frequently appearing as the authographic variant *Haplosporella*). However, since *Aplosporella* has been treated extensively by Petrak & Sydow (1927) who accepted 80 species, and the names have been frequently used in the literature since, it seems a good case for conservation. In anticipation of a fully documented proposal to this effect, I am continuing to use *Aplosporella*.

Hughes (1953b) used several *Aplosporella* names in the synonymy for the conidial state of *Phaeobotryosphaeria plicatula* (Berk. & Br.) Petch but these do not refer to the true *Aplosporella*. Conidia are smooth-walled, pale brown and when immature are enclosed in a thin mucilaginous sheath. Furthermore, the stromatic locules dehisce individually through separate ostioles rather than the single central common ostiole that is so typical of *Aplosporella*. Quite possibly *Cytosphaera* Died. is an immature state of conidial *P. plicatula*.

Since the monographic treatment by Petrak & Sydow (1927) many more species have been added to the genus. The situation now requires a revisionary treatment in which a reassessment of species limits is the most urgent problem. Most species have been introduced purely on a host substrate basis, so it seems likely that the results of any future revision will mean a drastic reduction in the number of accepted taxa.

A. BEAUMONTIANA Ahmad, *Biologia, Lahore* **8**: 144 (1962). (Fig. 99)
Pleosphaeropsis grewiae P.N. Mathur, *Indian Phytopath.* **22**: 492 (1970).

Conidiomata up to 700 µ diam.; ostiole 75 µ diam. Conidiogenous cells 15-18 × 2 µ. Conidia 16-21 × 11-11.5 µ. Paraphyses 25-60 × 1.5 µ.

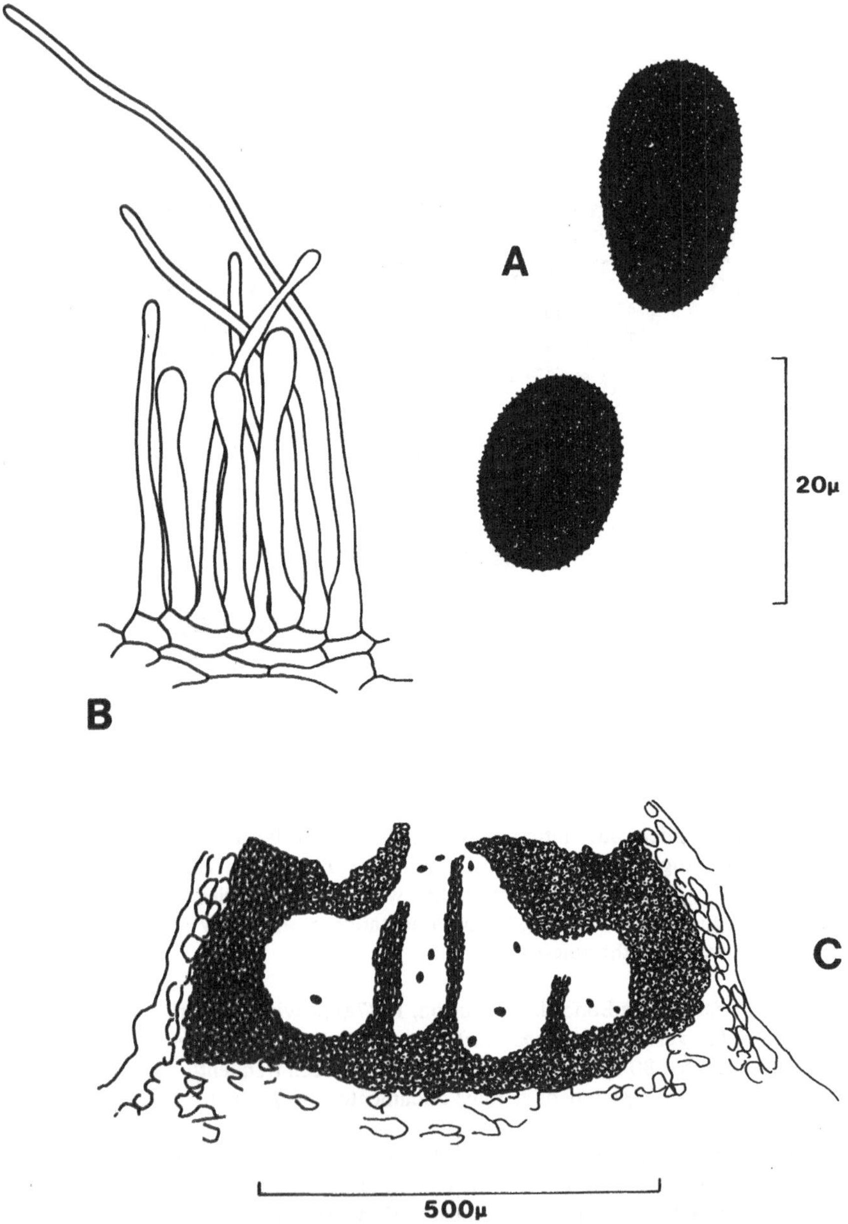

Fig. 99. *Aplosporella beaumontiana*, A, conidia; B, conidiogenous cells and paraphyses; C, vertical section of a conidioma.

On dead twigs and stems of *Artabotrys odoratissima*, *Beaumontia grandiflora*, *Dalbergia sissoo*, *Grewia asiatica*, *Mangifera indica*, *Pithecellobium dusea*, *P. dulce*, *Inga dulce*, Pakistan, India (150928, 91672, 151015, 163605, 198387, 214777, 197012, 136388 (holotype *P. grewiae*), 209354, 150929, 205615, 183707, 87046 (holotype *A. beaumontiana*), 183710, 163596).

41. LASIODIPLODIA Ell. & Ev. apud Clendenin, *Bot. Gaz.* **21**: 92 (1896).
Striodiplodia Zambettakis, *Bull. trimest. Soc. mycol. Fr.* **70**: 334 (1955).
Sp. typ.: *L. theobromae* (Pat.) Griff. & Maubl. (syn. *Botryodiplodia theobromae* Pat., *L. tubericola* Ell. & Ev.).
Illustration: Punithalingam (1976b).

Mycelium immersed or superficial, branched, septate, dark chocolate brown. *Conidiomata* eustromatic, immersed or superficial, separate or aggregated and confluent, globose, carbonous, dark brown, uni- or multilocular; wall of dark brown, thick-walled textura angularis, paler and thinner towards the conidiogenous region, often with dark brown superficial hyphae over the surface. *Ostiole* absent, dehiscence by irregular rupture. *Conidiophores* absent. *Conidiogenous cells* holoblastic, determinate, discrete, cylindrical, hyaline, smooth, with no percurrent or sympodial proliferation, formed from cells lining the inner pycnidial walls. *Conidia* acrogenous, hyaline when young, later becoming medianly 1 euseptate, dark brown, thick-walled, ellipsoid, base truncate, with longitudinal striations from apex to base. *Paraphyses* hyaline, cylindrical, septate.

Lasiodiplodia is clearly the correct generic name to adopt for the ubiquitous tropical to subtropical plant pathogen popularly known as *Botryodiplodia theobromae* Pat. The thick-walled, striate, slowly maturing conidia are diagnostic, in contrast to the vegetative hyphae that often give stromata a hirsute appearance but are of little taxonomic significance. The species is commonly referred to *Botryodiplodia* Sacc., typified by *B. juglandicola* (Schw.) Sacc., but if the differences in conidial morphology survive a reassessment of generic concepts in this group then *B. theobromae* will probably be maintained in *Lasiodiplodia* as Zambettakis (1955) has suggested.

Punithalingam (1976b) has fully reviewed the pathology of the species, but under the name *Botryodiplodia theobromae* Pat. He listed *Diplodia gossypina* Cke, 1879, as a synonym, which is an earlier name for the species. Considering the ubiquity of this taxon in tropical and subtropical regions, it would not be surprising if such earlier names were eventually found. For the time being, however, the epithet '*theobromae*' is adopted.

L. THEOBROMAE (Pat.) Griff. & Maubl., *Bull. trimest. Soc. Mycol. Fr.* **25**: 57 (1909).
Botryodiplodia theobromae Pat., *Bull. trimest. Soc. mycol. Fr.* **8**: 136 (1892).
Diplodia theobromae (Pat.) Nowell, *Diseases of Crop Plants in the Lesser Antilles*: 158 (1923).
Lasiodiplodia tubericola Ell. & Ev. apud Clendenin, *Bot. Gaz.* **21**: 92 (1896).
Diplodia tubericola (Ell. & Ev.) Taubenh., *Amer. J. Bot.* **2**: 328 (1915).
Botryodiplodia tubericola (Ell. & Ev.) Petrak, *Annls mycol.* **21**: 332 (1923).

Teleomorphosis: *Physalospora rhodina* Berk. & Curt. apud Cooke, *Grevillea* **17**: 92 (1889).

Conidiomata up to 5 (2-4) mm diam. Conidiogenous cells 5-15 × 3 µ. Conidia maturing slowly, for some while hyaline and thin-walled, then thick-walled, finally developing a median septum, dark brown pigmentation and longitudinal striations, 20-30 × 10-15 µ. Paraphyses up to 50 µ long.

Worldwide in distribution between 40°N and 40°S of equator, and represented in herb. IMI by 1100 collections on 280 different genera of host plants. Also found on a variety of other substrata.

42. PELLIONELLA Sacc. apud Penzig & Sacc., *Malpighia* **15**: 234 (1902).
? *Diplodiella* '(ut subgenus)' Karst. subgen. *Pellionella* Sacc., *Syll. fung.* **14**: 940 (1899).
Sp. typ.: *P. deformans* Penz. & Sacc.

Mycelium immersed, branched, septate, hyaline, pale brown or dark brown. *Conidiomata* pseudostromatic, peridermal to subperidermal, black, separate, globose, with an apical rostrum, multilocular, convoluted, locules confined to the upper part of the stroma and uniting in a common ostiolar channel; rostral tissue of textura intricata, hyphae and cells thick-walled with reduced lumina, darker and more sclerotioid towards the outside; stroma pseudoparenchymatic, cells thin-walled, pale brown, but dark brown, sclerotioid and thick-walled where the stroma is separated from unaffected host tissue. *Ostiole* single, circular, at the apex of an often excentric rostrum. *Conidiophores* absent. *Conidiogenous cells* holoblastic, discrete, cylindrical, thin-walled, smooth, hyaline, formed from the cells lining the locular walls. *Conidia* pale to medium brown, medianly 1 euseptate, septum thick, thin-walled, smooth, eguttulate, obtuse at the apex, truncate at the base, ellipsoid.

The genus is so far monotypic.

P. DEFORMANS Penz. & Sacc., *ibid.*: 234 (1902). (Fig. 100)

Conidiomata up to 2400 µ wide × 2900 µ high; rostrum up to 500 µ wide. Conidiogenous cells 13-26 × 2-3.5 µ. Conidia 19-21.5 × 8.5-10 µ.

On dead trunks of *Garcinia purpurea*, Java (IMI 206032 ex PAD).

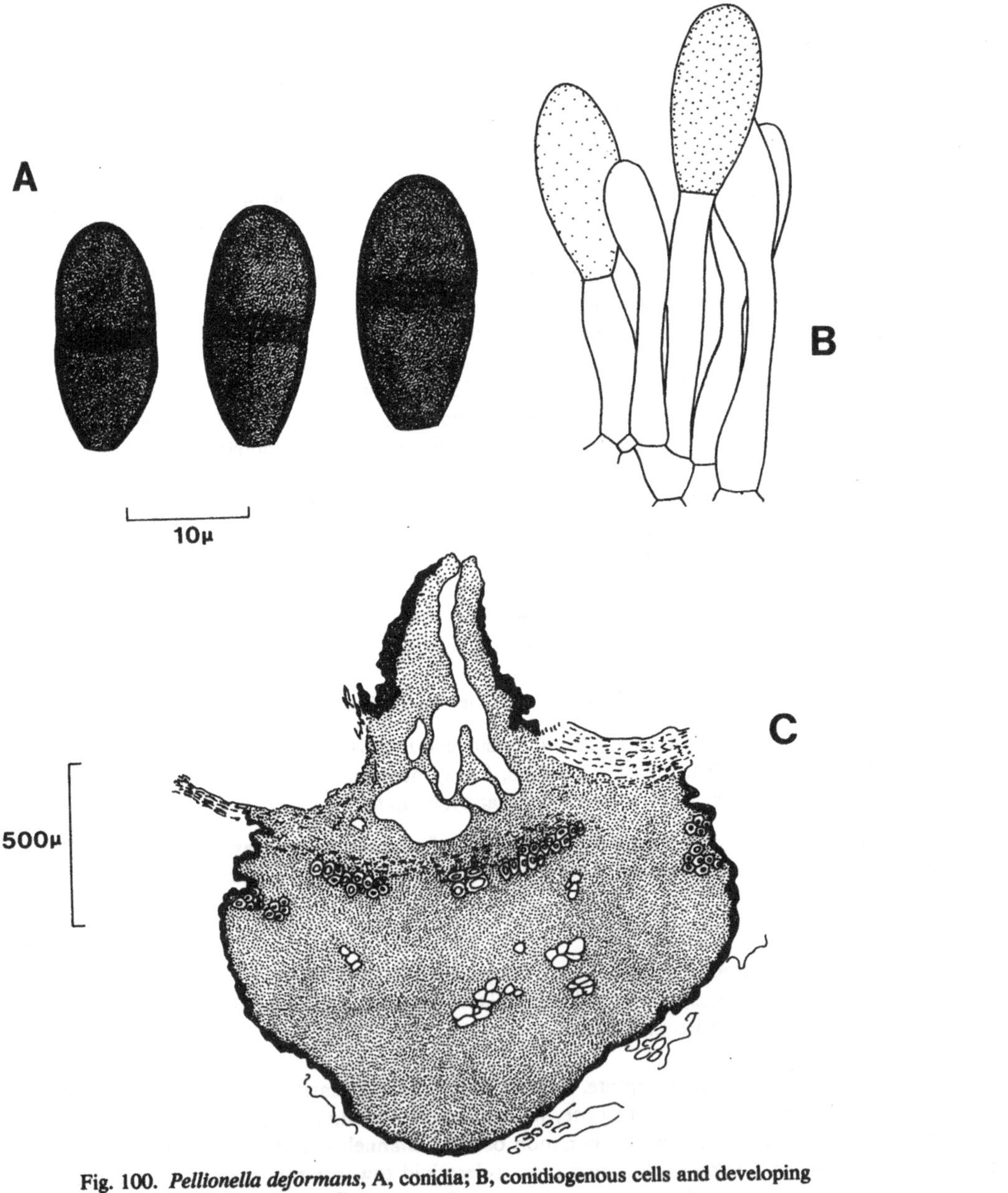

Fig. 100. *Pellionella deformans*, A, conidia; B, conidiogenous cells and developing conidia; vertical section of a conidioma.

43. ONCOSPORA Kalchbr. & Cke, *Grevillea* **9**: 19 (1880).
Original spp.: *O. bullata* Kalchbr. & Cke, *O. viridans* Kalchbr. & Cke
Lectotype sp.: *O. bullata* Kalchbr. & Cke, fide Clements & Shear (1931).

Mycelium immersed, of brown, branched, septate hyphae. *Conidiomata* initially immersed, later erumpent, pulvinate to bullate, solitary, closely aggregated, dark brown, epidermal to subepidermal, composed of a central and basal region of thin-walled but brown textura angularis, at the sides more sclerotioid and thick-walled. *Ostiole* absent, dehiscence by rupture of overlying host tissues, finally almost cupulate. *Conidiophores* cylindrical to irregular, branched at the base, septate, pale brown to hyaline, smooth, formed from the upper pseudoparenchyma. *Conidiogenous cells* holoblastic, integrated or discrete, determinate, cylindrical, pale brown to hyaline, smooth. *Conidia* pale brown, 1-euseptate, septum nearer the apex than the base, straight, curved or irregular, tapered to the truncate base, apex obtuse to bluntly acute, continuous, smooth, often irregularly guttulate in the lower cell.

The lectotype and *O. viridans* have been the subject of discussion by Höhnel (1910c). Three other species have been described in the genus but have not yet been revised.

O. BULLATA Kalchbr. & Cke, *ibid.*: 19 (1880). (Fig. 101)
? *Discella capparidis* Pat. & Har., *Bull. trimest. Soc. mycol. Fr.* **21**: 86 (1905).
? *Cryptosporium circinans* Welwitsch & Currey, *Trans. linn. Soc. Lond.* **26**: 286 (1870).

Conidiomata up to 650 µ wide × 350-400 µ deep. Conidiogenous cells 10 × 3-4 µ. Conidia 23-29 × 4.5-5.5 µ. Höhnel (1910) cited these two names as possible synonyms but there is a discrepancy in published measurements of conidia, so types should in this case be seen before possible transfers are made.

On lesions of leaves of *Capparis citrifolia*, Africa (IMI 196542, holotype of *O. bullata* ex K).

44. SCOLICOSPORIUM Lib. apud Roum., *Rev. Mycol.* **2**: 22 (1880).
Sp. typ.: *S. macrosporium* (Berk.) Sutton (syn. *Coryneum macrosporium* Berk., *S. fagi* Lib.).
Illustration: Allescher (1902).

Mycelium immersed, branched, septate, pale brown. *Conidiomata* acervular, subperidermal, brown, separate, composed of pale brown, thin-walled textura angularis. Dehiscence irregular. *Conidiophores* branched only at the base, 1-2 septate, cylindrical, hyaline, formed from the upper cells of the conidiomata. *Conidiogenous cells* holoblastic, determinate or indeterminate with 1-2 annellations, integrated or discrete, cylindrical, hyaline. *Conidia* smooth, thin-walled, fusiform, curved, pale brown with the end cells pale or hyaline, tapered to the obtuse apex and truncate base, with several transverse eusepta, often guttulate.

As pointed out by Sutton (1975a) the genus is heterogeneous. At least ten species have been described so far but it seems most unlikely that all of these are congeneric with *S. macrosporium*.

KEY

Conidia 39-45 × 4.5-5.5 µ, up to 8 euseptate .... *S. fusarioides*
Conidia 100-190 × 12-15 µ, 7-12 euseptate .... *S. macrosporium*

S. FUSARIOIDES (Sacc.) Sutton, *Mycol. Pap.* 138: 111 (1975). (Fig. 102)
Coryneum fusarioides Sacc., *Michelia* **2**: 120 (1880).

Conidiomata 150-200 µ diam. Conidiophores up to 3-7 × 1.5 µ, lacking percurrent proliferations. Conidia 39-45 × 4.5-5.5 µ, up to 8 euseptate.

On bark of *Populus fastigiata*, France (IMI 179078 ex PAD, holotype).

S. MACROSPORIUM (Berk.) Sutton, *Mycol. Pap.* 141: 185 (1977).
Coryneum macrosporium Berk., Hooker's *English Flora* **5**(2): 355 (1836).
Scolicosporium fagi Lib. apud Sacc., *Michelia* **2**: 355 (1881).
Sporidesmium vermiforme Riess in Fres., *Beitrage zur Myk.* 51 (1863).

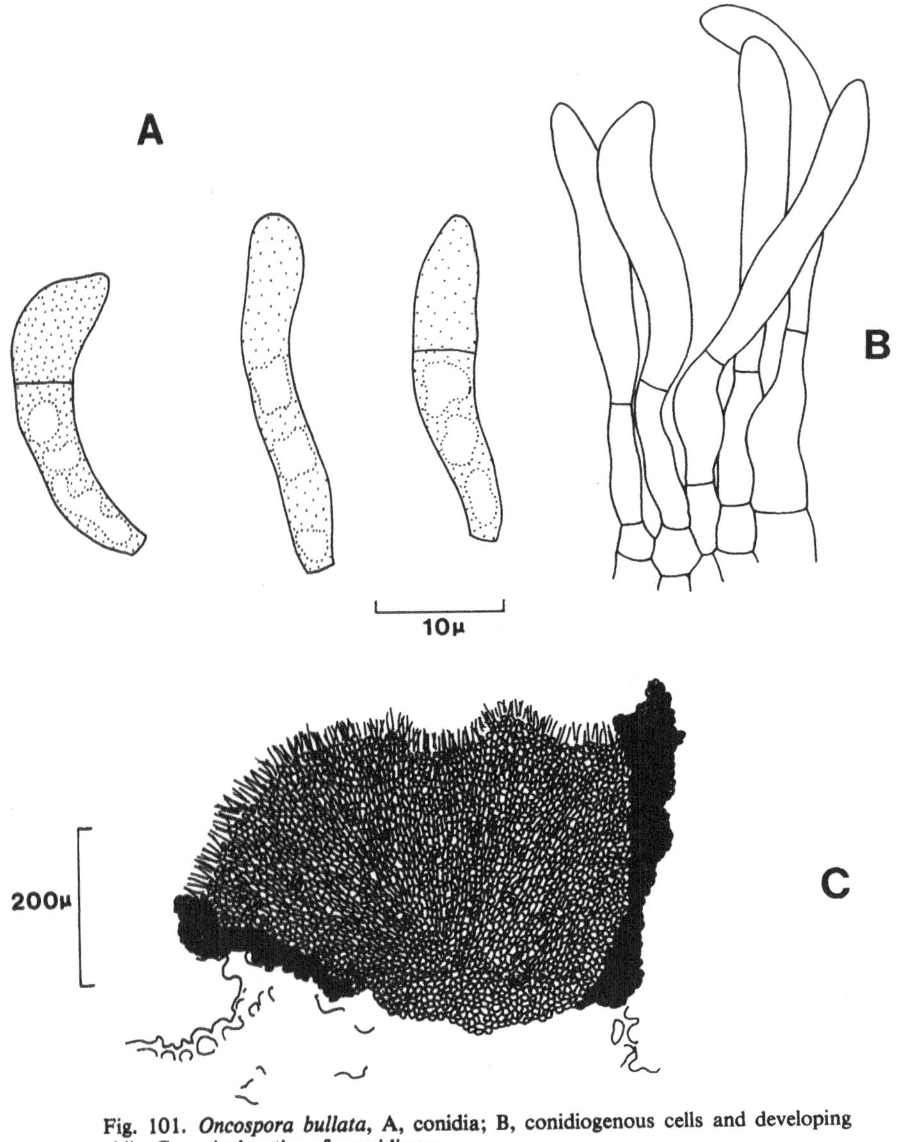

Fig. 101. *Oncospora bullata*, A, conidia; B, conidiogenous cells and developing conidia; C, vertical section of a conidioma.

Conidiomata up to 250 µ diam. Conidiophores up to 30 µ long × 3.5-5 µ wide. Conidia 100-190 × 12-15 µ, 7-12 euseptate.

Found mixed with or growing in conidiomata of *Asterosporium asterospermum* and *Stegonsporium pyriforme*.

On twigs of *Fagus* sp., Austria, U.K. (IMI 31095, 9890b).

45. SHAWIELLA Hansf., *Proc. linn. Soc., N.S.W.* **82**: 226 (1957).
Sp. typ.: *S. grevilleae* Hansf.
Illustration: Sutton (1975b).

Mycelium immersed or superficial, dark brown, branched, septate. *Conidiomata* infundibuliform, cupulate, superficial, separate or gregarious, medium brown, sessile; base of large, thin-walled textura angularis; periclinal wall 1 cell thick, of olivaceous, rounded to angular, thin-walled cells. *Conidiophores* absent. *Conidiogenous cells* holoblastic, discrete, determinate, papilliform, pale brown, smooth, formed from the base of the conidioma. *Conidia* nidulant, at first strongly

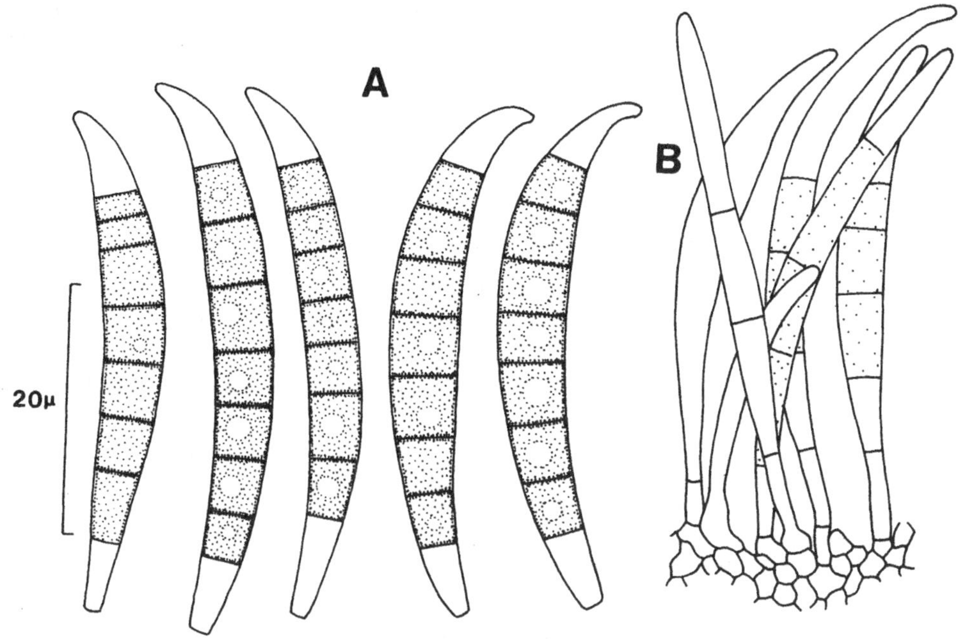

Fig. 102. *Scolicosporium fusarioides*, A, conidia; B, conidiogenous cells and developing conidia.

recurved in the conidioma, later ± straight or curved at the apex, pale brown, 8-13 distoseptate with reduced lumina, cylindrical to fusiform, attenuated towards the pale apex, base truncate.

Sutton (1975b) redescribed and illustrated the genus and species. So far it remains monotypic.

S. GREVILLEAE Hansf., *ibid.*: 226 (1975).

Conidiomata 30 μ high × 25 μ wide. Conidiogenous cells 4.5-6 × 2-4.5 μ. Conidia 55-80 × 5-7 μ.

On leaf hairs of *Grevillea robusta*, New Guinea (IMI 185278, holotype).

46. DICHOMERA Cke, *Nuovo G. bot. ital.* **10**: 24 (1878).

Original spp.: *D. saubinetii* (Mont.) Cke, *D. sphaerosperma* (Berk. & Curt.) Cke, *D. sarmenticia* (Sacc.) Cke, *D. stephensii* (Berk. & Br.) Cke, *D. oreades* (Dur. & Mont.) Cke, *D. camphorosmae* (Mont.) Cke, *D. minutula* (Dur. & Mont.) Cke

Lectotype sp.: *D. saubinetii* (Mont.) Cke (syn. *Hendersonia saubinetii* Mont.), fide Höhnel (1918a).

Mycelium immersed, branched, septate, pale to dark brown. *Conidiomata* eustromatic, immersed, subepidermal or subperidermal, erumpent, separate, dark brown, multilocular; walls of thick-walled, dark brown textura angularis on the outside, becoming paler, thin-walled, hyaline to subhyaline in the conidiogenous region. Occasionally unilocular stromata occur and show various gradations to the normal multilocular state. *Ostioles* 1 or many to each conidioma, sometimes papillate, circular. *Conidiophores* absent. *Conidiogenous cells* holoblastic, simple, determinate, some species indeterminate and annellidic, hyaline, doliiform to cylindrical, formed from the cells lining the locular walls. *Conidia* muriform, brown, euseptate, globose, pyriform or cylindrical, often variable and irregular in shape, constricted or not constricted at the septa, smooth, base truncate.

More than 40 species have been described in *Dichomera* which has usually been interpreted as the stromatic analogue of *Camarosporium* Schulz. Apart from recent accounts by Funk & Sutton (1972) and Sutton (1975a) there have been no modern illustrated treatments of species, so the genus requires revision.

Previously (Sutton, 1977a) I had considered *Sclerotheca* Bub. & Vleugel to be a later synonym for *Dichomera*, however subsequent re-examination of the only collection available has promoted a re-evaluation of this conclusion. The consistently percurrently proliferating conidiogenous cells and the often single pycnidia suggest *Camarosporium* rather than *Dichomera*. As in *D. gemmicola* the nature of the substrate in *S. strobilinum* must influence the development of pycnidial and stromatic conidiomata and it is difficult with such few collections to determine precise affinities. *D. gemmicola* is very frequently stromatic and only occasionally pycnidial. If *S. strobilinum* proves to behave in the same manner it can be placed in *Dichomera*, but if it is more consistently pycnidial, *Camarosporium* would appear to be a more preferable genus.

D. SAUBINETII (Mont.) Cke, *Nuovo G. bot. ital.* **10**: 24 (1878). (Fig. 103A)
Hendersonia saubinetii Mont., *Syll. gen. sp. crypt.* 263 (1856).

Conidiomata up to 2 mm diam.; locules up to 120 μ diam. Conidiogenous cells determinate, 7.5-14 × 3-4 μ. Conidia globose, subglobose, clavate or fusiform, with 1-5 transverse and 1 or more longitudinal or oblique septa, 11-13 × 7-10 μ.

On twigs and branches of *Rhamnus frangula*, *Corylus* sp., *Acer pseudoplatanus*, U.K. (IMI 21610, 21611, 60275, 79813).

D. EUCALYPTI (Wint.) Sutton, *Mycol. Pap.* 138: 182 (1975). (Fig. 104)
Camarosporium eucalypti Wint., *Rev. Mycol.* **8**: 212 (1886).
Camarosporellum eucalypti (Wint.) Tassi, *Bull. Lab. ort. Bot. Siena* **5**: 62 (1902).
Coryneum viminale Cke & Mass., *Grevillea* **20**: 36 (1891).

Conidiomata 100-265 μ diam., locules up to 75 μ diam. Conidiogenous cells with 1 annellation, 5-8 × 4-8.5 μ. Conidia globose, subglobose, obovoid or obpyriform, 10.5-12.5 × 7-9.5 μ.

On leaves of *Eucalyptus viminalis*, *E. rubida*, *E.* sp., Australia (IMI 59162, holotype *Camarosporium eucalypti* ex K, 75054, holotype *Coryneum viminale* ex K, 173942b).

D. GEMMICOLA Funk & Sutton, *Can. J. Bot.* **50**: 1514 (1972). (Fig. 103B)

Conidiomata diffuse, often simply pycnidial, then up to 350 μ diam. × 450 μ high. Conidiogenous cells with 1 annellation, 8-11 × 5.5-7.5 μ. Conidia long fusiform, with 11-17 transverse septa and several longitudinal septa, 70-85 × 12-17 μ.

On buds of *Picea engelmannii*, *P. glauca*, *P. sitchensis*, *Pseudotsuga menziesii*, B.C. & Sask., Canada (IMI 161667, holotype, 162698, 162699, 162700, 162701, 161668, 161670).

47. CAMAROSPORULA Petrak, *Sydowia* **8**: 99 (1954).

Sp. typ.: *C. persooniae* (P. Henn.) Petrak (syn. *Hendersonia persooniae* P. Henn.).

Mycelium immersed, brown, branched, septate. *Conidiomata* acervular, epidermal to subepidermal, separate or confluent, formed of brown, thick-walled textura globulosa or angularis. Dehiscence irregular. *Conidiophores* absent. *Conidiogenous cells* holoblastic, determinate, integrated, cylindrical to doliiform, hyaline, smooth, producing a single terminal conidium, formed from the upper pseudoparenchyma. *Conidia* medium to dark brown, muriform, distoseptate, thick-walled, lumina reduced, smooth, cylindrical, clavate, obclavate or irregular, base broadly truncate.

The genus is monotypic.

C. PERSOONIAE (P. Henn.) Petrak, *ibid.*: 99 (1954). (Fig. 105)
Hendersonia persooniae P. Henn., *Hedwigia* **40**: 97 (1901).
Dichomera persooniae (P. Henn.) P. Henn., *Hedwigia* **42**: (87) (1903).

Teleomorphosis: *Anthracostroma persooniae* (P. Henn.) Petrak, *Sydowia* **8**: 97 (1954).

Conidiomata up to 100 μ diam. Conidia with up to 3 transverse and several longitudinal or oblique distosepta, up to 15.5 μ long × 6.5-10 μ wide.

Causing lesions on leaves of *Persoonia elliptica*, *P. cornifolia*, *P. media*, *P. laevis* and *P.* sp., Australia (IMI 196495 (holotype), 94340, 148443, 161734, 161733, 153735).

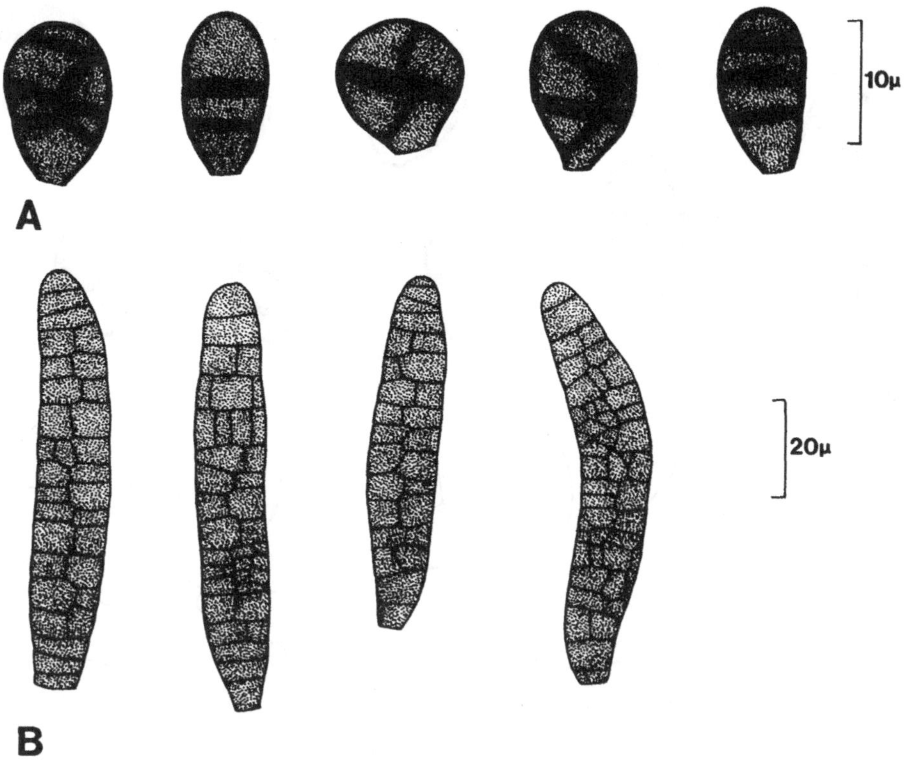

Fig. 103. A, *Dichomera saubinetii*, conidia; B, *D. gemmicola*, conidia.

48. MYXOCYCLUS Riess apud Fresenius, *Beiträg zur Mykologie*: 62 (1852).
Sp. typ.: *M. polycistis* (Berk. & Br.) Sacc. (syn. *Hendersonia polycistis* Berk. & Br., *M. confluens* Riess).

Mycelium immersed, branched, septate, brown. *Conidiomata* acervular, subperidermal, separate, punctiform, dark brown to black, of thick-walled, dark brown textura angularis at the base, thin-walled and pale brown in the conidiogenous region. Dehiscence irregular. *Conidiophores* cylindrical, flexuous, septate, branched at the base, pale brown, verruculose, becoming hyaline at the apex, enclosed in a gelatinous sheath, formed from the upper pseudoparenchyma. *Conidiogenous cells* holoblastic, integrated, determinate, cylindrical, pale brown to hyaline, becoming smooth. *Conidia* muriform, tapered to the base, apex obtuse, conidiogenous cells often attached, verruculose, 7-8 distoseptate with longitudinal or oblique distosepta, lumina reduced, constricted, enclosed in a gelatinous sheath.

Sutton (1975a) reviewed the history of the genus and species and gave a full description and illustration, together with an extensive synonymy. The genus contains two species, but *M. cenangioides* (Ell. & Bor.) Petrak is not congeneric with *M. polycistis*.

M. POLYCISTIS (Berk. & Br.) Sacc., *Annls mycol.* **6**: 559 (1908). (Fig. 106)

Conidiomata 400 µ wide at the base, 800 µ wide at the top. Conidiophores 90-220 × 4-6 µ. Conidia 50-66 × 20-26 µ.

On twigs and bark of *Betula papyrifera, B. occidentalis, B.* sp., Ala., U.S.A., Man., Sask., Canada (IMI 180178, holotype ex K, *C. irregulare*, WINF(M) 3197, 4413b, 4534, 4991a, 6049, 7031f, 7049, 7566, 8976, 11622, 6085b, 11454).

49. ENDOBOTRYELLA Höhn., *Sber. Akad. Wiss. Wien* **118**: 1536 (1909).
Sp. typ.: *E. oblonga* (Fckl) Höhn. (syn. *Myriocephalum oblongum* Fckl).

Mycelium immersed, branched, septate, pale brown. *Conidiomata* eustromatic, peridermal or subperidermal, separate, occasionally confluent, composed of thin-walled, paler brown textura intricata at the base and thicker-walled,

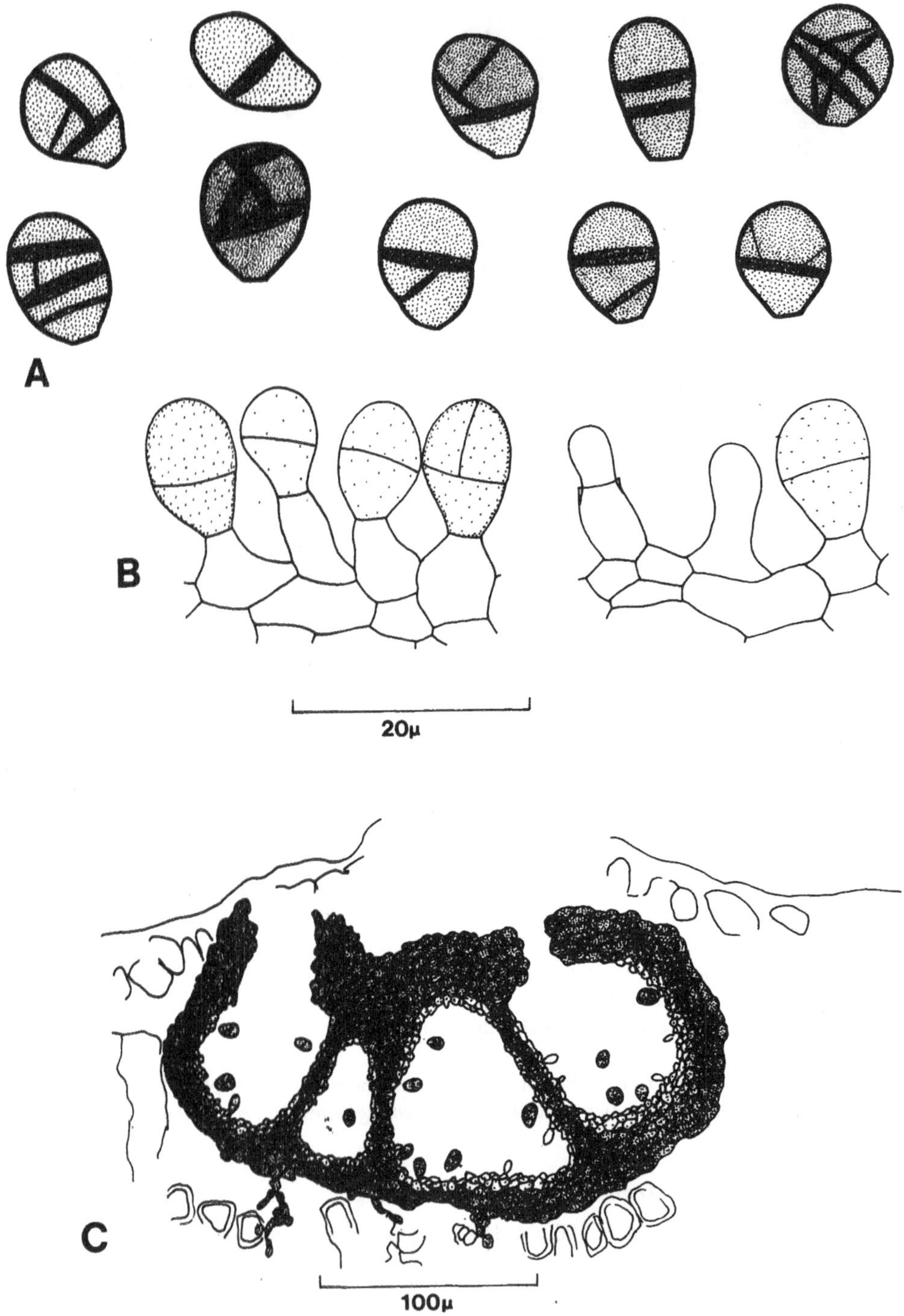

Fig. 104. *Dichomera eucalypti*, A, conidia; B, conidiogenous cells and developing conidia; C, vertical section of a conidioma.

brown textura porrecta at the sides, finally cupulate. Dehiscence irregular. *Conidiophores* hyaline, branched at the base, septate, straight, cylindrical, expanding towards the apices, smooth, formed from the basal and lateral walls. *Conidiogenous cells* holoblastic, determinate, integrated, short, cylindrical, hyaline, smooth, forming a single, apical, complex conidium. *Conidia* of up to 4 central cells each with several short lateral cells to form a cylindrical mass of small, thick-walled, brown, smooth, aseptate, eguttulate, spherical or cuneiform cells.

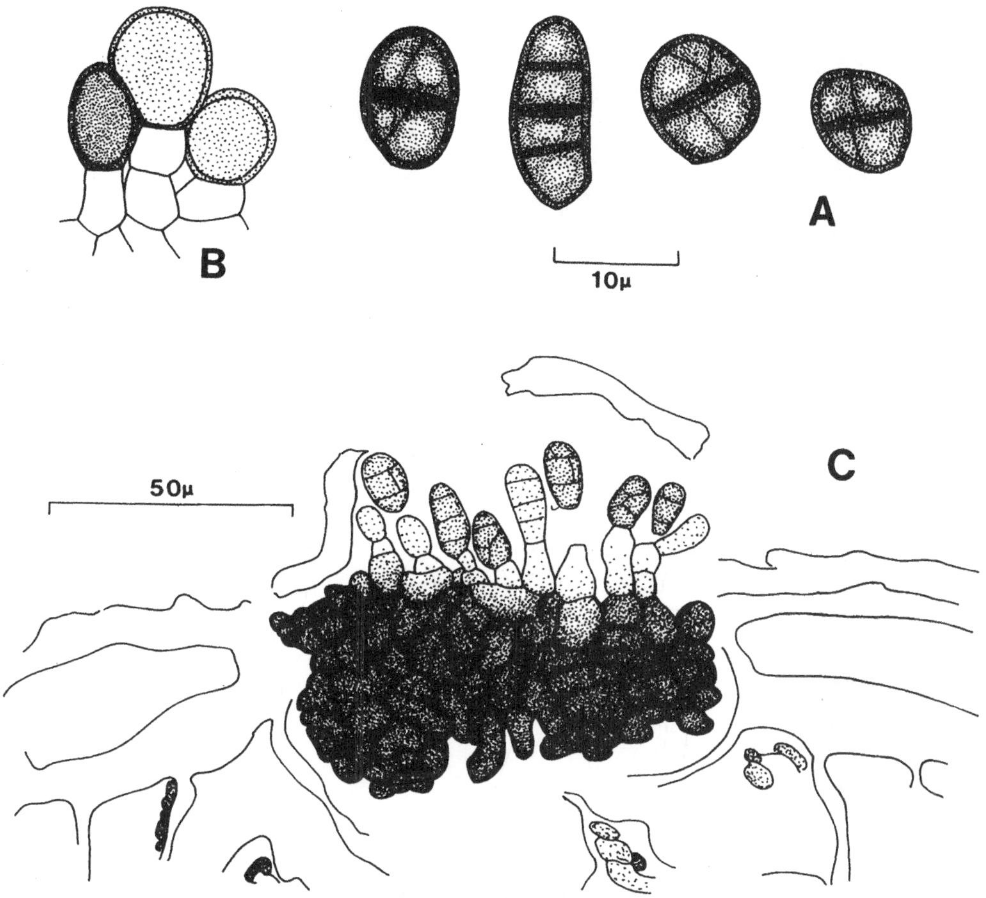

Fig. 105. *Camarosporula persooniae*, A, conidia; B, conidiogenous cells with developing conidia; C, vertical section of a conidioma.

Höhnel (1909) distinguished *Endobotryella* from *Thyrsidium* Mont. ex Mont. (= *Cheirospora* Moug. & Fr.) by the fact that the conidial masses and conidiophores were not enveloped in a gelatinous sheath. An additional difference is in the conidiomata. In *Cheirospora* they are acervular to sporodochial whereas in *Endobotryella* there is a distinct basal and lateral wall. Since conidiogenesis and conidial morphology are very similar the distinctions between the two genera are tenuous.

E. OBLONGA (Fckl) Höhn., *ibid.*: 1536 (1909). (Fig. 107)
Myriocephalum oblongum Fckl, *J. nassau. Ver. Naturk.* **23-24**: 351 (1870).
Thyrsidium oblongum (Fckl) Sacc., *Syll. fung.* **3**: 762 (1884).

Conidiomata up to 500 µ diam. Conidiophores up to 120 µ long × 2.5-3 µ wide. Composite conidia up to 27 µ long × 15 µ wide; individual cells up to 6.5 µ diam.

On bark of *Pinus*, Austria (IMI 200661, isotype *M. oblongum* ex K).

50. CHEIROSPORA Moug. & Fr. in Fr., *Syst. Orbis Veg.* 365 (1825).
Hyperomyxa Cda, *Icones fungorum* **3**: 34 (1839).
Moronopsis Delacr., *Bull. trimest. Soc. mycol. Fr.* **7**: 109 (1891).
Myriocephalum de Not., *Microm. ital. dec.* III: 13 (1842).
Rabdosporium Chev., *Flora generale des environs de Paris*: 428 (1826).
Thyrsidium Mont. ex Mont. in Durieu, *Expl. Sci. Alg. Bot.* **1**: 325 (1849).
Sp. typ.: *C. botryospora* (Mont.) Berk. & Br. (syn. *Stilbospora botryospora* Mont., *C. cheirospora* Fr.).

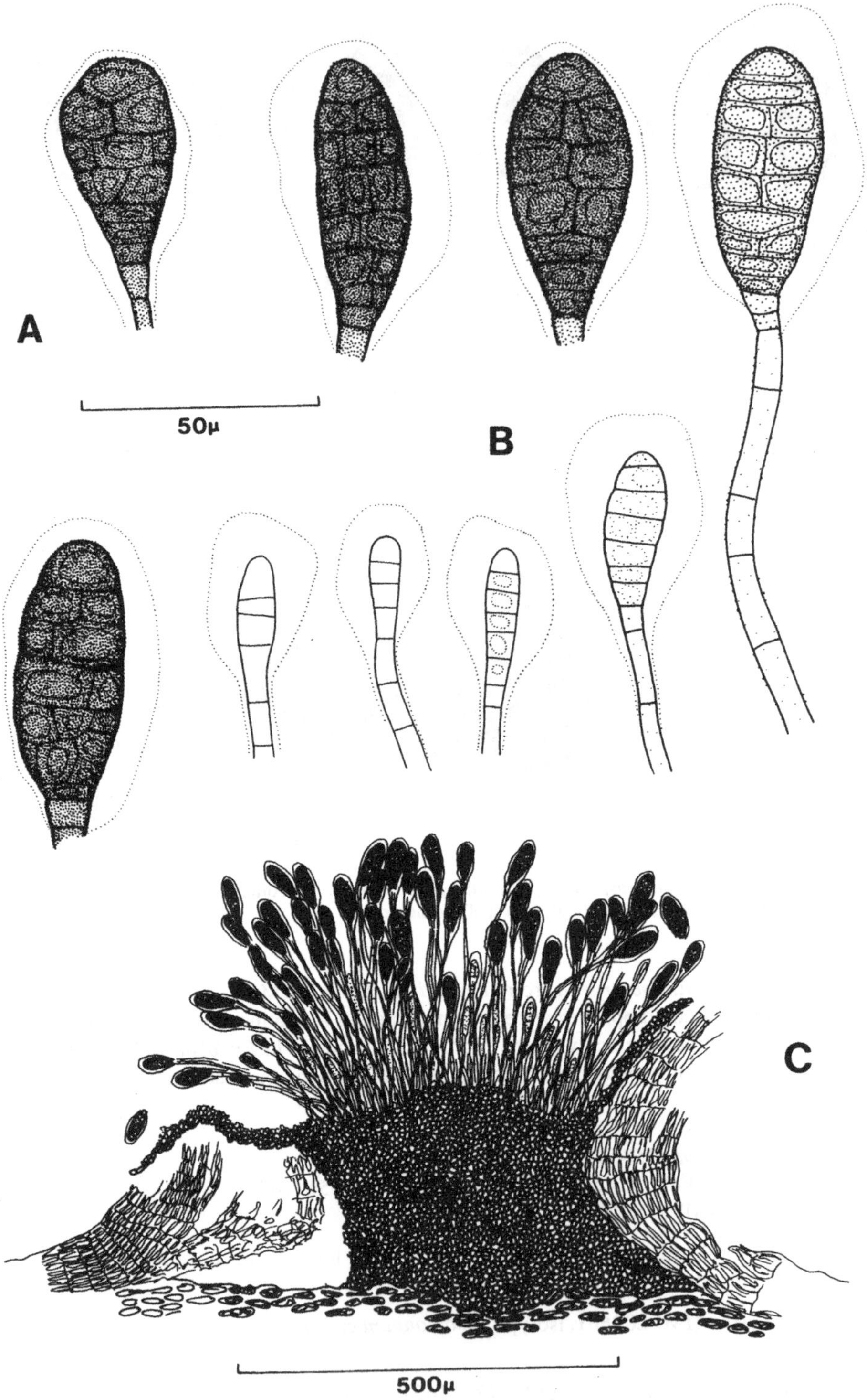

Fig. 106. *Myxocyclus polycistis*, A, conidia; B, developing conidia; C, vertical section of a conidioma.

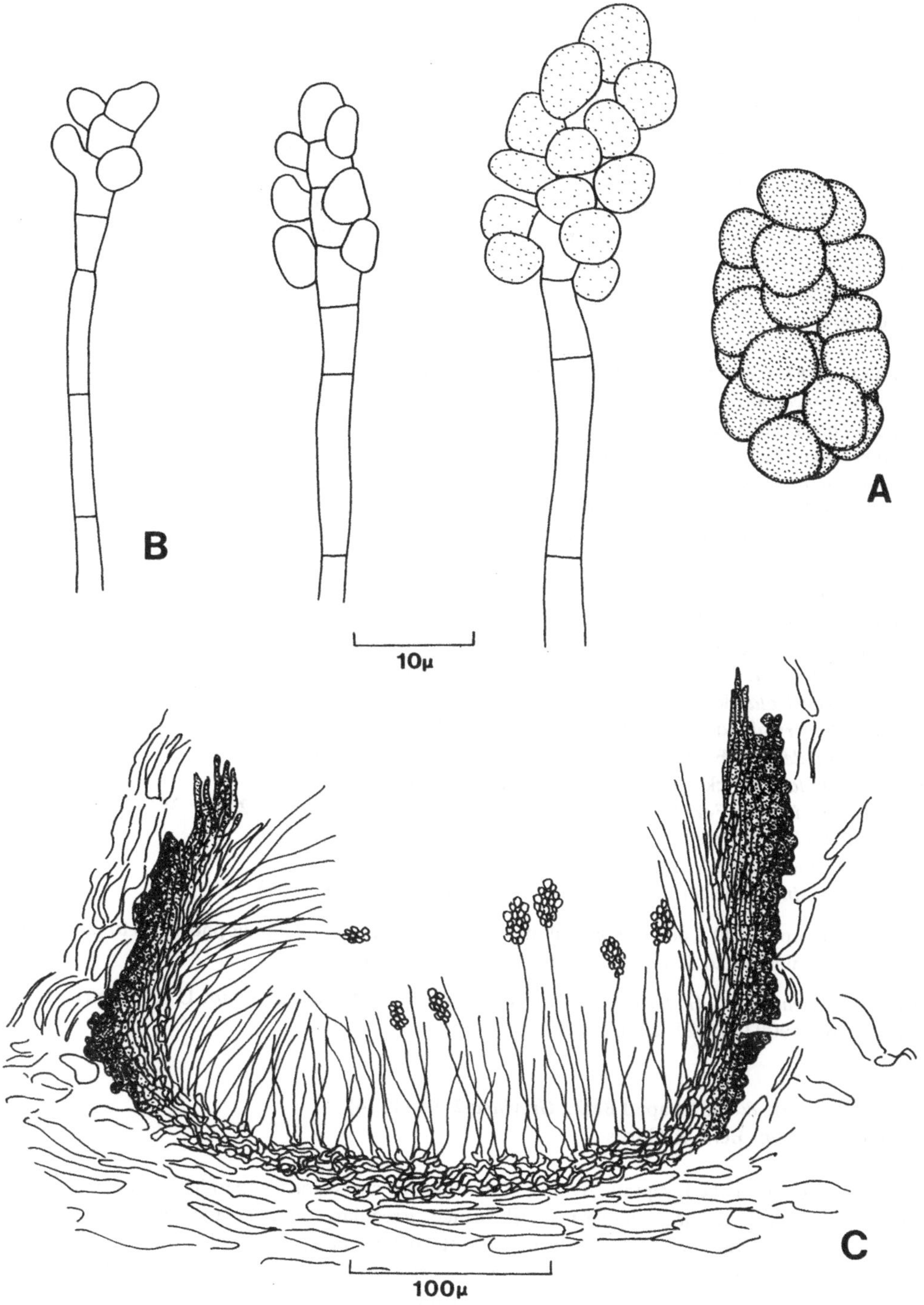

Fig. 107. *Endobotryella oblonga*, A, conidium; B, developing conidia; C, vertical section of a conidioma.

Mycelium immersed, branched, septate, hyaline. *Conidiomata* acervular, subepidermal or subperidermal, separate, formed of hyaline textura intricata. Dehiscence irregular, conidial masses strongly erumpent and adhering in a black pulvinate mass over the acervulus. *Conidiophores* hyaline, branched only at the base, septate, ± straight, cylindrical, expanding towards the apices, smooth, often enclosed in a gelatinous sheath, formed from the upper cells of the acervuli.

Conidiogenous cells holoblastic, determinate, integrated, short cylindrical, hyaline, smooth, forming a single apical complex conidium. *Conidia* of up to 6 central cells each with several short lateral acropetal branches which may in turn branch similarly to form a globose to cylindrical mass of small, thick-walled brown, smooth, aseptate, eguttulate, spherical cells, each complex conidium enclosed in a gelatinous sheath.

The several generic synonyms and widespread occurrence on many different substrates of the described species has resulted in many nomenclatural changes and synonyms. Hughes (1958) listed a comparatively brief synonymy for *C. botryospora* (Mont.) Hughes, but in doing so he omitted reference to earlier citations of this binomial by Berkeley & Broome (1850), Berkeley (1860), Cooke (1871) and Cooke (1886). Authority for the type species should be '(Mont.) Berk. & Br.' The following synonymy is largely surmised from the basic work of Hughes (1958) and by extrapolation to include earlier suggested synonymies, but it is clear that type studies are required to decide if there is a single composite species or several closely related ones.

CHEIROSPORA BOTRYOSPORA (Mont.) Berk. & Br., *Ann. Nat. Hist.* 2 ser., **5**: 455 (1850). (Fig. 108)
Cheirospora botryospora (Mont.) Hughes, *Can. J. Bot.* **36**: 748 (1958).
Stilbospora botryospora Mont., *Ann. Sci. nat.*, ser. 2, **6**: 338 (1836).
Thyrsidium botryosporum (Mont.) Mont. in Durieu, *Expl. Sci. Alg. Bot.* **1**: 325 (1849).
Myriocephalum botryosporum (Mont.) Fres., *Beitr. zur Mykol.* **2**: 40 (1852).
Rabdosporium diffusum Chev., *Flora generale des environs de Paris*: 428 (1826).
Stilbospora rhabdospora Fr., *Syst. Myc.* **3**: 485 (1832).
Stilbospora cheirospora Fr., *Syst. Myc.* **3**: 484 (1832).
Hyperomyxa stilbosporoides Cda, *Icones fungorum* **3**: 34 (1839).
Thyrsidium stilbosporoides (Cda) Sacc. *Syll. fung.* **3**: 762 (1884).
Myriocephalum hedericola de Not., *Mem. R. Accad. Sci. Torino* 2: **7**: 13 (1845).
Thyrsidium hedericola (de Not.) Durieu, *Expl. Sci. Alg. Bot.* **1**: 325 (1846).
Thyrsidium hedericola (de Not.) Durieu var. *carpini* Sacc., *Syll. fung.* **3**: 761 (1884).
Cheirospora hedericola (de Not.) Cke, *Grevillea* **14**: 127 (1886).
Myriocephalum densum Fckl, *Jahrb. Ver. Naturk. Nassau* **15**: 23 (1860).
Myriocephalum laxum Fckl, *Jahrb. Ver. Naturk. Nassau* **15**: 23 (1860).
Myriocephalum densum Fckl *β. hederae* Fckl, *Jahrb. nassau. Ver. Naturk.* **23-24**: 351 (1870).
Cheirospora micheneri Berk. & Curt., *North American Fungi* 492.
Thyrsidium micheneri (Berk. & Curt.) Sacc., *Syll. fung.* **3**: 762 (1884).
Moronopsis inquinans Delacr., *Bull. trimest. Soc. mycol. Fr.* **7**: 109 (1891).

Conidiomata up to 400 µ diam.; conidial mass up to 1050 µ diam. Conidiophores up to 200 µ long × 2-4 µ wide. Composite conidia up to 35 µ diam.; individual cells up to 5 µ diam.

On twigs of *Cornus alba, Hedera helix, Fagus grandifolia* and *Quercus robur*, U.K., Austria, N.Y., U.S.A. (IMI 123077, 31161, 49093, 57021, 22767, 42174).

51. ENDOBOTRYA Berk. & Curtis apud Berk., *Grevillea* 2: 98 (1874).
Sp. typ.: *E. elegans* Berk. & Curtis

Mycelium immersed, branched, septate, hyaline. *Conidiomata* acervular, subperidermal, separate, formed of hyaline to pale brown, thin-walled textura angularis. Dehiscence irregular, conidial masses effuse over the substrate. *Conidiophores* absent. *Conidiogenous cells* holoblastic, discrete, determinate, cylindrical, hyaline, smooth, formed from the upper pseudoparenchyma. *Conidia* muriform, consisting of a central column of 3-4 cells from which are formed columns of several lateral cells, ± cylindrical to doliiform, smooth, euseptate, markedly constricted, enclosed in a gelatinous sheath.

Höhnel (1909) reviewed *Endobotrya* and placed it in perspective with similar genera such as *Endobotryella* Höhn., *Hyperomyxa* Cda, *Thyrsidium* Mont. ex Mont. and *Thyrsidiella* Höhn. The genus remains monotypic.

E. ELEGANS Berk. & Curtis apud Berk., *ibid.*: 98 (1874). (Fig. 109)

Conidiogenous cells 7-10 × 4-7.5 µ. Conidia (excluding sheath) 32-46 × 20-25 µ. Sheath 2-8 µ thick.

On branches of *Fagus* sp. (IMI 192525, holotype ex K).

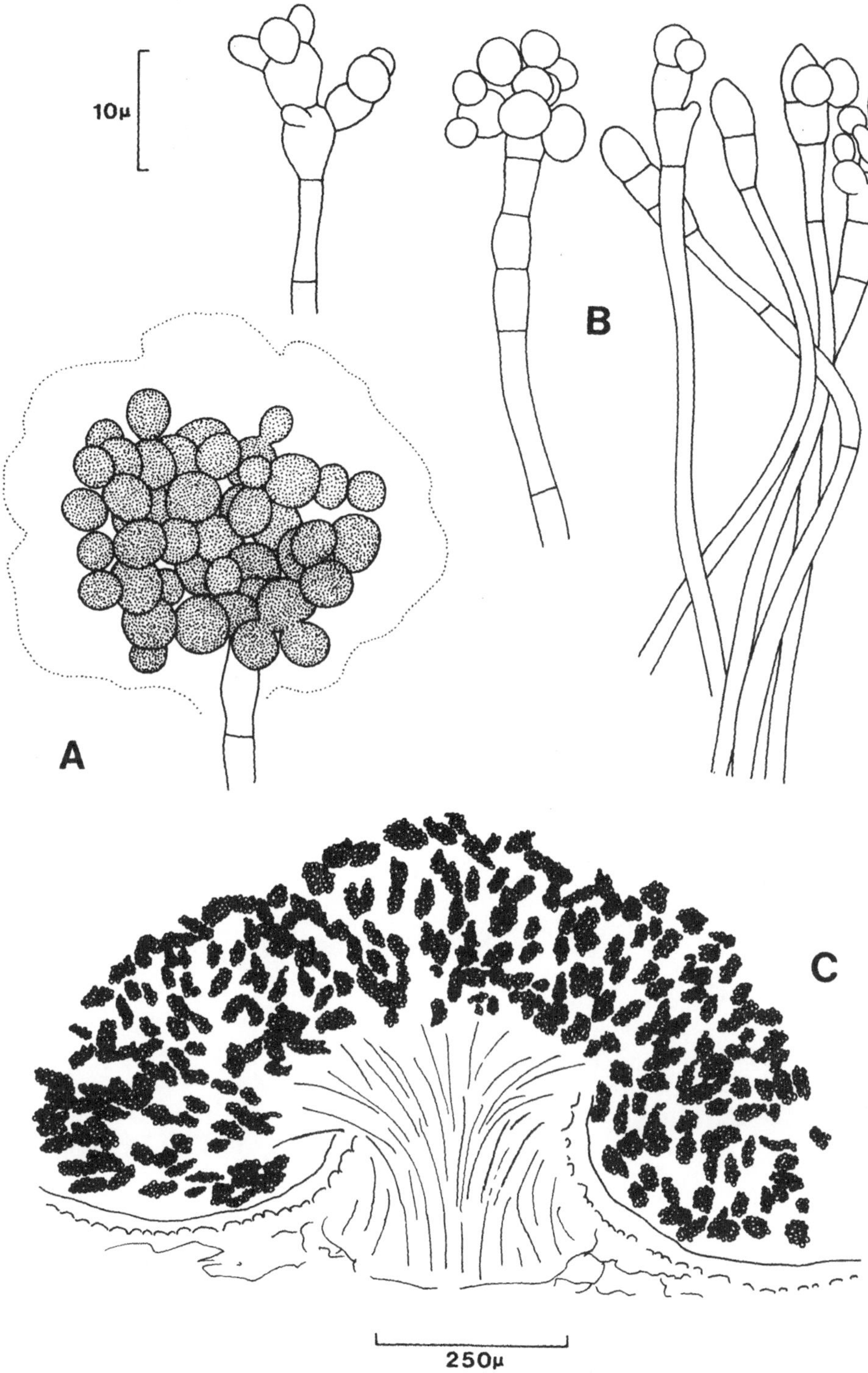

Fig. 108. *Cheirospora botryospora*, A, conidium; B, developing conidia; C, vertical section of a conidioma.

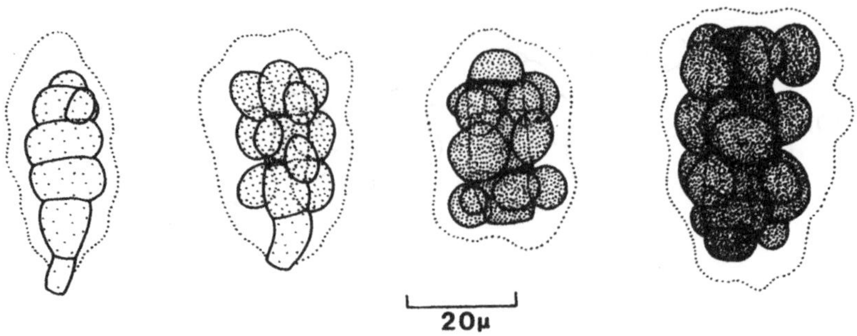

Fig. 109. *Endobotrya elegans*, Conidia.

BLASTOSTROMATINEAE, SYMPODIAL

The number of available criteria for distinguishing genera in this group is very limited. None of the conidia or conidiogenous cells are pigmented, all are smooth, hyaline and thin-walled. This limits the use of a synoptic key and places more reliance on the interpretation of conidiomatal type and conidial morphology in distinguishing taxa. Nevertheless, well-defined sections within the group can be recognised. Genera causing tar spots, discolouration of the cuticle, or which are obviously the anamorphoses of Hypodermataceae include *Thoracella, Crandallia, Aoria, Leptothyrina, Leptostroma, Fusamen* and *Conostroma*. Another group with hamate conidia includes *Libertella, Tryblidiopycnis* and *Polystigmina*. It would be interesting to speculate on the perfect-imperfect state connexions of such genera, bearing in mind the relatively small number of genera with sympodial or synchronously polyblastic conidiogenous cells.

DICHOTOMOUS KEY

(numbers in parentheses are species numbers)

1. Conidiomata associated with blackening of tissues (tar spots) or discolouring the cuticle; conidia small, unicellular, falcate, fusiform or ellipsoid 2
1. Conidiomata not associated with blackening of tissues 5
2. Conidiomata multilocular and divided 3
2. Conidiomata unilocular 4
3. Conidiogenous cells also showing percurrent proliferations, long filiform, branched; conidiomata effuse, locules irregular .... *Thoracella* (1)
3. Conidiogenous cells without proliferations, short, cylindrical; conidiomata circular, locules separated by distinct columns of tissue .... *Crandallia* (2)
3. Conidiogenous cells long, filiform, branched, without proliferations; conidiomata effuse, locules separated by distinct columns of tissue .... *Aoria* (3)
4. Conidiomata circular, basal wall pulvinate; conidiophores sparingly septate and branched; conidia curved at the base .... *Leptothyrina* (4)
4. Conidiomata circular to elongated, basal wall of several cell layers; conidiophores absent; conidia straight .... *Leptostroma* (5)
4. Conidiomata effuse and confluent, acervular; conidiogenous cells sympodial and/or annellate; conidia fusiform, straight or curved .... *Fusamen* (6)
5. Conidia branched 6
5. Conidia hamate 7
5. Conidia not so 8
6. Conidiomata superficial, setose, cupulate; conidia hyaline, consisting of a main aseptate axis and 2-4 divergent 1 septate arms at the apex .... *Belaina* (7)
6. Conidiomata eustromatic, unilocular; conidia hyaline, Y-shaped, each branch with a short setula .... *Furcaspora* (8)
6. Conidiomata acervular; conidia hyaline, consisting of a central angular cell and 4-6 small, ovoid, often curved peripheral cells .... *Vestigium* (9)
6. Conidiomata eustromatic, unilocular; conidia septate, filiform and often persistent on the conidiogenous cell, then appearing branched .... *Moralesia* (10)
7. Conidiomata acervular .... *Libertella* (11)
7. Conidiomata eustromatic, unilocular to multilocular, black .... *Tryblidiopycnis* (12)
7. Conidiomata pseudostromatic, yellow orange .... *Polystigmina* (13)
8. Conidia unicellular 9
8. Conidia either with appendages or with the apex prolonged and rostrate 10
8. Conidia septate, without appendages 11
9. Conidiomata acervular; conidiophores filiform, branched, septate; conidia small, cylindrical .... *Microgloeum* (14)
9. Conidiomata acervular; conidiophores filiform, branched, septate; conidia fusiform, straight or slightly curved .... *Rhabdogloeopsis* (15)
9. Conidiomata unilocular, multilocular or convoluted, locules situated around a central column of sterile tissue; conidiophores filiform, septate and branched .... *Conostroma* (16)
9. Conidiomata unilocular, longitudinal, pale brown; conidiophores septate, cylindrical .... *Hypocline* (17)
9. Conidiomata confluent, pulvinate, carbonous, unilocular; conidia fusiform; on other fungi .... *Anthracoderma* (18)

10. Conidiomata eustromatic, unilocular; conidia 2 septate, with 2 subapical setulae, the upper recurved .... *Gampsonema* (19)

10. Conidiomata superficial, cupulate, black, thick-walled; conidia attenuated towards the apices into a rostrum, base sometimes with an excentric cellular appendage .... *Heteropatella* (20)

10. Conidiomata acervular; conidia 1-2 septate, separating a small basal cell and a longer filiform apical cell, both devoid of contents .... *Monochaetiellopsis* (21)

10. Conidiomata navicular, setose; conidia with a single apical, and 2-6 suprabasal unbranched cellular appendages; conidiogenous cells sometimes with annellations .... *Ajrekarella* (22)

11. Conidiomata superficial 12

11. Conidiomata immersed 15

12. Conidiomata cupulate 13

12. Conidiomata not cupulate 14

13. Conidiomata black, multilocular, locules restricted to the base of the cup; conidia multiseptate, curved, released in small globose masses .... *Bothrodiscus* (23)

13. Conidiomata dark brown, unilocular, small, thin-walled; conidia multiseptate, straight or curved ...... *Oncosporella* (24)

13. Conidiomata pale brown; conidia straight or curved, septate, filiform .... *Septopatella* (25)

14. Conidiophores restricted to the upper wall, lower wall absent; conidia hyaline, cylindrical, 1 septate........................ *Leptochlamys* (26)

14. Conidiophores restricted to the lower wall, upper wall of separate cells in mucilage; conidia fusiform, hyaline, 3 septate .... *Asteromidium* (27)

15. Conidiomata acervular; conidiophores absent; conidia multiseptate, filiform .... *Phloeosporella* (28)

15. Conidiomata acervular; conidiophores branched and septate; conidia 1-2 septate, cylindrical to long pyriform .... *Diplosporonema* (29)

15. Conidiomata multilocular or convoluted; conidiogenous cells ampulliform to lageniform; conidia straight, hyaline, 1-3 septate .... *Septocyta* (31)

15. Conidiomata unilocular but convoluted; conidiophores much branched and anastomosed; conidia long, filiform, often curved, up to 7 septate .... *Stictosepta* (30)

SYNOPTIC KEY

Mycelial characters

Position in relation to substrate

a. superficial — 20

b. immersed — 1 2 3 4 5 6 7 8 9 10 11 12 13 14 15 16 17 18 19 20 21 22 23 24 25 26 27 28 29 30 31

Conidiomatal characters

Position in relation to substrate

a. superficial — 7 20 22 23 24 25 26

b. immersed — 1 2 3 4 5 6 8 9 10 11 12 13 14 15 16 17 18 19 21 27 28 29 30 31

c. semi-immersed — 7 22

Dehiscence

a. ostiolate, nonpapillate — 8 30

b. ostiolate, depressed — 13

c. linear-elongated — 2 10 17 22

d. furfuraceous — 8 30

e. irregular rupture — 1 2 3 4 5 6 7 9 11 12 14 15 16 18 19 20 21 23 24 25 26 27 28 29 31

Conidiomatal type

a. acervular — 6 9 14 15 21 28 29
b. cupulate — 7 20 23 24 25 26
c. simple eustromatic — 4 5 8 11 18 19 26 27 30 31
d. complex eustromatic — 1 2 3 16
e. complex eustromatic, pulvinate — 12
f. pseudostromatic — 13
g. flattened, pycnidial-like — 2 3 4 5
h. linear — 10 17 22

Locular configuration

a. unilocular — 4 5 6 7 8 9 10 11 14 15 17 18 19 20 21 22 24 25 26 27 28 29 30
b. multilocular — 2 3 12 18 23
c. multilocular, convoluted — 1 13 16 31

Other criteria

a. dark brown pigmentation — 1 2 3 4 5 7 8 9 10 11 12 16 18 19 20 22 23 24 25 26 27 30 31
b. hypocreaceous — 13 17
c. forming tar spots — 1 3
d. setose — 7 22
e. overlying wall composed of individual cells in a gelatinous matrix — 27

Conidiogenous cell characters

Insertion

a. discrete — 2 4 5 18 19 21 23 25 27 28 31
b. integrated — 1 2 3 4 6 7 8 9 10 11 12 13 14 15 16 17 18 19 20 21 22 24 25 26 28 29 30

Shape

a. ampulliform-doliiform — 19 31
b. lageniform — 2 4 18 21 26 27 28 30 31
c. cylindrical — 1 2 3 4 5 6 7 8 9 10 11 12 13 14 15 16 17 19 20 22 23 24 25 28 29 30

Conidiogenesis

a. occasionally showing annellations — 1 6 22

Conidiophore characters

Shape

a. irregularly branched and ramified — 18 30
b. long cylindrical to filiform, unbranched — 2 20 21 28
c. filiform, branched — 1 3 4 6 7 8 9 10 11 12 13 14 15 16 17 19 22 24 25 26 28 29

Conidial characters

Septation	
a. aseptate	1 2 3 4 5 6 8 11 12 13 14 15 16 17 18 22
b. medianly 1 septate	26 28 31
c. unequally 1 septate	29
d. 2-multiseptate	7 10 19 20 21 23 24 25 26 27 28 29 30 31
Shape	
a. oval-ellipsoid	3 5 14 16 17 18
b. cylindrical	1 2 14 16 17 21 24 25 26 28 29
c. fusiform	6 15 18 19 20 22 24 25 27 28
d. falcate	6 11 12 20 23 24 28
e. pyriform	3
f. clavate	29
g. filiform	10 25 30 31
h. hamate	11 12 13 21
i. branched	7 8 9 10 28
j. allantoid	4 18
k. Y-shaped	8
l. consisting of a large central cell and smaller peripheral cells	9
Guttulation	
a. eguttulate	4 5 6 7 8 9 11 12 13 14 15 16 17 18 19 20 21 22 23 26 27 28 29
b. guttulate	1 2 3 10 15 20 24 25 28 30 31
Appendages	
a. apical	8 19 20 21 22
b. basal	8 20 21 22 28
c. lateral	19
d. single	8 19 20 21 22 28
e. several at one point	22
f. cellular	20 21 22 28
g. simple	8 19 20 21 22 28
h. setulose	8 19
Other characters	
a. conidia exuded in globose masses	23
b. conidia exuded in tendrils	11

1. THORACELLA Oud., *Ned. Kruidk. Arch.* ser. 3, **2**: 267 (1901).
 Sp. typ.: *T. ledi* Oud.

Mycelium immersed, branched, septate, hyaline to pale brown. *Conidiomata* eustromatic, immersed, subcuticular, black, effuse, confluent, extensive, forming large black lesions, irregularly multilocular; sterile stromata composed of an upper black wall several cells thick, of thick-walled, dark brown textura angularis merging into a lower thicker layer of thick-walled, hyaline textura intricata; locules formed within this lower layer which at the base becomes darker with a few sclerotioid cells amongst the hyaline ones; epidermis distinct but filled with pale brown fungal cells; hypodermis and mesophyll penetrated by hyaline hyphae. *Ostiole* absent, dehiscence by irregular fissure. *Conidiophores* pale brown to hyaline, septate mainly at the base, tapered markedly towards the apex, branched at the base, less so above, flexuous, formed at the base of the locule. *Conidiogenous cells* holoblastic, sympodial or synchronous, indeterminate, integrated, filiform, hyaline, smooth, loci indistinct. (Some conidiogenous cells and conidiophores are annellate towards the base, but the diameter of annellations suggests that the conidia do not arise in this way. Perhaps annellations are ornamental incrustations although their regularity on the conidiophores does not favour this interpretation.) *Conidia* hyaline, aseptate, smooth, thin-walled, straight, apex obtuse, base ± acute, cylindrical, ± irregularly guttulate.

The genus is monotypic.

T. LEDI Oud., *ibid.*: 267 (1901). (Fig. 110)

Conidiophores up to 50 µ long × 2 µ wide at the base, narrowing to 1.5 µ wide at the apex. Conidia 6-9 × 1.5 µ.

Causing a tar spot on leaves of *Ledum palustre*, Netherlands (IMI 192312, holotype).

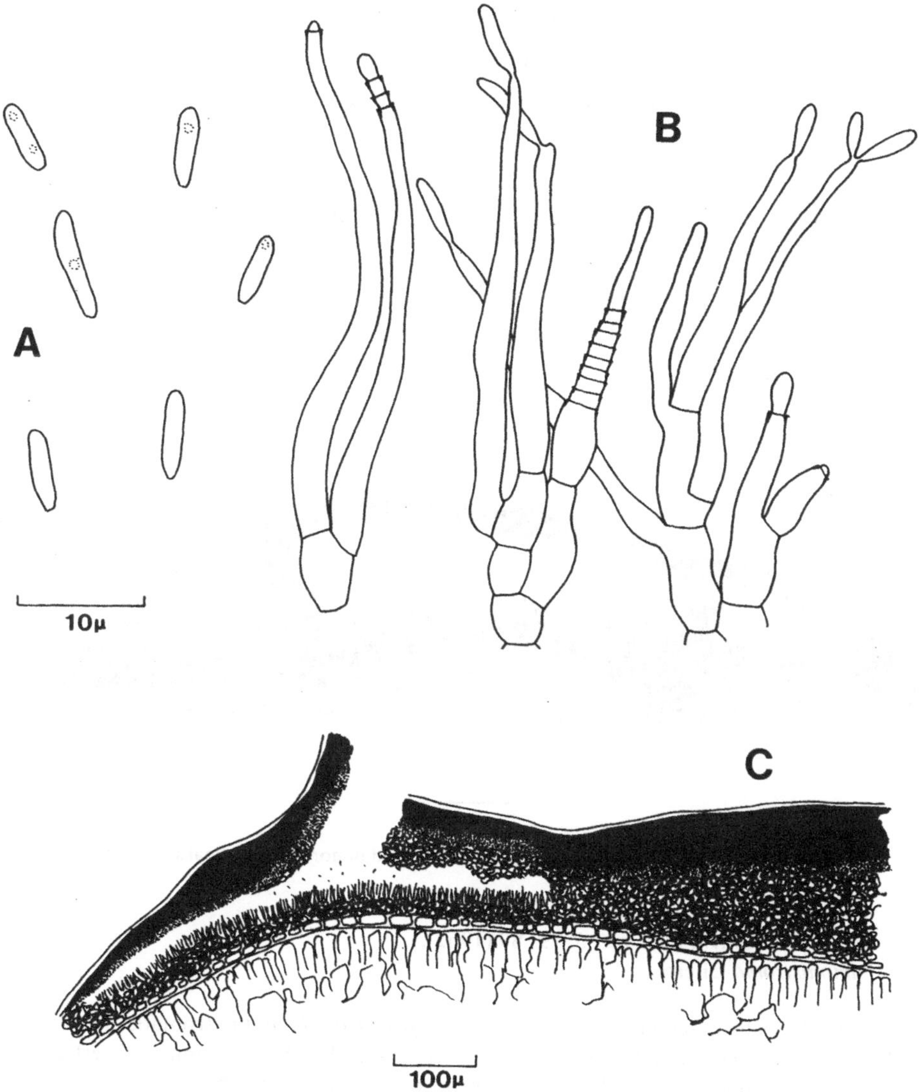

Fig. 110. *Thoracella ledi*, A, conidia, B, conidiophores and developing conidia; C, vertical section of a conidioma.

2. CRANDALLIA Ell. & Sacc., *Bull. Torrey bot. Cl.* **23**: 466 (1897).
Sp. typ.: *C. juncicola* Ell. & Sacc.

Mycelium immersed, branched, septate, hyaline. *Conidiomata* subcuticular, scutate, circular, separate, with a dark brown to black upper wall 1 cell thick composed of textura intricata, and a lower wall several cells thick, of hyaline thin-walled textura angularis; the lower wall of the cells adjacent to the upper epidermal wall are dark brown and thickened; upper and lower walls connected by 1 or more columns of dark brown sclerotioid vertically elongated pseudoparenchyma, so forming several locules; margin entire to irregular. *Ostiole* central, circular. *Conidiogenous cells* restricted to the lower wall, holoblastic, possibly sometimes sympodial, discrete or integrated on 1-septate conidiophores, determinate or indeterminate, hyaline, smooth, cylindrical, straight or somewhat irregular, abruptly tapered at the apices. *Conidia* hyaline cylindrical, aseptate, thin-walled, obtuse at each end, guttulate.

Previously (Sutton, 1973a) I interpreted conidial development as phialidic, but repeated examination by a variety of methods and stains has convinced me that conidia and conidiogenous cell walls are continuous until secession. As such interpretation is holoblastic rather than enteroblastic.

C. JUNCICOLA Ell. & Sacc., *ibid.*: 466 (1897). (Fig. 111)

Conidiomata up to 120 μ diam. Conidiogenous cells 8.5-14 × 2-2.5 μ. Conidia 9-10 × 2 μ.

On dead leaves and living stems of *Juncus drummondii*, Wyo., U.S.A. (IMI 97182, 97265).

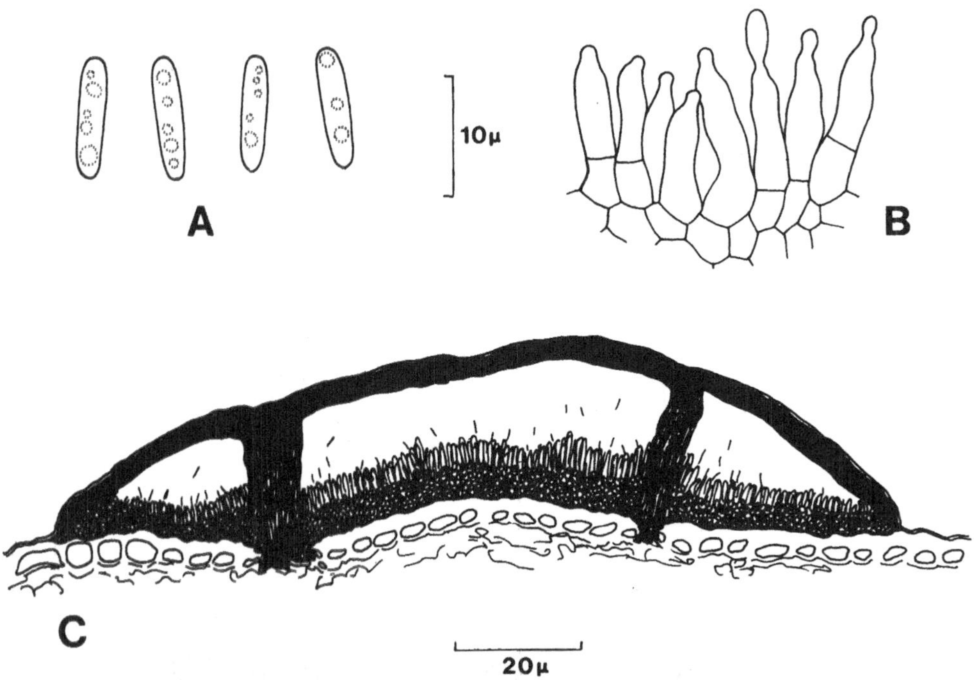

Fig. 111. *Crandallia juncicola*, A, conidia; B, conidiophores; C, vertical section of a conidioma.

3. AORIA Ciferri, *Atti Ist. Bot. Univ. Pavia*, ser. 5, **19**: 89 (1962).
Sp. typ.: *A. amphistroma* Ciferri

Mycelium immersed, branched, septate, hyaline to pale brown. *Conidiomata* stromatic, immersed, subepidermal, black, effuse, confluent, often extensive, multilocular, locules separated by thick columns of vertically elongated dark brown to black pseudoparenchyma; upper wall 1-2 cells thick, consisting of dark brown to black dead tissue which is closely appressed to the upper epidermis, itself packed with brown pseudoparenchyma and hyphae; lower wall consisting of a layer of subepidermal cells loosely packed with pale brown pseudoparenchyma (periodically with a few dark brown to black cells interspersed) from which conidiogenous cells arise. *Ostiole* absent, dehiscence by irregular fissures in the upper wall. *Conidiophores* hyaline, septate at the base and above, tapered to the apex, flexuous, branched immediately below transverse septa, formed from the base of the locules. *Conidiogenous cells* holoblastic, sympodial or synchronous, indeterminate, integrated, filiform, hyaline, smooth, loci indistinct. *Conidia* hyaline, aseptate, smooth, thin-walled, straight, apex obtuse, base ± acute, ellipsoid, minutely guttulate.

Considered by Ciferri (1962) to represent the conidial state of a *Linochora* or *Phyllachoropsis*, but *Linochora* is itself a conidial fungus and the generic name *Phyllachoropsis* cannot be traced. It seems more likely to be a conidial *Rhytisma*. At present the genus is monotypic.

A. AMPHISTROMA Ciferri, *ibid.*: 89 (1962). (Fig. 112)

Conidiophores up to 28 µ long × 3 µ wide at the base, tapering to 1 µ wide at the apex. Conidia 4-4.5 × 1.5-2 µ.

Causing a tar spot on leaves of *Lyonia* sp., Dominica (IMI 201376, holotype ex BPI).

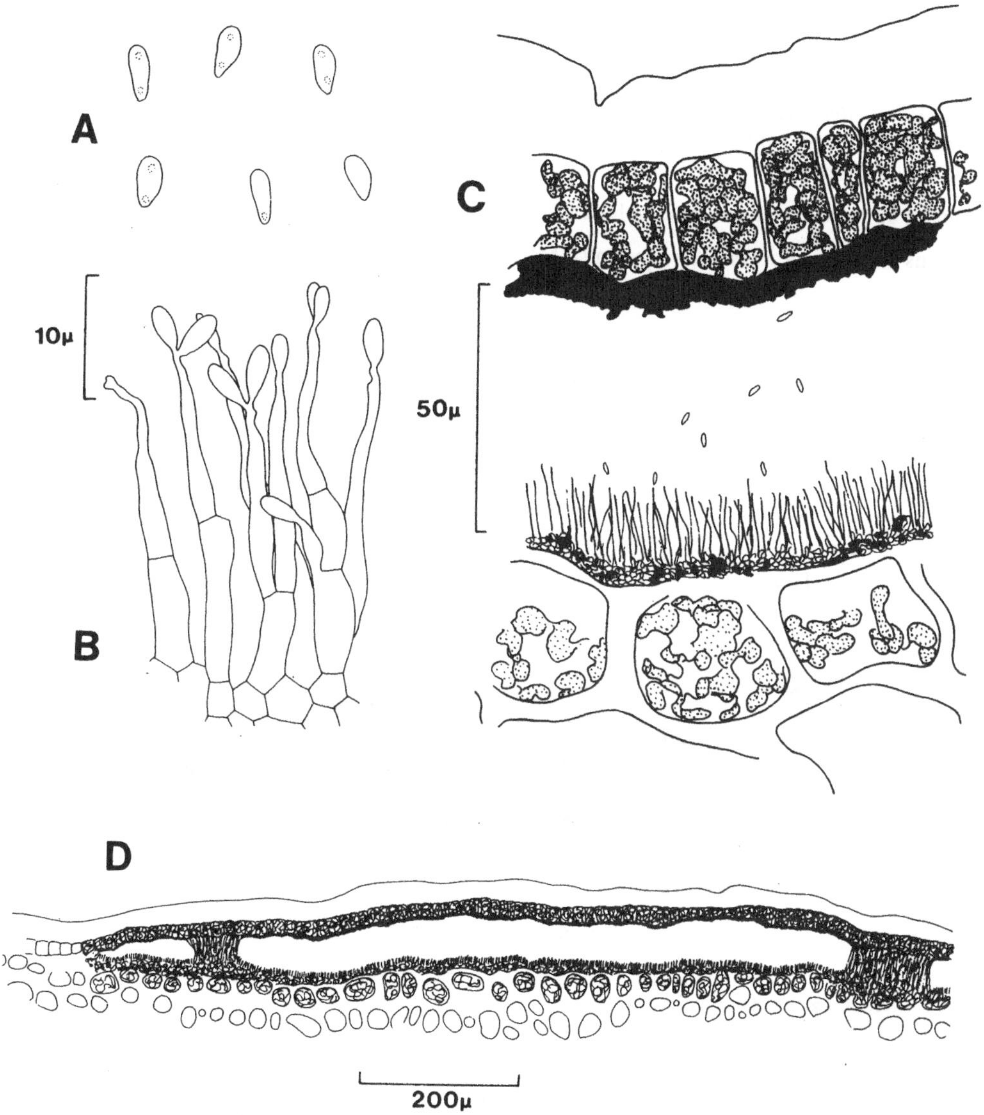

Fig. 112. *Aoria amphistroma*, A, conidia; B, conidiophores and developing conidia, C, detail of conidiomatal walls in relation to the substrate tissues; D, vertical section of part of a conidioma.

4. LEPTOTHYRINA Höhn., *Sber. Akad. Wiss. Wien* **124**: 123 (1915).
Sp. typ.: *L. rubi* (Duby) Höhn. (syn. *Sphaeria rubi* Duby).

Mycelium immersed, branched, septate, hyaline. *Conidiomata* subcuticular, amphigenous, black, shining, circular, consisting of an upper wall, 1 cell thick, of dark brown very thick-walled textura angularis and a two layered lower wall at the base of similar tissue to the upper wall, surmounted by a layer several cells thick of larger-celled, pale brown, thin-walled textura angularis. *Ostiole* absent, dehiscence by irregular rupture of overlying wall. *Conidiophores* hyaline, septate and branched only at the base, smooth, thin-walled, markedly attenuated towards the apices, formed only from the basal wall. *Conidiogenous cells* holoblastic, sympodial to synchronous, indeterminate, integrated or discrete, lageniform, hyaline, with 1-2 small, unthickened scars at the apices. *Conidia* hyaline, thin-walled, smooth, aseptate, eguttulate, tapered and often curved towards the base, apex obtuse.

Saccardo (1884) placed the type species in *Leptothyrium* Kze, a decision accepted by Clements & Shear (1931). However, *Leptothyrium* has falcate conidia formed from phialides, and conidiomata in which the lower wall merely consists of a one-celled layer of thick-walled textura angularis on which the phialides are directly inserted.

Apart from the type species, only two other taxa have been referred to *Leptothyrina* and these have not been recently revised.

L. RUBI (Duby) Höhn., *ibid.*: 123 (1915). (Fig. 113)
Sphaeria rubi Duby, *Botanicon Gall.* **2**: 712 (1830).
Leptothyrium rubi (Duby) Sacc., *Syll. fung.* **3**: 629 (1884).

Conidiomata up to 200 μ diam. Conidiophores 7-12 × 2 μ at the base. Conidia 3.5-4.5 × 1-1.5 μ.

Causing lesions on leaves of *Rubus*, Switzerland (IMI 203724 ex FH).

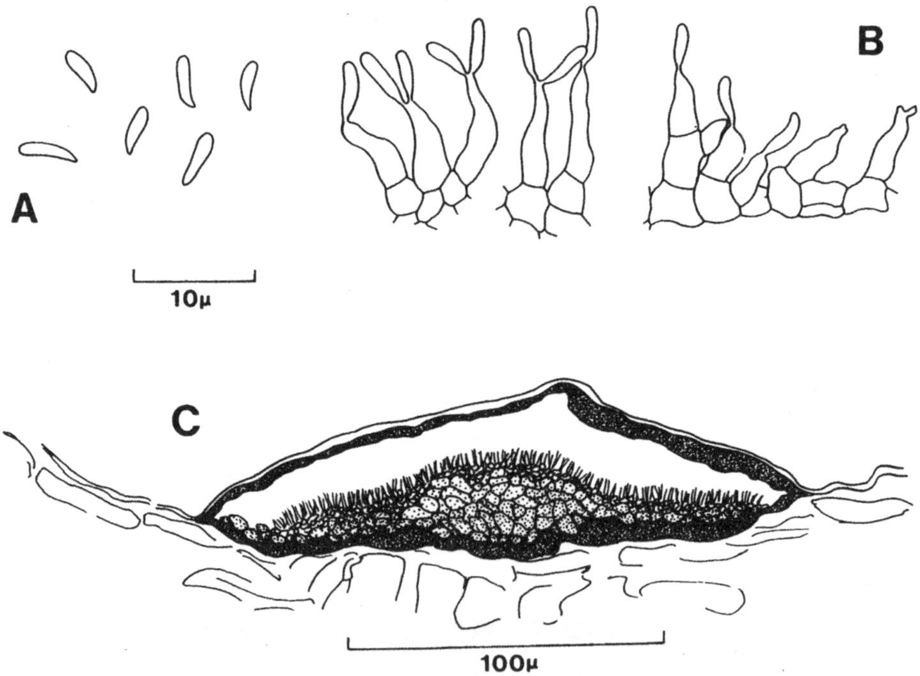

Fig. 113. ***Leptothyrina rubi*, A, conidia, B, conidiophores and developing conidia; C, vertical section of a conidioma.**

5. LEPTOSTROMA Fr. ex Fr., *Syst. Myc.* **2**: 597 (1923).
Leptostroma Fr., *Obs. Myc.* **1**: 196 (1815).
Original spp.: *L. scriptum* Fr. ex Fr., *L. scirpinum* Fr., *L. caricinum* Fr., *L. juncinum* Fr., *L. filicinum* Fr. ex Fr., *L. spiraeae* Fr., *L. vulgare* Fr., *L. hysterioides* Fr. ex Fr., *L. sphaeroides* Fr. ex Fr.
Lectotype sp.: *L. scirpinum* Fr., fide Höhnel (1915b), Clements & Shear (1931).

Mycelium immersed, branched, septate, dark brown. *Conidiomata* subcuticular, black, shining, circular, occasionally elongated, often collabent, consisting of an upper wall, 1 cell thick, of dark brown textura angularis or prismatica, and a lower wall which at the periphery is only of similar dark brown cells, mainly it consists of a layer many cells thick, of hyaline, thin-walled, small-celled textura angularis. *Ostiole* absent, dehiscence by irregular rupture of overlying wall. *Conidiophores* absent. *Conidiogenous cells* holoblastic, sympodial to synchronous, indeterminate, discrete, filiform, hyaline, with several slightly protuberant, unthickened scars towards the apices, formed from the upper cells of the lower wall. *Conidia* hyaline, thin-walled, aseptate, eguttulate, straight, ellipsoid.

Höhnel (1915b) gave extensive reasons for selecting *L. scirpinum* as the lectotype, but apart from this account the genus has not subsequently been studied. About 200 species have been described in the genus so far. Grove (1937) implied that *L. scirpinum* was the conidial state of *Hypoderma scirpinum* DC. ex Mérat but the species was not discussed in the revision of the Hypodermataceae by Darker (1967).

L. SCIRPINUM Fr., *ibid.*: 598 (1823). (Fig. 114)
L. xylomoides Fr., *Obs. Myc.* **1**: 197 (1815).

Teleomorphosis: *Hypoderma scirpinum* DC. ex Mérat, *Nouv. Fl. Paris* **1**: 152 (1821).

Conidiomata up to 220 μ diam. Conidiogenous cells 13-20 × 1-2 μ. Conidia 3-3.5 × 1-1.5 μ.

On dead culms of *Arundinaria macrosperma, Scirpus lacustris, S. validus, S.* sp., Miss., N. Dak., U.S.A., Hungary, Germany, Latvia, U.S.S.R., Sask., Canada (IMI 22861, 22862, 11958, 22859, 26829, 22860).

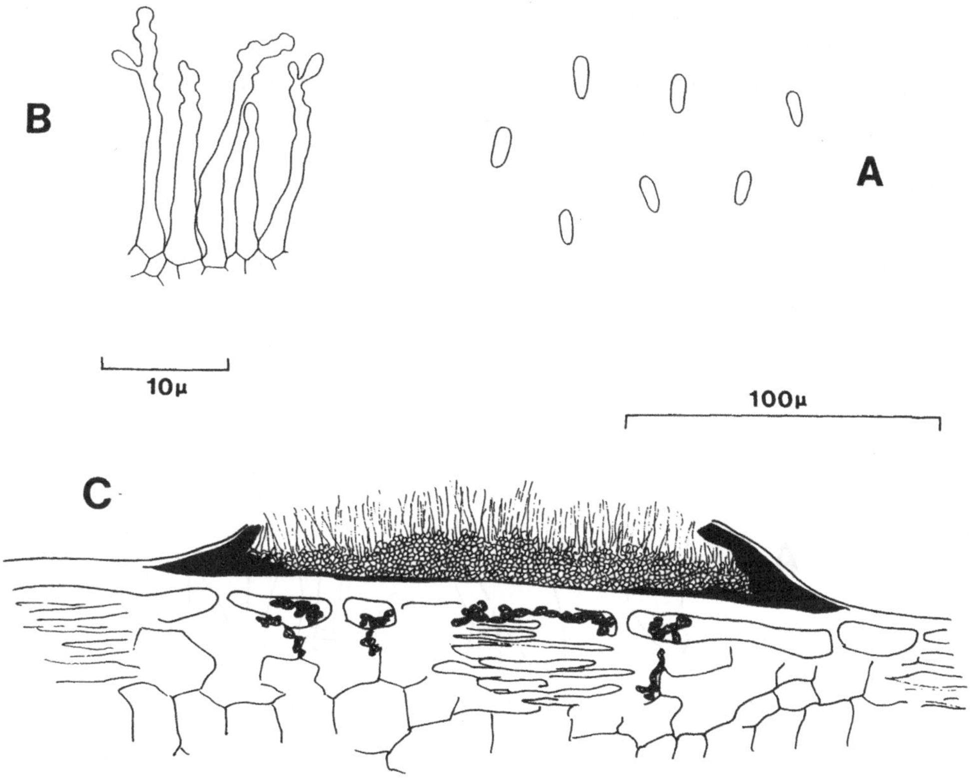

Fig. 114. *Leptostroma scirpinum*, A, conidia; B, conidiogenous cells; C, vertical section of a dehisced conidioma.

6. FUSAMEN (Sacc.) Karst., *Hyphomycetes fennici*: 143 (1892).
Fusarium Lk ex Fr. subgen. *Fusamen* Sacc., *Syll. fung.* **4**: 714 (1886).
Platycarpium Karst., *Acta Soc. Fauna Flora fenn.* **27**: 10 (1905).
Calogloeum H. Syd. apud Petrak & H. Syd., *Annls mycol.* **22**: 401 (1924).
Sp. typ.: *F. deformans* (Schröt.) Karst. (syn. *Fusarium deformans* Schröt.).

Mycelium immersed, branched, septate, hyaline to pale brown. *Conidiomata* acervular, separate or confluent, subcuticular, discolouring the cuticle, formed of hyaline to pale brown, thin-walled textura angularis. Dehiscence irregular. *Conidiophores* hyaline, sparingly branched and septate at the base, occasionally above, smooth, formed from the upper cells of the conidiomata. *Conidiogenous cells* holoblastic, indeterminate, integrated or discrete, cylindrical, hyaline, either sympodial with 1-2 broad flat unthickened denticles (in *F. fructigenum* and *F. amentorum*) or annellidic with 1-3 percurrent proliferations (in *F. deformans*). *Conidia* hyaline, aseptate, smooth, thin-walled, fusiform, straight or falcate, apex obtuse, base truncate.

Arx (1957a) accepted five species, all of which occur as parasites on catkins and leaves of *Salix*. He interpreted conidial development as phialidic (Arx, 1970) but this is erroneous. In the type species annellides are clearly present, but in *F. fructigenum* and *F. amentorum* the conidiogenous cells are sympodial with no percurrent proliferations. The combination of annellidic and sympodial conidiogenesis is documented for a few taxa (Sutton, 1964a; 1967a), and in *Fusamen* it would seem inadvisable to seperate taxa which have so many other characteristics in common. If, however, it is suggested, then *Platycarpium* would be a suitable genus for the sympodial species.

F. DEFORMANS (Schröt.) Karst., *ibid.*: 143 (1892). (Fig. 115)
synonymy in Arx (1957a; 1970).

Conidiophores 13.5-19.5 × 3.5-5 μ. Conidia falcate, 19-22 × 5.5-7 μ.

On female catkins of *Salix caprea*, Finland (IMI 21711).

F. AMENTORUM (de Lacr.) Arx, *Verh. K. Ned. Akad. Wet. Amst.* **51** (3): 57 (1957). (Fig. 116)
synonymy in Arx (1957a; 1970).

Conidiomata up to 250 μ diam. Conidiophores 12-25 × 2.5-3 μ. Conidia straight, 14-18 × 5.5-6.5 μ.

On female catkins of *Salix cinerea*, Latvia, Finland (IMI 30132, 21684).

F. FRUCTIGENUM (Karst.) Arx, *Verh. K. Ned. Akad. Wet. Amst.* **51** (3): 21 (1957). (Fig. 117)
synonymy in Arx (1957a; 1970).

Conidiomata up to 250 μ diam. Conidiophores 12-23 × 2.5-4.5 μ. Conidia falcate, 19-23 × 5-7 μ.

On *Salix myrtilloides*, Finland (IMI 153903, holotype *P. fructigenum* ex H).

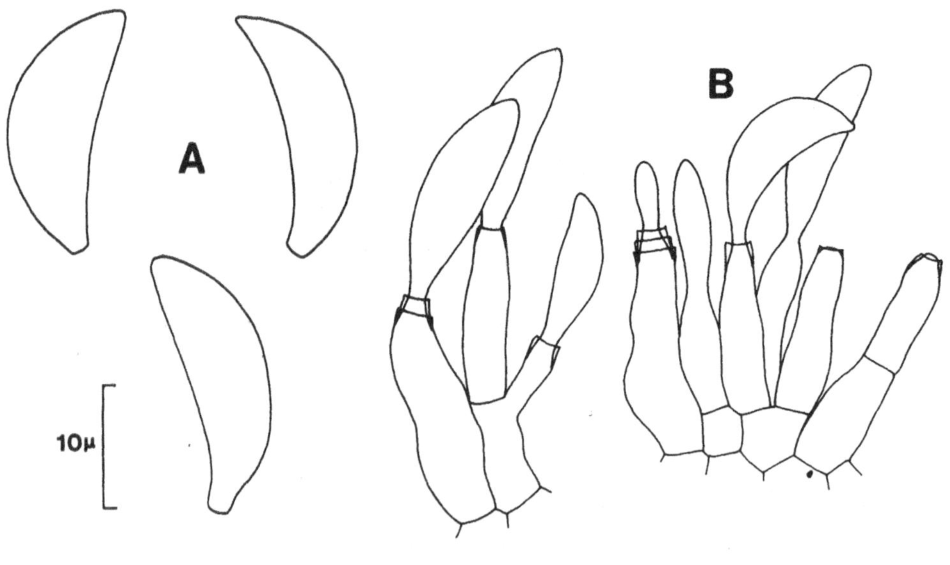

Fig. 115. *Fusamen deformans*, A, conidia; B, conidiophores and developing conidia.

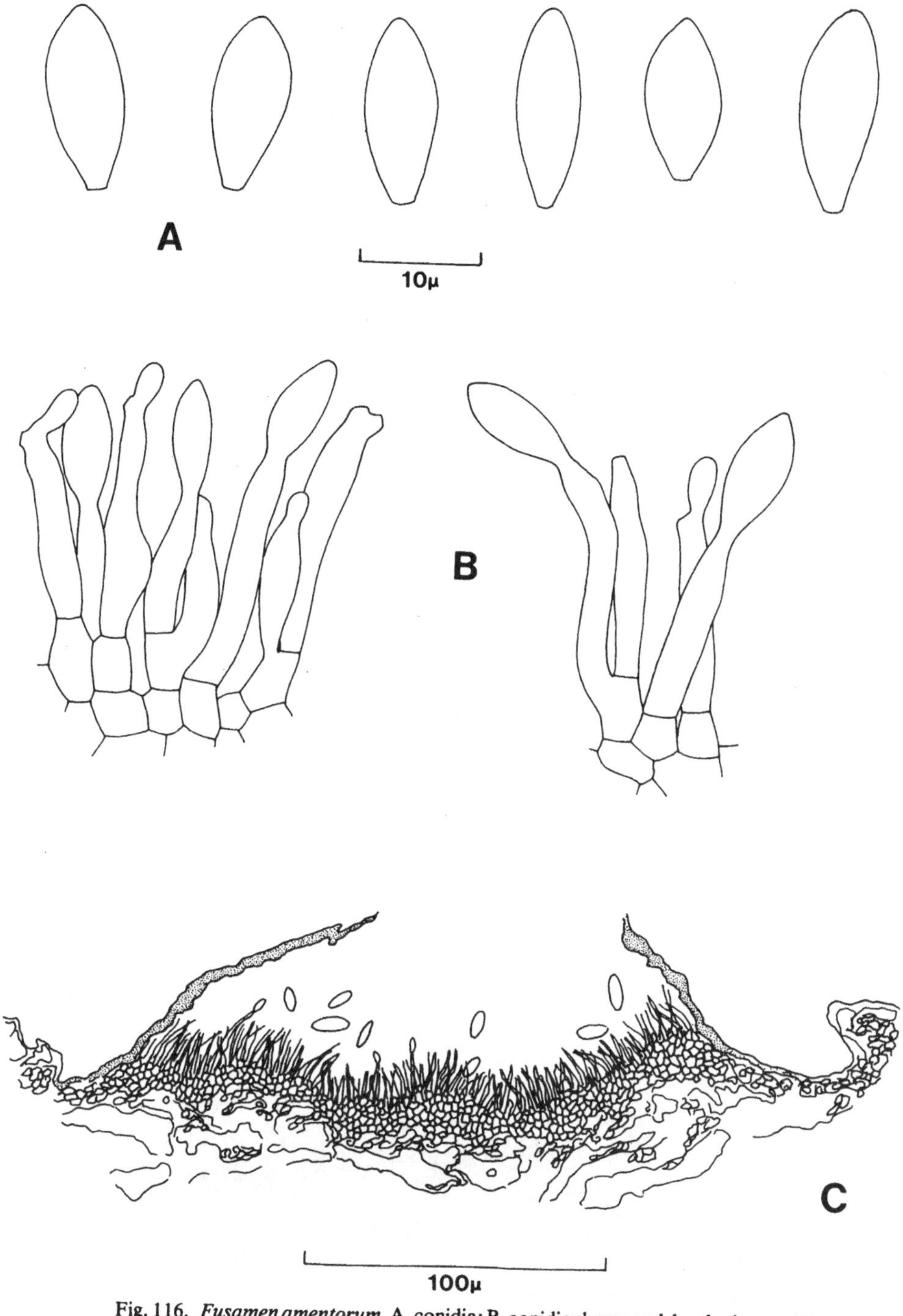

Fig. 116. *Fusamen amentorum*, A, conidia; B, conidiophores and developing conidia; C, vertical section of a conidioma.

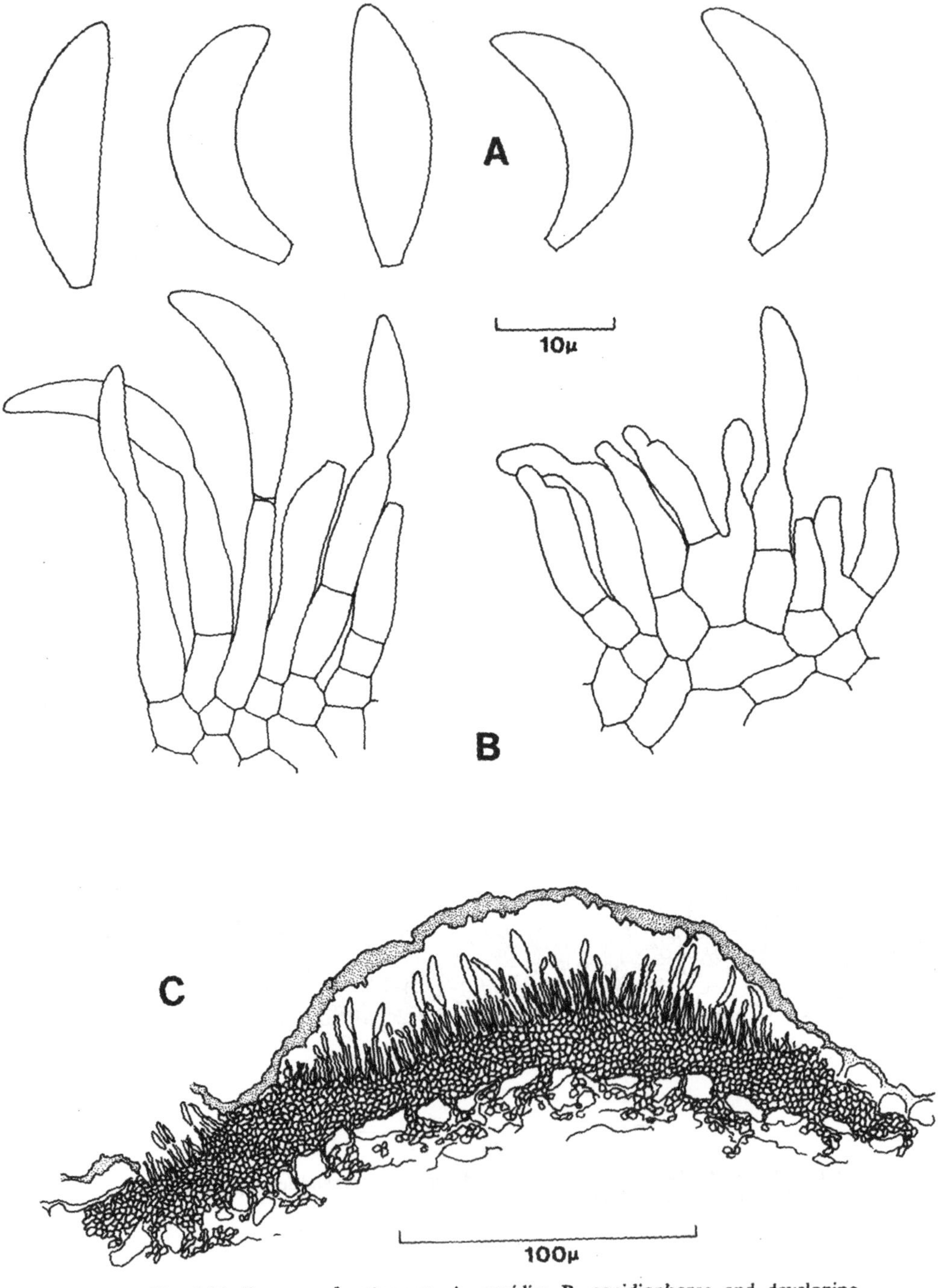

Fig. 117. *Fusamen fructigenum*, A, conidia; B, conidiophores and developing conidia; C, vertical section of a conidioma.

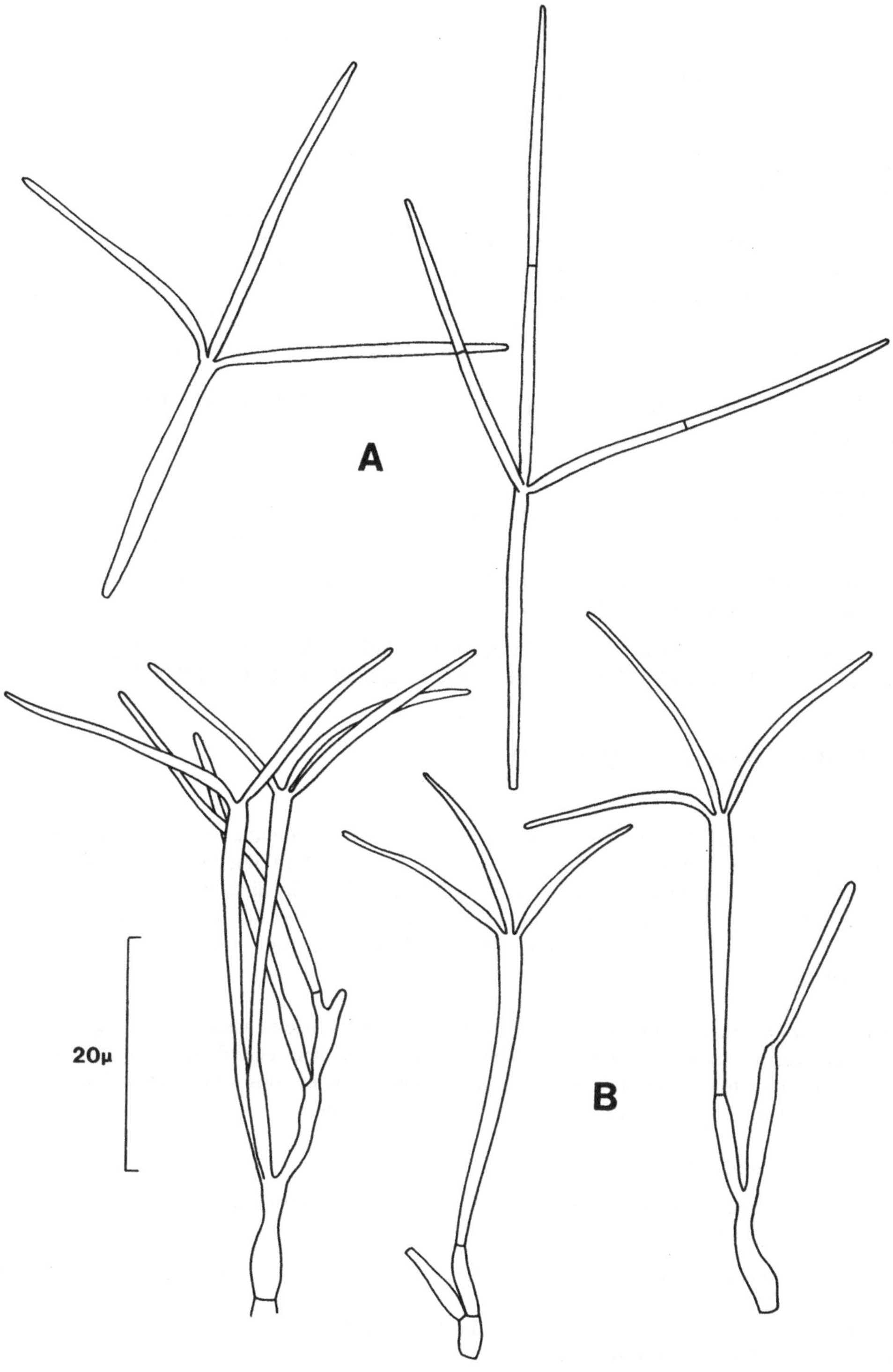

Fig. 118. *Belaina asclepiadis*, A, conidia; B, conidiophores and developing conidia.

7. BELAINA Bat. & Peres, *Mem. Soc. broteriana* **14**: 50 (1961).
Belainopsis Bat. & Maia, *Revta Biol. Lisboa* **5**: 84 (1965).
Crucellisporium Farr apud Farr & Horner, *Nova Hedwigia* **15**: 264 (1968).
Sp. typ.: *B. asclepiadis* Bat. & Peres.
Illustrations: Farr & Horner (1968); Batista & Peres (1961).

Mycelium immersed, brown, septate, branched. *Conidiomata* superficial to semi-immersed, separate, dark brown, sessile, at first ± closed, later opening widely to become cupulate; periclinal wall pseudoparenchymatous, composed of textura angularis towards the inside and textura porrecta extending longitudinally to setiform hyphae or setae on the outside. *Ostiole* absent. *Setae* restricted to the periclinal wall, acuminate or obtuse, septate, thick-walled, smooth, brown. *Conidiophores* hyaline, branched at the base and above, septate, smooth, ± cylindrical, restricted to the base of the conidioma. *Conidiogenous cells* holoblastic, sympodial or synchronous, integrated, hyaline, smooth, cylindrical, with 1-2 unthickened, slightly protuberant conidiogenous loci. *Conidia* consisting of a main axis the base of which is truncate at the point of secession, and 2-4, divergent, 1-septate, tapered arms arising from the apex, hyaline.

In a discussion of the generic names *Belainopsis* and *Crucellisporium*, Sutton (1977a) explained the synonymy for *Belaina*. The genus is monotypic.

B. ASCLEPIADIS Bat. & Peres, *ibid.*: 50 (1961). (Fig. 118)
Belainopsis malpighiacearum Bat. & Maia, *Revta Biol. Lisboa* **5**: 86 (1965).
Crucellisporium selaginellae Farr apud Farr & Horner, *Nova Hedwigia* **15**: 264 (1968).

Conidiomata 70-200 μ diam. Setae 35-95 × 1.5-3 μ. Conidiophores up to 30 μ long × 1.5-3 μ wide. Conidial main axis 15-30 × 1.5-2 μ; branches 21-40 μ long × 1-1.5 μ.

On leaves of *Asclepias curassavica*, Malpighiaceae indet., *Selaginella rupestris, Eucalyptus citriodora, E.* spp., Brazil and N. Carol., USA (IMI 175801, holotype of *C. selaginellae* ex BPI, 196480i, 186982b, 196481d, 88060, holotype of *B. asclepiadis*).

8. FURCASPORA Bonar, *Mycologia* **57**: 391 (1965).
Sp. typ.: *F. pinicola* Bonar

Mycelium immersed, branched, septate, pale brown. *Conidiomata* eustromatic, immersed, globose, unilocular, pale brown, thick-walled; walls of similar texture, of pale brown, thin-walled textura angularis. *Ostiole* central, circular, wide, unspecialized, furfuraceous. *Conidiophores* branched and septate at the base, hyaline, erect, formed from the inner cells of the conidiomatal wall. *Conidiogenous cells* holoblastic, sympodial, integrated or discrete, indeterminate, hyaline, with 1-3 small, scarcely protuberant loci at the apices. *Conidia* triradiate, hyaline, aseptate, each arm with an apical, extracellular, short, unbranched setula.

The type species was redescribed by Sutton (1972a). The only other species assigned to the genus is *F. indica* Talde (1975) but this is incorrectly placed in *Furcaspora*. The comparatively large conidiomata with a distinctive basal conidiogenous region and triradiate conidia lacking setulae agree fully in dimensions with *Suttoniella eriobotryae* (Ahmad) Ahmad. Consequently *F. indica* is regarded as a later synonym of this name.

F. PINICOLA Bonar, *ibid.*: 391 (1965). (Fig. 119)

Conidiomata up to 300 μ diam. Ostiole up to 250 μ diam. Conidia triradiate, main axis 18-20 × 3.5 μ, divergent arms 17-24 × 3.5 μ; setulae 10-14 μ long.

On dead needles of *Pinus monophylla, Abies concolor*, Calif., U.S.A. (WINF(M) 9790, holotype, 9791).

9. VESTIGIUM Pirozynski & Shoemaker, *Can. J. Bot.* **50**: 1163 (1972).
Sp. typ.: *V. felicis* Pirozynski & Shoemaker
Illustration: Pirozynski & Shoemaker (1972).

Mycelium immersed, branched, septate, pale brown to hyaline. *Conidiomata* acervular, subcuticular to epidermal and subepidermal, separate, formed of pale brown, thin-walled textura angularis. Dehiscence irregular. *Conidiophores* hyaline, branched towards the apices, septate, smooth, with acropleurogenous conidia, formed from the upper pseudoparenchyma

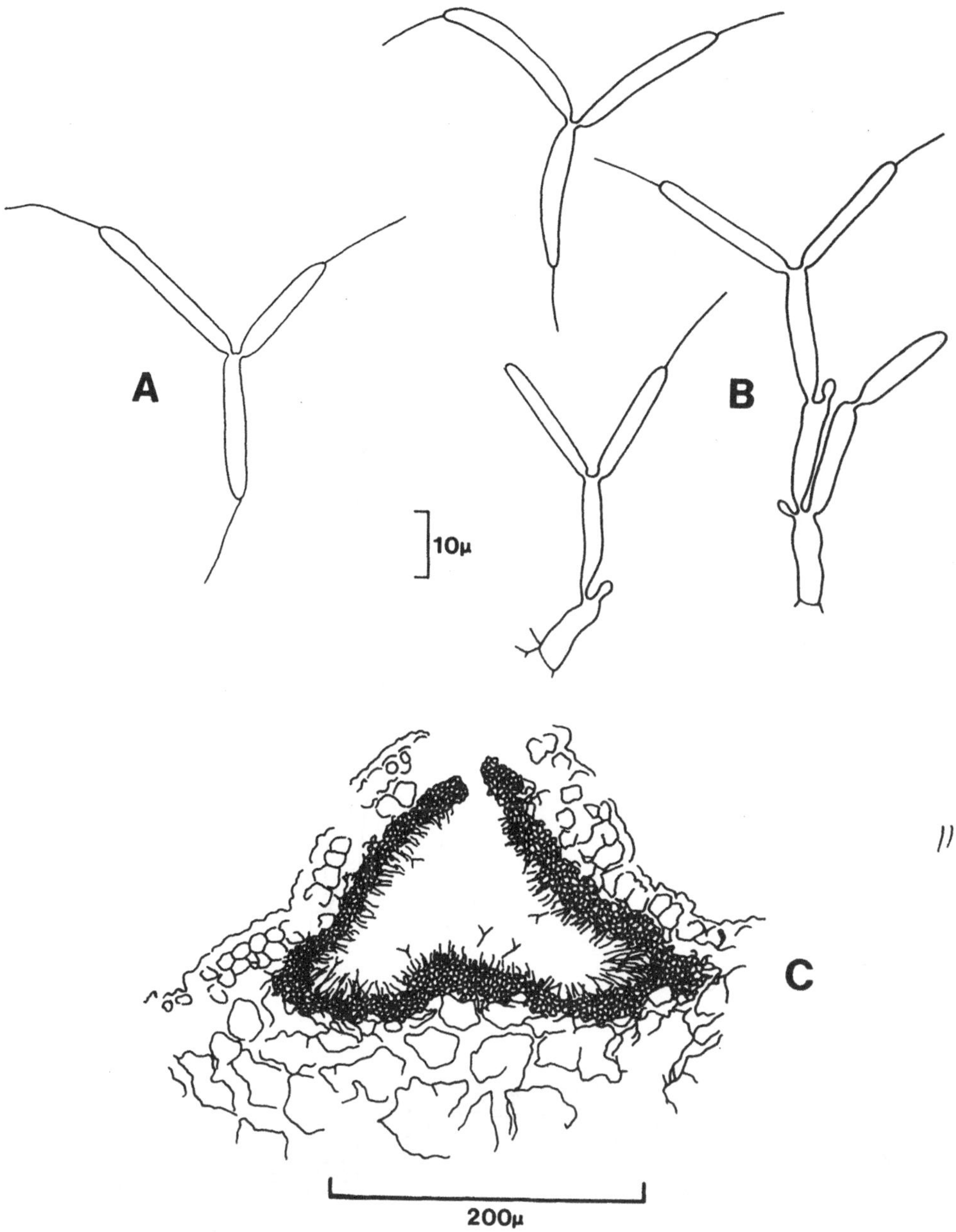

Fig. 119. *Furcaspora pinicola*, A, conidia; B, conidiophores and developing conidia; C, vertical section of a conidioma.

Conidiogenous cells holoblastic, sympodial, indeterminate, integrated, cylindrical, expanding towards the apices, hyaline, smooth, with 1-2 sympodial proliferations. *Conidia* hyaline, flat, smooth, thin-walled, eguttulate, consisting of a central angular cell and 4-6 small, ovoid, often curved, peripheral cells.

The genus is monotypic.

V. FELICIS Pirozynski & Shoemaker, *ibid.*: 1163 (1972).

Conidiomata up to 120 µ wide. Conidiophores up to 100 µ long expanding to 3 µ wide. Conidia 10-15 µ diam.

On young twigs of *Thuja plicata*, Canada.

10. MORALESIA Urries, *An. bot. A.J. Cavanilles* **14**: 167 (1955).
Sp. typ.: *M. arundinis* Urries
Illustrations: Urries (1955).

Mycelium immersed, branched, septate, pale brown to hyaline. *Conidiomata* eustromatic, immersed, separate, dark brown to black, elliptical to lenticular or elongated, unilocular, thick-walled, composed of an upper region of dark brown, thick-walled textura intricata, becoming hyaline towards the inside, and a basal region of pale brown to hyaline, thin-walled textura angularis. *Ostiole* absent, dehiscence by longitudinal rupture. *Conidiophores* hyaline, septate, branched dichotomously to subdichotomously, formed mainly from the basal and lateral walls. *Conidiogenous cells* holoblastic, sympodial, integrated, indeterminate, comprising the ultimate branches of the conidiophores with 2-4 small unthickened conidiogenous loci towards the apices. *Conidia* occasionally persistent, hyaline, filiform, minutely guttulate, multiseptate, thin-walled, smooth.

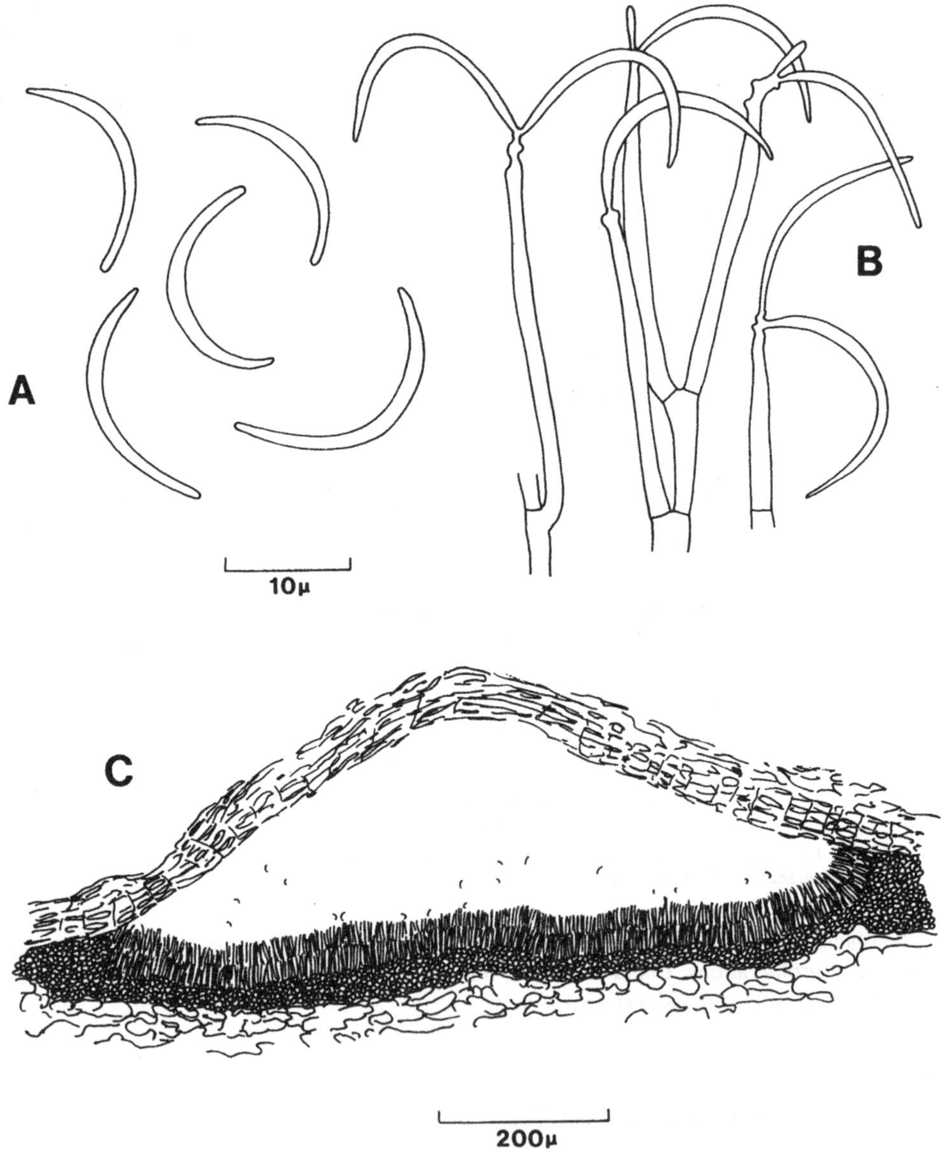

Fig. 120. *Libertella faginea*, A, conidia; B, conidiophores and developing conidia; C, vertical section of a conidioma.

These data are deduced from the excellent original account and illustration by Urries (1955). The genus resembles *Eriospora* Berk. & Br. but appears to differ mainly by the longitudinal line of dehiscence and structure of the conidioma. At present it remains monotypic.

M. ARUNDINIS Urries, *ibid.*: 167 (1955).

Conidiomata 500-900 µ long × 200-400 µ wide. Conidiophores 5-15 × 1.5 µ. Conidia 150-200 × 2.5 µ.

On dead culms of *Arundo donax*, Canary Islands.

11. LIBERTELLA Desm., *Ann. Sci. nat.*, ser. 1, **19**: 275 (1830).
Original spp.: *L. betulina* Desm., *L. rosae* Desm., *L. faginea* Desm.
Lectotype sp.: *L. faginea* Desm., fide Saccardo (1880a).

Mycelium immersed, branched, septate, pale brown to hyaline. *Conidiomata* acervular, subperidermal, separate or confluent, simple or irregular in shape, consisting of pale brown to hyaline, thin-walled textura angularis to intricata. Dehiscence irregular, conidia oozing in brightly coloured tendrils, globules or ligulate masses. *Conidiophores* repeatedly branched at the base and above, often fasciculate, septate in association with branches, hyaline, pale brown at the base, tapered towards the apices, formed from the upper cells of the pseudoparenchyma. *Conidiogenous cells* holoblastic, sympodial, integrated, indeterminate, hyaline, with 2 to several, small, indistinct, slightly protuberant conidiogenous loci towards the apices. *Conidia* hyaline, aseptate, thin-walled, smooth, eguttulate, curved (often markedly), filiform.

Although Saccardo (1880a) listed *L. faginea* as representative of the genus, Clements & Shear (1931) quoted *L. betulina* as lectotype species. Both species are congeneric but Saccardo's choice must be followed since it is earlier.

Croxall (1950) regarded *Libertella* species as the macro-conidial states of *Diatrypella* Ces. & de Not. He described conidia as meristem phialospores but this interpretation is at variance with my own observations. The several species in Herb. IMI show holoblastic sympodial development, there being no trace of phialidic apertures. *Libertella* species can be distinguished readily from the β-conidia of *Phomopsis* species because despite a strong similarity in conidial shape the conidiogenous cells are quite different. In *Phomopsis* they are truly phialidic.

About 70 species have been described. A revision with correllation to ascomycete states is required before any sense can be made of the genus.

L. FAGINEA Desm., *ibid.*: 276 (1830). (Fig. 120)

Conidia strongly curved, 14-17 × 1.5 µ.

On branches of *Fagus sylvatica*, G.B., Germany, France, Roumania (IMI 22075, 35329, 22082-91, 42058-61, 22410a, 46000, 60797, 79135, 79804, 103680, 129416, 115539).

L. BETULINA Desm., *ibid.*: 276 (1830). (Fig. 121B)

Conidia falcate, 14-15 × 1.5 µ.

On branches of *Betula* sp., France (Rabenh. Fungi europ. 354 ex K).

L. heveae (Petch) comb. nov. (Fig. 121A)
Phlyctema (as '*Phlyctaena*') *heveae* Petch, *Ann. R. bot. Gdn, Peradeniya* **6**: 237 (1917).

Conidia strongly curved, 17-24 × 1.5-2 µ.

From dry rot of *Hevea* wood, India (IMI 198097).

12. TRYBLIDIOPYCNIS Höhn., *Sber. Akad. Wiss. Wien* **127**: 562 (1918).
Sp. typ.: *T. pinastri* Höhn.

Mycelium immersed, branched, septate, hyaline. *Conidiomata* eustromatic, semi-immersed, becoming erumpent, circular, applanate to collabent, separate or aggregated and confluent, black, unilocular or multilocular; wall thick, composed of an outer layer one cell thick of dark brown, thick-walled textura angularis and an inner layer several cells thick, of similar thin-walled tissue which is pale brown at the periphery but becoming hyaline towards the conidiogenous region. *Ostiole* absent, dehiscence by irregular breakdown of upper wall. *Conidiophores* hyaline, septate, simply branched

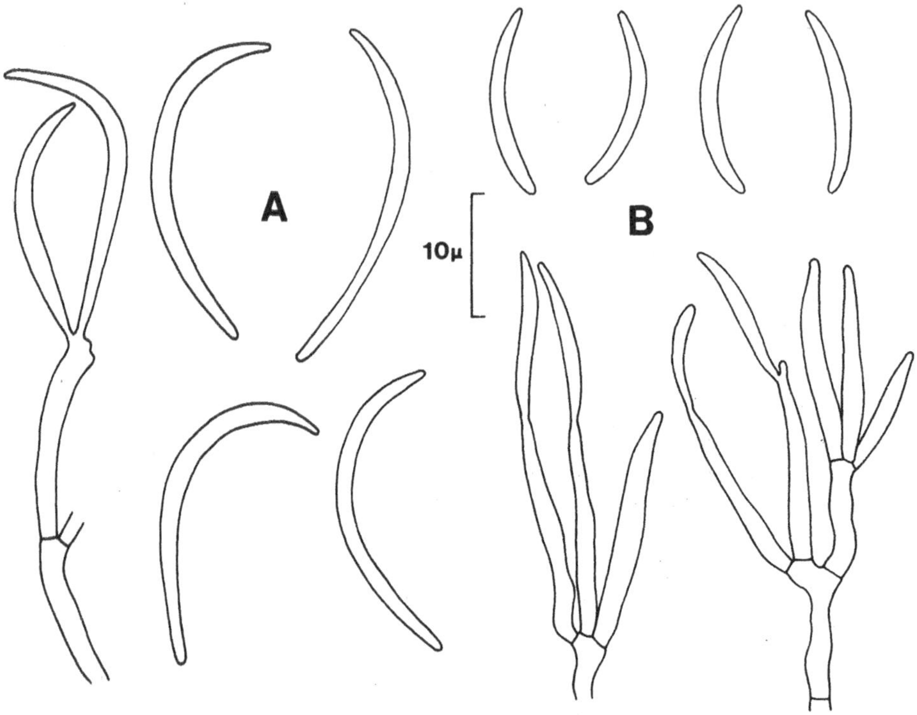

Fig. 121. A, *Libertella heveae*, conidia and conidiophore, B, *L. betulina*, conidia and conidiophores.

above, ± straight, tapered to the apices, formed all round the locular cavities. *Conidiogenous cells* holoblastic, sympodial or synchronous, indeterminate, mostly discrete, hyaline, with 1-2 apical, small, unthickened, scarcely protuberant scars. *Conidia* hyaline, aseptate, hamate or fusiform, thin-walled, eguttulate.

Höhnel (1918b) introduced this generic name for the conidial state of *Tryblidiopsis pinastri* (Pers.) Karst. and Gremmen (1957) provided the experimental proof of their connection. The conidia he figured however, are more irregular than was found in Höhnel's material.

T. PINASTRI Höhn., *ibid.*: 562 (1918). (Fig. 122)

Teleomorphosis: *Tryblidiopsis pinastri* (Pers.) Karst., *Mycol. Fenn.* **1**: 262 (1871).

Conidiomata up to 650 μ diam. × 200 μ high. Conidiophores 14-25 × 2.5-3 μ. Conidia 19-28 × 1 μ.

On dead branches of *Picea*, Austria (IMI 202957 ex FH).

13. POLYSTIGMINA Sacc., *Syll. fung.* **3**: 622 (1884).
Rhodoseptoria Naumov, *Bull. trimest. Soc. mycol. Fr.* **29**: 279 (1913).
Sp. typ.: *P. rubra* (Desm.) Sacc. (syn. *Septoria rubra* Desm.).

Mycelium immersed, branched, septate, hyaline. *Conidiomata* pseudostromatic, immersed, yellow-orange, occupying the whole of the lesion, composed of loosely interwoven hyphae and cells, multilocular; locules often convoluted; subepidermal tissue of small-celled, thin-walled, hyaline textura angularis. *Ostioles* single to each locule, circular, depressed. *Conidiophores* hyaline, branched profusely at the base and rather less so above, septate, smooth, with acrogenous conidia, formed from the inner cells of the locular walls. *Conidiogenous cells* holoblastic, sympodial to synchronous, indeterminate, integrated, straight or flexuous, hyaline, smooth, with several irregular nodulose conidiogenous loci at the apices. *Conidia* hyaline, aseptate, thin-walled, irregularly guttulate, smooth, hamate, tapered towards the apices.

Accounts of the genus have been given by Grove (1937) and Petch (1943).

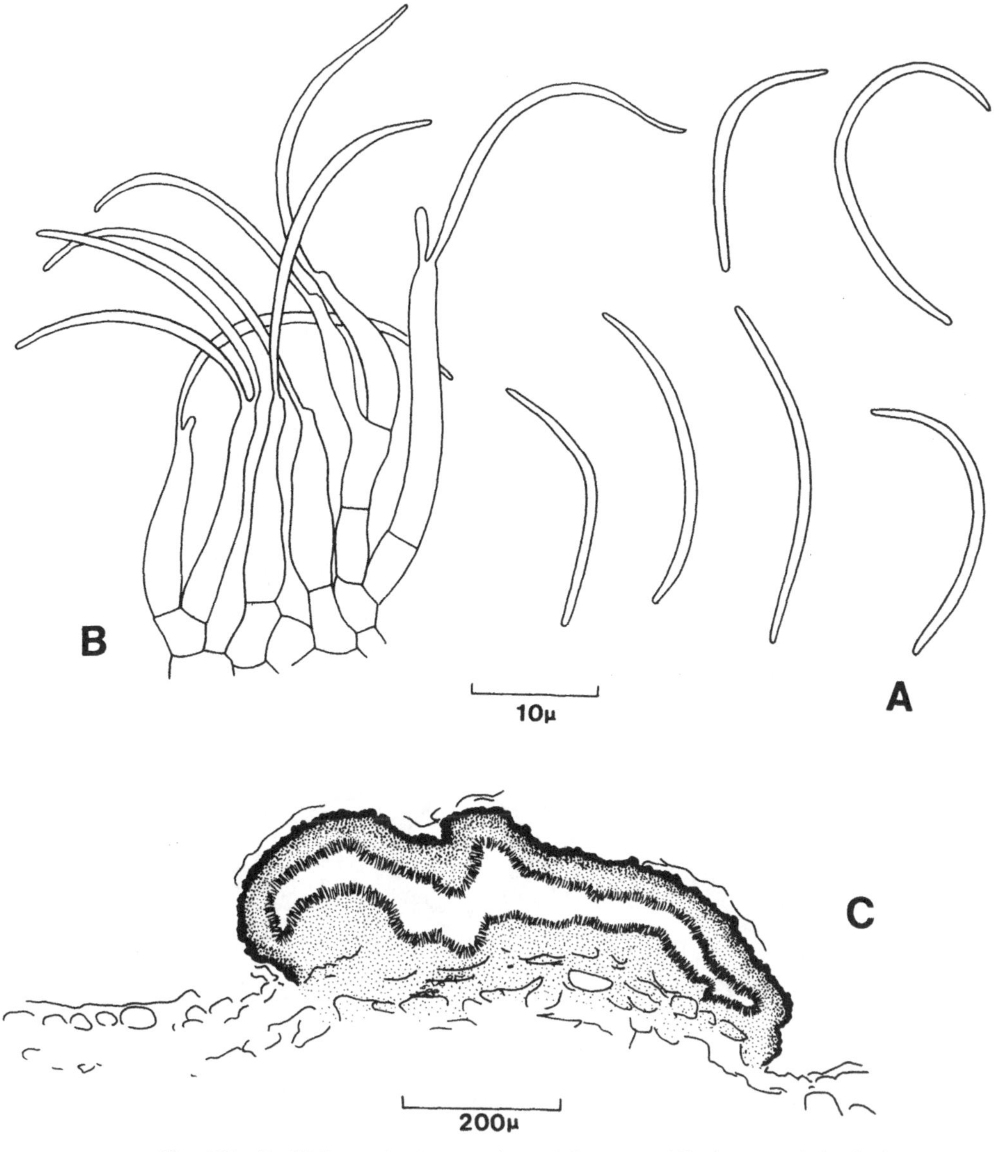

Fig. 122. *Tryblidiopycnis pinastri*, A, conidia; B, conidiophores and developing conidia; C, vertical section of a conidioma.

P. RUBRA (Desm.) Sacc., *Syll. fung.* **3**: 622 (1884). (Fig. 123)
Septoria rubra Desm., *Ann. Sci. nat.*, 2 ser., **19**: 342 (1843).
Libertella rubra (Desm.) Bonorden, *Handb. der Allegem. Mykologie* **1**: 55 (1851).
Rhodoseptoria ussuriensis Naumov, *Bull. trimest. Soc. mycol. Fr.* **29**: 278 (1913).

Teleomorphosis: *Polystigma rubrum* (Pers.) St. Amans, *Flore Agenaise*: 516 (1821).

Conidiomata occupying almost the whole of the leaf lesion, up to .7 cm diam. Conidiophores up to 26 μ long × 2-3.5 μ wide. Conidia 23-35 × 1 μ.

On leaf lesions of *Prunus domestica, P. spinosa, P. amygdalus, P. padus, P. cornuta*, Italy, Germany, Roumania, Czechoslovakia, G.B., Azores, Cyprus, Sweden, France, Turkey, U.S.S.R., India (IMI 16900-904, 16907-9, 16911-913, 120527, 31818, 12792, 30600, 42334, 46028, 46946, 69433, 135768, 185885, 75984).

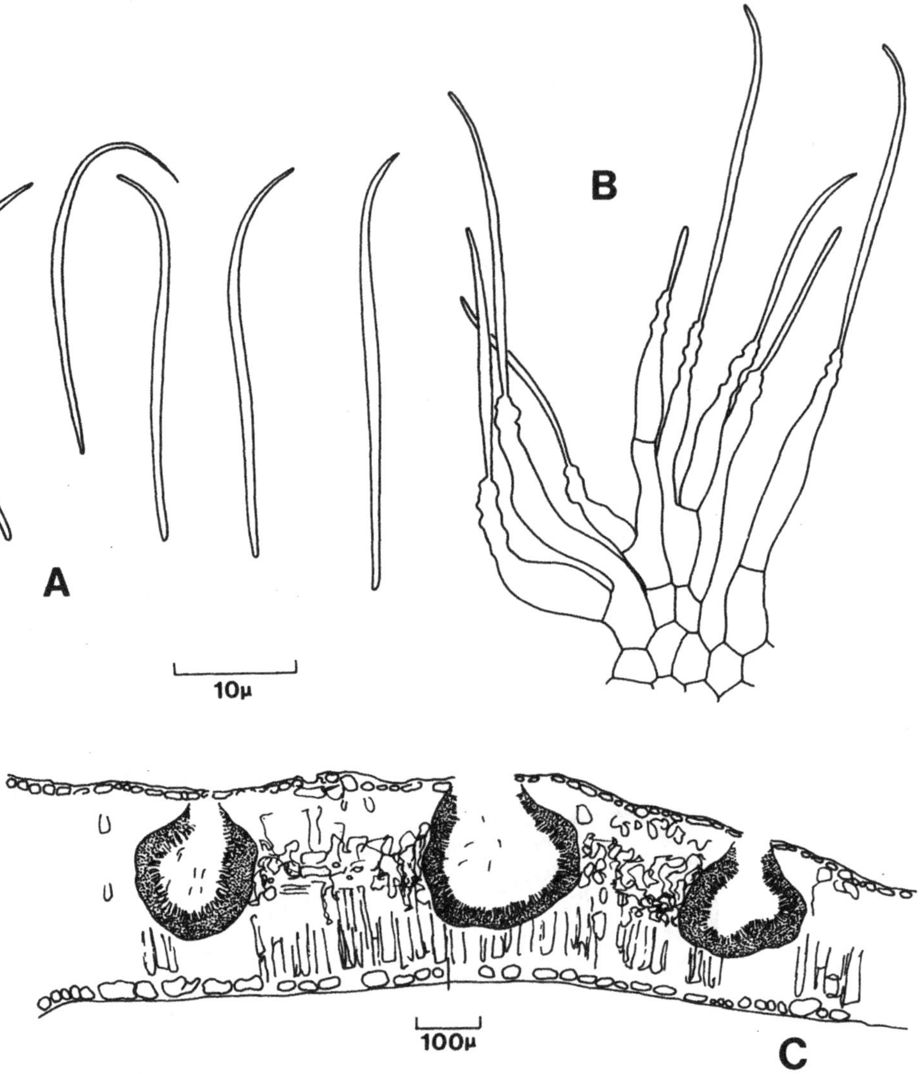

123. *Polystigmina rubra*, A, conidia; B, conidiophores and developing conidia; C, vertical section of conidiomata.

14. MICROGLOEUM Petrak apud Syd. & Petrak, *Annls mycol.* **20**: 215 (1922).
Sp. typ.: *M. pruni* Petrak

Mycelium immersed, branched, septate, hyaline. *Conidiomata* acervular, subepidermal to epidermal, ± circular, separate or eventually confluent, composed of hyaline to pale brown, thin-walled textura angularis. Dehiscence irregular. *Conidiophores* hyaline, septate, branched at the base and immediately below septa, fasciculate, tapered to the apices, formed from the upper pseudoparenchyma. *Conidiogenous cells* holoblastic, sympodial, discrete or more frequently integrated, indeterminate, terminal or as filiform lateral branches just below transverse septa, hyaline, smooth, gradually tapered to the apices, with 1-2 small unthickened loci. *Conidia* hyaline, aseptate, thin-walled, eguttulate, straight, cylindrical, slightly attenuated towards the truncate base, apex obtuse, mixed in the holotype with macroconidia.

Von Arx (1970) suggested *M. pruni* was possibly the microconidial state of *Blumeriella jaapii* (Rehm) von Arx, of which *Phloeosporella padi* (Lib.) von Arx is the macroconidial state. In the holotype of *M. pruni*, the conidia were mixed with larger, falcate, 1-septate conidia corresponding exactly with the macroconidia of *B. jaapii* collections in IMI. There seems no doubt that *M. pruni* is microconidial *B. jaapii*.

Bubák (1906a) described *Hainesia feurichii* from leaves of *Prunus padus*, mixed with *Cylindrosporium padi* (Lib.)

Karst. ex Sacc. (a later synonym of *P. padi*). Conidial size was 3-4 × 1-1.5 µ and conidiophores were 20 × 1-1.5 µ. Although Petrak apud Sydow & Petrak (1922) purposely excluded the use of *H. feurichii* when they proposed *M. pruni* because of a difference in conidial size it seems most unlikely that he was dealing with a different species if the macroconidia were present. If *H. feurichii* is the same as *M. pruni*, it certainly has priority, but confirmation of such a synonymy in a well-known plant pathogen must be based on a study of the type.

M. PRUNI Petrak, *ibid.*: 215 (1922). (Fig 124)

Teleomorphosis: *Blumeriella jaapii* (Rehm) von Arx, *Phytopath. Z.* **42**: 164 (1961).

Conidiomata up to 200 µ diam. Conidiophores up to 45 µ long × 2.5-3 µ at the base. Conidiogenous cells 8-16 µ long, narrowing to 1 µ wide at the apex. Conidia 7.5-9.5 × 1.5 µ.

Causing leaf lesions (shot-hole) on *Prunus*, Idaho, U.S.A. (IMI 204130, holotype ex BPI, addit. material with microconidia sub *Phloeosporella padi*).

15. RHABDOGLOEOPSIS Petrak, *Annls mycol.* **23**: 52 (1925).
Sp. typ.: *R. balsameae* (J.J. Davis) Petrak (syn. *Gloeosporium balsameae* J.J. Davis).
Illustrations: Nag Raj & Morgan-Jones (Fig. 2, 1973).

Mycelium immersed, hyaline, branched, septate. *Conidiomata* acervular, subepidermal, elongated, separate or confluent, formed of hyaline to pale brown, thin-walled textura angularis. Dehiscence irregular. *Conidiophores* hyaline, smooth, septate, sparsely branched, with acropleurogenous conidia, formed from the upper pseudoparenchyma. *Conidiogenous cells* holoblastic, sympodial, indeterminate, integrated, cylindrical or slightly curved, hyaline, smooth, with 1-3 sympodial, unthickened proliferations towards the apices. *Conidia* hyaline, aseptate, base obtuse to ± truncate, apex obtuse, thin-walled, smooth, guttulate, fusiform, straight or slightly curved.

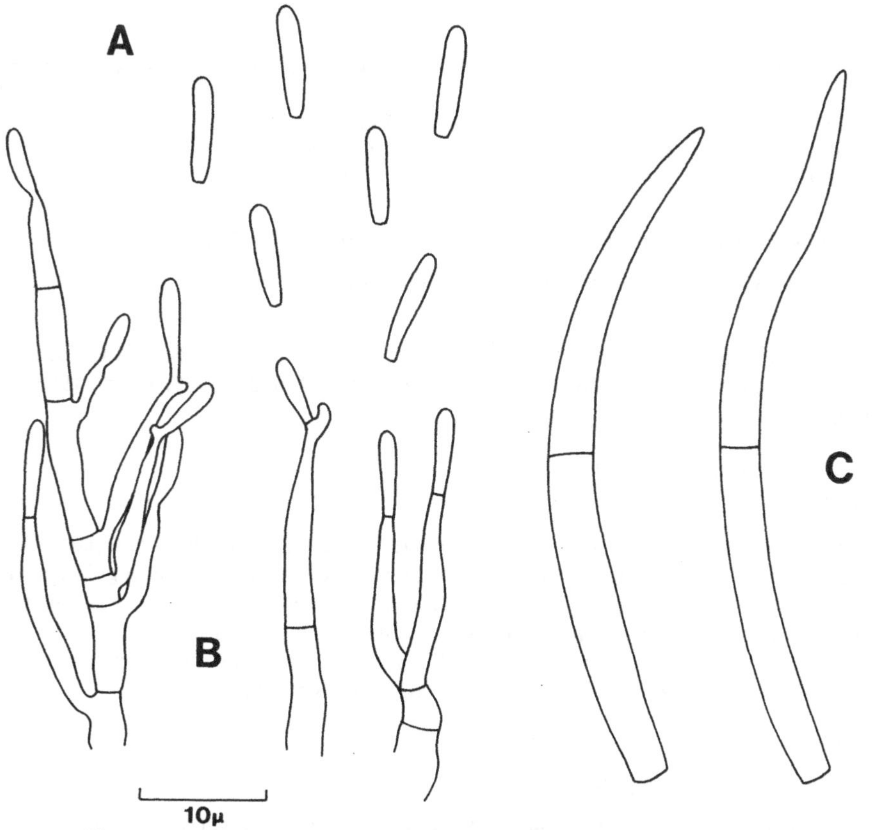

Fig. 124. *Microgloeum pruni*, A, conidia; B, conidiophores and developing conidia; C, macroconidia.

The genus was redescribed by Nag Raj & Morgan-Jones (1973) in an account of *Rhabdogloeopsis* and *Rhabdogloeum* Syd. Unfortunately the illustration purported to be of *Rhabdogloeopsis balsameae* is in fact of *Rhabdogloeum pseudotsugae* and vice versa, a situation that will not lessen confusion between the two genera. Arx (1957a; 1970) included *Rhabdogloeopsis* in the synonymy for *Kabatiella* Bub. but Nag Raj & Morgan-Jones (1973) have demonstrated no basis for this redisposition.

R. BALSAMEAE (J.J. Davis) Petrak, *ibid.*: 52 (1925).
Gloeosporium balsameae J.J. Davis, *Trans. Wisc. Acad. Sci. Arts Lett.* **20**: 409 (1922).
Kabatiella balsameae (J.J. Davis) Arx, *Verh. K. Ned. Akad. Wet. Amst.* **51**: 62 (1957).
Rhabdogloeum abietinum Dearn., *Mycologia* **20**: 241 (1928).

Conidiomata 250-400 μ diam. Conidiophores 28-40 × 3-4 μ. Conidia 19-27 × 3.5-5 μ.

On needles of *Abies balsamea, A. fraseri*, Wisc., N.C., U.S.A.

16. CONOSTROMA Moesz, *Bot. Közl.* **19**: 44 (1920-21).
Discosporella Höhn., *Mykol. Unters.* **1**: 340 (1923).
Sp. typ.: *C. didymum* (Fautrey & Roum.) Moesz (syn. *Dendrophoma didyma* Fautrey & Roum.).

Mycelium immersed, septate, pale brown to hyaline, branched. *Conidiomata* eustromatic, separate, pulvinate, dark brown, subperidermal, unilocular, multilocular or irregular, situated around a central column of sterile tissue composed of thin-walled, pale brown textura angularis with somewhat larger cells in the centre; locular wall tissue and basal stroma of smaller pseudoparenchyma. Dehiscence initially by the central column rupturing the bark, then opening widely. *Conidiophores* hyaline, branched at the base and along their length, septate, smooth, with acrogenous conidia, tapered markedly towards the apices, formed from the inner cells of the locular walls. *Conidiogenous cells* holoblastic, sympodial, indeterminate, integrated, straight or flexuous, hyaline, smooth, with up to 3 conidia formed sympodially in a very restricted area at the apices. *Conidia* hyaline, aseptate, thin-walled, eguttulate, smooth, cylindrical, apex obtuse, base ± acute.

A full account of *C. didymum*, the conidial state of *Colpoma quercinum* (Fr.) Wallr., was given by Twyman (1947).

C. DIDYMUM (Fautrey & Roum.) Moesz, *ibid.*: 45 (1920-21). (Fig. 125)
Discosporella didyma (Fautrey & Roum.) Höhn., *Mitt. bot. Inst. tech. Hochsch., Wien* **4**: 80 (1927).
Dendrophoma didyma Fautrey & Roum., *Rev. Mycol.* **14**: 9 (1892).

Teleomorphosis: *Colpoma quercinum* (Pers.) Wallr., *Fl. Crypt. Germ.* **2**: 423 (1823).

Conidiomata up to 650 μ diam. × 200 μ deep. Conidiophores up to 35 × 2.5-3.5 μ, tapering to 1 μ at the apices. Conidia 5-7 × 1.5 μ.

On branches of *Quercus* spp., Germany, Roumania, U.K., U.S.A., Czechoslovakia, Austria, France (IMI 13793, 13794, 13795, 13796, 11316, 30424, 32621, 32765, 11224, 38455, 41633, 46058, 46909, 19038b, 73447, 80729c, 80714, 80737b, 86850, 95223a, 82746, 19059b).

17. HYPOCLINE Syd., *Annls mycol.* **37**: 245 (1939).
Sp. typ.: *H. penniseti* Syd.

Mycelium immersed, hyaline, branched, septate. *Conidiomata* eustromatic, pale brown, separate, rarely aggregated, longitudinal, unilocular, opening widely; upper wall several cells thick of brown textura angularis; basal wall hyaline, thinner. *Conidiophores* hyaline, branched only at the base, 1-2 septate, smooth, cylindrical, ± straight, formed from the upper cells of the basal wall. *Conidiogenous cells* holoblastic, sympodial, indeterminate, integrated, cylindrical, hyaline, smooth, with 1-3 apical conidiogenous loci, scars unthickened, slightly protruding. *Conidia* hyaline, aseptate, thin-walled, smooth, eguttulate, apex obtuse, base ± truncate, cylindrical, straight or slightly bent.

The genus is monotypic.

H. PENNISETI Sydow, *ibid.*: 245 (1939). (Fig. 126)

Conidiomata 200-400 μ long × 100-300 μ wide. Conidiophores 21-24 × 2.5-3.5 μ. Conidia 9.5-12 × 3-4 μ.

Causing lesions on *Pennisetum* ? *trachyphyllum*, Uganda (IMI 51540, holotype).

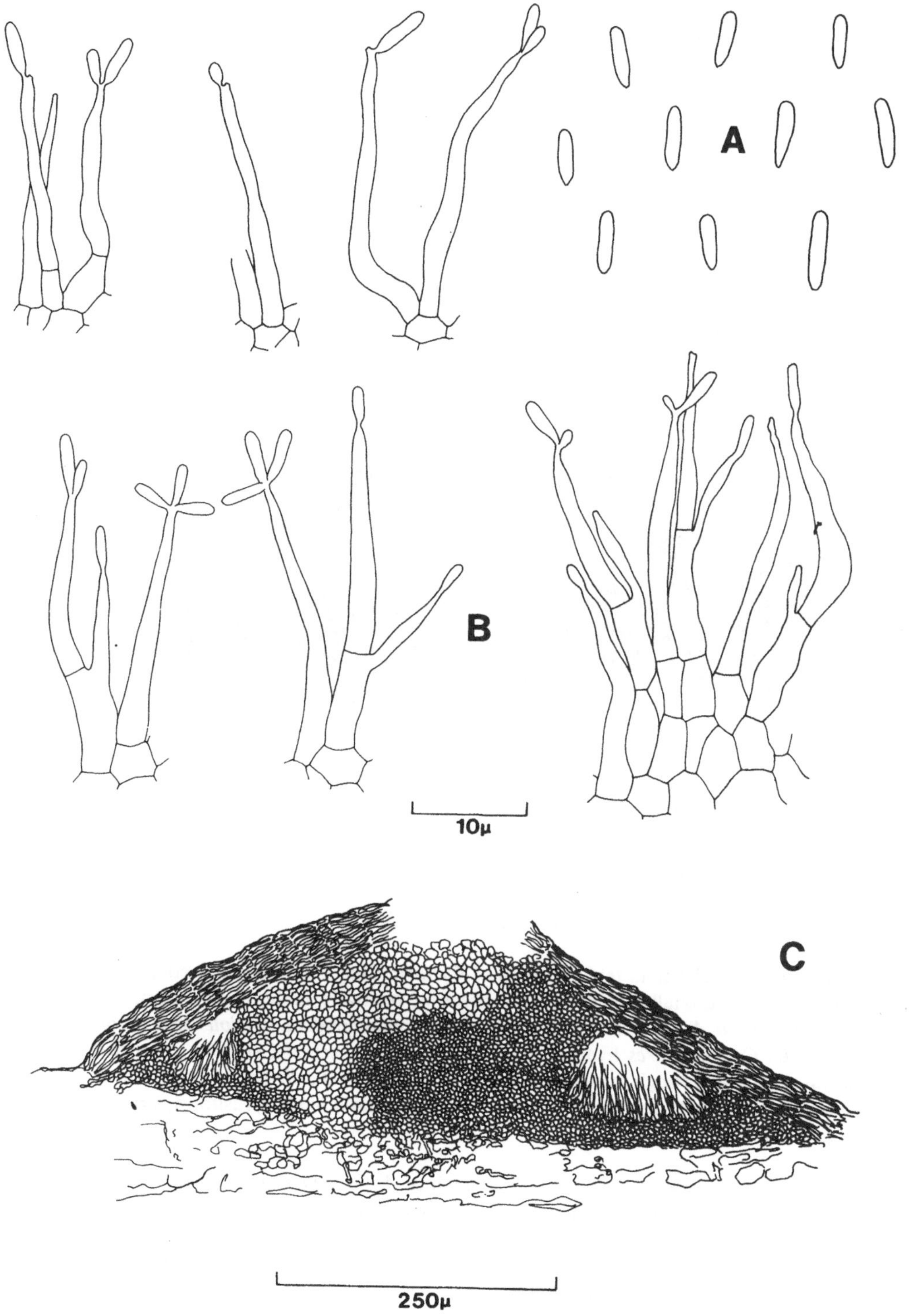

Fig. 125. *Conostroma didymum*, A, conidia; B, conidiophores and developing conidia; C, vertical section of a conidioma.

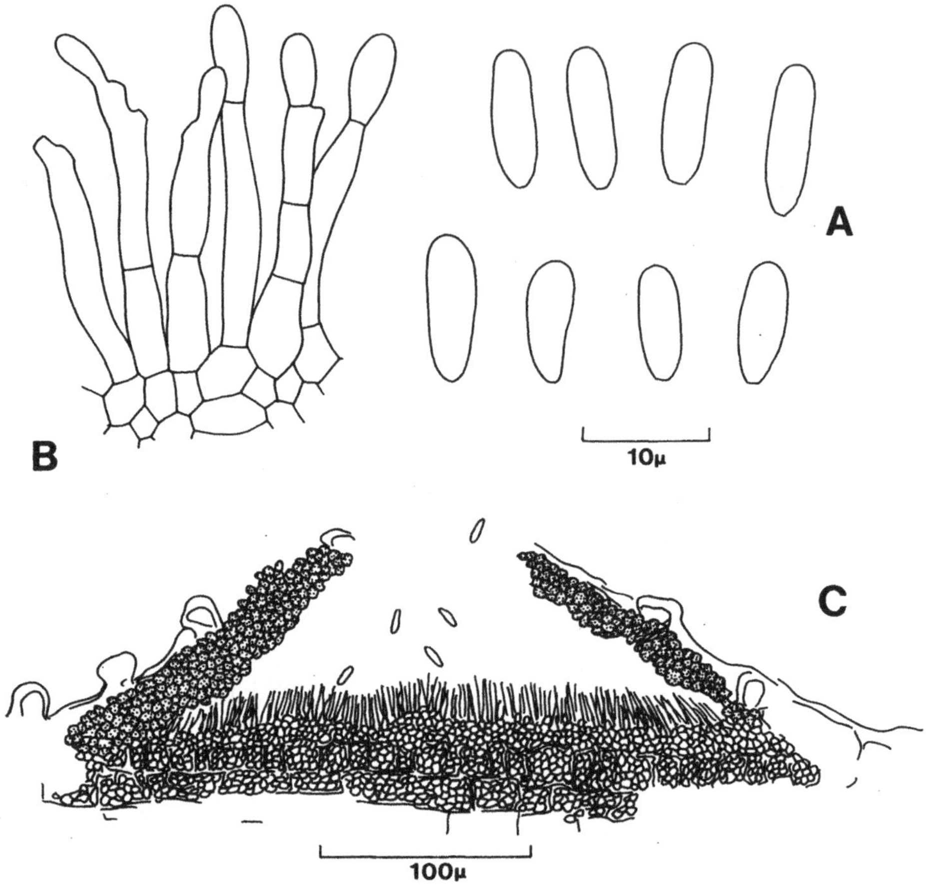

Fig. 126. *Hypocline penniseti*, A, conidia; B, conidiophores and developing conidia; C, vertical section of a conidioma.

18. ANTHRACODERMA Speg., *Bol. Acad. Nac. Cienc. Cordoba* **11**: 154 (1887).
Original spp.: *A. hookeri* Speg., *A. selenospermum* Speg.
Lectotype sp.: *A. hookeri* Speg., fide Petrak & Sydow (1935).

Mycelium immersed, intracellular in textura oblita of *Cyttaria* tissue. *Conidiomata* confluent, pulvinate, carbonous, dark brown, unilocular or occasionally divided; basal tissue of relatively large, thick-walled, brown textura angularis, lateral and upper walls up to 4 cells thick, of similar pseudoparenchyma. *Ostiole* absent, dehiscence by irregular breakdown of upper wall. *Conidiophores* hyaline, branched irregularly, septate, smooth, contorted, originating from the walls of the locule but ramifying through the locular cavity. *Conidiogenous cells* holoblastic, occasionally sympodial, with flat, unthickened scars, indeterminate, integrated, hyaline smooth. *Conidia* hyaline, fusiform, straight or irregular, aseptate, smooth, thin-walled, eguttulate.

Petrak & Sydow (1935) studied the holotype and maintained *Anthracoderma* as an acceptable genus. Apart from the two species described by Spegazzini (1887) the only further addition has been *A. duvidosum* Viégas (1945). From the excellent account of the species it is clear that it is not congeneric with *A. hookeri*. The species is not a hyperparasite but is corticolous. Differences in structure of the stromata and conidiogenesis exclude it from *Anthracoderma*, but until the type is examined it would be premature to place it elsewhere.

A. HOOKERI Speg., *ibid.*: 154 (1881). (Fig. 127)

Conidiomata up to 250 µ diam. Conidia 9-19 × 3-3.5 µ.

On fructification of *Cyttaria hookeri*, Chile (IMI 196762, holotype ex LPS).

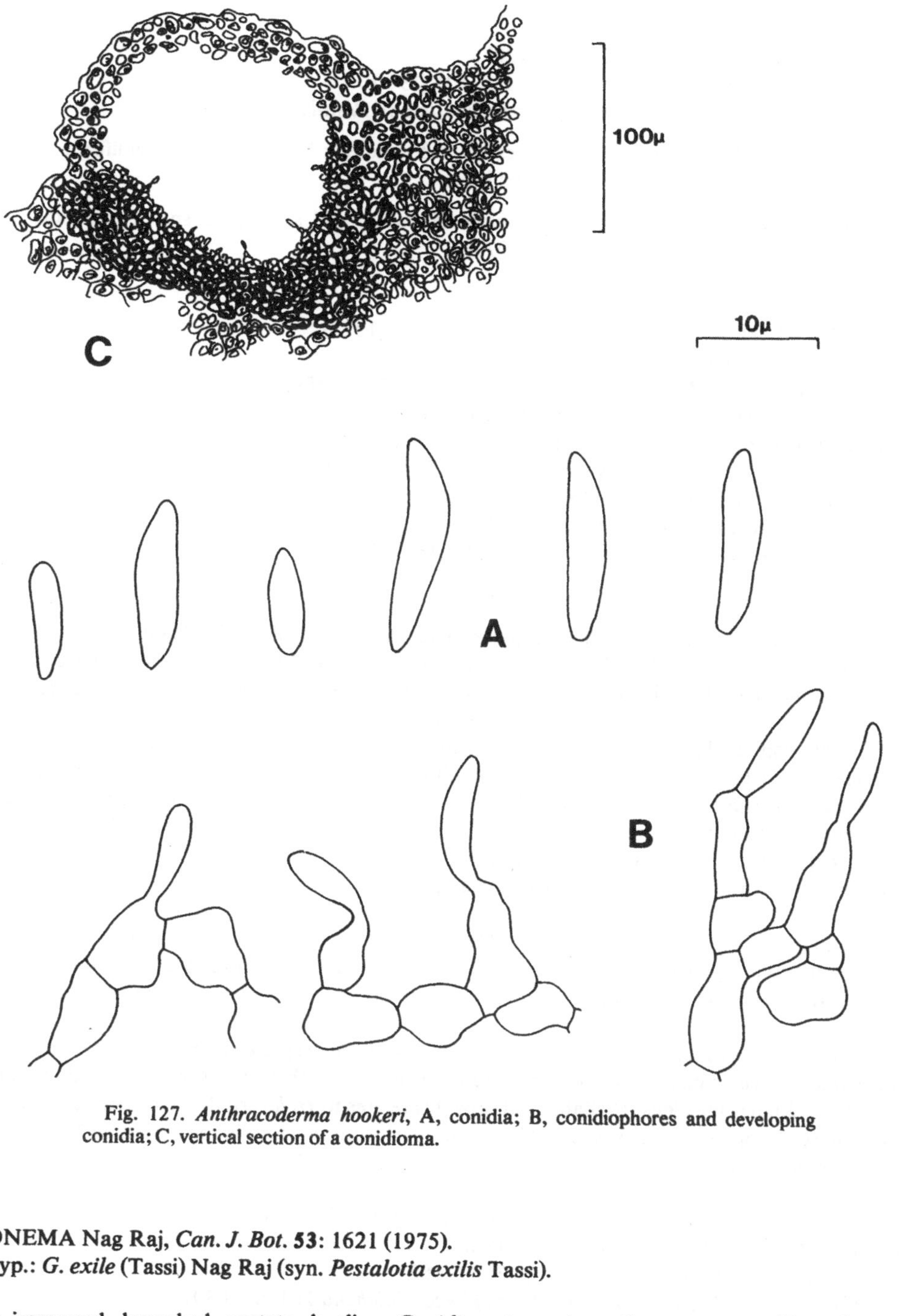

Fig. 127. *Anthracoderma hookeri*, A, conidia; B, conidiophores and developing conidia; C, vertical section of a conidioma.

19. GAMPSONEMA Nag Raj, *Can. J. Bot.* **53**: 1621 (1975).
Sp. typ.: *G. exile* (Tassi) Nag Raj (syn. *Pestalotia exilis* Tassi).

Mycelium immersed, branched, septate, hyaline. *Conidiomata* eustromatic, separate, epidermal to subepidermal, depressed, ± flattened, pale brown, unilocular; upper wall thick, composed of relatively large-celled, thin-walled, pale brown textura angularis, lower wall thin, of small-celled, similar hyaline tissue. *Ostiole* absent, dehiscence by irregular rupture, often opening widely. *Conidiophores* cylindrical, branched above, fasciculate, sparingly septate, hyaline, formed from the inner cells of both upper and lower walls. *Conidiogenous cells* holoblastic, sympodial, either integrated on conidiophores and then cylindrical, or discrete and then ampulliform, indeterminate, hyaline, with 1-2 apical conidiogenous loci, scars not thickened, not protuberant, but closely spaced. *Conidia* slightly curved, 2-euseptate, hyaline, apex and base obtuse, thin-walled, eguttulate, with 2 subapical filiform unbranched divergent setulae.

The genus is monotypic.

G. EXILE (Tassi) Nag Raj, *ibid.*: 76 (1975). (Fig. 128)
Pestalotia exilis Tassi, *Bull. Lab. Ort. Bot. Siena* **3**: 101 (1900).
Hyalotia exilis (Tassi) Guba, *Monograph of Monochaetia and Pestalotia*: 283 (1961).

Conidiomata up to 220 µ wide. Conidiophores up to 30 µ long × 1.5-2 µ wide. Ampulliform conidiogenous cells 4 × 4 µ. Conidia 19-23 × 2 µ; setulae up to 20 µ long.

On dead leaves of *Eucalyptus grandis, E. paniculata, E. robusta, E. saligna, E.* sp., Brazil, India, Hawaii, U.S.A. (IMI 194560*g*, 186981*d*, 209217, 227974*a*, 186978*g*).

20. HETEROPATELLA Fckl, *Jb. Nassau. Ver. Naturk.* **27-28**: 54 (1873).
Excipulina Sacc., *Syll. fung.* **3**: 688 (1884).
? *Cucurbitariopsis* Massal., *Mem. Accad. Verona* **65**, ser. 3: 133 (1889).
Excipulella Höhn., *Sber. Akad. Wiss. Wien* **124**: 109 (1915).
Pseudodiscosia Hosterm. & Laub., *Gartenwelt* **25**: 66 (1921).
Darlucis Clements apud Clements & Shear, *Genera of Fungi*: 364 (1931).
Falcipatella Gucevicz, *Bot. Mater.* **8**: 145 (1952).
Falcipatellina Gucevicz, *Bot. Mater.* **8**: 147 (1952).
Viennotiella Negru, *Bull. trimest. Soc. mycol. Fr.* **80**: 228 (1964).
Sp. typ.: *H. lacera* Fckl

Mycelium immersed or superficial, branched, septate, pale brown or hyaline. *Conidiomata* eustromatic, superficial, separate, dark brown to black, unilocular, depressed in the centre at maturity and then appearing cupulate; wall thick, outer layer at the base and sides 2-3 cells thick, of brown, thin-walled textura globulosa or angularis, at the top of textura porrecta which curves over the conidiogenous region, inner layer of hyaline textura oblita which becomes more textura angularis towards the conidiophores. *Ostiole* absent, dehiscence by breakdown of the central textura porrecta. *Conidiophores* hyaline, branched at the base, often septate above, smooth, cylindrical, straight or curved, formed mostly from the base, also from the sides of the locule. *Conidiogenous cells* holoblastic, sympodial, integrated, indeterminate with 1-3 apical, unthickened, scarcely projecting conidiogenous loci, hyaline. *Conidia* hyaline, up to 3 euseptate, continuous or constricted at the septa, thin-walled, smooth, often guttulate, fusiform, apical cell prolonged with an unbranched, filiform appendage, base truncate, often with a short, excentric appendage.

The generic synonymy has been discussed by Sutton (1977a) under the individual generic names. More than 20 species have been described in *Heteropatella* but their relationships to the species described under the synonymous generic names has not been determined. Consequently considerable work is required before the number of accepted taxa can be realized.

It is quite possible that the genus consists of no more than two or three species that can occur on a wide range of substrates. Revision based on type studies is urgently required. Of the names that have been proposed in *Heteropatella, H. umbilicata* (Pers. ex Fr.) Jaap is the oldest available, but how this relates to *H. lacera* the type species is unknown.

The teleomorphoses belong in *Heterosphaeria* Grev. but apart from the correlations listed by Nannfeldt (1932) these have not been extensively investigated.

H. UMBILICATA (Pers.) Jaap, *Annls mycol.* **5**: 266 (1907). (Fig. 129C, D, E)
Peziza umbilicata Pers., *Mycol. Europ.* **1**: 323 (1822).
Hymenula umbilicata (Pers.) Fr., *Elench. fung.* **2**: 37 (1828).
Heteropatella lacera Fckl forma *umbilicata* (Pers.) Sacc., *Michelia* **2**: 116 (1880).

Conidiomata up to 300 µ diam. Conidiophores up to 30 µ long × 2-2.5 µ wide. Conidia 3 septate, 17-22 × 2-2.5 µ, basal appendage 1-2.5 µ, apical appendage 17-29 µ.

On indet. wood, *Oxyria digyna, Campanula uniflora, Rhinanthus minor, Bartsia alpina, Arcangeliella officinalis, Epilobium alsinifolium, Gnaphalium norvegicum, Daucus carota, Pastinaca sativa, Angelica silvestris,* U.K., Italy, Austria, Sweden, Germany, Greenland, Denmark, Yugoslavia (IMI 95564, 109976, 109977, 109912, 109913, 109914, 109915, 109979, 109978 ex GRO, 164298, 29810, 61799, 29808, 11292, 30336, 29261, 194017, 184266, 29809).

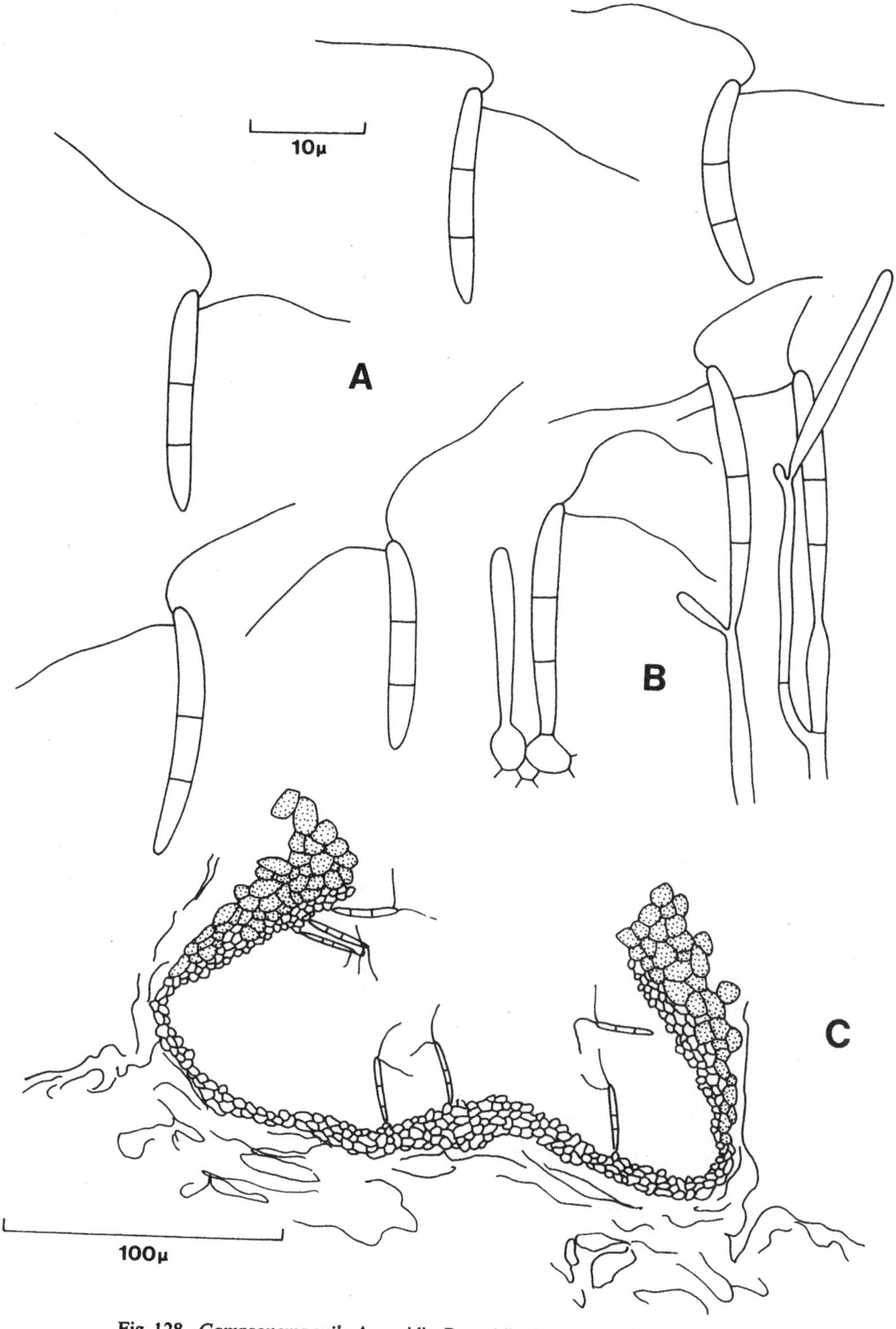

Fig. 128. *Gampsonema exile*, A, conidia; B, conidiophores and developing conidia; C, vertical section of a conidioma.

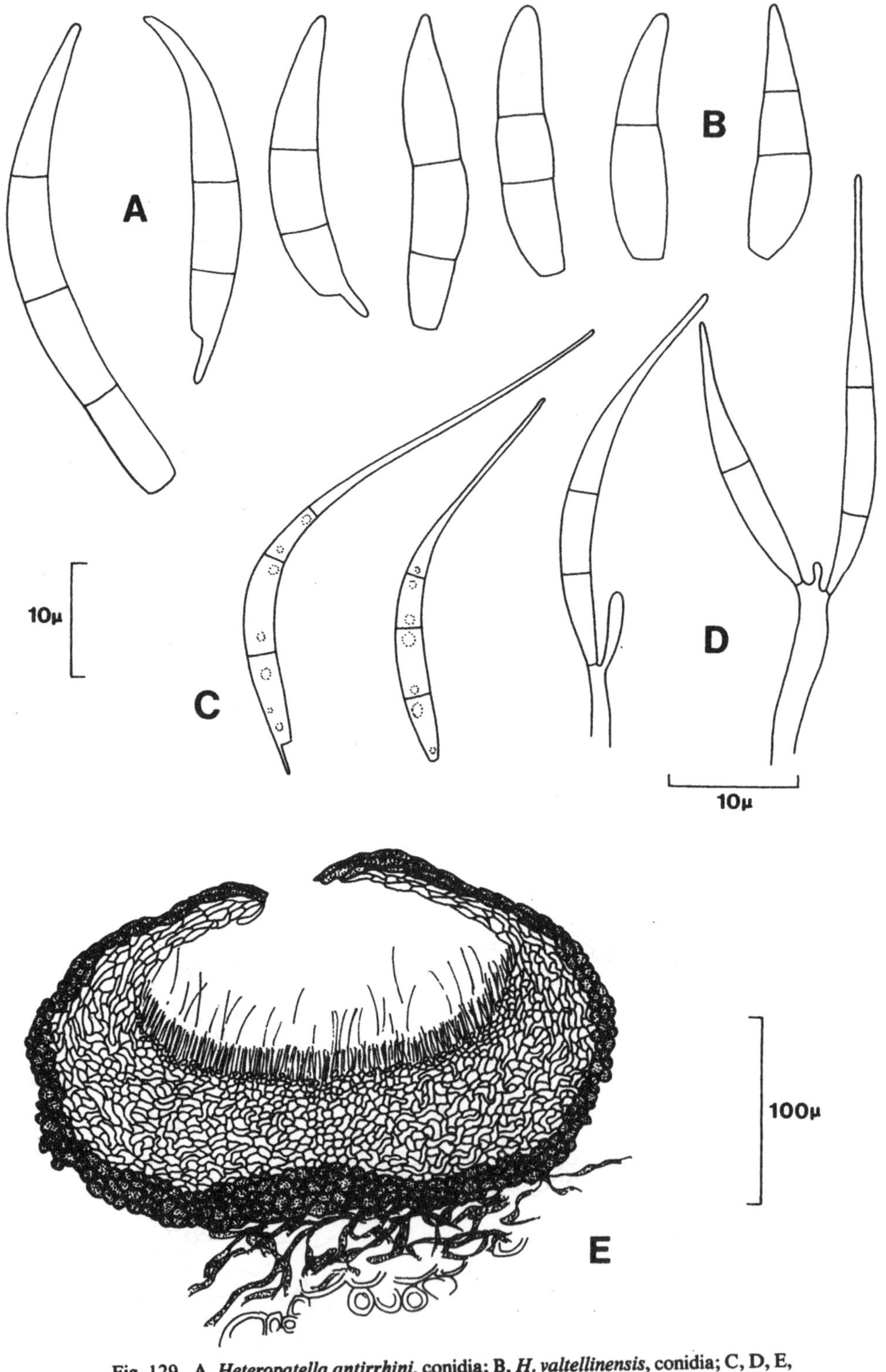

Fig. 129. A, *Heteropatella antirrhini*, conidia; B, *H. valtellinensis*, conidia; C, D, E, *H. umbilicata*, conidia, developing conidia, vertical section of a conidioma.

H. ANTIRRHINI Buddin & Wakefield, *Trans. Br. mycol. Soc.* **11**: 188 (1926). (Fig. 129A)
Cercosporella antirrhini Wakefield, *Kew Bull. 1918*: 233 (1918).
Pseudodiscosia antirrhini (Wakefield) Buddin & Wakefield, *Trans. Br. mycol. Soc.* **14**: 215 (1929).

Conidiomata acervular, foliicolous, up to 150 µ diam. Conidia 24-40 × 3.5-4.5 µ, 1-3 septate, apex shortly rostrate, base sometimes with an appendage 3-4 µ long.

On *Antirrhinum orontium*, *A.* sp., U.K., Germany (IMI 21840, 21839, 35230, 77145, 50876).

H. VALTELLINENSIS (Trav.) Wollenw., *Z.f.Parasit.* **111**(3): 507 (1931). (Fig. 129B)
Excipulina valtellinensis Trav., *Annls mycol.* **1**: 316 (1903).
Heteropatella dianthi Buddin & Wakefield, *Trans. Br. mycol. Soc.* **14**: 220 (1929).
Pseudodiscosia dianthi Hosterm. & Laub., *Gartenw.* **25**: 65 (1921).

Conidiomata acervular, foliicolous, up to 200 µ diam. Conidia 22-28 × 5-5.5 µ, 1-2 septate, apex shortly rostrate.

On *Dianthus caryophyllus*, Australia, Kenya (IMI 205350, 63662).

21. MONOCHAETIELLOPSIS Sutton & Di Cosmo, *Can. J. Bot.* **55**: 2536 (1977).
Sp. typ.: *M. themedae* (Kandasw. & Sundar.) Sutton & Di Cosmo (syn. *Monochaetiella themedae* Kandasw. & Sundar.).

Mycelium immersed, branched, septate, pale brown to hyaline. *Conidiomata* acervular, epidermal to subepidermal, immersed, separate or confluent, linear, formed of pale brown, thin-walled textura angularis. Dehiscence irregular. *Conidiophores* short, hyaline, smooth, cylindrical, 1-2 septate, branched at the base, formed from the upper pseudoparenchyma. *Conidiogenous cells* holoblastic, synchronous or sympodial, with up to 5, ± protuberant, unthickened conidiogenous loci towards the apices, discrete or integrated, indeterminate, cylindrical to lageniform or irregular, hyaline, smooth. *Conidia* hyaline, smooth, thin-walled, straight or slightly curved, 1-2 euseptate separating a very small basal cell, a long median cell and a long apical cellular, flexuous, unbranched appendage, the appendage and basal cell when present devoid of contents.

The two *Monochaetiella* species added after the type species differ considerably from it. The conidia are septate with a small basal cell and a long apical cell or appendage devoid of cytoplasmic contents. The conidiogenous cells in *Monochaetiella* are simple phialides whereas those in *M. themedae* and *M. cymbopogonis* are polyblastic. Such differences are sufficient to separate these two taxa into *Monochaetiellopsis*.

KEY

Conidia 19-30 × 2-3 µ; apical appendage 12-29 µ long .... *M. themedae*
Conidia 12-15.5 × 1.5 µ; apical appendage 7-16 µ long .... *M. cymbopogonis*

M. THEMEDAE (Kandasw. & Sundar.) Sutton & Di Cosmo, *ibid.*: 2541 (1977). (Fig. 130A, C, D)
Monochaetiella themedae Kandasw. & Sundar., *Ind. Phytopath.* **9**: 202 (1957).

Teleomorphosis: *Hypnotheca graminis* Tommerup, *Trans. Br. mycol. Soc.* **55**: 467 (1970).

Conidiomata up to 1 mm long. Conidiophores 9-12.5 × 2-3 µ. Conidia 19-30 × 2-3 µ; apical appendage 12-29 µ long.

Causing leaf lesions on *Themeda tremula* and *Heteropogon contortus*, India, Australia, Tanzania (IMI 78370a, 151159a, 62200).

M. CYMBOPOGONIS (Punith. & Sarwar) Sutton & Di Cosmo, *ibid.*: 2541 (1977). (Fig. 130B)
Monochaetiella cymbopogonis Punith. & Sarwar apud Punithalingam, *Trans. Br. mycol. Soc.* **53**: 311 (1969).

Conidiomata .5 mm long. Conidiophores 10-15 × 2-3 µ. Conidia 12-15.5 × 1.5 µ; apical appendage 7-16 µ long.

Causing leaf lesions on *Cymbopogon nardus*, India (IMI 132844, holotype).

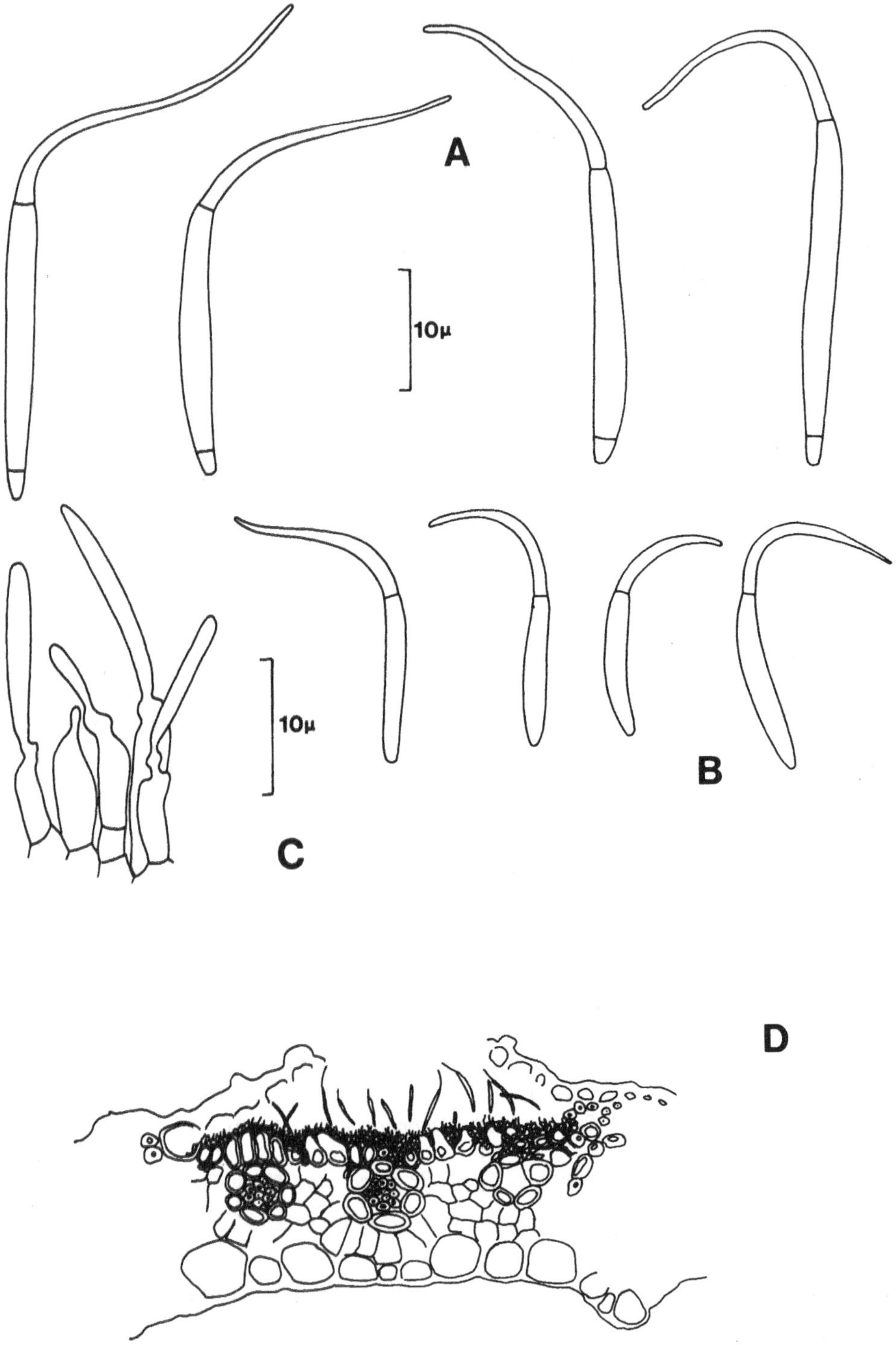

Fig. 130. A, C, D, *Monochaetiellopsis themedae*, conidia, conidiogenous cells, vertical section of a conidioma; B, *M. cymbopogonis*, conidia.

22. AJREKARELLA Kamat & Kalani, *Mycopath. Mycol. appl.* **24**: 300 (1964).
Sp. typ.: *A. polychaetriae* Kamat & Kalani
Illustration: Sutton (1967a).

Mycelium immersed, brown to pale brown, branched, septate. *Conidiomata* initially immersed, then erumpent and finally appearing superficial, separate, dark brown, sessile, navicular and elongated, rarely discoid, unilocular, initially closed, later dehiscing by a raphe; wall pseudoparenchymatous, of textura angularis, cells dark brown, thick-walled on the outside, becoming pale brown to hyaline and thin-walled towards the inside. *Setae* sparse, straight or curved, unbranched, dark brown, smooth, thick-walled, septate, tapered to acute apices. *Conidiophores* hyaline, branched irregularly, septate, smooth, ± cylindrical, formed from the inner layer of the wall. *Conidiogenous* cells holoblastic, annellidic, less rarely sympodial, indeterminate, integrated, cylindrical, hyaline, smooth, with 1-2 percurrent and/or 1 sympodial proliferations. *Conidia* hyaline, aseptate, navicular, apex acute, base truncate, thin-walled, smooth, eguttulate, with 1(-2) apical unbranched cellular appendages, and 2-6 basal, unbranched cellular appendages inserted above the truncate base.

The genus and species were fully redescribed and illustrated by Sutton (1967a), and so far it remains monotypic.

A. POLYCHAETRIAE Kamat & Kalani, *ibid.*: 300 (1964). (Fig. 131)

Conidiomata 650-950 µ long × 500 µ wide. Setae 3-7 septate, 175-385 × 4-7 µ. Conidiophores 12-19.5 × 2-3 µ. Conidia 12-14.5 × 4-5 µ; apical appendages 10-13 µ long; basal appendages 8-16 µ long.

On dead twigs of *Jasminum calophyllum* and *Wigandia urens*, India, Zambia (IMI 101970, holotype, 93916*e*).

23. BOTHRODISCUS Shear, *Bull. Torrey bot. Cl.* **34**: 312 (1907).
Pycnocalyx Naumov, *Bull. Soc. Oural Sci. Nat. Trud. Bur. Mykol.* **35**: 34 (1915).
Sp. typ.: *B. berenice* (Berk. & Curt.) Groves (syn. *Fusisporium berenice* Berk. & Curt., *B. pinicola* Shear).
Illustration: Groves (1968a).

Mycelium immersed, branched, septate, pale brown. *Conidiomata* stromatic, cupulate, superficial, separate, dark brown, at first closed, later opening widely, broadly stipitate; composed of thick-walled, pale brown plectenchymatic textura intricata, becoming pseudoparenchymatic textura angularis and darker at the periphery, extending above into a periclinal, usually incurved margin; multilocular, the globose locules regularly arranged at the flattened base of the cup and concealed by the incurved margin. Each locule dehisces separately by irregular rupture at the apex, the conidial mass of a single locule being discharged in toto on the surface of the stroma. *Conidiophores* absent. *Conidiogenous cells* holoblastic, sympodial, discrete, indeterminate, hyaline, smooth cylindrical, with 1-3 flattened, unthickened scars towards the apices, formed from the inner cells of the locular wall. *Conidia* hyaline, 0-7 euseptate, continuous, base truncate, apex obtuse, thin-walled, smooth, eguttulate, falcate.

Groves (1968a) has established the synonymy of the type species, and has also described two additional *Ascocalyx* species with *Bothrodiscus* conidial states. These are differentiated by septation and conidial size but have received no formal names.

B. BERENICE (Berk. & Curt.) Groves, *Can. J. Bot.* **46**: 1273 (1968). (Fig. 132)
synonymy in Groves (1968a).

Conidiomata up to 850 µ diam. Conidiogenous cells 9-16 × 3-4.5 µ. Conidia 25-28 × 3-3.5 µ, 3-5 septate.

On branches of *Abies balsamea*, Ont., Que., Can., Mich., U.S.A. (IMI 125729, 105177, 122895, 20896, 43758, 73273, 29676).

24. ONCOSPORELLA Karst., *Medd. Soc. Fauna Flora fenn.* **14**: 105 (1887).
Sp. typ.: *O. punctiformis* Karst.

Mycelium immersed, branched, septate, pale brown. *Conidiomata* eustromatic, superficial, separate, dark brown to black, opening widely to become cupulate; periclinal wall pseudoparenchymatous, composed of thick-walled, brown textura porrecta; at the base composed of hyaline to pale brown textura angularis. *Ostiole* absent. *Conidiophores* hyaline or thicker-walled, and pale brown, branched and septate only at the base, smooth, cylindrical, formed from the base of the conidioma. *Conidiogenous cells* holoblastic, sympodial, integrated, less often discrete, indeterminate, with 1-3 flat,

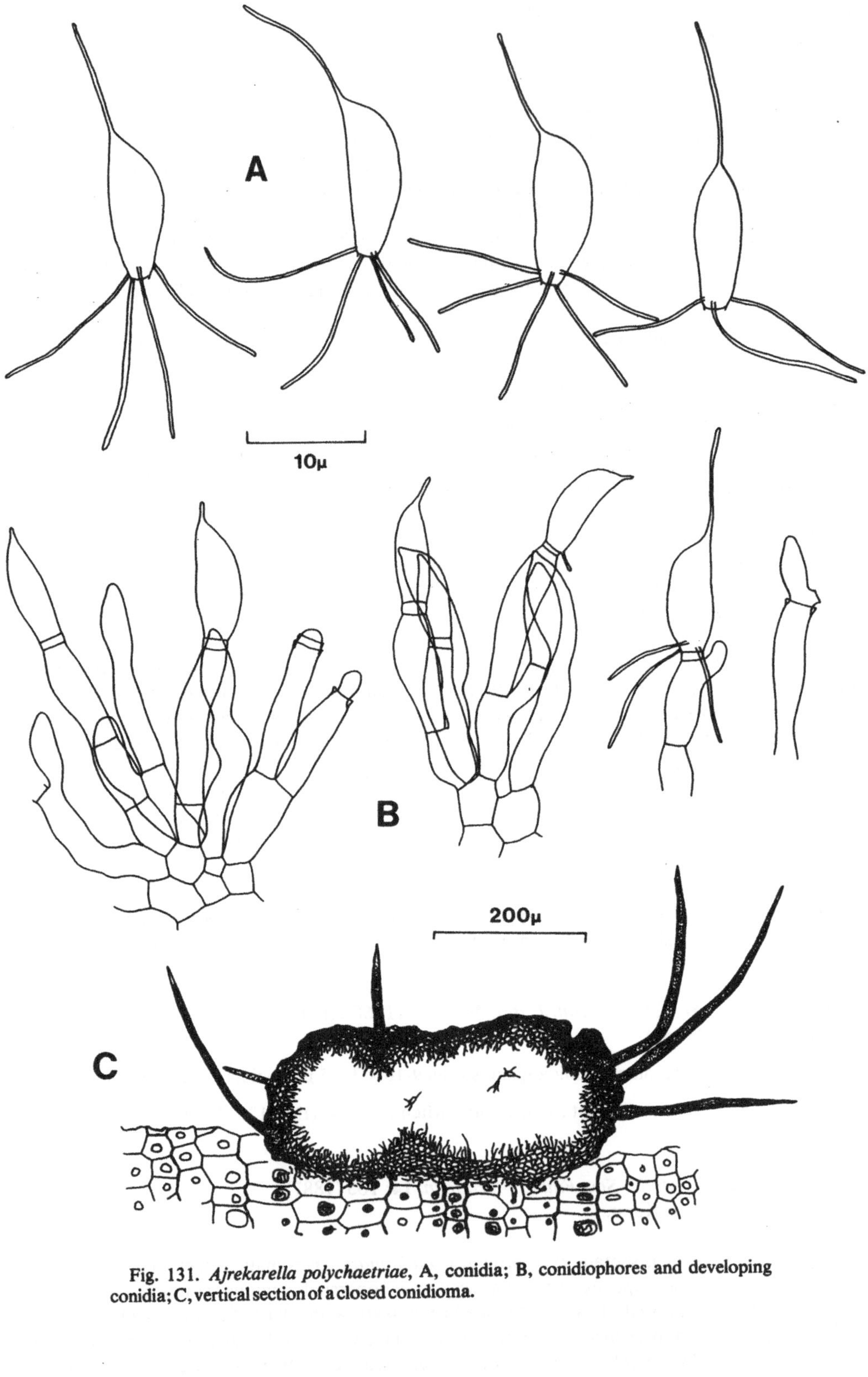

Fig. 131. *Ajrekarella polychaetriae*, A, conidia; B, conidiophores and developing conidia; C, vertical section of a closed conidioma.

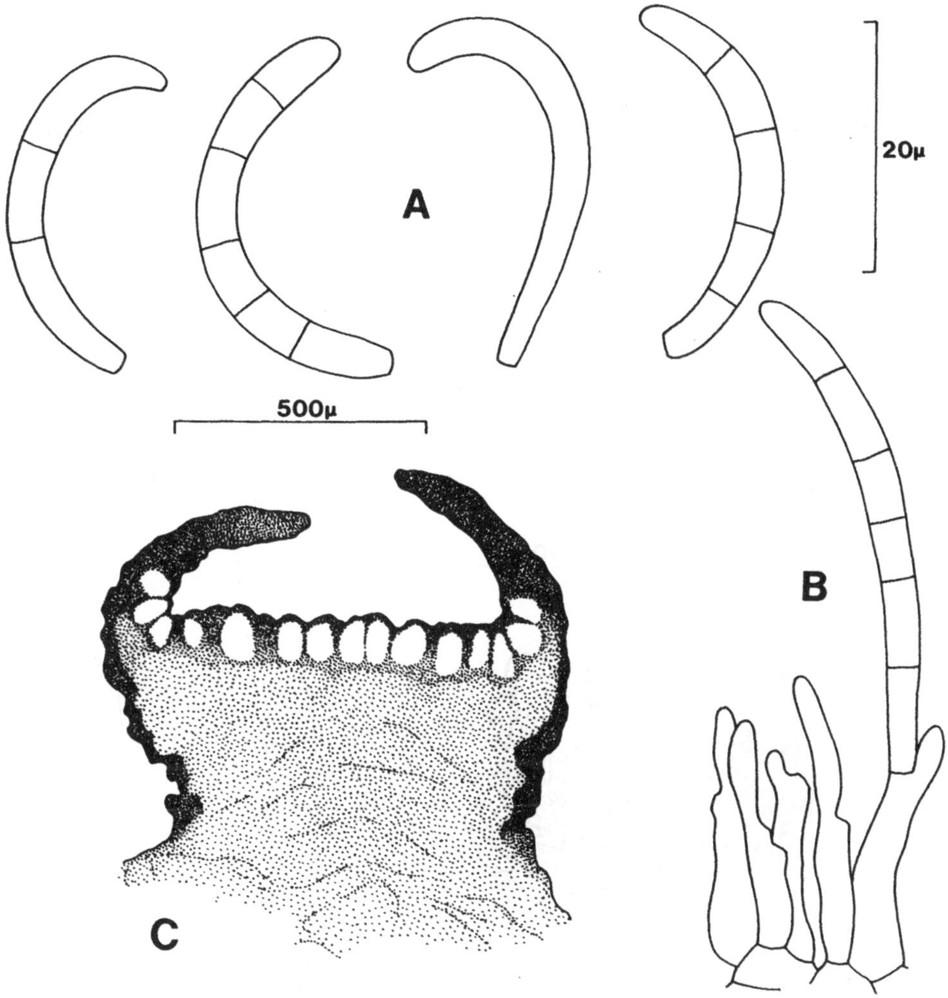

Fig. 132. *Bothrodiscus berenice*, A, conidia; B, conidiogenous cells and developing conidia; C, vertical section of a conidioma.

unthickened, slightly protuberant scars towards the apices. *Conidia* hyaline, up to 7 euseptate, continuous or slightly constricted at septa, apex obtuse, base truncate, thin-walled, smooth, often guttulate, straight or strongly curved.

Saccardo (1892) transferred *O. punctiformis* to *Collonema* Grove but the latter genus has been shown to be a Sphaeriaceous ascomycete by Sutton (1977a) so the synonymy is untenable. At present there is only a single species in the genus.

O. PUNCTIFORMIS Karst., *ibid.*: 105 (1887). (Fig. 133)
Collonema punctiformis (Karst.) Sacc., *Syll. fung.* **10**: 397 (1892).

Conidiomata up to 110 μ wide and 80 μ deep. Conidiophores 13-20 × 2-4 μ. Conidia 28-47 × 4-4.5 μ.

On dead wood of *Populus*, Finland (IMI 200425, holotype).

25. SEPTOPATELLA Petrak, *Annls mycol.* **23**: 128 (1925).
Sp. typ.: *S. septata* (Jaap) Petrak (syn. *Pseudocenangium septatum* Jaap).
Illustration: Dyko & Sutton (1979b).

Mycelium immersed, branched, septate, hyaline to subhyaline. *Conidiomata* superficial, often subtended by a superficial, pale brown, septate, branched mycelium, pulvinate, separate to occasionally aggregated, dark brown to black,

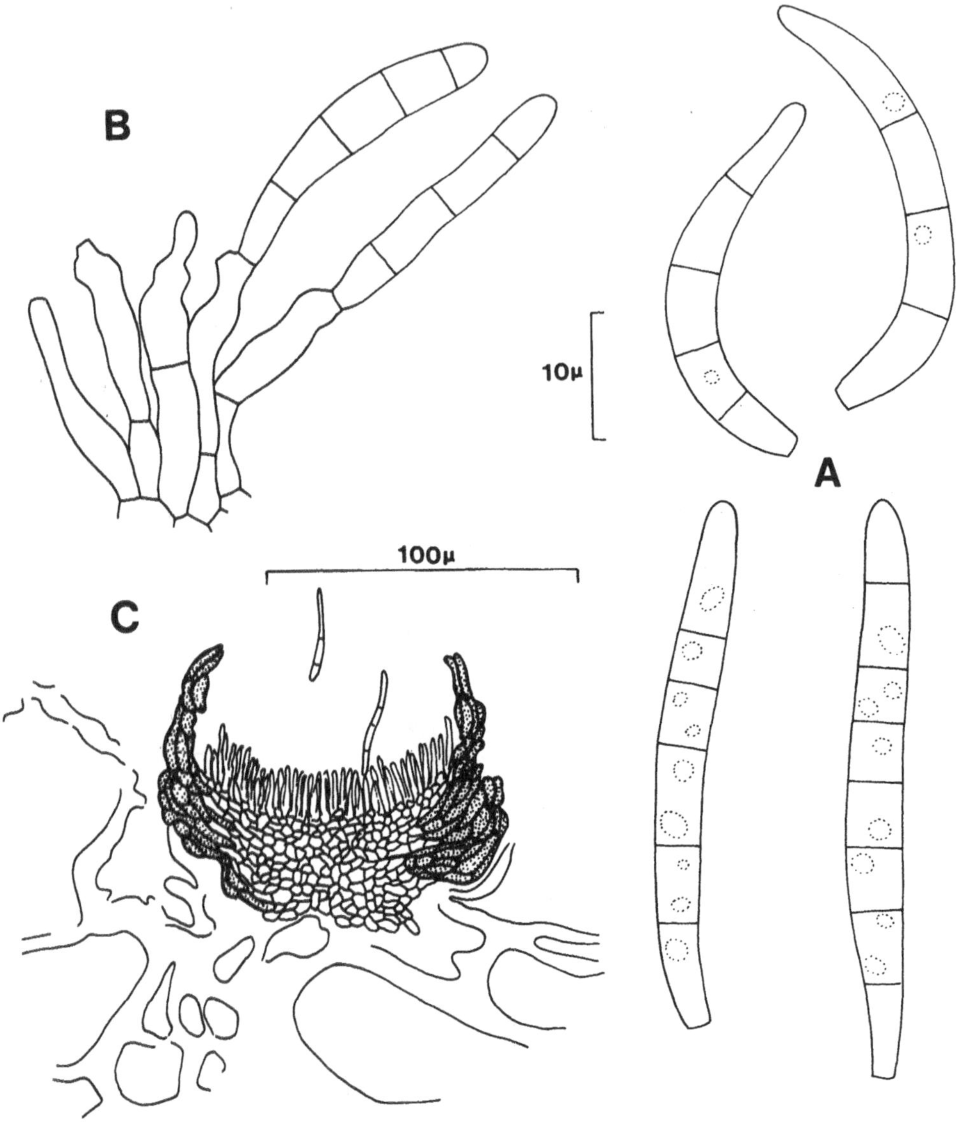

Fig. 133. *Oncosporella punctiformis*, A, conidia; B, conidiophores and developing conidia, C, vertical section of a conidioma.

finally opening widely, cupulate; basal wall of small-celled, brown, thin-walled textura angularis becoming textura porrecta as it merges into the periclinal walls; a hypostroma attaches the conidioma to the substrate. *Ostiole* absent. *Conidiophores* hyaline, septate and branched only at the base, thin-walled, ± cylindrical, formed from the gelatinized basal wall or the conidioma. *Conidiogenous cells* holoblastic, sympodial, integrated, indeterminate, cylindrical, hyaline, smooth, produced as 2-3 branches from the apex of the conidiophores. *Conidia* hyaline, 3-4 euseptate, thin-walled, smooth, minutely guttulate, straight or curved. occasionally irregularly filiform.

The genus is monotypic and has been revised by Dyko & Sutton (1979b). Contrary to a previous published account (Morgan-Jones & Kendrick, 1972) conidiogenesis is sympodially holoblastic not phialidic.

S. SEPTATA (Jaap) Petrak, *ibid.*: 128 (1925).

Pseudocenangium septatum Jaap, *Annls mycol.* **6**: 219 (1908).

Conidiomata 500-900 µ diam. Conidiogenous cells 9-28 × 2-3 µ. Conidia 30-68 × 1.5-2 µ.

On needles of *Pinus pumila*, *P. montana*, *P. ponderosa*, *P. nigra*, Czechoslovakia, Greece, Austria, Calif., U.S.A. (IMI 225740, 225733, 228411).

26. LEPTOCHLAMYS Died., *Annls mycol.* **19**: 299 (1921).
Sp. typ.: *L. thecicola* (Berk. & Br.) A. Racovitza (syn. *Sphaeropsis thecicola* Berk. & Br., *L. scapicola* (Berk. & Br.) Died.).

Mycelium immersed, branched, septate, hyaline to pale brown. *Conidiomata* subcuticular, separate, rarely confluent, dark brown, scutate to pulvinate, with a 1-3 celled upper wall of dark brown, thick-walled textura angularis, basal wall absent, ± circular, irregular (not radiate); margin entire to irregular; definite *ostiole* absent, dehiscence by breakdown of the upper wall at the centre. *Conidiophores* hyaline, smooth, sparsely septate and branched at the base and above, restricted to the upper wall. *Conidiogenous cells* holoblastic, occasionally sympodial, discrete or more rarely integrated, indeterminate, hyaline, smooth, forming 1-2 conidia from unthickened apical to subapical loci. *Conidia* hyaline, medianly 1 euseptate, straight, cylindrical, thin-walled, eguttulate, base truncate, apex obtuse.

Clements & Shear (1931) placed *Leptochlamys* in synonymy with *Megaloseptoria* Naumov but Shoemaker (1967) did not accept this conclusion. Apart from differences in conidiomata (eustromata compared with pycnidia), the method of conidial ontogeny and type of conidial morphology are quite distinct.

L. THECICOLA (Berk. & Br.) Racovitza, *Bull. Sec. Sci. Acad. Rom.* **30**: figs 1-2, pl. 1 (1947). (Fig.134)
Leptochlamys scapicola (Karst.) Died., *Annls mycol.* **19**: 299 (1921).
Septoria thecicola Berk. & Br. var. *scapicola* Karst., *Medd. Soc. Fauna Flora fenn.* **16**: 11 (1888).
Expanded synonymy in Racovitza (1959).

Conidiomata up to 70 μ diam. × 30 μ high. Conidiogenous cells 4-8.5 × 2 μ. Conidia 15-17.5 × 1.5 μ.

On capsular stalks of *Polytrichum juniperinum*, Germany (IMI 199146, LE).

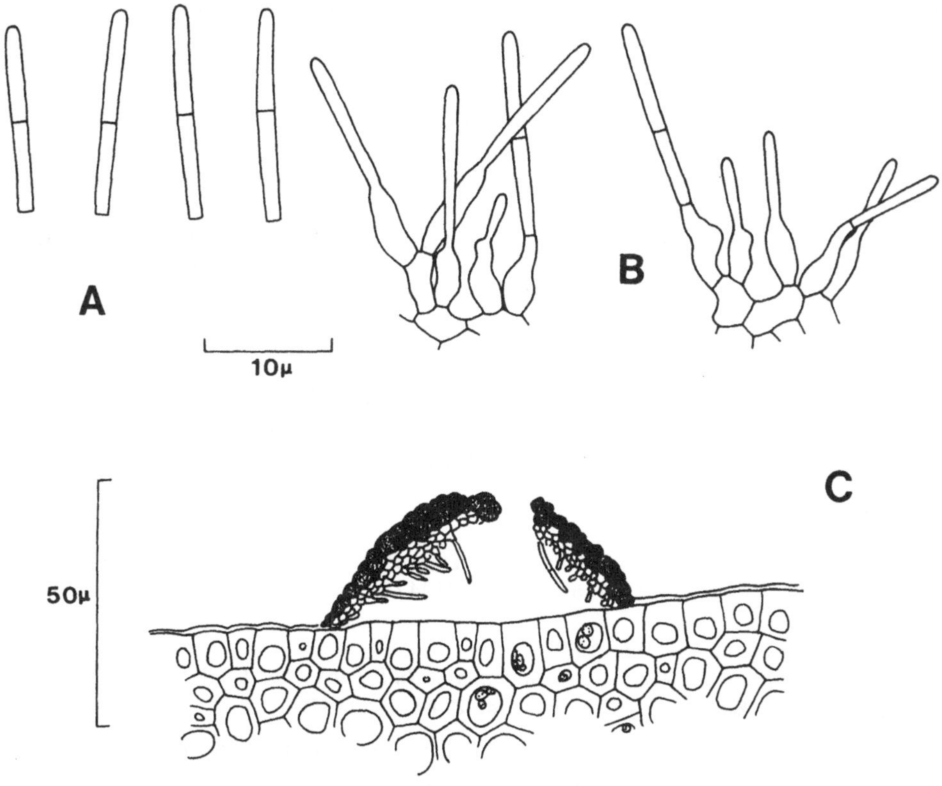

Fig. 134. *Leptochlamys scapicola*, A, conidia; B, conidiophores and developing conidia; C, vertical section of a conidioma.

27. ASTEROMIDIUM Speg., *An. Soc. Cient. Arg.* **26**: 66 (1888).
Sp. typ.: *A. imperspicuum* Speg.

Mycelium immersed, branched, septate, hyaline. *Conidiomata* acervular, subcuticular, separate or confluent, pulvinate to doliiform, at the base composed of hyaline to pale brown, thin-walled textura angularis which extends laterally, finally with separate cells dispersed in a mucilaginous matrix to form the overlying wall; cuticle discoloured and occasionally pseudoparenchymatous walls adjacent to the upper epidermal wall also discoloured. Dehiscence irregular. *Conidiophores* absent. *Conidiogenous cells* holoblastic, discrete, indeterminate, ± cylindrical, hyaline, smooth, with 1-2 sympodial proliferations, scars unthickened, flat, formed from the basal and lateral walls. *Conidia* cylindrical to fusiform, gently tapered at each end, apex obtuse, base truncate, thin-walled, eguttulate, hyaline, 3 septate.

The name *Asteromidium* first appeared for a subgeneric taxon of *Actinonema* Pers. by Saccardo (1884). Two species were included, *A. guepinii* (Mont.) Sacc. and *A. uvariae* (Lév.) Sacc., but neither was listed when Spegazzini (1888) introduced *Asteromidium* as a generic name.

Clements & Shear (1931) listed *Dearnessia* Bub. as a synonym but this genus is truly pycnidial. Petrak & Sydow (1936) gave an account of *Asteromidium* concluding that it was an acceptable genus. A second species, *A. secundum* was added by Batista, Peres & Maia (1959).

A. IMPERSPICUUM Speg., *ibid.*: 66 (1888). (Fig. 135)

Conidiomata up to 150 µ diam. × 80 µ high. Conidiogenous cells 8-16 × 2.5-4.5 µ. Conidia 21-25 × 3 µ.

Hypophyllous on living leaves of Sapindaceae, Paraguay (IMI 199540, isotype of *A. imperspicuum* ex K).

28. PHLOEOSPORELLA Höhn., *Annls mycol.* **22**: 201 (1924).
Sp. typ.: *P. ceanothi* (Ell. & Ev.) Höhn. (syn. *Cylindrosporium ceanothi* Ell. & Ev.).

Mycelium immersed, branched, septate, hyaline. *Conidiomata* acervular, subepidermal, ± circular, discrete, composed of hyaline to pale brown, thin-walled textura angularis. Dehiscence irregular. *Conidiophores* absent. *Conidiogenous cells* holoblastic, sympodial, discrete, indeterminate, hyaline, smooth, lageniform to cylindrical, with 1-2 broad, flat unthickened apical scars, formed from the upper pseudoparenchyma. *Conidia* hyaline, 2 euseptate, thin-walled, smooth, guttulate, straight, curved or irregular, tapered gradually to the obtuse apex and abruptly to the truncate base.

At least 5 taxa have been referred to *Phloeosporella* including *P. maculans* (Bereng.) Höhn., the anamorphosis of *Mycosphaerella mori* (Fckl) Wolf. Conidiogenesis in this species is holoblastic and annellidic so it should be referred to *Phloeospora* Wallr. rather than *Phloeosporella* as suggested by Boerema & Verhoeven (1973).

P. CEANOTHI (Ell. & Ev.) Höhn., *ibid.*: 201 (1924). (Fig. 136)
Cylindrosporium ceanothi Ell. & Ev., *Proc. Acad. Phil. 1891*: 84 (1891).

Lesions circular, often confluent, brown, margin distinct but irregular, up to .5 cm diam. Conidiomata amphigenous, up to 250 µ diam. Conidiogenous cells 12-16 × 3.5-4 µ, with 1-2 distinct flat apical scars. Conidia 62-67 × 4-6 µ, 2 septate, guttulate.

Causing leaf lesions on *Ceanothus velutinus, C. sanguineus, C. integerrimus*, Colo., Wash., Iowa, Ore., Calif., U.S.A. (IMI 21352, 21353, 25928, 142186, 101343, 101342 ex K).

P. ariaefoliae (Ell. & Ev.) comb. nov. (Fig. 137)
Cylindrosporium ariaefoliae Ell. & Ev. (as '*ariaefolium*'), *Bull. Torrey bot. Cl.* **27**: 575 (1900).
Phloeospora ariaefoliae (Ell. & Ev.) Höhn., *Öst. Bot. Z.* **66**: 106 (1916).

Lesions small, brown, margin diffuse, up to 500 µ diam. Conidiomata epiphyllous, up to 200 µ diam. Conidiophores 11-21 × 3-3.5 µ, with 1-2 distinct flat apical scars. Conidia, fusiform, falcate, eguttulate, 2 septate, 32-38 × 5-5.5 µ.

On leaves of *Spiraea discolor*, Calif., U.S.A. (IMI 92430).

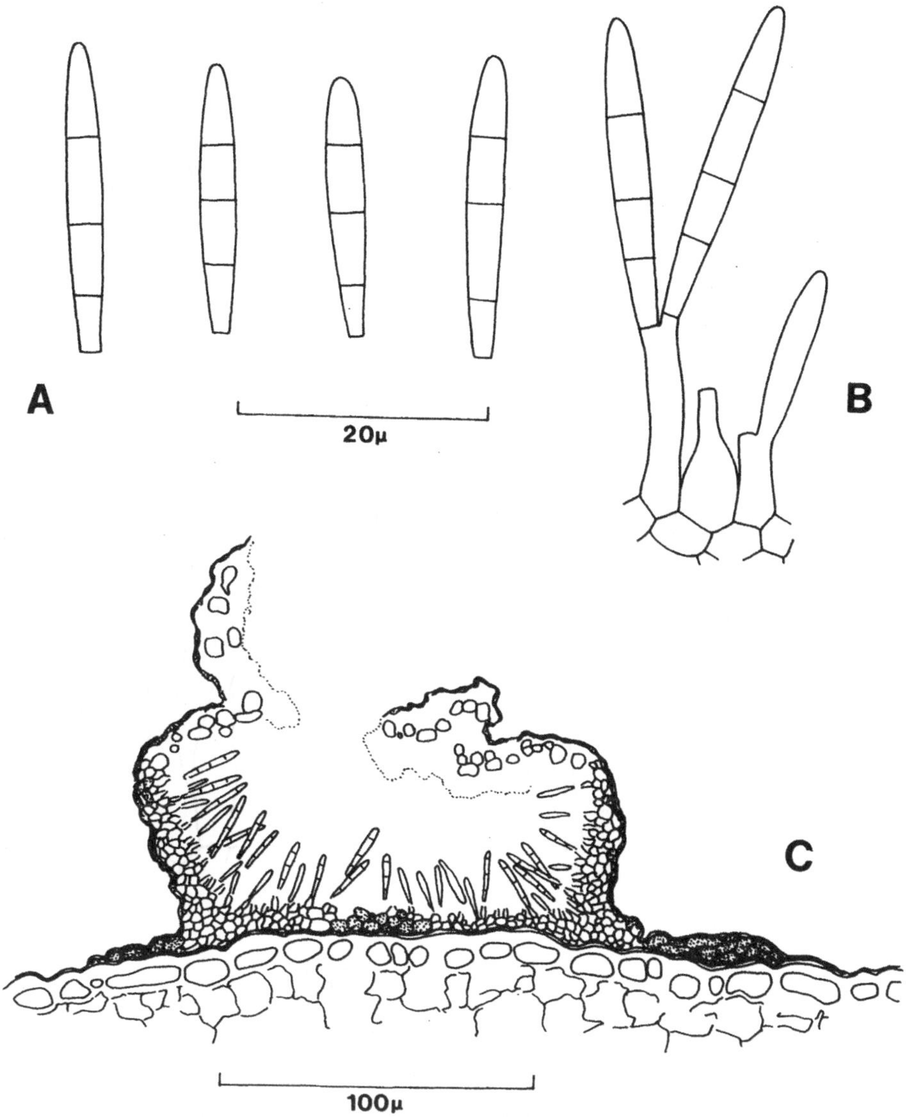

Fig. 135. *Asteromidium imperspicuum*, **A**, conidia; **B**, conidiogenous cells and developing conidia; **C**, vertical section of a conidioma.

P. hedysari (Solheim) comb. nov. (Fig. 138)
Septogloeum hedysari Solheim, *Univ. Wyo. Publ.* **10**: 42 (1943).

Lesions diffuse, large, spreading, pale brown, margin diffuse, .5-1 cm diam. Conidiomata hypophyllous, immersed, almost cupulate, up to 220 μ diam. Conidiogenous cells doliiform to cylindrical, 6-17 × 4.5-7 μ. Conidia medianly 1 septate, cylindrical, apex obtuse, eguttulate, 41-55 × 5-6.5 μ.

On leaves of *Hedysarum marginatum*, Wyo., U.S.A. (IMI 97269, isotype).

P. leucosceptri (Keissl.) comb. nov. (Fig. 139)
Cylindrosporium leucosceptri Keissl., *Öst. Bot. Z.* 124 (1924).
Cercoseptoria leucosceptri (Keissl.) Petrak, *Sydowia* **1**: 347 (1947).

Lesions small, medium brown, margin diffuse, up to .1 cm diam. Conidiomata, white, amphigenous but mainly epiphyllous, up to 100 μ diam. Conidiophores 23-35 × 3-3.5 μ, with 2-3 distinct, flat, apical scars. Conidia

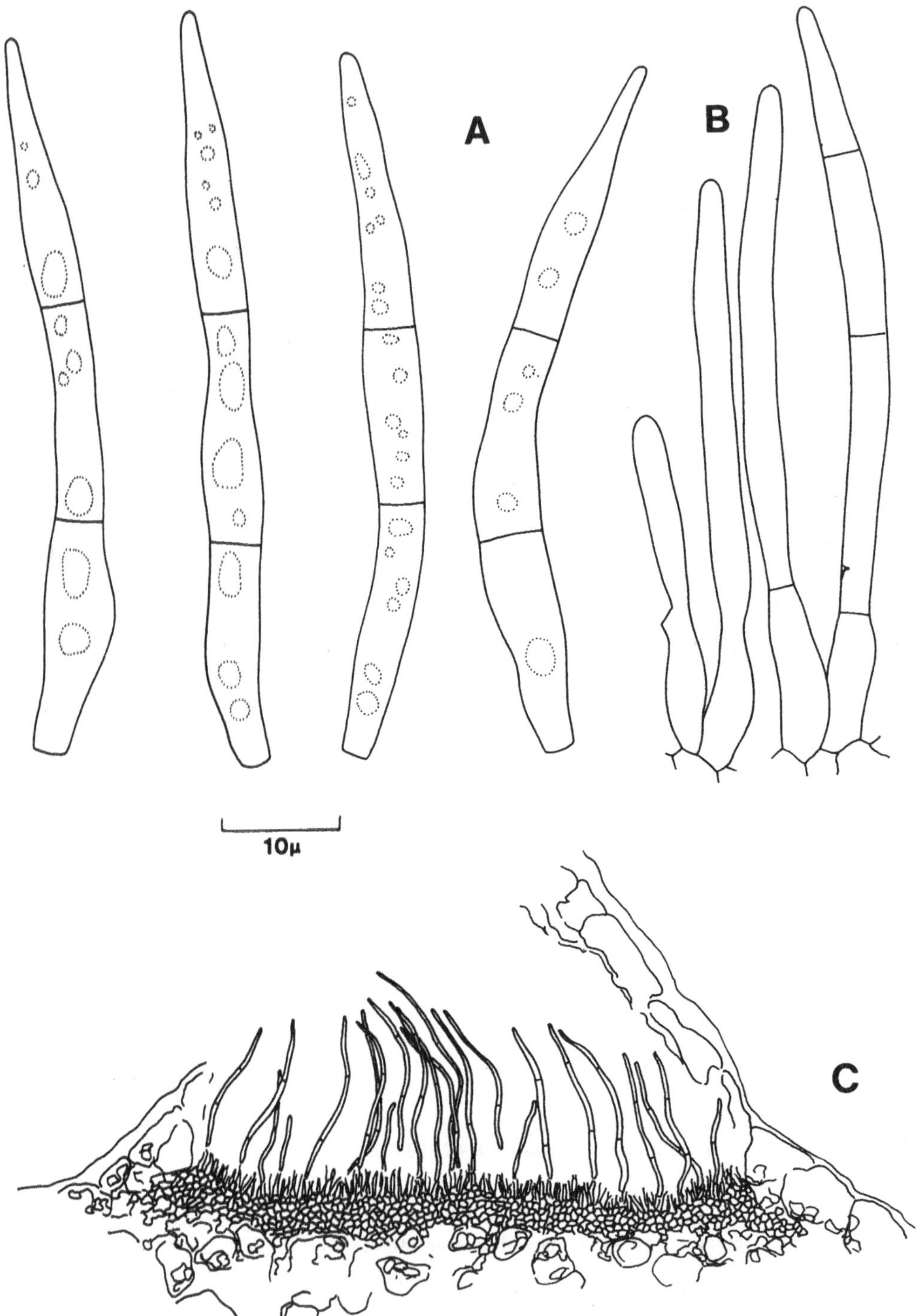

Fig. 136. *Phloeosporella ceanothi*, A, conidia; B, conidiogenous cells and developing conidia; C, vertical section of a conidioma.

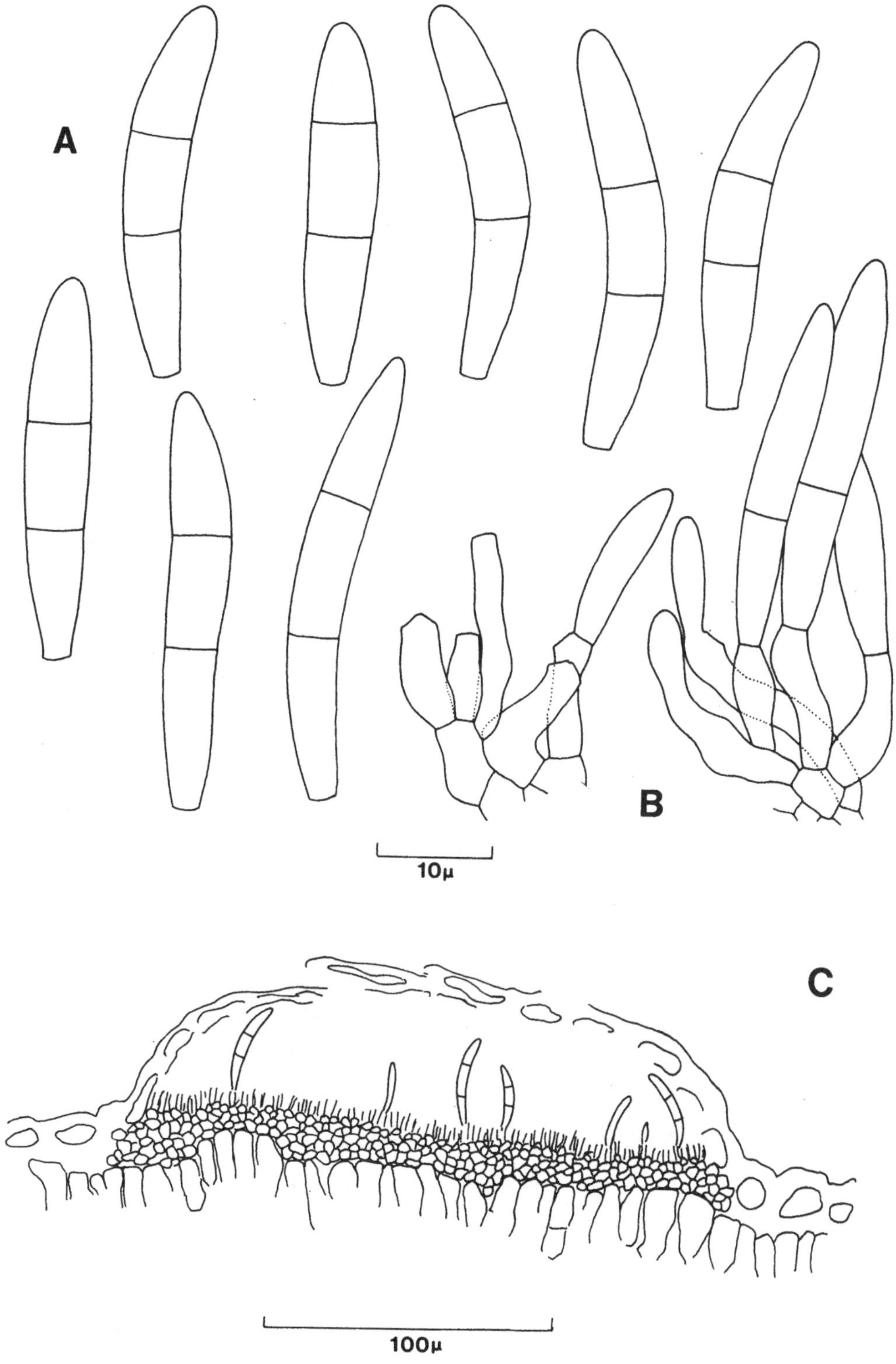

Fig. 137. *Phloeosporella ariaefoliae*, A, conidia; B, conidiophores and developing conidia; C, vertical section of a conidioma.

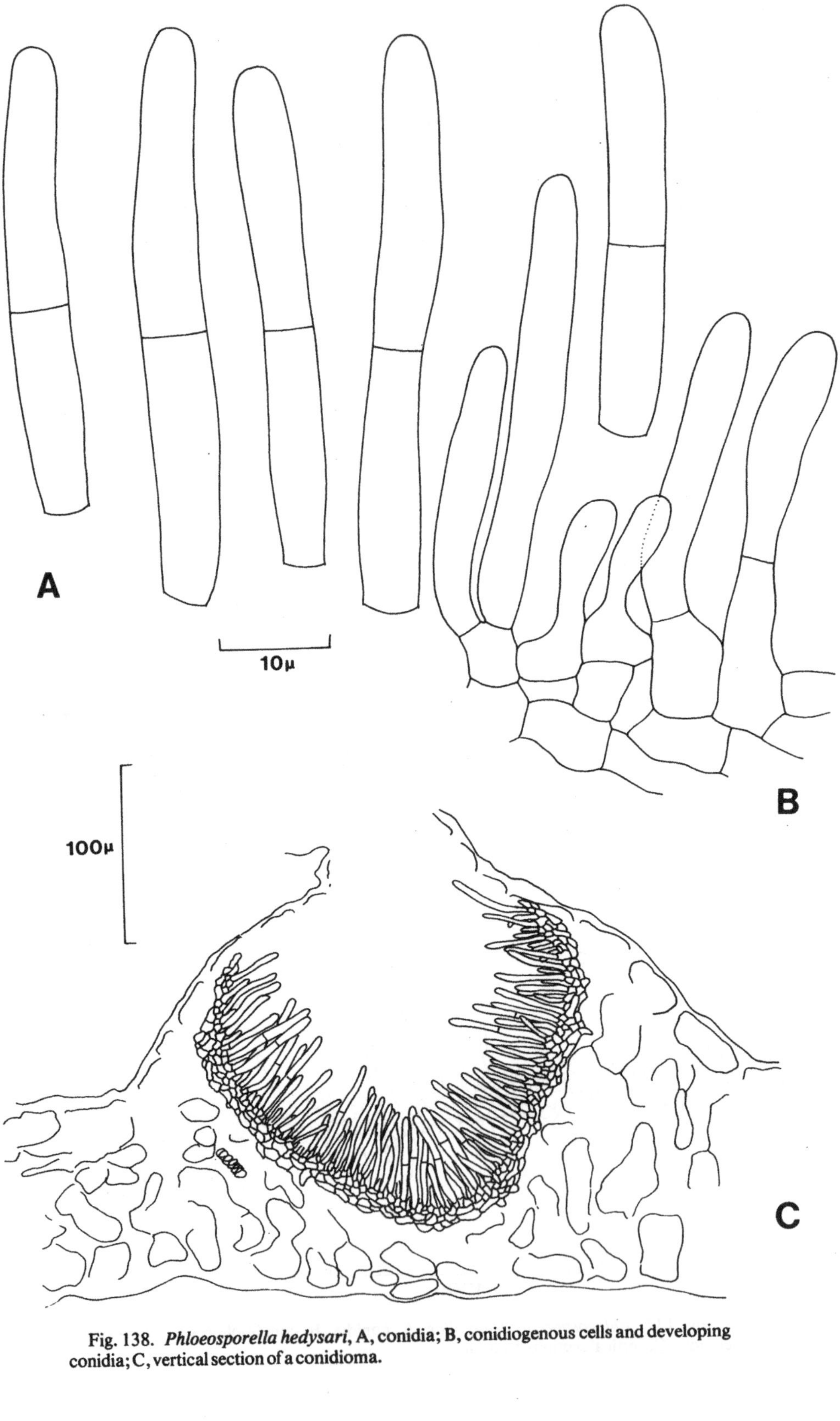

Fig. 138. *Phloeosporella hedysari*, A, conidia; B, conidiogenous cells and developing conidia; C, vertical section of a conidioma.

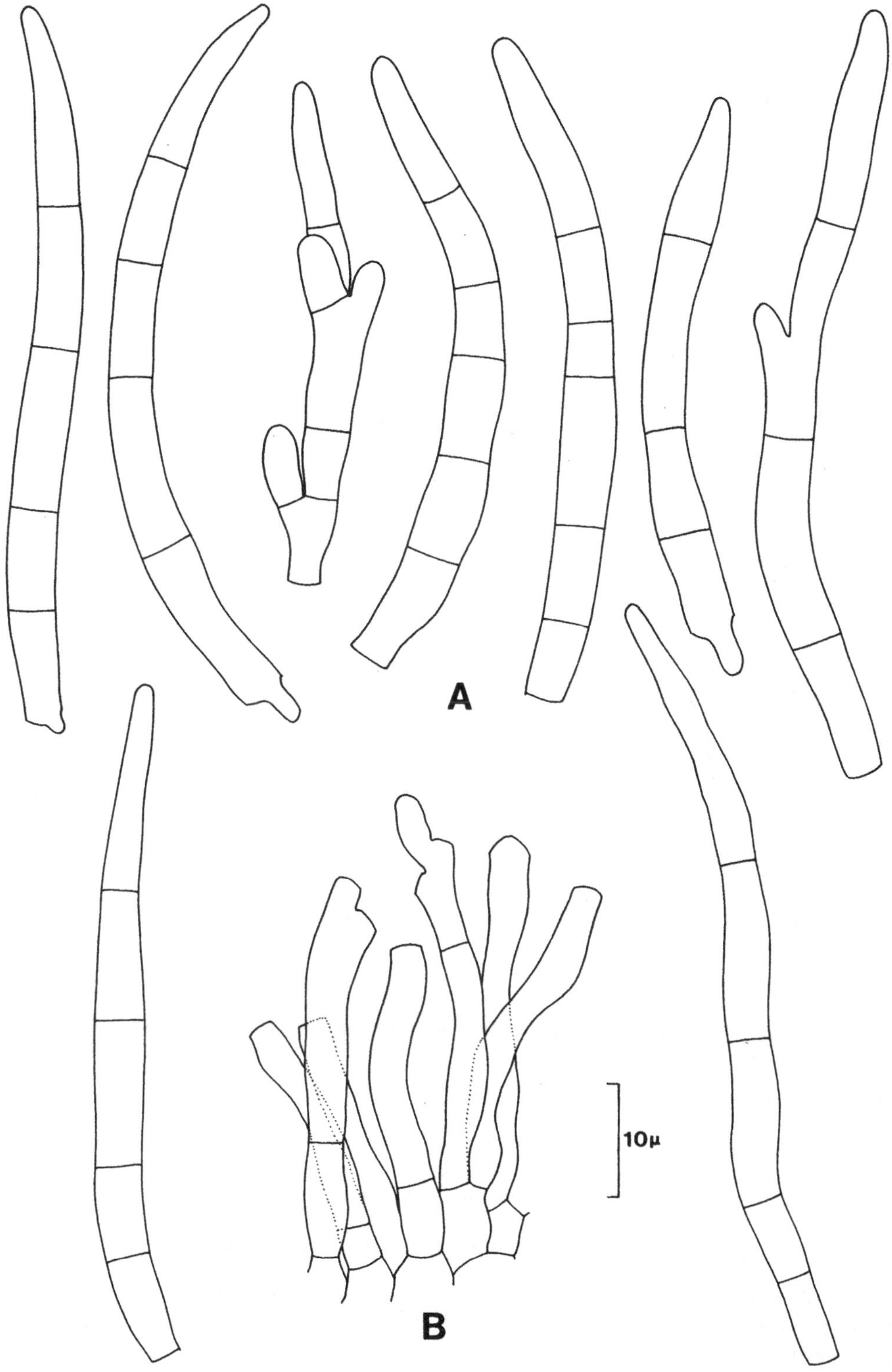

Fig. 139. *Phloeosporella leucosceptri*, A, conidia; B, conidiophores.

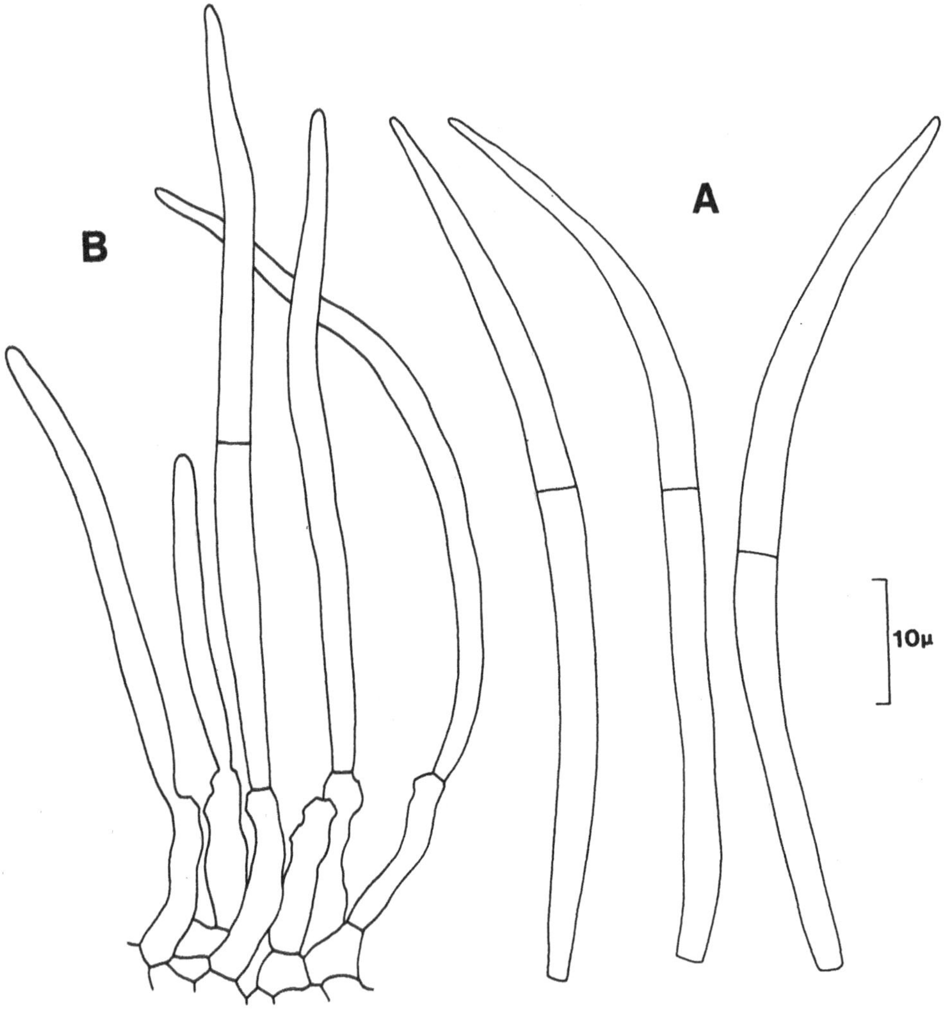

Fig. 140. *Phloeosporella padi*, A, conidia; B, conidiogenous cells and developing conidia.

occasionally shortly branched, 3-5 septate, base truncate, sometimes with a short central appendage or protuberance, eguttulate, 45-70 × 4-5 μ.

On leaves of *Leucosceptrum canum*, India (IMI 113593).

P. PADI (Lib.) von Arx, *Phytopath. Z.* **42**: 163 (1961). (Fig. 140)
Ascochyta padi Lib., *Pl. Crypt. Ard.* fasc. 2, No. 153 (1832).
Cylindrosporium padi (Lib.) Karst. ex Sacc., *Syll. fung.* **3**: 738 (1884)

Teleomorphosis: *Blumeriella jaapii* (Rehm) von Arx, *Phytopath. Z.* **42**: 164 (1961).

Lesions small, brown, margin distinct, raised, eventually centres dropping out leaving a shot-hole effect. Conidiomata up to 250 μ diam. Conidiogenous cells 10-15 × 2-3 μ with up to 3 indistinct scars. Conidia 69-74 × 3 μ, 1 septate, eguttulate.

Causing leaf shot-hole of *Prunus mahaleb*, *P. cerasus*, *P. melanocarpa*, *P. pennsylvanica*, *P. domestica*, *P. virginiana* var. *demissa*, Sask., Ont., Canada, N.Y., Calif., Ore., U.S.A., Latvia, Roumania (IMI 172078, 125701, 154350, 137320, 26620, 21455, 21456, 25894, 26621, 172079).

29. DIPLOSPORONEMA Höhn., *Sber. Akad. Wiss. Wien* **126**: 335 (1917).
Phragmosporonema Moesz, *Mag. Bot. Lap.* **33**: 52 (1934).
Sp. typ.: *D. delastrei* (Delacr.) Petrak (syn. *Gloeosporium delastrei* Delacr.).

Mycelium immersed, hyaline, branched, septate. *Conidiomata* acervular, epidermal to subepidermal, separate, formed of hyaline to very pale brown, thin-walled textura angularis. Dehiscence irregular. *Conidiophores* hyaline, branched at the base and above, either penicillately or irregularly, septate, smooth, formed from the upper pseudoparenchyma. *Conidiogenous cells* holoblastic, sympodial, indeterminate, integrated or discrete, ± straight, hyaline, smooth, with 1-2 conidia formed sympodially from broad, flat, unthickened scarcely protuberant scars. *Conidia* hyaline, 1-2 euseptate, thin walled, eguttulate, smooth, base truncate, apex obtuse, clavate.

Höhnel (1917a) introduced this generic name for conidial *Pyrenopeziza agrostemmatis* Fckl (*Diplocarpon agrostemmatis* (Fckl) Nannf.) but did not formally ascribe a species to it, even though it is understood to be based on *Gloeosporium delastrei* Delacr. from the discussion. The binomial *Diplosporonema delastrei* was not used until Petrak (1947a) pointed out that *Phragmosporonema* Moesz was based on the same species. Since he placed *P. delastrei* in synonymy with *D. delastrei* it is clear that Petrak should be ascribed as the validating authority.

D. DELASTREI (Delacr.) Petrak, *Sydowia* **1**: 74 (1947). (Fig. 141)
Gloeosporium delastrei Delacr. apud Mont., *Ann. Sci. nat.*, ser. 4, **5**: 345 (1840).
Marssonia delastrei (Delacr.) Sacc., *Michelia* **2**: 119 (1880).
Marssonina delastrei (Delacr.) Magn., *Hedwigia* **45**: 88 (1906).
Phragmosporonema delastrei (Delacr.) Moesz & Smarods apud Moesz, *Mag. Bot. Lap.* **33**: 52 (1934).

Conidiomata up to 150 µ diam. Conidiophores up to 35 µ long × 3-4 µ wide. Conidia 19.5-31.5 × 5-7 µ.

On leaves, rarely stems, of *Agrostemma githage, Silene* spp., *Melandrium* sp., and other Caryophyllaceae, N. Dak., Wisc., U.S.A., Austria, G.B., Canada, Latvia, Czechoslovakia (IMI 22172, 12243, 86690, 95172, 95310b, 102186, 168493, 25910, 25756, 33486, 31077).

30. STICTOSEPTA Petrak, *Sydowia* **17**: 230 (1964).
Sp. typ.: *S. cupularis* Petrak

Mycelium immersed, branched, septate, hyaline. *Conidiomata* eustromatic, immersed, globose to collabent, within the substrate papillate, unilocular, often convoluted, hyaline; walls thick, of hyaline, thin-walled textura intricata. *Ostiole* central, circular, single, furfuraceous. *Conidiophores* hyaline, septate, branched, anastomosing, formed from the inner cells of the locular wall. *Conidiogenous cells* holoblastic, sympodial or synchronous, integrated, indeterminate (occasionally instead of becoming a conidium, a lateral branch will develop into a conidiogenous cell with loci at a higher level), hyaline, thin-walled, with usually 2 small, unthickened, apical, slightly protuberant conidiogenous loci. *Conidia* hyaline, thin-walled, smooth, with several eusepta, slightly constricted at the septa, each cell medianly guttulate, straight or curved, base truncate, apex obtuse.

The genus is monotypic.

S. CUPULARIS Petrak, *ibid.*: 230 (1964). (Fig. 142)

Conidiomata up to 700 µ diam × 330 µ high. Conidiogenous cells 4-14 µ long × 1.5-2 µ wide. Conidia 7 septate, 52-54 × 2 µ.

On dead branches of *Fraxinus*, Czechoslovakia (IMI 204093, isotype ex BPI).

31. SEPTOCYTA Petrak, *Annls mycol.* **25**: 330 (1927).
Sp. typ.: *S. ruborum* (Lib.) Petrak (syn. *Ascochyta ruborum* Lib., *S. ramealis* (Rob.) Petrak).

Mycelium immersed, branched, septate, hyaline to pale brown. *Conidiomata* eustromatic, immersed, separate, erumpent, dark brown to black, finally opening widely, unilocular, multilocular or convoluted, thick-walled; wall of pale brown, thin-walled textura angularis except in the dehiscent region which is darker brown and more thick-walled. *Ostiole* absent, dehiscence by breakdown of the upper wall. *Conidiophores* absent. *Conidiogenous cells* holoblastic, sympodial with 1-3 apical, scarcely protruding, unthickened denticles, indeterminate, discrete, ampulliform to lageniform, hyaline,

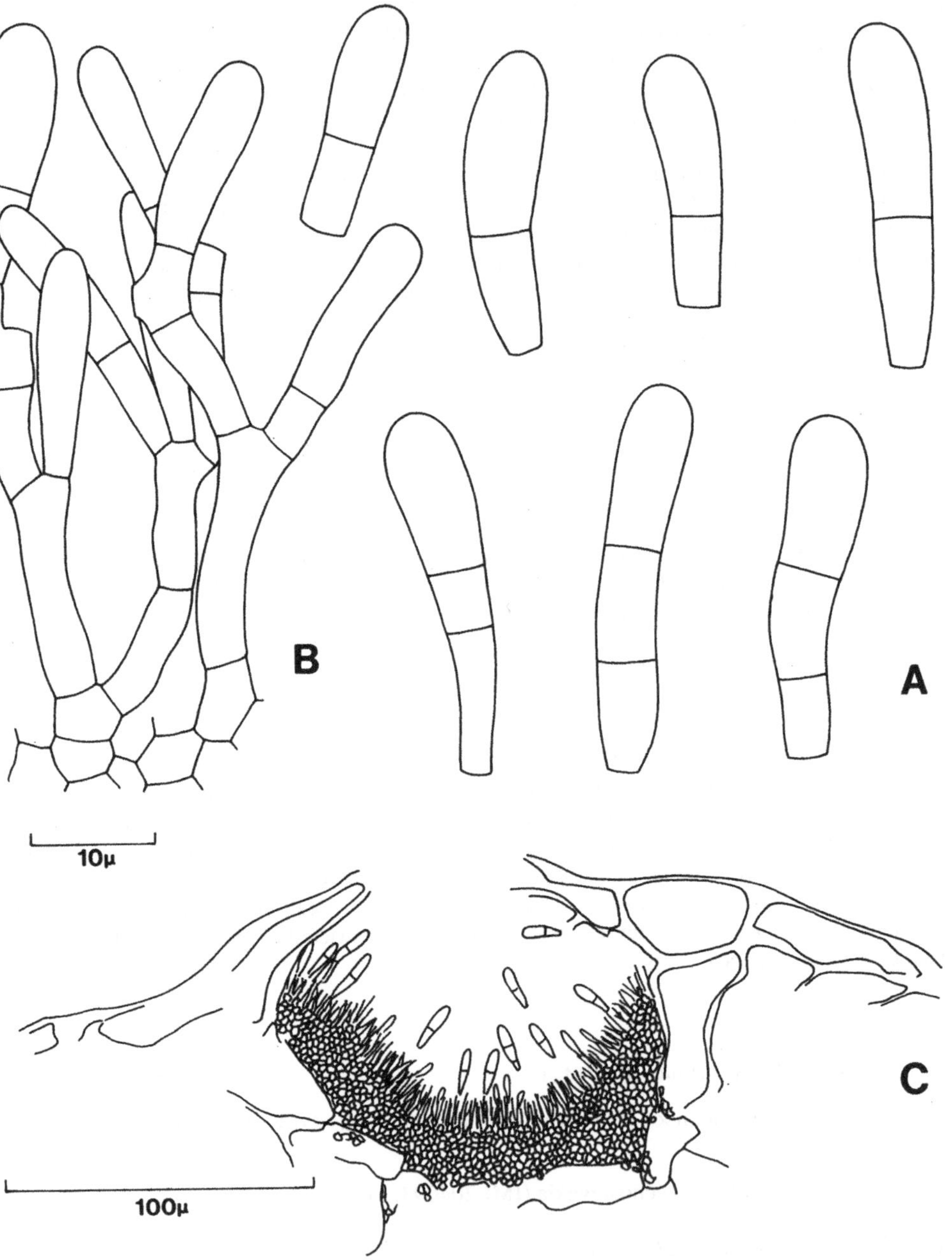

Fig. 141. *Diplosporonema delastrei*, A, conidia; B, conidiophores and developing conidia; C, vertical section of a conidioma.

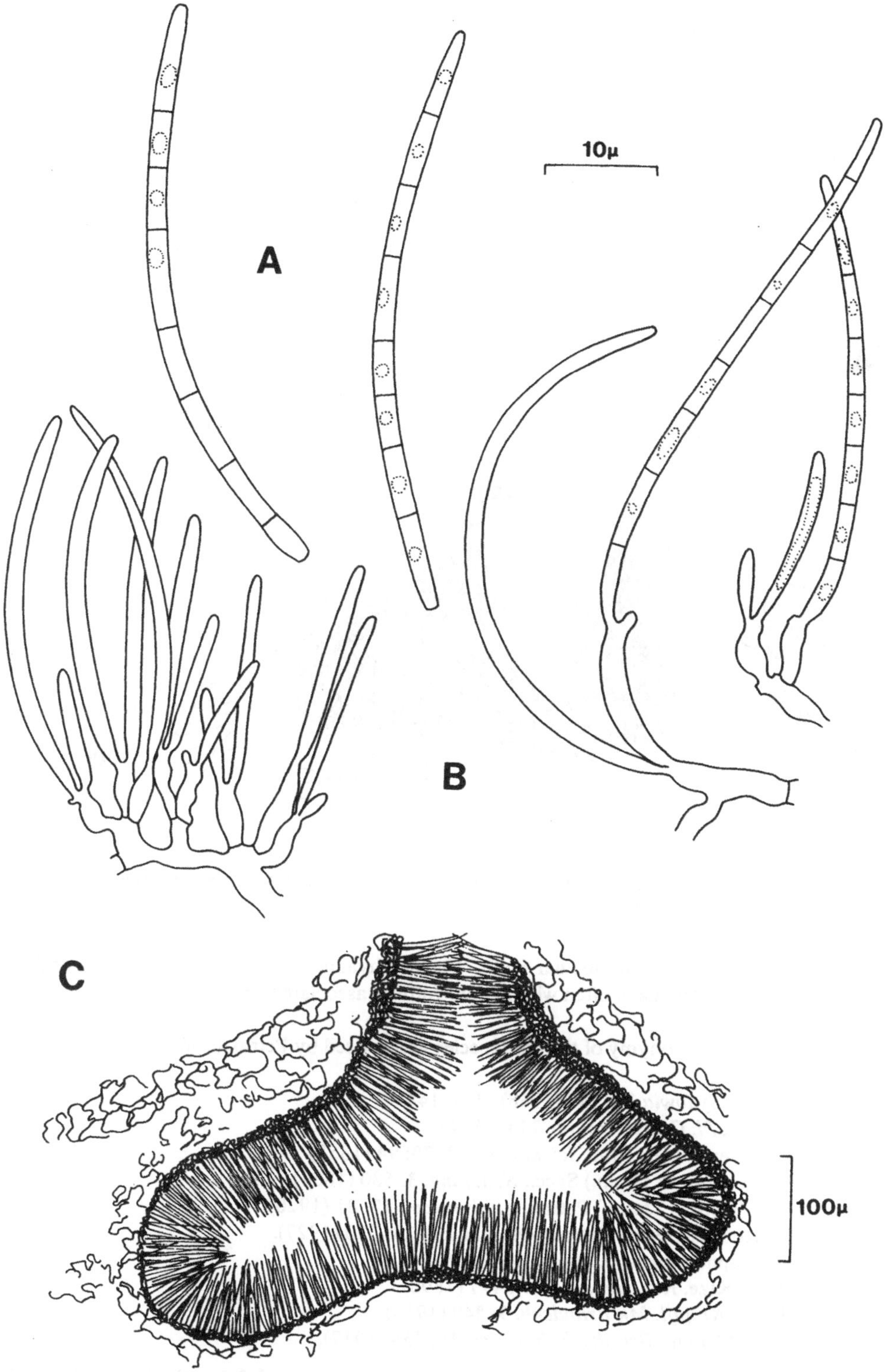

Fig. 142. ***Stictosepta cupularis***, A, conidia; B, conidiophores and developing conidia; C, vertical section of a conidioma.

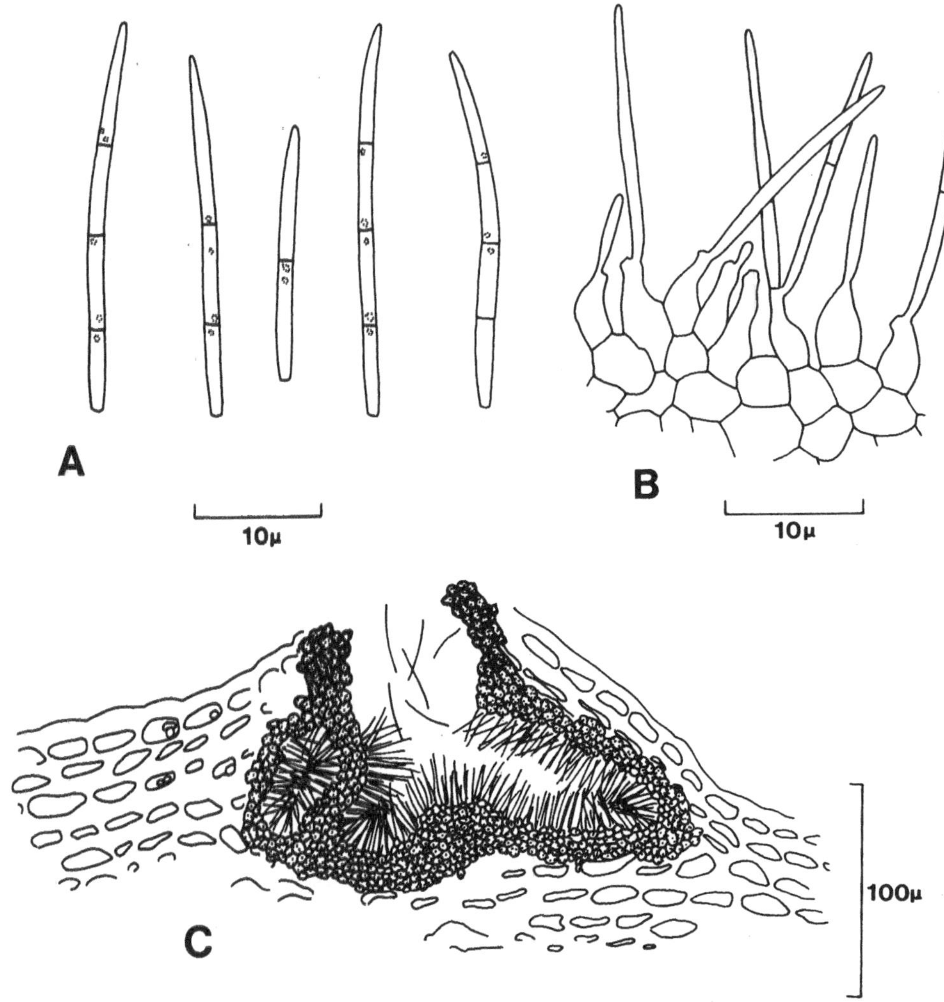

Fig. 143. *Septocyta ruborum*, A, conidia; B, conidiogenous cells and developing conidia; C, vertical section of a conidioma.

smooth, formed from the inner cells of the locular walls. *Conidia* hyaline, 1-3 euseptate, smooth, straight or slightly curved, acicular, apex obtuse, base truncate, often with minute guttules associated with septa.

The genus is monotypic. The synonymy of the type species was revised by Petrak (1968a).

S. RUBORUM (Lib.) Petrak, *Sydowia* **21**: 245 (1968). (Fig. 143)
Ascochyta ruborum Lib., *Pl. Crypt. Ard.* no. 214 (1834).
Septoria ramealis Rob. ex Desm., *Ann. Sci. nat.*, ser. 3, **20**: 94 (1853).
Rhabdospora ramealis (Rob. ex Desm.) Sacc., *Syll. fung.* **3**: 580 (1884).
Cytosporina ramealis (Rob. ex Desm.) Petrak, *Annls mycol.* **17**: 81 (1920).
Septocyta ramealis (Rob. ex Desm.) Petrak, *Annls mycol.* **25**: 330 (1927).
Rhabdospora rubi Ell. apud Martin, *J. Mycol.* **3**: 90 (1887).
Filaspora rubi (Ell.) O. Ktze, *Rev. Gen. Pl.* **3**(2): 477 (1898).
Cytosporina rubi Died., *Kryptfl. Mk Brandenb.* **9**: 549 (1914).
Hemidothis rubi (Died.) Höhn., *Ber. dtsch. Bot. Ges.* **35**: 354 (1917).

Conidiomata up to 300 µ diam × 140 µ deep. Conidiogenous cells 5-8 × 2-3.5 µ. Conidia 21-32 × 1.5 µ.

On stems of *Rubus plicatus*, *R. sulcatus*, *R. fruticosus*, *R.* sp., Austria, U.K. (IMI 33621, 13153, 18293, 113752, 21539, 21540, 31103, 100668).

BLASTOSTROMATINEAE, ANNELLIDIC

The majority of genera placed in this section are acervular, and a high proportion have pigmented conidia, thus providing an unexpected correlation with the annellidic type of conidiogenous cell. Furthermore, several fit into fairly well-defined groups. The most readily identifiable is the *Pestalotia*, *Pestalotiopsis*, *Truncatella*, *Monochaetia*, *Seiridium*, *Seimatosporium*, *Morinia* complex. The genera are conveniently separated by conidial septation and appendage number, except for *Seimatosporium* which is distinguished from all the others by an excentric, exogenous basal appendage. In fact, whereas those taxa with central endogenous basal appendages are split into 5 separate genera, in *Seimatosporium* they are included in a single genus. Several of the latter are known to have teleomorphoses, and a reasonable dismemberment of *Seimatosporium* in the future may follow a reappraisal of perfect-imperfect state connections. *Pestalotia* is also an oddity for it is the only genus in the complex with distoseptate conidia. Another recognisable group is comprised of *Phloeospora*, *Colletogloeum* and *Ciferriella*. Since there are a large number of described species, especially in *Phloeospora* and genera with which it has been confused, it is quite likely that the generic boundaries as presently defined will change in the future.

The many pigmented genera in this section can be grouped within a recognised series from the simple aseptate conidium to the multiseptate one. Those with smooth conidia belong in the *Melanconium*, *Lamproconium*, *Melanconiopsis* complex while those with verruculose conidia are in the *Leptomelanconium*, *Gloeocoryneum*, *Lecanosticta* complex. Various subordinate characters like pigmentation and ornamentation of the conidiogenous cells certainly indicate natural groupings. *Coryneum* and *Stegonsporium*, although clearly belonging together, are unrelated to other genera in the section. The distoseptate conidia with very reduced cell lumina suggest an affinity with *Asterosporium* (Blastostromatineae, monoblastic).

DICHOTOMOUS KEY

(numbers in parentheses are species numbers)

1. Conidia with cellular, setulose, or mucilaginous appendages (not including persistent conidiogenous cells) 2
1. Conidia lacking appendages 12
 2. Appendages mucilaginous 3
 2. Appendages cellular 4
 2. Appendage single, subapical; conidia unequally septate, pale brown; conidiomata acervular .... *Obstipipilus* (1)
3. Conidiomata eustromatic, subcuticular; conidia hyaline, 1 euseptate, mucilaginous sheath becoming everted to form up to 9 tentacular filiform apical appendages .... *Comatospora* (2)
3. Conidiomata eustromatic, uredinicolous; conidia hyaline, 1 septate with a mucilaginous apical cap; conidiogenous cells with periclinal thickening .... *Sphaerellopsis* (see p. 470)
3. Conidiomata acervular to immersed cupulate; conidia fusiform, brown, aseptate, verruculose with a mucilaginous appendage at the apex and base .... *Hymenopsis* (31) (see also key 25)
3. Conidiomata pseudostromatic, immersed; conidia fusiform, muriform, distoseptate, with a basal mucilaginous sheath.... .... *Shearia* (3)
 4. Conidia concolourous 5
 4. Conidia versicolourous 6
5. Conidiomata acervular, setose; conidia falcate fusiform, pale brown, with a single apical and a single exogenous basal appendage .... *Annellolacinia* (4)
5. Conidiomata eustromatic, conic, unilocular; conidia fusiform, pale brown, medianly 1 septate, with a single apical appendage .... *Uniseta* (5)
5. Conidiomata acervular; conidia multiseptate, with a single apical and a single exogenous basal appendage.... .... *Seimatosporium* (13) (see also key 8, 32)
 6. Conidiomata eustromatic; conidia 1 septate with a small paler basal cell bearing 3 lateral appendages, and a much longer, thicker-walled, darker upper cell with a single apical appendage .... *Coma* (6)
 6. Conidiomata acervular to cupulate, 2-many septate 7
7. Conidiomata cupulate; conidia distoseptate, apical cell with many dichotomously branched appendages, basal cell with an endogenous much branched appendage .... *Pestalotia* (7)
7. Conidiomata acervular; conidia euseptate 8
 8. Basal appendage, if present, exogenous .... *Seimatosporium* (13) (see also key 5, 32)

8. Basal appendage, if present, endogenous 9
9. Conidia with a single apical appendage (but occasionally branched) 10
9. Conidia with several apical appendages 11
10. Conidia 3-4 septate .... *Monochaetia* (11)
10. Conidia 5 septate .... *Seiridium* (12)
11. Conidia 3 septate .... *Truncatella*(9)
11. Conidia 4 septate .... *Pestalotiopsis* (8)
11. Conidia muriform .... *Morinia* (10)
12. Conidia hyaline 13
12. Conidia pale to dark brown 17
13. Conidia aseptate 14
13. Conidia 1 septate 15
13. Conidia multiseptate 16
14. Conidiomata acervular to eustromatic, immersed; conidia falcate, fusiform, hyaline .... *Disculina* (15)
14. Conidiomata eustromatic, multilocular to convoluted; conidia fusiform, straight, hyaline ..... *Macrophomopsis* (14)
14. Conidiomata acervular; conidia obpyriform, occasionally pale brown; conidiogenous cells often with marked periclinal thickening .... *Discosporium* (see p. 573)
15. Conidiomata eustromatic, linear; conidia fusiform, slightly curved, medianly 1 septate .... *Scaphidium* (16)
15. Conidiomata acervular, subcuticular, sometimes associated with a dendroid subicle; conidia often unequally 1 septate, basal cell cylindrical, truncate, apical cell ellipsoid to globose or attenuated and laterally inclined .... *Marssonina* (17)
15. Conidiomata cupulate, dark or pale brown; conidia straight, cylindrical or filiform .... *Pseudocenangium* (18)
15. Conidiomata acervular, dark brown; conidiophores broadly cylindrical, brown, smooth, apices wide; conidia 1 septate, strongly curved .... *Anaphysmene* (19)
16. Conidiomata acervular, subcuticular .... *Ahmadia* (20)
16. Conidiomata acervular, subepidermal or subperidermal .... *Phloeospora* (21)
17. Conidia aseptate 18
17. Conidia distoseptate, with reduced cell lumina 26
17. Conidia euseptate 27
18. Conidia smooth 19
18. Conidia verruculose or verrucose 25
19. Conidiomata stromatic 20
19. Conidiomata acervular 21
20. Conidiomata pseudostromatic, convoluted, multilocular, immersed; conidia pale brown, ellipsoid, with a long persistent conidiogenous cell .... *Bellulicauda* (24)
20. Conidiomata eustromatic, strongly erumpent, multilocular; conidia large, pale brown, ellipsoid .... *Phaeodomus* (25)
20. Conidiomata eustromatic, immersed, hemispherical, multilocular; conidia circular in outline and with a central hyaline pore .... *Lasmeniella* (27)
20. Conidiomata eustromatic, convoluted, immersed; conidia large, dark brown; conidiogenous cells often with marked periclinal thickening .... *Neomelanconium* (see p. 626)
21. Conidiomata with paraphyses; conidia fusiform .... *Lamproconium* (28)
21. Conidiomata lacking paraphyses 22
22. Conidia thin-walled 23
22. Conidia thick-walled 24
23. Conidiomata discolouring the cuticle; conidiophores, straight, cylindrical; conidia short cylindrical to cuneiform .... *Piggotia* (34)
23. Conidiomata epidermal to subepidermal; conidiophores filiform, tapered to the apex; conidia ellipsoid, guttulate, with a thickened basal scar .... *Enerthidium* (33)
24. Conidiophores septate, branched at the base, annellations many, distinct; conidia comparatively small .... *Melanconium* (29)
24. Conidiophores absent, annellations few; conidia comparatively large .... *Melanconiopsis* (30)
25. Conidiomata acervular; conidiogenous cells brown, verrucose; conidia verrucose .... *Leptomelanconium* (36) (see also key 28)
25. Conidiomata acervular to cupulate, occasionally setose; conidia and conidiophores verruculose .... *Hymenopsis* (31) (see also key 3)

25. Conidiomata eustromatic, subcuticular; conidia pale brown, verruculose ..... *Gaubaea* (32)
   26. Conidiomata with paraphyses; conidia with longitudinal distosepta ..... *Stegonsporium* (43)
   26. Conidiomata lacking paraphyses; conidia lacking longitudinal distosepta ..... *Coryneum* (42)
27. Conidia 1 septate 28
27. Conidia 2-many septate 29
   28. Conidiomata eustromatic, multilocular, superficial; conidia 0-2 septate, mostly 1 septate, globose to pyriform, smooth ..... *Dothideodiplodia* (26)
   28. Conidiomata acervular, epidermal; conidia 1 septate, smooth, short cylindrical to cuneiform ..... *Didymosporina* (35)
   28. Conidiomata acervular; conidia 0-1 septate, verrucose ..... *Leptomelanconium* (36) (see also key 25)
29. Conidia pale brown 30
29. Conidia dark brown 31
   30. Conidiogenous cells abruptly tapered at the apices with 1-4 annellations; conidia verruculose ..... *Ciferriella* (23)
   30. Conidiogenous cells cylindrical, not abruptly tapered; conidia smooth to verruculose ..... *Colletogloeum* (22)
31. Conidiomata acervular 32
31. Conidiomata eustromatic 33
   32. Conidiomata with paraphyses; conidia smooth, often with an excentric basal scar ..... *Stilbospora* (39)
   32. Conidiomata lacking paraphyses; conidia smooth ..... *Seimatosporium* (13) (see also key 5, 8)
   32. Conidiomata lacking paraphyses; conidiogenous cells with 1 verrucose annellation; conidia 2-4 septate, broad at the base, straight ..... *Gloeocoryneum* (37)
   32. Conidiomata lacking paraphyses; conidiogenous cells verruculose with 1-2 annellations; conidia 1-4 septate, straight or curved ..... *Lecanosticta* (38)
33. Conidiomata subcuticular, multilocular and convoluted; conidia regularly 3 septate, pale brown ..... *Hendersonula* (40)
33. Conidiomata subperidermal; multilocular to convoluted but with a single ostiole; conidia versicoloured, 2-7 septate ..... *Neohendersonia* (41)
33. Conidiomata subperidermal, convoluted; conidia variably septate; conidiogenous cells often with marked periclinal thickening ..... *Hendersoniopsis* (see p. 635)

SYNOPTIC KEY

Conidiomatal characters

Position in relation to substrate

a. superficial: 7 18 25 26

b. immersed, epidermal to subepidermal or peridermal: 1 3 5 6 8 9 10 11 12 13 14 15 16 19 21 22 23 24 27 28 29 30 31 33 35 36 37 38 39 41 42 43

c. immersed, subcuticular: 2 4 6 17 20 32 34 40

d. semi-immersed: 25 31

Dehiscence

a. ostiolate, non-papillate: 5 14 16 26

b. ostiolate, papillate: 3 24 41

c. lifting a flap of host tissue: 38

d. furfuraceous: 5

e. irregular rupture: 1 2 4 6 7 8 9 10 11 12 13 15 17 18 19 20 21 22 23 25 27 28 29 30 31 32 33 34 35 36 37 38 39 40 42 43

Pigmentation

a. dark brown (non acervular) 2 3 5 7 13 14 15 16 18 19 24 25 26 27 31 32 40 41

Other characters

a. setose 4 31
b. subiculate 17

Type of stroma

a. acervular 1 4 6 8 9 10 11 12 13 17 19 20 21 22 23 28 29 30 31 33 34 35 36 37 38 39 42 43
b. cupulate 7 13 18 31
c. simple eustromatic 2 5 14 15 16 31 32 41
d. complex eustromatic, irregular 27 40
e. complex eustromatic, pulvinate 25
f. complex eustromatic, effuse 26
g. pseudostromatic 3 24
h. linear 16

Locular configuration

a. unilocular 1 2 3 4 5 6 7 8 9 10 11 12 13 15 16 17 18 19 20 21 22 23 24 28 29 30 31 33 34 35 36 37 38 39 42 43
b. multilocular 25 26 27 40 41
c. multilocular, convoluted 14 27

Conidiogenous cell characters

Insertion

a. discrete 1 2 3 4 5 6 13 14 15 16 20 21 23 24 25 26 27 30 31 32 33 34 35 36 38 39 40 41
b. integrated 7 8 9 10 11 12 13 14 15 17 18 19 22 23 28 29 31 33 34 36 37 38 39 42 43

Shape

a. ampulliform-doliiform 2 3 6 17 21 23 25 32 39 41
b. lageniform 1 2 6 13 14 16 20 21 23 24 25 32 41
c. cylindrical 3 4 5 7 8 9 10 11 12 13 14 15 16 17 18 19 20 21 22 24 25 26 27 28 29 30 31 33 34 35 36 37 38 39 40 41 42 43

Pigmentation

a. hyaline 1 2 3 4 5 6 7 8 9 10 11 12 13 14 15 16 17 18 20 21 22 24 25 26 27 28 29 30 33 39 40 41 42 43
b. pale brown 19 20 22 23 31 32 34 35 36 37 38

Ornamentation

a. smooth 1 2 3 4 5 6 7 8 9 10 11 12 13 14 15 16 17 18 19 21 23 24 25 26 27 28 29 30 31 33 34 35 39 40 41 42 43
b. verruculose 20 22 32 34 36 37 38

Annellide morphology

a. ragged margin — 34
b. prominently thickened — 31 32 33 36 37
c. with a distinct narrow 'phialidic' channel — 2

Conidiophore characters

Shape

a. irregularly branched and ramified — 17
b. long cylindrical to filiform, unbranched or only branched at the base — 14 15 19 22 23 28 29 31 34 37 38 39
c. filiform, branched — 7 8 9 10 11 12 13 14 18 22 33 36 42 43

Pigmentation

a. hyaline — 7 8 9 10 11 12 13 14 15 17 18 22 28 29 33 39 42 43
b. pale brown — 19 22 23 31 34 36 37 38 42 43

Ornamentation

a. smooth — 7 8 9 10 11 12 13 14 17 18 19 23 28 29 31 33 34 39 42 43
b. verruculose — 22 34 36 37 38

Conidial characters

Pigmentation

a. hyaline — 2 14 15 16 17 18 19 20 21 22
b. brown — 1 3 4 5 13 22 23 24 25 26 27 28 29 30 31 32 33 34 35 36 37 38 39 40 42 43
c. versicoloured — 6 7 8 9 10 11 12 13 41

Septation

a. distoseptate — 3 7 42 43
b. euseptate — 1 2 5 6 8 9 10 11 12 13 16 17 18 19 20 21 22 23 26 36 37 38 39 40 41
c. aseptate — 4 14 15 18 19 20 22 23 24 25 27 28 29 30 31 32 33 34 36
d. medianly 1 septate — 2 5 16 18 19 20 22 26 35 36 38 40
e. unequally 1 septate — 1 6 17 26 35
f. 2 septate — 13 20 21 22 23 26 37 38 41
g. 3 septate — 9 13 20 21 22 23 37 38 39 40 41
h. 4 septate — 8 11 13 21 22 39 41
i. 5 septate — 7 12 13 21 22 41
j. variably transversely septate — 21 41 42
k. muriform — 3 10 43

Shape	
a. ellipsoid-oval	1 2 13 24 25 29 31 33 36 42 43
b. cylindrical	6 18 20 21 22 37 39 40 42
c. fusiform	3 4 5 6 7 8 9 10 11 12 13 14 15 16 17 18 19 20 21 22 28 31 32 36 38 39 40 41 42
d. falcate	4 15 16 17 19 21 22
e. pyriform or obpyriform	17 25 26 27 29 30 31 36 42 43
f. clavate	1 8 13 17 24 26 32 37 39 41 42 43
g. globose	27 29
h. filiform	21 23 38
i. cuneiform	34 35
Ornamentation	
a. smooth	1 2 3 4 5 6 7 8 9 10 11 12 13 14 15 16 17 18 19 20 21 22 24 25 26 27 28 29 30 33 34 35 39 40 41 42 43
b. verrucose	8 11 13 22 23 31 32 36 37 38
c. longitudinally striate	12
Guttulation	
a. guttulate	2 4 8 10 17 18 22 30 31 33 38
b. eguttulate	1 3 5 6 7 8 9 11 12 13 14 15 16 18 19 20 21 22 23 24 25 26 27 28 29 31 32 34 35 36 37 39 40 41 42 43
Appendages	
a. apical	1 2 4 5 6 7 8 9 10 11 12 13 31
b. basal, exogenous	3 4 13 31
c. basal, endogenous or as a persistent conidiogenous cell	7 8 9 11 12 24 27
d. lateral, ± median	6
e. single	1 3 4 5 6 9 11 12 13 24 27 31
f. several at one point	2 7 8 9 10 13
g. cellular	4 5 6 7 8 9 10 11 12 13 24 27
h. setulose	1
i. mucilaginous	2 3 31
j. simple	1 2 3 4 5 6 8 9 10 11 12 13 24 27 31
k. irregularly branched	8 9 11 12 13
l. dichotomously branched	7
m. tentacular	2
n. infundibular	31
o. cap-like	3 31
p. spathulate	8
q. espathulate	1 2 3 4 5 6 7 8 9 10 11 12 13
Wall thickness	
a. thin-walled	1 2 4 5 14 15 16 17 19 20 21 22 23 24 25 26 31 32 33 34 35 40
b. thick-walled	3 6 7 8 9 10 11 12 13 22 27 28 29 30 36 37 38 39 41 42 43
c. basal scar comparatively prominently thickened	33 41

Paraphyses characters

a. present	28	39	43
b. enclosed in a mucilaginous sheath	28		
c. septate	28	39	43
d. branched	28	43	

1. OBSTIPIPILUS Sutton, *Can. J. Bot.* **46**: 187 (1968).
Sp. typ.: *O. malabaricus* (T.S. & K. Ramakr.) Sutton (syn. *Kellermania malabarica* T.S. & K. Ramakr.).

Mycelium immersed, branched, septate, pale brown to hyaline. *Conidiomata* acervular, subepidermal, separate or confluent, formed of pale brown, thin-walled textura angularis. Dehiscence irregular. *Conidiophores* absent. *Conidiogenous cells* holoblastic, annellidic, discrete, indeterminate, lageniform, hyaline or pale brown, smooth, with 1-4 percurrent proliferations, formed from the upper pseudoparenchyma. *Conidia* obovate, 1 euseptate, upper cell larger, slightly constricted, olivaceous to pale brown, thin-walled, smooth, base truncate; apical cell with an unbranched hyaline setula inclined to one side.

The genus is monotypic.

O. MALABARICUS (T.S. & K. Ramakr.) Sutton, *ibid.*: 188 (1968). (Fig. 144)
Kellermania malabarica T.S. & K. Ramakr., *Proc. Ind. Acad. Sci.* Sect. B, **32**: 211 (1950).
Marssonina poonensis Patwardhan & Pande, *Sydowia* **23**: 100 (1970).

Conidiomata up to 100 µ diam. Conidiogenous cells 5.5-12 × 4-8 µ. Conidia 11.5-14 × 4-5.5 µ, appendage 4-5.5 µ long.

Causing lesions on leaves of *Anogeissus latifolia*, *A. pendula* (IMI 110502, holotype *K. malabarica*, 122278).

2. COMATOSPORA Pirozynski & Shoemaker, *Can. J. Bot.* **49**: 539 (1971).
Sp. typ.: *C. suttonii* Pirozynski & Shoemaker

Mycelium immersed, septate, pale brown, branched. *Conidiomata* eustromatic, separate, flattened-hemispherical, dark brown, subcuticular, unilocular, thick-walled, composed of an outer layer of irregular dark brown textura angularis restricted to the upper and lateral regions and an inner layer of hyaline, thin-walled tissue. Dehiscence irregular. *Conidiophores* absent. *Conidiogenous cells* holoblastic, annellidic, indeterminate, discrete, ampulliform to lageniform, hyaline, smooth, with 1-4 percurrent proliferations, formed from the inner cells of the stromatal wall. *Conidia* medianly 1 euseptate, hyaline, continuous, tapered abruptly to the truncate base, apex obtuse, thin-walled, guttulate, ellipsoid to ovoid, with up to 9 apical, extracellular, filiform, gelatinous appendages formed by the splitting and eversion from the base of a gelatinous envelope covering immature conidia.

The genus is monotypic.

C. SUTTONII Pirozynski & Shoemaker, *ibid.*: 539 (1971). (Fig. 145)

Conidiomata 150-300 µ diam. Conidiogenous cells 7-11 × 2-4.5 µ. Conidia 11.5-16 × 5-7 µ; appendages 15-20 µ long.

On dead branches of *Picea mariana*, Canada (IMI 144358, holotype).

3. SHEARIA Petrak, *Annls mycol.* **22**: 180 (1924).
Sp. typ.: *S. formosa* (Ell. & Ev.) Petrak (syn. *Stegonsporium formosum* Ell. & Ev., *S. magnoliae* (Shear) Petrak). Petrak).

Mycelium immersed, branched, septate, brown. *Conidiomata* pseudostromatic, separate or aggregated, immersed, peridermal to subperidermal, globose to conical, papillate, dark brown, unilocular, thick-walled; lateral and basal walls composed of peridermal cells and thick-walled, brown textura angularis, outer layers consisting of very dark brown, thick-

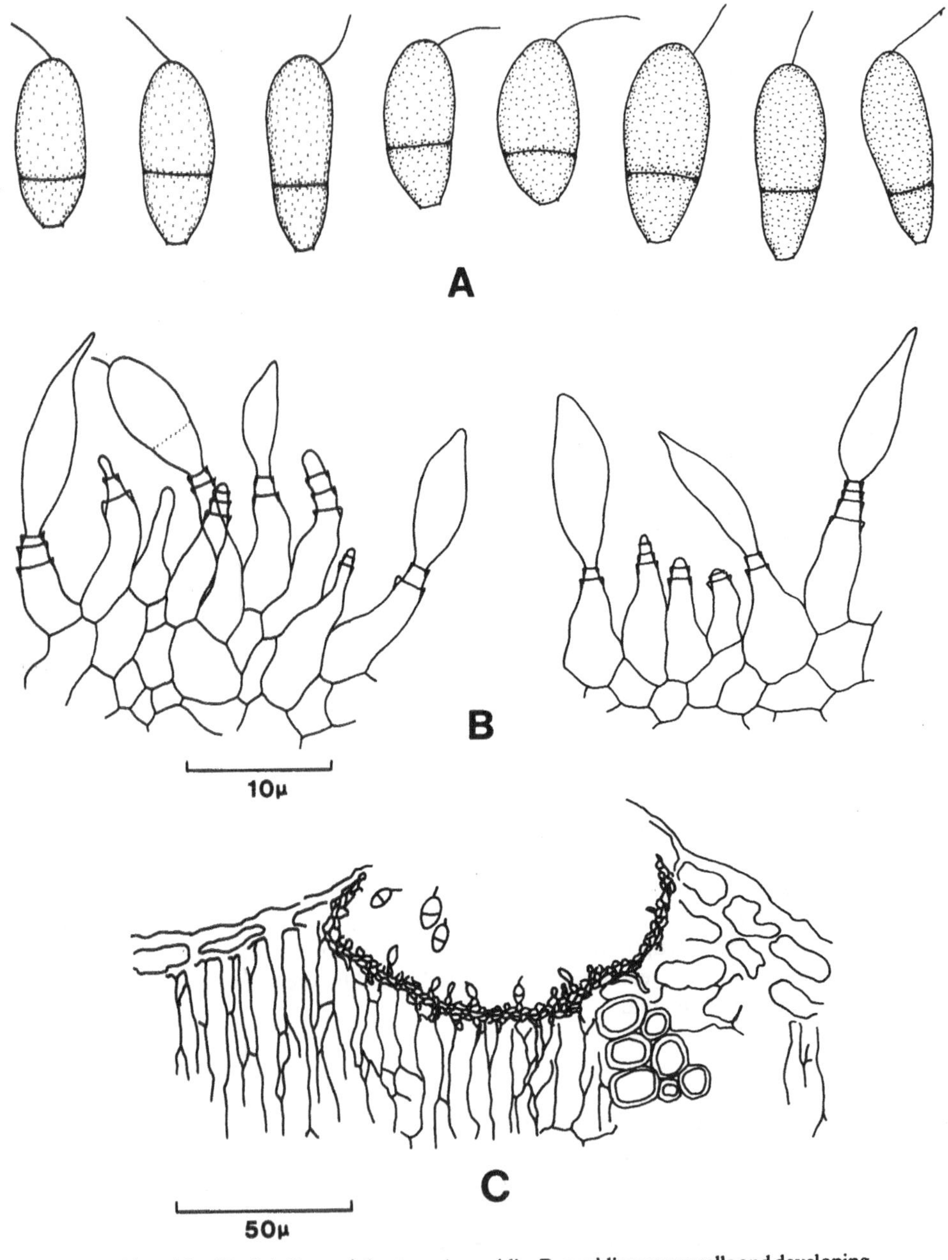

Fig. 144. *Obstipipilus malabaricus*, A, conidia; B, conidiogenous cells and developing conidia; C, vertical section of a conidioma.

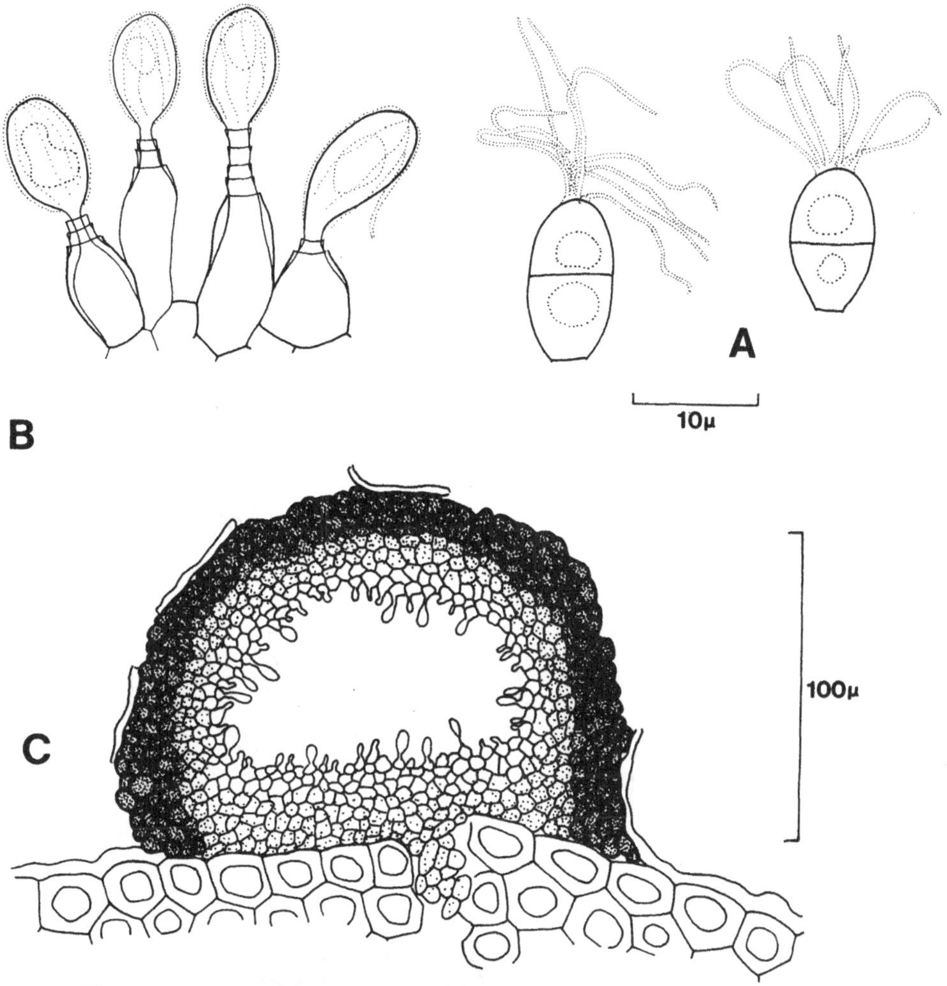

Fig. 145. *Comatospora suttonii*, A, conidia; B, conidiogenous cells and developing conidia; C, vertical section of a closed conidioma.

walled, occluded cells; layers lining the locule of thin-walled, hyaline tissue. *Ostiole* central, circular, papillate, with a long ostiolar channel lined with textura porrecta. *Conidiophores* absent. *Conidiogenous cells* holoblastic, annellidic, discrete, indeterminate, ampulliform, doliiform or cylindrical, hyaline, smooth, often thick-walled, with 1-2 percurrent proliferations, formed from the inner cells of the locular wall. *Conidia* fusiform, with several transverse and lateral distosepta, lumina reduced, continuous, thick-walled, smooth, base truncate, apex obtuse, initially enveloped in a gelatinous sheath which is depressed at the apex, at maturity persisting as a basal lateral sheath.

This most distinctive genus is so far monotypic. Petrak (1962) revised the synonymy for the species.

S. FORMOSA (Ell. & Ev.) Petrak, *Sydowia* **15**: 217 (1962). (Fig. 146).
Stegonsporium formosum Ell. & Ev., *Bull. Torrey bot. Cl.* **10**: 76 (1843).
Shearia magnoliae (Shear) Petrak, *Annls mycol.* **22**: 180 (1924).
Camarosporium magnoliae Shear, *Bull. Torrey bot. Cl.* 455 (1902).

Teleomorphosis: *Pleomassaria magnoliae* Shear, *ibid.*: 455 (1902).

Conidiomata up to 750 μ high × 600 μ wide at the base. Ostiolar channel up to 300 μ long. Conidiogenous cells 7-15 × 5-10 μ. Conidia 90 × 30 μ.

On branches of *Magnolia grandiflora, M.* sp., Wash., Miss., U.S.A. (IMI 200273, 200150 ex K).

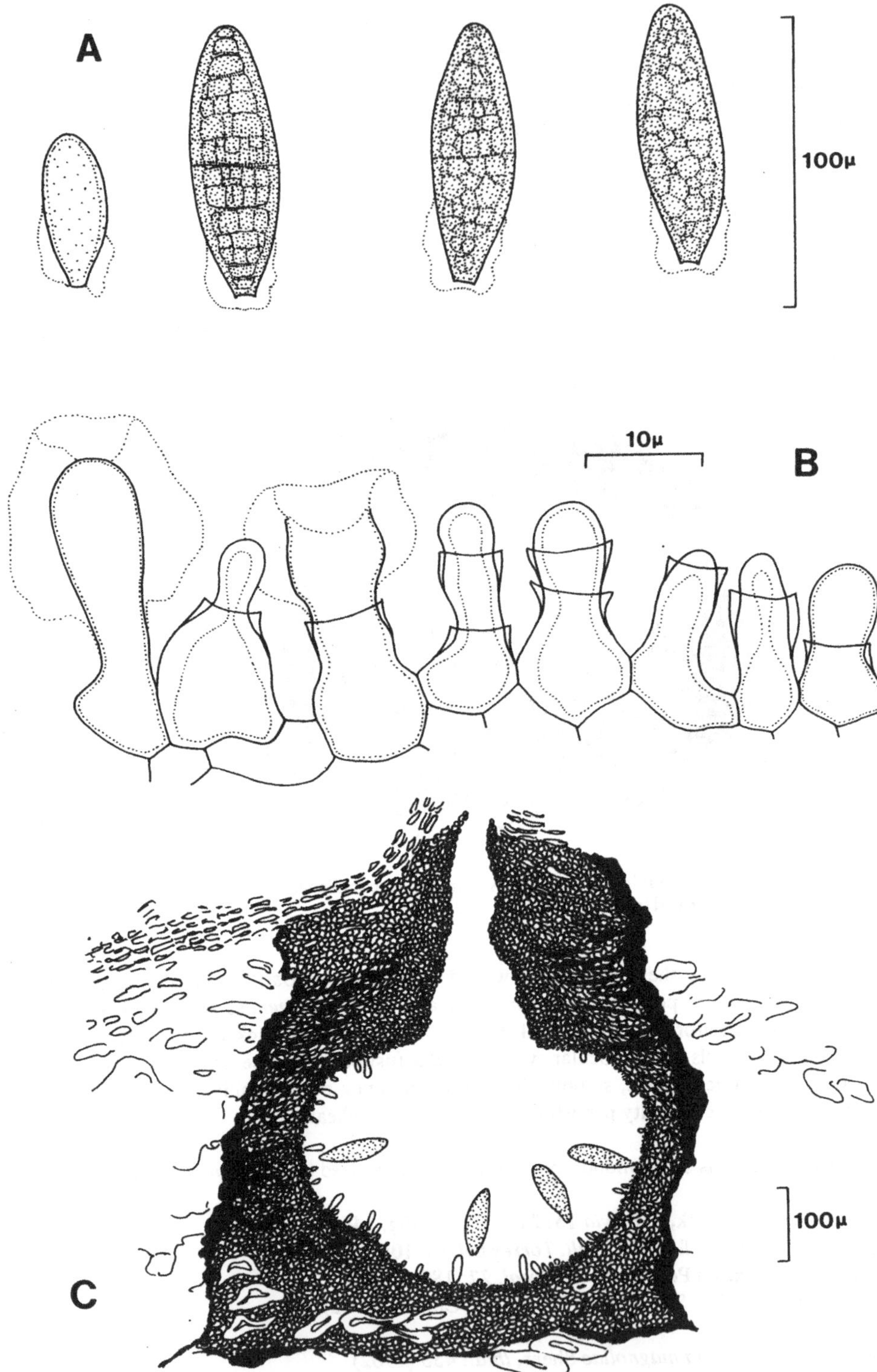

Fig. 146. *Shearia formosa*, A, conidia; B, conidiogenous cells and developing conidia; C, vertical section of a conidioma.

4. ANNELLOLACINIA Sutton, *Mycol. Pap.* 97: 31 (1964).
Sp. typ.: *A. dinemasporioides* Sutton
Illustration: Sutton (1964a).

Immersed mycelium pale brown to hyaline, branched, septate. *Conidiomata* acervular, subcuticular, separate, amphigenous, formed of pale brown, thin-walled textura angularis. Dehiscence irregular. *Setae* mostly marginal, unbranched, straight, tapered to acute apices, thick-walled, septate at the base, smooth, dark brown, at first recurved over the acervulus. *Conidiophores* absent. *Conidiogenous cells* holoblastic, annellidic, discrete, indeterminate, cylindrical, hyaline to pale brown, smooth, with 1-8 percurrent proliferations, formed from the upper pseudoparenchyma. *Conidia* falcate, fusiform, aseptate, pale brown, thin-walled, smooth, guttulate, base truncate, apex tapered into a single, cellular unbranched appendage; at the base with a single, lateral, cellular, unbranched appendage.

The genus is monotypic.

A. DINEMASPORIOIDES Sutton, *ibid.*: 33 (1964). (Fig. 147)

Conidiomata 100 µ diam. Setae 45-120 × 3-5 µ. Conidiogenous cells 7-20 × 2-3.5 µ. Conidia 15-19 × 4-5 µ; appendages 12-23 µ long.

Causing lesions on leaves of *Ananas comosus*, *A. sativus*, Ghana, Sierra Leone, Trinidad, India (IMI 37897c (holotype), 37872p, 21619, 58508, 58509a, 87938a, 101325, 112859).

5. UNISETA Ciccarone, *Nuovo G. bot. ital., N.S.* **54**: 15 (1947).
Sp. typ.: *U. flagellifera* (Ell. & Ev.) Ciccarone (syn. *Pestalotia flagellifera* Ell. & Ev.).
Illustration: Nag Raj (1974).

Mycelium immersed, branched, septate, brown. *Conidiomata* eustromatic, gregarious, conic, brown, immersed, unilocular, thick-walled; wall composed of an outer layer of thicker-walled, dark brown textura angularis, and an inner layer of thin-walled, pale brown tissue. *Ostiole* single, circular, furfuraceous. *Conidiophores* absent. *Conidiogenous cells* holoblastic, annellidic, indeterminate, discrete, cylindrical to flexuous, hyaline, smooth, with 1-2 percurrent proliferations, formed from the inner cells of the stromatal wall. *Conidia* pale brown, medianly 1 euseptate, continuous, base truncate, ± thick-walled, smooth, eguttulate, fusiform, upper cells slightly wider, straight or slightly curved, with a cellular, unbranched, filiform appendage at the apex.

Ciccarone (1947) gave a full synonymy for *U. flagellifera*. Guba (1961) asserted that *Pestalotia flagellifera*, the basionym, was identical with *Mastigosporella hyalina* (Ell. & Ev.) Höhn., but it is clear from the illustrations by Nag Raj (1974) that *Mastigosporella* represents quite a different genus from *Uniseta*.

U. FLAGELLIFERA (Ell. & Ev.) Ciccarone, *ibid.*: 15 (1947).
Pestalotia flagellifera Ell. & Ev., *J. Mycol.* **5**: 156 (1889).
Barklayella flagellifera (Ell. & Ev.) Sacc., *Syll. fung.* **10**: 476 (1892).
Neobarclaya flagellifera (Ell. & Ev.) Sacc., *Syll. fung.* **14**: 46 (1895).
Neobarclaya flagellifera (Ell. & Ev.) O. Kze, *Rev. gen. Pl.* III(1): 177 (1893).

Conidiomata up to 550 µ diam. × 200 µ deep. Conidiogenous cells 11-15.5 µ long. Conidia 11-15 × 4 µ; appendage 32-38 µ long.

On branches of *Comptonia aspleniifolia*, N.J., U.S.A.

6. COMA Nag Raj & Kendrick, *Can. J. Bot.* **50**: 614 (1972).
Sp. typ.: *C. circularis* (Cke & Massee) Nag Raj & Kendrick (syn. *Pestalozziella circularis* Cke & Massee).
Illustrations: Nag Raj & Kendrick (1972a); Sutton (1974).

Mycelium immersed, dark brown, branched, septate. *Conidiomata* eustromatic, hypophyllous, aggregated, often confluent, subcuticular to epidermal, flattened, lenticular, oval, circular or irregular, dark brown to black, unilocular; tissue of dark brown, thick-walled outer and thinner paler inner textura angularis. Dehiscence by irregular rupture. *Conidiophores* absent. *Conidiogenous cells* holoblastic, annellidic, indeterminate, discrete, doliiform, hyaline, smooth, with 1 percurrent proliferation, formed from the inner cells of the stromatal wall. *Conidia* 1 euseptate, the upper cell 3-4 times

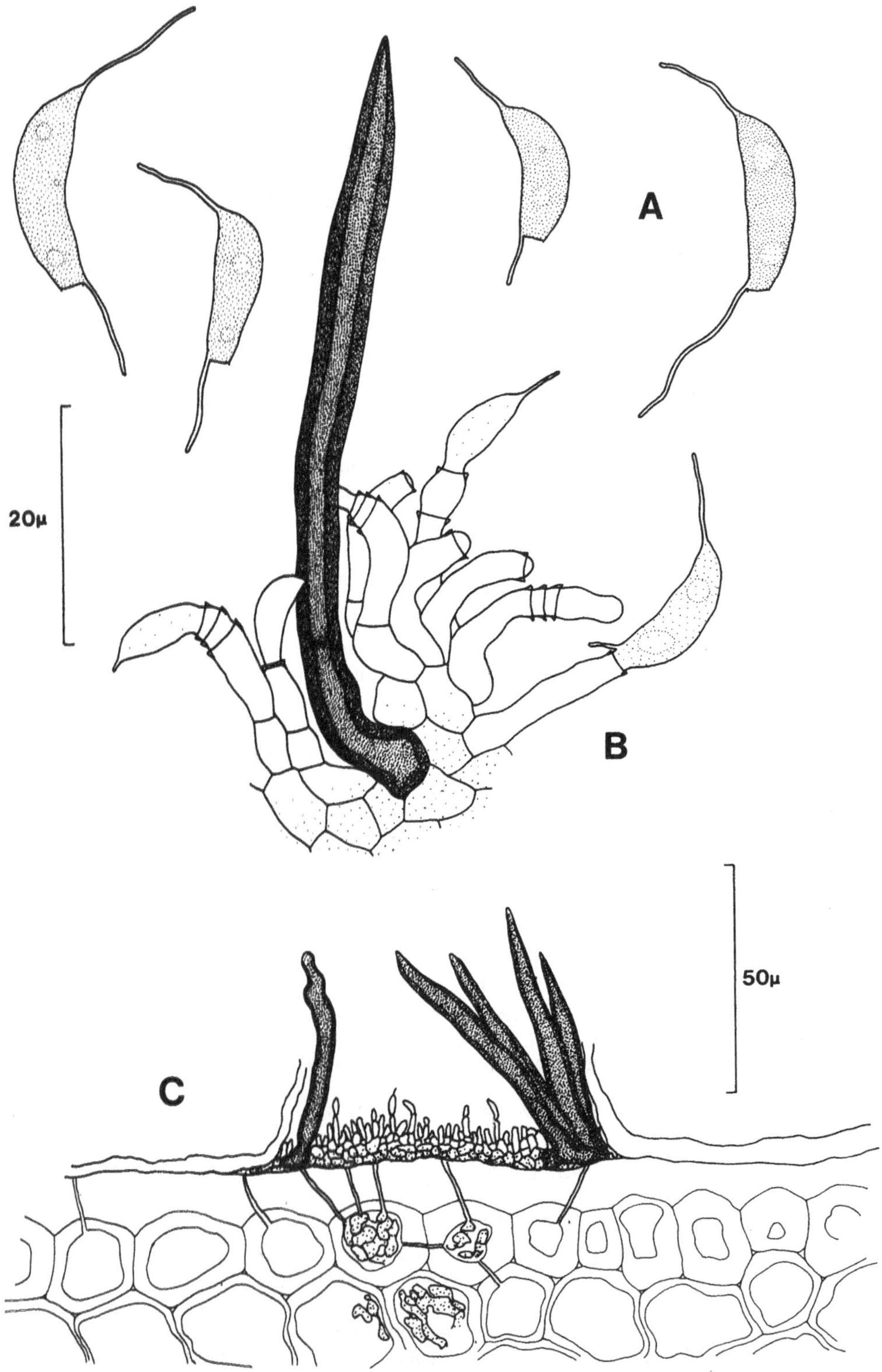

Fig. 147. ***Annellolacinia dinemasporioides***, A conidia; B, conidiophores with developing conidia and a seta; C, vertical section of a conidioma.

as long as the lower cell, constricted, upper cell pale brown, thick-walled, base truncate, ± guttulate, cylindrical to fusiform, with a single cellular unbranched apical appendage delimited by a septum and 1-3 lateral, unbranched, cellular appendages inserted just below the septum.

Although described by Nag Raj & Kendrick (1972a) the account was modified by Sutton (1974) who enlarged the synonymy.

C. CIRCULARIS (Cke & Massee) Nag Raj & Kendrick, *ibid.*: 614 (1972).
synonymy in Sutton (1974).

Teleomorphosis: *Pseudopeziza eucalypti* Hansf., *Proc. linn. Soc. N.S.W.* **81**: 10 (1956).

Conidiomata up to 380 µ diam. Conidia 24-34 × 5.5-9.5 µ; apical appendage 16.5-22.5 × 1.5-3 µ; lateral appendages 14.5-28.5 µ long.

On leaves of *Eucalyptus parviflora*, *E.* sp., Australia (IMI 57992 (holotype of *Gloeosporiella eucalypti*), holotype of *Pestalozziella circularis*, K).

7. PESTALOTIA de Not., *Mem. R. Accad. Sci., Torino* 2, **3**: 80 (1839).
Sp. typ.: *P. pezizoides* de Not.

Mycelium immersed, branched, septate, brown. *Conidiomata* eustromatic, cupulate, separate or confluent, black or dark brown, at first immersed, then erumpent, composed of thick-walled, dark brown textura angularis, giving rise to thinner-walled, paler tissue in the conidiogenous region. Dehiscence irregular. *Conidiophores* hyaline, branched irregularly, septate, smooth, formed from the upper cells of the pseudoparenchyma. *Conidiogenous cells* holoblastic, annellidic, indeterminate, integrated, cylindrical, hyaline, smooth, with 1-3 percurrent proliferations. *Conidia* fusiform, straight or slightly curved, 6-celled; basal cell hyaline, thin-walled, truncate, with an endogenous, cellular, simple or dichotomously branched appendage; apical cell conic, thin-walled, hyaline, with 3-9 apical, simple or dichotomously branched appendages; median cells separated from end cells by eusepta but from one another by distosepta; lumina reduced, medium brown.

Sutton (1969a) has fully reviewed the history of this genus and its relationships with *Pestalotiopsis* Stey. and *Truncatella* Stey., including pertinent references to the arguments for and against restricting the name to a single species. Guba (1961) accepted 220 species in *Pestalotia* sensu lato but Sutton (1969a) preferred to regard the genus as monotypic. Until an appraisal of the variability of 5-celled *Pestalotia* species in culture is made, no further redispositions of names in *Pestalotiopsis* seems advisable.

P. PEZIZOIDES de Not., *ibid.*: 80 (1839). (Fig. 148)

Conidiomata 200-600 µ diam. Conidiophores 20-50 × 2-3 µ. Conidia 29-37 × 6-9 µ; apical appendages 20 µ long; basal appendage 10 µ long; median cells 24-29 µ long.

On twigs of *Vitis vinifera*, *V. aestivalis*, *V.* sp., Italy, U.S.A. (IMI 83642, holotype, 83643, 83533, 83534, all ex K).

8. PESTALOTIOPSIS Stey., *Bull. Jard. bot. Brux.* **19**: 300 (1949).
Robillarda Cast., *Catalogue des plantes qui croissant naturellement aux environs de Marseilles*: 205 (1845), nom. rej. prop.
Discosiopsis Edward et al., *Sydowia* **26**: 268 (1974).
Sp. typ.: *P. guepinii* (Desm.) Stey. (syn. *Pestalotia guepinii* Desm.).

Mycelium immersed, branched, septate, hyaline to pale brown. *Conidiomata* acervular, epidermal to subepidermal, separate or confluent, formed of brown, thin-walled textura angularis. Dehiscence irregular. *Conidiophores* hyaline, branched and septate at the base and above, cylindrical or lageniform, formed from the upper cells of the pseudoparenchyma. *Conidiogenous cells* holoblastic, annellidic, indeterminate, integrated, cylindrical, hyaline, smooth, with several percurrent proliferations. *Conidia* fusiform, straight or slightly curved, 4 euseptate; basal cell hyaline, truncate, with an endogenous, cellular, simple or rarely branched appendage; apical cell conic, hyaline, with 2 or more apical, simple or branched, spathulate or espathulate appendages; median cells brown, sometimes versicoloured, thicker-walled, smooth or verruculose.

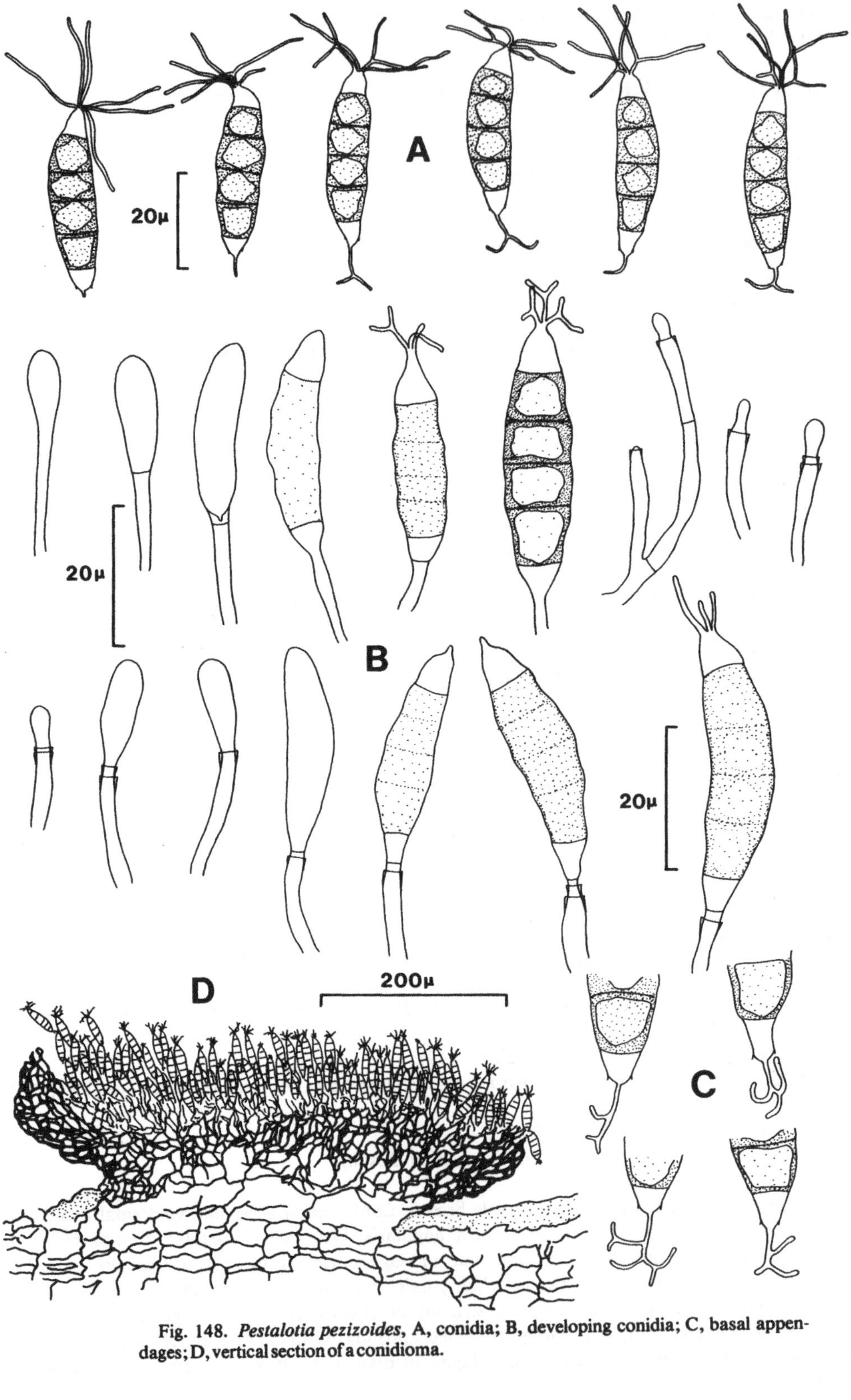

Fig. 148. *Pestalotia pezizoides*, A, conidia; B, developing conidia; C, basal appendages; D, vertical section of a conidioma.

By restricting *Pestalotia* to a single species, Steyaert (1949) found it necessary to reassign the many species described in *Pestalotia* to other genera. Several of these were placed in two new genera, *Pestalotiopsis* and *Truncatella*, but the majority of the more than 600 species described in the genus remained unexamined and therefore stayed in *Pestalotia*. Guba (1961) preferred to maintain *Pestalotia* with a wider generic concept than that advanced by Steyaert (1949), consequently in his monograph, 220 species were accepted in the genus, with *Truncatella*, *Pestalotiopsis* and *Labridella* reduced to synonymy.

There are three major problems concerning the taxonomy of *Pestalotia* auct. or *Pestalotiopsis*.

The first is a basic decision in favour of either the approach by Guba (1961) or that of Steyaert (1949). Sutton (1969a) has fully discussed the arguments for and against each approach and has presented evidence favouring the generic rearrangement by Steyaert (1949). The value of differences in conidial wall structure between *Pestalotia* and *Pestalotiopsis* has been substantiated by electron microscopy studies (Griffiths & Swart, 1974a, b).

The second problem in the group involves nomenclature. By opting for Steyaert's system, the names of a large number of taxa which should be referred to *Pestalotiopsis* have not yet been combined into that genus. Consequently erroneous names in *Pestalotia* are still in use.

The last problem concerns species differentiation. The studies by Guba (1961), Steyaert (1949, 1953a, b, 1954, 1955, 1961) and most other workers in the group have been based on dried herbarium material. It is clear, however, that when taxa are grown in artificial culture, they are shown to be far more variable. In consequence the species limits often overlap. So far identification in culture and the application of names based on herbarium taxonomy have not been reconciled into any suitable modern system.

P. GUEPINII (Desm.) Stey., *ibid.*: 312 (1949). (Fig. 149)
Pestalotia guepinii Desm., *Ann. Sci. nat.*, ser. 2, **13**: 181 (1840).

Conidiomata up to 200 µ diam. Conidiophores 10-15 × 1-3 µ, with 1-2 proliferations. Conidia 21-27 × 6.5-8.5 µ, smooth; median cells concolourous, 15-19 µ long; apical appendages 2-5, mostly 3, unbranched, apices obtuse, 16-33 µ long; basal appendage 4-12 µ long.

On *Camellia sinensis, C.* sp., *Laurus sassafras, Azalea* sp., *Erica* sp., *Hippocratea* sp., *Jatropha curcas, Pinus patula, Rhododendron* sp., *Terminalia bellerica*, France, Netherlands, U.K., New Zealand, Kenya, Germany, Ghana, India, U.S.A. (IMI 89441b, 22483, 51698, 121040, 45130b, 44045b, 83708a, b, 120337a, 123227, 149787, 215627, 83913 ex PC, 83728b ex BR, K, 84979 ex BM, 84982 ex BM, 85127, 85168b, 178800 ex K).

9. TRUNCATELLA Stey., *Bull. Jard. bot. Brux.* **19**: 293 (1949).
Pseudopestalotia Elenkin & Chi, *Zh. Bol. Rast.* **6**: 94 (1912), nom. tent.
Ahmadinula Petrak, *Sydowia* **7**: 375 (1953).
Sp. typ.: *T. angustata* (Pers. ex Lk) Hughes (syn. *Stilbospora angustata* Pers. ex Lk, *T. truncata* (Lév.) Stey.).

Mycelium immersed, branched, septate, hyaline to pale brown. *Conidiomata* acervular, epidermal to subepidermal, separate or confluent, formed of brown, thin-walled textura angularis. Dehiscence irregular. *Conidiophores* hyaline, branched and septate at the base and above, cylindrical, formed from the upper cells of the pseudoparenchyma. *Conidiogenous cells* holoblastic, annellidic, indeterminate, integrated, cylindrical, hyaline, smooth, with several percurrent proliferations. *Conidia* fusiform, straight, 3 euseptate; basal cell hyaline, truncate, sometimes with an endogenous cellular, simple or rarely branched appendage; apical cell conic, hyaline, with 1 or more apical, simple or branched cellular appendages; two median cells brown, smooth, thick-walled, often constricted at the septa.

In his reassessment of *Pestalotia* species, Steyaert (1949) proposed the generic name *Truncatella* for the group of species known as *Pestalotia* sect. Quadriloculatae by Guba (1961). Sections within *Truncatella* were separated according to the number of apical appendages — Monosetulatae, Bisetulatae, Trisetulatae and Multisetulatae. Five species were assigned to the genus, all based on existing species of *Pestalotia*, and subsequently other authors added four more.

There has been no revision of the 33 species placed in *Pestalotia* sect. Quadriloculatae by Guba (1961), nor of the eight species of *Monochaetia* sect. Quadriloculatae which should be considered in any reorganisation of potential *Truncatella* species. *Monochaetia paeoniae* (Maubl.) Sacc. & D. Sacc., *M. excipuliformis* Bub. and *M. syringae* (Oud.) Allesch. in the latter group have been examined and should more correctly be referred to *Seimatosporium* Cda.

Of the species accepted by Steyaert (1949), *T. angustata* (Pers. ex Lk) Hughes (syn. *T. truncata* (Lév.) Stey.) seems to have a wide host range, *T. laurocerasi* (Westd.) Stey. seems to be common on the Rosaceae, and *T. hartigii* (Tubeuf) Stey. confined to the Coniferae. However Guba (1961), although listing separate synonymies for the species, was really

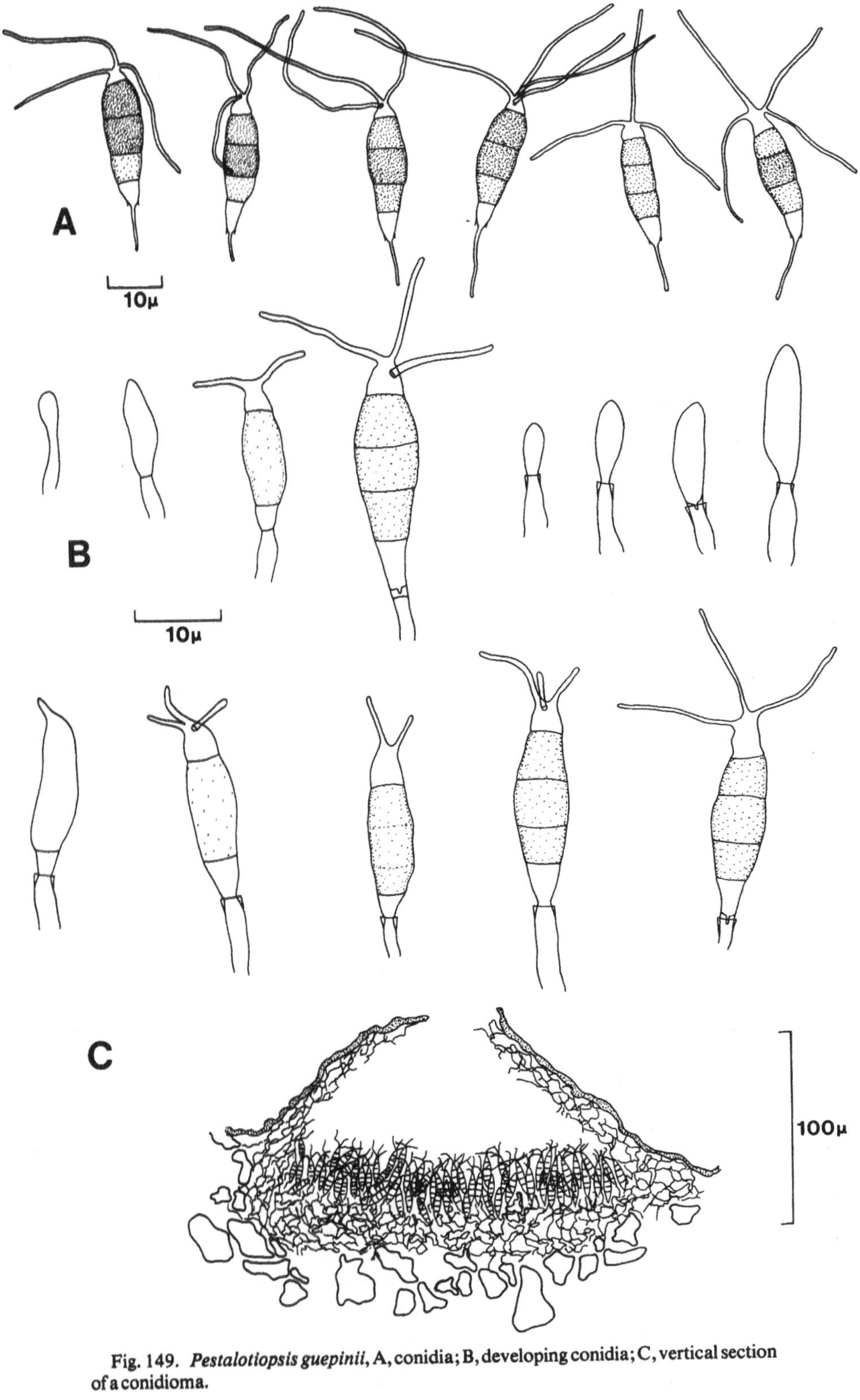

Fig. 149. *Pestalotiopsis guepinii*, A, conidia; B, developing conidia; C, vertical section of a conidioma.

unable to list sufficient grounds for keeping them distinct. Until species separation is clarified by type studies in the genus it would be pointless to attempt an account of the species based on published descriptions.

T. ANGUSTATA (Pers. ex Lk) Hughes, *Can. J. Bot.* **36**: 822 (1958). (Fig. 150)
Stilbospora angustata Pers. ex Lk, *Linn. Spec. Pl.* IV, **6**(2): 95 (1824).
Stilbospora angustata Pers., *Syn. Meth. fung.* 96 (1801).
Uredo angustata (Pers.) Strauss, *Ann. Wetterausich, Ges.* **2**: 112 (1810).
Sporidesmium angustatum (Pers. ex Lk) Cda in Sturm's *Deutschl. Fl.* III *(Pilze)*, **2**(7): 49 (1829).
Dicoccum truncatum Cda, *Icon. fung.* **1**: 6 (1837).
Phragmotrichum lignicola Cda, *Icon. fung.* **2**: 9 (1838).
Seiridium lignicola (Cda) Sacc., *Syll. fung.* **3**: 783 (1884).
Pestalotia truncata Lév., *Ann. Sci. nat.*, ser. 3, **5**: 285 (1846).
Truncatella truncata (Lév.) Stey., *Bull. Jard. bot. Brux.* **19**: 295 (1949).
Coryneum affine Sacc., *Michelia* **2**: 634 (1882).

Conidiomata up to 350 µ diam. Conidiophores up to 24 µ long × 2 µ wide, with 1-3 annellations. Conidia 15-20 × 6-8 µ; median cells 11-14 µ long; basal appendage absent; apical appendages variable, single or 2-4 µ, sometimes branched irregularly or dichotomously, 11-29 µ long.

On and from *Rumex* sp., *Bertholletia excelsa, Fagus sylvatica, Cotoneaster bacillaris, Malus* sp., *Dactylis glomerata, Phleum pratense, Prunus persica, P.* sp., *Rhododendron* sp., *Rubus ursinus, Triticum* sp., *Vicia faba, Ribes* sp., *Rosa* sp., soil, Brazil, Australia, New Zealand, India, Eire, UK., Sweden, Norway, France, represented by 30 collections including IMI 179075 holotype *C. affine* ex PAD.

10. MORINIA Berl. & Bres., *XIV Ann. Soc. Alp. Trident.* 82 (1889).
Rinomia Nieuwland, *Amer. Midl. Nat.* **4**: 381 (1916).
Pestalotia de Not. subgen. *Pestalozziana* Pass., *Rend. R. Accad. Lincei Roma*, 4 ser., **7**: 51 (1891).
Sp. typ.: *M. pestalozzioides* Berl. & Bres.
Illustration: Allescher (1903).

Mycelium immersed, branched, septate, pale brown to hyaline. *Conidiomata* acervular, subepidermal, separate, applanate, ± circular, dark brown to black, composed of pale brown, thin-walled textura angularis. Dehiscence irregular. *Conidiophores* cylindrical, flexuous, septate, branched at the base, hyaline, smooth, formed from the upper pseudoparenchyma. *Conidiogenous cells* holoblastic, integrated, ± indeterminate, cylindrical, hyaline, smooth, probably with percurrent proliferations. *Conidia* muriform with 5 transverse eusepta and 1-3 vertical eusepta, end cells hyaline, median cells medium brown, fusiform, smooth, basal cell truncate with the remains of the conidiogenous cell attached, apical cell conic with 2-4 cellular, unbranched, filiform appendages.

The genus was briefly discussed by Guba (1961) and well-illustrated by Allescher (1903).

M. PESTALOZZIOIDES Berl. & Bres., *ibid.*: 82 (1889).
Rinomia pestalozzioides (Berl. & Bres.) Nieuwland, *Amer. Midl. Nat.* **4**: 381 (1916).
Pestalotia artemisiae Pass., *Rend. R. Accad. Lincei Roma*, 4 ser., **7**: 51 (1891).

Conidia 16-22 × 5-7 µ (8-10 µ fide Allescher, 1903); intermediate cells 12-15 µ long; apical appendages 12-19 µ long.

On dead stems of *Artemisia camphorata*, Italy.

11. MONOCHAETIA (Sacc.) Allesch. in Rabenh., *Krypt. Flora* **7**: 665 (1902).
Pestalotia de Not. subgen. *Monochaetia* Sacc., *Syll. fung.* **3**: 797 (1884).
Lennisia Nieuwland nom. nov., *Amer. Midl. Nat.* **4**: 380 (1916).
Original spp.: 23 species listed, all of which were combinations from existing genera.
Lectotype sp.: *M. monochaeta* (Desm.) Allesch. (syn. *Pestalotia monochaeta* Desm.), fide Clements & Shear (1931).

Mycelium immersed, branched, septate, pale brown to hyaline. *Conidiomata* acervular, separate or confluent, ± circular or linear, composed of brown, thin-walled textura angularis. Dehiscence irregular. *Conidiophores* hyaline, cylin-

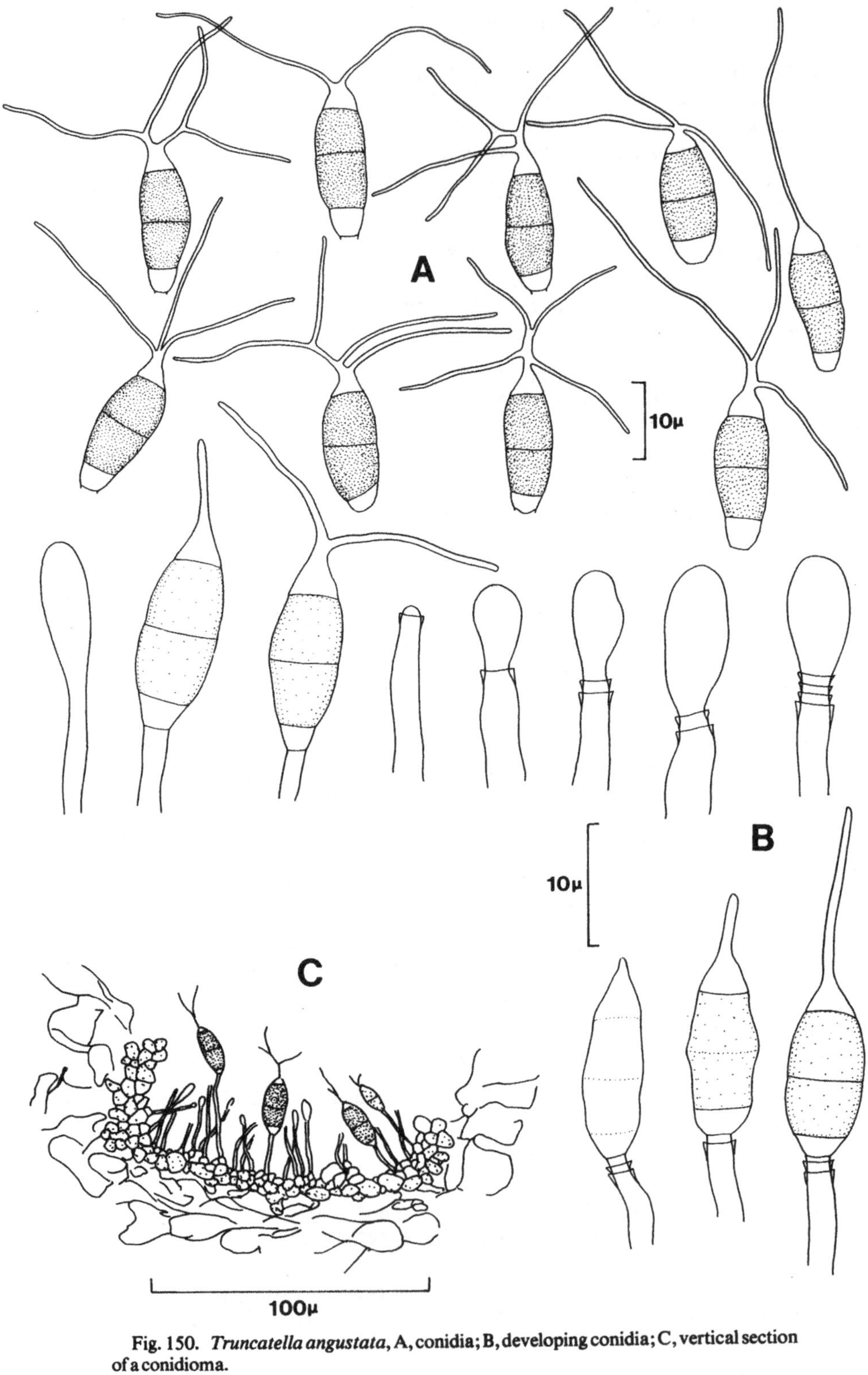

Fig. 150. *Truncatella angustata*, A, conidia; B, developing conidia; C, vertical section of a conidioma.

drical, straight or curved, sparsely branched and septate only at the base, formed from the upper cells of the conidiomata. *Conidiogenous cells* holoblastic, annellidic, with several apical annellations, indeterminate, discrete, more rarely integrated, hyaline, cylindrical. *Conidia* 4-euseptate, apical and basal cell hyaline, median cells thick-walled, brown, smooth, continuous or constricted at the septa, at the base with an unbranched, cellular, usually short, central appendage, at the apex with a simple or branched, cellular, filiform appendage.

So far more than 120 taxa have been described in *Monochaetia*, and these mostly comprise both 4 septate and 5 septate species. Guba (1961) adopted *Monochaetia* for 3-, 4- and 5- septate species and accepted 41 species in three sections based on septation of conidia. His section Quadriloculatae contains several species now referable to *Truncatella* Stey. or *Seimatosporium* Cda, the Quinqueloculatae is the true *Monochaetia*, and the Sexloculatae contains species referable to *Seiridium* Nees ex Fr. and other genera. Fourteen species were accepted in the Quinqueloculatae including the type species *M. monochaeta*. The distinctions between genera in the *Monochaetia–Seiridium* and *Pestalotia–Pestalotiopsis* complexes have been fully discussed by Sutton (1969a).

KEY

1. Apical conidial appendages often branched 2
1. Apical conidial appendage simple 3
   2. Conidia 26-37 × 7-8.5 µ .... *M. karstenii*
   2. Conidia 18.5-21 × 5.5-6 µ .... *M. karstenii* var. *gallica*
   2. Conidia 15-17 × 5-5.5 µ .... *M. nattrassii*
3. Apical and basal appendage spathulate .... *M. dimorphospora*
3. Appendages simple 4
   4. Conidia verrucose 5
   4. Conidia smooth 6
5. Conidia 8.5-9.5 µ wide; basal appendage 6-11 µ long .... *M. carissae*
5. Conidia 7.5-8 µ wide; basal appendage 1-3 µ long .... *M. concentrica*
   6. Conidia 21-30 × 7-8 µ; basal appendage absent .... *M. kansensis*
   6. Conidia rarely exceeding 20 µ long 7
7. Conidia thick-walled; apical and basal appendage 1-2 µ long .... *M. saccardiana*
7. Conidia curved; apical appendage 5-6 µ long; basal appendage 2-3.5 µ long .... *M. ceratoniae*
7. Conidia straight; apical appendage 6-9 µ long; basal appendage 1-2 µ long .... *M. monochaeta*

M. CARISSAE Munjal & Kapoor, *Ind. Phytopath.* **15**: 259 (1963). (Fig. 151B)

Conidia 24-26 × 8.5-9.5 µ; median cells thick-walled, verrucose, 17-19 µ long; apical appendage simple, 12-16 µ long; basal appendage 6-11 µ long.

On pods and leaves of *Carissa carandas, C. spinarum, Cassia* sp., India (IMI 108218, 160056c and HCIO 27159, holotype).

M. CERATONIAE (Sousa da Camera) Sutton, *Mycol. Pap.* 88: 42 (1963). (Fig. 151A)
Amphichaeta ceratoniae Sousa da Camara, *Agron. lusit.* **13**: 147 (1951).

Conidia 17.5-20 × 5.5-6.5 µ; median cells thin-walled, smooth, 13-13.5 µ long; apical appendage 5-6 µ long; basal appendage 2-3.5 µ long.

On leaves and twigs of *Ceratonia siliqua, Diospyros kaki, Dodonea* sp., India, Portugal (IMI 86969, holotype ex LISE, 147955, 113516).

M. CONCENTRICA (Berk. & Br.) Sacc., *Syll. fung.* **18**: 485 (1906). (Fig. 152C)
Pestalotia concentrica Berk. & Br., *Grevillea* **2**: 156 (1874).
Synonymy in Guba (1961).

Conidia 23-28 × 7.5-8 µ; median cells thick-walled 16.5-18.5 µ long, verrucose; apical appendage 13-16 µ long; basal appendage 1-3 µ long.

On leaves of *Castanea* sp., Ala, U.S.A. (IMI 86701, holotype ex K).

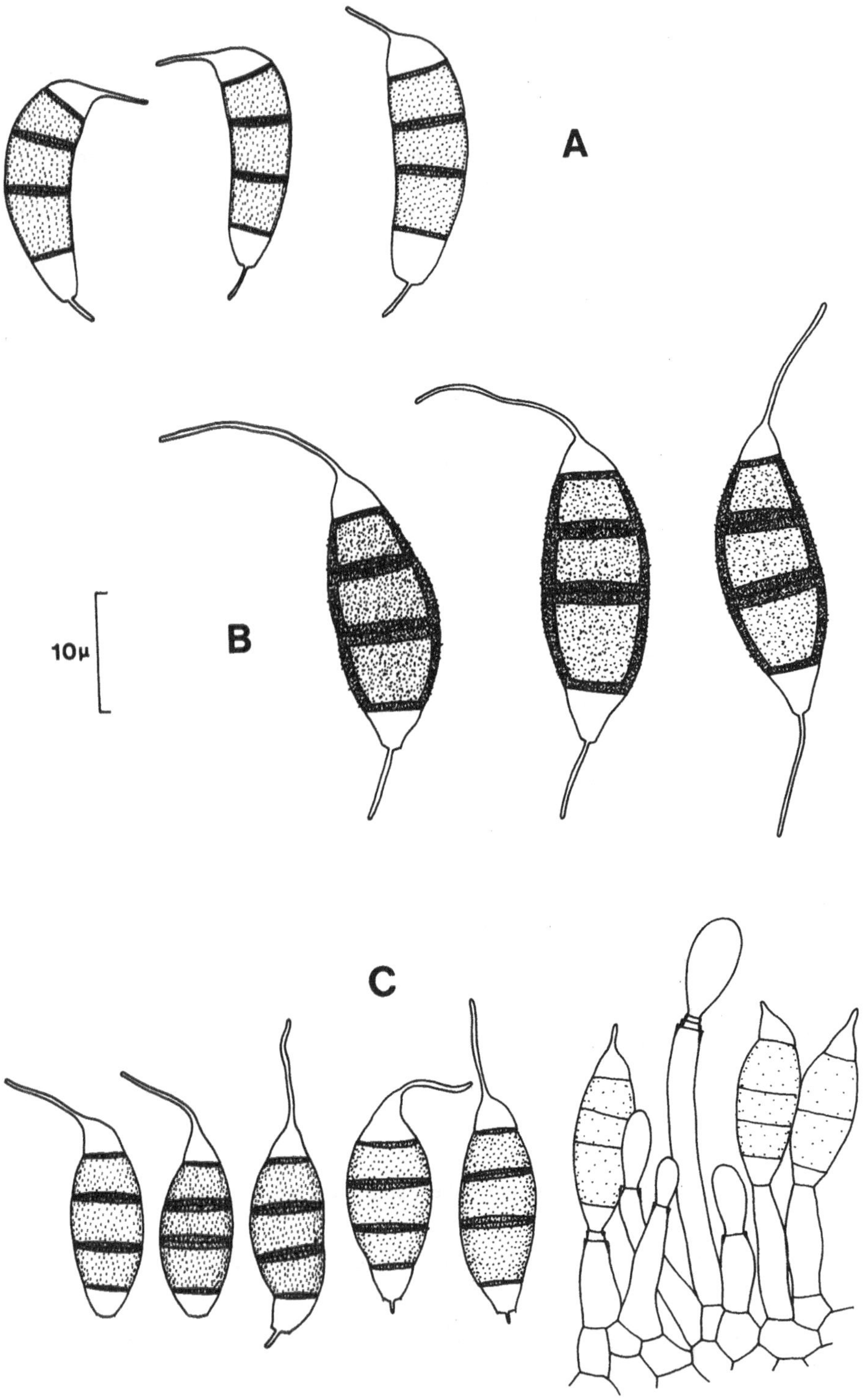

Fig. 151. A, *Monochaetia ceratoniae*, conidia; B, *M. carissae*, conidia; C, *M. monochaeta*, conidia, conidiogenous cells and developing conidia.

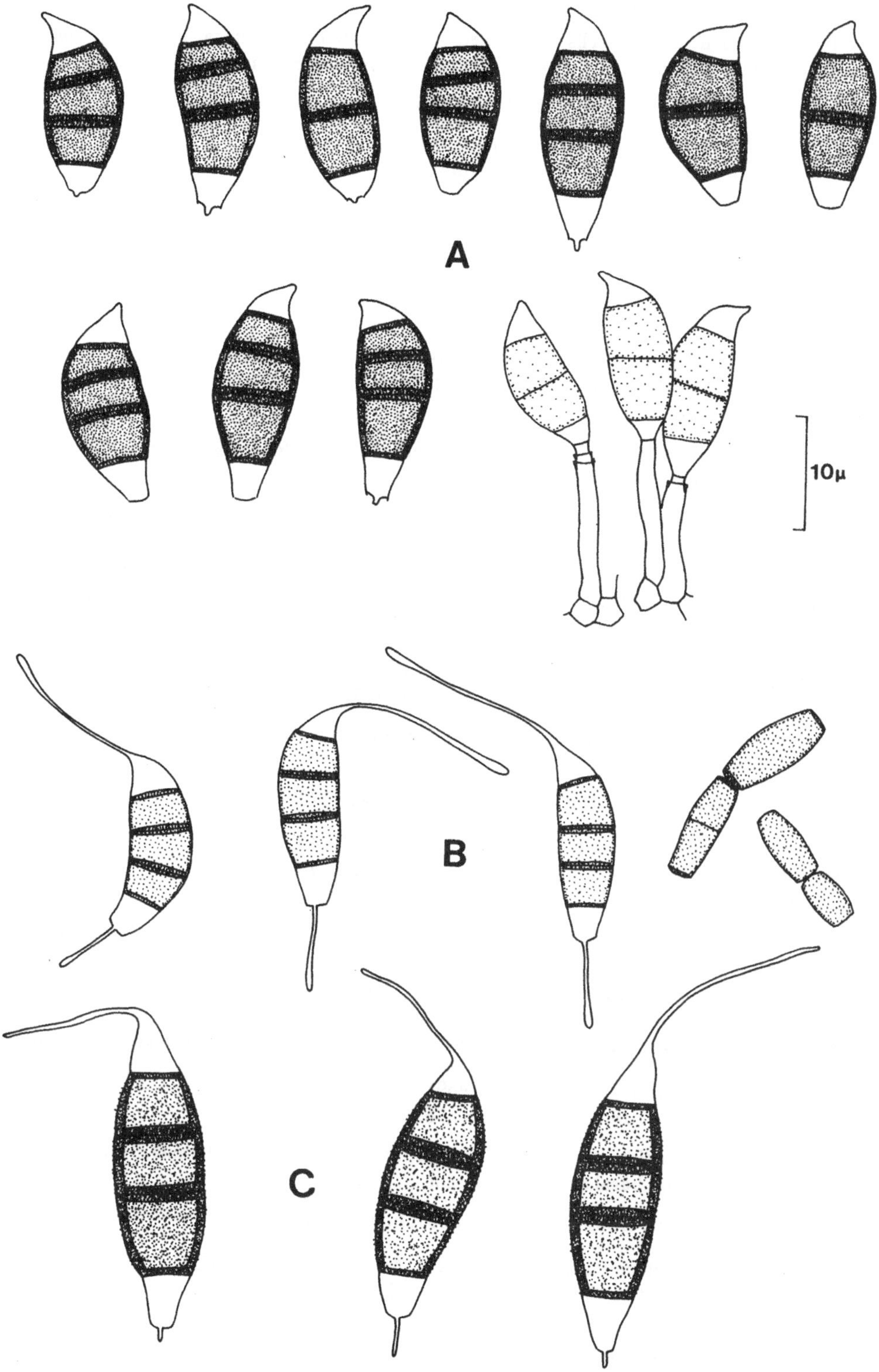

Fig. 152. A, *Monochaetia saccardiana*, conidia, conidiogenous cells and developing conidia; B, *M. dimorphospora*, conidia and chlamydospores, C, *M. concentrica*, conidia.

M. DIMORPHOSPORA Yokoyama, *Trans. Br. mycol. Soc.* **65**: 500 (1975). (Fig. 152B)

Conidia 16-18 × 5-5.5 µ; median cells thin-walled, smooth, 10-11.5 µ; apical appendage spathulate, 14-17 µ long; basal appendage spathulate, 6-8 µ long. In culture brown thallic chlamydospores are produced.

From leaves of *Castanea pubinervis*, Japan (IMI 187550, isoholotype).

M. KANSENSIS (Ell. & Barth.) Sacc., *Syll. fung.* **18**: 485 (1906). (Fig. 153C)
Pestalotia kansensis Ell. & Barth., *Erythea* **4**: 26 (1896).
Synonymy in Guba (1961).

Conidia 21-30 × 7-8 µ; median cells thick-walled, smooth, 17-18 µ long; apical appendage 6-9 µ long; basal appendage absent.

On leaves of *Quercus* sp., Austria (IMI 87094, holotype ex NY).

M. KARSTENII (Sacc. & Syd.) Sutton, *Can. J. Bot.* **47**: 2091 (1969). (Fig. 153A)
Pestalotia karstenii Sacc. & Syd. in Sacc., *Syll. fung.* **14**: 1030 (1899).
Synonymy in Sutton (1969a).

Conidia 15-17 × 5-5.5 µ; median cells thin-walled, smooth, 10-11.5 µ long; apical appendage often branched, 12-20 µ long; basal appendage 1-3 µ long.

On and from leaves of *Camellia japonica, C. caudata, Camellia* sp., *Allium sativum, Cassia obtusifolia, Curcuma decipiens, Daucus carota, Peltophorum ferrugineum, Rhododendron* sp., and from bird feathers, New Zealand, U.K., North Borneo, India, Dominica, Italy, France, Czechoslovakia, Miss., U.S.A. (IMI 84980, 84975, 84976, 84977 ex BM, 83729 ex K, 22481, 22479, 57020, 123805, 22482, 217376, 83728a, 163677, 205323, 99728, 73431, 60580, 102486, 102484, 134460, 86639, 93725, 209880).

M. KARSTENII var. GALLICA (Stey.) Sutton, *Can. J. Bot.* **47**: 2091 (1969). (Fig. 153B).
Pestalotiopsis gallica Stey., *Bull. Jard. Bot. Brux.* **19**: 308 (1949).

Conidia 18.5-21 × 5.5-6 µ; median cells thin-walled, smooth, 13.5-15.5 µ; apical appendage 12-15 µ long; basal appendage 1-3 µ long.

On leaves of *Camellia* sp., France (IMI 83728c ex BR, holotype, 83728c ex PC, 83728c ex K).

M. SACCARDIANA (Vogl.) Sacc., *Syll. fung.* **22**: 1229 (1913). (Fig. 152A)
Pestalotia saccardiana Vogl., *Atti Soc. Ven.–Trent. Sci. Nat. Padova* **9**: 233 (1885).
Synonymy in Guba (1961).

Conidia 15-19 × 6.5-7.5 µ; median cells 2-3, thick-walled, smooth, 11-13.5 µ long; apical appendage 1-2 µ long; basal appendage 1-2 µ long.

Reported by Sutton (1975a) as *M. mucronata* (Massal.) Maire. This species was included in section Quadriloculatae by Guba (1961) but many conidia develop 4 septa.

On leaves of *Quercus pubescens*, Italy (IMI 179686 ex VER, holotype of *Coryneum mucronatum* Massal.).

M. NATTRASII (Stey.) Sutton, *Can. J. Bot.* **47**: 2091 (1969). (Fig. 153D)
Pestalotiopsis nattrassii Stey., *Trans. Br. mycol. Soc.* **36**: 82 (1953).

Conidia 26-37 × 7-8.5 µ; median cells thin-walled, smooth, but with reduced lumina, 18-24 µ long; apical appendage branched, 18-33 µ long; basal appendage absent.

From *Camellia sinensis*, Kenya, Hong Kong (IMI 38807, holotype, 122004).

M. MONOCHAETA (Desm.) Allesch. in Rabenh., *Krypt. Fl.* **7**: 667 (1902). (Fig. 151C)
Pestalotia monochaeta Desm., *Ann. Sci. nat.*, ser. 3, **10**: 355 (1848).
Synonymy in Guba (1961).

Conidia 16-18.5 × 6.5-7 µ; median cells smooth, thin-walled, 11-13 µ long; apical appendage 6-9 µ long; basal appendage 1-2 µ long.

On leaves of *Quercus* spp., Italy, Austria (IMI 181240, holotype *Coryneum rostratum* ex K, 87156, holotype *Pestalotia saccardoi* ex LPS).

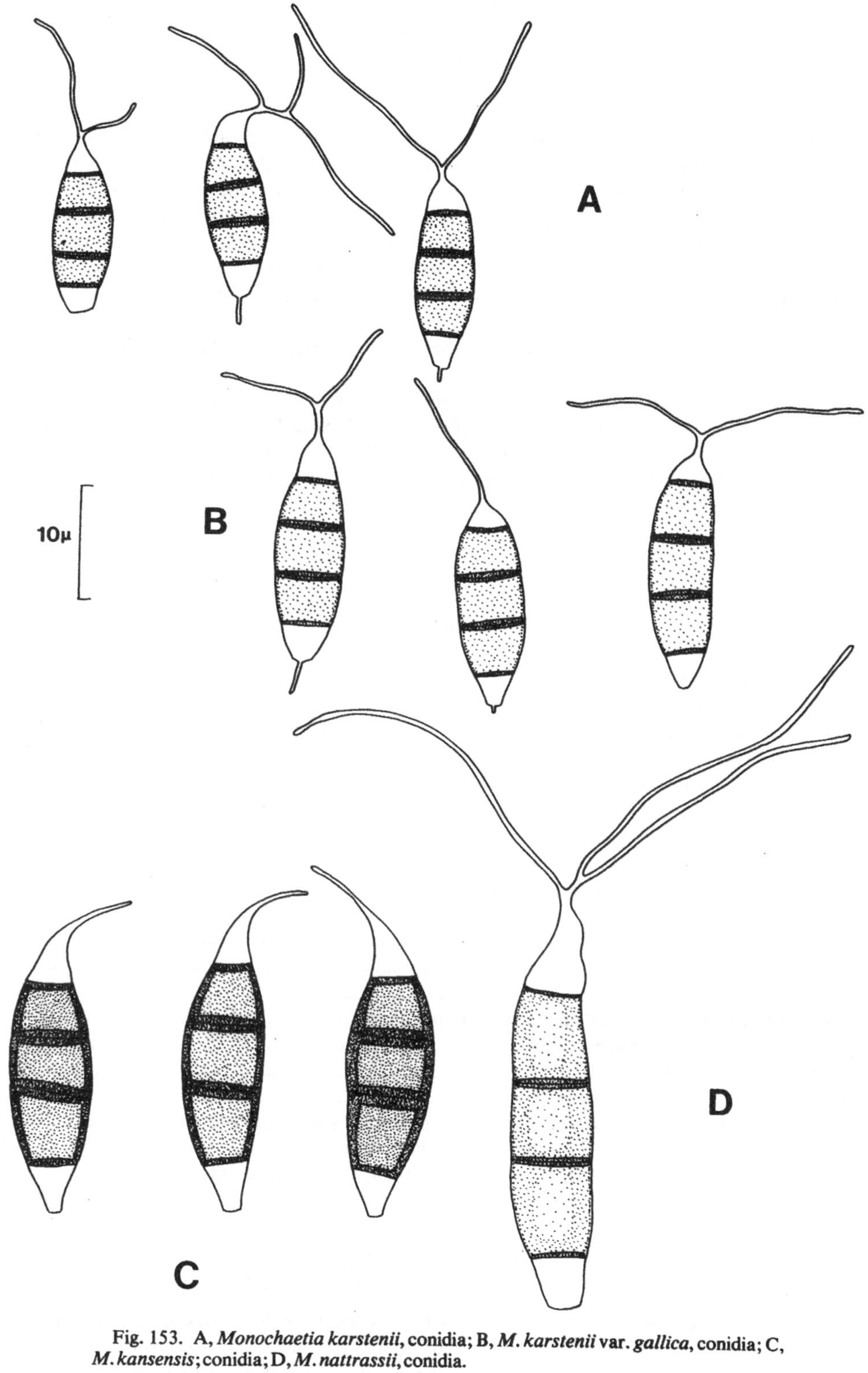

Fig. 153. A, *Monochaetia karstenii*, conidia; B, *M. karstenii* var. *gallica*, conidia; C, *M. kansensis*; conidia; D, *M. nattrassii*, conidia.

12. SEIRIDIUM Nees ex Fr., *Syst. Myc.* **1**: XL (1821).
Seiridium Nees, *Das System der Pilze und Schwämme*: 22 (1817).
Hyaloceras Dur. & Mont. in Dur., *Flore d'Algerie Crypt.* **1**: 587 (1846).
? *Adea* Petrak, Engler's *Bot. Jahrb. Beibl.* **142**: 144 (1929).
Sp. typ.: *S. marginatum* Nees ex Steudel

Mycelium immersed, branched, septate, pale brown to hyaline. *Conidiomata* acervular, separate or confluent, circular or linear, composed of brown, thin-walled textura angularis. Dehiscence irregular. *Conidiophores* hyaline, cylindrical, straight or curved, sparsely branched and septate at the base, formed from the upper cells of the conidiomata. *Conidiogenous cells* holoblastic, annellidic, with several apical annellations, indeterminate, discrete or integrated, hyaline, cylindrical. *Conidia* 5- euseptate, fusiform, apical and basal cells hyaline, 4 median cells thick-walled, smooth, verruculose, striate or punctate, brown, constricted or continuous at the septa which may be equidistant or not; basal cell with or without an endogenous, cellular, simple appendage; apical cell with a single, cellular, simple or branched appendage.

The generic synonymy has been fully discussed by Guba (1961), Shoemaker, Müller & Morgan-Jones (1966), and Sutton (1969a, 1975a).

More than 20 taxa have now been placed in *Seiridium* and although there has been no recent revision of the group, Guba (1961) accepted 15 species in *Monochaetia* (Sacc.) Allesch. sect. Sexloculatae, which is the equivalent of *Seiridium* sensu stricto. There are probably many more species referable to *Seiridium* than have actually been described in the genus, because for many years it had been misinterpreted, and the majority of species placed by Guba in *Monochaetia* sect. Sexloculatae have names only in *Monochaetia*.

KEY

1. Conidial appendages longer than 5 µ 2
1. Conidial appendages shorter than 5 µ 3
    2. Conidia smooth; apical appendage frequently branched, up to 30 µ long .... *S. corni*
    2. Conidia striate; apical appendage unbranched, 20-50 µ long .... *S. marginatum*
    2. Conidia smooth; apical appendage unbranched, up to 20 µ long .... *S. unicorne*
3. Conidia striate 4
3. Conidia smooth 5
    4. Median coloured cells of equal size .... *S. indicum*
    4. Lowest of median coloured cells larger than the rest .... *S. intermedium*
5. Conidia 10-13 µ wide 6
5. Conidia less than 10 µ wide on average 7
    6. Conidia 36-42 × 11-13 µ .... *S. abietinum*
    6. Conidia 26-36 × 10-12 µ .... *S. viburni*
7. Conidia with a distinct though short apical appendage up to 3 µ long 8
7. Conidia with conical, slightly projecting apical cells 9
    8. Conidia 19-24 × 8-8.5 µ; median cells 15.5-19 µ .... *S. anceps*
    8. Conidia 25-29.5 × 9.5-10.5 µ; median cells 19-23 µ .... *S. tecomae*
9. Conidia 21-30 × 8-9 µ; median cells 17-21.5 µ; short basal appendage sometimes present .... *S. cardinale*
9. Conidia 24-27 × 10-11 µ; median cells 20-22 µ; basal appendage absent .... *S. castaneae*
9. Conidia 23-34 × 8.5-10 µ; median cells 18.5-26.5 µ; basal appendage usually present .... *S. juniperi*

S. ABIETINUM (Ell. & Ev.) Sutton, *Mycol. Pap.* 138: 51 (1975). (Fig. 154A)
Coryneum abietinum Ell. & Ev., *Proc. Acad. nat. Sci., Phil. 1894*: 375 (1894).

Conidiomata subperidermal, 200-400 µ diam. Conidiophores up to 60 µ long × 2-2.5 µ with 0-1 annellations. Conidia 36-42 × 11-13 µ, straight or slightly curved, smooth; median cells dark brown, 25-30 µ long; apical cell conic projecting into a short appendage 1-3 µ long; basal cell often with an appendage 2-3 µ long.

On bark of *Abies balsamea*, *A.* sp., N.F., Canada (IMI 178765, holotype, 178763, 178764 ex NY).

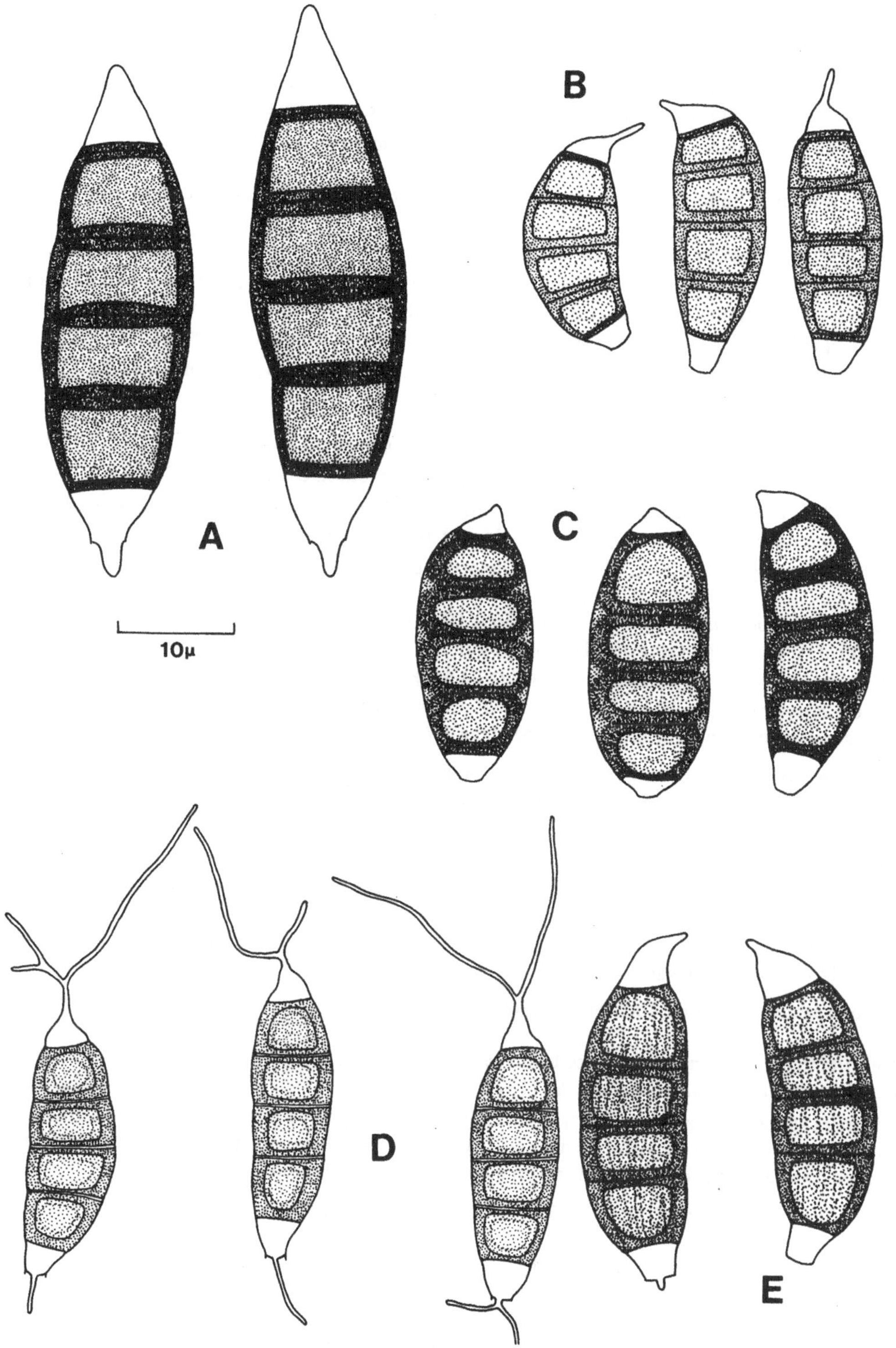

Fig. 154. A, *Seiridium abietinum*, conidia; B, *S. anceps*, conidia; C, *S. castaneae*, conidia; D, *S. corni*, conidia; E, *S. indicum*, conidia.

S. ANCEPS (Sacc.) Sutton, *Mycol. Pap.* 138: 58 (1975). (Fig. 154B)
Coryneum anceps Sacc., *Grevillea* **21**: 68 (1893).

Conidiomata subepidermal, up to 150 µ diam. Conidiophores up to 15 µ long × 1.5-2 µ wide, with 0-1 annellations. Conidia 19-24 × 8-8.5 µ, straight or occasionally curved, smooth; median cells brown, 15.5-19 µ long; apical cells with a corniform appendage 3 µ long; basal cell lacking an appendage.

On leaves of *Euonymus alatus*, Siberia, U.S.S.R. (IMI 179083, holotype, ex PAD).

S. CARDINALE (Wagener) Sutton & Gibson, *CMI Descriptions of pathogenic fungi and bacteria*: 326 (1972). (Fig. 156B)
Coryneum cardinale Wagener, *J. Agr. Res.* **58**: 8 (1939).

Conidiomata subperidermal, associated with cankers, 200-300 µ diam. Conidiophores up to 25 µ long × 1.5-2 µ wide, with 0-2 annellations. Conidia 21-30 × 8-9 µ, straight or slightly curved, smooth; apical cell conic, projecting into an appendage 1 µ long; basal cell sometimes with a short appendage 1 µ long.

On and from *Cupressus macrocarpa*, *C.* sp., U.K., Eire, U.S.A. (IMI 75045, holotype ex K, 137371, 159269, 141216, 158985, 137368, 8894).

S. CASTANEAE (Berk. & Curt. ex Sacc.) Sutton, *Mycol. Pap.* 138: 92 (1975). (Fig. 154C)
Coryneum decipiens Berk. & Curt. β *Castaneae* Berk. & Curt. ex Sacc., *Syll. fung.* **3**: 779 (1884).

Conidia 24-27 × 10-11 µ, fusiform, straight or slightly curved; median cells 20-22 µ long, smooth, brown; apical cell conic, prolonged with a short appendage 1 µ long; basal cell lacking any appendages.

On bark of *Castanea* sp., Va, U.S.A. (IMI 74983, holotype ex K).

S. CORNI (Allesch.) Sutton, *Can. J. Bot.* **47**: 2091 (1969). (Fig. 154D)
Pestalotia corni Allesch., *Bot. Centrabl.* **42**: 106 (1890).

Conidiomata subperidermal, 200-400 µ diam. Conidiophores 8-33 × 2-3 µ; with up to 2 annellations. Conidia 22-27 × 6.5-8 µ, straight, fusiform, smooth; apical cell with a single, simple or branched appendage up to 30 µ long; basal cell with a simple or branched appendage 3-7 µ long.

On twigs of *Cornus alba*, *C. stolonifera*, Germany, Man., Sask., Canada (WINF(M) 6348, ex HBG, IMI 132456 (WINF(M) 4216), WINF(M) 7251a, 7249a, 10380, 10309a, 8515, 2333b, 6696b, 6744b).

S. INDICUM Pavgi & U.P. Singh, *Sydowia* **24**: 116 (1971). (Fig. 154E)

Conidiomata subcuticular to epidermal, up to 250 µ diam. Conidiophores up to 15 µ long, with 1-2 annellations. Conidia straight or slightly curved, fusiform, 27-29 × 9-10 µ (22-31(-33) × 7.5-9 µ in original description); median cells striate, 19-23 µ long; apical cell conical, projecting with a short appendage up to 2 µ long; basal cell occasionally with an appendage 1 µ long.

On leaves of *Spiraea micrantha*, India (IMI 113602 holotype).

S. INTERMEDIUM (Sacc.) Sutton, *Mycol. Pap.* 138: 119 (1975). (Fig. 155A)
Coryneum intermedium Sacc., *Michelia* **1**: 91 (1877).

Conidiomata subperidermal, up to 200 µ diam. Conidiophores up to 45 µ long with 0-1 annellations. Conidia 27-31 × 10.5-11.5 µ, fusiform, straight; median cells striate, 22-24 µ long; apical cells conic, projecting into a short appendage up to 3 µ long; basal cell often with a short appendage 1 µ long.

On bark of *Ulmus* sp., Italy, U.K. (IMI 179074, holotype ex PAD, 6707a).

S. JUNIPERI (Allesch.) Sutton, *Mycol. Pap.* 138: 135 (1974). (Fig. 155B)
Pestalotia juniperi Allesch., *Ber. bot. Ver. Landshut*: 84 (1892).
Coryneum juniperi (Allesch.) Allesch., *Krypt. Fl. Dtsch* **7**: 650 (1902).

Conidiomata subperidermal, 300-500 µ diam. Conidiophores up to 30 µ long × 2-4 µ wide with up to 3 annellations. Conidia 23-34 × 8.5-10 µ, fusiform, straight, smooth; median cells 18.5-26.5 µ long; apical cell conic, projecting into a short appendage 1 µ long; basal cell sometimes with an appendage 1 µ long.

On galls caused by *Podisoma clavariiforme* on *Juniperus communis*, Germany (IMI 180072 ex Staatsherbarium München, holotype).

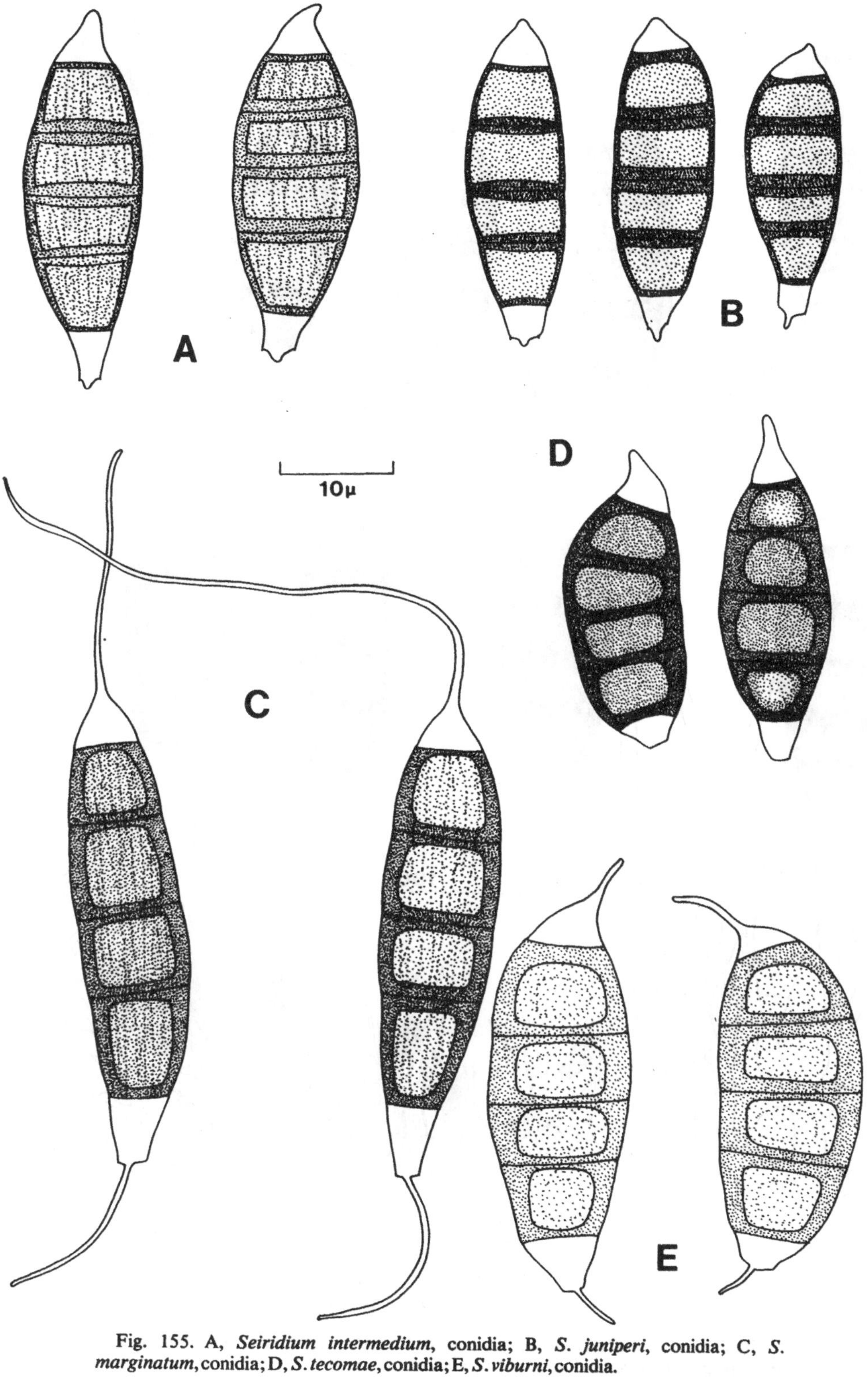

Fig. 155. A, *Seiridium intermedium*, conidia; B, *S. juniperi*, conidia; C, *S. marginatum*, conidia; D, *S. tecomae*, conidia; E, *S. viburni*, conidia.

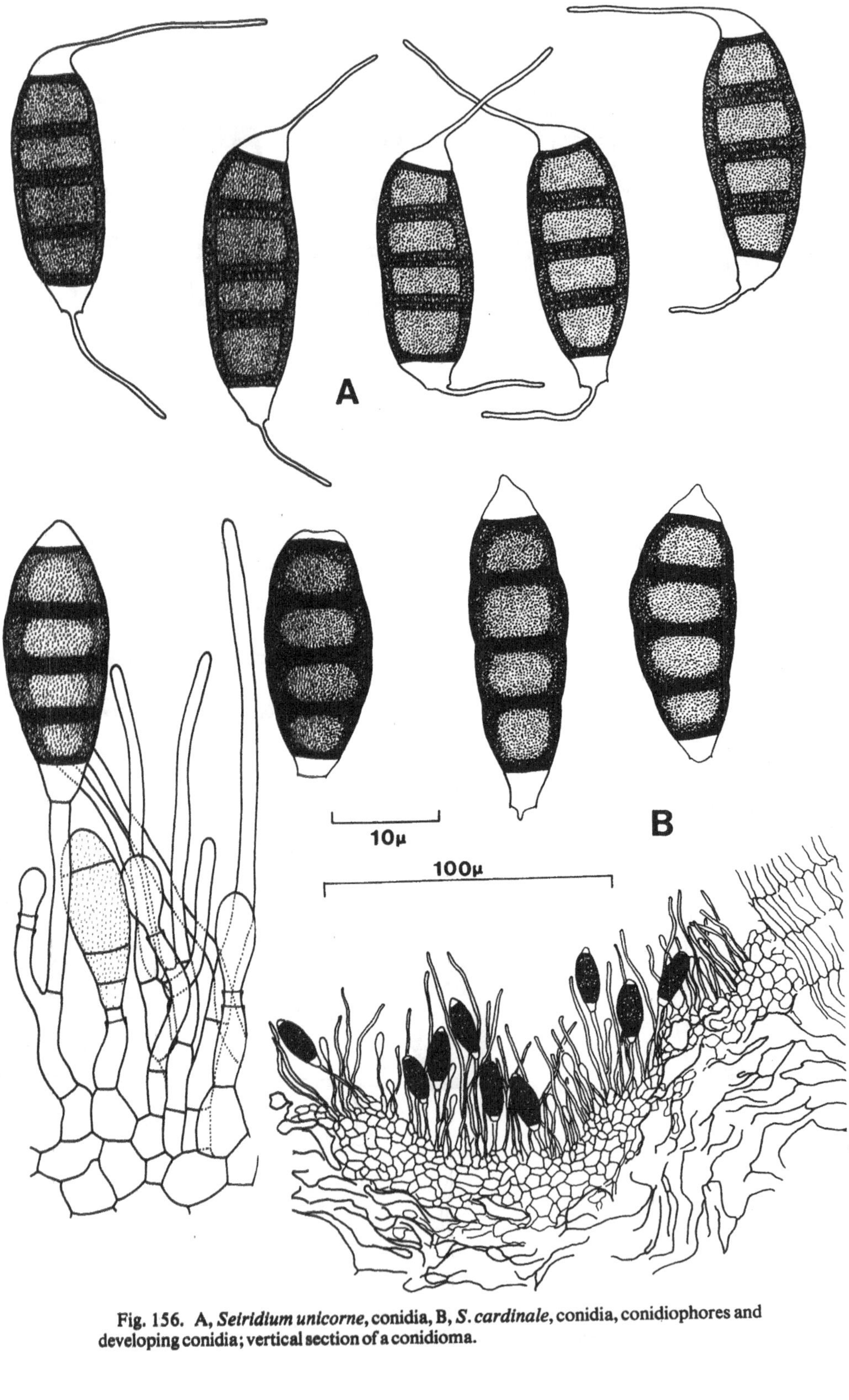

Fig. 156. A, *Seiridium unicorne*, conidia, B, *S. cardinale*, conidia, conidiophores and developing conidia; vertical section of a conidioma.

S. MARGINATUM Nees ex Steudel, *Nomenclator Botanicus*: 375 (1824). (Fig. 155C)
S. marginatum Nees, *Das System der Pilze und Schwämme*: 22 (1817).

Teleomorphosis: *Blogiascospora marginata* (Fckl) Shoemaker et al., *Can. J. Bot.* **44**: 248 (1966).

Conidiomata subcuticular, 300-500 μ diam. Conidiophores 20-50 μ × 6 μ wide, with up to 3 annellations. Conidia fusiform, straight, 35-45 × 9-10.5 μ; median cells striate, 32-33 μ long; apical cell with a filiform unbranched appendage up to 40 μ long; basal cells also with an appendage measuring up to 15 μ long.

On twigs of *Rosa canina*, *R.* sp., Switzerland, France, Germany (IMI 102908, 102909 ex B, 93432a ex G).

S. TECOMAE (Sacc.) Sutton, *Mycol. Pap.* 138: 174 (1975). (Fig. 155D)
Coryneum tecomae Sacc., *Atti Accad. scient. veneto-trent.-istriana* **2**(2): 190 (1873).

Conidia 25-29.5 × 9.5-10.5 μ, straight or slightly curved, fusiform, smooth; median cells brown, 19-23 μ long; apical cell conic projecting into a short appendage 2-3 μ long; basal cell lacking an appendage.

On twigs of *Bignonia grandiflora*, Italy (IMI 179082 ex PAD, holotype).

S. UNICORNE (Cke & Ell.) Sutton, *Mycol. Pap.* 138: 74 (1975). (Fig. 156A)
Pestalotia unicornis Cke & Ell., *Grevillea* **7**: 6 (1878).
Monochaetia unicornis (Cke & Ell.) Sacc., *Syll. fung.* **18**: 485 (1906).
Cryptostictis cupressi Guba, *Monograph of Monochaetia and Pestalotia*: 47 (1961).

Teleomorphosis: *Lepteutypa cupressi* (Nattrass, Booth & Sutton) Swart, *Trans. Br. mycol. Soc.* **61**: 79 (1973).

Conidiomata subperidermal, associated with cankers, up to 650 μ diam. Conidiophores 15-35 × 2-4 μ with 0-3 annellations. Conidia 24-28 × 8.5-9.5 μ, straight, fusiform, smooth; median cells 18.5-22 μ long, brown; apical cell with an unbranched appendage 11-19 μ long; basal cells with an endogenous unbranched appendage 8-13 μ long.

On and from *Cupressus forbesii*, *C. lusitanica*, *C. macrocarpa*, *C.* sp., *Chamaecyparis thyoides*, *C. lawsoniana*, *Juniperus procera*, Kenya, Swaziland, Uganda, India, U.S.A. (IMI 5816, type of *Pestalotia unicornis*, 52255, 136771, 165510, 52257, 37158, 63666, 52258, 52254, 52256, 40096, 56917, 31199, 5817, 8895).

S. VIBURNI (Sousa da Camara & da Luz) Sutton, *Mycol. Pap.* 88: 48 (1963). (Fig. 155E)
Amphichaeta viburni Sousa da Camara & da Luz, *Agron. lusit.* **1**: 57 (1939).
Amphichaeta viburni Sousa da Camara, *Agron. lusit.* **11**: 187 (1949).

Conidiomata subperidermal, 500-700 μ diam. Conidiophores 15-25 × 2-3 μ with 0-1 annellations. Conidia 26-36 × 10-12 μ, straight curved or sigmoid, smooth; median cells 22-29 μ long; apical cell prolonged into a short appendage 1.5-5 μ long; basal cell often with an appendage .5-3 μ long.

On branches of *Viburnum tinus*, Portugal (IMI 86077, holotype ex LISE).

13. SEIMATOSPORIUM Cda in Sturm, *Deutschl. Flora* **3**: 79 (1833).
Sporocadus Cda, *Icon. fung.* **3**: 24 (1839).
Cryptostictis Fckl, *Fungi rhenani* 1838, suppl. fasc. IV (1866).
Sarcostroma Cke, *J. Queckett Microsc. Cl.* **2**: 13 (1872).
Dochmolopha Cke, *Nuovo G. bot. ital.* **10**: 25 (1878).
Labridium Vestergr., *Öfv. K. Vet. Akad. Förhandl.* **1**: 43 (1897).
Amphichaeta McAlp., *Proc. linn. Soc. N.S.W.* **29**: 118 (1904).
Diploceras (Sacc.) Died., *Krypt. Fl. M. Brandenb.* **9**: 887 (1915).
Hyaloceras Dur. & Mont. subgen. *Diploceras* Sacc., *Syll. fung.* **10**: 48 (1892).
Leptocoryneum Petrak, *Hedwigia* **65**: 278 (1925).
Disaeta Bonar, *Mycologia* **20**: 299 (1928).
Seiridina Höhn., *Mitt. bot. Inst. tech. Hochsch. Wien* **7**: 31 (1930).
Coryneopsis Grove, *J. Bot., Lond.* **70**: 33 (1932).
Allelochaeta Petrak, *Sydowia* **9**: 464 (1955).
Basipilus Subram., *Proc. natn. Inst. Sci., India, B, Biol. Sci.* **27**: 243 (1961).
Monoceras Guba, *A monograph of Monochaetia and Pestalotia*: 290 (1961).
Sp. typ.: *S. rosae* Cda

Mycelium immersed, branched, septate, hyaline to pale brown. *Conidiomata* acervular, separate, immersed, foliicolous or corticolous, brown, composed of brown, thin- or thick-walled textura angularis. *Conidiophores* cylindrical, septate, branched, hyaline, filiform, formed from the upper cells of the conidiomata. *Conidiogenous cells* holoblastic, annellidic, integrated or discrete, hyaline, indeterminate, with 0-many simple annellations. *Conidia* variable, cylindrical, fusiform or clavate, 2-5 euseptate, sometimes constricted at the septa but in 4-celled species often typically with the thin periclinal walls between thick septa collapsing, septa equidistant or not, smooth, eguttulate, evenly pigmented or with the apical and basal cell, or the basal one only, hyaline, and the rest various degrees of brown, lacking appendages entirely, or with the apical cell provided with a single, cellular, simple or branched appendage, basal cell truncate with a marginal frill, appendage when present exogenous, cellular, simple or branched; some conidia with only an apical or a basal appendage.

The generic synonymy has been demonstrated by Sutton (1963, 1964a, 1973a, 1975a, c, 1977a), Shoemaker (1964a), Shoemaker & Müller (1964) and Nag Raj (1978b). At present the genus is quite heterogeneous with respect to conidial morphology and it is conceivable that in the future it will be desirable to split it into smaller genera. The basis on which this would occur is conjectural, but it may be by conidial septation, pigmentation, perfect state relationships, or it may be by a combination of such criteria. There are validly published generic names already available to accommodate such segregations and it would be a relatively simple matter to re-instate generic names for this purpose. Meanwhile, descriptive studies of *Seimatosporium* species have been presented by Sutton (1963, 1964a, 1975a, c), Swart & Griffiths (1974), Pirozynski & Shoemaker (1970), Shoemaker (1964a), and Shoemaker & Müller (1964).

KEY

1. Conidia 2 septate 2
1. Conidia 3 septate 4
1. Conidia 4 septate 18
1. Conidia 5 septate 23
   2. Conidia with a basal appendage only .... *S. rosarum*
   2. Conidia lacking any appendages 3
3. Conidia 17.5-24 × 8-9.5µ .... *S. acerinum*
3. Conidia 9.5-12 × 4.5-5.5 µ .... *S. caninum*
   4. Conidia lacking any appendages 5
   4. Conidia with appendages, sometimes only basal 10
5. Conidia wider than 6.5 µ on average 6
5. Conidia less than 6.5 µ wide on average 7
   6. Conidia 18.5-23 × 7.5-8.5 µ; median cells darker brown .... *S. cassiopes*
   6. Conidia 22-31 × 8.5-10.5 µ; cells evenly dark brown, very thick-walled .... *S. pezizoides*
   6. Conidia 15.5-20 × 6.5-8.5 µ; clavate, markedly tapered to the base; subterminal cell often darker brown .... *S. rhododendri*
7. Conidia distinctly fusiform 8
7. Conidia ellipsoid 9
   8. Conidia 15-18 × 5-6.5 µ; median cells pale brown, and cells subhyaline .... *S. glandigenum*
   8. Conidia 13-18 × 4.5-5.5 µ; median or upper three cells medium brown; periclinal wall often collapsed .... *S. vaccinii*
9. Conidia 13-15 × 5.5-6.5 µ; transverse septa thick and prominent, periclinal walls collapsed .... *S. lichenicola*
9. Conidia 11-15 × 4.5-5.5 µ; periclinal walls not collapsed .... *S. pestalozzioides*
   10. Appendages branched or more than one at each end 11
   10. Appendages unbranched 12
11. Conidia 13-15.5 × 5-6.5 µ; appendages basal only, 10-20 µ long .... *S. effusum*
11. Conidia 15-19 × 5-6 µ; single at each end, 10-20 µ long .... *S. anomalum*
11. Conidia 15-18.5 × 4.5-5.5 µ; 2 appendages at each end, 7-17 µ long .... *S. hypericinum*
   12. Conidia with thin, indistinct septa 13
   12. Conidia with dark brown, thick, prominent septa 16
13. Conidia less than 20 µ long 14
13. Conidia more than 20 µ long 15
   14. Conidia 11.5-18 × 3.5-5 µ; appendages 2-7 µ long .... *S. discosioides*
   14. Conidia 16-21 × 6-7 µ; appendages 7-18 µ long .... *S. fusisporum*
15. Conidia 19.5-30 × 4-5 µ, appendages 5-16 µ long .... *S. kriegerianum*

15. Conidia 28-37.5 × 4-5 μ, appendages 6-10 μ long .... *S. falcatum*
15. Conidia 36-157 × 3-6 μ, appendages 3-86μ long .... *S. leptospermi*
16. Conidia distinctly fusiform, 11-19 × 3-5 μ .... *S. lonicerae*
16. Conidia ellipsoid to long doliiform 17
17. Conidia 15.5-18 × 5.5-6.5 μ; upper of median cells longer than lower; appendages 1-3 μ long .... *S. monochaetioides*
17. Conidia 11.5-17 × 4-5.5 μ; apical and/or basal appendages sometimes absent, 15-35 μ long .... *S. caudatum*
17. Conidia 10-15 × 3-4 μ; apical and/or basal appendage sometimes absent, 1-8 μ long .... *S. rosae*
18. Appendages branched .... *S. dilophosporum*
18. Appendages unbranched 19
19. Conidia less than 20 μ long on average 20
19. Conidia more than 20 μ long on average 22
20. Appendages less than 15 μ long .... *S. arbuti*
20. Appendages more than 15 μ long 21
21. Conidia 13-15.5 × 6-7 μ, dark brown; appendages 12-29 μ long .... *S. cadicola*
21. Conidia 16.5-21 × 5.5-6.5 μ, medium brown; appendages 15-34 μ long .... *S. hollosii*
21. Conidia 16.5-19.5 × 6-7 μ, medium brown; appendages 15-29 μ long .... *S. kennedyae*
21. Conidia 16-20.5 × 7-9 μ, medium brown; appendages 14-30 μ long .... *S. hakeae*
22. Conidia 18-23 × 5-7 μ, pale brown; appendages 11-25 μ long .... *S. grevilleae*
22. Conidia 20-25.5 × 7-8 μ, strongly curved, reddish brown; appendages 6-11.5 μ long .... *S. sublunatum*
22. Conidia 21.5-29.5 × 5.5-8 μ, pale brown; appendages 5-15.5 μ long .... *S. daviesiae*
23. Conidia lacking appendages 24
23. Conidia with appendages 25
24. Conidia 29-32 × 9-10 μ, fusiform; septa prominent, periclinal walls collapsed .... *S. berckmansii*
24. Conidiomata cupulate; conidia 28-39 × 9-12.5 μ, smooth .... *S. macrospermum*
24. Conidia 26-32.5 × 7-9 μ, verruculose .... *S. missionum*
25. Conidia more than 25 μ long .... *S. parasiticum*
25. Conidia less than 25 μ long 26
26. Conidia 15-22.5 × 5-6 μ, appendages 2-18 μ long, often the apical one absent .... *S. consocium*
26. Conidia 18.5-22.5 × 7.5-9 μ, appendages 3-9 μ long .... *S. foliicola*
26. Conidia 19-23 × 6-7 μ, appendages 3-30 μ long .... *S. mariae*

S. ANOMALUM (Harkn.) Shoemaker, *Can. J. Bot.* **42**: 416 (1964). (Fig. 157B)
Pestalotia anomala Harkn., *Bull. Calif. Acad. Sci.* 13 (1884).
Diploceras anomala (Harkn.) Höhn., *Sber. Akad. Wiss. Wien* **125**: 93 (1916).
Cryptostictis anomala (Harkn.) Guba, *Monograph of Monochaetia and Pestalotia*: 270 (1961).

Conidiomata 200-300 μ diam. Conidiophores 15 × 2-3 μ, with 1-3 annellations. Conidia 15-19 × 5-6 μ, 3 septate, straight or slightly curved, median cells pale brown, 9.5-13 μ long; apical cell hyaline with a simple or irregularly branched appendage 13-20 μ long; basal cell hyaline with a simple or branched appendage 10-17 μ long.

On twigs of *Eriogonum virgatum, E. tomentosum*, Calif., U.S.A. (IMI 86130, 86128, holotype ex K).

S. ACERINUM (Bauml.) Sutton, *Mycol. Pap.* 138: 56 (1975). (Fig. 157A)
Coryneum acerinum Bauml., *Verh. f. Natur-Heilk, Presburg* 23, ser. 14: 52 (1902).

Conidiophores up to 45 × 3-4 μ with 1-4 annellations. Conidia 17.5-24 × 8-9.5 μ, 2 septate, straight, cells evenly medium brown, transverse septa prominent, periclinal walls collapsed; appendages absent.

On leaves of *Acer campestris*, Czechoslovakia (holotype in BP).

S. ARBUTI (Bonar) Shoemaker, *Can. J. Bot.* **42**: 416 (1964). (Fig. 157C)
Disaeta arbuti Bonar, *Mycologia* **20**: 299 (1928).
Cryptostictis arbuti (Bonar) Zeller, *Phytopath.* **21**: 972 (1931).
Amphichaeta hispanica Bausá Alcalde, *An. Jard. bot. Madrid* **7**: 498 (1948).
Amphichaeta rhododendri Camara, *Agron. lusit.* **11**: 187 (1949).
Cryptostictis bisetulata Guba, *Monograph of Monochaetia and Pestalotia*: 272 (1961).

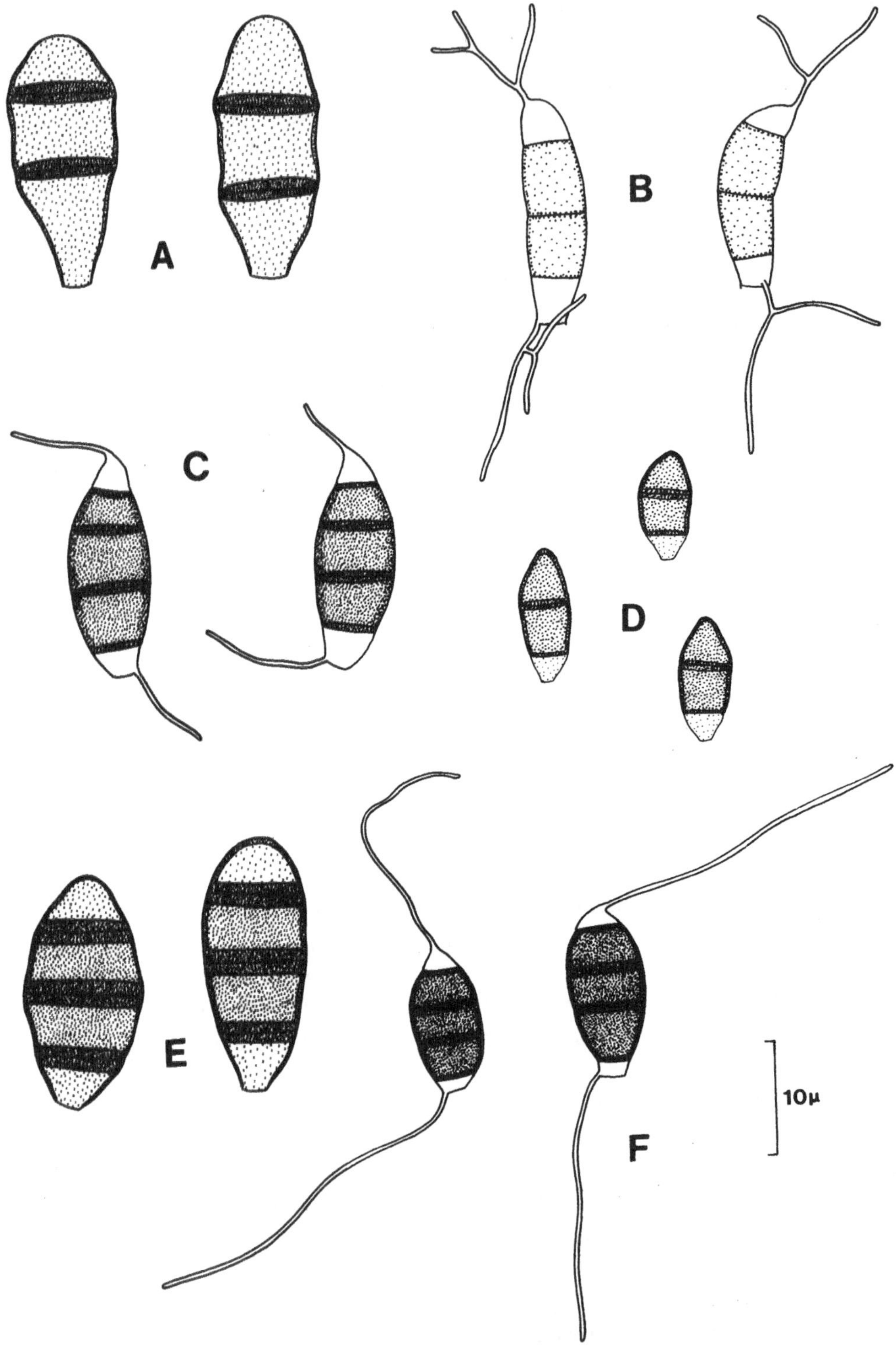

Fig. 157. A, *Seimatosporium acerinum*, conidia; B, *S. anomalum*, conidia; C, *S. arbuti*, conidia; D, *S. caninum*, conidia; E, *S. cassiopes*, conidia; F, *S. cadicola* conidia.

Conidiomata 150-200 µ diam. Conidiophores 10-30 × 2 µ, with 0-1 annellations. Conidia 14.5-21 × 4.5-5.5 µ, 4 septate, straight; median cells medium brown, 10-14 µ long; apical cell hyaline with an unbranched appendage 6-12 µ long; basal cell hyaline with an unbranched appendage 5-11 µ long.

On leaves of *Arbutus menziesii, Cistus* sp., *Ledum glandulosum, Phillyrea media, Rhododendron* sp., Calif., U.S.A., Portugal, Spain, Algeria (IMI 86040 ex UC, holotype *D. arbuti*, 87151 ex LISE, holotype *A. rhododendri*, 92425 ex BPI, 92884 ex DAOM, 93435 ex PC, 93434 ex MA, holotype of *A. hispanica*).

S. BERCKMANSII (Milbrath) Sutton, *Mycol. Pap.* 138: 64 (1975). (Fig. 158A)
Coryneum berckmansii Milbrath, *Phytopath.* **30**: 594 (1940).

Conidiomata up to 300 µ diam. Conidiophores 10-30 × 2-3 µ, with 0-1 annellations. Conidia 29-32 × 9-10 µ, 5 septate, straight, transverse septa prominent, periclinal walls collapsed, evenly pale brown; appendages absent.

On leaves of *Thuja orientalis* var. *conspicua*, Ore., U.S.A. (IMI 178772 ex NY, holotype).

S. CADICOLA (Sutton) Shoemaker, *Can. J. Bot.* **42**: 416 (1964). (Fig. 157F)
Cryptostictis cadicola Sutton, *Mycol. Pap.* 88: 14 (1963).

Conidiomata 175-250 µ diam. Conidiophores 10-20 × 2-3 µ with 1-2 annellations. Conidia 13-15.5 × 6-7 µ, 4 septate, straight; median cells medium brown, 10.5-12.5 µ long; apical cell hyaline with an unbranched appendage 18-29 µ long; basal cell hyaline with an unbranched appendage 12-29 µ long.

On leaves of *Acrotricha* sp. aff. *ovalifolia*, Vict., Australia (IMI 86131 ex K, holotype).

S. CANINUM (Brun.) Sutton, *Mycol. Pap.* 138: 69 (1975). (Fig. 157D)
Hendersonia canina Brun., *Act. Soc. linn. Bordeaux* **52**: 146 (1897).
Coryneopsis canina (Brun.) Grove, *J. Bot. Lond.* **70**: 34 (1932).
Coryneum biseptatum Kapoor & Munjal, *Indian Phytopath.* **21**: 107 (1968).

Conidiomata 350-500 µ diam. Conidiophores 4-30 × 1-1.5 µ, with 0-1 annellations. Conidia 9.5-12 × 4.5-5.5 µ, 2 septate, straight; upper two cells medium brown, basal cell paler, transverse septa prominent; appendages absent.

On acorns of *Quercus incana*, India (IMI 177628 ex HCIO, holotype *C. biseptatum*).

S. CASSIOPES (Rostrup) Sutton, *Mycol. Pap.* 138: 76 (1975). (Fig. 157E)
Coryneum cassiopes Rostrup, *Rept. Second Norwegian arctic Exped. Fram.* 9 (1906).
Coryneum sorbi Pk, *N.Y. St. Mus. Bull. 1911*: 109 (1912).
Coryneopsis cisticola Bausá Alcalde, *An. Jard. bot. Madr.* **7**: 497 (1948).

Conidiomata up to 400 µ diam. Conidiophores 12-16 × 2.5-3.5 µ, with 1-3 annellations. Conidia 18.5-23 × 7.5-8.5 µ, 3 septate, septa prominent, periclinal walls often collapsed, straight; 2 median cells or the upper 3 cells slightly darker brown, basal and sometimes apical cell paler; appendages lacking.

On leaves of *Cassiope tetragona, Sorbus californica, Cistus populifolius*, Norway, Spain, Calif., U.S.A. (IMI 176983 ex C, holotype *C. cassiopes*, 181952 ex NYS, holotype *C. sorbi*, 180481 ex MA, holotype *C. cisticola*).

S. CAUDATUM (Preuss) Shoemaker, *Can. J. Bot.* **52**: 414 (1964). (Fig. 158C)
Sporocadus caudatus Preuss, *Linnaea* **24**: 146 (1851).
Pestalotia caudata (Preuss) Cke, *Nuovo G. bot. ital.* **10**: 27 (1878).
Cryptostictis caudata (Preuss) Sacc., *Syll. fung.* **3**: 444 (1884).
Cryptostictis ludibunda Vestergr., *Öfv. K. Vet.-Akad. Förh.* **10**: 41 (1897).
Coryneum confusum Kabát & Bub. apud Bub. & Kabát, *Hedwigia* **52**: 361 (1912).
Coryneum foliorum (Fckl) Petrak, *Fungi Polonici exsiccati Fasc.* II, nr. 26 (1920).
Hendersonia foliorum Fckl, *Jb. nassau. Ver. Naturk.* **23-24**: 392 (1870).
Leptocoryneum foliorum (Fckl) Petrak, *Hedwigia* **64**: 278 (1925).

Conidiomata 150-250 µ diam. Conidiophores 5-15 × 2-3 µ, with 0-1 annellations. Conidia 11.5-17 × 4-5.5 µ, 3 septate, straight; 2 median cells pale brown, with prominent septa, 6-9 µ long; apical cell hyaline, with or without an unbranched appendage 15-35 µ long; basal cell hyaline, with or without an unbranched appendage 13-30 µ long.

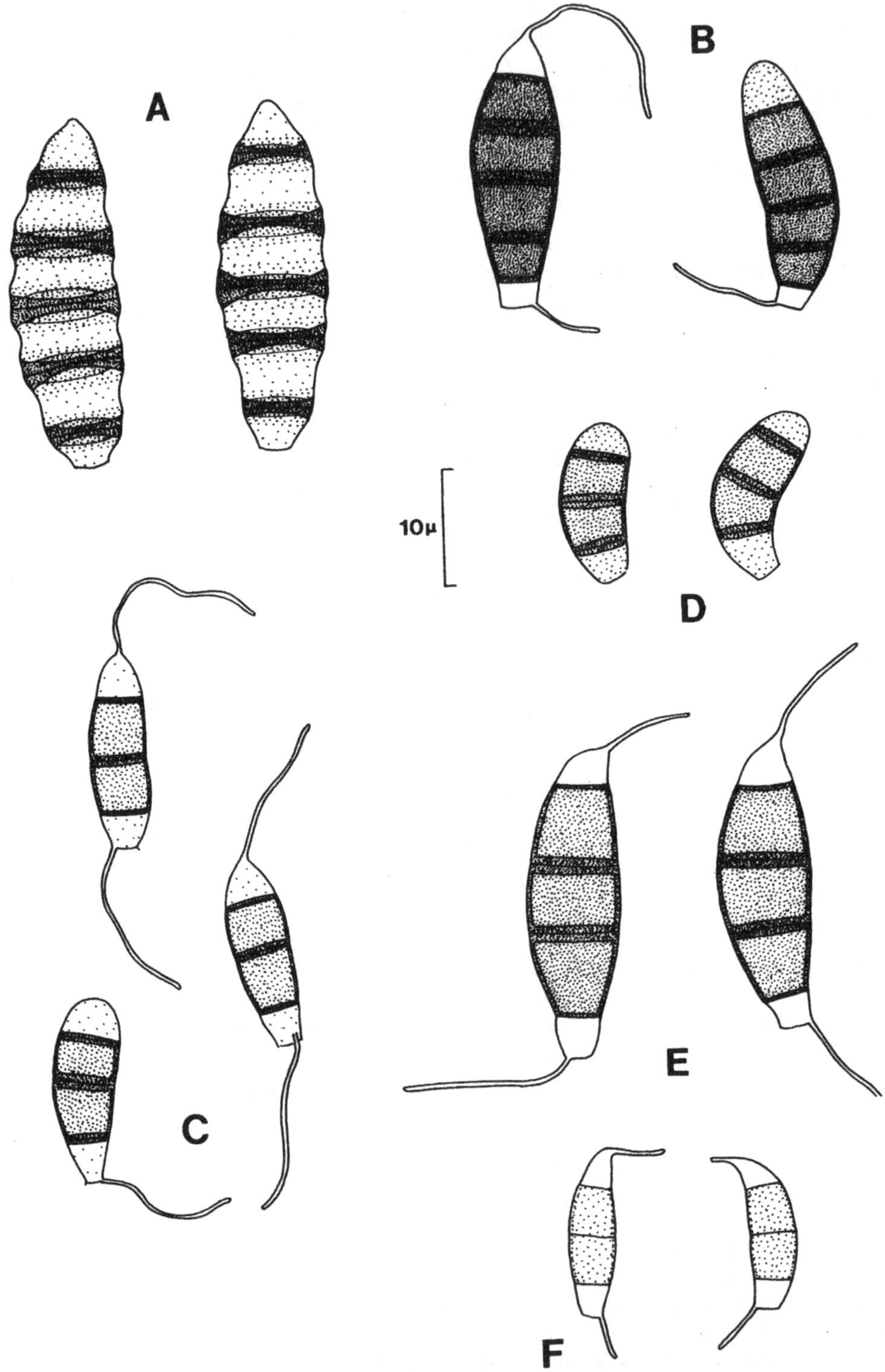

Fig. 158. A, *Seimatosporium berckmansii*, conidia; B, *S. consocium*, conidia; C, *S. caudatum*, conidia; D, *S. dacicum*, conidia; E, *S. daviesiae*, conidia; F, *S. discosioides*, conidia.

On leaves and stems of *Rosa indica*, *R.* sp., *Rubus occidentalis*, *Salix capraea*, *Thuja occidentalis*, *Rumex* sp., Pakistan, Germany, France, Sweden, Austria, N.Y., U.S.A. (IMI 85600 ex B, holotype *Sporocadus caudatus*, 93476 ex B, 93478 ex B, 85611 ex UPS, holotype *C. ludibunda*, 182381 ex G, holotype *H. foliorum*, 180331 ex K, 119455, 103540).

S. CONSOCIUM (Pk) Shoemaker, *Can. J. Bot.* **42**: 416 (1964). (Fig. 158B)
Pestalotia consocia Pk, *Rep. N.Y. St. Mus.* **39**: 48 (1886).
Monochaetia consocia (Pk) Sacc., *Syll. fung.* **18**: 485 (1906).
Cryptostictis consocia (Pk) Guba, *Monograph of Monochaetia and Pestalotia*: 280 (1961).

Conidiomata 100-150 µ diam. Conidiophores 5-10 × 2-3 µ with 0-1 annellations. Conidia 15-22.5 × 5-6 µ, 5 septate, straight or curved; median cells brown, 10-16 µ long with distinct septa; apical cell often hyaline, sometimes pale brown, lacking an appendage or with an unbranched appendage 3-18 µ long; basal cell hyaline with an unbranched appendage 2-13 µ long.

On leaves of *Hamamelis virginiana*, Ont., Canada (IMI 22467).

S. DAVIESIAE (McAlp.) Shoemaker, *Can. J. Bot.* **42**: 416 (1964). (Fig. 158E)
Amphichaeta daviesiae McAlp., *Proc. linn. Soc. N.S.W.* **29**: 118 (1904).
Cryptostictis daviesiae (McAlp.) Subram. & Ramakr., *J. Ind. bot. Soc.* **35**: 227 (1956).

Conidiomata 200-300 µ diam. Conidiophores 15-25 × 2-3 µ with 0-1 annellations. Conidia 21.5-29.5 × 5.5-8 µ, 4 septate, straight or slightly curved; median cells pale brown, 16.5-20 µ long; apical cell hyaline with an unbranched appendage 7-15.5 µ long; basal cell hyaline with an unbranched appendage 5-14 µ long.

On stems of *Daviesia latifolia*, Vict., Australia (IMI 71291 ex DAVFP, holotype).

S. DILOPHOSPORUM (Cke) Sutton, *Mycol. Pap.* 97: 35 (1964). (Fig. 159A)
Hyaloceras dilophospora Cke, *Grevillea* **19**: 5 (1890).
Diploceras dilophospora (Cke) Guba, *Monograph of Monochaetia and Pestalotia*: 287 (1961).

Conidiomata up to 300 µ diam. Conidiophores 7-17.5 × 1.5-4 µ with 1-4 annellations. Conidia 25-32 × 4-5 µ, 4 septate, straight or slightly curved; median cells pale brown, 14-20 µ long; apical cell hyaline with an appendage branched once, 10-19 µ long; basal cell with a simple or branched appendage 9-21 µ long.

On leaves of *Melaleuca squarrosa*, N.S.W., Australia (IMI 58841 ex K, holotype, 144626, 144628, 144634, 142126).

S. DISCOSIOIDES (Ell. & Ev.) Shoemaker, *Can. J. Bot.* **42**: 415 (1964). (Fig. 158F)
Pestalotia discosioides Ell. & Ev., *J. Mycol.* **4**: 51 (1888).
Monochaetia discosioides (Ell. & Ev.) Sacc., *Syll. fung.* **18**: 485 (1906).
Amphichaeta rosicola H.C. Greene, *Trans. Wisc. Sci. Arts Lett.* **47**: 127 (1959).
Coryneum microstictum Berk. & Br. var. *foliae* Dearn. & Overh., *Mycologia* **20**: 243 (1928).

Conidiomata 100-200 µ diam. Conidiophores 10-15 × 2-3 µ, with 1-2 annellations. Conidia 11.5-18 × 3.5-5 µ, 3 septate, curved; median cells pale brown, 7.5-10 µ long; apical cell with a short appendage 2-7 µ long; basal cell with a short appendage 2-6 µ long.

On leaves of *Rosa heliophila*, *R.* sp., Del., Wisc., U.S.A. (IMI 84988 ex BM, isotype *P. discosioides*, 92919 ex WIS, holotype *A. rosicola*).

S. EFFUSUM (Vestergr.) Shoemaker, *Can. J. Bot.* **42**: 416 (1964). (Fig. 159B)
Pestalotia effusa Vestergr., *Öfv. K. Vet.-Akad. Förh.* **10**: 45 (1897).
Cryptostictis effusa (Vestergr.) Guba, *Monograph of Monochaetia and Pestalotia*: 282 (1961).

Conidiomata 200-300 µ long × 100 µ wide. Conidiophores 5-15 × 1.5-2.5 µ with 0-1 annellations. Conidia 13-15.5 × 5-6.5 µ, straight, clavate, 3 septate; median cells pale brown 6.5-8 µ long; apical cell hyaline, without an appendage; basal cell hyaline with a single, irregularly branched appendage 10-20 µ long.

On twigs of *Lonicera caerulea*, Sweden (IMI 92705 ex UPS holotype).

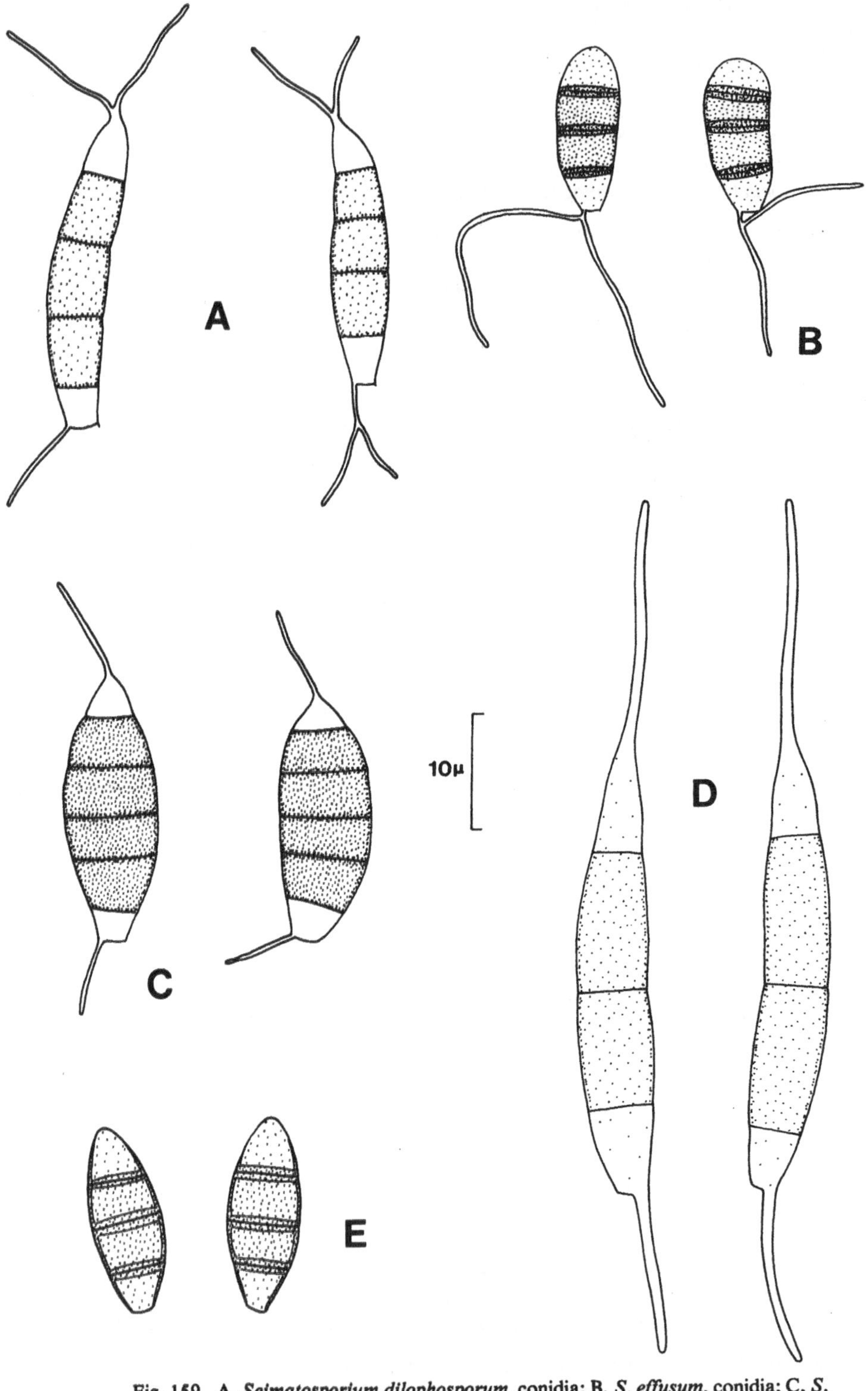

Fig. 159. A, *Seimatosporium dilophosporum*, conidia; B, *S. effusum*, conidia; C, *S. foliicola*, conidia; D, *S. falcatum*, conidia; E, *S. glandigenum*, conidia.

S. FALCATUM (Sutton) Shoemaker, *Can. J. Bot.* **42**: 416 (1964). (Fig. 159D)
Cryptostictis falcata Sutton, *Mycol. Pap.* 88: 25 (1963).

Conidiomata up to 200 µ diam. Conidiophores 5-10 × 1-2 µ, usually without any annellations. Conidia 28-37.5 × 4-5 µ, 3 septate, falcate; median cells pale brown, 15.5-29 µ long; apical cell long conic, very pale brown, with a short appendage 6-7 µ long, basal cell again very pale brown with a short appendage 6.5-10 µ long.

On leaves of *Eucalyptus radiata, E. perriniana, E.* sp., Vict., N.S.W., Australia (IMI 59166 ex K, holotype, 168731, 173941b).

S. FOLIICOLA (Berk.) Shoemaker, *Can. J. Bot.* **42**: 416 (1964). (Fig. 159C)
Podisoma foliicola Berk. in Smith, *English Flora* **5**: 362 (1836).
Hendersonia foliicola (Berk.) Fckl, *Jb. nassau. Ver. Naturk.* **23-24**: 391 (1870).
Pestalotia juniperi Rostrup, *Bot. Tidsskr.* **19**: 211 (1895).
Monochaetia juniperi (Rostrup) Allesch. in Rabenh., *Kryptogamen Fl.* **7**: 670 (1902).

Conidiomata sporodochial, punctiform, 150 µ diam. Conidiophores 5-15 × 2-3 µ, with 0-1 annellations. Conidia 18.5-22.5 × 7.5-9 µ, straight, fusiform, 5 septate; median cells pale brown, septa prominent, 13.5-17.5 µ long; apical cell hyaline with a single unbranched appendage 4-9 µ long; basal cell hyaline with an unbranched appendage 3-9 µ long.

On needles of *Juniperus communis*, U.K., ? Denmark (IMI 70725 ex K, holotype *P. foliicola*; 71567 ex C, holotype *P. juniperi*).

S. FUSISPORUM Swart & D.A. Griffiths, *Trans. Br. mycol. Soc.* **62**: 360 (1974). (Fig. 160A)

Conidiomata up to 200 µ diam. Conidiophores 10-20 × 2-3 µ, with 1-2 annellations. Conidia 16-21 × 6-7 µ, 3 septate, fusiform, straight; median cells 11-14 µ long, pale brown; apical cell very pale brown with an unbranched appendage 9-18 µ long; basal cell also very pale brown with an unbranched appendage 7-16 µ long.

From and on leaves of *Eucalyptus polyanthemos*, Vict., Australia (IMI 163446, holotype).

S. GLANDIGENUM (Bub. & Frag.) Sutton, *Mycol. Pap.* 138: 112 (1975). (Fig. 159E)
Coryneum glandigenum Bub. & Frag., *Hedwigia* **57**: 12 (1915).

Conidiomata up to 250 µ diam. Conidia 15-18 × 5-6.5 µ, fusiform, straight, 3 septate; median cells 7.5-10 µ, slightly darker brown than the end cells; appendages absent.

On acorns of *Quercus ballota*, Spain (IMI 180484 ex BPI, holotype).

S. GREVILLEAE (Loos) Shoemaker, *Can. J. Bot.* **42**: 416 (1964). (Fig. 160B)
Amphichaeta grevilleae Looş, *Trans. Br. mycol. Soc.* **33**: 41 (1950).
Cryptostictis grevilleae (Loos) Subram. & Ramakr., *J. Indian bot. Soc.* **35**: 231 (1956).

Conidiomata 150-300 µ diam. Conidiophores 10-15 × 1.5 µ, with 0-1 annellations. Conidia 18-23 × 5-7 µ, 4 septate, straight or slightly curved; median cells pale brown, 12-16 µ long; apical cell hyaline with an unbranched appendage 14-25 µ long; basal cell hyaline with an unbranched appendage 11-19 µ long.

On and from leaves of *Grevillea robusta, G. chrysophaea, G. rosmarinifolia, G.* sp., *Pinus radiata*, decaying leaves, N.C., U.S.A., Australia, New Zealand, Sri Lanka (IMI 85604 ex Tea Research Institute, Sri Lanka, holotype, 211586, 163442, 91799, 151544, 205737, 223613).

S. HAKEAE (Sutton) Shoemaker, *Can. J. Bot.* **42**: 416 (1964). (Fig. 160C)
Cryptostictis hakeae Sutton, *Mycol. Pap.* 88: 30 (1963).

Conidiomata 50-150 µ diam. Conidiophores 5-15 × 2-3 µ, with 0-1 annellations. Conidia 16-20.5 × 7-9 µ, 4 septate, straight or curved, fusiform to clavate; median cells medium brown, 13.5-16.5 µ long; apical cell hyaline with an unbranched appendage 14-31 µ long; basal cell hyaline with an unbranched appendage 20-30 µ long.

On and from leaves of *Hakea saligna, H.* sp., *Grevillea aquilifolium, Pteridium* litter, Kenya, U.K., W. Australia, S. Australia (IMI 94283 ex K, holotype, 84966 ex BM, 84972 ex BM, 216913, 163440, 96703).

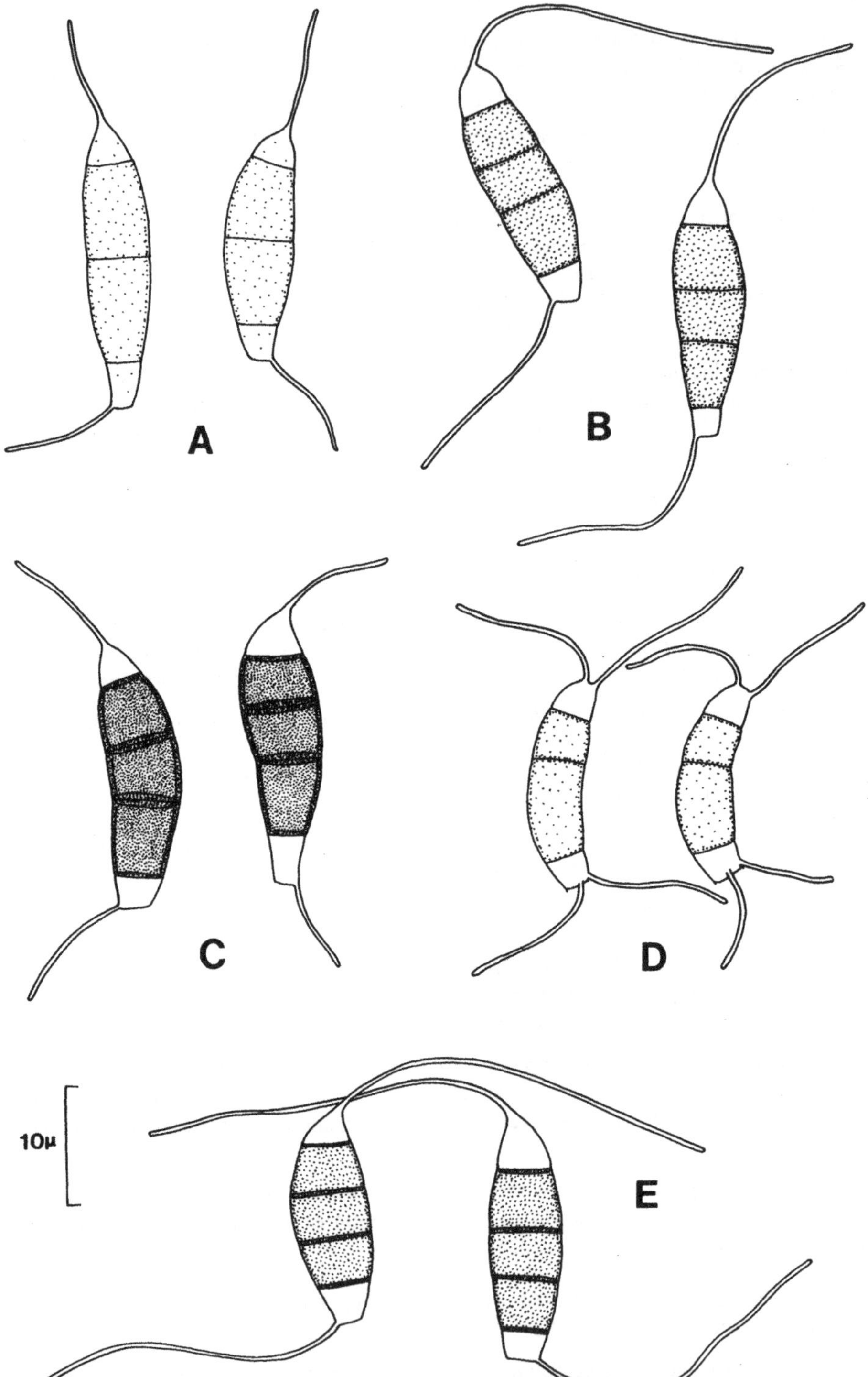

Fig. 160. A, *Seimatosporium fusisporum*, conidia; B, *S. grevilleae*, conidia; C, *S. hakeae*, conidia; D, *S. hypericinum*, conidia; E, *S. hollosii*, conidia.

S. HOLLOSII (Tóth) Shoemaker, *Can. J. Bot.* **42**: 416 (1964). (Fig. 160E)
Cryptostictis hollosii Tóth, *Ann. hist.-nat. Mus. Hung.* **52**: 103 (1960).

Conidiomata up to 250 μ diam. Conidiophores 5-15 × 2-3 μ, with 1-2 annellations. Conidia 16.5-21 × 5.5-6.5 μ, straight or curved, 4 septate; median cells medium brown, 11-14 μ long; apical cells hyaline with an unbranched appendage 17-34 μ long; basal cell hyaline with an unbranched appendage 15-30 μ long.

On branches of *Fumana procumbens*, Hungary (IMI 86605, holotype, 86606 ex BP).

S. HYPERICINUM (Ces.) Sutton, *Trans. Br. mycol. Soc.* **64**: 483 (1975). (Fig. 160D)
Pestalotia hypericina Ces., Rabenh., *Herb. Myc. II*, no. 64 (1855).
Hyaloceras hypericina (Ces.) Sacc., *Syll. fung.* **10**: 485 (1892).
Diploceras hypericina (Ces.) Died., *Krypt. Fl. M. Brandenb.* **9**: 877 (1915).

Conidiomata up to 200 μ diam. Conidiophores 4-12 × 2.5-3 μ, with 1-2 annellations. Conidia 15-18.5 × 4.5-5.5 μ, 3 septate, curved, median cells pale brown, 10-13 μ long; apical cell hyaline with two unbranched appendages 12-17 μ long; basal cell hyaline, with two unbranched appendages 7-13 μ long.

On stems of *Hypericum perforatum, H. quadrangulum*, Italy, Sweden, Germany, U.K., France (IMI 30455, 31189).

S. KENNEDYAE (McAlp.) Shoemaker, *Can. J. Bot.* **42**: 416 (1964). (Fig. 161A)
Amphichaeta kennedyae McAlp., *Proc. Linn. Soc. N.S.W.* **29**: 119 (1904).
Cryptostictis macalpinei Subram. & Ramakr., *J. Indian bot. Soc.* **35**: 231 (1956).

Conidiomata 100-200 μ diam. Conidiophores 5-15 × 2-3 μ, with 0-1 annellations. Conidia 16.5-19.5 × 6-7 μ, 4 septate, straight or curved; median cells medium brown, 11-14 μ long; apical cell hyaline, with an unbranched appendage 19-29 μ long; basal cell hyaline, with an unbranched appendage 15-22 μ long.

On *Kennedya prostrata, Hardenbergia monophylla*, Vict., Australia (IMI 71292 ex DAVFP, lectotype *A. kennedyae*, 71293 ex DAVFP, holotype *C. macalpinei*).

S. KRIEGERIANUM (Bres.) Morgan-Jones & Sutton apud Sutton, *Mycol. Pap.* 97: 37 (1964). (Fig. 161C)
Pestalotia kriegeriana Bres., *Hedwigia* **33**: 209 (1894).
Monochaetia kriegeriana (Bres.) Allesch. in Rabenh., *Krypt. Fl.* **1**(7): 669 (1902).
Hyaloceras kriegeriana (Bres.) Died., *Krypt. Fl. M. Brandenb.* **7**: 878 (1915).
Monoceras kriegeriana (Bres.) Guba, *Monograph of Monochaetia and Pestalotia*: 290 (1961).

Teleomorphosis: *Paradidymella tosta* (Berk. & Br.) Petrak, *Annls mycol.* **25**: 238 (1927).

Conidiomata 80-110 μ diam. Conidiophores 10-20 × 1-3.5 μ, with up to 3 annellations. Conidia 19.5-30 × 4-5 μ, straight or curved, 3 septate; median cells 11.5-16.5 μ long, pale brown; apical cell hyaline, with an unbranched appendage 5-16 μ long, basal cell with an unbranched appendage 7-15 μ long.

On *Chamaenerion angustifolium, Epilobium hirsutum, E. fleischeri, E.* sp., *Callistemon indicum, C. paludosum*, U.K., India, Vict., Australia, Switzerland, Germany (IMI 104570 ex B, holotype *P. kriegeriana*, 104519, 105420 ex S, 104401 ex K, 33498, 30224, 104400, 110881, 163445, 189296, 188376).

S. LEPTOSPERMI Bagnall & Sheridan, *N.Z. Jl. Bot.* **10**: 73 (1972). (Fig. 161B)

Conidiomata 350 μ diam. × 520 μ high. Conidiophores up to 18 μ long × 3 μ wide, apparently lacking annellations. Conidia extremely variable, dimensions quoted by the original authors are 36-157(92) × 3-6 (3.5) μ, 3 septate; median cells slightly darker than the end cells; apical cell pale brown, projecting into an unbranched appendage 9-86(27) μ long; basal cell of similar pigmentation, with an unbranched appendage 3-56(15) μ long.

On leaves of *Leptospermum scoparium*, N.Z. (IMI 116381, holotype).

S. LICHENICOLA (Cda) Shoemaker & Müller, *Can. J. Bot.* **42**: 405 (1964). (Fig. 161E)
Sporocadus lichenicola Cda, *Icon. fung.* **3**: 24 (1839).
Hendersonia lichenicola (Cda) Fr., *Summa Veg. Scan.* **2**: 416 (1849).
Hendersonia hysterioides (Fckl) Fckl, *Jb. nassau. Ver. Naturk.* **23-24**: 392 (1970).
Cryptostictis hysterioides Fckl, *Fungi rhenani* no. 1838 (1866).
Coryneum microstictum Berk. & Br., *Ann. Mag. nat. Hist.* **5**: 458 (1850).

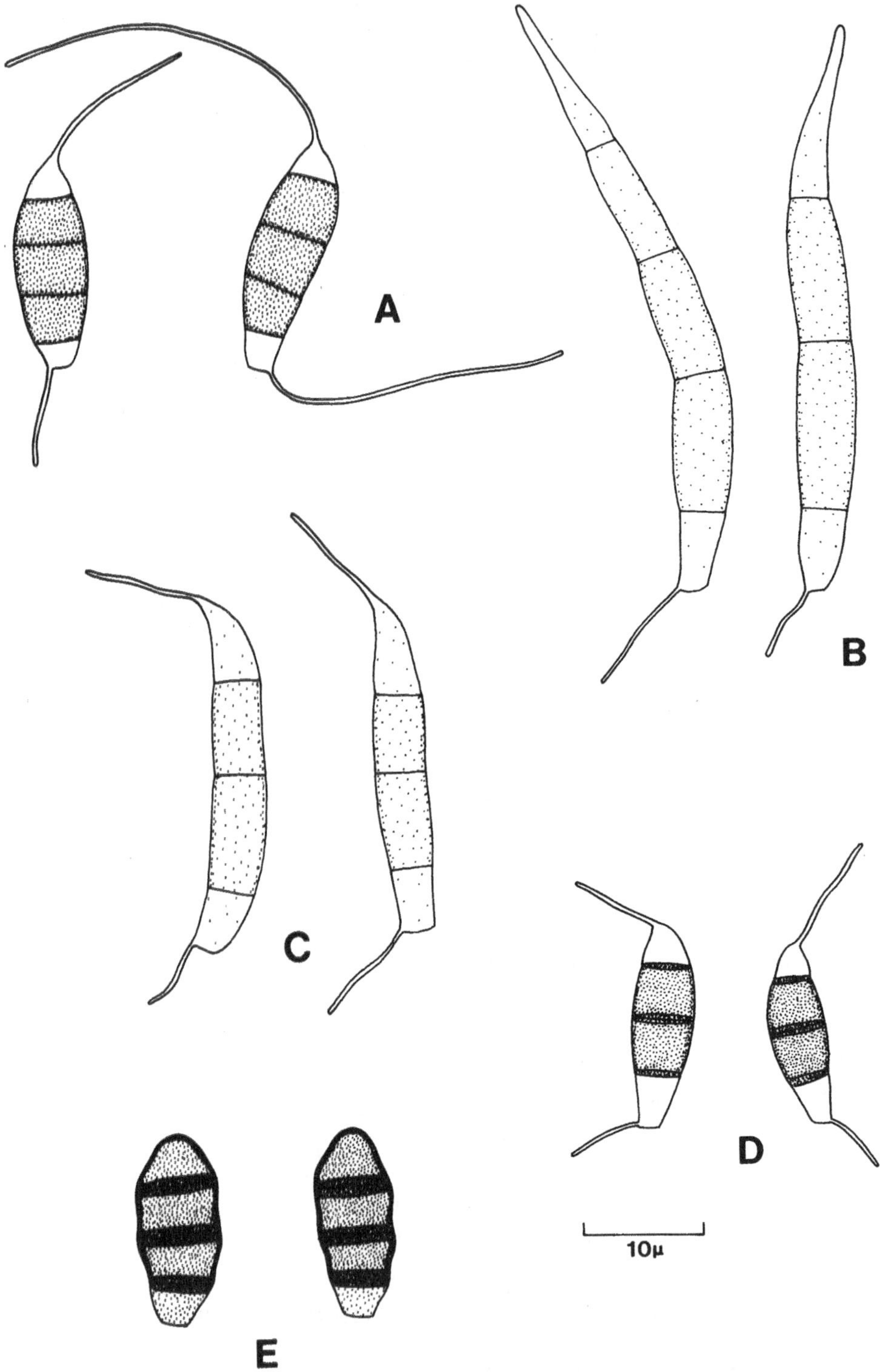

Fig. 161. A, *Seimatosporium kennedyae*, conidia; B, *S. leptospermi*, conidia; C, *S. kriegerianum*, conidia; D, *S. lonicerae*, conidia; E, *S. lichenicola*, conidia.

Leptocoryneum microstictum (Berk. & Br.) Petrak, *Hedwigia* **65**: 278 (1925).
Stilbospora microsticta (Berk. & Br.) Höhn., *Mitt. bot. Inst. tech. Hochsch. Wien* **7**: 29 (1930).
Coryneopsis microsticta (Berk. & Br.) Grove, *J. Bot. Lond.* **70**: 34 (1932).
Coryneum trotterianum Massalongo apud Sacc., *Annls mycol.* **12**: 293 (1914).
Coryneum foliicola Fckl, *Jb. nassau. Ver. Naturk.* **23-24**: 372 (1870).
Coryneopsis foliicola (Fckl) Grove, *J. Bot. Lond.* **70**: 34 (1932).
Coryneum corni-asperiifoliae Frag., *Fungi Hort. Matrit.* 86 (1917).
Coryneum microstictoides Sacc. & Penz. subsp. *epilobii* Karst., *Meddn. Soc. Fauna Flora fenn.* **13**: 164 (1888).
Coryneum epilobii (Karst.) Oud., *Enum. Syst. fung.* **4**: 142 (1923).
Coryneum microstictoides Sacc. & Penz., *Michelia* **2**: 633 (1882).
Coryneum maculicola Fckl, *Jb. nassau. Ver. Naturk.* **23-24**: 372 (1870).
Coryneum ruborum Oud., *Ned. kruidk. Archf.*, ser. 2, **6**: 295 (1894).
Hendersonia crataegi Brenckle, *Mycologia* **10**: 217 (1918).
Coryneum carbonaceum Kabát & Bub., *Annls mycol.* **14**: 346 (1916).
Coryneum sorbi Migula, *Crypt. Germ. Austr. Helv. exsicc. Fasc.* 56-57, Pilze 381 (1934).
Amphichaeta europaea Grove, *J. Bot. Lond.* **55**: 134 (1917).
Cryptostictis moravica Petrak, *Annls mycol.* **21**: 17 (1923).
Monochaetia excipuliformis Bub., *Annls mycol.* **4**: 120 (1906).
Hendersonia rubi (Westd.) Sacc., *Michelia* **1**: 214 (1878).
Coryneopsis rubi (Westd.) Grove, *British Stem- and Leaf-Fungi* **2**: 330 (1937).

Teleomorphosis: *Clethridium corticola* (Fckl) Shoemaker & Müller, *Can. J. Bot.* **42**: 404 (1964).

Conidiomata up to 350 μ diam. Conidiophores up to 27 μ long × 1.5-2.5 μ wide, with 1-3 annellations. Conidia 13-15 × 5.5-6.5 μ, 3 septate, fusiform, septa prominent, periclinal walls collapsed; median two or upper three cells darker than the basal one; appendages lacking.

Variability within the species has been discussed by Sutton (1975a) and Shoemaker & Müller (1964).

On *Kerria* sp., *Rosa canina, R.* sp., *Crataegus mollis, C.* sp., *Rubus idaeus, Laurus nobilis, Salix longifolia, Vitis* sp., *Juniperus communis, Quercus* sp., Germany, France, U.K., Eire, Turkey, Austria, N. Dak., U.S.A. (IMI 181486 ex K, lectotype *C. microstictum,* 21834, 93432b, 68246, 79706b, 118978, 6425, 21820, syntype *H. crataegi,* 92931, 80503e, 94681, 127666, 181485 ex K, 180332 ex K, 182379 ex G, 182380 ex G and more than 50 collections in herb. IMI).

S. LONICERAE (Cke) Shoemaker, *Can. J. Bot.* **42**: 417 (1964). (Fig. 161D)
Dochmolopha lonicerae Cke, *Nuovo G. bot. ital.* **10**: 25 (1878).
Cryptostictis lonicerae (Cke) Sacc., *Syll. fung.* **3**: 444 (1884).
Hendersonia lonicerae Thüm., *Mycotheca universalis* no. 578, non *H. lonicerae* Fr., *Summa Veg. Scand.* 146 (1849).
Cryptostictis physocarpi Vestergr., nom. nov., *Bot. Notiser*: 166 (1899).
Pestalotia syringae Oud., *Nederl. Kruidk. Arch.* **5**, ser. 2, 510 (1889).
Monochaetia syringae (Oud.) Allesch. in Rabenh., *Krypt. Fl.* **1**(7): 675 (1902).

Conidiomata up to 150 μ diam. Conidiophores 5-10 × 1-3 μ, with 0-1 annellations. Conidia 11-19 × 3-5 μ, 3 septate, straight; median cells medium brown, septa distinct, 6.5-10 μ long; apical cell hyaline, with a short unbranched appendage 4-10 μ long; basal cell hyaline with a similar appendage 2-9 μ long.

On twigs of *Physocarpus opulifolius, P. amurensis, Lonicera tatarica, Syringa vulgaris,* Germany, Sweden, Netherlands (IMI 84917 ex BM, 84968 ex BM and B, 99727 ex GRO, holotype *P. syringae*).

S. MACROSPERMUM (Berk. & Br.) Sutton, *Mycol. Pap.* 138: 130 (1975). (Fig. 162B)
Coryneum macrospermum Berk. & Br., *Ann. Mag. nat. Hist.* 9 (1861).
Stilbospora ulmi Grove nom. nov., *British Stem- and Leaf-Fungi* **2**: 326 (1937).
Coryneum discolor Fautrey, *Rev. mycol.* **11**: 152 (1889).

Conidiomata almost cupulate, 350 μ diam. × 175 μ deep. Conidiophores up to 37 μ long × 3μ wide with 0-1 annellations. Conidia 28-39 × 9-12.5 μ, 5 septate, straight or slightly curved; median cells 19-27 μ long, medium brown, with prominent septa; end cells paler brown; appendages absent.

On poles of *Ulmus* sp., stems of *Vitis* sp., France, U.K. (IMI 75052 ex K, holotype *C. discolor,* 74902 ex K, holotype *C. macrospermum*).

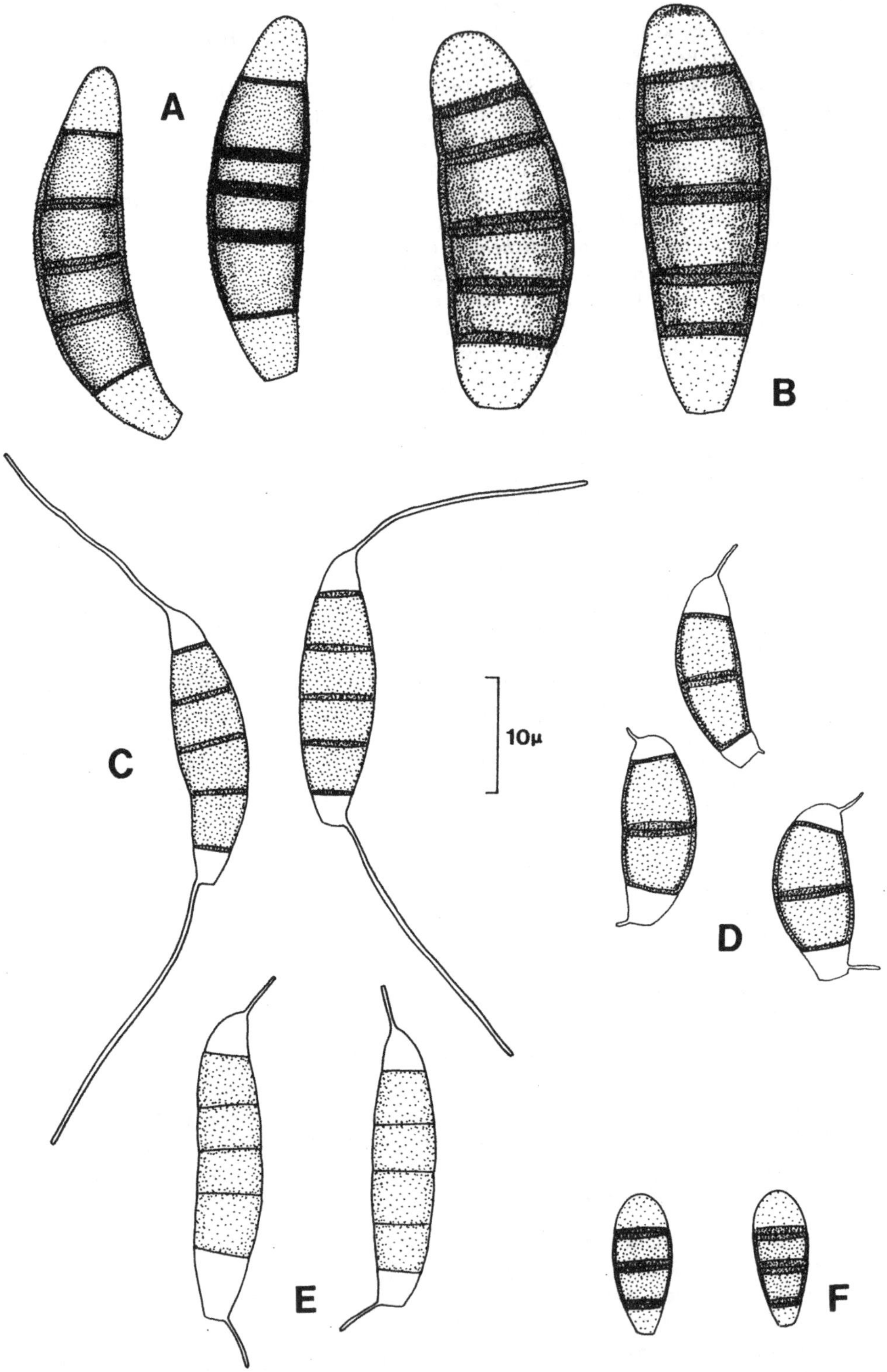

Fig. 162. A, *Seimatosporium missionum*, conidia; B, *S. macrospermum*, conidia; C, *S. mariae*, conidia; D, *S. monochaetioides*, conidia; E, *S. parasiticum*, conidia; F, *S. pestalozzioides*, conidia.

S. MARIAE (Clinton) Shoemaker, *Can. J. Bot.* **42**: 417 (1964). (Fig. 162C)
Pestalotia mariae Clinton apud Pk, *Rep. N.Y. St. Mus.* **27**: 102 (1875).
Dochmolopha clintonii Cke, *Nuovo G. bot. ital.* **10**: 25 (1878).

Conidiomata 175-250 µ diam. Conidiophores 15-30 × 2-3 µ, with 0-1 annellations. Conidia 19-23 × 6-7 µ, straight, fusiform, 5 septate; median cells 14-17 µ long, pale brown; apical cell hyaline, with an unbranched appendage 3-25 µ long; basal cell hyaline, with a similar appendage 2-30 µ long.

On *Rhododendron californicum, R.* sp., *Correa reflexa*, Calif., U.S.A., Vict., Australia (IMI 92426, 85098 ex K, 168734).

S. MISSIONUM (Speg.) Sutton, *Mycol. Pap.* 138: 139 (1975). (Fig. 162A)
Coryneum missionum Speg., *An. Mus. nac. Hist. B. Aires* **20**: 410 (1910).

Conidiomata up to 200 µ diam. Conidiophores up to 14 µ long × 2.5-5 µ, with 0-2 annellations. Conidia 26-32.5 × 7-9 µ, curved, 5 septate; median cells 17-21.5 µ long, medium brown, verruculose; apical and basal cell paler brown; appendages absent.

On leaves of *Pithecellobium contortisiliquum*, Argentina (IMI 176869 ex LPS, holotype).

S. MONOCHAETIOIDES (Speg.) Sutton, *Mycol. Pap.* 138: 140 (1975). (Fig. 162D)
Coryneum monochaetioides Speg., *Revta chil. Hist. nat.* **28**: 28 (1924).

Conidiomata up to 120 µ diam. Conidiophores 5-22 × 1.5-3 µ, with 0-1 annellations. Conidia 15.5-18 × 5.5-6.5 µ, 3 septate, straight or slightly curved; median cells 11-13 µ long, medium brown, septa prominent; apical cell hyaline, with an appendage 1-3 µ long; basal cell hyaline, with an appendage 1-2 µ long.

On leaves of *Aristotelia maqui*, Chile (IMI 176868 ex LPS, holotype).

S. PARASITICUM (Dearn. & House) Shoemaker, *Can. J. Bot.* **42**: 417 (1964). (Fig. 162E)
Pestalotia monochaetioidea Sacc. & Ell. var. *parasitica* Dearn. & House, *Bull. N.Y. St. Mus.* 76: 242 (1923).
Cryptostictis parasitica (Dearn. & House) Guba, *Monograph of Monochaetia and Pestalotia*: 300 (1961).

Conidiomata up to 100 µ diam. Conidiophores 5-10 × 2-4 µ, with 0-1 annellations. Conidia 24-31 × 4.5-6.5 µ, 5 septate, often irregularly curved; median cells 17-24 µ long, pale brown; apical cell hyaline, sometimes with an appendage 1-8.5 µ long; basal cell hyaline, with an appendage 1.5-9 µ long.

On leaves of *Physocarpus opulifolius*, branches of *Vitis vinifera*, Germany, Pakistan (IMI 93477 ex B, 133451, 133450).

S. PESTALOZZIOIDES (Sacc.) Sutton, *Mycol. Pap.* 138: 155 (1975). (Fig. 162F)
Coryneum pestalozzioides Sacc., *Michelia* **1**: 91 (1877).
Coryneum tumoricola Pk, *40th Ann. Rept St. Mus. N.Y. 1886*: 61 (1920).

Conidiomata up to 150 µ diam. Conidiophores 8-25 × 1.5-2 µ, with 0-1 annellations. Conidia 11-15 × 4.5-5.5 µ, 3 septate, septa prominent, straight, ellipsoid to fusiform; median cells pale brown, 5-8 µ long; apical and basal cell paler brown; appendages absent.

On leaves of *Crataegus oxyacantha*, *Ulmus* sp., *Salix aurita*, Italy, Austria, N.Y., U.S.A. (IMI 179085 ex PAD, holotype *C. pestalozzioides*, 181671 ex NYS, holotype *C. tumoricola*, 182374b ex G).

S. PEZIZOIDES (Ell. & Ev.) Sutton, *Mycol. Pap.* 138: 157 (1975). (Fig. 163A)
Coryneum pezizoides Ell. & Ev., *Bull. Torrey bot. Cl.* **24**: 292 (1897).

Conidiomata sporodochial to disciform, up to 400 µ wide × 100 µ deep. Conidiophores 12-23 × 2.5-3 µ. Conidia 22-31 × 8.5-10.5 µ, dark brown, clavate to fusiform, 3 septate, septa prominent and very dark brown, periclinal walls collapsed; appendages absent.

On *Salix* sp., indet. bark, Col., U.S.A., Pakistan (IMI 75077 ex K holotype, 139948).

S. RHODODENDRI (Schw.) Pirozynski & Shoemaker, *Can. J. Bot.* **48**: 2202 (1970). (Fig. 163C)
Coryneum rhododendri Schw., *Trans. Amer. Phil. Soc.* **2A**: 307 (1832).
Coryneum triseptatum Pk, *Ann. Rep. N.Y. St. Mus.* **27**: 102 (1875).

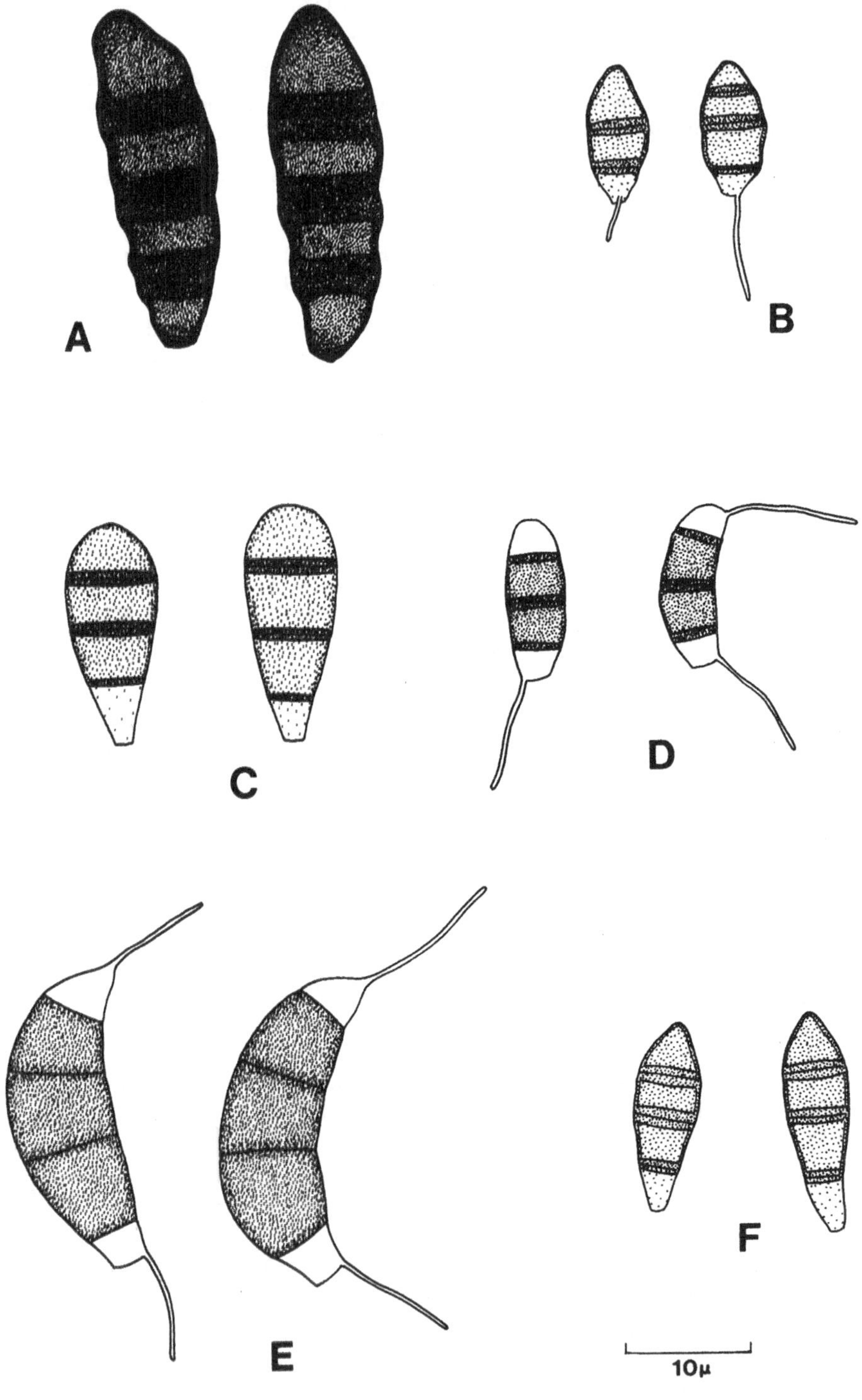

Fig. 163. A, *Seimatosporium pezizoides*, conidia; B, *S. rosarum*, conidia; C, *S. rhododendri*, conidia; D, *S. rosae*, conidia; E, *S. sublunatum*, conidia; F, *S. vaccinii*, conidia.

Conidiomata up to 250 µ diam. Conidiophores 12-25 × 1-2 µ, with up to 5 annellations. Conidia 15.5-20 × 6.5-8.5 µ, 3 septate, clavate; subapical cell thick-walled, brown, apical and lower median cell paler, basal cell sub-hyaline; appendages absent.

From leaves of *Pyrus communis*, N.Z. (IMI 151958).

S. ROSAE Cda in Sturm's *Deutschl. Fl.* III, **3**: 79 (1833). (Fig. 163D)
Hendersonia cynosbati Fckl, *Jb. nassau. Ver. Naturk.* **23-24**: 392 (1870).
Cryptostictis cynosbati (Fckl) Sacc., *Syll. fung.* **3**: 443 (1884).

Conidiomata 250-400 µ diam. Conidiophores 5-15 × 1.5-3 µ, with 0-1 annellations. Conidia 10-15 × 3-4 µ, 3 septate, straight or curved; median cells medium brown with distinct septa, 5-9 µ long; apical cell hyaline or pale brown, with or without an unbranched appendage 1-6.5 µ long; basal cell hyaline, with an unbranched appendage 1-8 µ long.

On *Rosa canina*, *R. pimpinellifolia*, *R. moschata*, *R.* sp., *Rhododendron* sp., *Sedum reflexum*, *Rubus* sp., *Paeonia moutan*, Czechoslovakia, France, Germany, Switzerland, Pakistan (IMI 83916 ex P, 84978 ex BM, 93433 ex P, 96260 ex GRO, holotype *H. cynosbati*, 141925, 137028, 139912, 139903).

S. ROSARUM (P. Henn.) Sutton, *Mycol. Pap.* 138: 168 (1975). (Fig. 163B)
Coryneum rosarum P. Henn., *Nyt. Mag. Naturvid.* **42**: 32 (1904).

Conidiomata 175-200 µ diam. Conidia 9.5-12 × 3-5 µ, 2-3 septate, septa prominent or not, fusiform, straight or curved; two upper cells pale brown, lower cell paler; apical cell without an appendage; basal with or without an unbranched appendage 2-10 µ long.

On leaves of *Rosa canina*, *R.* sp., Norway, U.K. (IMI 179738 ex B, holotype *C. rosarum*, 181485).

S. SUBLUNATUM Sutton, *Mycol. Pap.* 97: 39 (1964). (Fig. 163E)

Conidiomata up to 200 µ diam. Conidiophores 5-15 × 1-2 µ, with 1-3 annellations. Conidia 20-25.5 × 7-8 µ, 4 septate, strongly curved; median cells reddish brown, 15-19 µ long; apical cell hyaline with an unbranched appendage 8-11.5 µ long; basal cell hyaline, with an unbranched appendage 6-11 µ long.

From leaves of *Carissa carandas*, India (IMI 104265, holotype).

S. VACCINII (Fckl) Eriksson, *Svensk bot. Tidskr.* **68**: 242 (1974). (Fig. 163F)
Coryneum vaccinii Fckl, *Jb. nassau. Ver. Naturk.* **27-28**: 81 (1873).
Coryneum microstictum Berk. & Br. var. *staphyleae* W. Davis, *Phytopath.* **21**: 1172 (1931).
Coryneum rhododendri Schw. var. *fusoideum* Dearn., *Mycologia* **33**: 364 (1941).

Conidiomata up to 100 µ diam. Conidiophores up to 28 µ long × 1.5-2 µ wide, with 1-2 annellations. Conidia 13-18 × 4.5-5.5 µ, 3 septate, septa prominent, periclinal wall often collapsed, fusiform; median or the 3 upper cells darker than the basal cell; appendages absent.

On stems of *Vaccinium myrtillus*, *V.* sp., *Rhododendron catawbiense*, *R.* sp., *Staphylea trifolia*, *Crataegus oxyacantha*, Switzerland, N.Z., Tenn., Mass., U.S.A., U.K. (IMI 182373 ex G, holotype *C. vaccinii*, 170068, 181440 ex TENN, holotype *C. rhododendri* var. *fusoideum*, 181438, 181439 ex TENN, 181173, 181172, 73466a).

14. MACROPHOMOPSIS Petrak, *Annls mycol.* **22**: 108 (1924).

Sp. typ.: *M. coronillae* (Desm.) Petrak (syn. *Sphaeria coronillae* Desm.).

Mycelium immersed, branched, septate, pale brown. *Conidiomata* eustromatic, immersed, subperidermal, separate, dark brown, multilocular to convoluted; wall thick, composed of pale brown thin-walled textura angularis, darker and more thick-walled near the ostiole. *Ostiole* single, circular, central. *Conidiophores* hyaline, branched and septate only at the base, cylindrical, straight to irregular, formed from the inner cells of the wall. *Conidiogenous cells* holoblastic, indeterminate, discrete or integrated, hyaline, smooth, with 1-2 apical annellations. *Conidia* hyaline, thin-walled, aseptate, fusiform, straight, eguttulate, apex obtuse, base truncate.

Petrak & Sydow (1927) quickly removed *Macrophomopsis* and placed it as a synonym of *Dothiorella* Sacc. but this redisposition cannot be upheld because of their erroneous interpretation of the typification of *Dothiorella*. Clements & Shear (1931) regarded *Macrophomopsis* as a synonym of *Phomopsis* but this is quite clearly incorrect. *M. coronillae* is most closely related to the fungus called *Fusicoccum aesculi* Cda by Saccardo (1880a) but differs in having annellidic conidiogenous cells.

M. CORONILLAE (Desm.) Petrak, *ibid.*: 108 (1924). (Fig. 164)
Sphaeria coronillae Desm., *Ann. Sci. nat.*, 2 ser., **13**: 188 (1840).
Macrophoma coronillae (Desm.) Neger, *Ber. dtsch. Bot. Ges.* **28**: 479 (1910).
Dothiorella coronillae (Desm.) Petrak apud Petrak & Sydow, *Beih. Rep. spec. nov. regn. veg.* **42**: 217 (1927).
Macrophoma leguminarum Jaap, *Annls mycol.* **14**: 31 (1916).

Conidiomata up to 1100 µ diam. × 800 µ deep. Conidiophores 12-27.5 × 3-4 µ. Conidia 24-28 × 6.5-7.5 µ.

On galls of *Asphondylia* on pods of *Sarothamnus scoparius*, Germany (IMI 201967 ex K).

15. DISCULINA Höhn., *Sber. Akad. Wiss. Wien* **125**: 104 (1916).
Macmillanina O. Kuntze, *Rev. Gen. Pl.* **3**: 490 (1898).
Sp. typ.: *D. vulgaris* (Fr.) Sutton (syn. *Cryptosporium vulgare* Fr., *D. neesii* (Cda) Höhn.).

Mycelium immersed, hyaline, branched, septate. *Conidiomata* eustromatic, subperidermal, separate, rarely confluent, at the base formed of hyaline, thin-walled textura angularis, at the sides and above composed of pale brown, thin-walled cells. Dehiscence irregular. *Conidiophores* hyaline, septate, branched at the base, irregular, tapered towards the apices, formed mainly from the basal wall but sometimes from the upper walls. *Conidiogenous cells* holoblastic, annellidic, integrated or discrete, indeterminate, hyaline, smooth, with 1-2 percurrent proliferations. *Conidia* falcate, hyaline, aseptate, thin-walled, smooth, base truncate, apex obtuse, eguttulate.

In taking up the name *Cryptosporum* Kze, Fries (1832) decided to change the name *C. neesii* Cda to *C. vulgare* because he believed the species to occur on more than one host. *C. vulgare* has legislated priority and should be used instead of *C. neesii* as a basionym for combinations into other genera. Petrak (1925) regarded *Disculina* as a 'melanconioid Phomopsideae' and placed it erroneously as a synonym of *Ascochytopsis*.

The genus contains 4 species but only the type species has been examined.

D. VULGARIS (Fr.) Sutton, *Mycol. Pap.* 141: 75 (1977). (Fig. 165)
Cryptosporium vulgare Fr., *Syst. myc.* **3**: 482 (1832).
C. neesii Cda apud Sturm, *Flora Deutschlands* **3**: 109 (1837).
Disculina neesii (Cda) Höhn., *Sber. Akad. Wiss. Wien* **125**: 104 (1916).

Teleomorphosis: *Ophiovalsa suffusa* (Fr.) Petrak, *Sydowia* **19**: 272 (1966).

Conidiomata up to 900 µ diam. Conidiophores up to 30 µ long × 4 µ wide at the base. Conidia 25-45 × 4-4.5 µ.

On dead twigs and branches of *Alnus incana*, *A. glutinosa*, *A.* spp., U.K., Latvia, Hungary, Austria (IMI 37200, 33487, 17256, 105286, 173497, 341, 90628, 33979, 35338, 14074).

16. SCAPHIDIUM Clem., *Bot. Survey Nebr.*: 5 (1901).
Sp. typ.: *S. boutelouae* Clem.

Mycelium immersed, branched, septate, pale brown, concentrated in palisade tissue surrounding vascular bundles. *Conidiomata* like solitary pycnidia or loosely aggregated into linear eustromata; when pycnidial, separate, conical, brown, immersed, scarcely papillate, subepidermal, unilocular or convoluted, thick-walled; composed of brown, thick-walled textura angularis, somewhat thinner and paler at the base; when stromatic, linear between veins, consisting of brown pseudoparenchyma in the palisade tissue connecting the locules. *Ostioles* single to each locule or pycnidium, central, circular. *Conidiophores* absent. *Conidiogenous cells* holoblastic, annellidic, determinate, discrete, cylindrical to lageniform, hyaline to pale brown, smooth, formed from the inner cells of the locular wall. *Conidia* hyaline, medianly 1-euseptate, not constricted, apex obtuse, base ± truncate, thin-walled, smooth, eguttulate, fusiform, straight or slightly curved.

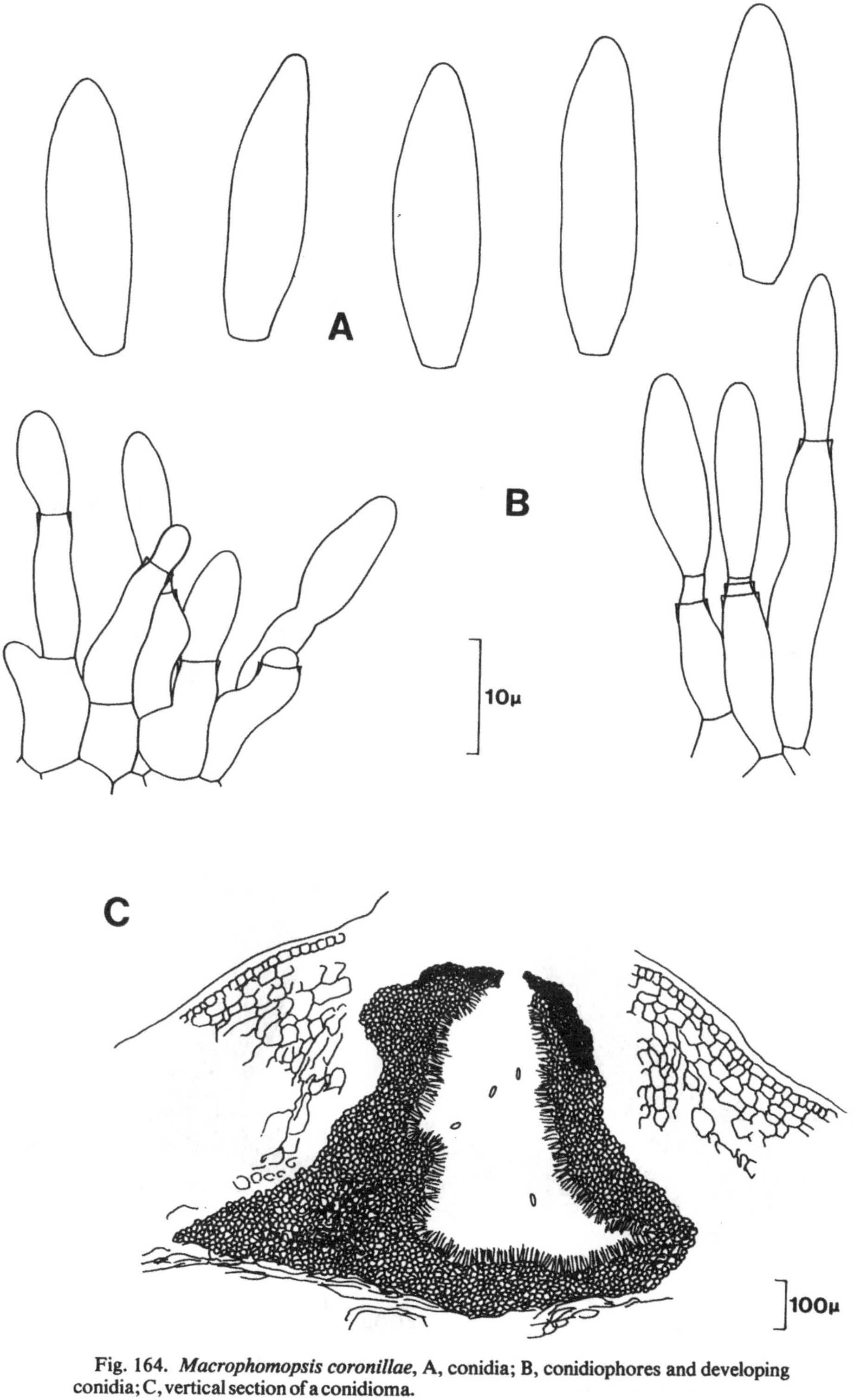

Fig. 164. ***Macrophomopsis coronillae***, A, conidia; B, conidiophores and developing conidia; C, vertical section of a conidioma.

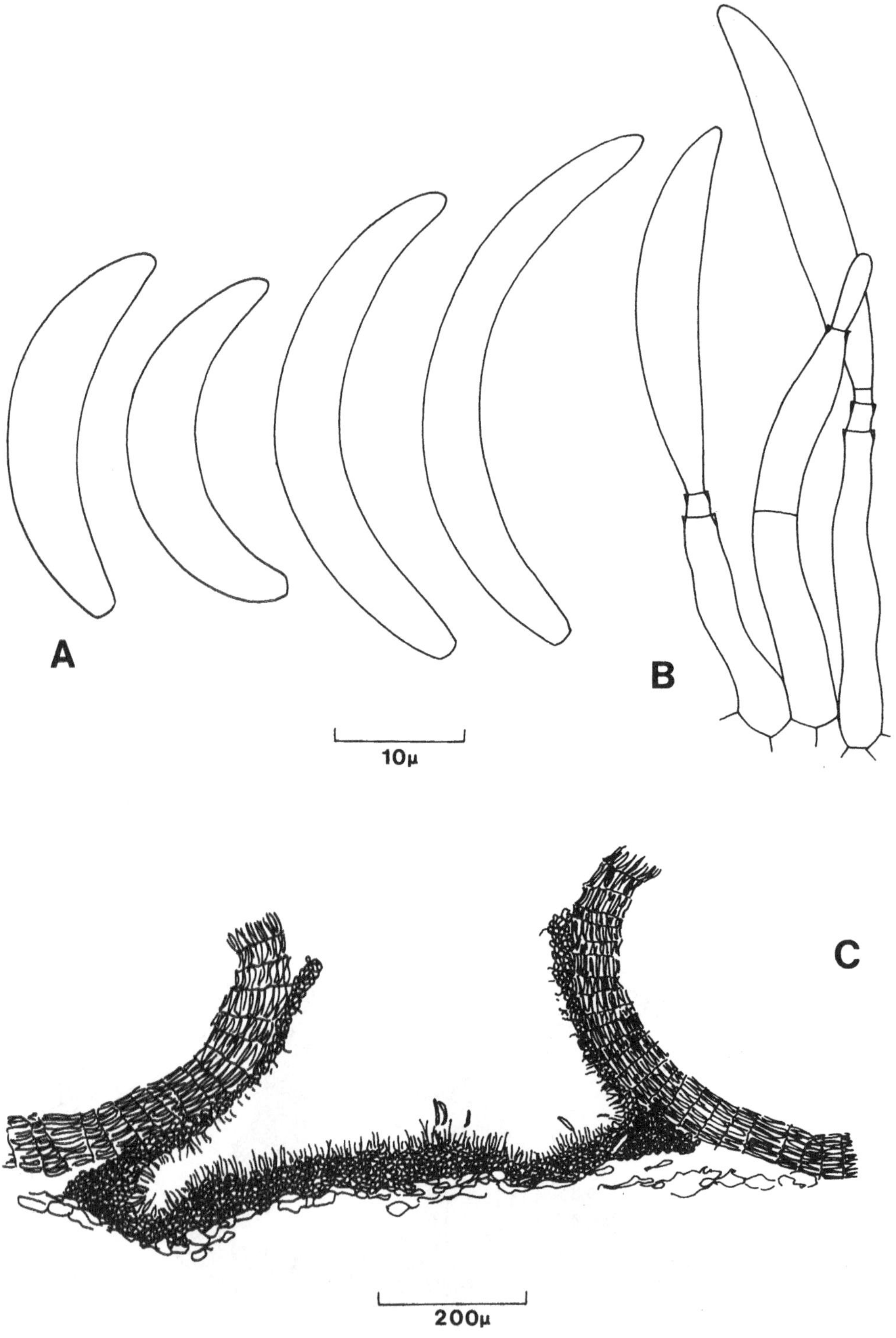

Fig. 165. ***Disculina vulgaris***, A, conidia; B, conidiophores and developing conidia; C, vertical section of a conidioma.

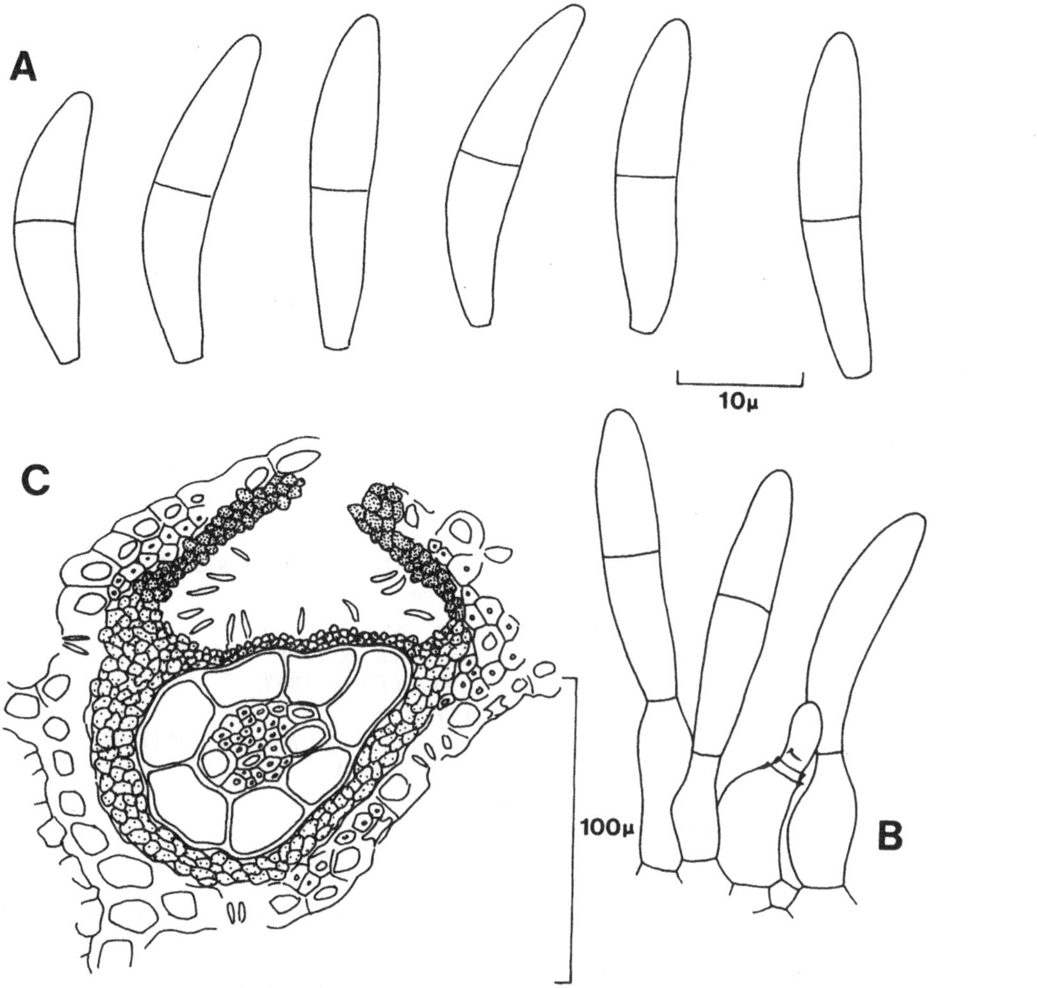

Fig. 166. *Scaphidium boutelouae*, A, conidia; B, conidiogenous cells and developing conidia; C, vertical section of a conidioma.

The genus is monotypic.

S. BOUTELOUAE Clem., *ibid.*: 5 (1901). (Fig. 166)

Individual locules up to 100 µ wide. Conidiogenous cells 9-13 × 4-5.5 µ. Conidia 22-28.5 × 4.5-5 µ.

Causing lesions on leaves of *Bouteloua curtipendula*, N. Mex., U.S.A. (IMI 182785).

17. MARSSONINA Magnus nom. nov., *Hedwigia* **45**: 2 (1906), nom. cons. prop.
Gloeosporium Desm. & Mont., *Ann. Sci. nat.* **12**: 295 (1849), nom. rej. prop.
Marssonia Fischer, *Fungi europaei exsiccati cent.* XIX, no. 1857 (1874).
Marssoniella Höhn., *Sber. Akad. Wiss. Wien* **125**: 108 (1916).
Sp. typ.: *M. fragariae* (Lib.) Kleb. (syn. *Leptothyrium fragariae* Lib., *M. potentillae* (Desm.) Magnus).

Mycelium immersed, hyaline to pale brown, branched, septate. *Conidiomata* acervular, subcuticular, dark brown to black, amphigenous, separate, rarely confluent, sometimes associated with a dendroid subcuticular subicle, pseudoparenchymatic. Dehiscence irregular. *Conidiophores* hyaline, branched irregularly, mostly lateral, septate, smooth, comprising the subcuticular acervuli. *Conidiogenous cells* holoblastic, annellidic, sympodial and indeterminate, or monoblastic and determinate, integrated or discrete, doliiform or cylindrical, hyaline, smooth. *Conidia* hyaline, 1-septate, separating equal or unequal cells, smooth, base truncate, guttulate, apical cells obtuse or subacute, sometimes straight or curved, ± constricted at the septum.

Höhnel (1910d, 1916b) showed that the type species of *Gloeosporium, G. castagnei* Desm. & Mont., has 1-septate conidia and that it had been referred to *Marssonina* by Magnus (1906). *Gloeosporium* antedates *Marssonina* by several years, and if the respective type species are congeneric, as seems likely, the fungi at present placed in *Marssonina* should be transferred to *Gloeosporium* in accordance with nomenclatural priority. Arx (1957a) considered the two genera to be identical but argued for the conservation of *Marssonina* against *Gloeosporium* to preserve current usage. The fungi hitherto placed in *Gloeosporium*, the majority of which are not congeneric with *G. castagnei*, have been removed to other genera, notably by Höhnel (1916b) and Arx (1957a). There has not been any modern revision of *Marssonina* or the more than 150 taxa described in it. There is wide variation in conidial morphology, structure of conidiomata and their relationships with the substrates in which they occur. It therefore seems probable that *Marssonina* will eventually be fragmented into smaller generic units particulary if differences in conidiogenesis are correlated with such morphological distinctions.

Some species have been correlated with *Diplocarpon* states (Wolf 1912, 1924) others with *Drepanopeziza* (Rimpau, 1962).

M. FRAGARIAE (Lib.) Kleb., *Ascomyceten*: 288 (1918). (Fig. 167)
Leptothyrium fragariae Lib., *Crypt. Ard. exs.* 163 (1832).
Gloeosporium fragariae (Lib.) Mont., *Ann. Sci. nat.* **12**: 296 (1849).
Marssonina potentillae (Desm.) Magnus forma *fragariae* (Lib.) Ohl, *Bot. Rast.* **4**: 15 (1910).
Phyllosticta potentillae Desm., *Ann. Sci. nat.* **8**: 31 (1847).
Marssonina potentillae (Desm.) Magnus, *Hedwigia* **45**: 88 (1906).
Gloeosporium potentillae (Desm.) Oud., *Ned. Kruidk. Arch.*, ser. 2, **1**: 259 (1873).
Marssonia potentillae (Desm.) Fischer in Rabenh., *Fungi Europ.* 1857.
Marssonina fragariae Sacc., *Malpighia* **10**: 276 (1896).
Marssonina potentillae (Desm.) Fischer var. *fragariae* Sacc., *Syll. fung.* **13**: 496 (1898).
Leptothyrium dryadearum Desm., *Ann. Sci. nat.* **9**: 277 (1848).
Ascochyta fragariae Sacc., *Michelia* **1**: 169 (1877).
Ascochyta colorata Pk, *Rept N.Y. St. Mus.* **38**: 94 (1884).
Septoria potentillarum Fckl, *Jb. nassau. Ver. Naturk.* **23-24**: 196 (1870).

Teleomorphosis: *Diplocarpon earlianum* (Ell. & Ev.) Wolf, *J. Elisha Mitchell Sci. Soc.* **39**: 158 (1924).

Conidia curved, with a truncate base and an acute apex, 16.5-29 × 5.5-8 µ.

Although Wolf (1924) regarded the species on *Fragaria* to be distinct from that on *Potentilla*, neither Vassiljevsky & Karakulin (1950) nor von Arx (1957a; 1970) considered them to be so.

On leaves of *Potentilla anserina, P. procumbens, P. fissa, P. reptans, P. palustris, P.* sp., *Rubus saxatilis, Fragaria vesca, F. virginiana, F. californica, F. grandiflora, F. moschata, F.* sp., Jamaica, Malaysia, S. Rhodesia, Hong Kong, Hungary, U.S.S.R., Roumania, Ont., Canada, U.K., Wyo., U.S.A., Germany, Austria, Sweden, Zambia (IMI 22321, 22319, 17243, 30222, 13030, 30027, 30564, 182794, 22320, 98279, 159741, 178342, 168524, 159740, 24818, 24692, 57098, 59534, 100801, 101170, 114824, 16284, 159768, 129731, 33563, 63029, 97173, 154419, 172090, 172091).

M. ROSAE (Lib.) Died., *Krypt. Fl. Mk Brandenb.* **9**: 830 (1915). (Fig. 168)
Asteroma rosae Lib., *Mem. Soc. Linn.* **5**: 404 (1827).
Actinonema rosae (Lib.) Fr., *Summa Veg. Scand.* 424 (1849).
Marssonia rosae Trail, *Fung. Inverar.*: 46 (1889).
Dicoccum rosae Bonord., *Bot. Zeit.* 282 (1853).
Marssonia rosae (Bonord.) Briosi & Cav., *Funghi parassiti della coltivate od utile exs.* 97 (1889).

Teleomorphosis: *Diplocarpon rosae* Wolf, *Bot. Gaz.* **54**: 231 (1912).

Conidiomata often associated with subcuticular fibrils. Conidia slightly curved, base truncate, apex obtuse, 13.5-16.5 × 4.5-5.5 µ.

This species is more commonly known under the *Actinonema* name, but Sutton (1977a) has shown that it is inapplicable to the fungus causing black spot of rose leaves.

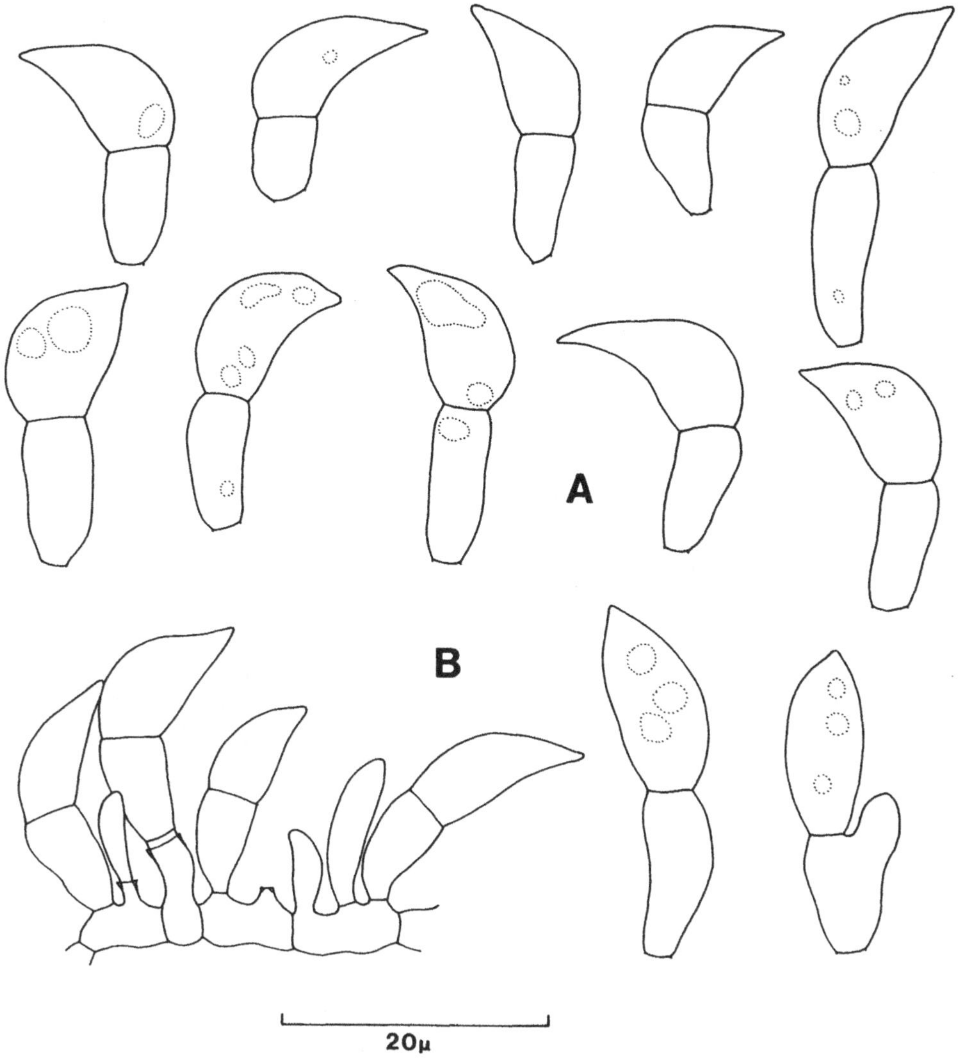

Fig. 167. *Marssonina fragariae*, A, conidia; B, conidiogenous cells and developing conidia.

On leaves of *Rosa canina, R. damascena, R. indica, R.* sp., U.K., Germany, Venezuela, India, Pakistan, Hong Kong, S. Rhodesia, Thailand, Austria, Hungary, Roumania, Italy, Sierra Leone, Nigeria, Zambia, Ghana, Papua New Guinea, Cuba, Jamaica, U.S.A., U.S.S.R., Gambia, Libya, New Zealand, Ethiopia, Bolivia, Turkey, Malawi, Nepal, Australia, Uruguay, Malaysia, represented in herb. IMI by 78 collections.

18. PSEUDOCENANGIUM Karst., *Meddn. Soc. Fauna Flora fenn.* **13**: 163 (1886), non *Pseudocenangium* Knapp, *Schw. Z. Pilzk.* **2**: 52 (1924).

Siroscyphella Höhn., *Sber. Akad. Wiss. Wien* **119**: 650 (1910).

Sp. typ.: *P. pinastri* Karst.

Illustrations: Dyko & Sutton (1979b).

Mycelium immersed, branched, septate, hyaline to pale brown. *Conidiomata* superficial, separate, dark or pale brown, at first closed, later opening widely to become cupulate; basal wall consisting of a central mass of brown textura angularis which merges into the periclinal, often incurved wall of pale brown to hyaline textura intricata in gelatin. *Ostiole* absent. *Conidiophores* hyaline, repeatedly branched penicillately or irregularly at the base and above, septate, smooth, cylindrical

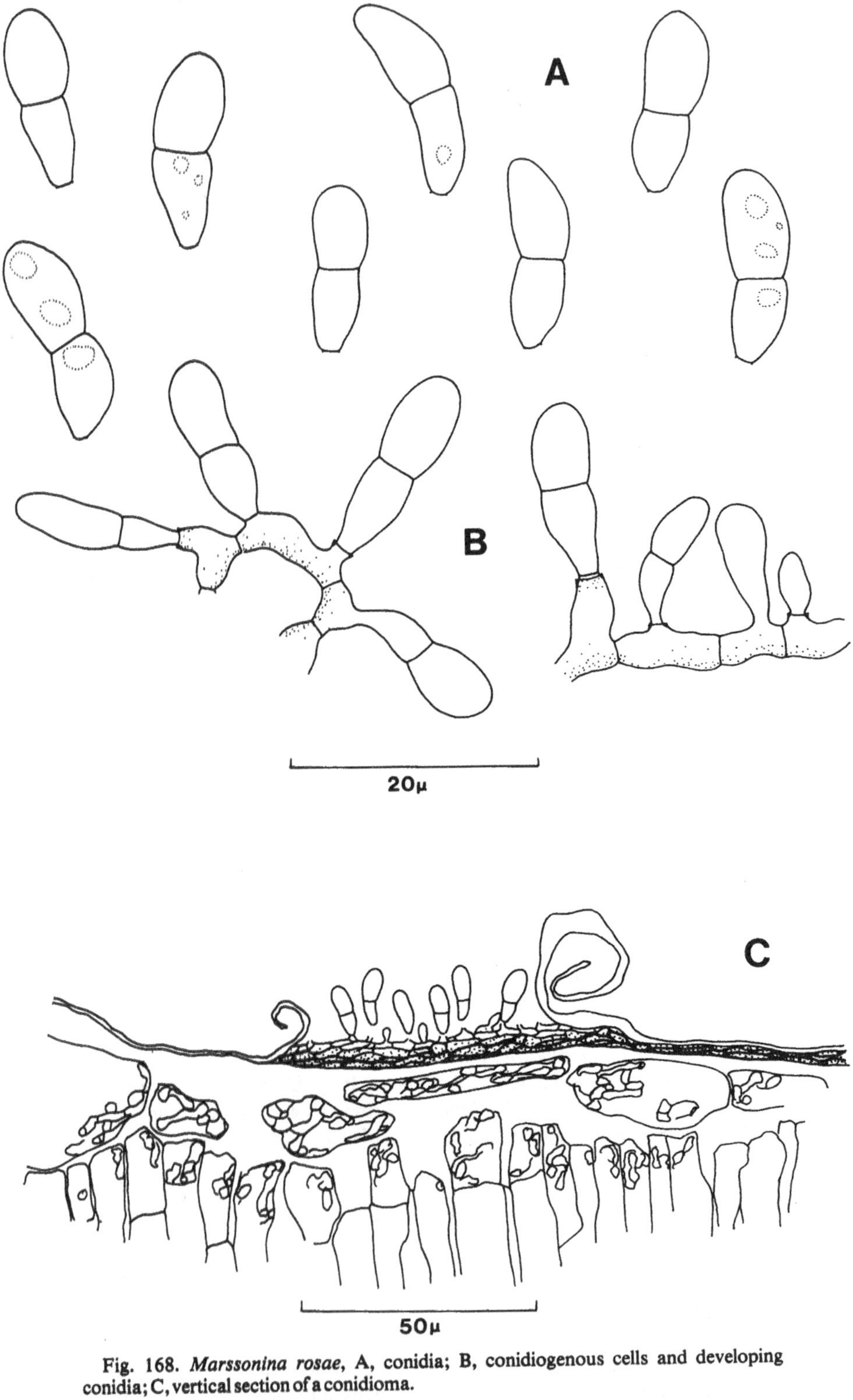

Fig. 168. *Marssonina rosae*, A, conidia; B, conidiogenous cells and developing conidia; C, vertical section of a conidioma.

or curved, formed from loose, irregularly branched cells at the base and sides of the conidiomata. *Conidiogenous cells* holoblastic, annellidic, with up to 3 indistinct apical percurrent proliferations, integrated, indeterminate, hyaline, terminal or lateral, cylindrical. *Conidia* hyaline, septate, thin-walled, ± guttulate, straight, cylindrical or filiform, truncate at the base, obtuse and occasionally tapered at the apex.

Pseudocenangium and related genera such as *Linodochium* and *Septopatella* have been revised by Dyko & Sutton (1979b) leading to the acceptance of only two species in the genus.

KEY

Conidia 0-1 septate, 26-33 × 1 µ, with acute apices .... *P. pinastri*
Conidia 0-1 septate, 12-18 × 2 µ, apices obtuse .... *P. succineum*

P. PINASTRI Karst., *ibid.*: 164 (1886).

Conidiomata up to 50 µ diam. Conidia 0-1 septate, 26-33 × 1 µ.

On needles of *Pinus sylvestris*, Finland (IMI 209762, holotype ex H).

P. SUCCINEUM (Sprée) Dyko & Sutton, *Can. J. Bot.* **57**: 372 (1979).
Dacrymyces succineus Sprée, *Bot. Zeit.* **22**: 175 (1864).
? *Calloria succinea* Fr., *Summa Veg. Scand.*: 359 (1849).
Dacrymyces succineus Fr., *Hymen. europ.*: 699 (1874).
Hymenula fumosellina Stärb., *Bih. K. svensk. Akad. Handl.* **21**(3): 35 (1895).
Siroscyphella fumosellina (Stärb.) Höhn., *Sber. Akad. Wiss. Wien* **119**: 650 (1910).
Siroscyphella succinea (Fr.) Höhn., *Sber. Akad. Wiss. Wien* **127**: 374 (1918).
Hymenula fumosa Cke & Ell., *Grevillea* **6**: 87 (1878).
Pseudocenangium alpinum Petrak, *Annls mycol.* **23**: 124 (1925).

Teleomorphosis: *Phialea fumosella* (Cke & Ell.) Sacc., *Syll. fung.* **8**: 269 (1889).

Conidiomata pale brown, up to 300 µ diam. Conidia medianly 1 septate, cylindrical, apices obtuse, 12-18 × 2 µ.

On needles of *Pinus pumilio*, *P. sylvestris*, *P. rigida*, *P.* sp., Czechoslovakia, Sweden, Switzerland, France, U.K., N.B., Canada, N.J. and Ohio, U.S.A., represented by 17 collections in herb. IMI including isotypes of *P. alpinum*, *H. fumosellina* and *D. succineus* Sprée ex K.

19. ANAPHYSMENE Bub., *Annls mycol.* **4**: 124 (1906).
Wakefieldia Arnaud, *Bull. trimest. Soc. mycol. Fr.* **69**: 303 (1954).
Sp. typ.: *A. heraclei* (Lib.) Bub. (syn. *Cheilaria heraclei* Lib.).
Illustrations: Bubák (1906a); Arnaud (1954); Sutton (1972a).

Mycelium immersed, branched, septate, brown. *Conidiomata* acervular, caulicolous, epidermal to subepidermal, separate or often confluent, formed of dark brown, thick-walled textura angularis. Dehiscence irregular. *Conidiophores* erect, cylindrical to subulate, parallel, thick-walled, septate, dark brown, formed from the upper pseudoparenchyma. *Conidiogenous cells* holoblastic, annellidic, integrated, indeterminate, cylindrical, pale brown, smooth, with 1 percurrent proliferation. *Conidia* hyaline, medianly 1-euseptate, thin-walled, smooth, eguttulate, base truncate, apex acute, falcate, fusiform.

Höhnel (1916c) established the typification of *Cheilaria* Lib. which excluded *C. heraclei*. He accepted the disposition by Bubák (1906a), as did Petrak (1959b). Sutton (1972a) redescribed the genus and listed the synonymy. At present the genus remains monotypic.

A. HERACLEI (Lib.) Bub., *ibid.*: 124 (1906). (Fig. 169)
Synonymy as in Sutton (1972a).

Conidiomata 200-300 µ diam. Conidiophores 17-25 × 7-9 µ. Conidia 18-38 × 4.5-6.5 µ.

On stems of *Heracleum sphondylium*, France, Germany (IMI 99095 holotype ex PC, 100363).

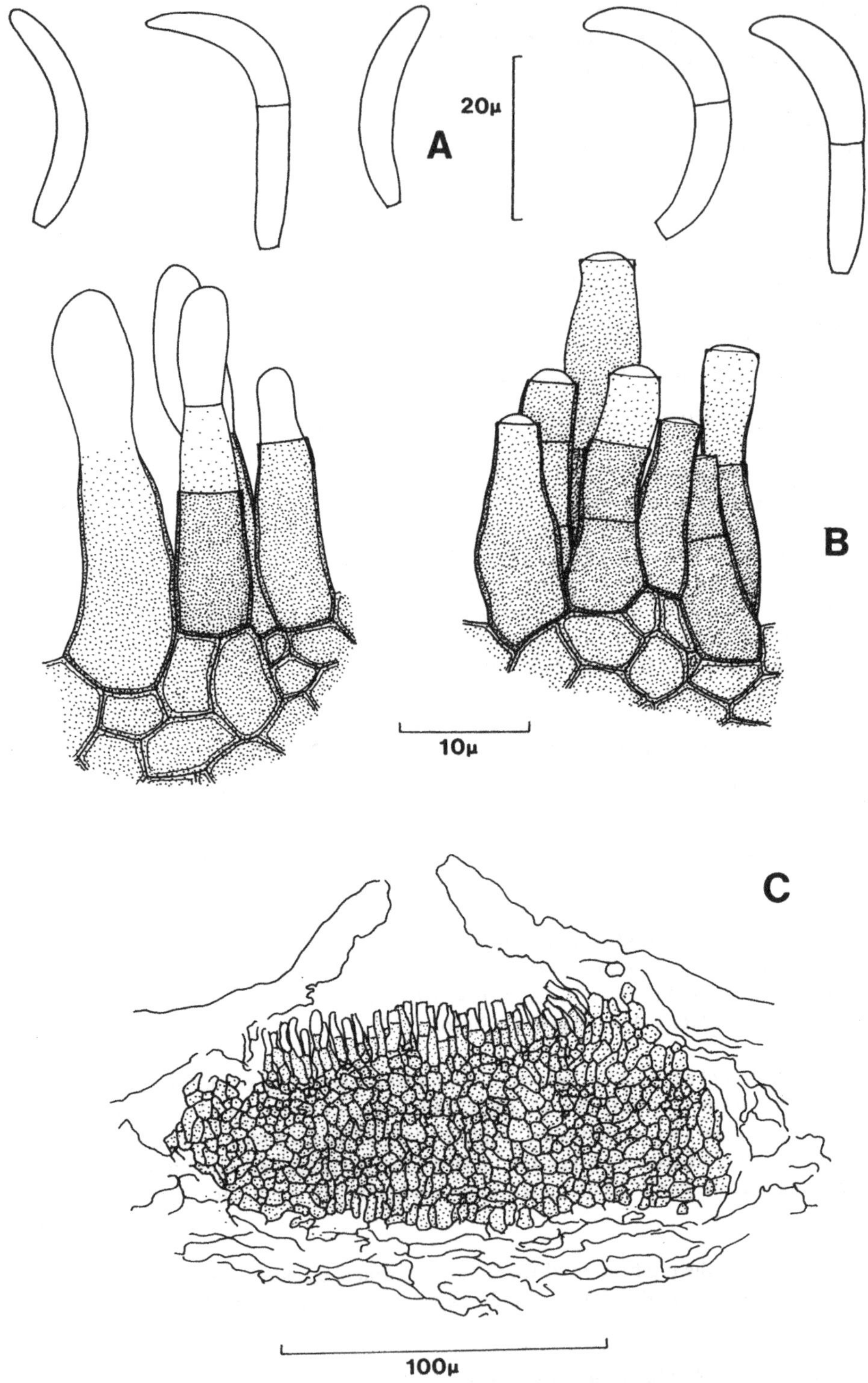

Fig. 169. *Anaphysmene heraclei*, A, conidia; B, conidiogenous cells; C, vertical section of a conidioma.

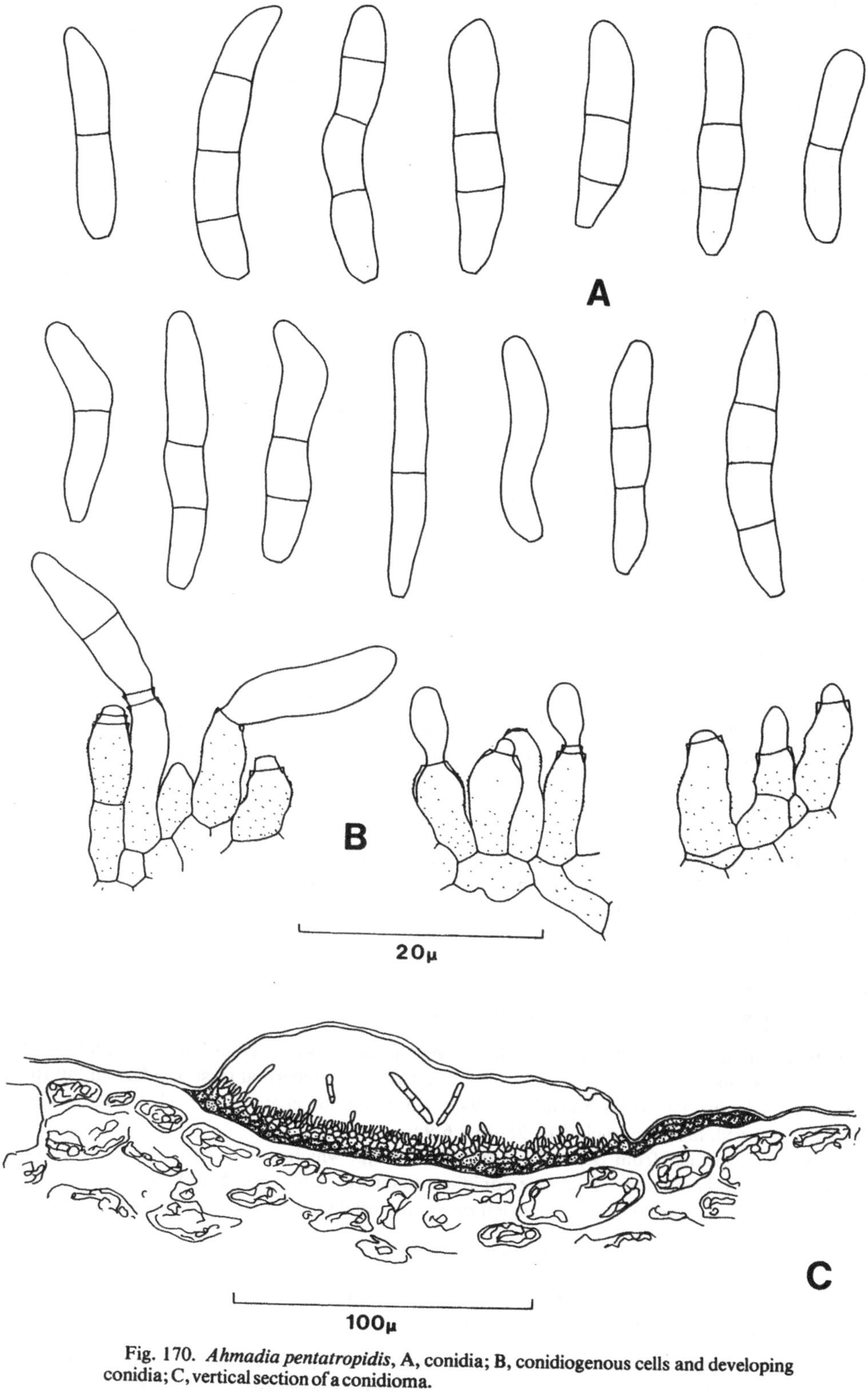

Fig. 170. *Ahmadia pentatropidis*, A, conidia; B, conidiogenous cells and developing conidia; C, vertical section of a conidioma.

20. AHMADIA Syd., *Annls mycol.* **37**: 445 (1939).
Sp. typ.: *A. pentatropidis* Syd.

Mycelium immersed, branched, septate, hyaline. *Conidiomata* acervular, subcuticular, separate, occasionally confluent, composed of pale brown, thin-walled textura angularis. Dehiscence irregular. *Conidiophores* absent. *Conidiogenous cells* holoblastic, annellidic, discrete, indeterminate, cylindrical or doliiform, hyaline to pale brown, roughened towards the base, with 1-2 percurrent proliferations, formed from the upper pseudoparenchyma. *Conidia* hyaline, 0-3 euseptate, straight or bent irregularly, truncate at the base, obtuse at the apex, thin-walled, smooth, eguttulate.

This genus is distinguished from *Colletogloeum* Petrak, which it resembles in ontogeny and conidial morphology, by the subcuticular rather than subepidermal conidiomata. If in the future sufficient grounds are found for merging the two genera, *Ahmadia* will take precedence over the more well-known *Colletogloeum* for it was published earlier. At the moment it remains monotypic.

A. PENTATROPIDIS Syd., *ibid.*: 445 (1939). (Fig. 170)

Conidiomata up to 150 µ diam. Conidiogenous cells 8-15 × 3.5-4 µ. Conidia 18-25 × 3.5-4 µ.

On leaves of *Pentatropis cynanchoides*, Pakistan (IMI 92123 (holotype), 92284, 113837).

21. PHLOEOSPORA Wallr., *Fl. Crypt. Germ.* **2**: 176 (1833).
Helicobolus Wallr., *Fl. Crypt. Germ.* **2**: 751 (1833).
Phloeochora Höhn., *Ber. dtsch. Bot. Ges.* **35**: 253 (1917).
Original spp.: *P. ulmi* (Fr. ex Kze) Wallr., *P. oxyacanthae* (Schmidt ex Kze) Wallr.
Lectotype sp.: *P. ulmi* (Fr. ex Kze) Wallr. (syn. *Septoria ulmi* Fr. ex Kze), fide Wallroth (1883), p. 752), Donk (1964).
Illustrations: Sutton & Pollack (1974).

Mycelium immersed, branched, septate, hyaline. *Conidiomata* acervular, subepidermal, circular, discrete or confluent, composed of hyaline to pale brown, thin-walled textura angularis. Dehiscence irregular. *Conidiophores* absent. *Conidiogenous cells* holoblastic, annellidic, occasionally also sympodial, discrete, indeterminate hyaline, smooth, cylindrical, with several apical inconspicuous annellations, formed from the upper cells of the acervuli. *Conidia* hyaline, septate, smooth, eguttulate, apex obtuse, base truncate, cylindrical, curved, attenuated towards the apices.

The generic synonymy has been discussed by Sutton (1977a) and the type species described and illustrated by Sutton & Pollack (1974). So far more than 160 species have been described in *Phloeospora*. Any revision of the genus requires a consideration at the same time of *Septoria, Septogloeum, Cylindrosporium* auct., and the segregates from *Cercospora* which have acervular-sporodochial conidiomata.

P. bullata (H. & P. Syd.) comb. nov. (Figs. 171, 172)
Septogloeum bullatum H. & P. Syd., *Annls mycol.* **10**: 44 (1912).

Lesions initially small and discrete, later confluent, very dark brown to black, margin vein-limited, sometimes raised, up to .5 cm diam. Conidiomata acervular, amphigenous, epidermal to subepidermal, up to 400 µ diam. Conidiogenous cells discrete, short, cylindrical, with 1-3 apical annellations, 8.5-31 × 5-6.5 µ. Conidia (2) 3-4 septate, slightly curved, fusiform to clavate, 28-46 × 6-9 µ.

This species is distinct from *Septoria combretiana* Cejp, which was described with very small pycnidial conidiomata, 70-90 µ diam., and 3-5 septate conidia measuring 25-65 × 3-3.5 µ. Although the isotype (ex PRE) was examined there appeared to be no fungus left on the material so these differences could not be verified.

On leaves of *Combretum gazense*, *C.* sp., Sudan, S. Africa (IMI 59111, 38046).

P. CERCOCARPI (Bonar) Sutton & Pollack, *Mycopath. Mycol. appl.* **52**: 349 (1974).
Septogloeum cercocarpi Bonar, *Mycologia* **38**: 345 (1946).

Conidiomata up to 120 µ diam. Conidiogenous cells 8-16 × 2.5-4 µ, with up to 3 apical annellations, occasionally sympodial. Conidia falcate, 1-3 septate, fusiform, apex obtuse, 25-29 × 5-6 µ.

On leaves of *Cercocarpus betuloides*, Calif., U.S.A. (IMI 160456, holotype).

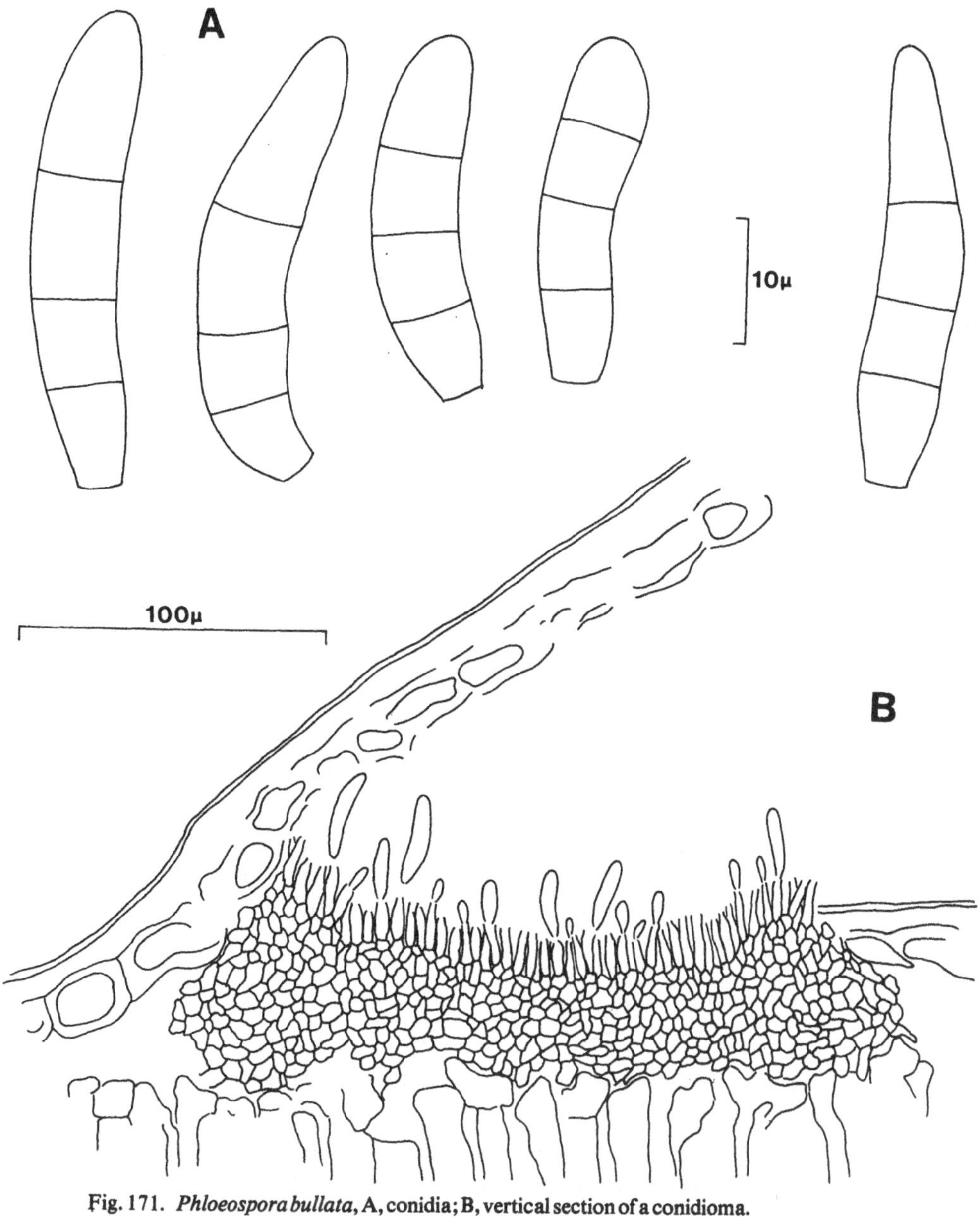

Fig. 171. *Phloeospora bullata*, A, conidia; B, vertical section of a conidioma.

P. ehretiae sp. nov. (Fig. 173)

Laesiones plus vel minusve circulares sed margine saepe irregulare, atro brunneae et colubrinae, .1-.8 cm diam. Conidiomata acervularia, hypophylla, separata vel confluentia, epidermalia, ex textura angulari pallide brunnea, hyalina, usque ad 180 μ diam. Cellulae conidiogenae holoblasticae, annellidicae, raro sympodiales, cum 1-2 annellationibus apicalibus, discretae, hyalinae, cylindricae vel leniter curvatae, ex cellulis superioribus conidiomatibus oriundae, 7-11 × 3-3.5 μ. Conidia hyalina, 2(-3) septata, laevia, eguttulata, falcata, apicem versus deminuta, basim truncata, 45-66 × 3-4 μ.

In foliis vivis *Ehretiae laevis*, Udaipur, India, Babu Singh 33, Dec. 1960, IMI 86496 holotypus.

Lesions more or less circular, discrete or confluent and then causing marginal paler brown leaf necrosis, dark brown but mottled due to angular groups of pale epidermal cells, margin often irregular, .1-.8 cm diam. *Conidiomata* acervular, hypophyllous, abundant, separate or confluent, epidermal, composed of pale brown to hyaline textura angularis, up to 180 μ diam. *Conidiogenous cells* holoblastic, annellidic, rarely sympodial with 1-

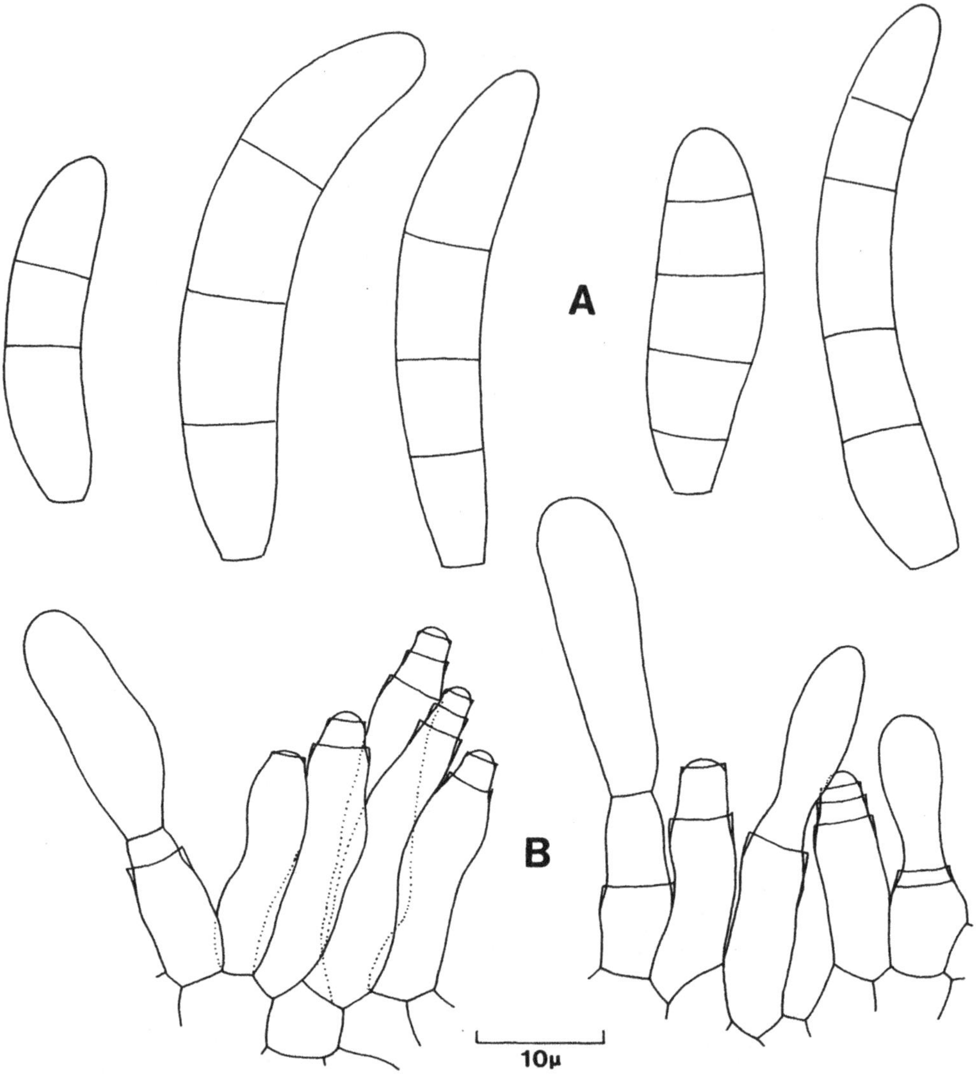

Fig. 172. *Phloeospora bullata*, A, conidia; B, conidiogenous cells and developing conidia.

2 apical annellations, discrete, hyaline, cylindrical or slightly curved, formed from the upper cells of the acervuli, 7-11 × 3-3.5 μ. Conidia hyaline, 2(-3), septate, smooth, eguttulate, falcate, gradually tapered to the apex, truncate at the base, 45-66 × 3-4 μ.

No species in the genera *Septogloeum, Cylindrosporium* or *Phloeospora* have been described from this host. *Cercospora ehretiae* Togashi & Katsuki and *C. poonensis* Viswanathan, both described from *Ehretia*, appear to be satisfactorily maintained in *Cercospora*. Although *Cercosporella ehretiae* Sawada was not validly published, it was described with conidia 4-12 septate and 55-183 × 4-5 μ which exludes it from comparison with *P. ehretiae*.

P. koenigii (Thirum.) comb. nov. (Fig. 174)
Cylindrosporium koenigii Thirum., *Bull. Torrey bot. Cl.* **75**: 177 (1948).

Lesions circular, sometimes confluent, slightly raised, medium brown, margin distinct, up to .2 cm diam. Conidiomata acervular, epiphyllous, epidermal, orange, up to 200 μ diam. Conidiogenous cells often integrated, cylindrical to filiform, with 1-3 apical annellations, 25-40 × 4-5 μ. Conidia 4-12 septate, curved, falcate, apex obtuse, 87-112 × 5.5 μ.

On leaves of *Murraya koenigii*, India (IMI 4306, holotype, 84652).

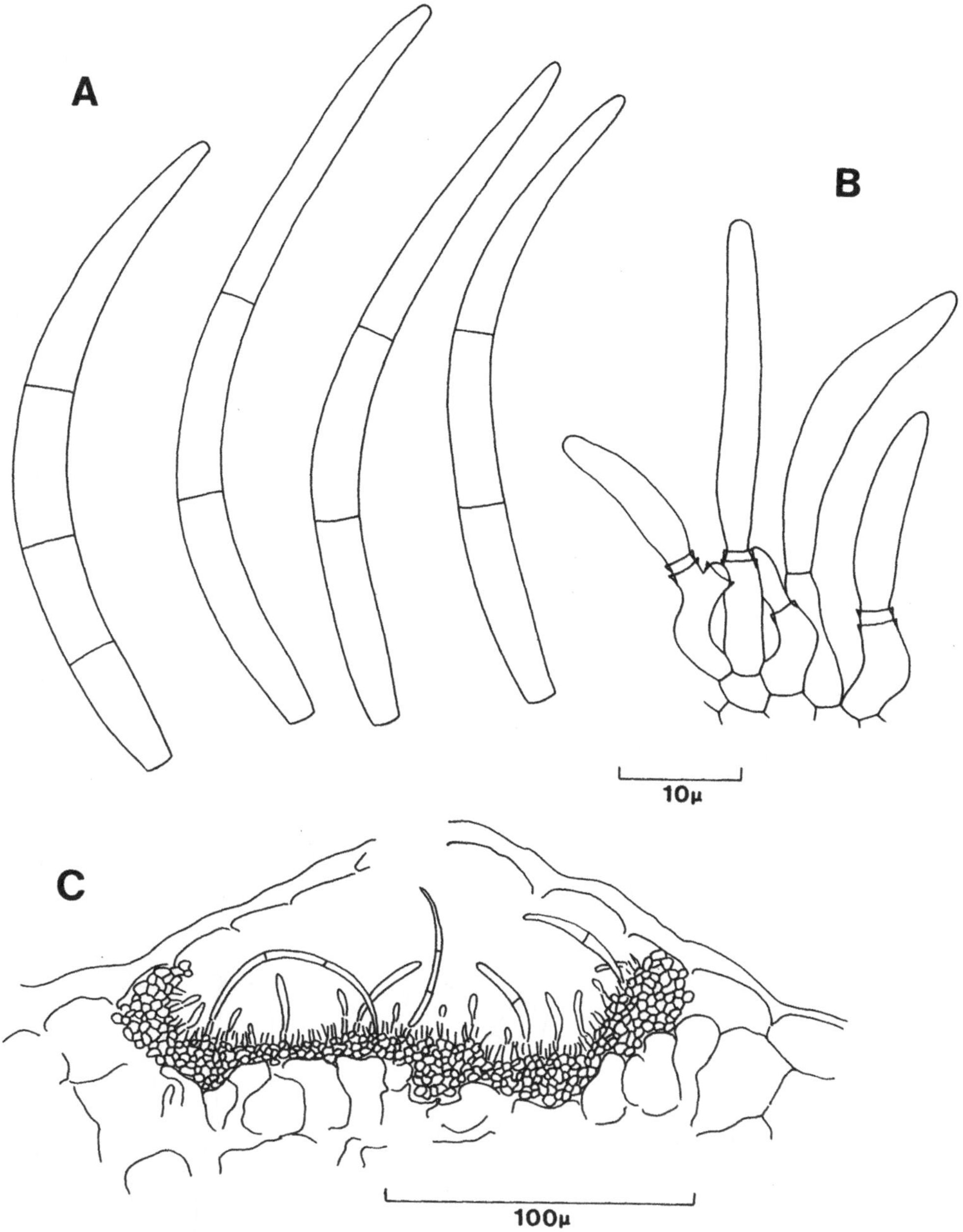

Fig. 173. *Phloeospora ehretiae*, A, conidia; B, conidiogenous cells and developing conidia; C, vertical section of a conidioma.

P. MACULANS (Béreng.) Allesch. in Rabenh., *Krypt. Fl.* **6**: 935 (1900). (Fig. 175)
Fusarium maculans Béreng., *Atti Congr. Milano 1844*: 474 (1844).
Phloeosporella maculans (Béreng.) Höhn., *Mitt. bot. Inst. tech. Hochsch. Wien* **4**: 77 (1927).
Cercosporella maculans (Béreng.) Wolf, *J. Elisha Mitchell Sci. Soc.* **51**: 165 (1935).
Septoria mori Lév., *Ann. Sci. nat.*, ser. 3, **5**: 279 (1846).
Phloeospora mori (Lév.) Sacc., *Michelia* **1**: 175 (1878).
Septogloeum mori (Lév.) Briosi & Cav., *I Funghi Parassiti della Piante coltivate od utili.* Fasc. I. No. 21 (1888).
Cylindrosporium mori (Lév.) Berl., *Riv. Pat. veg. Padova* **5**: 205 (1896).
Cercospora pulvinulata Sacc. & Wint. forma *angulosa* Savulescu & Sandu, *Herb. Myc. Rom. Fasc.* IV, No. 188 (1930).

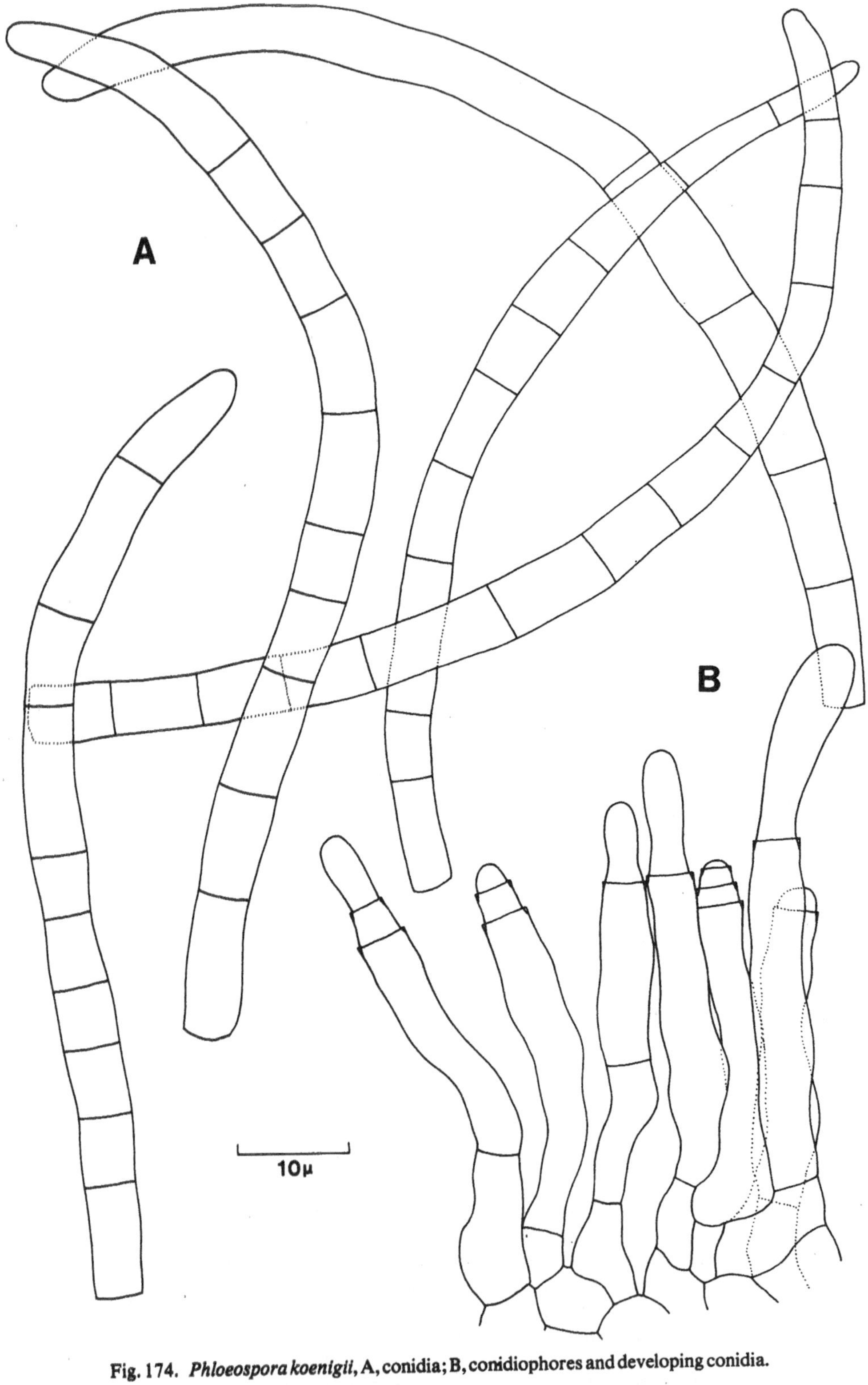

Fig. 174. *Phloeospora koenigii*, A, conidia; B, conidiophores and developing conidia.

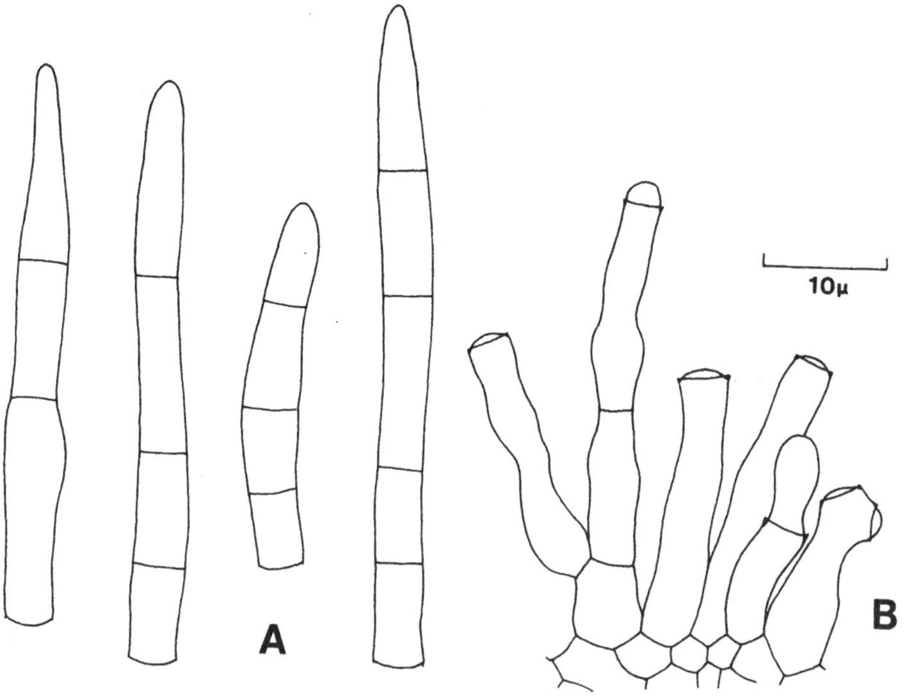

Fig. 175. *Phloeospora maculans*, A, conidia; B, conidiophores and developing conidia.

Teleomorphosis: *Mycosphaerella mori* (Fckl) Wolf, *J. Elisha Mitchell Sci. Soc.* **51**: 165 (1935).

Conidiomata up to 200 μ diam. Conidiogenous cells 9-25 × 3.5-4 μ, with 1 apical annellation. Conidia usually straight, 2-4 septate, tapered towards the apex, or cylindrical, obtuse, 29-55 × 4-5 μ.

On leaves of *Morus alba, M. nigra, M. indica, M.* sp., U.K., Roumania, Italy, Hungary, Turkey, Austria, Cyprus, Libya, Tanzania, Kenya, Malawi, S. Rhodesia, Iraq, Nepal, India, Australia (IMI 22531, 22530, 31114, 76240, 80972, 82849, 59543, 115846, 123550, 139004, 135766, 132503, 145619, 148731, 178052, 192508, 205477, 210287, 8346, isotype of *C. pulvinulata* f. *angulosa*, 22727, 22528, 11982, 11520, 22529).

P. ULMI (Fr. ex Kze) Wallr., *Fl. Germ.* **2**: 177 (1833).
Septoria ulmi Fr., *Nov. Fl. Suec.* **5**: 78 (1819).
Septoria ulmi Fr. ex Kze in Kze & Schm., *Mykol. Hefte* **2**: 107 (1823).
Cylindrosporium ulmi (Fr. ex Kze) Vassiljevsky apud Vassiljevsky & Karakulin, Fungi Imperf. Paras. II. Melanconiales: 580 (1950).
Septogloeum ulmi (Fr. ex Kze) Briosi & Cav., *I Funghi Parassiti della Piante coltivate od utili* Fasc. III & IV, No. 98 (1890).
Septogloeum ulmi (Fr. ex Kze) Died., *Krypt. Fl. M. Brandenb.* **9**: 836 (1916).

Teleomorphosis: *Mycosphaerella ulmi* Kleb., *Zeitschr. Pfl.* **12**: 259 (1902).

Conidiomata up to 200 μ diam. Conidiogenous cells 10-14 × 4-5 μ, with 1-4 apical annellations. Conidia falcate, 3-5 septate, fusiform, apex tapered, obtuse, 30-60 × 4.5-6 μ.

On leaves of *Ulmus campestris, U. glabra* f. *pendula, U. laevis, U. scabra, U.* sp., Hungary, Germany, Austria, Roumania, Italy, U.K. (IMI 11190, 11191, 22536, 22537, 22538, 22539, 11837, 22721, 172096, 186514, 29991, 33415, 41846, 46977, 53290, 57014).

22. COLLETOGLOEUM Petrak, *Sydowia* **7**: 368 (1953).
Sp. typ: *C. sissoo* (Syd.) Sutton (syn. *Cercospora sissoo* Syd., *Colletogloeum dalbergiae* (Ahmad) Petrak).
Illustration: Sutton (1964a).

Mycelium immersed, branched, septate, hyaline to pale brown. *Conidiomata* acervular, epidermal to subepidermal, separate, occasionally confluent, composed of pale brown to hyaline, thin-walled textura angularis. Dehiscence irregular. *Conidiophores* hyaline or very pale brown, sparsely branched, septate, smooth, cylindrical or slightly irregular, formed from the upper cells of the acervulus. *Conidiogenous cells* holoblastic, annellidic, integrated or discrete, indeterminate, cylindrical or doliiform, with several percurrent proliferations. *Conidia* hyaline or pale brown, 0-multiseptate, straight, curved or irregular, truncate at the base, obtuse at the apex, usually thin-walled, smooth, guttulate or eguttulate.

The genus and, at that time, only species, were described by Sutton (1964a). It differs from *Ahmadia* Syd. only in having epidermal to subepidermal conidiomata as opposed to subcuticular conidiomata. A small number of additions have subsequently been made to the genus (Pollack & Sutton, 1972; Sutton, 1975a; Singh, 1977).

KEY

1. Conidia <30 µ long 2
1. Conidia >30 µ long 5
   2. Conidia up to 1 septate 3
   2. Conidia up to 3-4 septate 4
3. Conidia 13.5-20 × 2.5-3 µ, acute at the apex .... *C. artocarpi*
3. Conidia 11.5-14.5 × 2.5-3 µ, obtuse at the apex .... *C. obtusum*
   4. Conidiophores 20-40 µ long; conidia 1-3 septate, obtuse at the apex .... *C. olacis*
   4. Conidiophores up to 18 µ long; conidia 0-2 septate, tapered to the apex .... *C. rhois*
   4. Conidiophores 10-20 µ long; conidia 1-4 septate, obtuse at the apex .... *C. sissoo*
5. Conidiophores up to 50 µ long .... *C. feroniae*
5. Conidiophores <20 µ long 6
   6. Conidia thick-walled, 5-8 septate .... *C. glochidiicola*
   6. Conidia thin-walled 7
7. Conidia 28-72 × 4-6 µ; on *Dovyalis* .... *C. dovyalidis*
7. Conidia 34-80 × 3.5-4 µ; on *Fouquieria* .... *C. fouquieriae*

C. ARTOCARPI S.M. Singh, *Indian Phytopath.* **29**: 71, 1976 (1977). (Fig. 176)

Lesions originally discrete and circular, later coalescing, confluent and spreading. Conidiomata amphigenous, up to 175 µ diam. Conidiophores frequently branched, 10-20 × 2-4 µ, with 1-4 percurrent proliferations. Conidia pale brown, smooth, 0-1 euseptate (0-4 in original account), straight or strongly curved, tapered to the apices, 13.5-20 × 2.5-3 µ.

On leaves of *Artocarpus heterophyllus*, *A. integrifolia*, India, Nepal (IMI 158138b (not 158137b as in original account), holotype of *C. artocarpi*, 182431).

C. DOVYALIDIS (van der Byl) Sutton, *Mycol. Pap.* 138: 97 (1975).
Coryneum dovyalidis van der Byl, *S. Afr. J. Sci.* **26**: 318 (1929).

Lesion 6 cm diam., concentrically zoned. Conidiomata amphigenous, up to 150 µ diam. Conidiogenous cells 9-18 × 5-7 µ, with one percurrent proliferation. Conidia pale brown, finely verruculose, cylindrical, curved, 3 euseptate, 28-72 × 4-6 µ.

On leaf of *Dovyalis*, S. Africa (IMI 177762, holotype ex Universiteit van Stellenbosch).

C. feroniae (T.S. & K. Ramakr.) comb nov. (Figs. 177, 178)
Septogloeum feroniae T.S. & K. Ramakr., *Proc. Ind. Acad. Sci.* sect. B, 32: 78 (1950).

Lesions either small and scarcely larger than conidiomata, or large and spreading. Conidiomata up to 1500 µ diam., black, hypophyllous. Conidiophores pale brown, multiseptate, minutely verruculose, up to 50 µ long × 4-6.5 µ wide, with 1-6 percurrent proliferations. Conidia very pale brown, finely verruculose, curved, irregular or straight and cylindrical, 1-4 euseptate, 34-50 × 4-5.5 µ.

On leaves of *Feronia elephantum*, *F. limonia*, *F.* sp., India, Burma (IMI 101865, holotype, 86504, 161586, 164829).

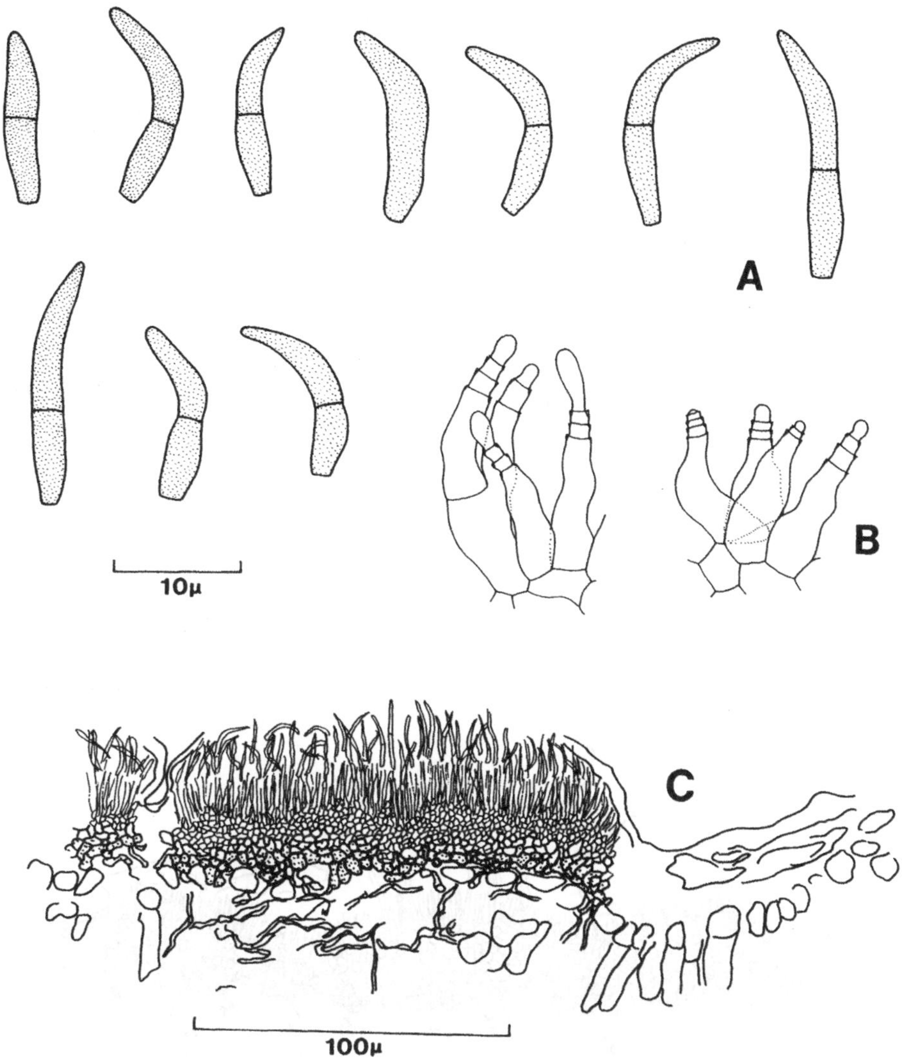

Fig. 176. *Colletogloeum artocarpi*, A, conidia; B, conidiophores; C, vertical section of a conidioma.

C. FOUQUIERIAE Pollack & Sutton, *Mycologia* **64**: 703 (1972). (Fig. 179)

Lesions up to 1 cm often fully occupied by conidiomata that are amphigenous, up to 250 μ diam. and black. Conidiophores very pale brown, smooth, sparsely septate, 4-10 × 4 μ, with 1-3 percurrent proliferations. Conidia very pale brown, smooth, straight or slightly curved, 0-4 euseptate, 34-80 × 3.5-4 μ.

On leaves of *Fouquieria splendens*, Ariz., U.S.A. (IMI 154116, isotype).

C. GLOCHIDIICOLA (Seshadri) Sutton, *Mycol. Pap.* 138: 113 (1975). (Fig. 180)
Coryneum glochidiicola Seshadri, *Mycopath. Mycol. Appl.* **30**: 182 (1966).

Lesions circular, sometimes confluent, up to .5 cm diam. Conidiophores pale brown, smooth, branched or simple, up to 15 μ long × 3.5-4.5 μ wide, with 1-2 percurrent proliferations. Conidia straight or curved irregularly, sometimes sigmoid, tapered markedly to the apex, smooth, thicker-walled, 5-8 euseptate, 27-46 × 4.5-5.5 μ.

On leaves of *Glochidion hohenackeri*, India (IMI 137302, holotype).

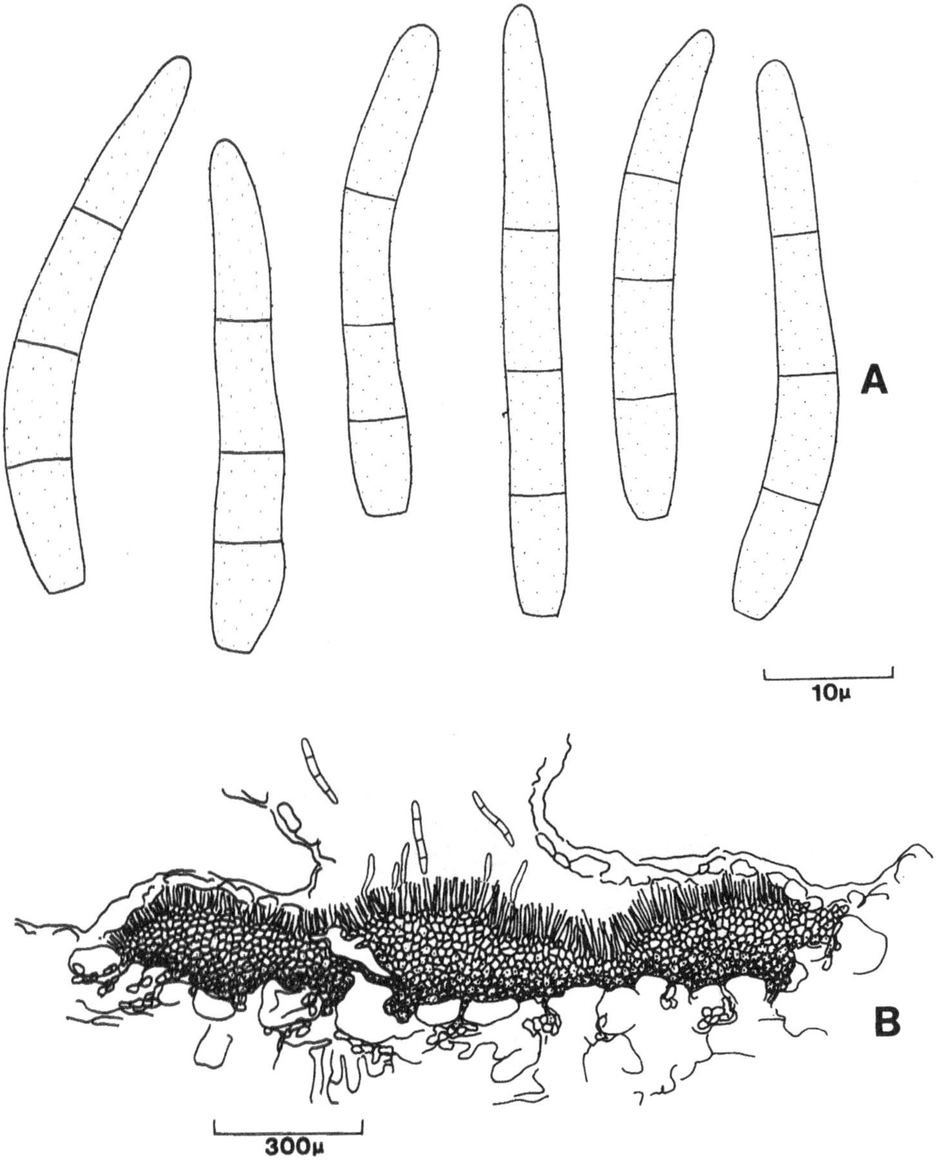

Fig. 177. *Colletogloeum feroniae*, A, conidia; B, vertical section of a conidioma.

C. obtusum sp. nov. (Fig. 181)

Conidiomata acervularia, epiphylla, raro hypophylla, brunnea, abundantia, ex texturis angularibus pallide brunneis, 150 μ diam. Conidiophora hyalina, 1-2 septata, basim versus ramosa, cylindrica vel lageniformia, 10-17 × 2-4 μ, ex cellulis superioribus acervulorum formata. Cellulae conidiogenae holoblasticae, annellidicae, in conidiophoris incorporate vel discretae, indeterminatae, cum 1-6 proliferationibus percurrentibus, 6-13.5 × 2-4 μ. Conidia pallide brunnea, laevia, 0-1 euseptata, recta vel irregulariter curvata, basim truncatam versus attenuata, apice obtusa, 11.5-14.5 × 2.5-3 μ.

In foliis vivis *Artocarpi integrifoliae*, Pokhara, Nepal, K.L. Tuladhar 999, 29 Mar. 1970, IMI 152717, holotypus.

Lesions at first circular with a prominent dark brown irregular margin, central tissue white to cream white .5-1.0 cm diam., later becoming confluent producing extensive necrosis. Conidiomata acervular, epiphyllous, rarely hypophyllous, brown, abundant, epidermal, of pale brown, thin-walled textura angularis, up to 150 μ diam. Con-

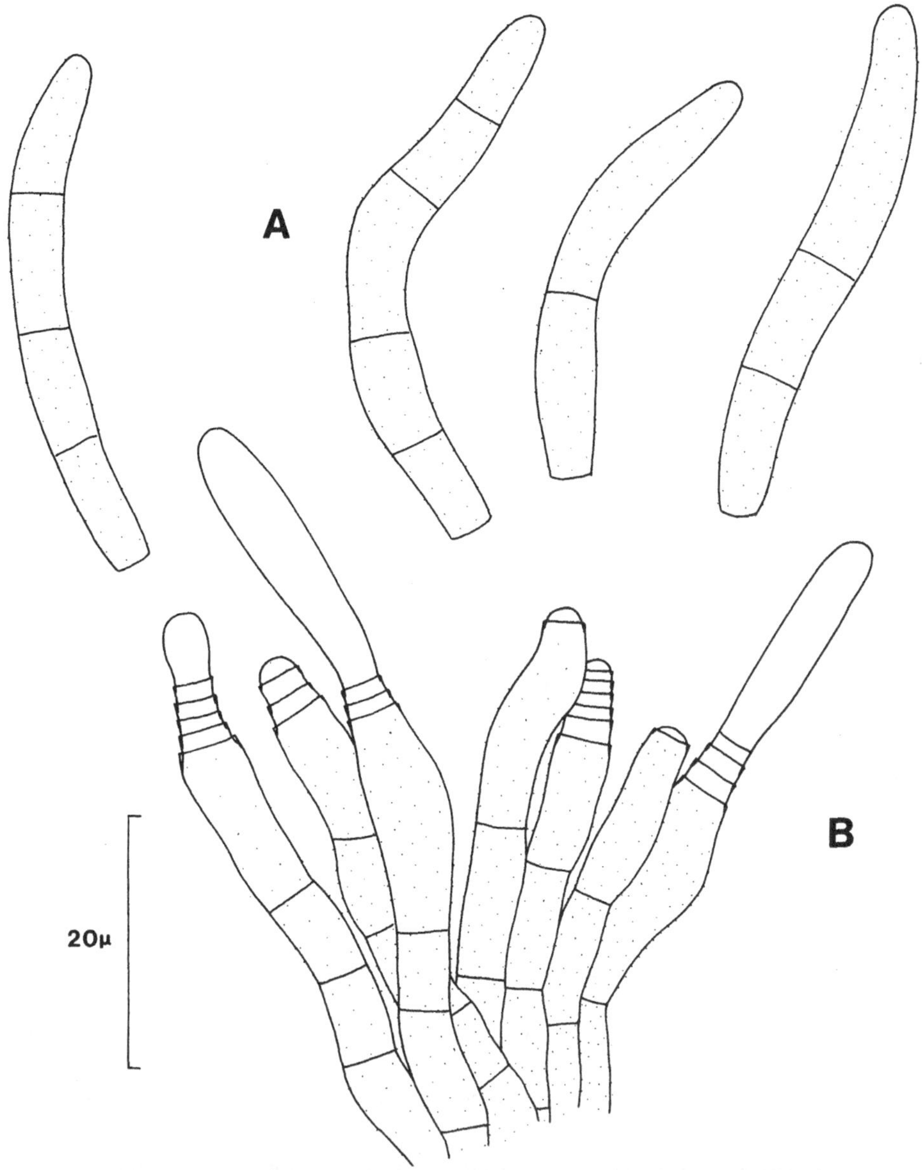

Fig. 178. *Colletogloeum feroniae*, A, conidia; B, conidiophores and developing conidia.

idiophores relatively short, hyaline, 1-2 septate, branched near the base, cylindrical to lageniform, 10-17 × 2-4 µ, formed from the upper cells of the acervulus. Conidiogenous cells holoblastic, annellidic, integrated or discrete, indeterminate, with 1-6 percurrent proliferations, 6-13.5 × 2-4 µ. Conidia pale brown, smooth, 0-1 euseptate, straight or irregularly curved, tapered to a truncate base, apex obtuse, 11.5-14.5 × 2.5-3 µ.

This species is distinguished from *C. artocarpi* by the smaller conidia that are obtuse at the apices rather than tapered and acute.

C. olacis sp. nov. (Fig. 182)

Conidiomata acervularia, amphigena, punctiformia, brunnea, abundantia, epidermalia vel subepidermalia, ex texturis angularibus brunneis formata, 80 µ diam. Conidiophora hyalina, 2-5 septata, laevia, ad basim ramosa, filiformia, 20-40 × 3-4 µ, ex cellulis superioribus acervulorum formata. Cellulae conidiogenae holoblasticae, annellidicae, in conidiophoris incorporatae, indeterminatae, cum 1-4 proliferationibus percurrentibus, 7-17 × 3-4 µ.

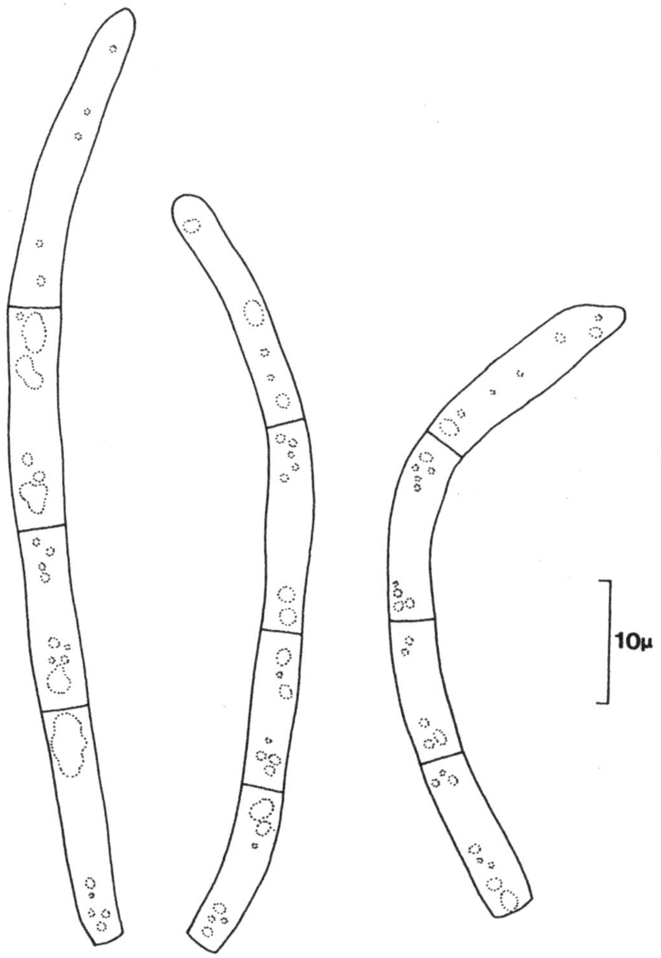

Fig. 179. *Colletogloeum fouquieriae*, conidia.

Conidia pallide brunnea, laevia, 1-3 euseptata, recta vel leniter curvata, basim truncatam versus abrupte deminuta, apice obtusa, 15.5-25.5 × 3.5-4.5 µ.

In foliis vivis *Olacis scandentis*, Balarampur, India, M. Mandal C-729, 23 Feb. 1966, IMI 136010, holotypus.

Lesions circular with a rather diffuse brown to dark brown irregular margin, central tissue paler, not exceeding 1 cm diam. Conidiomata acervular, amphigenous, punctiform, brown, abundant, epidermal to subepidermal, of brown, thin-walled textura angularis, up to 80 µ diam. Conidiophores hyaline, 2-5 septate, smooth, branched only at the base, filiform, 20-40 × 3-4 µ, formed from the upper cells of the acervulus. Conidiogenous cells holoblastic, annellidic, integrated, indeterminate, with 1-4 percurrent proliferations, 7-17 × 3-4 µ. Conidia pale brown, smooth, 1-3 euseptate, straight or slightly curved, tapered abruptly to a truncate base, apex obtuse, 15.5-25.5 × 3.5-4.5 µ.

C. RHOIS (Halst.) Sutton, *Mycol. Pap.* 138: 165 (1975). (Fig. 183)
Coryneum rhois Halst., *Bull. Torrey bot. Cl.* **20**: 251 (1893).

Lesions angular, brown, mottled with patches of grey tissue, often confluent, up to .5 cm diam. Conidiomata amphigenous, sparse, up to 150 µ diam. Conidiophores sparsely septate and branched, 9-18 × 2.5-3.5 µ, with 1-4 percurrent proliferations. Conidia pale brown, smooth, 0-2 euseptate, straight or irregularly curved, tapered to the apices, 16-27 × 3.5-4.5 µ.

On leaves of *Rhus toxicodendron*, N.J., U.S.A. (IMI 178770 ex NY, isotype).

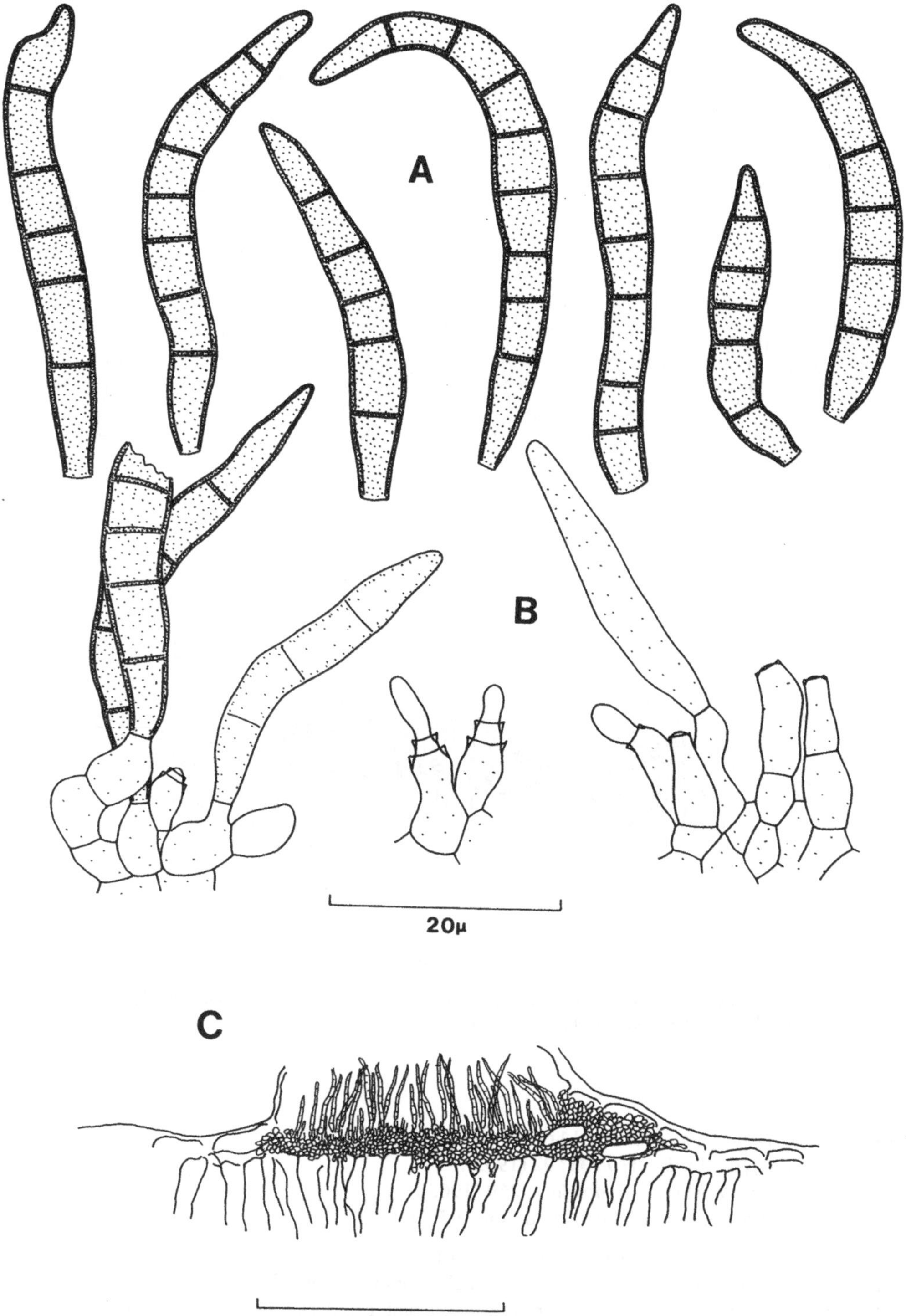

Fig. 180. *Colletogloeum glochidiicola*, A, conidia; B, conidiophores and developing conidia; C, vertical section of a conidioma.

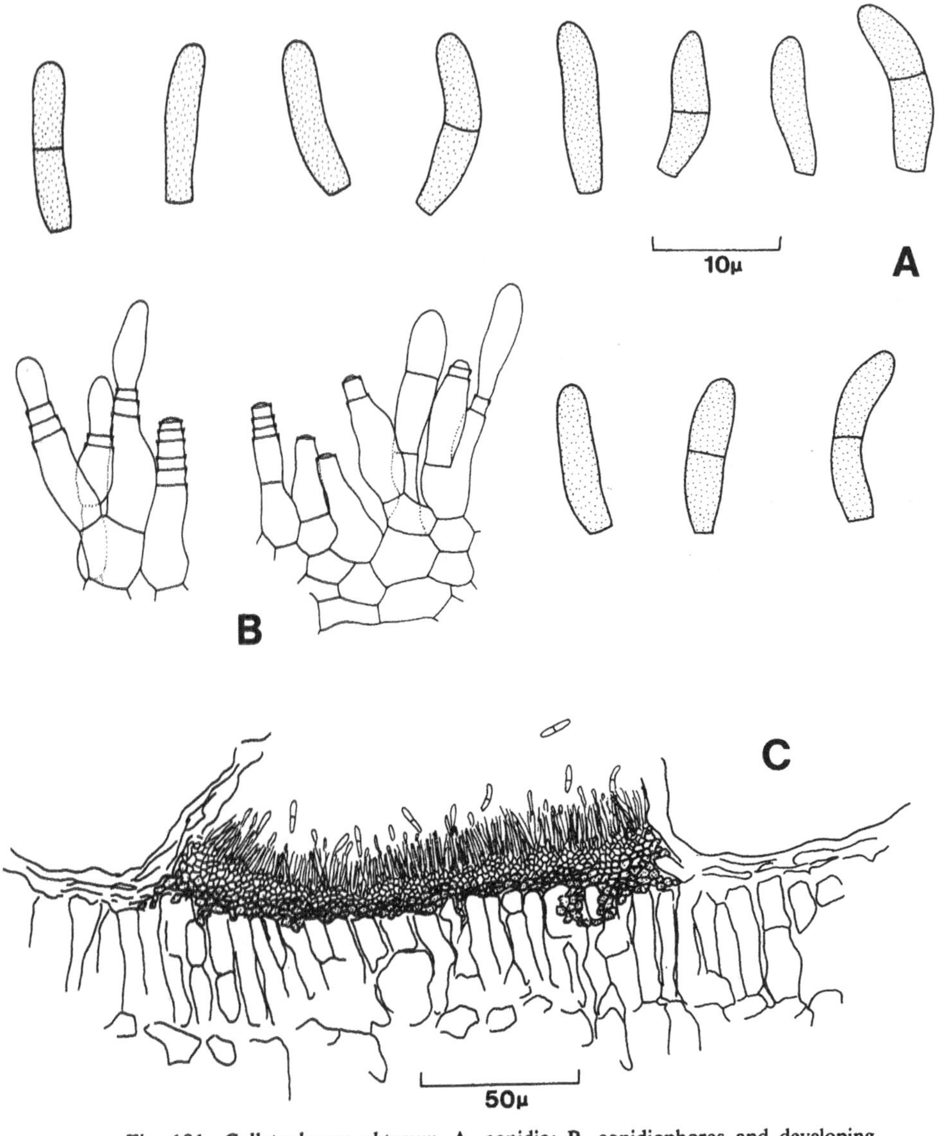

Fig. 181. *Colletogloeum obtusum*, A, conidia; B, conidiophores and developing conidia; C, vertical section of a conidioma.

C. SISSOO (Syd.) Sutton, *Mycol. Pap.* 97: 14 (1964). (Fig. 184)
Cercospora sissoo Syd., *Annls mycol.* **31**: 92 (1933).
Septogloeum dalbergiae Ahmad, *Sydowia* **7**: 269 (1963).
Colletogloeum dalbergiae (Ahmad) Petrak, *Sydowia* **7**: 367 (1953).

Lesions on leaves and pods, evenly brown, almost fully occupied by conidiomata, margin irregular, up to .5 cm diam. Conidiomata mostly hyphophyllous, black, confluent, 50-200 μ diam. Conidiophores branched and septate at the base, 10-20 × 3-4 μ, with 1-3 percurrent proliferations. Conidia hyaline or very pale brown, smooth, 1-4 euseptate, straight, curved, sigmoid or irregular, apex obtuse, 15-27 × 3-4 μ.

On leaves and pods of *Dalbergia sissoo*, India, Pakistan (IMI 8196, isotype *C. sissoo*, 119162, 165197, 136184, 115053, 132064, 139365, 139367, 80188, 62846, 115053, 113577, 90789b, 90825, 112874, 86029f, 89415, 89416, 89417, 88668, 85950, 85119, 86035, 98220, 85416).

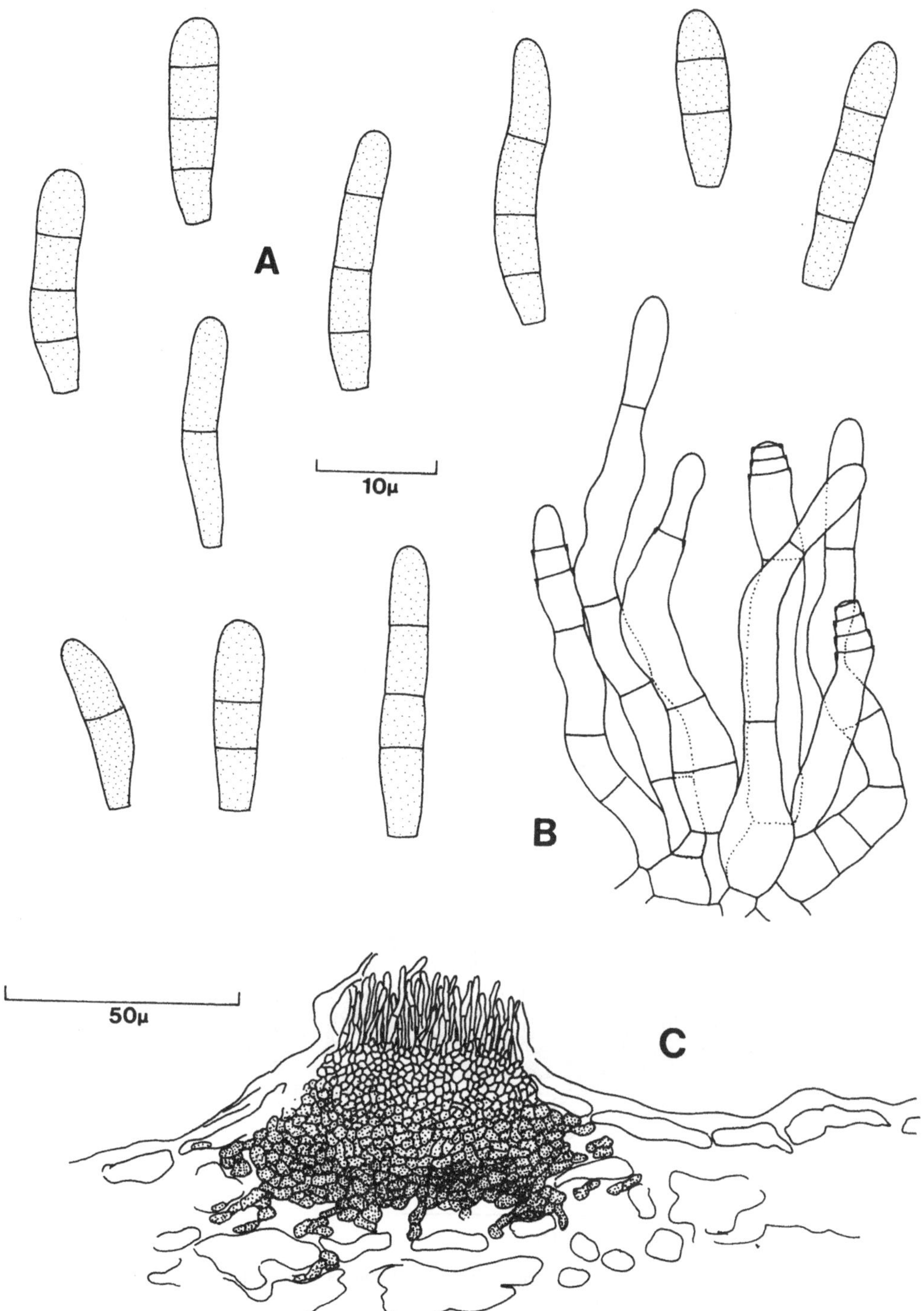

Fig. 182. *Colletogloeum olacis*, A, conidia; B, conidiophores and developing conidia; C, vertical section of a conidioma.

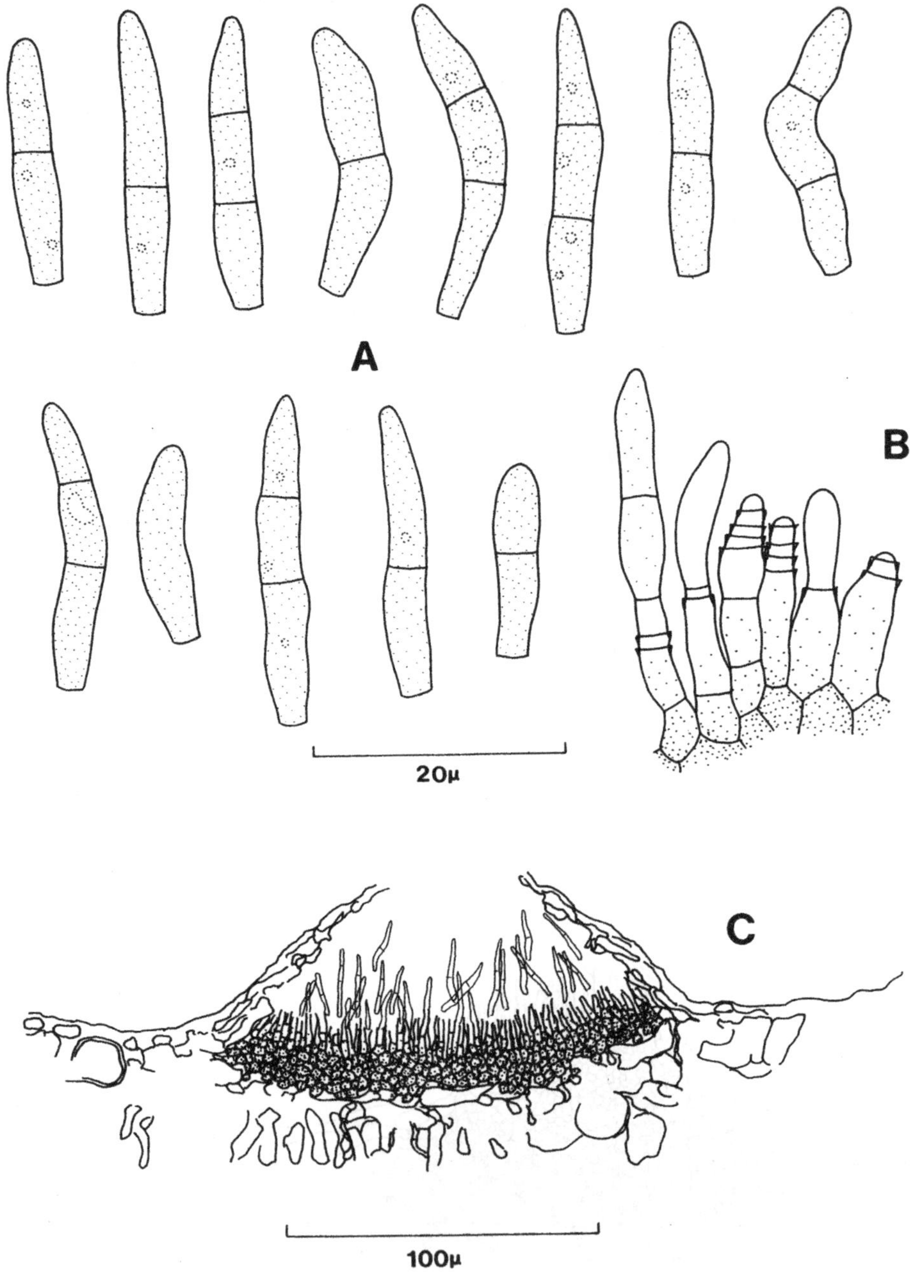

Fig. 183. *Colletogloeum rhois*, A, conidia; B, conidiophores and developing conidia; C, vertical section of a conidioma.

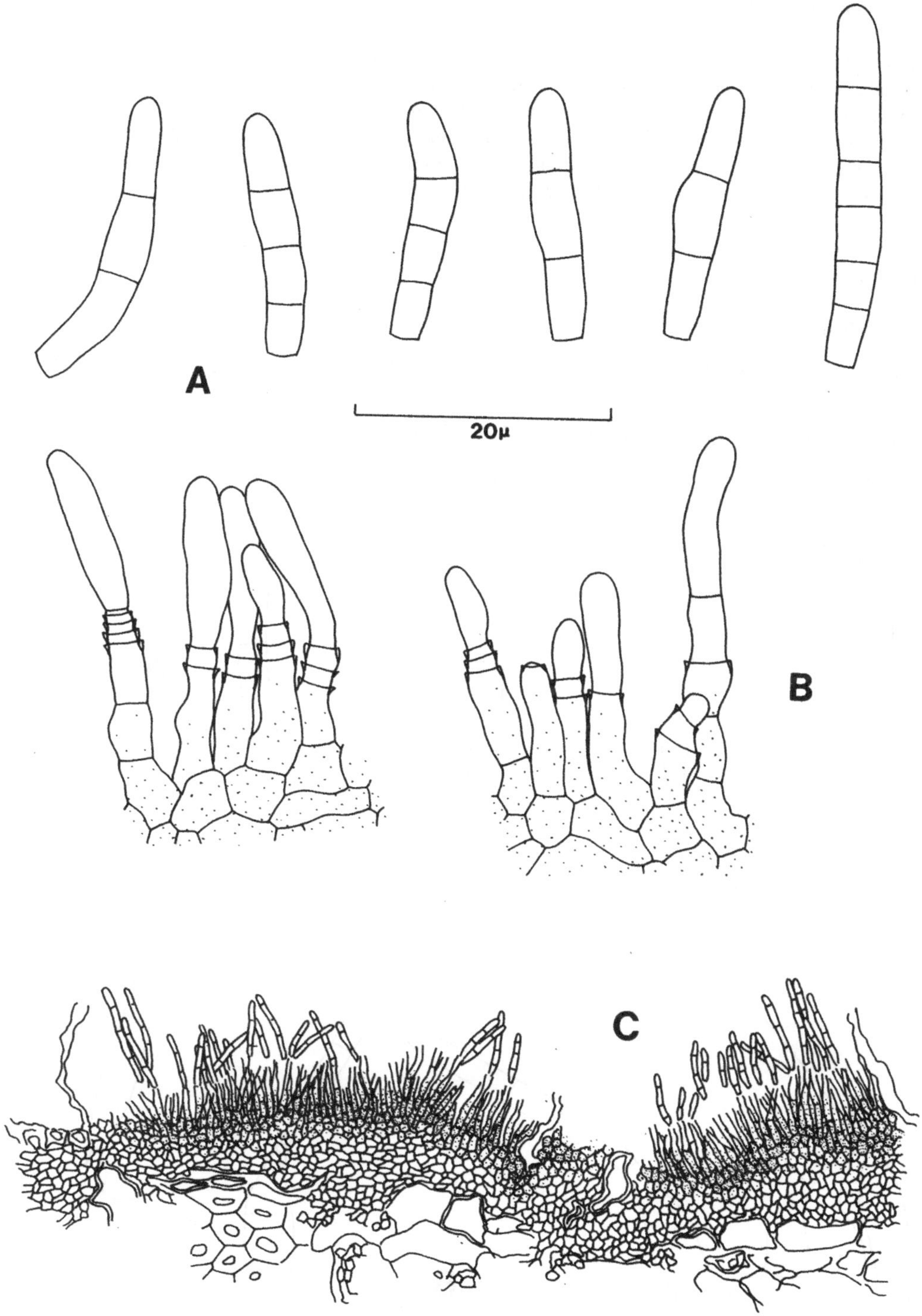

Fig. 184. *Colletogloeum sissoo*, A, conidia; B, conidiophores and developing conidia; C, vertical section of a conidioma.

23. CIFERRIELLA Petrak apud Petrak & Ciferri, *Annls mycol.* **28**: 409 (1930).

Sp. typ.: *C. domingensis* Petrak

Mycelium immersed, branched, septate, hyaline to pale brown. *Conidiomata* acervular, epiphyllous, subepidermal, separate, brown, small, composed of brown, ± thick-walled textura angularis. Dehiscence irregular, conidia produced above the substrate. *Conidiophores* pale brown, septate, sparingly branched at the base, broadly cylindrical, smooth, formed from the upper acervular cells. *Conidiogenous cells* holoblastic, annellidic, indeterminate, integrated or discrete, pale brown, smooth, markedly tapered at the apices with 1-4 barely discernible annellations. *Conidia* pale brown, verruculose, 0-3 euseptate, markedly tapered to a truncate base, obtuse at the apex, straight or slightly curved, eguttulate.

Material was too fragmentary to take a vertical section of the conidiomata, but they appear to be simple acervuli. So far the genus remains monotypic.

C. DOMINGENSIS Petrak, *ibid.*: 409 (1930). (Fig. 185)

Conidiomata 50-80 μ diam. Conidiophores 19-26 μ long × 6 μ wide at the base, tapered to 1.5 μ wide at the apex. Conidia 26-32 × 2.5 μ.

Causing leaf lesions on *Vitex umbrosa*, Dominica (IMI 201890, isotype ex ZT).

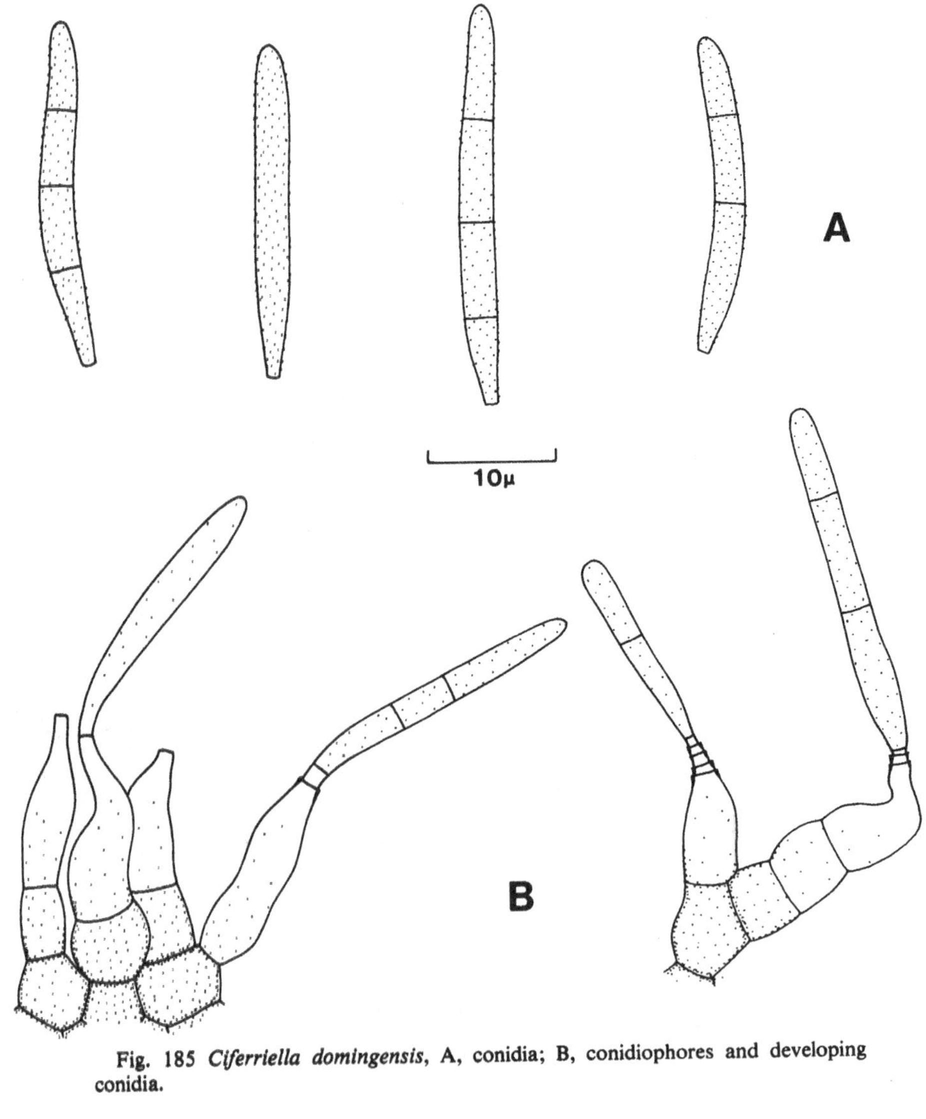

Fig. 185 *Ciferriella domingensis*, A, conidia; B, conidiophores and developing conidia.

24. BELLULICAUDA Sutton, *Can. J. Bot.* **45**: 1254 (1967).
Sp. typ.: *B. dialii* (Syd.) Sutton (syn. *Diachorella dialii* Syd.).

Mycelium immersed, branched, septate, pale brown to hyaline. *Conidiomata* pseudostromatic, immersed, black, vein-limited, angular, single, consisting of dark brown, thick-walled textura angularis concentrated in the upper and lower epidermises and immediate palisade tissue, between which is a single convoluted locule, the wall composed of small-celled, hyaline, thin-walled pseudoparenchyma, *Ostiole* single, circular, papillate, epiphyllous, often excentric. *Conidiophores* absent. *Conidiogenous cells* holoblastic, annellidic, indeterminate, discrete, cylindrical, hyaline, smooth, with 1-3 percurrent proliferations, formed from the inner cells of the pycnidial wall. *Conidia* pale brown, eguttulate, smooth, aseptate, clavate to ellipsoid, base truncate, often with the remains of the conidiogenous cell attached.

The genus is monotypic.

B. DIALII (Syd.) Sutton, *ibid.*: 1254 (1967). (Fig. 186)
Diachorella dialii Syd., *Annls mycol.* **36**: 193 (1938).

Conidiomata 400-750 µ diam. Conidiogenous cells 8.5-17 × 3-4.5 µ. Conidia 9-13 × 4.5 µ; basal appendage 6-10 µ long.

Causing leaf lesions on *Dialium guineense* and *D. angolense*, Sierra Leone, Zambia (IMI 21609, holotype, 110335).

25. PHAEODOMUS Höhn., *Sber. Akad. Wiss. Wien* **118**: 69 (1909).
Sp. typ.: *P. erumpens* (Berk. & Curt.) Petrak & Syd. (syn. *Sphaeropsis erumpens* Berk. & Curt., *P. lauracearum* Höhn.).

Mycelium immersed, branched, septate, brown. *Conidiomata* eustromatic, separate, occasionally confluent, erumpent, irregularly pulvinate, dark brown to black, uni- to multi-locular, locules simple, effuse, peripheral in the upper part of the stroma; wall tissue of thick-walled, brown textura angularis, darker and more sclerotioid at the periphery. *Ostiole* absent, dehiscence by irregular breakdown or rupture of the overlying wall. *Conidiophores* absent. *Conidiogenous cells* holoblastic, annellidic, indeterminate, discrete, straight, hyaline, smooth, cylindrical, doliiform or conical, with 1-2 apical annellations, formed from the inner cells of the locular wall. *Conidia* aseptate, pale brown, eguttulate, smooth, base truncate, sometimes with a frill, apex obtuse, ellipsoid to obpyriform.

Petrak (1925) considered *Phaeodomus* to be intermediate between *Cyclodomus* Höhn. and *Hemidothis* Syd. Later, Petrak & Sydow (1927) maintained the genus as distinct. It is monotypic.

P. ERUMPENS (Berk. & Curt.) Petrak & Syd., *Beih. Rep. spec. nov. regn. veg.* **42**: 179 (1927). (Fig. 187)
Sphaeropsis erumpens Berk. & Curt., *J. Linn. Soc. Bot.* **10**: 353 (1896).
Phoma erumpens (Berk. & Curt.) Sacc., *Syll. fung.* **3**: 117 (1884).
Macrophoma erumpens (Berk. & Curt.) Berl. & Vogl., *Atti Soc. Venet.-Trent. Sci. Nat.* **10**: 188 (1886).
Phaeodomus lauracearum Höhn., *Sber. Akad. Wiss. Wien* **118**: 69 (1909).

Conidiomata up to 900 µ wide × 700 µ deep. Conidiogenous cells 9.5-19 × 7-13 µ. Conidia 24-28.5 × 10-13 µ.

On indet. leaves of Lauraceae, Brazil (IMI 202952, holotype *P. lauracearum* ex FH).

26. DOTHIDEODIPLODIA Murashkinsky, *Trans. Siberian Inst. Agr. For.* **9**: 27 (1928).
Sp. typ.: *D. agropyri* Murashkinsky

Mycelium immersed, branched, septate, pale brown to hyaline. *Conidiomata* eustromatic, originating from an immersed pseudoparenchymatic hypostroma which ruptures the host tissues between the veins, finally becoming longitudinally lenticular, applanate at the edges, dark brown to black and multilocular, the interlocular tissues often breaking down to give a furfuraceous aspect with single locules appearing like aggregated pycnidia. Insufficient material available to determine locular or stromatic tissue structure. *Ostioles* single to each locule, circular. *Conidiophores* absent. *Conidiogenous cells* holoblastic, annellidic, discrete, indeterminate, hyaline, smooth, cylindrical, with 1-2 apical annellations, formed from the inner cells of the locular walls. *Conidia* for a long time hyaline, unicellular and thin-walled, but finally thicker-walled and pale brown, 1(-2) euseptate, eguttulate, smooth, ellipsoid to obovoid, truncate at the base, obtuse at the apex.

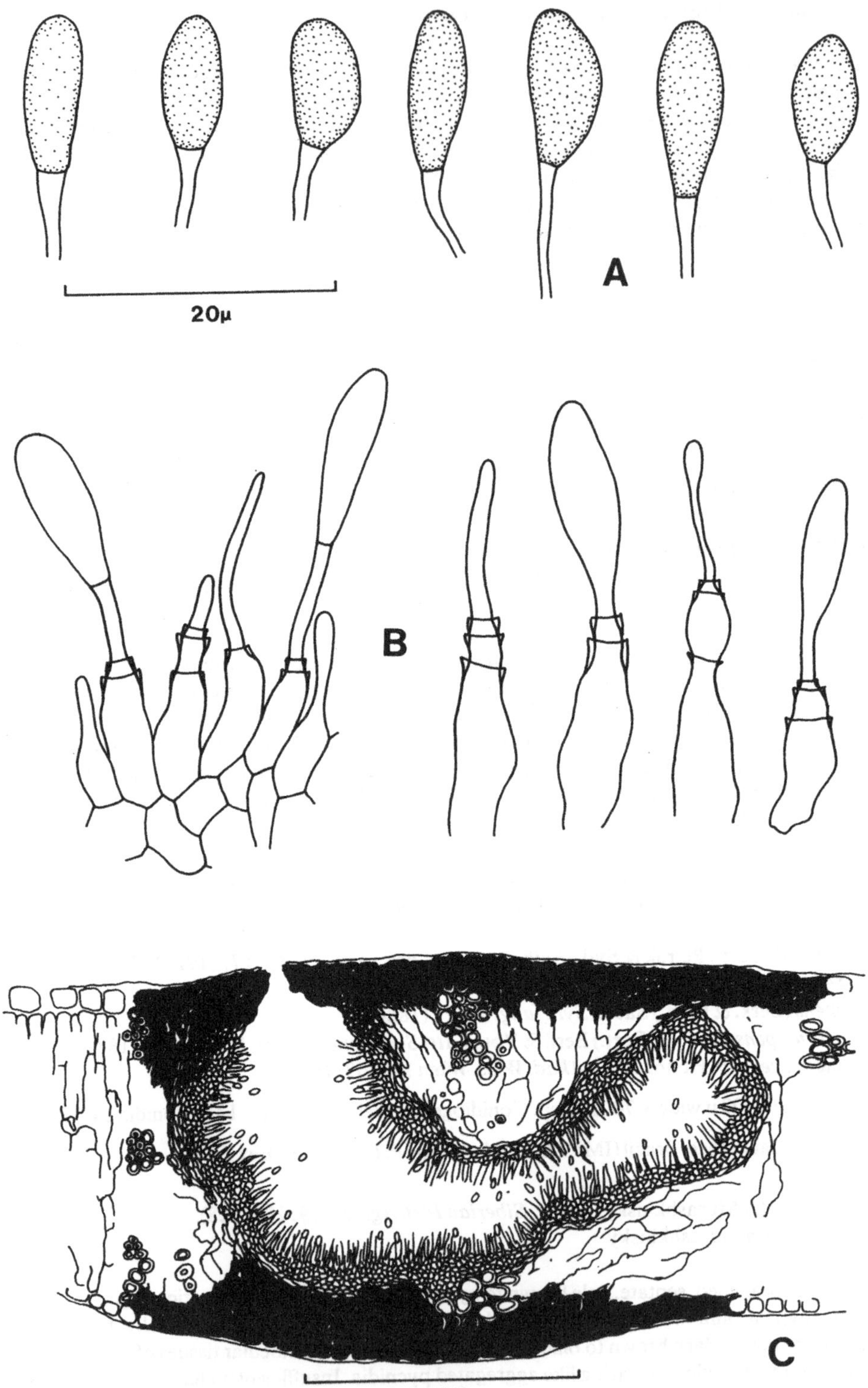

Fig. 186. ***Bellulicauda dialii***, A, conidia; B, conidiogenous cells and developing conidia; C, vertical section of a conidioma.

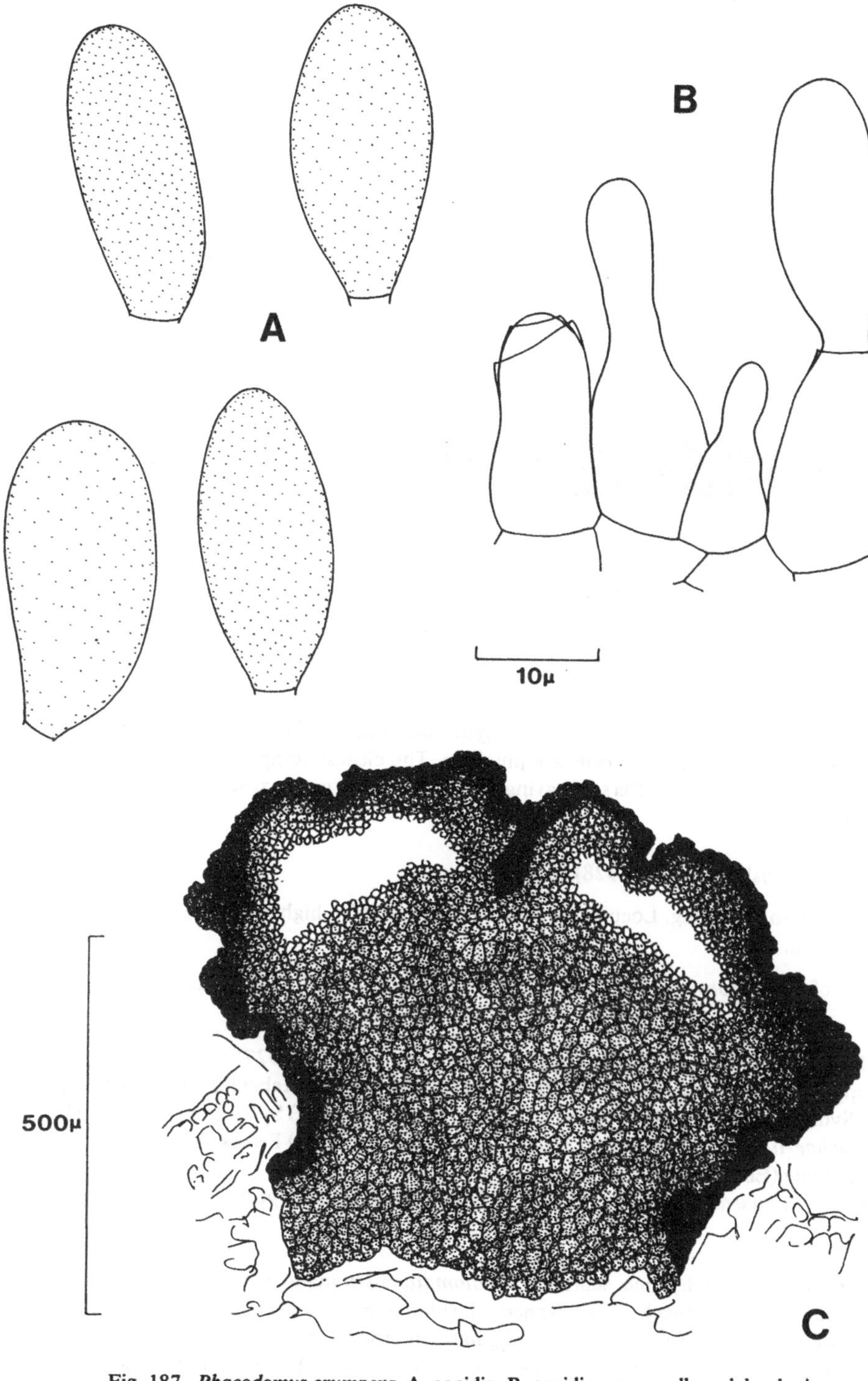

Fig. 187. ***Phaeodomus erumpens***, A, conidia; B, conidiogenous cells and developing conidia; C, vertical section of a conidioma.

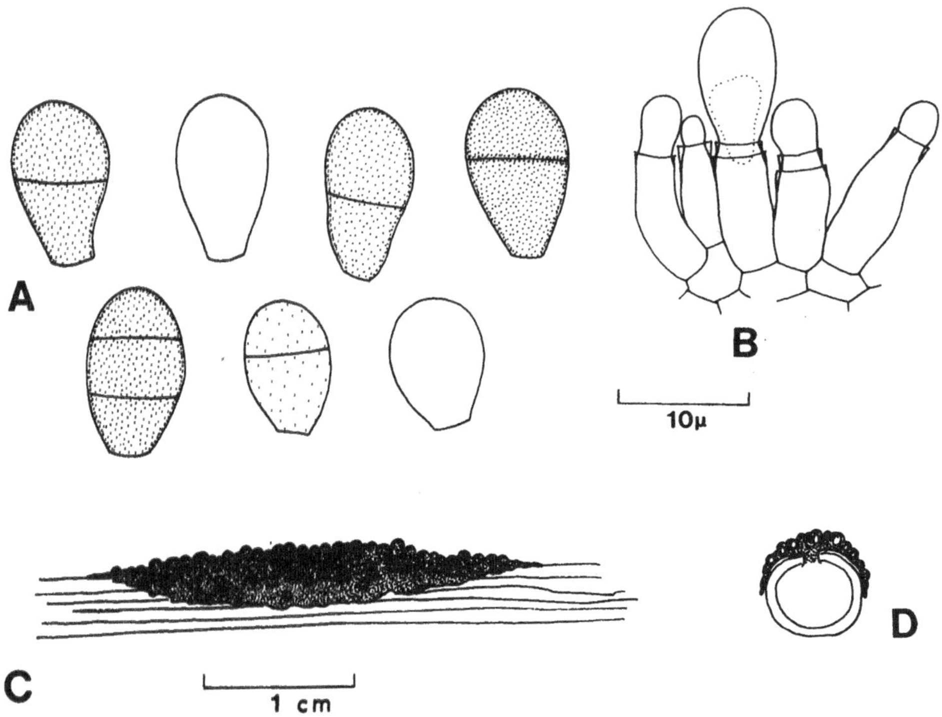

Fig. 188. *Dothideodiplodia agropyri*, A, conidia; B, conidiogenous cells and developing conidia; C, habit sketch of conidioma; D, diagrammatic sketch of vertical section of a conidioma.

Zambettakis (1955) placed *Dothideodiplodia* in synonymy with *Placodiplodia* Bub. but the stromata in the latter genus are completely immersed and conidiogenous cells are phialidic. The closest comparison is with *Dichomera* Cke where *D. gemmicola* Funk & Sutton (1972) only differs in having conidia with longitudinal septa. A second species, *D. lespedezae* Ziling, has not been examined.

D. AGROPYRI Murashkinsky, *ibid.*: 27 (1928). (Fig. 188)

Conidiomata up to 3 cm long. Locules up to 200 µ diam. × 150 µ high. Conidiogenous cells 8-12 × 3-5 µ. Conidia 10.5-14 × 7-8.5 µ.

On dead culms of *Agropyron repens*, Siberia, U.S.S.R. (IMI 199147, holotype ex LE).

27. LASMENIELLA Petrak & Syd., *Beih. Rep. spec. nov. regn. veg.* **42**: 301 (1927).

Original spp.: *L. guaranitica* (Speg.) Petrak & Syd., *L. globulifera* (Rabenh.) Petrak & Syd., *L. eucalypti* (Frag.) Petrak & Syd., *L. dendritica* (Racib.) Petrak & Syd., *L. ficina* (Syd.) Syd., *L. congoensis* (Har. & Pat.) Petrak & Syd., *L. machaerii* (P. Henn.) Petrak & Syd., *L. vochysiae* (P. Henn.) Petrak & Syd., *L. cocoes* (P. Henn.) Petrak & Syd., *L. alpina* (Ell. & Ev.) Petrak & Syd.

Lectotype sp.: *L. guaranitica* (Speg.) Petrak & Syd. (syn. *Lasmenia guaranitica* Speg.), fide Clements & Shear (1931).

Mycelium immersed, branched, hyaline, septate. *Conidiomata* eustromatic, subepidermal, separate or aggregated and confluent, forming circular colonies, black, ± hemispherical, shining, multilocular, the locules simple or convoluted; upper and lower walls of dark brown, thick-walled textura angularis, interlocular tissue of brown, thinner walled cells; locules lined with a layer of hyaline cells. *Ostioles* absent, dehiscence by irregular breakdown of upper walls at different points over the stromata. *Conidiophores* absent or only occasionally septate at the base. *Conidiogenous cells* holoblastic, determinate or indeterminate, discrete, hyaline to pale brown, swollen at the base, cylindrical to tapered above, often 1-3 flared percurrent proliferations, formed from the inner cells of the locular walls. *Conidia* brown, thick-walled, smooth, aseptate, flat, circular in outline, truncate at the base, often with a central pore.

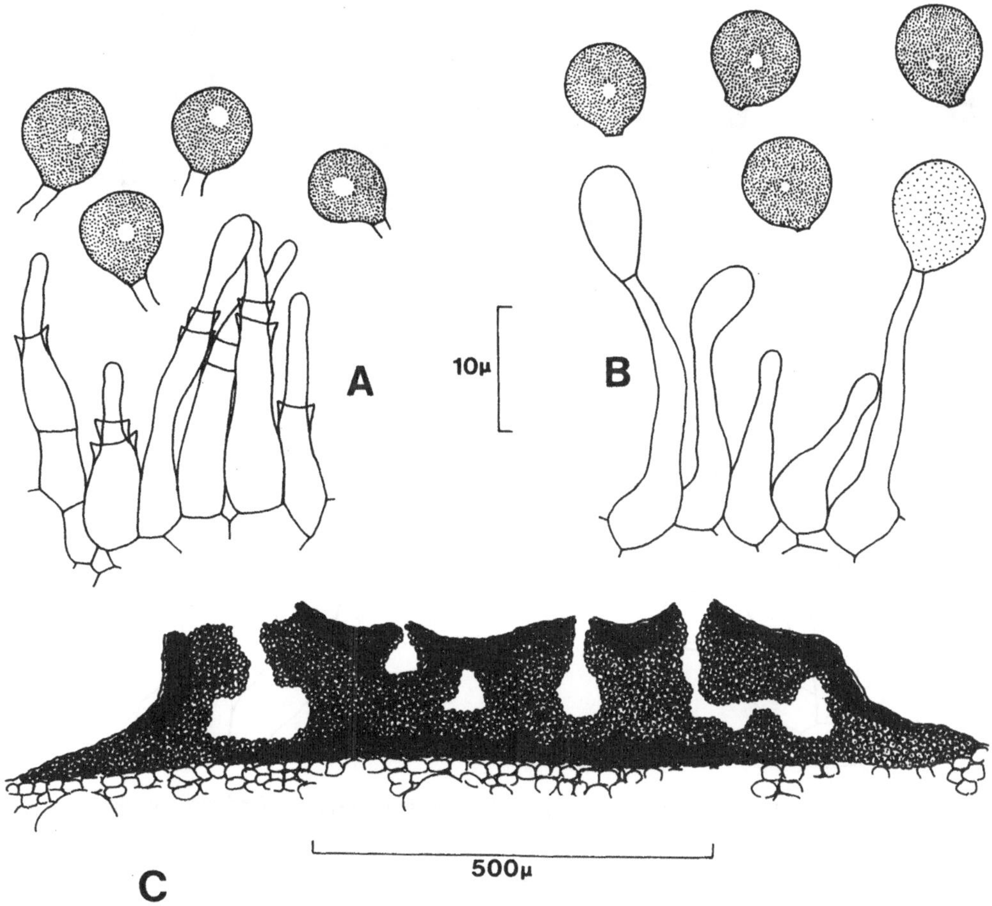

Fig. 189. A, *Lasmeniella congoensis*, conidia, conidiophores, B, C, *L. guaranitica*, conidia and conidiogenous cells, vertical section of a conidioma.

Petrak & Sydow (1927) accepted 10 species in the genus, and subsequent additions have brought the total to 13.

L. GUARANITICA (Speg.) Petrak & Syd., *ibid.*: 302 (1927). (Fig. 189B, C)
Lasmenia guaranitica Speg., *An. Soc. Cient. Arg.* **22**: 199 (1889).

Conidiomata up to 1200 µ diam., epidermal, multilocular. Conidiogenous cells 13-24 µ long × 1-1.5 µ wide at the apex, lacking annellations. Conidia 7-9 µ diam., medium brown, with a central clear spot.

On dead leaves of *Achatocarpus* sp., Paraguay (IMI 200650, isotype *L. guaranitica* ex K).

L. COCOES (P. Henn.) Petrak & Syd., *ibid.*: 312 (1927). (Fig. 190A)
Coniothyrium cocoes P. Henn., Engler's *Bot. Jahrb.* **34**: 53 (1927).

Conidiomata up to 320 µ diam., subepidermal, simple. Conidiogenous cells 10-29 µ long × 1.5-2 µ wide at the apex. Conidia 9.5-12 µ diam., thick-walled, very dark brown, lacking a central clear spot.

From leaves of *Cocos nucifera*, Jamaica, British Solomon Isl., Fla, U.S.A. (IMI 153680, 155435, 195452).

L. DENDRITICA (Racib.) Petrak & Syd., *ibid.*: 306 (1927). (Fig. 190B)
Aplosporella dendritica Racib., *Parasit. Alg. und Pilze Java's* **2**: 38 (1900).

Conidiomata up to 520 µ diam., subepidermal, multilocular. Conidiogenous cells 15-28 µ long × 1-1.5 µ wide at the apex, with up to 3 annellations. Conidia 15-16 µ diam., medium brown with a large central clear spot.

On leaves of *Eugenia jambolana*, *E. aquea*, Java, India (IMI 21872, 103999).

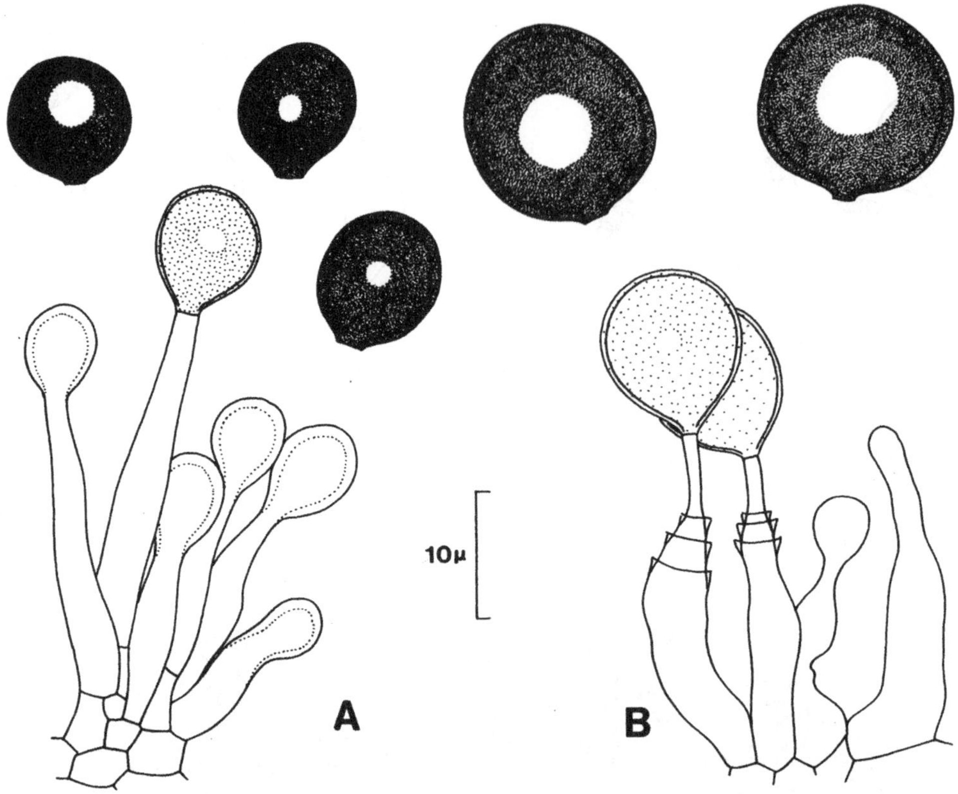

Fig. 190. A, *Lasmeniella cocoes*, conidia, conidiophores; B, *L. dendritica*, conidia, conidiogenous cells.

L. CONGOENSIS (Har. & Pat.) Petrak & Syd., *ibid.*: 308 (1927). (Fig. 189A)
Aplosporella congoensis Har. & Pat., *Bull. Mus. hist. nat. Paris* **5**: 369 (1911).

Conidiomata up to 600 µ diam., subepidermal, multilocular. Conidiogenous cells 11-18 µ long × 1-1.5 µ wide at the apex, with 1-2 flared annellations. Conidia 6.5-7.5 µ diam., medium brown, with a small central spot.

On leaves of *Albizzia gummifera*, Ghana (IMI 36904a).

28. LAMPROCONIUM (Grove) Grove, ***British Stem- and Leaf-fungi*** **2**: 321 (1937).
Melanconium Lk ex Fr. subgen. *Lamproconium* Grove, *Kew Bull. 1918*: 170 (1918).
Sp. typ.: *L. desmazieresii* (Berk. & Br.) Grove (syn. *Discella desmazieresii* Berk. & Br.).

Mycelium immersed, branched, septate, pale brown to hyaline. *Conidiomata* acervular, subperidermal, separate, formed of pale brown to hyaline, thin-walled textura angularis. Dehiscence irregular. *Conidiophores* hyaline, septate, branched at the base, filiform, cylindrical, formed from the upper pseudoparenchyma. *Conidiogenous cells* holoblastic, annellidic, integrated, indeterminate, cylindrical, hyaline, smooth, with 1-3 flared percurrent proliferations. *Conidia* ellipsoid, aseptate, thick-walled, smooth, base truncate, apex obtuse to bluntly apiculate, bluish. *Paraphyses* filiform, hyaline, branched, septate, longer than conidiophores, often with a gelatinous sheath.

The genus is monotypic.

L. DESMAZIERESII (Berk. & Br.) Grove, *ibid.*: 321 (1937). (Fig. 191)
Discella desmazieresii Berk. & Br., *Ann. nat. Hist.* **5**: 377 (1850).
Melanconium desmazieresii (Berk. & Br.) Sacc., *Michelia* **2**: 355 (1880).
Epidochium maertensii Westd., *Bull. Acad. Sci. Brux.* **21**: 238 (1854).

Conidiomata up to 900 µ diam. Conidiophores up to 40 µ long × 2 µ wide. Paraphyses up to 100 µ long × 2 µ wide. Conidia 29-32.5 × 8-9 µ.

On branches of *Tilia grandifolia*, *T.* sp., Austria, G.B. (IMI 721, 725, 3065, 180535).

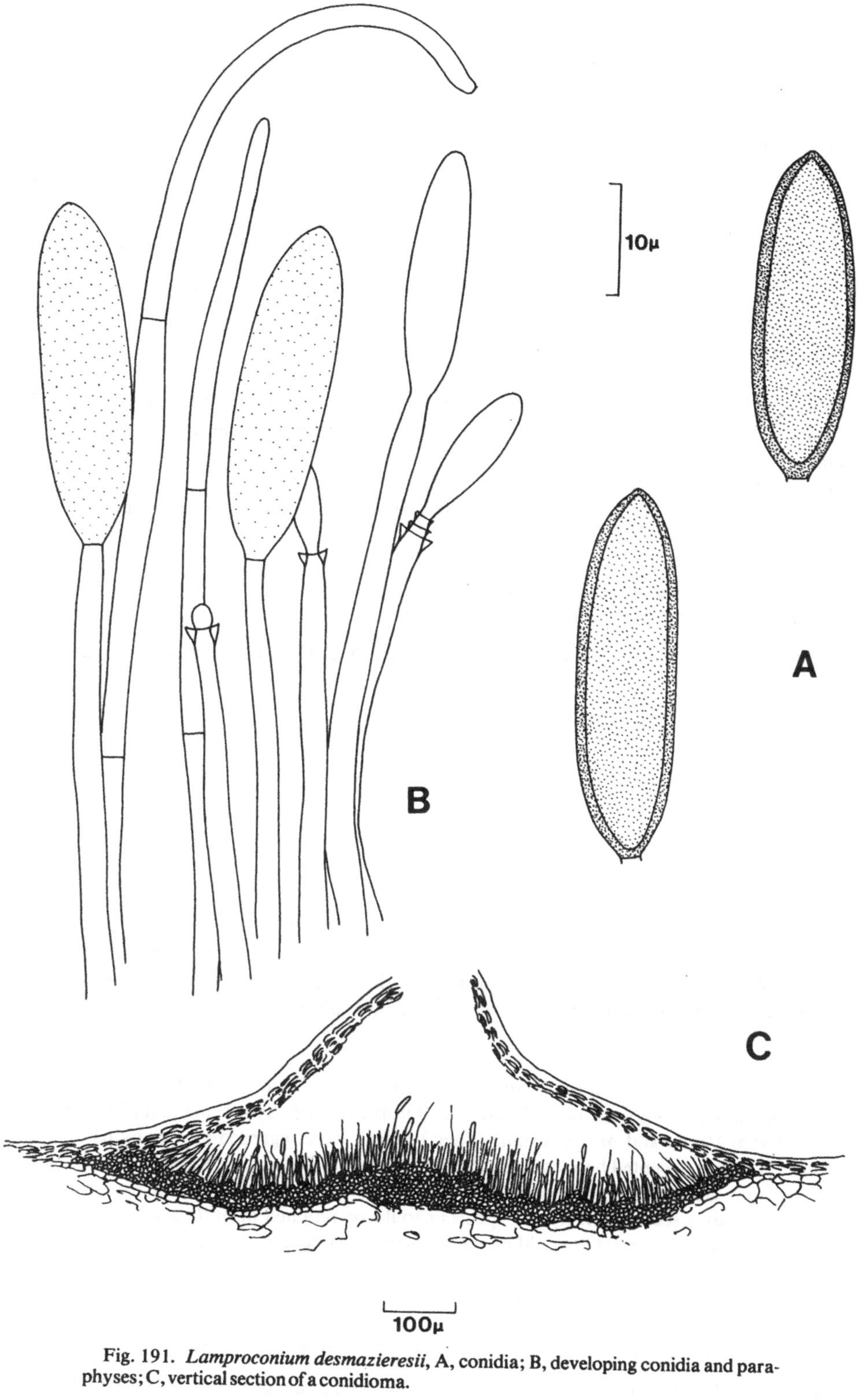

Fig. 191. *Lamproconium desmazieresii*, A, conidia; B, developing conidia and paraphyses; C, vertical section of a conidioma.

29. MELANCONIUM Lk ex Fr., *Syst. Myc.* **1**: XL (1821).
Melanconium Lk, *Mag. Ges. naturf. Fr. Berl.* **3**: 9 (1809).
Dapsilosporium Cda in Sturm, *Deutschlands Flora* **3**: 75 (1837).
Trichoconium Cda in Sturm, *Deutschlands Flora* **3**: 73 (1837).
Callosisperma Preuss, *Linnaea* **26**: 721, *1853* (1855).
Original spp.: None listed.
Lectotype sp.: *M. atrum* Lk ex Schlecht.
Illustration: Sutton (1964c); Wehmeyer (1941).

Mycelium immersed, branched, septate, hyaline to subhyaline. *Conidiomata* acervular, peridermal or subperidermal, lenticular, separate or occasionally confluent, composed of pale brown, thin-walled textura angularis. Dehiscence irregular, black conidial masses often spreading. *Conidiophores* septate and branched only at the base, hyaline, smooth, filiform, formed from the upper cells of the conidiomata. *Conidiogenous cells* holoblastic, annellidic with up to 5 apical annellations, indeterminate, discrete, less often integrated, hyaline, smooth. *Conidia* aseptate, brown, smooth, guttulate or eguttulate, globose to ellipsoid, base truncate.

The status of the type species has been reviewed by Sutton (1964c). More than 200 taxa have been described in the genus, and apart from the removal by Petrak & Sydow (1927) of a few species to other genera there has been no revision of the group. Wehmeyer (1941) included several taxa as the anamorphoses of *Melanconis* species but the names applied and their synonymies largely originated from the work of Tulasne & Tulasne (1863).

Without extensive type studies related to correlations with teleomorphoses there can be no meaningful account of the genus.

M. ATRUM Lk ex Schlecht., *Flora berol.* **2**, *Cryptogamia*: 136 (1824).
Melanconium atrum Lk, *Mag. Ges. naturf. Fr. Berl.* **3**: 9 (1809).

Conidiomata .5-1.5 mm diam. Conidiophores 10-30 × 2.5-4 μ. Conidia elliptical, with a central guttule, 10-12 × 6.5-7.5 μ.

On bark of *Fagus sylvatica* (IMI 102914, lectotype ex L).

30. MELANCONIOPSIS Ell. & Ev., *Bull. Torrey bot. Cl.* **27**: 275 (1900).
Sp. typ.: *M. inquinans* Ell. & Ev.

Mycelium immersed, branched, septate, hyaline. *Conidiomata* acervular, peridermal to subperidermal, separate, irregular in shape, convoluted, composed of hyaline to pale brown textura angularis. Dehiscence irregular. *Conidiophores* absent. *Conidiogenous cells* holoblastic, annellidic, indeterminate, discrete, cylindrical, hyaline, smooth, formed from the upper pseudoparenchyma. *Conidia* aseptate, thick-walled, dark brown, often guttulate, smooth, spherical, truncate at the base.

So far, seven taxa have been described in *Melanconiopsis.* Petrak & Sydow (1927) accepted two species, the type being included in subgen. *Eumelanconiopsis*, and *M. olivacea* (Otth) Petrak & Syd. being placed in subgen. *Otthinula. M. ulmigena* Höhn., *M. ailanthi* Höhn. and *M. incrustans* (Sacc.) Petrak were regarded as synonyms of *Cytoplea juglandis* (Schum.) Petrak, which in turn is accepted here as *Cyclothyrium juglandis* (Schum. ex Rabenh.) Sutton. *M. elzoi* Speg. has not been re-investigated, but Butin (1968) records it as common in Chile. In *M. africana* Luc, the conidia eventually develop 1-2 transverse septa so it seems unlikely the species should be retained in *Melanconiopsis*. Structure of the conidiomata, conidiophores and conidia suggest it would be more correctly assigned to either *Camarosporium* or *Dichomera.*

M. INQUINANS Ell. & Ev., *ibid.*: 275 (1900). (Fig. 192)

Teleomorphosis: *Massariovalsa sudans* (Berk. & Curt.) Sacc. *Syll. fung.* **2**: LV (1883).

Conidiomata up to 750 μ diam. Conidiogenous cells 8-19 × 4-6.5 μ. Conidia 24-27 × 16-17.5 μ.

On branches of *Acer macrophyllum, Carya* sp., Calif. and Fla, U.S.A. (IMI 154427, 94617a, 94616).

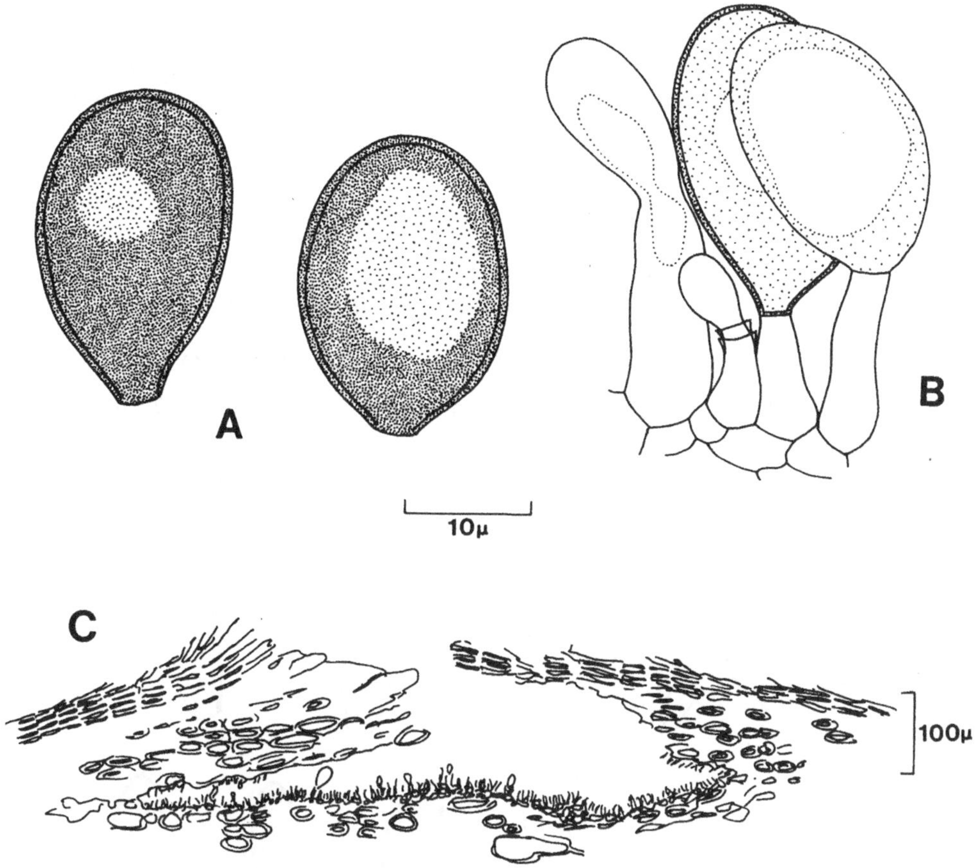

Fig. 192. *Melanconiopsis inquinans*, A, conidia; B, conidiogenous cells and developing conidia; C, vertical section of a conidioma.

31. HYMENOPSIS Sacc., *Syll. fung.* **4**: 744 (1886).
Schoenbornia Bub., *Bull. l'Herb. Boissier*, 2 ser., **6**(5): 483 (1906).
Chaetodiscula Bub. & Kabát, *Hedwigia* **50**: 44 (1910).
Phaeopolynema Speg., *An. Mus. nac. Buenos Aires* **23**: 117 (1912).
Original spp.: 16 spp., all of which were transfers from existing genera.
Lectotype sp.: *H. trochiloides* (Sacc.) Sacc. (syn. *Myrothecium trochiloides* (Sacc.) Sacc.) fide Clements & Shear (1931).

Mycelium immersed, hyaline, branched, septate. *Conidiomata* eustromatic, immersed, dark greenish brown, separate, circular to longitudinally elongated, unilocular, at first closed, later opening out to become shallow disciform to acervular, composed of a thick basal wall of hyaline, thin-walled textura angularis extending at the margin and at first above into brown, thicker-walled textura porrecta, occasionally in the upper part running out into short sterile hyphae. *Setae* sometimes present, brown, septate. *Ostiole* absent, dehiscence by breakdown of the upper wall and its recurvature. *Conidiophores* hyaline, branched at the base, 1-2 septate, smooth, cylindrical, formed from the upper cells of the basal wall, also from the lateral walls. *Conidiogenous cells* holoblastic, annellidic, indeterminate, integrated, cylindrical, expanding and pale to dark brown towards the apices, hyaline elsewhere, smooth below, slightly rough above, with up to 3 flared, thickened, percurrent proliferations. *Conidia* fusiform, occasionally pyriform, apex obtuse, base truncate, brown, thick-walled, ± guttulate, aseptate, verruculose or smooth, straight, sometimes with an apical and/or basal gelatinuous appendage.

Petrak (1921) regarded *Schoenbornia* as a synonym of *Trullula* but this was based on his own collections, not the type. I have studied the type of *S. basidio-annulata* and *Cesatia spartii* Ces. and they are entirely different species.

Clements & Shear (1931) placed both *Schoenbornia* and *Phaeopolynema* Speg. as synonyms of *Coniothyris* Clem. & Shear (=*Coniothyriella* Speg., nom. confus.). The basic generic synonymy has been outlined by Sutton (1977a). It is difficult to assess the number of taxa referable to *Hymenopsis*. Tulloch (1972) lists many species placed in *Myrothecium* and related genera as belonging here. Apart from the taxa listed below it seems quite probable that *Hymenopsis* will eventually be known as a comparatively large genus.

H. saccardoana (Speg.) comb. nov. (Fig. 193)
Diplodia saccardoana Speg., *Atti Soc. Critt. Ital.* ser. 2, **3**: 70 (1881).
Schoenbornia saccardoana (Speg.)., Bub., *Annls mycol.* **14**: 153 (1916).
Sphaeropsis saccardoana (Speg.) Sacc., *Syll. fung.* **3**: 292 (1884).
Coniothyrium saccardoanum (Speg.) Tassi, *Bull. Lab. Ort. Bot. Siena* **5**: 25 (1902).
Coniothyrium saccardoanum (Speg.) Allesch. in Rabenh., *Krypt. Fl.* **7**: 55 (1903).
Schoenbornia basidio-annulata Bub., *Bull. l'Herb. Boissier*, 2 ser., **6**(5): 484 (1906).

Conidiomata up to 300 µ diam., ± cupulate. Conidiophores up to 23 µ long × 1.5-2.5 µ wide, with 1-2 proliferations. Conidia fusiform to clavate, 10.5-12 × 3.5-4 µ.

On dead stems of *Spartium junceum*, Yugoslavia, Austria (IMI 194249, holotype of *S. basidio-annulata* ex FH, 194250).

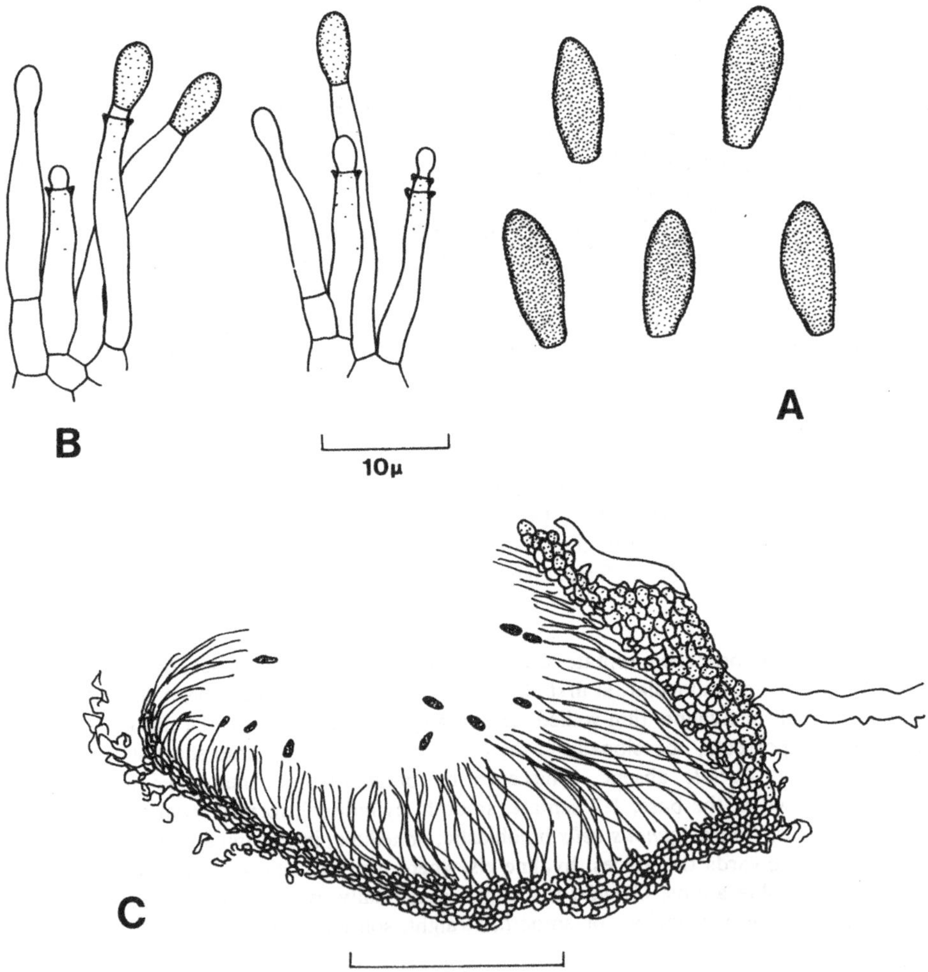

Fig. 193. *Hymenopsis saccardoana*, A, conidia; B, conidiophores and developing conidia; C, vertical section of a conidioma.

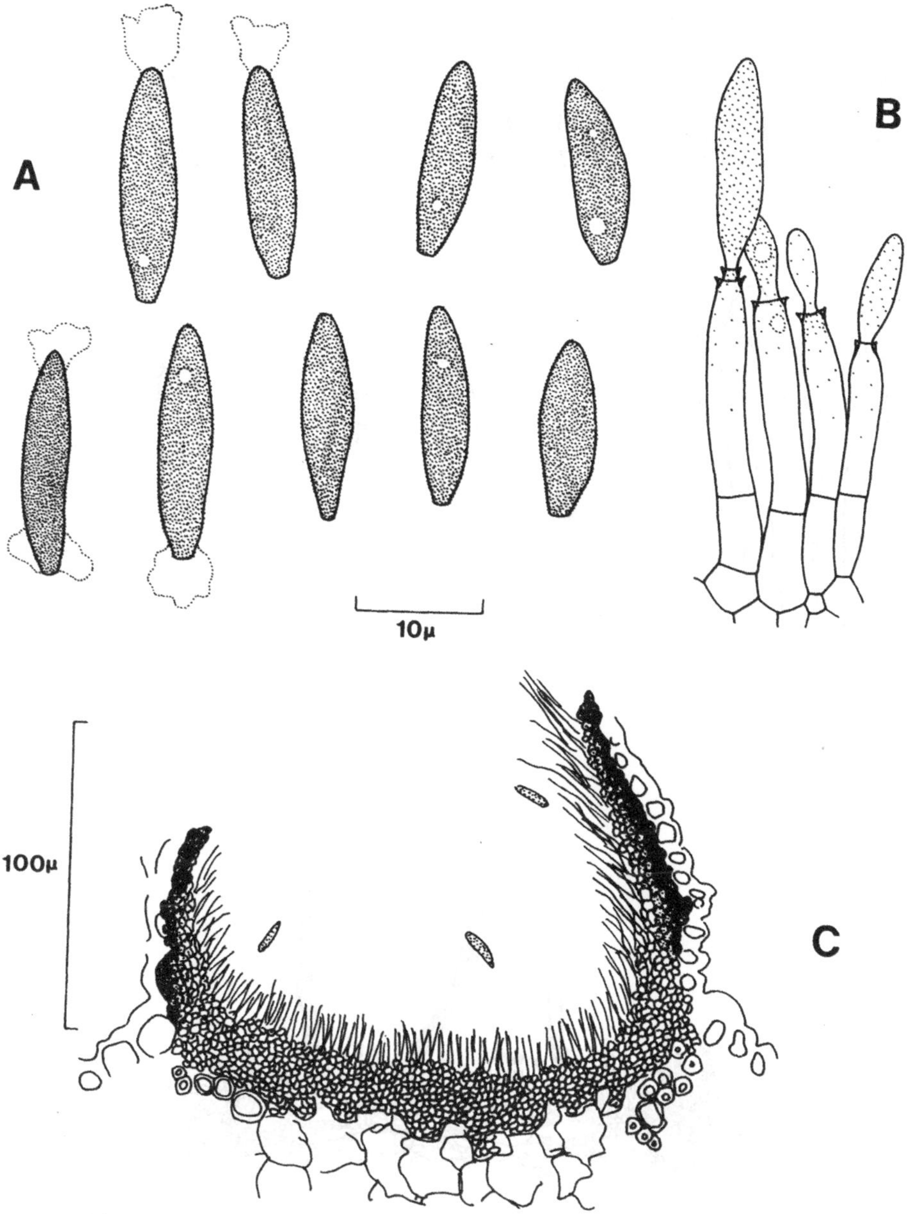

Fig. 194. *Hymenopsis typhae*, A, conidia; B, conidiophores and developing conidia; C, vertical section of a conidioma.

H. typhae (Pk) comb. nov. (Fig. 194)

Melanconium typhae Pk, *Bot. Gaz.* **6**: 275 (1881).
Cryptomela typhae (Pk) Died., *Krypt. Fl. Mk Brandenb.* **9**(7): 865 (1915).
Chaetodiscula hysteriformis Bub. & Kabát, *Hedwigia* **50**: 44 (1910).
Fusella typhae Lindau, Rabenh., *Krypt. Fl.* **1**(8): 566 (1907).

Conidiomata longitudinally elongated, up to 350 μ long × 150 μ wide. Conidiophores up to 25 μ long × 2.5-3 μ wide, 1-2 proliferations, not pigmented. Conidia fusiform, 14-19 × 3.5-4.5 μ, sometimes with gelatinous appendages.

This species is popularly known as *Cryptomela typhae* but *Cryptomela* is a later synonym of *Cryptosporium* Kze ex Cda and is therefore untenable.

On dead stems and leaves of *Typha latifolia*, U.K., Czechoslovakia (IMI 194177, holotype of *Chaetodiscula hysteriformis* ex FH, 164985, 17861, 34285, 86856a, 34406, 89623a, 86855a).

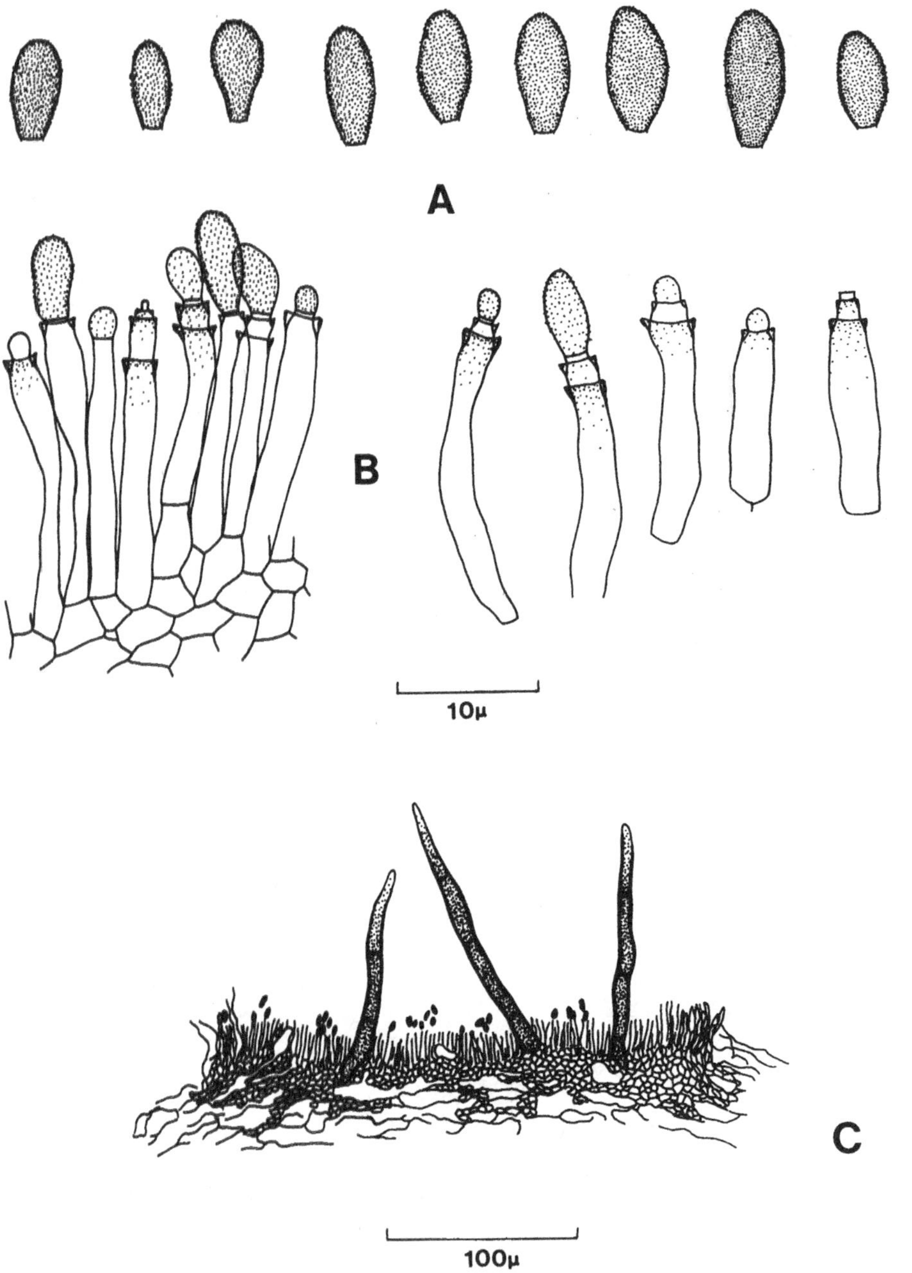

Fig. 195. *Hymenopsis argentinensis*, A, conidia; B, conidiophores and developing conidia; C, vertical sections of a conidioma.

H. argentinensis (Speg.) comb. nov. (Fig. 195)
Phaeopolynema argentinense Speg., *An. Mus. nac. Buenos Aires* **23**: 117 (1912).
Amerosporium argentinense (Speg.) Petrak & Syd., *Annls mycol.* **35**: 335 (1937).

Conidiomata applanate, occasionally setose, up to 300 µ diam. Conidiophores 12-20 × 2-3.5 µ with 1-3 brown annellations. Conidia often clavate, occasionally fusiform, 7-9 × 3-4.5 µ.

On dead stems of *Boehmeria candida*, Argentina (IMI 110813, holotype of *Phaeopolynema argentinense* ex LPS, 110812).

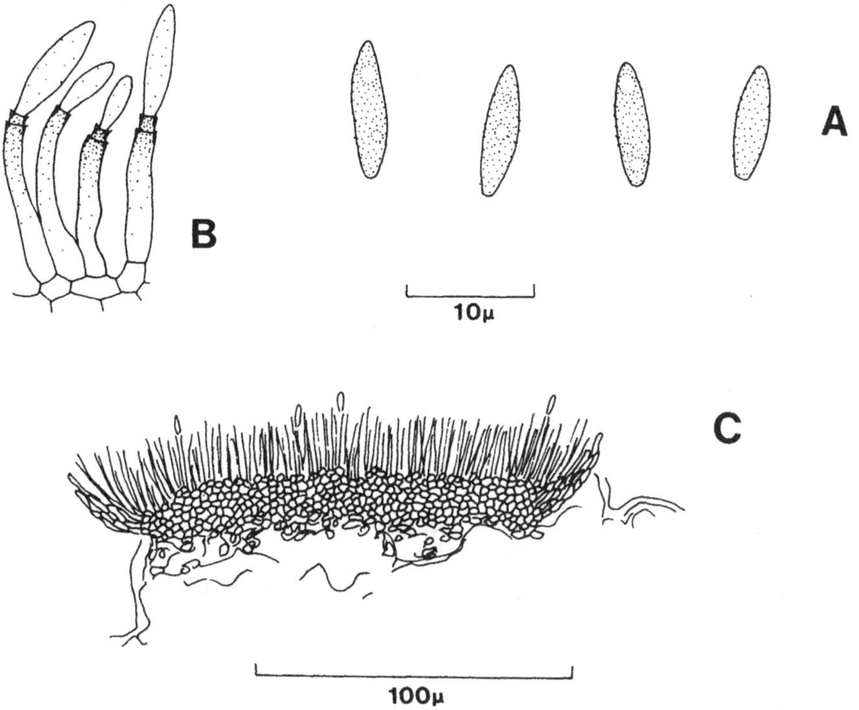

Fig. 196. *Hymenopsis trochiliodes*, A, conidia; B, conidiogenous cells and developing conidia; C, vertical section of a conidioma.

H. TROCHILOIDES (Sacc.) Sacc., *Syll. fung.* **4**: 144 (1886). (Fig. 196)
Myrothecium trochiloides Sacc., *Michelia* **2**: 367 (1881).

Conidiomata up to 250 μ diam., effuse cupulate. Conidiophores 12-16 × 1.5-2 μ, with up to 3 proliferations. Setae not observed in the type, but present in an otherwise conspecific isolate (IMI 206198). Conidia fusiform, pale greenish brown, often smooth, 9.5-11 × 2.5-3 μ.

On leaves and stems of *Phragmites communis*, and from *Lens culinaris*, France, India (IMI 152287, holotype ex PAD, 33868, 33867, ex K, 147046, 152288, 148786 ex PAD, 206198).

32. GAUBAEA Petrak, *Bot. Arch.* **43**: 89 (1941).
Sp. typ.: *G. bornmuelleri* (P. Magn.) Petrak (syn. *Leptothyrium bornmuelleri* P. Magn., *G. insignis* Petrak).

Mycelium immersed, septate, branched, pale brown to hyaline. *Conidiomata* eustromatic, subcuticular, applanate, circular, separate, dark brown to black; thin upper wall consisting of small, dark brown, thick-walled textura angularis, lower wall of pale brown, discrete textura angularis or textura intricata. *Ostiole* absent, dehiscence by irregular rupture of cuticle and upper wall. *Conidiophores* absent. *Conidiogenous cells* holoblastic, annellidic, indeterminate, discrete, cylindrical or lageniform, pale brown, verruculose with a single ragged percurrent proliferation, formed from the upper cells of the lower wall. *Conidia* pale brown, thin-walled aseptate, fusiform to obpyriform, apex obtuse, base broadly truncate, verruculose.

Petrak (1947a) found an earlier epithet for the type species, and Speer (1972) added a second species, *G. rechingeri* Speer on *Atraphaxis* which differs in having longer conidiophores (15-25 × 3-6 μ) and larger conidia (18-21 × 6-7 μ).

G. BORNMUELLERI (P. Magn.) Petrak, *Sydowia* **1**: 67 (1947). (Fig. 197)
Leptothyrium bornmuelleri P. Magn., *Ber. dtsch. Bot. Ges.* **19**: 447 (1901).
Gaubaea insignis Petrak, *Bot. Arch.* **43**: 89 (1941).

Conidiomata up to 400 μ diam. Conidiogenous cells 10-14 × 4-9 μ. Conidia 12.5-15.5 × 5.5-6.5 μ.

On dead stems of *Calligonum arborescens*, Turkmenia (IMI 202258 ex ZT).

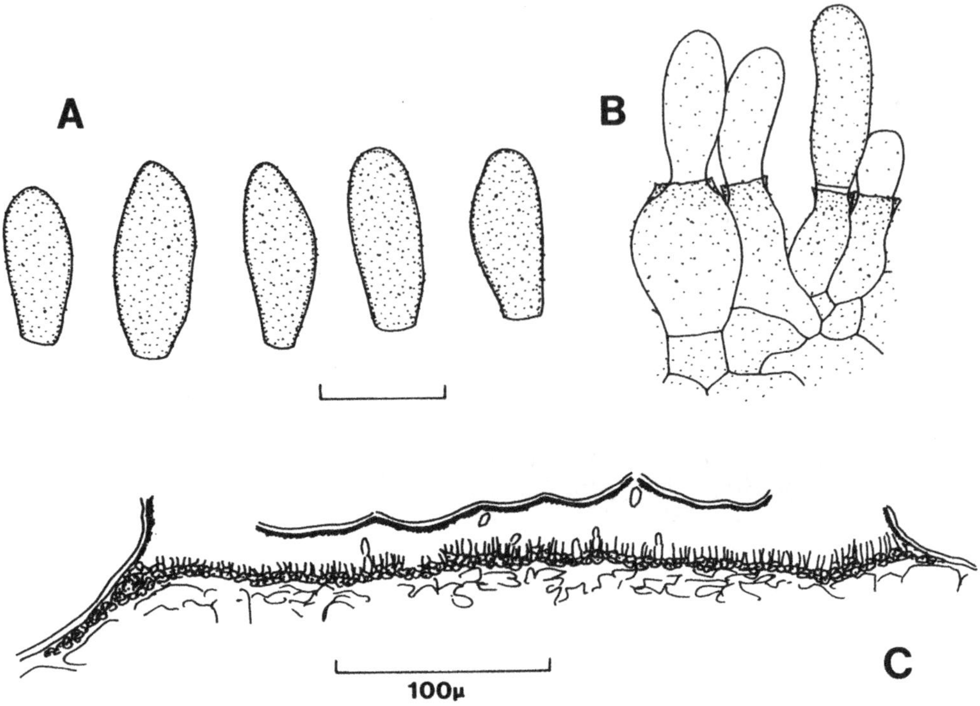

Fig. 197. *Gaubaea bornmuelleri*, A, conidia; B, conidiogenous cells with developing conidia; C, vertical section of a conidioma.

33. ENERTHIDIUM Syd., *Annls mycol.* **37**: 244 (1939).
Sp. typ.: *E. canarii* Syd.

Mycelium immersed, branched, septate, pale brown to hyaline. *Conidiomata* acervular, amphigenous, epidermal to subepidermal, separate, rarely confluent, formed of pale brown, thin-walled textura angularis. Dehiscence irregular. *Conidiophores* absent. *Conidiogenous cells* holoblastic, annellidic, discrete, indeterminate, cylindrical, straight or flexuous, hyaline to pale brown, smooth, with 2-5 darker percurrent proliferations, formed from the upper pseudoparenchyma. *Conidia* pale brown, aseptate, thin-walled, smooth, base truncate with a frill, apex obtuse, with a guttule at each end of the conidium, straight, cylindrical.

The genus is monotypic.

E. CANARII Syd., *ibid.*: 244 (1939). (Fig. 198)

Conidiomata up to 180 µ diam. Conidiogenous cells 12-20 × 2-2.5 µ. Conidia 6-7.5 × 3 µ.

Causing lesions on leaves of *Canarium schweinfurthii*, Sierra Leone (IMI 21654, holotype).

34. PIGGOTIA Berk. & Br., *Ann. Mag. Natur. Hist.* ser. 2, **7**: 95 (1851).
Basiascella Bub., *Ann. Naturh. Hofmus., Wien* **28**: 216 (1914).
Sp. typ.: *P. ulmi* (Grev.) Keissl. (syn. *Asteroma ulmi* Grev., *P. astroidea* (Berk.) Berk. & Br.).

Mycelium immersed, branched, septate, pale brown. *Conidiomata* acervular, subcuticular, applanate or irregularly pulvinate with copious conidial masses, separate or frequently coalescing to form large composite conidiomata, black, cuticle dark brown, lower wall of pale to medium brown almost cuboid textura prismatica. *Ostiole* absent, dehiscence by irregular rupture of the cuticle. *Conidiophores* branched and septate only near the base, pale brown, smooth, cylindrical, usually straight, formed from the upper cells of the acervulus. *Conidiogenous cells* holoblastic, annellidic, indeterminate, discrete, cylindrical, pale brown, smooth or verruculose above, with 1-2 simple or ragged percurrent proliferations. *Conidia* pale brown, thin-walled, aseptate, smooth, cylindrical to almost cuneiform, apex obtuse, base truncate.

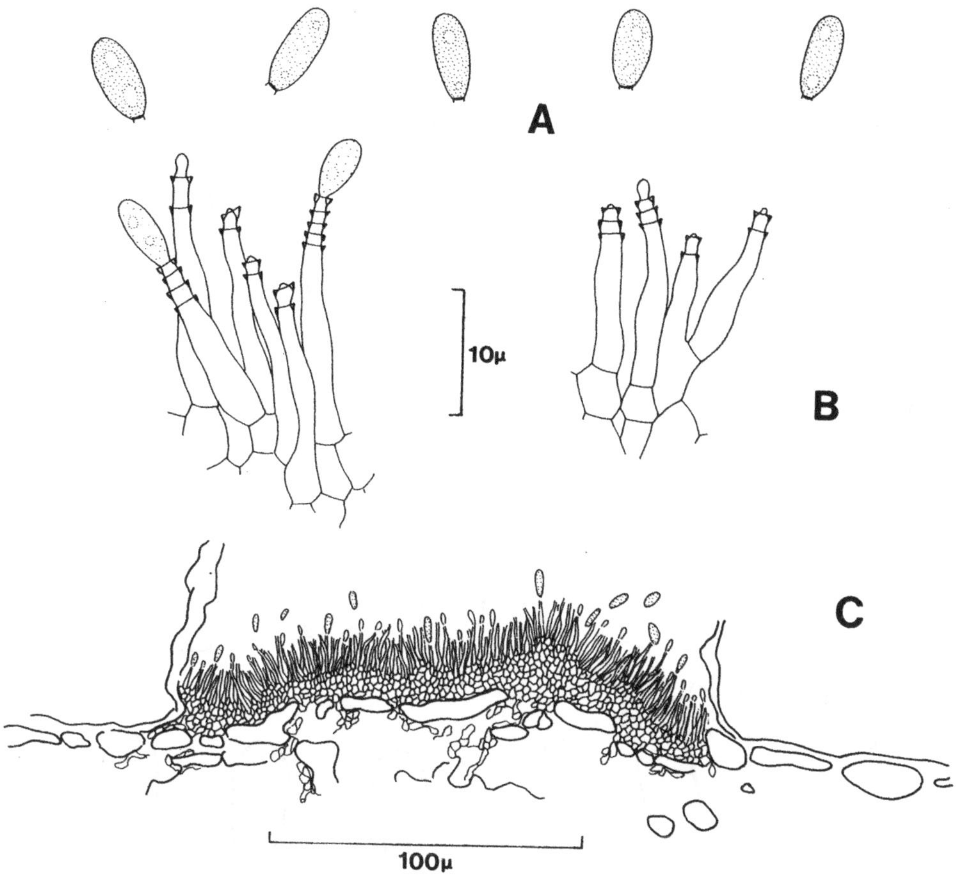

Fig. 198. *Enerthidium canarii*, A, conidia; B, conidiophores and developing conidia; C, vertical section of a conidioma.

This is a comparatively small genus with only 10 taxa described in it. *P. cinnamomi* P. Henn. and *P. gneti* P. Henn. were both regarded as synonyms of conidial *Glomerella cingulata* (Stonem.) Spauld. & Schrenk by von Arx (1957a, 1970), and *P. fraxini* Berk. & Curt. was transferred to *Asteromella* by Petrak (1923). The remainder still need to be revised.

P. ULMI (Grev.) Keissler, *Ann. Naturh. Mus. Wien* 1923-1933: 207 (1933). (Fig. 199B, C)
Asteroma ulmi Grev., *Flora edin.*: 368 (1824).
Dothidea astroidea Berk. in Smith, *Engl. Flora* **5**(2): 287 (1836).
Piggotia astroidea (Berk.) Berk. & Br., *Ann. Mag. Natur. Hist.* ser. 2, **7**: 95 (1851).
Basiascella gallarum Bub., *Ann. Naturh. Hofmus., Wien* **28**: 216 (1914).

Teleomorphosis: *Platychora ulmi* (Schleich.) Petrak, *Annls mycol.* **23**: 103 (1925).

Conidiomata up to 1000 μ diam. Conidiogenous cells 10-12 × 4-4.5 μ, with ragged annellations. Conidia 8-10 × 4.5-5 μ.

Causing leaf lesions on *Ulmus montana*, *U. campestris*, *U. procera*, G.B., Austria (IMI 12584, 22669, 111060, 202558, holotype *D. astroidea* ex K).

P. coryli (Desm.) comb. nov. (Fig. 199A)
Cheilaria coryli Desm., *Ann. Sci. nat.* ser. 3, **20**: 226 (1853).
Gloeosporium coryli (Desm.) Sacc., *Michelia* **2**: 117 (1880).
Labrella coryli (Desm.) Sacc., *Syll. fung.* **3**: 648 (1884).
Monostichella coryli (Desm.) Höhn., *Sber. Akad. Wiss. Wien* **125**: 95 (1916).

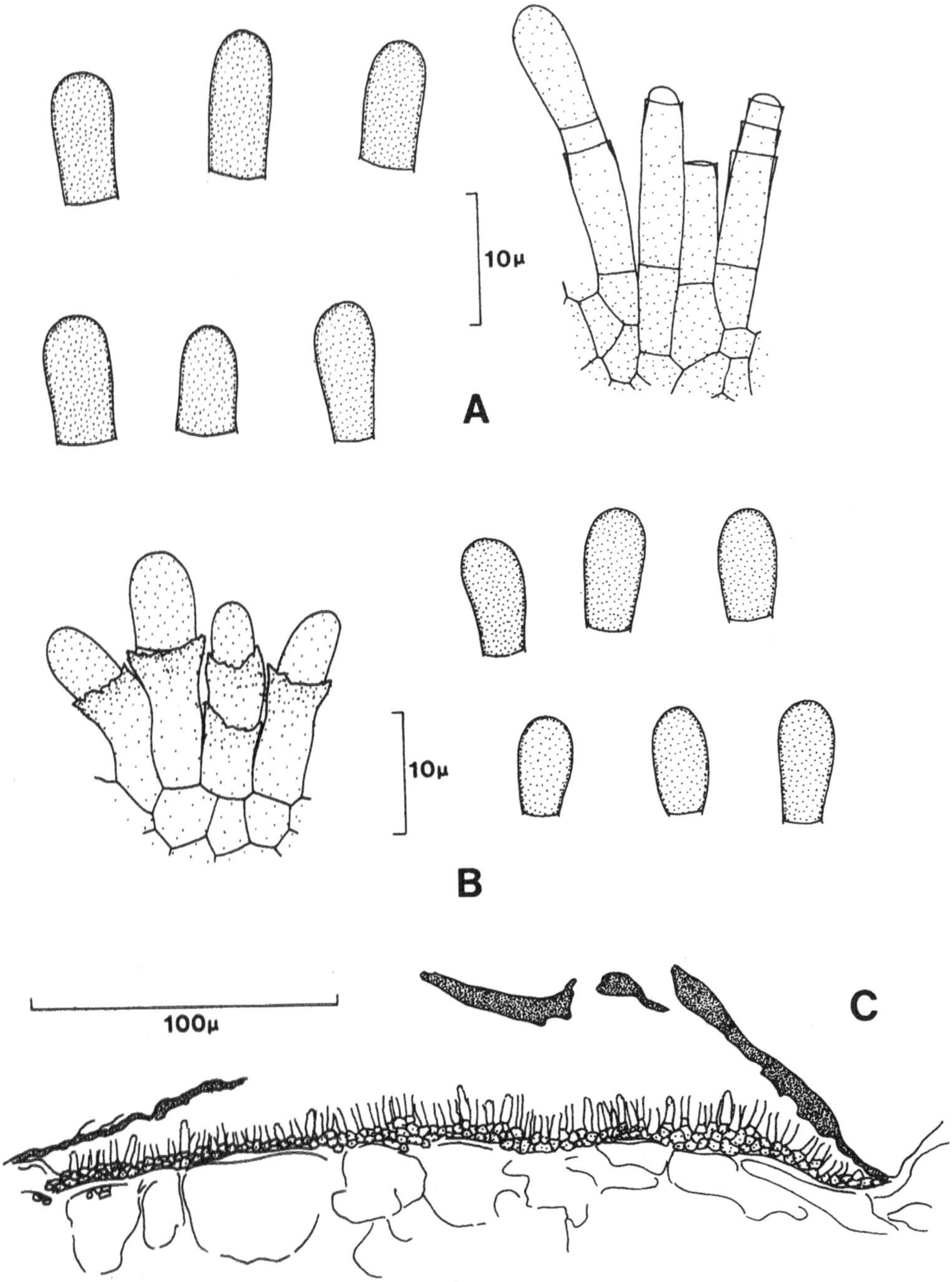

Fig. 199. A, *Piggotia coryli*, conidia and conidiogenous cells, B, C, *P. ulmi*, conidia and conidiogenous cells, vertical section of a conidioma.

Conidiomata up to 150 μ diam. Conidiogenous cells 9-13 × 3.5-4 μ, with smooth annellations. Conidia 8.5-11 × 4-5 μ.

Overholts (1926), von Arx (1970) and Vassiljevsky & Karakulin (1950) give a wider range in conidial size (10-15 × 4.5-6.5 μ, 12-16 × 5-7 μ, and 7-17 × 5-7 μ respectively), which substantiates the separation of this species from *P. ulmi* somewhat more convincingly than the material I have examined.

On leaves of *Corylus avellana*, *C. rostrata*, *C. californica*, *C.* sp., U.K., Germany, Latvia, U.S.S.R., Italy, Ont. Canada & Wash., U.S.A. (IMI 21865, 33572, 13146, 21703, 21704, 21705, 31171, 101824).

35. DIDYMOSPORINA Höhn., *Sber. Akad. Wiss. Wien* **125**: 83 (1916).
Phaeomarssonia Bubák, *Bot. Közl.* **14**: (75), (1915).
Sp. typ.: *D. aceris* (Lib.) Höhn. (syn. *Cheilaria aceris* Lib.).

Mycelium immersed, branched, septate, pale brown. *Conidiomata* acervular, epidermal, epiphyllous, separate, rarely confluent, formed of pale brown, thin-walled textura angularis. Dehiscence irregular. *Conidiophores* absent. *Conidiogenous cells* holoblastic, annellidic, discrete, indeterminate, short cylindrical or doliiform, pale brown, smooth, with 1-3 percurrent proliferations, formed from the upper pseudoparenchyma. *Conidia* cuneiform to cylindrical, medianly 1- euseptate or lower cells slightly smaller, continuous, olivaceous to pale brown, thin-walled, smooth, base wide, truncate, apex obtuse.

Although *Phaeomarssonia* antedates *Didymosporina* it is regarded as a later homonym of *Phaeomarssonia* Speg. (published as *Phaeomarsonia*). Synonymies have been listed for the species by Höhnel (1916d) and Vassiljevsky & Karakulin (1950). Apart from the type species, only *D. africana* H. Syd. has been described in the genus. There has been no reassessment of this species which was reported from leaves of *Rhus viminalis* in S. Africa.

D. ACERIS (Lib.) Höhn., *ibid.*: 83 (1916). (Fig. 200)
Synonymy in Höhnel (1916d).

Conidiomata up to 115 µ diam. Conidiogenous cells 6-8 × 3-5 µ. Conidia 7-9.5 × 3-4 µ.

Causing leaf lesions on *Acer campestris, A. negundo*, Roumania, Italy, Germany, France (IMI 164169b ex PAD, 21612, 12346, 22153, 61602).

36. LEPTOMELANCONIUM Petrak apud Syd., *Annls mycol.* **21**: 179 (1923).
Sp. typ.: *L. allescheri* (Schnabl) Petrak (syn. *Cryptomela allescheri* Schnabl, *L. asperulum* (Moesz) Petrak).

Mycelium immersed, branched, septate, brown. *Conidiomata* acervular, subepidermal, separate or confluent, formed of pale brown, thin-walled textura angularis. Dehiscence irregular or regular. *Conidiophores* hyaline, branched at the base, septate, smooth, formed from the upper pseudoparenchyma. *Conidiogenous cells* holoblastic, annellidic, integrated or discrete, indeterminate, cylindrical, pale brown, verruculose, with up to 4 verruculose percurrent proliferations. *Conidia* navicular to clavate, 0-1 euseptate, continuous, brown, thick-walled, verruculose, base truncate.

Sutton & Chao (1970) redefined the genus and Sutton (1974) added a third species. Morgan-Jones (1971d) also placed *Gloeocoryneum cinereum* (Dearn.) Weindlmayr in *Leptomelanconium* but the long conidiogenous cells lacking annellations and 3-septate conidia tend to extend the limits of this genus unnecessarily.

L. ALLESCHERI (Schnabl) Petrak, *Sydowia* **16**: 358 (1963). (Fig. 201C, D, E)
Synonymy in Sutton & Chao (1970).

Conidiomata up to 400 µ diam. Conidiophores rarely septate, basally branched, 1 annellation, 13-17 × 3.5 µ. Conidia 8-12 × 3.5-5 µ, aseptate, navicular.

On needles of *Pinus mugho, P. montana, P. pumilio, P. monticola, P.* sp., Germany, Czechoslovakia, Canada (IMI 144360, 144361, 144362, 144363 ex PR, 24741, 132454).

L. PICEAE Sutton & Chao, *Trans. Br. mycol. Soc.* **55**: 40 (1970). (Fig. 201A)

Conidiomata up to 500 µ diam. Conidiophores branched and septate, 1 annellation, 10-15 × 3-4 µ. Conidia 6-9 × 3-4 µ, medianly 1 septate, elliptical to clavate.

On cones of *Picea mariana*, Ont., Man., Sask., Canada (IMI 142894, holotype, 142897, 142899, 142898, 142896).

L. AUSTRALIENSE Sutton, *Nova Hedwigia* **25**: 163 (1974). (Fig. 201B)

Conidiomata 150-300 µ diam. Conidiogenous cells cylindrical or doliiform, up to 2 annellations, 7-11 × 3.5-7 µ. Conidia 7-11 × 4.5-6 µ, aseptate, elliptical to pyriform.

Causing lesions on leaves of *Eucalyptus ficifolia*, Australia (IMI 159079a, holotype, 190003, 190004).

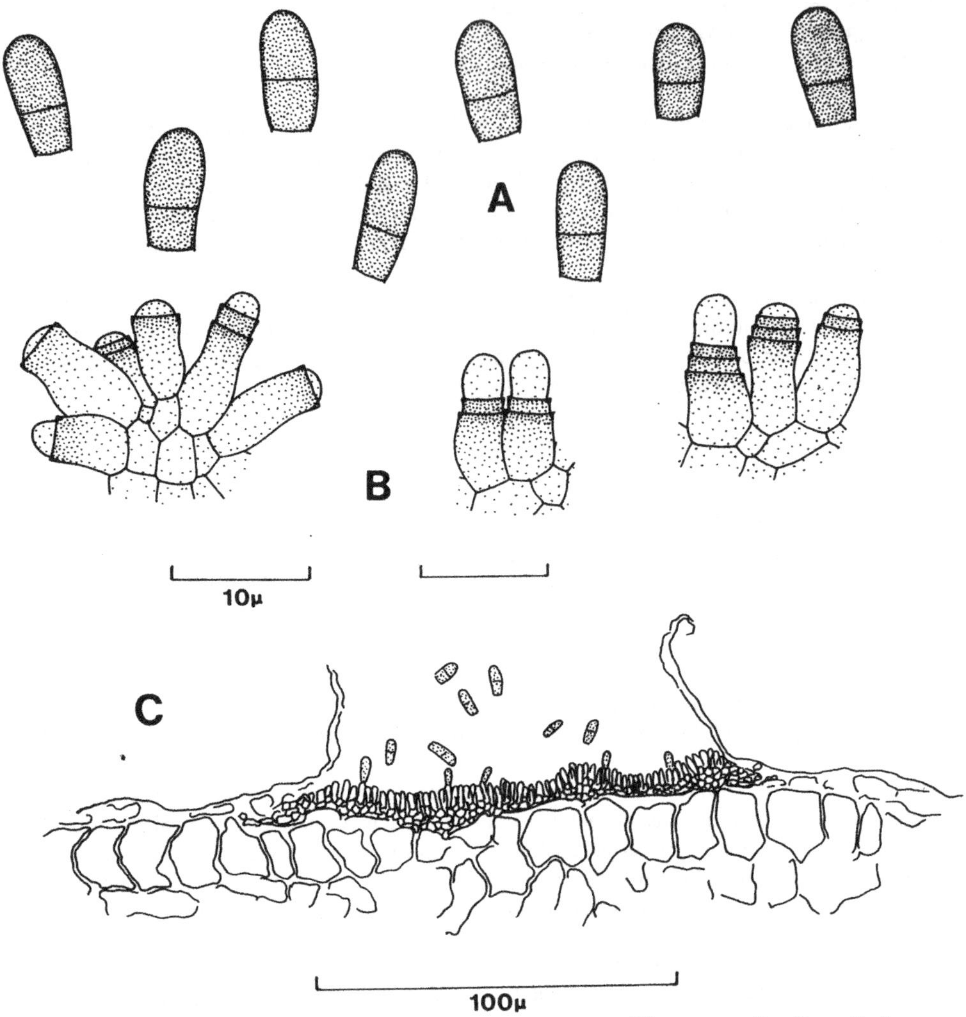

Fig. 200. *Didymosporina aceris*, A, conidia; B, conidiogenous cells; C, vertical section of a conidioma.

37. GLOEOCORYNEUM Weindlmayr, *Sydowia* **17**: 100 (1964).
Sp. typ.: *G. cinereum* (Dearn.) Weindlmayr (syn. *Coryneum cinereum* Dearn.).

Mycelium immersed, branched, septate, hyaline to pale brown. *Conidiomata* acervular, subepidermal, separate, formed of pale brown, thin-walled textura angularis. Dehiscence irregular. *Conidiophores* hyaline, branched at the base, septate, smooth, formed from the upper pseudoparenchyma. *Conidiogenous cells* holoblastic, integrated or discrete, determinate, cylindrical, hyaline to very pale brown, smooth, often lacking contents after spore formation. *Conidia* acrogenous, ovoid to ellipsoidal, 3-5 euseptate, continuous, brown, thick-walled, verruculose, base truncate, apex obtuse.

Morgan-Jones (1971d) regarded *G. cinereum* as congeneric with *Leptomelanconium* Petrak if the limits of the genus were enlarged slightly. Sutton (1975a) preferred to maintain *Gloeocoryneum* as a distinct genus on the basis of differences in the conidiogenous cells and conidia.

G. CINEREUM (Dearn.) Weindlmayr, *ibid.*: 101 (1964). (Fig. 202)
Coryneum cinereum Dearn., *Mycologia* **16**: 171 (1924).
Leptomelanconium cinereum (Dearn.) Morgan-Jones, *Can. J. Bot.* **49**: 1012 (1971).

Conidiomata up to 600 μ diam. Conidiophores 15-35 × 3-4 μ. Conidia 20-25 × 6-12 μ.

On needles of *Pinus cembroides* var. *monophylla*, *P. contorta*, *P. jeffreyi*, Calif., Ore., U.S.A. (IMI 75046, holotype ex K, 107969, 142214).

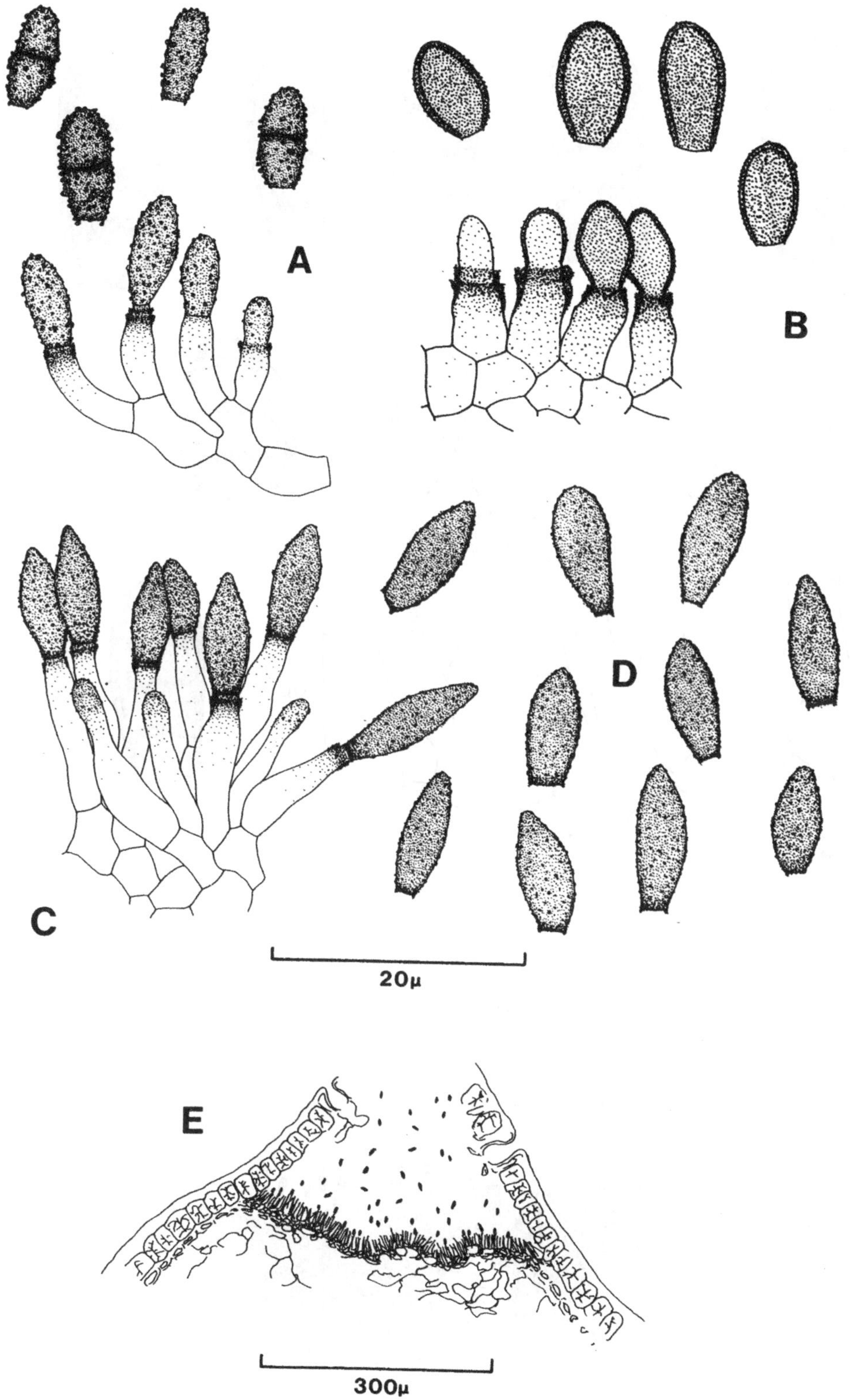

Fig. 201. A, *Leptomelanconium piceae*, conidia and conidiophores; B, *L. australiense*, conidia and conidiogenous cells; C, D, E, *L. allescheri*, conidiophores, conidia, vertical section of a conidioma.

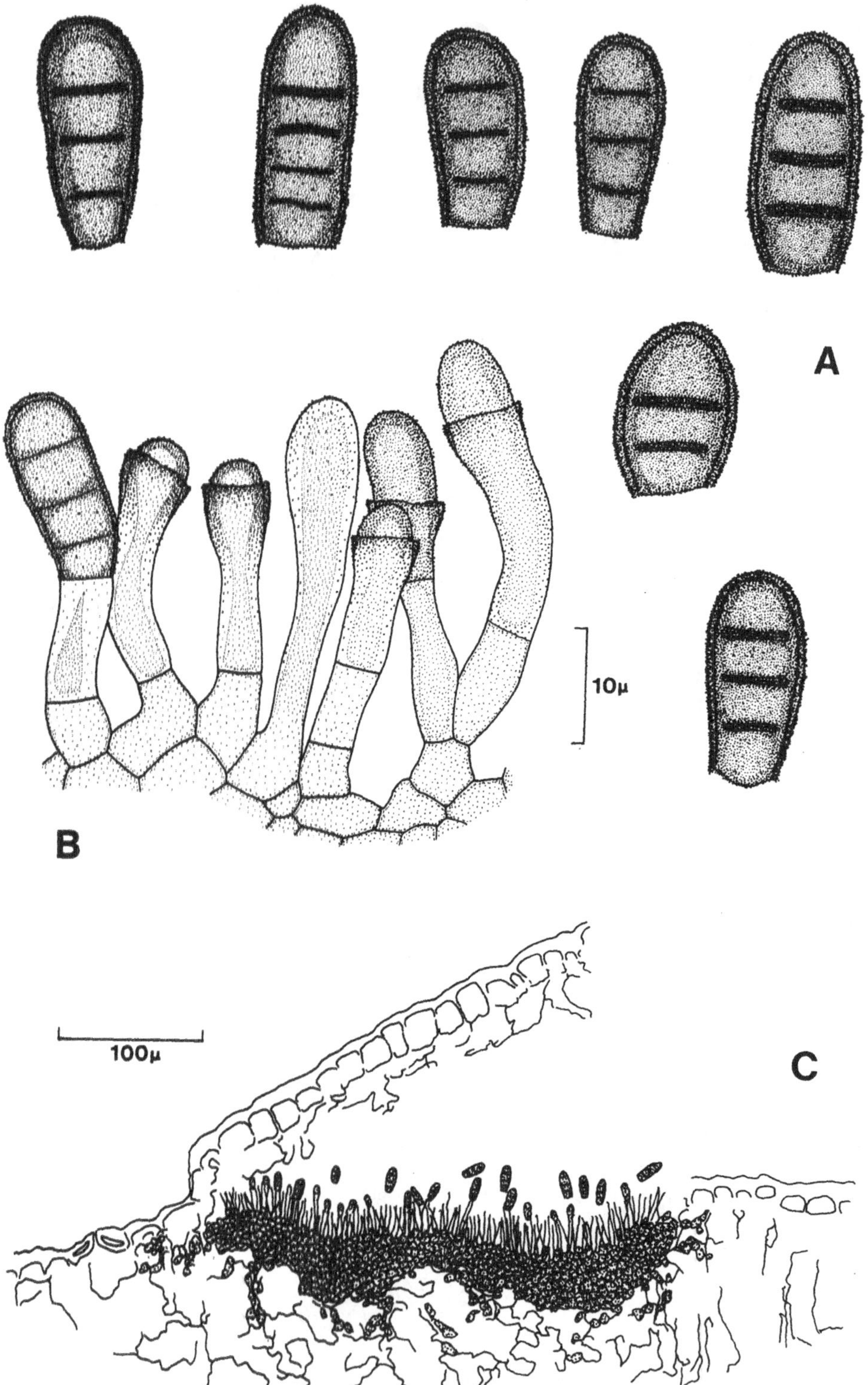

Fig. 202. *Gloeocoryneum cinereum*, A, conidia; B, conidiophores and developing conidia; C, vertical section of a conidioma.

38. LECANOSTICTA Syd. apud Syd. & Petrak, *Annls mycol.* **20**: 211 (1922).
Sp. typ.: *L. acicola* (Thüm.) Syd. (syn. *Cryptosporium acicola* Thüm., *L. pini* Syd.).

Mycelium immersed, branched, septate, pale brown. *Conidiomata* acervular, subepidermal, separate, formed of brown, thin- or thick-walled textura angularis. Dehiscence by pushing back a flap of epidermis that remains attached. *Conidiophores* hyaline to pale brown, branched, septate, smooth, formed from the upper cells of the pseudoparenchyma. *Conidiogenous cells* holoblastic, integrated or discrete, indeterminate, cylindrical, hyaline, with 1-2 often widely spaced percurrent proliferations. *Conidia* acrogenous, straight or curved, fusiform, tapered to the rounded apex and truncate base, 1-3 euseptate, continuous, pale brown, verrucose.

The single species, which causes brown spot needle blight has been fully described and illustrated by Punithalingam & Gibson (1973) who also list its synonymy.

L. ACICOLA (Thüm.) Syd. apud Syd. & Petrak, *Annls mycol.* **22**: 400 (1924). (Fig. 203).
Additional syn. *Dothistroma acicola* (Thüm.) Schischkina & Tzanava, *Nov. Sist. niz. Rast., 1967*: 277 (1967).

Teleomorphosis: *Scirrhia acicola* (Desm.) Siggers, *Phytopathology* **29**: 1076 (1939).

Conidiomata up to 120 µ diam. Conidiophores up to 30 µ long. Conidia 15-35 × 3-4 µ.

On needles of *Pinus variabilis, P. caribaea, P. contorta* var. *latifolia, P. palustris, P. pinaster*, Man., Canada, U.S.A. (IMI 114862, 104312, 91061, 215781 ex BPI).

39. STILBOSPORA Pers. ex Mérat, *Nouv. Fl. Env. Paris.*, Ed. 2, **1**: 147 (1821).
Stilbospora Pers., *Neues Magazin fur die Botanik* (J.J. Romer): 93 (1797).
Stilbospora Pers., *Syn. Meth. Fung.* 69 (1801).
Janospora (Stärb.) Höhn., *Mykol. Unters.* **3**: 319 (1923).
Hendersonia Berk. subgen. *Janospora* Starb., *Bih. Svensk. Akad. Handl.* **19**: 86 (1894).
Original spp.: *S. asterospora* Pers. ex Mérat, *S. sphaerosperma* Pers. ex Mérat, *S. ovata* Pers. ex Mérat, *S. macrosperma* Pers. ex Mérat, *S. microsperma* Pers. ex Mérat, *S. uredo* DC. ex Mérat.
Lectotype sp.: *S. macrosperma* Pers. ex Mérat, fide Clements & Shear (1931).

Mycelium immersed, branched, septate, pale brown to hyaline. *Conidiomata* acervular, separate or confluent, circular, composed of pale brown to dark brown textura angularis. Dehiscence irregular. *Conidiophores* cylindrical, branched and septate mainly at the base, hyaline, smooth, formed from the upper cells of the conidiomata. *Conidiogenous cells* holoblastic, annellidic, integrated or discrete, indeterminate, hyaline, smooth, cylindrical, with 0-4 percurrent proliferations. *Conidia* cylindrical, fusiform to clavate, 3-4 euseptate, continuous or sometimes constricted at the equidistant septa, dark brown, finely punctate, eguttulate, obtuse at the apex, truncate at the base. *Paraphyses* filiform, hyaline, cylindrical, septate.

Sutton (1975a) explained the synonymy and lectotypification of the genus and added two species. Hughes (1958) has listed some nomina excludenda. However more than 60 taxa have been described in the genus so far and the majority of these have not been re-examined. Even with those species currently accepted in the genus there is a need for studies to determine the species limits.

S. CISTINA (Cke) Sutton, *Mycol. Pap.* 138: 79 (1975). (Fig. 205B, C)
Coryneum cistinum Cke, *Grevillea* **14**: 5 (1885).
Hendersonia cisti da Camara, *Agron. lusit.* **13**: 145 (1951).

Conidiophores 12-40 × 3-5 µ, with 0-1 annellations. Paraphyses up to 55 µ long. Conidia 35-47 × 14-18 µ, 3 septate, fusiform.

On branches of *Cistus laurifolius, C. ladaniferus*, U.K., Portugal (IMI 75049 ex K, holotype *C. cistinum*, 177020 ex LISE, holotype *H. cisti*).

S. MACROSPERMA Pers. ex Mérat, *Nouv. Fl. Env. Paris*, Ed. 2, **1**: 147 (1821). (Fig. 205A)
Stilbospora macrosperma Pers., *Syn. Meth. Fung.*: 96 (1801).
Uredo macrosperma (Pers.) Strauss, *Ann. Wetterauisch. Ges.* **2**: 112 (1810).

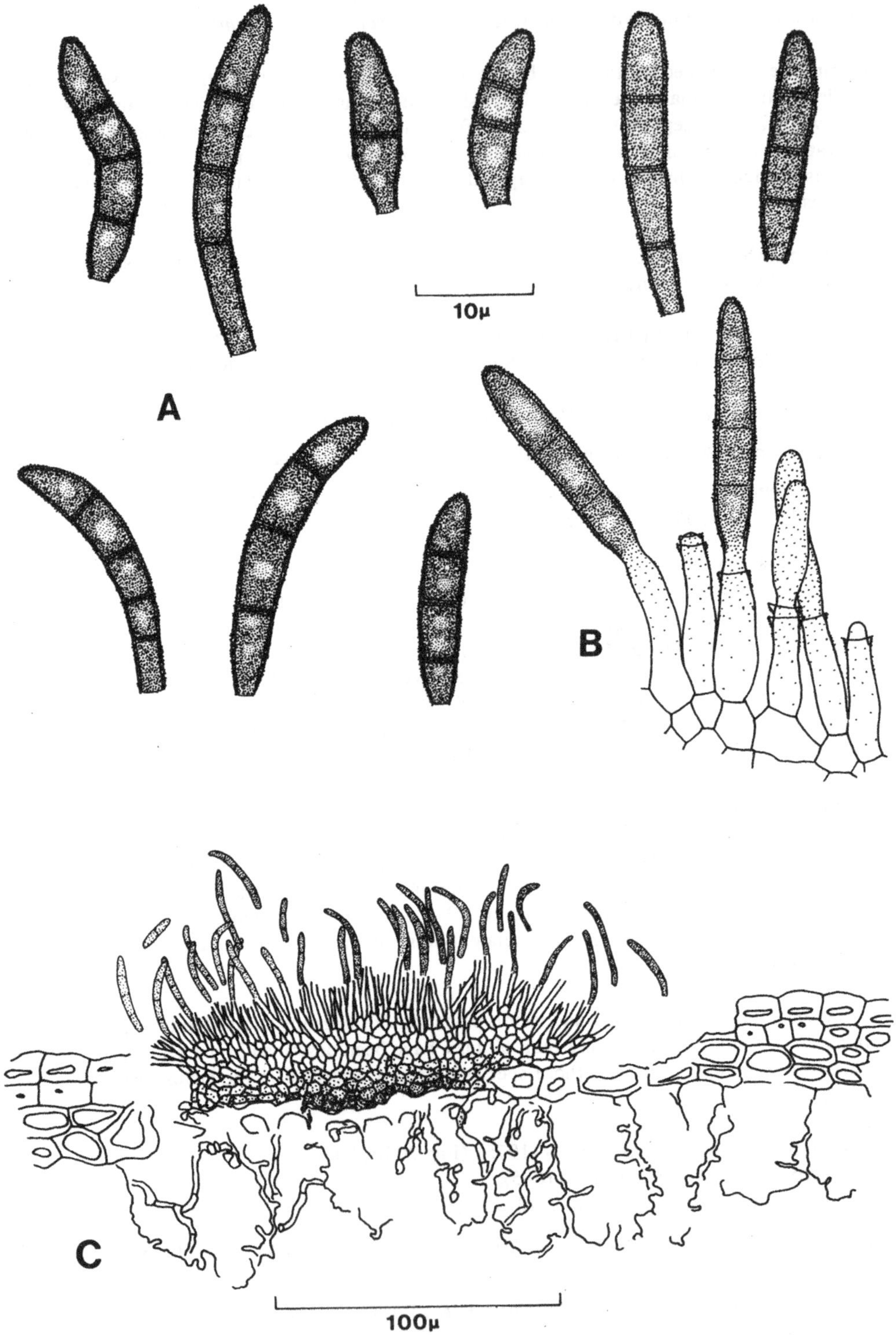

Fig. 203. *Lecanosticta acicola*, A, conidia; B, conidiogenous cells and developing conidia; C, vertical section of a conidioma.

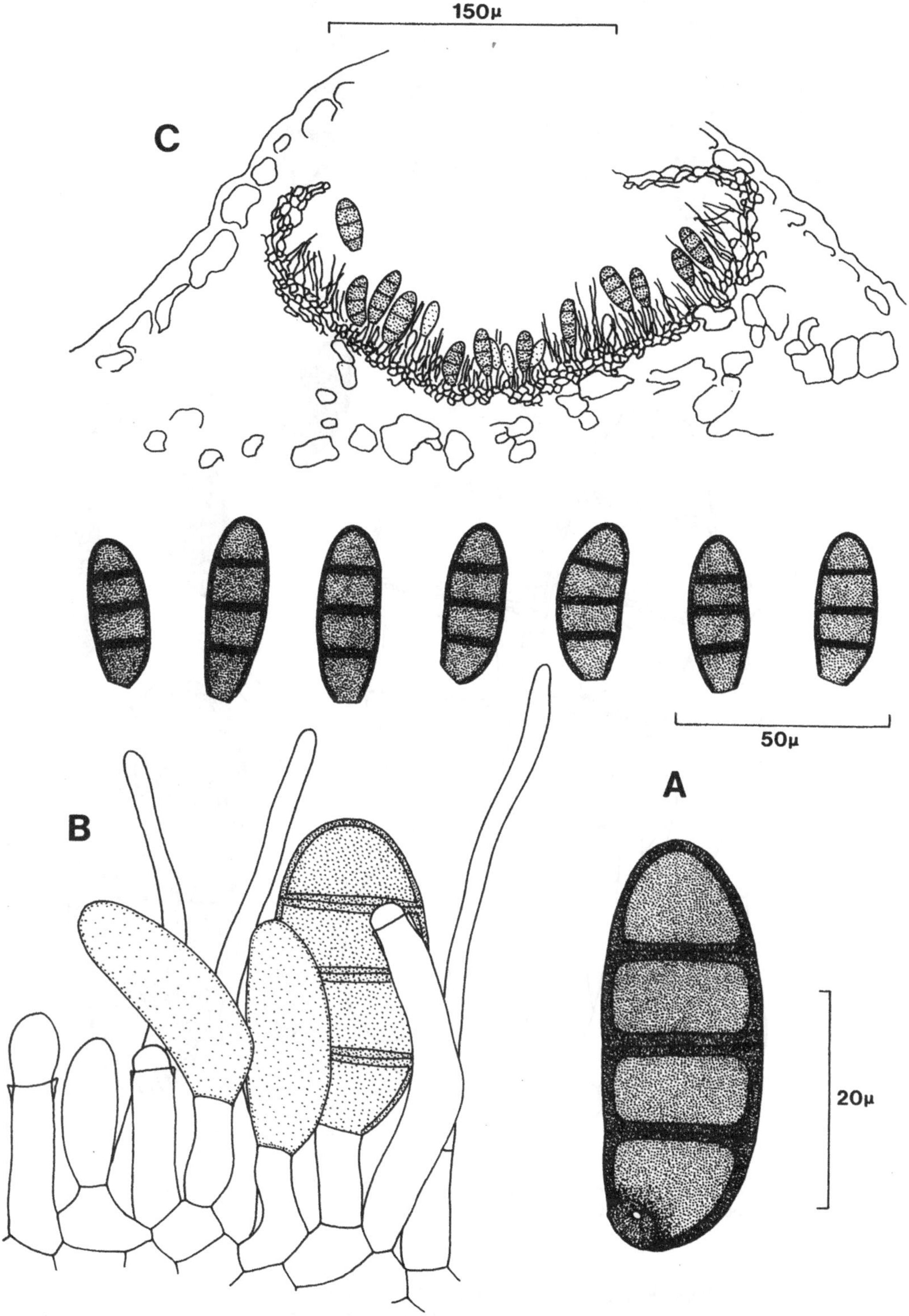

Fig. 204. *Stilbospora pistaciae*, A, conidia; B, conidiogenous cells and developing conidia, paraphyses, C, vertical section of a conidioma.

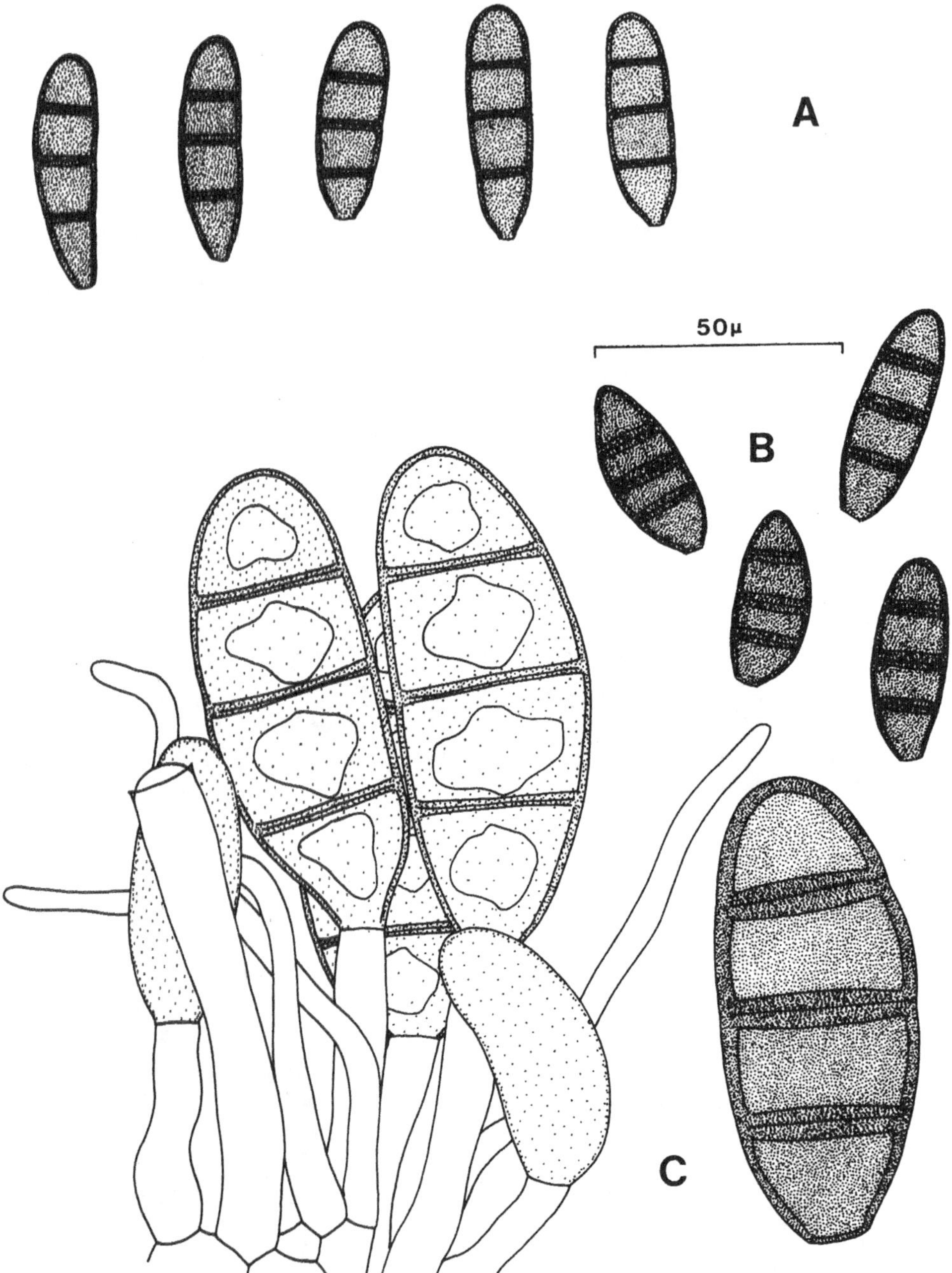

Fig. 205. A, *Stilbospora macrosperma*, conidia; B, C, *S. cistina*, Conidia, conidiogenous cells and paraphyses.

Stilbospora macrospora Pers., Romer's *Neues Mag. Bot.* **1**: 94 (1794).
Puccinia macrospora (Pers.) Sprengel, *Linn. Syst. Veg.* XVI **4**(1): 569 (1827).
Sporidesmium macrosporum (Pers.) Cda, Sturm's *Deutschl. Flora* 3 (Pilze), **2**(7): 47 (1829).

Conidiomata up to 1300 μ diam. Conidiophores 15-40 × 5-6.5 μ, with 1-4 annellations. Conidia 36-48 × 11.5-12.5 μ, 3 septate, cylindrical to long obovoid. Paraphyses up to 100 μ long.

On branches of *Carpinus betulus*, *Fagus sylvatica*, Czechoslovakia, Austria (IMI 33642, 13048, 94581).

S. PISTACIAE (Pat.) Sutton, *Mycol. Pap.* 138: 159 (1975). (Fig. 204)
Coryneum pistaciae Pat., *Cat. rais. pl. cell. Tunisie* 128 (1897).

Foliicolous. Conidiomata up to 280 µ diam. Conidiophores up to 30 µ long × 3-5 µ wide, with 0-1 annellations. Paraphyses up to 60 µ long. Conidia 33-45 × 13-15 µ, 3 septate, cylindrical to ellipsoid, basal cell often inclined to one side.

On leaves of *Pistacia lentiscum*, Tunisia (IMI 178504 ex FH, holotype).

40. HENDERSONULA Speg., *An. Soc. Cient. Arg.* **10**: 160 (1880).
Sp. typ.: *H. australis* Speg.

Mycelium immersed, branched, septate, hyaline. *Conidiomata* eustromatic, subcuticular, becoming epidermal, irregular in shape but generally circular in outline, becoming erumpent particularly at the periphery, irregularly pulvinate, carbonous, blackish brown, irregularly multilocular; upper, lateral and interlocular tissue of dark brown, thick-walled textura angularis, basal tissue hyaline to pale brown, thin-walled. Dehiscence by irregular breakdown of the upper wall. *Conidiophores* absent. *Conidiogenous cells* holoblastic, annellidic, indeterminate, discrete, hyaline, cylindrical, smooth, with occasionally a single percurrent proliferation, cytoplasm retracted after conidial delimitation. *Conidia* pale brown, smooth, thin-walled, 1-3 (mostly 3) euseptate, septa thick and prominent, cylindrical to fusiform, apex obtuse, base truncate, eguttulate.

Petrak & Sydow (1935) accepted *Hendersonula* and later, Petrak (1949) redisposed several species into other genera. Finally Petrak (1951a) reduced *Hendersonula* to synonymy with *Hendersonia* auct. He regarded the original account of the fungus to be erroneous because the stromatic tissue referred to by Spegazzini (1880) consisted not only of the *Hendersonula* but also of the *Dothidella* beneath which it grew. Petrak (1951a) acknowledged a difference in growth for *H. australis* but preferred not to maintain the genus for stromatic *Hendersonia* species. Since *Hendersonia* is a nomen rejiciendum for *Stagonospora* (Sacc.) Sacc. and a generic name is still required for species like *H. australis*, it seems preferable to restrict that part of Spegazzini's description and type to the imperfect fungus alone.

The common plant pathogen, *H. toruloidea* Nattrass, is not strictly congeneric with *H. australis*. It is not a hyperparasite and although conidiomata are stromatic, the conidia are formed from phialides. Conidia are versicoloured with a central pigmented cell and a hyaline cell at each end. Furthermore there is an associated *Scytalidium* hyphomycete state which is more frequently encountered than the pycnidial state. For these reasons it will be necessary to separate *H. toruloidea* into a different genus.

So far, *Hendersonula* contains about 30 published taxa. A revision of these species resulting in the acceptance of perhaps only two taxa in the genus is currently in progress (Dyko & Sutton, unpublished.).

H. AUSTRALIS Speg., *ibid.*: 160 (1880). (Fig. 206)
Hendersonia dothidellae Petrak, *Sydowia* **5**: 421 (1951).

Conidiomata up to 2 mm diam. × 250 µ high. Conidiogenous cells 8-13 × 4-4.5 µ. Conidia 8-17 × 4-5 µ.

Growing beneath stromata of *Placostroma australe* (Speg.) Theiss. & Syd., Argentina (IMI 12278b, isotype ex LPS).

41. NEOHENDERSONIA Petrak, *Annls mycol.* **19**: 190 (1921).
Sp. typ.: *N. kickxii* (Westd.) Sutton & Pollack (syn. *Stilbospora kickxii* Westd., *N. pyriformis* (Otth) Petrak).

Mycelium immersed, branched, septate, pale brown. *Conidiomata* eustromatic, separate or gregarious, immersed, subperidermal, globose or collabent, papillate, dark brown to black, unilocular or multilocular, thick-walled; wall of two layers, the outer of dark brown, thick-walled textura porrecta or intricata, the inner of hyaline, thin-walled textura angularis. *Ostiole* single, circular, distinctly papillate. *Conidiophores* absent. *Conidiogenous cells* holoblastic, single or annellidic, discrete, determinate or indeterminate, cylindrical, lageniform, doliiform or ampulliform, hyaline, smooth, with 0-3 percurrent proliferations, formed from the inner cells of the locular walls. *Conidia* obovoid, cylindrical to clavate or fusiform, 2-7 euseptate or distoseptate and then with reduced lumina, continuous or slightly constricted, versicoloured, basal and sometimes apical cells paler than the median cells, thick-walled, smooth, base truncate, sometimes thickened, apex obtuse, eguttulate.

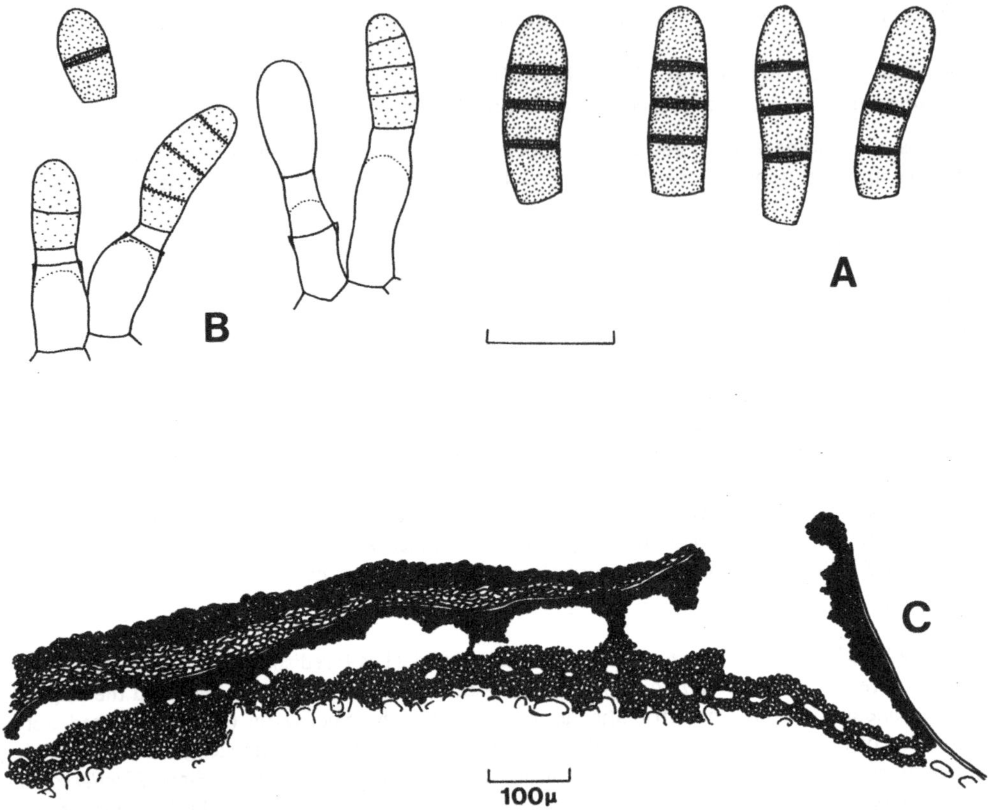

Fig. 206. *Hendersonula australis*, A, conidia; B, conidiogenous cells and developing conidia; C, vertical section of a conidioma.

Sutton & Pollack (1974) described and illustrated the type species and subsequently Sutton (1975a) expanded the concept of the genus to include *N. congoensis* (Torrend) Sutton. These are the only species assigned to the genus.

N. KICKXII (Westd.) Sutton & Pollack, *Mycopath. Mycol. appl.* **52**: 334 (1974). (Fig. 207A, C)
Stilbospora kickxii Westd., *Bull. Acad. r. Belg. Cl. Sc.* **18**: 409 (1851).
Coryneum kickxii (Westd.) Trav., *Bull. Soc. bot. Ital.* 215 (1904).
Neohendersonia pyriformis (Otth) Petrak, *Annls mycol.* **19**: 191 (1921).
Hendersonia pyriformis Otth, *Mitt. nat. Ger. Bern* 164 (1866).
Hendersonia loricata Sacc. & Roum., *Michelia* **2**: 629 (1882).

Conidiomata 1300 µ diam. × 750 µ high. Ostiole 250 µ diam. Conidia 2(-3) distoseptate, apical cell larger and darker than lower ones, base thickened, 17.5-30 × 8.5-14 µ.

On branches and bark of *Fagus sylvatica*, Belgium, Austria, G.B., U.S.A. (IMI 168433, holotype *S. kickxii* ex BR, 30358, 31104, 12841b, 22764).

N. CONGOENSIS (Torrend) Sutton, *Mycol. Pap.* 138: 84 (1975). (Fig. 207B)
Coryneum congoense Torrend apud Wild., *Bull. Jard. bot. État Brux.* **4**: 10 (1914).

Conidiomata 950 µ diam. × 425 µ high. Ostiole 125 µ diam. Conidia 3-7 (mostly 6-7) euseptate, apical and basal cells pale brown, penultimate cells medium brown, median cells dark brown, 51-69 × 17.5-21 µ.

On stems of *Aloe* or *Agave*, Congo (IMI 180829 ex BR, holotype).

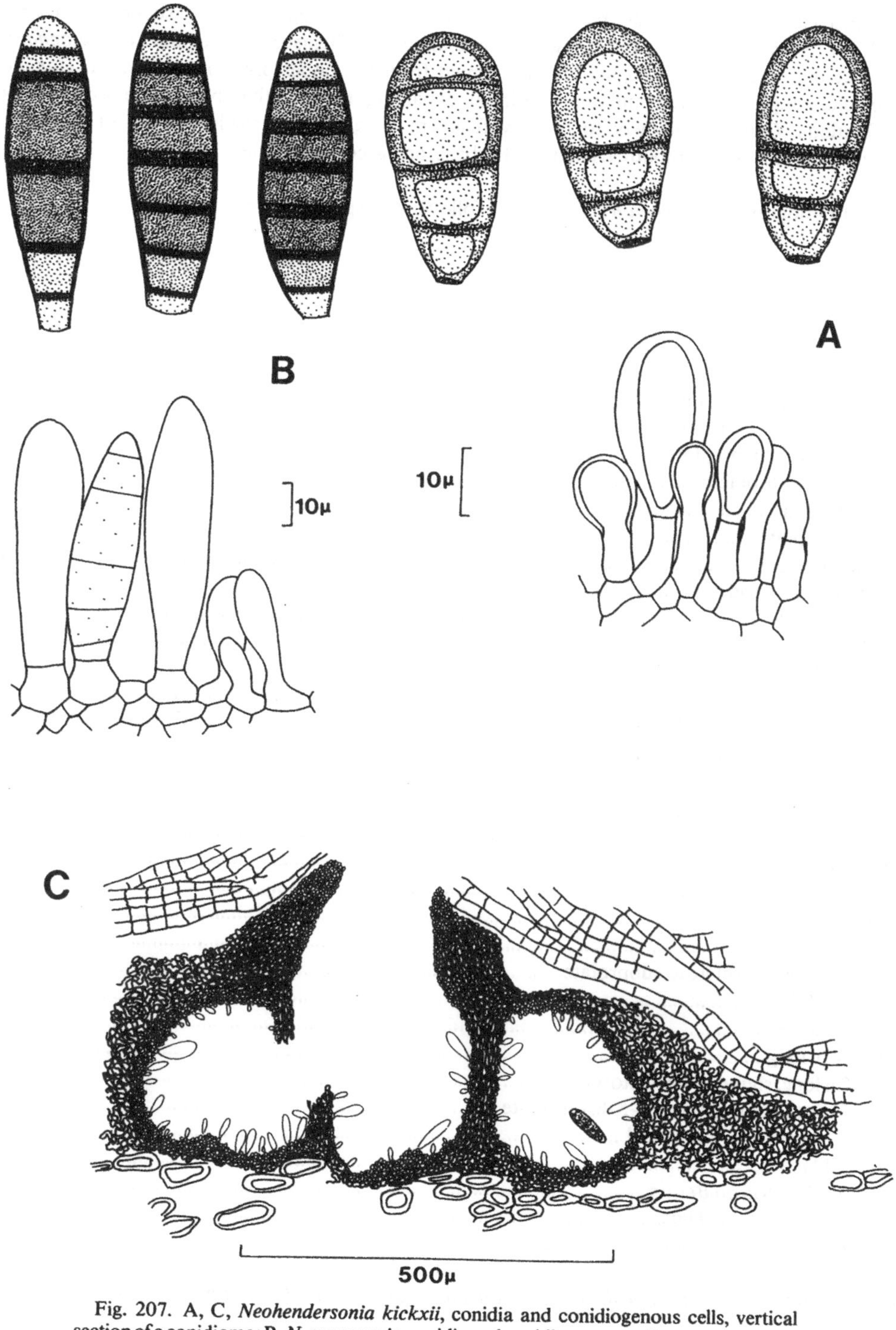

Fig. 207. A, C, *Neohendersonia kickxii*, conidia and conidiogenous cells, vertical section of a conidioma; B, *N. congoensis*, conidia and conidiogenous cells.

42. CORYNEUM Nees ex Schw., *Schr. d. Naturf.–Ges. Leipzig* **1**: 76 (1822).
Coryneum Nees, *Das System der Pilze und Schwamme* 34 (1817).
Sp. typ.: *C. umbonatum* Nees ex Steudel

Mycelium immersed, branched, septate, pale brown. *Conidiomata* acervular, or rarely subepidermal when foliicolous, separate or confluent, formed of brown, thin-walled textura angularis. Dehiscence irregular. *Conidiophores* cylindrical, septate, branched at the base, hyaline to pale brown, formed from the upper pseudoparenchyma. *Conidiogenous cells* holoblastic, annellidic, integrated, indeterminate, cylindrical, expanding towards the apices, pale brown, smooth, with 0-1 percurrent proliferations. *Conidia* narrowly to broadly fusiform, or globose, straight, curved or sigmoid, transversely distoseptate, lumina reduced, brown, smooth, truncate at the base, apical cell often paler.

Sutton (1975a) monographed *Coryneum* accepting the following 19 species and one variety.

KEY

1. Conidia narrowly fusiform to cylindrical, l/b ratio >4/1	2
1. Conidia broadly fusiform, l/b ratio <4/1	8
2. Conidia more than 8 distoseptate	3
2. Conidia 8 distoseptate or less	4
3. Conidia 11 distoseptate	*C. megaspermum*
3. Conidia 9-17 distoseptate	*C. stromatoideum*
3. Conidia 7-11 distoseptate	*C. carpinicola*
4. Conidia less than 13 µ wide	5
4. Conidia 13-16 µ wide	6
4. Conidia 14-22 µ wide	7
5. Conidia 45-60 × 11-12 µ	*C. japonicum*
5. Conidia 57-80 × 10-13 µ, apex recurved	*C. castaneicola*
5. Conidia 100-125 × 10-13 µ, cylindrical	*C. megaspermum* var. *cylindricum*
6. Conidia 57-72 × 13-16 µ, slightly curved	*C. umbonatum*
6. Conidia 80-90 × 13-15 µ, straight	*C. cesatii*
7. Conidia 68-82 × 18-22 µ	*C. neesii*
7. Conidia 50-71 × 14-19 µ	*C. modonium*
8. Conidia >20 µ wide	9
8. Conidia <20 µ wide	10
9. Conidia 56-70 × 24-32 µ	*C. elevatum*
9. Conidia 44-53 × 19-23 µ	*C. depressum*
9. Conidia 30-40 × 20-30 µ, prolate to subprolate	*C. clusiae*
9. Conidia 61-70 × 24-32 µ	*C. pyricola*
10. Conidia <40 µ long	11
10. Conidia >40 µ long	12
11. Conidia 25-40 µ long, sigmoid, 5-6 distoseptate	*C. psidii*
11. Conidia 31-36 µ long, symmetrical, 4-5 distoseptate	*C. betulinum*
12. Conidia <14 µ wide, symmetrical	*C. calophylli*
12. Conidia >14 µ wide	13
13. Conidia 45-53 × 16-18 µ, on *Betula*	*C. brachyurum*
13. Conidia 50-58 × 14-17 µ, on *Alnus*	*C. sydowianum*
13. Conidia 40-58 × 15-21 µ on *Ulmus*	*C. compactum*

C. BETULINUM Schulzer, *Rada jugoslavenske akademije zuanosti i umjetnosti mat. – prirod. raz.* **64**: 3 (1882). (Fig. 208A)

Conidia 31-36 × 14-17 µ, broadly fusiform, straight, 4-5 distoseptate.

On twigs of *Betula rubrum*, *B.* sp., N. Carol. & N.J., U.S.A. (IMI 74906, 180798).

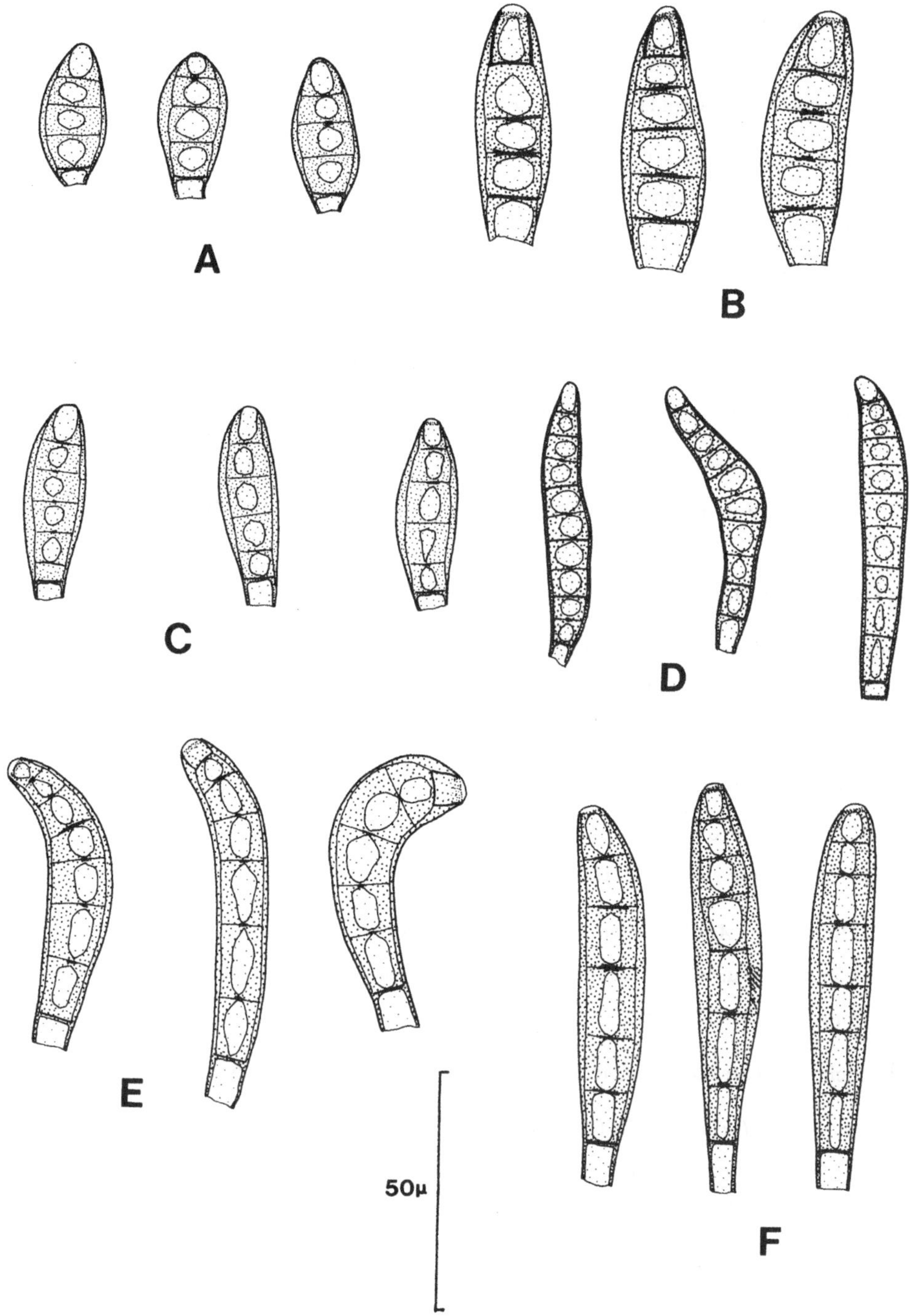

Fig. 208. A, *Coryneum betulinum*, conidia; B, *C. brachyurum*, conidia; C, *C. calophylli*, conidia; D, *C. carpinicola*, conidia; E, *C. castaneicola*, conidia; F, *C. cesatii*, conidia.

C. BRACHYURUM Lk, *Linné Spec. Pl.* **6**(2): 124 (1825). (Fig. 208B)

Teleomorphosis: *Pseudovalsa lanciformis* (Fr.) Ces. & de Not., *Comm. Soc. Critt. ital.* **1**: 206 (1861).

Conidia 45-53 × 16-18 µ, narrowly fusiform, straight, tapered each end, 4-5(-6) distoseptate.

On twigs of *Betula* spp., Austria, Finland, Germany, U.K., Italy, Sweden, 41 collections in IMI.

C. CALOPHYLLI (Syd.) Morgan-Jones, *Can. J. Bot.* **49**: 1849 (1971). (Fig. 208C)

Conidia 38-48 × 12.5-14.5 µ, narrowly fusiform, occasionally clavate, straight, 5-6 distoseptate, apical and 2 basal septa prominent.

On bark of *Calophyllum inophyllum*, Philippines (IMI 183293, holotype).

C. CARPINICOLA Sutton, *Mycol. Pap.* 138: 20 (1975). (Fig. 208D)

Conidia 50-68 × 8-11 µ, narrowly fusiform, bent, 7-11 distoseptate, alternate ones prominent.

On twigs of *Carpinus caroliana*, *C.* sp., Ont., Canada (IMI 180709, holotype, 180630).

C. CASTANEICOLA Berk. & Curt. apud Berk., *Grevillea* **2**: 154 (1874). (Fig. 208E)

Conidia 57-80 × 10-13 µ, narrowly fusiform, often curved at the abruptly tapered apices, 6-7 distoseptate, often prominent.

On bark of *Castanea dentata*, *C.* sp., Penn. & Ga, U.S.A. (IMI 180179 (holotype), 180703).

C. CESATII Sutton, *Mycol. Pap.* 138: 23 (1975). (Fig. 208F)

Conidia 80-90 × 13-15 µ, narrowly fusiform, straight, 6-7 distoseptate, often prominent.

On bark of *Aesculus* sp., Italy (IMI 184830, holotype).

C. CLUSIAE Sutton, *Kew Bull.* **17**: 311 (1963). (Fig. 209A)

Conidia 30-40 × 20-30 µ, oval to pyriform, 3-5 distoseptate.

On dead leaves of *Clusia* sp., Venezuela (IMI 75572, holotype).

C. COMPACTUM Berk. & Br., *Ann. Mag. nat. Hist.* 2 ser. **5**: 458 (1850). (Fig. 209B)

Conidia 40-55 × 18-22 µ, broadly fusiform or clavate, variable in shape, straight, 4-6 distoseptate.

On bark of *Ulmus* sp., U.K. (IMI 49512, lectotype).

C. DEPRESSUM Schmidt ex Steudel, *Nomenclator botanicus*: 139 (1824). (Fig. 209C)

Teleomorphosis: *Pseudovalsa umbonata* (Tul.) Sacc., *Fungi veneti* **4**: 20 (1875).

Conidia 44-53 × 19-23 µ, broadly fusiform, straight, 4-5(-6) distoseptate.

On bark and twigs of *Quercus* spp., U.K., France, Belgium, Germany, Austria, Hungary, Italy, Russia, 20 collections in IMI.

C. ELEVATUM (Riess) Sutton, *Mycol. Pap.* 138: 31 (1975). (Fig. 209D)

Conidia 56-59 × 24-28 µ, broadly fusiform, straight, 5-7 distoseptate, occasionally with a longitudinal septum.

On bark of *Quercus* spp., U.K., Pakistan, Colo., U.S.A. (IMI 86875, 177066, 108530, 180582, 5808, 80729a).

C. JAPONICUM (Sacc.) Sutton, *Mycol. Pap.* 138: 33 (1975). (Fig. 209E)

Conidia 45-60 × 11-12 µ, narrowly fusiform, straight or curved, 5-7 distoseptate, sometimes prominent.

On bark of *Quercus* spp., Ont., Canada, Japan, Wash., U.S.A. (IMI 180599, 179080 (holotype), 180710, 180712b).

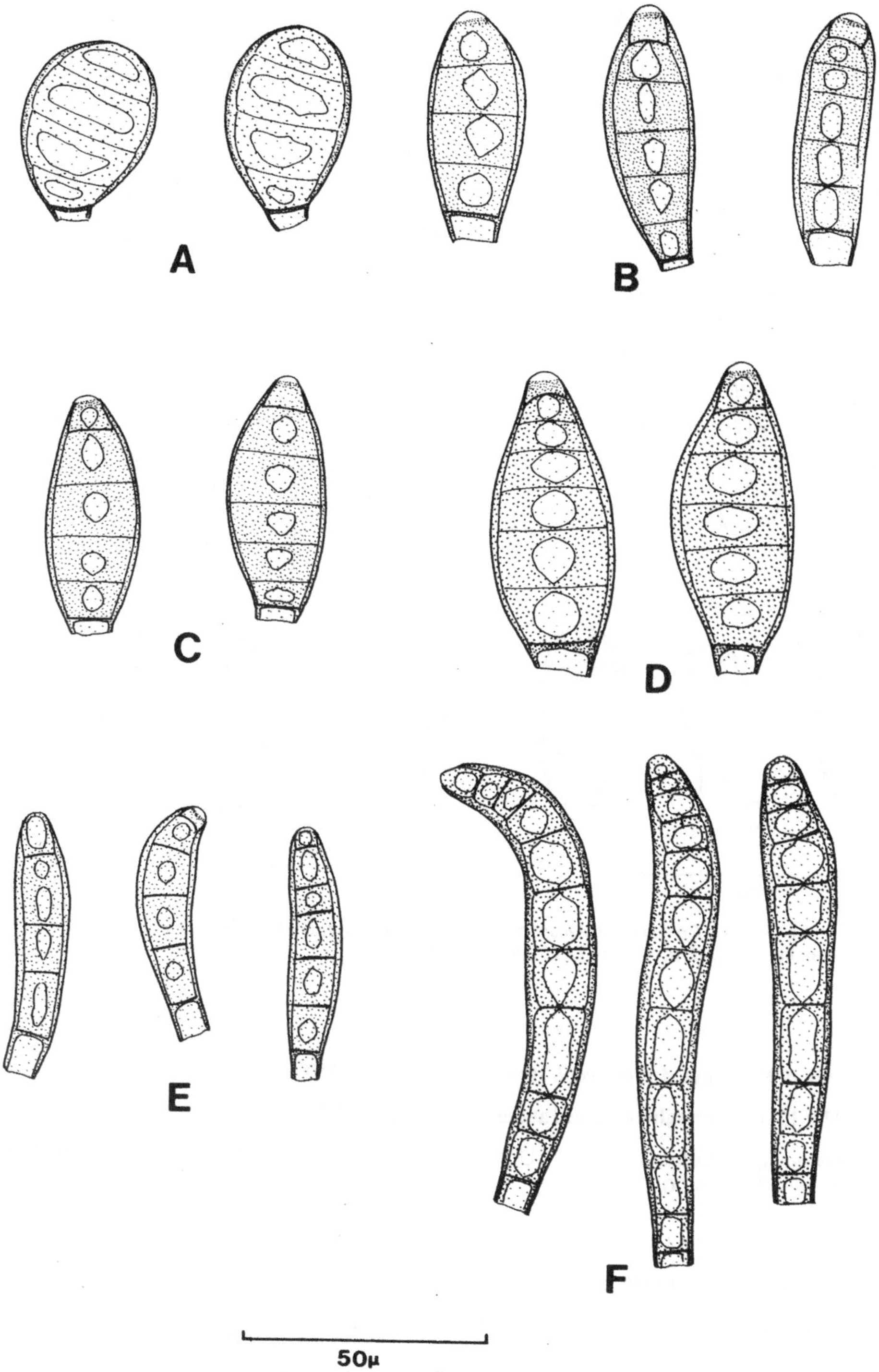

Fig. 209. A, *Coryneum clusiae*, conidia; B, *C. compactum*, conidia; C, *C. depressum*, conidia; D, *C. elevatum*, conidia; E, *C. japonicum*, conidia; F, *C. megaspermum*, conidia.

M

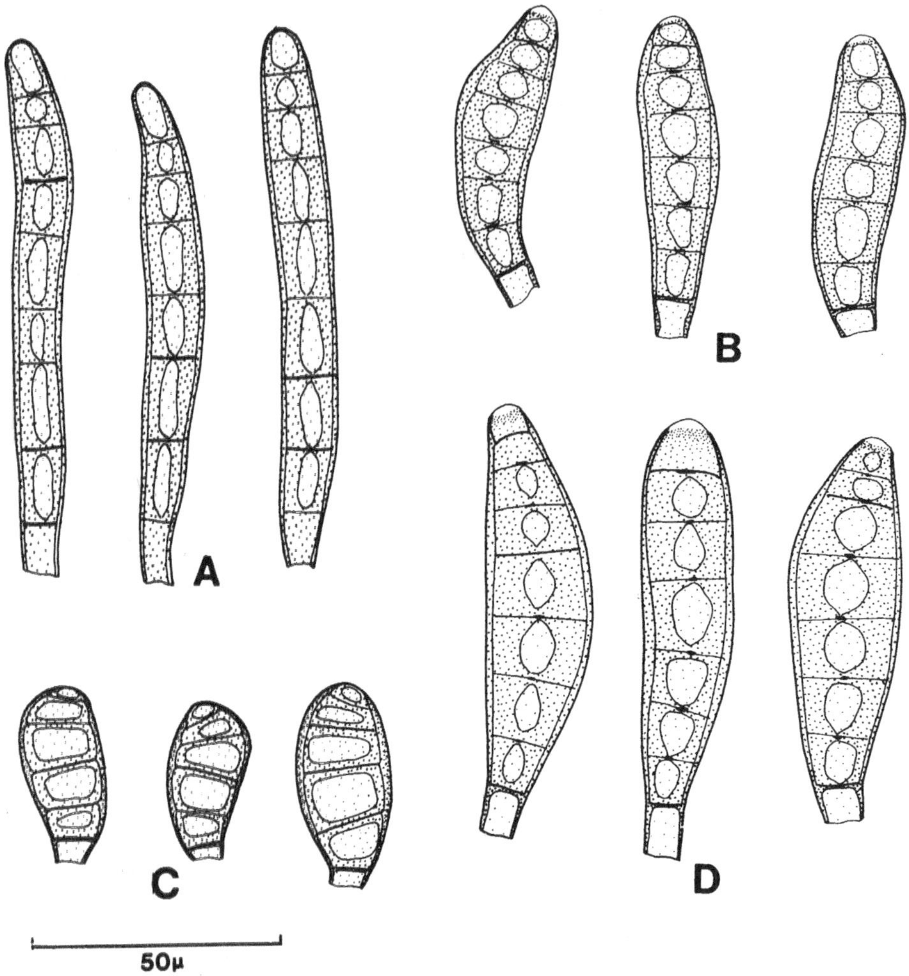

Fig. 210. A, *Coryneum megaspermum* var. *cylindricum*, conidia; B, *C. modonium* conidia; C, *C. psidii*, conidia; D, *C. neesii*, conidia.

C. MEGASPERMUM H. & P. Sydow, *Annls mycol.* **10**: 84 (1912). (Fig. 209F)

Conidia 73-97 × 13-16 µ, straight or curved at the apex, fusiform, 7-11 distoseptate.

On bark of *Quercus* spp., Colo., Tex., N.J., U.S.A., Man., Canada, Pakistan (IMI 178505 (holotype), 180702, 102444, 80197b, 180474, 180400, 180712a).

C. MEGASPERMUM var. CYLINDRICUM Sutton, *Mycol. Pap.* 138: 36 (1975). (Fig. 210A)

Conidia 100-125 × 10-13 µ, straight, cylindrical, 7-8 distoseptate.

On branches of *Quercus* sp., Ont., Canada (IMI 180701, holotype).

C. MODONIUM (Sacc.) Griffon & Maublanc, *Bull. trimestr. Soc. mycol. Fr.* **26**: 379 (1910). (Fig. 210B)

Teleomorphosis: *Pseudovalsa modonia* (Tul.) Höhn., *Annls mycol.* **16**: 125 (1918)

Conidia 50-71 × 14-19 µ, variable, clavate, narrowly fusiform, straight or curved, 5-8 distoseptate.

On bark of *Castanea* spp., Belgium, Roumania, France (IMI 180397 (isotype), 22763, 179079).

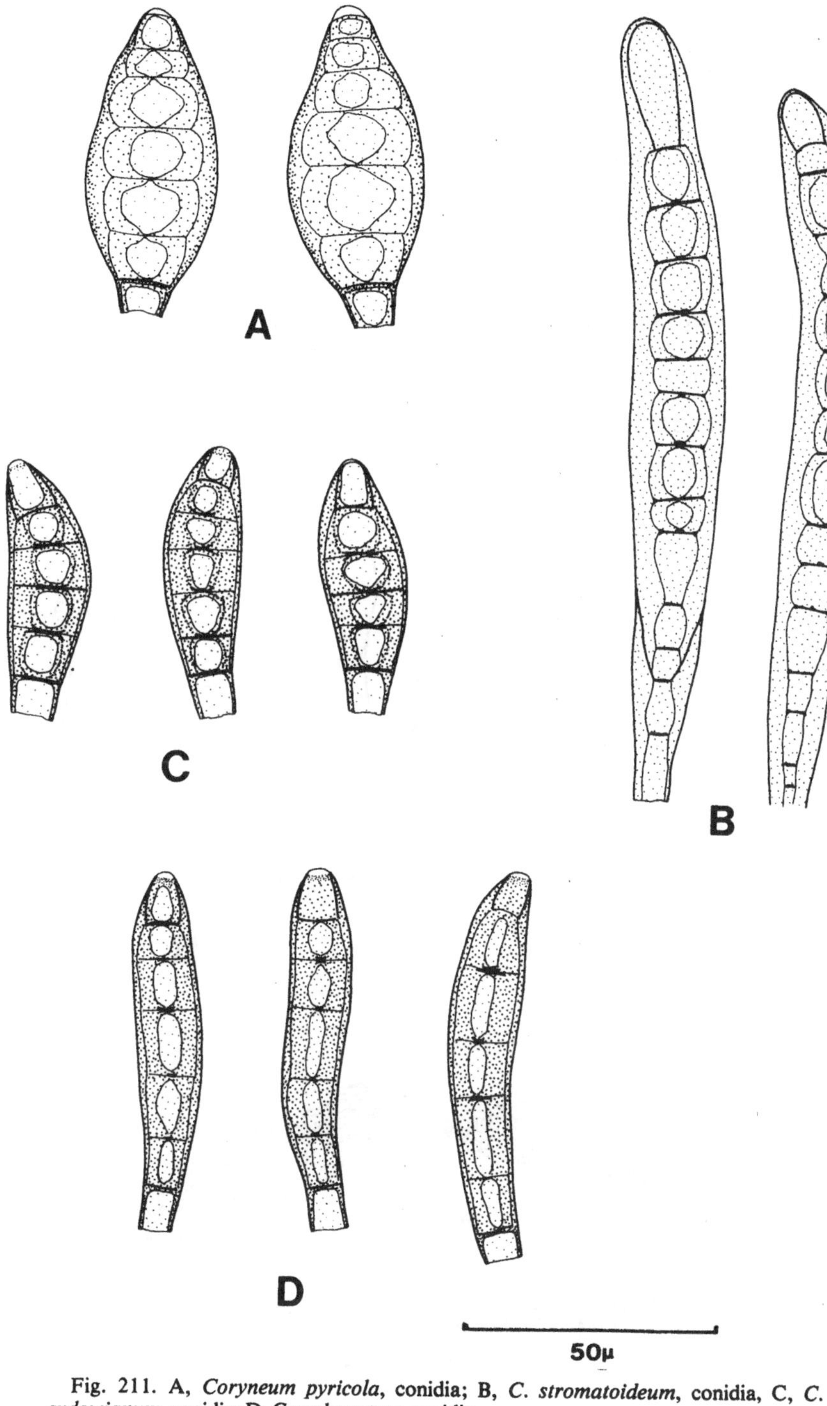

Fig. 211. A, *Coryneum pyricola*, conidia; B, *C. stromatoideum*, conidia, C, *C. sydowianum*, conidia; D, *C. umbonatum*, conidia.

C. NEESII Sutton, *Mycol. Pap.* 138: 40 (1975). (Fig. 210D)

Conidia 68-82 × 18-22 µ, narrow to broadly fusiform, straight, less often curved, 6-8 distoseptate.

On branches of *Quercus* spp., Austria, Malta, U.K. (IMI 15650 (isotype), 70478, 13942).

C. PSIDII Sutton, *Kew Bull.* **17**: 311 (1963). (Fig. 210C)

Conidia 25-40 × 14-17 µ, clavate, 5-6 distoseptate.

On branches of *Psidium guajava*, Sierra Leone, India (IMI 137497, holotype, 217534).

C. PYRICOLA Ahmad, *Biologia, Lahore* **6**: 127 (1961). (Fig. 211A)

Conidia 61-70 × 24-32 µ, broadly fusiform, straight, 5-7 distoseptate.

On branches of *Pyrus*, Pakistan (IMI 80233b, holotype)

C. STROMATOIDEUM (Dearn.) Sutton, *Mycol. Pap.* 138: 44 (1975). (Fig. 211B)

Conidia 105-180 × 16-20 µ, cylindrical, 9-17 distoseptate.

On bark of *Tsuga canadensis*, N.Y., U.S.A. (IMI 80280, holotype).

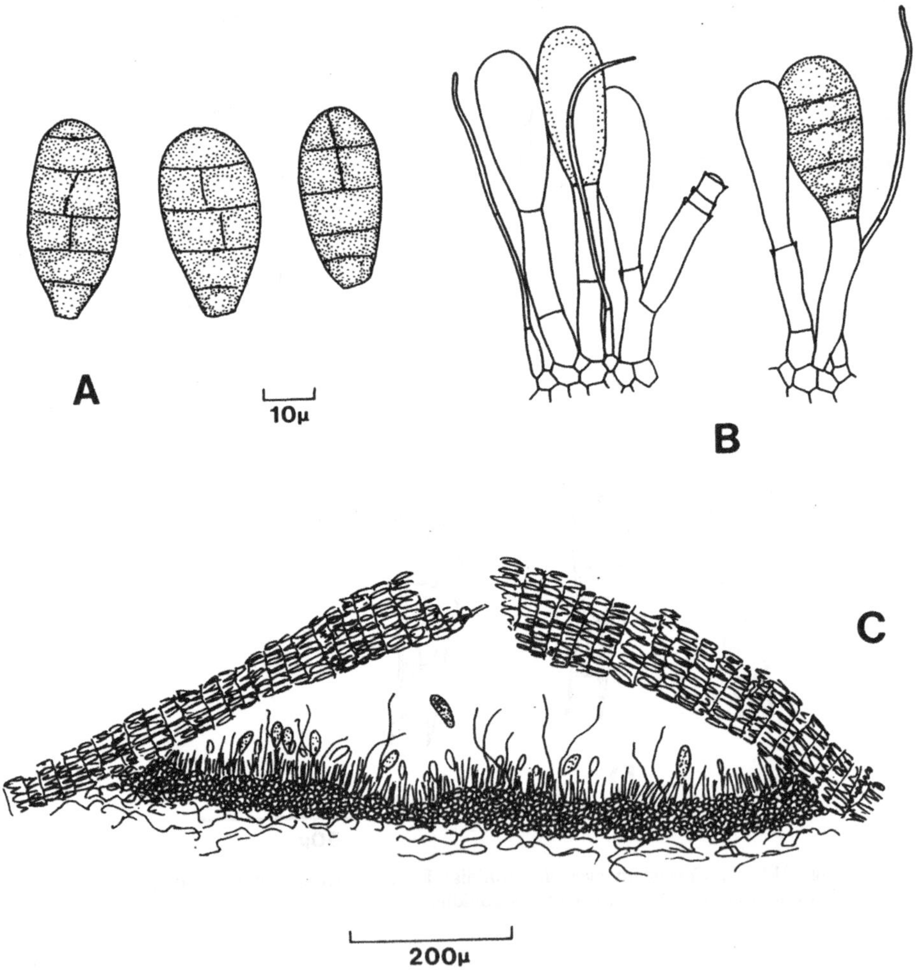

Fig. 212. *Stegonsporium ovatum*, A, conidia; B, conidiophores and paraphyses, C, vertical section of a conidioma.

C. SYDOWIANUM Allescher, *Hedwigia* **35**: [33], (1896). (Fig. 211C)

Conidia 50-58 × 14-17 µ, straight or curved, narrowly fusiform, 5-6 distoseptate.

On bark of *Alnus incana*, Germany (IMI 180073, holotype).

C. UMBONATUM Nees ex Steudel, *Nomenclator botanicus*: 139 (1824). (Fig. 211D)

Teleomorphosis: *Pseudovalsa longipes* (Tul.) Sacc., *Fungi veneti* **4**: 20 (1875).

Conidia 57-72 × 13-16 µ, narrowly fusiform, straight or slightly curved, 5-7 distoseptate.

On bark and twigs of *Quercus* spp., N.Y., Penn. & N.J., U.S.A., Ont., Canada, U.K., Germany, Czechoslovakia, France, The Netherlands, Austria, Poland, 34 collections in IMI.

43. STEGONSPORIUM Cda in Opiz, *Naturalientausch* **11**: 458 (1826).
Original spp.: *S. curvatum* Cda, *S. rotundum* Cda
Lectotype sp.: *S. pyriforme* (Hoffm. ex Fr.) Cda (syn. *Stilbospora pyriformis* Hoffm. ex Fr., *Stegonsporium curvatum* Cda).

Mycelium immersed, branched, septate, hyaline to pale brown. *Conidiomata* acervular, separate or confluent, immersed, circular, dark brown to black, corticolous, formed of small-celled, hyaline textura angularis. *Conidiophores* cylindrical, septate and branched only towards the base, hyaline, smooth, formed from the upper cells of the conidioma. *Conidiogenous cells* holoblastic, annellidic, with up to 3 percurrent proliferations, integrated or discrete, hyaline. *Conidia* obovoid to clavate, with several transverse and longitudinal distosepta, cell lumina considerably reduced, brown, smooth, obtuse at the apex, truncate at the base. *Paraphyses* filiform, septate, hyaline, flexuous.

The genus was briefly described and discussed by Sutton (1975a). So far more than 60 species have been described in *Stègonsporium* and a révision, similar to that published for *Coryneum* by Sutton (1975a) is in press (van Warmelo & Sutton, 1980).

S. PYRIFORME (Hoffm. ex Fr.) Cda, *Icon. fung.* **3**: 23 (1839). (Fig. 212)
Synonymy in van Warmelo & Sutton (1980).
Conidiomata up to 1000 µ diam. Conidiophores up to 40 × 5-6 µ. Conidia transversely 4-5 distoseptate, with upper cells often with longitudinal distosepta, 35-40 × 17-20 µ.

On branches of *Acer pseudoplatanus, A. platanoides* var. *brilliantissima, A.* sp., Germany, U.K., Eire, Roumania (IMI 61607, 198106, 168546, 173227, 1378, 1377, 22754, 82731, 82746, 92704, 61537, 61750, 27245, 22761, 22760, 22759, 22758, 22756, 22755, 38571, 27717).

PHIALOPYCNIDIINEAE subord. nov.

Cellulae conidiogenae phialidicae; conidiomata pycnidialia
Conidiogenous cells phialidic; conidiomata pycnidial

The majority of genera in this suborder have simple, thin-walled, immersed, papillate, globose pycnidia, ampulliform phialides or separate conidiophores, and aseptate, ellipsoid, cylindrical or fusiform, hyaline, smooth, thin-walled conidia. Such genera have been considered characteristic of the Sphaeropsidales in the past. Allowing for such similarity in pycnidial, conidiogenous cell and conidial characters, the suborder still contains a wide selection of genera, several with unusual or distinctive features. Such is the variation in structure of pycnidia particularly that it would be presumptuous to designate natural groups with any degree of certainty. There appears to be a continuum from the simplest form with thin-walled pycnidia, ampulliform phialides and hyaline aseptate conidia (e.g. in *Phoma*) through genera with thicker-walled pycnidia (occasionally with a very distinct structure), cylindrical and then integrated conidiogenous cells, and pigmented septate conidia. However this does not imply that *Wojnowicia*, with large, septate brown conidia and thick-walled rostrate pycnidia also has septate filiform conidiophores. It does not, they are simple ampulliform phialides. Nor does it imply that a genus like *Dendrodomus*, with thick-walled pycnidia and filiform conidiophores, is likely to have brown septate conidia. It does not, they are hyaline aseptate and thin-walled. Thus complexities in pycnidial structure, conidiophores and conidia are not necessarily correlated characters. Each genus may show complexities in one or more features. Consequently there remains a body of genera in which complexity in any of these characters is not achieved, and several other genera which show diversity and complexity to a variable degree.

The simplest group comprises *Ampelomyces, Chaetophoma, Rhizosphaera, Hapalosphaeria* and *Phoma*. Other genera which, by a summation of criteria, obviously should be placed together in any classification are a) *Pseudodiplodia, Ascochytulina* and *Chaetodiplodia*, b) *Microsphaeropsis* and *Readeriella*, c) *Pseudorobillarda* and *Neottiospora*, d) *Leptodothiorella* and *Asteromella*, and e) *Pyrenochaeta, Pleurophoma, Dendrodomus, Phialophorophoma* and *Sirophoma*.

DICHOTOMOUS KEY

(numbers in parentheses are species numbers)

1. Conidia hyaline, aseptate 2
1. Conidia hyaline, 1-septate 17
1. Conidia hyaline, multiseptate 19
1. Conidia brown 22
   2. Conidiogenous cells discrete, ampulliform, doliiform, lageniform or cylindrical 3
   2. Conidiogenous cells integrated on septate conidiophores 10
3. Pycnidia setose 4
3. Pycnidia lacking setae 6
   4. Pycnidia columnar, superficial, setae restricted to the apex, aseptate; phialides sometimes integrated; conidia bacillar .... *Cylindroxyphium* (1)
   4. Pycnidia globose, superficial 5
5. Setae hyaline, long, filiform, septate; conidia cylindrical, often 1 guttulate .... *Dasysticta* (2)
5. Setae stellate, brown; conidia ovoid .... *Staurophoma* (3)
5. Setae dark brown, verrucose; conidia ellipsoid .... *Chaetasbolisia* (4)
5. Setae pale brown, smooth; conidia ellipsoid to cylindrical .... *Chaetosticta* (5)
   6. Pycnidia superficial 7
   6. Pycnidia immersed 8
7. Pycnidia thin-walled, pale brown; conidia hyaline to very pale brown; hyperparasitic on Erysiphaceae .... *Ampelomyces* (6)
7. Pycnidia thin-walled, dark brown, with a repent superficial mycelium; conidia bacillar .... *Chaetophoma* (7)
7. Pycnidia thin-walled, brown, attached by a hypostroma to the substrate; conidia comparatively large, guttulate .... *Rhizosphaera* (8)
7. Pycnidia thick-walled, dark brown, base immersed in substrate; phialides percurrently proliferating; conidia large, unigut-tulate .... *Trematophoma* (9)
   8. Phialides cylindrical; conidia fusiform .... *Amerosporiopsis* (10)

8. Phialides with several conidiogenous loci (polyphialides); conidia fusiform, falcate (in some species becoming brown and verruculose) .... *Selenophoma* (11)
8. Phialides ampulliform, lageniform or doliiform 9
9. Pycnidia pale brown; conidia globose; on anthers of *Rubus* .... *Hapalosphaeria* (12)
9. Pycnidia pale brown, often aggregated; conidia fusiform, minutely guttulate; hyperparasitic on Uredinales .... *Metazythia* (13)
9. Pycnidia brown, thin-walled (in most species); conidia ellipsoid to subglobose, sometimes becoming 1-septate .... *Phoma* (14)
9. Pycnidia, brown, thick-walled; conidia cylindrical to fusiform, comparatively large; in culture usually only sclerotia are formed .... *Macrophomina* (15)
9. Pycnidia brown, thick-walled; conidia globose to pyriform, apparently thick-walled .... *Melanophoma* (16)
10. Conidiophores distinctly filiform, septate and branched; conidiogenous loci on very small lateral branches immediately below septa 11
10. Conidiophores comparatively short, septate, irregularly branched; conidiogenous loci on longer lateral branches 13
11. Pycnidia setose, thin-walled, brown; conidia ellipsoid .... *Pyrenochaeta* (20)
11. Pycnidia very pale brown, without setae; conidia with an apical and basal short cellular appendage; hyperparasitic on hymenomycetes .... *Eleutheromyces* (18)
11. Pycnidia dark brown, without setae; conidia with a short basal and a long apical cellular appendage; hyperparasitic on hymenomycetes .... *Eleutheromycella* (19)
11. Pycnidia rostrate, without setae, brown, superficial; conidia ellipsoid .... *Conidioxyphium* (17)
11. Pycnidia neither setose, nor conidia with appendages 12
12. Pycnidia thin-walled, brown, immersed; conidia oval; in marine environments .... *Phialophorophoma* (21)
12. Pycnidia thick-walled, dark brown, immersed; conidia globose to subglobose .... *Sirophoma* (22)
12. Pycnidia thick-walled, dark brown, globose to collabent, semi-immersed; conidia straight, ellipsoid .... *Pleurophoma* (23)
12. Pycnidia thick-walled, dark brown, globose to collabent, superficial; paraphyses near the ostiole; conidia ellipsoid to irregular .... *Dendrodomus* (24)
13. Conidia with an appendage(s) 14
13. Conidia lacking an appendage(s) 15
14. Pycnidia immersed; conidia fusiform, with the apex drawn out into a short appendage .... *Ciliosporella* (25)
14. Pycnidia immersed; conidia navicular with an apical and basal branched cellular appendage .... *Brycekendrickia* (26)
15. Pycnidia setose, thick-walled; conidia falcate, guttulate .... *Chaetophiophoma* (27)
15. Pycnidia without setae, dark brown, almost lacking an upper wall; conidia straight, cylindrical, guttulate .... *Coleophoma* (28)
15. Pycnidia not so 16
16. Pycnidia thick-walled, brown, immersed; conidia slightly invaginated in the middle; microconidial states of *Guignardia* .... *Leptodothiorella* (29)
16. Pycnidia thick-walled, brown, immersed; conidia ellipsoid; microconidial state of *Mycosphaerella* .... *Asteromella* (30)
16. Pycnidia thin-walled, brown, superficial; conidia ellipsoid .... *Aposphaeria* (31)
16. Pycnidia thick-walled, dark brown, superficial; conidia globose to subglobose .... *Pleurophomopsis* (32)
17. Pycnidia setose; conidiogenous cells cylindrical to integrated; conidia cylindrical or slightly curved .... *Chaetosphaeronema* (33)
17. Pycnidia lacking setae 18
18. Pycnidia immersed, pale brown, thin-walled; conidia medianly 1 septate, ellipsoid .... *Ascochyta* (34)
18. Pycnidia superficial, dark brown, thick-walled; conidia medianly 0-1(-2) septate, ellipsoid .... *Didymochaeta* (35)
18. Pycnidia immersed, dark brown, thick-walled; conidiogenous cells sometimes integrated; conidia fusiform, 0-1 septate .... *Clypeopycnis* (36)
19. Conidia with an appendage(s) 20
19. Conidia lacking appendages 21
20. Pycnidia thick-walled, dark brown, occasionally divided; conidia 2-septate, apex extended into a filiform appendage .... *Chaetoconis* (37)
20. Pycnidia thin-walled, brown, uniloculate; conidia 1-3 septate with 3 diverging mucilaginous appendages derived from the splitting of a sheath; paraphyses present .... *Pseudorobillarda* (38)

21. Pycnidia superficial; conidiophores filiform, septate, with conidiogenous loci immediately below septa; conidia 3 septate, falcate, fusiform .... *Sphaerographium* (39)
21. Pycnidia superficial; conidiophores filiform, septate, irregularly branched, loci apical; conidia 3 septate, fusiform, straight .... *Topospora* (40)
21. Pycnidia superficial, often aggregated; conidiogenous cells cylindrical; conidia hyaline or very pale brown, multiseptate, filiform .... *Megaloseptoria* (41)
    22. Conidia aseptate 23
    22. Conidia 1 septate 26
    22. Conidia multiseptate 28
23. Pycnidia rostrate; conidiogenous cells ampulliform; conidia ellipsoid, pale brown .... *Capitorostrum* (42)
23. Pycnidia globose 24
    24. Pycnidia with a pulvinate cushion at the base, thin-walled; conidia fusiform to falcate .... *Coniella* (43)
    24. Pycnidial base not pulvinate 25
25. Conidia globose, pyriform, cylindrical or ellipsoid; conidiogenous cells ampulliform, monophialidic .... *Microsphaeropsis* (44)
25. Conidia globose, pyriform or ellipsoid; conidiogenous cells ampulliform, percurrently proliferating........ *Coniothyrium* (see p. 122)
25. Conidia deltoid, with up to 3 short projections at the apex; conidiogenous cells polyphialidic and also percurrent .... *Readeriella* (45)
25. Conidia fusiform, with an infundibular mucilaginous appendage at the base formed by eversion of the sheath; phialides cylindrical .... *Neottiospora* (46)
    26. Pycnidia setose, superficial; phialides ampulliform; conidia ellipsoid, straight, medianly 1 septate, smooth........ *Chaetodiplodia* (47)
    26. Pycnidia immersed, clypeate; phialides ampulliform; conidia ellipsoid, straight, medianly 1 septate, verruculose........ *Ascochytulina* (48)
    26. Pycnidia neither setose nor clypeate 27
27. Pycnidia thin-walled, pale brown, immersed; phialides ampulliform; conidia medianly 1 septate, smooth, ellipsoid .... *Pseudodiplodia* (49)
27. Pycnidia thick-walled, dark brown, immersed; phialides lageniform; conidia 1-2 septate, smooth, cylindrical .... *Stenocarpella* (50)
27. Pycnidia thin-walled, brown, immersed; conidiophores irregularly branched and septate; conidia medianly 1 septate, ellipsoid, surrounded by a rigid granular sheath .... *Tunicago* (51)
    28. Pycnidia with hyaline setae, conidia 3 septate, thick-walled verruculose, fusiform to ellipsoid . *Angiopomopsis* (52)
    28. Pycnidia rostrate, without setae; conidia 3 septate, thick-walled, verrucose, ellipsoid .... *Ceratopycnis* (53)
    28. Pycnidia excentrically rostrate, setose; conidia fusiform to cylindrical, smooth, 2-8 septate .... *Wojnowicia* (54)
    28. Pycnidia neither setose nor rostrate; conidia muriform; conidiogenous cells percurrently proliferating........ *Camarosporium* (see p. 123)

SYNOPTIC KEY

Mycelial Characters

Position in relation to substrate

a. superficial	1 3 4 5 6 7 8 9 15 17 18 19 27 42
b. immersed	6 10 11 12 13 14 15 18 19 20 21 22 23 24 25 26 28 29 30 31 32 33 34 35 36 37 38 39 40 41 43 44 45 46 47 48 49 50 51 52 53 54

Other characters

a. subiculate	3
b. hyphopodiate	27

Conidiomatal characters

Position in relation to substrate

a. superficial or finally appearing to be superficial	1	2	3	4	5	6	7	8	9	17	18	23	24	27	31
	32	35	39	40	41	42	47	53							
b. immersed	10	11	12	13	14	15	19	20	21	22	25	26	28	29	30
	33	34	36	37	38	43	44	45	46	48	49	50	51	52	54
c. semi-immersed	9	13	14	19	39	53	54								

Tissue type

a. textura angularis	1	2	3	4	5	6	7	8	9	10	11	12	13	14	15
	16	18	19	20	21	22	23	24	25	26	27	28	29	30	31
	32	34	35	36	37	38	39	41	42	43	44	45	46	47	48
	49	50	51	52	53	54									
b. textura globulosa	35	53													
c. textura intricata	3	40	48												
d. textura oblita	39														
e. textura porrecta	17	39													
f. textura prismatica	33	42													

Dehiscence

a. ostiolate, non papillate	1	4	8	9	14	15	17	18	19	20	22	28	29	32	39
	45	47	49	51	52										
b. ostiolate, papillate	5	13	16	20	21	24	30	33	34	35	36	37	38	40	41
	42	43	44	48	50	53	54								
c. ostiolate, depressed	10	16	23	25	31										
d. linear, longitudinal	26														
e. fimbriate	1	18													
f. operculate	46														
g. irregular	2	3	6	7	11	12	27								

Shape

a. globose	2	3	4	5	6	7	8	9	10	11	12	13	14	15	16
	19	20	21	22	23	25	26	27	28	29	30	31	32	33	34
	35	36	37	38	40	41	43	44	45	46	47	48	49	50	51
	52	54													
b. cylindrical, corniform or rostrate	1	17	18	39	42	53	54								
c. branched	17														
d. collabent	24	28	31	38	47										
e. basal wall pulvinate	43														
f. clypeate	48														

Wall thickness

a. thin-walled (1-3 cells wide)	1	2	4	5	6	7	8	11	12	13	14	16	17	20	27
	34	38	42	43	44	47	49								
b. thick-walled (5 cells wide or more)	3	9	10	14	15	18	19	20	21	22	23	24	25	26	28
	29	30	31	32	33	35	36	37	39	40	41	44	45	46	48
	50	51	52	53	54										

Pigmentation

a. hypocreaceous	18														
b. brown to dark brown	1	2	3	4	5	6	7	8	9	10	11	12	13	14	15
	16	17	19	20	21	22	23	24	25	26	27	28	29	30	31
	32	33	34	35	36	37	38	39	40	41	42	43	44	45	46
	47	48	49	50	51	52	53	54							

Setal characters

a. present	1	2	3	4	5	20	33	47	52	54

Pigmentation

a. brown	1	3	4	5	20	33	47	54
b. hyaline	2	52						

Branching

a. branched	3								
b. unbranched	1	2	4	5	20	33	47	52	54

Conidiogenous cell characters

Insertion

a. discrete	1	2	3	4	5	6	7	8	9	10	11	12	13	14	15
	16	24	28	29	31	33	34	35	36	38	41	42	43	44	45
	47	48	49	50	51	52	53	54							
b. integrated	1	9	13	14	17	18	19	20	21	22	23	24	25	26	27
	28	29	30	31	32	33	36	37	38	39	40	41	43	46	51

Shape

a. ampulliform-doliiform	2	3	4	5	6	7	8	11	12	13	14	15	16	31	34
	35	36	38	42	44	45	47	48	49	50	53	54			
b. lageniform	9	15	24	28	29	30	32	41	44	45	50	51	52		
c. cylindrical	1	9	10	17	18	19	20	21	22	23	24	25	26	27	28
	29	30	31	32	33	37	39	40	41	43	44	45	46	51	52

Pigmentation

a. hyaline	1	2	3	4	5	6	7	9	10	11	12	13	14	15	16
	17	18	19	20	21	22	23	24	25	26	27	28	29	30	31
	32	33	34	35	36	37	38	39	40	41	42	43	44	45	46
	47	48	49	50	51	52	53	54							
b. brown	8														

Other characters

a. polyphialidic	11	45
b. occasionally showing annellations	45	

Conidiophore characters

Shape

a. irregularly branched and ramified (to produce aggregates of cells)	13	16

b. cylindrical to filiform, unbranched or branched sparingly at the base	1 9 14 26 30 31 32 33 41 43 46
c. filiform, branched	17 18 19 20 21 22 23 24 25 27 28 29 37 38 39 40 51

Conidial characters

Septation	
a. distoseptate	52
b. euseptate	5 33 34 35 36 37 38 39 40 41 47 48 49 53 54
c. aseptate	1 2 3 4 5 6 7 8 9 10 11 12 13 14 15 16 17 18 19 20 21 22 23 24 25 26 27 28 29 30 31 32 36 42 43 44 45 46
d. medianly 1 septate	5 14 33 34 35 36 38 47 48 49 50 51
e. 2 septate	35 37
f. 2-3 septate	5 38 50
g. 3 septate	39 40 52 53
h. multiseptate	41 54
Shape	
a. oval-ellipsoid	1 2 3 4 5 6 7 8 14 16 17 18 20 21 23 24 30 31 34 35 40 44 47 48 49 51 53
b. cylindrical	1 2 6 14 15 28 29 33 40 47 53
c. fusiform	10 11 13 14 15 19 24 25 26 27 36 37 38 39 43 44 46 50 52 54
d. falcate	11 25 26 27 39 43
e. pyriform or obpyriform	8 9 14 16 17 32 42 44
f. clavate	9
g. filiform	41
h. globose	12 14 22 32 44
i. deltoid	45
Guttulation	
a. guttulate	2 4 5 6 8 9 10 13 14 15 17 20 21 24 27 28 29 32 33 36 37 38 41 46 51 54
b. eguttulate	1 2 3 7 11 12 14 15 16 17 18 19 20 21 22 23 24 25 26 29 30 31 34 35 36 39 40 42 43 44 45 47 48 49 50 52 53 54
Wall thickness	
a. thin-walled	1 2 3 4 5 6 7 8 9 10 11 12 13 14 15 17 18 19 20 21 22 23 24 25 26 27 28 29 30 31 32 33 34 35 36 37 38 39 40 41 42 43 44 46 47 48 49 54
b. thick-walled	16 43 44 45 50 52 53
c. tunicate	51
Ornamentation	
a. smooth	1 2 3 4 5 6 7 8 9 10 11 12 13 14 15 17 18 19 20 21 22 23 24 25 26 27 28 29 30 31 32 33 34 35 36 37 38 39 40 41 42 43 44 45 46 47 49 50 51 54
b. verruculose	11 16 44 48 52 53

Pigmentation															
a. hyaline	1	2	3	4	5	7	8	9	10	11	12	13	14	15	17
	18	19	20	21	22	23	24	25	26	27	28	29	30	31	32
	33	34	35	36	37	38	39	40	41						
b. brown	6	11	16	42	43	44	45	46	47	48	49	50	51	52	53
	54														

Appendages						
a. apical	18	19	25	26	37	
b. basal	18	19	26	38	46	
c. single	18	19	25	26	37	46
d. several inserted at one point	38					
e. cellular	18	19	25	26	37	
f. mucilaginous	38	46				
g. simple	18	19	25	37	38	46
h. branched	26					
i. infundibular or tentacular	46					

Paraphyses characters

a. present	28	38
b. inflated	28	
c. filiform	38	
d. septate	38	
e. aseptate	28	

Sclerotia

a. present	14	15

Chlamydospores

a. present	14

1. **CYLINDROXYPHIUM** Batista & Ciferri, *Quad. Inst. Univ. Pavia* 31: 77 (1963).
Sp. typ.: *C. virginianum* Batista & Ciferri

Mycelium superficial, branched, septate, brown. *Conidiomata* pycnidial, superficial, separate, dark brown, cylindrical to doliiform, slightly tapered towards the apex, unilocular; wall of thick-walled, brown textura angularis. *Setae* apical, short, 0-2 septate, dark brown, acuminate. *Ostiole* single, circular, central, slightly fimbriate. *Conidiophores* short, hyaline, 1-2 septate, branched only at the base, cylindrical, formed from the inner cells of the pycnidial wall. *Conidiogenous cells* enteroblastic, phialidic, determinate, integrated or discrete, hyaline, smooth, apertures apical or lateral immediately below transverse septa, channel and collarette minute, periclinal wall thickened. *Conidia* hyaline, aseptate, thin-walled, eguttulate, cylindrical to ellipsoid.

The genus is monotypic.

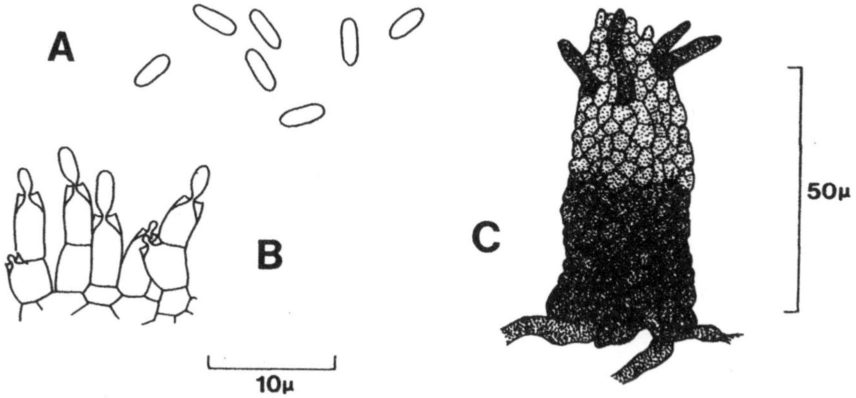

Fig. 213. *Cylindroxyphium virginianum*, A, conidia; B, conidiophores and developing conidia; C, side view of a conidioma.

C. VIRGINIANUM Batista & Ciferri, *ibid.*: 77 (1963). (Fig. 213)

Conidiomata up to 70 µ high × 30 µ wide at the base; setae up to 20 µ long × 3 µ wide. Conidiogenous cells 3-6 × 2.5-3 µ. Conidia 3.5-4 × 1.5 µ.

On leaves of *Quercus virginiana*, Fla., U.S.A. (IMI 201362, holotype ex BPI).

2. DASYSTICTA Speg., *An. Mus. Nac. Buenos Aires* **23**: 108 (1912).
Sp. typ.: *D. sapindophila* Speg.

Mycelium apparently absent. *Conidiomata* pycnidial, superficial, amongst leaf hairs, separate, globose, pale brown, unilocular, wall 1-2 cells thick, of textura angularis. *Setae* hyaline, thick-walled, septate, tapered, flexuous. *Ostiole* apparently absent. *Conidiophores* absent. *Conidiogenous cells* enteroblastic, phialidic, determinate, discrete, doliiform to ampulliform, hyaline, smooth, channel and collarette minute, formed directly from the cells comprising the pycnidial wall. *Conidia* hyaline, aseptate, thin-walled, smooth, straight, cylindrical, often with a large central guttule.

The genus is monotypic.

D. SAPINDOPHILA Speg., *ibid.*: 108 (1912). (Fig. 214)

Conidiomata up to 150 µ diam.; setae up to 650 µ long. Conidia 5.5-8 × 2.5 µ.

On leaves of *Serjania caracasana*, Argentina (IMI 196498, holotype).

3. STAUROPHOMA Höhn., *Denk. Akad., Wien* **83**: 34 (1907).
Sp. typ.: *S. panici* Höhn.
Illustration: Morgan-Jones, Nag Raj & Kendrick (1972f).

Mycelium superficial, branched, septate, pale brown, anastomosing to form a subicle. *Conidiomata* pycnidial, superficial, separate, globose, brown, unilocular, thick-walled; walls composed of brown, thick-walled textura angularis to intricata. *Ostiole* absent, dehiscence by irregular rupture. *Setae* scattered over the pycnidial wall, straight, branched stellately at the apex, dark brown, smooth, thick-walled, aseptate. *Conidiophores* absent. *Conidiogenous cells* enteroblastic, phialidic, determinate, discrete, doliiform to ampulliform, hyaline, smooth, formed from the inner cells of the pycnidial wall. *Conidia* hyaline, aseptate, ovoid, smooth, thin-walled, both ends obtuse, eguttulate.

The genus is monotypic.

S. PANICI Höhn., *ibid.*: 34 (1907).

Conidiomata up to 75 µ diam. Setae 17 µ long × 4.5 µ wide, 18 µ wide at the apex. Conidiogenous cells 5-6 × 3-4 µ. Conidia 4 × 2 µ.

On leaves of *Panicum sulcatum*, Brazil.

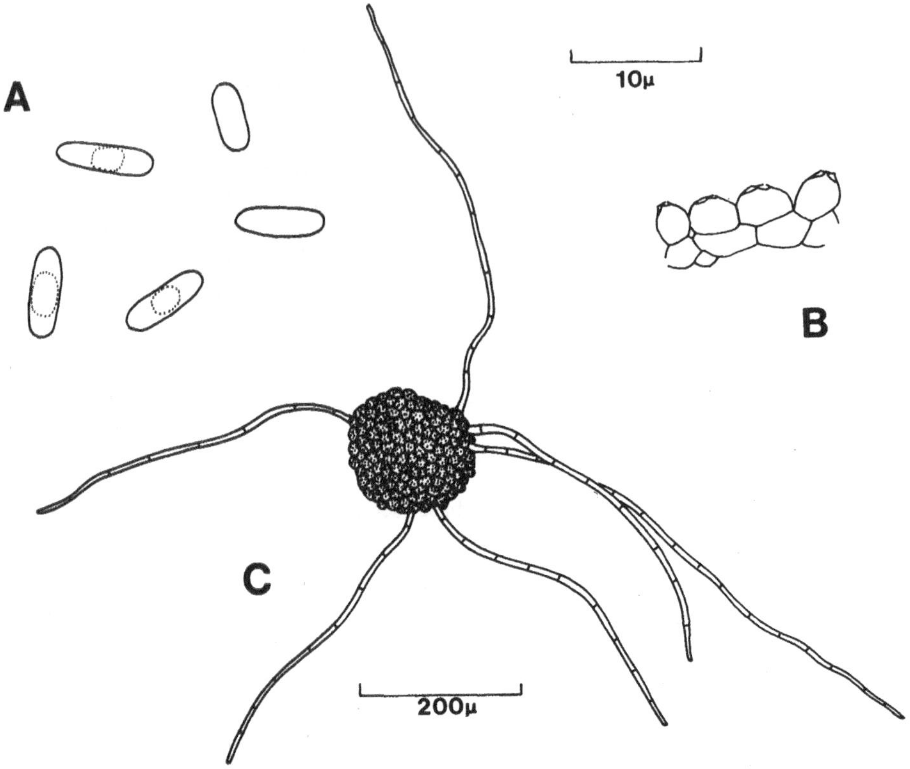

Fig. 214. *Dasysticta sapindophila*, A, conidia; B, conidiogenous cells; C, surface view of a conidioma.

4. CHAETASBOLISIA Speg., *Physis* **4**: 293 (1918).
Sp. typ.: *C. erysiphoides* (Griff. & Maubl.) Speg. (syn. *Chaetophoma erysiphoides* Griff. & Maubl.).

Mycelium superficial, branched, septate, brown. *Conidiomata* pycnidial, superficial, separate, dark brown, globose, unilocular, thin-walled, wall 1-2 cells thick, of brown, thin-walled textura angularis; setae several, scattered over the pycnidia, straight or curved, septate, thick-walled, irregularly verrucose. *Ostiole* central, circular. *Conidiophores* absent. *Conidiogenous cells* enteroblastic, phialidic, discrete, doliiform to ampulliform, thin-walled, hyaline, channel and collarette minute, formed from the inner cells of the pycnidial wall. *Conidia* hyaline, aseptate, thin-walled, smooth, minutely guttulate, smooth, both ends obtuse, oval.

The type species was redescribed by Batista & Ciferri (1963) but not through an examination of the type collection. There are 7 taxa described in the genus so far but no revision has been made.

C. ERYSIPHOIDES (Griff. & Maubl.) Speg., *ibid.*: 293 (1918). (Fig. 215)
Chaetophoma erysiphoides Griff. & Maubl., *Bull. trimest. Soc. mycol. Fr.* **25**: 60 (1909).

Conidiomata up to 85 µ diam.; setae up to 60 µ long × 3-4 µ wide. Conidiogenous cells 3.5-7 × 2.5-5 µ. Conidia 5-6 × 2 µ.

Hypophyllous on leaves of *Photinia loriformis*, Yunnan, China (IMI 12578).

5. CHAETOSTICTA Petrak & H. Syd., *Annls mycol.* **23**: 270 (1925).
Trichocicinnus (Sacc.) Höhn., *Mitt. bot. Inst. tech. Hochsch. Wien* **3**: 115 (1926).
Pyrenochaeta deNot. subgen. *Trichocicinnus* Sacc., *Annls mycol.* **3**: 512 (1905).
Sp. typ.: *C. perforata* (Ell. & Ev.) Petrak & Syd. (syn. *Chaetomella perforata* Ell. & Ev.).

Mycelium mostly superficial, tomentose, pale brown, branched, septate. *Conidiomata* pycnidial, hypophyllous, separate, globose, brown, superficial, unilocular, thin-walled, of textura angularis. *Ostiole* central, circular, papillate. *Setae* pale brown, obtuse, septate. *Conidiophores* absent. *Conidiogenous cells* enteroblastic, monophialidic, discrete, doliiform-ampulliform, hyaline, smooth, channel wide, collarette minimal, formed from cells lining the inner pycnidial wall. *Conidia* hyaline, with 0-several eusepta, not constricted, cylindrical, obtuse at each end, thin-walled, smooth, eguttulate. Paraphyses absent.

Considered by Clements & Shear (1931) to be a later name for *Dasypyrena* Speg. which was shown by Petrak (1950) to be synonymous with *Actinopeltis* Höhn. However, Crane (1971) has fully reviewed the confusion surrounding typification of

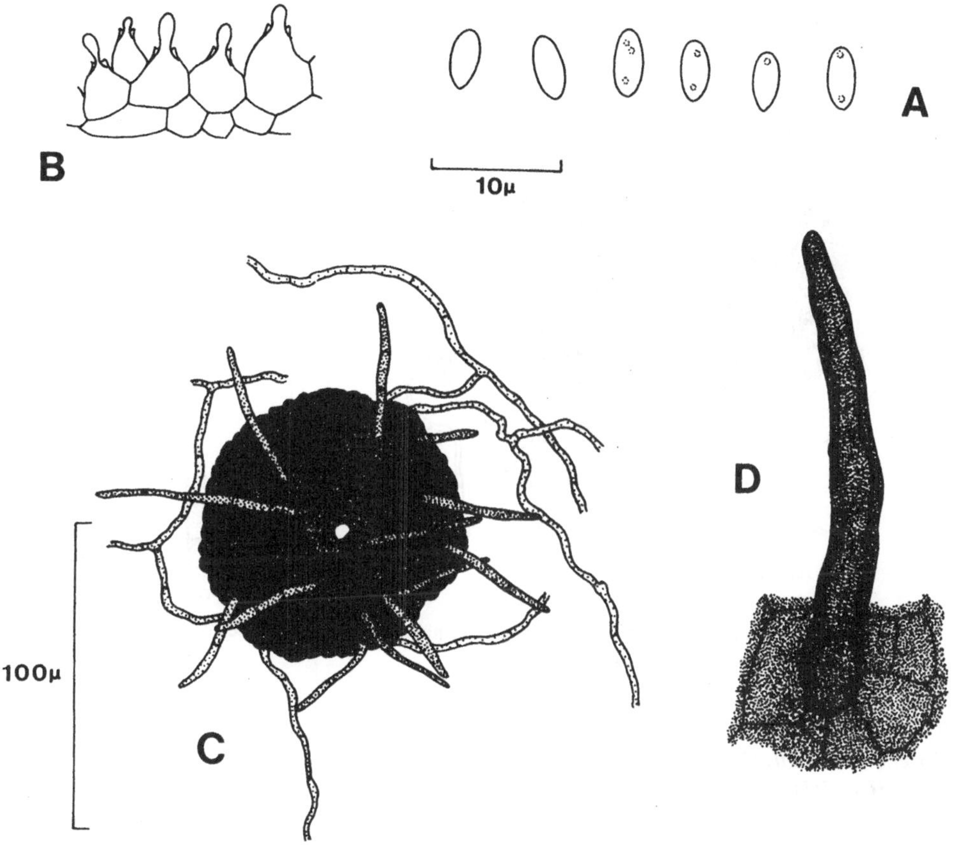

Fig. 215. *Chaetasbolisia erysiphoides*, A, conidia; B, conidiogenous cells and developing conidia; C, surface view of a conidioma and superficial mycelium; D, seta.

C. perforata and the relationships with *Nematostroma occidentale* (Ell. & Ev.) Barr, the presumed teleomorphosis. In addition to aseptate conidia (which are the only type I have hitherto seen), Crane (1971) described 1-3 septate conidia. Evidently septation in the species is variable and some collections have consistently aseptate conidia whereas others are 1 septate or 2-3 septate.

C. PERFORATA (Ell. & Ev.) Petrak & H. Syd., *ibid.*: 270 (1925). (Fig. 216)
Chaetomella perforata Ell. & Ev., *J. Mycol.* **1**: 153 (1885).
Pyrenochaeta erysiphoides Sacc., *Annls mycol.* **3**: 512 (1905).
Trichocicinnus erysiphoides (Sacc.) Höhn., *Mitt. bot. Inst. tech. Hochsch. Wien* **3**: 115 (1926).

Conidiomata 105-300 µ diam.; ostiole 31-42 µ diam.; setae 1-9 septate, 100-300 µ long × 3.5-10 µ wide. Conidiogenous cells 8-10 × 4-9 µ. Conidia 3-septate, 17.5-31 × 2.5-4.5 µ.

Hypophyllous on living leaves of *Cirsium discolor, C. altissimum, C. arvense,* and *Artemisia ludoviciana*, Italy, U.S.A. (IMI 205841, holotype of *P. erysiphoides* ex PAD).

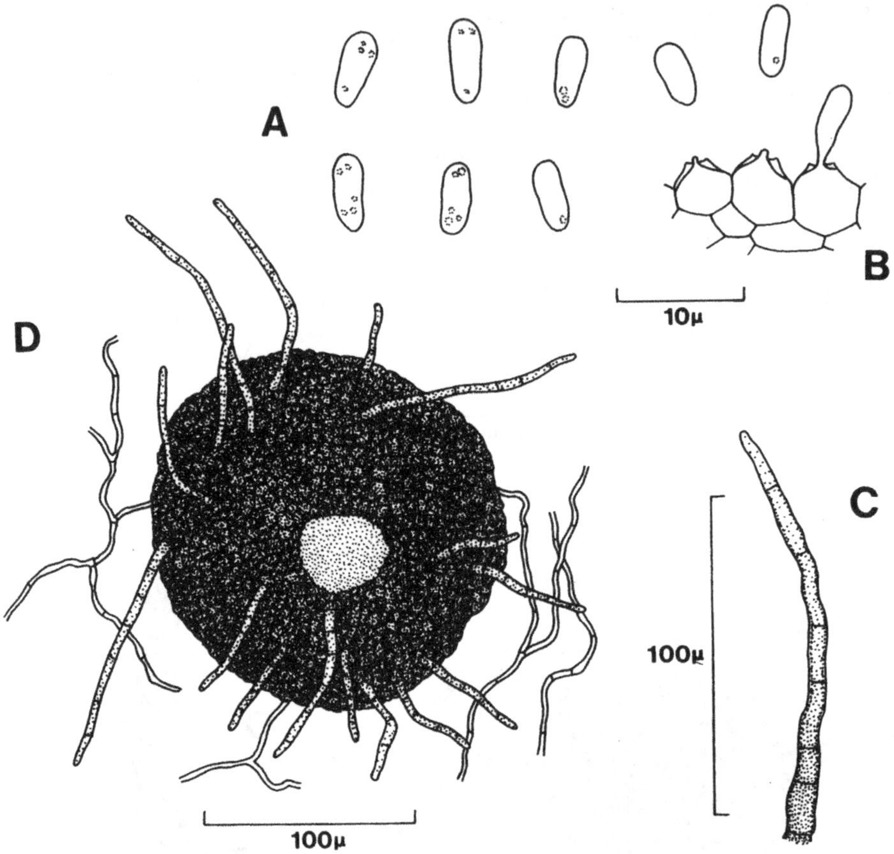

Fig. 216. *Chaetosticta perforata*, A, conidia; B, conidiogenous cells and developing conidia; C, seta; D, surface view of a conidioma.

6. AMPELOMYCES Ces. ex Schlecht., *Bot. Zeit.* **10**: 302 (1852).
Cicinobolus Ehrenb., *Bot. Zeit.* **11**: 16 (1853).
Byssocystis Riess, *Hedwigia* **1**: 23 (1853).
Sp. typ.: *A. quisqualis* Ces.

Mycelium superficial or immersed, hyaline to pale brown, branched, septate. *Conidiomata* pycnidial, superficial, in, on or around hyphae of Erysiphales, separate, globose or elongated to pyriform, sometimes papillate, pale brown, unilocular, wall 1 cell thick, of pale brown textura angularis. No distinct ostiole present, apex sometimes papillate, but not of different texture from rest of pycnidial wall. Dehiscence by apical rupture of pycnidium. *Conidiophores* absent. *Conidiogenous cells* enteroblastic, phialidic, determinate, discrete, doliiform to ampulliform, hyaline, smooth, channel and collarette minute, periclinal wall towards apex thickened, formed directly from the pycnidial wall cells. *Conidia* very pale brown, aseptate, thin-walled, smooth, guttulate, straight or curved, cylindrical to fusiform.

More than 40 binomials have been described in '*Cicinnobolus*' but apart from the generic synonymy elucidated by Rogers (1959) there has been no definitive reappraisal of either species names or species delimitation in the genus. The following dimensions are from a wide range of material on a variety of hosts in IMI.

Donk (1966) did not agree with the interpretation of Art. 70 by Rogers (1959) and preferred to adopt the well-known and almost universally used '*Cicinnobolus cesatii* deBary' for this common hyperparasite of Erysiphales. It had been previously assumed that conidia were hyaline, but they are in fact slightly thick-walled and pale brown. The name *Cicinnobella* P. Henn. had been used for 'brown-spored *Cicinnobolus* species', although Höhnel (1911b) showed *Cicinnobella* to be based on an immature ascomycete. Since *Asbolisia* Speg. and *Diblastospermella* Speg. would therefore be considered as alternatives, their adoption should be viewed with caution since *Ampelomyces* is now known to have pigmented conidia. The genus *Cicinnobolus* contains more than 40 described taxa and these have been recognised mainly according to the dif-

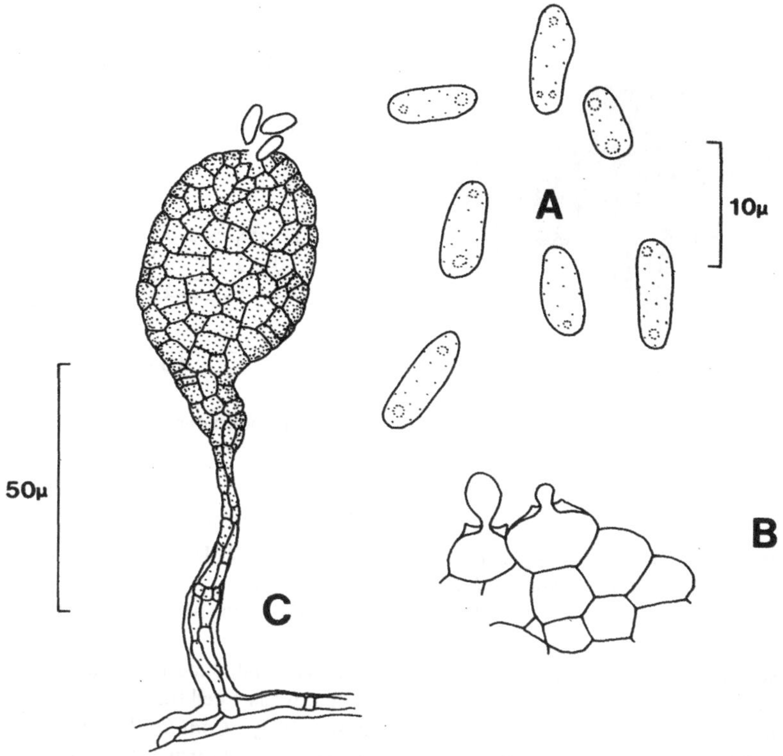

Fig. 217. *Ampelomyces quisqualis*, A, conidia; B, conidiogenous cells and developing conidia; C, side view of a conidioma.

ferent species of Erysiphales on which they occur. However I am unaware of any modern revision of the genus, and clearly a reassessment is required.

A. QUISQUALIS Ces. ex Schlecht., *ibid.*: 301 (1852). (Fig. 217)

Conidiomata 40-105 μ long × 30-50 μ wide. Conidiogenous cells 4.5-5.5 μ wide. Conidia 4-6.5 × 2-2.5 μ.

Hyperparasitic on Erysiphales, world wide in distribution, represented by about 100 collections in IMI.

7. CHAETOPHOMA Cke, *Grevillea* 7: 25 (1878).

Original spp.: *C. ilicifolia* Cke, *C. quercifolia* Cke, *C. catesbeyi* (Thüm.) Cke, *C. anthelmintica* Cke, *C. catalpae* Cke, *C. cycadis* Cke

Lectotype sp.: *C. quercifolia* Cke, fide Clements & Shear (1931).

Mycelium superficial, brown, irregularly branched, anastomosing, septate, of varying width. *Conidiomata* pycnidial, superficial, globose, separate or aggregated, medium brown, unilocular, thin-walled, of textura angularis. *Ostiole* absent. *Conidiophores* absent. *Conidiogenous cells* enteroblastic, phialidic, discrete, determinate, ampulliform, collarette and channel minute, formed from the inner cells of the pycnidial wall. *Conidia* hyaline, aseptate, thin-walled, eguttulate, smooth, cylindrical.

More than 60 taxa have been described in this genus but none have been reassessed in the light of modern taxonomic developments.

C. QUERCIFOLIA Cke, *ibid.*: 25 (1878). (Fig. 218)

Conidiomata up to 30 μ diam. Conidiogenous cells 2.5-4 × 2-4 μ. Conidia 2.5-3 × 1 μ.

Hypophyllous on leaves of *Quercus nigra*, N. Carol., U.S.A. (IMI 5974, designated lectotype of *C. quercifolia*, ex K).

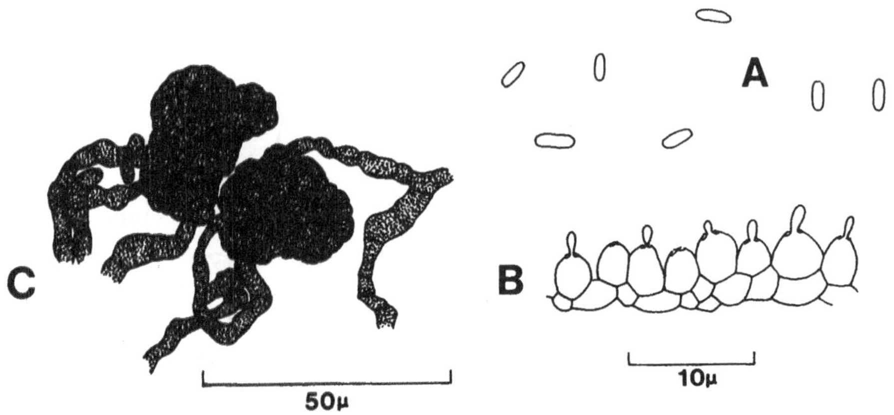

Fig. 218. *Chaetophoma quercifolia*, A, conidia; B, conidiogenous cells and developing conidia; C, surface view of conidiomata.

8. RHIZOSPHAERA Mangin & Hariot, *Bull. trimest. Soc. mycol. Fr.* **23**: 56 (1907).
Rhizophoma Petrak & H. Syd., *Beih. Rep. spec. nov. regni veg.* **42**: 472 (1927).
Gelatosphaera Batista & Maia, *Ann. Soc. biol. Pern.* **16**: 144 (1959).
Sp. typ.: *R. pini* (Cda) Maubl. (syn. *Coniothyrium pini* Cda, *R. abietis* Mangin & Hariot).

Mycelium immersed, branched, septate, brown. *Conidiomata* pycnidial, superficial, globose, black, produced above stomata and connected to the substrate by a hypostroma, thin-walled; walls 1 cell thick, of large-celled, brown, thin-walled textura angularis. *Ostiole* central, circular, non-papillate, opening widely. *Conidiophores* absent. *Conidiogenous cells* enteroblastic, phialidic, determinate, discrete, pale brown, smooth, ampulliform, channel wide, periclinal thickening marked, collarette minimal, comprising the cells of the conidiomatal wall. *Conidia* hyaline to subhyaline, smooth, thin-walled, aseptate, guttulate, cylindrical to ellipsoid, obtuse at the apex, occasionally truncate at the base.

The most recent account of the genus is by Kobayashi (1967) who reviewed its history and the synonymy of *Rhizophoma* Petrak & H. Syd. Four species were accepted, including one which was at that time unnamed. Morelet (1973a) later described it as *R. kobayashii* (Morelet) Morelet. Morelet (1973b) added another species, *R. media*, but Sutton (1977a) suggested that two specific names in *Gelatosphaera* Batista & Maia had precedence. It is evident from the account by Kobayashi (1967) that all are probably the same as *R. oudemansii* Maubl.

KEY

1. Conidia 7-12.5 μ wide 2
1. Conidia 3-8 μ wide 3
   2. Conidia 15-21 μ long .... *R. pini*
   2. Conidia 20-37.5 μ long .... *R. kobayashii*
3. Conidia 5-10 μ long .... *R. kalkoffii*
3. Conidia 9-13 μ long .... *R. oudemansii*

R. KALKOFFII Bub., nom. nov., *Ber. dtsch. Bot. Ges.* **32**: 188 (1914). (Fig. 219A, B, C)
Sphaeronaema pini Desm., *Ann. Sci. nat.*, ser. 10, **3**: 347 (1948), non *R. pini* (Cda) Maubl. (1907).
Sclerophoma pini (Desm.) Höhn., *Sber. Akad. Wiss. Wien* **118**: 1234 (1916).
Rhizophoma pini (Desm.) Petrak & H. Syd., *Beih. Rep. spec. nov. regni veg.* **42**: 472 (1927).
Rhizosphaera radicata Naumov, *Bull. trimest. Soc. mycol. Fr.* **30**: 385 (1914).

Conidiomata 50-95 × 115 μ. Conidia 5-10 × 3-5 μ.

On dead needles of *Picea abies*, *P. pungens*, *P.* sp., U.K., Que, Canada (IMI 225393, 61842, 25962, 8444).

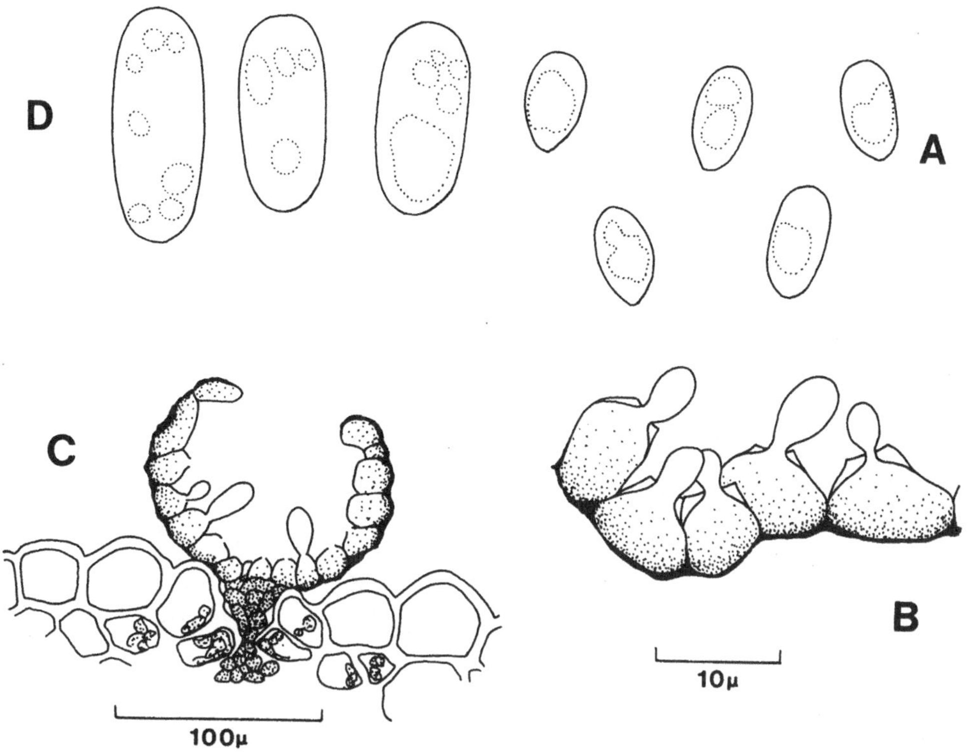

Fig. 219. A, B, C, *Rhizosphaera kalkoffii*, conidia, conidiogenous cells, vertical section of a conidioma; D, *R. pini*, conidia.

R. KOBAYASHII (Morelet) Morelet, *Bull. Soc. Sci. nat. Archéol. Toulon Var* **204**: 8 (1973).
Rhizosphaera pini var. *kobayashii* Morelet, *ibid.*: **180**: 4 (1969).

Conidiomata 65-85 × 70-100 µ. Conidia 20-37.5 × 7.5-12.5 µ.

On dead needles of *Abies mariesii, Pinus pumila*, Japan.

R. OUDEMANSII Maubl., nom. nov., *Bull. trimest. Soc. mycol. Fr.* **23**: 173 (1907).
Sacidium abietis Oud., *Contr. Fl. Myc. Pay Bas* **17**: 333 (1900), non *R. abietis* Mangin & Hariot (1907).
Rhizosphaera media Morelet, *Bull. Soc. Sci. nat. Archéol. Toulon Var* **203**: 12 (1973).
Gelatosphaera abietina Batista & Maia, *Ann. Soc. Biol. Pernambuco* **16**: 144 (1959).
Gelatosphaera tsugae Batista & Maia, *Publ. Inst. Mic. Univ. Recife* 214: 14 (1959).

Conidiomata 60-100 µ diam. Conidia 9-16 × 5.5-9 µ.

On needles of *Picea glauca, Abies grandis, A.* sp., *Tsuga heterophylla*, France, Holland, Germany, Alaska, U.S.A.

R. PINI (Cda) Maubl., *Bull. trimest. Soc. mycol. Fr.* **23**: 171 (1907). (Fig. 219D)
Coniothyrium pini Cda, *Icon. fung.* **4**: 38 (1840).
Sacidium pini (Cda) Fr., *Summa Veg. Scand.* 420 (1849).
Leptothyrium pini (Cda) Sacc., *Syll. fung.* **3**: 627 (1884).
Rhizosphaera abietis Mangin & Hariot, *Bull. trimest. Soc. mycol. Fr.* **23**: 56 (1907).

Conidiomata 50-65 × 125 µ. Conidia 13-25 × 7-12.5 µ.

On needles of *Picea* sp., *Abies pectinata*, Pakistan, Austria (IMI 33617, 30670, 37216, 81159).

9. TREMATOPHOMA Petrak, *Annls mycol.* **22**: 152 (1924).
Sp. typ.: *T. lignicola* Petrak

Mycelium immersed, branched, septate, pale brown. *Conidiomata* pycnidial, semi-immersed, separate, elongated,

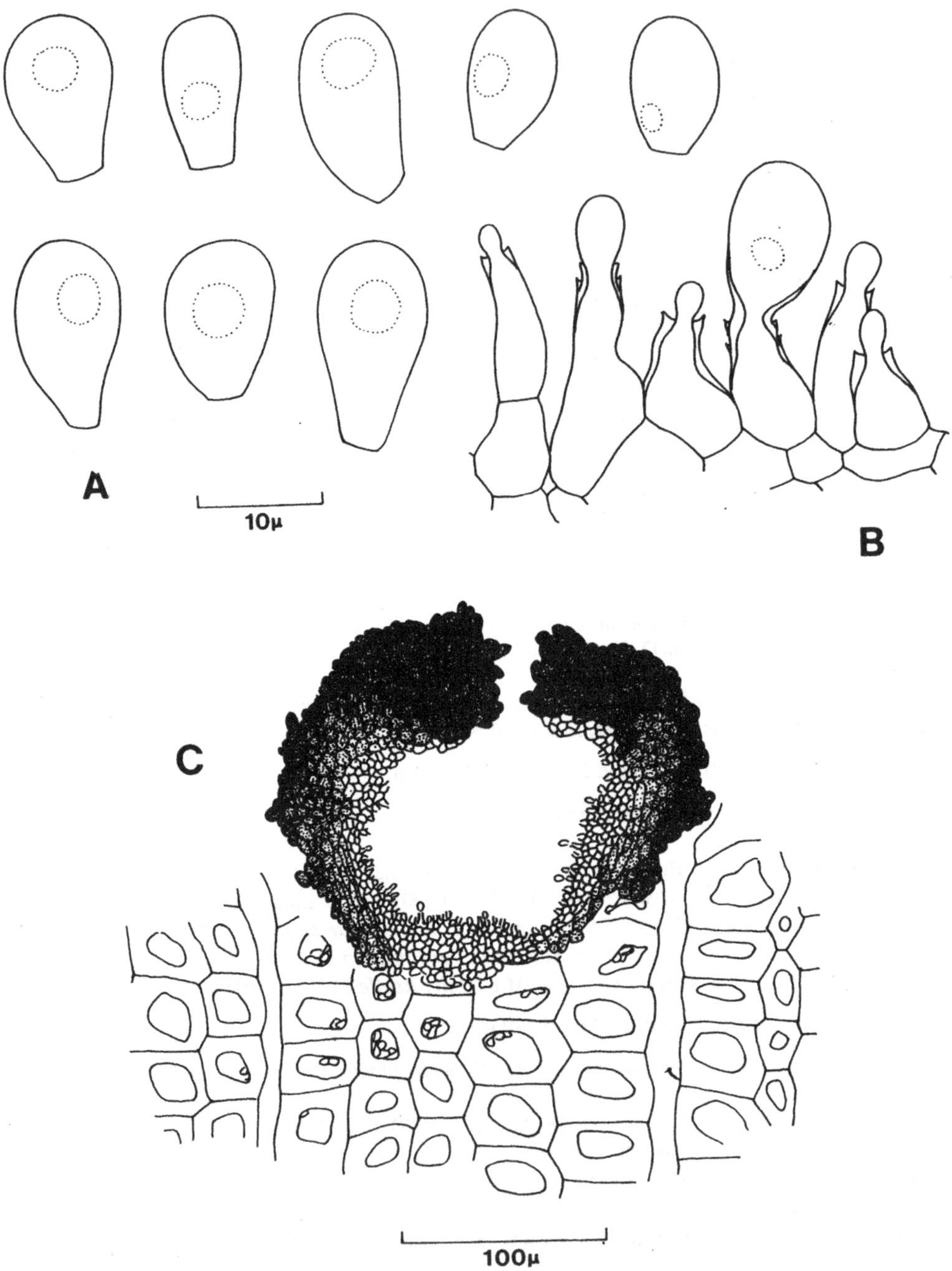

Fig. 220. . *Trematophoma lignicola*, A, conidia; B, conidiogenous cells and developing conidia; C, vertical section of a conidioma.

dark brown to black, unilocular; wall thick, consisting of a hyaline to pale brown base and thick lateral and upper walls of dark brown textura angularis. *Ostiole* single, circular, central. *Conidiophores* occasionally present, hyaline, 1-2 septate, branched only at the base, smooth, formed from the inner cells of the pycnidial wall. *Conidiogenous cells* enteroblastic, phialidic, discrete or occasionally integrated, hyaline, smooth, doliiform to cylindrical, indeterminate, often percurrently proliferating to form 2 collarettes, channel wide, collarettes minute, periclinal wall considerably thickened. *Conidia* hyaline, thin-walled, aseptate, smooth, with one central guttule, base broad and truncate, broadly obovoid.

Petrak & Sydow (1927) gave an additional account of *Trematophoma* where it was keyed out from related genera and regarded as conidial *Trematosphaeria* Fckl. The genus remains monotypic.

T. LIGNICOLA Petrak, *ibid.*: 152 (1924). (Fig. 220)

Conidiomata up to 250 µ diam. Conidiogenous cells 7.5-20 × 4-6 µ. Conidia 11.5-17.5 × 6.5-9 µ, base up to 4 µ wide.

On dead wood lying in water, Austria (IMI 197515 ex K, isotype).

10. AMEROSPORIOPSIS Petrak, *Bot. Arch.* **43**: 84 (1941).
Sp. typ.: *A. gaubae* Petrak

Mycelium immersed, pale brown to hyaline, branched, septate. *Conidiomata* pycnidial, separate, globose, greenish black, immersed, unilocular, thick-walled; wall several cells thick, in the upper part of dark brown, thick-walled, textura angularis, at the sides of paler, thinner-walled, elongated cells, at the base of pale brown, almost isodiametric cells. *Ostiole* central, circular, depressed. *Conidiophores* absent. *Conidiogenous cells* enteroblastic, phialidic, determinate, discrete, cylindrical, straight, hyaline, smooth, abruptly tapered at the apex to a minute aperture, formed mostly from the base of the pycnidium. *Conidia* hyaline, aseptate, fusiform to limoniform, apex obtuse to acute, base protuberant, ± truncate, thin-walled, smooth, often guttulate.

The genus only contains the type species.

A. GAUBAE Petrak, *ibid.*: 84 (1941). (Fig. 221)

Conidiomata up to 180 µ diam. Conidiogenous cells 11-14 × 1.5-2 µ. Conidia 8-10.5 × 2.5-3.5 µ.

On dead leaves of *Sesleria* sp., Iran (IMI 201893, holotype ex ZT).

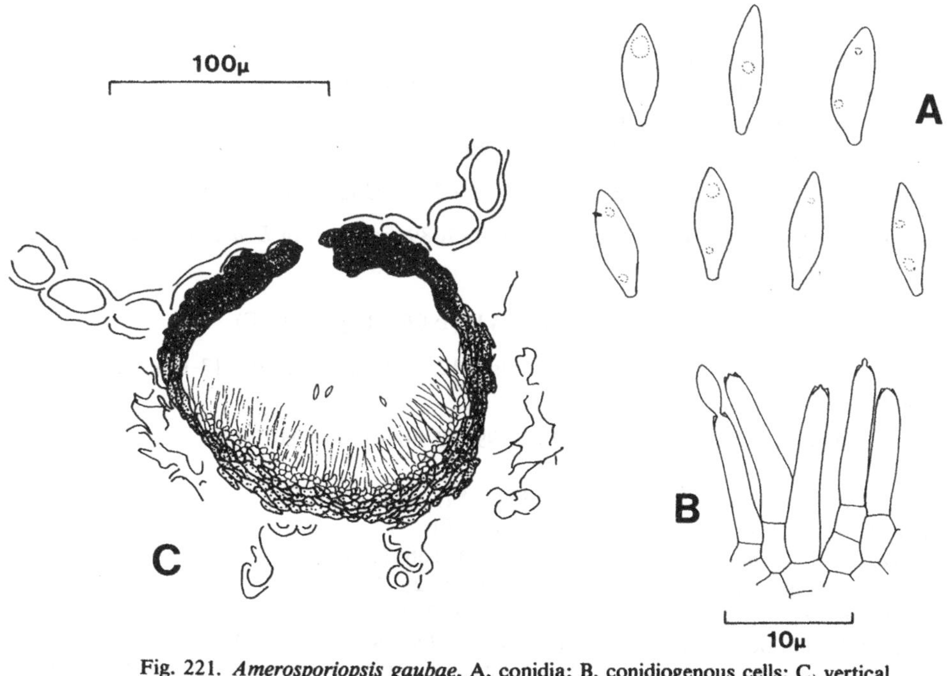

Fig. 221. *Amerosporiopsis gaubae*, A, conidia; B, conidiogenous cells; C, vertical section of a conidioma.

11. SELENOPHOMA Maire, *Bull. Soc. Bot. Fr.* **53**: CLXXXVII (1906).
Neopatella Sacc. apud H. & P. Syd., *Annls mycol.* **6**: 530 (1908).
Ludwigiella Petrak, *Annls mycol.* **20**: 319 (1922).
Selenophomopsis Petrak, *Annls mycol.* **22**: 182 (1924).
Falcispora Bub. & Serebr., *Hedwigia* **52**: 269 (1912).
Sp. typ.: *S. catananches* Maire

Mycelium immersed, branched, septate, pale brown. *Conidiomata* pycnidial, separate or aggregated, immersed but often appearing superficial in old substrata, globose, dark brown, unilocular, thin-walled; walls 2-3 cells thick, composed of brown, thick-walled, large-celled textura angularis. *Ostiole* absent, dehiscence by irregular rupture of upper wall. *Conidiophores* absent. *Conidiogenous cells* enteroblastic, phialidic, discrete, determinate, hyaline to pale brown, smooth, with 1-3 apertures, collarette and channel minute, periclinal wall thickened, formed from the inner cells of the pycnidial wall. *Conidia* aseptate, falcate, fusiform, eguttulate, smooth, at first thin-walled, hyaline, later pale brown with thicker walls (at least in the type species) and verruculose.

The popular concept of *Selenophoma*, as exemplified in the work of Sprague (1950) on graminicolous species, is at variance with data obtained from a study of the type species. Consequently the generic synonymy which has included *Neopatella* Sacc., *Falcispora* Bub. & Serebr., *Lunospora* Frandsen and *Pseudoseptoria* Speg. must be discarded in favour of the inclusion of *Selenophomopsis* Petrak, *Neopatella* Sacc. and *Ludwigiella* Petrak which are truly congeneric with *Selenophoma*. Therefore the graminicolous *Selenophoma* species have been accommodated in *Pseudoseptoria* which is the earliest available generic name for them.

The type species does differ from those subsequently included in the genus inasmuch that a proportion of the conidia become pale brown with thicker walls and develop verruculose ornamentation.

KEY

1. Conidia exceeding 20 µ in length .... *S. juncea*
1. Conidia less than 20 µ long 2
   2. Conidia 15-19 × 5-6 µ .... *S. asterina*
   2. Conidia 13-16 × 4-5 µ .... *S. catananches*
   2. Conidia 13.5-17 × 3-4 µ .... *S. scrophulariae*
   2. Conidia 17-19 × 3 µ .... *S. straussiana*

S. asterina (Berk. & Br.) comb. nov. (Fig. 222B)
Leptothyrium asterinum Berk. & Br., *Ann. mag. nat. Hist.*, ser. 5, **7**: 129 (1881).
Macrophoma asterina (Berk. & Br.) Syd., *Annls. mycol.* **19**: 141 (1921).
Ludwigiella asterina (Berk. & Br.) Petrak, *Annls mycol.* **20**: 319 (1922).

Conidiomata 175-200 µ diam. Conidiogenous cells 7-11 × 3-8 µ. Conidia 15-19 × 5-6 µ.

On leaf lesions of *Aster tripolium*, G.B., Germany (IMI 147089, holotype of *L. asterinum* ex K, 147088, 105571, 30556).

S. CATANANCHES Maire, *Bull. Soc. Bot. Fr.* **53**: CLXXXVII (1906). (Fig. 222D, E)

Conidiomata up to 100 µ diam. Conidiogenous cells 8-12 × 5.5-12 µ. Conidia 13-16 × 4-5 µ, often pale brown and verruculose.

On dead stems of *Catananche caerulea*, Algeria (IMI 199598, holotype ex MPU).

S. JUNCEA (Mont.) Arx, *Verh. Akad. Wet. Amst.* **51**: 139 (1957). (Fig. 222A, F)
Sacidium junceum Mont., *Ann. Sci. nat.*, ser. 4, **5**: 341 (1856).
Phlyctema juncea (Mont.) Höhn., *Sber. Akad. Wiss. Wien* **119**: 661 (1910).
Sarcophoma juncea (Mont.) Höhn., *Sber. Akad. Wiss. Wien* **125**: 110 (1916).
Selenophomopsis juncea (Mont.) Petrak, *Annls mycol.* **22**: 182 (1924).
Gloeosporium subfalcatum Bomm., Rouss. & Sacc., *Contr. Myc. Belg.* **4**: 90 (1891).
Myxosporium subfalcatum (Bomm., Rouss. & Sacc.) Allesch. in Rabenh., *Krypt. Fl. Dtsch* **7**: 531 (1903).
Cryptosporium lunulatum Bauml., *Beitr. Crypt. Fl. Pressb. Comit.* 26 (1887).

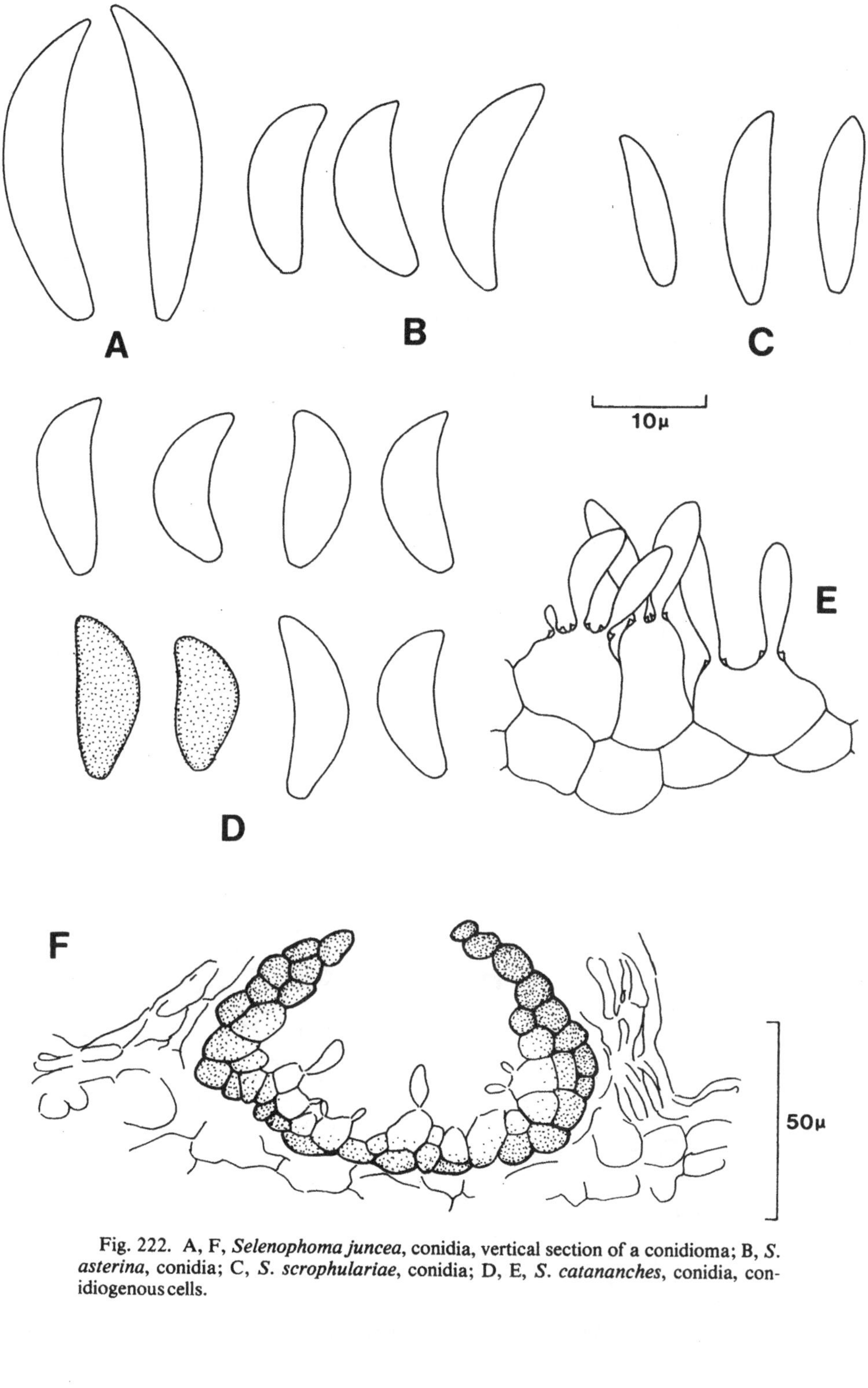

Fig. 222. A, F, *Selenophoma juncea*, conidia, vertical section of a conidioma; B, *S. asterina*, conidia; C, *S. scrophulariae*, conidia; D, E, *S. catananches*, conidia, conidiogenous cells.

Cryptosporium ludwigii Sacc., *Annls mycol.* **11**: 18 (1913).
Falcispora androssowii Bub. & Serebr., *Hedwigia* **52**: 269 (1912).

Teleomorphosis: *Guignardia cytisi* (Fckl) Arx & Müller, *Beitr. Krypt. Fl. Schw.* **11**: 58 (1954).

Conidiomata 100-200 µ diam. Conidiogenous cells 15-20 × 4-7 µ. Conidia 25-29 × 5.5 µ.

On dead stems of *Spartium junceum, Sarothamnus scoparius, Glycyrrhiza glandulifera*, Austria, Germany, Hungary, France, G.B., U.S.S.R. (IMI 147382, isotype *Sacidium junceum* ex PC, 147101, 147097, 147099, 147100, 9528b, 199709, isotype of *Falcispora androssowii* ex K).

S. scrophulariae sp. nov. (Fig. 222C)

Conidiomata immersa, sphaerica, globosa, atro-brunnea vel nigra, interdum aggregata, 100-150 µ diam., paries cum latitudine 1-3 cellularum atrobrunnearum sclerotioideum pseudoparenchymaticarum et 15 µ crassus; ostiola singula primo orbiculata, dein irregulariter rumpentes, 40 µ diam. Cellulae conidiogenae polyphialidicae, variabiles, discretae, determinatae, doliiformes, obpyriformes vel ovatae, hyalinae vel pallide brunneae, 8-15 µ longae × 5-10 µ crassae, cum 1-3 aperturis apicalibus vel lateralibus, ex cellulis interioribus pycnidiorum formatae. Conidia primo ex locis cellulorum conidiogenorum et dein phialibus successivis oriunda, hyalina, septata, eguttulata, falcata, ad apices obtusa et ad basim leniter truncata, 13.5-17 × 3-4 µ.
In caulis emortuis *Scrophulariae* sp., Swat, Bahrain, Pakistan, S. Ahmad, Apr. 1954, IMI 98391, holotypus.

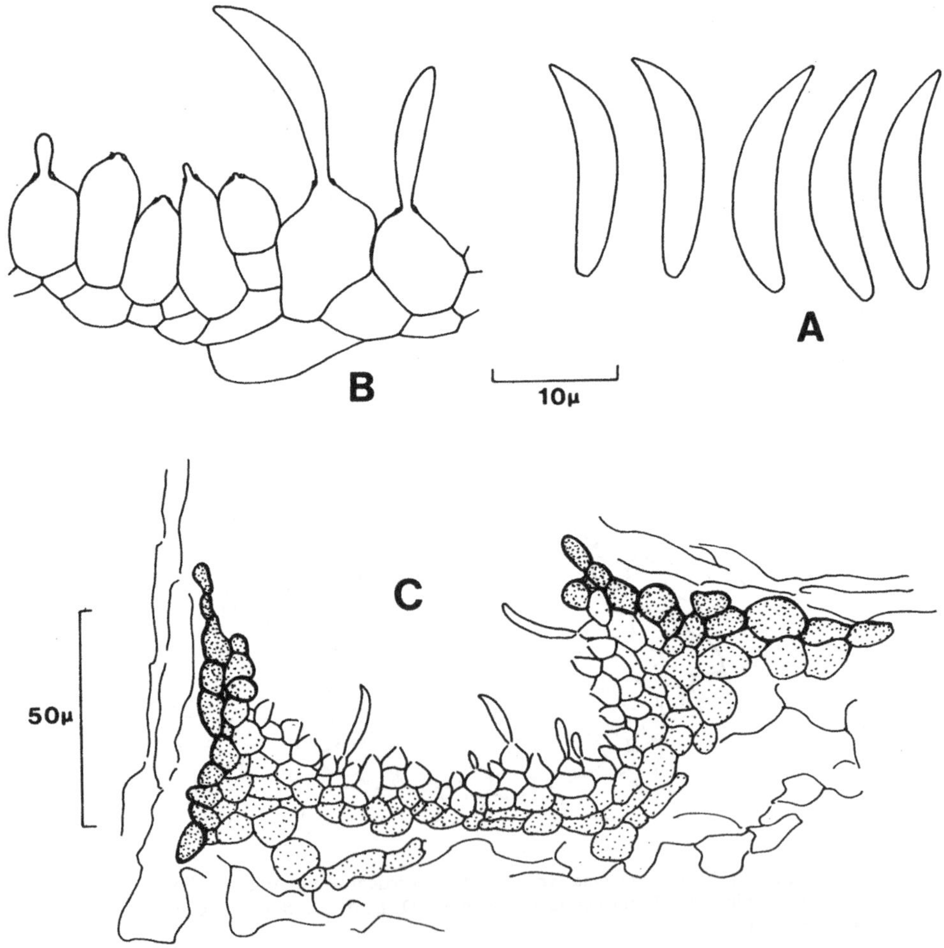

Fig. 223. *Selenophoma straussiana*, A, conidia; B, conidiogenous cells and developing conidia; C, vertical section of a conidioma.

S. STRAUSSIANA (Sacc.) Petrak, *Ann. naturh. Mus. Wien* **50**: 504 (1940). (Fig. 223)
Neopatella straussiana Sacc. apud H. & P. Syd., *Annls mycol.* **6**: 530 (1908).
Selenophoma straussiana (Sacc.) Luc, *Rev. Mycol.* **18**, *suppl. colon.* 2: 100 (1953).
Dothichiza straussiana (Sacc.) Höhn., *Sber. Akad. Wiss. Wien* **118**: 75 (1909).
Leptothyrium thymi Bub., *Ann. naturh. Mus. Wien* **28**: 216 (1914).

Conidiomata up to 100 µ diam. Conidiogenous cells 6-12 × 5-7.5 µ. Conidia 17-19 × 3 µ.

On dead stems of *Dianthus orientalis, D.* sp., Bulgaria, Iran (IMI 30685, 94555, 31099).

12. HAPALOSPHAERIA Syd., *Annls mycol.* **6**: 305 (1908).
Sp. typ.: *H. deformans* (Syd.) Syd. (syn. *Paepalopsis deformans* Syd.).

Mycelium immersed, branched, septate, hyaline to pale brown. *Conidiomata* pycnidial, immersed, separate or aggregated, globose, pale brown, unilocular; wall thin, formed of pale brown, thin-walled textura angularis. Distinct ostiole absent, dehiscence by a relatively wide apical opening. *Conidiophores* absent. *Conidiogenous cells* enteroblastic, phialidic, discrete, determinate, ampulliform, hyaline, smooth, formed from the inner cells of the pycnidial wall, collarette and aperture minute. *Conidia* globose, hyaline, aseptate, smooth, thin-walled, eguttulate.

The genus is monotypic.

H. DEFORMANS (Syd.) Syd., *Annls mycol.* **6**: 305 (1908). (Fig. 224)
Paepalopsis deformans Syd., *Annls mycol.* **5**: 398 (1907).

Conidiomata up to 80 µ diam. Conidiogenous cells 2.5-3 × 2.5-4 µ wide. Conidia 3.5-4 µ diam.

On anthers of *Rubus fruticosus, R. balfourianus, R. idaeus, R.* sp., U.K. (IMI 19181, 194434, 92193, 92194).

13. METAZYTHIA Petrak, *Sydowia* **4**: 373 (1950).
Sp. typ.: *M. caespitosa* Petrak

Mycelium immersed, branched, septate, hyaline. *Conidiomata* pycnidial, immersed, becoming erumpent, unilocular or occasionally divided, dark brown, separate or aggregated, thin-walled; walls of thin-walled, pale brown, textura angularis. *Ostiole* single to each pycnidium, circular, central, slightly papillate. *Conidiophores* hyaline, branched and septate to form aggregations of cells rather than linear conidiophores, formed from the inner cells of the pycnidial walls. *Conidiogenous cells* enteroblastic, phialidic, determinate, integrated, doliiform to ampulliform, hyaline, smooth, collarette and aperture minute, periclinal wall markedly thickened. *Conidia* hyaline, aseptate, thin-walled, often biguttulate, cylindrical to fusiform, occasionally irregular.

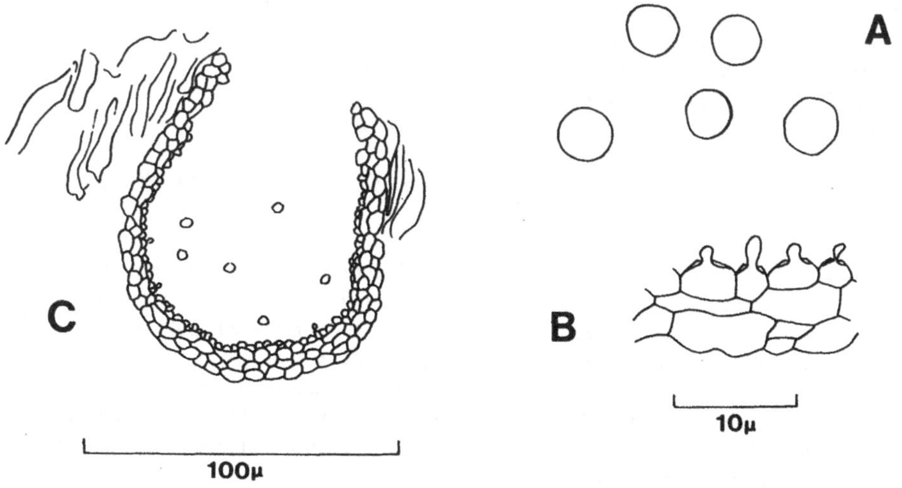

Fig. 224. *Hapalosphaeria deformans*, A, conidia; B, conidiogenous cells; C, vertical section of a conidioma.

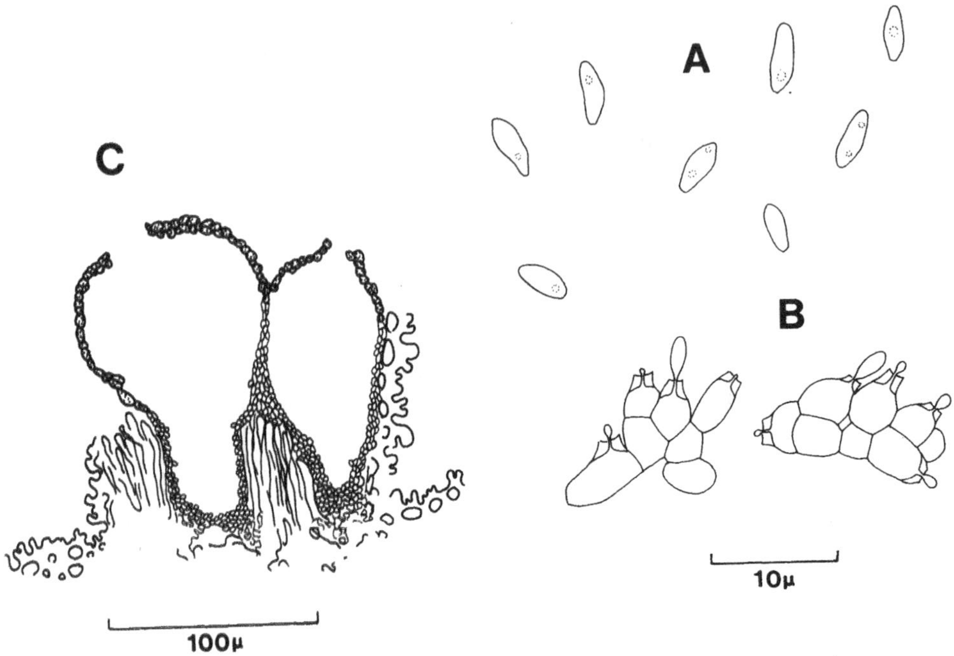

Fig. 225. *Metazythia caespitosa*, A, conidia; B, conidiophores with developing conidia; C, vertical section of conidiomata.

The genus is monotypic.

M. CAESPITOSA Petrak, *ibid.*: 373 (1950). (Fig. 225)

Conidiomata up to 100 µ diam. Conidiogenous cells 3-5 × 3-6.5 µ. Conidia 4-6 × 1.5-2 µ.

Hyperparasitic on indet. fungal stroma on *Chusquea serrulata*, Ecuador (IMI 201886, isotype ex ZT).

14. PHOMA Sacc., nom. cons., *Michelia* **2**: 4 (1880).
Phoma Fr. ex Fr., *Syst. Myc.* **1**: LII (1821); **2**: 546 (1823), nom. rej.
Plenodomus Preuss, *Linnaea* **24**: 145 (1851).
Leptophoma Höhn., *Sber. Akad. Wiss. Wien* **124**: 73 (1915).
Sclerophomella Höhn., *Hedwigia* **59**: 237 (1917).
Polyopeus Horne, *J. Bot. Lond.* **58**: 239 (1920).
Deuterophoma Petri, *Boll. R. Staz. Pat. Veg. Roma*, n.s. 9, **4**: 396 (1929).
? *Vialina* Curzi, *Boll. R. Staz. Pat. Veg. Roma*, n.s., **15**: 252 (1935).
Peyronellaea Goid. ex Goid. apud Togliani, *Ann. Sper. agr.*, N.S., **6**: 92 (1952).
Peyronellaea Goid., *R.C. Accad. Lincei*, ser. 8, **1**: 451 (1946).
Ascochyta Lib. sect. *Phyllostictoides* Zherbele, *Trudy vses. Inst. Zashch. Rast.* **29**: 20 (1971).
Sp. typ.: *P. herbarum* Westd.

Mycelium immersed, branched, septate, hyaline or pale brown. *Conidiomata* pycnidial, immersed, or semi-immersed, sometimes becoming erumpent, unilocular, brown, globose, separate or aggregated, occasionally confluent, thin-walled (in *P. lingam* becoming thick-walled and pseudosclerenchymatous); walls of thin-walled, pale to medium brown textura angularis. *Ostioles* single or several to each pycnidium, central, not papillate. *Conidiophores* only present in *P. cava* and *P. tracheiphila* and then either filiform, septate, and branched, or short, irregularly branched, and ramified respectively. *Conidiogenous cells* enteroblastic, phialidic, integrated or discrete, ampulliform to doliiform, hyaline, smooth, collarette and aperture minute, periclinal wall markedly thickened. *Conidia* hyaline, aseptate or occasionally 1 septate, thin-walled, often guttulate, ellipsoid, cylindrical, fusiform, pyriform or globose.

The generic synonymy for *Phoma* has been discussed by Sutton (1977a).

So far many more than 2000 species have been described in the genus which has traditionally been used for simple stem inhabiting pycnidial fungi with small, hyaline, unicellular conidia. Species separation has been by a combination of slight morphological differences and host substrate. This has led to a multiplicity of described species which neither reflects the fundamental relationships between taxa nor in a practical sense is of much value to mycologists, pathologists and ecologists. For these reasons, and the fact that the number of available taxonomic criteria on the natural substrates is very limited, Boerema and his co-workers have, over several years, evolved a system for the identification and naming of *Phoma* species which is based on behaviour of isolates in vitro. By analysing under standardized conditions colony habit, pigmentation, chlamydospore formation, crystal formation, features of the pycnidia, conidia and conidiogenous cells, and relating these to type studies of described taxa, they are gradually achieving a scheme for identification of *Phoma* species that has a well-known parallel in *Fusarium.*

In spite of this well-documented work, Boerema & Griffin (1974) in a discussion of *Phoma* species on *Viburnum* were moved to state 'It is only with much experience that *P. viburni* can be distinguished with certainty from *P. exigua* var. *inoxydabilis* and *P. macrostoma* var. *incolorata.*' This statement sums up much of the hesitancy anyone might feel in attempting the identification of *Phoma* species in culture. For many taxa gross cultural characteristics are so variable and dependent on the nature of the medium, light regime, C/N ratio etc. that the problems of identification seem unsurmountable. The number of key characters is relatively small but fortunately many of these are stable. It is general however, that identification is by a combination of different characters and that these are linked to one or two key characters.

The following key is based on the data published by Boerema over several years. In the majority of cases my own observations have been based on isolates named by Boerema or cited by him in his work. The standard conditions for growth are oat agar, malt agar, 20-22°C (in the dark for the first week for determination of growth rate), and alternating 12 hr periods of near U.V. (black light) and dark.

The majority of species documented by Boerema are included in the key, but some have been omitted because it has not been possible to verify the data due to lack of material. These include *P. valerianellae* Boerema & de Jong (1968) on *Valerianella* species, *P. viburni* (Roum. & Sacc.) Boerema & Griffin (1974) on *Viburnum* species, *P. hoehnelii* van Kesteren (1972), a ubiquitous saprophyte frequently recorded on *Angelica* species, *P. hoehnelii* subsp. *amplior* (Sacc. & Roum.) Boerema & van Kesteren apud Boerema (1976), a ubiquitous saprophyte especially on *Foeniculum* species, *P. hoehnelii* var. *urticae* Boerema & van Kesteren apud Boerema (1976), common on dead *Urtica* species, *P. urticicola* van der Aa & Boerema apud Boerema (1976) on leaves and stems of *Urtica dioica*, *P. arachidicola* Marasas, Pauer & Boerema (1974) on *Arachis hypogaea*, *P. aubrietiae* (Moesz) Boerema apud Boerema & Valckx (1970) on *Aubrietia* sp., *P. macdonaldii* Boerema (1970b) from *Helianthus annuus* and *P. telephii* (Vestergr.) van Kesteren (1971, 1972) from *Sedum.*

KEY

1. Conidia formed from filiform conidiophores .... *P. cava*
1. Conidia formed from ampulliform to doliiform phialides 2
   2. Cultures producing either chlamydospores, sclerotia or an *Epicoccum* state 3
   2. Cultures not producing accessory vegetative or spore states 6
3. *Epicoccum* state present; cultures with a similar range of pigmentation as found in *Epicoccum purpurascens* .... *P. epicoccina*
3. Chlamydospores multicellular, alternarioid (some unicellular chlamydospores also formed) 4
3. Chlamydospores multicellular, aggregating to form sclerotia .... *P. chrysanthemicola*
3. Chlamydospores unicellular, no alternarioid ones formed 5
   4. Colonies consistently with salmon pink patches; conidia 4-5 × 2-2.5 µ .... *P. sorghina*
   4. Chlamydospores catenate; conidia 5-9 × 2.5-3 µ .... *P. glomerata*
   4. Chlamydospores lateral; conidia 6-7.5 × 3-4 µ .... *P. jolyana*
   4. Chlamydospores solitary, not lateral; conidia 4.5-6 × 2-2.5 µ .... *P. pomorum*
5. Conidia 3-5 × 1.5-2 µ, biguttulate, aseptate .... *P. eupyrena*
5. Colonies uniform, sparse formation of crystals on MA; conidia 6-11.5 × 2-3.5 µ, often becoming 1-septate .... *P. medicaginis*
5. Colonies variable, abundant formation of crystals on MA; conidia 7-10 × 2.5-4 µ, often becoming 1-septate (*P. multirostrata* may key out here if it forms chlamydospores, but is separated by the smaller aseptate conidia) .... *P. medicaginis* var. *pinodella*
   6. Pycnidia usually with setae round the ostiole .... *P. leveillei*
   6. Pycnidia lacking setae, not rostrate 7

6. Pycnidia lacking setae, rostrate; cultures variable, occasionally with unicellular chlamydospores; conidia 5-6.5 × 2-2.5 µ (*P. nebulosa* may key out here if pycnidia become rostrate, but it does not form chlamydospores and conidia are 4-6 × 1.5-2 µ) .... *P. multirostrata*

7. Colonies with vinaceous to red pigmentation 8

7. Colonies lacking such pigmentation 9

8. Aerial mycelium sparse, reverse vinaceous; conidia 3.5-5 × 2 µ, biguttulate .... *P. destructiva*

8. Aerial mycelium sparse, reverse red, turning blue with NaOH; conidia 4-5 × 1.5-2 µ .... *P. herbarum*

8. Aerial mycelium felty to woolly, reverse livid red, turning purple-blue with NaOH; conidia 5.5-10.5 × 2.5-3.5 µ, often becoming 1-septate .... *P. macrostoma*

8. Aerial mycelium dense, felty, reverse rosy buff; conidia 5-7 × 2-3 µ, sometimes with a single septum and then 15.5-22 × 4-6 µ .... *P. dennisii*

9. Conidia often becoming 1-septate 10

9. Conidia unicellular 11

10. Colonies scalloped or lobed at the margin, NaOH turns agar blue-green; conidia 5.5-10 × 2.5-3.5 µ, often becoming 1-septate .... *P. exigua* and its varieties

10. Colonies saffron to brown in reverse, margin not scalloped; pycnidia very thick-walled; conidia 5.5-9 × 3-3.5 µ, occasionally 1-septate .... *P. complanata*

10. Colonies olivaceous grey, not scalloped, stilboid tufts present; pycnidia sparse; conidia 4.5-9 × 2.5-3.5 µ, becoming 1-septate .... *P. lycopersici*

11. Conidial width not exceeding 1.5 µ 12

11. Conidia ▷2 µ wide 13

12. Colonies floccose, reverse olivaceous to black; pycnidia pseudoparenchymatous; conidia 3-4 × 1.5 µ, irregularly guttulate .... *P. tropica*

12. Colonies felty, white to grey; pycnidia pseudoparenchymatous; conidia 4.5-6.5 × 1.5 µ, often formed from the mycelium. On *Citrus* .... *P. tracheiphila*

12. Colonies variable; pycnidia pseudoparenchymatous or pseudosclerenchymatous, often sterile and then sclerotial; conidia 3.5-4.5 × 1.5 µ. Mainly on Cruciferae .... *P. lingam*

13. Conidial l/b ratio 1-1.5/1 14

13. Conidial l/b ratio ▷2/1 15

14. Colonies mostly lacking aerial mycelium; pycnidia formed in concentric zones; conidia 3.5-4.5 × 2.5-3 µ, almost globose, often with a single guttule .... *P. capitulum*

14. Colonies with dense to moderately dense aerial mycelium; reverse yellow-saffron; conidia 3-3.5 × 2-2.5 µ, often with a single guttule .... *P. fimeti*

14. Colonies with little aerial mycelium; reverse buff to apricot; pycnidia often with brown hyphae over their surfaces; conidia 2.5-4.5 × 2-2.5 µ, larger ones biguttulate .... *P. putaminum*

15. Colonies variable but NaOH gives blue-green pigmentation to agar; conidia 4-6.5 × 2.5-3.5 µ, eguttulate .... *P. hedericola*

15. Colonies usually green to greenish brown, some whitish green; conidia 4-6 × 1.5-2 µ, straight or slightly curved, biguttulate .... *P. nebulosa*

15. Colonies olivaceous brown; pycnidia abundant in areas lacking aerial mycelium; conidia 6-9 × 3.5-4.5 µ, biguttulate .... *P. betae*

P. BETAE Frank, *Z. Ver. Rubenz. Indus. Dtsch. Reichs.* **42**: 904 (1892). (Fig. 226A)
Phyllosticta betae Oud., *Ned. Kruidk. Arch.*, 2 ser., **2**: 181 (1876).

Teleomorphosis: *Pleospora betae* (Berl.) Nevodovsky, *Griby rossi exs.* 247 (1915).

Colonies with olivaceous brown felty aerial mycelium bearing irregular patches of grey to white, reverse greenish brown. Pycnidia formed in areas comparatively clear of aerial mycelium. Conidia 6-9 × 3.5-4.5 µ, straight or slightly curved, ellipsoid, usually with two large guttules.

Additional data in Booth (1967) and Booth & Cornford (1959).

On and from *Beta vulgaris*, *B. trigyna*, *B. cicla*, *B.* spp., represented by 30 collections in IMI from New Zealand, U.K., Canada, Malaysia, Cyprus, France, Barbados, Netherlands, Australia, Egypt, Czechoslovakia, Iraq.

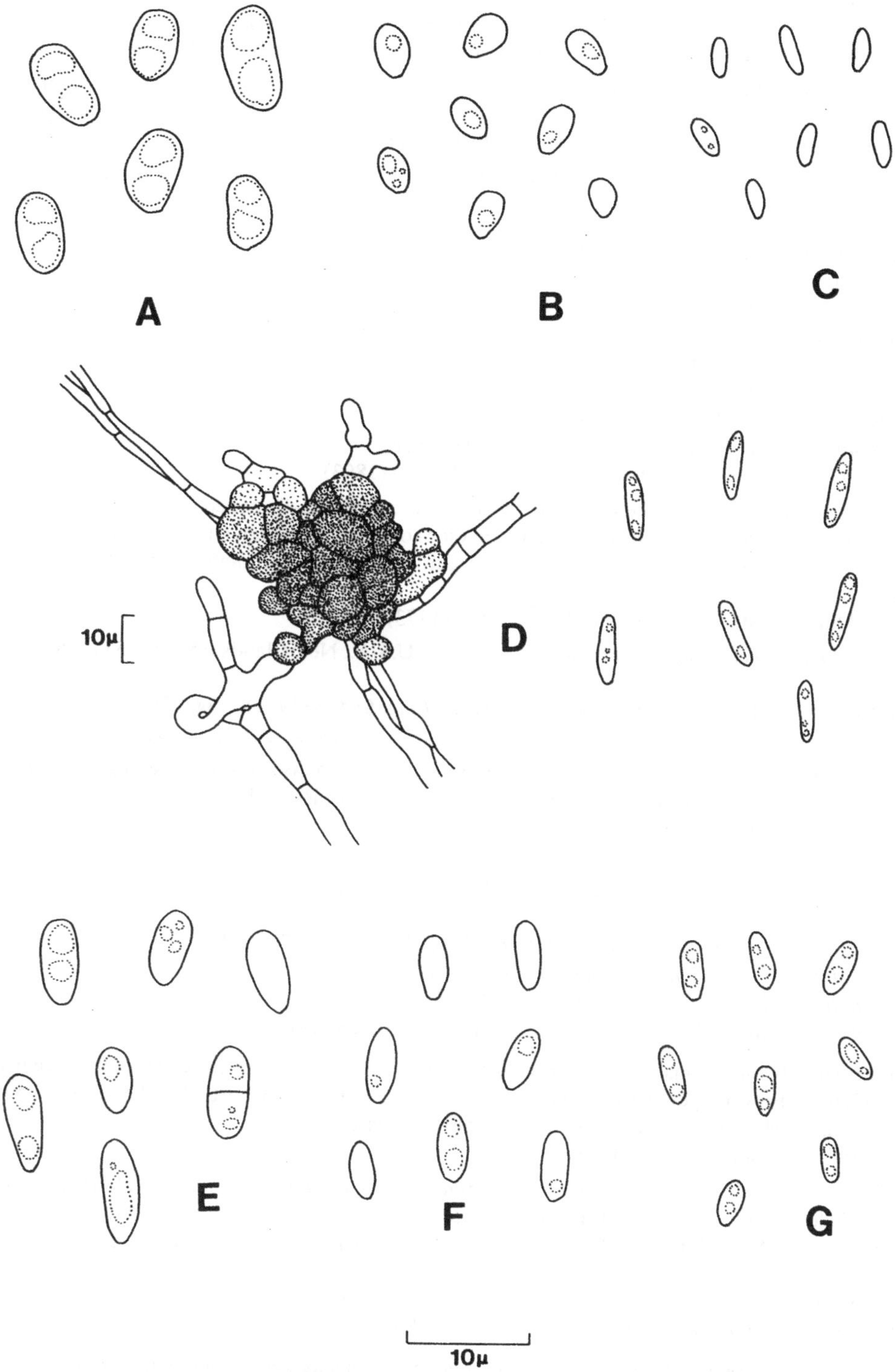

Fig. 226. A, *Phoma betae*, conidia; B, *P. capitulum*, conidia; C, *P. cava*, conidia; D, *P. chrysanthemicola*, pseudosclerotium and conidia; E, *P. complanata*, conidia; F, *P. dennisii*, conidia; G, *P. destructiva*, conidia.

P. CAPITULUM Pawar, Mathur & Thirum., *Trans. Br. mycol. Soc.* **50**: 261 (1967). (Fig. 226B)
P. ostiolata Pawar, Mathur & Thirum., *Trans. Br. mycol. Soc.* **50**: 262 (1967).
P. ostiolata var. *brunnea* Pawar, Mathur & Thirum., *Trans. Br. mycol. Soc.* **50**: 263 (1967).

Colonies with very little aerial mycelium, when present white and extremely thin. Pycnidia abundant, formed in concentric dark brown zones. Conidia 3.5-4.5 × 2.5-3 µ, subglobose to ellipsoid, often with a single guttule.
Additional data and synonymy provided by Boerema & Dorenbosch (1968).

From saline soil, marine environment and oak forest soil, India (IMI 113693, holotype *P. capitulum*, 210984, 229519).

P. CAVA Schulz., *Verh. zool.-bot. Ges. Wien* **21**: 1248 (1871). (Fig. 226C)
Aposphaeria cava (Schulz.) Sacc. & Schulz., *Syll. fung.* **3**: 174 (1884).
Coniothyrium cavum (Schulz.) O. Ktze, *Rev. Gen. Pl.* III^{II}: 459 (1893).
Aposphaeria collabescens Schulz. & Sacc., *Syll. fung.* **3**: 170 (1883).
Coniothyrium collabescens (Schulz. & Sacc.) O. Ktze, *Rev. Gen. Pl.* III^{II}: 459 (1893).
Phoma aposphaerioides Briard & Hariot, *Rev. myc.* **12**: 178 (1890).
Phoma collabens Schulz. & Sacc., *Hedwigia* **23**: 109 (1884).
Aposphaeria minuta Berl., *Fungi moricolae*, fasc. VI: 18 (1888).
Coniothyrium minutum (Berl.) O. Ktze, *Rev. Gen. Pl.* III^{II}: 459 (1893).

Colonies olivaceous grey, dense felty aerial mycelium, not concentrically zoned, reverse saffron to apricot. Conidiogenous cells integrated on filiform, septate conidiophores, conidiogenous loci occurring immediately below transverse septa. Conidia 3-4 × 1-2 µ, occasionally guttulate, straight or slightly curved, cylindrical to slightly irregular.
For additional data see Boerema & Dorenbosch (1973).

From leaves of *Hedera helix*, ear of white-tailed deer, N.Y., U.S.A., Netherlands (IMI 163515, 184366).

P. CHRYSANTHEMICOLA Hollós, *Ann. hist.-nat. Mus. natn. Hung.* **5**: 456 (1907). (Fig. 226D)

Colonies irregularly olivaceous grey with darker or lighter patches, felty aerial mycelium, occasionally with indistinct zonation, reverse dark brown. Reputedly with orange-red discolouration fading by addition of NaOH. Conidia 5-6.5 × 1-1.5 µ, straight, cylindrical with 1-4 guttules. Chlamydospores aggregated to form pseudosclerotia, pale to dark brown.
For additional data see Dorenbosch (1970).

From rhizosphere of *Triticum*, W. Australia (IMI 175673, 175680).

P. COMPLANATA (Tode ex Fr.) Desm., *Ann. Sci. nat.*, ser. 3, **16**: 299 (1851). (Fig. 226E)
Sphaeria complanata Tode, *Fungi Mecklenb. Fasc.* 2, 21 (1791).
Sphaeria complanata Tode ex Fr., *Syst. Myc.* **2**: 508 (1823).
Sclerophomella complanata (Tode ex Fr.) Höhn., *Hedwigia* **59**: 238 (1917).

Colonies irregularly olivaceous grey, cream grey to white, dense felty aerial mycelium, not concentrically zoned, reverse saffron to brown. Pycnidia very thick-walled in culture and on the host. Conidia 5.5-9 × 3.5 µ, irregularly guttulate, straight, ellipsoid to cylindrical, occasionally 1 septate.
For additional data see Boerema (1976).

On and from stems of *Angelica sylvestris*, *Boerhavia* sp., *Conopodium majus*, *Heracleum sphondylium*, Umbelliferae indet., U.K., Australia, Finland (IMI 60896, 194764, 22555, 22554, 202700, 202705, 202706 202701, 196573, 191852, 24770, 22553, 22556, 368, 56126, 226011, 77908, 69743b, 22552).

P. DENNISII Boerema nom. nov., *Trans. Br. mycol. Soc.* **67**: 307 (1976). (Fig. 226F)
Phoma oleracea var. *solidaginis* Sacc., *Syll. fung.* **3**: 135 (1884).

Colonies cream grey, dense felty aerial mycelium, not concentrically zoned. Boerema (1976) reports a rosy buff discolouration in reverse, but not evident in his named isolate at CMI. Conidia 5-7 × 2-3 µ, irregularly guttulate, straight, ellipsoid to cylindrical; occasionally with a single septum, then reported by Boerema & Bollen (1975) with conidia 15.5-22 × 4-6 µ.
For additional data see Boerema (1976) and Boerema & Bollen (1975).

From *Solidago* sp., Netherlands (IMI 194765).

P. DESTRUCTIVA Plowr., *Gard. Chron.* **16**: 621 (1881). (Fig. 226G)
Diplodina destructiva (Plowr.) Petrak, *Annls mycol.* **10**: 19 (1921).
Phyllosticta lycopersici Pk, *Bull. N.Y. St. Mus.* 219-220: 55 (1887).

Colonies variable, usually with sparse aerial mycelium, but when present floccose and dark brown with patches of grey to whitish grey, often sectoring; reverse dark brown, often with a vinaceous tint. Conidia 3.5-5 × 2 µ, consistently biguttulate, as in the α-conidia of *Phomopsis* species, straight, cylindrical, ellipsoid or slightly irregular, reputedly produced in saffron tendrils.

For additional data see Boerema & Dorenbosch (1973).

This species and its relationship to *P. lycopersici* Cooke (anamorphosis of *Didymella lycopersici* Kleb.) have been the subject of discussion by Brookes & Searle (1921), Dennis (1946), Knight-Fisher (1960), Laundon (1971) and Boerema & Dorenbosch (1973). The consensus of opinion is that the two species can only be distinguished by cultural characters. *P. lycopersici* differs from *P. destructiva* by limited pycnidial production in colonies showing a well-developed aerial mycelium and stilboid aggregations of hyphae.

On and from *Lycopersicon esculentum*, U.S.S.R., India, Papua New Guinea, Guadeloupe, Nigeria, Sudan, Zambia, Mauritius, Libya (IMI 129003, 188639, 123242, 75370, 58540, 74643b, 22557, 184082, 173142, 124596, 138126, 206868, 202963, 229015, 62996).

P. EPICOCCINA Punithalingam, Tulloch & Leach, *Trans. Br. mycol. Soc.* **59**: 341 (1972). (Fig. 227A)

Colonies extremely variable in pigmentation, showing a similar range to that found in *Epicoccum purpurascens*, aerial mycelium present or absent, zonate or not. Conidia 5-8.5 × 2.5-3 µ, eguttulate or with minute guttules, straight or slightly curved, cylindrical. *Epicoccum* state present, conidia 15-25 µ diam., verrucose, muriform.

From seed of *Dactylis glomerata, Avena* sp., and from *Beta vulgaris, Malus* sp., *Triticum* sp., *Zea mays*, U.K., Eire, India, N.Z., Ore., U.S.A. (IMI 164070, holotype, 183207, 165573b, 170490, 226928, 172767, 200403, 167908).

P. EUPYRENA Sacc., *Michelia* **1**: 525 (1879). (Fig. 227B)

Colonies dark brown, quite variable, often with dense felty olivaceous green to brown aerial mycelium, rarely grey, at the margin mycelium usually submerged. Conidia 3-5 × 1.5-2 µ, cylindrical or ellipsoid, straight or slightly curved, often biguttulate. Chlamydospores pale, medium or dark brown, smooth, terminal or intercalary, often catenate, abundant.

Additional data from Dorenbosch (1970), Boerema (1976).

Represented in herb. IMI by about 100 collections from 30 different host genera, soil, sewage, sand and veterinary sources from Sri Lanka, Pakistan, India, Kuwait, Malaysia, Australia, New Caledonia, Norway, Eire, U.K., Canada, Nigeria, Malawi, Sierra Leone, South Africa.

P. EXIGUA Desm., *Ann. Sci. nat.*, ser. 3, **11**: 282 (1849). (Fig. 227C)
Synonymy in Boerema & Höweler (1967), Boerema (1967a), Boerema (1976), Boerema & Dorenbosch (1973).

Colonies very variable with a scalloped or lobed margin, usually with dense felty white, black or dark olivaceous aerial mycelium, not concentrically zoned. Application of NaOH gives a blue-green pigmentation to the agar, changing to brown-red. Conidia 5.5-10 × 2.5-3.5 µ, straight or slightly curved, ellipsoid or cylindrical, often biguttulate and becoming 1 septate.

This species has been separated into five varieties by Boerema (1976) and the diagnostic features tabulated clearly.

P. exigua var. *exigua* is recognised by the above-listed features for the species.
P. exigua var. *linicola* (Naumov & Vass.) Maas (1965) causes foot rot and damping off of *Linum usitatissimum* and occurs on other plant substrates. It has a slower growth rate than var. *exigua*, produces more uniform colonies which are dominated by black tinges, and also shows the colour reaction by addition of NaOH.
P. exigua var. *sambuci-nigrae* (Sacc.) Boerema & Höweler (1967) is associated with leaf lesions and dead shoots of *Sambucus nigra*. It has a similar growth rate to var. *exigua* and shows a similar colour reaction to addition of NaOH, but is separated by the relatively uniform olivaceous mycelial mat and a colony margin that is entire and not scalloped.

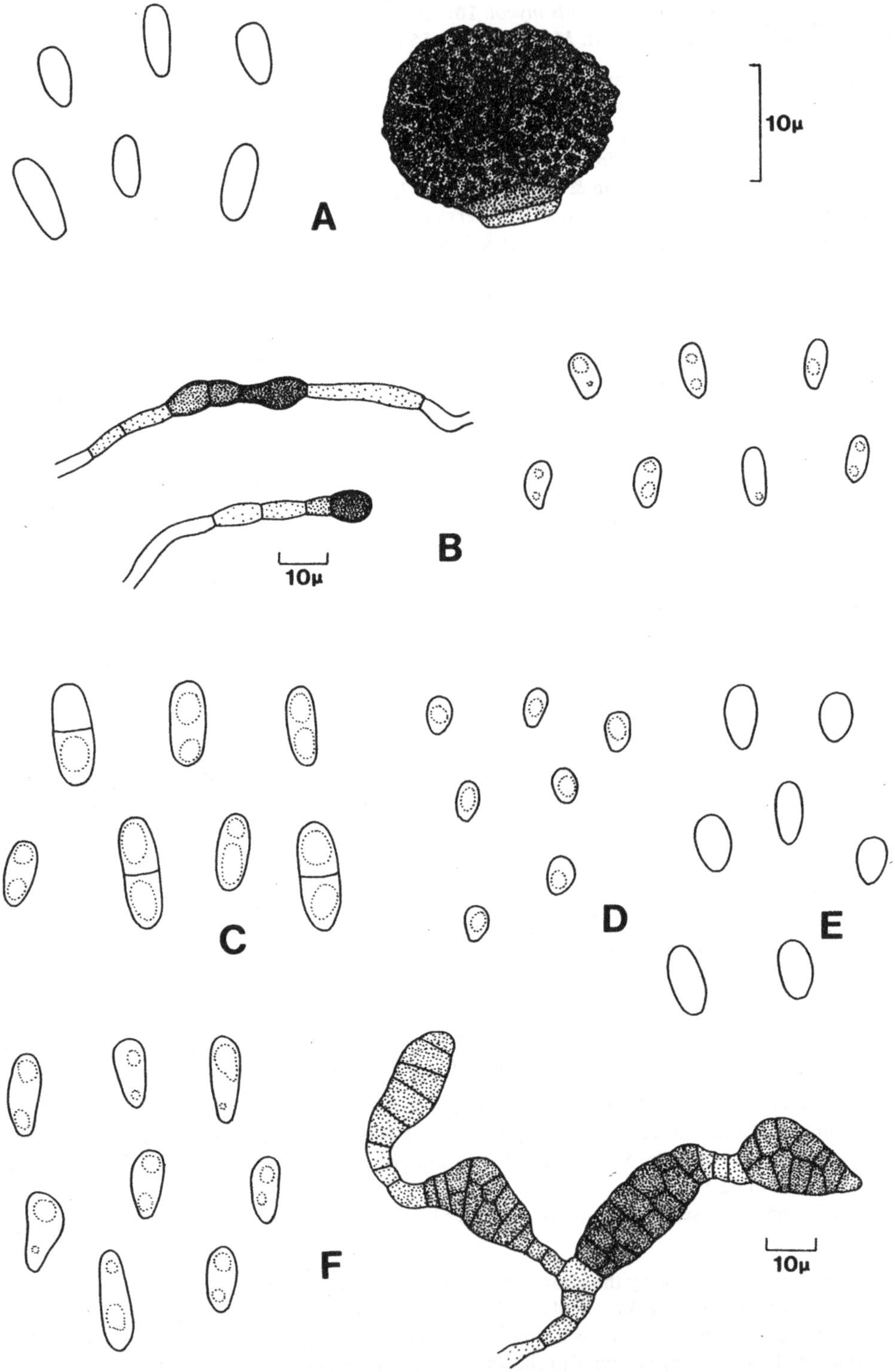

Fig. 227. A, *Phoma epicoccina*, conidia and *Epicoccum* conidium; B, *P. eupyrena*, chlamydospores and conidia; C, *P. exigua*, conidia; D, *P. fimeti*, conidia; E, *P. hedericola*, conidia, F, *P. glomerata*, conidia and chlamydospores.

P. exigua var. *inoxydabilis* Boerema & Vegh apud Vegh et al. (1974) is associated with leaf lesions and dead shoots of *Vinca* spp., and also occurs on other plant substrates. It is precisely similar to var. *exigua* but differs in that it shows no colour reaction when NaOH is added.
P. exigua var. *foveata* (Foister) Boerema (1967a) is known to cause gangrene of *Solanum tuberosum* and is occasionally found on other plant substrates in or near potato fields. Again the growth rate is similar to var. *exigua* but the colonies are relatively uniform with a white-grey mycelial mat and the margin is entire not lobed. In addition the medium is coloured yellowish brown by anthraquinone pigments which turn red in alkaline conditions; older cultures develop yellow-green crystals.
P. exigua has been recorded from a large number of substrates by Boerema & Dorenbosch (1973) and Boerema & Höweler (1967). It is known in herb. IMI from almost 200 collections and cultures from 74 different host genera, soil, dung and litter from U.K., France, Eire, Cyprus, Netherlands, Norway, Turkey, Israel, India, Indonesia, Sabah, Brunei, Sri Lanka, Nepal, Mauritius, Pakistan, Iran, Solomon Is., Australia, New Zealand, Togo, Ethiopia, Malawi, Nigeria, Sierra Leone, Zambia, Zaire, Kenya, Yemen, Libya, Venezuela, Colombia, Honduras.

P. FIMETI Brun., *Bull. Soc. bot. Fr.* **36**: 338 (1889). (Fig. 227D)

Colonies olivaceous brown or occasionally whiter, aerial mycelium moderately dense to dense, not concentrically zoned, reverse yellow to saffron on OA. Conidia 3-3.5 × 2-2.5 µ, frequently with a large guttule, ellipsoid.
For additional data see Dorenbosch (1970) and Boerema & Dorenbosch (1973).

On and from *Hibiscus* sp., *Apium graveolens, Carica papaya, Bouea burmanica, Eucalyptus pauciflora, Lycopersicon esculentum, Pterocarpus angolensis*, soil, oyster shell, wood, sand, degraded wool, U.K., Netherlands, India, Burma, Zambia, Windward Isl., N.S.W., Australia, N. Zealand (IMI 170404, 223968, 197122, 165736h, 164570, 168146, 222992, 166083, 174037, 193288, 177860, 166457, 166356, 178661, 163514, 168406).

P. GLOMERATA (Cda) Wollenw. & Hochapf., *Z. Parasitkde* **8**: 592 (1936). (Fig. 227F)
Synonymy in Boerema, Dorenbosch & van Kesteren (1965, 1968, 1971, 1977) and Boerema & Dorenbosch (1973).

Colonies extremely variable, inasmuch as formation of pycnidia, aerial mycelium and chlamydospores and size and pigmentation of pycnidia, conidia and chlamydospores is influenced by the C/N ratio of the medium and the age of the culture. Aerial mycelium sparse, although cultures often sectoring to produce olivaceous grey, grey white or brown strains of denser aerial mycelium. Conidia 5-9 × 2.5-3 µ, often slightly curved, otherwise straight, cylindrical or tapered towards the base, usually biguttulate. Chlamydospores brown, catenate, *Alternaria*-like, with transverse and longitudinal septa.
Additional data from Dorenbosch (1970), Boerema & Dorenbosch (1973), Boerema, Dorenbosch & van Kesteren (1965, 1968, 1971, 1973, 1977) and Morgan-Jones (1967).

This species is known in herb. IMI from more than 300 collections on 94 different host genera, from the air, seawater, cement, litter, paint, paper, wool, soil and various medical and veterinary sources from U.K., Yugoslavia, Portugal, Cyprus, Turkey, U.S.S.R., Israel, Kuwait, Iraq, Pakistan, India, Bangladesh, Malaysia, Burma, Hong Kong, Selangor, Zambia, Malawi, Nigeria, Ethiopia, Sudan, Kenya, S. Rhodesia, Yemen, Senegal, Sierra Leone, Gambia, Tanzania, Egypt, S. Africa, Australia, N. Zealand, Togo, British Solomon Isl., Papua New Guinea, Antigua, Canada, U.S.A., Venezuela, Brazil.

P. HEDERICOLA (Dur. & Mont.) Boerema, *Trans. Br. mycol. Soc.* **67**: 295 (1976). (Fig. 227E)
Phyllosticta hedericola Dur. & Mont., *Flore d'Algerie Crypt.* **1**: 611 (1849).
Phyllosticta destructiva var. *hederae* (Dur. & Mont.) Oud., *Ned. Kruidk. Arch.* **1**: 257 (1873).

Colonies with a short dense aerial mycelium with whitish, black-greenish olivaceous tinges, margin regular. Addition of NaOH gives a blue-green pigmentation to the agar with most isolates. Conidia (from host) 4-6.5 × 2.5-3.5 µ (Boerema, 1976, quotes conidia from culture as 2.5-6 × 1.5-3, mostly 3.5-5 × 1.5-2.5 µ), ellipsoid, eguttulate.

On and from leaves and stems of *Hedera helix, H. himalaica,* Austria, Czechoslovakia, Portugal, U.K., Italy, Malta, India, Pakistan (IMI 69218, 61264, 34697, 34693, 34694, 101519, 213792, 199774, 103560, 82736, 80207, 147957, 141853b, 35331, 11673, 30185, 228867).

P. HERBARUM Westd., *Bull. Acad. Belg. Cl. Sci.* **19**: 118 (1852). (Fig. 228B)
Synonymy in Boerema (1963, 1970b).

Colonies comparatively slow-growing with sparse grey-green aerial mycelium but abundant pycnidium formation. Some strains and sectors in some cultures however develop dense aerial mycelium lacking pycnidia. Pigmentation is typically reddish, which, by the application of NaOH, turns blue. Conidia 4-5 × 1.5-2 µ, cylindrical or ellipsoid, straight, eguttulate.

Additional data from Dorenbosch (1970) and Boerema (1976).

Represented in herb. IMI by more than 100 collections on 35 different host genera, from soil, air, dung, sewage, spoil tips and other sources, from U.K., Eire, Germany, Czechoslovakia, Cyprus, Netherlands, Turkey, Yugoslavia, Latvia, U.S.S.R., Kenya, Malawi, Egypt, Mozambique, Tanzania, India, Sri Lanka, Pakistan, Nepal, Sarawak, Brunei, Papua New Guinea, Australia, Barbados, Trinidad.

P. JOLYANA Pirozynski & Morgan-Jones nom. nov., *Trans. Br. mycol. Soc.* **51**: 200 (June 1968). (Fig. 228A)
Peyronellaea musae Joly, *Rev. Myc.* **26**: 97 (July, 1961).
Peyronellaea nainensis Tandon & Bilgrami, *Curr. Sci.* **30**: 344 (Dec. 1961).
Phoma musae (Joly) Boerema, Dorenbosch & van Kesteren, *Persoonia* **4**: 63 (1965), non *P. musae* (Cke) Sacc., 1884, nec Carp., 1919.
Phoma jolyi Morelet nom. nov., *Bull. Soc. Sci. nat. Archéol. Toulon Var* **177**: 9 (July 1968).

Colonies variable, greenish black, with or without a felty aerial mycelium, in lighter isolates occasionally sectoring to produce the darker strains. Pycnidia abundant. Reported by Boerema, Dorenbosch & van Kesteren (1977) to have an olivaceous buff to amber colour in reverse. Conidia 6-7.5 × 3-4 µ, ellipsoid or slightly irregular, often biguttulate. Chlamydospores *Alternaria*-like, usually single and although initially terminal eventually appearing lateral.

Boerema, Dorenbosch & van Kesteren (1977) separated a var. *circinata* (Kuznetzova) Boerema, Dorenbosch & van Kesteren based on *Peyronellaea circinata* Kuznetzova. This had a similar growth rate but more abundant production of chlamydospores and irregular conidia. Such differences are somewhat difficult to apply in identification, for the whole complex is quite variable.

On and from *Carica papaya, Citrus* sp., *Coffea arabica, Eriobotrya japonica, Musa* sp., *Sterculia urens*, soil, air, India, Kenya (IMI 227612, 227602, 155819, 225236, 80891, 156978, 201642, 201648, 201650, 201647, 226558, 182341, 209743).

P. LEVEILLEI Boerema & Bollen nom. nov., *Persoonia* **8**: 115 (1975). (Fig. 228C)
Pyrenochaeta acicola (Lév.) Sacc., *Syll. fung.* **3**: 220 (1884).
Vermicularia acicola Lév., *Ann. Sci. nat.*, ser. 3, **9**: 259 (1848).
Pyrenochaeta spinaciae Verma & Negru, *Mycopath. Mycol. appl.* **30**: 310 (1966).

Colonies slow-growing, grey brown to olivaceous brown, with or without felty aerial mycelium, often in concentric zones. Pycnidia abundant in non-felty cultures, setose. Conidia 2.5-3.5 × 1.5 µ, cylindrical, straight, biguttulate.

Additional data from Dorenbosch (1970).

Bearing in mind that the majority of *Pyrenochaeta* species have been described with simple doliiform or ampulliform conidiogenous cells, it seems likely that many more synonyms for this species will eventually be confirmed.

On and from *Capsicum annuum, Chrysanthemum* sp., *Coffea arabica, Combretum* sp., *Fragaria vesca, Hevea* sp., *Juniperus depressa, Malus* sp., *Pothos* sp., *Rhododendron* sp., *Scabiosa ochroleuca, Vaucheria filamentosa*, soil, fresh water, U.K., Oman, India, Ghana, Malaysia, N. Zealand, Kuwait, Sudan, Egypt, Canada, Yugoslavia, Singapore (IMI 195157, 198476, 198577, 201101, 204663, 210986, 210985, 228266, 168144, 193561, 187632, 187631, 187623, 200047, 203460c, 211624, 195945, 227669, 172808, 203967, 184514, 172542, 190149, 210085, 199831, 172612).

P. LINGAM (Tode ex Schw.) Desm., *Ann. Sci. nat.*, ser. 3, **11**: 281 (1849). (Fig. 228D)
Sphaeria lingam Tode, *Fungi Mecklenb. Sel.*: 51 (1791).
Sphaeria lingam Tode ex Schw., *Schr. Naturforsch. Ges. (Leipzig)* **1**: 45 (1822).
Plenodomus lingam (Tode ex Schw.) Höhn., *Sber. Akad. Wiss. Wien* **120**: 463 (1911).

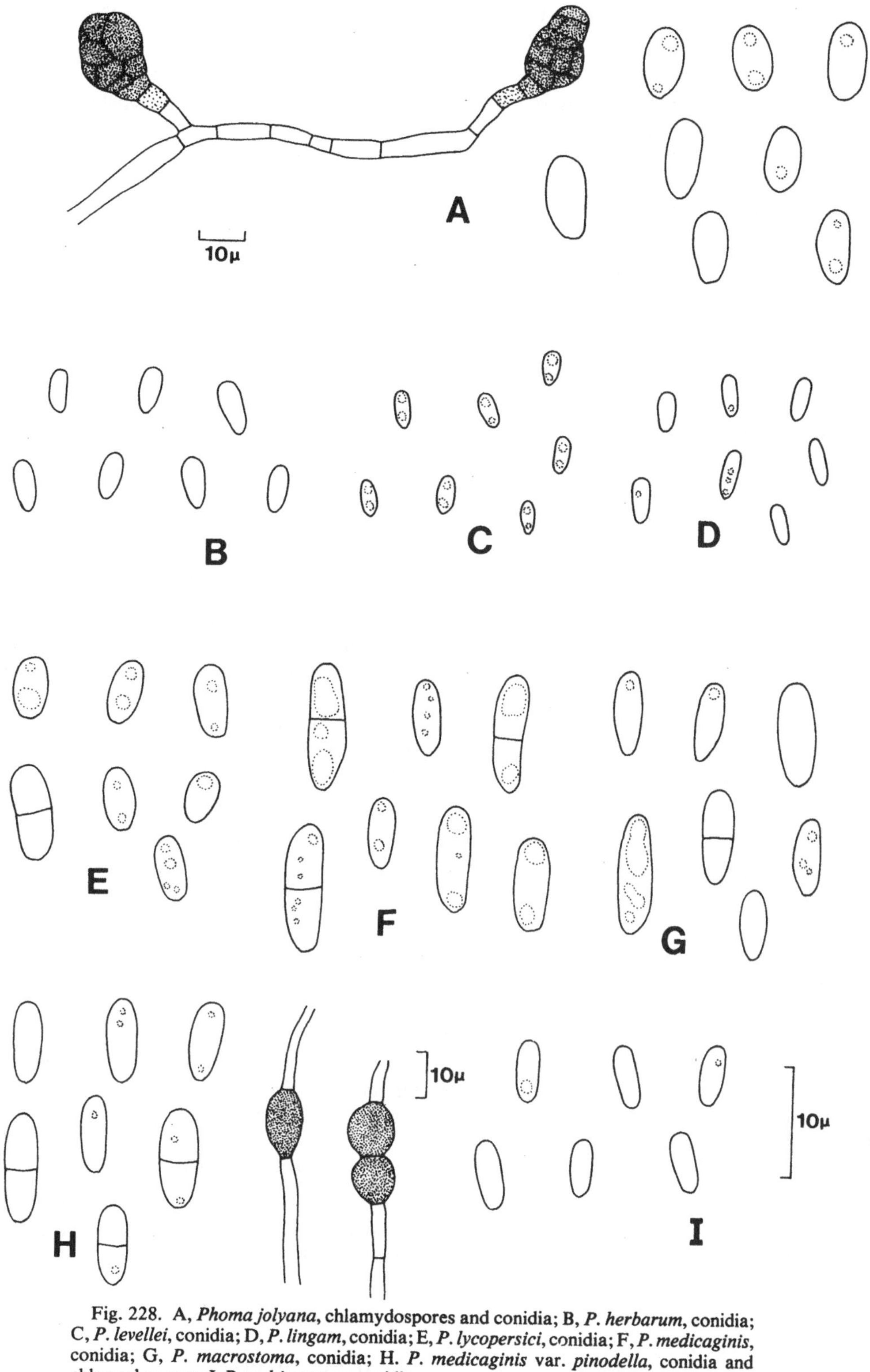

Fig. 228. A, *Phoma jolyana*, chlamydospores and conidia; B, *P. herbarum*, conidia; C, *P. levellei*, conidia; D, *P. lingam*, conidia; E, *P. lycopersici*, conidia; F, *P. medicaginis*, conidia; G, *P. macrostoma*, conidia; H. *P. medicaginis* var. *pinodella*, conidia and chlamydospores; I, *P. multirostrata*, conidia.

Teleomorphosis: *Leptosphaeria maculans* (Desm.) Ces. & de Not., *Comm. Soc. critt. ital.* 235 (1863).
Synonymy in Boerema & van Kesteren (1963).

Colonies variable in form and rate of growth, pycnidial production and pigmentation. Aerial mycelium felty or woolly, grey, cream or white or mixtures of different pigmentations. Pycnidia pseudoparenchymatous in culture but thick-walled and pseudosclerenchymatous during overwintering on the host substrate. Frequently sterile sclerotia (immature pycnidia) also develop. Conidia 3.5-4.5 × 1.5 μ, cylindrical, occasionally with a minute guttule, straight.

Additional data from Boerema (1976).

On and from *Brassica campestris*, *B. oleracea*, *B. chinensis*, *B. napus* var. *napobrassica*, *B. napus* var. *oleifera*, *B. oleracea* var. *capitata*, *B. oleracea* var. *ramosa*, *B. rapa*, *Artemisia* sp., *Berteroa incana*, *Beta vulgaris*, *Chrysanthemum* sp., *Clematis* sp., *Echinops* sp., *Erysimum* sp., *Erythrina indica*, *Ficus palmata*, *Hesperia matronalis*, *Linaria genistifolia*, *Matthiola incana*, *Raphanus maritimus*, *R. sativus* and represented by 74 collections from U.K., Eire, France, Roumania, Netherlands, Czechoslovakia, Hungary, Malta, Pakistan, India, Kenya, S. Rhodesia, Zambia, Ethiopia, Australia, New Zealand, North Borneo.

P. LYCOPERSICI Cke, *Grevillea* **13**: 94 (1885). (Fig. 228E)
Sphaeronaema lycopersici Plowr., *Gard. Chron.* **16**: 260 (1881).
Ascochyta lycopersici (Plowr.) Brun., *Bull. Soc. bot. Fr.* **34**: 430 (1887).
Phoma lycopersici (Plowr.) Jacz., *Nouv. Mem. Soc. Imp. Nat. Moscou* **15**: 350 (1898).
Diplodina lycopersici Hollós, *Ann. Mus. Nat. Hung.* **5**: 461 (1907).

Teleomorphosis: *Didymella lycopersici* Kleb., *Z. PflKr. PflPath. PflSchutz* **31**: 12 (1921).

Colonies olivaceous grey, dense felty aerial mycelium often in stilboid tufts, not concentrically zoned; reverse dark brown. Pycnidia sparse. Conidia 4.5-9 × 2.5-3.5 μ, often becoming medianly 1-septate, variably guttulate but usually with a guttule at each end, as in α-conidia of *Phomopsis*, straight, cylindrical or ellipsoid.

For additional data see Boerema & Dorenbosch (1973).

The distinction between this species and *P. destructiva* is marginal. *P. destructiva* differs in having smaller aseptate conidia and cultures in which aerial mycelium is usually lacking.

P. lycopersici was described in detail by Holliday & Punithalingam (1970).

On and from *Lycopersicon esculentum*, U.K., Malaysia, Tonga, Nigeria, Papua New Guinea, Poland, Netherlands, Brunei, Togo, India, Tahiti, S. Rhodesia (IMI 210426, 202048, 203562, 192649, 195696, 184500, 198847, 177843, 93441, 150063, 152809, 159519, 130267, 173141, 103216, 101664, 150065, 132415, 166380, 147191, 141625, 132414, 132413, 60074, 61313, 61314, 143948).

P. MEDICAGINIS Malbr. & Roum., *Rev. Mycol.* **8**: 91 (1886). (Fig. 228F)
P. herbarum Westd. forma *medicaginis* Westd. in Rabenh., *Fungi europ.* exs. n. ser. 2, no. 455b (1862).
Diplodina medicaginis Oud., *Ned. Kruidk. Arch.* ser. 3, **2**: 884 (1903).
Ascochyta imperfecta Pk, *Bull. N.Y. St. Mus.* 157: 21 (1912).

Colonies consisting of a flat grey felt of aerial mycelium, uniform in appearance on most media, usually concentrically zoned; chlamydospore formation only in older cultures; on MA small crystals appear in the medium. Pycnidia large, scattered. Conidia 6-11.5 × 2-3.5 μ, straight or slightly irregular, cylindrical, often becoming 1-septate, variably guttulate.

Additional data in Boerema, Dorenbosch & Leffring (1965) and Boerema (1976).

On and from *Brassica oleracea*, *Cicer arietinum*, *Glycine soja*, *Lathyrus odoratus*, *Medicago sativa*, *Solanum tuberosum*, *Trifolium subterraneum*, soil, cattle feed, India, Zambia, Canada, U.K., Australia (IMI 224794, 213420, 213426, 202187, 224691, 230682, 116999, 225036, 229080, 199818, 215973).

P. MEDICAGINIS var. PINODELLA (L.K. Jones) Boerema apud Boerema, Dorenbosch & Leffring, *Neth. J. Pl. Path.* **71**: 88 (1965). (Fig. 228H)
Ascochyta pinodella L.K. Jones, *Bull. N.Y. St. agric. Exp. St.* 547: 10 (1927).
Phoma trifolii E.M. Johnston & Valleau, *Bull. Ky agric. Exp. St.* 339: 73 (1933).
Additional synonyms in Boerema (1976).

Colonies extremely variable, often forming different sectors with dense felty dark or pale aerial mycelium and rare small pycnidia, or sparse aerial mycelium and abundant pycnidia; chlamydospore formation in young cultures; on MA abundant, fan-shaped crystals formed in the medium. Growth rate faster than in *P. medicaginis* var. *medicaginis*. Conidia 7-10 × 2.5-4 µ, straight, cylindrical to ellipsoid, often becoming 1-septate, minutely guttulate.

Additional data in Dorenbosch (1970) and Boerema (1976).

On and from *Beta vulgaris, Casuarina* sp., *Coffea arabica, Galanthus* sp., *Gossypium* sp., *Oryza*, sp., *Petroselinum sativum, Phaseolus aureus, Pisum sativum, P.*sp., *Trifolium pratense, T. repens, T.* sp., *Vicia faba, Vigna sinensis*, straw, hay, U.K., Tanzania, India, Nigeria, Canada, Netherlands, N. Zealand (IMI 54861, 182028, 182036, 201943, 171846, 202253, 202254, 116998, 226732, 220640, 183351, 68457, 68429, 135520, 137795, 160179, 165710, 176668, 231281, 231285, 193335, 195694, 179342, 181726, 199830, 199827, 192609, 224360).

P. MACROSTOMA Mont., *Ann. Sci. nat.*, ser. 3, **11**: 52 (1849). (Fig. 228G)
Synonymy in Boerema & Dorenbosch (1970, 1973).

Colonies fast-growing, variable; aerial mycelium felty or woolly with white, grey or olivaceous areas; colonies coloured livid red, changing to purple-blue with addition of NaOH. Conidia 5.5-10.5 × 2.5-3.5 µ, very variable, occasionally becoming 1 septate, cylindrical, ellipsoid or pyriform, variably guttulate.

Boerema & Dorenbosch (1970) have separated a var. *incolorata* (Horne) Boerema & Dorenbosch which can be distinguished only by the absence of red pigment in the hyphae.

Represented in herb. IMI by almost 100 collections on and from 41 different host genera from U.K., Germany, Netherlands, Cyprus, Sweden, Turkey, Venezuela, India, Pakistan, Nepal, Sri Lanka, S. Rhodesia, Egypt, Nigeria, Kenya, Tanzania, Ethiopia, Hong Kong, New Zealand.

P. MULTIROSTRATA (Mathur et al.) Dorenbosch & Boerema, *Mycopath. Mycol. appl.* **50**: 256 (1973). (Fig. 228I)
Synonymy in Dorenbosch & Boerema (1973).

Cultures thermophilic, growing more rapidly at 30°C than 20°C, similar to *P. medicaginis* var. *pinodella* in colony morphology and occasionally with chlamydospores. Pycnidia often aggregated and typically rostrate. Conidia 5-6.5 × 2-2.5 µ, straight, cylindrical, occasionally ellipsoid, eguttulate or with a single guttule.

Additional data in Dorenbosch & Höweler (1968).

On and from *Allium cepa, Coriandrum sativum, Solanum tuberosum, Sorghum* sp., soil, air, leaf litter, India, Sri Lanka (IMI 222937, 224036, 230678, 209528, 81598, 206943, 211985, 211966, 212995, 212371, 212378, 213390, 226541, 225701, 226690, 226689, 113681, 213470, 213434, 213457, 213461, 231039, 213014).

P. NEBULOSA (Pers. ex S.F. Gray) Berk., *Outl. Br. Fung.*: 314 (1860). (Fig. 229A)
Synonymy in Boerema (1976).

Colonies variable, usually dull green to green olivaceous brown, some strains mostly lacking aerial mycelium, others whitish to green and moderately felty. Conidia 4-6 × 1.5-2 µ, straight or slightly curved, cylindrical to ellipsoid, biguttulate.

Pycnidia reported to be often rostrate by Boerema (1976).

On and from *Onobrychis viciifolia, Althaea rosea, Brassica oleracea, Lamium* sp., *Petroselinum sativum, Urtica dioica*, U.K., Austria, Belgium (IMI 222912, 194766, 134790, 134791, 227915, 46900, 225857, 195947, 104188b).

P. POMORUM Thüm., *Fungi pomicoli*: 105 (1879). (Fig. 229F)
Synonymy in Boerema, Dorenbosch & van Kesteren (1965, 1968) under *P. prunicola* (Opiz) Wollenw. & Hochapf., and in Boerema, Dorenbosch & van Kesteren (1971, 1977).

Colonies again extremely variable, often with different sectors showing abundant pycnidial production and little aerial mycelium or dense felty aerial mycelium which may be olivaceous green, white, grey or black. Conidia 4.5-6 × 2-2.5 µ, straight or slightly curved, cylindrical or ellipsoid, eguttulate or with the occasional guttule. Chlamydospores of two types, either single and formed in chains, or *Alternaria*-like and solitary.

Additional data from Dorenbosch (1970), Boerema & Dorenbosch (1973), Boerema, Dorenbosch & van Kesteren (1965, 1968, 1971, 1977) and Morgan-Jones (1967).

On and from *Arachis hypogaea*, *Capsicum* sp., *Ceanothus dentatus*, *Gossypium* sp., *Lithospermum* sp., *Malus sylvestris*, *Pinus caribaea*, *Polygonum tataricum*, *Prunus amygdalus*, *P. persica*, *Pyrus communis*, *P. malus*, *Rhododendron* sp., *Salvinia modesta*, *Solanum tuberosum*, *Trichosanthes dioica*, from soil, U.K., Netherlands, Nigeria, Libya, India, Nepal, Sri Lanka, Papua New Guinea, Sabah (IMI 124911, 202068, 198483, 205626, 215645, 196270a, 204783, 122266, 130123, 195943b, 145624b, 146716, 130366, 215529, 211626, 211625, 216997, 207567, 207568, 216995, 178500, 183776, 183771, 205878, 205875, 216822, 216866, 216867, 132163, 199799, 202150, 202151, 213425, 201782b, 202190).

P. PUTAMINUM Speg., *Atti Soc. Critt. Ital.*, ser. 2, **3**: 66 (1881). (Fig. 229B)
Aposphaeria putaminum (Speg.) Sacc., *Syll. fung.* **3**: 177 (1884).
Coniothyrium putaminum (Speg.) O. Ktze, *Rev. Gen. Pl.* III[II]: 459 (1893).

Colonies comparatively fast-growing, with little aerial mycelium, although when present olivaceous to grey; reverse buff to apricot. Pycnidia often with brown hyphae over their surfaces. Conidia 2.5-4.5 × 2-2.5 μ, ellipsoid or almost globose, smaller conidia with a single guttule, larger ones often biguttulate.

Additional data in Boerema & Dorenbosch (1973).

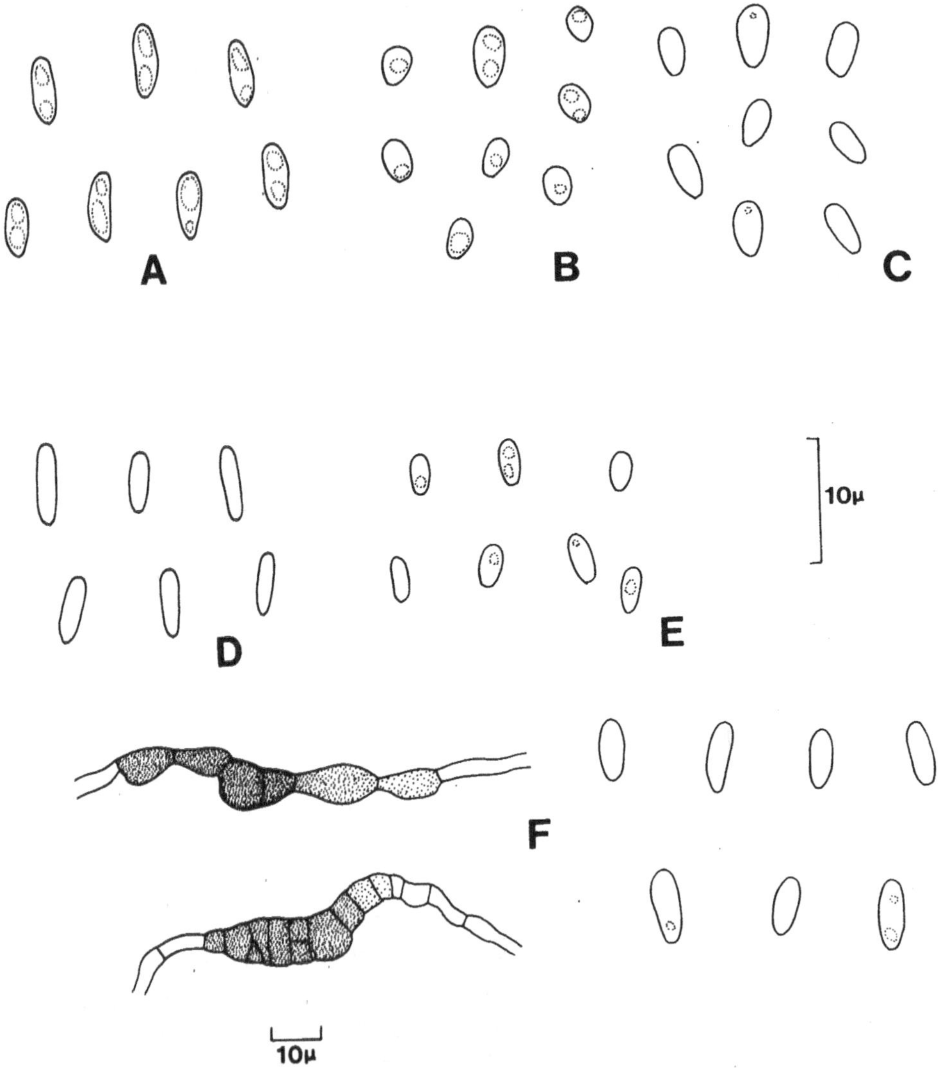

Fig. 229. A, *Phoma nebulosa*, conidia; B, *P. putaminum*, conidia; C, *P. sorghina*, conidia; D, *P. tracheiphila*, conidia; E, *P. tropica*, conidia; F, *P. pomorum*, chlamydospores and conidia.

On and from *Achras sapota*, *Artabotrys japonica*, *Citrullus vulgaris*, *Convallaria majalis*, *Cyperus* sp., *Ficus* sp., *Madhuca indica*, *Vandellia multiflora*, soil, air, wood, leaf litter and fibreglass, Australia, India, Burma, U.K., Nigeria, Malaysia, Kuwait, Venezuela, Netherlands (IMI 203490, 189303, 198750, 163513, 208414, 210671, 209303, 209061, 178979, 188668, 193492, 195787, 202490, 212385, 230056, 178829, 179751, 179750, 226031, 190895, 190273, 187215, 192958, 192956, 213455).

P. SORGHINA (Sacc.) Boerema, Dorenbosch & van Kesteren, *Persoonia* **7**: 134 (1973). (Fig. 229C)
Synonymy in Boerema, Dorenbosch & van Kesteren (1968) under *P. indianensis* (Deshpande & Mantri) Boerema, Dorenbosch & van Kesteren, in Boerema, Dorenbosch & van Kesteren (1971) under *P. glumarum* Ell. & Tracy, and in Boerema, Dorenbosch & van Kesteren (1973, 1977).

Colonies extremely variable, usually with a fluffy to dense aerial mycelium which is basically grey-green to olivaceous or darker, but with very characteristic white to salmon pink tinges or areas; reverse often reddish. Conidia 4-5 × 2-2.5 µ, ellipsoid, eguttulate. Chlamydospores single-celled -celled or *Alternaria*-like but formation very variable and not entirely necessary for identification, the cultural characteristics being quite diagnostic.
Additional data are found in Punithalingam & Holliday (1972).

Represented in herb. IMI by more than 500 collections from 76 different host genera, soil, cattle and poultry feed and several other substrates, from at least 50 different countries.

P. TRACHEIPHILA (Petri) Kantachveli & Gikachvili, *Lav. Ist. prot. piante Georgia* **5**: 20 (1948). (Fig. 229D)
Deuterophoma tracheiphila Petri, *Boll. R. Staz. Pat. Veg. Roma, N.S.* **9**: 396 (1929).
Bakerophoma tracheiphila (Petri) Cif., *Atti Ist. bot. Univ. Pavia*, ser. 5, **5**: 307 (1946).

Colonies with a fairly even grey to white aerial mycelial felt. Pycnidial formation variable. Conidia formed in pycnidia from septate conidiophores which are short, and irregularly branched, and also directly from the mycelium, 4.5-6.5 × 1.5 µ, cylindrical, straight or slightly curved, eguttulate.
For additional data see Punithalingam & Holliday (1973).

On and from *Citrus limonia*, *C. medica*, *C. medica* var. *limonum*, *C.* spp., Cyprus, Italy, Turkey, Uganda, Iraq (IMI 21608, 114369, 149896, 151684, 230150, 141601, 73027, 182961, 201277, 201281, 207209).

P. TROPICA Schneider & Boerema, *Phytopath. Z.* **83**: 361 (1975). (Fig. 229E)

Colonies with floccose aerial mycelium at first, later becoming woolly, pale olivaceous grey to chocolate brown, with an olivaceous to black reverse. Conidia 3-4 × 1.5 µ, straight, ellipsoid, indistinctly guttulate.

On and from *Aerides falcatum*, *Amorphophallus* sp., *Barleria* sp., *Borreria hispida*, *Bougainvillea spectabilis*, *B.* sp., *Coffea arabica*, *Cordia colococca*, *Croton bonplandianum*, *Dioscorea alata*, *D. belophylla*, *Heliconia* sp., *Manihot utilissima*, *Mimusops* sp., *Ricinus communis*, *Solanum tuberosum*, *Syzygium jambolana*, *Thespesia populnea*, India, Australia, Pakistan, Fiji, Tahiti, Venezuela, Guadalcanal, Togo, U.K. (IMI 226447, 208424, 210487, 204259a, 211597, 199786, 227829, 199828, 208423, 206234, 209243, 200996c, 206490, 213428, 205897, 198436, 199892, 207556, 214815c, 216923b, 209007, 208624, 208631a).

15. MACROPHOMINA Petrak, *Annls mycol.* **21**: 314 (1909).

Sp. typ.: *M. phaseolina* (Tassi) Goid. (syn. *Macrophoma phaseolina* Tassi, *M. philippines* Petrak).
Illustration: Holliday & Punithalingam (1970).

Mycelium superficial or immersed, brown to hyaline, branched, septate, often dendroid in culture. *Conidiomata* pycnidial, separate, globose, dark brown, immersed, unilocular, thick-walled; wall consisting of an outer layer of dark brown thick-walled textura angularis, becoming hyaline towards the inside. *Ostiole* central, circular, papillate. *Conidiophores* absent. *Conidiogenous cells* enteroblastic, phialidic, determinate, discrete, lageniform to doliiform, hyaline, smooth, with wide aperture and minute collarette, formed from the inner cells of the pycnidial wall. *Conidia* hyaline, aseptate, obtuse at each end, straight, cylindrical to fusiform, thin-walled, smooth, ± guttulate. *Sclerotia* black, smooth, hard, formed of dark brown, thick-walled cells.

The full synonymy for the type species was given by Holliday & Punithalingam (1970) together with notes on diseases, distribution, physiological specialization and transmission. In culture only the sclerotia are usually produced.

M. PHASEOLINA (Tassi) Goid., *Ann. Sper. agr.* N.S. **1**: 457 (1947).

Conidiomata 100-200 μ diam. Conidiogenous cells 5-13 × 4-6 μ. Conidia 14-30 × 5-10 μ. Sclerotia 50-300 μ diam.

Plurivorous, represented on a wide range of hosts in Herb. IMI by approximately 500 collections.

16. MELANOPHOMA Papendorf & Du Toit, *Trans. Br. mycol. Soc.* **50**: 503 (1967).
Sp. typ.: *M. karroo* Papendorf & du Toit
Illustration: Papendorf & du Toit (1967).

Described only from culture. *Mycelium* white or buff, olivaceous, branched, septate, smooth. *Conidiomata* pycnidial, separate, globose to ampulliform, medium brown, immersed to semi-immersed, unilocular, thin-walled; wall 2-4 cells thick, of pale brown textura angularis, darker around the ostiole. *Ostiole* papillate or slightly depressed, single, circular. *Conidiophores* absent. *Conidiogenous cells* enteroblastic, phialidic, determinate, discrete, ampulliform to doliiform, hyaline, smooth, channel wide, collarette minimal, formed from the inner cells of the pycnidial wall. *Conidia* dark smoky brown, aseptate, base ± rounded, apex obtuse, thick-walled, verruculose, eguttulate, ovoid, broadly elliptical or subglobose, with a hyaline, membranous, persistent, epispore.

The genus is monotypic.

M. KARROO Papendorf & du Toit, *ibid.*: 503 (1967).

Conidiomata 100-400 × 90-250 μ. Conidiogenous cells 5-10 × 6-8 μ. Conidia 7-14.5 × 5-8.5 μ.

Isolated from leaf litter of *Acacia karroo*, S. Africa.

17. CONIDIOXYPHIUM Batista & Ciferri, *Quad. Ist. bot. Univ. Pavia* 31: 72 (1963).
Sp. typ.: *C. gardeniorum* Batista & Ciferri

Mycelium superficial, branched, septate, verrucose, brown. *Conidiomata* pycnidial, superficial, separate, elongated, unilocular, produced on and merging into a basal sterile stalk, expanding slightly in the locular region, finally with a long tapered rostrate apex which is obtuse and not fimbriate; stalk of ± parallel columns of dark brown, thick-walled longitudinally elongated textura porrecta, pycnidial wall of ± cuboid brown pseudoparenchyma, rostrate apex of parallel columns of pale brown thin-walled, longitudinally elongated pseudoparenchyma. Pycnidial structures often branch irregularly to form up to 5 fertile regions from one stalk. *Ostiole* central, circular, apical. *Conidiophores* long, irregularly filiform, hyaline, septate, unbranched, consisting of several ± cuboid conidiogenous cells, formed from the base and sides of the pycnidial wall. *Conidiogenous cells* enteroblastic, phialidic, determinate, integrated, hyaline, smooth, with apical or lateral apertures immediately below septa, occasionally polyphialidic. *Conidia* hyaline, aseptate, thin-walled, smooth, guttulate, ellipsoid to cylindrical.

The genus contains two species.

C. GARDENIORUM Batista & Ciferri, *ibid.*: 73 (1963). (Fig. 230)

Pycnidial conidiomata up to 600 μ long × 50 μ wide. Conidiophores up to 4 μ wide. Conidia 4.5-8 × 2-2.5 μ.

On leaves of *Gardenia jasminoides*, La, U.S.A. (IMI 201352, isotype ex BPI).

18. ELEUTHEROMYCES Fckl, *Jahrb. nassau. Ver. Naturk.* **23-24**: 183 (1870).
Sp. typ.: *E. subulatus* (Fr.) Fckl (syn. *Sphaeronaema subulatum* Tode ex Fr.).
Illustration: Seeler (1943).

Mycelium immersed or superficial, hyaline, branched, septate. *Conidiomata* pycnidial, separate, rarely confluent, long conical, very pale brown, translucent, superficial, unilocular, thick-walled; wall composed of thick-walled, pale to very pale brown textura angularis, somewhat fimbriate at the apex. *Ostiole* central, single, circular, at the tip of the neck.

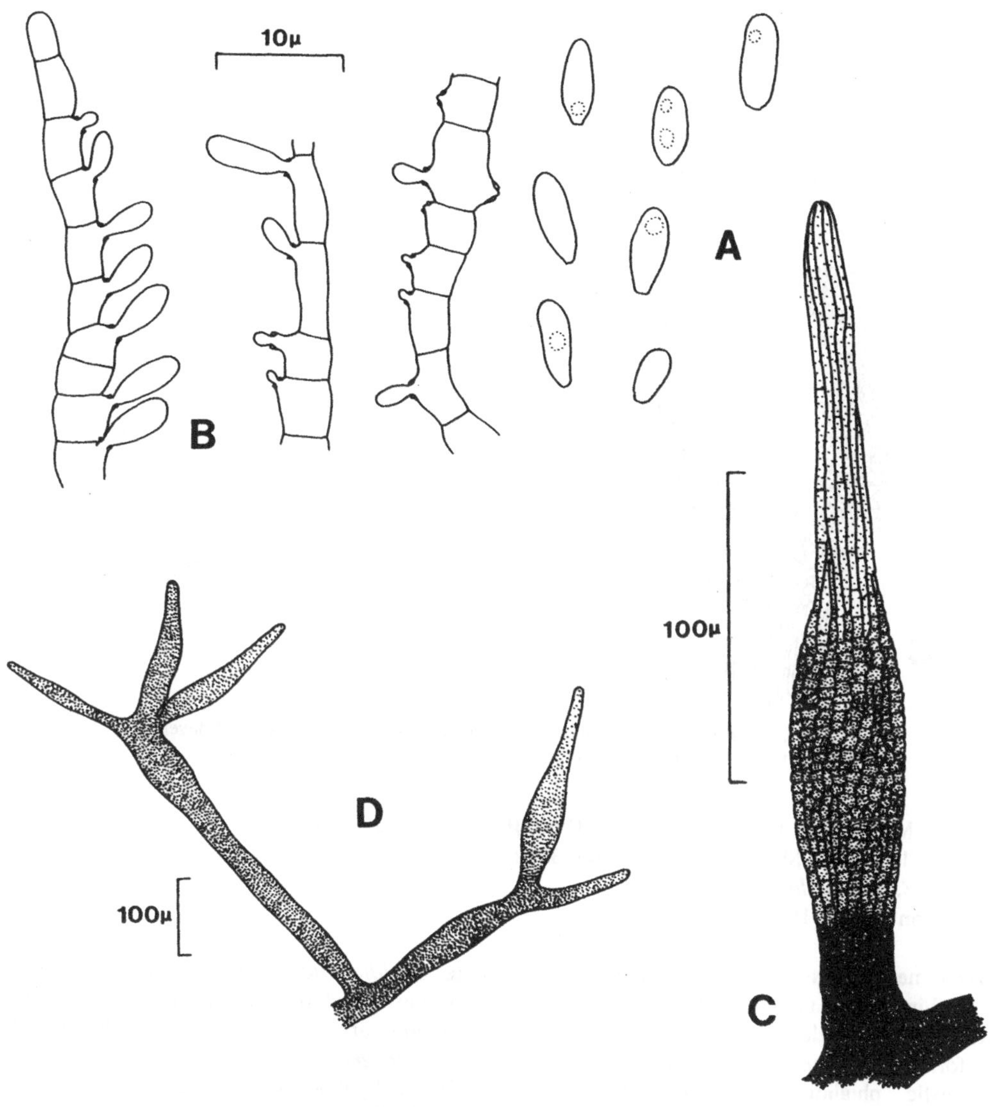

Fig. 230. *Conidioxyphium gardeniorum*, A, conidia; B, conidiophores and developing conidia; C, side view of a conidioma; D, branched conidiomata.

Conidiophores hyaline, branched mainly at the base, smooth, cylindrical or flexuous, with many septa, formed from the inner cells of the pycnidial wall. *Conidiogenous cells* enteroblastic, phialidic, integrated, determinate, cylindrical, hyaline, smooth, each with an aperture immediately below a transverse septum, collarette and channel minute. *Conidia* hyaline, aseptate, lenticular, apex and base drawn out into short, unbranched cellular appendages, smooth, thin-walled, eguttulate, abstriction occurring at the tip of the basal appendage.

Seeler (1943) described conidiogenesis as phialidic. The genus is monotypic.

E. SUBULATUS (Fr.) Fckl, *ibid.*: 183 (1870). (Fig. 231)

Conidiomata 130-350 µ diam. at the base, up to 3300 µ high. Conidiophores up to 60 µ long × 2.5-3.5 µ wide. Conidia 4-5.5 × 1.5-2 µ (excluding appendages); appendages 2-7 µ long.

On *Polyporus picipes* and hymenomycete indet., Canada, Latvia (IMI 26792, 12791, 125697a).

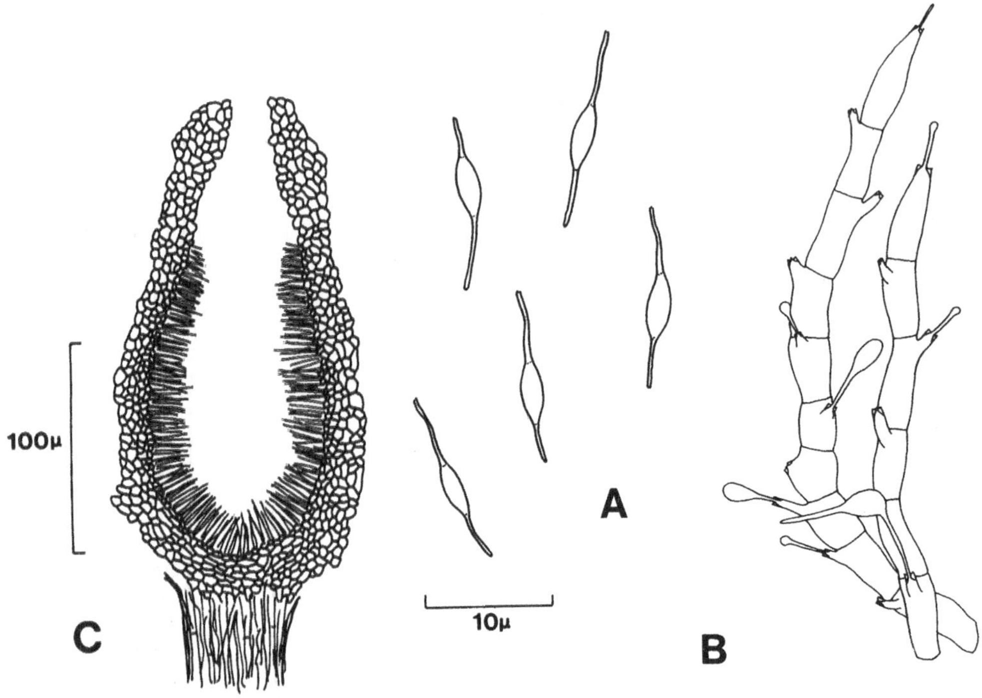

Fig. 231. *Eleutheromyces subulatus*, A, conidia; B, conidiophores and developing conidia; C, vertical section of a conidioma.

19. ELEUTHEROMYCELLA Höhn., *Sber. Akad. Wiss. Wien* **117**: 1023 (1908).
Eleutheris Clements & Shear, *Genera of fungi*: 371 (1931).
Sp. typ.: *E. mycophila* Höhn.
Illustration: Seeler (1943).

Mycelium immersed or superficial, hyaline, branched, septate. *Conidiomata* pycnidial, separate, globose, dark brown, immersed or semi-immersed, unilocular, thick-walled, composed of thin-walled, brown textura angularis, with a short neck. *Ostiole* central, circular, single. *Conidiophores* hyaline, branched mainly at the base, smooth, cylndrical or flexuous, with several septa, formed from the inner cells of the pycnidial wall. *Conidiogenous cells* figured as holoblastic but almost certainly enteroblastic, phialidic, integrated, determinate, cylindrical, hyaline, smooth, each producing a conidium immediately below a transverse septum. *Conidia* hyaline, aseptate, fusiform, apex and base drawn out into a short basal and relatively long apical unbranched, cellular appendage, smooth, thin-walled, eguttulate, abstriction occurring at the tip of the basal appendage.

Seeler (1943) has redescribed this genus which is separated from *Eleutheromyces* Fckl by the dark pigmentation of the pycnidia. The genus is monotypic.

E. MYCOPHILA Höhn., *ibid.*: 1023 (1908).

Conidiomata 300-900 µ diam. Conidiophores up to 60 µ long × 1.5-2.5 µ wide. Conidia 5-6.5 × 1.5 µ (excluding appendages); apical appendage 12-30 µ long; basal appendage 2-3 µ long.

On *Polystictus versicolor*, Austria.

20. PYRENOCHAETA de Not., *Mem. R. Accad. Torino ser.* 2, **10**: 347 (1849).
Herpotrichiopsis Höhn., *Sber. Akad. Wiss. Wien* **113**: 115 (1914).
Lasiophoma Naumov, *Bull. Soc. Oural. d'Amis Sci. nat.* **35**: 30 (1915).
Pyrenochaetella Karst. ex Höhn., *Hedwigia* **59**: 256 (1917).
Sp. typ.: *P. nobilis* de Not.

Mycelium immersed or superficial, pale brown to hyaline, branched, septate. *Conidiomata* pycnidial, separate, rarely aggregated, globose, immersed or superficial, brown, unilocular, thin- to thick-walled; wall of pale to medium brown textura angularis. *Ostiole* single, central, circular. *Setae* abundant around the ostiole, less so over the rest of the pycnidium, erect or lax, dark brown, thick-walled, smooth, septate, usually tapered to the apices. *Conidiophores* long, filiform, branched at the base, multiseptate, hyaline, formed from the inner cells of the conidiomatal wall. *Conidiogenous cells* enteroblastic, phialidic, determinate, integrated, arising as very short lateral branches immediately below transverse septa, hyaline, smooth, channel, collarette and periclinal thickening minute. *Conidia* hyaline, aseptate, smooth, ± guttulate, straight, cylindrical to ellipsoid.

The generic synonymy has been explained by Sutton (1977a) who agreed with Schneider (1976) in placing *Pyrenochaetella* as a synonym. Schneider (1976) designated a neotype for *P. nobilis*, and accepted only 11 species in the genus and later gave a descriptive account (Schneider, 1979). More than 120 species have been assigned to *Pyrenochaeta* and the majority will have to be removed to other genera. Most of these species resemble *Pyrenochaeta* sensu stricto in having simple setose pycnidia and small unicellular conidia. They differ in having simple ampulliform to lageniform phialides instead of septate conidiophores and acropleurogenous phialides. Such species could possibly be accommodated in *Phoma* Sacc. at subgeneric level where the presence of setae on the pycnidia differentiates them from the species at present accepted in *Phoma*. If, however, the setae are regarded as a valuable taxonomic criterion at generic level, then it will be necessary to introduce a new generic name to take such species. So far, no described genus appears to be available.

KEY

1. Conidia less than 3.5 µ long .... *P. unguis-hominis*
1. Conidia more than 3.5 µ long 2
   2. Conidia 2-2.5 µ wide .... *P. rubi-idaei*
   2. Conidia 1.5 µ wide 3
3. Conidiophores long, filiform, more than 30 µ long; setae long, multiseptate, up to 250 µ long, abundant 4
3. Conidiophores less than 30 µ long; setae less than 150 µ long; conidia 5-7 µ long; microsclerotia formed in culture .... *P. lycopersici*
   4. Conidia 8-9 µ long .... *P. ilicis*
   4. Conidia 5-6 µ long .... *P. fallax*

P. FALLAX Bres. apud Strasser, *Verh. zool. bot. Ges. Wien*: 615 (1905). (Fig. 232D)

Conidiomata up to 325 µ diam.; setae abundant, up to 200 µ long × 8 µ wide, up to 6 septate. Conidiophores up to 45 µ long × 2-2.5 µ wide, 4 septate. Conidia 5-6 × 1.5 µ.

On dead stems of *Urtica dioica*, U.K. (IMI 95239a).

P. ILICIS Wilson, *Scot. Bot. Rev.* **1**: 161 (1912). (Fig. 232C)

Conidiomata up to 800 µ diam.; setae abundant, up to 250 µ long × 10 µ wide, tapered, thick-walled at the base, up to 6 septate. Conidiophores up to 60 µ long × 2-2.5 µ wide, 4-5 septate. Conidia 8-9 × 1.5 µ.

On dead leaves of *Ilex aquifolium*, U.K., Eire (IMI 214346, 214345, 182245b, 193364b, 188542, 169634, 169405).

P. LYCOPERSICI Schneider & Gerlach, *Phytopath. Z.* **56**: 121 (1966). (Fig. 232A, B)

Conidiomata up to 175 µ diam.; setae sparse, up to 150 µ long × 10 µ wide, not markedly tapered, 2-3 septate. Conidiophores 20-35 µ long × 3 µ wide, 3-5 septate. Conidia 5-7 × 1.5 µ. Microsclerotia formed in culture.

From roots of *Lycopersicon esculentum*, Germany, U.K., New Zealand (IMI 173997, holotype, 174300, 211266, 211265, 136604, 174306, 174304, 225043).

P. RUBI-IDAEI Cavara, *Rev. Myc.* **11**: 188 (1889). (Fig. 232F)

Conidiomata up to 200 µ diam.; setae sparse, up to 200 µ long × 6 µ wide, markedly tapered. Conidiophores up to 20 µ long × 3-4.5 µ wide, 1-2 septate. Conidia 5.5-8 × 2-2.5 µ.

Causing lesions on leaves of *Rubus idaeus*, Italy (IMI 22692).

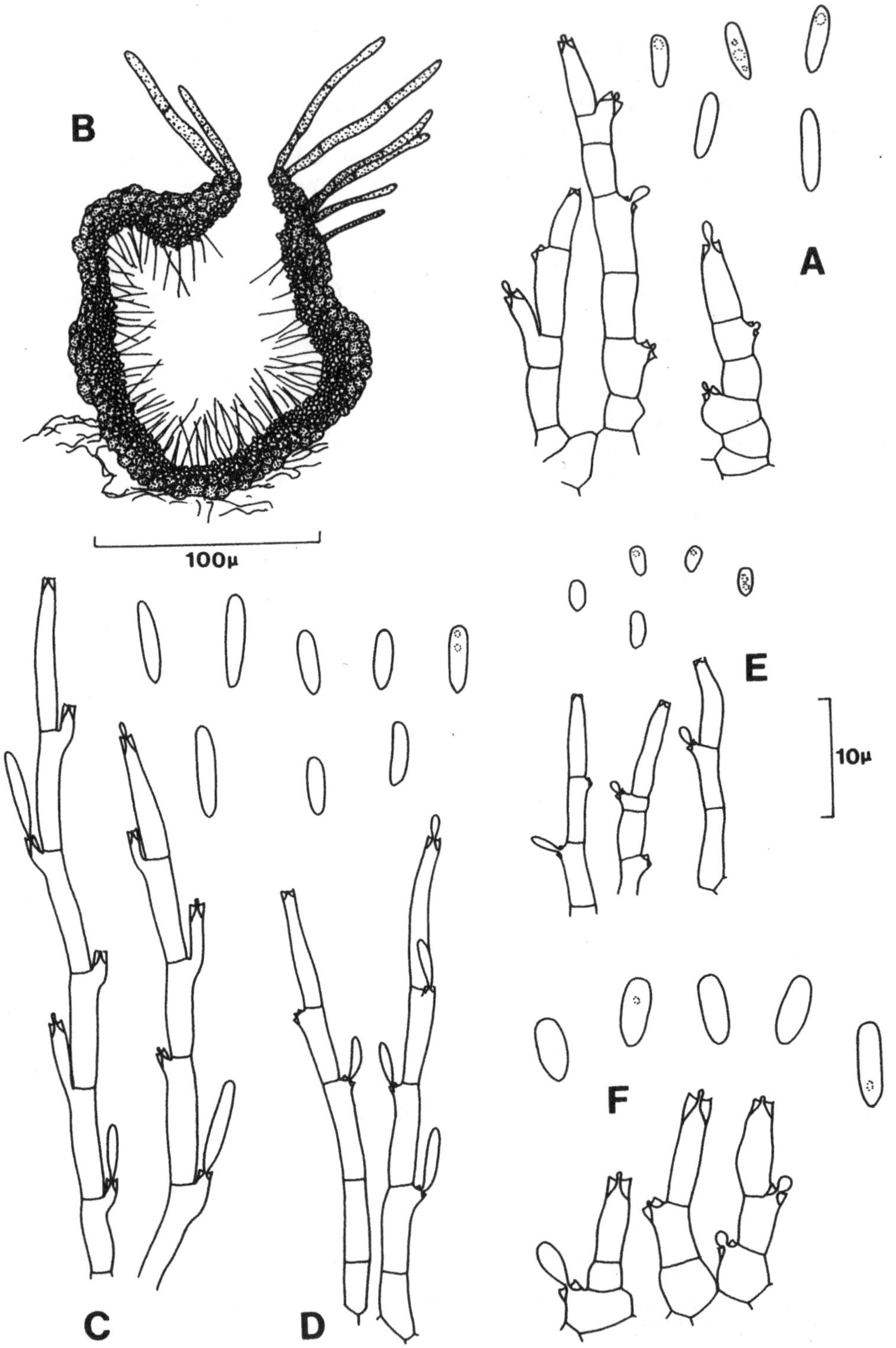

Fig. 232. A, B, *Pyrenochaeta lycopersici*, conidia, conidiophores and vertical section of a conidioma; C, *P. ilicis*, conidia and conidiophores; D, *P. fallax*, conidia and conidiophores; E, *P. unguis-hominis*, conidia and conidiophores; F, *P. rubi-idaei*, conidia and conidiophores

P. UNGUIS-HOMINIS Punithalingam & English, *Trans. Br. mycol. Soc.* **64**: 539 (1975). (Fig. 232E)

Conidiomata up to 500 µ diam.; setae sparse, up to 45 µ long × 4 µ wide, obtuse at the apices. Conidiophores 15-20 µ long × 2 µ wide, 2-3 septate. Conidia 3-3.5 µ × 1-1.5 µ.

From toe nails of *Homo sapiens*, U.K. (IMI 184840, holotype, 171211).

21. PHIALOPHOROPHOMA Linder, *Farlowia* **1**: 402 (1944).
Sp. typ.: *P. litoralis* Linder
Illustrations: Linder (1944); Sutton & Pirozynski (1963).

Mycelium immersed, branched, septate, brown. *Conidiomata* pycnidial, separate, globose with a short neck, dark brown, immersed, unilocular, thick-walled; wall composed of dark brown, thick-walled textura angularis. *Ostiole* single, circular, papillate. *Conidiophores* hyaline, branched at the base, septate, smooth, with acropleurogenous conidia, formed from the inner cells of the pycnidial wall. *Conidiogenous cells* enteroblastic, phialidic, indeterminate, integrated, cylindrical to lageniform, hyaline, smooth, with terminal or lateral apertures immediately below transverse septa, channel minute, collarette prominent. *Conidia* hyaline, aseptate, thin-walled, smooth, ± guttulate, oval.

Sutton & Pirozynski (1963) redescribed the species from British material but noted minor differences from the original account by Linder (1944). The genus remains monotypic.

P. LITORALIS Linder, *ibid.*: 403 (1944).

Conidiomata 200 µ diam. Conidiophores 0-4 septate, 10-25 × 2-2.5 µ. Conidia 3-4.5 × 1.5-2 µ.

On driftwood, from sea water, from *Spartina* sp., G.B., Yugoslavia (IMI 189757, 187622, 95240a, 206433).

22. SIROPHOMA Höhn., *Hedwigia* **59**: 257 (1917).
Sp. typ.: *S. singularis* Höhn.

Mycelium immersed, branched, septate, pale brown. *Conidiomata* pycnidial, dark brown, separate, globose, unilocular, immersed, thick-walled; wall uniformly of brown, ± thick-walled textura angularis. *Ostiole* single, circular, central. *Conidiophores* long, filiform, branched only at the base, hyaline, septate, formed from the inner cells of the pycnidial wall. *Conidiogenous cells* enteroblastic, phialidic, integrated, determinate, apertures apical or lateral and immediately below transverse septa, hyaline, clindrical, channel wide, periclinal wall thickened. *Conidia* hyaline, aseptate, thin-walled, eguttulate, smooth, globose to ellipsoid or pyriform.

The genus contains at least six species, none of which have been recently revised.

S. SINGULARIS Höhn., *ibid.*: 258 (1917). (Fig. 233)

Conidiomata up to 300 µ diam. Conidiophores up to 40 µ long × 2.5-3 µ wide. Conidia 3-4.5 × 2.5 µ.

On bare wood of *Viburnum opulus*, Austria (holotype ex FH).

23. PLEUROPHOMA Höhn., *Sber. Akad. Wiss. Wien* **123**: 117 (1914).
Sp. typ.: *P. pleurospora* (Sacc.) Höhn. (syn. *Phoma pleurospora* Sacc.).

Mycelium immersed, branched, septate, hyaline. *Conidiomata* pycnidial, separate, superficial, black, globose to collabent, unilocular, thick-walled; wall consisting of a thin outer layer of dark brown, thick-walled textura angularis, middle layer of thin-walled hyaline cells amongst which are isolated dark brown thick-walled cells, and an inner layer of hyaline to very pale brown, small-celled textura angularis. *Ostiole* single, circular, central, depressed. *Conidiophores* hyaline, branched at the base, more rarely above, septate, filiform, long, formed from the inner cells of the pycnidial wall. *Conidiogenous cells* enteroblastic, phialidic, integrated, determinate, hyaline, thin-walled, with apertures immediately below transverse septa, or occasionally apical, collarette and channel minute. *Conidia* hyaline, cylindrical to ellipsoid, eguttulate, thin-walled, smooth, aseptate.

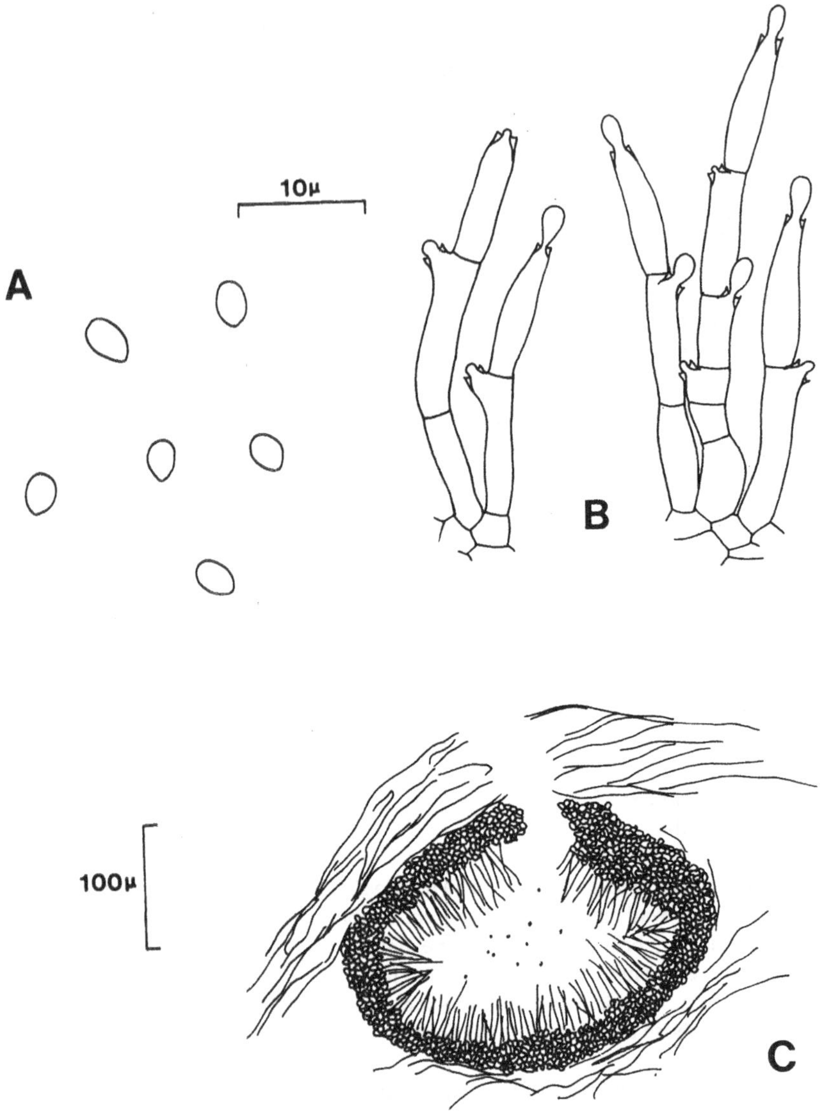

Fig. 233. *Sirophoma singularis*, A, conidia; B, conidiophores; C, vertical section of a conidioma.

Conceivably little different from the later-described *Phialophorophoma* Linder, although the two genera could be separated on the basis of differences in the pycnidial wall. Although the detailed structure of *Pleurophoma* is known, such details in *Phialophorophoma* have not yet been elucidated. If pycnidia are of similar structure, then *Phialophorophoma* will be reduced to synonymy.

The genus contains at least 8 described species, of which only *P. frangulae* Höhn. has been reassigned (to *Pleurostromella* Petrak by Petrak, 1923). It therefore requires no modern revision.

P. PLEUROSPORA (Sacc.) Höhn., *ibid.*: 117 (1914). (Fig. 234)
Dendrophoma pleurospora (Sacc.) Sacc., *Syll. fung.* **3**: 178 (1884).
Phoma pleurospora Sacc., *Michelia* **2**: 97 (1880).

Conidiomata up to 300 µ diam. Conidiophores up to 55 µ long × 2 µ wide. Conidia 3.5-4 × 1.5 µ.

On bare wood of *Salix fragilis*, G.B. (IMI 202437 ex K).

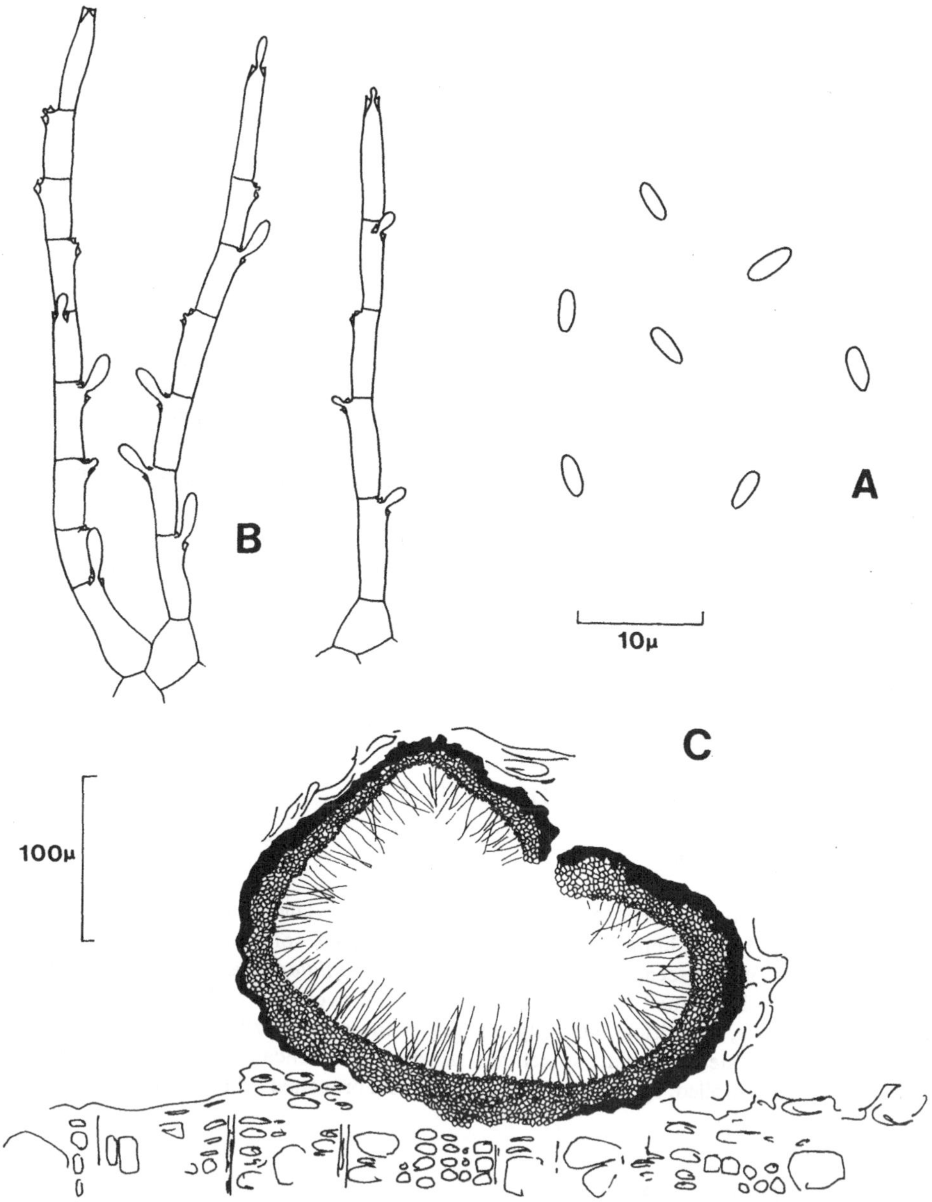

Fig. 234. *Pleurophoma pleurospora*, A, conidia; B, conidiophores; C, vertical section of a conidioma.

24. DENDRODOMUS Bub., *Bot. Közl.* **14**: (63) (1915).
Sp. typ.: *D. annullatus* Bub.

Mycelium immersed, brown, branched, septate. *Conidiomata* pycnidial, separate or aggregated, globose to collabent, dark brown, unilocular, thick-walled, outer layer of dark brown, thick-walled textura angularis, becoming larger but tbin-walled and paler and finally with a region of thin-walled, small-celled, hyaline pseudoparenchyma from which conidiogenous cells arise; towards the ostiole sterile hyphae like paraphyses extend into the pycnidial cavity. *Ostiole* central, circular, ± papillate. *Conidiophores* hyaline, branched at the base, less along the length, smooth, flexuous, with several septa. *Conidiogenous cells* enteroblastic, phialidic, integrated, rarely discrete, determinate, cylindrical or ampulliform, hyaline, smooth, on conidiophores formed with apertures immediately below transverse septa, when discrete apertures apical, collarette and channel minute, periclinal wall comparatively thick. *Conidia* hyaline, aseptate, smooth, thin-walled, irregularly guttulate, ellipsoid to cylindrical.

The genus is monotypic.

D. ANNULLATUS Bub., *ibid.*: (63) (1915). (Fig. 235)

Conidiomata up to 300 μ diam. Conidiophores up to 40 μ long × 2.5-3.5 μ wide. Conidia 4-5 × 1.5 μ.

On dead stems of *Scrophularia bosniaca*, Yugoslavia (IMI 194178, holotype).

25. CILIOSPORELLA Petrak, *Annls mycol.* **25**: 217 (1927).
Sp. typ.: *C. selenospora* Petrak
Illustration: Nag Raj (1974).

Mycelium immersed, branched, septate, brown. *Conidiomata* pycnidial, separate or aggregated, globose, pale brown, immersed, unilocular or occasionally multilocular, thick-walled; wall of several layers of brown, thin-walled textura angularis. Locules dehiscing by a single, circular, depressed ostiole. *Conidiophores* hyaline, branched, septate, smooth, formed from the inner cells of pycnidial wall. *Conidiogenous cells* enteroblastic, phialidic, determinate, integrated, cylindrical or tapered to the apex, hyaline, smooth, channel and collarette minute. *Conidia* hyaline, aseptate, base truncate, apex prolonged into a short cellular, unbranched appendage, thin-walled, smooth, eguttulate, fusiform or slightly falcate, straight.

I have not examined this species but in addition to the protologue there is a brief generic description and illustration in Nag Raj (1974). The genus is monotypic.

C. SELENOSPORA Petrak, *ibid.*: 218 (1927).
Conidiomata up to 200 μ diam. Conidiophores up to 17 μ long × 2-2.5 μ. Conidia 20.5-27.5 × 2.5-3 μ.

On dead stems of *Trifolium alpinum*, Austria.

26. BRYCEKENDRICKIA Nag Raj, *Can. J. Bot.* **51**: 1337 (1973).
Sp. typ.: *B. indica* Nag Raj
Illustration: Nag Raj (1973b).

Mycelium immersed, pale brown, branched, septate. *Conidiomata* pycnidial, epiphyllous, separate or gregarious to confluent, globose or irregular, brown, immersed, unilocular or convoluted, sometimes divided, outer layer of dark brown thick-walled cells. *Ostiole* absent, dehiscence by longitudinal rupture. *Conidiophores* hyaline, branched sparingly, septate, smooth, with acrogenous conidia, formed from the inner cells of the pycnidial wall. *Conidiogenous cells* enteroblastic, phialidic, integrated, determinate, cylindrical, abruptly tapered at the apex, hyaline, smooth, channel narrow, collarette minimal. *Conidia* hyaline, aseptate, navicular, sometimes falcate, base truncate, apex obtuse with a single, branched, cellular appendage, smooth, thin-walled, eguttulate, with a single basal branched cellular appendage inserted laterally.

The genus is monotypic.

B. INDICA Nag Raj, *ibid.*: 1337 (1973). (Fig. 236)

Conidiomata 190-450 × 130-230 μ. Conidiogenous cells 6.5-12 × 1.5-2.5 μ. Conidia 11-16 × 2-3.5 μ; appendage branches 3.5-16.5 μ long.

On rotten leaves of *Bambusa* and *Sporobolus*, India.

27. CHAETOPHIOPHOMA Speg., *An. Mus. Nac. Buenos Aires* **20**: 388 (1910).
Sp. typ.: *C. trematis* Speg.

Mycelium superficial, branched, septate, closely appressed to the substrate, fibrillose, pale brown, with alternate obovate hyphopodia. *Conidiomata* pycnidial, superficial, separate, ± globose, pale brown, unilocular, thin-walled, membranous, walls 1 cell thick, of large irregular-celled textura angularis. *Ostiole* absent, dehiscence by the breakdown of the upper wall. *Conidiophores* hyaline, branched at the base and above, septate, smooth, with acropleurogenous conidia, formed from the inner cells of the pycnidial wall. *Conidiogenous cells* enteroblastic, phialidic, determinate, integrated or discrete, cylindrical, hyaline, smooth, terminal or lateral just below septa, collarette and channel minute. *Conidia* hyaline, falcate, often with 2 guttules, thin-walled, aseptate, smooth.

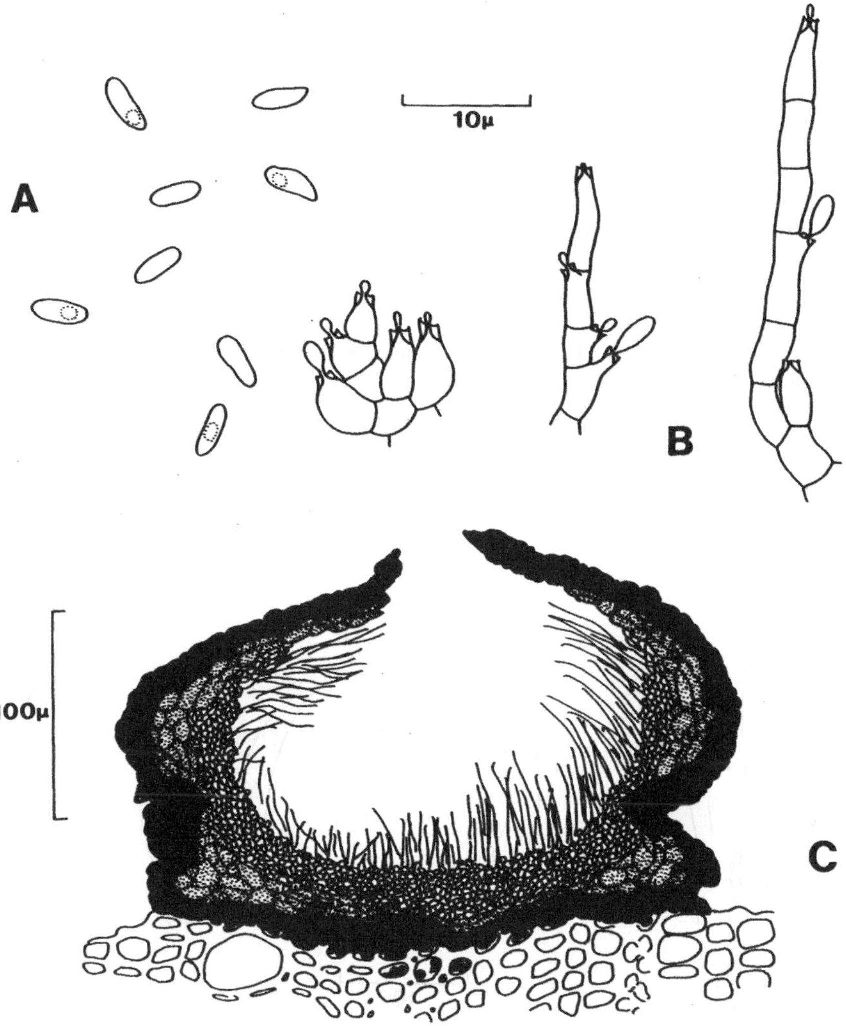

Fig. 235. *Dendrodomus annullatus*, A, conidia; B, conidiophores; C, vertical section of a conidioma.

The genus is monotypic.

C. TREMATIS Speg., *ibid.*: 388 (1910). (Fig. 237)

Pycnidia 50-75 μ diam. Conidiophores up to 25 μ long × 2-3 μ wide. Conidia 16-24 × 2 μ.

On leaves of *Trema micrantha*, Argentina (IMI 196609, holotype).

28. COLEOPHOMA Höhn., *Sber. Akad. Wiss. Wien* **116**: 637 (1907).
Coleonaema Höhn., *Mitt. Bot. Inst. tech. Hochsch. Wien* **1**: 95 (1924).
Bactropycnis Höhn., *Hedwigia* **62**: 65 (1920).
Rhabdostromina Died., *Annls mycol.* **19**: 297 (1921).
Sp. typ.: *C. crateriformis* (Dur. & Mont.) Höhn. (syn. *Septoria crateriformis* Dur. & Mont.).

Mycelium immersed, branched, septate, hyaline to pale brown. *Conidiomata* pycnidial, separate, globose or flattened at the base, black, immersed, unilocular, thick-walled, composed of brown, thick-walled textura angularis; lower wall often thicker. *Ostiole* single, circular, not papillate. *Conidiophores* pale brown at the base, hyaline above, branched, septate,

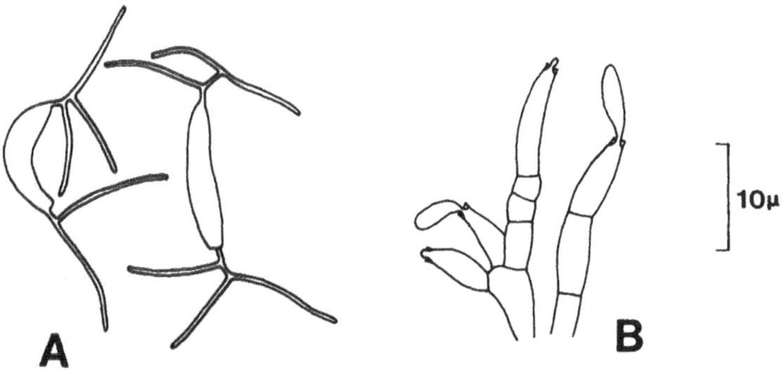

Fig. 236. *Brycekendrickia indica*, A, conidia; B, conidiophores and developing conidia.

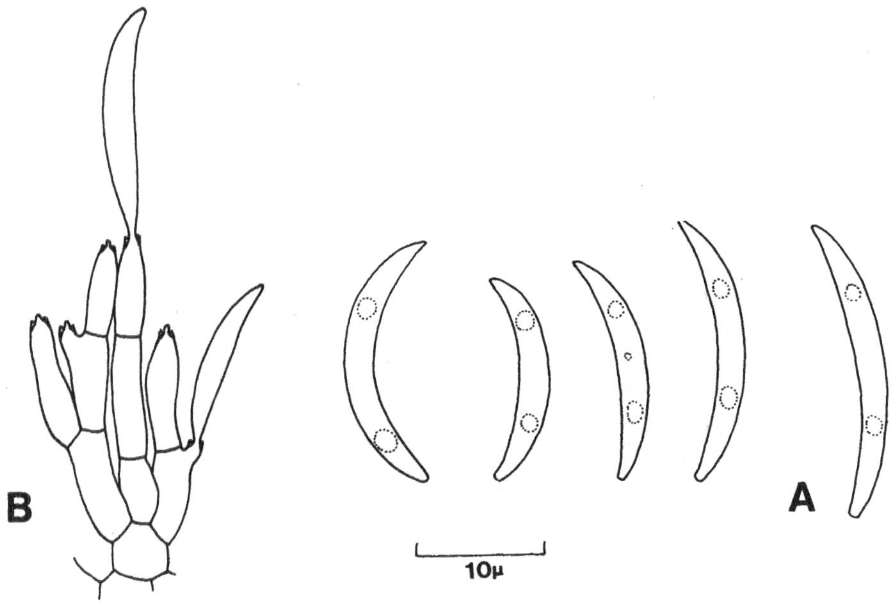

Fig. 237. *Chaetophiophoma trematis*, A, conidia; B, conidiophores and developing conidia.

smooth, formed from the inner cells of the pycnidial wall, sometimes confined to the base. *Conidiogenous cells* enteroblastic, phialidic, determinate, discrete and ampulliform to lageniform, or integrated and then ± cylindrical, hyaline, smooth, channel and collarette minute. *Conidia* hyaline, aseptate, cylindrical, apex obtuse, base acute, thin-walled, smooth, guttulate, straight. *Paraphyses* hyaline, septate at the base, formed amongst and on conidiophores, cylindrical or long clavate, collapsing.

About 20 species have been described in *Coleophoma* but the regularity with which members of the genus occur on a variety of hosts suggests that there are only a few plurivorous species. These can be separated by conidial size and morphology as the collections on 17 hosts in Herb. IMI indicate.

KEY

1. Conidia 3-4.5 µ wide .... *C. oleae*
1. Conidia 2-3 µ wide 2
   2. Conidia 11-13 µ long .... *C. crateriformis*
   2. Conidia 12.5-18 µ long .... *C. empetri*
   2. Conidia 16.5-24 µ long .... *C. cylindrospora*

C. CRATERIFORMIS (Dur. & Mont.) Höhn., *ibid.*: 637 (1907). (Fig. 238B)
Septoria crateriformis Dur. & Mont., *Fl. Alger.* **1**: 590 (1846-49).
Ascospora crateriformis (Dur. & Mont.) Dur. & Mont. apud Mont., *Syll. Crypt.* 275 (1856).
Phoma crateriformis (Dur. & Mont.) Sacc., *Michelia* **2**: 90 (1880).
Macrophoma crateriformis (Dur. & Mont.) Berl. & Vogl., *Atti Soc. Venet.-Trent. Sc. Nat.* **10**: 199 (1886).
Septoria oleae var. *phillyreae* Mont., *Ann. Sci. nat.* 2 ser., **1**: 342 (1834).

Conidia with small guttules at each end, 11-13 × 2-2.5 µ.

On leaves of *Photinia serulata*, France (IMI 31026).

C. CYLINDROSPORA (Desm.) Höhn., *Ber. dtsch. Bot. Ges.* **37**: 14 (1919). (Fig. 238A, E, F)
Sphaeropsis cylindrospora Desm., *Ann. Sci. nat.* 3 ser., **11**: 277 (1849).
Phoma cylindrospora (Desm.) Sacc., *Michelia* **1**: 527 (1879).
Macrophoma cylindrospora (Desm.) Berl. & Vogl., *Atti Soc. Venet.-Trent. Sc. Nat.* **10**: 197 (1886).
Bactropycnis concentrica Höhn., *Hedwigia* **62**: 65 (1920).

Conidia with several large irregular or globose guttules, 16.5-24 × 2.5-3 µ.

On or from *Sedum, Ilex aquifolium, Laurus, Hedera helix, H. himalaica, H. canariensis, Juniperus, Malus, Picea* and *Mahonia*, U.K. and Pakistan (IMI 188550, 188577a, 188495a, 86894b, 86868c, 89645b, 175909, 168658a, 168655, 168372, 93768, 105565, 90116a, 107235, 104574, 188560a and holotype of *B. concentrica*, FH).

C. EMPETRI (Rostr.) Petrak, *Annls mycol.* **27**: 331 (1929). (Fig. 238C)
Septoria empetri Rostr., *Medd. Groenland* **3**: 573 (1888).
Rhabdostromina empetri (Rostr.) Died., *Annls mycol.* **19**: 296 (1921).
Sporonema oxycocci Shear, *Bull. Torrey bot. Cl.* **37**: 308 (1907).
Coleophoma rhododendri Syd., *Annls mycol.* **34**: 397 (1936).
? *Coleophoma ericae* Höhn., *Hedwigia* **62**: 62 (1920).

Conidia sparsely guttulate, usually 1 or more at each end, only occasionally small guttules throughout, 12.5-18 × 2-3 µ.

On or from *Empetrum nigrum, Ficus, Eucalyptus, Rhododendron* cf. *ponticum, Fraxinus, Betula, Lonicera periclymenum, Taxus baccata* and from the air, U.K., Germany, India (IMI 50748 (holotype *C. rhododendri*), 183999, 183998, 183992, 182350, 182338, 105243, 105244, 105245, 91145, 90495, 123649, 86864, 188571a 20055, 1027, 31034, 33463, 20055).

C. OLEAE (DC. ex Dur. & Mont.) Petrak & Syd., *Beih. rep. spec. nov. regn. veg.* **42**: 469 (1927). (Fig. 238D)
Sphaeria oleae DC., *Fl. franç.* **6**: 136 (1815).
Septoria oleae DC. ex Dur. & Mont., *Fl. Alger.* **1**: 590 (1846-49).
Diplodia oleae (DC. ex Dur. & Mont.) de Not., *Klotzsch Herb. myc.* 1056 (1846).
Hendersonia oleae (DC. ex Dur. & Mont.) Desm., *Pl. Crypt. du Nord Fr.* ed 1. no. 1888.
Ascospora oleae (DC. ex Dur. & Mont.) Dur. & Mont. apud Mont., *Syll. Crypt.* 275 (1856).
Septoria olivae Pass. & Thüm., *Rev. Inst. Coimbra* **28**: 49 (1881).
Phoma oleae (DC. ex Dur. & Mont.) Sacc., *Michelia* **2**: 91 (1880).
Macrophoma oleae (DC. ex Dur. & Mont.) Berl. & Vogl., *Atti Soc. Venet.-Trent. Sc. nat.* **10**: 202 (1886).
Coleonaema oleae (DC. ex Dur. & Mont.) Höhn., *Mitt. Bot. Inst. tech. Hochsch. Wien* **1**: 95 (1924).
Macrophoma nerii Jaap, *Annls mycol.* **14**: 31 (1916).

Conidia with a few relatively large guttules, 15.5-22.5 × 3-4.5 µ.

On *Olea europaea* and *O. cuspidata*, and from indet. host, Crete, Israel, India, Pakistan, Calif., U.S.A. (IMI 33462, 171851, 171849, 154335, 142028, 142455, 133457, 22109, 142010, 151753, 186804).

29. LEPTODOTHIORELLA Höhn., *Mykol. Unters.* **1**: 328 (1923).
? *Leptasteromella* Petrak, *Sydowia* **20**: 235 (1968).
Leptodothiorella van der Aa, *Stud. Mycol., Baarn* **5**: 13 (1973).
Original spp.: None listed.
Type sp.: *L. concinna* Syd., first species ascribed to the genus.
Illustration: van der Aa (1973).

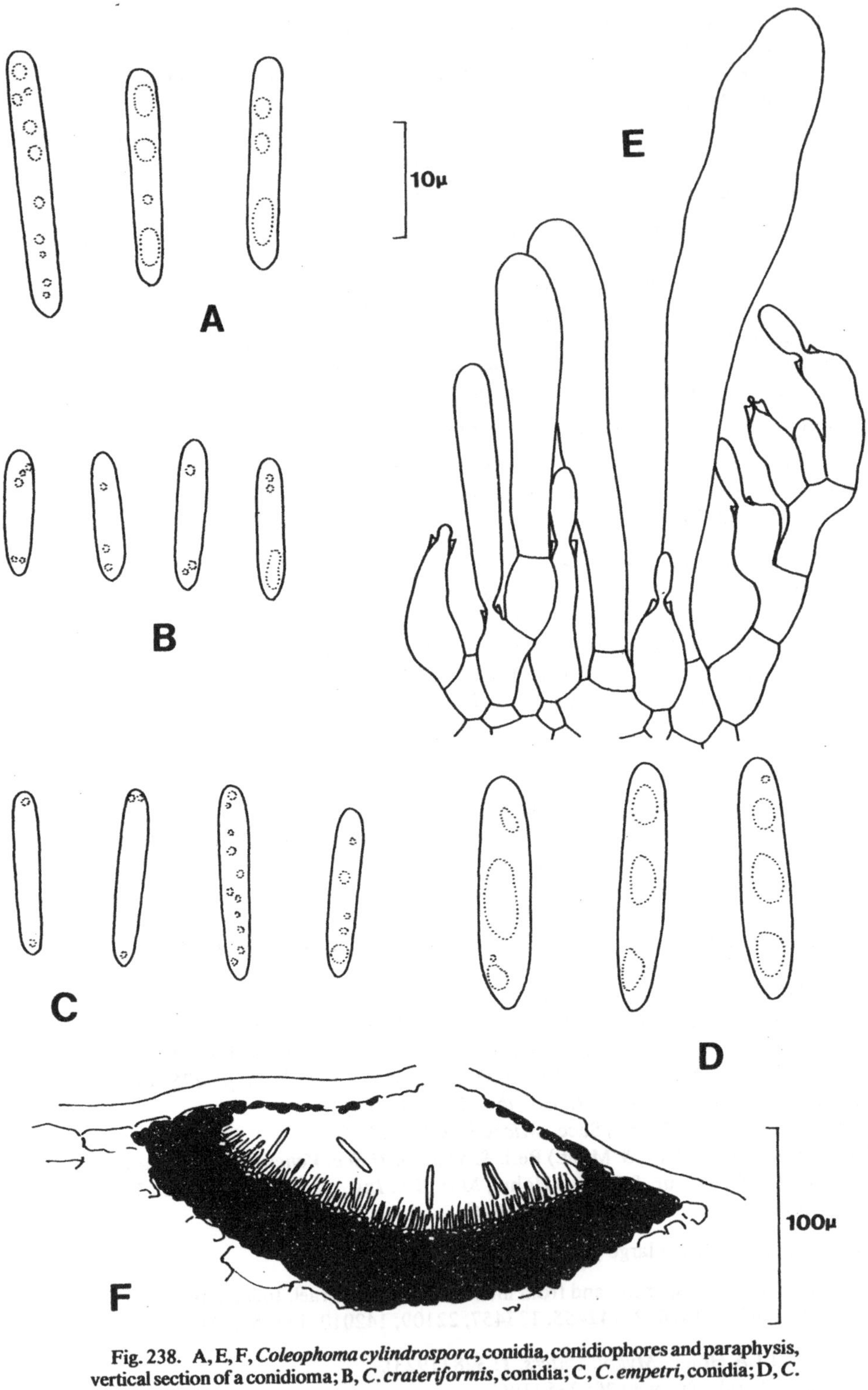

Fig. 238. A, E, F, *Coleophoma cylindrospora*, conidia, conidiophores and paraphysis, vertical section of a conidioma; B, *C. crateriformis*, conidia; C, *C. empetri*, conidia; D, *C. oleae*, conidia.

Mycelium immersed, branched, septate, brown or pale brown. *Conidiomata* pycnidial, globose to subglobose, dark brown to black, separate or aggregated, immersed, thick-walled; wall of hyaline to pale or dark brown, thick-walled textura angularis, becoming more hyaline and thinner-walled towards the conidiogenous region. *Ostiole* single, circular, central. *Conidiophores* hyaline, septate, sparingly branched, above and from the base, smooth, formed from the inner cells of the conidiomatal wall. *Conidiogenous cells* enteroblastic, phialidic, determinate, discrete or integrated, channel and collarette minute, periclinal thickening very distinct. *Conidia* hyaline, aseptate, smooth, guttulate at each end, straight, cylindrical or slightly swollen towards each end.

Leptodothiorella accommodates the imperfect spermatial states of the ascomycetes *Botryosphaeria* and *Guignardia* (van der Aa, 1973; Punithalingam, 1974a). Although many *Leptodothiorella* states are known for *Guignardia* species and various *Phyllosticta* species, less than ten have received formal binomials.

Van der Aa (1973) interpreted spermatia as blastoconidia developing in basipetal succession and separating from the spermatiogenous cells by a septum. Between two and three conidia were illustrated in each chain. My observations do not confirm this opinion. The conidiogenous cells are distinct phialides with a very characteristic and easily discernible apical structure. The formation of successive conidia through the phialidic aperture gives rise to short chains of cylindrical spores.

Van der Aa (1973) and Punithalingam (1974a) have provided accounts of individual *Guignardia* and *Phyllosticta* species together with their *Leptodothiorella* states, and Sutton (1977a) has summarized the nomenclatural status of *Phyllosticta* and *Phyllostictina*, the latter being the generic name under which the macroconidial states of *Guignardia* species had been known prior to the work of van der Aa (1973).

LEPTODOTHIORELLA state of GUIGNARDIA SANSEVIERIAE Punithalingam, *Mycol. Pap.* 136: 45 (1974). (Fig. 239)

Conidiomata up to 250 µ diam. Conidiogenous cells cylindrical to lageniform, 4.5-8 × 3-4 µ. Conidia 7-9 × 2 µ.

Causing lesions on *Sansevieria* sp., U.K. (IMI 147696, holotype, 139282).

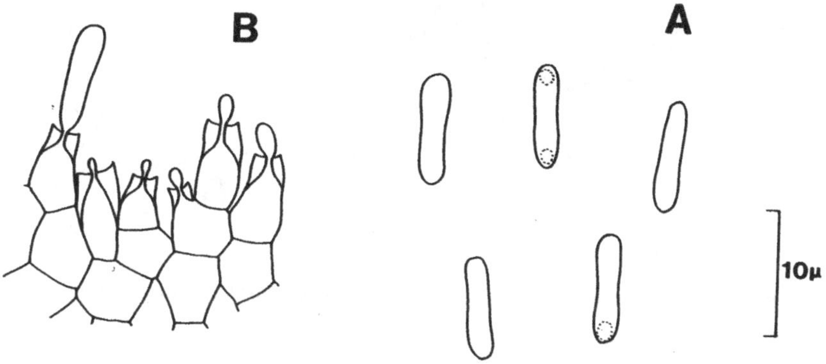

Fig. 239. *Leptodothiorella* state of *Guignardia sansevieriae*, A, conidia; B, conidiophores and developing conidia.

30. ASTEROMELLA Pass. & Thüm., de Thümen *Mycotheca universalis* 1689 (1880).

Stictochorella Höhn., *Ber. dtsch. Bot. Ges.* **35**: 253 (1917).
Stictochorellina Petrak, *Annls mycol.* **20**: 337 (1922).
Apiosporella Speg., *An. Mus. Nac. Hist. Nat. Buenos Aires* **23**: 106 (1912), non Speg., 1910, nec Höhn., 1909.
Haplosporidium Speg., *An. Mus. Nac. Hist. Nat. Buenos Aires* **23**: 130 (1912).
Sp. typ.: *A. ovata* Thüm.
Illustrations: Higgins (1920, 1929, 1936); Dring (1961).

Mycelium immersed, branched, septate, hyaline or pale brown. *Conidiomata* pycnidial, globose, separate or more frequently aggregated, dark brown, immersed, unilocular, thick-walled, wall of brown, ± thick walled textura angularis. *Ostiole* central, circular, ± papillate. *Conidiophores* hyaline, smooth, 1-3 septate, tapered at the apex, branched only at the base, formed from the inner cells of the pycnidial wall. *Conidiogenous cells* enteroblastic, phialidic, integrated or less often discrete, determinate, hyaline, apertures apical or lateral on short branches produced immediately below transverse septa, collarette and channel minute. *Conidia* hyaline, aseptate, thin-walled, eguttulate, cylindrical to oval.

Asteromella species are generally regarded as the microconidial or spermogonial states of *Mycosphaerella* species. Since Higgins (1920, 1929, 1936) established the connections between perfect and imperfect states of *Mycosphaerella* a number of later authors have extended our knowledge of additional correlations or described more species (Arx, 1949; Rupprecht, 1959; Boerema & Kesteren, 1963; Batista, Peres & Maia, 1960). However, separation of the imperfect taxa is usually on a host basis rather than on morphological differences. So far there are more than 140 taxa described in the genus.

A. OVATA Thüm., *ibid.*: 1689 (1880). (Fig. 240)

Conidiomata up to 70 μ diam. Conidiophores 4-15 × 3-4 μ. Conidia 2.5-3 × 1 μ.

Hypophyllous on living leaves of *Menispermum canadense*, Italy (IMI 199822, isotype of *A. ovata* ex K).

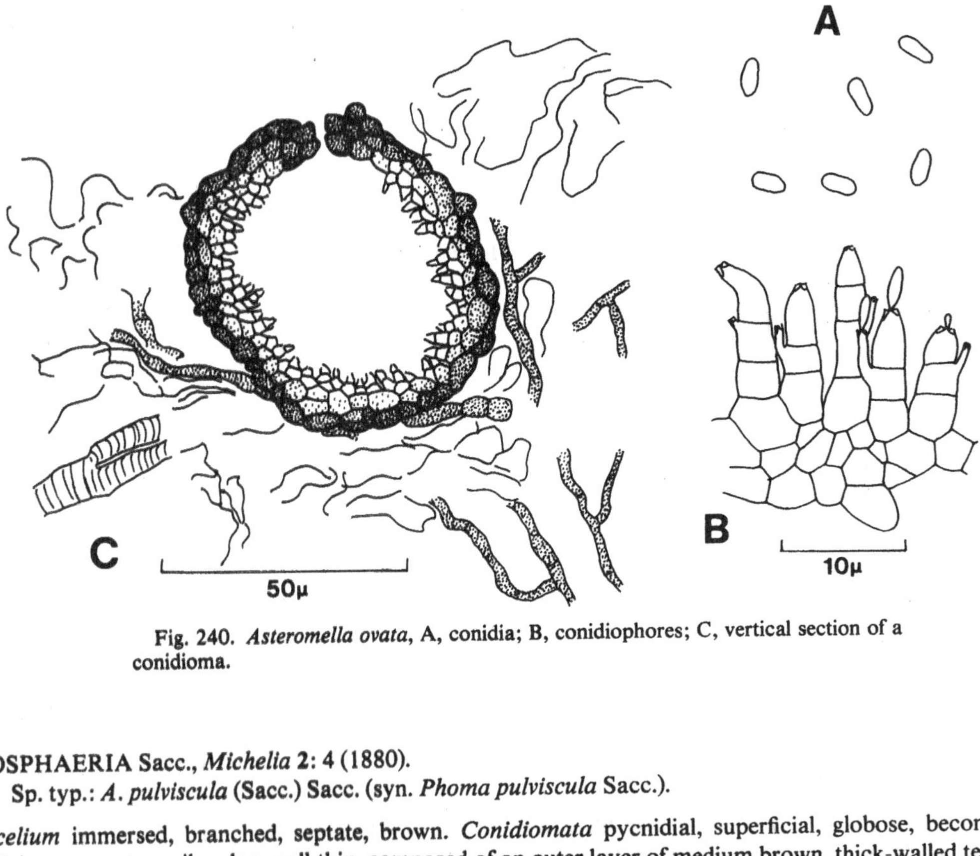

Fig. 240. *Asteromella ovata*, A, conidia; B, conidiophores; C, vertical section of a conidioma.

31. APOSPHAERIA Sacc., *Michelia* 2: 4 (1880).
Sp. typ.: *A. pulviscula* (Sacc.) Sacc. (syn. *Phoma pulviscula* Sacc.).

Mycelium immersed, branched, septate, brown. *Conidiomata* pycnidial, superficial, globose, becoming collabent, black, shining, separate, unilocular; wall thin, composed of an outer layer of medium brown, thick-walled textura angularis and an inner layer of smaller-celled, thin-walled hyaline tissue. *Ostiole* single, circular, central, depressed. *Conidiophores* hyaline, branched at the base, septate above, short, cylindrical, formed from the inner cells of the pycnidial wall. *Conidiogenous cells* enteroblastic, phialidic, integrated or discrete, determinate, hyaline, smooth, apertures apical or lateral just below septa, collarette occasionally prominent, channel minute, periclinal wall thickened. *Conidia* hyaline, aseptate, thin-walled, eguttulate, cylindrical or ellipsoidal, smooth.

Aposphaeria was introduced for the spermogonial states of *Melanomma* (Nits.) Fckl and *Cucurbitaria* Gray ex Grev., and Saccardo (1884) thought the type species was probably the conidial state of *Melanomma pulviscula* (Currey) Massee & Crossl. However Booth (1957) showed that *Chaetosphaeria pulviscula* (Currey) Booth developed a hyphomycete state assignable to *Menispora caesia* Preuss. Chesters (1938) however, has shown that the conidial state of *M. pulvis-*

pyrius (Pers.) Fckl can be assigned to *Aposphaeria agminalis* Sacc., and of *M. fuscidulum* Sacc. to *A. fuscidula* Sacc. Both agree quite satisfactorily with the generic concept exemplified by *A. pulviscula*. About 200 taxa have been placed in *Aposphaeria* and so far there has been no revisionary treatment of them.

A. PULVISCULA (Sacc.) Sacc., *ibid.*: 4 (1880). (Fig. 241)
Phoma pulviscula Sacc., *Michelia* **1**: 259 (1878).

Conidiomata up to 80 μ diam. Conidiophores up to 12.5 μ long × 2-5 μ wide. Conidiogenous cells 4-6.5 × 2-5.5 μ. Conidia 3-4.5 × 1.5 μ.

On bare wood of *Salix*, indet. wood, Italy, France (IMI 202557, isotype *P. pulviscula* ex K, 201770 ex PAD).

32. PLEUROPHOMOPSIS Petrak, *Annls mycol.* **22**: 156 (1924).
Original spp.: *P. salicicola* Petrak, *P. salicina* Petrak
Lectotype sp.: *P. salicicola* Petrak, fide Clements & Shear (1931).

Mycelium immersed, branched, septate, hyaline to pale brown. *Conidiomata* pycnidial, superficial, separate, globose, dark brown; lateral and upper walls thick, composed of dark brown, thick-walled textura angularis becoming hyaline towards the inner conidiogenous region; basal wall hyaline to pale brown, of thin-walled textura angularis. *Ostiole* central, circular, scarcely papillate. *Conidiophores* hyaline, 1-2 septate, cylindrical, branched at the base, with acropleurogenous conidia, formed all round the locular cavity. *Conidiogenous cells* enteroblastic, phialidic, determinate, integrated or discrete, cylindrical, hyaline, smooth, loci terminal or on short lateral branches immediately below septa, collarette and channel minute, but periclinal wall thickened. *Conidia* hyaline, ± spherical, aseptate, smooth, thin-walled, often with a single guttule.

This generic description is based on *P. salicina* Petrak which is not the lectotype species. If it is subsequently shown that *P. salicina* and *P. salicicola* are not congeneric then this generic account will have to be modified. Apart from the two original species, there have been four subsequent additions, *P. lignicola* Petrak, *P. eucalypti* Petrak, *P. valentina* de Urries and *P. strictae* Ahmad, none of which have been examined.

P. SALICINA Petrak, *ibid.*: 156 (1924). (Fig. 242)

Conidiomata up to 400 μ diam. Conidiophores 9-16 × 2-2.5 μ. Conidia 2-3 μ diam.

On wood of *Salix* sp., Czechoslovakia (IMI 84821, isotype ex K).

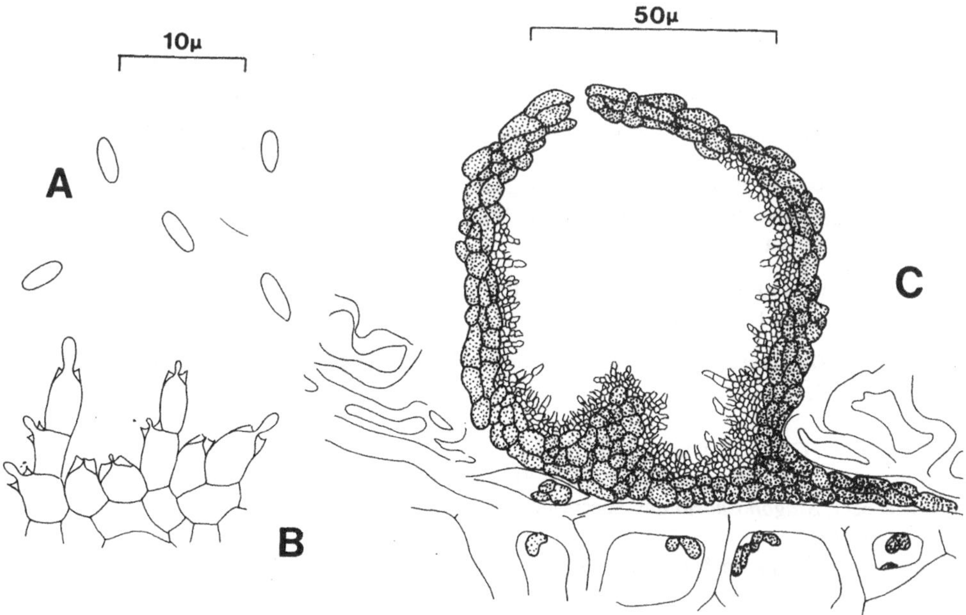

Fig. 241. *Aposphaeria pulviscula*, A, conidia; B, conidiophores; C, vertical section of a conidioma.

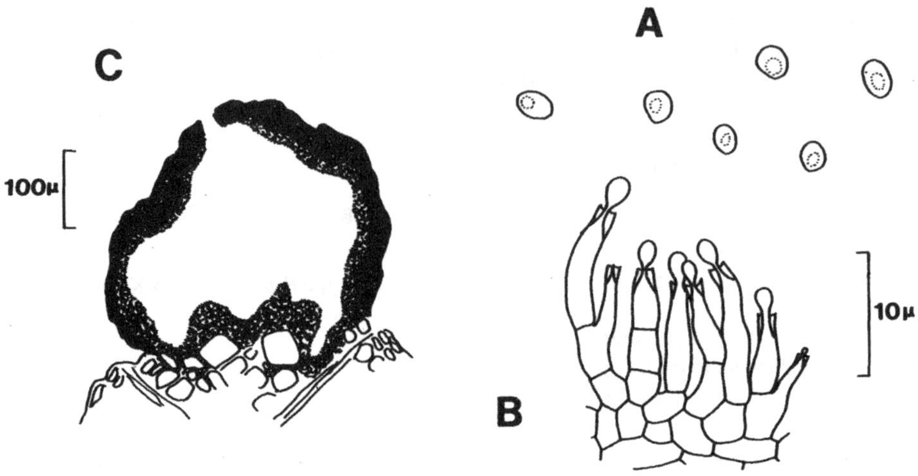

Fig. 242. *Pleurophomopsis salicina*, A, conidia; B, conidiophores and developing conidia; C, vertical section of a conidioma.

33. CHAETOSPHAERONEMA Moesz, *Bot. Közl.* **14**: 152 (1915).
Pseudophoma Höhn., *Sber. Akad. Wiss. Wien* **125**: 73 (1916).
Original spp.: *C. hispidulum* (Cda) Moesz, *C. herbarum* (Hollós) Moesz
Lectotype sp.: *C. hispidulum* (Cda) Moesz (syn. *Sphaeronaema hispidulum* Cda).

Mycelium immersed, branched, septate, hyaline. *Conidiomata* pycnidial, separate, rarely aggregated, globose, dark brown, immersed, epidermal to subepidermal, rostrate, unilocular, thick-walled; walls composed of thin-walled, pale brown textura prismatica, the outer layer of cells and the tissue comprising the beak dark brown and thicker-walled. *Setae* restricted to the beak, numerous, dark brown, septate, ± straight, acuminate. *Ostiole* single, circular, central. *Conidiophores* hyaline, branched irregularly and only sparingly, septate, smooth, formed from the inner cells of the pycnidial wall. *Conidiogenous cells* enteroblastic, phialidic, determinate, integrated or discrete, cylindrical, hyaline, smooth, channel and collarette minute, periclinal wall thickened. *Conidia* hyaline, medianly 1 euseptate, straight or slightly curved, slightly tapered to the base, apex and base obtuse, thin-walled, smooth, often guttulate.

Clements & Shear (1931) selected *C. hispidulum* as the lectotype species, and Petrak (1944) placed *C. herbarum* as a synonym of it. Petrak (1944) accepted three species in the genus, *C. hispidulum*, *C. galii* (Petrak) Petrak and *C. collivagum* (Petrak) Petrak but only the type has been studied. An extensive synonymy was proposed for *C. hispidulum*.

C. HISPIDULUM (Cda) Moesz, *ibid.*: 152 (1915). (Fig. 243)
Sphaeronaema hispidulum Cda, *Icon. Fung.* **4**: 39 (1840).
synonymy in Petrak (1944).

Conidiomata up to 450 µ diam.; setae up to 100 µ long; beak up to 200 µ long. Conidiogenous cells 9.5-11.5 × 2.5-3 µ. Conidia 12-13 × 2 µ.

On dead stems of *Dictamnus albus*, *Adonis vernalis*, *Echium vulgare* and from the air, Australia, India (IMI 30308, 94352, 33460, 182342).

34. ASCOCHYTA Lib., *Pl. Crypt. Ard.* fasc. 1: 8 (1830).
Ascoxyta Lib., *Mem. Soc. Sci. Agric. Lille* 1829-30: 175 (1831).
Ascochyta Lib. sect. *Stagonosporioides* Zherbele, *Trudy vses Inst. Zashch. Rast.* **29**: 20 (1971).
Sp. typ.: *A. pisi* Lib.
Illustrations: Punithalingam & Holliday (1972); Boerema & Bollen (1975).

Mycelium immersed, branched, septate, hyaline to pale brown. *Conidiomata* pycnidial, amphigenous, separate, globose, brown, immersed, unilocular, thin-walled, of textura angularis. *Ostiole* central, circular, slightly papillate.

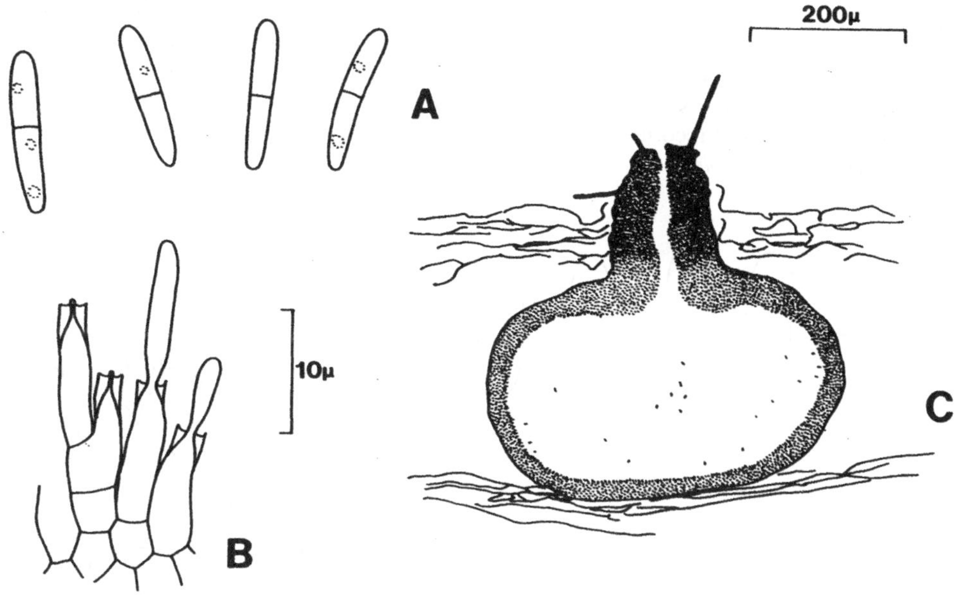

Fig. 243. *Chaetosphaeronema hispidulum*, A, conidia; B, conidiophores and developing conidia; C, vertical section of a conidioma.

Conidiophores absent. *Conidiogenous cells* enteroblastic, phialidic, determinate, discrete, doliiform to lageniform, hyaline, smooth, formed from the inner cells of the pycnidial wall. *Conidia* hyaline, medianly 1 septate, continuous or constricted, thin-walled, smooth, ± guttulate, cylindrical to irregular.

The type species has been fully described and illustrated by Punithalingam & Holliday (1972) and a detailed study of its conidiogenesis has been reported by Boerema & Bollen (1975). Melnik (1977) has provided the only modern compilation of the genus since the Saccardoan era in which 328 species are accepted as occurring in the U.S.S.R. Punithalingam (1979) has revised the species in Gramineae, but apart from this account there is no satisfactory comprehensive treatment of the genus or groups of species within it.

Several hundred binomials have been proposed for taxa in *Ascochyta* and it is clear that detailed revisionary work is required, for many species have been described simply because they have been found on hosts for which previous reports of *Ascochyta* infection are lacking. There is considerable variation in conidial shape, septation and size, and it is conceivable that a taxonomy based on host families rather than host genera or species may provide the circumscription by which some order may evolve out of the present apparent confusion.

Boerema & Bollen (1975) have attempted to separate *Ascochyta* from *Phoma* Sacc. using electron microscopy data in combination with established taxonomic criteria. The major distinction is phialidic conidiogenesis in *Phoma* in contrast to annellidic conidiogenesis in *Ascochyta*. In addition, when septa occur in *Phoma* they are secondary, whereas in *Ascochyta* they were interpreted as distoseptate. These concepts have been fully discussed by Punithalingam (1979) who rejects the septa in *Ascochyta* as distosepta. As for conidiogenesis, although there may be a fundamental distinction between monopolar repetitive budding (phialidic) development in *Phoma* and secession of successive conidia at the same level in *Ascochyta* (annellidic) as envisaged by Boerema & Bollen (1975), for practical purposes, by light microscopy the two types of conidiogenous cell (if they are indeed distinct) cannot be distinguished. For the purposes of identification therefore, such conidiogenous cells are referred to as phialides.

A. PISI Lib., *Pl. Crypt. Ard. fasc.* 1: 59 (1830). (Fig. 244)
Sphaeria concava Berk., *Annls Nat. Hist.*, ser. 1, **6**: 363 (1841).
Ascochyta pisicola (Berk.) Sacc., *Syll. fung.* **3**: 397 (1884).
Depazea pisicola Berk. apud Currey, *Trans. Linn. Soc. Lond.* **22**: 334 (1859).
Diplodina macrophomoides Camara, *An. Inst. Super. Agron. Lisboa* **3**: 111 (1930).
Ascochyta orobi Sacc., *Michelia* **1**: 161 (1878).
Additional varietal synonyms given by Melnik (1977).

Conidiomata 100-250 µ diam. Conidiogenous cells 7-10 × 5-8 µ. Conidia 1-2 septate, 11.5-14.5 × 3.5-4.5 µ.

Causing leaf, stem and pod spot on *Lathyrus ochrus, L. gorgoni, L. sativus, L. odoratus, Medicago arabica, Phaseolus vulgaris, Pisum sativum, Trifolium subterraneum, Vicia sativa, V. villosa, V. faba*, G.B., Belgium, Romania, Czechoslovakia, Poland, Italy, Eire, Cyprus, Egypt, Libya, Mauritius, N. Iran, India, Malaysia, Burma, New Guinea, Hong Kong, Nepal, Malawi, Kenya, Uganda, Ethiopia, Zambia, Australia, Jamaica, Canada, Bolivia. Additional records in Punithalingam & Holliday (1972). 72 Collections in IMI, including IMI 163629 ex K, isotype of *A. pisi*.

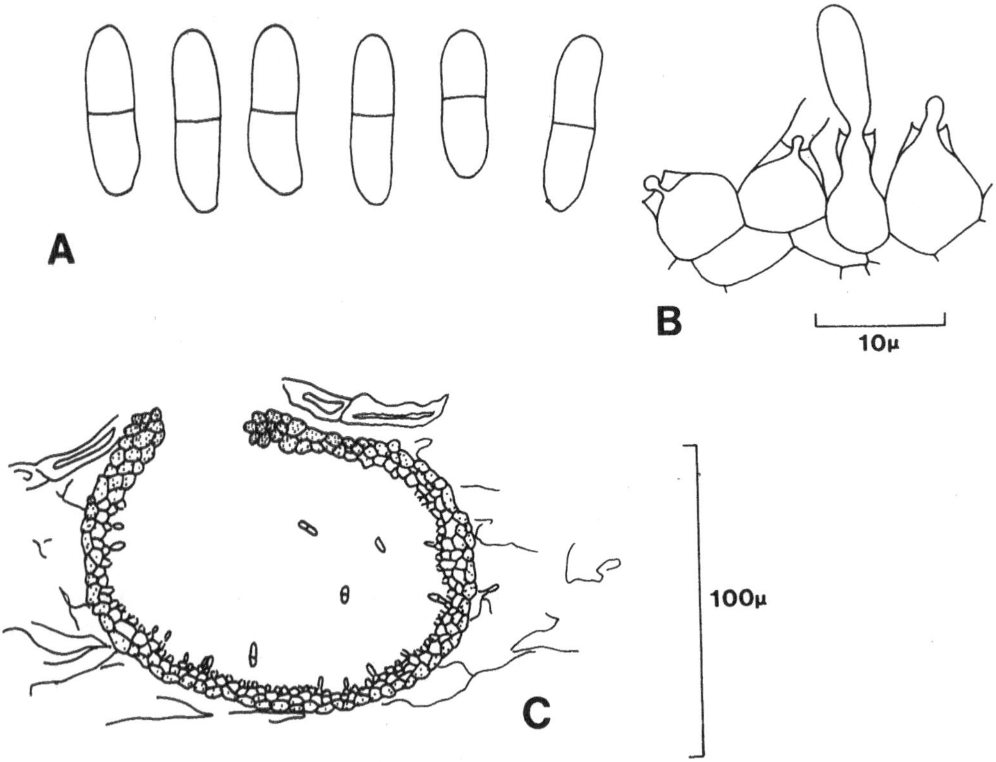

Fig. 244. *Ascochyta pisi*, A, conidia; B, conidiogenous cells and developing conidia; C, vertical section of a conidioma.

35. DIDYMOCHAETA Sacc. & Ell., *Bull. Torrey bot. Cl.* **25**: 510 (1898).
Sp. typ.: *D. fraserae* (Ell. & Ev.) Sutton (syn. *Ascochyta fraserae* Ell. & Ev., *D. americana* Sacc. & Ell.).

Mycelium immersed, brown, branched, septate. *Conidiomata* pycnidial, finally becoming superficial, separate or aggregated, globose, dark brown, unilocular, thick-walled, outer layer of dark brown, thick-walled textura globulosa, often pruinose, becoming smaller, thinner-walled and paler towards the locule, finally small-celled hyaline textura angularis. *Ostiole* central, circular, ± papillate. *Conidiophores* absent. *Conidiogenous cells* enteroblastic, phialidic, discrete, determinate, doliiform, hyaline, smooth, with minute collarette but prominent channel and thickened periclinal walls, formed from inner cells of the pycnidial wall. *Conidia* cylindrical, 1(-2) euseptate, hyaline, thin-walled, obtuse, occasionally slightly constricted at the septum, straight.

Although Sydow & Sydow (1900) transferred the name of the type species to *Vermiculariella* Oud., Petrak (1944) found that *V. elymi* Oud., upon which the latter genus was based, consisted of several discordant elements.

Apart from the type species, three other taxa have been described in *Didymochaeta, D. opuntiicola* Speg., *D. columbiana* Fairm. and *D. australiana* McAlp. Without determining the modes of conidiogenesis it is not possible to say whether they are correctly placed. *D. opuntiicola* was reported with conidiophores 10-15 × 1.5-2 µ which tends to indicate that this species is not congeneric with *D. fraserae*.

D. FRASERAE (Ell. & Ev.) Sutton, *Mycol. Pap.* 141: 65 (1977). (Fig. 245)
Ascochyta fraserae Ell. & Ev., *Bull. Torrey bot. Cl.* **24**: 289 (1897).
Diplodina fraserae (Ell. & Ev.) Tracy & Earle in Green, *Plantae Bakerianae* **1**: 36 (1901).
Didymochaeta americana Sacc. & Ell., *Bull. Torrey bot. Cl.* **25**: 510 (1898).
Vermiculariella americana (Sacc. & Ell.) Syd. & Syd., *Hedwigia Beibl.* **39**: (3) (1900).

Conidiomata up to 250 µ diam. Conidiogenous cells 7-10 × 4-7 µ. Conidia 10.5-16.5 × 3.5-4.5 µ.

On dead stems of *Frasera speciosa*, Colo., U.S.A. (IMI 191331, isotype *A. fraserae*, 191332, 191333 ex NY).

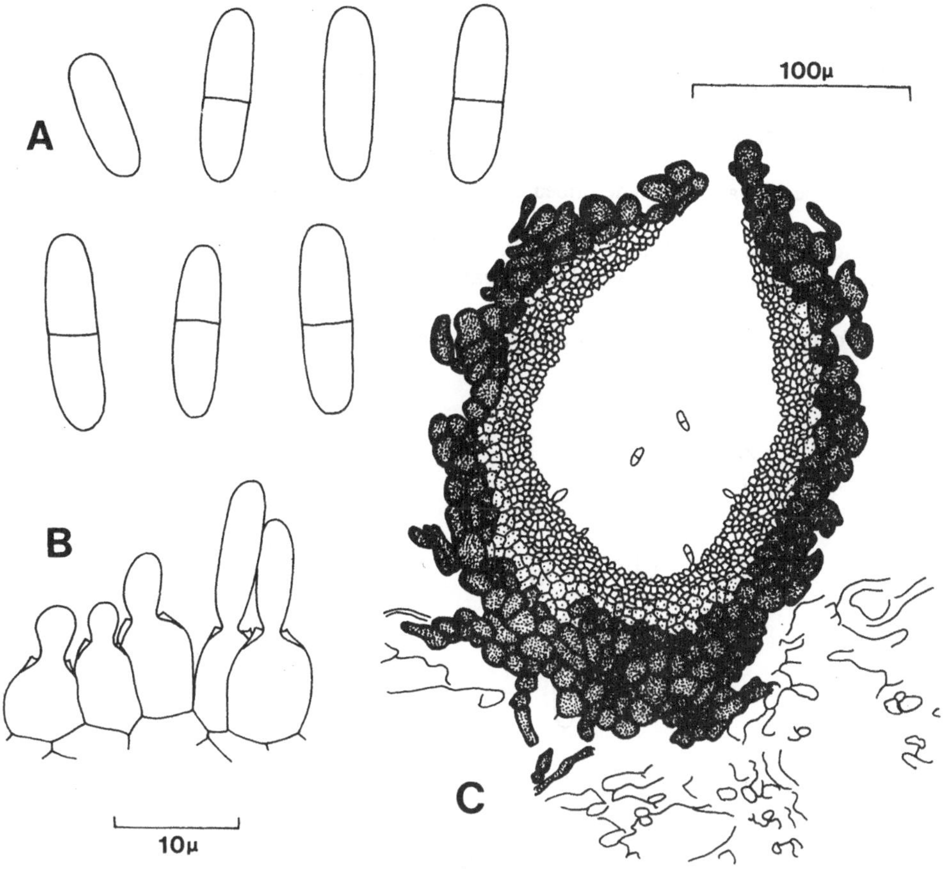

Fig. 245. *Didymochaeta fraserae*, A, conidia; B, conidiogenous cells and developing conidia; C, vertical section of a conidioma.

36. CLYPEOPYCNIS Petrak, *Annls mycol.* **23**: 76 (1925).
Sp. typ.: *C. aeruginascens* Petrak

Mycelium immersed, branched, septate, hyaline. *Conidiomata* pycnidial, separate, globose, dark brown, immersed, epidermal to subepidermal, unilocular, thick-walled, composed of a basal region of small-celled, thin-walled, hyaline textura angularis, the lateral walls becoming larger-celled and pale brown, surmounted by dark brown, thick-walled ostiolar tissue. *Ostiole* central, circular, papillate. *Conidiophores* hyaline, branched irregularly, septate, smooth, hyaline, formed from the inner cells of the pycnidial wall. *Conidiogenous cells* enteroblastic, phialidic, determinate, integrated or discrete, ampulliform to doliiform, hyaline, smooth, channel and collarette minute. *Conidia* hyaline, medianly 1-euseptate or aseptate, straight, fusiform, apex and base obtuse, thin-walled, smooth, often guttulate.

The genus is monotypic.

C. AERUGINASCENS Petrak, *ibid.*: 86 (1925). (Fig. 246)

Conidiomata up to 110 μ wide. Conidiogenous cells up to 5.5 × 3 μ. Conidia 8-11.5 × 2 μ.

On dead stems of *Ribes missouriensis*, S. Dak., U.S.A. (IMI 20956).

37. CHAETOCONIS Clements, *Genera of fungi*: 125 (1909).
Amphorula Grove, *J. Bot. Lond.* **60**: 82 (1922).
Sp. typ.: *C. polygoni* (Ell. & Ev.) Clements (syn. *Kellermania polygoni* Ell. & Ev.).

Mycelium immersed, branched, septate, brown. *Conidiomata* pycnidial, depressed, dark brown, immersed, unilocular or multilocular, thick-walled; wall several layers thick, of compact outer dark brown and inner, paler, thin-walled textura angularis. *Ostioles* 1 or more, circular, papillate. *Conidiophores* hyaline, branched, septate, smooth, with acropleurogenous conidia, formed from the inner pycnidial wall cells. *Conidiogenous cells* enteroblastic, phialidic, determinate, integrated, cylindrical, hyaline, smooth, channel and collarette minute. *Conidia* hyaline, 2-euseptate, continuous, base obtuse, apex prolonged into a filiform, cellular, unbranched appendage, thin-walled, smooth, guttulate, obclavate.

The genus and single species have been redescribed by Sutton (1968b) and Pirozynski & Morgan-Jones (1968).

C. POLYGONI (Ell. & Ev.) Clements, *ibid.*: 176 (1909). (Fig. 247)
synonymy in Sutton (1968), Pirozynski & Morgan-Jones (1968).

Teleomorphosis: *Ceriospora polygonacearum* (Petrak) Pirozynski & Morgan-Jones, *Trans. Br. mycol. Soc.* **51**: 189 (1968).

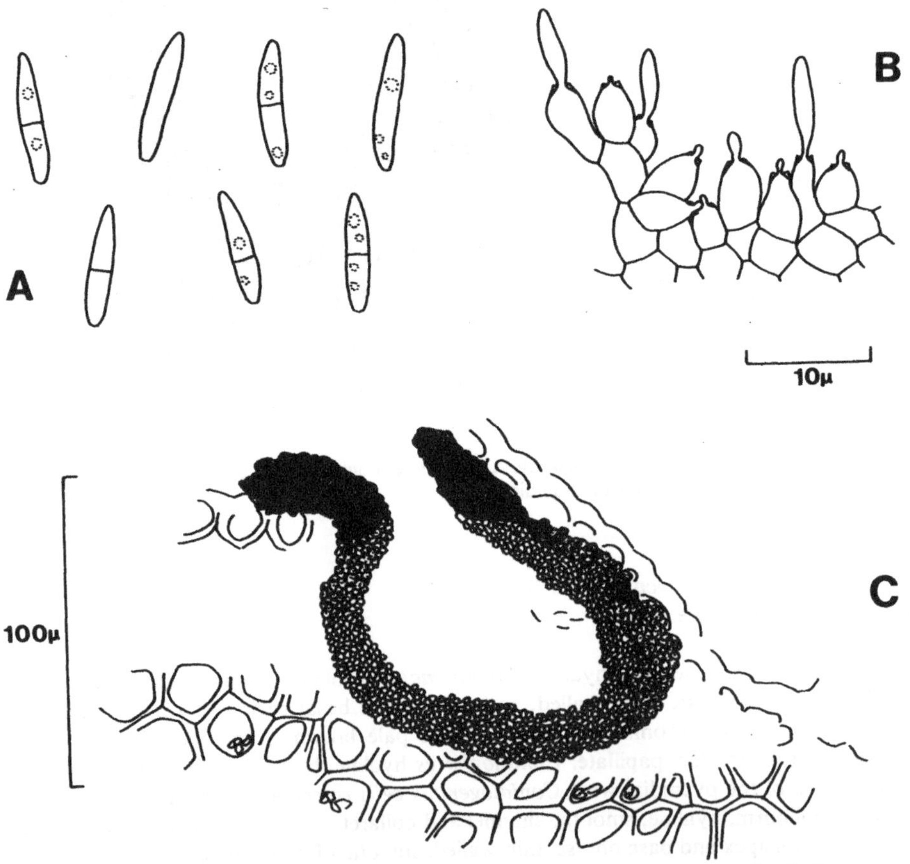

Fig. 246. *Clypeopycnis aeruginascens*, A, conidia; B, conidiophores and developing conidia; C, vertical section of a conidioma.

Conidiomata 250 µ diam., wall 20-30 µ thick. Conidiophores 12-25 × 2-3.5 µ. Conidia 20-25 × 3.5-4 µ (excluding apical cell); appendage 13-17 µ long.

On stems of *Polygonum* spp., *Rumex acetosa*, U.K., Sweden, Russia (IMI 120926 (holotype *Robillarda discosioides*), 124601 (holotype *Amphorula sachalinensis*), 16634, 107950, 110301).

38. PSEUDOROBILLARDA Morelet, *Bull. Soc. Sci. Nat. Archéol. Toulon Var* **175**: 5 (1968).
Pseudorobillarda Nag Raj, Morgan-Jones & Kendrick, *Can. J. Bot.* **50**: 862 (1972).
Sp. typ.: *P. phragmitis* (Cunnell) Morelet (syn. *Robillarda phragmitis* Cunnell).
Illustrations: Nag Raj, Morgan-Jones & Kendrick (1972); Cunnell (1958).

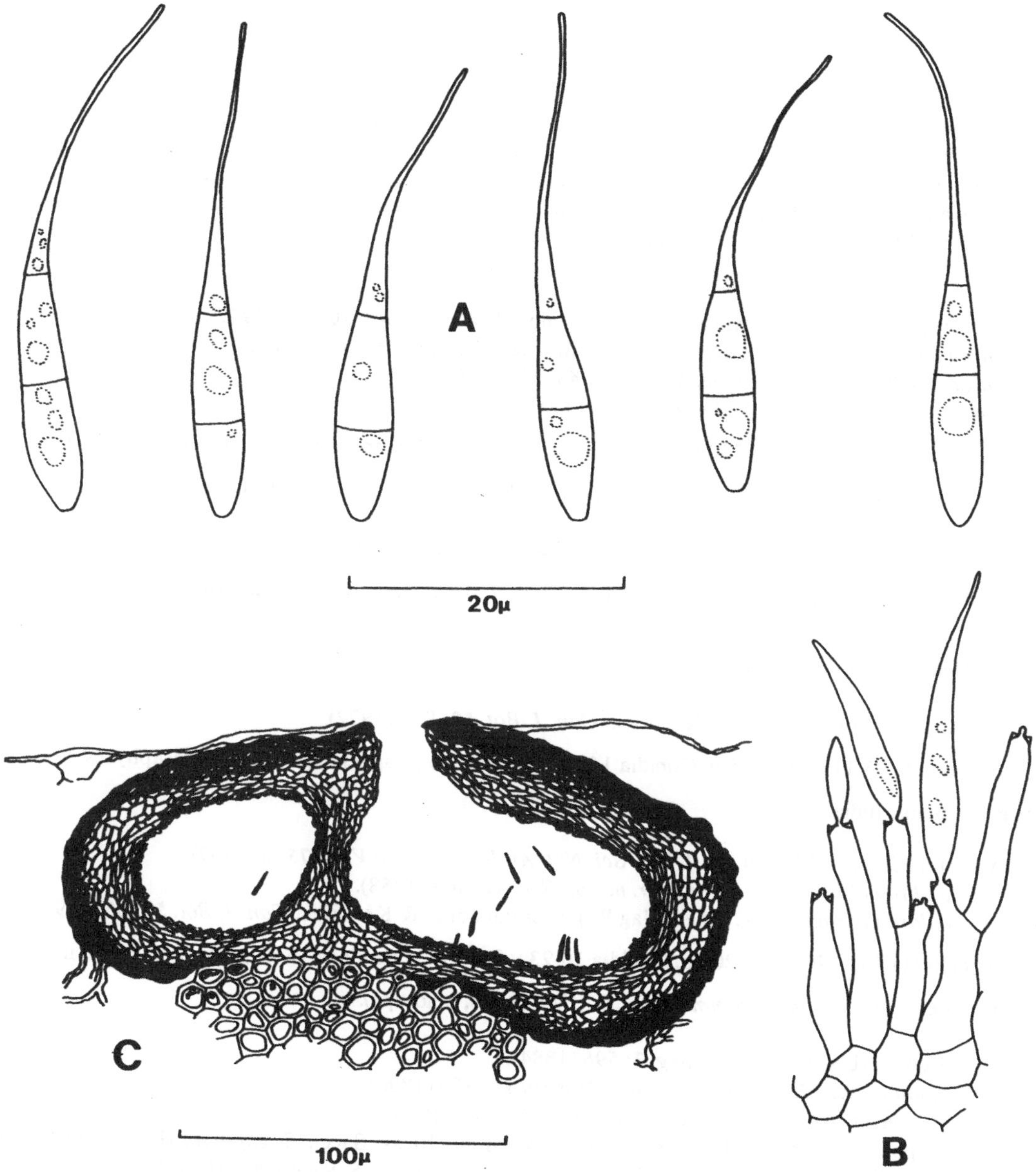

Fig. 247. *Chaetoconis polygoni*, A, conidia; B, conidiophores and developing conidia; C, vertical section of a conidioma.

Mycelium immersed, pale brown to hyaline, branched, septate. *Conidiomata* pycnidial, separate or aggregated, globose or flattened, brown, immersed, unilocular, thin-walled, of textura angularis. *Ostiole* central, circular, papillate. *Conidiophores* occasionally present, short, 1-2 septate, hyaline, branched at the base, formed from the inner cells of the pycnidial wall. *Conidiogenous cells* enteroblastic, phialidic, determinate, integrated or discrete, ampulliform to doliiform, hyaline, smooth, channel minute, collarette tubular, conspicuous. *Conidia* hyaline, 1-3 euseptate, constricted, apex obtuse to ± truncate, thin-walled, smooth, guttulate, fusiform, base ± acute with one extracellular flexuous appendage with 3 diverging branches, arising by splitting of the conidial sheath. *Paraphyses* filiform, simple or branched, hyaline, ± septate.

Following the treatment by Nag Raj, Morgan-Jones & Kendrick (1972) it was realised that *Pseudorobillarda* Nag Raj et al. was a later homonym of *Pseudorobillarda* Morelet (1968b). Consequently Nag Raj, Morgan-Jones & Kendrick (1973) placed their generic name into synonymy with that of Morelet and issued a revised list of the correct species names to adopt.

KEY

1. Conidia 1-3 septate .... *P. indica*
1. Conidia 1 septate 2
   2. Appendage less than 16 µ long .... *P. agrostis*
   2. Appendage more than 16 µ long 3
3. Conidia 17-22.5 (19) × 2.5-3.5 µ, eguttulate .... *P. bambusae*
3. Conidia 16-23 × 2.5-4 µ, guttulate .... *P. phragmitis*

P. AGROSTIS (Sprague) Nag Raj, Morgan-Jones & Kendrick, *Can. J. Bot.* **50**: 863 (1972).
Robillarda agrostis Sprague apud Sprague & W.B. Cke, *Mycologia* **31**: 47 (1939).
Robillarda muehlenbergiae Sprague, *Mycologia* **43**: 551 (1951).
Robillarda lathyri Rupprecht, *Sydowia* **11**: 433 (1957).

Conidiomata 260-360 × 140-230 µ. Conidia 13-20 (16.5) × 2-3 µ; appendage 10-16.5 µ. Paraphyses 28-52(34) µ.

On leaves of *Agrostis*, *Muhlenbergia*, and stems of *Lathyrus*, Ore., Calif., U.S.A., Germany.

P. BAMBUSAE Nag Raj, Morgan-Jones & Kendrick, *Can. J. Bot.* **50**: 864 (1972).

Conidiomata 190-460 × 85-160 µ. Conidia 17-22.5(19.5) × 2.5-3.5 µ; appendage 16.5-22.5 µ. Paraphyses 45-52 (50) µ.

On dead leaves of *Bambusa* sp., India.

P. INDICA Nag Raj, Morgan-Jones & Kendrick, *Can. J. Bot.* **50**: 865 (1972).

Conidiomata 95-115 × 65-105 µ. Conidia 14.5-21.5 × 3.5-5.5 µ; appendage 16-19 µ. Paraphyses 29-33 µ.

On indet. stems, India.

P. PHRAGMITIS (Cunnell) Morelet, *Bull. Soc. Sci. Nat. Archéol. Toulon Var* **175**: 6 (1969).
Robillarda phragmitis Cunnell, *Trans. Br. mycol. Soc.* **41**: 405 (1958).
Pseudorobillarda phragmitis (Cunnell) Nag Raj, Morgan-Jones & Kendrick, *Can. J. Bot.* **50**: 866 (1972).

Conidiomata 150-320 × 100-200 µ. Conidia 16-23 × 2.5-4 µ; appendage 15-23 µ. Paraphyses 58 µ.

On dead stems of *Phragmites communis*, U.K. (IMI 70768, holotype).

39. SPHAEROGRAPHIUM Sacc., *Syll. fung.* **3**: 596 (1884).
Pseudographium Jacz., *Nouv. Mem. soc. nat. Moscou* **15**: 370 (1898).
? *Cryptorhynchella* Höhn., *Sber. Akad. Wiss. Wien* **124**: 88 (1915).
Original spp.: *S. lonicerae* (Fckl) Sacc., *S. squarrosum* (Riess) Sacc., *S. capillare* (Ell. & Harkn.) Sacc., *S. hystricinum* (Ell.) Sacc., *S. echinatum* Sacc., *S. fraxini* (Pk) Sacc., *S. microperae* (Cke) Sacc.
Lectotype sp.: *S. lonicerae* (Fckl.) Sacc. = *S. squarrosom* (Riess) Sacc. (syn. *Sphaeronaema squarrosom* Riess), fide Höhnel (1915c), Clements & Shear (1931).

Mycelium immersed, branched, septate, hyaline to pale brown. *Conidiomata* superficial to semi-immersed, pycnidial, separate, rostrate, ± globose at the base, dark brown, unilocular, thick-walled; wall consisting of an outer layer of brown textura angularis and an inner region of textura oblita, cells becoming small and more like textura porrecta towards the apex. *Ostiole* single, circular, apical. *Conidiophores* hyaline, septate, filiform, branched mainly from the base, formed at the sides and the base of the pycnidial wall. *Conidiogenous cells* enteroblastic, phialidic, indeterminate, integrated, apertures produced immediately below transverse septa, hyaline, smooth, aperture, collarette and channel minute. *Conidia* hyaline, smooth, 3 septate (1-2 septate in immature conidia), eguttulate, fusiform, falcate, attenuated more towards the apices.

The generic synonymy has been discussed by Sutton (1977a). At least 17 species have been described in the genus but there has been no comprehensive revision so far. di Cosmo (1978) in a treatment of species placed in *Corniculariella* Karst., accepted *S. hystricinum* (Ell.) Sacc. as a synonym of *C. hystricina* (Ell.) di Cosmo.

S. LONICERAE (Fckl) Sacc., *ibid.*: 597 (1884). (Fig. 248)
Sphaeronaema lonicerae Fckl, *Jahrb. Nassau. Ver. Naturk.* **23-24**: 275 (1870).
Sphaeronaema squarrosum Riess, *Bot. Z.* 133 (1853).
Pseudographium squarrosum (Riess) Jacz., *Nouv. Mem. Soc. Imp. Nat. Moscou* **15**: 371 (1898).
Sphaerographium squarrosum (Riess) Sacc., *Syll. fung.* **3**: 597 (1884).

Conidiomata up to 400 µ high × 200 µ wide at the base. Conidiophores up to 40 µ long × 2-2.5 µ wide. Conidia 23-28.5 × 1.5-2 µ.

On dead twigs of *Lonicera xylostei*, Austria (IMI 12163).

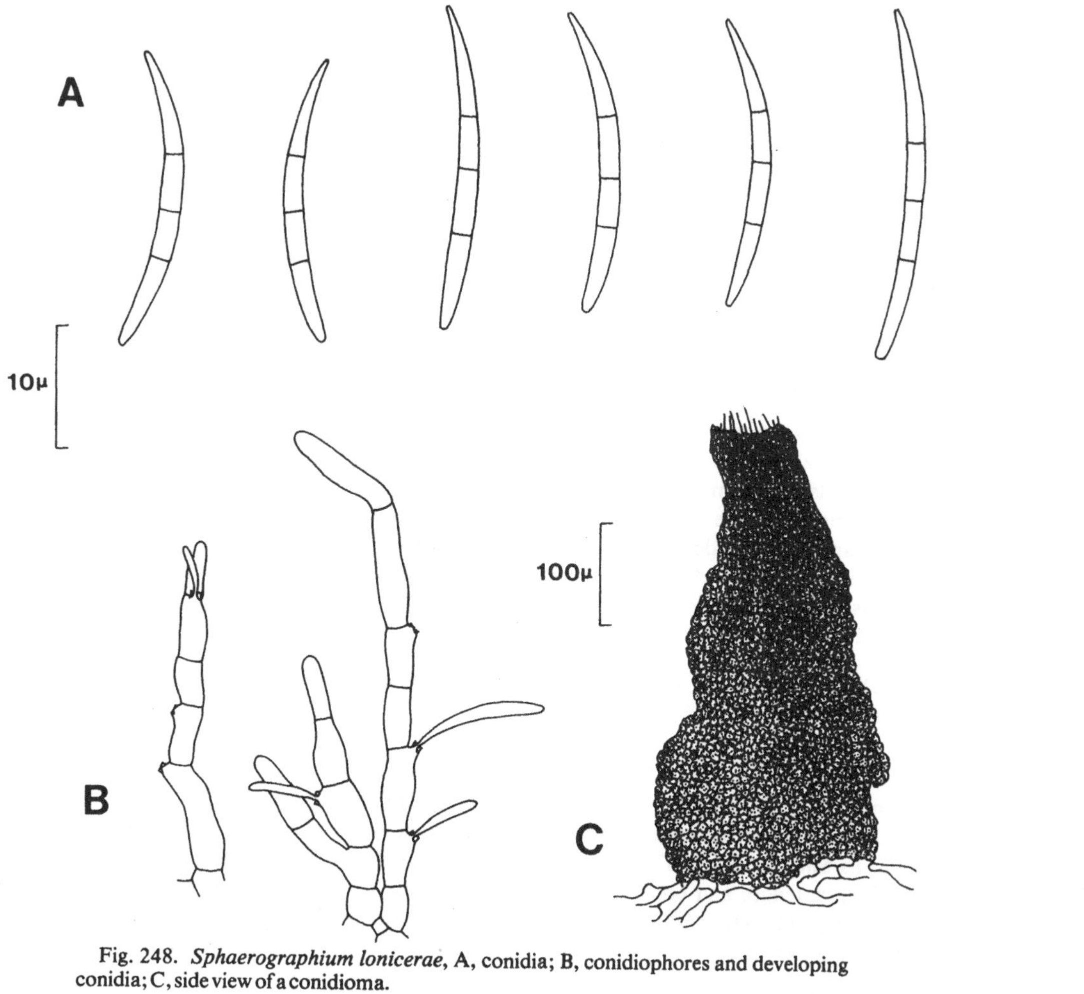

Fig. 248. *Sphaerographium lonicerae*, A, conidia; B, conidiophores and developing conidia; C, side view of a conidioma.

40. TOPOSPORA Fr., *Corpus Flor. Prov. Sueciae* I. *Fl. Scan.*: 347 (1835).
Mastomyces Mont., *Ann. Sci. nat.*, ser. 3, **10**: 134 (1848).
Clinterium Fr., *Summa Veg. Scand.* 418 (1849).
Sphaerocista Preuss, *Linnaea* **25**: 734 (1852).
Hypocenia Berk. & Curt., *Grevillea* **3**: 4 (1874).
Sirodiplospora Naumov, *Mater. Mikol. Fitop.* **1**: 22 (1915).
? *Sirexcipulina* Petrak, *Annls mycol.* **21**: 278 (1923).
Sp. typ.: *T. uberiformis* (Fr.) Fr. (syn. *Sphaeria uberiformis* Fr.).

Mycelium immersed, branched, septate, pale brown. *Conidiomata* pycnidial, initially immersed, later strongly erumpent and often apparently superficial, separate or aggregated, globose to obpyriform, black or very dark brown, unilocular with a thick wall; walls several cells thick, consisting of an outer layer of brown textura oblita and an inner layer of textura intricata giving rise to very thin-walled hyaline tissue on the inside. *Ostiole* single, circular, central, papillate. *Conidiophores* abundant, hyaline, branched at the base and above and often lying parallel to the main branch, cylindrical, smooth, formed from the inner cells of the pycnidial wall. *Conidiogenous cells* enteroblastic, phialidic, determinate, integrated, apertures terminal on branches, collarette conspicuous, periclinal thickening and channel minute. *Conidia* hyaline, 3 euseptate, smooth, straight, fusiform, apex obtuse, base truncate.

The generic synonymy of *Topospora* has been outlined by Sutton (1977a) under each separate generic name. In the seven genera, more than thirty separate species have been described. Several of these have been dealt with in the treatment of the teleomorphosis, *Godronia* Moug. & Lév., by Groves (1965) but apparently not based on type studies.

T. UBERIFORMIS (Fr.) Fr., *K. Vet. Akad. Handl.* 1848: 153 (1848). (Fig. 249)
Sphaeria uberiformis Fr., *Syst. Myc.* **2**: 491 (1823).
Mastomyces uberiformis (Fr.) Karst., *Acta Soc. Fauna Fl. fenn.* **6**: 34 (1890).
Mastomyces friesii Mont., *Ann. Sci. nat.*, ser. 3, **10**: 135 (1848).

Teleomorphosis: *Godronia uberiformis* Groves, *Can. J. Bot.* **43**: 1245 (1965).

Conidiomata up to 800 μ wide × 1350 μ high. Conidiophores up to 35 μ long × 2.5-3 μ wide. Conidia 17-22.5 × 3-3.5 μ.

On twigs of *Ribes floridum*, *R.* sp., U.K., N. Dak., U.S.A., Ont., Canada (IMI 22330, 26882, 29804, 29803, 29806, 29805).

41. MEGALOSEPTORIA Naumov, *Bolezn. Rast.* **14**: 144 (1925).
Sp. typ.: *M. mirabilis* Naumov

Mycelium immersed, branched, septate, brown. *Conidiomata* pycnidial, separate, globose, dark brown to black, superficial, sessile, often aggregated in groups, unilocular, thick-walled; wall several cells thick, of brown, thin-walled textura angularis, more darkly pigmented on the outside. *Ostiole* single, circular, slightly papillate. *Conidiophores* hyaline, branched and septate mainly at the base, smooth, straight or irregular, formed from the inner cells of the pycnidial wall. *Conidiogenous cells* enteroblastic, determinate, discrete or integrated, doliiform, ampulliform or irregularly cylindrical, hyaline, smooth, collarette evident, channel wide, periclinal thickening marked. *Conidia* hyaline to very pale brown, with several transverse eusepta, continuous, tapered near the obtuse apex and truncate base, thin-walled, smooth, cylindrical, straight or slightly curved, often with 2 guttules in each cell.

Petrak (1963b) and Shoemaker (1967) have provided modern accounts of the species and its relationship with *Cucurbitaria piceae* Borthwick.

M. MIRABILIS Naumov, *ibid.*: 144 (1925). (Fig. 250)

Conidiomata 400-600 μ diam.; ostiole 40-60 μ diam. Conidia 5-7 septate, 155-195 × 7-9 μ.

Examination of fresh material has shown that rather than each conidiogenous cell producing only a single conidium as Shoemaker (1967) has suggested, a succession is formed from a phialidic aperture. My measurements of septa, length and breadth also differ from those of Shoemaker who reported conidia as 31-46 septate and measuring 220-315 × 5-7 μ. Evidently the British material is somewhat less mature than the collections examined by Shoemaker.

On buds of *Picea abies*, *P. engelmannii*, U.K. (IMI 176627b, 176628).

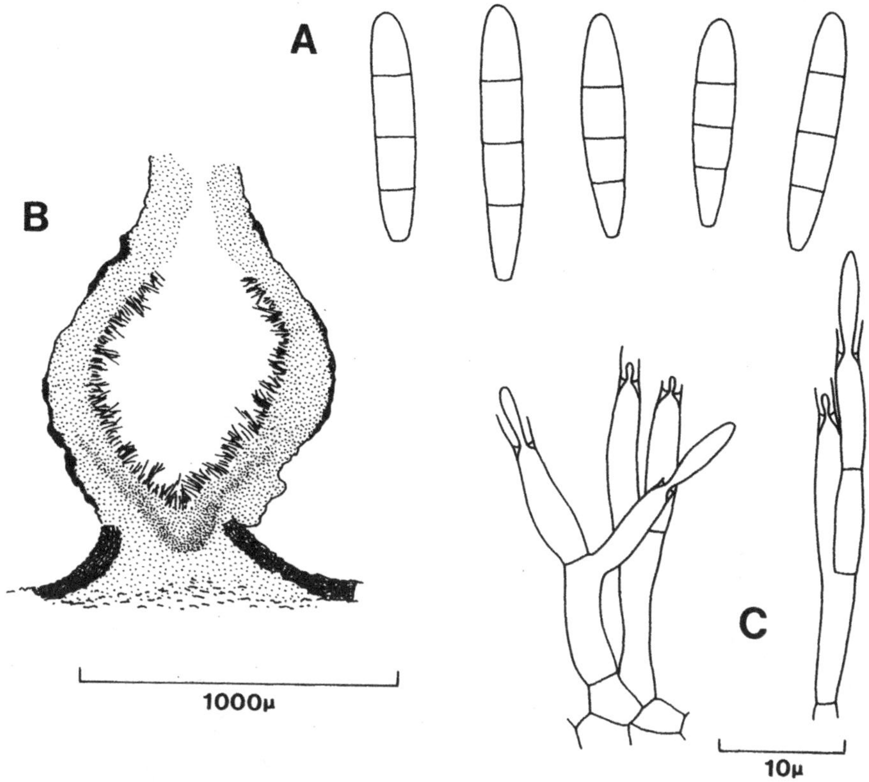

Fig. 249. *Topospora uberiformis*, A, conidia; B, vertical section of a conidioma; C, conidiophores.

42. CAPITOROSTRUM Batista, *Rev. Biol., Lisboa* **1**: 140 (1957).
Sp. typ.: *C. asteridiellae* Batista

Mycelium superficial, branched, septate, pale brown. *Conidiomata* pycnidial, superficial, solitary, separate, globose at the base but with a long neck, thin-walled, base of thin-walled, pale brown textura angularis, neck of parallel textura prismatica. *Ostiole* single, circular. *Conidiophores* absent. *Conidiogenous cells* enteroblastic, phialidic, ampulliform, hyaline, smooth, channel and collarette minute, formed from the inner cells of the pycnidial wall. *Conidia* pale brown, smooth, aseptate, eguttulate, ellipsoid.

The genus is monotypic.

C. ASTERIDIELLAE Batista, *ibid.*: 140 (1957). (Fig. 251)

Conidiomata up to 50 µ diam. at the base, neck up to 100 µ long. Conidiogenous cells 3.5 × 3 µ. Conidia 3-4 × 1.5-2 µ.

On colonies of *Asteridiella fraseriana* on *Cryptocarya meissneri*, N.S.W., Australia (IMI 23788b, isotype).

43. CONIELLA Höhn., *Ber. dtsch. Bot. Ges.* **36**: (316) (1918).
Anthasthoopa Subram. & Ramakr., *Proc. Ind. Acad. Sci.* **43**: 173 (1956).
Cyclodomella Mathur, Bhatt & Thirum., *Sydowia* **13**: 144 (1959).
Embolidium Batista, *Broteria* N.S., **33**: 194 (1964).
Pilidiella Petrak & Syd., *Beih. rep. spec. nov. regni veg.* **42**: 462 (1927).
Sp. typ.: *C. fragariae* (Oud.) Sutton (syn. *Coniothyrium fragariae* Oud., *C. pulchella* Höhn.).
Illustrations: Sutton (1969b; 1975b).

Mycelium immersed, hyaline, branched, septate. *Conidiomata* pycnidial, separate, globose, pale brown, immersed to

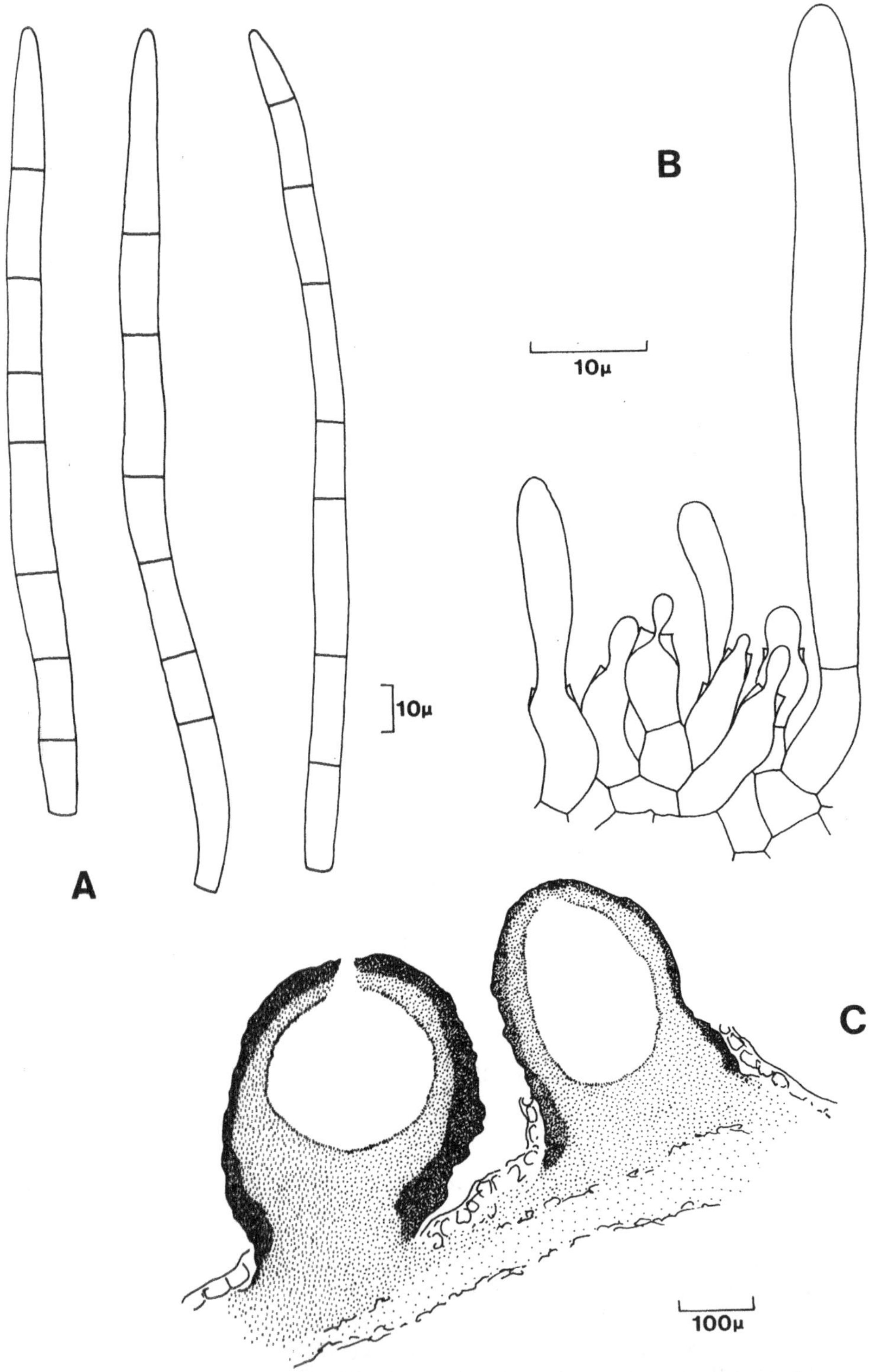

Fig. 250. *Megaloseptoria mirabilis*, A, conidia; B, conidiogenous cells and developing conidia; C, vertical section of conidiomata.

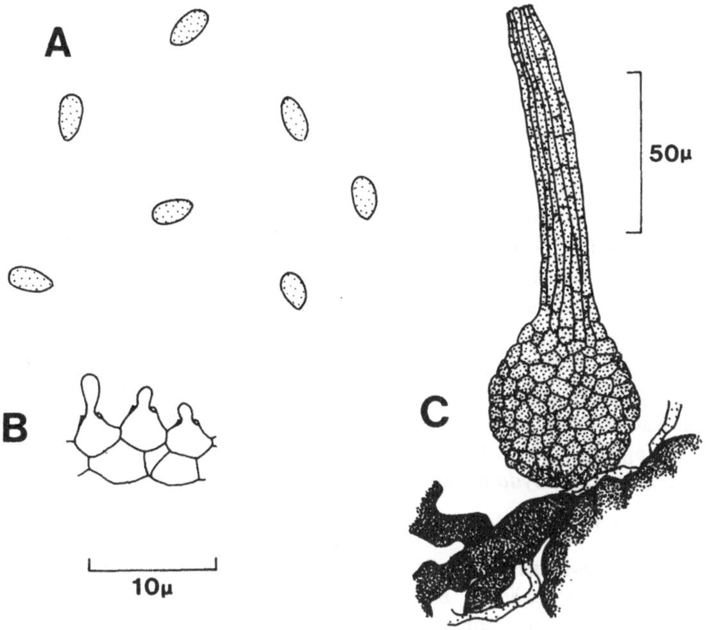

Fig. 251. *Capitorostrum asteridiellae*, A, conidia; B, conidiogenous cells; C, side view of a conidioma.

semi-immersed, unilocular, thin-walled, composed of thin-walled, pale brown textura angularis except at the base where is situated a pulvinate convex conidiogenous region of small-celled hyaline pseudoparenchyma. *Ostiole* simple, circular, central, not protruding. *Conidiophores* usually absent. *Conidiogenous cells* enteroblastic, phialidic, discrete, very rarely aggregated on 1-2-celled conidiophores branched at the base, determinate, cylindrical, hyaline, smooth, collarette and channel minute, periclinal thickening distinct, formed from the pulvinate basal pseudoparenchyma. *Conidia* olivaceous to brown, aseptate, base truncate, apex obtuse to almost apiculate, smooth, thin or thick-walled, eguttulate, sometimes with a longitudinal raphe.

Previous treatments have been by Petrak & Sydow (1927) and Sutton (1969b). By making new combinations into *Pilidiella* Petrak & Sydow, Arx (1957a) and van der Aa (1972) have implied that this genus is distinct from *Coniella*. The only difference is in the pigmentation of conidia. In *C. pulchella* conidia are dark brown whereas in *P. quercicola* they are pale brown. Pycnidia, conidiogenesis, and orientation of conidiophores are identical, consequently I regard *Pilidiella* as a synonym of *Coniella*.

KEY

1. Conidia more or less globose, l/b ratio <2/1 2
1. Conidia more or less ellipsoid or falcate to cylindrical, l/b ratio >2/1 3
   2. Conidia 6.5-7.5 × 3.5-4.5 μ .... *C. minima*
   2. Conidia 10-14.5 × 8-10.5 μ .... *C. australiensis*
   2. Conidia 7.5-11 × 5.5-7.5 μ .... *C. fragariae*
3. Conidia <10 μ long .... *C. delicata*
3. Conidia >10 μ long 4
   4. Conidia <3.5 μ wide 5
   4. Conidia >3.5 μ wide 6
5. Conidia 10-15 μ long .... *C. granati*
5. Conidia 15-29 μ long .... *C. castaneicola*
   6. Conidia 15.5-22 × 4.5-5 μ .... *C. musaiaensis*
   6. Conidia 11-16 × 3.5-5 μ .... *C. musaiaensis* var. *hibisci*
   6. Conidia 10-15.5 × 4.5-7 μ .... *C. petrakii*

C. AUSTRALIENSIS Petrak, *Sydowia* **9**: 567 (1955). (Fig. 252 I)

Conidia 10-14 × 7-11 µ, dark brown, globose to napiform.

On *Geranium, Citrus, Eucalyptus deglupta, Viola tricolor* and *Macaranga indica*, India, France, Israel, W. New Britain, Netherlands (IMI 190100, 34645, 21227, 159341, 1531, 116140, 136228, 112973).

C. castaneicola (Ell. & Ev.) comb. nov. (Fig. 252A)
Gloeosporium castaneicola Ell. & Ev., *Proc. Acad. nat. Sci. Philad.* 435 (1895).
Pilidiella castaneicola (Ell. & Ev.) Arx, *Verh. K. ned. Akad. Wet. Natuurk.* **51**: 69 (1957).
Sphaeropsis quercicola Oud., *Ned. kruidk. Arch.* 3 ser., **2**: 752 (1902).
Pilidiella quercicola (Oud.) Petrak, *Beih. rep. spec. nov. regni veg.* **42**: 462 (1927).
Dendrophoma corticalis Woronichin, *Moniteur Jard. Bot. Tiflis* **28**: 21 (1913).
Selenophoma kamatii Kalani, *Sydowia* **23**: 203 (1970).
Selenophoma kamatii Kalani, *Mycopath. mycol. appl.* **22**: 216 (1964), based on a different type.
Embolidium eucalypti Bat. & Peres apud Bat., Peres & Bezerra, *Broteria*, N.S., **33**: 194 (1964).
Coniella eucalypticola Nag Raj, *Can. J. Bot.* **54**: 1370 (1976).
Coniella citri Agarwal & Sharma, *Sydowia* **26**: 261 (1974).

Conidia 15-29 × 2.5-3.5 µ, fusiform, falcate, pale brown.

On *Castanea sativa, C.* sp., *Quercus robur, Tilia, Eucalyptus viminalis, E. grandis, E. saligna, E. citriodora, E. tereticornis, E.* spp., *Agropyron scabrum, Betula* sp., *Syzygium jambolanum, Mangifera indica, Citrus media, Shorea robusta*, Brazil, Cuba, U.S.A., India, Australia, U.K., represented in herb. IMI by 29 collections.

C. GRANATI (Sacc.) Petrak & Sydow, *Beih. rep. spec. nov. regni veg.* **42**: 461 (1927). (Fig. 252B)
Phoma granati Sacc., *Nuovo G. bot. ital.* **8**: 200 (1876).
Macrophoma granati (Sacc.) Berl. & Vogl., *Atti Soc. Venet. Trent. Sc. nat.* **10**: 202 (1886).
Pilidiella granati (Sacc.) van der Aa apud Arx, *Verh. K. ned. Akad Wet.* ser. 2, **61**: 51 (1972).
Phoma versoniana Sacc., *Michelia* **2**: 272 (1881).
Zythia versoniana (Sacc.) Sacc., *Syll. fung.* **3**: 614 (1884).
Anathasthoopa simba Subram. & Ramakr., *Proc. Ind. Acad. Sci.* **43**: 174 (1956).
Coniella simba (Subram. & Ramakr.) Sutton, *Can. J. Bot.* **47**: 607 (1969).

Conidia 10-15 × 2.5-3.5 µ, straight or slightly curved, fusiform, pale brown.
For a discussion of the synonymy and relationship with *Nectriella versoniana* Sacc., see Dodge (1936).

On *Punica granatum, Hevea* sp., *Anogeissus acuminata, Caesalpinia pulcherrima, Vitis vinifera*, Jamaica, Nigeria, Cyprus, U.K., India, Greece, Burma (IMI 15241b, 90700, 15731, 15732, 185873, 37658, 124026, 173839, 182675, 183240, 110496).

C. MINIMA Sutton & Thaung apud Sutton, *Nova Hedwigia* **26**: 10 (1975). (Fig. 252F)

Conidia 6.5-7.5 × 3.5-4.5 µ, globose to subglobose, pale brown.

On *Eucalyptus camaldulensis*, Burma (IMI 179300, holotype).

C. MUSAIAENSIS Sutton, *Can. J. Bot.* **47**: 607 (1969). (Fig. 252D)

Conidia 15.5-22 × 4.5-5 µ, fusiform, straight or falcate.

On *Piliostigma thonningii, Acacia arabica* and from soil, Sierra Leone, Pakistan, India (IMI 103345, holotype, 139784b, 174356).

C. musaiaensis Sutton var. **hibisci** var. nov. (Fig. 252C)
A *Coniella musaiaensis* differt conidia minore, 11-16 × 3.5-5 µ, pallide brunnea, fusiformia, recta vel curvata.
In foliis *Hibisci esculenti*, Eruwa, W. Nigeria, leg. Arny, 25 Jul. 1967, IMI 129200, holotypus.

Causing large spreading necrotic lesions on *Hibiscus* spp., especially in Nigeria. Other collections on *Hibiscus* spp., *Anogeissus pendula, Bauhinia malabarica, Vitis vinifera*, India, Australia, Togo, Nigeria (IMI 129200 holotype, 195541, 188348, 188347, 187885, 187880, 169312, 168797, 167514, 167511, 162914, 151487, 148081, 110963).

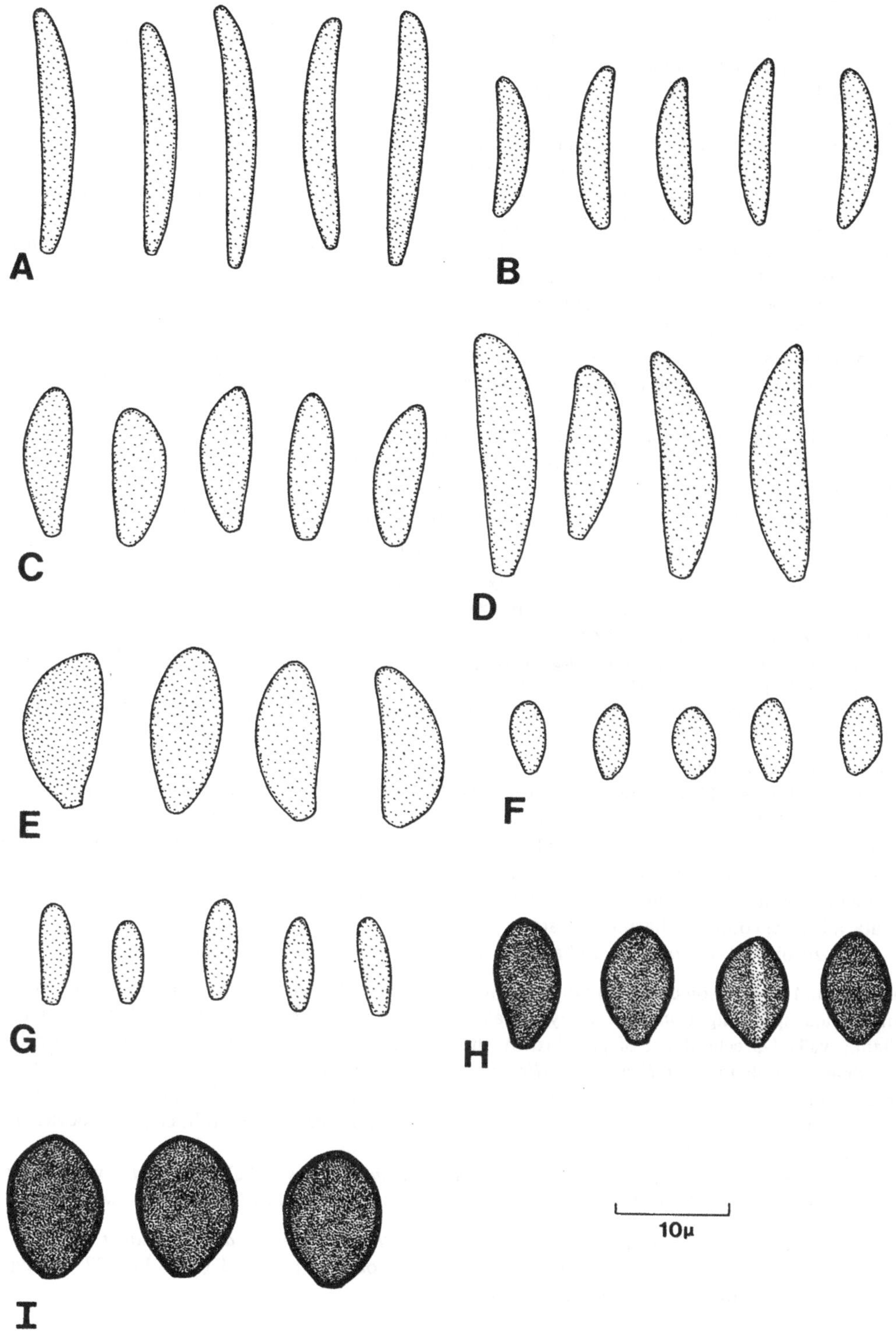

Fig. 252. A, *Coniella castaneicola*, conidia; B, *C. granati*, conidia; C, *C. musaiaensis* var. *hibisci*, conidia; D, *C. musaiaensis*, conidia; E, *C. petrakii*, conidia; F, *C. minima*, conidia; G, *C. delicata*, conidia; H, *C. fragariae*, conidia; I, *C. australiensis*, conidia.

C. delicata sp. nov. (Fig. 252G)

Coloniae in PDA et MA celeriter crescentes, roseae sed mycelio aerio sparso. Conidiomata pycnidialia, immerse vel superficialia, pallide brunnea, globosa, laevia, usque ad 150 µ diam.; paries 1-3 cellulis crassus ex textura angulari pallide brunnea. Ostiolum singulum, circulare, papillatum, 10 µ diam. Cellulae conidiogenae phialidicae, discretae, hyalinae, cylindricae, 7-10 × 2 µ, ex cellulis superioribus pulvini convexi formatae et restrictae. Conidia 7-9 × 2.5-3 µ, aseptata, pallide brunnea, laevia, ellipsoidea, ad apicem obtusa et basim truncata.

Ex *Aeridis crassifoliis*, Thailand, G.F. Laundon 9055, 16 Jan. 1975, IMI 191546, holotypus.

On PDA and MA, colonies are rapidly growing, roseous in colour, with very little aerial mycelium. *Conidiomata* pycnidial, immersed or superficial, pale brown, globose, smooth, up to 150 µ diam.; walls 1-3 cells thick, of pale brown, thin-walled textura angularis. *Ostiole* single, circular, papillate, protruding, 10 µ diam. *Conidiogenous cells* enteroblastic, phialidic, discrete, hyaline, cylindrical, 7-10 × 2 µ, formed from and restricted to the upper cells of a convex cushion of small-celled textura angularis at the base of the pycnidium. *Conidia* 7-9 × 2.5-3 µ, aseptate, pale brown, smooth, ellipsoid, apex obtuse, base truncate.

The only described species with conidia comparable in size to *C. delicata* is *C. minima* Sutton & Thaung. However conidia of both species are of quite different shapes and this is reflected in their l/b ratios.

C. FRAGARIAE (Oud.) Sutton, *Mycol. Pap.* 141: 47 (1977). (Fig. 252H)

Coniothyrium fragariae Oud., *Ver. Ned. K. Ak. Wet. ser.* 2, **18**: 37 (1883).

Coniella pulchella Höhn., *Ber. dtsch. bot. Ges.* **36**: (316) (1918).

Cyclodomella nigra Mathur, Bhatt & Thirum. apud Mathur & Thirum., *Sydowia* **13**: 145 (1959).

Conidia 7.5-11 × 5.5-7.5 µ, brown, thick-walled, globose to napiform.

C. diplodiella (Speg.) Petrak & Sydow may be an even earlier name for the type species, but Sutton (1969b) has indicated the unsatisfactory nature of the holotype.

On *Rosa laxa*, *R.* spp., *Triticum* sp., *Fragaria vesca*, *F.* spp., *Ananas comosus*, *Pisum sativum*, *P.* spp., *Phyllanthus* sp., *Anogeissus latifolia*, *Pinus patula*, *Cupressus* sp., and from soil, Tanzania, S. Africa, Egypt, India, Australia, New Caledonia, Canada, U.K. (IMI 201764, 201762, 181372, 176871, 175422, 171399, 161687, 159000, 133811, 124759, 122337, 108648, 104866, 104661, 100414, 81599 (holotype *C. nigra*), 61255, 34644, 21228).

C. petrakii sp. nov. (Fig. 252E)

Coloniae in PDA et MA celeriter crescentes, mycelio aerio sparso, cum massis nigris conidiis abundantibus. idiomata pycnidialia, immersa vel superficialia, pallide brunnea, globosa, laevia, usque ad 300 µ diam.; paries 1-3 cellulis crassus ex textura angulare pallide brunnea. Ostiolum singulum, circulare, papillatum, 30 µ diam. Cellulae conidiogenae phialidicae, discretae, hyalinae, cylindricae, 8-14 × 2-3 µ, ex cellulis superioribus pulvini convexi formatae et restrictae. Conidia 10-15.5 × 4.5-7 µ, aseptata, pallide vel medio brunnea, laevia, ellipsoidea, raro curvata, ad apicem obtusa et basim truncata.

Ex rhizosphera *Vitis viniferae*, India, M.A. Salam, 29 Jan. 1965, IMI 112027, holotypus.

On PDA and MA, colonies are rapidly growing, with very little aerial mycelium, but black conidial masses abundant and spreading. *Conidiomata* pycnidial, immersed or superficial, pale brown, globose, smooth, up to 300 µ diam.; walls 1-3 cells thick, of pale brown, thin-walled textura angularis. *Ostiole* single, circular, papillate, protruding, 30 µ diam. *Conidiogenous cells* enteroblastic, phialidic, discrete, hyaline, cylindrical, 8-14 × 2-3 µ, formed from and restricted to the upper cells of a convex cushion of small-celled textura angularis at the base of the pycnidium. *Conidia* 10-15.5 × 4.5-7 µ, aseptate, pale to medium brown, smooth, ellipsoid, occasionally slightly curved, apex obtuse, base truncate.

C. petrakii is distinguished from the closely related *C. musaiaensis* and *C. musaiaensis* var. *hibisci* by conidial size, having conidia up to 7 µ wide. Neither *C. musaiaensis* nor its var. *hibisci* exceed 5 µ in width.

On or from rhizosphere of *Vitis vinifera*, *Vitis* sp., *Hymenocardia acida*, *Parinaria excelsa*, *Eucalyptus camaldulensis*, Burma, India, Nigeria, Tanzania, Sierra Leone, Switzerland (IMI 112027 holotype, 107889, 107791a, 104660, 100482, 45087a, 21225, 162826, 179301).

44. MICROSPHAEROPSIS Höhn., *Hedwigia* **59**: 267 (1917).

? *Monoplodia* Westd., *Bull. Acad. R. sci. Belg.*, 2 ser., **7**: 91 (1859).

? *Phyllostictella* Tassi, *Bull. Lab. ort. Bot. Siena* **4**: 4 (1901).

? *Coniothyrium* Cda subgen. *Phyllostictella* (Tassi) Sacc. & D. Sacc., *Syll. fung.* **18**: 308 (1906).

Coniothyriopsis Speg., *An. Mus. Nac. Buenos Aires* **20**: 361 (1910).

Asteropsis G. Frag., *Trab. Mus. Nac. Cienc. Nat. Madrid*, ser. bot., **12**: 50 (1917).
Cryptophaeella Höhn., *Sber. Akad. Wiss. Wien* **126**: 360 (1917).
Microsporella Höhn., *Hedwigia* **60**: 146 (1918).
Dothisphaeropsis Höhn., *Ber. dtsch. Bot. Ges.* **36**: 214 (1918).
Sclerothyrium Höhn., *Hedwigia* **60**: 181 (1918).
? *Coniothyrinula* Petrak, *Annls mycol.* **21**: 2 (1923).
Sp. typ.: *M. olivacea* (Bonord.) Höhn. (syn. *Coniothyrium olivaceum* Bonord.).

Mycelium immersed, branched, septate, hyaline to pale brown. *Conidiomata* pycnidial, globose, separate, brown, unilocular, immersed; wall thin, composed of brown, thin-walled textura angularis. *Ostiole* single, circular, central, sometimes papillate. *Conidiophores* absent. *Conidiogenous cells* enteroblastic, phialidic, determinate, discrete, doliiform, ampulliform or cylindrical, hyaline, smooth, collarette minute, channel and periclinal thickening prominent, formed from the inner cells of the pycnidial wall. *Conidia* brown, thin or thick-walled, aseptate, smooth or ornamented, globose, subglobose, pyriform or short cylindrical.

The generic synonymy has been discussed under the individual generic names by Sutton (1977a), and it is quite clear from this that there are probably earlier generic names than *Microsphaeropsis* for the species which would previously have been placed in *Coniothyrium*. However, so many species are involved, that the ad hoc redisposition of taxa into *Microsphaeropsis*, when that name is itself not necessarily the earliest available, would seem to be an inadvisable policy to adopt. Type studies correlated with behaviour in culture will lead to a more useful reassessment of the group. Meanwhile, the following species which cannot be retained in *Coniothyrium* (Sutton, 1977a), have been documented in *Microsphaeropsis*, although this is very much an interim measure.

KEY

1. Conidia smooth 2
1. Conidia ornamented, or apparently so 5
    2. Conidia thick-walled 3
    2. Conidia thin-walled 4
3. Conidiomata <100 µ diam.; conidia globose .... *M. globulosa*
3. Conidiomata up to 250 µ diam.; conidia fusiform, with a central guttule .... *M. callista*
    4. Conidia fusiform .... *M. conielloides*
    4. Conidia elliptical .... *M. olivacea*
    4. Conidia cylindrical .... *M. arundinis*
5. Conidia ornamentation not prominent, walls uneven and wrinkled rather than verrucose .... *M. pseudaspera*
5. Conidia ornamentation prominent 6
    6. Conidiomata 350-420 µ diam. Conidia asperate, globose .... *M. centaureae*
    6. Conidiomata <150 µ diam. Conidia with distinct wall protuberances .... *M. onychiuri*
    6. Conidiomata up to 250 µ diam. Conidia verrucose .... *M. sarcinellae*

M. arundinis (Ahmad) comb. nov. (Fig. 253B)
Alveophoma arundinis Ahmad, *Biologia, Lahore* **17**: 1 (1971).

Conidiomata up to 175 µ diam. Conidiogenous cells 3-5.5 × 2.5-3 µ. Conidia cylindrical, thin-walled, guttulate, smooth, 4-4.5 × 1.5 µ.

On *Arundo donax*, Pakistan (IMI 137064, holotype).

M. CALLISTA (H. Syd.) Sutton, *Mycol. Pap.* 123: 35 (1971). (Fig. 254A)
Coniothyrium callistum H. Syd., *Annls mycol.* **35**: 358 (1937).

Conidiomata 150-200 µ diam. Conidiogenous cells 4-7 × 2.5-4 µ. Conidia elliptical to fusiform, thick-walled, with a central guttule, 7-8.5 × 4.5-5.5 µ, brown, but darker at each end.

On leaves of *Eucalyptus haemastoma*, N.S.W., Australia (IMI 21230, holotype).

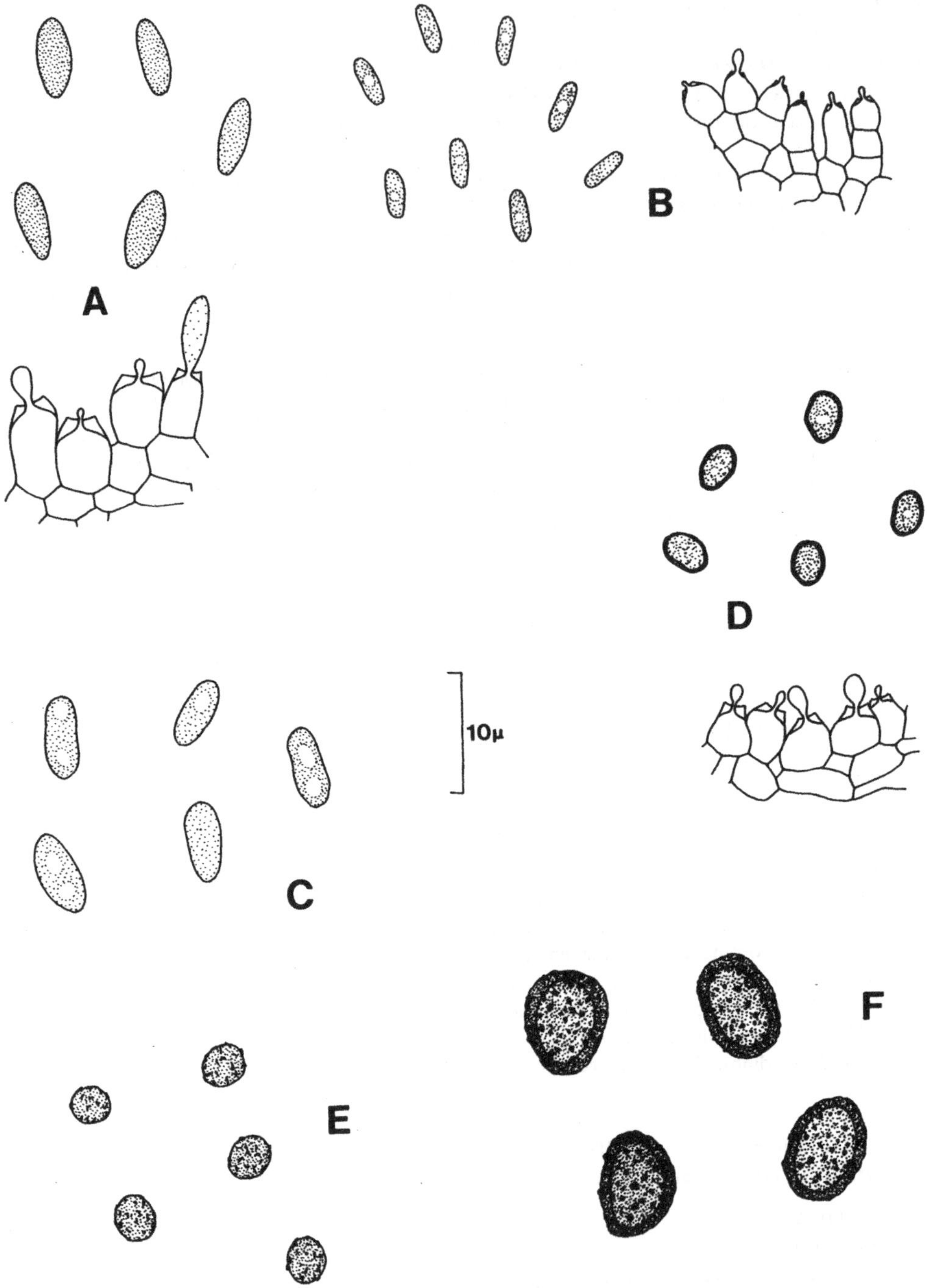

Fig. 253. A, *Microsphaeropsis conielloides*, conidia and conidiogenous cells; B, *M. arundinis*, conidia and conidiogenous cells; C, *M. olivacea*, conidia; D, *M. pseudaspera*, conidia and conidiogenous cells; E, *M. onychiuri* conidia; F, *M. sarcinellae*, conidia.

M. CENTAUREAE Morgan-Jones, *Can. J. Bot.* **52**: 2576 (1974).

Conidiomata 350-420 μ diam. Conidiogenous cells 3-6 × 2.5-3 μ. Conidia ellipsoid, pale to yellowish brown, asperate, 5-6 × 4.5 μ.

From *Centaurea diffusa*, B.C., Canada.

M. CONIELLOIDES Sutton, *Nova Hedwigia* **25**: 168 (1974). (Fig. 253A)

Conidiomata up to 250 μ diam. Conidiogenous cells 3.5-7.5 × 3-6.5 μ. Conidia *Coniella*-like, pale brown, thin-walled, smooth, 5.5-7.5 × 3-3.5 μ.

From leaves of *Eucalyptus pauciflora* and isolated from the air, Australia, India (IMI 160566, holotype, 160529, 160530, 182340).

M. GLOBULOSA (Camara) Sutton, *Nova Hedwigia* **25**: 170 (1974). (Fig. 254B, C)
Coniothyrium globulosum Camara, *Agron. lusit.* **10**: 305 (1948).
Coniothyrium olivaceum Bonord. var. *eucalypti* G. Frag., *Bol. Soc. Brot.* ser., 2, **2**: 59 (1924).
Microsphaeropsis eucalypti (G. Frag.) Sutton, *Mycol. Pap.* 123: 37 (1971).
? *Coniothyrium eucalypti* G. Frag., *Bol. Soc. Brot.* ser. 2, **2**: 59 (1924).

Conidiomata up to 100 μ diam. Conidiogenous cells 4-9 × 3-5 μ. Conidia globose, subglobose or pyriform, dark brown, thick-walled, smooth, 6-7 × 5-6 μ.

On leaves of *Eucalyptus globulus*, Portugal (IMI 165749, holotype *C. globulosum* ex LISE, 151356, holotype of *C. olivaceum* var. *eucalypti* ex MA, 151357, holotype of *C. eucalypti* ex MA).

M. OLIVACEA (Bonord.) Höhn., *Hedwigia* **59**: 267 (1917). (Fig. 253C)
Coniothyrium olivaceum Bonord. apud Fckl, *Jahrb. nassau. Ver. Naturk.* **23-24**: 377 (1870).

Conidiomata 200-300 μ diam. Conidiogenous cells 4-6 × 3-4 μ. Conidia oval to ellipsoid, pale brown, smooth, thin-walled, 4-8 × 2.5-5 μ.

This species is distinguished from *Coniothyrium fuckelii* Sacc. (which is also congeneric with *M. olivaceum*) by the larger conidia.

From and on *Citrus* sp., *Pyrus malus, Capparis decidua, Crescentia cujeta, Chenopodium* sp., *Dalbergia sissoo, Laburnum* sp., *Eucalyptus ficifolia, Cytisus scoparius, Ficus carica, Hyphaene parvula, Juniperus macropoda, J. horizontalis, Morus alba, Pinus wallichiana, Salvadora oleoides, Sapindus mukorossi, Cydonia vulgaris, Diospyros* sp., *Macadamia* sp., *Paspalum distichum, Prunus persica, Pseudotsuga menziesii, Zizyphus jujuba, Z. oxyphylla* and from air, grass debris, water and soil, U.K., France, Iraq, Pakistan, India, Kenya, Tanzania, Singapore, Papua New Guinea, Australia, Venezuela, U.S.A. (IMI 207211, 4084, 139839c, 139665a, 151510a, 157355b, 140214, 140232b, 102671b, 159079b, 187556b, 140168b, 176852, 142466b, 140191b, 142024, 140920b, 132808, 210425, 181863, 214148, 213858, 195543, 197848, 199471, 149689, 138452a, 171396, 171720, 195651, 187155, 223985).

M. ONYCHIURI (Punith.) Morgan-Jones, *Can. J. Bot.* **52**: 2578 (1974). (Fig. 253E)
Coniothyrium onychiuri Punith., *Trans. Br. mycol. Soc.* **55**: 497 (1970).

Conidiomata 100-140 μ diam. Conidiogenous cells 3-5 × 2-3 μ. Conidia globose to ellipsoid, thick-walled, pale brown, with distinct protuberances on the wall, 3-4 × 2-3.5 μ.

From *Onychiurus pseudofimetarius*, N.Z. (IMI 137464, holotype, 137447).

M. PSEUDASPERA Sutton, *Nova Hedwigia* **25**: 170 (1974). (Fig. 253D)

Conidiomata up to 225 μ diam. Conidiogenous cells 3.5-5 × 2-4 μ. Conidia elliptical to subglobose, pale brown, smooth but with uneven walls, 3-4 × 2.5-3 μ.

On dead branches of *Eucalyptus* sp., Portugal (IMI 165750 ex LISE, holotype).

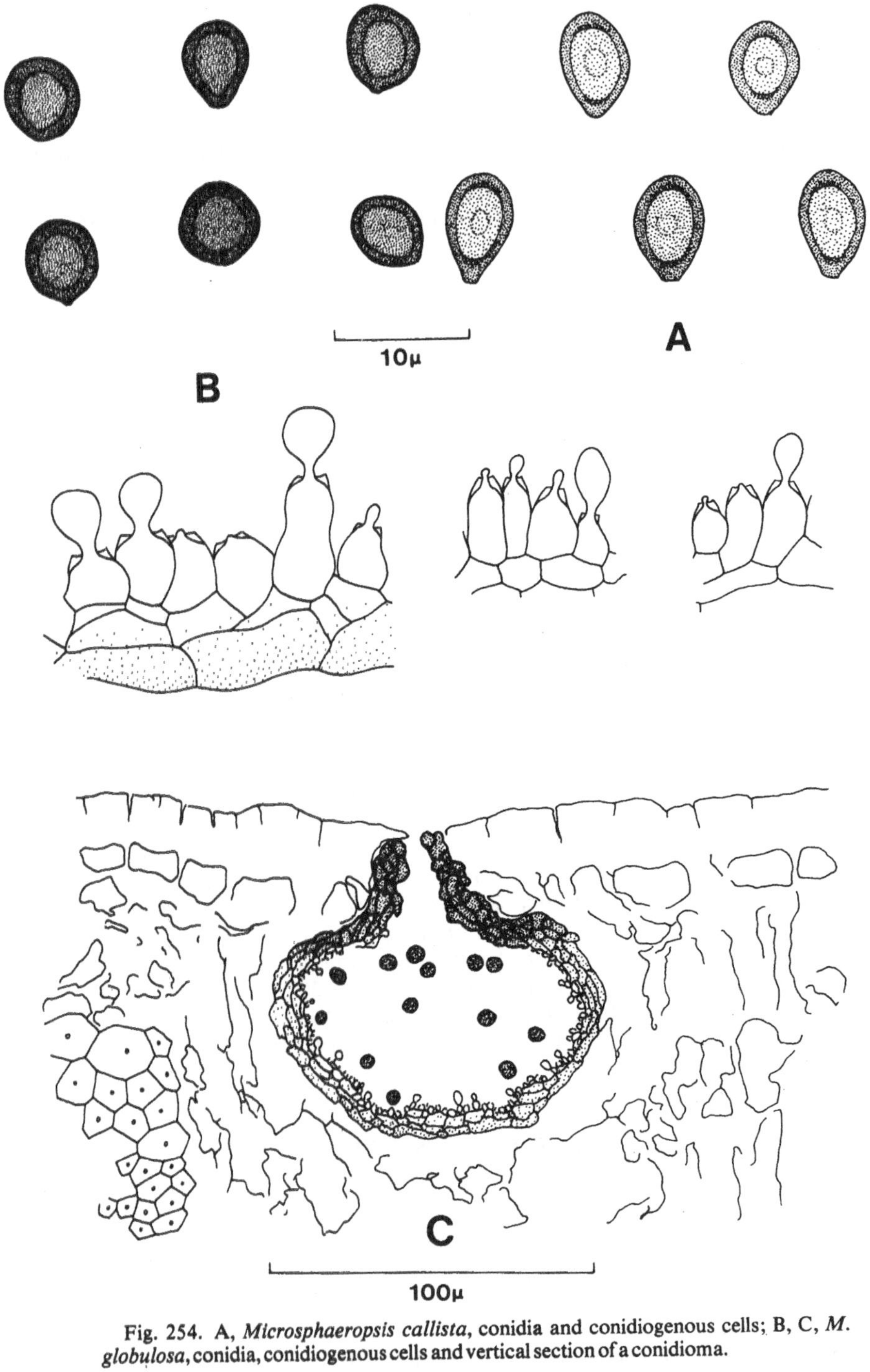

Fig. 254. A, *Microsphaeropsis callista*, conidia and conidiogenous cells; B, C, *M. globulosa*, conidia, conidiogenous cells and vertical section of a conidioma.

M. SARCINELLAE (Sahni) Morgan-Jones, *Can. J. Bot.* **52**: 2578 (1974). (Fig. 253F)
Coniothyrium sarcinellae Sahni, *Mycopath. Mycol. appl.* **36**: 276 (1968).

Conidiomata 150-250 µ diam. Conidiogenous cells 4-7 × 3-4 µ. Conidia ellipsoid to globose, brown, thick-walled, verrucose, 5-9.5 × 6-7 µ.

On *Sarcinella palawanensis* on *Celastrus paniculatus*, India (IMI 112817b, holotype).

45. READERIELLA H. & P. Syd., *Annls mycol.* **6**: 484 (1908).
Sp. typ.: *R. mirabilis* H. & P. Syd.

Mycelium immersed, branched, septate, brown. *Conidiomata* pycnidial, separate, globose, dark brown, immersed, unilocular, thick-walled; wall several cells thick, of outer, elongated, dark brown, thick-walled textura angularis and inner paler tissue. *Ostiole* single, circular, not papillate. *Conidiophores* absent. *Conidiogenous cells* enteroblastic, monophialidic, occasionally polyphialidic, indeterminate, discrete, short lageniform to ampulliform, hyaline, smooth, channel proliferating percurrently, producing 3 collarettes. *Conidia* pale brown, aseptate, base truncate, apex ± flattened with 3 lateral obtuse projections, deltoid, thick-walled, smooth, guttulate.

The genus and species have been redescribed by Sutton (1971b).

R. MIRABILIS H. & P. Syd., *ibid.*: 484 (1908). (Fig. 255)

Conidiomata 150 µ diam.; ostiole 40 µ diam. Conidiogenous cells 7.5-13.5 × 4.5-5 µ. Conidia 7-9.5 µ long × 6.5-9 µ wide at the apex.

On leaves of *Eucalyptus capitellata*, *E. regnans*, *E. viminalis*, *E.* sp., Australia, Brazil, U.K. (IMI 59143b, ex K, 147802, holotype ex S, 108602, 184644c, 229566).

46. NEOTTIOSPORA Desm., *Ann. Sci. nat.*, **19**: 346 (1843).
Samukuta Subram. & Ramakr., *J. Ind. bot. Soc.* **36**: 75 (1957).
Spilomyces Petrak & Syd., *Beih. nov. spec. regn. veg.* **42**: 293 (1927).
Sp. typ.: *N. caricina* (Desm.) Höhn. (syn. *Sphaeria caricina* Desm., *N. caricum* Desm.).

Mycelium immersed, brown, septate, branched. *Conidiomata* pycnidial, gregarious, globose, collabent, pale to dark brown, immersed, subepidermal, unilocular, thin-walled, composed of small, pale brown, thin-walled textura angularis. *Ostiole* absent, dehiscence operculate, by a circular ring of tissue between the pycnidial wall and the dark brown thicker operculum. *Conidiophores* hyaline, branched mainly at the base, septate, smooth, formed from the inner pycnidial wall cells. *Conidiogenous cells* enteroblastic, phialidic, determinate, integrated, cylindrical, hyaline, smooth, channel and collarette minute. *Conidia* hyaline to olivaceous, aseptate, fusiform, apex obtuse to subacute, base flattened, smooth, guttulate, at first enclosed in a gelatinous sheath which later becomes everted towards the base to produce a basal funnel-shaped appendage which then splits into tentacular processes.

The modern concept of structure and ontogeny in this genus is derived from the work of Cunnell (1957a), Pirozynski & Shoemaker (1971) and Nag Raj (1973a). At least 15 taxa have been described in *Neottiospora* and several have already been removed to other genera.

N. CARICINA (Desm.) Höhn., *Ber. dtsch. Bot. Ges.* **37**: 158 (1919). (Fig. 256)
Sphaeria caricina Desm., *Ann. Sci. nat.* **6**: 246 (1836).
Neottiospora caricum Desm., *Ann. Sci. nat.* **19**: 346 (1843).
Darluca caricum (Desm.) Fckl, *Jb. nassau Ver. Naturk.* **23-24**: 380 (1870).
Samukuta berkeleyi Subram. & Ramakr., *J. Ind. Bot. Soc.* **36**: 75 (1957).
Spilomyces atramentarius (Schröt.) Petrak & Syd., *Beih. nov. spec. regn. veg.* **42**: 293 (1927).
Hendersonia atramentaria Schröt., *Hedwigia* **17**: 173 (1878).

Conidiomata 200-450 µ diam. Operculum 70-170 µ diam. Conidiophores 15-22 × 1.5-2.5 µ. Conidia 10-16 × 2-4 µ; appendage 9-15 × 3-5 µ.

On dead leaves of *Iris pseudacorus*, *Carex riparia*, *C. elata*, *C. acutiformis*, *C. pendula*, *C. appropinquata*, *Panicum maximum*, *Typha latifolia*, G.B., Germany, Austria, Canada, Uganda (IMI 21202, 21271, 23212, 23220, 27777, 28539, 28564, 28569, 21070, 30577, 12554, 34853d, 34898, 34893e, 47145, 49854, 17672b, 67901, 125760, 91753c, 184303, 67900).

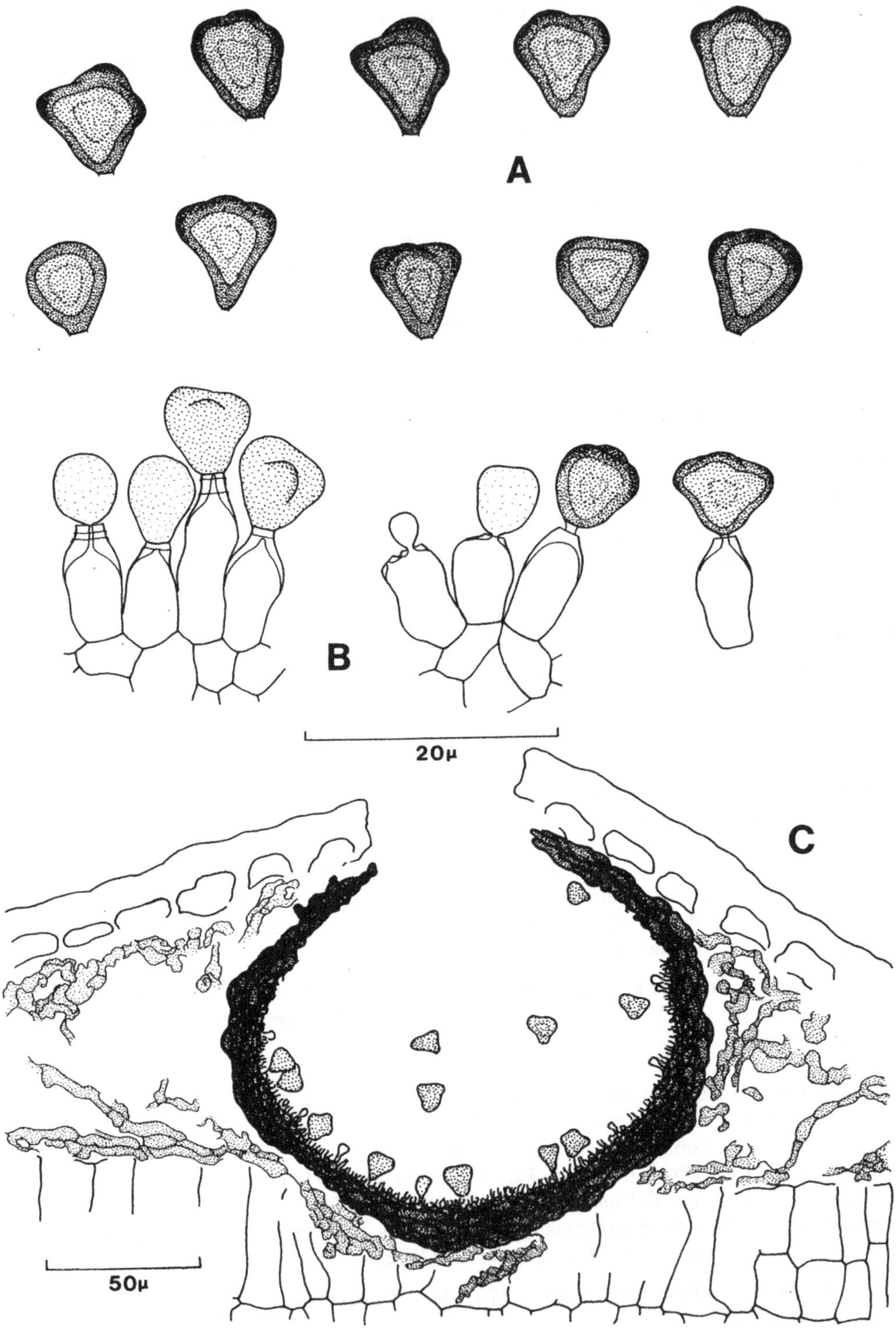

Fig. 255. *Readeriella mirabilis*, A, conidia; B, conidiogenous cells and developing conidia; C, vertical section of a conidioma.

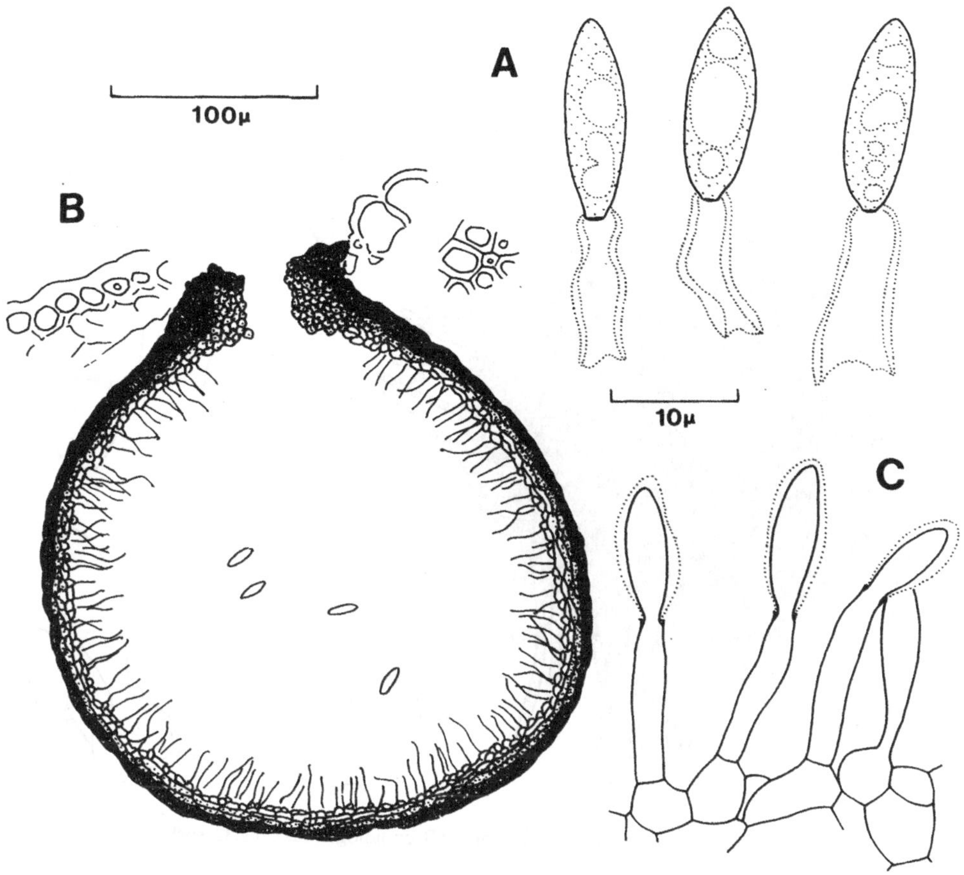

Fig. 256. *Neottiospora caricina*, A, conidia; B, vertical section of a conidioma; C, conidiogenous cells with developing conidia.

47. CHAETODIPLODIA Karst., *Hedwigia* **23**: 62 (1884).
Sp. typ.: *C. caulina* Karst.

Mycelium immersed, branched, septate, pale brown to hyaline. *Conidiomata* pycnidial, superficial on cortex from which epidermis has been removed, dark brown to black, globose, finally collabent, separate, less often aggregated, unilocular, wall thin, composed of thin-walled, brown textura angularis. *Ostiole* single, circular, central. *Setae* numerous over whole pycnidium, straight or flexuous, smooth, pale brown, generally aseptate, becoming septate with age, apices obtuse, not markedly tapered. *Conidiophores* absent. *Conidiogenous cells* enteroblastic, phialidic, discrete, determinate, hyaline, smooth, ampulliform to doliiform, channel narrow, collarette minute, periclinal wall thickened, formed from the inner cells of the pycnidial wall. *Conidia* medianly 1-euseptate, continuous or occasionally slightly constricted, pale brown, thin-walled, septum prominent, smooth, eguttulate, cylindrical to ellipsoid, apex obtuse, base sometimes ± truncate.

More than 20 species have been described in the genus so far but no revisionary treatment has been published.

C. CAULINA Karst., *ibid.*: 62 (1884). (Fig. 257)

Conidiomata up to 200 µ diam. Setae 40-150 µ long × 10 µ wide. Conidiogenous cells 6-9 × 6-7 µ. Conidia 13.5-16 × 5-5.5 µ.

On dead stems of *Chenopodium album*, Finland (IMI 200035, holotype ex H).

48. ASCOCHYTULINA Petrak, *Annls mycol.* **20**: 342 (1922).
Sp. typ.: *A. deflectens* (Karst.) Petrak (syn. *Diplodia deflectens* Karst.).

Mycelium immersed, branched, septate, pale brown. *Conidiomata* pycnidial, separate, dark brown to black, globose

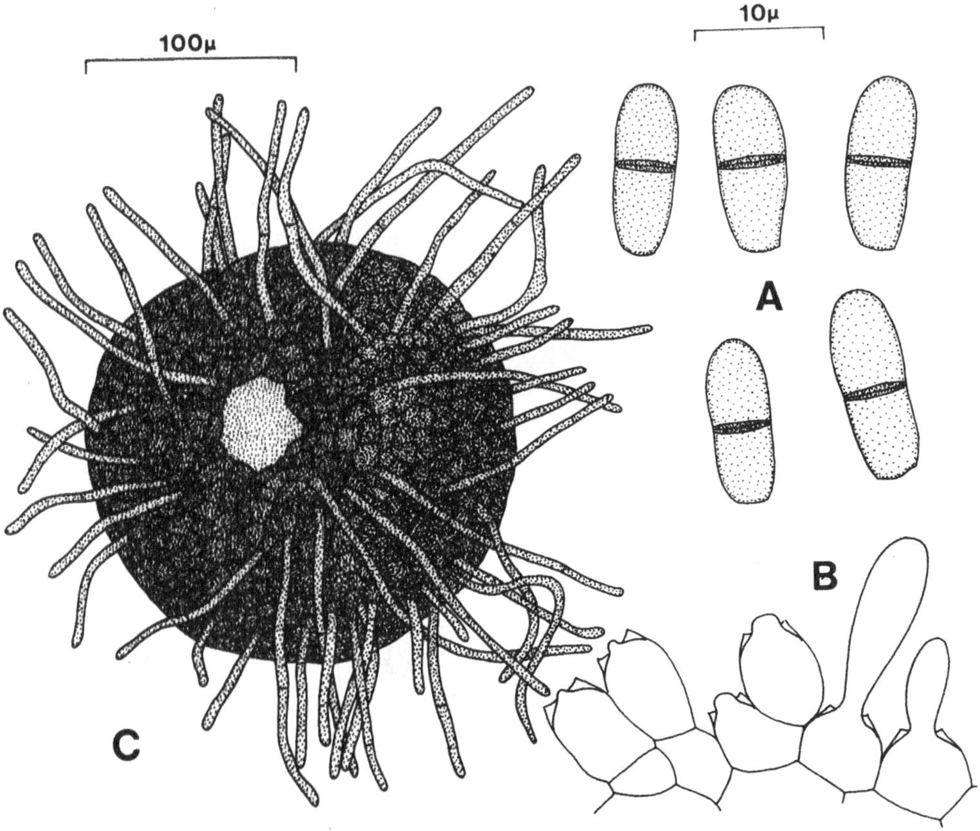

Fig. 257. *Chaetodiplodia caulina*, A, conidia; B, conidiogenous cells; C, surface view of a conidioma.

to flattened and collabent, immersed, unilocular, thick-walled, clypeate; walls 5-6 cells thick of brown textura angularis on the outside, becoming paler and thinner-walled towards the conidiogenous layer; clypeus extending just beyond the diameter of the pycnidia, composed of loosely anastomosing dark brown textura intricata. *Ostiole* central, circular, ± papillate. *Conidiophores* absent. *Conidiogenous cells* enteroblastic, phialidic, doliiform to ampulliform, discrete, determinate, hyaline, smooth, formed from cells lining the inner wall of the pycnidium. *Conidia* ellipsoid, pale brown, minutely verruculose, medianly 1 euseptate, base ± truncate, apex obtuse.

The genus is monotypic.

A. DEFLECTENS (Karst.) Petrak, *ibid.*: 342 (1922). (Fig. 258)
Diplodia deflectens Karst., *Hedwigia* **23**: 18 (1884).
Phoma sempervirentis Oud., *Hedwigia* **37**: 314 (1898).
Pseudodiplodia lonicerae Höhn., *Sber. Akad. Wiss. Wien* **111**: 1021 (1902).
Ascochytula lonicerae (Höhn.) Höhn., *Sber. Akad. Wiss. Wien* **124**: 53 (1915).

Conidiomata up to 200 µ diam. Conidiogenous cells 8-10.5 × 7.5-9.5 µ. Conidia 15.5-19 × 7-8 µ.

On dead stems of *Lonicera sempervirens*, *L. tatarica*, France, Romania, Netherlands, Finland (IMI 90822a, holotype *Phoma sempervirentis*, 6375, 12395, 33435, Karsten 1486 ex H, holotype of *Diplodia deflectens*).

49. PSEUDODIPLODIA (Karst.) Sacc., *Syll. fung.* **3**: 621 (1884).
Diplodia Fr. subgen. *Pseudodiplodia* Karst., *Medd. Soc. Fauna flora fenn.* **11**: 156 (1884).
Sp. typ.: *P. ligniaria* (Karst.) Sacc. (syn. *Diplodia ligniaria* Karst.).

Mycelium immersed, branched, septate, hyaline. *Conidiomata* pycnidial, separate, dark brown, globose, immersed, unilocular, thin-walled; walls 2-4 cells thick, of brown textura angularis on the outside, thinner-walled and pale towards the

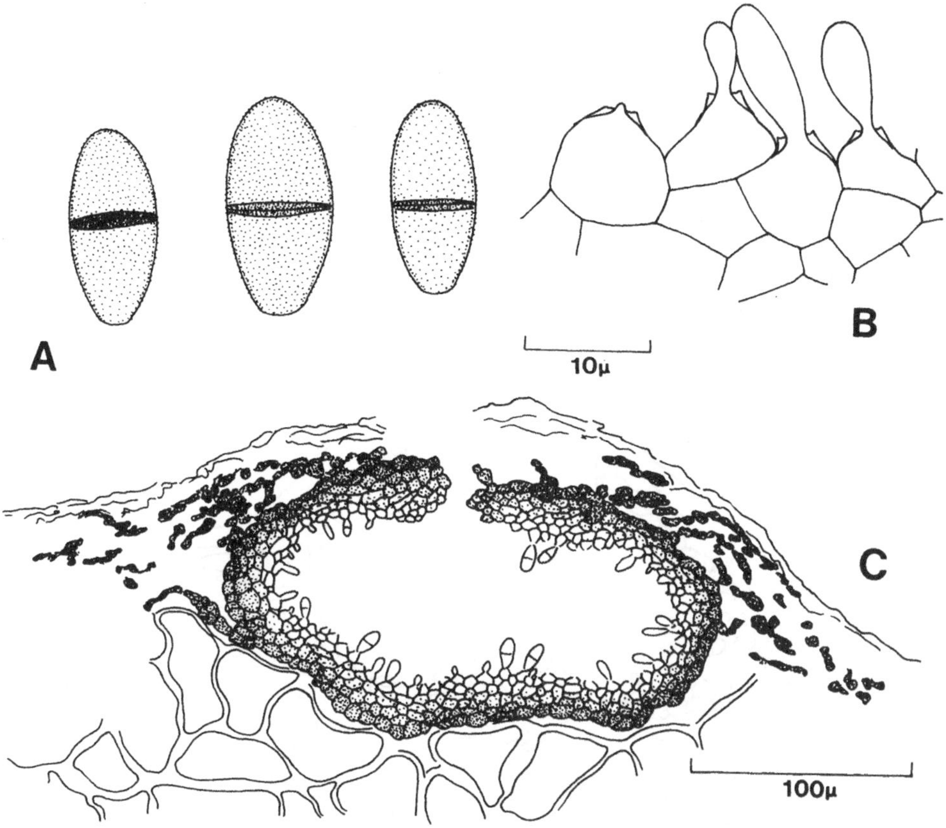

Fig. 258. *Ascochytulina deflectens*, A, conidia; B, conidiogenous cells and developing conidia; C, vertical section of a conidioma.

conidiogenous layer, hyaline at the base. *Ostiole* single, central, circular. *Conidiophores* absent. *Conidiogenous cells* enteroblastic, phialidic, doliiform to ampulliform, discrete, determinate, hyaline, smooth, formed from the inner cells of the pycnidial wall, collarette and channel minute. *Conidia* ellipsoid, pale brown, smooth, medianly 1 euseptate, often slightly constricted at the septum, apex and base obtuse.

Petch (1943) redescribed the genus but gave no details of conidiogenesis. *P. corticis* Grove which was included is rather doubtful as an acceptable species because pycnidia were described as nearly superficial and 'sporophores' (conidiophores) as branched and fasciculate. Petrak (1953b) also accepted *Pseudodiplodia* and made many new combinations from *Ascochytella* Tassi and *Ascochytula* (Poteb.) Died. so that now, about 100 taxa have been referred to the genus. Apart from the absence of a clypeus and the smooth conidia, *Pseudodiplodia* is very similar to *Ascochytulina* Petrak.

P. LIGNIARIA (Karst.) Sacc., *ibid.*: 621 (1884). (Fig. 259)
Diplodia ligniaria Karst., *Medd. Soc. Fauna Flora fenn.* **11**: 156 (1884).

Conidiomata up to 150 µ diam. Conidiogenous cells 5-6.5 × 4.5-7.5 µ. Conidia 12.5-13 × 6-7 µ.

On dead wood, Finland (IMI 200426, holotype ex H).

50. STENOCARPELLA Syd., *Annls mycol.* **15**: 258 (1917).
Phaeostagonosporopsis Woronichin, *Defense des plantes Leningrad* **2**: 333 (1925).
Hendersoniopsis Woronichin, *Fungal and bacterial diseases of agricultural Plants*: 255 (1922).
Sp. typ.: *S. macrospora* (Earle) Sutton (syn. *Diplodia macrospora* Earle, *S. zeae* Syd.).

Mycelium immersed, brown, branched, septate. *Conidiomata* pycnidial, separate or sometimes confluent, globose or elongated, dark brown, subepidermal, unilocular, thick-walled; walls composed of dark brown, thick-walled textura

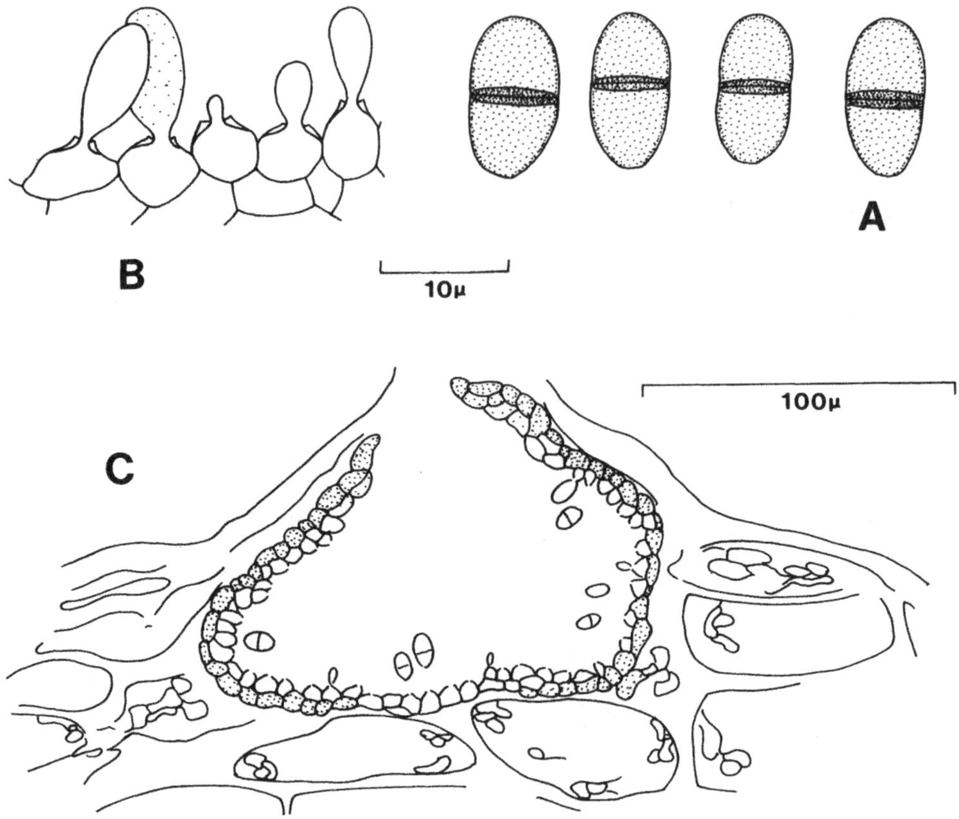

Fig. 259. *Pseudodiplodia ligniaria*, A, conidia; B, conidiogenous cells and developing conidia; C, vertical section of a conidioma.

angularis. *Ostiole* single, circular, papillate, protruding. *Conidiophores* usually absent. *Conidiogenous cells* enteroblastic, phialidic, determinate, discrete (only very rarely integrated on 1-septate conidiophores), cylindrical, collarette and channel minute, periclinal wall thickened, formed from the inner cells of the pycnidial wall. *Conidia* pale brown, 0-3 euseptate, continuous or constricted, cylindrical, straight or bent, fusiform, apex obtuse, base ± tapered and truncate, thick-walled, smooth, eguttulate.

The genus contains two species, previously dealt with by Sutton (1964a) in *Diplodia* and by Petrak & Sydow (1927) in *Macrodiplodia*. Shear & Stevens (1935) and Sutton (1964a) suggested the species be accommodated in a genus distinct from *Diplodia*. Typification of *Diplodia* and the other genera involved means that *Stenocarpella* becomes the acceptable generic name to use in this context.

Stenocarpella maydis (Berk.) comb. nov. (Fig. 260A)
Sphaeria maydis Berk., *Hooker's London J. Bot.* **6**: 15 (1847).
Sphaeria zeae Schw., *Syn. Fung. Amer. bor.* 207 (1832), non *S. zeae* Schw., 1822.
Diplodia zeae (Schw.) Lév., *Ann. Sci. nat.* **3**: 258 (1848).
Diplodia ? maydicola Speg., *An. Mus. Nac. B. Aires* **20**: 370 (1910).
Macrodiplodia zeae (Schw.) Petrak & H. Syd. apud Petrak, *Annls mycol.* **21**: 189 (1923).
Phaeostagonosporopsis zeae (Schw.) Woronichin, *Defense des Plantes Leningrad* **2**: 333 (1925).
Hendersonia zeae (Schw.) Haszl., *Math. term. Közl.* **10**: 48 (1872).
Diplodia zeae-maydis Mechtijeva, *Bot. Mater.* **15**: 156 (1962).

Conidiomata 150-300 μ diam. Conidiogenous cells 10-20 × 2-3 μ. Conidia 0-2 septate, 15-34 × 5-8 μ.

On culms and seeds of *Zea mays* and culms of *Arundinaria*, U.S.A., Swaziland, Tanzania, S. Rhodesia, Kenya, Canada, Pakistan, Nepal, S. Africa (IMI 96546 (isotype), 96582, 96581, 96575, 96577, 96579, 96578, 96580,

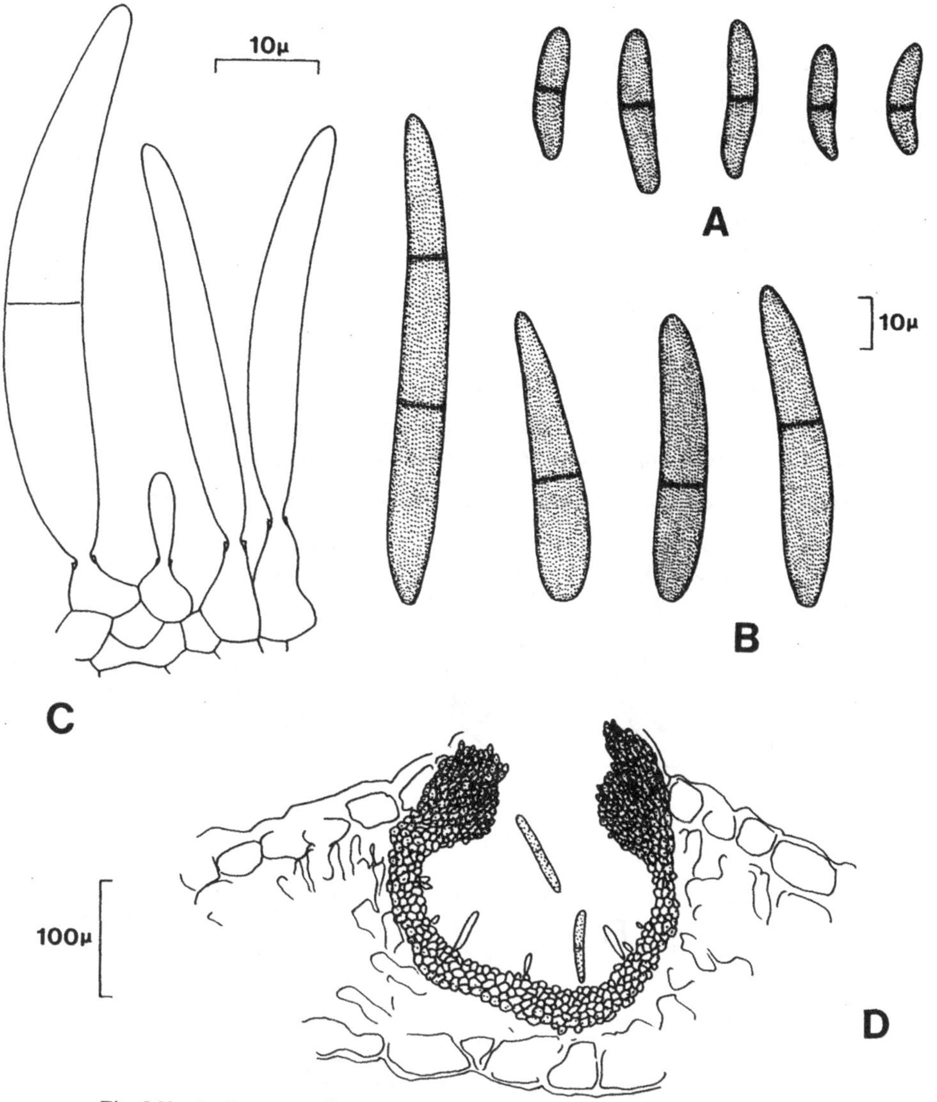

Fig. 260. A, *Stenocarpella maydis*, conidia, B, C, D, *S. macrospora*, conidia, conidiogenous cells and developing conidia, vertical section of a conidioma.

160678, 160647, 96576, 4091b, 6692, 6693, 68117, 68490, 95576b, 95934, 124682, 160759, 160259, 160657, 160652, 160694, 160681, 160682, 160686, 160689, 160657a, 189174, 141704, 142706, 141695, 141701, 141699, 141708).

S. MACROSPORA (Earle) Sutton, *Mycol. Pap.* 141: 202 (1977). (Fig. 260B, C, D)
Diplodia macrospora Earle, *Bull. Torrey bot. Cl.* **24**: 29 (1897).
Macrodiplodia zeae (Schw.) Petrak & H. Syd. var. *macrospora* (Earle) Petrak & H. Syd., *Annls mycol.* **21**: 189 (1923).
Stenocarpella zeae Syd., *Annls mycol.* **15**: 258 (1917).

Conidiomata 200-300 μ diam. Conidiogenous cells 8-15 × 3-4 μ. Conidia 0-3 septate, 44-82 × 7.5-11.5 μ.

On stems, seeds and leaves of *Zea mays*, Ghana, Malawi, Sierra Leone, Tanzania, U.S.A., Zambia, Nigeria, Sabah, Guinea, Cuba, Nepal, Brazil (IMI 12790 (isotype), 38605a, 40985, 6724a, 6397, 6398, 6396, 56925, 59802, 96612, 59547, 60333, 62099b, 69725, 4091a, 101946, 103914 110781b, 119681, 119679, 119667, 119731, 119732, 119735a, 119737, 119736, 119721, 120965a, 122716, 123818a, 139028, 142502a, 162393, 163426, 145639, 162938b, 173480).

51. TUNICAGO Sutton & Pollack, *Can. J. Bot.* **55**: 326 (1977).
Sp. typ.: *T. uniolae* Sutton & Pollack

Mycelium immersed, branched, septate, pale brown to hyaline. *Conidiomata* pycnidial, immersed, separate, dark brown to black, globose, unilocular, thick-walled; walls of dark brown, thick-walled textura angularis. *Ostiole* single, circular, central. *Conidiophores* hyaline, smooth, septate, irregularly branched at the base and above, formed from the inner cells of the conidiomatal wall. *Conidiogenous cells* enteroblastic, phialidic, integrated or discrete, determinate, apical or lateral, cylindrical or lageniform, channel wide, collarette and periclinal thickening minute. *Conidia* perprolate, brown, smooth, medianly 1 euseptate, guttulate; enclosed by a thick, brown, granular, rigid sheath at maturity.

The genus is monotypic. A distinctive feature of the developing conidia is the very irregular hyaline sheath which later becomes granular, brown and rigidly perprolate.

T. UNIOLAE Sutton & Pollack, *ibid.*: 327 (1977). (Fig. 261)

Conidiomata 155 µ wide × 130 µ thick. Ostiole 20 µ diam. Conidiophores up to 20 µ long × 3-4.5 µ wide. Conidia 10-14 × 5.5-8 µ (excluding sheath), 14-19 × 9.5-11 µ (including sheath).

On leaves and inflorescences of *Uniola paniculata*, Fla., U.S.A. (IMI 187331, holotype, BPI 71860, paratype).

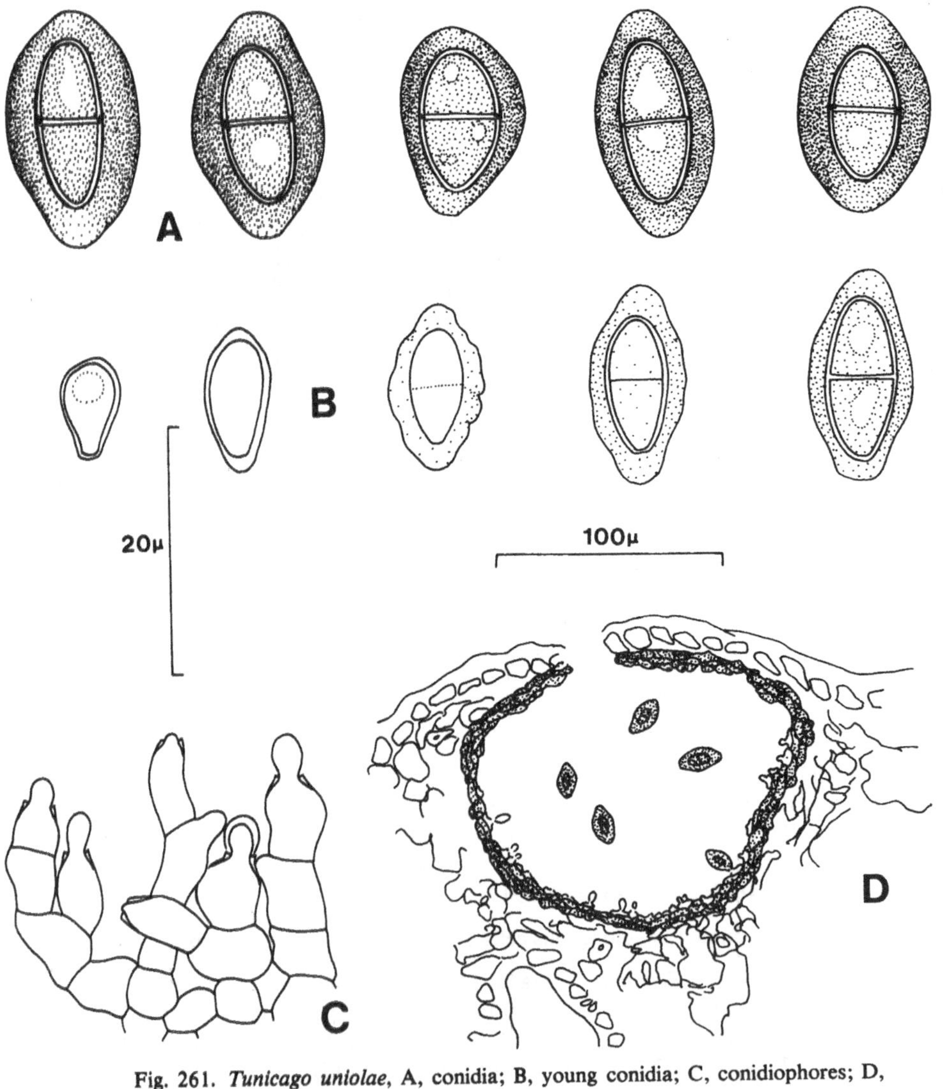

Fig. 261. *Tunicago uniolae*, A, conidia; B, young conidia; C, conidiophores; D, vertical section of a conidioma.

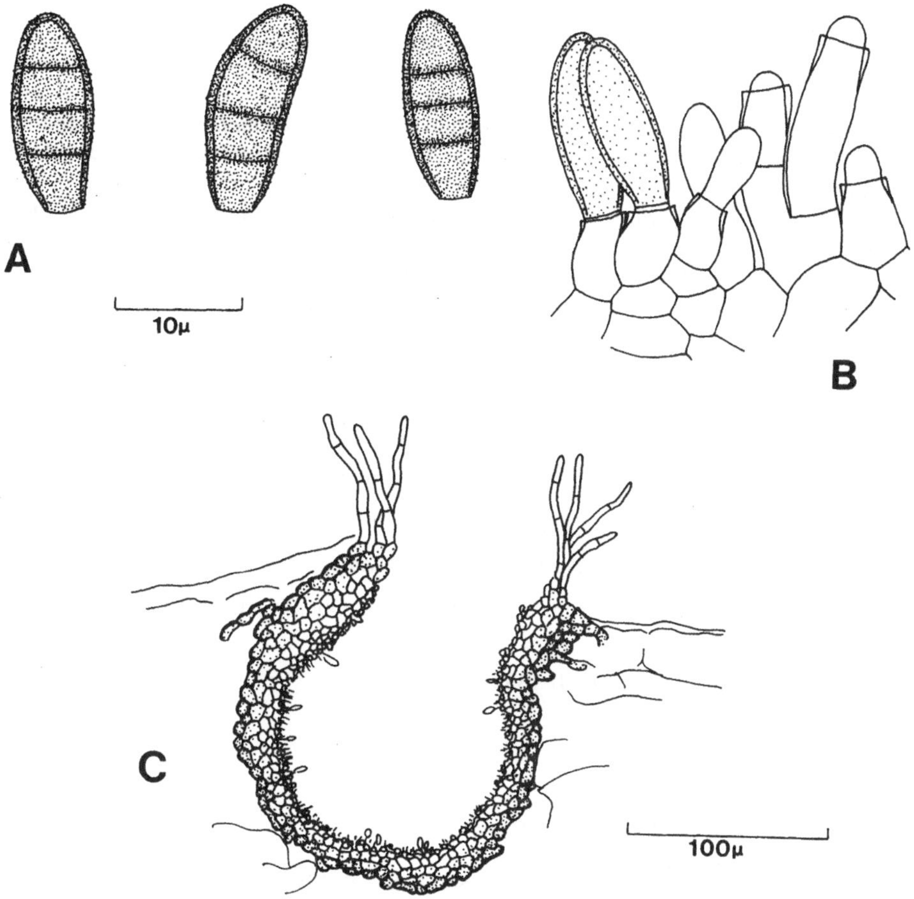

Fig. 262. *Angiopomopsis lophostoma*, A, conidia; B, conidiogenous cells and developing conidia; C, vertical section of a conidioma.

52. ANGIOPOMOPSIS Höhn., *Sber. Akad. Wiss. Wien* **121**: 407 (1912).
Sp. typ.: *A. lophostoma* Höhn.
Illustration: Sutton (1975d).

Immersed mycelium branched, septate, pale brown to hyaline. *Conidiomata* pycnidial, immersed, subepidermal, separate, globose to subglobose, dark brown; walls thick, composed of dark brown, thick-walled textura angularis, becoming hyaline and thinner-walled towards the inner conidiogenous region. *Ostiole* central, scarcely papillate, circular. *Setae* ostiolar, hyaline to subhyaline, 1-4 septate, straight or flexuous, unbranched, smooth. *Conidiophores* absent. *Conidiogenous cells* enteroblastic, phialidic, indeterminate, discrete, doliiform or short cylindrical, hyaline, smooth, with up to 1 percurrent proliferation, formed from the inner pycnidial wall cells. *Conidia* medium brown, 3 distoseptate, continuous, straight or slightly curved, broadly cymbiform, apex obtuse, base truncate, thick-walled, verrucose.

Sutton (1975d) provided a modern description of this monotypic genus which previously had been placed in synonymy with *Wojnowicia* Sacc. by Saccardo in Trotter (1931) and Clements & Shear (1931).

A. LOPHOSTOMA Höhn., *ibid.*: 407 (1912). (Fig. 262)

Conidiomata up to 200 μ diam.; walls 14.5-42 μ thick; ostiole 50-75 μ diam. Setae up to 85 × 10 μ. Conidiogenous cells 4.5-9.5 × 3-4.5 μ. Conidia 14-16 × 5.5-6.5 μ.

On dead leaves of *Phragmites*, Java (IMI 164118, holotype ex FH).

53. CERATOPYCNIS Höhn., *Sber. Akad. Wiss. Wien* **124**: 86 (1915).
Rhynchophorus Hollós, *Mat. Term. Közl. Mag. Tudom. Akad.* **35**: 54 (1926).
Sp. typ.: *C. clematidis* Höhn.

Mycelium immersed, branched, septate, pale to medium brown. *Conidiomata* pycnidial, separate or gregarious, rostrate, globose at the base, dark brown, immersed, unilocular, thick-walled; wall several cells thick, of dark brown textura angularis to globosa, paler towards the inside, consisting of thicker-walled textura prismatica in the rostrum. *Ostiole* single, circular, central, at the apex of the rostrum. *Conidiophores* absent. *Conidiogenous cells* enteroblastic, phialidic, determinate, discrete, doliiform to cylindrical, hyaline, smooth, channel relatively wide, collarette and periclinal wall thickening minimal, formed from the inner cells of the basal conidiomatal wall but not the rostrum. *Conidia* dark brown, 3-euseptate, continuous, base truncate, apex obtuse, thick-walled, verrucose, straight, cylindrical, often slightly narrower in the middle.

Morgan-Jones, Nag Raj & Kendrick (1972f) describe yellowish brown conidia as formed from proliferating phialides on septate, branched conidiophores, details which are at variance with my observations. Two bryophilous species were described by Racovitza (1959) but the structure of the conidiomata and conidial morphology are quite different from *C. clematidis*. Neither *C. muscorum* nor *C. bryophila* are congeneric with the type species. To which of the several pale brown spored pycnidial genera they belong is difficult to assess without examining conidiogenesis in the types.

C. CLEMATIDIS Höhn., *ibid.*: 86 (1915). (Fig. 263)
Rhynchophorus clematidis Hollós, *Mat. Term. Közl. Mag. Tudom. Akad.* **35**: 54 (1926).

Conidiomata up to 140 µ diam.; rostrum up to 150 µ long. Conidiogenous cells 6.5-10 × 5-7 µ. Conidia 18.5-28.5 × 8.5-9.5 µ.

On stems of *Clematis vitalba*, Pakistan (IMI 133347).

54. WOJNOWICIA Sacc., *Syll. fung.* **10**: 328 (1892).
Adella Petrak, *Annls mycol.* **34**: 228 (1936).
Guceviczia Glazer, *Not. Sist. crypt. Inst. bot. acad. Sci. U.S.S.R.* **12**: 167 (1959).
Sp. typ.: *W. hirta* Sacc.
Illustrations: Sutton (1975d).

Mycelium immersed, branched, septate, pale brown to hyaline. *Conidiomata* pycnidial, at first immersed, later appearing superficial by decay of host tissues, separate, globose, often markedly papillate, dark brown; walls thick, composed of dark brown, thick-walled textura angularis becoming hyaline and thin-walled towards the inner conidiogenous region. *Ostiole* central or displaced to one side, ± papillate, circular. *Setae* formed around the ostiole or from the lateral pycnidial walls, straight or flexuous, unbranched, brown, septate, smooth. *Conidiophores* absent. *Conidiogenous cells* enteroblastic, phialidic, determinate, discrete, doliiform to ampulliform, hyaline, smooth, channel and collarette minute, formed from the inner pycnidial wall cells. *Conidia* pale brown, with several transverse eusepta, continuous, straight or curved, fusiform or cylindrical, apex and base obtuse, thin-walled, smooth, ± guttulate.

Sutton (1975d) reviewed the genus and accepted the following two species.

W. HIRTA Sacc., *ibid.*: 328 (1892).
Hendersonia hirta Schröt., *Hedwigia* **29**: 61 (1890).
Wojnowicia tenella Pat., *Cat. rais. Pl. cell. Tunisie*: 122 (1897).
Hendersonia graminis McAlp., *Bull. Dept. Agr. Vict.* **9**: 9 (1904).
Wojnowicia graminis (McAlp.) Sacc. & D. Sacc., *Syll. fung.* **18**: 367 (1906).

Conidiomata up to 220 µ diam.; walls 25-40 µ thick; ostiole often subapical or lateral. Setae up to 150 × 5 µ. Conidiogenous cells 3.5-6 × 3-5 µ. Conidia 7-8 euseptate, falcate, 31-44 × 3-4 µ.

On culms of *Triticum sativum*, *T. aestivum*, *T.* sp., indet. Gramineae, Tunisia, Italy, Vict., S. Australia, Australia, Canada, Turkey, Eire, U.K. (IMI 167079, holotype *W. tenella* ex FH, 154294 ex PAD, 154293, holotype *H. graminis* ex Vict. Pl. Res. Inst. Burnley, 22773-776, 177192, 92920, 102667, 16858b, 95196, 95177, 223283).

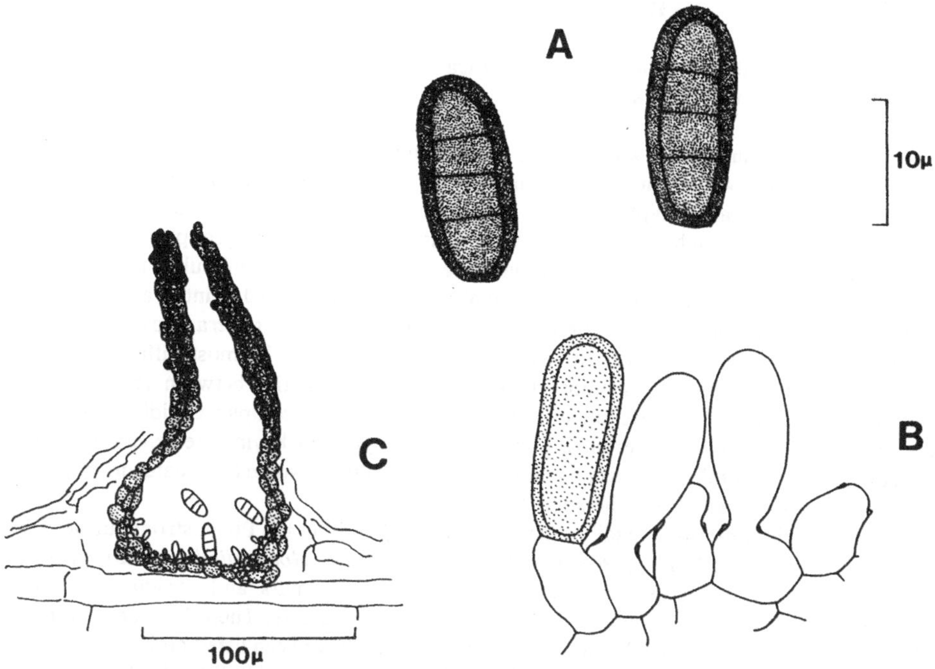

Fig. 263. *Ceratopycnis clematidis*, A, conidia; B, conidiogenous cells and developing conidia; C, vertical section of a conidioma.

W. EPHEDRAE Hollós, *Ann. Hist. nat. Hung.* **4**: 364 (1906).
Adella olympica Petrak, *Annls mycol.* **34**: 228 (1936).
Guceviczia setosa Glazer, *Not. Sist. crypt. Inst. bot. acad. Sci. U.S.S.R.* **12**: 167 (1959).

Conidiomata up to 300 µ diam.; walls 60 µ thick; ostiole apical. Setae up to 235 × 10 µ. Conidiogenous cells 6.5-11 × 3-5.5 µ. Conidia straight, 2-euseptate, 20.5-23 × 2.5-3 µ.

On leaves of *Buxus* sp. and stems of *Ephedra distachya*, France, Hungary (IMI 152606 ex K, holotype *A. olympica*, 155871, 155844, 155870, 155845 ex BP, 240962a).

PHIALOSTROMATINEAE subord. nov.

Cellulae conidiogenae phialidicae; conidiomata stromatica
Conidiogenesis phialidic; conidiomata stromatic

Whereas in the Blastostromatineae it was possible to separate a large number of genera into three groups based on the ability or inability to form a plurality of conidia from a single conidiogenous cell, this is not possible in the Phialostromatineae. A few genera are polyphialidic and some show percurrent proliferation, but since these only number 8 genera out of the 128 placed in the suborder such characters are of little value in grouping genera. Any other means of further subdividing the genera within the Phialostromatineae are fraught with difficulty, for the characters are rarely absolute and a gradation from one to another is quite normal. Discrete, as opposed to integrated conidiogenous cells could have been perhaps the most useful, although there would be difficulties with genera showing both types of insertion. Similarly any separation based on differences in stromatic characters would be most difficult to apply because of the inherent variability of the conidiomata in this suborder. Not only is the distinction between acervulus and simple eustroma sometimes unclear but the recognition of different stromatic types in the more complex conidiomata would lead to the fragmentation of the suborder into many small groups. I have therefore opted for keying the genera in the Phialostromatineae as a whole and clearly, in the synoptic key, this has unfortunately resulted in some large categories. In addition, a synoptic key dealing with 128 genera is somewhat cumbersome to use.

There are several well-circumscribed readily recognisable groups within the Phialostromatineae. The most obvious is the *Dinemasporium, Diarimella, Polynema, Pseudolachnea, Stauronema, Hoehneliella, Dimastigosporium* group, all of which genera are characterized by setose, cupulate to acervular conidiomata and conidia with setulose appendages. Genera are mainly separated by the septation and arrangement of the setulae. There is a large complex of genera with acervular conidiomata and hyaline, unicellular conidia, many of which have been listed and compiled by von Arx (1957a, 1970). In separating these it is important to stress the relationships of the acervuli to the substrate (subcuticular, epidermal, subepidermal), the insertion and arrangement of the conidiogenous cells, the phialidic apparatus and the precise nature of conidial morphology. In a group where there is a comparatively small number of criteria available for identification it is the more subtle criteria that are eventually employed in distinguishing taxa. This group includes *Asteroma, Rhodesia Discosporina, Cylindrosporium, Discula, Discogloeum, Monostichella, Gloeosporidina, Cryptocline* and *Cryptosporiopsis*. There is also a multiplicity of genera with complex, mainly immersed, multilocular to convoluted eustromata, filiform, branched conidiophores and minute hyaline aseptate conidia. These include *Pleuroplaconema, Cytogloeum, Zythiostroma, Endothiella, Hainesia, Waydora, Stilbophoma, Kendrickomyces, Pragmopycnis, Sirodothis, Kaskaskia, Cytosporella, Cytonaema,*. *Pleurocytospora, Sporonema, Microdiscula* and *Amphicytostroma*. In this group the conidiophore morphology and the structure of the conidiomata are of prime importance. Some stromata are highly distinctive, as in *Waydora, Stilbophoma, Kendrickomyces* and *Pragmopycnis*, and as such are more important for identification than other characters. Another well-defined group is comprised of *Corniculariella, Gelatinosporium* and *Foveostroma* and here again, primary distinctions between genera are in tissue structure of the conidiomata.

DICHOTOMOUS KEY

(numbers in parentheses are species numbers)

1. Conidia branched 2
1. Conidia with appendages or setulae 3
1. Conidia neither branched nor with appendages 14
    2. Conidiomata eustromatic, flat, unilocular; conidia Y-shaped; conidiogenous cells polyphialidic .... *Suttoniella* (1)
    2. Conidiomata eustromatic, applanate to pulvinate, uniloculate; conidia consisting of a 2-celled base and 3 septate, divergent arms .... *Eriosporella* (2)
3. Conidiomata setose 4
3. Conidiomata not setose 8
    4. Conidiomata hysteriform; phialides inclined to one side at the apex; conidia allantoid, guttulate, with 3-6 setulae at the apex and base .... *Diarimella* (3)
    4. Conidiomata cupulate to flat-cupulate 5
5. Conidia septate 6
5. Conidia aseptate 7
    6. Conidia hyaline or pale brown and verruculose, 1 septate with a setula at the apex and 1-3 just above the base .... *Polynema* (7)

6. Conidia 1-many septate, with one or more simple or branched setulae at apex and base .... *Pseudolachnea* (5)
6. Conidia 1 septate, hyaline to pale brown, each end with an unbranched or branched appendage .... *Hoehneliella* (8)
7. Conidia with a single setula at the apex and base, sometimes pale brown .... *Dinemasporium* (4)
7. Conidia with a setula at each end and 1-2 medianly inserted setulae .... *Stauronema* (6)
7. Conidia pyriform with 2 divergent cellular appendages at the apices .... *Dimastigosporium* (9)
8. Appendage mucilaginous 9
8. Appendage cellular 10
9. Conidiomata acervular; conidia cylindrical, aseptate, hyaline, slightly constricted in the middle; appendage apical .... *Rhabdogloeum* (10)
9. Conidiomata eustromatic; conidia ellipsoid, medianly 1-septate, hyaline, with an apical mucilaginous cap; hyperparasitic on Uredinales .... *Sphaerellopsis* (11)
9. Conidiomata eustromatic; conidia ellipsoid, medianly 1-septate, hyaline, with filiform apical mucilaginous appendages; conidiogenous cells often percurrently proliferating .... *Comatospora* (see p. 257)
9. Conidiomata pseudostromatic; conidia cylindrical, aseptate, hyaline, with a small everted apical mucilaginous cap .... *Ceuthospora* (12)
10. Appendage(s) apical 11
10. Appendage basal 12
10. Appendages apical and basal 13
11. Conidiomata acervular; conidia aseptate, falcate to fusiform, apex prolonged into a short appendage.. *Monochaetiella* (13)
11. Conidiomata eustromatic, effuse, black; conidiogenous cells often polyphialidic; conidia aseptate, very irregular, apex prolonged into an irregular appendage .... *Diachorella* (14)
11. Conidiomata pseudostromatic; conidia hyaline, aseptate, ellipsoid, with an apical, filiform cellular appendage separated by a septum.. *Neoplaconema* (15)
12. Conidiomata eustromatic, immersed, convoluted and irregularly multilocular; conidia allantoid, aseptate, with a basal filiform cellular appendage not separated by a septum .... *Strasseria* (16)
12. Conidiomata eustromatic, superficial, multilocular; conidia fusiform, basal cell modified as an appendage and separated by a septum .... *Strasseriopsis* (17)
13. Conidiomata eustromatic, flattened, unilocular; conidia pale brown, 1 septate, with two apical cellular appendages and one basal cellular appendage .... *Ciliochorella* (18)
13. Conidiomata eustromatic, multilocular, circular to longitudinal; conidia hyaline, aseptate, fusiform, with an apical and an exogenous basal unbranched setula .... *Placonema* (19)
14. Conidia hyaline, aseptate 15
14. Conidia brown and/or septate 50
15. Conidiomata eustromatic, superficial, multilocular, attached to the substrate by a hypostroma; conidiogenous cells lageniform; conidia stellate, with up to 6 short protuberances .... *Tribolospora* (20)
15. Conidia lacking any protuberances 16
16. Conidiogenous cells ampulliform, doliiform to lageniform 17
16. Conidiogenous cells cylindrical, only occasionally on septate conidiophores 25
16. Conidiogenous cells integrated on septate conidiophores which are distinct from, and raised above the wall tissue on which they arise 34
17. Conidiomata eustromatic, lifting a flap of host tissue at dehiscence; conidia falcate, fusiform .... *Pseudostegia* (50)
17. Conidiomata infundibuliform to nidulariaceous, superficial; conidia cylindrical .... *Satchmopsis* (21)
17. Conidiomata not so 18
18. Conidiomata subcuticular 19
18. Conidiomata epidermal to subepidermal although at maturity some appearing superficial 20
18. Conidiomata superficial 24
19. Conidiomatal surface cerebriform, with an upper and lower wall, uni- or multilocular; phialides over both walls; conidia ellipsoid, multiguttulate .... *Myxothyrium* (34)
19. Conidiomata with upper and lower wall; phialides restricted to lower wall; conidia cylindrical to ellipsoid, variable, eguttulate .... *Pseudothyrium* (35)
19. Conidiomata with only a lower wall (acervular), discolouring the cuticle; conidia variable, straight or curved, ellipsoid to fusiform.. *Asteroma* (36)
20. Conidiomata acervular to sporodochial; phialides discrete or terminal on closely appressed conidiophores; conidia ellipsoid .... *Kabatina* (22)
20. Conidiomata eustromatic 21

21. Conidiomata with a central column of tissue surrounded by a periclinal locule; conidia cylindrical, uniguttulate .... *Cyclodomus* (23)
21. Conidiomata regularly multilocular 22
21. Conidiomata unilocular 23
    22. Conidiomata remaining immersed, discrete; conidia straight or slightly curved; conidial *Omphalospora* .... *Podoplaconema* (24)
    22. Conidiomata remaining immersed, dothideoid; conidia bacilliform; spermatial *Dothidea* .... *Asteromellopsis* (25)
    22. Conidiomata strongly erumpent, becoming superficial; conidia pyriform .... *Sclerophoma* (26)
23. Conidiogenous cells polyphialidic; conidia comparatively large, ellipsoid .... *Sarcophoma* (27)
23. Conidiomata pale brown to hyaline, appearing violet-purple under low power; monophialidic; conidia ellipsoid to cylindrical .... *Calocline* (28)
23. Conidiomata golden brown, occasionally irregularly multilocular; conidia ellipsoid to pyriform .... *Pseudosclerophoma* (29)
23. Conidiomata dark brown, linear; conidia short cylindrical .... *Sphaeriothyrium* (31)
23. Conidiomata dark brown to black, pulvinate; conidia cylindrical to oblong .... *Paradiscula* (30)
    24. Conidiomata sessile or short stipitate, widely cupulate; conidia spherical .... *Acleistia* (32)
    24. Conidiomata attached to substrate by a central column or foot, black, applanate to pulvinate or stellate, multilocular, cylindrical; entomogenous .... *Mattirolella* (33)
25. Conidiomata acervular 26
25. Conidiomata eustromatic 31
    26. Conidiomata usually setose (some species lacking setae); conidia cylindrical or falcate; appressoria formed readily in culture .... *Colletotrichum* (51)
    26. Conidiomata not setose 27
27. Conidiomata subcuticular, discolouring the cuticle; phialides lageniform to cylindrical, on short conidiophores; conidia ellipsoid, straight or slightly curved, comparatively large .... *Monostichella* (46)
27. Conidiomata epidermal, subepidermal or subperidermal 28
    28. Conidia fusiform, usually eguttulate; conidiogenous cells monophialidic with comparatively long phialidic channel .... *Rhodesia* (37)
    28. Conidia ellipsoid, doliiform, cylindrical or pyriform 29
29. Conidia cylindrical to ellipsoid 30
29. Conidia pyriform, often in chains; conidiophores long cylindrical .... *Gloeosporidina* (47)
29. Conidia broadly doliiform, solitary; conidiogenous cells wide, with distinct collarettes .... *Cryptocline* (48)
    30. Conidia minute; conidiogenous cells often polyphialidic; associated with scab and anthracnose damage .... *Sphaceloma* (52)
    30. Conidiomata subperidermal, with a sterile pulvinate cushion in the centre; conidia cylindrical, multiguttulate .... *Discosporina* (39)
    30. Conidiomata white, often pushing up a flap of tissue at dehiscence; conidia cylindrical .... *Cylindrosporium* (40)
    30. Conidiomata pale brown; conidia ellipsoid to pyriform, fairly regular .... *Discula* (41)
    30. Conidiomata pale brown; conidia ellipsoid, cylindrical, curved, pyriform, irregular .... *Discogloeum* (42)
31. Conidiomata very dark brown to black 32
31. Conidiomata pale brown 33
    32. Stromata widely effuse, confluent, multilocular; conidia irregular in shape, often curved, multiguttulate .... *Apomelasmia* (43)
    32. Conidiomata discrete, flattened, circular, unilocular; conidia ellipsoid to cylindrical or slightly curved .... *Cryptomycella* (44)
33. Conidiomata extremely variable, from simple acervuli to multilocular, rostrate eustromata; conidia straight, ellipsoid to pyriform, large, with a protuberant scar .... *Cryptosporiopsis* (45)
33. Conidiomata reddish brown, unilocular; conidia ellipsoid .... *Erythrogloeum* (38)
33. Conidiomata immersed cupulate; phialides with wide channel at the apex, little periclinal thickening; conidia catenate, guttulate, cylindrical, broad at the base .... *Sirexcipula* (49)
    34. Conidiomata acervular 35
    34. Conidiomata superficial, or at least with the locules elevated above the substrate 36
    34. Conidiomata neither superficial nor simple acervuli 44
35. Conidia fusiform, either straight or falcate, often with accompanying microconidia .... *Gloeosporidiella* (53)
35. Conidia small, ellipsoid; conidiophores very irregularly branched .... *Pleuroplaconema* (54)

35. Conidia ellipsoid, straight or slightly curved; conidiophores long, filiform .... *Cytogloeum* (55)
  36. Conidiomata hypocreaceous (reddish, yellowish) 37
  36. Conidiomata brown to black 38
37. Conidiomata entomogenous on scale insects, extremely variable in shape and pigmentation; conidia fusiform, straight; paraphyses present .... *Aschersonia* (56)
37. Conidiomata globose to subglobose, of textura intricata; conidia spherical; resinicolous .... *Pycnidiella* (57)
37. Conidiomata reddish brown, multilocular and convoluted but with a single ostiole; conidia minute, bacilliform .... *Zythiostroma* (58)
37. Conidiomata reddish to orange brown, multilocular, often with a single rostrate ostiole; conidia ellipsoid .... *Endothiella* (59)
  38. Conidiomata cupulate 39
  38. Conidiomata spherical to reniform, with a longitudinal raphe, setose; conidia fusiform, guttulate, sometimes curved .... *Chaetomella* (60)
  38. Conidiomata cylindrical, of textura porrecta; conidia ellipsoid, minute .... *Cornucopiella* (61)
  38. Conidiomata not cupulate 40
39. Conidiomata dark brown, thick-walled; conidia fusiform, straight to curved .... *Ascochytopsis* (62)
39. Conidiomata pale brown, thin-walled; conidia fusiform, often curved, guttulate .... *Hainesia* (63)
39. Conidiomata pulvinate to verruciform, becoming excipuliform, multilocular; conidia ellipsoid, minute .... *Agyriellopsis* (64)
  40. Conidiomata consisting of a thick-walled, dark brown to black upper wall, unilocular; conidiophores either ramified or synnematous 41
  40. Conidiomata consisting of several locules but appearing like botryose pycnidia at low power 42
  40. Conidiomata with locules arranged like orange segments but with a single ostiole; conidia fusiform, straight or allantoid .... *Waydora* (65)
  40. Conidiomata comprised of a single multilocular or convoluted stroma 43
41. Conidiophores synnematous; conidia ellipsoid .... *Stilbophoma* (66)
41. Conidiophores irregularly ramified; conidia allantoid .... *Pragmopycnis* (67)
  42. Conidiophores with conidiogenous loci immediately below transverse septa; conidia bacillar .... *Sirodothis* (68)
  42. Conidiophores irregularly branched, mostly with terminal conidiogenous loci; conidia ellipsoid .... *Kaskaskia* (69)
43. Conidioma like a large pycnidium but convoluted and multilocular inside and with a single ostiole; conidia cylindrical, straight .... *Camaropycnis* (70)
43. Conidiomata more or less pulvinate but convoluted and multilocular; dehiscence by breakdown of upper layers; conidia ellipsoid to oval .... *Cytosporella* (71)
43. Conidiomata spherical, pulvinate or moriform, multilocular and convoluted; ostioles numerous; conidia biguttulate, cylindrical .... *Dothiorina* (72)
  44. Conidiomata associated with tar-spots; conidia cylindrical, tapered to the base .... *Melasmia* (73)
  44. Conidiomata flat, hemispherical, circular; conidia falcate, fusiform .... *Pilidium* (74)
  44. Conidiomata applanate, subcuticular, thin-walled; collarettes distinct; conidia cylindrical, straight .... *Siroplacodium* (75)
  44. Conidiomata immersed 45
45. Conidiomata rostrate, convoluted; conidia straight or slightly curved .... *Cytonaema* (76)
45. Conidiomata pseudostromatic, unilocular, yellowish brown; conidia pyriform .... *Rhodosticta* (77)
45. Conidiomata complex, consisting of several different layers, locules initiated in upper layers and merging into 1-2 furfuraceous ostioles; conidia allantoid, fusiform .... *Kendrickomyces* (78)
45. Conidiomata not so 46
  46. Conidia fusiform, usually biguttulate, sometimes with, or replaced by, hamate β-conidia .... *Phomopsis* (79)
  46. Conidia pyriform to globose; phialides consistently percurrently proliferating .... *Discosporium* (80)
  46. Conidia ellipsoid or variable in shape; monophialidic .... *Phacidiopycnis* (81)
  46. Conidia distinctly allantoid, falcate and fusiform 47
  46. Conidia straight or only slightly curved 48
47. Conidiomata convoluted and multilocular but with a single ostiole; conidia small, allantoid .... *Cytospora* (82)
47. Conidiomata comparatively simple; conidia larger, fusiform .... *Phlyctema* (83)
  48. Conidiomata pulvinate, regularly multilocular; conidia straight, cylindrical .... *Blennoria* (84)
  48. Conidiomata not pulvinate 49
49. Conidiomata irregularly multilocular; conidiophores sparingly branched; paraphyses present; conidia cylindrical .... *Plectophomella* (85)

49. Conidiomata irregularly multilocular; conidiophores filiform, much branched; conidia straight or slightly curved .... *Pleurocytospora* (86)
49. Conidiomata unilocular to convoluted; conidiophores branched irregularly; conidia straight, fusiform *Sporonema* (87)
49. Conidiomata multilocular, convoluted; conidiophores filiform, much branched; conidia ellipsoid, minute .... *Microdiscula* (88)
49. Conidiomata formed in association with *Amphiporthe* state, convoluted; conidiophores sparingly branched; conidia straight or slightly curved .... *Amphicytostroma* (89)
 50. Conidia hyaline, 1-septate 51
 50. Conidia hyaline, multiseptate 59
 50. Conidia brown 66
51. Conidiomata acervular to cupulate, setose; conidia medianly 1 septate, cylindrical, with 2-3 apical to subapical and basal, unbranched, cellular appendages .... *Dwayalomella* (90)
51. Conidia lacking appendages 52
 52. Conidiomata superficial, tubular, of textura porrecta; conidiophores branched; conidiogenous loci apical; conidia ellipsoid, medianly 1 septate .... *Nanoschema* (91)
 52. Conidiomata strongly erumpent and appearing to be superficial 53
 52. Conidiomata immersed 54
53. Conidiomata consisting of a thick pseudostromatic base surmounted by globose conidiogenous structures that are multilocular; conidiophores irregularly branched; conidia medianly 1 euseptate, ellipsoid, guttulate .... *Fuckelia* (92)
53. Conidiomata eustromatic, conical to rostrate, of textura angularis, unilocular; conidiophores irregularly branched; conidia curved to sigmoid, 1 septate .... *Cornicularíella* (93)
53. Conidiomata eustromatic, pulvinate, of textura oblita, uni- to multilocular; conidiophores irregularly branched; conidia curved to sigmoid, 1 septate .... *Gelatinosporium* (94)
53. Conidiomata eustromatic, pulvinate, becoming furfuraceous, of textura intricata to angularis, multilocular; conidiophores irregularly branched; conidia curved to sigmoid, 1 septate, guttulate .... *Foveostroma* (95)
 54. Conidiomata acervular; conidiogenous cells lageniform or integrated; conidia 0-1 septate, fusiform, straight or slightly curved .... *Cylindrogloeum* (96)
 54. Conidiogenous cells discrete 55
 54. Conidiogenous cells integrated on filiform or short septate conidiophores 56
55. Conidiomata eustromatic, very pale brown, unilocular; phialides lageniform to cylindrical; conidia medianly 1 septate, guttulate, fusiform .... *Diplozythiella* (97)
55. Conidiomata eustromatic, dark brown to black, multilocular; phialides ampulliform; conidia ellipsoid, 0-1 septate .... *Ceuthodiplospora* (98)
55. Conidiomata pseudostromatic, linear, dark brown, unilocular and convoluted; phialides ampulliform; conidia 0-1 septate, fusiform .... *Placonemina* (99)
 56. Conidiophores short 57
 56. Conidiophores long, filiform 58
57. Conidiomata thin-walled, globose, hyperparasitic in *Phyllachora*; conidiophores branched very irregularly; conidia straight or slightly curved, medianly 1 septate, variably guttulate .... *Davisiella* (100)
57. Conidiomata pale brown, thin-walled, unilocular; conidiophores short, sparingly septate, unbranched; conidia medianly 1 septate, guttulate, straight .... *Helhonia* (101)
 58. Conidiophores filiform; conidiogenous apertures apical on main axes and short lateral branches arising below septa; conidia fusiform, markedly tapered to the apex, less so to the base .... *Sirococcus* (102)
 58. Conidiophores filiform, irregularly septate and branched, conidiogenous apertures apical or immediately below septa; conidia fusiform to ellipsoid, not markedly tapered to the apex .... *Diplodina* (103)
59. Conidia bearing a foot-cell (as in *Fusarium*) 60
59. Conidia anastomosing after formation, then appearing like branched complex conidia .... *Titaeospora* (105)
59. Conidia cylindrical, with several dichotomously or irregularly branched setulae at each end .... *Dilophospora* (104)
59. Conidia not as above 61
 60. Conidiomata acervular; conidia 1-2 septate, fusiform, falcate .... *Pycnofusarium* (106)
 60. Conidiomata superficial, globose, reddish brown, with a grey-white tomentum; conidiophores irregularly branched; conidia 3 septate, fusiform, falcate .... *Botryocrea* (107)
61. Conidiomata acervular 62
61. Conidiomata eustromatic 63
 62. Conidiomata formed in association with extensive dark brown pseudostroma; phialides lageniform to cylindrical .... *Cheilaria* (108)

62. Conidiophores septate, branched; conidiogenous loci apical and on short branches immediately below septa; conidia variable in shape, very rarely branched, 1-3 septate .... *Myriellina* (109)
62. Conidiophores sparingly septate and branched; phialides broad with a wide channel and marked periclinal thickening; conidia 1-3 septate, straight or curved, obovoid .... *Septogloeum* (110)
63. Conidiomata unilocular 64
63. Conidiomata multilocular 65
64. Conidiomata eustromatic; conidiophores branched, septate; conidiogenous loci apical and lateral just below septa; conidia long filiform, multiseptate, guttulate .... *Phlyctaeniella* (111)
64. Conidiomata eustromatic, depressed, flattened, circular; phialides discrete or integrated on short septate conidiophores; conidia 2-3 septate, fusiform, falcate .... *Periperidium* (112)
65. Conidiomata initially immersed, later strongly erumpent and appearing superficial; conidiophores branched, septate, filiform; conidia 2-5 septate, fusiform, falcate .... *Brunchorstia* (113)
65. Conidiomata immersed, linear; phialides ampulliform, discrete or on short, septate, branched conidiophores; conidia 1-2 septate, fusiform but irregular in shape .... *Cytoplacosphaeria* (114)
66. Conidia aseptate 67
66. Conidia septate 72
67. Conidiomata setose, superficial, globose; phialides cylindrical; paraphyses branched; conidia pale brown, smooth, fusiform .... *Amerosporium* (115)
67. Conidiomata not setose 68
68. Conidiomata acervular 69
68. Conidiomata eustromatic 70
69. Conidia rhomboid .... *Vanderystiella* (116)
69. Conidia fusiform or oval; conidiophores branched .... *Greeneria* (117)
69. Conidia globose to pyriform, thick-walled, verruculose; phialides ampulliform .... *Fairmaniella* (118)
70. Phialides ampulliform; conidia verruculose; conidiomata multilocular .... *Cytoplea* (119)
70. Phialides cylindrical 71
71. Conidiomata superficial to semi-immersed, unilocular, convoluted or multilocular; conidia globose, smooth; resinicolous .... *Epithyrium* (120)
71. Conidiomata immersed, multilocular; phialides large, percurrently proliferating; conidia globose, thick-walled, smooth, dark brown .... *Neomelanconium* (121)
71. Conidiomata immersed, multilocular; phialides occasionally integrated; conidia pale brown, smooth, short cylindrical, often with a central guttule .... *Cyclothyrium* (122)
71. Conidiomata immersed, multilocular; paraphyses present; conidia cylindrical to fusiform, smooth, thin- or thick-walled .... *Phaeocytostroma* (123)
72. Conidia 1 septate 73
72. Conidia multiseptate 74
73. Conidiomata eustromatic, unilocular to multilocular; phialides cylindrical; conidia smooth, fusiform, pale brown, with a thickened scar at each end .... *Hendersonina* (124)
73. Conidiomata eustromatic, linear, unilocular; phialides ampulliform; conidia ellipsoid, medianly 1 septate .... *Placodiplodia* (125)
74. Conidia euseptate 75
74. Conidia distoseptate 76
75. Conidiomata eustromatic, multilocular, immersed; conidia 3-4 septate, verruculose .... *Endocoryneum* (126)
75. Conidiomata eustromatic, multilocular, immersed; phialides percurrently proliferating; conidia 2-7 septate, smooth; microconidia phialidic, allantoid .... *Hendersoniopsis* (127)
76. Conidia transversely septate, thickened at each end .... *Massariothea* (128)
76. Conidia muriform .... *Camarographium* (129)
76. Conidia muriform, with a mucilaginous basal cap; conidiogenous cells often percurrent........ *Shearia* (see p. 257)

SYNOPTIC KEY

Mycelial characters

Position in relation to substrate

a. immersed 1 2 3 4 5 6 7 8 10 11 12 13 14 15 16 17 18 19 20 21 22 23 24 25 26 27 28 29 30 31 32 33 34 35 36 37 38 39 40 41 42 43 44 45 46 47 48 49 50 51 52 53 54 55 57 58 59 60 61 62 63 64 65 66 67 68 69 70 71 72 73 74 75 76 77 78 79 80 81 82 83 84 85 86 87 88 89 90 91 92 93 94 95 96 97 98 99 100 101 102 103 104 105 106 107 108 109 110 111 112 113 114 115 116 117 118 119 120 121 122 123 124 125 126 127 128

b. superficial 32 56 57 60 64 107 120

c. haustorial 33

d. subiculate 56

e. fibrillar 36

Conidiomatal characters

Position in relation to substrate

a. immersed 1 2 7 10 11 12 13 14 15 16 18 19 22 23 24 25 26 27 28 29 30 31 34 35 36 37 38 39 40 41 42 43 44 45 46 47 48 49 50 51 52 53 54 55 71 73 74 75 76 77 78 79 80 81 82 83 84 85 86 87 88 89 90 95 96 97 98 99 100 101 102 103 104 105 106 108 109 110 111 112 113 114 116 117 118 119 121 122 123 124 125 126 127 128

b. subcuticular 5 7 18 34 35 36 46 51 75

c. semi-immersed 11 30 58 59 63 68 69 71 92 93 94 113 120

d. superficial 3 4 5 6 7 8 9 17 20 21 26 32 33 56 57 58 59 60 61 62 63 64 65 66 67 68 69 70 72 81 91 92 93 94 95 107 115 120

Dehiscence

a. ostiolate, papillate 12 16 58 59 72 78 79 82 85 98 99 104 114 123 128

b. ostiolate, non-papillate 11 17 18 20 28 45 57 65 70 77 88 93 100 126 127

c. ostiolate, depressed 15

d. ostiolate, rostrate 59 76 93 128

e. ostiolate, furfuraceous 12 49 78 82 95 127

f. linear-longitudinal 3 19 30 44 60 75 125

g. stellate 74

h. lifting a flap of host and/or fungal tissue 40 50 66 81

i. irregular rupture 1 2 4 5 6 7 8 9 10 13 14 19 21 22 23 24 25 26 27 29 31 32 33 34 35 36 37 38 39 40 41 42 43 44 45 46 47 48 49 51 52 53 54 55 56 61 62 63 64 65 67 68 69 71 73 81 83 84 85 87 89 90 91 92 94 95 96 97 101 102 103 105 106 107 108 109 110 111 112 113 115 116 117 118 119 120 121 122 123 124

Stromatal type

a. acervular — 5 7 10 13 22 36 37 38 39 40 41 42 45 46 47 48 51 52 53 55 80 90 96 105 106 108 109 110 116 117 118

b. cupulate — 3 4 5 6 7 8 9 32 62 63 64 90

c. simple eustromatic — 1 2 11 18 27 29 30 43 44 45 49 57 60 75 76 79 80 81 83 87 93 97 100 101 111 115 120 125 128

d. complex eustromatic — 16 23 24 25 26 54 73 78 79 82 85 86 88 89 98 103 104 114 119 121 122 123 124 126 127 128

e. complex eustromatic, pulvinate — 17 20 26 30 33 56 58 59 65 66 67 68 69 70 71 72 81 84 92 94 95 102 107 113

f. tubular — 21 61 91 93

g. linear — 3 19 28 31 114 125

h. flattened — 1 14 18 34 35 43 50 74 112

i. pseudostromatic — 12 15 77 99

j. locules arranged like orange segments — 65

k. pyriform to conic — 70

l. clypeate — 16 104

Locular configuration

a. unilocular — 1 2 3 4 5 6 7 8 9 10 11 13 15 18 21 24 27 28 29 30 31 32 34 35 36 37 38 39 40 41 42 44 45 46 47 48 49 50 51 52 53 55 57 59 60 61 62 63 66 67 74 75 76 79 80 81 83 87 90 91 93 96 97 99 100 101 105 106 107 108 109 110 111 112 115 116 117 118 120 125 128

b. multilocular — 11 17 19 20 23 33 59 64 65 73 77 79 80 81 89 98 104 114 119 120 128

c. irregularly multilocular, convoluted — 12 14 16 24 25 26 34 45 54 56 57 58 59 68 69 70 71 72 77 78 79 82 84 85 86 87 88 92 94 95 99 102 103 113 120 121 122 123 124 126 127

d. with a central sterile area — 39 45

Tissue type

a. textura angularis — 1 2 3 4 5 6 7 8 9 10 11 12 13 14 15 16 17 19 21 22 23 25 26 27 28 29 30 31 33 34 35 36 37 38 39 40 41 42 43 45 46 47 48 49 51 52 53 54 58 59 60 62 63 64 65 66 67 68 69 70 71 73 74 75 76 78 79 80 81 82 83 84 85 87 91 92 93 95 96 97 98 99 100 101 102 103 104 105 107 108 109 110 111 112 113 114 115 116 117 119 120 121 122 123 124 125 126 127 128

b. textura epidermoidea — 108 118

c. textura globulosa — 19 24 29 50 62 86

d. textura intricata — 2 4 12 15 17 20 32 55 56 57 63 65 71 73 76 77 80 82 88 89 90 92 95 104 106 120 126 128

e. textura oblita — 65 72 78 93 94

f. textura porrecta — 3 4 5 7 8 59 61 63 91

g. textura prismatica — 1 18 21 33 35 44 107 118

Pigmentation (not including acervular types)	
a. dark brown	1 2 3 4 5 6 7 8 9 11 12 15 16 17 18 19 20 21 23 24 25 26 27 29 30 31 33 34 35 43 44 45 50 60 61 62 64 65 66 67 68 69 70 71 72 73 74 75 76 78 79 80 81 82 83 84 85 86 87 88 89 90 91 92 93 94 95 97 98 99 100 101 102 103 104 111 112 113 114 115 119 120 121 122 123 124 125 126 127 128
b. hypocreaceous	28 32 56 57 63 77 106 107
c. white	40
d. red	58 59
e. tar spot or black and effuse	14 43 73 108
f. discolouring the cuticle	36 46
g. pruinose	59 68 95

Setal characters

a. present	3 4 5 6 7 8 9 51 60 63 90 115
Pigmentation	
a. brown	3 4 5 6 7 8 9 51 60 90 115
b. hyaline	63
Septation	
a. septate	3 4 5 6 7 8 9 51 60 63 115
b. aseptate	90
Branching	
a. unbranched	3 4 5 6 7 8 51 60 63 90 115
b. branched at the base	9
Ornamentation	
a. smooth	3 4 5 6 7 8 9 51 60 63 90
b. verrucose	115
c. coiled	60
d. acuminate, straight	3 4 5 6 7 8 9 51 60 90 115

Conidiogenous cell characters

Insertion	
a. integrated	2 4 5 6 7 8 9 11 12 14 15 16 19 21 22 37 39 40 41 42 43 44 47 49 51 52 53 54 55 56 57 58 59 60 61 62 63 64 65 66 67 68 69 70 71 72 73 74 75 76 77 78 79 80 81 82 83 84 85 86 87 88 89 91 92 93 94 95 96 100 101 102 103 105 106 107 109 110 111 112 113 117 120 122 123 124 127
b. discrete	1 2 3 10 11 13 14 15 16 17 18 20 21 23 24 26 27 28 29 30 31 32 33 34 35 36 37 38 39 40 41 42 43 44 45 46 47 48 49 50 51 52 62 77 79 81 90 96 97 98 99 101 104 108 110 112 114 115 116 118 119 121 122 123 124 125 126 127 128

Shape

a. ampulliform-doliiform — 1 11 13 17 20 21 23 24 26 27 28 29 31 35 36 46 50 52 85 98 99 100 103 104 110 114 118 119 125 126 128

b. lageniform — 1 10 11 13 15 18 19 20 21 30 34 35 36 38 44 46 51 53 57 64 73 77 79 81 85 92 96 97 100 103 108 111 112 119 124

c. cylindrical — 2 4 5 6 7 8 9 10 12 14 16 19 22 28 32 33 35 36 37 38 39 40 41 42 43 45 46 47 48 49 50 51 53 54 55 56 57 58 59 60 61 62 63 64 65 66 67 68 69 70 71 72 73 74 75 76 77 78 79 80 81 82 83 84 86 87 88 89 90 91 93 94 95 97 100 101 102 103 105 106 107 108 109 110 113 115 116 117 120 121 122 123 124 126 127 128

d. cylindrical but curved at the apices — 3

Other characters

a. polyphialidic — 1 14 27 52 67

b. percurrently proliferating — 80 121 127

c. enclosed in mucilage — 115

d. brown — 52 127

e. collarettes long — 75 127

f. forming catenate conidia — 47 49

Conidiophore characters

Shape

a. irregularly branched and ramified, or branches very short and almost indistinguishable from stromatal wall — 54 67 85 100

b. long cylindrical to filiform, unbranched or branched sparingly at the base — 2 11 14 15 16 19 37 39 41 42 44 47 49 51 52 62 77 79 80 81 84 94 96 101 110 112 122 123 124 127

c. filiform, much branched — 4 5 6 7 8 9 12 21 22 40 43 47 53 55 56 57 58 59 60 61 62 63 64 65 66 68 69 70 71 72 73 74 75 76 78 79 80 81 82 83 84 86 87 88 89 91 92 93 94 95 100 102 103 105 106 107 109 113 117

d. synnematous — 66

Conidiogenous cell or conidiophore characters

Distribution in the stroma

a. restricted to the base — 1 7 8 9 10 13 14 18 21 22 23 25 32 33 35 36 37 39 40 41 42 46 47 48 50 51 52 53 55 67 73 74 75 91 96 105 106 108 109 110 112 115 116 117 118

b. produced at the base and sides or all round the stromatal wall

2 3 4 5 6 11 12 15 16 17 19 20 24 26 27
28 29 30 31 34 38 43 44 45 49 54 56 57 58 59
60 61 62 63 64 65 66 68 69 70 71 72 76 77 78
79 80 81 82 83 84 85 86 87 88 89 90 92 93 94
95 97 98 99 100 101 102 103 104 107 111 113 114 119 120
121 122 123 124 125 126 127 128

Conidial characters

Pigmentation

a. hyaline

1 2 3 4 5 6 7 9 10 11 12 13 14 15 16
17 19 20 21 22 23 24 25 26 27 28 29 30 31 32
33 34 35 36 37 38 39 40 41 42 43 44 45 46 47
48 49 50 51 52 53 54 55 56 57 58 59 60 61 62
63 64 65 66 67 68 69 70 71 72 73 74 75 76 77
78 79 80 81 82 83 84 85 86 87 88 89 90 91 92
93 94 95 96 97 98 99 100 101 102 103 104 105 106 107
108 109 110 111 112 113 114

b. brown

4 7 8 11 18 49 80 115 116 117 118 119 120 121 122
123 124 125 126 127 128

Size

a. greater than 10 µ long

1 2 5 6 7 8 10 11 12 13 15 16 17 18 19
21 23 27 35 36 39 40 41 42 43 44 45 46 48 50
51 53 56 62 70 74 79 80 81 83 84 90 91 93 94
95 96 97 99 100 101 102 103 104 105 106 107 108 109 110
111 112 113 114 115 116 119 121 122 123 124 125 126 127 128

b. less than 10 µ long

3 4 9 14 20 22 24 25 26 28 29 30 31 32 33
34 35 36 37 38 40 47 49 52 54 55 56 57 58 59
60 61 63 64 65 66 67 68 69 71 72 73 74 75 76
77 78 79 82 85 86 87 88 89 91 92 98 99 115 117
118 119 120 123

Septation

a. distoseptate

128

b. euseptate

2 5 7 8 11 18 90 91 92 93 94 95 96 97 98
99 100 101 102 103 104 105 106 107 108 109 110 111 112 113
114 124 125 126 127

c. aseptate

1 3 4 6 7 9 10 12 13 14 15 16 19 20 21
22 23 24 25 26 27 28 29 30 31 32 33 34 35 36
37 38 39 40 41 42 43 44 45 46 47 48 49 50 51
52 53 54 55 56 57 58 59 60 61 62 63 64 65 66
67 68 69 70 71 72 73 74 75 76 77 78 79 80 81
82 83 84 85 86 87 88 89 93 96 98 99 100 103 104
115 116 117 118 119 120 121 122 123

d. medianly 1 septate

5 7 8 11 90 91 92 93 94 95 96 97 98 99 100
101 102 103 104 105 106 109 114 124 125

e. unequally 1 septate

17 110

f. 2-3 septate (rarely more)

103 106 108 109 110 112 114

g. multiseptate

2 5 18 104 107 111 113 126 127 128

Shape

a. oval-ellipsoid: 4 11 15 22 24 26 27 28 31 34 36 38 41 45 51 52 54 55 58 59 60 61 64 66 69 71 75 79 80 81 85 87 88 91 92 98 117 118 119 123 125 126 127

b. cylindrical: 5 8 10 12 18 21 23 25 30 33 36 39 40 45 51 55 58 68 70 72 73 75 76 84 85 86 88 89 90 100 104 123 124 126 127

c. fusiform: 4 5 6 7 8 11 13 14 15 17 18 19 21 24 35 36 37 39 43 44 50 51 53 56 60 62 63 73 74 79 83 87 88 93 94 95 96 97 99 100 101 102 103 105 106 107 108 112 113 114 115 117 123 124 127 128

d. falcate: 3 5 6 13 36 50 51 53 62 63 74 83 89 93 94 95 96 100 105 106 107 108 112 113 128

e. pyriform or obpyriform: 9 22 26 27 29 41 46 47 69 77 80 87 110 118 126

f. filiform or acicular: 5 62 93 94 95 101 111

g. spherical: 32 57 120 121

h. hamate: 79

i. branched: 1 2

j. allantoid-reniform: 3 4 16 65 67 76 78 82 86

k. Y-shaped: 1

l. trifurcate: 2

m. protuberate: 20

n. turbinate: 48 49 122

o. biconic-rhomboid: 116

p. anastomosing: 105

q. *Fusarium*-like (with a foot cell): 106 107

r. thickened at each end: 124 128

s. irregular in shape: 42 81 109 114

Ornamentation

a. verruculose: 7 11 118 119 126

b. smooth: 1 2 3 4 5 6 7 8 9 10 11 12 13 14 15 16 17 18 19 20 21 22 23 24 25 26 27 28 29 30 31 32 33 34 35 36 37 38 39 40 41 42 43 44 45 46 47 48 49 50 51 52 53 54 55 56 57 58 59 60 61 62 63 64 65 66 67 68 69 70 71 72 73 74 75 76 77 78 79 80 81 82 83 84 85 86 87 88 89 90 91 92 93 94 95 96 97 98 99 100 101 102 103 104 105 106 107 108 109 110 111 112 113 114 115 116 117 120 121 122 123 124 125 127 128

Appendages

a. apical: 3 4 5 6 7 8 9 10 11 12 13 14 15 18 19 51 90 104

b. basal: 3 4 5 6 7 8 16 17 18 19 90 104 106 107

c. lateral: 6 7

d. cellular: 9 13 14 15 16 17 18 51 90 106 107

e. mucilaginous: 10 11 12

f. setulose: 3 4 5 6 7 8 19 104

g. single: 3 4 5 6 7 8 9 10 11 12 13 14 15 16 17 18 19 51 104 106 107

h. several inserted at one point: 3 5 18 90

i. simple	3	4	5	6	7	9	10	11	13	14	15	16	17	18	19
	51	90	106	107											
j. branched	5	8													
k. dichotomously branched	104														
l. cap-like	10	11													
m. infundibular	12														
Wall thickness															
a. thin-walled	1	2	3	4	5	6	7	8	9	10	11	12	13	14	15
	16	17	18	19	20	21	22	23	24	25	26	27	28	29	30
	31	32	33	34	35	36	37	38	39	40	41	42	43	44	45
	46	47	48	49	50	51	52	53	54	55	56	57	58	59	60
	61	62	63	64	65	66	67	68	69	70	71	72	73	74	75
	76	77	78	79	80	81	82	83	84	85	86	87	88	89	90
	91	92	93	94	95	96	97	98	99	100	101	102	103	104	105
	106	107	108	109	110	111	112	113	114	115	116	117	119	120	122
	123	125													
b. thick-walled	118	119	120	121	123	124	126	127	128						
Guttulation															
a. guttulate	3	4	5	8	10	21	22	23	26	30	34	37	39	43	46
	48	49	51	54	56	59	60	63	67	72	75	79	81	92	95
	97	100	101	111	119	122									
b. eguttulate	1	2	4	5	6	7	9	11	12	13	14	15	16	17	18
	19	20	21	22	24	25	26	27	28	29	30	31	32	33	35
	36	37	38	40	41	42	44	45	46	47	50	51	52	53	54
	55	57	58	59	61	62	64	65	66	68	69	70	71	73	74
	75	76	77	78	79	80	81	82	83	84	85	86	87	88	89
	91	93	94	96	98	99	100	102	103	104	105	106	107	108	109
	110	112	113	114	115	116	117	118	119	120	121	123	124	125	126
	127	128													
c. consistently multiguttulate	34	39	49	56	79	97									
d. consistently with a more or less central large guttule	23	81													
e. consistently biguttulate	49	63	79	92	101										

Paraphysis characters

a. present	85	115	123	128
b. simple	85			
c. branched	85	115	123	128
d. filiform	85	115	123	128

1. SUTTONIELLA Ahmad, *Biologia, Lahore* **6**: 257 (1961).
   Pseuderiospora Petrak, *Sydowia* **13**: 109 (1959).
   Suttonia Ahmad, *Biologia, Lahore* **6**: 128 (1960).
   Sp. typ.: *S. gaubae* (Petrak) Ahmad (syn. *Pseuderiospora gaubae* Petrak).

Mycelium immersed, brown, branched, septate. *Conidiomata* eustromatic, immersed, epidermal to subepidermal, dark brown, flattened, ± circular, often confluent; upper wall of dark brown thick-walled textura angularis, lower wall of light brown to hyaline tissue, sometimes of textura prismatica. *Ostiole* absent, dehiscence by irregular fissures in the upper wall. *Conidiophores* absent. *Conidiogenous cells* enteroblastic, phialidic, indeterminate, discrete, doliiform to irregular, hyaline, smooth, with 1-3 apertures over the apex, channel and collarettes minute, restricted to the lower conidiomatal wall. *Conidia* hyaline, aseptate, Y-shaped, the two upper arms constricted where they meet the basal arm, branches tapered towards the apices, thin-walled, smooth, eguttulate.

Sutton (1972a) has fully illustrated and described the genus which only contains two species.

KEY

Upper arms 30-37 µ long .... *S. gaubae*
Upper arms 12-14 µ long .... *S. eriobotryae*

S. ERIOBOTRYAE (Ahmad) Ahmad, *ibid.*: 258 (1961). (Fig. 264)
Suttonia eriobotryae Ahmad, *Biologia, Lahore* **6**: 128 (1960).
Furcaspora indica Talde, *Indian Phytopath.* **27**: 630 (1975).

Conidiomata up to 900 µ diam. Conidiogenous cells 8-12 µ × 4-6.5 µ. Conidia: upper arms 12-14 × 2-2.5 µ, basal arm 11-12 × 2.5 µ.

On dead leaves of *Eriobotrya japonica*, Pakistan (IMI 80680, holotype).

S. GAUBAE (Petrak) Ahmad, *ibid.*: 258 (1961). (Fig. 265)
Pseuderiospora gaubae Petrak, *Sydowia* **13**: 109 (1959).
Suttonia gaubae (Petrak) Ahmad, *Biologia, Lahore* **6**: 128 (1960).

Conidiomata up to 300 µ diam. Conidiogenous cells 10-17.5 × 5-7.5 µ. Conidia: upper arms 30-37 × 2.5 µ, basal arm 18-20 × 2.5 µ.

On dead leaves of *Alyxia buxifolia, Mangifera indica*, Australia, India (IMI 57983b, 227721).

2. ERIOSPORELLA Höhn., *Sber. Akad. Wiss. Wien* **125**: 109 (1916).
Pseuderiospora Keissl., *Anz. Akad. Wiss. Wien* **60**: 75 (1923).
Sp. typ.: *E. calami* (Neissl) Höhn. (syn. *Cryptosporium calami* Neissl).

Mycelium immersed, branched, septate, hyaline. *Conidiomata* eustromatic, epidermal to subepidermal, separate, sessile, applanate to pulvinate at maturity, consisting of a central locule, a thin basal wall and thicker lateral walls extending into the disorganised epidermal tissue around the conidioma; walls composed of hyaline, thin-walled textura angularis to intricata. *Ostiole* absent, dehiscence by breakdown of overlying tissues. *Conidiophores* hyaline, branched and septate at the base, smooth, irregular, formed from the inner cells of the locular wall. *Conidiogenous cells* enteroblastic, phialidic, integrated or discrete, hyaline, smooth, cylindrical, collarette absent, channel minute, periclinal wall thickened. *Conidia* consisting of a short 0-1 septate main axis which is obtuse at the point of secession, and 3 divergent, septate, guttulate, tapered arms arising from the apex, hyaline, smooth.

The genus is monotypic.

E. CALAMI (Neissl) Höhn., *ibid.*: 109 (1916). (Figs 266, 267)
Cryptosporium calami Neissl in Rabenh., *Hedwigia* **17**: 176 (1878).
Pseuderiospora castanopsidis Keissl., *Anz. Akad. Wiss. Wien* **60**: 75 (1923).

Conidiomata up to 150 µ diam. Conidiophores 7-17 × 2 µ. Conidia: basal cell up to 17 × 1.5-2 µ; arms 3-5 septate, up to 65 µ long.

On dead leaves of *Calamus*, India (IMI 200662 isotype of *C. calami* ex K, 209213, 209214, holotype *P. castanopsidis* ex W).

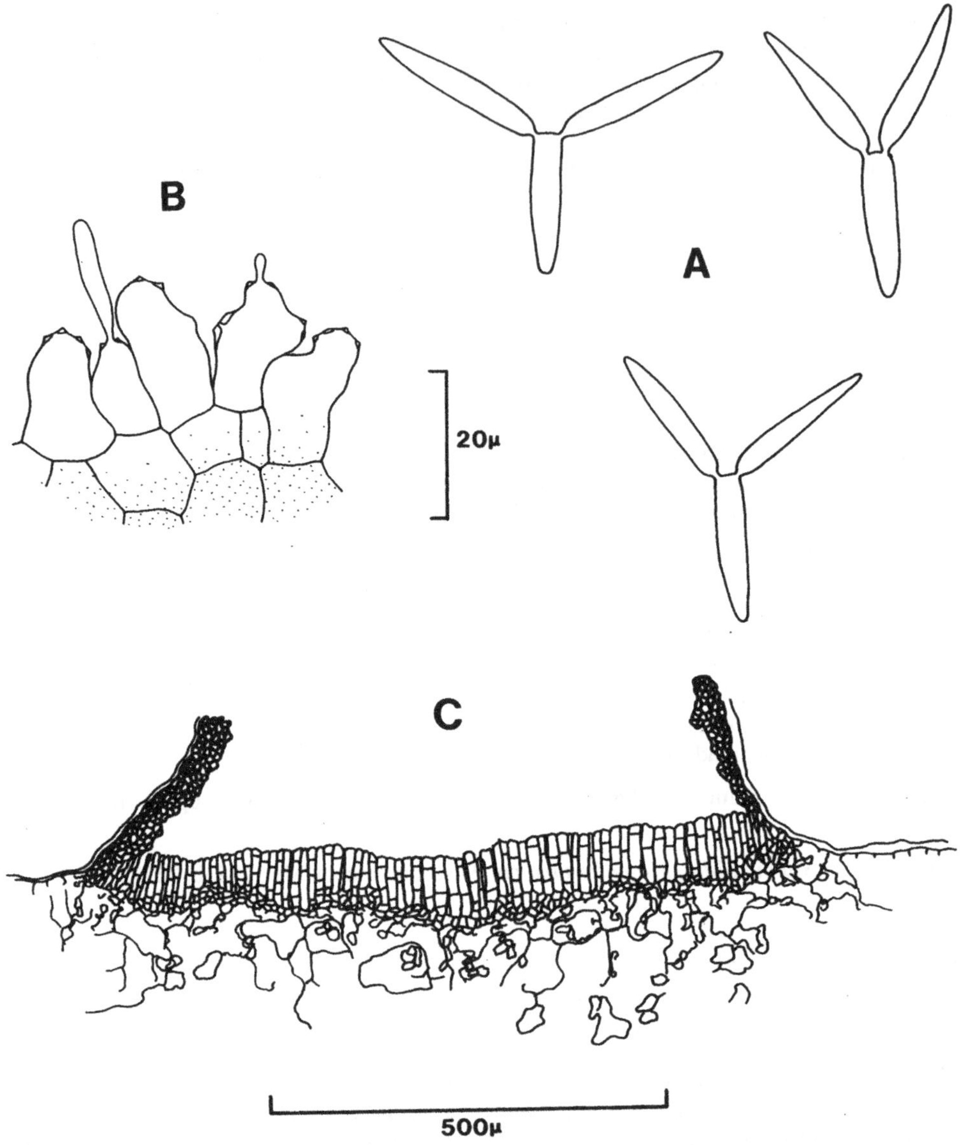

Fig. 264. *Suttoniella eriobotryae*, A, conidia; B, conidiogenous cells; C, vertical section of a conidioma.

3. **DIARIMELLA** gen. nov.
(etym. dia-(G. pref.), through, rim-(L, noun), fissure, et-ella (L, suff.), diminutive).

Mycelium immersum. Conidiomata eustromatica, superficialia, separata, navicularia vel hysterioidea, unilocularia; parietes crassi, ex textura angulari formati, ad marginum versus ex textura porrecta. Per raphe longitudinalem dehiscens. Setae conspicuae, rigidae, atro brunnea acuminatae, et inconspicuae, pallide brunneae, flexuosae. Conidiophora absentia. Cellulae conidiogenae enteroblasticae, phialidicae, determinatae, discretae, apicem versus leniter curvatae, parietibus basalibus et lateralibus restrictae. Conidia hyalina, allantoidea vel fusiformia, aseptata, biguttulata, apicem acuta, basim truncta utrinque cum 1-3 setulis non ramosis vestita.

Sp. typ.: *D. setulosa* Sutton

Mycelium immersed. Conidiomata eustromatic, superficial, separate, navicular to hysteriform, unilocular; walls thick, of textura angularis at the margin merging into textura porrecta which eventually produces setae.

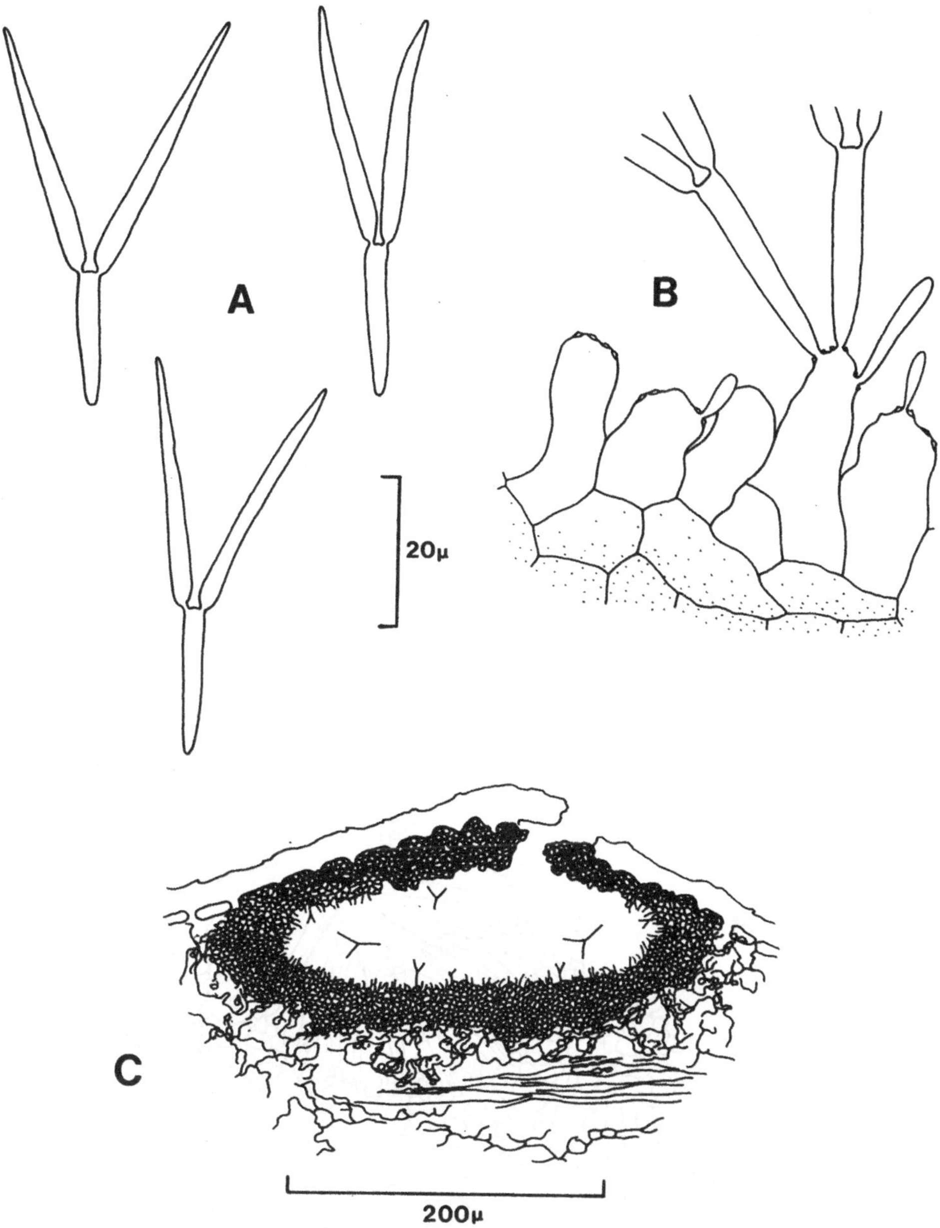

Fig. 265. *Suttoniella gaubae*, A, conidia; B, conidiogenous cells with developing conidia; C, vertical section of a conidioma.

Dehiscence by a longitudinal raphe. Setae of two types, conspicuous, rigid, dark brown, acuminate, and inconspicuous, pale brown, flexuous. Conidiophores absent. Conidiogenous cells enteroblastic, phialidic, determinate, discrete, curved at the apices, confined to the base and lower regions of the lateral walls. Conidia hyaline, allantoid to fusiform, aseptate, biguttulate, apex acute, base truncate, each with 3-6 unbranched setulae.

Hysteriform conidiomata are uncommon in the coelomycetes, being represented only by *Tetranacrium* Hudson & Sutton (1964), *Chaetomella* Fckl (Stolk, 1963; Sutton & Sarbhoy, 1976), and *Polymorphum* Chev. (Hawksworth & Punithalingam, 1973). Of these genera only in *Chaetomella* are the conidiomata setose, and in none are the conidia setulose. *Chaetomella* differs from *Diarimella* in the simple conidia and filiform septate conidiophores.

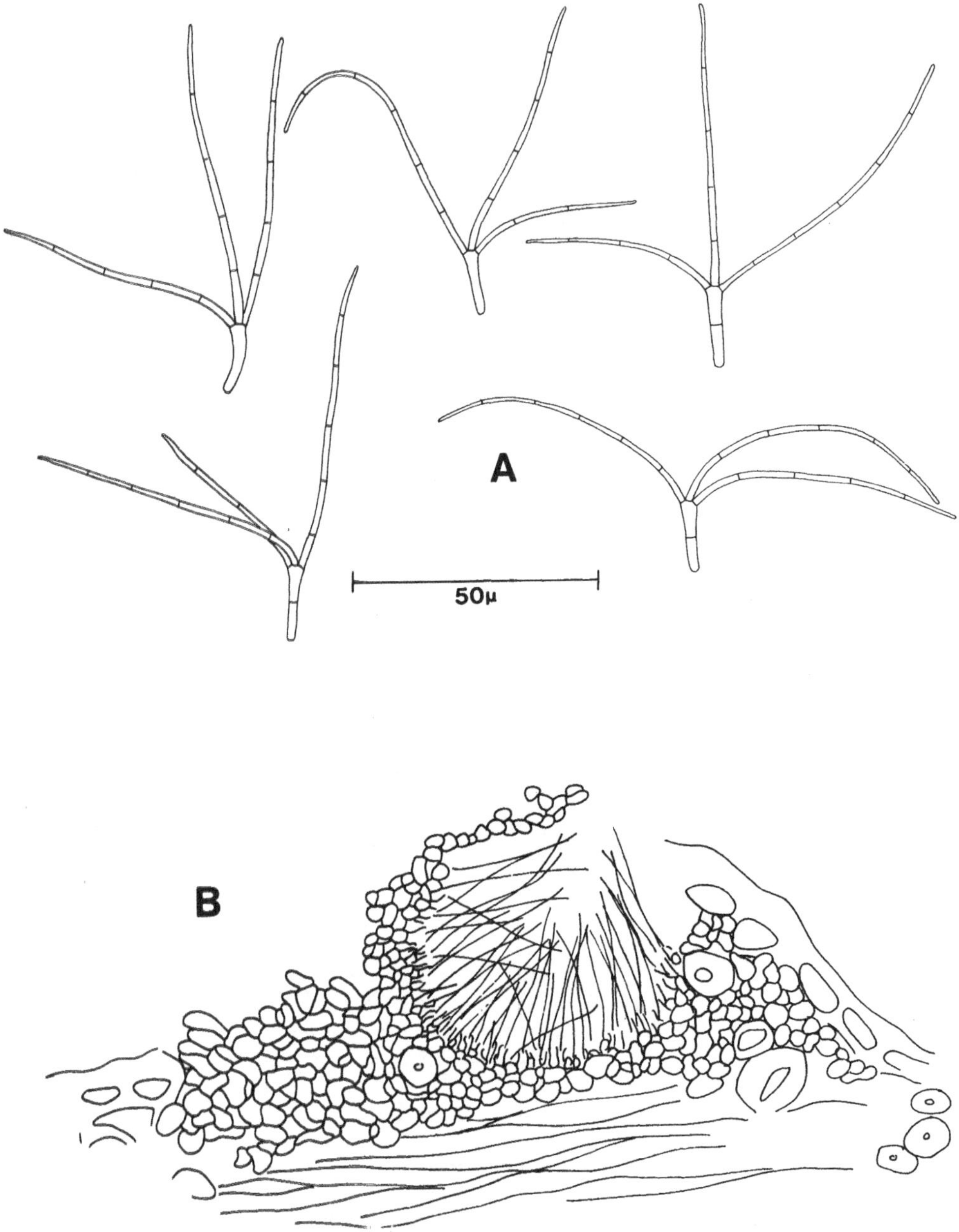

Fig. 266. *Eriosporella calami*, A, conidia; B, vertical section of a conidioma.

Thus apart from *Dinemasporium* Lév., *Stauronema* (Sacc.) H. & P. Syd. & Butl. and *Pseudolachnea* Ranoj., which all have cupulate, setose conidiomata, and setulose conidia, there are really no taxa closely related to *Diarimella*. These genera differ from *Diarimella* by the erect phialides which do not curve at the apices and the method of conidiomatal dehiscence which is not longitudinal. The only other related genus with apically curved phialides is *Oramasia* Urries (Sutton, 1978b) but here the conidia lack any setulae and the conidiomata are sporodochial.

Diarimella setulosa sp. nov. (Fig. 268)

Mycelium immersum, septatum, ramosum, hyalinum vel pallide brunneum. *Conidiomata* eustromatica, superficialia, separata, navicularia vel hysterioidea, atro brunnea vel nigra, unilocularia, setosa, 700-900 μ

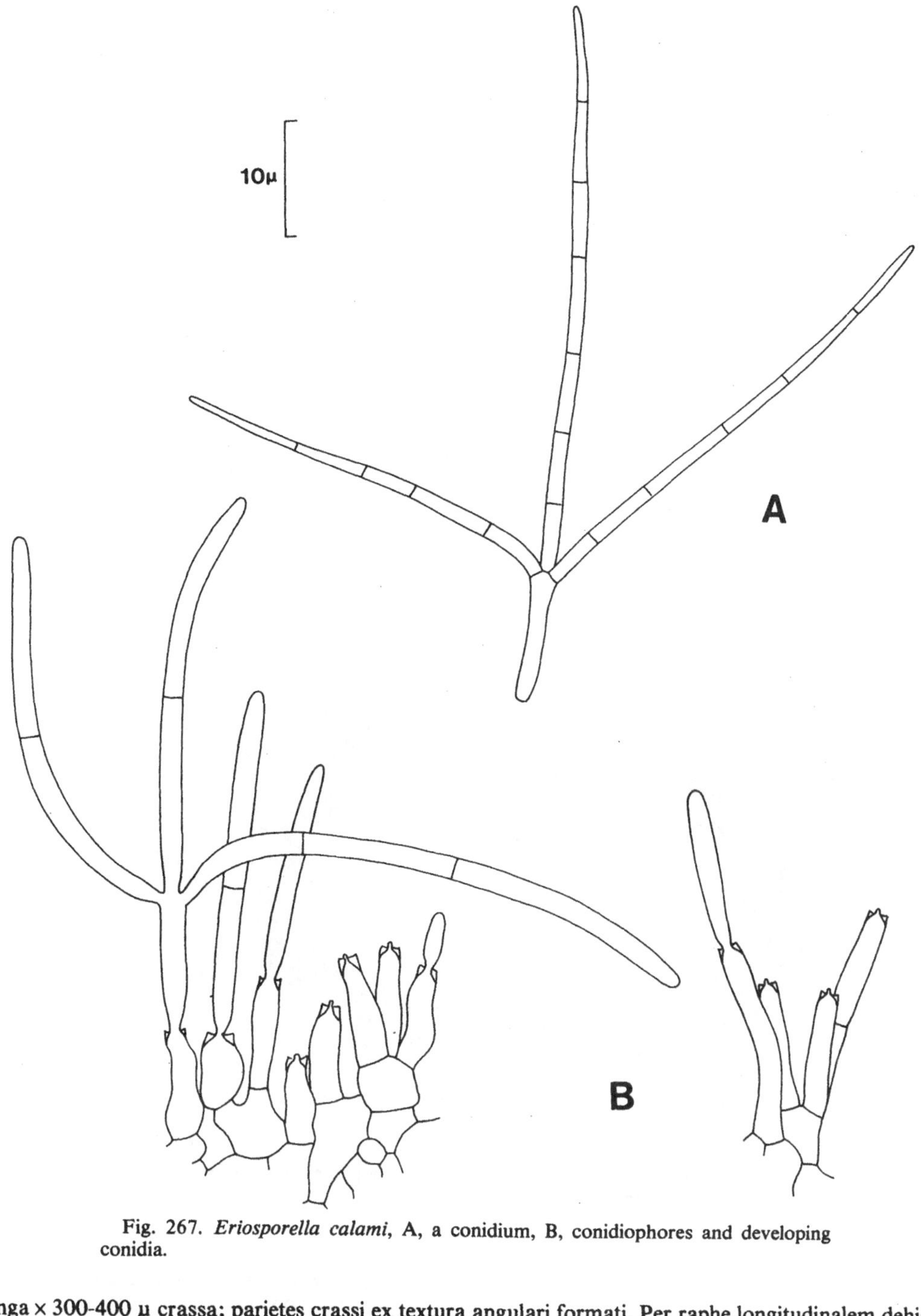

Fig. 267. *Eriosporella calami*, A, a conidium, B, conidiophores and developing conidia.

longa × 300-400 μ crassa; parietes crassi ex textura angulari formati. Per raphe longitudinalem dehiscens. *Setae* conspicuae, rigidae, atro brunneae, laeves, apicem versus deminutae et pallidiores, 3-10 septatae, 90-240 × 5-10 μ; setae inconspicuae pallide brunneae, flexuosae, aseptatae, laeves, apicem deminutae, usque ad 40 μ longae. Cellulae conidiogenae enteroblasticae, phialidicae, determinatae, discretae, hyalinae, filiformes, apicem curvatum versus deminutae, 8-17 × 2-3 μ, parietibus basalibus et lateralibus restrictae. *Conidia* hyalina, allantoidea vel fusiformia, aseptata, biguttulata, 4-5.5 × 2 μ, apicem acuta, basim obtusa, cum 1-3 setulis nonramosis, 3-6 μ longis vestita.

In ramiculis emortuis *Bougainvilleae spectabilis*, Poona, India, I.K. Kalani, 4 Nov. 1963, IMI 103785, holotypus.

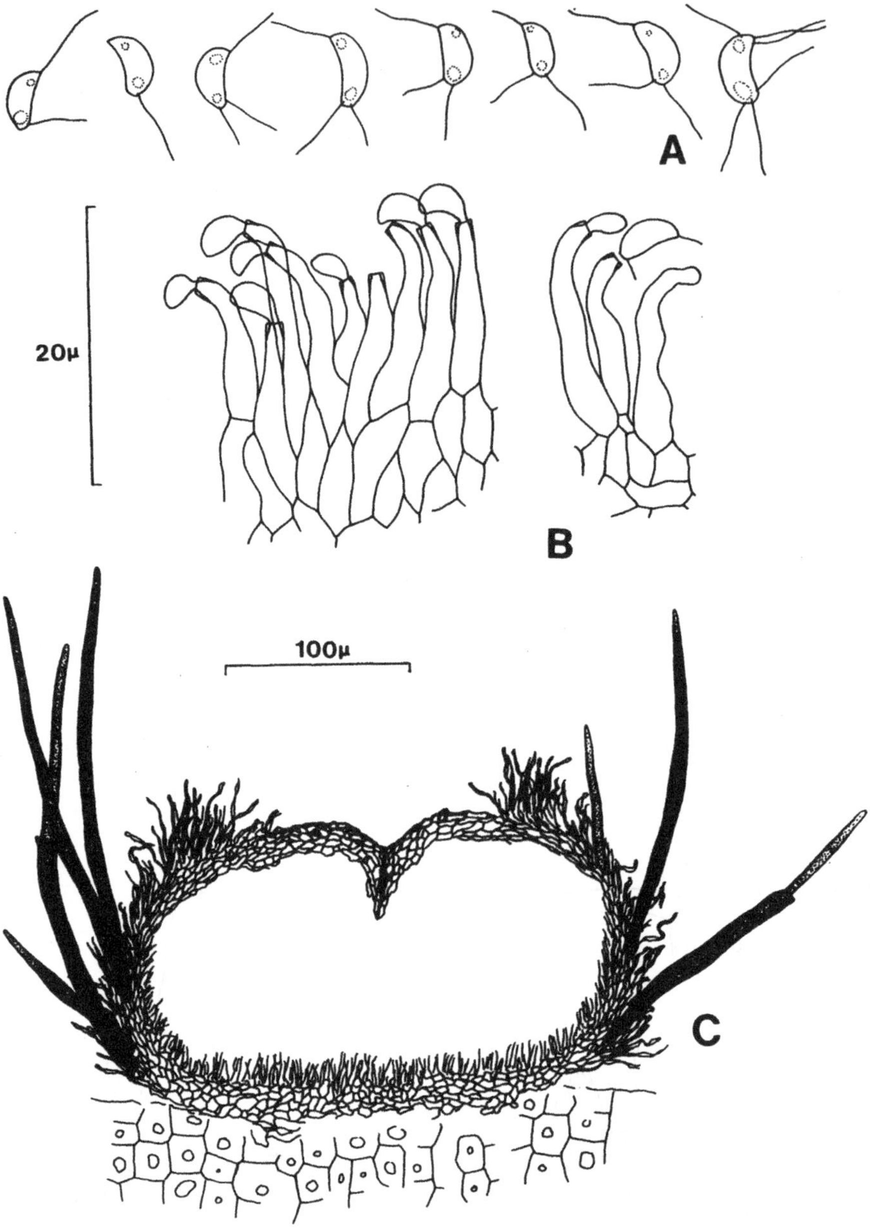

Fig. 268. *Diarimella setulosa*, A, conidia; B, conidiophores and developing conidia; C, vertical section of a closed conidioma.

Mycelium immersed, septate, branched, hyaline to pale brown. *Conidiomata* eustromatic, superficial, separate, navicular to hysteriform, dark brown to black, unilocular, 700-900 μ long × 300-400 μ wide; wall thick, consisting at the base of thin-walled, hyaline to pale brown textura angularis, at the sides of similar, darker tissue which develops into textura porrecta at the periphery and finally runs out into two types of setae. Dehiscence by a longitudinal raphe composed of brown, thick-walled textura angularis which lacks any covering setae. *Setae* of two types: conspicuous setae, rigid, dark brown, smooth, tapered and paler towards the subacute apices, 3-10 septate, 90-240 × 5-10 μ, confined to the base and sides of the conidiomata: inconspicuous setae pale brown, flexuous, aseptate, smooth, tapered to the apices, up to 40 μ long; particularly dense near the glabrous raphe, less abundant on the lateral walls. *Conidiophores* absent. *Conidiogenous cells* enteroblastic, phialidic, determinate, discrete, hyaline, filiform, tapered towards the slightly curved apices, 8-17 × 2-3 μ, confined to the base and the lower regions of the lateral walls. *Conidia* hyaline, allantoid to fusiform, aseptate, biguttulate, apex acute, base obtuse, 4-5.5 × 2 µ; apex and base each with 1-3 unbranched, setulae, 3-6 µ long.

4. DINEMASPORIUM Lév., *Ann. Sci. nat.*, ser. 3, **5**: 274 (1846).
Dendrophoma Sacc., *Michelia* **2**: 4 (1880).
Pycnidiochaeta Camara, *Agron. lusit.* **12**: 109 (1950).
Amphitiarospora Agnihothrudu, *Sydowia* **16**: 75, *1962* (1963).
Sp. typ.: *D. strigosum* (Pers. ex Fr.) Sacc. (syn. *Peziza strigosa* Pers. ex Fr., *D. graminum* (Lib.) Lév.).

Mycelium immersed, branched, septate, hyaline or brown. *Conidiomata* separate or aggregated, superficial, black, at first globose and closed, later opening out to become cupulate; walls at the base of brown textura angularis, at the periphery of either textura angularis, textura porrecta or textura intricata. Dehiscence by exposure of conidiogenous region as the conidiomata become cupulate. *Setae* abundant or sparse, brown, septate, acuminate, smooth, originating from the periphery, less commonly from the conidiogenous region. *Conidiophores* hyaline, simple or branched irregularly above, septate, smooth, cylindrical or tapered to the apices, formed from the basal and peripheral walls. *Conidiogenous cells* enteroblastic, phialidic, integrated, determinate, cylindrical or tapered, hyaline, smooth, with a minute channel and collarette, but periclinal thickening conspicuous. *Conidia* hyaline or very pale brown, aseptate, guttulate or eguttulate, thin-walled, smooth, straight or allantoid, fusiform, with a single unbranched setula at each end.

Dinemasporium is a comparatively large heterogeneous genus in which more than 50 taxa have been described. Although there have been no modern reassessments of the taxa, isolated accounts of accepted species have more recently appeared (Webster, 1955; Sutton, 1965b, 1969c; Morgan-Jones, 1971e). The generic synonymy has been discussed by Sutton (1977a).

In the past the genus has included species with aseptate and septate conidia, but in the following treatment *Dinemasporium* is being restricted to those taxa with aseptate conidia. The species with septate conidia are referred to *Pseudolachnea* Ranoj. Septation in this complex is relatively invariable within species and is considered to be a useful means of separating taxa at the generic level.

KEY

Conidia 9.5-11 × 1.5-2 µ; setulae 5-7 µ long; on Gramineae .... *D. strigosum*
Conidia 3.5-5 × 1 µ; setulae 1 µ long .... *D. cytosporoides*
Conidia 4-6 × 2-2.5 µ, pale brown; setulae 4-6 µ long .... *D. aberrans*

D. ABERRANS Sutton, *Can. J. Bot.* **47**: 2098 (1969). (Fig. 269)

Conidiomata 90-150 µ diam. Setae 75-120 × 3-6 µ, 5-11 septate. Conidiophores 16-22.5 × 1.5-2 µ. Conidia fusiform, 1-2 guttulate, 4-6 × 2-2.5 µ; setulae 4-6 µ long.

On rotten leaves, Nigeria (IMI 99367b, holotype).

D. CYTOSPOROIDES (Sacc.) Sutton, *Trans. Br. mycol. Soc.* **48**: 613 (1965). (Fig. 270)
Phoma cytosporoides Sacc., *Michelia* **1**(5): 522 (1879).
Dendrophoma cytosporoides (Sacc.) Sacc., *Michelia* **2**(6): 4 (1880).

Conidiomata 120-150 µ diam. Setae sparse, 80-120 × 4-8 µ, 2-6 septate. Conidiophores up to 40 µ long. Conidia allantoid, hyaline, 3.5-5 × 1 µ; setulae 1 µ long.

This species differs from *D. minutum* Batista et al. which has similar conidia which measure 2.5-4 × 1.5-2 µ and setulae up to 3 µ long. *D. canadense* Morgan-Jones also differs in the larger conidiomata (200-450 µ diam.) and conidia measuring 5-6 × 2-2.5 µ with setulae 3-4 µ long.

On wood of *Ulmus* sp., *Deutzia scabra*, France (IMI 111879, lectotype, 111880 ex PAD).

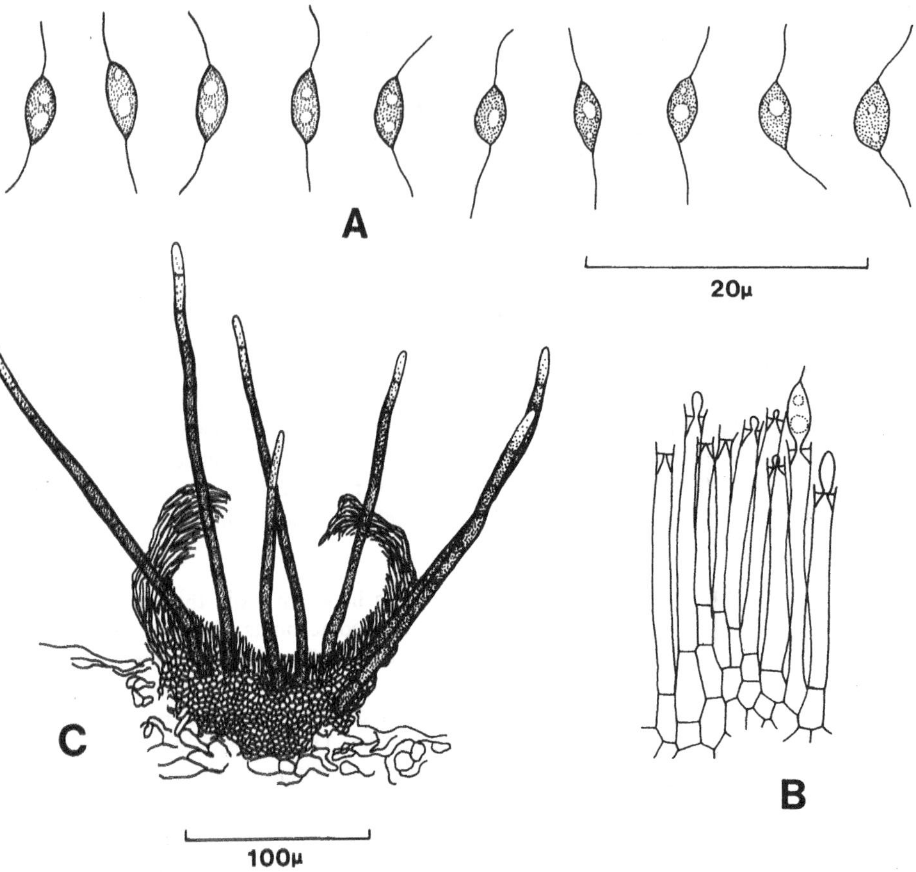

Fig. 269. *Dinemasporium aberrans*, A, conidia; B, conidiophores and developing conidia; C, vertical section of a conidioma.

D. STRIGOSUM (Pers. ex Fr.) Sacc., *Michelia* **2**: 281 (1881). (Fig. 271)
Peziza strigosa Pers., *Syn. Meth. Fung.*: 648 (1801).
Peziza strigosa Pers. ex Fr., *Syst. Myc.* **2**: 103 (1822).
Polynema strigosum (Pers. ex Fr.) Fckl, *Jb. nassau. Ver. Naturk.* **23-24**: 367 (1870).
Vermicularia graminum Lib., *Pl. Crypt. Ard. exs.* 348 (1837).
Excipula graminum (Lib.) Cda, *Icon. fung.* **3**: 29 (1839).
Dinemasporium graminum (Lib.) Lév., *Ann. Sci. nat.*, 3 ser., **5**: 274 (1846).
Excipula betulae Fckl, *Jb. nassau. Ver. Naturk.* **15**: 64 (1860).
Polynema betulae (Fckl) Fckl, *Jb. nassau. Ver. Naturk.* **23-24**: 367 (1870)
Dinemasporium longisetum Syd., *Annls mycol.* **34**: 400 (1936).

Teleomorphosis: *Phomatospora dinemasporium* Webster, *Trans. Br. mycol. Soc.* **38**: 364 (1955).

Conidiomata up to 200 μ diam. Setae 60-220 × 4-6 μ, 2-7 septate. Conidiophores up to 25 × 1.5-2.5 μ. Conidia straight or slightly curved, 9.5-11 × 1.5-2 μ, setulae 5-7 μ long.

This species has been known in the literature as *D. graminum*, but examination of relevant types has shown that the basionym upon which this is based is antedated by *Peziza strigosa* Pers. ex Fr., so the name *D. strigosum* is used instead. It is a common species most frequently found on graminaceous substrates but not exclusively so.

It is known from 95 herb. IMI collections on 46 different host genera and is world-wide in distribution (including IMI 94858, isotype *E. betulae* ex K, 94746, isotype *D. longisetum* ex K 94741, isotype *P. strigosa* ex K, 94827, isotype *V. graminum* ex K).

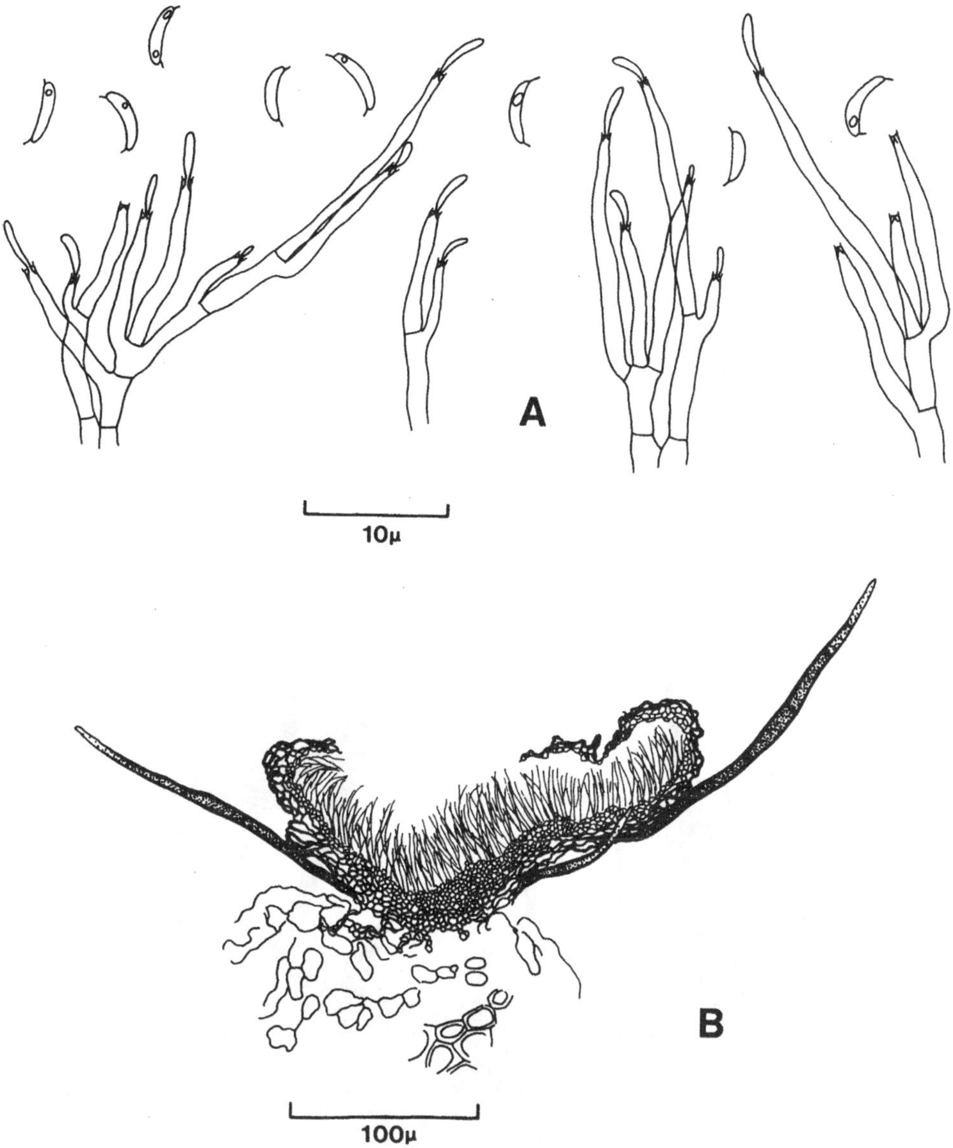

Fig. 270. *Dinemasporium cytosporoides*, A, conidia and conidiophores; B, vertical section of a conidioma.

5. PSEUDOLACHNEA Ranojević, *Annls mycol.* **8**: 393 (1910).
Dinemasporiella Bub. & Kabát, *Hedwigia* **52**: 358 (1912).
Dinemasporiopsis Bub. & Kabát, nom. nov. in Died., *Krypt. Fl. Mk Brandenb.* **9**: 750 (1915).
Pseudolachnella Teng, *Sinensia* **7**: 775 (1936).
Chaetopatella Hino & Katumoto, *J. Jap. Bot.* **33**: 238 (1958).
Sp. typ.: *P. hispidula* (Schrad. ex Fr.) Sutton (syn. *Peziza hispidula* Schrad. ex Fr., *Pseudolachnea bubakii* Ranojević).

Mycelium immersed, pale to dark brown, septate, branched. *Conidiomata* subcuticular or superficial, separate, dark brown to black, at first closed, later opening to become globose or flattened cupulate, or conical; periclinal wall pseudoparenchymatous, composed of brown textura angularis towards the inside, extending into textura porrecta at the periphery. *Ostiole* absent. *Setae* restricted to the outer wall, at first incurved over the conidiogenous region, later divergent, acuminate, thick-walled, dark brown. *Conidiophores* hyaline, branched at the base, sparingly septate at the base, smooth,

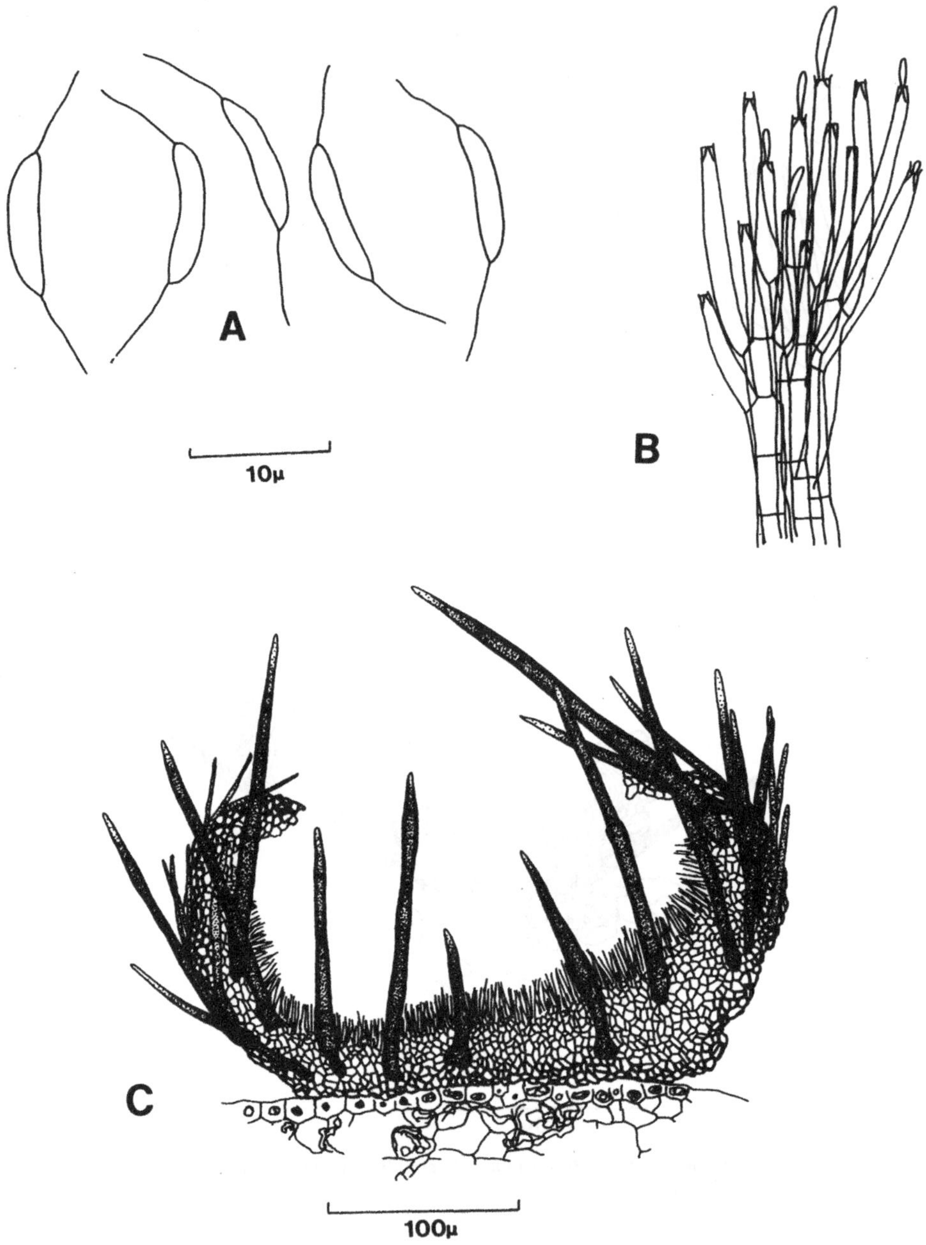

Fig. 271. *Dinemasporium strigosum*, A, conidia; B, conidiophores; C, vertical section of a conidioma.

cylindrical, formed from the base and sometimes from the periclinal wall. *Conidiogenous cells* enteroblastic, phialidic, integrated or discrete, hyaline, smooth, cylindrical, collarette and channel minute. *Conidia* hyaline, 1- many euseptate, continuous, apices and bases obtuse, thin-walled, smooth, ± guttulate, fusiform to cylindrical, straight or curved irregularly, with 1 or more apical, simple or branched setulae and 1 or more basal, simple or branched setulae inserted above the basal scar.

P. hispidula is the typical representative with cupulate conidiomata on stems, bark and wood of a wide variety of substrates. The other species all occur on culms or leaf-sheaths of various bamboos and have flattened conidiomata that are regarded as reduced from the cupulate type owing to the hard substrate upon which they are formed.

KEY

1. Conidia 1- septate .... *P. hispidula*
1. Conidia 2- septate or more 2
   2. Setulae branched 3
   2. Setulae simple 4
3. Conidia 8- septate, 75-105 × 3 µ .... *P. coronata*
3. Conidia 5- septate, 32-39 × 2.5-3 µ .... *P. ryukyuensis*
   4. Conidia 20-70 × 3-4 µ, 7-12 septate .... *P. scolecospora*
   4. Conidia 62-81 × 2.5-3 µ, 15 septate .... *P. longiciliata*

P. coronata (Hino & Katumoto) comb. nov. (Fig. 272)
Chaetopatella coronata Hino & Katumoto, *J. Jap. Bot.* **33**: 239 (1958).

Conidiomata conical, 150-240 × 3-5 µ. Setae 150-260 × 3-5 µ. Conidiophores 13-19 × 1.5-2 µ. Conidia guttulate, cylindrical, 8-septate, 75-105 × 3 µ; apical setulae 1-2, dichotomously branched, 5-11 µ long; basal setula branched, 3-6 µ long.

On leaf sheaths of *Pleioblastus hindsii*, Japan (IMI 145716, 110663, holotype).

P. HISPIDULA (Schrad. ex Fr.) Sutton, *Mycol. Pap.* 141: 167 (1977). (Fig. 273)
Peziza hispidula Schrad. ex Fr., *Syst. Myc.* **2**: 98 (1823).
Excipula hispidula (Schrad. ex Fr.) Cke & Ell., *Grevillea* **4**: 179 (1876).
Dinemasporium hispidulum (Schrad. ex Fr.) Curtis, *Geological and National History Survey of North Carolina* **3**: 120 (1867).
Dinemasporiella hispidula (Schrad. ex Fr.) Bub. & Kabát, *Hedwigia* **52**: 358 (1912).
Dinemasporiopsis hispidula (Schrad. ex Fr.) Bub. & Kabát in Diedicke, *Krypt. Fl. Mk Brandenb.* **9**: 750 (1915).
Polynema hispidulum (Schrad. ex Fr.) Fries, *Summa Veg. Scand.*: 367 (1849).
Pseudolachnea bubakii Ranojević, *Annls mycol.* **8**: 393 (1910).
Dinemasporium graminum var. *herbarum* Cke, *Handbook of British Fungi* **1**: 459 (1871).
Dinemasporium herbarum (Cke) Grove, *British Stem- and Leaf-fungi* **2**: 137 (1937).
Dinemasporium epixylon Fautrey, *Rev. mycol.* **11**: 152 (1889).
Dinemasporium patellum Cke & Ell., *Grevillea* **7**: 38 (1878).
Dinemasporium hispidulum (Schrad. ex Fr.) Sacc. var. *herbarum* (Cke) Allesch. in Rabenh., *Kryptogamenflora* **7**: 425 (1903).

Conidiomata cupulate 260-850 µ diam. Setae up to 400 × 3-5 µ. Conidiophores 20-30 × 1.5-2 µ. Conidia medianly 1 septate, eguttulate, 15.5-18.5 × 2-3 µ; apical and basal setulae single, unbranched, 2-3 µ long.

On *Urtica dioica*, *U.* sp., *Fraxinus*, *Ballota nigra*, *Festuca*, *Abies*, *Picea excelsa*, *Robinia pseudacacia*, *Lunaria rediviva*, *Quercus ilex*, *Quercus*, *Salix*, *Solanum dulcamara*, *Sambucus nigra*, *Sarothamnus scoparius*, G.B., Eire, Czechoslovakia, Germany, U.S.A., U.S.S.R., Pakistan (IMI 94856, holotype *D. patellum* ex K, 94868, holotype *D. graminum* var. *herbarum* ex K, 94860, holotype *D. epixylon* ex K, 94468, 94488, 94489, 94486, 94490, 94519, 94520, 94522, 94524-26, 96547-9, 21001, 81106, 83709, 83138, 165393, 33974, 102571a, 22879, 6619a, 21622, 122442, 33978, 34705, 44944, 57018, 30071).

P. longiciliata (Hino & Katumoto) comb. nov. (Fig. 274B, E)
Chaetopatella longiciliata Hino & Katumoto, *J. Jap. Bot.* **33**: 238 (1958).

Conidiomata conical, 350-500 µ diam. Setae 130-210 × 3-4 µ. Conidiophores 16.5-32.5 × 2.5-3 µ. Conidia cylindrical or slightly curved, 15 septate, guttulate, 62-81.5 × 2.5-3 µ; apical setula single, unbranched, 32-47 µ long; basal setula single unbranched, 8-13 µ long.

On dead culms of *Sasa kurilensis*, Japan (IMI 110665, holotype).

P. ryukyuensis (Hino & Katumoto) comb. nov. (Fig. 275)
Chaetopatella ryukyuensis Hino & Katumoto, *Bull. Fac. Agr. Yamaguti Univ.* 11: 10 (1960).

Conidiomata conical, 200-300 µ diam. Setae 110-260 µ long. Conidiophores 10-15 × 2-3 µ. Conidia cylindrical

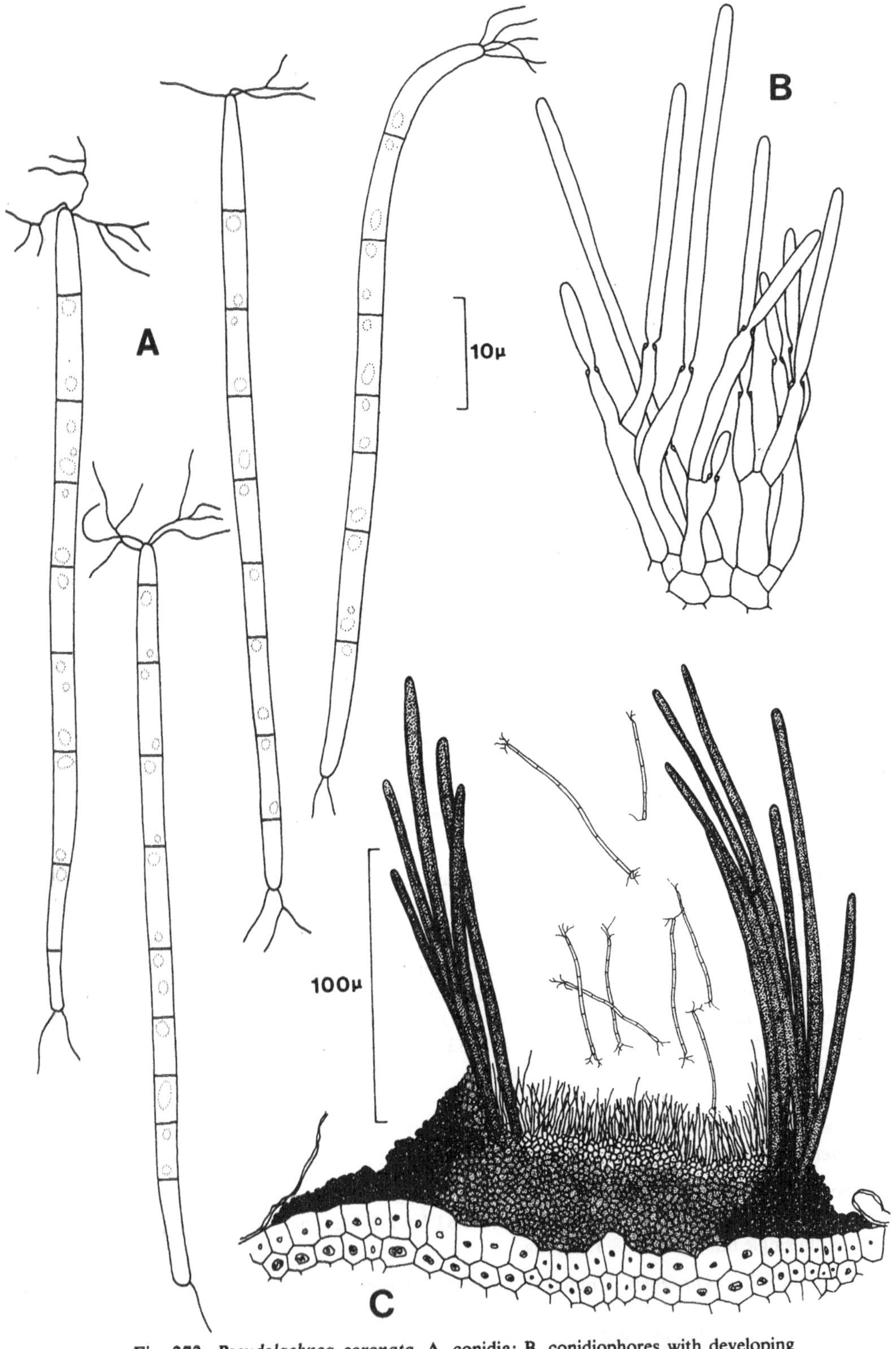

Fig. 272. *Pseudolachnea coronata*, A, conidia; B, conidiophores with developing conidia; C, vertical section of a conidioma.

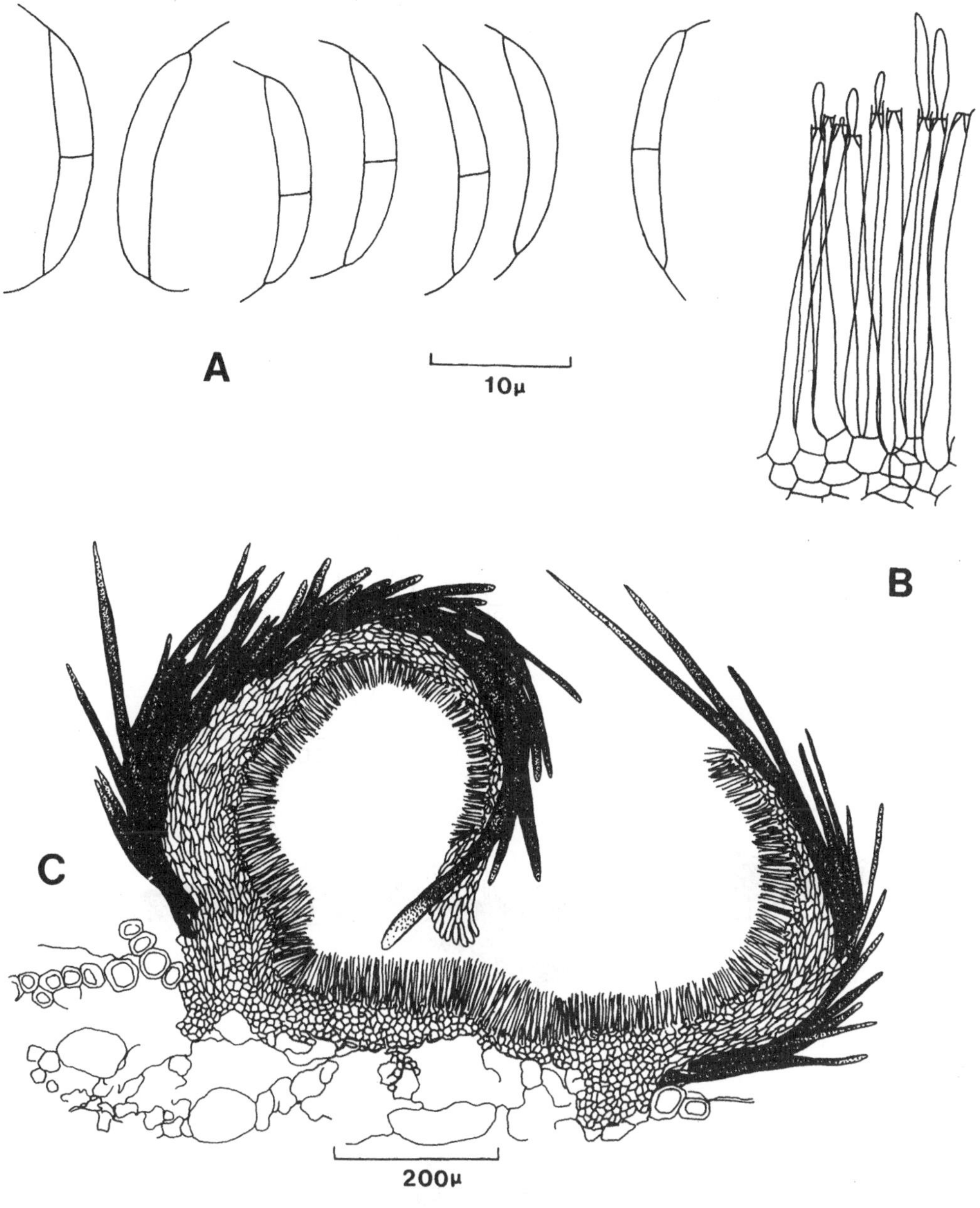

Fig. 273. *Pseudolachnea hispidula*, A, conidia; B, conidiophores; C, vertical section of a conidioma.

to fusiform, 5 septate, guttulate, 32-39 × 2.5-3 μ; apical setulae 1-2, dichotomously branched, 5-6 μ long; basal setula 1, branched once at the base, 3-4 μ long.

On dead leaf sheaths of *Pleioblastus linearis*, Japan (IMI 145717, 110667, holotype).

P. SCOLECOSPORA Teng & Shen, *Contr. Biol. Lab., Sci. Soc. China Bot.* **8**: 277 (1933). (Fig. 274A, C, D)
Pseudolachnella scolecospora (Teng & Shen) Teng, *Sinensia* **7**: 775 (1936).
Heteropatella setulosa Hino & Katumoto, *Bull. Fac. Agr. Yamaguti Univ.* **5**: 234 (1954).
Chaetopatella setulosa (Hino & Katumoto) Hino & Katumoto, *Icones fungorum bambusicolorum Japonicorum* 252 (1961).

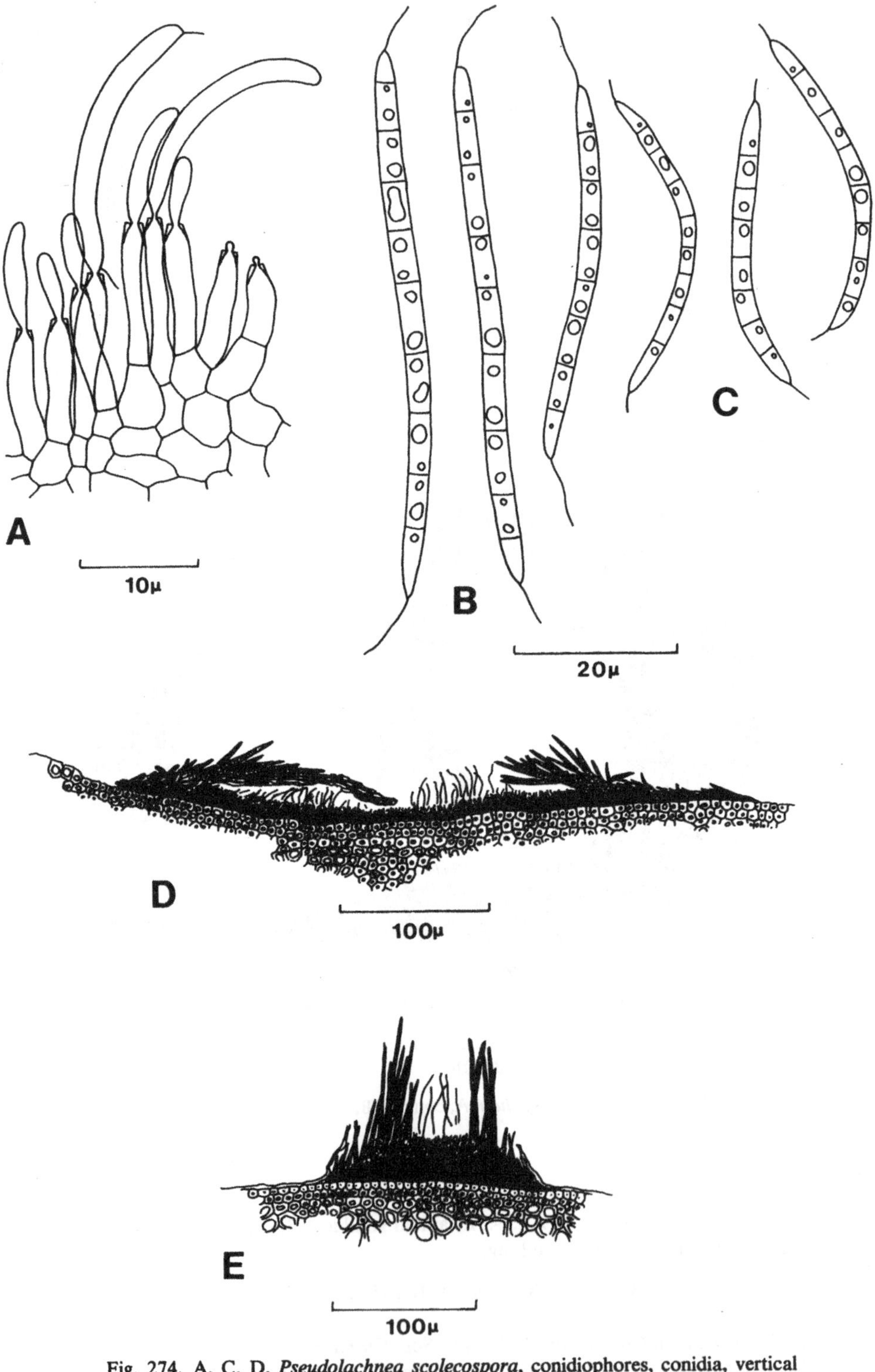

Fig. 274. A, C, D, *Pseudolachnea scolecospora*, conidiophores, conidia, vertical section of a conidioma; B, E, *P. longiciliata*, conidia, vertical section of a conidioma.

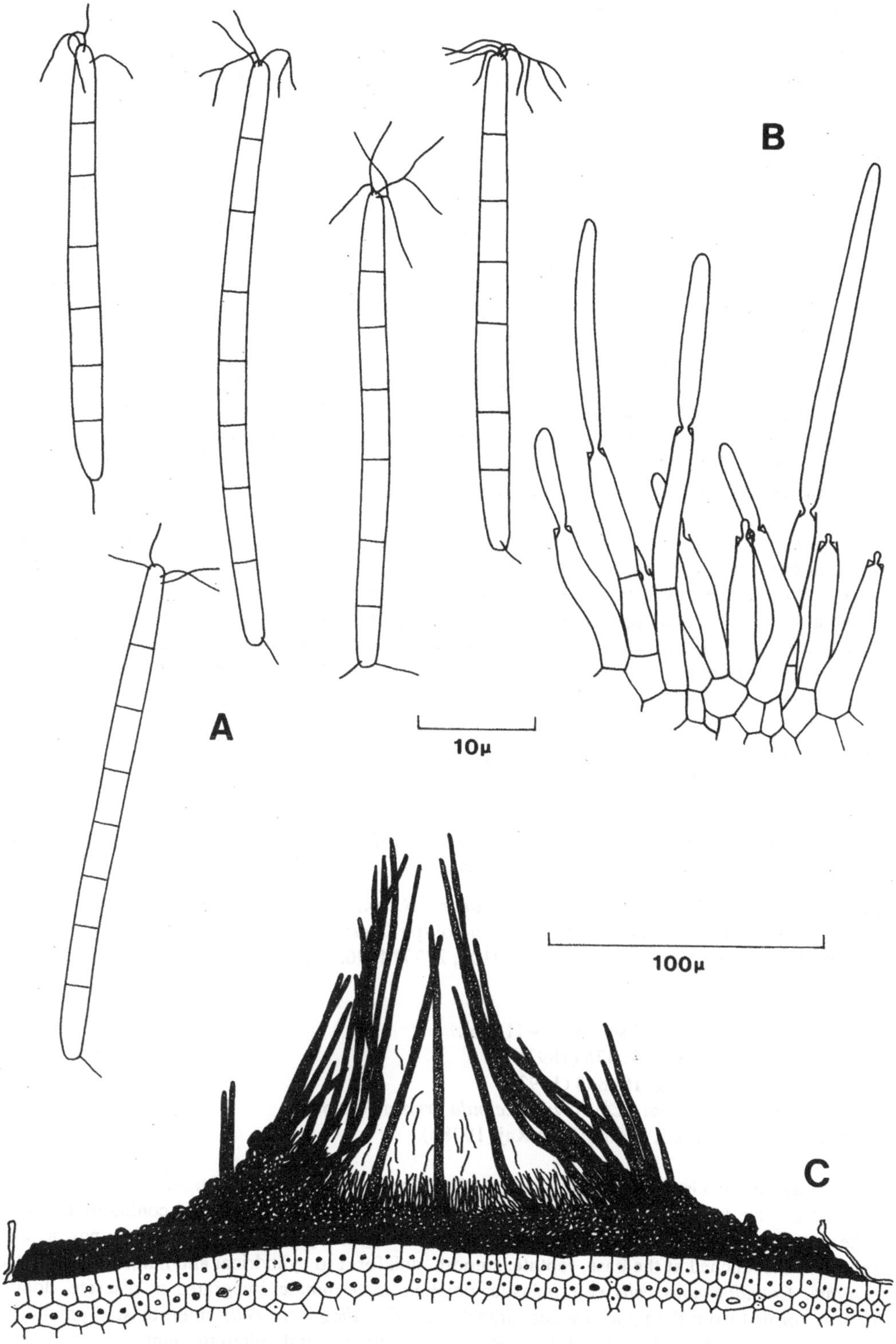

Fig. 275. *Pseudolachnea ryukyuensis*, A, conidia; B, conidiophores; C, vertical section of a conidioma.

Conidiomata conical or flattened, .8-2 mm long × 90-120 µ high. Setae 100-200 × 3.5-5 µ. Conidiophores 16.5-18 × 3-3.5 µ. Conidia falcate, 20-70 × 3-4 µ, 7-12 septate; apical and basal setulae single, unbranched, 3-4 µ long.

On dead culms of *Pleioblastus simonii*, *Sasa borealis* var. *purpurascens*, *S. lasioclada*, *Phyllostachys* sp., *Dendrocalamus*, India, Japan (IMI 170466, 110668 holotype *H. setulosa*).

6. STAURONEMA (Sacc.) H. & P. Syd. & Butler, *Annls mycol.* **14**: 217 (1916).
Dinemasporium Lév. subgen. *Stauronema* Sacc., *Syll. fung.* **3**: 686 (1884).
Sp. typ.: *S. cruciferum* (Ell.) H. & P. Syd. & Butler (syn. *Dinemasporium cruciferum* Ell.).

Mycelium immersed, branched, septate, hyaline or brown. *Conidiomata* separate or aggregated, superficial, black, at first globose and closed, later opening out to become cupulate; walls at the base and sides of brown textura angularis, at the sides darker brown and thicker-walled. Dehiscence by exposure of the conidiogenous region as the conidiomata become cupulate. *Setae* abundant, brown, septate, acuminate, smooth, originating from the side walls, occasionally also from the conidiogenous region. *Conidiophores* hyaline, simple or branched at base, septate, smooth, cylindrical, formed from the base and the side walls. *Conidiogenous cells* enteroblastic, phialidic, integrated, occasionally discrete, determinate, cylindrical, hyaline, smooth, with a minute channel and collarette, but periclinal thickening conspicuous. *Conidia* hyaline, aseptate, eguttulate, thin-walled, smooth, straight or slightly curved, fusiform, with a single, unbranched setula at each end and 1-2 originating in the middle of the conidium.

So far, five species have been described in *Stauronema* but there has not been any revision, particularly with respect to species of *Dinemasporium* (e.g. *D. sacchari* P. Henn.) which are better placed in *Stauronema*. *S. platense* (Speg.) H. & P. Syd. & Butler (1916) was reported with conidia 15 × 4-5 µ and appendages 15 × 1 µ, *S. sacchari* H. & P. Syd. & Butler (1916) with conidia 10-12 × 3 µ and appendages 10-12 µ long, *S. indicum* Kalani (1964) with conidia 10.5-21 × 4.5-5 µ and appendages 17-25 µ long, and *S. spinificis* Subhedar & V.G. Rao (1976) with conidia 6-8.5 × 3-4 µ and setulae 6-8 µ long. No types have been examined, but since some of these measurements overlap in range, it seems probable that there are in fact fewer species than the five described.

S. CRUCIFERUM (Ell.) H. & P. Syd. & Butler, *ibid.*: 217 (1916). (Fig. 276)
Dinemasporium cruciferum Ell., *Bull. Torrey bot. Cl.* **9**: 20 (1882).
Dinemasporium bicristatum Cke, *Grevillea* **6**: 135 (1878).

Conidiomata up to 300 µ diam. Setae 1-4 septate, 125-240 µ × 5-8 µ. Conidiophores up to 20 µ long × 2 µ wide. Conidia 10.5-13.5 × 3-4 µ; 1-3 setulae, 10-19 µ long.

On *Bambusa* sp., *Saccharum munja*, *S. officinarum*, *S.* sp., *Sorghum vulgare*, *Triticum* sp., *Zea mays*, indet. Gramineae, U.S.A., India, Pakistan (IMI 94859, holotype *D. cruciferum* ex K, 214792a, 82285, 227713, 227732b, 227715, 98414, 227732b, 215859, 227704, 199103, 227714, 96362, 183841, 98416, 96358, 227716a).

7. POLYNEMA Lév., *Ann. Sci. nat.*, 3 ser., **5**: 274 (1846).
Barklayella Sacc., *Syll. fung.* **10**: 475 (1892).
Neobarclaya Sacc., *Syll. fung.* **10**: 843 (1892).
Sp. typ.: *P. ornatum* (de Not.) Lév. (syn. *Excipula ornata* de Not.).
Illustrations: Sutton (1963, 1968c); Nag Raj (1978b).

Mycelium immersed, hyaline, branched, septate. *Conidiomata* acervular to cupulate, subcuticular, separate or gregarious, formed of pale brown to hyaline, thin-walled textura angularis, in cupulate conidiomata becoming textura porrecta at the margin. Dehiscence irregular. *Setae* erect, divergent, pale brown, smooth, apex acute, thin-walled. *Conidiophores* hyaline, branched at the base, septate, smooth, doliiform to cylindrical and filiform, formed from the upper pseudoparenchyma. *Conidiogenous cells* enteroblastic, phialidic, determinate, integrated or discrete, doliiform to cylindrical and filiform, formed from the upper pseudoparenchyma. *Conidiogenous cells* enteroblastic, phialidic, determinate, integrated or discrete, doliiform to cylindrical, hyaline, smooth, channel and collarette minute. *Conidia* pale brown, 0-2 euseptate, continuous, base obtuse to truncate, apex obtuse, thin-walled, verruculose, eguttulate, fusiform, lower cells sometimes slightly wider, with one simple or branched, cellular, apical appendage and 1-3 unbranched, cellular appendages inserted laterally above the basal scar.

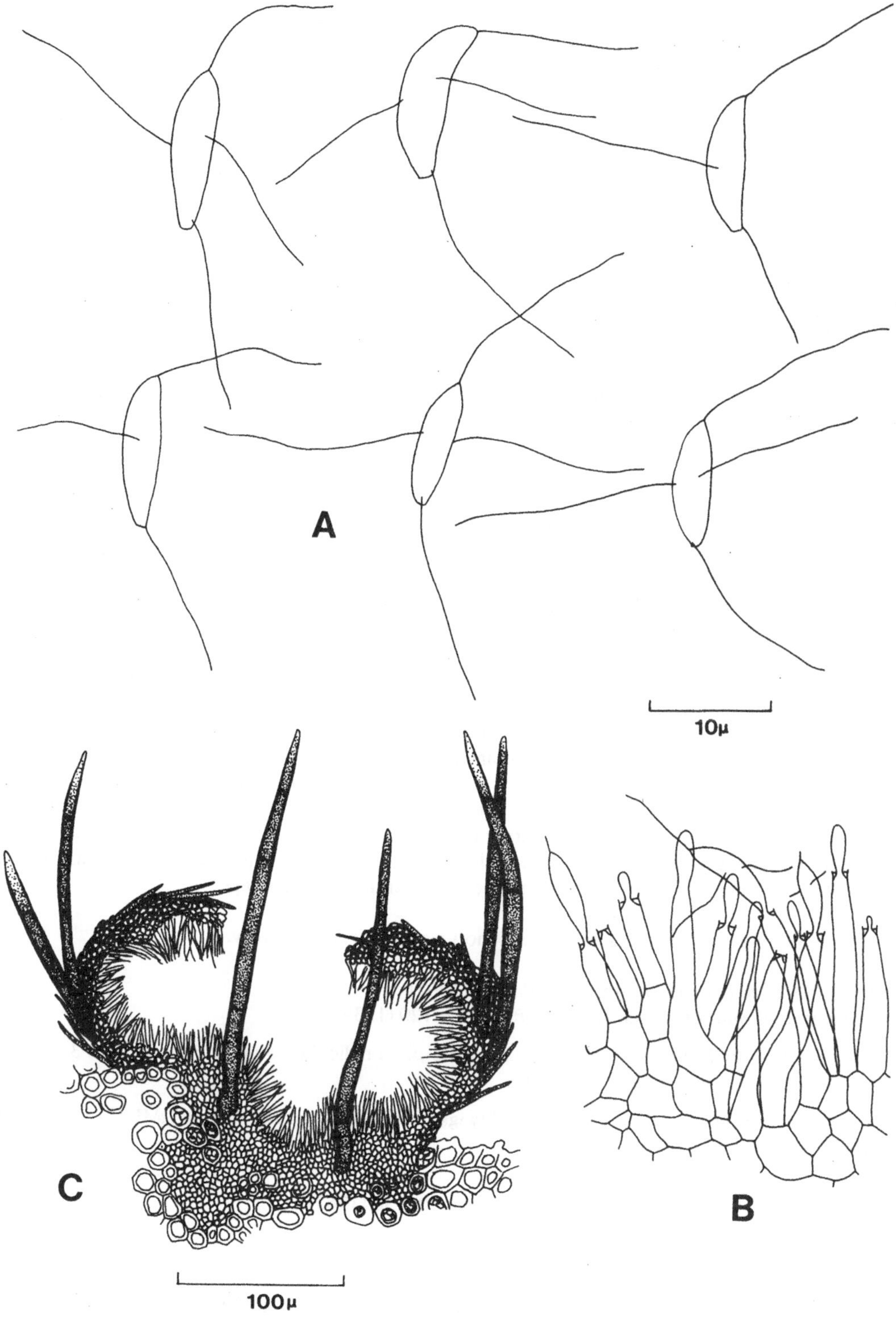

Fig. 276. *Stauronema cruciferum*, A, conidia; B, conidiophores and developing conidia; C, vertical section of a conidioma.

Sutton (1963, 1968c) has illustrated and described this genus and the history of its synonyms. Apart from the type species there is *P. careyae* (Patil & Thirum.) Morelet, a species of doubtful retention in *Polynema*. No acervular setae were mentioned in the original description and the conidial morphology seems quite at variance with that of *P. ornatum*.

Nag Raj (1978b) has recently presented evidence suggesting that *Neobarclaya* Sacc. be reinstated as a distinct genus based on the single original species, and that *Polynema* can be reserved for two taxa, *P. ornatum* (de Not.) Lév. and *P. radiatum* (Ell. & Ev.) Nag Raj. For species in which conidiogenesis and conidial morphology are comparable, the most marked differences being in structure of the conidiomata (possibly substrate induced), I still prefer to keep them in a single genus, *Polynema*. The re-examination by Nag Raj (1978b) of *Dinemasporium radiatum* Ell. & Ev., shows there to be another species in *Polynema* differing from *P. ornatum* by the shorter conidia.

P. ORNATUM (de Not.) Lév., *ibid.*: 274 (1846). (Fig. 277)
Excipula ornata de Not., *Mem. Accad. Torino* **7**: 3 (1845).
Pestalotia primaria Ell. & Ev., *J. mycol.* **2**: 103 (1886).
Barclayella primaria (Ell. & Ev.) Sacc., *Syll. fung.* **10**: 475 (1892).
Neobarclaya primaria (Ell. & Ev.) O. Kuntze, *Rev. Gen. Pl.* III: 177 (1893).

Conidiomata 100-300 µ diam. Conidiophores 5-33 × 2-3 µ. Setae up to 150 µ long. Conidia 10-14 × 3.5-4.5 µ; apical appendage 6-8 µ long; basal appendages 4-9 µ long.

On stems of *Scirpus fluviatilis* and wood of *Castanea* sp., Italy, U.S.A. (IMI 104663, holotype *E. ornata* ex PR, 91259, holotype *P. primaria* ex FH).

8. HOEHNELIELLA Bres. & Sacc. apud Strasser, *Verh. zool.-bot. Ges., Wien* **52**: 437 (1902).
Klebahnopycnis Kirschst., *Annls mycol.* **37**: 120 (1939).
Sp. typ.: *H. perplexa* Bres. & Sacc.
Illustration: Vasant Rao & Sutton (1976).

Mycelium immersed, pale brown to hyaline, septate, branched. *Conidiomata* superficial, separate, dark brown, at first closed, later cupulate; periclinal and basal wall of hyaline, thin-walled textura angularis which becomes darker, thicker-walled and textura porrecta at the periphery. *Setae* peripheral, incurved, later divergent, thick-walled, dark brown, septate, rigid or flexuous, acuminate. *Conidiophores* hyaline, branched and septate at the base, more rarely above, smooth, cylindrical, straight or slightly curved, formed from the base, less so from the periclinal wall. *Conidiogenous cells* enteroblastic, phialidic, integrated, hyaline, smooth, apical, or as lateral branches immediately below septa, channel and collarette minute, apical periclinal wall thickened. *Conidia* solitary or catenate, pale brown, medianly 1-euseptate, continuous, apex and base obtuse to truncate, slightly thickened, thin-walled, smooth, guttulate, cylindrical to fusiform, straight, each end with an extracellular, unbranched or dichotomously to irregularly branched appendage.

Vasant Rao & Sutton (1976) fully described the monotypic genus and its history.

H. PERPLEXA Sacc., & Berl. apud Strasser, *ibid.*: 437 (1902).
Klebahnopycnis clematidis Kirschst., *Annls mycol.* **37**: 120 (1939).

Conidiomata up to 220 µ diam. Setae 75-210 × 3-5 µ. Conidiophores up to 50 µ long × 2-3.5 µ wide. Conidia 10.5-12.5 × 3-3.5 µ; appendages 2-6 µ long.

On decorticated runners of *Clematis vitalba* and twigs of *Berberis vulgaris*, Austria (IMI 33169, WINF(M) 10700, holotype *H. perplexa* ex PAD).

9. DIMASTIGOSPORIUM Faurel & Schotter, *Rev. Mycol.* **30**: 156 (1965).
Sp. typ.: *D. musimonum* Faurel & Schotter
Illustration: Faurel & Schotter (1965).

Mycelium not described. *Conidiomata* superficial, separate, black or blackish green, opening widely to become cupulate; periclinal wall pseudoparenchymatous, extending into setae at the periphery. *Ostiole* absent. *Setae* restricted to the outer wall, divergent, thick-walled, dark brown, rigid, jointed and sometimes branched at the base. *Conidiophores* hyaline, smooth, cylindrical, branched sparingly at the base, septate, often unequally bifurcate at the apex, formed from the base and sometimes from the periclinal wall. *Conidiogenous cells* integrated, hyaline, cylindrical; conidiogenesis not

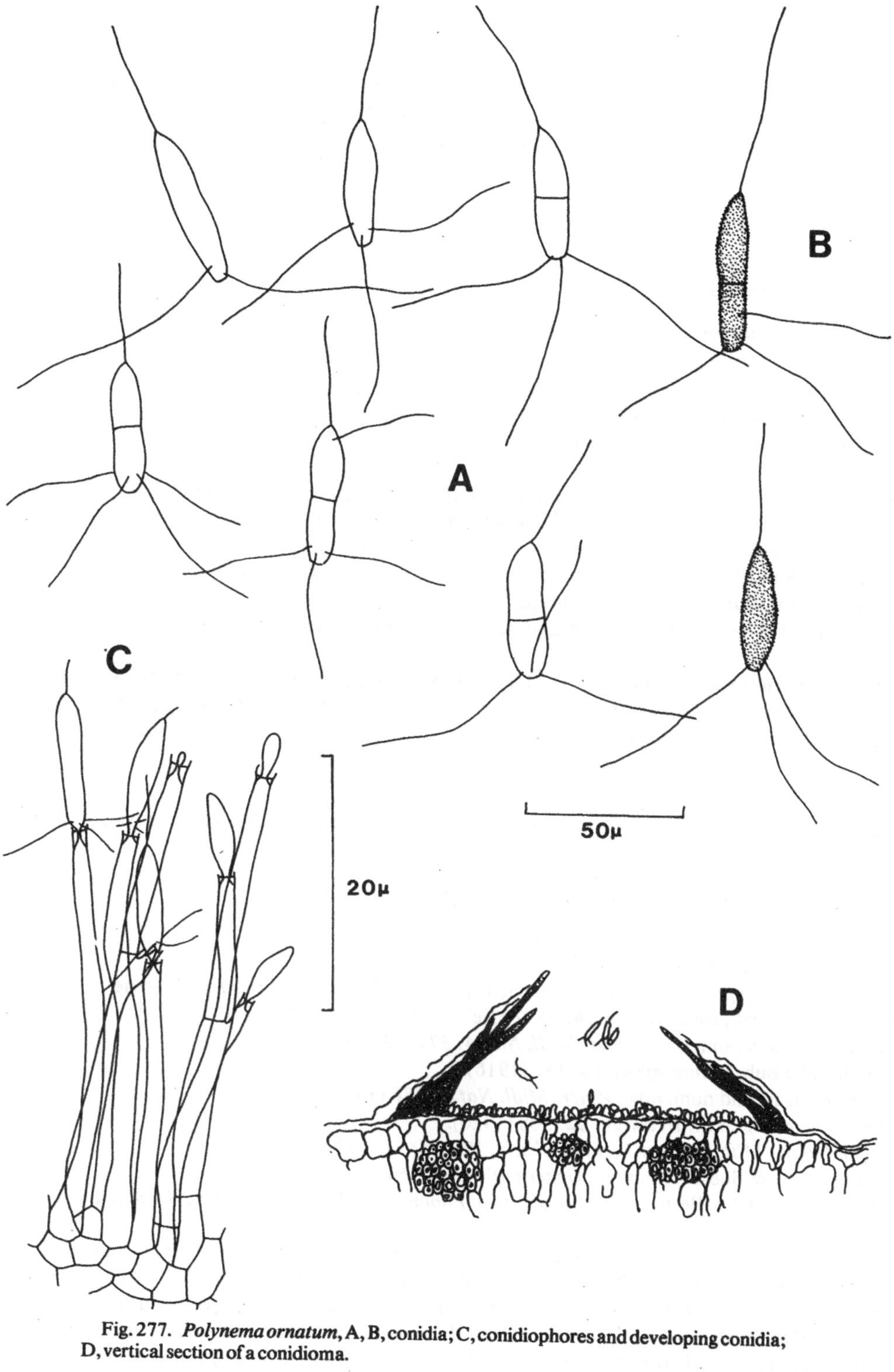

Fig. 277. *Polynema ornatum*, A, B, conidia; C, conidiophores and developing conidia; D, vertical section of a conidioma.

described. *Conidia* hyaline, aseptate, thin-walled, pyriform, eguttulate, the apex with two divergent, unbranched, equal, cellular appendages. During conidiogenesis the body of the conidium is formed first and is attached to the conidiogenous cell by one of the appendages.

The original material of the type species has been destroyed (Mme J. Nicot, pers. comm.), consequently without details of conidiogenesis the affinities of *D. musimonum* cannot be determined precisely. Other morphological evidence suggests that it is related to *Dinemasporium* Lév., *Stauronema* (Sacc.) Syd. & Butl., *Polynema* Lév. and *Pseudolachnea* Ranoj.

D. MUSIMONUM Faurel & Schotter, *ibid.*: 156 (1965).

Conidiomata up to 250 µ diam. × 75-100 µ high. Setae 150-200 × 3-4 µ. Conidiophores 15-20 × 1.5-2 µ. Conidia 8-10 × 3-4 µ; appendages 15-25 µ long.

On dung of *Ammotragus lervia*, Central Sahara.

10. RHABDOGLOEUM H. Syd. apud Sydow & Petrak, *Annls mycol.* **20**: 215 (1922).
Sp. typ.: *R. pseudotsugae* H. Syd.
Illustrations: Nag Raj & Morgan-Jones (Fig. 1, 1973).

Mycelium immersed, hyaline, branched, septate. *Conidiomata* acervular, subepidermal, elongated, separate or confluent, formed of hyaline to pale brown, isodiametric, thin-walled pseudoparenchyma. Dehiscence irregular. *Conidiophores* absent. *Conidiogenous cells* enteroblastic, phialidic, determinate or indeterminate, discrete, cylindrical to lageniform, hyaline, smooth, channel wide, collarette minute, rarely proliferating percurrently, formed from the upper cells of the pseudoparenchyma. *Conidia* hyaline, aseptate, base truncate, apex obtuse, thin-walled, smooth, guttulate, cylindrical, slightly constricted in the middle, with a gelatinous, unbranched, extracellular appendage at the apex.

The genus was redescribed by Nag Raj & Morgan-Jones (1973) in an account of *Rhabdogloeopsis* Petrak and *Rhabdogloeum*. Unfortunately the illustration purported to be of *Rhabdogloeum pseudotsugae* is in fact of *Rhabdogloeopsis balsameae* and vice versa.

R. PSEUDOTSUGAE H. Syd., *ibid.*: 215 (1922).

Conidiomata 200-250 µ diam. Conidiogenous cells 9-120 × 4-6.5 µ. Conidia 14-17 × 4-6 µ; appendage 3-5.5 × 1.5-2 µ.

On needles of *Pseudotsuga taxifolia*, *P.* sp., Mont., U.S.A., B.C., Canada.

11. SPHAERELLOPSIS Cke, *Grevillea* **12**: 23 (1883).
Darluca Castagne, *Suppl. Cat. Pl. Marseille*: 53 (1851), non Rafinesque, 1820, Rubiaceae
Botryella H. & P. Syd., *Annls mycol.* **14**: 95 (20 Feb., 1916).
Kabathia Nieuwland nom. nov., *Amer. Midl. Nat.* **4**: 377 (13 Mar., 1916).
Diplodothiorella Bub., *Annls mycol.* **14**: 151 (1916).
Mycepimyce Nieuwland nom. nov., *Amer. Midl. Nat.* **4**: 386 (1916).
Diploplacosphaeria Petrak, *Hedwigia* **62**: 308 (1920).
Darlucella Höhn., *Mitt. bot. inst. tech. Hochsch., Wien* **4**: 57 (1927).
Diploplacis Clem. & Shear nom. nov., *Genera of Fungi*: 364 (1931).
Sp. typ.: *S. filum* (Biv.-Bern. ex Fr.) Sutton (syn. *Phoma filum* Biv.-Bern. ex Fr., *S. quercuum* Cke).

Mycelium immersed, branched, septate, hyaline to pale brown. *Conidiomata* eustromatic, immersed, deep-seated but becoming erumpent, unilocular or multilocular, the locules often almost superficial and then appearing like separate pycnidia, each locule with a separate simple ostiole; basal wall composed of pale brown textura angularis, locular wall of dark brown, thick-walled textura angularis. *Conidiophores* hyaline to pale brown, septate, branched only at the base and then infrequently, smooth, formed from the inner cells of the locular walls. *Conidiogenous cells* enteroblastic, phialidic, indeterminate, integrated or discrete, cylindrical to doliiform, hyaline to pale brown, smooth, often with 1-3 percurrent proliferations. *Conidia* initially hyaline, finally very pale brown and irregularly verruculose, 0-1 euseptate, not constricted, apex obtuse, base truncate, straight, ellipsoid, the apex with a gelatinous cap that occasionally becomes everted.

Sphaerellopsis was made a synonym of *Ascochyta* Lib. by Saccardo (1884), but Petrak (1968b) regarded it as a satisfactory genus in its own right and this conclusion was re-iterated by Melnik (1976). Re-examination of the type has shown that the hypophyllous, single or aggregated erumpent pycnidia are uredinicolous. Consequently it provides the next available generic name for the ubiquitous rust hyperparasite after *Darluca* Cast., which is a later homonym of *Darluca* Raf., 1820, Rubiaceae. Neither *Sphaerellopsis* nor its later synonyms were included in the discussion of *Eudarluca australis* (Fr.) O. Eriksson by Eriksson (1966).

S. FILUM (Biv.-Bern. ex Fr.) Sutton, *Mycol. Pap.* 141: 196 (1977). (Fig. 278).
Phoma filum Biv.-Bern. ex Fr., *Syst. myc.* **2**: 547 (1823).
Sphaeria filum Biv.-Bern., *Stirp. Sicilia, Man.* **3**: 12 (1813-1816).
Darluca filum (Biv.-Bern. ex Fr.) Berk., *Outline of British Fungology*: 318 (1860).
Kabathia filum (Biv.-Bern. ex Fr.) Nieuwland, *Amer. Midl. Nat.* **4**: 377 (1916).
Sphaerellopsis quercuum Cke, *Grevillea* **12**: 23 (1883).
Ascochyta quercuum (Cke) Sacc., *Syll. fung.* **3**: 393 (1884).
Diplodina quercuum (Cke) Tracy & Earle, *Bull. Torrey bot. Cl.* **26**: 494 (1899).
Diplodia punctata Lév., *d'Orb. Dict. d'Hist. nat.* **12**: 779 (1840).
Ascochyta androsaces Achund., *Izv. Akad. Nauk azerb. SSR* 1971(3): 6 (1971).
Ascochyta cerinthes Lobik, *Morbi Pl. Leningr.* **17**: 173 (1928).
Ascochyta contubernalis Oud., *Ned. Kruidk. Arch.*, ser. 2, **5**: 170 (1888).
Ascochyta densiuscula Sacc. & Malbr., *Michelia* **2**: 621 (1881).
Ascochyta dianthi Lib., *Pl. Crypt. Ard.* 158 (1832).
Ascochyta graminicola Sacc. var. *ciliolata* Sacc., *Michelia* **1**: 127 (1878).
Ascochyta graminicola Sacc. var. *coeruleae* Br. & Har., *Rev. mycol.* **13**: 17 (1891).
Ascochyta pucciniophila Starb., *Bot. Centralbl.* **64**: 382 (1895).
Ascochyta quercus-ilicis Güssow, *J. Bot.* **46**: 123 (1908).
Ascochyta solidaginis (Schw. ex Fr.) Starb., *Bih. Kongl. Svensk, Vetensk.-Akad. Handl.* **19**(2): 84 (1893).
Ascochyta solidaginis (Thüm.) Keissl., *Beih. Bot. Centralbl.* **29**: 427 (1912).
Ascochyta staticis Nagorny, *Bol. Rast.* **7**: 120 (1913).
Hendersonia uredinicola Desm., *Ann. Sci. nat.*, 3 ser., **11**: 345 (1849).
Darluca arcuata Ell. & Ev., *Bull. Torrey bot. Cl.* **22**: 437 (1895).
Darlucella arcuata (Ell. & Ev.) Höhn., *Mitt. bot. Inst. tech. Hochsch., Wien* **4**: 57 (1927).
Diplodothiorella sadurneri Bub., *Annls mycol.* **14**: 151 (1916).
Diploplacosphaeria ruthenica Petrak, *Hedwigia* **62**: 308 (1920).
Diploplacis ruthenica (Petrak) Clem. & Shear, *Genera of Fungi*: 364 (1931).
Botryella nitidula H. & P. Syd., *Annls mycol.* **14**: 95 (1916).

Teleomorphosis: *Eudarluca caricis* (Fr.) Eriksson, *Bot. Not.* **119**: 35 (1966).

Pycnidial conidiomata up to 200 µ diam. Conidiophores 7.5-20 × 5-7 µ. Conidia 17-20 × 5.5-6.5 µ.

Hyperparasitic on Uredinales on various hosts. World wide in distribution and represented in IMI by 273 collections including (on leaves of *Quercus virens*, Ga, U.S.A., IMI 197712, isotype *S. quercuum*).

12. CEUTHOSPORA Grev., *Scot. Crypt. Flora* **5**: 253-254 (1827), nom. cons. against *Ceuthospora* Fries, 1825, Ascomycetes.
Siroplaconema Petrak, *Annls mycol.* **20**: 331 (1922).
Lectotype sp.: *C. lauri* (Grev.) Grev. (syn. *Cryptosphaeria lauri* Grev.).

Mycelium immersed, branched, septate, hyaline to dark brown. *Conidiomata* pseudostromatic, immersed, black, circular to oval, occupying the upper and lower epidermises and the parenchyma inbetween, the total part of which is separated from the uncolonised tissue by dark brown sclerotial walls. Locules single or aggregated within the stromata, hyaline to pale brown, with separate or communal epiphyllous, hypophyllous or amphigenous furfuraceous ostioles. *Conidiophores* hyaline, branched irregularly, septate, smooth, with acropleurogenous conidia, formed from the inner cells of the locular walls. *Conidiogenous cells* enteroblastic, phialidic, determinate, integrated or discrete, cylindrical, hyaline, smooth, channel and collarette minute, periclinal wall thickened. *Conidia* hyaline, cylindrical, aseptate, smooth, thin-walled, base obtuse, apex abruptly tapered, eguttulate, with an apical, fan-shaped, gelatinous appendage.

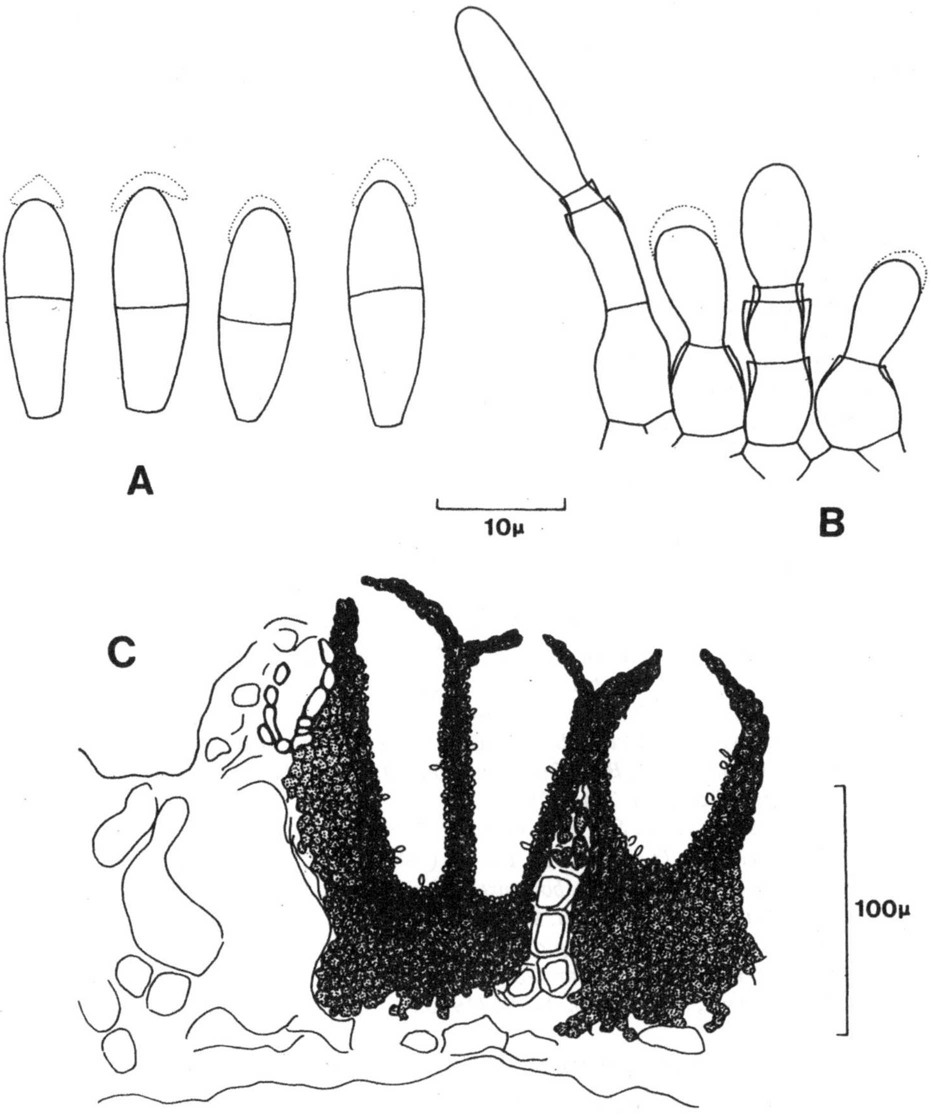

Fig. 278. *Sphaerellopsis filum*, A, conidia, B, conidiophores and developing conidia; C, vertical section of conidiomata.

The nomenclatural complexities of this generic name have been discussed by Sutton (1972b), leading to the formal conservation of *Ceuthospora* Greville, 1827 against Fries, 1825, with *C. lauri* (Grev.) Grev. as the lectotype species.

The genus at present contains in excess of 100 described taxa and no modern revision has yet been attempted. A provisional assessment of some temperate species and the fact that they invariably colonize dead or moribund leaf tissue in litter habitats suggests that many of these names could possibly refer to a rather small number of plurivorous species. For instance, *C. buxi* (Fr.) Petrak (on *Buxus sempervirens*, U.K., Switzerland, IMI 49362, 49361, 127975, 105572, 214192), *C. feurichii* Bubák (on *Vinca minor*, U.K., IMI 178356, 107232) and *C. melaleuca* Ferd. & Winge (on *Ginkgo biloba*, Eire, IMI 199438, 199439) do not have conidia exceeding 12.5 μ × 2.5 μ. On the other hand *C. euonymi* Grove (on *E. pendulus*, *E. radicans*, Pakistan, India, U.K., IMI 136954, 136953, 136947, 175000, 174986, 93333a, 82725a, 108585), *C. hederae* Grove (on *Hedera helix*, *H. himalaica*, Pakistan, U.K., IMI 117403, 107953, 91671, 81243, 90903), *C. phacidioides* Grev. (commonly on *Ilex aquifolium*, Germany, Eire, U.K., IMI 33456, 214193, 202379e, 193368a, 80077, 117643, 20933, 35343, 95740a, 107233, 107952, 100413, 95573, 95568, 95567, 80713a, 80337), and *C. rhododendri* Grove (on *Rhododendron*, U.K., IMI 107234) cannot be distinguished from one another by conidial size or morphology of conidiomata and fall well within the range of variation quoted by Sutton (1972b) for *C. lauri*. Type studies on these taxa have not yet been undertaken but it seems most probable that such names will eventually be reduced to synonymy with *C. lauri*.

C. LAURI (Grev.) Grev., *ibid.*: 254 (1827). (Fig. 279)
Cryptosphaeria lauri Grev., *Flora edinensis*: 361 (1824).
Sphaeria lauri Sowerby, *Coloured figures of English fungi* **3**: 371 (1803).

Conidiomata up to 900 µ diam.; locules 150-250 µ diam. Conidiophores up to 20 × 3.5 µ. Conidia 11.5-14 × 2.5 µ.

On dead leaves of *Prunus laurocerasus*, U.K. (IMI 153020, lectotype *S. lauri* ex E, 107954, 105046).

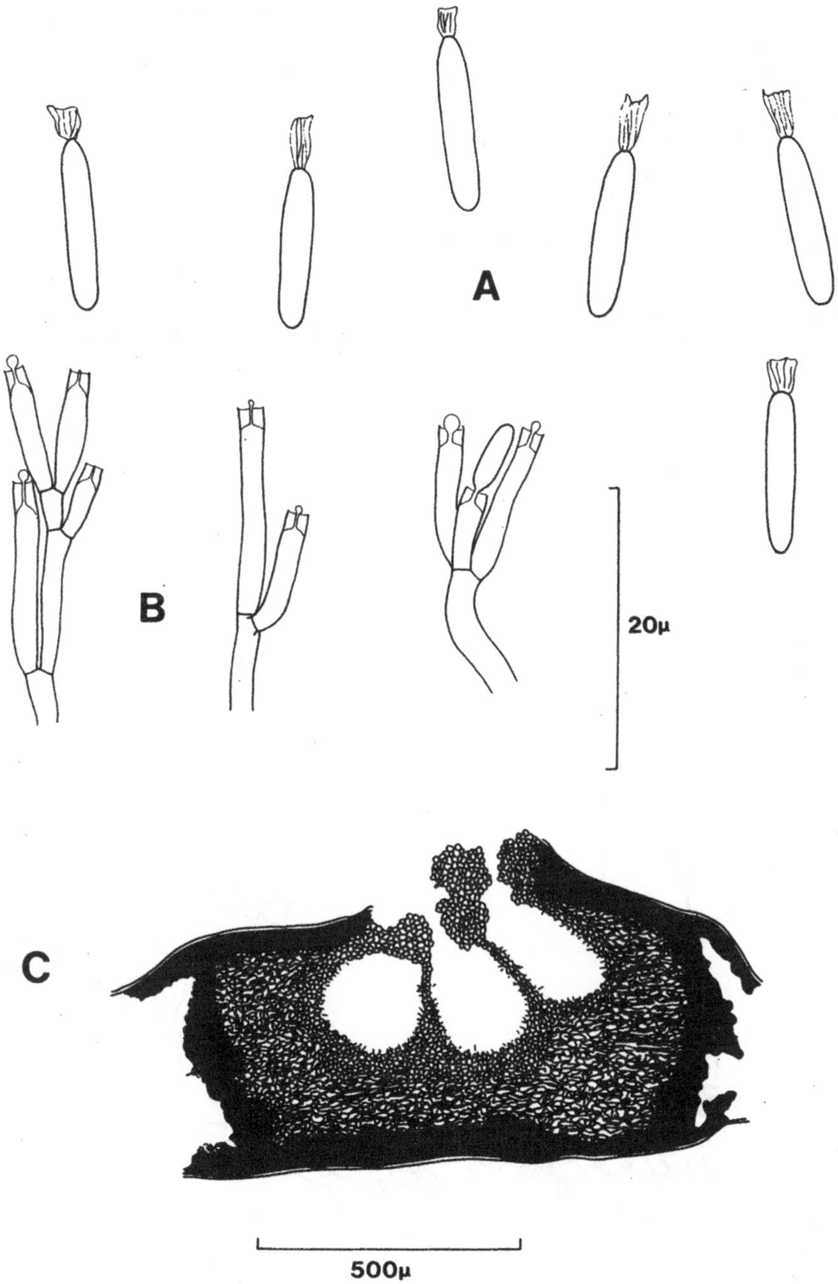

Fig. 279. *Ceuthospora lauri*, A, conidia; B, conidiophores; C, vertical section of a conidioma.

13. MONOCHAETIELLA Castellani, *Nuovo G. bot. ital.* n.s., **49**: 487 (1943).
Sp. typ.: *M. hyparrheniae* Castellani

Mycelium immersed, branched, septate, brown. *Conidiomata* acervular, subepidermal, separate or confluent, formed of brown, thin-walled textura angularis. Dehiscence irregular. *Conidiophores* absent. *Conidiogenous cells* enteroblastic, phialidic, discrete, determinate, cylindrical, doliiform to lageniform, hyaline or pale brown, smooth, channel and collarette minute, periclinal wall thickened, formed from the upper pseudoparenchyma. *Conidia* fusiform, straight or slightly curved, base obtuse, apex prolonged into a short, unbranched appendage, inclined to one side, thin-walled, hyaline, aseptate, smooth.

The absence of acervular setae and appressoria at present can be used to distinguish this genus from *Colletotrichum* Cda. Nag Raj (1973a) reduced *Rostrospora* Subram. & Ramakr. and *Ellisiella* Sacc. to synonymy with *Colletotrichum* and these show considerable similarity with *Monochaetiella* in having appendaged conidia.

It may be necessary to reduce *Monochaetiella* to synonymy with *Colletotrichum* if the type species is studied in culture and found to have setae and appressoria.

M. HYPARRHENIAE Castellani, *ibid.*: 487 (1943). (Fig. 280)

Conidiomata up to 400 μ diam. Conidiogenous cells 6-15 × 5-6 μ. Conidia 21-31 × 6.5-7 μ; appendage 3-7 μ long.

Causing lesions on leaves of *Hyparrhenia rufa*, *H. variabilis*, *H. cymbaria* forma *macra*, *H. cymbaria*, *H. filipendula*, *H. umbrosa*, Sudan, Uganda, Malawi, Zambia (IMI 35545, 59042, 80498, 164010, 112446, 112459, 61940, 112491).

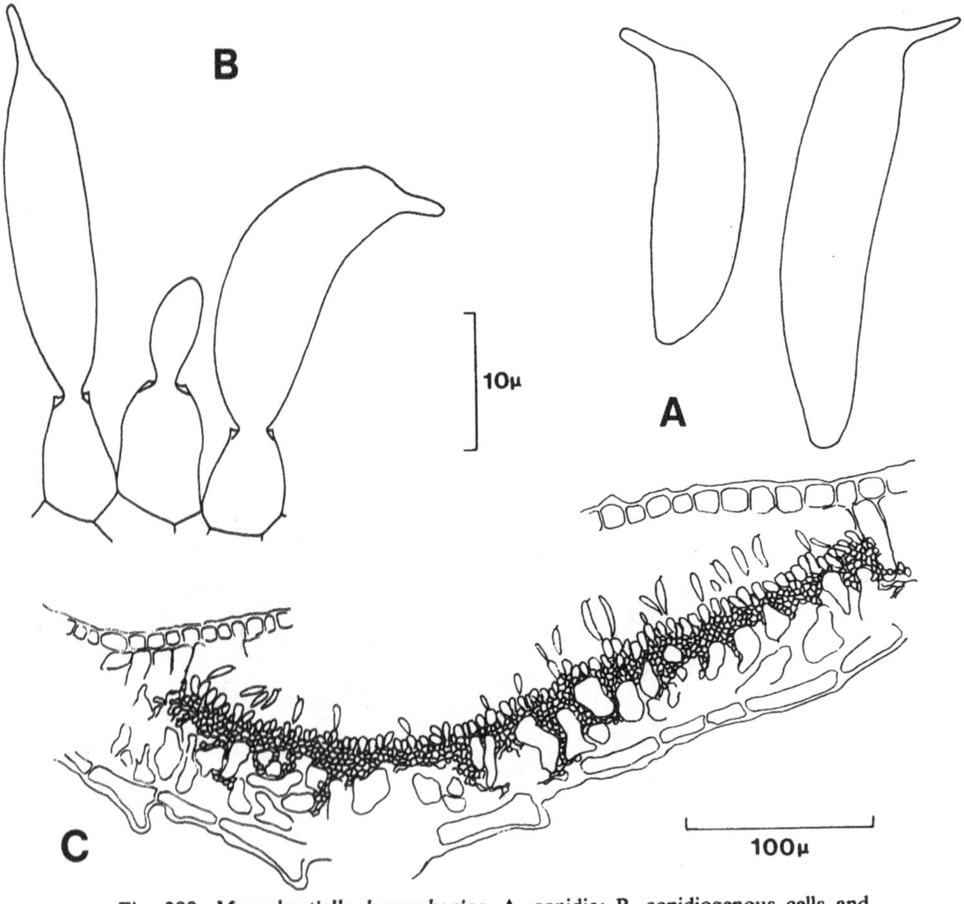

Fig. 280. *Monochaetiella hyparrheniae*, A, conidia; B, conidiogenous cells and developing conidia; C, vertical section of a conidioma.

14. DIACHORELLA Höhn., *Hedwigia* **60**: 192 (1918).
Chaetomelasmia Danilova apud Naumov & Danilova, *Bot. Mater.* **7**: 139 (1951).
Sp. typ.: *D. onobrychidis* (DC. ex Fr.) Höhn. (syn. *Rhytisma onobrychidis* DC. ex Fr.).

Mycelium immersed, branched, septate, hyaline. *Conidiomata* epidermal, black, large and irregular, often confluent, applanate, composed of a basal layer of hyaline, thin-walled textura angularis, the walls of the cells adjacent to the lower epidermal wall thickened and dark brown to black; upper wall consisting of a thin layer of dark brown pseudoparenchyma and black discoloured upper epidermal wall and cuticle. *Ostiole* absent, dehiscence by irregular longitudinal fissures. *Conidiophores* branched only at the base, hyaline, thin-walled, cylindrical, tapered or inflated towards the apices, formed from the upper cells of the lower wall. *Conidiogenous cells* enteroblastic, polyphialidic, integrated or discrete, indeterminate, hyaline, smooth, with 1-2 apertures near the apex, collarette and periclinal thickening marked, channel minute. *Conidia* hyaline, aseptate, smooth, thin-walled, narrowly obpyriform to fusiform or often irregular, base obtuse, apex prolonged into a filiform, irregular, unbranched, cellular appendage.

Ciccarone (1963) recognised three formae within *D. onobrychidis*, and Sutton (1967b) provided details of conidiogenesis in what was then called *D. onobrychidis* forma *lathyri* (Fckl) Ciccarone.

KEY

Conidia 3-5 × 1.5-3 µ; appendages 4.5-8.5 µ long .... *D. onobrychidis*
Conidia 5-12 × 1.5-2.5 µ; appendages 10-30 µ long .... *D. lathyri*
Conidia 12-20 × 1-2.5 µ; appendages 13-42 µ long .... *D. caraganae*

D. ONOBRYCHIDIS (DC. ex Fr.) Höhn., *ibid.*: 192 (1918). (Fig. 281A)
Xyloma onobrychidis DC., *Fl. française* **5**(6): 159 (1815).
Rhytisma onobrychidis DC. ex Fr., *Syst. myc.* **2**: 569 (1823).
Placosphaeria onobrychidis (DC. ex Fr.) Sacc., *Michelia* **2**: 283 (1881).
Placosphaeria onobrychidis (DC. ex Fr.) Sacc. var. *minuta* Brun., Liste Sphaerops.: 24 (1886).
Placosphaeria onobrychidis (DC. ex Fr.) Sacc. var. *exappendiculata* Brun., Liste Sphaerops.: 24 (1886).
Placosphaeria onobrychidis (DC. ex Fr.) Sacc. subsp. *hedysari* Scalia, *Prima contrib. conosc. fl. micol. Catanie* 19 (1899).

Conidiomata 150-250 µ diam. × 100 µ high. Conidia 3-5 × 1.5-3 µ; appendages 4.5-8.5 µ long.

Causing small leaf lesions on *Hedysarum coronarium, H. americanum, Onobrychis* sp., Iraq, Malta, Sask., Canada, Italy (IMI 70559, 22679, 15465, 128662).

D. lathyri (Fckl) comb. nov. (Fig. 281B, C, D)
Rhytisma onobrychidis DC. ex Fr. b *lathyri* Fckl, *Jb. nassau. Ver. Naturk.* **23-24**: 265 (1870).

Conidiomata 80-100 µ diam. Conidia 5-12 × 1.5-2.5 µ; appendages 10-30 µ long.

Causing large leaf lesions on *Lathyrus tuberosus*, Roumania (IMI 22678).

D. caraganae (Danilova) comb. nov.
Diachorella onobrychidis (DC. ex Fr.) Höhn. forma *caraganae* Ciccarone, *Phytopath. Medit.* **2**: 248 (1963).
Chaetomelasmia caraganae Danilova apud Naumov & Danilova, *Bot. Mater.* **7**: 139 (1951).

Conidia 12-20 × 1-2.5 µ; appendages 13-42 µ long.

Causing leaf lesions on *Caragana arborescens*, U.S.S.R.

15. NEOPLACONEMA Sutton, *Kew Bull.* **31**: 463 (1977).
Sp. typ.: *N. napelli* (Maire & Sacc.) Sutton (syn. *Placosphaeria napelli* Maire & Sacc.).

Mycelium immersed, branched, septate, hyaline to brown. *Conidiomata* pseudostromatic, immersed, black, circular, epidermal to subepidermal; mycelium concentrated in the epidermis and subepidermis, of dark brown textura intricata, main body of the stroma hypodermal, consisting of a central undivided locule bounded by thin-walled, small textura

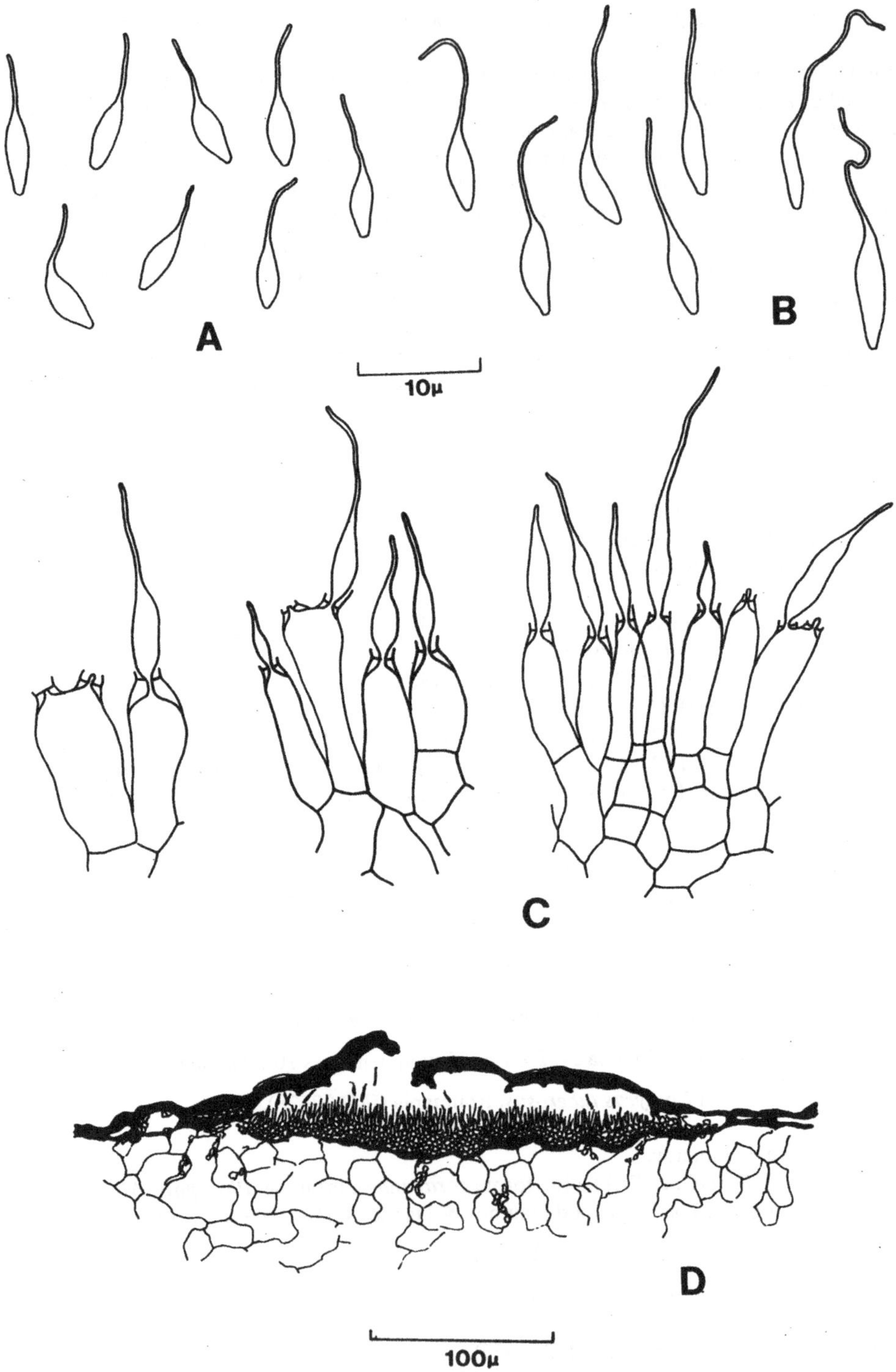

Fig. 281. A, *Diachorella onobrychidis*, conidia; B, C, D, *D. lathyri*, conidia, conidiophores, vertical section of a conidioma.

angularis constituting a wall, the base of which is seated almost directly on the substrate. The locule is surrounded by pseudostromatic pseudoparenchyma of textura intricata intermixed with host tissue which becomes progressively more sclerotioid towards the periphery. *Ostiole* central, circular, collabent. *Conidiophores* hyaline, branched only at the base, 1-3 septate, smooth, formed from the locular wall. *Conidiogenous cells* enteroblastic, phialidic, determinate, integrated or discrete, lageniform, hyaline, smooth, channel and collarette minute, periclinal wall thickened. *Conidia* hyaline, aseptate,

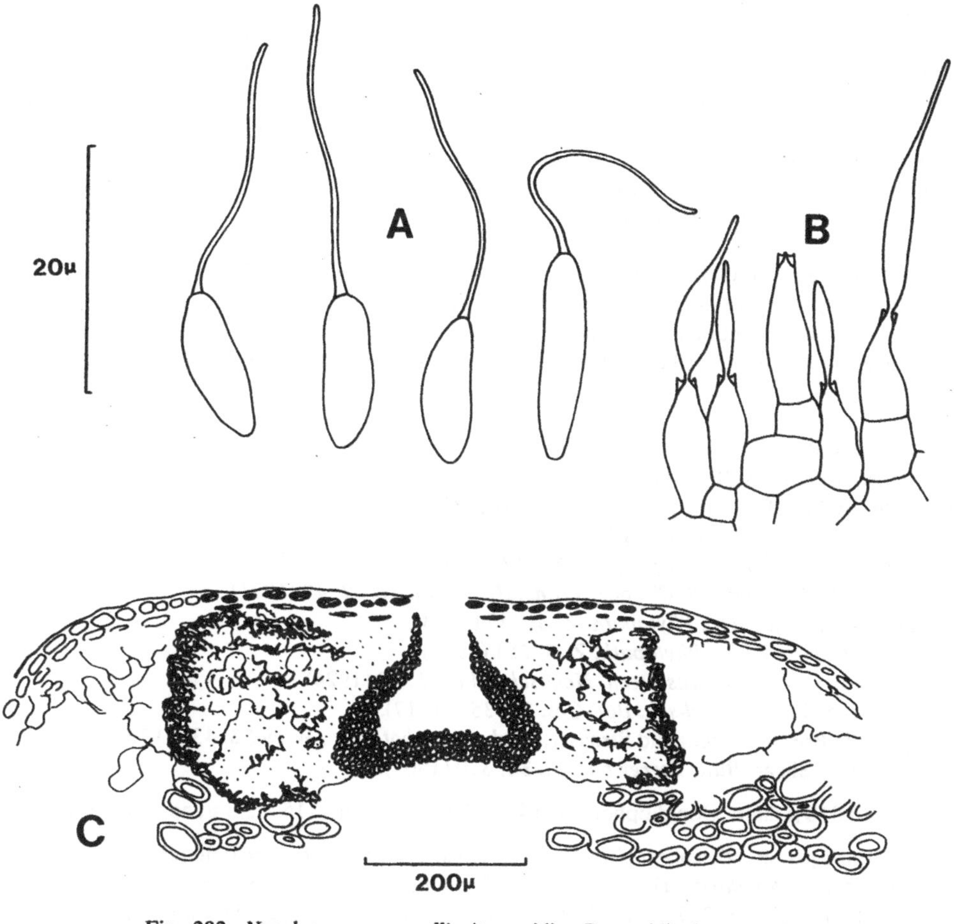

Fig. 282. *Neoplaconema napelli*, A, conidia; B, conidiophores and developing conidia; C, vertical section of a conidioma.

thin-walled, straight or slightly curved, obtuse at the apex and base, the apex prolonged into a flexuous, curved, unbranched, cellular appendage separated at the base by a septum.

The reasons necessitating the erection of *Neoplaconema* rather than the acceptance of *Placonema* for this single species on *Aconitum napellum* have been fully discussed by Sutton (1977b).

N. NAPELLI (Maire & Sacc.) Sutton, *ibid.*: 463 (1977). (Fig. 282)
Placosphaeria napelli Maire & Sacc. apud Sacc. & Syd., *Syll. fung.* **16**: 898 (1902).
Placonema napelli (Maire & Sacc.) Petrak, *Annls mycol.* **19**: 60 (1921).

Conidiomata 650 μ diam. × 200 μ high; locules 250 μ diam. Conidiophores 9-16 × 2.5-4 μ. Conidia 11.5-16.5 × 5 μ. Appendage up to 24 μ long.

On stems of *Aconitum napellum*, Germany (IMI 24787).

16. STRASSERIA Bres. & Sacc. apud Strasser, *Verh. zool.-bot. Ges., Wien* **52**: 436 (1902).
Plagiorhabdus Shear, *Bull. Torrey bot. Cl.* **34**: 310 (1907).
Cytotriplospora Bayliss-Elliott & Chance, *Trans. Br. mycol. Soc.* **7**: 47 (1921).
Haplostromella Höhn., *Ber. dtsch. Bot. Ges.* **35**: 355 (1917).
Sp. typ.: *S. geniculata* (Berk. & Br.) Höhn. (syn. *Sphaeropsis geniculata* Berk. & Br., *Strasseria carpophila* Bres. & Sacc.).

Mycelium immersed, branched, septate, pale brown to hyaline. *Conidiomata* eustromatic, immersed, peridermal to subperidermal, globose to subglobose, black or dark brown, separate or aggregated, unilocular, lacunose or multilocular, thick-walled; walls of dark brown, thick-walled textura angularis, thinner at the base, darker near the ostiole, locular walls pale brown, of thin-walled tissue. *Ostiole* single, circular, ± papillate, often clypeate. *Conidiophores* cylindrical or lageniform, hyaline, 1-2 septate, branched at the base, formed from the inner cells of the locular walls. *Conidiogenous cells* enteroblastic, phialidic, determinate, integrated, less often discrete, hyaline, smooth, channel and collarette minute. *Conidia* hyaline, aseptate, smooth, botuliform, apex obtuse, base truncate with a filiform, flexuous, unbranched appendage by which the conidium is attached to the conidiogenous cell.

Recent accounts of this genus have been by Dennis (1943), Rupprecht (1958), Guba (1961), and Sutton (1967b). Rupprecht (1958) confined his study to coniferous species and gave a brief synonymy for *S. geniculata* (Berk. & Br.) Höhn. Comparison of coniferous collections with those from non-coniferous substrates indicates that a single species is involved. At least 8 species have been described in the genus but none have been the subject of revision apart from that listed below.

S. GENICULATA (Berk. & Br.) Höhn., *Ber. dtsch. Bot. Ges.* **37**: 158 (1919). (Fig. 283)
Sphaeropsis geniculata Berk. & Br., *Ann. Mag. nat. Hist.* **5**, ser. 2, 375 (1919).
Phoma geniculata (Berk. & Br.) Sacc., *Syll. fung.* **3**: 102 (1884).
Pestalozzina rollandii Fautr., *Rev. mycol.* **17**: 71 (1895).
Monochaetia rollandii (Fautr.) Sacc., *Syll. fung.* **18**: 485 (1906).
Cytotriplospora pini Bayliss-Elliott & Chance, *Trans. Br. mycol. Soc.* **7**: 47 (1921).
Allantophoma exogenospora Kleb., *Phytopath. Z.* **6**: 247 (1933).
Allantophoma nematospora Kleb., *Phytopath. Z.* **6**: 251 (1933).
Phlyctema pithya Höhn., *Mitt. bot. Inst. tech. Hochsch., Wien* **1**: 45 (1924).
Haplostromella pithya Höhn., *Ber. dtsch. Bot. Ges.* **35**: 355 (1917).
Strasseria carpophila Bres. & Sacc. apud Strasser, *Verh. Zool.-bot. Ges., Wien* **52**: 436 (1902).
Plagiorhabdus crataegi Shear, *Bull. Torrey bot. Cl.* **34**: 310 (1907).

Conidiomata 100-350 µ diam. Conidiophores 6-14 × 2-2.5 µ. Conidia 8-15 × 2-3 µ; appendage 12-15 µ long.

On *Malus* sp., *Crataegus* sp., *Picea excelsa*, *Pinus strobus*, *P. nigra* var. *maritima*, *P. radiata*, *P. sylvestris*, *P. halepensis*, *Pseudotsuga douglasii*, *Taxus baccata*, *Rhododendron* sp., U.K., Eire, Wash., U.S.A., Austria, U.S.S.R., Australia (IMI 22766, 22765, 194148, 112184 (holotype of *P. crataegi*), holotype of *P. pithya* ex FH, 223857a, ex FH (Höhn.), 208976a, 231784, 215129b, 231783, 22586, 225827, 231782, 201199b, 225364, 214835, 211627, 195310b, 194148).

17. STRASSERIOPSIS Sutton & Kobayashi, *Mycologia* **61**: 1066 (1969).
Sp. typ.: *S. tsugae* (Kobayashi) Sutton & Kobayashi (syn. *Phellostroma tsugae* Kobayashi).
Illustrations: Kobayashi (1964).

Mycelium immersed, branched, septate, brown. *Conidiomata* eustromatic, sessile, superficial, separate, pulvinate, carbonous, blackish brown, thick-walled; tissue of hyaline, thick-walled textura intricata, at the periphery of small-celled, dark brown, thick-walled textura angularis, locules peripheral, pycnidial-like, each with a papillate, circular ostiole. *Conidiophores* absent. *Conidiogenous cells* enteroblastic, phialidic, discrete, very rarely integrated on 1-septate conidiophores, determinate, ampulliform to cylindrical, hyaline, smooth, channel minute, collarette minimal, formed directly from the inner cells of the locular walls. *Conidia* 1 euseptate, upper cells broadly fusiform, apex obtuse to papillate, basal cells (appendage) filiform, hyaline, smooth.

The genus is monotypic.

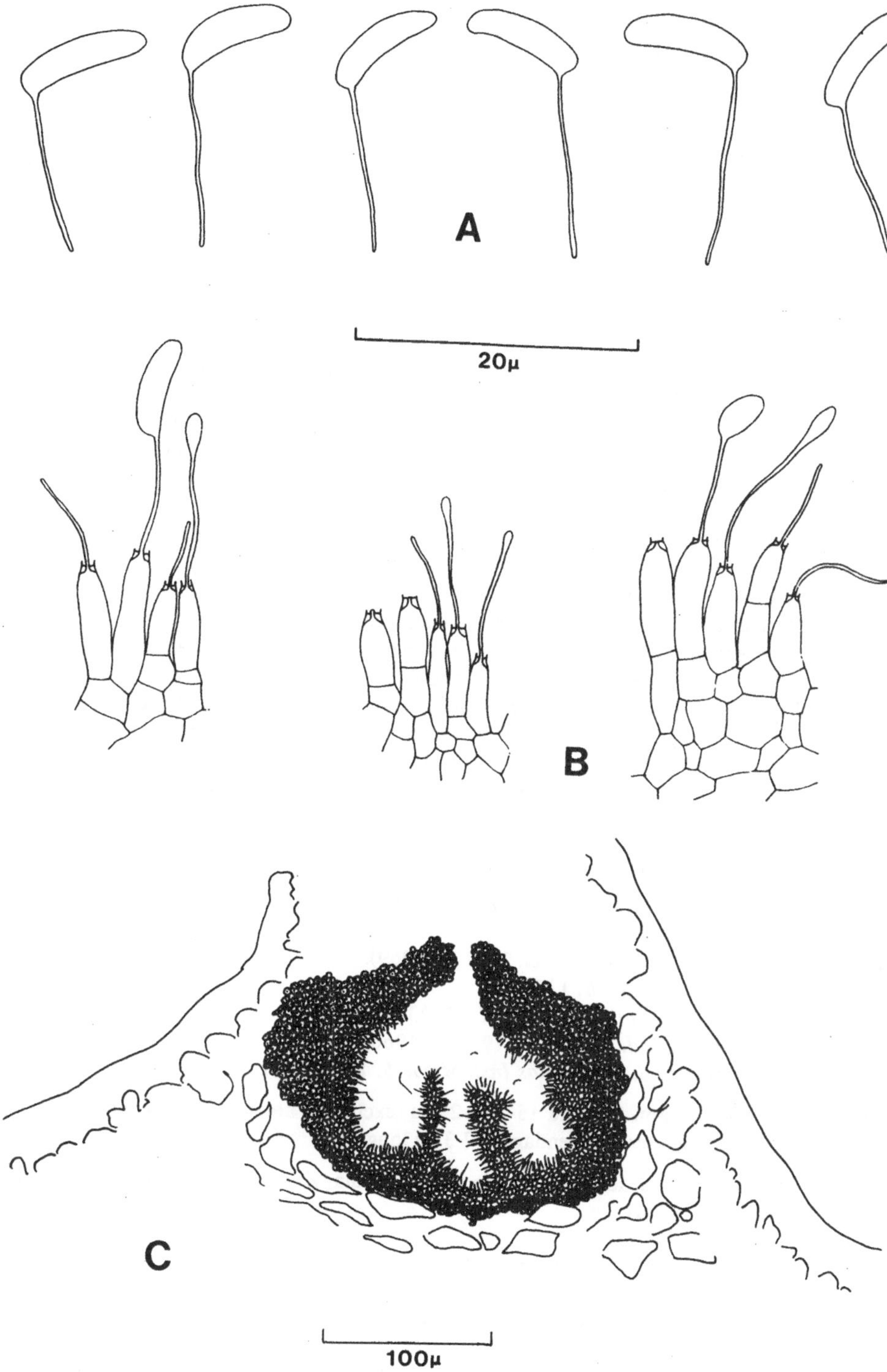

Fig. 283. *Strasseria geniculata*, A, conidia; B, conidiophores and developing conidia; C, vertical section of a conidioma.

S. TSUGAE (Kobayashi) Sutton & Kobayashi, *ibid.*: 1069 (1969).
Phellostroma tsugae Kobayashi, *Trans. mycol. Soc. Japan* **5**: 8 (1964).

Conidiomata 1-3 mm wide; cavities up to 250 µ diam. Conidiogenous cells 8-15 × 4-6 µ. Conidia 26-31 µ long (incl. app.), upper cell 3.5-4 µ wide, basal cells (appendage) 14-18 µ long × 1 µ wide.

Causing stem and twig blight of *Tsuga sieboldii*, Japan (WINF(M) 9742 (holotype), 9745, 9744, 9743).

18. CILIOCHORELLA Syd. apud Syd. & Mitter, *Annls mycol.* **33**: 63 (1935).
Plagionema Subram. & Ramakr., *J. Indian bot. Soc.* **32**: 135 (1953).
Phaeociliospora Batista & Peres, *Atas Inst. Micol.* **5**: 152 (1967).
Sp. typ.: *C. mangiferae* Syd.

Mycelium immersed, dark brown, branched, septate. *Conidiomata* eustromatic, amphigenous, separate or aggregated, subcuticular to epidermal, flattened, lenticular, oval, circular or irregular, dark brown to black, unilocular, thick-walled; upper and lower walls of dark reddish brown, thick-walled textura prismatica. *Ostiole* central, circular, rarely papillate. *Conidiophores* absent. *Conidiogenous cells* enteroblastic, phialidic, determinate, discrete, ampulliform, hyaline, smooth, with a relatively long narrow neck, channel and collarette minute, formed only from the upper cells of the basal wall. *Conidia* 3-euseptate, two median cells very pale brown, end cells hyaline, all thin-walled, continuous or constricted, straight or slightly curved, cylindrical, with an eccentric, short, cellular, basal appendage and at the apex with one ± straight and one lateral cellular, unbranched appendage.

Sutton (1964a) has briefly reviewed the history of the genus. In addition to *C. mangiferae*, another 4 species have been described in *Ciliochorella*. Viswanathan (1959) added *C. eucalypti* with apparently very small conidia, and Kalani (1963) separated *C. indica* with conidia intermediate between *C. mangiferae* and *C. eucalypti*. Munjal (1966) reduced *C. indica* to synonymy with *C. mangiferae* and also thought *C. eucalypti* belonged to this species, but he had not seen the type. Apparent differences between dimensions he attributed to varied methods of measurement of conidia by the different authors. He added *C. castaneae* with apparently much smaller conidia.

It is not clear from the account by Subramanian & Ramakrishnan (1953) whether the length dimensions include or exclude the apical and basal appendages. The measurements of conidial length by Sydow & Mitter (1935) excluded the appendages, and on this basis it would appear that *C. castaneae* differs little from *C. mangiferae*.

The single acceptable species is a characteristic inhabitant of angiosperm leaf litter and occupies a niche in the sub-tropical to tropical ecosystem that *Discosia artocreas* does in temperate regions.

C. MANGIFERAE Syd. apud Syd. & Mitter, *ibid.*: 63 (1935). (Fig. 284)
Plagionema indicum Subram. & Ramakr., *J. Ind. bot. Soc.* **32**: 135 (1953).
Discosia nobilis Petrak, *Sydowia* **5**: 232 (1951).
Ciliochorella eucalypti Viswanathan, *Curr. Sci.* **28**: 171 (1959).
Discosia jambolanae Ahmad, *Biologia* **6**: 135 (1961).
Ciliochorella indica Kalani, *Mycopath. Mycol. appl.* **19**: 239 (1963).
Ciliochorella castaneae Munjal, *Ind. Phytopath.* **19**: 199 (1966).
Phaeociliospora eugeniae Batista & Peres, *Atas Inst. Micol.* **5**: 152 (1967).

Conidiomata up to 800 µ diam. Conidia 15-20 × 2-4 µ (excluding apical and basal appendages); basal appendage 1-12 µ long; apical and lateral appendages 17-33 µ long.

Occurring on a wide variety of dead leaves in the tropics, India, Zambia, Pakistan, Sierra Leone, Nigeria, and represented by 33 collections in IMI.

19. PLACONEMA (Sacc.) Petrak, *Annls mycol.* **19**: 60 (1921).
Placosphaeria Sacc. subgen. *Placonema* Sacc. apud Sacc. & D. Sacc., *Syll. fung.* **18**: 293 (1906).
Shanoria Subram. & Ramakr., *Trans. Br. mycol. Soc.* **39**: 317 (1956).
Stigmateopsis Batista apud Batista, Peres & Maia, *Nova Hedwigia* **2**: 473 (1960).
Sp. typ.: *P. bambusacearum* (Sacc. & Syd.) Petrak (syn. *Placosphaeria bambusacearum* Sacc. & Syd.).
Illustrations: Subramanian & Ramakrishnan (1956); Ciccarone (1966).

Mycelium immersed, branched, septate, hyaline to pale brown. *Conidiomata* eustromatic, immersed, subepidermal, erumpent, separate or confluent, circular to longitudinal, blackish brown; walls of dark brown, thick-walled textura

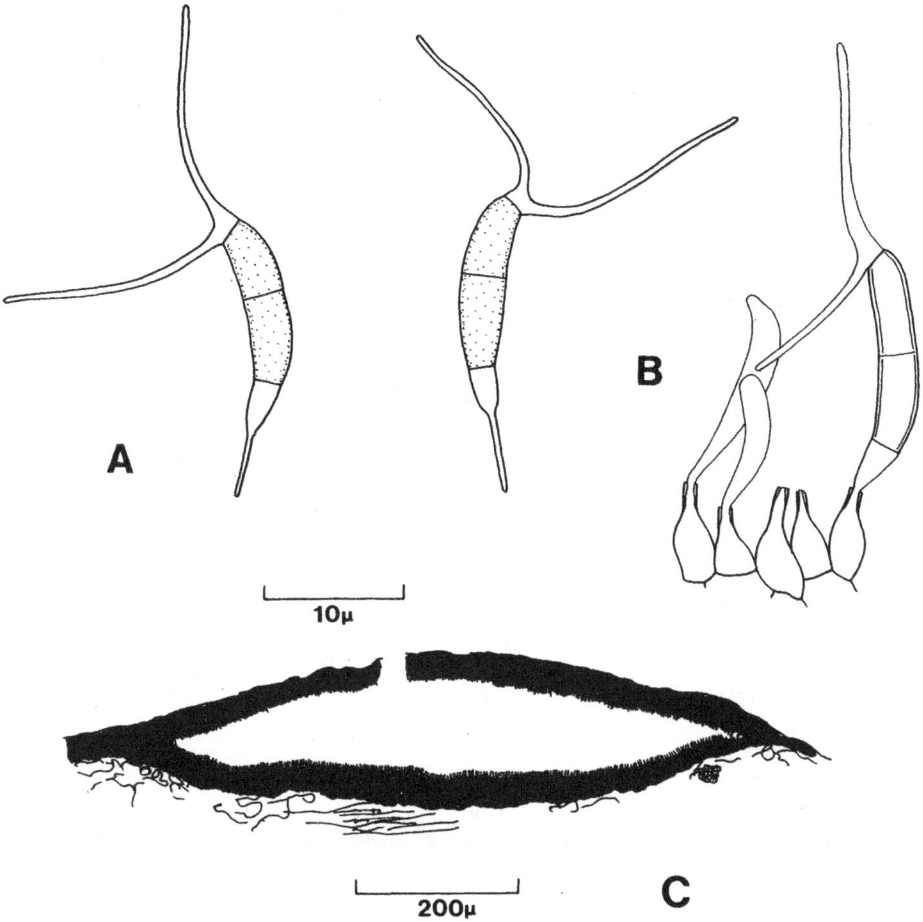

Fig. 284. *Ciliochorella mangiferae*, A, conidia, B, conidiogenous cells and developing conidia; C, vertical section of a conidioma.

globulosa or textura angularis, the upper thicker than the lower one, locules separated by vertical walls of thinner-celled pseudoparenchyma. Dehiscence by irregular longitudinal rupture of the upper wall. *Conidiogenous cells* hyaline, smooth, reputedly holoblastic, but quite possibly enteroblastic and phialidic, determinate, cylindrical to long lageniform, discrete, formed all round the locular walls. *Conidia* hyaline, aseptate, thin-walled, fusiform to navicular, with an acute àpex and extracellular, short, unbranched apical setula, and a truncate base with an exogenous, lateral, extracellular, short, unbranched basal setula.

The generic synonymy and nomenclatural implications have been fully explained by Sutton (1977b), and illustrated accounts have appeared by Subramanian & Ramakrishnan (1956), Ciccarone (1966) and Farr (1968).

P. BAMBUSACEARUM (Sacc. & Syd.) Petrak, *ibid.*: 60 (1921).
Placosphaeria bambusacearum Sacc. & Syd., *Annls mycol.* **2**: 172 (1904).
Shanoria bambusacearum (Sacc. & Syd.) Ciccarone, *Sydowia* **19**: 179 (1966).
Ciliochorella bambusarum Shanor, *Mycologia* **38**: 335 (1946).
Shanoria bambusarum (Shanor) Subram. & Ramakr., *Trans. Br. mycol. Soc.* **39**: 317 (1956).

Teleomorphosis: *Kiehlia bambusina* (Speg.) Farr var. *bambusarum* (Shanor) Farr, *Mycologia* **60**: 928 (1968).

Conidiomata up to 2.5 mm long × .5 mm wide. Conidiophores 10-26 × 2-2.5 µ. Conidia 14-16.5 × 5 µ; setulae 9-15 µ long.

On leaves of *Arthrostylidium racemiflorum*, *Merostachys speciosa*, Brazil, El Salvador.

P. BAMBUSACEARUM (Sacc. & Syd.) Petrak var. BAMBUSAE (Batista) Sutton, *Kew Bull.* **31**: 462 (1977).
Stigmateopsis bambusae Batista, *Nova Hedwigia* **2**: 474 (1960).
Shanoria bambusacearum (Sacc. & Syd.) Cicc. var. *bambusae* (Batista) Farr, *Mycologia* **60**: 928 (1968).

Conidiomata 420-500 × 240-280 µ. Conidia 10-12 × 5-7 µ; setulae 7-10 µ long.

On leaves of *Bambusa*, Brazil.

20. TRIBOLOSPORA D.A. Reid, *Aust. J. Bot.* **14**: 31 (1966).
Sp. typ.: *T. sycopsidis* D.A. Reid
Illustrations: Reid (1966).

Mycelium immersed, sparse, branched, septate, brown. *Conidiomata*, eustromatic, with a small central point of attachment, separate, pulvinate or flattened, carbonous, black; tissue of loosely interwoven, dark brown, thick-walled textura intricata, giving rise in the upper part to separate pycnidial-like locules, the walls of which are of pale brown to

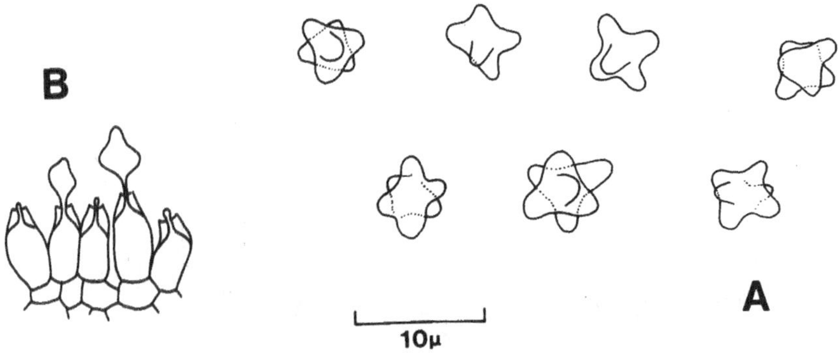

Fig. 285. *Tribolospora sycopsidis*, A, conidia; B, conidiogenous cells with developing conidia.

hyaline, loosely arranged hyphae; unilocular, each with a separate, circular ostiole. *Conidiophores* absent. *Conidiogenous cells* enteroblastic, phialidic, discrete, determinate, cylindrical, hyaline, smooth, channel and collarette minute but periclinal wall considerably thickened, formed directly from the inner cells of the locular walls. *Conidia* hyaline, aseptate, smooth, thin-walled, eguttulate, stellate with up to 6 short protuberances. Point of secession of conidia has not been determined.

The genus is monotypic.

T. SYCOPSIDIS D.A. Reid, *ibid.*: 31 (1966). (Fig. 285)

Conidiomata up to 1 mm diam.; situated in hypophyllous vein axils in association with acarodomatia, with up to 50 pycnidial-like locules. Conidiogenous cells 6-7 × 2.5-3 µ. Conidia 6-7.5 µ between protuberances.

On leaves of *Sycopsis dunnii*, Australia (IMI 188150, holotype).

21. SATCHMOPSIS Sutton & Hodges apud Sutton, *Nova Hedwigia* **26**: 1 (1975).
Sp. typ.: *S. brasiliensis* Sutton & Hodges
Illustration: Sutton (1975b).

Mycelium immersed, dark brown, branched, septate. *Conidiomata* infundibuliform to nidulariaceous, amphigenous, superficial, separate, dark brown to black, sessile; basal wall several cells thick, of dark brown large-celled textura angularis below and smaller-celled hyaline, thin-walled cells above, periclinal wall 1-cell thick, of vertically elongated thick-walled, brown textura prismatica. *Conidiophores* sparse, 1-2 septate, hyaline, sparingly branched, restricted to the base of the conidioma. *Conidiogenous* cells enteroblastic, phialidic, discrete, rarely integrated on 1-2 septate conidiophores, determinate, ampulliform to lageniform, hyaline, smooth, channel and collarette minute. *Conidia* nidulant, hyaline, aseptate, ± guttulate, cylindrical, both ends obtuse.

S. BRASILIENSIS Sutton & Hodges apud Sutton, *ibid.*: 3 (1975).

Conidiomata 175 µ wide × 70-135 µ deep. Conidiogenous cells 4-7 × 2.5-3 µ. Conidia 11.5-15.5 × 1-1.5 µ.

On dead leaves of *Eucalyptus camaldulensis*, *E. grandis*, *E. paniculata*, *E. robusta*, *E. saligna*, *E. tereticornis*, *E. viminalis*, *E.* sp., Brazil, India, Australia, Jamaica, Hawaii, U.S.A., Venezuela. Represented in herb. IMI now by more than 30 collections.

22. KABATINA Schneider & von Arx, *Phytopath. Z.* **57**: 179 (1966).
Sp. typ.: *K. thujae* Schneider & von Arx

Mycelium immersed, branched, septate, hyaline to pale brown. *Conidiomata* acervular, epidermal to subepidermal, separate, rarely confluent, pulvinate, formed of pale brown, thick-walled textura angularis. Dehiscence irregular. *Conidiophores* hyaline to pale brown, rarely branched, 1-septate, smooth, formed from the upper pseudoparenchyma. *Conidiogenous cells* enteroblastic, phialidic, integrated or discrete, determinate, cylindrical to doliiform, smooth, channel and collarette minute, periclinal wall thickened. *Conidia* hyaline, aseptate, ± guttulate, thin-walled, smooth, ellipsoid.

The two species of *Kabatina* are separated by behaviour in culture and the range of hosts attacked. Differences are slender, and Morelet (1973b) suggested *K. juniperi* Schneider & von Arx be reduced to varietal rank within *K. thujae*. *Kabatina* species have been recorded from a range of coniferous substrates at C.M.I. including *Chamaecyparis pisifera plumosa aurea*, *C. plumosa durea*, *C.* sp., *Cupressocyparis leylandii*, *Cupressus macrocarpa* and *C.* sp. and further work on species differentiation is clearly required. A third species, *K. populi*, was described by Butin & Schneider (1976) from *Populus euramericana*. It is characterised by conidiomata 150-300 µ diam., conidiophores 10-18 µ long and conidia 6-9 × 2.5-3.5 µ. In culture it produces discrete conidiophores. It is not pathogenic.

K. THUJAE Schneider & von Arx, *ibid.*: 180 (1966). (Fig. 286A, C)

Conidiomata 50-170 µ diam. Conidiophores 5-8 × 3-4 µ. Conidia 5-8 × 2.5-3.5 µ.

From and on *Thuja occidentalis ericoides*, *T. plicata* and *T.* sp., U.K., Germany (IMI 151271, isotype ex CBS, 176629, 178614).

K. THUJAE Schneider & von Arx var. JUNIPERI (Schneider & von Arx) Morelet, *Bull. Soc. Sci. nat. Archéol. Toulon Var* **204**: 8 (1973). (Fig. 286B)
Kabatina juniperi Schneider & von Arx, *Phytopath. Z.* **57**: 181 (1966).

Conidiomata up to 100 µ diam. Conidiophores 4-12 × 2-4 µ. Conidia 3-5 × 2 µ.

From and on *Juniperus chinensis*, *J.* sp., U.K., Germany, Netherlands (IMI 151272, isotype ex CBS, 151273, 177271, 173099, 177270).

23. CYCLODOMUS Höhn., *Sber. Akad. Wiss. Wien* **118**: 1527 (1909).
Sp. typ.: *C. umbellulariae* Höhn.

Mycelium immersed, branched, septate, dark brown. *Conidiomata* eustromatic, collabent, black, epidermal, thick-walled, with a central column of vertically arranged brown textura angularis surrounded by a periclinal locule; wall of large-celled, thick-walled, dark brown pseudoparenchyma, darker in the upper part. *Ostiole* absent, dehiscence by irregular breakdown of upper wall. *Conidiogenous cells* enteroblastic, phialidic, determinate, discrete, ampulliform to doliiform, pale brown, smooth, periclinal wall thickened, formed from the inner cells of the pycnidial wall and the central column. *Conidia* hyaline, aseptate, cylindrical, apex obtuse, base acute, thin-walled, smooth, straight, with a prominent central guttule.

The genus is monotypic.

C. UMBELLULARIAE Höhn., *ibid.*: 1528 (1909). (Fig. 287)

Conidiomata up to 250 µ diam. Conidiogenous cells 4-6 × 5-8 µ. Conidia 15-17 × 2-2.5 µ.

On dead leaves of *Umbellularia californica*, Calif., U.S.A. (IMI 194353, holotype).

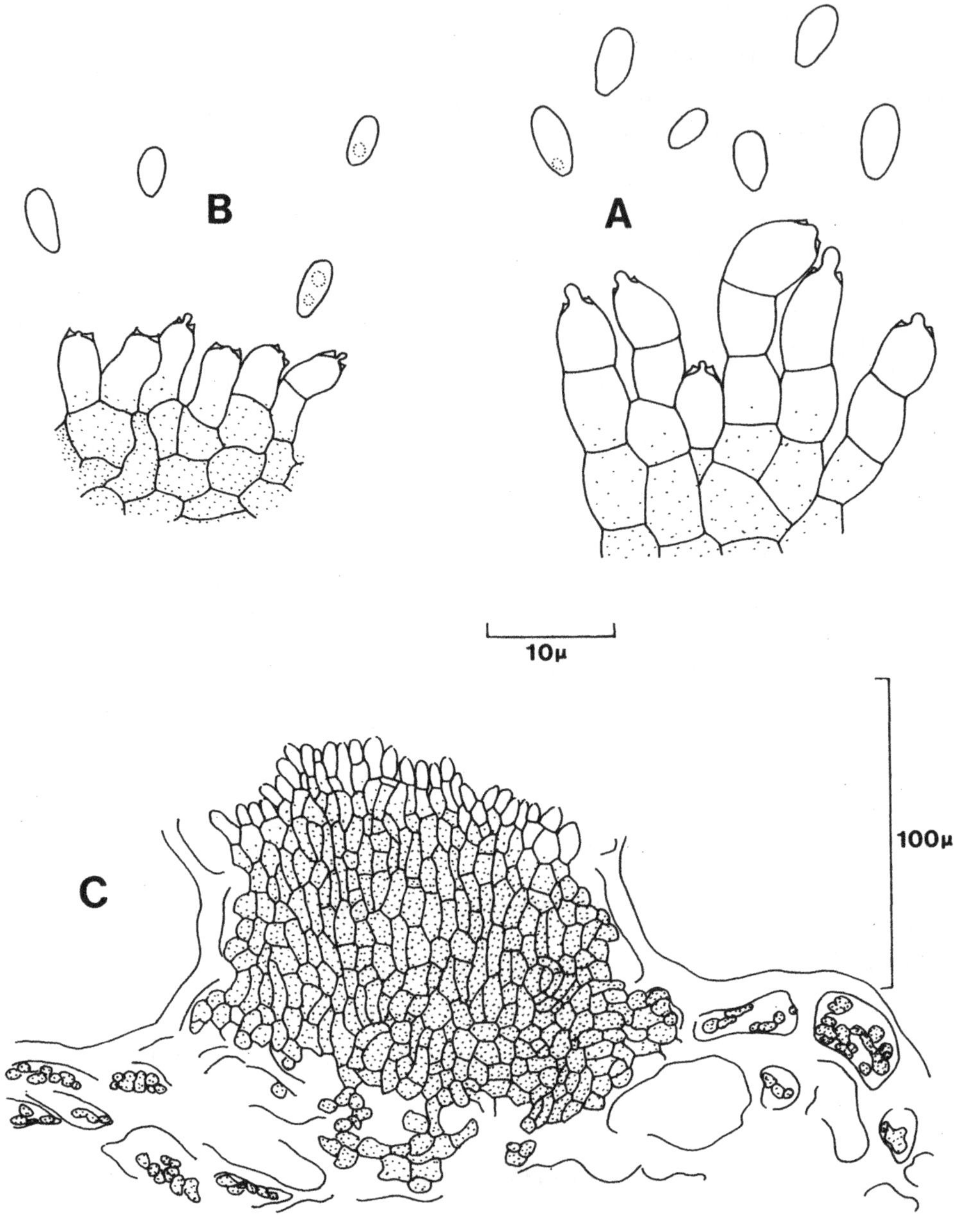

Fig. 286. A, C, *Kabatina thujae*, conidiophores, vertical section of a conidioma; B, *K. thujae* var. *juniperi*, conidia and conidiogenous cells.

24. PODOPLACONEMA Petrak, *Annls mycol.* **19**: 83 (1921).
Sp. typ.: *P. melaenum* (Fr.) Petrak (syn. *Sphaeria melaena* Fr.).

Mycelium immersed, hyaline to brown, branched, septate. *Conidiomata* eustromatic, separate like pycnidia or aggregated, often indistinguishable from immature ascocarps, globose-depressed or globose, dark brown, immersed, unilocular or multilocular, epidermal to subepidermal, thick-walled; walls of dark brown, thick-walled textura globulosa. *Ostiole* absent, dehiscence by breakdown of upper wall. *Conidiophores* absent. *Conidiogenous cells* enteroblastic, phialidic, determinate, discrete, ampulliform, hyaline or pale brown, smooth, formed all round the locular wall. *Conidia* hyaline, aseptate, thin-walled, smooth, eguttulate, straight or slightly curved, cylindrical or tapered towards the base.

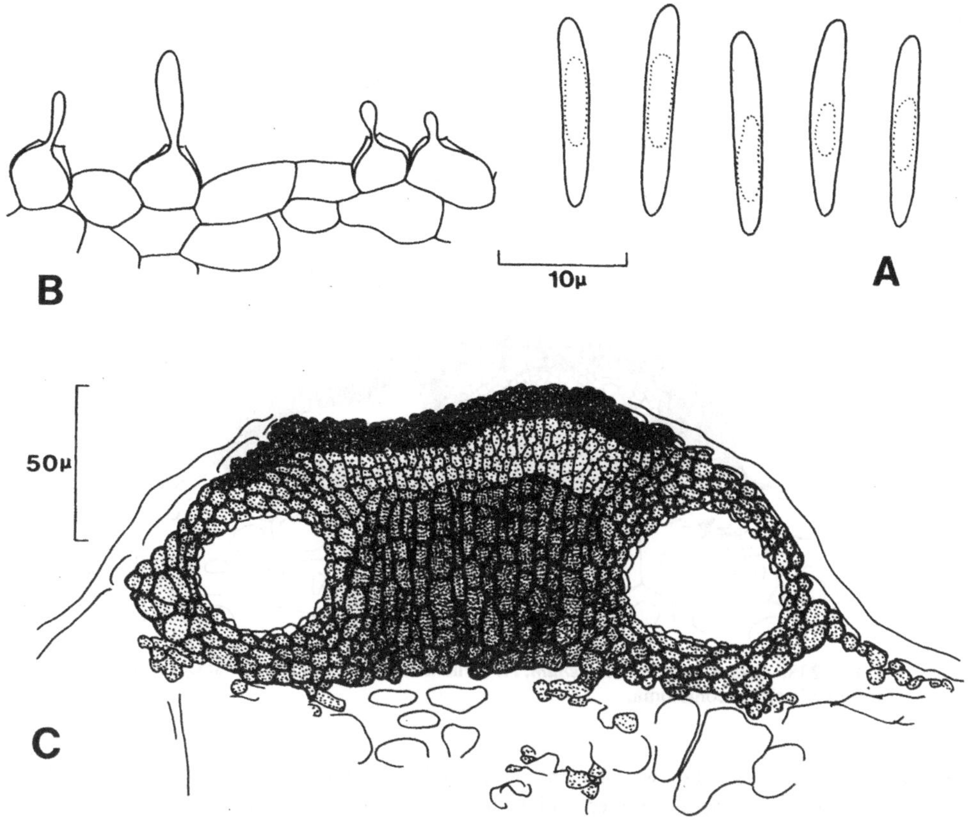

Fig. 287. *Cyclodomus umbellulariae* A, conidia; B, conidiogenous cells and developing conidia; C, vertical section of a closed conidioma.

Apart from the type, there are only three other species described in the genus, *P. jahnii* Chardon, *P. lauracearum* Bat., Bez. & Cif. and *P. manaosense* Bat., Garnier & Bez. None of these have been reexamined since their original description.

P. MELAENUM (Fr.) Petrak, *ibid.*: 84 (1921). (Fig. 288)
Sphaeria melaena Fr., *Systema mycologicum* **2**: 431 (1832).
Phoma melaena (Fr.) Mont. & Dur. in Kickx, *Fl. crypt. Fland.* **1**: 437 (1867).

Teleomorphosis: *Omphalospora melaena* (Fr.) Höhn., *Sber. Akad. Wiss. Wien* **128**: 603 (1919).

Conidiomata up to 80 μ diam. Conidiogenous cells 5-7 × 3-4.5 μ. Conidia 2.5-4.5 × 1 μ.

On dry stems of *Silene nutans*, Austria (IMI 11614).

25. ASTEROMELLOPSIS Hess & Müller, *Ber. Schw. Bot. Ges.* **61**: 18 (1951).
Sp. typ.: *A. insculpta* Hess & Müller
Illustration: Hess & Müller (1951).

Mycelium immersed, branched, septate, brown. *Conidiomata* stromatic, dothideoid, subepidermal to erumpent, composed of large-celled, angular pseudoparenchyma, at the periphery and base dark brown, multilocular, the locules separate, sparse, convoluted to lacunose. *Ostiole* absent, dehiscence by breakdown of upper wall. *Conidiophores* absent. *Conidiogenous cells*? enterothallic. *Conidia* hyaline, aseptate, thin-walled, eguttulate, cylindrical.

Asteromellopsis was introduced by Hess & Müller (1951) to accommodate what is probably the spermatial or microconidial state of *Dothidella insculpta* (Wallr.) Theiss. & Syd. (later referred to *Dothidea insculpta* Wallr. by Müller &

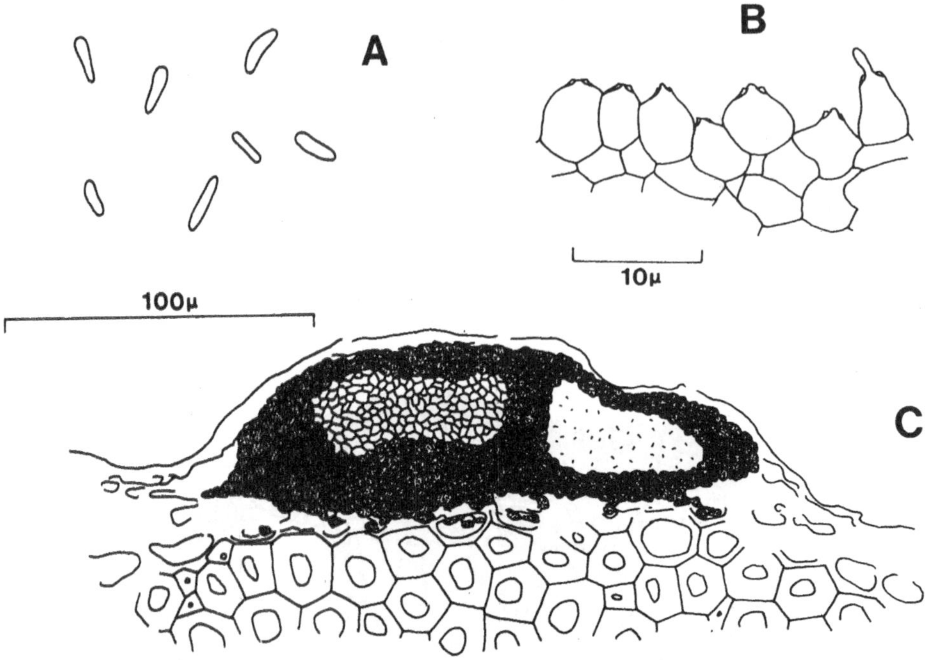

Fig. 288. *Podoplaconema melaenum*, A, conidia; B, conidiogenous cells; C, vertical section of a closed conidioma.

Arx, 1962). Precise details of conidiogenesis were lacking, and a slide from the holotype (ZT) was little help in determining ontogeny. The smaller cells (or protoplasmic entities) which stained heavily and appeared to be in small groups surrounded by the walls of pseudoparenchymatic cells may be analogous to the results of cytoplasmic division prior to spermatial secession reported by Higgins (1920, 1929, 1936) and Dring (1961) for *Mycosphaerella* spp. However, no papillae were found on the internal faces of the pseudoparenchymatic cells, so confirmation of this apparent similarity is not yet forthcoming.

A. INSCULPTA Hess & Müller, *ibid.*: 18 (1951).

Spermatial *Dothidea insculpta* Wallr., *Fl. Crypt. Germ.* 2: 864 (1833).

Locules 15-25 μ diam. Conidia 1.5-2 × 1 μ.
On dead stems of *Clematis vitalba*, Switzerland (holotype ex ZT).

26. SCLEROPHOMA Höhn., *Sber. Akad. Wiss. Wien* **118**: 1234 (1909).
? *Blastophoma* Kleb., *Phytopath. Z.* **6**: 273 (1933).
? *Schizophoma* Kleb., *Phytopath Z.* **6**: 281 (1933).
? *Stichophoma* Kleb., *Phytopath. Z.* **6**: 279 (1933).
Original spp.: *S. pythiophila* (Cda) Höhn., *S. pithya* (Thüm.) Höhn., *S. piceae* (Fiedl.) Höhn., *S. pini* (Desm.) Höhn.
Lectotype sp.: *S. pythiophila* (Cda) Höhn. (syn. *Sphaeronaema pythiophilum* Cda), fide Sutton (1977b).

Mycelium immersed, branched, septate, pale to dark brown. *Conidiomata* eustromatic, immersed to semi-immersed, strongly erumpent, black, globose or irregular, separate or aggregated, unilocular, multilocular or convoluted, thick-walled; walls composed of dark brown, thick-walled textura angularis, somewhat darker at the periphery. *Ostiole* absent, dehiscence by breakdown of the overlying wall. *Conidiophores* absent. *Conidiogenous cells* enteroblastic, phialidic, determinate, discrete, hyaline to pale brown, doliiform to ampulliform, channel wide, periclinal thickening and collarette minimal, formed from the inner cells of the locular walls. *Conidia* hyaline, aseptate, ellipsoid, sometimes tapered to the base, smooth, a proportion typically 1 guttulate.

Sutton & Waterston (1970) described the type species fully with details of its host range, transmission and distribution etc., and Butin (1964) proved the correlation between it and *Sydowia polyspora* (Bref. & Tav.) E. Müller. More recently, Hermanides-Nijhof (1977) has described and illustrated the *Hormonema dematioides* Lagerberg & Melin state that frequently develops in culture.

More than 30 taxa have been described in *Sclerophoma* and several have subsequently been segregated into genera of the stromatic *Phoma* complex. This has been done sporadically in the past so no full treatment of the genus has yet appeared.

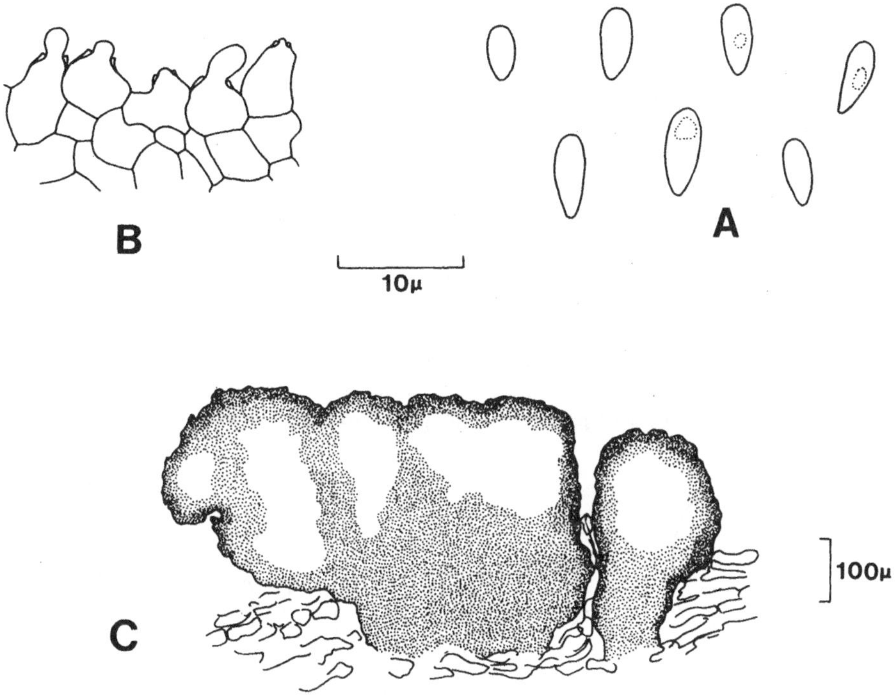

Fig. 289. *Sclerophoma pythiophila*, A, conidia; B, conidiogenous cells; C, vertical section of conidiomata.

S. PYTHIOPHILA (Cda) Höhn., *ibid.*: 1234 (1909). (Fig. 289)
Sphaeronaema pythiophilum Cda, *Icon. fung.* **4**: 40 (1840).
Phoma pythiophila (Cda) Sacc., *Syll. fung.* **3**: 101 (1884).
Sclerotiopsis pythiophila (Cda) Oud., *Ned. Kruidk. Arch.*, ser. 3, **2**: 247 (1897).
Dothichiza pythiophila (Cda) Petrak, *Annls mycol.* **21**: 267 (1923)
Sphaeropsis acicola Lév., *Ann. Sci. nat.*, 3 ser., **9**: 256 (1848).
Phoma acicola (Lév.) Sacc., *Michelia* **2**: 272 (1881).
Stichophoma acicola (Lév.) Kleb., *Phytopath. Z.* **6**: 279 (1933).
Pycnis pinicola Zopf, *Hedwigia* **20**: 146 (1881).
Phoma pinicola (Zopf) Sacc., *Syll. fung.* **3**: 100 (1884).
Phoma strobiligena Desm., *Ann. Sci. nat.*, 3 ser., **11**: 280 (1849).
Dothichiza ferruginosa Sacc., *Syll. fung.* **3**: 673 (1884).

Teleomorphosis: *Sydowia polyspora* (Bref. & Tav.) E. Müller, *Sydowia* **7**: 342 (1953).

Conidiomata up to 200 µ when single, then up to 750 µ diam when multilocular. Conidiogenous cells 4-7 × 4-5 µ. Conidia 4-8 × 2-3 µ.

The epithet for this species has consistently appeared as '*pithyophila*', even though Corda (1840) originally published it as '*pythiophilum*'. He gave no etymology for the epithet which however was clearly derived from pyth-o (G) meaning rot, decay and — philo (G) meaning love. As such, the epithet makes sense and should be retained in its original form.

Represented by more than 60 collections in herb. IMI on species of *Abies, Cupressus, Juniperus, Picea, Pinus* from Eire, U.K., France, Germany, Czechoslovakia, Latvia, Kenya, Tanzania, Uganda, S. Africa, Canada, Calif., U.S.A. and Australia.

27. SARCOPHOMA Höhn., *Sber. Akad. Wiss. Wien* **125**: 75 (1916).
Sp. typ.: *S. miribelii* (Fr.) Höhn. (syn. *Sphaeria miribelii* Fr., *S. endogenospora* Höhn.).

Mycelium immersed, septate, branched, pale brown to hyaline. *Conidiomata* eustromatic, dark brown, globose to flattened globose, separate, epidermal to subepidermal, unilocular, thick-walled; wall composed of pale brown, thin-walled textura angularis, somewhat thicker at the base than the upper and side walls. *Ostiole* absent, dehiscence by irregular rupture of the overlying wall. *Conidiophores* absent. *Conidiogenous cells* enteroblastic, polyphialidic, discrete, determinate, ampulliform to doliiform, hyaline, smooth, with 1-3 conidiogenous loci, collarette and periclinal thickening minute, channel comparatively wide, formed from the inner cells of the conidiomatal wall. *Conidia* hyaline, aseptate, smooth, eguttulate, thin-walled, ellipsoid to obpyriform, apex obtuse, base often truncate.

The most comprehensive treatment of *S. miribelii* has been by van der Aa (1975) who proved the correlation with *Guignardia miribelii* van der Aa.

Apart from the type species, which has had two names in *Sarcophoma*, there are three other taxa, *S. eriogoni* (Ell. & Ev.) Arx, *S. garganica* (Sacc.) Arx (von Arx, 1957a) and *S. juncea* (Mont.) Höhn. The latter proves to be more correctly placed in *Selenophoma* Maire, but it has not been possible to verify the two other species.

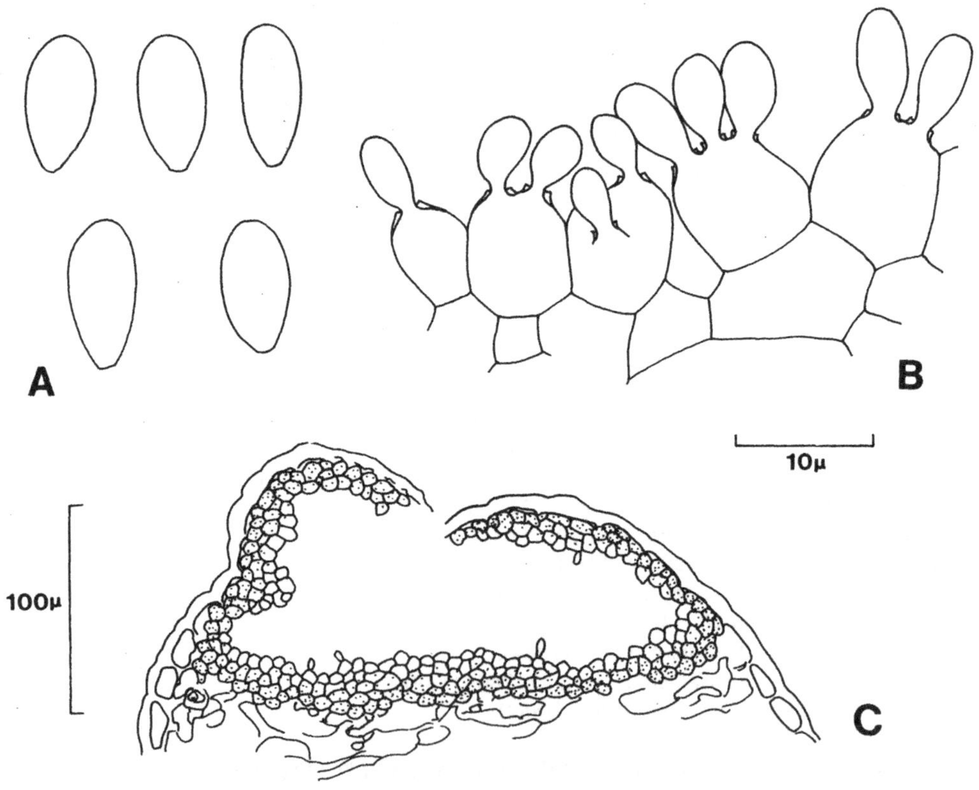

Fig. 290. *Sarcophoma miribelii*, conidia; B, conidiogenous cells and developing conidia; C, vertical section of a conidioma.

S. MIRIBELII (Fr.) Höhn., *Hedwigia* **60**: 133 (1918). (Fig. 290).
Sphaeria miribelii Fr., *Linnaea* **5**: 548 (1830).
Sphaeropsis miribelii (Fr.) Lév., *Ann. Sci. nat.*, ser. 3, **5**: 296 (1846).
Phoma miribelii (Fr.) Sacc., *Michelia* **2**: 90 (1880).
Macrophoma miribelii (Fr.) Berl. & Vogl., *Atti Soc. ven. – trent. Sci. nat.* **10**: 179 (1886).
Sphaeria delitescens Wallr., *Fl. Crypt. Germ.* **2**: 777 (1833).
Phoma delitescens (Wallr.) Sacc., *Syll. fung.* **3**: 105 (1884).
Phacidium buxi Lasch in Klotzsch, *Herb. viv. myc.* **1**: 1154 (1847).
Gloeosporium pachybasium Sacc., *Michelia* **2**: 117 (1880).
Phoma phacidioides Sacc., *Michelia* **2**: 274 (1881).
Phyllosticta phacidioides (Sacc.) Allesch. in Rabenh., *Krypt. Fl.* **1** (6): 25 (1898).
Gloeosporium louisiae Bauml., *Verh. Ver. Natur-Heilk, Pressburg N.F.* **9**: 100 (1896).
Macrophoma miribelii (Fr.) Berl. & Vogl. forma *ramicola* Oud., *Ned. kruidk. Arch.*, ser. 3, **2**: 734 (1902).
Sarcophoma endogenospora Höhn., *Sber. Akad. Wiss. Wien* **125**: 75 (1916).
Sclerophoma confusa Petrak, *Annls mycol.* **20**: 23 (1922).
Phyllostictina confusa (Petrak) Petrak apud Petrak & Syd., *Beih. rep. spec. nov. regni veg.* **42**: 191 (1927).

Teleomorphosis: *Guignardia miribelii* van der Aa, *Persoonia* **8**: 283 (1975).

Conidiomata up to 300 μ diam. × 150 μ high. Conidiogenous cells 7.5-10 × 5-9 μ. Conidia 9.5-11 × 4.5-5.5 μ.

On leaves of *Buxus sempervirens*, *B. arborescens*, *B.* sp., Italy, Austria, Hungary, Roumania, Germany, Latvia, U.K., Australia (IMI 21916e, 31089a, 115204, 29760, 105566, 50071, 11196, 11775a, 33618a, 12687a, 52228).

28. CALOCLINE H. Syd., *Annls mycol.* **37**: 417 (1939).
Sp. typ.: *C. chusqueae* H. Syd.

Mycelium immersed, hyaline to very pale brown, branched, septate. *Conidiomata* hypophyllous, eustromatic, loosely aggregated into a linear row between veins, separated by immersed, hyaline to very pale brown textura angularis, the cells of which become more elongated in the upper region; conidiomata separate, globose to cornute, violet-purple, semi-immersed to superficial, unilocular, walls composed of thick-walled, hyaline textura angularis on the outside and thin-walled, much smaller hyaline cells on the inside. *Ostioles* single to each locule, circular, central. *Conidiophores* absent. *Conidiogenous cells* enteroblastic, phialidic, determinate, discrete, doliiform to lageniform, hyaline, smooth, channel and collarette minute, periclinal wall thick, formed from the inner cells of the pycnidial walls. *Conidia* hyaline, aseptate, continuous, apex obtuse, base slightly attenuated, thin-walled, eguttulate, smooth, oval.

The structure of the conidiomatal wall is unusual, for the constituent cells are thick-walled and large in comparison with those of the conidiogenous cell region, yet there is a characteristic violet-purple pigmentation evident under low power. The genus is so far monotypic.

C. CHUSQUEAE H. Syd., *ibid.*: 417 (1939). (Fig. 291)

Conidiomata up to 150 μ high × 130 μ wide. Conidiogenous cells 5-7 × 2.5-4 μ. Conidia 4.5-5.5 × 1.5-2 μ.

Causing lenticular lesions on leaves of *Chusquea serrulata*, Ecuador (IMI 31013, holotype).

29. PSEUDOSCLEROPHOMA Petrak, *Annls mycol.* **21**: 283 (1923).
Sp. typ.: *P. negundinis* Petrak

Mycelium immersed, branched, septate, pale brown. *Conidiomata* eustromatic, immersed, subepidermal, finally strongly erumpent, separate or aggregated, unilocular or irregularly multilocular, globose to pulvinate, golden brown; walls thick, composed of an outer layer of thin-walled golden brown textura angularis to globulosa with an inner layer of smaller-celled, thin-walled, hyaline tissue. *Ostiole* absent, dehiscence by irregular rupture of upper wall. *Conidiophores* absent. *Conidiogenous cells* enteroblastic, phialidic, determinate, discrete, ampulliform to doliiform, hyaline, smooth, collarette and channel minute, periclinal wall thickened, formed from the inner cells of the locular walls. *Conidia* hyaline, aseptate, straight, smooth, eguttulate, thin-walled, ellipsoid.

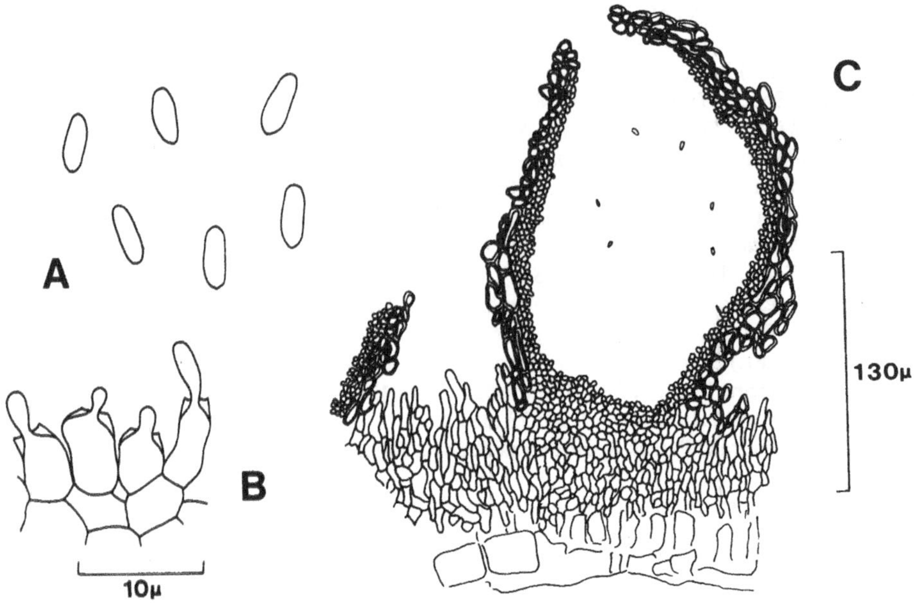

Fig. 291. *Calocline chusqueae*, A, conidia; B, conidiogenous cells and developing conidia; C, vertical section of a conidioma.

Clements & Shear (1931) placed this name in synonymy with *Plenozythia* Syd., a genus which was referred to the Zythiaceae. No data on *Plenozythia* are presently available. Sutton (1977a) thought the genus may be the same as *Phoma* Sacc., but the stromatic conidiomata evident on the type clearly exclude this possibility.

P. NEGUNDINIS Petrak, *ibid.*: 283 (1923). (Fig. 292)

Conidiomata up to 220 μ diam. Conidiogenous cells 6-7 × 5-6 μ. Conidia 4-6.5 × 3-3.5 μ.

On dead twigs of *Acer negundo*, Czechoslovakia (IMI 204131, isotype ex BPI).

30. PARADISCULA Petrak, *Annls mycol.* **39**: 307 (1941).
Sp. typ.: *P. spuria* (Vestergr.) Petrak (syn. *Phoma spuria* Vestergr.).
Illustration: Morgan-Jones (1975b).

Mycelium immersed, branched, septate, hyaline to pale brown. *Conidiomata* eustromatic, immersed, subepidermal, separate, occasionally aggregated, black, elliptical to elongated, finally pulvinate; wall thick, composed of pale to medium brown, thin-walled textura angularis, thicker at the base than the sides and above. *Ostiole* absent, dehiscence by a longitudinal fissure opening widely. *Conidiophores* absent. *Conidiogenous cells* enteroblastic, phialidic, discrete, determinate, hyaline, smooth, lageniform, channel wide, collarette and periclinal thickening minimal, formed from the basal and lateral walls. *Conidia* hyaline, aseptate, thin-walled, smooth, occasionally guttulate, straight, cylindrical to oblong.

Morgan-Jones (1975b) has fully illustrated and described the type and only species.

P. SPURIA (Vestergr.) Petrak, *ibid.*: 308 (1941).
Phoma spuria Vestergr., *Jahreskat. Wien Krypt. Tauschanst.* 4 (1879).

Conidiomata 200-500 μ diam. Conidiogenous cells 5-14 × 2-3.5 μ. Conidia 4-6 × 1.5-2.5 μ.

On stems of *Potentilla argentea*, Czechoslovakia.

31. SPHAERIOTHYRIUM Bub., *Ber. dtsch. Bot. Ges.* **34**: 299 (1916).
Neoplacosphaeria Petrak, *Annls mycol.* **19**: 74 (1921).
Sp. typ.: *S. filicinum* Bub.

Mycelium immersed, branched, septate, brown. *Conidiomata* eustromatic, separate, linear, dark brown, epidermal to subepidermal, unilocular, thick-walled; wall of dark brown, often large-celled, thick-walled textura angularis, basal wall narrower. *Ostiole* absent, dehiscence by breakdown of the upper wall. *Conidiophores* absent. *Conidiogenous cells* enteroblastic, phialidic, discrete, smooth, hyaline, ampulliform to doliiform, collarette and channel minute, periclinal wall thickened, formed from the inner cells lining the locular cavity. *Conidia* hyaline, aseptate, thin-walled, straight, ellipsoid to cylindrical.

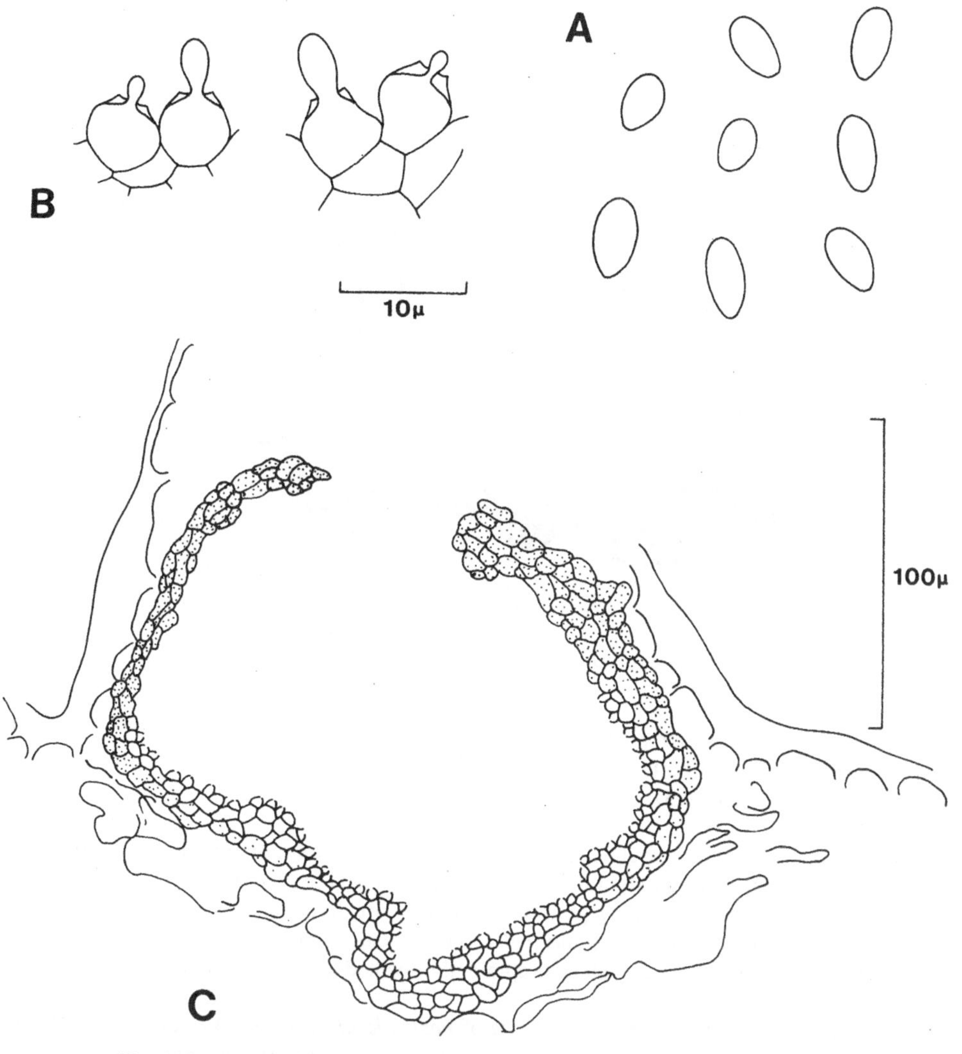

Fig. 292. *Pseudosclerophoma negundinis*, A, conidia; B, conidiogenous cells and developing conidia; C, vertical section of a conidioma.

Apart from the type species the only other to be described in *Sphaeriothyrium* is *S. praecastrense* (C. Massal.) Bub. This has not been reexamined.

S. FILICINUM Bub., *ibid.*: 299 (1916). (Fig. 293)
Neoplacosphaeria polonica Petrak, *Annls mycol.* **19**: 75 (1921).

Conidiomata up to .15 cm long × 150 μ wide. Conidiogenous cells 7 × 4 μ. Conidia 4.5-6 × 2.5 μ.

On petioles of *Struthiopteris germanica*, *Pteridium aquilinum*, Czechoslovakia, Germany (IMI 194245, lectotype, 31129).

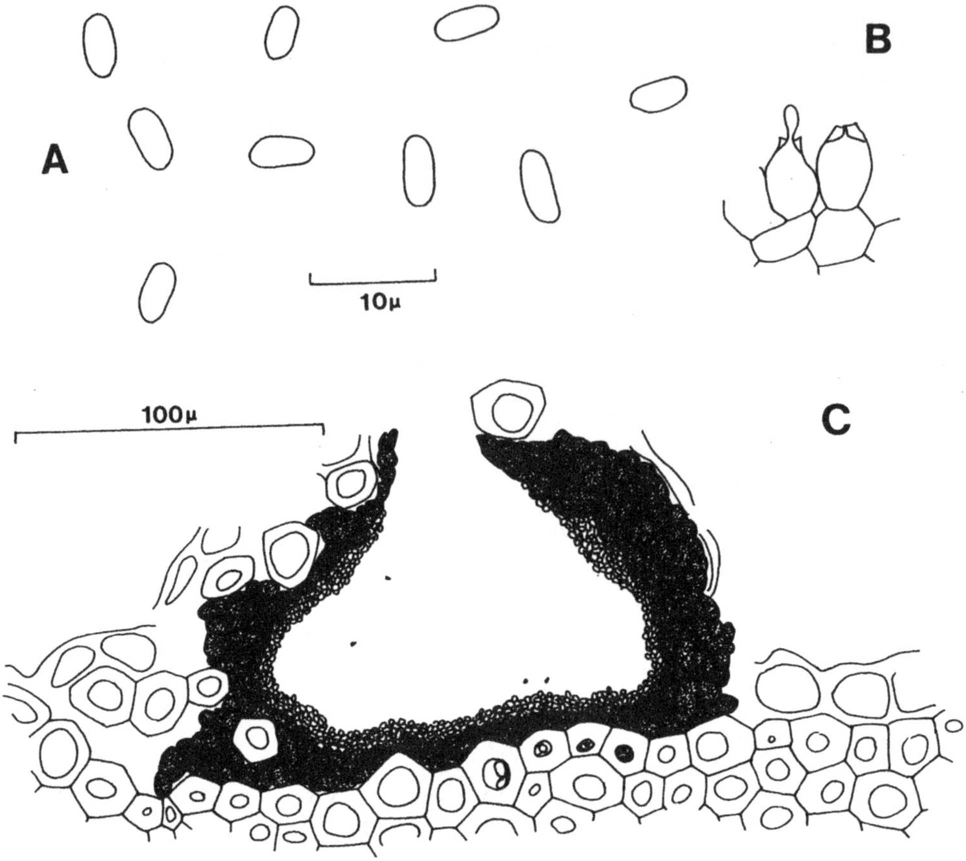

Fig. 293. *Sphaeriothyrium filicinum*, A, conidia; B, conidiogenous cells; C, vertical section of a conidioma.

32. ACLEISTIA Bayliss-Elliott, *Trans. Br. mycol. Soc.* **5**: 420 (1915).
Sp. typ.: *A. alniella* Bayliss-Elliott
Illustration: Bayliss-Elliott (1915).

Mycelium superficial and immersed, hyaline, branched, septate. *Conidiomata* superficial, separate, gregarious, pale brown to subhyaline, sessile or very shortly stipitate, widely cupulate; basal tissue textura intricata, of loosely aggregated, interwoven, branched, septate hyphae extending peripherally into short, hyaline, upwardly curved, septate, thin-walled, smooth hairs. *Conidiophores* absent. *Conidiogenous cells* enteroblastic, phialidic, hyaline, smooth, cylindrical, erect, formed from the interwoven hyphae. *Conidia* hyaline, aseptate, thin-walled, eguttulate, spherical.

The generic name was introduced for the conidial state of *Pezizella alniella* (Nyl.) Dennis. From the original account it seems likely that conidial formation probably precedes ascus development in the same fructification. The genus is monotypic.

A. ALNIELLA Bayliss-Elliott, *ibid.*: 420 (1915).

Teleomorphosis: *Pezizella alniella* (Nyl.) Dennis, *Mycol. Pap.* 62: 56 (1956).

Conidiomata up to 250 µ diam. Conidiogenous cells 12-15 µ long. Conidia 2-2.5 µ diam.

On female catkins of *Alnus glutinosa*, U.K. (IMI 35426a (holotype), 35427, 182850).

33. MATTIROLELLA Colla, *Boll. Lab. Zool. Portici* **22**: 44 (1928).
Sp. typ.: *M. silvestrii* Colla
Illustrations: Khan & Kimbrough (1974).

Entomogenous. *Mycelium* internal, haustorial. *Conidiomata* eustromatic, superficial, attached to the substrate by a small central column or foot, separate, black, crustose, applanate to pulvinate or stellate, consisting of a basal layer of thick-walled, haustorial mother cells upon which is situated a layer of hyaline pseudoparenchyma; vertical columns of sterile hyphae connect this layer to the dark brown, thick-walled, reticulate overlying upper wall (epihymenium) so forming several separate locules. *Ostiole* absent, dehiscence by breakdown of the upper wall. *Conidiophores* absent. *Conidiogenous cells* enteroblastic, phialidic, hyaline, smooth, discrete, determinate, collarette long, meristem basal, formed from the basal pseudoparenchyma. *Conidia* hyaline, thin-walled, eguttulate, aseptate, cylindrical.

Khan & Kimbrough (1974) have fully reviewed the morphology and development of *Mattirolella crustosa* Khan & Kimbrough and placed the genus in a newly described order Termitariales, together with *Termitaria* Thaxter. The genus is distinguished from *Termitaria* Thaxter which lacks the longitudinal hyphae that make *Mattirolella* multilocular.

M. SILVESTRII Colla, *ibid.*: 44 (1928).

Conidiomata stellate, with 8-10 locules, 60-70 µ diam.

Parasitizing integument of *Rhinotermes marginalis*, Guyana.

M. CRUSTOSA Khan & Kimbrough, *Amer. J. Bot.* **61**: 398 (1974).

Conidiomata 85-90 µ diam., multilocular. Conidia 3 × 1.5 µ.

On termites, Panama.

34. MYXOTHYRIUM Bub. & Kabát, *Svensk Bot. Tidskr.* **9**: 379 (1915).
Sp. typ.: *M. leptideum* (Fr.) Bub. & Kabát (syn. *Sphaeria leptidea* Fr.).

Mycelium immersed, branched, septate, brown. *Conidiomata* eustromatic, hypophyllous, subcuticular, separate or aggregated and confluent, dark brown, surface ± cerebriform, uni- or multilocular, consisting of a basal and an upper wall of dark brown, thick-walled textura angularis 1-3 cells thick. *Ostiole* absent, dehiscence by breakdown of upper wall. *Conidiophores* absent. *Conidiogenous cells* enteroblastic, phialidic, determinate, discrete, doliiform to ampulliform, hyaline, formed from the conidiomatal walls. *Conidia* thin-walled, smooth, with several large guttules, aseptate, obtuse at each end, straight, cylindrical to fusiform.

The genus is monotypic.

M. LEPTIDEUM (Fr.) Bub. & Kabát, *ibid.*: 379 (1915). (Fig. 294)
Sphaeria leptidea Fr., *Syst. Myc.* **2**: 522 (1823).
Sphaerella leptidea (Fr.) Fckl, *Jb. nassau. Ver. Naturk.* **23-24**: 100 (1870).
Phoma leptidea (Fr.) Sacc., *Syll. fung.* **3**: 111 (1884).
Phyllosticta leptidea (Fr.) Allesch. in Rabenh., *Kryptogamen Flora* **1**(6): 94 (1901).

Conidiomata up to 200 µ diam. Conidiogenous cells 4-6 × 2-3 µ. Conidia 5-6 × 1.5 µ.

On dead leaves of *Vaccinium vitis-idaea*, Czechoslovakia, Sweden (IMI 194256, 194257).

35. PSEUDOTHYRIUM Höhn., *Mitt. Bot. Inst. tech. Hochsch., Wien* **4**: 109 (1927).
Sp. typ.: *P. polygonati* (Tassi) Höhn. (syn. *Leptothyrium polygonati* Tassi).

Mycelium immersed, branched, septate, hyaline. *Conidiomata* subcuticular, separate, rarely confluent, formed of an upper wall, 1 cell thick, of very dark brown, thick-walled textura angularis and a basal wall of similar structure, above which is a layer of hyaline thin-walled textura prismatica. *Conidiophores* absent. *Conidiogenous cells* enteroblastic, phialidic, discrete, determinate, ampulliform, hyaline, smooth, collarette and channel extremely small, formed from the upper pseudoparenchyma. *Conidia* hyaline, thin-walled, eguttulate, smooth, straight, cylindrical, acicular.

Arx (1957a) reduced this generic name to synonymy with *Cylindrosporella* Höhn., and transferred the name of the type species to that genus. However, whereas *Asteroma* (syn. *Cylindrosporella*) is characterised by subcuticular acervuli which give the disorganised cuticle a pigmented appearance, in *Pseudothyrium* the entire subcuticular conidioma is enclosed in a dark brown pigmented wall 1 cell thick, the cuticle remaining unpigmented. In addition *P. polygonati* occurs

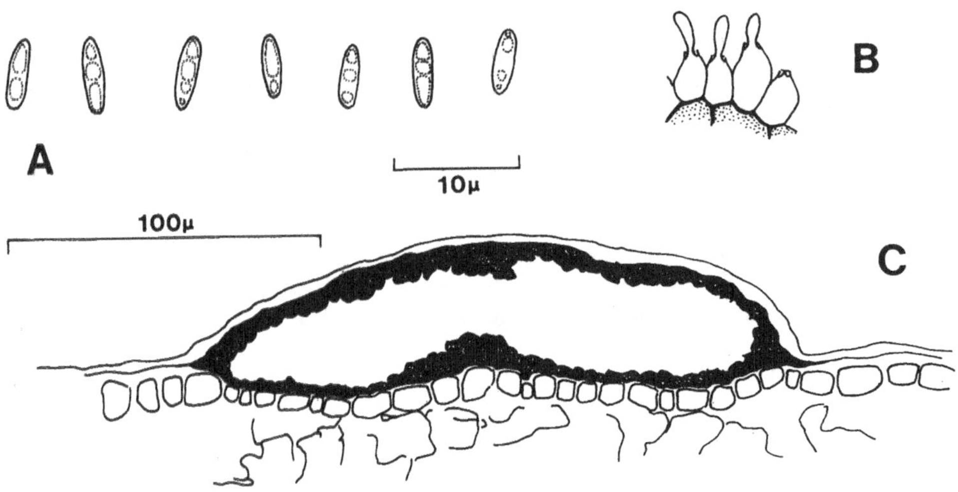

Fig. 294. *Myxothyrium leptideum*, A, conidia; B, conidiogenous cells; C, vertical section of a closed conidioma.

on stems whereas *Asteroma* species are foliicolous. For these reasons *Pseudothyrium* is maintained as a genus distinct from *Asteroma*.

P. POLYGONATI (Tassi) Höhn., *ibid.*: 109 (1927). (Fig. 295)
Leptothyrium polygonati Tassi, *Rev. mycol.* **18**: 171 (1896).
Cylindrosporella polygonati (Tassi) Arx, *Verh. K. Ned. Akad.* **51**: 15 (1957).

Conidiomata up to 400 µ diam. Conidiogenous cells 3.5-7 × 2-3 µ. Conidia 5.5-14 × 1-1.5 µ.

On stems of *Polygonatum officinale*, Latvia (IMI 30266).

36. ASTEROMA DC. ex St. Amans, *Flore Agenaise* 516 (1821).
Asteroma DC., *Fl. française* **5**: 162 (1815).
Depazea Fr. ex Ficinus & Schubert, *Flora Gegend um Dresden* **2** *Cryptogamie*: 311 (1823).
Depazea Fr., *Obs. Myc.*: 365 (1818).
Cylindrosporella Höhn., *Sber. Akad. Wiss. Wien* **125**: 94 (1916).
Gloeosporina Höhn., *Sber. Akad. Wiss. Wien* **125**: 94 (1916).
Titaeosporina van Luyk, *Annls mycol.* **17**: 112 (1919).
Actinonemella Höhn., *Mykol. Unters.* **1**: 338 (1923).
Ribaldia Ciferri, *Atti Ist. bot. Univ. Pavia*, ser. 5, **10**: 243 (1954).
Sp. typ.: *A. padi* DC. ex Fr.

Immersed mycelium branched, septate, pale brown to hyaline. *Conidiomata* acervular, subcuticular, separate or confluent, formed of pale brown to hyaline, thin-walled textura angularis; cuticle discoloured; some species associated with fibrils. *Conidiophores* absent. *Conidiogenous cells* enteroblastic, phialidic, discrete, determinate, ampulliform to lageniform, not cylindrical, hyaline, smooth, collarette and channel minute, formed from the upper cells of the pseudoparenchyma. *Conidia* hyaline, thin-walled, eguttulate, smooth, straight or curved, cylindrical to fusiform and acicular or even more broadly fusiform.

Asteroma was first introduced by De Candolle with 6 species, *A. phyteumatis* (as '*phyteumae*'), *A. dentariae*, *A. polygonati*, *A. violae*, *A. fraxini* and *A. padi*. The name was taken up by St. Amans (Apr., 1821) for *A. fraxini* and by Mérat (June, 1821) for *A. phyteumatis* and *A. polygonati*, and Fries (1823) adopted the genus with a single species, *A. padi*.

Höhnel (1916e) selected *A. phyteumatis* as lectotype species and this was followed by Clements & Shear (1931), but Schüepp (1959) pointed out that since *A. phyteumatis* was referred by Fries (1823) as a synonym of *Dothidea stellaris* Pers. ex Fr. (= *Leptotrochila phyteumatis* (Fckl) Schüepp), it cannot serve as a lectotype.

A. polygonati was placed by Fries (1823) as a synonym of *Dothidea asteroma* Fr. and *A. fraxini* was referred by Fries (1828) to *Septoria fraxini* DC. ex Fr, Even though St. Amans (1821) used *Asteroma fraxini*, since Fries (1828)

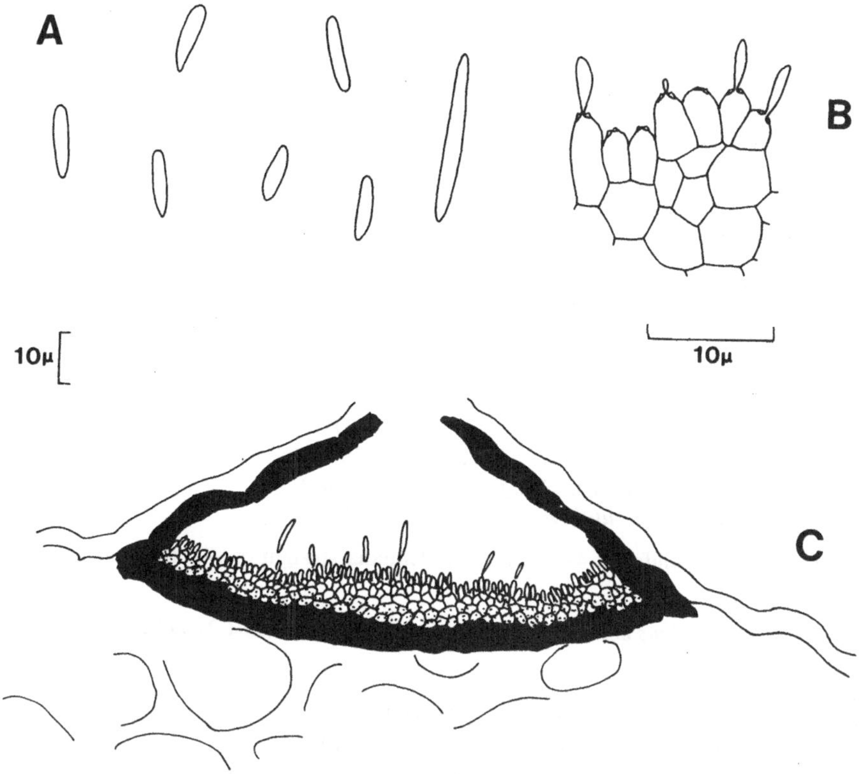

Fig. 295. *Pseudothyrium polygonati*, A, conidia; B, conidiogenous cells; C, vertical section of a conidioma.

placed the name in synonymy with *Septoria fraxini* it cannot be selected as lectotype species for to do so would be expressly against the intent of Fries (*International Code* Art. 13), consequently the genus is typified by *A. padi*. To adopt *A. fraxini* as lectotype would effectively reduce *Asteroma* to synonymy with *Septoria*, leaving the adoption of *Depazea* Fr. ex Ficinus & Schubert as obligatory for these fungi.

As a result of his conclusion over typification of *Asteroma*, Höhnel (1916e, 1923a) proposed *Actinonemella* to accommodate *A. padi*. Arx (1957a) reduced *Actinonemella* to synonymy with *Cylindrosporella*, but since *Actinonemella* and *Asteroma* are based on the same type species, *Asteroma* becomes the earliest generic name for the species formerly placed in *Cylindrosporella*. Arx (1957a, 1970) accepted 13 species in *Cylindrosporella* but maintained *Titaeosporina* separately.

At present there are in excess of 150 taxa described in *Asteroma* and it is extremely doubtful that the majority are congeneric with *A. padi*. Evidently more revisionary work is required.

A. alneum (Pers. ex Fr.) comb. nov. (Fig. 296D)
Xyloma alneum Pers., *Syn. Meth. fung.* 108 (1801).
Dothidea alnea Pers. ex Fr., *Syst. mycol.* **2**: 264 (1823).
Cylindrosporella alnea (Pers. ex Fr.) Höhn., *Hedwigia* **62**: 49 (1921).
Gloeosporium alneum Westd., *Prodr. Fl. Batav.* 2: **4**: 15 (1866).
Leptothyrium cylindrospermum Bon. in Rabenh., *Fungi europ.* 678 (1863).
Gloeosporium cylindrospermum (Bon.) Sacc., *Syll. fung.* **3**: 715 (1884).
Gloeosporium alnicola Dearn. & Hse, *N.Y. St. Mus. Bull.* 188: 32 (1916).

Teleomorphosis: *Gnomoniella tubiformis* (Tode) Sacc., *Michelia* **2**: 312 (1881).

Conidiogenous cells ampulliform, 4-6 × 3-4 µ. Conidia fusiform, straight or slightly curved, 10.5-13.5 × 3 µ.

Causing leaf lesions on *Alnus glutinosa*, *A. incana*, *A. virida*, Latvia, Hungary, Germany, U.K. (IMI 11134, 21709, 33305, 31170, 50771).

A. coryli (Fckl) comb. nov. (Fig. 298)
Leptothyrium coryli Fckl, *Jb. nassau. Ver. Naturk.* **23-24**: 120 (1870) non Lib. in Cke, 1879.
Ribaldia coryli (Fckl) Ciferri, *Atti Ist. bot. Univ. Pavia* **10**: 243 (1953).
Leptothyrium corylinum Sacc., nom. nov., *Syll. fung.* **3**: 626 (1884).
Septoria avellanae Berk. & Br. in Rabenh., *Fungi europ.* 1958 (1874), and *Grevillea* **5**: 56 (1876).
Titaeosporina avellanae (Berk. & Br.) Petrak, *Annls mycol.* **32**: 352 (1934).

Conidiogenous cells ampulliform to lageniform, 5-11 × 3.5 µ. Conidia straight, falcate or strongly curved, fusiform to acicular, 11.5-18.5 × 1-1.5 µ.

Causing leaf lesions of *Corylus avellana*, Austria, Germany, U.K. (IMI 30692, 31163, 31090, 33198, 13049, 54500, 12510).

A carpini (Lib.) comb. nov. (Fig. 297)
Leptothyrium carpini Lib., *Pl. Crypt. Ard.* 3: no. 256 (1834).
Gloeosporium carpini (Lib.) Desm., *Ann. sci. nat.*, ser. 3, **20**: 214 (1853).
Cylindrosporella carpini (Lib.) Höhn., *Sber. Akad. Wiss. Wien* **125**: 96 (1916).
Leptothyrium carpinicola Sacc. & Syd., *Syll. fung.* **14**: 933 (1899).
Leptothyrium betuli Oud., *Ned. Kruidk. Arch.* 3 ser., **2**: 347 (1901).

Conidiogenous cells ampulliform, 4-6 × 2.5-3.5 µ. Conidia straight or slightly curved, fusiform to acicular, 7.5-13.5 × 1-1.5 µ.

Causing leaf lesions of *Carpinus betulus*, Germany, Czechoslovakia (IMI 21697-21700).

A. caryae (Pk) comb. nov. (Fig. 296C)
Phyllosticta caryae Pk, *40th Rep. N.Y. St. Mus.* 57 (1887).
Cylindrosporella caryae (Pk) Petrak, *Annls mycol.* **22**: 162 (1924).
Gloeosporium caryae Ell. & Dearn., *Proc. Acad. nat. Sci. Phil.* 82 (1891).
Gloeosporium caryae Ell. & Ev., *Proc. Acad. nat. Sci. Phil.* 167 (1893).

Conidiogenous cells ampulliform, 6-7 × 3.5-4.5 µ. Conidia straight, curved or irregular, fusiform, 7-10 × 1.5-2 µ.

Causing leaf lesions on *Carya ovata* and *Hicoria glabra*, Canada, U.S.A. (IMI 21701, 33156, 21702).

A. inconspicuum (Cav.) comb. nov. (Fig. 296B)
Gloeosporium inconspicuum Cav., *Fungi Longob.* 5: 249 (1895).
Gloeosporina inconspicua (Cav.) Höhn., *Sber. Akad. Wiss. Wien* **125**: 94 (1916).
Cylindrosporella inconspicua (Cav.) Arx, *Verh. k. Ned. Akad. Wetench. Afd. Natuurk.* ser. 3, **51** (3): 95 (1957).

Conidiogenous cells ampulliform, 4-5 × 2.5-3.5 µ. Conidia straight or slightly curved, cylindrical or fusiform, 4-6 × 1-1.5 µ.

Causing leaf lesions on *Ulmus americana*, Italy, Germany (IMI 21715, 21716).

A. microspermum (Pk) comb. nov. (Fig. 296A)
Septoria microsperma Pk, *34th Rep. N.Y. St. Mus.* 44 (1883).
Sacidium microspermum (Pk) Davis, *Trans. Wisc. Acad. Sci. Arts Lett.* **18**: 88 (1915).
Cylindrosporella microsperma (Pk) Petrak, *Annls mycol.* **22**: 42 (1924).
Gloeosporium betulae-albae Sacc. & Dearn., *Annls mycol.* **13**: 125 (1915).

Conidiogenous cells lageniform, 6-8 × 2-3 µ. Conidia fusiform, straight or slightly curved, 6.5-9.5 × 2-2.5 µ.

Causing leaf lesions on *Betula allegheniensis*, *B.* sp., Canada, U.K. (IMI 159751, 188502).

A. PADI DC. ex Fr., *Syst. mycol.* **2**: 602 (1823). (Fig. 299)
Actinonema padi (DC. ex Fr.) Fr., *Summa Veg. Scand.* 424 (1849).
Actinonemella padi (DC. ex Fr.) Höhn., *Mykol. Unters.* **1**: 301 (1923).
Dothidea padi (DC. ex Fr.) Sprengel, *Syst. Veg.* **4** (1): 417 (1827).
Phlyctidium padi (DC. ex Fr.) Wallr., *Flora Germanica* **2**: 417 (1833).
Cylindrosporella padi (DC. ex Fr.) Arx, *Verh. k. Ned. Akad. Wet. Afd. Natuurk*, ser. 3, **51** (3): 114 (1957).

Teleomorphosis: *Gnomonia padicola* (Lib.) Kleb., *Z. f. Pflanzenkr.* **18**: 129 (1908).

Conidiogenous cells lageniform to cylindrical, up to 16 µ long × 2.5-3.5 µ wide. Conidia straight or slightly curved, fusiform, 11-15 × 2.5-3.5 µ.

Causing leaf lesions on *Prunus padus*, Germany, Czechoslovakia (IMI 20865, 30606, 13019, 16320).

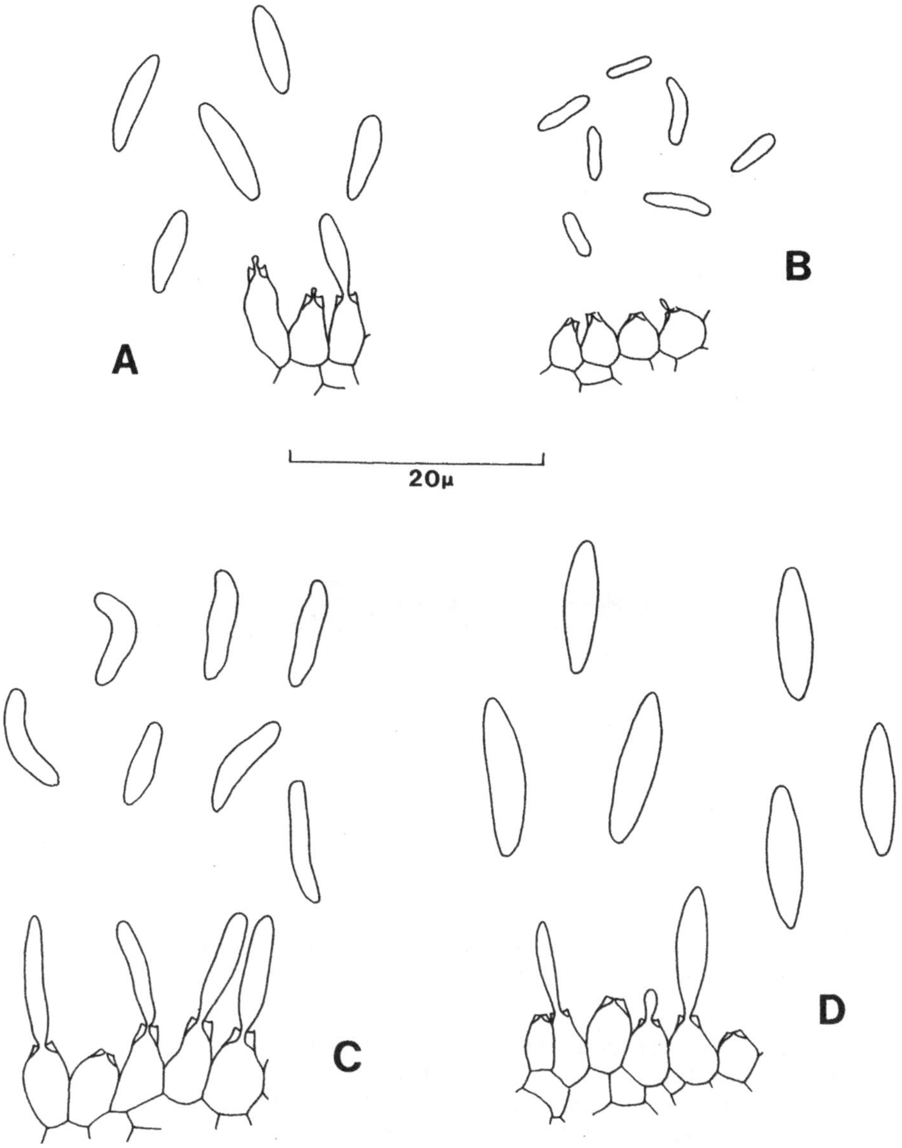

Fig. 296. A, *Asteroma microspermum*, conidia and conidiogenous cells; B, *A. inconspicuum*, conidia and conidiogenous cells; C, *A. caryae*, conidia and conidiogenous cells; D, *A. alneum*, conidia and conidiogenous cells.

A. FRONDICOLA (Fr. ex Ficinus & Schubert) Morelet, *Bull. Soc. Sci. nat. Arch. Toulon Var* 34 (221): 15 (1978).
Depazea frondicola Fr. ex Ficinus & Schubert, *Flora Gegend um Dresden* **2** *Cryptogamie*: 311 (1823).
Depazea frondicola Fr., *Obs. Myc.*: 365 (1818).
Leptothyrium tremulae Lib., *Pl. Crypt. Ard.* 161 (1823).
Gloeosporium tremulae (Lib.) Pass., *Hedwigia* **13**: 187 (1874).

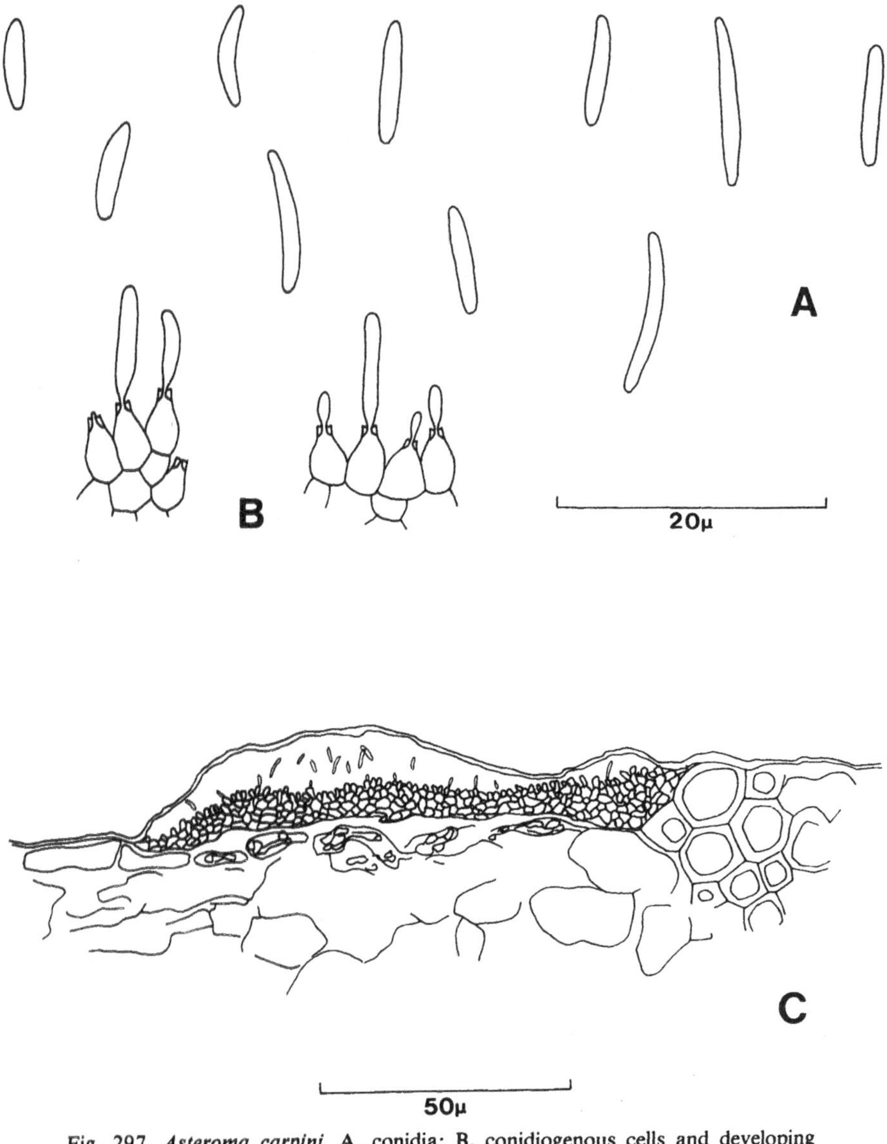

Fig. 297. *Asteroma carpini*, A, conidia; B, conidiogenous cells and developing conidia; C, vertical section of a conidioma.

Titaeosporina tremulae (Lib.) van Luyk, *Annls mycol.* **17**: 112 (1919).
Gloeosporium populi-albae Desm., *Bull. Soc. bot. Fr.* **4**: 799 (1857).
Leptothyrium circinans Fckl, *Jb. nassau. Ver. Naturk.* **23-24**: 383 (1870).
Gloeosporium circinans (Fckl) Sacc., *Syll. fung.* **3**: 712 (1884).
Gloeosporium populi-albae β tremulae Sacc., *Michelia* **1**: 219 (1878).
Gloeosporidium tremulae (Lib.) Höhn., *Sber. Akad. Wiss. Wien.* **125**: 95 (1916).

Conidiogenous cells lageniform to cylindrical, 7-12 × 2-3 µ. Conidia straight, fusiform, 5-9 × 2-2.5 µ, often larger, up to 24 µ long and anastomosing.

Causing leaf lesions on *Populus tremula*, *P. alba*, *P. ciliata*, Czechoslovakia, Sweden, Italy, Germany, Roumania, Austria, France, Finland, Latvia, U.S.S.R., Pakistan (IMI 30693, 31178, 33197, 13050, 17249, 21748, 21749, 29075, 11526, 46060, 21803, 21804, 50777, 60869, 62999, 80213).

The following additional taxa have not been examined but are included in the treatment by von Arx (1957a).

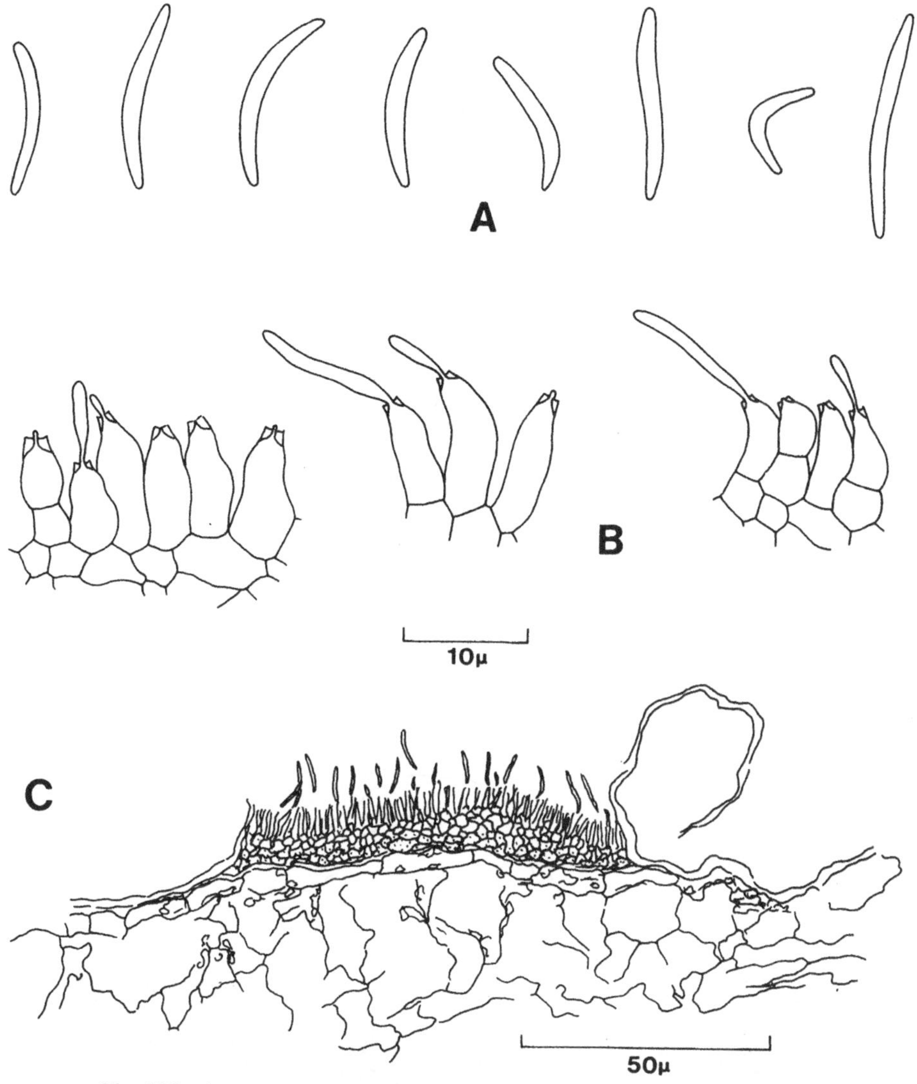

Fig. 298. *Asteroma coryli*, A, conidia; B, conidiogenous cells and developing conidia; C, vertical section of a conidioma.

A. ostryae (Massal.) comb. nov.
Leptothyrium ostryae Massal., *Atti Ist. Veneta Sci.* **59**: 686 (1900).

A. ulmeum (Miles) comb. nov.
Gloeosporium ulmeum Miles, *Bot. Gaz.* **71**: 161 (1921).

A. leptothyrioides (Kabát & Bub.) comb. nov.
Gloeosporium leptothyrioides Kabát & Bub., *Hedwigia* **46**: 296 (1907).

A. vleugelianum (Bub.) comb. nov.
Gloeosporium vleugelianum Bub., *Svensk. bot. Tid.* **5**: 345 (1911).

A. smilacinum (Bub.) comb. nov.
Gloeosporium smilacinum Bub., *Annls mycol.* **14**: 39 (1916).

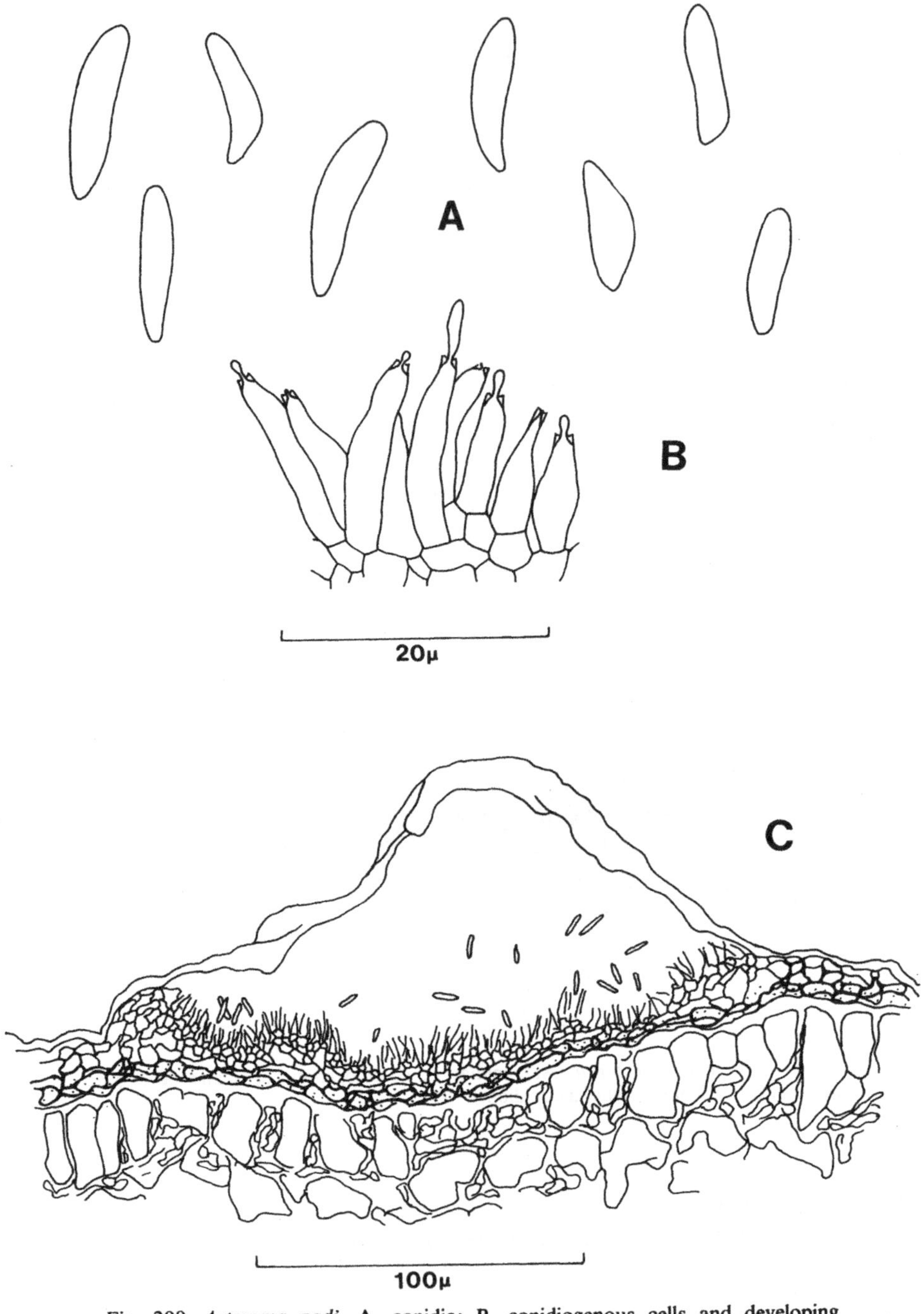

Fig. 299. *Asteroma padi*, A, conidia; B, conidiogenous cells and developing conidia; C, vertical section of a conidioma.

37. RHODESIA Grove, *British Stem- and Leaf- Fungi* **2**: 364 and 205 (1937).
Myxosporina Höhn., *Ber. dtsch. bot. Ges.* **37**: 115 (1919), not validly published.
Myxosporina Höhn. ex Arx, *Verh. K. Ned. Akad. Wet. Natuurk.* **51**: 31 (1957).
Sp. typ.: *R. subtecta* (Rob.) Grove (syn. *Fusarium subtectum* Rob.).

Mycelium immersed, hyaline, branched, septate. *Conidiomata* acervular, epidermal, separate or aggregated, formed of pale brown pseudoparenchyma. Dehiscence irregular, acervulus opening widely. *Conidiophores* hyaline, sparingly branched and septate towards the base, hyaline, cylindrical, smooth, with acrogenous conidia, formed from the upper acervular cells. *Conidiogenous cells* enteroblastic, phialidic, determinate, integrated or discrete, cylindrical, hyaline, smooth, channel minute, collarette evident, periclinal wall thickened. *Conidia* rose-coloured in mass, hyaline, aseptate, thin-walled, smooth, ± guttulate, abruptly tapered to a ± truncate base, oval.

The generic synonymy was explained by Sutton (1977a) and so far the genus remains monotypic.

R. SUBTECTA (Rob.) Grove, *ibid.*: 205 (1937). (Fig. 300)
Fusarium subtectum Rob., *Ann. Sci. nat.*, ser. 3, **3**: 358 (1845).
Hainesia subtecta (Rob.) Grove, *J. Bot. Lond.* 4 (1932).
Myxosporina subtecta (Rob.) Höhn., *Ber. dtsch. bot. Ges.* **37**: 115 (1919).
Gloeosporidium subtectum (Rob.) Höhn., *Mitt. Bot. Inst. tech. Hochsch., Wien* **4**: 74 (1927).
Hymenula psammae Oud., *Ned. Kruidk. Arch.* 3, **1**: 532 (1898).

Conidiomata 200-600 µ diam. Conidiophores 13-22 × 2 µ. Conidia 5-7 × 2.5 µ.

On leaves and stems of *Ammophila arenaria*, Germany (IMI 16618).

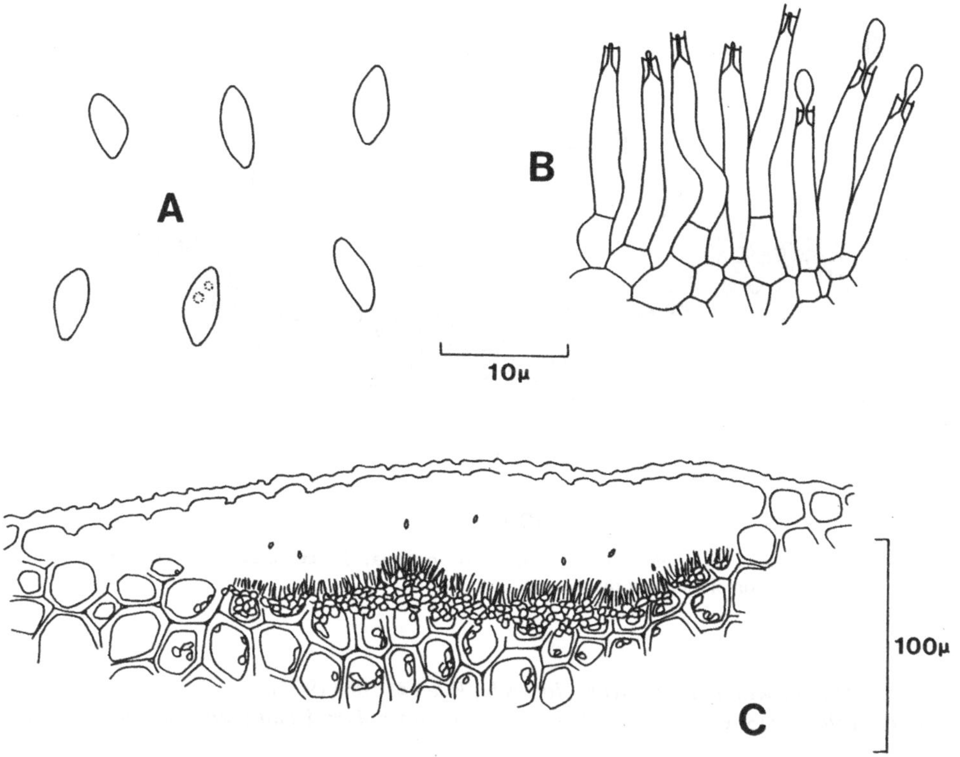

Fig. 300. *Rhodesia subtecta*, A, conidia; B, conidiophores and developing conidia; C, vertical section of a conidioma.

38. ERYTHROGLOEUM Petrak, *Sydowia* **7**: 378 (1953).
Sp. typ.: *E. hymenaeae* Frag. & Cif. ex Petrak

Mycelium immersed, branched, septate, pale brown to hyaline. *Conidiomata* eustromatic, epidermal, separate, immersed, formed at the base of pale brown, thin-walled textura angularis, and at the sides by slightly darker thicker-walled tissue. Dehiscence irregular. *Conidiophores* absent. *Conidiogenous cells* enteroblastic, phialidic, discrete, determinate, lageniform to cylindrical, hyaline, smooth, channel and thickened periclinal wall prominent, collarette absent, formed from the basal, lateral and upper wall. *Conidia* hyaline, aseptate, thin-walled, smooth, ± guttulate, oval.

The genus is monotypic.

E. HYMENAEAE Frag. & Cif. ex Petrak, *Sydowia* 7: 379 (1953). (Fig. 301)
Phyllosticta hymenaeae Frag. & Cif., nom. nud.

Conidiomata up to 250 µ diam. Conidiogenous cells 6.5-9.5 × 2.5-3.5 µ. Conidia 5-7.5 × 2.5-3 µ.

Causing lesions on leaves of *Hymenaea courbaril*, Costa Rica (IMI 94666, authentic for the name.)

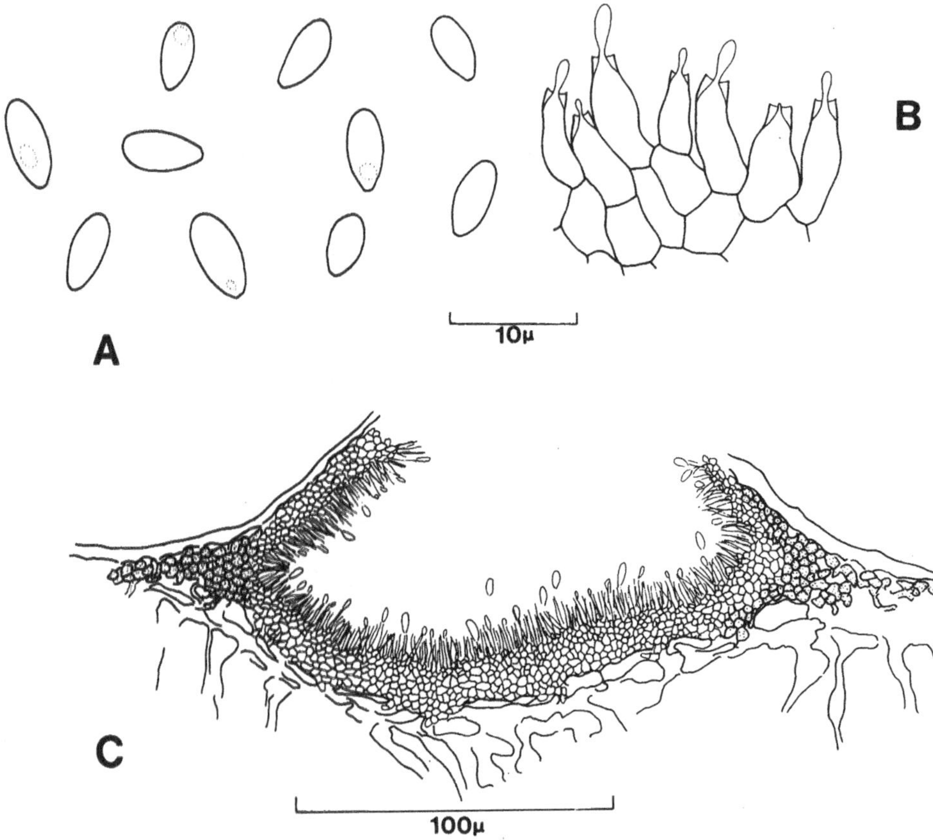

Fig. 301. *Erythrogloeum hymenaeae*, A, conidia; B, conidiogenous cells; C, vertical section of a conidioma.

39. DISCOSPORINA Höhn., *Mitt. bot. Inst. tech. Hochsch., Wien* 4: 106 (1927).

Sp. typ.: *D. deplanata* (Speg. & Roum.) Höhn. (syn. *Melanconium deplanatum* Speg. & Roum.).

Mycelium immersed, branched, septate, hyaline to pale brown. *Conidiomata* acervular, subperidermal, separate, consisting of a central pulvinate mass of textura angularis which is composed of large, thin-walled, brown cells, surrounded by a thinner region of hyaline to pale brown, smaller-celled tissue; both arise from a basal pseudostroma of mixed fungus-host tissue. *Ostiole* absent, dehiscence irregular. *Conidiophores* hyaline, branched and septate only at the base, smooth, formed mainly at the periphery, usually absent from the central pulvinate tissue. *Conidiogenous cells* enteroblastic, phialidic, determinate (or occasionally indeterminate by percurrent proliferations), integrated or discrete, cylindrical, hyaline, smooth, channel and collarette minute, periclinal wall thickened. *Conidia* hyaline, cylindrical, straight or slightly curved, apex obtuse, tapered to a truncate to obtuse base, thin-walled, multiguttulate, smooth.

Arx (1957a, 1970) included *Discosporina* as a synonym of *Cryptosporiopsis* Bub. & Kabát but neither conidiogenous cells nor conidial morphology warrant this. In *Cryptosporiopsis*, conidiogenous cells are relatively broad with very marked periclinal thickening at the apices, with a wide channel. Consequently conidia have a relatively broad base.

Apart from the type, three additional taxa have been placed in *Discosporina* but these have not recently been revised.

D. DEPLANATA (Speg. & Roum.) Höhn. (as 'Lib.'), *ibid.*: 106 (1927). (Fig. 302)
Melanconium deplanatum Speg. & Roum. apud Roum., *Rev. mycol.* **2**: 17 (1880).
Myxosporium deplanatum (Speg. & Roum.) Sacc., *Michelia* **2**: 354 (1881).
Myxofusicoccum deplanatum (Speg. & Roum.) Died., *Annls mycol.* **10**: 71 (1912).
Discosporium deplanatum (Speg. & Roum.) Höhn., *Mitt. bot. Lab. tech. Hochsch., Wien* **6**: 25 (1929).
Fusicoccum carpini Sacc., *Syll. fung.* **3**: 250 (1884).

Stromata up to 1200 μ diam. Conidiophores 8-16 × 2-2.5 μ. Conidia 19.5-23 × 2.5-3 μ.

On branches of *Carpinus betulus*, Austria (IMI 30316, 30439, 31080).

40. CYLINDROSPORIUM Grev., *Scottish Crypt. Flora* **1**: 27 (1823).
Cylindrodochium Bonorden, *Handb. der Allgem. Myk.*: 132 (1851).
Sp. typ.: *C. concentricum* Grev.
Illustration: Sutton & Gibson (1977).

Mycelium immersed, branched, septate, hyaline. *Conidiomata* acervular, white, slimy, subcuticular, separate or confluent, formed of pale brown to hyaline, thin-walled textura angularis. Dehiscence irregular but often pushing up a flap of host tissue. *Conidiophores* hyaline, parallel, branched only at the base, 1-2 septate, smooth, formed from the upper pseudoparenchyma. *Conidiogenous cells* enteroblastic, phialidic, integrated, cylindrical, hyaline, smooth. *Conidia* straight or slightly curved, aseptate, cylindrical, thin-walled, smooth, hyaline, eguttulate.

Few genera have been the subject of more confusion than *Cylindrosporium*, mainly because of the interpretation by Saccardo (1884) and his exclusion of the type species. The nomenclatural status of *Cylindrosporium* and the genera confused with it have been summarised by Wakefield (1940). Höhnel (1924) was first to point out that *Cylindrosporium* was monotypic, subsequently confirmed as such by von Arx (1957a, 1970), Sutton (1967c) and Petrak (1972). Thomson (1936) provided a full account of the species based on the type and fresh material. Thomson (1936) and von Arx (1957a, 1970) are in favour of regarding conidia as thallic, seceding by fragmentation of a conidiogenous cell, but Thomson (1936) illustrates acropleurogenous development of conidia in culture which suggests holoblastic or enteroblastic rather than holothallic ontogeny. Sutton & Gibson (1977) and Rawlinson, Sutton & Muthyalu (1978) have clearly demonstrated development to be phialidic. This places *Cylindrosporium* close to *Asteroma* DC. ex St. Amans. The genera only differ by the cuticular discolouration shown in *Asteroma* but which is absent in *Cylindrosporium*. Many hundreds of species have been described in *Cylindrosporium*, but they need to be revised with similar studies on *Phloeospora, Septogloeum, Septoria* and the many segregates from these genera.

C. CONCENTRICUM Grev., *ibid.*: 27 (1823).
Gloeosporium concentricum (Grev.) Berk. & Br., *Ann. mag. Nat. Hist.* 2, **5**: 455 (1850).
Cylindrodochium concentricum (Grev.) Bonord., *Handb. der Allgem. Myk.*: 132 (1851).

Teleomorphosis: *Pyrenopeziza brassicae* Sutton & Rawlinson, *Trans. Br. mycol. Soc.* **71**: 426 (1978).

Conidiomata 100-200 μ diam. Conidia 8-15 × 3-4.5 μ.

Causing light leaf spot of *Brassica oleracea, B. oleracea* var. *botrytis, B. napus, B. napus* var. *oleifera, B. rapa, B.* spp., New Zealand, U.K. (IMI 101696, holotype of *C. concentricum* ex E, 217176, 217177, 217178, 217179, 201689, 81824, 136630, 98130, 203945, 203944, 204289, 204288, 204287, 203943, 204383, 204284, 204282, 81823, 204285, 204286, 204290).

41. DISCULA Sacc., *Syll. fung.* **3**: 674 (1884).
Gloeosporidium Höhn., *Sber. Akad. Wiss. Wien* **125**: 95 (1916).
Original spp.: 20 species listed, all transferred from existing genera.
Lectotype sp.: *D. quercina* (Westd.) Arx (syn. *Gloeosporium quercinum* Westd., *D. platani* (Pk) Sacc.), fide Höhnel (1915d) and Arx (1957a); = *D. umbrinella* (Berk. & Br.) Sutton.

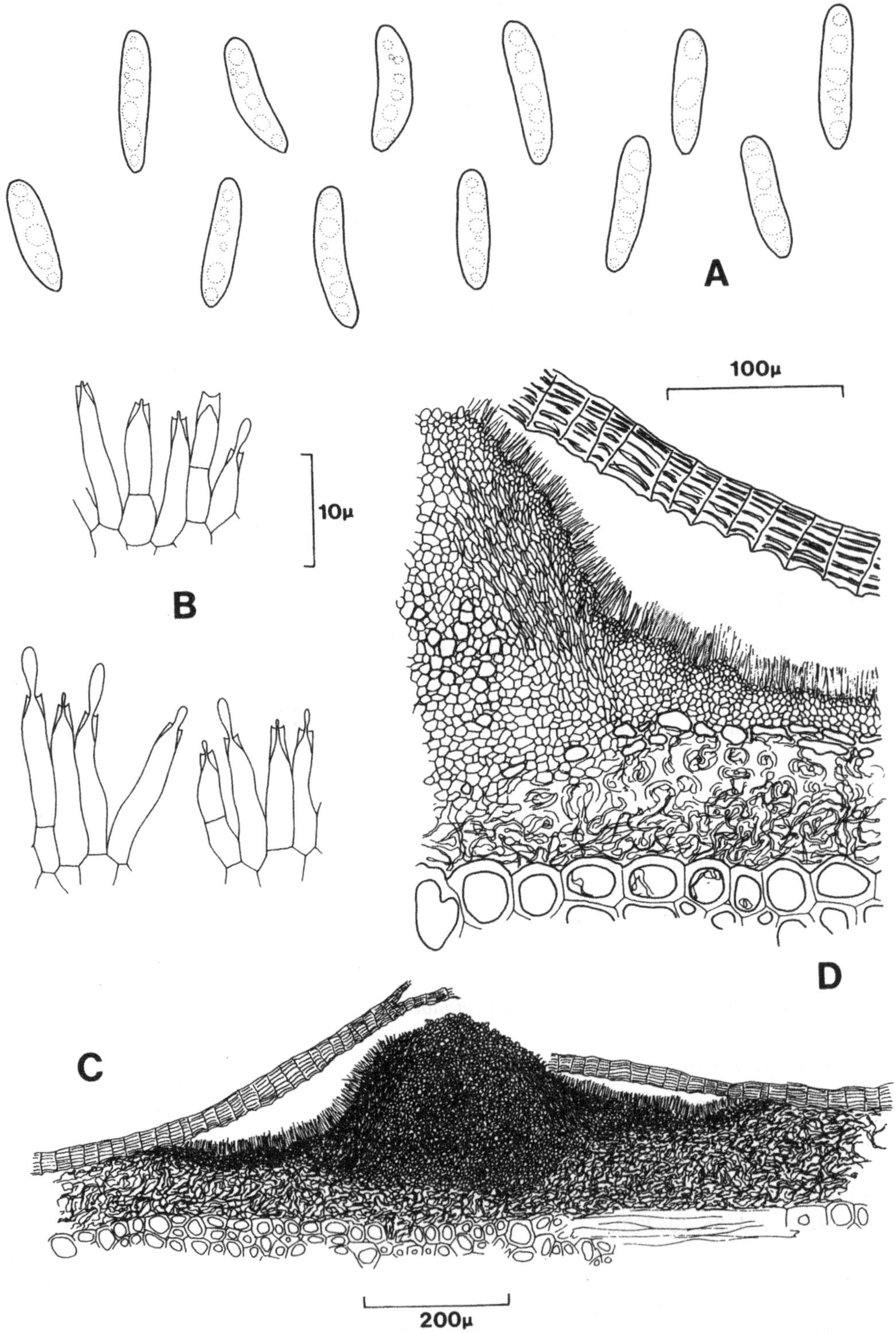

Fig. 302. *Discosporium deplanatum*, A, conidia; B, conidiophores; C, vertical section of a conidioma; D, detailed structure of pseudoparenchyma.

Mycelium immersed, branched, septate, hyaline or pale brown. *Conidiomata* acervular, epidermal, separate or confluent, formed of pale brown, thin-walled textura angularis. Dehiscence irregular. *Conidiophores* hyaline, septate and branched only at the base, straight or curved to irregular, tapered towards the apices, formed from the upper cells of the conidiomata. *Conidiogenous cells* enteroblastic, phialidic, integrated or discrete, determinate, hyaline, smooth, channel and collarette minute, periclinal wall distinctly thickened. *Conidia* hyaline, aseptate, smooth, thin-walled, straight or slightly curved, apex obtuse, base ± truncate, ellipsoid or clavate.

Arx (1957a) accepted 12 taxa in this genus and in doing so adopted a very broad concept for the type species. This was stated to occur on a wide variety of temperate deciduous leaves, and 46 synonyms were cited for the species name. Few of these have subsequently been checked and so far it has not been possible to examine the rest of the species placed in the genus. The teleomorphosis belongs in *Apiognomonia* Höhn.

Discula umbrinella (Berk. & Br.) comb. nov. (Fig. 303)
Gloeosporium umbrinellum Berk. & Br., *Ann. Mag. Nat. Hist.*, 3 ser., **18**: 121 (1886).
Gloeosporidium umbrinellum (Berk. & Br.) Petrak, *Annls mycol.* **20**: 15 (1922).
Myxosporium platanicola Ell. & Ev., *Proc. Acad. Phil. 1894*: 372 (1894).
synonymy as cited in Arx (1957a).

Teleomorphosis: *Apiognomonia errabunda* (Rob.) Höhn., *Annls mycol.* **16**: 51 (1918).

Conidiomata up to 150 µ diam. Conidiophores 13-21 × 3-5 µ, tapering to 2-3 µ at the apices. Conidia 9.5-12 × 3.5-5 µ.

The concept of *A. errabunda* propounded by Arx (1951) embraces both *Gnomonia platani* Kleb. and *G. quercina* Kleb. (Boerema & Verhoeven, 1972). Arx (1957a) accepted *D. quercina* (Westd.) Arx as the correct name of the anamorphosis but unfortunately this is later homonym of *D. quercina* (Cke) Sacc., 1884. The next available epithets *Hymenula platani* Lév., 1848, and *Fusarium platani* Mont., 1849, would also be pre-empted in *Discula* by *D. platani* (Pk) Sacc., 1884, and *D. platani* (Oud.) Höhn. ex Petrak & Sydow, 1927 respectively. The same applies to *Labrella fagi* Desm. & Rob., 1853 and *D. fagi* Oud., 1889. Consequently, if the synonymy suggested by Arx (1957a) is accepted, then the next available epithet for this fungus is *Gloeosporium umbrinellum* Berk. & Br.

On leaves of *Platanus occidentalis*, *P. acerifolia*, *P. orientalis*, *P.* x *hybrida*, *P.* sp., *Acer palmata*, *Quercus pedunculata*, *Q. alba*, *Q. palustris*, *Q.* sp., U.K., Eire, Italy, Hungary, Romania, Germany, N.S.W., Australia, New Zealand, Ont., Canada (IMI 123720, 205479, 21738, 11424, 50711, 33494, 21741, 16574, 23313, 12466, 185084, 194416, 203949, 16653, 12797, 21740, 21739, 25903, 190464, 175588, 50773, 25906, 102910, 101740, 94843, 80522).

42. DISCOGLOEUM Petrak, *Annls mycol.* **21**: 285 (1923).
Sp. typ.: *D. veronicae* (Lib.) Petrak (syn. *Leptothyrium veronicae* Lib., *Discogloeum veronicarum* (Ces.) Petrak).

Mycelium immersed, branched, septate, hyaline. *Conidiomata* acervular, separate or confluent, circular, composed of hyaline to pale brown textura angularis. *Conidiophores* hyaline, smooth, cylindrical, branched and septate only at the base, formed from the upper cells of the acervuli. *Conidiogenous cells* enteroblastic, phialidic, hyaline, smooth, cylindrical, discrete or integrated, determinate, collarette prominent, channel narrow, periclinal wall thickened. *Conidia* hyaline, aseptate, thin-walled, eguttulate, smooth, apex obtuse, base truncate, cylindrical to obovoid, sometimes slightly curved, occasionally irregular.

Arx (1957a, 1970) has redescribed the genus, and Morgan-Jones (1971f) gave more precise details of conidial ontogeny. The genus is not very distinct from *Cylindrosporium* Grev. and only appears to differ by the slightly more irregularly shaped conidia and prominent phialidic apparatus. Three other taxa in addition to the type have been referred to *Discogloeum* but they have not yet been re-examined.

D. VERONICAE (Lib.) Petrak, *Annls mycol.* **27**: 370 (1929). (Fig. 304)
synonymy as in Arx (1970).

Conidiomata up to 175 µ diam. Conidiophores 13-17 × 3-4.5 µ. Conidia 10-17 × 3.5-4.5 µ.

Causing leaf lesions on *Veronica tournefortii*, Germany (IMI 21810).

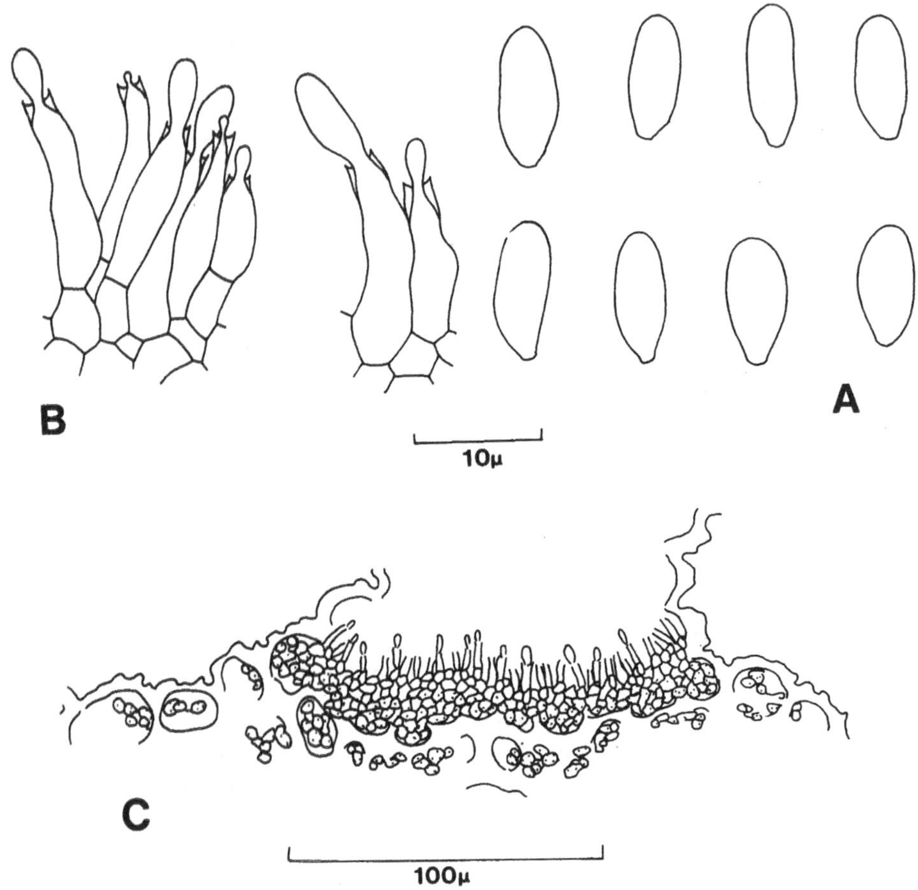

Fig. 303. *Discula umbrinella*, A, conidia; B, conidiophores and developing conidia; C, vertical section of a conidioma.

43. APOMELASMIA Grove, *British Stem- and Leaf- Fungi* **2**: 363 (1937).
Cheilariopsis Petrak, *Sydowia* **13**: 63 (1959).
Sp. typ.: *A. urticae* (Lib.) Grove (syn. *Cheilaria urticae* Lib.).

Mycelium immersed, branched, septate, brown to dark brown. *Conidiomata* stromatic, immersed, black, effuse, separate or often confluent and then extensive, and lateral regions of pale brown, thin-walled textura angularis, upper region of similar tissue but the whole covered by a layer of dark brown, larger, thick-walled pseudoparenchyma which extends beyond the locules and often completely covers the substrate. *Ostiole* absent, dehiscence by an irregular fissure. *Conidiophores* hyaline, 1-2 septate, cylindrical, branched sparingly at the base, with acrogenous conidia, formed all round the locular cavity. *Conidiogenous cells* enteroblastic, phialidic, determinate, integrated or discrete, cylindrical or long lageniform, hyaline, smooth, collarette and channel minute. *Conidia* hyaline, ± cylindrical to fusiform, aseptate, smooth, thin-walled, straight or irregularly curved, apex and base obtuse, often abruptly tapered at one or both ends, guttulate.

The genus is monotypic.

A. URTICAE (Fr.) Grove, *ibid.*: 189 (1937). (Fig. 305)
Rhytisma urticae Fr., *Syst. mycol.* 2: 370 (1823).
Xyloma urticae (Fr.) Wallr., *Fl. Germ.* **2**: 409 (1833).
Cheilaria urticae (Fr.) Lib., *Pl. Crypt. Ard.* 1: no. 62 (1830).
Placosphaeria urticae (Fr.) Sacc., *Malpighia* **2**: 19 (1888).
Cheilariopsis urticae (Fr.) Petrak, *Sydowia* **13**: 63 (1959), not validly published.
Melasmia urticae Grove, *J. Bot., Lond.* **56**: 319 (1918).

Teleomorphosis: *Aporhytisma urticae* (Wallr.) Höhn., *Ber. dtsch. Bot. Ges.* **35**: 421 (1917).

Conidiomata extensive and confluent, locule up to 200 µ diam. Conidiogenous cells 9-16 × 3.5-4 µ. Conidia 17-25 × 3.5-4.5 µ.

On dead stems of *Urtica dioica*, Germany, Latvia, Poland, Austria, U.K. (IMI 12203, 50863, 50735, 30420, 12678, 73353b, 94235, 164984).

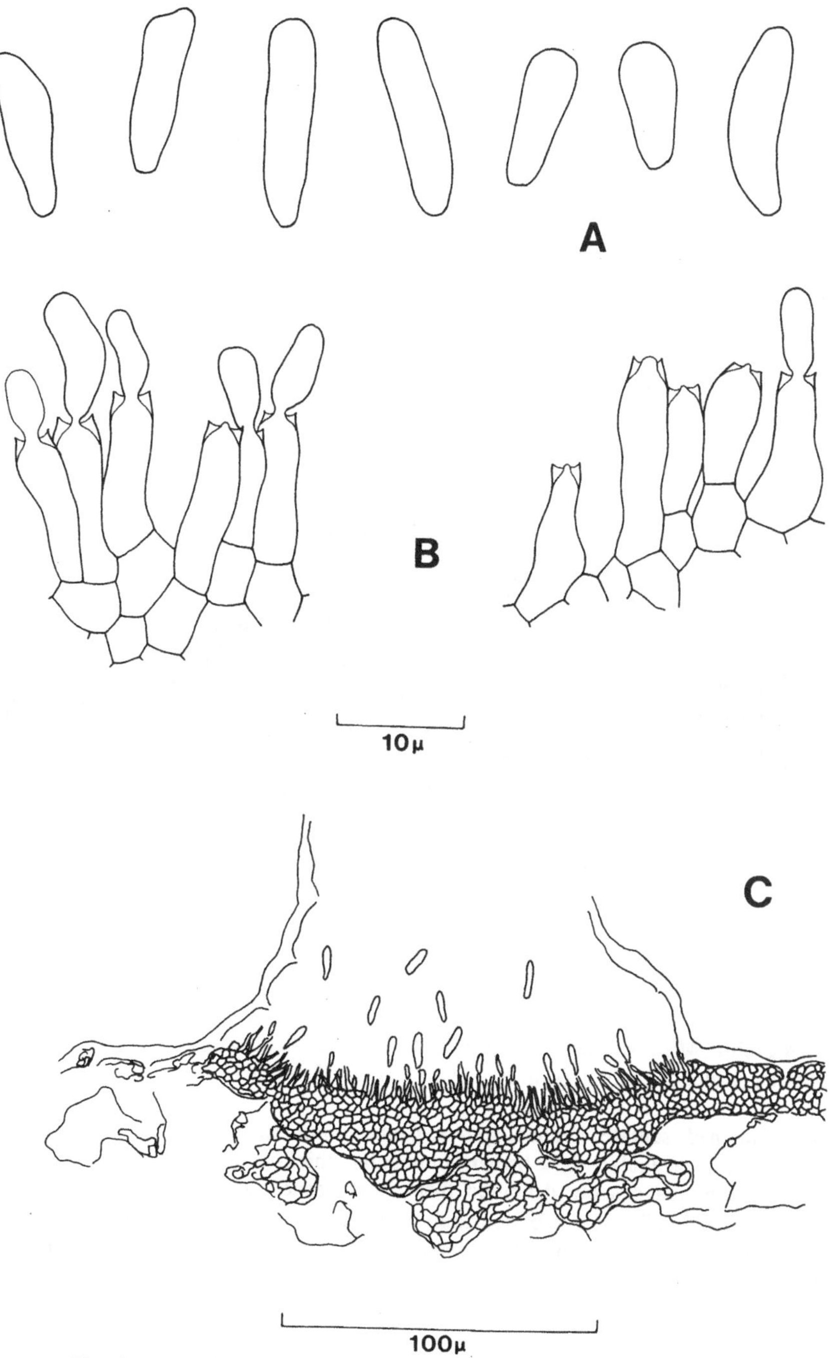

Fig. 304. *Discogloeum veronicae*, A, conidia; B, conidiophores and developing conidia; C, vertical section of a conidioma.

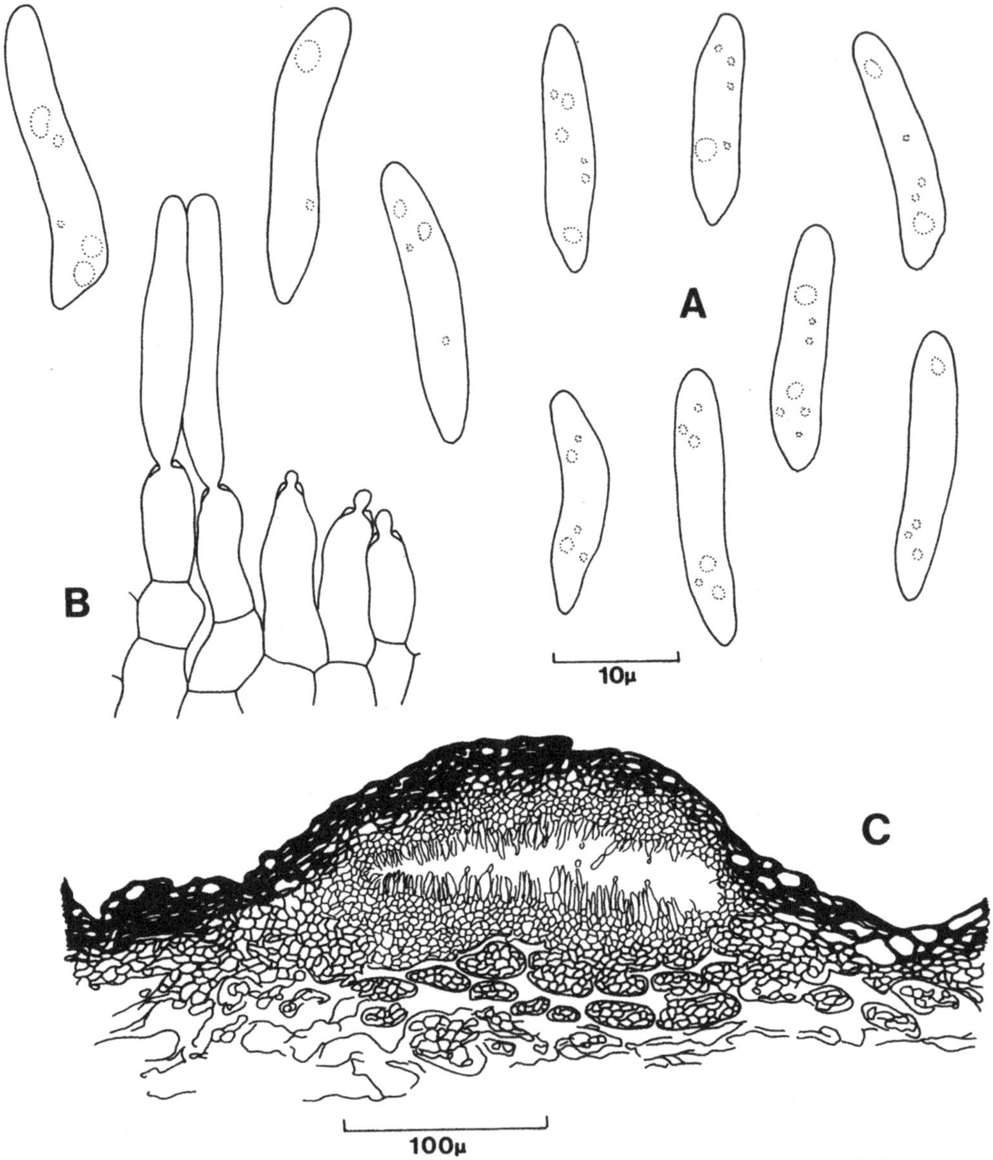

Fig. 305. *Apomelasmia urticae*, A, conidia; B, conidiophores and developing conidia; C, vertical section of a closed conidioma.

44. CRYPTOMYCELLA Höhn., *Mitt. bot. Inst. tech. Hochsch., Wien* **2**: 48 (1926).
Sp. typ.: *C. pteridis* (Kalchbr.) Höhn. (syn. *Fusidium pteridis* Kalchbr.).

Mycelium immersed, branched, septate, hyaline, sparse. *Conidiomata* stromatic, immersed, separate or aggregated, subepidermal, flattened, circular to irregular, dark brown, unilocular, thick-walled, above and at the base; walls of dark brown, thin-walled textura prismatica sharply delimited from host tissue. *Ostiole* absent, dehiscence by irregular to longitudinal fissure. *Conidiophores* hyaline, 1-2 septate, rarely branched, smooth, cylindrical, formed from the upper and lower walls. *Conidiogenous cells* enteroblastic, phialidic, determinate, integrated or discrete, cylindrical or tapered at the apex, sometimes abruptly, hyaline, smooth, apertures and collarettes minute. *Conidia* hyaline, aseptate, eguttulate, thin-walled, straight or slightly curved, fusiform.

Two species, *C. pteridis* and *C. maxima* Höhn., were originally placed in the genus. Bache-Wigg (1940) gave a full account of the pathology and taxonomy of *C. pteridis* on *Pteridium latiusculum*.

C. PTERIDIS (Kalchbr.) Höhn., *ibid.*: 48 (1926). (Fig. 306)
Fusidium pteridis Kalchbr., *Bot. Z.* **19**: 296 (1861).
Gloeosporium pteridis (Kalchbr.) Bub. & Kabát, *Ber. naturw. med. Ver. Innsbruck* **30**: 31 (1906).
Gloeosporium pteridis Harkn., *Bull. Calif. Acad. Sci.* **1**: 12 (1884).
Gloeosporium pteridii Karakulin nom. nov. apud Vassiljevsky & Karakulin, *Fungi Imperfecti Parasitici* Pars II. Melanconiales: 42 (1950).
Gloeosporium obtegens Syd., *Annls mycol.* **2**: 172 (1904).

Teleomorphosis: *Cryptomycina pteridis* (Rebent. ex Fr.) Höhn., *Ber. dtsch. Bot. Ges.* **35**: 422 (1917).

Conidiomata up to 450 µ diam. Conidiogenous cells 10-16 × 3.5-4 µ. Conidia 16.5-20.5 × 3.5-5 µ.

On fronds of *Pteridium aquilinum, P. pubescens, P. latiusculum, P. lanuginosum*, Germany, Canada, U.S.A., Roumania, Austria, Latvia, Finland, (IMI 23154, 23155, 23156, 34010, 13123, 32769, 167004, 25955, 11261, 12405, 50500, 50499, 73261, 22683, 122926, 33154, 154390).

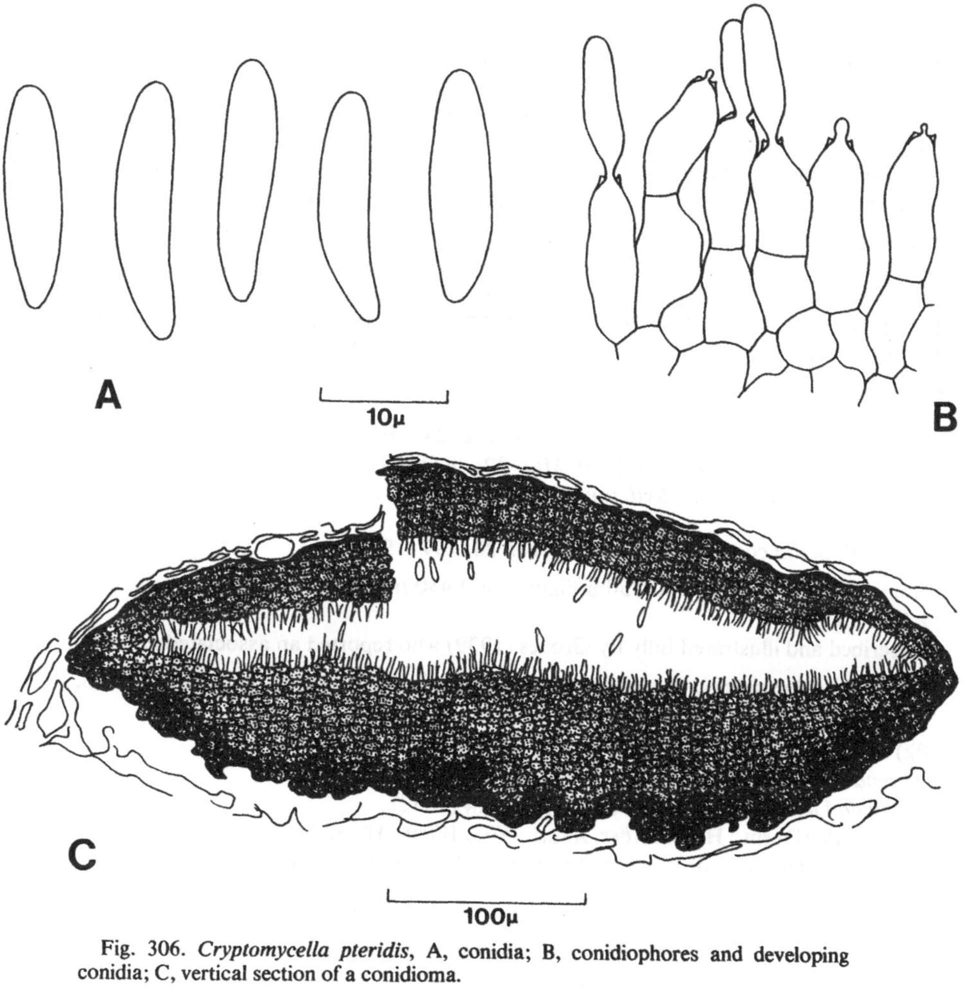

Fig. 306. *Cryptomycella pteridis*, A, conidia; B, conidiophores and developing conidia; C, vertical section of a conidioma.

45. CRYPTOSPORIOPSIS Bub. & Kabát, *Hedwigia* **52**: 360 (1912).
Pachydiscula Höhn., *Z.f. Garungsph.* **5**: 210 (1914).
Tuberculariella Höhn., *Z.f. Garungsph.* **5**: 209 (1914).
Lagynodella Petrak, *Annls mycol.* **20**: 207 (1922).
Discosporiella Petrak, *Annls mycol.* **21**: 14 (1923).
Sp. typ.: *C. scutellata* (Otth) Petrak (syn. *Sphaeropsis scutellata* Otth, *C. nigra* Bub. & Kabát).

Mycelium immersed, branched, septate, hyaline or pale brown. *Conidiomata* acervular to stromatic, peridermal to subperidermal, separate or confluent, formed of pale brown to hyaline, thin-walled textura angularis. Dehiscence irregular, but in some species forming a massive erumpent stroma which may be multilocular and dehisce by a large ostiole (*C. pruinosa*). *Conidiophores* absent. *Conidiogenous cells* enteroblastic, phialidic, discrete, indeterminate, cylindrical, hyaline, smooth, with 1-several proliferations that may or may not be percurrent. *Conidia* comparatively large, hyaline, thin-walled, guttulate or eguttulate, smooth, straight, ellipsoid, apex obtuse, base abruptly tapered to a distinct truncate scar.

Cryptosporiopsis species are the anamorphoses of *Pezicula* species and as such, much of the published information on them is documented in articles dealing with *Pezicula* (Dennis, 1974; Nannfeldt, 1932; Groves, 1938, 1939, 1940, 1941; Seaver, 1951). There are little differences between *Cryptosporiopsis* species and the characters used to distinguish taxa are few and unreliable. Occasionally major differences in conidial size are useful, as are the conical stromatic conidiomata in a few species.

C. ABIETINA Petrak, *Annls mycol.* **23**: 125 (1925). (Fig. 307A)
Myxosporium abietinum Rostr., *Tids. Skow.* **13**: 89 (1901).

Teleomorphosis: *Pezicula livida* (Berk. & Br.) Rehm, *Ascomycetes*: 462 (1878).

Conidiomata immersed, not specially erumpent, .5-2 mm long, reddish brown. Conidiogenous cells 12-15 × 4-5 µ. Conidia 25-33.5 × 9.5-10 µ.

Described briefly by Seaver (1951) and Nannfeldt (1932). Grove (1937) recorded microconidia 4-5 × 1.5-2 µ formed from conidiophores 10-17 µ long, associated with the *Cryptosporiopsis* state.

On and from twigs, bark and needles of *Abies alba, A. pectinata, A.* sp., *Larix* sp., *Picea excelsa, P. sitchensis, P.* sp., *Pseudotsuga douglasii, Pinus contorta, P. sylvestris, P. strobus, P. radiata, Sequoia sempervirens, Taxus baccata, Thuja occidentalis*, Austria, Switzerland, U.K., France, N.Y., U.S.A., Jamaica, Kenya (IMI 11293, 145134, 184533, 94535, 146512, 146518, 146508, 146509, 58048, 51227, 29724, 188937, 146504, 193726, 155138, 58046, 145214, 145213, 145212, 68739, 201964, 29722, 29720, 29723, 29721, 223361, 223362, 224305, 190241, 184537, 184536, 108507, 108506).

C. CORNINA (Pk) Petrak & Syd., *Annls mycol.* **22**: 370 (1924). (Fig. 307B)
Sphaeropsis cornina Pk, *Ann. Rep. N.Y. St. Mus.* 32: 38 (1879).
Macrophoma cornina (Pk) Sacc., *Syll. fung.* **10**: 192 (1892).

Teleomorphosis: *Pezicula corni* Petrak, *Annls mycol.* **20**: 197 (1922).

Conidiomata black, conical, 200-350 µ diam. × 300-450 µ high. Conidiogenous cells 10-30 × 3-6 µ. Conidia 24-32.5 × 11.5-13.5 µ.

Described and illustrated fully by Groves (1939) who reported an associated microconidial state.

On twigs of *Cornus instolonea, C. circinata*, Sask., Ont., Canada (IMI 21338, 21339, 29716, 29715).

C. coryli (Pk) comb. nov. (Fig. 307C)
Sphaeronaema coryli Pk, *Ann. Rep. N.Y. St. Mus.* 24: 85 (1872).
Catinula turgida Desm., *Ann. Sci. nat.*, ser. 3, **18**: 374 (1852).
Dothichiza turgida (Desm.) Höhn., *Sber. Akad. Wiss. Wien* **118**: 887 (1909).
Psilospora turgida (Desm.) Höhn., *Sber. Akad. Wiss. Wien* **124**: 103 (1915).

Teleomorphosis: *Pezicula corylina* Groves, *Mycologia* **30**: 47 (1938).

Conidiomata cylindrical to conic, separate, opening widely at maturity, .2-.5 mm diam. × .3-.05 mm high. Conidiophores 10-40 × 2.5-5 µ. Conidia 19-23.5 × 7.5-8.5 µ.

The earliest name applied to this *Cryptosporiopsis* species is *Catinula turgida* Pk, but since Pirozynski & Morgan-Jones (1968) used *Cryptosporium turgidum* Berk. & Br. as the basionym for their combination into *Cryptosporiopsis*, the epithet is pre-empted. *C. turgida* (Berk. & Br.) Pirozynski & Morgan-Jones is a species on *Fraxinus* with curved conidia like *C. nobilis* (Sacc.) Pirozynski & Morgan-Jones. Neither species comes within my concept of *Cryptosporiopsis*. They would more satisfactorily be placed in *Gloeosporidiella* Petrak.

On twigs of *Corylus avellana, C. rostrata, C.* sp., U.K., Ont., Canada (IMI 20929, 20928, 123031, 21646, 73216, 73215, 20930, 73193, 20926, 20927, 20925, 20924, 20923, 20922).

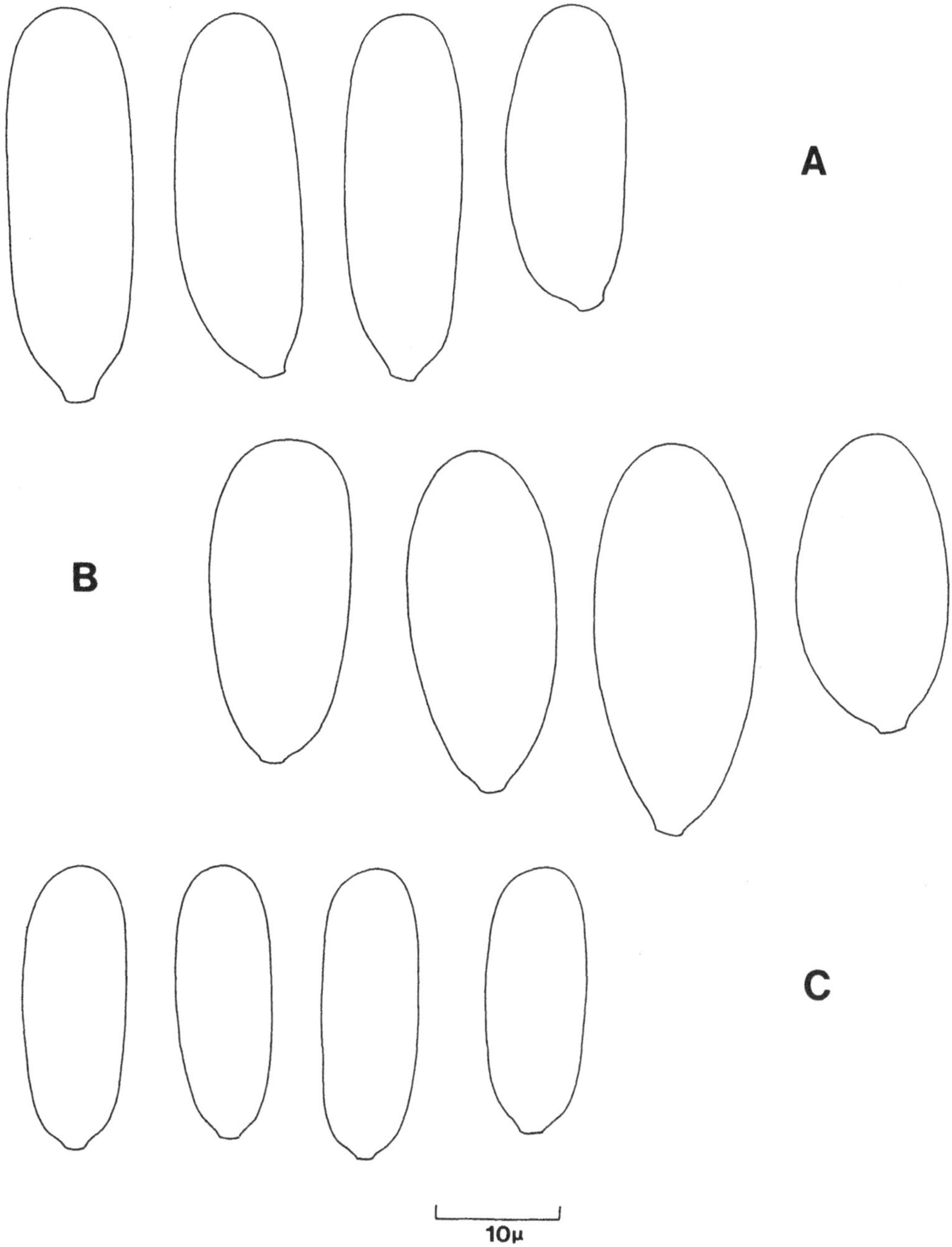

Fig. 307. A, *Cryptosporiopsis abietina*, conidia; B, *C. cornina*, conidia; C, *C. coryli*, conidia.

C. CORTICOLA (Edgerton) Nannfeldt, *Nova Acta Reg. Soc. Scient. Upsal.* ser. 4, **8**: 94 (1932). (Fig. 308A)
Myxosporium corticola Edgerton, *Annls mycol.* **6**: 51 (1908).

Teleomorphosis: *Pezicula corticola* (Jorg.) Nannfeldt, *Nova Acta Reg. Soc. Scient. Upsal.* ser. 4, **8**: 94 (1932).

Conidiomata 1-2 mm diam., forming white conidial masses. Conidiophores up to 15 μ long × 4-5 μ wide. Conidia 30-32.5 × 9.5-10.5 μ.

There is some controversy over the synonymy for the *Cryptosporiopsis* name, and this cannot be resolved without recourse to examination of the types. Nannfeldt (1932) included several synonyms for the species based

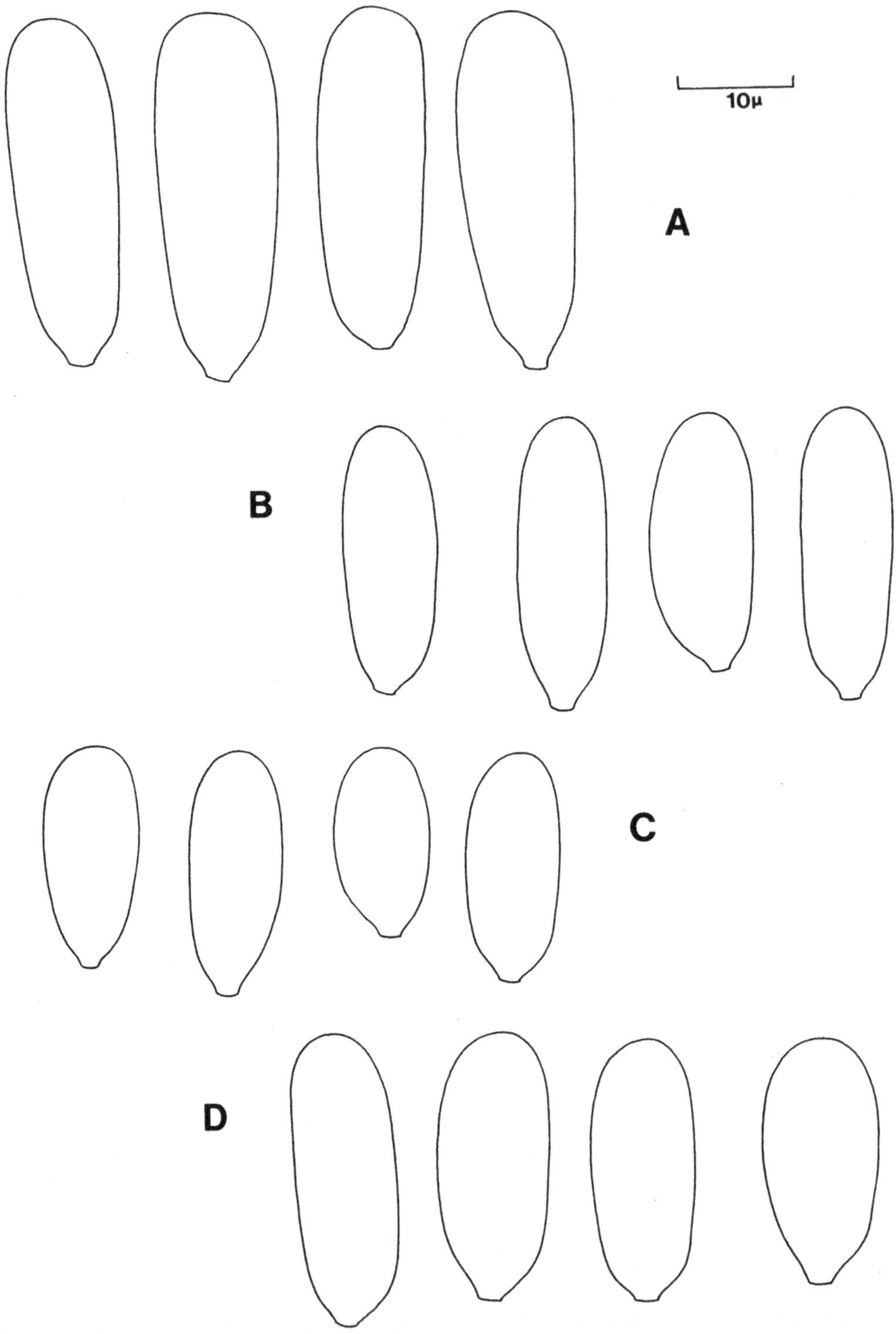

Fig. 308. A, *Cryptosporiopsis corticola*, conidia; B, *C. diplodioides*, conidia; C, *C. fasciculata*, conidia; D, *C. grisea*, conidia.

on *Cytospora pyri* Fckl and Petrak (1923) and von Arx (1957a, 1970) were of the same opinion. Weindlmayr (1965) however used *C. pyri* as the earliest available epithet for the type species of *Phacidiopycnis* Potebnia. Depending on which point of view is adopted either the synonymy for *Cryptosporiopsis corticola* is greatly enlarged, or, the names based on *Cytospora pyri* are referable to a species belonging in *Phacidiopycnis*.

Causing bark canker of *Malus* sp., *Pyrus* sp., U.K., Cyprus (IMI 22387, 57671, 62470, 206459).

C. DIPLODIOIDES (Allesch.) Petrak, *Annls mycol.* **21**: 272 (1923). (Fig. 308B)
Myxosporium diplodioides Allesch., *Ber. Bayer. Bot. Ges.* **4**: 37 (1896).
Pachydiscula diplodioides (Allesch.) Höhn., *Z.f. Garungsph.* **5**: 210 (1915).

Conidiomata black to dark brown, irregular in texture, up to 1 mm diam. Conidiogenous cells 12-16 × 3-4 µ. Conidia 22.5-26.5 × 7.5-9.5 µ.

Nannfeldt (1932) does not record any *Pezicula* state for this species

On twigs and bark of *Sorbus aria*, Austria (IMI 31036, 30427).

C. FASCICULATA (Tode ex Tul.) Petrak, *Annls mycol.* **21**: 186 (1923). (Fig. 308C)
Tubercularia fasciculata Tode, *Fungi Mecklenb. Sel.* 1: 20 (1790).
Tubercularia fasciculata Tode ex Tul., *Ann. Sci. nat.*, ser. 3, **20**: 144 (1853).

Teleomorphosis: *Pezicula carpinea* (Pers. ex Pers.) Tul. ex Fckl, *Jb. nassau. Ver. Naturk.* **23-24**: 279 (1870).

Conidiomata pulvinate, 1-3 mm diam., dark reddish brown. Conidiogenous cells up to 100 µ long × 2.5-3 µ wide (5-7 µ wide at the apices). Conidia 16.5-20.5 × 8-9 µ.

Fully described by Groves (1939), Seaver (1951) and Dennis (1974) who give a much larger range in conidial size.

On twigs of *Carpinus betulus, C. caroliniana, Fagus sylvatica, F.* sp., *Castanea sativa,* U.K., Sweden, Austria (IMI 11605, 32778, 12176, 16654, 23381, 73218, 167025, 29714, 214084, 205587, 188553, 188548).

C. GRISEA (Pers. ex Cda) Petrak, *Annls mycol.* **23**: 186 (1925). (Fig. 308D)
Naemaspora grisea Pers. ex Cda, *Icon. fung.* **3**: 26 (1839).
Naemaspora grisea Pers., *Syn. Meth. Fung.*: 110 (1801).
Myxosporium griseum (Pers. ex Cda) Sacc., *Syll. fung.* **3**: 726 (1884).
Discosporium griseum (Pers. ex Cda) Höhn., *Z.f. Garungsph.* **5**: 199 (1915).
Discosporiopsis grisea (Pers. ex Cda) Petrak, *Annls mycol.* **20**: 311 (1922).
Myxosporium coryleum (Sacc.) Died., *Krypt. Fl. Mk Brandenb.* **9**: 794 (1915).
Myxosporium incarnatum (Desm.) Bonord. subsp. *roumeguerei* Sacc. forma *coryleum* Sacc., *Annls mycol.* **7**: 436 (1909).

Teleomorphosis: *Pezicula coryli* (Tul.) Tul., *Sel. Fung. Carp.* **3**: 183 (1865).

Conidiomata pulvinate, conidial masses pale orange, 1-1.5 mm diam. Conidiogenous cells 15-20 × 2-3 µ. Conidia 21.5-28 × 9-10.5 µ.

This species can be distinguished from the *C. coryli* state of *P. corylina* by the narrower conidia and pulvinate rather than conical conidiomata.

On bark and twigs of *Corylus avellana*, U.K., W. Germany (IMI 22388, 27775, 33471, 50756, 39105, 22389, 22395, 22392, 22393, 22394, 22396, 22397, 29719, 29718, 29717, 63568, 23382).

C. MALICORTICIS (Cordl.) Nannfeldt, *Nova Acta Reg. Soc. Scient. Upsal.* ser. 4, **8**: 91 (1932). (Fig. 309A)
Gloeosporium malicorticis Cordl., *Bot. Gaz.* **30**: 57 (1900).
Gloeosporium perennans Zeller & Childs, *Oregon Agr. Coll. Exp. St. Bull.* 217: 14 (1925).
Cryptosporiopsis perennans (Zeller & Childs) Wollenw., *Arb. biol. Abt. (Anst.-Reichsanst.), Berl.* **22**: 549 (1939).

Teleomorphosis: *Pezicula malicorticis* (Jacks.) Nannfeldt, *Nova Acta Reg. Soc. Scient. Upsal.* ser. 4, **8**: 91 (1932).

Conidiomata pulvinate, yellowish, gelatinous, large and often confluent, 1 mm diam. Conidiophores 10-15 × 3 µ. Conidia 11.5-16 × 3-4 µ. Associated microconidia 5-6 × 1-1.5 µ.

R

The anamorphosis is nowhere near typical of the genus or the species accepted in it. For many years the teleomorphosis has been known as *Neofabraea malicorticis* Jacks. and it is possible that differences in the anamorphosis will confirm the disposition of the teleomorphosis in *Neofabraea* rather than *Pezicula*.

Causing perennial canker and fruit rot of *Malus* spp., *Pyrus* spp., U.K., Eire, New Zealand, W. Australia, Tasmania, Australia, B.C., Canada, represented by 33 collections in herb. IMI.

C. PRUINOSA (Pk) Wollenw., *Arb. Biol. Abt. (Anst.-Reichsanst.) Berl.* **22**: 527 (1939). (Fig. 310)
Sphaeronaema pruinosum Pk, *Ann. Rep. N.Y. St. Mus.* 24: 85 (1872).
Lagynodella pruinosa (Pk) Petrak, *Annls mycol.* **20**: 207 (1922).

Teleomorphosis: *Pezicula pruinosa* Farlow, *Mycologia* **14**: 102 (1922).

Conidiomata cylindrical to conical (stromatic), pruinose, .5-2.0 mm high, up to 500 μ diam. at the base, tapering to 250 μ diam. at the apex. Conidiogenous cells 17-22 × 4-4.5 μ. Conidia 24-26 × 9-10 μ.

Fully described by Groves (1939) and Seaver (1951). A distinctive species on account of the conical conidiomata.

On twigs of *Amelanchier alnifolia, A.* sp., N. Dak., N.Y., Idaho, U.S.A. (IMI 197178, holotype *S. pruinosum* ex NYS, 22729, 29725, 22730).

C. QUERCINA Petrak, *Annls mycol.* **22**: 159 (1924). (Fig. 309B)

Teleomorphosis: *Pezicula cinnamomea* (DC. ex Pers.) Sacc., *Michelia* **2**: 332 (1881).

Conidiomata up to 1 mm diam., remaining mostly depressed. Conidiogenous cells up to 30 μ long × 3-4 μ wide. Conidia 18.5-27 × 9.5-11.5 μ.

Dennis (1974) has provided a full account of the *Pezicula* state but gave no details of the *Cryptosporiopsis*.

On twigs of *Quercus rubra, Q. robur, Q.* sp., *Ulmus* sp., W. Germany, U.K., Ont., Canada (IMI 30419, 30374, 43785, 32279, 23380, 102904).

C. SCUTELLATA (Otth) Petrak, *Hedwigia* **62**: 315 (1921). (Fig. 309C)
Sphaeropsis scutellata Otth, *Mittheil. Bern.* 60 (1886).
Macrophoma scutellata (Otth) Sacc., *Syll. fung.* **11**: 496 (1895).
Cryptosporiopsis nigra Bubák & Kabát, *Hedwigia* **62**: 361 (1921).

Teleomorphosis: *Pezicula ocellata* (Pers. ex Fr.) Seaver, *The North American Cup-Fungi (Inoperculates)*: 345 (1951).

Conidiomata subperidermal, acervular, not distinctly erumpent, up to .5 mm diam. Conidiogenous cells 20-30 × 3-4 μ. Conidia 20.5-25 × 7-8 μ.

Described by Nannfeldt (1932) and later by Seaver (1951) who provided a brief account of the conidial state.

On twigs and branches of *Salix grandifolia, S. pentandra, S. incana, S. repens, S.* sp., *Populus tremuloides, P. deltoides, P. tremula, P.* sp., Austria, Latvia, Ont., Canada, Colo., U.S.A. (IMI 26994, 29757, 50834, 12083, 12010, 73203, 25984, 73223).

C. TILIAE Petrak, *Annls mycol.* **42**: 54 (1944). (Fig. 309D)

Conidiomata often aggregated, pulvinate to conical, 1-2 mm diam. Conidiogenous cells 20-35 × 4-6 μ. Conidia 26.5-29 × 9-10 μ.

No teleomorphosis has been described for this species and apart from the original description there has been no subsequent account.

On bark of *Tilia* sp., Austria (IMI 30033).

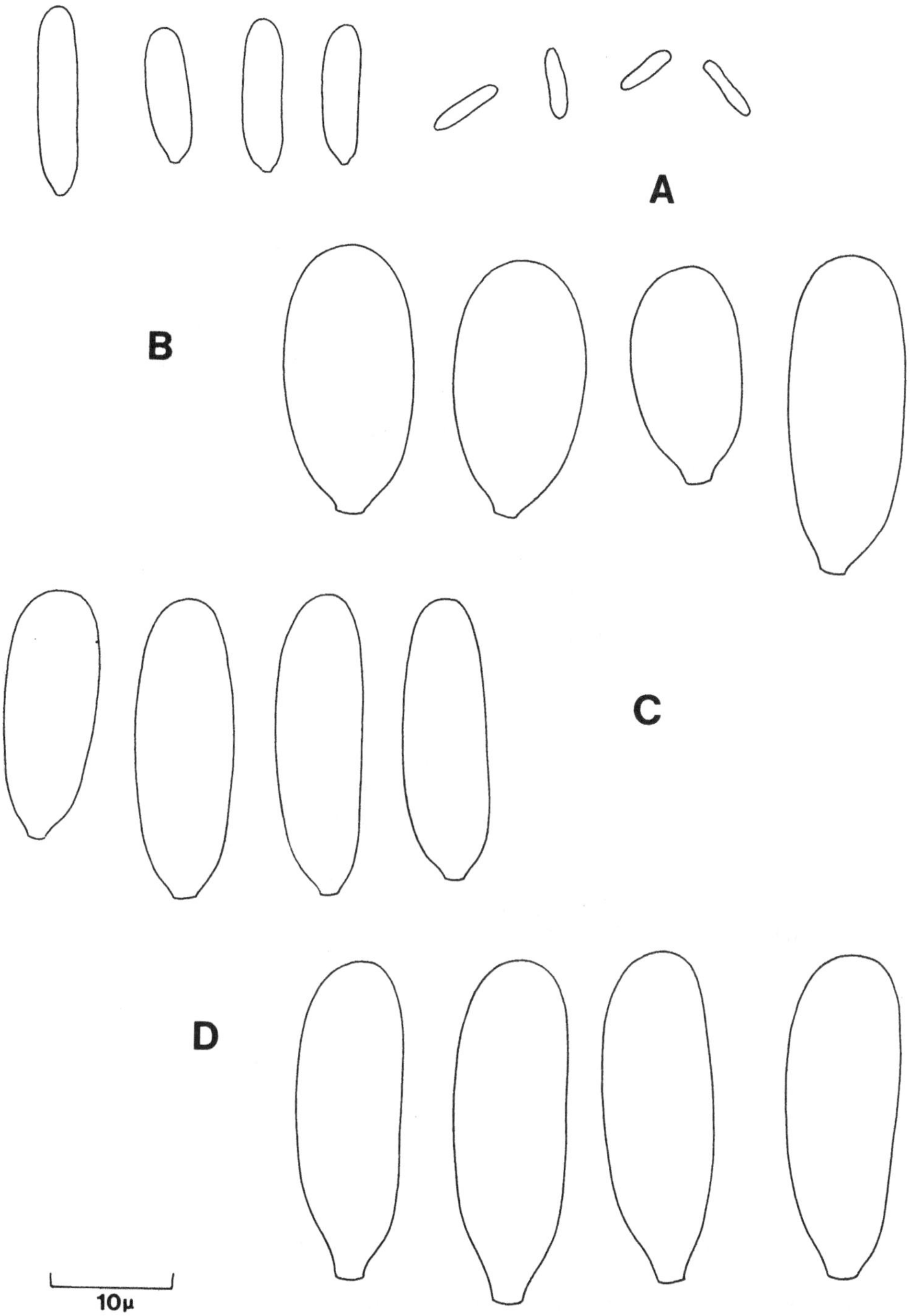

Fig. 309. A, *Cryptosporiopsis malicorticis*, macroconidia and microconidia; B, *C. quercina*, conidia; C, *C. scutellata*, conidia; D, *C. tiliae*, conidia.

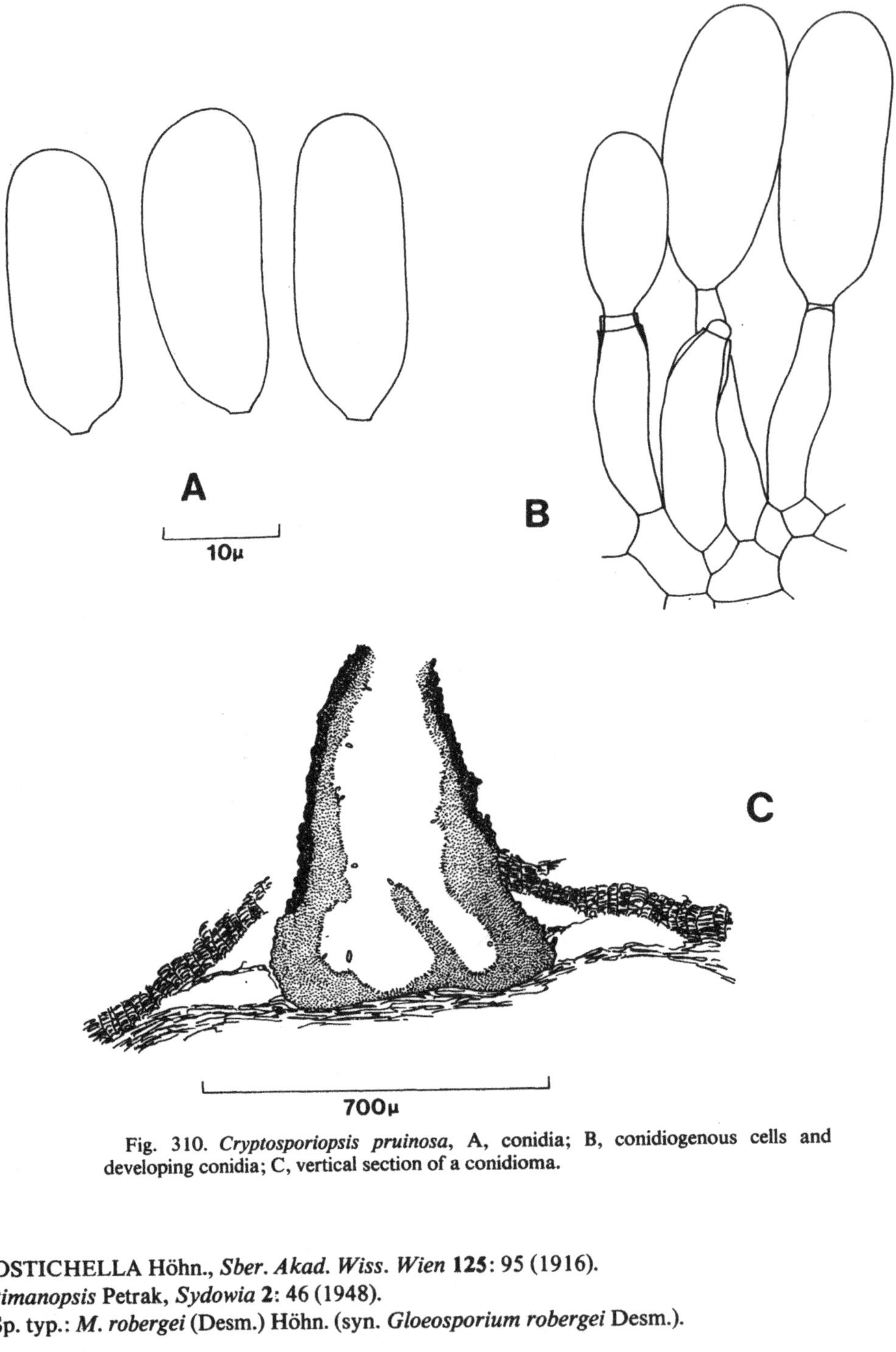

Fig. 310. *Cryptosporiopsis pruinosa*, A, conidia; B, conidiogenous cells and developing conidia; C, vertical section of a conidioma.

46. MONOSTICHELLA Höhn., *Sber. Akad. Wiss. Wien* **125**: 95 (1916).
Antimanopsis Petrak, *Sydowia* **2**: 46 (1948).
Sp. typ.: *M. robergei* (Desm.) Höhn. (syn. *Gloeosporium robergei* Desm.).

Mycelium immersed, branched, septate, hyaline to pale brown. *Conidiomata* acervular, subcuticular, eventually sometimes epidermal, discolouring the cuticle, separate or confluent, usually circular, formed of pale brown to hyaline, thin-walled textura angularis. Dehiscence irregular. *Conidiophores* absent or just occasionally there is a septum near the base. *Conidiogenous cells* enteroblastic, phialidic, determinate, discrete, very rarely integrated, doliiform to cylindrical, hyaline, smooth, with a wide or narrow channel, minute collarette and either prominent or minimal periclinal thickening, formed from the upper cells of the conidiomata. *Conidia* hyaline, aseptate, thin-walled, smooth, guttulate or eguttulate, truncate at the base but scarcely protruding, apex obtuse, ellipsoid, pyriform, sometimes slightly curved.

Arx (1957a, 1970) accepted 10 species in the genus but it is evident from conidial shape alone that his concept was heterogeneous. Morgan-Jones (1971b) illustrated the type species and described conidiogenesis as phialidic, but interpreted development in *M. falcata* (Dearn. & House) Arx as annellidic (Morgan-Jones, 1971a). *M. coryli* (Desm.) Höhn. also has distinct annellides and in addition the conidia are pale brown and thicker-walled. Consequently the species is referred to *Piggotia* Berk. & Br.

M. ROBERGEI (Desm.) Höhn., *ibid.*: 95 (1916). (Fig. 311E)
Gloeosporium robergei Desm., *Ann. Sci. nat.*, ser. 3, **20**: 214 (1853).

Conidiomata up to 150 µ diam. Conidiogenous cells 6-11 × 4-6.5 µ. Conidia straight, eguttulate, 16-18 × 7.5-9 µ.

On leaves of *Carpinus betulus*, Austria, Italy, Germany (IMI 21788, 21789, 13033).

M. SALICIS (Westd.) Arx, *Verh. K. Ned. Akad. Wet.* **51**(3): 131 (1957). (Fig. 311D)
Gloeosporium salicis Westd., *Herb. Crypt. Belg.* exs. 1269.
Gloeosporidiella salicis (Westd.) Nannf., *Svensk bot. Tidskr.* **25**: 20 (1931).
Gloeosporium aterrimum Fckl, *Jb. nassau. Ver. Naturk.* **23-24**: 277 (1870).
Gloeosporium mesopotamicum Bub., *Ann. naturhist. Hofmus. Wien* **28**: 217 (1914).

Conidiomata up to 200 µ diam. Conidiogenous cells 4.5-17 × 4.5-6 µ. Conidia often slightly curved, guttulate, 14-17 × 7.5-8.5 µ.

On leaves of *Salix triandra, S. fragilis, S. alba, S. amygdalina, S. viminalis, S. alba × fragilis, S.* spp., Iraq, Roumania, Hungary, Austria, Czechoslovakia, Germany, Italy, Ont., Canada, Wisc., U.S.A. (IMI 21791, 21792, 21794, 21796, 21795, 21793, 50775, 43774, 159754, 94244, 11197, 37204, 30543, 50081, 50082, 224577).

M. indica sp. nov. (Fig. 311A, B, C)
Mycelium immersum, ramosum, septatum, hyalinum vel pallide brunneum. *Conidiomata* acervularia, subcuticularia, separata, circularia, usque ad 150 µ diam., ex textura angulari pallide brunnea vel hyalina. Dehiscentia irregularis. *Conidiophora* sparsa, basim versus 1-septata, hyalina, laevia. *Cellulae conidiogenae* enteroblasticae, phialidicae, determinatae, semper discretae, raro in conidiophoris integratae, cylindricae, apicem versus attenuatae, hyalinae, laeves, 7-20 × 2-3.5µ, ex cellulis superioribus conidiomatum formatae. *Conidia* hyalina, laevia, aseptata, guttulata, recta vel leviter curvata, ad apicem obtusa, ad basim truncata, 9-11.5 × 4.5-5.5 µ.

In foliis vivis *Ficus bengalense*, India, R.B. Somani VII, 30 Oct. 1975, IMI 198385, holotypus.

Although this species has longer conidiogenous cells it resembles *M. salicis* quite closely but has much smaller conidia though they are of comparable shape. It is associated with distinct leaf lesions 1-2 cm diam., and as in other species of the genus, discolours the cuticle.

47. GLOEOSPORIDINA Petrak, *Annls mycol.* **19**: 214 (1921).
Sp. typ.: *G. moravica* Petrak

Mycelium immersed, branched, septate, hyaline. *Conidiomata* acervular, subepidermal to epidermal, separate or confluent, formed of hyaline, thin-walled textura angularis. Dehiscence irregular. *Conidiophores* hyaline to pale brown, branched sparingly, septate, smooth, formed from the upper pseudoparenchyma. *Conidiogenous cells* enteroblastic, phialidic, determinate, integrated or discrete, cylindrical, hyaline to pale brown, smooth, channel narrow or wide, periclinal wall thickened or not, collarette distinct. *Conidia* hyaline, aseptate, base ± truncate, apex obtuse, thin-walled, smooth, eguttulate, pyriform.

An illustrated account of the genus was given by Sutton & Pollack (1973).

G. cecidii (Kohlmeyer) comb. nov. (Fig. 312C)
Sphaceloma cecidii Kohlmeyer, *J. Elisha Mitchell scient. Soc.* **88**(4): 255 (1972).

Conidiomata 200-700 µ diam. Conidiogenous cells 20-30 × 1.5-2.5 µ, with a relatively long phialidic channel. Conidia 3-4.5 × 1.8-2.5 µ excluding the basal frill, at the base with a frill 2.5 µ long.

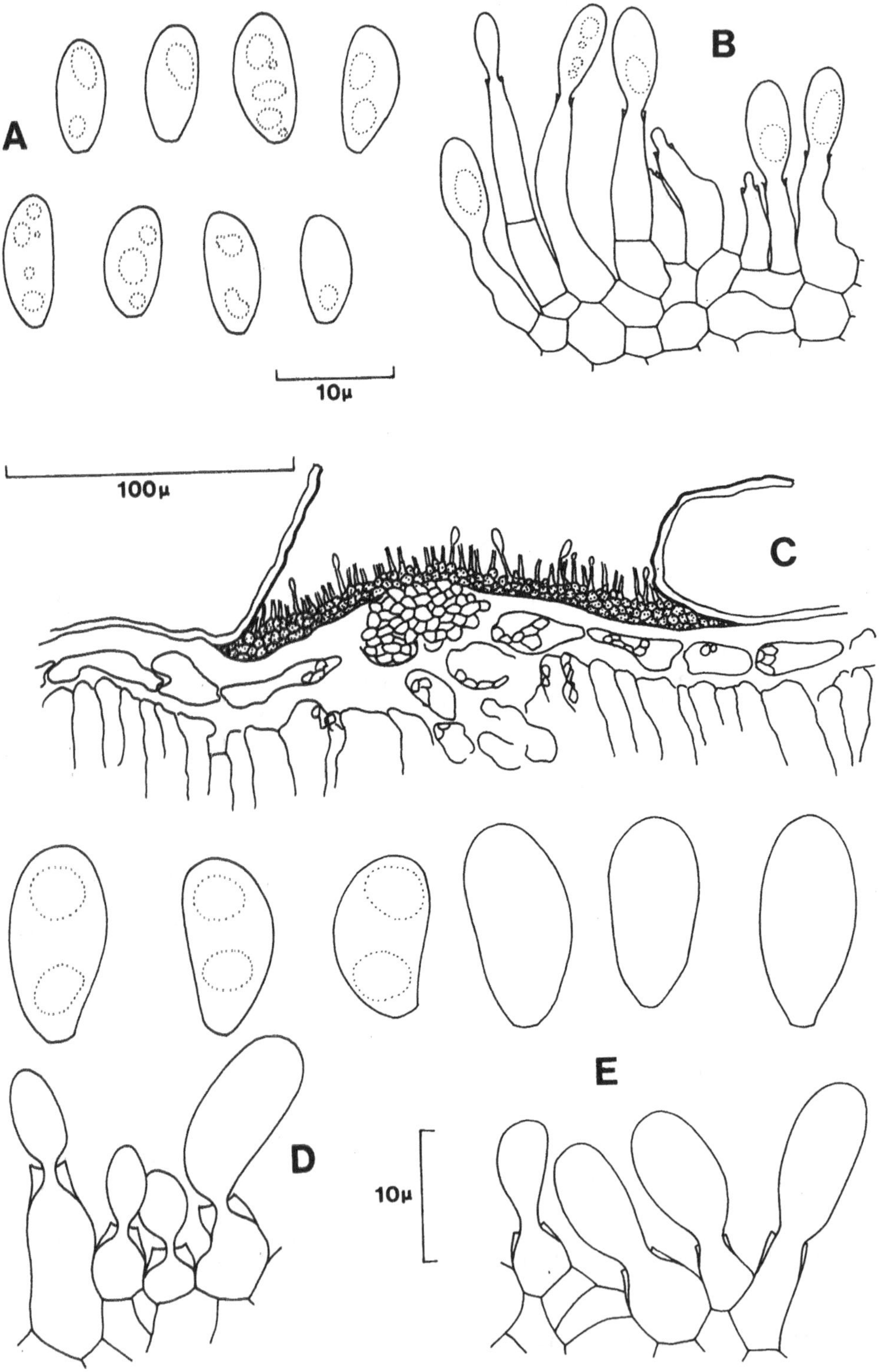

Fig. 311. A, B, C, *Monostichella indica*, conidia, conidiophores, vertical section of a conidioma; D, *M. salicis*, conidia and conidiogenous cells; E, *M. robergei*, conidia and conidiogenous cells.

The monophialidic, hyaline, cylindrical conidiogenous cells exclude this fungus from *Sphaceloma* de Bary. Although the basal frill is present, in other respects the species is congeneric with *G. moravica*. The frill could be indicative of percurrently proliferating phialides but no successive collarettes were found. Instead of the point of conidial delimitation being within the collarette, as in *G. moravica* and the other species, in *G. cecidii* it occurs above the collarette.

On galls of *Haloguignardia oceanica* in *Sargassum natans* (IMI 178401 ex Herb. J. Kohlmeyer).

G. CERCOCARPI (Ell. & Ev.) Sutton & Pollack, *Mycologia* **65**: 1133 (1973). (Fig. 312B)

Conidiomata confluent, coalescing. Conidiogenous cells 7-18 × 1.5-2 μ, with a narrow phialidic channel. Conidia 2.5-5.5 × 1.5-2.5 μ.

Causing lesions on leaves of *Cercocarpus betuloides*, Calif., U.S.A. (IMI 107244 ex K, isotype).

G. MORAVICA Petrak, *Annls mycol.* **19**: 214 (1921). (Fig. 312A)

Conidiomata up to 210 μ diam. Conidiogenous cells 6.5-13.5 × 2 μ, with a wide phialidic channel. Conidia 2.5-3.5 × 2 μ, often produced in chains.

On leaves of *Quercus robur*, Austria (IMI 154473 ex K).

G. canthiicola sp. nov. (Fig. 312D, E)

Lesiones distinctae, usque ad .8 cm diam., circulares sed margine irregulari diffuso pallide aurantio brunneo et centro leniter atro brunniore. Conidiomata acervularia, circularia, epiphylla, sparsa, usque ad 400 μ diam., ex textura angulari pallide brunnea vel hyalina parietibus tenuibus. Conidiophora hyalina, raro septata et ramosa. Cellulae conidiogenae enteroblasticae, phialidicae, cylindricae, laeves, discretae vel in conidiophoris incorporatae, 10-12 × 2.5-3.5 μ. Conidia hyalina, aseptata, pyriformia, laevia, parietibus tenuibus, eguttulata, 6.5-8 × 3-4 μ.

In foliis vivis *Canthii* ? *venosi*, Kamaila Forest Reserve, Chisamba, Zambia, A. Angus M2496, 12 Sept. 1963, IMI 115397, holotypus.

Lesions distinct, up to .8 cm diam., more or less circular but with an irregular, diffuse, pale orange margin and a slightly darker brown centre. Conidiomata acervular, circular, epiphyllous, sparse, scattered, up to 400 μ diam., epidermal, composed of pale brown to hyaline, thin-walled textura angularis, the upper walls of the ruptured epidermal cells retaining a little fungal tissue. Dehiscence irregular. Conidiophores hyaline, rarely septate or branched. Conidiogenous cells enteroblastic, phialidic, cylindrical, smooth, collarette and periclinal thickening distinct, 10-12 × 2.5-3.5 μ. Conidia hyaline, aseptate, pyriform, base often truncate, apex obtuse, smooth, thin-walled, eguttulate, 6.5-8 × 3-4 μ.

This species is readily distinguished from those already described by the larger conidia.

48. CRYPTOCLINE Petrak apud H. Syd. & Petrak, *Annls mycol.* **22**: 402 (1924).
Gloeotrochila Petrak, *Sydowia* **1**: 50 (1947).
Original spp.: *C. effusa* Petrak, *C. abietina* Petrak
Lectotype sp.: *C. effusa* Petrak, fide Clements & Shear (1931).
Illustrations: Morgan-Jones (1973).

Mycelium immersed, branched, septate, hyaline. *Conidiomata* acervular, epidermal to subepidermal, separate or confluent, formed of pale brown to hyaline, thin-walled textura angularis. Dehiscence irregular. *Conidiophores* absent. *Conidiogenous cells* enteroblastic, phialidic, discrete, indeterminate, cylindrical, hyaline, smooth, with 1-2 percurrent proliferations. *Conidia* hyaline or very pale brown, thin-walled, eguttulate or guttulate, smooth, cylindrical to doliiform or ellipsoid, with a broad, flat base.

Despite the accounts by Morgan-Jones (1973), and von Arx (1957a), *Cryptocline* remains a very heterogeneous entity. The only common feature shared by recently illustrated species is the epidermal to subepidermal location of the conidiomata. As the genus is presently conceived conidiogenous cells may be discrete or integrated on branched septate conidiophores; conidiogenesis may be typically phialidic, percurrently phialidic or annellidic; conidia may be cylindrical, fusiform or doliiform and have either a small area of abscission or a relatively wide basal scar. Separation of the taxa on the basis of conidial morphology and nature of the conidiogenous process would fragment the genus into several distinct groups.

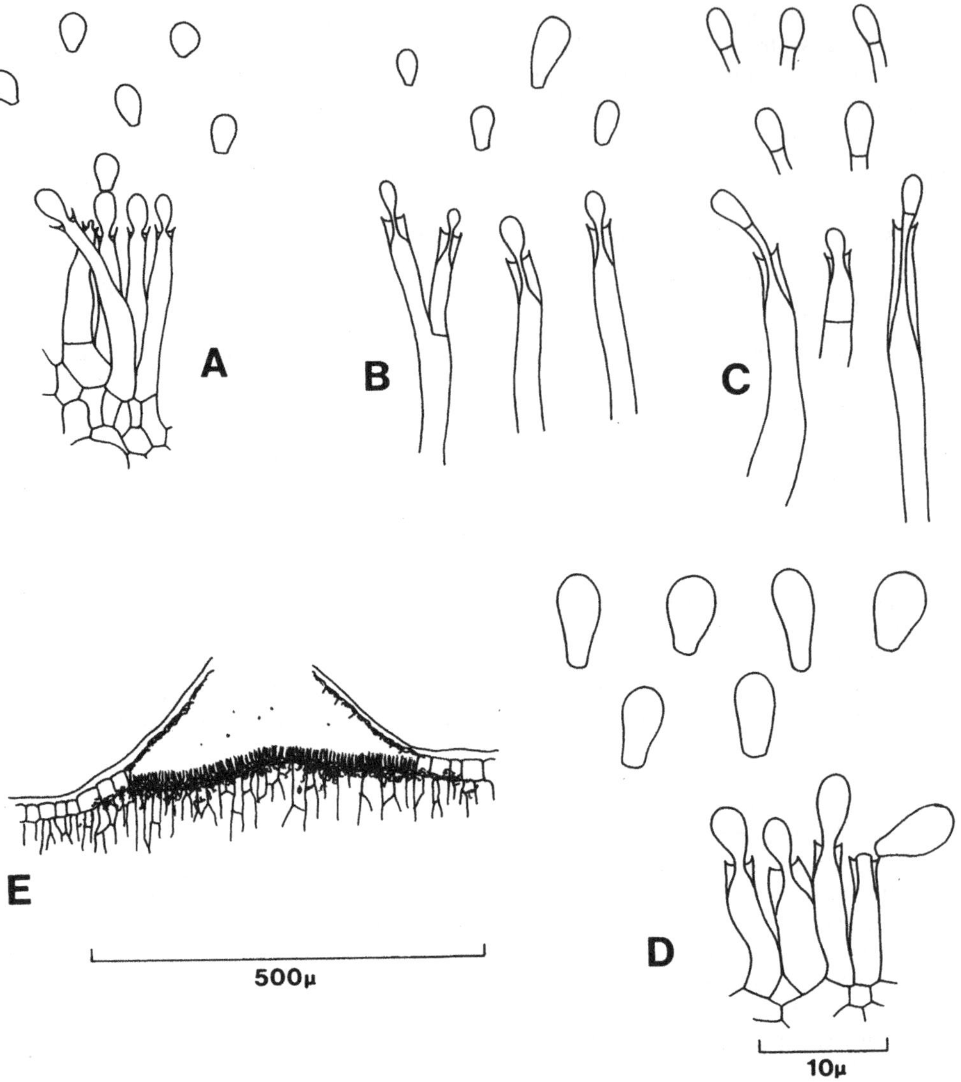

Fig. 312. A, *Gloeosporidina moravica*, conidia and conidiogenous cells; B, *G. cercocarpi*, conidia and conidiophores; C, *G. cecidii*, conidia and conidiophores; D, E, *G. canthiicola*, conidia, conidiogenous cells, vertical section of a conidioma.

Congeneric with the type species, *C. effusa*, are *C. paradoxa* (de Not.) von Arx, *C. taxicola* (Allesch.) Petrak and *C. betularum* (Ell. & Mart.) von Arx. *C. dubia* (Bauml.) von Arx is probably referable to *Cryptosporiopsis* Bub. & Kabát, whilst *C. conigena* (Sacc. & Roum.) von Arx, *C. nigricans* (Cke & Mass.) Sutton and *C. tremuloides* (Ell. & Ev.) Sutton should be referred to three separate distinct genera. However, until the limits of genera in this area have stabilized and become rationalized with modern taxonomic concepts, it would be premature to propose new genera at the present time. The remaining species, *C. abietina* Petrak, *C. ceanothi* (H.C. Greene) von Arx, *C. cinerescens* (Bub.) von Arx, *C. cyclaminis* (Sibilia) von Arx, *C. hypophyllum* (D.E. Ell. & Gill) Morgan-Jones & Nag Raj, and *C. phacidiella* (Grove) von Arx have more similarity with *Discogloeum* Petrak than *Cryptocline*, although in *D. veronicae* the conidia tend to be slightly more irregular than in these species assigned to *Cryptocline*.

C. PARADOXA (de Not.) von Arx, *Verh. K. Ned. Akad. Wet.*, sect. 2, **51**: 115 (1957). (Fig. 313)
synonymy as in Morgan-Jones (1973).

Conidiomata 150-180 μ diam. Conidiogenous cells 13-23 × 3-4.5 μ. Conidia 7.5-9.5 × 4-5 μ.

On leaves of *Hedera helix*, Germany, U.K., Eire (IMI 33164, 105570, 120679, 122864).

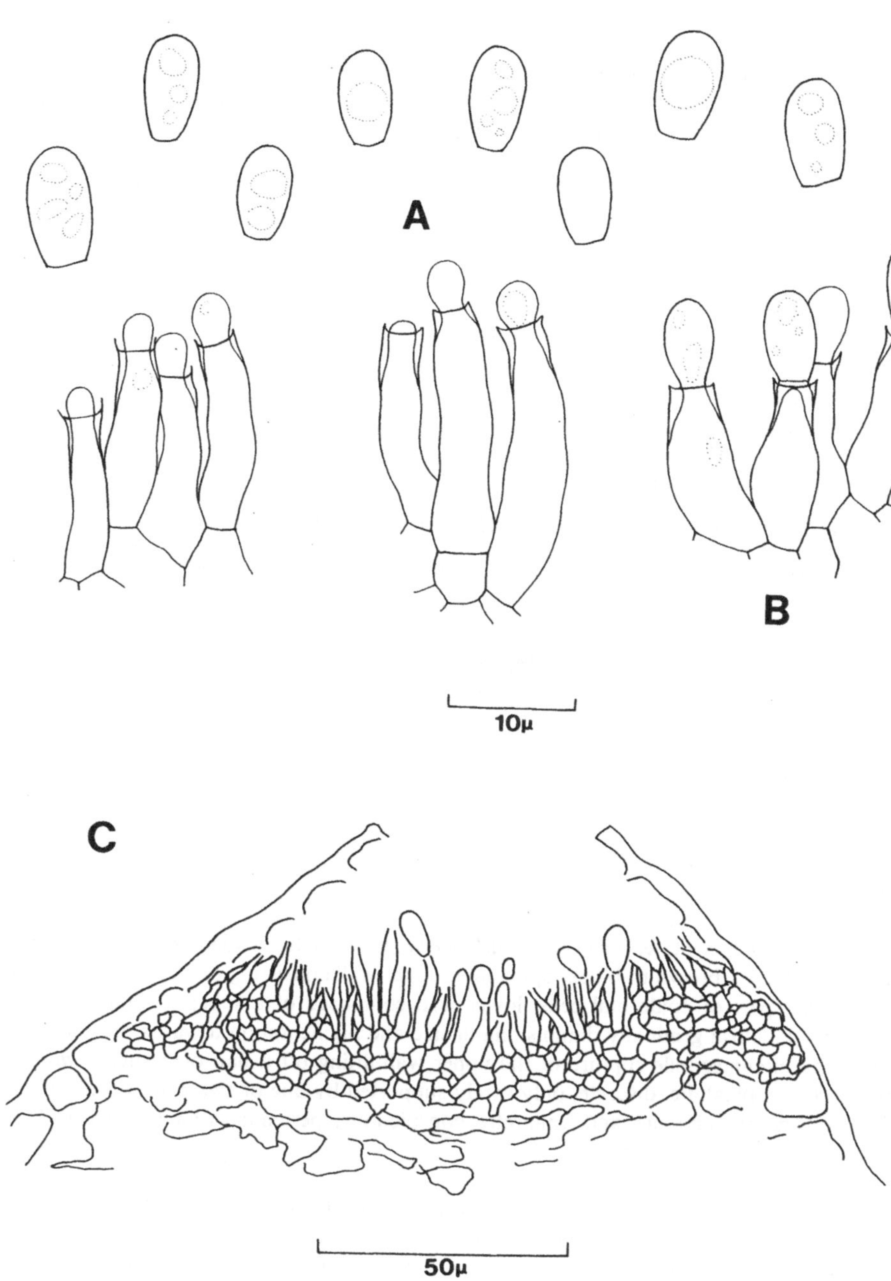

Fig. 313. *Cryptocline paradoxa*, A, conidia; B, conidiogenous cells and developing conidia; C, vertical section of a conidioma.

49. SIREXCIPULA Bub., *Hedwigia* **46**: 296 (1907).
Sirodomus Petrak, *Sydowia* **1**: 157 (1947).
Sp. typ.: *S. kabatiana* Bub.

Mycelium immersed, branched, septate, pale brown to hyaline. *Conidiomata* eustromatic, immersed, globose to collabent or finally shallow cupulate, unilocular, pale brown, appearing pruinose, thick-walled; walls of even thickness, composed of medium brown, large-celled, thin-walled textura angularis. *Ostiole* absent, dehiscence by irregular breakdown of the upper wall, finally widely open. *Conidiophores* branched and septate mainly at the base, hyaline, cylindrical, straight, formed from the inner cells of the conidiomatal wall. *Conidiogenous cells* enteroblastic, phialidic, integrated or discrete, determinate, hyaline, cylindrical, smooth, channel comparatively wide, collarette minute, periclinal thickening almost absent. *Conidia* catenate but formed in acropetal succession, cylindrical, straight, hyaline to subhyaline, thin-walled, occasionally guttulate, apex obtuse, base truncate, smooth.

Sutton (1977a) has explained the generic synonymy. Until the addition of *S. montana*, the genus was monotypic. Previous authors (Bubák & Kabát, 1907; Petrak, 1947c) had regarded conidiogenesis as thallic, but the conidia, though formed in chains, actually originate from a phialidic aperture.

KEY

Conidia 6 × 2.5 µ .... *S. kabatiana*
Conidia 7-8.5 × 3.5 µ .... *S. montana*

S. KABATIANA Bub., *ibid.*: 295 (1907). (Fig. 314B, C)

Conidiomata up to 300 µ diam. Conidiophores 13-22 × 2 µ. Conidia 6 × 2.5 µ.

On dead leaves of *Funkia sieboldii*, Austria, Czechoslovakia (IMI 194246, holotype ex BPI, 194247 ex BPI).

S. montana (Höhn.) comb. nov. (Fig. 314A)
Peckia montana Höhn., *Sber. Akad. Wiss. Wien* **123**: 127 (1914).
Sirodomus montanus (Höhn.) Petrak, *Sydowia* **1**: 157 (1947).

Conidiomata up to 250 µ diam. Conidiophores 13-20 × 2.5-3.5 µ. Conidia 7-8.5 × 3.5 µ.

On dead leaves of *Veratrum* sp., Austria (IMI 203247, holotype ex FH).

50. PSEUDOSTEGIA Bub., *J. Mycol.* **12**: 56 (1906).
Sp. typ.: *P. nubilosa* Bub.

Mycelium immersed, branched, septate, brown. *Conidiomata* eustromatic, subepidermal, ± circular, disciform, separate, occasionally confluent, dark brown to black, applanate; upper wall consisting of small, dark brown, thick-walled cells concentrated in the upper part of epidermal cells, lower wall composed of a subepidermal basal layer of brown, relatively large-celled textura globulosa gradually diminishing in size and pigmentation towards the conidiogenous region. *Ostiole* absent, dehiscence by circumscissile rupture of the upper wall lifting off a flap of tissue. *Conidiophores* absent. *Conidiogenous cells* enteroblastic, phialidic, discrete, only rarely supported by one or two cells, determinate, ampulliform to cylindrical, pale brown, smooth, channel and collarette conspicuous, periclinal wall thickened. *Conidia* hyaline, aseptate, falcate, thin-walled, smooth, eguttulate.

Bubák (1906b) compared *Pseudostegia* with *Anaphysmene* Bub. but the latter does not have the specialized method of dehiscence as in *Pseudostegia* and furthermore the conidiogenous cells are holoblastic and annellidic. In the original account, Bubák (1906b) described peripheral setae. These were figured as of similar size to conidia. The type was not found to be setose so it is possible that attached conidia were interpreted as setae.

Pseudostegia closely resembles *Colletotrichum* Cda except for the specialized method of dehiscence and it is on the latter basis that the two genera are kept apart.

Bubák (1906b) suggested that *Cryptosporium nubilosum* Ell. & Ev. was the same species although there was discrepancy in conidial width. If this was proved to be the case, the name *P. nubilosa* is already preoccupied so the *P. nubilosa* (Ell. & Ev.) Bub. listed by Cash (1952) is a later homonym.

P. NUBILOSA Bub., *ibid.*: 56 (1906). (Fig. 315)
Macmillanina nubilosa (Ell. & Ev.) Kze, *Rev. Gen. Pl.* **3**(2): 491 (1898).
Cryptosporium nubilosum Ell. & Ev., *J. Mycol.* **5**: 166 (1889).

Conidiomata up to 300 µ diam. Conidiogenous cells 5-8 × 3.5-5 µ. Conidia 18-22 × 3-3.5 µ.

On dead leaves of *Carex* sp., Ky, U.S.A. (IMI 194255, holotype of *P. nubilosa* ex BPI).

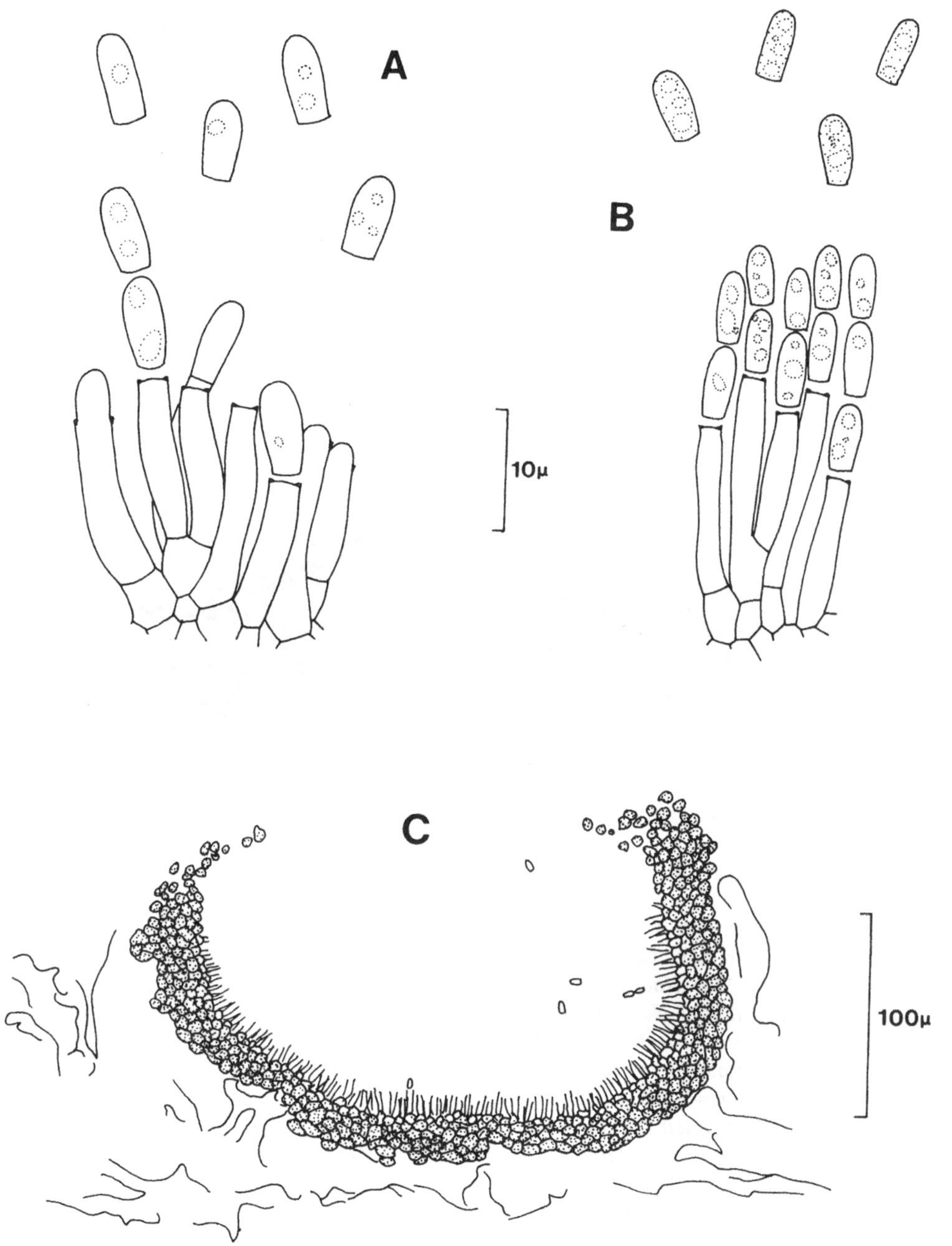

Fig. 314. A, *Sirexcipula montana*, conidia and conidiophores; B, C, *S. kabatiana*, conidia, conidiophores, vertical section of a conidioma.

51. COLLETOTRICHUM Cda in Sturm, *Deutschlands Flora* **3**: 41 (1831-1832).
Vermicularia Tode, *Fungi Mecklenb. Sel.* **1**: 31 (1790).
Vermicularia Tode ex Fr., *Syst. Orb. Veg.*: 111 (1825).
Dicladium Ces., *Flora* **35**: 398 (1852).
Steirochaete Braun & Casp., *Krankh. Pfl.*: 28 (1854).
Fellneria Fckl, *Fungi rhenani* 1923 (1869).
Ellisiella Sacc., *Michelia* **2**: 147 (1880).
? *Coniothyriella* Speg., *Bol. Acad. Nac. Cienc. Cordoba* **11**: 605 (1889).

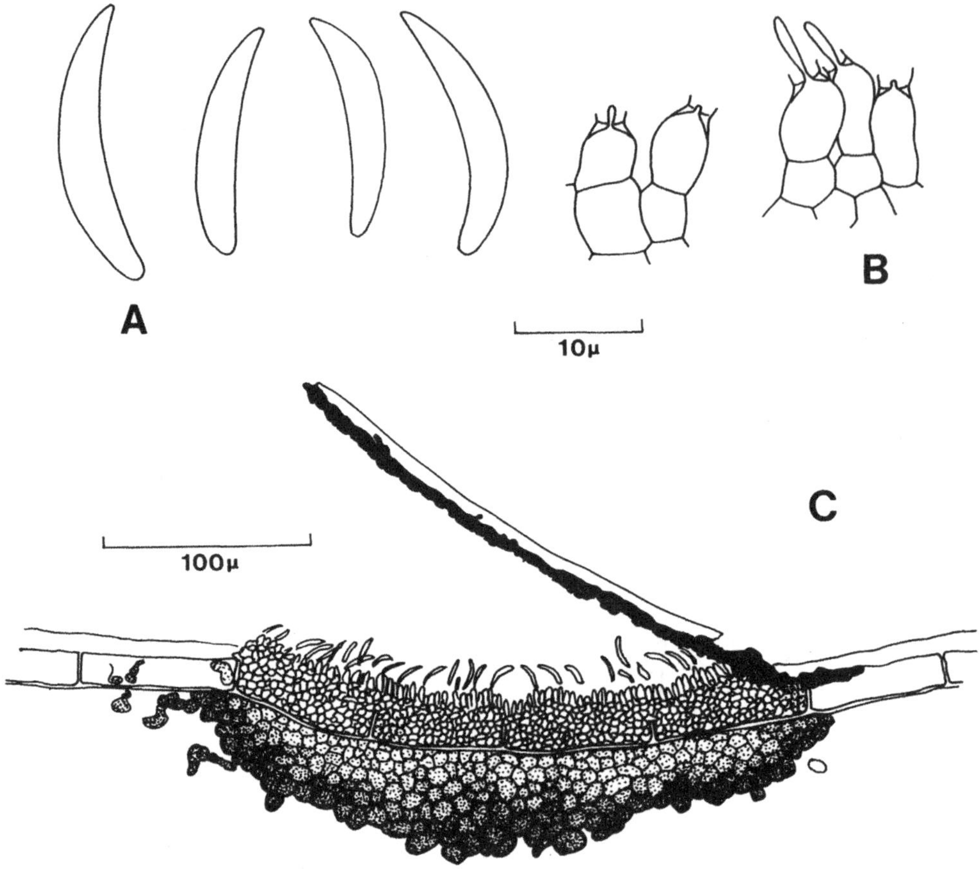

Fig. 315. *Pseudostegia nubilosa*, A, conidia; B, conidiogenous cells, C, vertical section of a conidioma.

Colletotrichopsis Bub., *Öst. Bot. Z.* **54**: 184 (1904).
Coniothyris Clements, nom. nov., *Genera of Fungi*: 133 (1909).
Didymariopsis Speg., *An. Mus. Nac. Buenos Aires* **20**: 424 (1910).
Gloeosporiopsis Speg., *An. Mus. Nac. Buenos Aires* **20**: 404 (1910).
Fominia Girzitska, *Bull. Jard. bot. Kiev* **5-6**: 168 (1927).
Lophodiscella Tehon, *Mycologia* **25**: 253 (1933).
? *Blennorella* Kirschst., *Hedwigia* **81**: 201 (1944).
Ellisiellina Camara, *Agron. Lusit.* **11**: 72 (1949).
Rostrospora Subram. & Ramakr., *J. Madras Univ.* B, **22**: 67 (1952).
Colletostroma Petrak, *Sydowia* **7**: 346 (1953).
Schizotrichella Morris, *Mycologia* **48**: 733 (1956).
Peresia Maia, *Publ. Inst. Mic. Univ. Recife* 267: 28 (1960).

Sp. typ.: *C. dematium* (Pers. ex Fr.) Grove (syn. *Sphaeria dematia* Pers. ex Fr., *C. lineola* Cda).

Mycelium immersed, branched, septate, hyaline, pale brown or dark brown. *Conidiomata* acervular, subcuticular, epidermal, subepidermal or peridermal, separate or confluent, composed of hyaline to dark brown, thin- or thick-walled textura angularis; dehiscence irregular. *Sclerotia* sometimes present in culture, dark brown to black, often confluent, occasionally setose. *Setae* in conidiomata or sclerotia, brown, smooth, septate, tapered to the apices. *Conidiophores* hyaline to brown, septate, branched only at the base, smooth, formed from the upper cells of the conidiomata. *Conidiogenous cells* enteroblastic, phialidic, hyaline, smooth, determinate, cylindrical, integrated or discrete, channel minute but occasionally collarette and periclinal thickening quite prominent. *Conidia* hyaline, aseptate (except prior to germination), straight or falcate, smooth, thin-walled, sometimes guttulate, muticate or with the apex prolonged into a simple cellular appendage. *Appressoria* brown, entire or with crenate to irregular margins, simple or repeatedly germinating to produce complex columns of several closely connected appressoria.

The taxonomy of *Colletotrichum* has been largely concerned with classical descriptive criteria such as conidial shape and size, and presence, absence and morphology of setae. Supplementary criteria have included type of damage to the host caused by the species, and the identity of the host itself. Several hundred taxa have been described in *Colletotrichum* on this basis. Arx (1957b) reduced these to eleven accepted species, in which nine forms were recognized within *Glomerella cingulata* (Stonem.) Spauld. & Schrenk (*C. gloeosporioides* (Penz.) Sacc.), and three within *C. dematium* (Pers. ex Fr.) Grove. Exceptionally large synonymies were suggested for *C. gloeosporioides*, *C. graminicola* (Ces.) Wils. and *C. dematium*, and some names were confirmed as such by examination of type specimens. As these synonymies are investigated and species concepts reappraised, the number of clearly-defined acceptable taxa is gradually increasing (Sutton, 1962; Simmonds, 1965; Nag Raj, 1973a).

However there are still many common taxa already described that require investigation in vitro to determine the nature of setae, sclerotia, appressoria and cultural behaviour. In addition it is necessary to relate morphology in culture to the slight morphological differences in vivo, particularly in conidia, that have been noted by original authors to determine if such differences are consistent. Quite apart from such basic mycological investigations it is also necessary to determine host ranges by cross inoculation so that ultimately a system for the classification and identification of *Colletotrichum* species can be evolved similar to that now known for *Fusarium* Lk ex Fr. (Booth, 1971).

In the following descriptions, cultural characters have been taken from PDA cultures grown on alternating 12 hrs near UV/12 hr dark at 25°C, and mycelial rather than conidial appressorial characters from slide cultures on PCA under the same conditions.

KEY

1. Conidia straight 2
1. Conidia falcate 8
    2. Conidia fusiform (attenuated at each end) 3
    2. Conidia cylindrical 4
3. Colonies lacking sclerotia; medium with pink pigmentation; wide host range .... *C. acutatum*
3. Colonies with abundant setose sclerotia; conidia medianly constricted; frequently on Solanaceae .... *C. coccodes*
3. Conidia sometimes slightly curved, 15-19 × 4 µ; on Brassicae .... *C. higginsianum*
3. Conidia sometimes slightly curved, 14-17 × 3.5-4 µ; mycelial chlamydosphores present; appressoria absent; on Scrophulariaceae .... *C. fuscum*
    4. Conidia rarely more than 12 µ long; appressoria small; spherical sclerotia present; on Leguminosae .... *C. lindemuthianum*
    4. Conidia more than 12 µ long 5
5. Conidia 4.5-6 µ wide 6
5. Conidia usually not more than 4.5 µ wide 7
    6. Conidia 10-15 × 4.5-6.5 µ; setae present; appressoria typically crenate; often on Orchidae .... *C. crassipes*
    6. Conidia 12-17 × 4.5-5.5 µ; sclerotia and setae absent; appressoria irregular but with large or deep lobes; on *Musa* .... *C. musae*
    6. Conidia 14-15 × 4.5-6 µ; sclerotia and setae present; appressoria long clavate, entire; on Cucurbitaceae .... *C. orbiculare*
7. Colonies dense to floccose, pale chocolate brown; sclerotia absent; setae usually absent. Conidia 12.5-19 × 4 µ, formed from the mycelium; coffee berry disease (CBD) pathogen .... '*C. coffeanum*'
7. Colonies very variable; sclerotia occasionally present; conidia 9-24 × 3-4.5 µ; appressoria 6-20 × 4-12 µ; host range very wide .... *C. gloeosporioides* group
    8. On Gramineae 9
    8. On other substrates 10
9. Conidial apex prolonged into an appendage .... *C. caudatum*
9. Sclerotia absent; conidia 15.5-26.5 × 4-5 µ; appressoria 12.5-14.5 × 9.5-12 µ, circular; on *Saccharum* .... *C. falcatum*
9. Sclerotia present; conidia 18.5-27.5 × 3-4.5 µ; appressoria 11.5-15 × 8.5-9 µ, margin irregular; on *Sorghum* .... *C. sublineolum*
9. Sclerotia present; colonies vinaceous lilac; conidia 23.5-29 × 3.5-5 µ; appressoria 17.5-20 × 12.5-14 µ, very irregular; on *Zea* .... *C. graminicola*
    10. Conidia not strongly falcate 11
    10. Conidia strongly falcate 12

11. Sclerotia absent; conidia straight or only slightly curved, 16-19 × 3-4.5 µ; appressoria 10-13.5 × 6.5-13.5 µ, very variable; on *Linum* .... *C. lini*
11. Sclerotia present; conidia moderately curved, 14.5-19.5 × 3.5-4 µ; appressoria clavate or circular, 6.5-11 × 5.5-8.5 µ; on *Corchorus* .... *C. corchori*
11. Sclerotia present; conidia moderately curved, 14-24 × 4-6 µ, appressoria chocolate brown, crenate, 12-18 × 8.5-12 µ; on *Hedera* .... *C. trichellum*
11. Conidia 14.5-18.5 × 5-5.5 µ, widest nearer the base; on *Artemisia* .... *C. phyllachoroides*
 12. Conidia 2-2.5 µ wide; wide host range, usually saprophytic .... *C. dematium* group
 12. Conidia 3.5-4 µ wide 13
13. On *Allium* .... *C. circinans*
13. On Leguminosae .... *C. truncatum*
13. Wide host range, usually pathogenic .... *C. capsici* group

The *C. gloeosporioides, C. dematium* and *C. capsici* group species (there should also be a group species for graminicolous taxa and this may eventually prove to be *C. sublineolum*) are taxa in which separation of subspecific elements cannot be made satisfactorily at the moment.

C. ACUTATUM Simmonds, *Qd Jl. Agr. Anim. Sci.* **22**: 458 (1965). (Fig. 316A)

Colonies of dense, white aerial mycelium, becoming pinkish grey, conidial masses salmon pink, colony reverse pink. Sclerotia absent. Setae absent. Conidia fusiform, occasionally medianly constricted, 8.5-16.5 × 2.5-4 µ. Appressoria sparse, pale to dark brown, clavate or slightly irregular, 8.5-10 × 4.5-6 µ, occasionally becoming complex.

A full account of the species and its relationships with the *C. gloeosporioides* complex is to be found in Simmonds (1965). Dingley & Gilmour (1972) recognised *C. acutatum* f. sp. *pinea* on *Pinus* seedlings in New Zealand causing terminal crook disease.

From *Capsicum, Delphinium, Carica papaya, Lycopersicon esculentum, Pinus elliottii*, Australia, New Zealand (IMI 199172, 214465, 117617-117623, authentic for the name).

C. COCCODES (Wallr.) Hughes, *Can. J. Bot.* **36**: 754 (1958). (Fig. 316B)
Chaetomium coccodes Wallr., *Fl. Crypt. Germ.* **2**: 265 (1833).
Vermicularia atramentaria Berk. & Br., *Ann. Mag. Nat. Hist.* **5**: 378 (1850).
Colletotrichum atramentarium (Berk. & Br.) Taubenh., *Mem. N.Y. Bot. Gdn* **6**: 549 (1916).
An additional 13 synonyms are listed by von Arx (1957b) under the name *C. atramentarium*.

Colonies mainly consisting of black sclerotia, only rarely with diffuse white aerial mycelium, conidial masses honey. Sclerotia abundant, black, setose, globose, separate or confluent. Setae present. Conidia fusiform, straight, consistently medianly constricted, 16-22 × 3-4 µ. Appressoria common, medium brown, long clavate, occasionally irregular and edge almost crenate, 11-16.5 × 6-9.5 µ, rarely becoming complex.

The species has been fully illustrated and described by Mordue (1967).

From and on *Lycopersicon esculentum, Solanum tuberosum, S. dulcamara, Mnium undulatum, Medicago tribuloides, M. sativa, Zizyphus*, soil, U.K., Poland, S. Rhodesia, Nigeria, Kenya, Ethiopia, Sudan, Turkey, Cyprus, Jordan, Pakistan, India, Burma, Sabah, Australia, New Zealand, P.Q. and Sask., Canada, represented in herb. IMI by 57 collections.

C. HIGGINSIANUM Sacc. apud Higgins, *J. Agr. Res.* **10**: 161 (1917). (Fig. 316D)

Not known at C.M.I. in culture so details of sclerotia and appressoria not available. Setae present, but very sparse. Conidia straight or slightly curved, fusiform, often with large guttules, 16.5-19 × 4 µ.

Placed in synonymy with *C. gloeosporioides* (Penz.) Sacc. by von Arx (1957b) but distinguished from this species by its consistent association with leaf lesions on *Brassica* spp. and its different conidial shape. Scheffer (1950) provides a detailed account of the host range and growth in culture for American material.

On leaves of *Brassica campestris, B. chinensis, B. chinensis* var. *communis, B. rapa, B. juncea* var. *rugosa, B. oleracea* var. *capitata*, from North Borneo, Netherlands, New Guinea, Sabah, Hong Kong, Sarawak, Brunei and Jamaica (IMI 88773, 94806, 179217, 187488, 183488, 106300, 21053, 106298, 86958, 36700, 78807, 81401, 81469).

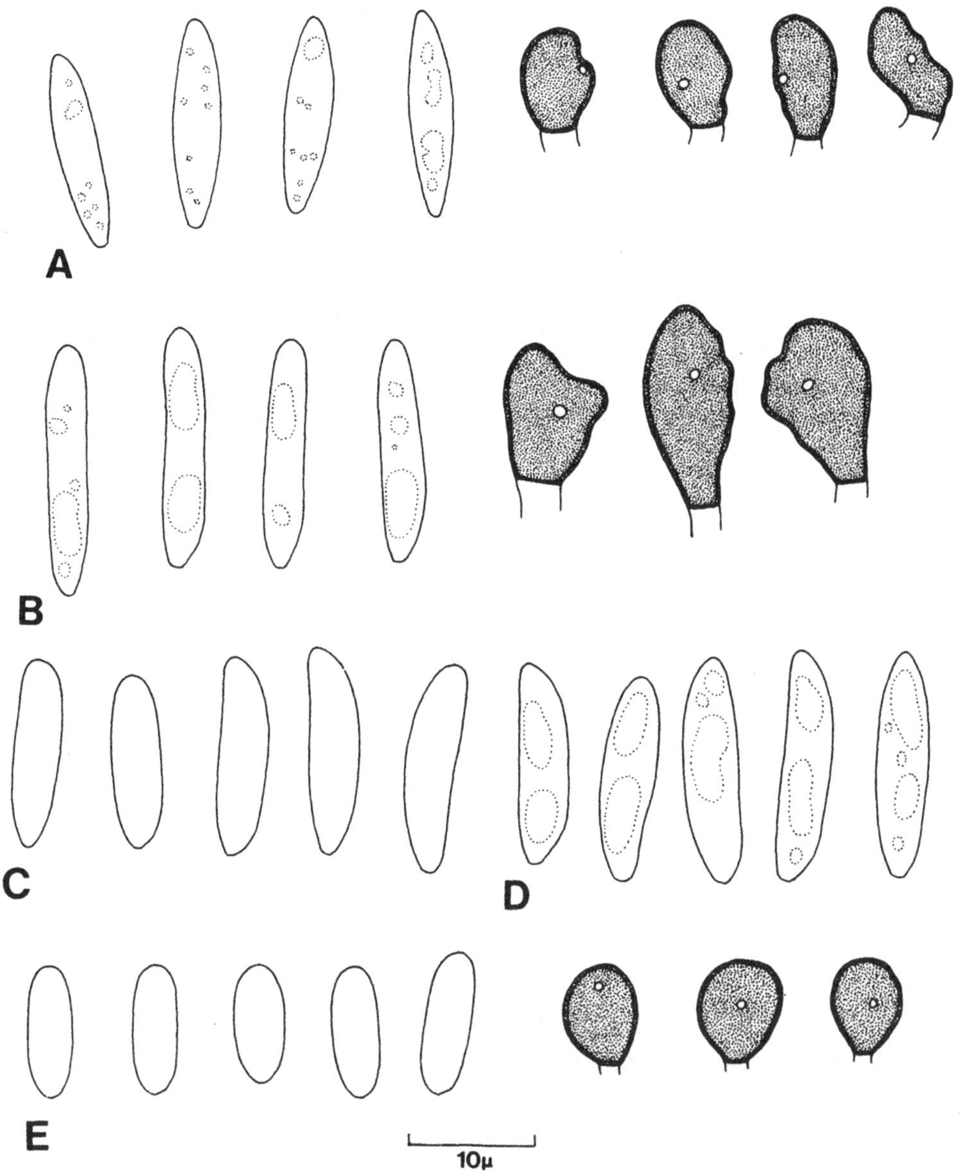

Fig. 316. A, *Colletotrichum acutatum*, conidia and appressoria; B, *C. coccodes*, conidia and appressoria; C, *C. fuscum*, conidia; D, *C. higginsianum*, conidia; E, *C. lindemuthianum*, conidia and appressoria.

C. FUSCUM Laub., *Gartenwelt* **31**: 674 (1927)1927). (fig. 316C)
Two additional synonyms have been listed by Arx (1957b).

Colonies black, of dense, very dark brown aerial mycelium, with irregular patches of greyish mycelium or covered in a diffuse grey felt, colony reverse black. Sclerotia often present, globose to columnar, separate or confluent. Setae present. Conidia straight or slightly curved, obtuse at the apices, 14-17 × 3.5-4 μ. Appressoria absent, but the vegetative mycelium is thick-walled and dark brown, cells occasionally with an appressorial germination pore.

Limited details of growth in culture and pathogenicity are provided by Kowalski (1968).

On and from *Antirrhinum majus*, *Digitalis purpurea*, New Zealand, U.K. (IMI 197877, 197675, 190613, 179993, 168497).

C. LINDEMUTHIANUM (Sacc. & Magn.) Br. & Cav., *Funghi Parass.* 50 (1889). (Fig. 316E)
Gloeosporium lindemuthianum Sacc. & Magn., *Michelia* **1**: 128 (1880).
Gloeosporium socium Sacc., *Bull. trimest. Soc. mycol. Fr.* **12**: 71 (1896).

Colonies slow growing, dark brown to black, with abundant brown aerial mycelium, margin regular and even, immersed, conidial masses honey or paler. Sclerotia occasionally present on PCA but sparse and scattered, globose. Setae present. Conidia straight, cylindrical, obtuse at the apices, 9.5-11.5 × 3.5-4.5 µ. Appressoria sparse, pale to dark brown, clavate or circular in outline, regular, 8 × 6-7 µ, not becoming complex.

A full account of the species has been provided by Mordue (1971).

On or from *Cassia fistula, C. occidentalis, C. tora, Lablab niger, Dolichos, Glycine max, Lathyrus odoratus, Medicago sativa, Mucuna pruriens* var. *utilis, Phaseolus acutifolius, P. atropurpureus, P. nanus, P. trilobus, P. vulgaris, P.* sp., *Pisum sativum, Vigna sesquipedalis, V. sinensis, V. unguiculata, V. vexillata*, from Austria, Poland, Italy, Germany, France, U.K., Jamaica, Trinidad, Cuba, Venezuela, Nigeria, Liberia, Sierra Leone, Zambia, Tanzania, Uganda, S. Rhodesia, Malawi, Ethiopia, Oman, Mauritius, India, Burma, Malaysia, Hong Kong, British Solomon Isl., Sarawak, Brunei, Papua New Guinea, New Zealand, represented in herb. IMI by 92 collections.

C. CRASSIPES (Speg.) Arx, *Verh. Akad. Wet. Amst.* **51**: 77 (1957). (Fig. 317A)
Gloeosporium crassipes Speg., *Ampel. ital.* 405 (1878).
Colletotrichum guajaci Syd., *Annls mycol.* **28**: 197 (1930).
Colletotrichum eranthemi Vasant Rao, *Sydowia* **16**: 104 (1963).
An additional 17 synonyms are listed by von Arx (1957b).

Colonies with variable development of dark chocolate brown aerial mycelium, sometimes absent, sometimes abundant, margin usually distinct, immersed, and regular, colony reverse dark brown, conidial masses honey. Sclerotia absent. Setae present. Conidia straight, cylindrical, obtuse at the apices, 10-15 × 4.5-6.5 µ. Appressoria abundant, medium to dark brown, long clavate or circular but with crenate or deeply divided edges, 10.5-14 × 7-9.5 µ, often becoming complex.

Ram, Ram & Rocha (1973) compared cultural characters of *C. crassipes* with *C. gloeosporioides* from *Theobroma cacao*. The two species can be readily distinguished on conidial size.

On or from *Aerangis kotschyana, Alocasia macrorhiza, Amaryllis* sp., *Bulbophylum cylindrum, Cattleya* sp., *Coelogyne cristata, Croton oblongifolium, Cymbidium* spp., *Dendrobium* sp., *Eranthemum bicolor, Euphorbia abyssinica, Guajacum officinale, Oncidium* spp., *Orchid* (various), *Hevea brasiliensis, H.* sp., *Masdervallia harryana* var. *lilacina, M.* sp., *Murraya* sp., *Persea americana, Piper* sp., *Saccharum officinarum, Theobroma cacao, Vanda* sp., from Uganda, Ethiopia, Egypt, India, Nepal, Brunei, British Solomon Isl., Hong Kong, Papua New Guinea, Burma, Malaysia, Sabah, New Zealand, Cuba, Brazil, Venezuela, Trinidad, U.K., represented by 37 collections in herb. IMI, including IMI 103771, holotype of *C. eranthemi* Vasant Rao.

C. MUSAE (Berk. & Curt.) Arx, *Verh. Akad. Wet. Amst.* **51**: 107 (1957). (Fig. 317B)
Myxosporium musae Berk. & Curt., *Grevillea* **3**: 13 (1874).
Gloeosporium musarum Cke & Mass., *Grevillea* **16**: 3 (1887).

Colonies often with abundant white aerial mycelium beoming grey with age, with abundant ochreous to cinnamon conidial masses, usually coalescing. Sclerotia absent. Setae absent. Conidia straight, cylindrical, obtuse at the apices, 12-17 × 4.5-5.5 µ. Appressoria common, medium to dark brown, irregular in shape, often with large or deep lobes, 9-13 × 9-11.5 µ, often becoming complex.

Plata et al. (1974) gave an account of the morphology, life history and symptoms of the species.

On and from *Musa sapientum, M. cavendishii, M. paradisiaca*, Cyprus, Egypt, Yemen, Kenya, Tanzania, Nigeria, Zambia, Malawi, Mauritius, Pakistan, India, Sri Lanka, New Zealand, British Solomon Isl, Dominica, Windward Isl, St. Lucia, Cuba, Venezuela, Colombia, represented by 63 collections in herb. IMI.

C. ORBICULARE (Berk. & Mont.) Arx, *Verh. Akad. Wet. Amst.* **51**: 112 (1957). (Fig. 317C)
Gloeosporium orbiculare Berk. & Mont., *Pl. Portug. Welw.* 7 (1853).
Colletotrichum lagenarium (Pass.) Ell. & Halst., *Bull. Torrey bot. Cl.* **20**: 250 (1893).
An additional 10 synonyms are listed by von Arx (1957b).

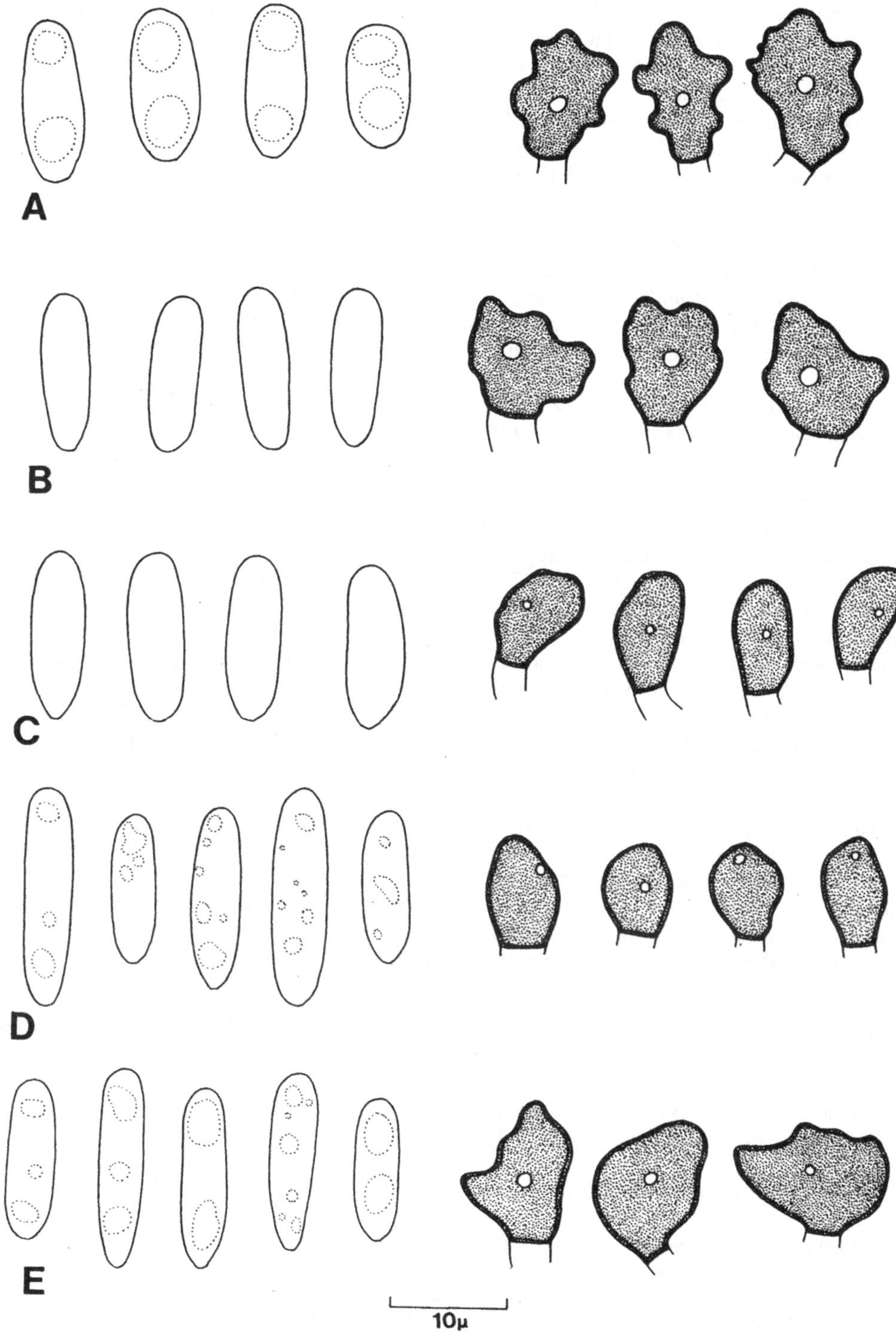

Fig. 317. A, *Colletotrichum crassipes*, conidia and appressoria; B, *C. musae*, conidia and appressoria; C, *C. orbiculare*, conidia and appressoria; D, *C. coffeanum*, conidia and appressoria; E, *C. gloeosporioides*, conidia and appressoria.

Colonies irregular in morphology, with abundant white-greyish aerial mycelium (occurring in tufts and patches on OA), reverse dark brown, conidial masses salmon pink. Sclerotia abundant, scattered, separate (except on PDA where they are aggregated into complex large masses). Setae present. Conidia straight, cylindrical, obtuse at the apices, 14-15 × 4.5-6 µ. Appressoria abundant, medium brown, long clavate or slightly irregular, 9-10 × 5-6 µ, often becoming complex.

Arx & Van der Velden (1961) studied various strains in culture and cross-inoculated them on Cucurbitaceous and other hosts. Jenkins, Winstead & McCombs (1962) reported a perfect state *G. cingulata* var. *orbicularis* but did not formally publish the name.

On or from *Benincasa corifera, B. hispida, Colocynthus vulgaris* var. *fistulosus, C. citrullus, Cucumis melo, C. sativus, C. citrullus, C. vulgaris, C.* sp., *Citrullus vulgaris, Cucurbita pepo, C. maxima, C.* sp., *Echinocystis macrocarpa, Lagenaria vulgaris, L. ciceraria, L. leucantha, L. vulgaris, L.* sp., *Luffa acutangula, Momordica charantia, Telfairia occidentalis, T. pedata*, from Germany, Roumania, Italy, U.K., S. Rhodesia, Zambia, Ethiopia, Tanzania, Uganda, Nigeria, S. Africa, Mauritius, India, Nepal, Burma, S. Cameroons, N. Borneo, Sabah, Hong Kong, Brunei, Sarawak, Malaysia, Tahiti, Fiji, U.S.A., represented by 57 collections in herb. IMI.

"COLLETOTRICHUM COFFEANUM" Noack, *Z.f. Pflanzenkr.* **11**: 202 (1901). (Fig. 317D)
Gloeosporium coffeanum Delacr., *Bull. trimest. Soc. mycol. Fr.* **13**: 110 (1897).

Colonies of dense or floccose pale chocolate brown aerial mycelium, sometimes greyish with a lighter centre, reverse greenish grey, lacking acervuli. Sclerotia absent. Setae usually absent. Conidia straight, cylindrical, 12.5-19 × 4 µ, formed from individual hyphae rather than acervuli. Appressoria moderately abundant, pale to medium brown, circular, or slightly irregular, 8-9.5 × 5.5-6.5 µ, often becoming complex.

The relationships of the Coffee Berry Disease pathogen to *C. gloeosporioides* (Penz.) Sacc. have been fully discussed by Firman & Waller (1977), and although the name *C. coffeanum* Noack has been applied in this context, the species was not originally described from berries. Arx (1957b) included the name in synonymy with *C. gloeosporioides* together with 7 other taxa specifically described from *Coffea*. All except *G. coffeigenum* Averna – Sacca (1926) were originally described from leaves and the synonymy suggested by Arx (1957b) is probably correct. *G. coffeigenum* is the exception in being described from berries affected by *Stephanoderes* in Brazil, and although the fungus produced acervuli on the natural substrate, it only formed conidia from separate hyphae in culture. No setae were reported, hence its description in *Gloeosporium* auct. rather than *Colletotrichum*. Conidia were described as ellipsoid, cylindrical or ovate-cylindrical, straight or very slightly curved, measuring 10-21.6 × 5.4-8 µ. The width measurements are in excess of those known for the CBD pathogen but length is identical. Length/width ratios calculated from the original illustrations indicate that the width dimensions given in the description are incorrect and should be approximately 4-5 µ instead of 5.5-8 µ. The geographical distribution of CBD only includes records within Africa, although there are reports of a similar disease in Guatemala and Costa Rica (Scheiber et al., 1970). It seems unlikely that *G. coffeigenum* can be taken up for the CBD pathogen because the original data strongly suggest it is the saprophytic CCM form of *C. gloeosporioides* (Gibbs, 1969).

On and from berries of *Coffea arabica*, Kenya, Tanzania, Ethiopia (IMI 95757, 131279, 108201, 190857, 170660, 191257, 119256, 191255, 191247, 191261, 191252, 191247).

C. GLOEOSPORIOIDES (Penz.) Sacc., *Fung. Agrum.* **2**: 6 (1882). (Fig. 317E)
Vermicularia gloeosporioides Penz., *Michelia* **2**: 450 (1880).

Teleomorphosis: *Glomerella cingulata* (Stonem.) Spauld. & Schrenk, *Science* ser. 2, **17**: 751 (1903).

Arx (1957b) gives more than 600 synonyms for the name in *Colletotrichum*, and Arx & Müller (1954) give 120 synonyms for the name in *Glomerella*.

This group species is one of the commonest to occur in the tropics and subtropics and is world-wide in distribution. Arx (1957b) recognised nine forms within the species but retained the names in *Colletotrichum*. There are probably many more to be differentiated by a combination of cultural characters, morphology, host range and pathogenicity.

The most comprehensive modern treatments of the species have been by Arx (1957b) and Mordue (1971), and readers are referred to the latter for a full description of its pathology together with illustrations of perfect and imperfect states.

In all the criteria that have been used for differentiating taxa in *Colletotrichum*, it is clear that *C. gloeosporioides* shows excessively wide variation, and for this reason it is meaningless to provide a standardized description. Conidia are straight, obtuse at the apex, 9-24 × 3-4.5 µ, and appressoria are 6-20 × 4-12 µ, clavate or irregular, sometimes becoming complex.

Represented in herb. IMI on 470 different host genera by 2200 collections.

C. CAUDATUM (Sacc.) Pk, *Bull. N.Y. St. Mus.* 131: 81 (1909). (Fig. 318A)
Ellisiella caudata Sacc., *Michelia* **2**: 147 (1880).
Ellisiellina caudata (Sacc.) Camara, *Agron. Lusit.* **11**: 72 (1949).
Rostrospora ciliata (T.S. & K. Ramakr.) Subram. & Ramakr., *J. Madras Univ.* B, **22**: 67 (1952).
Colletotrichum ciliatum T.S. & K. Ramakr., *Proc. Ind. Acad. Sci.* sect. B, **25**: 186 (1947).

Colonies of dense, pale to vinaceous brown, aerial mycelium, reverse greenish brown. Sclerotia absent. Setae abundant. Conidia 18.5-24 × 3.5-4 µ (excluding appendage), falcate, fusiform, apices prolonged into a filiform, unbranched, cellular appendage 10-16 µ long. Appressoria moderately abundant, clavate or circular, edge entire, 10-14 × 9-10 µ, rarely becoming complex.

This species has been the source of confusion both nomenclaturally and taxonomically since its inception. Arx (1957b) included it as a synonym of *C. graminicola*, but the apical appendage separates it readily from this species.

If the conidial appendage is considered to justify segregation of the taxon into a distinct genus, then *Ellisiella* Sacc. could be resurrected, but as Sutton (1966) suggested and Nag Raj (1973a) demonstrated, the species is biologically indistinguishable from other *Colletotrichum* species.

On and from *Sorghum nutans*, *Imperata cylindrica*, *I. cylindrica* var. *major*, *I.* sp., *Bothriochloa bladhii*, *Cymbopogon martinii*, *C.* sp., *Digitaria* sp., *Eulaliopsis binata*, Burma, Brunei, Sarawak, India, Australia, N.J., U.S.A. (IMI 79947, 176621, 176620, 176618, 149152, 176619, 176617, 116332, 196464, 79777, 182552, 162855, 190421, 211619, 197166).

C. FALCATUM Went, *Arch. Java Suikerindustrie* **1**: 265 (1893). (Fig. 319A)

Teleomorphosis: *Glomerella tucumanensis* (Speg.) Arx & Müller, *Beitr. Krypt.fl. Schw.* **11**: 195 (1954).

Colonies greyish white with sparse aerial mycelium and small dense felty patches, elsewhere reverse white to grey, conidial masses salmon pink (light race): some cultures have abundant greyish white aerial mycelium with poor sporulation and no distinct acervuli (dark race). Sclerotia absent from both races. Setae sparse. Conidia falcate (but not markedly so), fusiform, apices obtuse, 15.5-26.5 × 4-5 µ. Appressoria sparse, medium brown, clavate or circular, edge entire, 12.5-14.5 × 9.5-12 µ (6-21 × 6-17 µ).

Arx (1957b) included most of the described *Colletotrichum* species on Gramineae as synonyms of *C. graminicola* (Ces.) Wils. and accepted *Glomerella tucumanensis* (Speg.) Arx & Müller as the name for the perfect state of this species. However, it is known that *C. falcatum* is only strongly pathogenic on *Saccharum* species (Mordue, 1967), and Sutton (1968a) has demonstrated a morphological difference in appressoria between *C. falcatum* and taxa described from *Zea* and *Sorghum*. Previous studies (Sutton, 1965a) showed slight differences in conidial size between four graminicolous taxa and a less obvious distinction in conidial shape.

On and from *Saccharum officinarum*, associated with red rot, from Australia, Cuba, Sri Lanka, India, Malaysia, Nepal, New Caledonia, New Hebrides, Bangladesh, Reunion, Western Samoa, British Solomon Isl, Tonga, Trinidad, Venezuela, Mauritius, Ghana, Nigeria. Represented in herb. IMI by 53 collections.

C. SUBLINEOLUM P. Henn. apud Kabát & Bub., *Fungi imp. exs.* 186 (1905). Saccardo, *Syll. fung.* **22**: 1206 (1913). (Fig. 318C)
Colletotrichum graminicola (Ces.) Wils. f. sp. *sorghi* Messaien, Lafon & Malot, *Ann. Epiph.*, ser. C, **10**: 456 (1959).

Colonies with a sharply delimited advancing edge, with grey, compact, felty or occasionally woolly and undulate aerial mycelium, grey to greenish grey in reverse, conidial masses grey to greenish grey. Sclerotia usually absent, when present immersed to superficial, black, lenticular, dense. Setae abundant. Conidia falcate, fusiform, apices acute, 18.5-27.5 × 3-4.5 µ. Appressoria abundant, medium brown, edge moderately irregular (but not as much as in *C. graminicola*), 11.5-15 × 8.5-9 µ (6-25 × 4-16 µ).

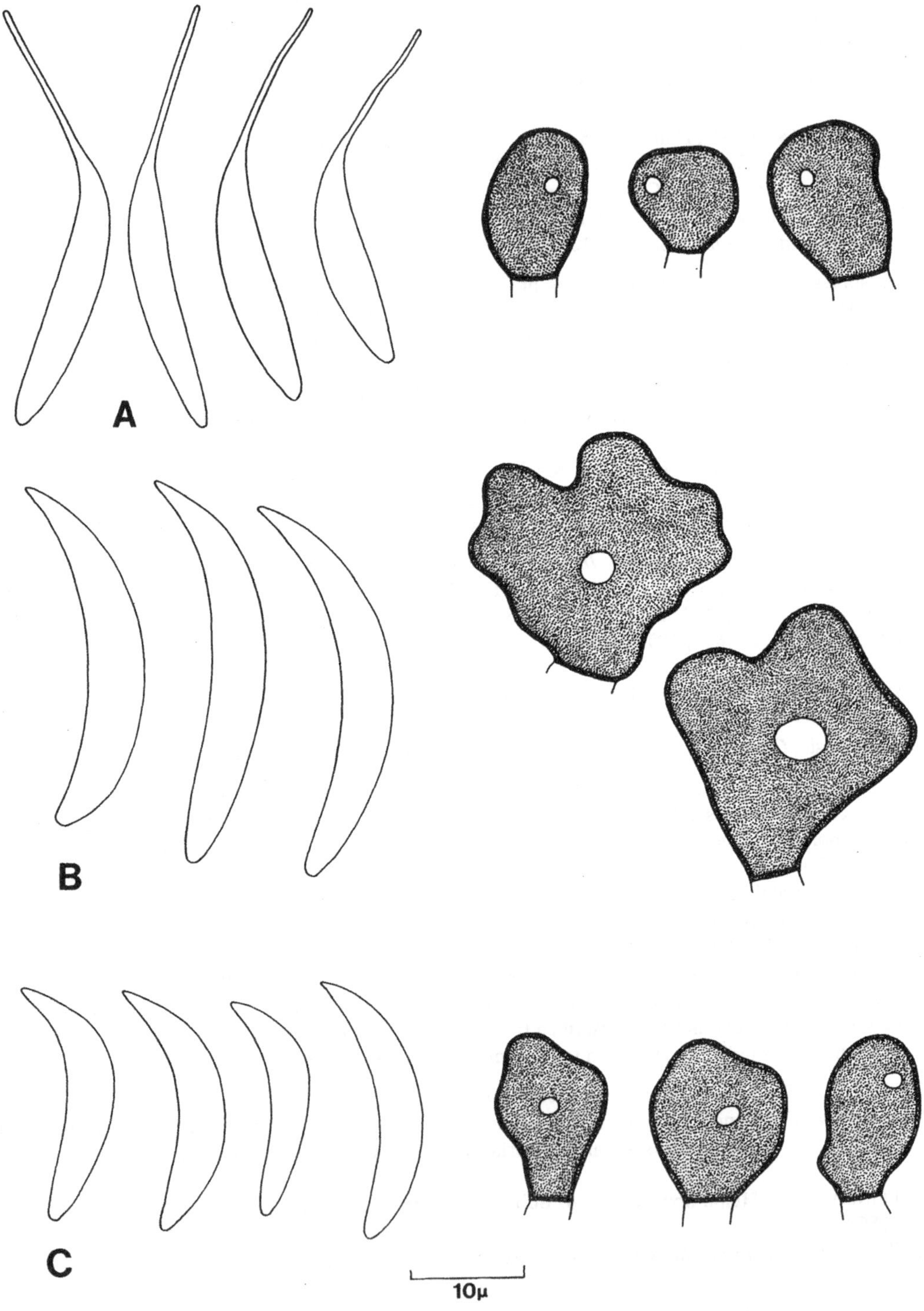

Fig. 318. A, *Colletotrichum caudatum*, conidia and appressoria; B, *C. graminicola*, conidia and appressoria; C, *C. sublineolum*, conidia and appressoria.

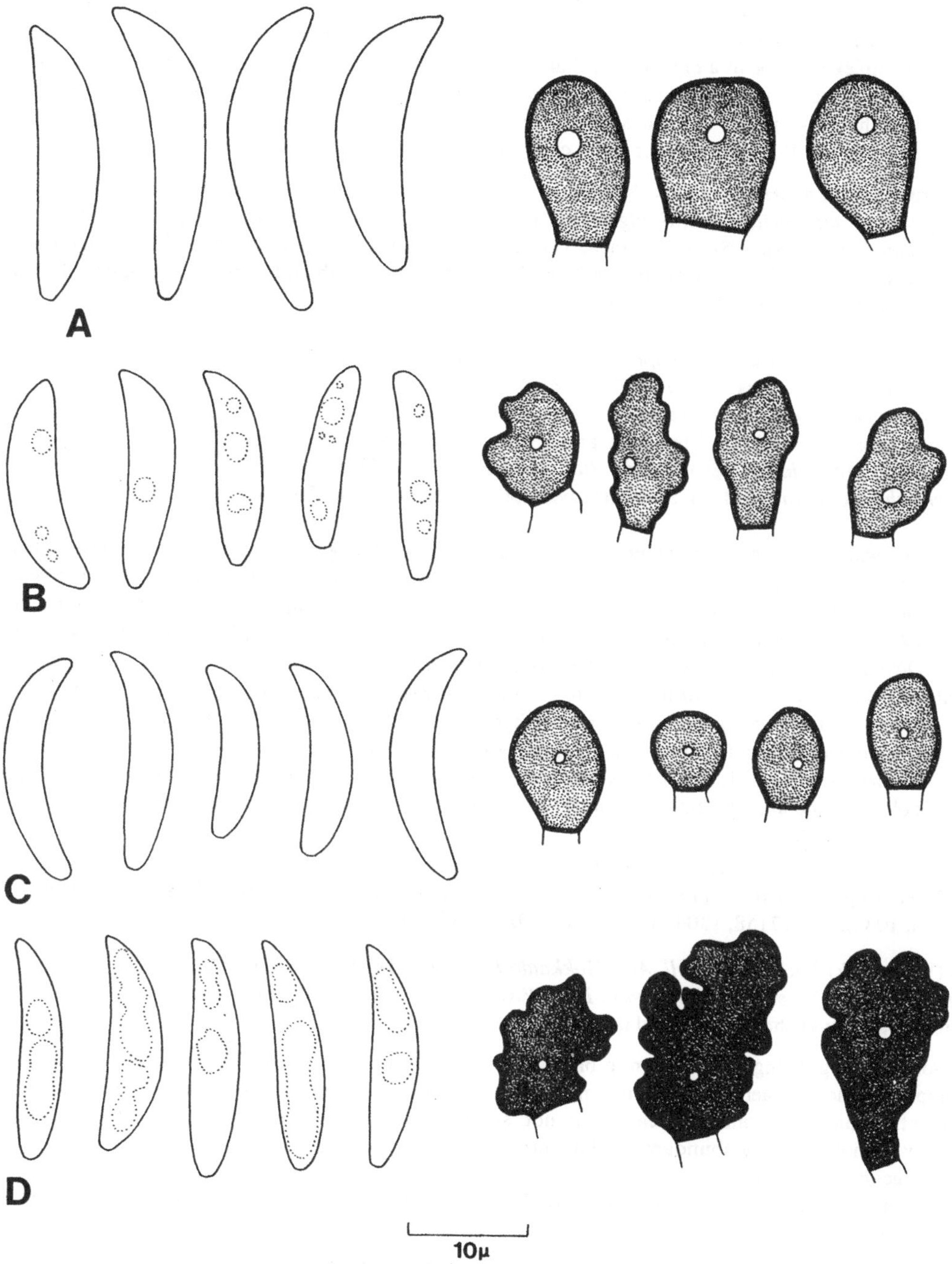

Fig. 319. A, *Colletotrichum falcatum*, conidia and appressoria, B, *C. lini*, conidia and appressoria, C, *C. corchori*, conidia and appressoria; D, *C. trichellum*, conidia and appressoria.

Since the work of Wilson (1914) and its confirmation by Arx (1957b) virtually all *Colletotrichum* species on Gramineae have been assigned to *C. graminicola*. Messaien, Lafon & Malot (1959) however, suggested formae speciales on different hosts such as *Zea*, *Sorghum*, *Saccharum* and *Secale*, but this treatment implies an absence of morphological differences between the taxa. Sutton (1966) clearly showed differences in adhesive area between appressoria derived from *Zea*, *Sorghum* and *Saccharum* isolates, consequently separation of taxa in this complex must be at a higher level than the rank of forma specialis. *C. falcatum* is well-established for the *Saccharum* fungus, and typification of *Dicladium graminicola* allows the adoption of *C. graminicola* for the *Zea* material. The only species to be described specifically from *Sorghum* is *C. sublineolum* from Togo, and it is proposed here to adopt the name for those collections and isolates from *Sorghum*.

On or from *Sorghum arundinaceum*, *S. dochna*, *S. dochna* var. *obovatum*, *S. gambicum* var. *celsum*, *S. guineense*, *S. halepense*, *S. margaritifolium*, *S. nigricans*, *S. sudanense*, *S. verticiflorum*, *S. propinquum*, *S. vulgare*, *S.* spp., from Ghana, Sierra Leone, Sudan, Guinea, Tanzania, Ethiopia, Nigeria, Zambia, Malawi, Uganda, S. Rhodesia, India, Pakistan, Nepal, Hong Kong, Papua New Guinea, Malaysia, Sabah, Burma, China, Cuba, Trinidad, U.S.A., represented in herb. IMI by 108 collections.

C. GRAMINICOLA (Ces.) Wilson, *Phytopathology* **4**: 110 (1914). (Fig. 318B)
Dicladium graminicola Ces., *Flora* **35**: 398 (1852).
Steirochaete graminicola (Ces.) Sacc., *Syll. fung.* **4**: 316 (1886).
Colletotrichopsis graminicola (Ces.) Muntañola, *Rev. argent. Agron.* **19**: 220 (1952).
Vermicularia graminicola (Ces.) Westd., *Bull. Acad. Roy. Brux.* **12**: n. 7 (1861).
Colletotrichum graminicola f. sp. *zeae* Messaien, Lafon & Malot, *Ann. Epiphyt.*, ser. C, **10**: 454 (1959).

Colonies with a diffuse, irregular margin, aerial mycelium fluffy, grey, immersed mycelium nodular, dendroid, dark brown, reverse vinaceous lilac, conidial masses salmon orange. Sclerotia formed in association with dendroid mycelium, semi-immersed to superficial, black. Setae abundant. Conidia falcate, fusiform, apices acute, 23.5-29 × 3.5-5 µ. Appressoria abundant, medium brown, edge very irregular, 17.5-20 × 12.5-14 µ (9-33 × 6-25 µ).

Dicladium graminicola was described from *Echinochloa crus-galli* and *Zea mays* and potential lectotype material was reported by Sutton (1966) to be probably *Zea*. Consequently the name *C. graminicola* should be strictly applied to the fungus on *Zea*, and if other graminicolous collections are found to differ, as seems likely from the work of Sutton (1968a), it is necessary to apply different names to them. *C. graminicola* sensu stricto on *Zea* can be distinguished from species on *Sorghum* and other Gramineae by conidial shape and size, morphology of the appressoria and cultural characteristics. An account of *C. graminicola* auct. has been given by Mordue (1967).

On and from *Zea mays*, Italy, S. Rhodesia, Malawi, Ghana, Sierra Leone, Zambia, Nigeria, S. Africa, Nepal (IMI 83255, lectotype *D. graminicola*, 84302, 53253, 59856b, 84773, 84301, 80024, 90824, 91798, 91797, 95576a, 110781*a*, 103423a, 117158, 130444, 134754, 160228, 182481, 38604b).

C. LINI (Westerdijk) Tochinai, *J. Coll. Agr. Hokkaido Imp. Univ.* **14**: 176 (1926). (Fig. 319B)
Gloeosporium lini Westerdijk, *Phytopath. Lab. W.C.S. Baarn, Jaersversl.* (1915).
Additional synonyms have been listed in von Arx (1957b).

Colonies slow growing, dark brown to black, often with immersed, dendroid aggregations of dark hyphae, often completely lacking aerial mycelium, margin distinct, conidial masses not discrete. Sclerotia absent. Setae present, abundant. Conidia falcate, but not strongly so, apices obtuse, often guttulate, 16-19 × 3-4.5 µ. Appressoria moderately abundant, medium brown, long clavate to irregular, quite variable, 10-13.5 × 6.5-13.5 µ, often becoming complex.

A number of races were recognised by Zarycka (1976), some of which showed significant morphological differences.

On and from *Linum usitatissimum*, *L. catharticum*, Holland, U.K., Ethiopia, Canada (IMI 104722, 104721, 88370, 103843, 92892b, 68481, 103844, 103842, 103841, 103840, 116676).

C. CORCHORI Ikata & Tanaka, *Ann. Phytopath. Soc. Japan* **10**: 141 (1940), non *C. corchori* Pavgi & U.P. Singh, *Mycopath. Mycol. appl.* **27**: 84 (1965). (Fig. 319C)

Cultures variable, with sparse development of thin, white or grey aerial mycelium, or lacking aerial mycelium completely; conidial masses honey yellow. Sclerotia only formed at periphery in old cultures. Setae frequent on small dark brown conidiomata scattered abundantly throughout the medium. Conidia falcate, fusiform, strongly curved towards the apices, 14.5-19.5 × 3.5-4 μ. Appressoria abundant in some isolates, sparse in others, clavate or circular, edge entire, 6.5-11 × 5.5-8.5 μ, occasionally becoming complex.

The physiology of parasitism of *C. corchori* has been studied by Purkayastha & Ray (1975) and Khan & Strange (1975) and the differentiation of 6 physiologic races was established by Choudhury & Ahmed (1969).

On or from *Corchorus capsularis*, India, Bangladesh, Malaysia (IMI 50541, 77637, 61677, 146596, 194570).

C. TRICHELLUM (Fr. ex Fr.) Duke, *Trans. Br. mycol. Soc.* **13**: 173 (1928). (Fig. 319D)
Sphaeria trichella Fr. ex Fr., *Syst. Myc.* **2**: 515 (1823).
Vermicularia trichella (Fr. ex Fr.) Fr. in Greville, *Scott. Crypt. Fl.* **6**: 345 (1828).
Amerosporium trichellum (Fr. ex Fr.) Lind, *Rostrup's Danish Fungi* 471 (1913).
Colletotrichum gloeosporioides (Penz.) Sacc. var. *hederae* Pass., *Atti R. Acc. Linc. Roma* **6**: 469 (1889).
Colletotrichum hedericola Laub., *Arb. kais. biol. Anst. Land- u. Forstw.* **5**: 503 (1907).
Colletotrichum helicis (Desm.) Morgan-Jones, *Can. J. Bot.* **49**: 1931 (1971).
Cheilaria helicis Desm., *Ann. Sci. nat.* **8**: 27 (1847).
Monostichella helicis (Desm.) Höhn., *Sber. Akad. Wiss. Wien* **125**: 56 (1916).

Colonies variable, of very sparse aerial mycelium and abundant distinct black acervuli with cinnamon buff to drab grey conidial masses, or with abundant white brownish vinaceous to olive grey aerial mycelium and small indistinct acervuli, colony reverse grey to vinaceous. Sclerotia occasionally present, but sparse, globose, separate or confluent, covered by aerial mycelium. Setae present. Conidia falcate, not strongly curved, fusiform with large guttules, 14-24 × 4-6 μ. Appressoria common, chocolate brown, irregular, edge crenate and lobed, 12-18 × 8.5-12 μ, occasionally becoming complex.

A full account has been given by Sutton (1962).

On or from *Hedera helix, H. canariensis*, Fla, U.S.A., Germany, Italy, Austria, France, Eire, Hungary, U.K., Turkey, India, Australia, represented in herb. IMI by 30 collections including IMI 107240 ex PC holotype of *C. helicis*.

C. PHYLLACHOROIDES (Ell. & Ev.) Arx, *Verh. Akad. Wet. Amst.* **51**: 119 (1957). (Fig. 320A)
Gloeosporium phyllachoroides Ell. & Ev., *North Amer. Fungi* 3071 (1893).
Gloeosporium heterophyllae Ell. & Ev., *J. Mycol.* **10**: 169 (1904).

Not known in culture at C.M.I. so details of sclerotia and appressoria not available. Conidiomata occupying almost all the lesion tissue, black, shining. Setae absent or very sparse. Conidia fusiform, falcate, apices obtuse, 14.5-18.5 × 5-5.5 μ. There appear to be no studies on behaviour in culture or host range of this distinctive species.

On leaves of *Artemisia vulgaris* var. *californica, A. heterophylla*, Calif., U.S.A. (IMI 107241, holotype *C. phyllachoroides* ex NY, 154389).

C. DEMATIUM (Pers. ex Fr.) Grove, *J. Bot., Lond.* **56**: 341 (1918). (Fig. 320B)
Sphaeria dematium Pers., *Syn. Meth. Fung.* 88 (1801).
Vermicularia dematium Pers. ex Fr., *Summa Veg. Scand.*: 420 (1849).
An additional 84 synonyms were listed by Arx (1957b).

Colonies extremely variable, with white to pale mouse grey or pale grey vinaceous patches at the centre or elsewhere, reverse dark brown, conidial masses olive grey to light vinaceous salmon. Sclerotia abundant, black, conical. Setae abundant. Conidia falcate, fusiform, apices acute, 19.5-24 × 2-2.5 (-3.5) μ. Appressoria abundant, medium brown, clavate to circular, edge usually entire, 8-11.5 × 6.5-8 μ, often becoming complex and forming long, closely branched chains.

This is the falcate-spored group species accepted by Arx (1957b), but the suggested synonymy contains some inaccuracies (Sutton, 1962). It is here interpreted as mainly a temperate species, distinguishable from *C. capsici* by the narrower conidia, cultures which often develop sclerotia, and the fact that it is usually saprophytic. Such a distinction however is not entirely satisfactory because of the wide variability shown by the non-graminicolous falcate spored isolates that have been examined. A more reasonable basis for separating taxa in the complex would be similar to that initiated by Arx (1957b) who recognised 3 formae within *C. dematium*. Until the whole complex can be examined in this light any piecemeal separation will be inadequate.

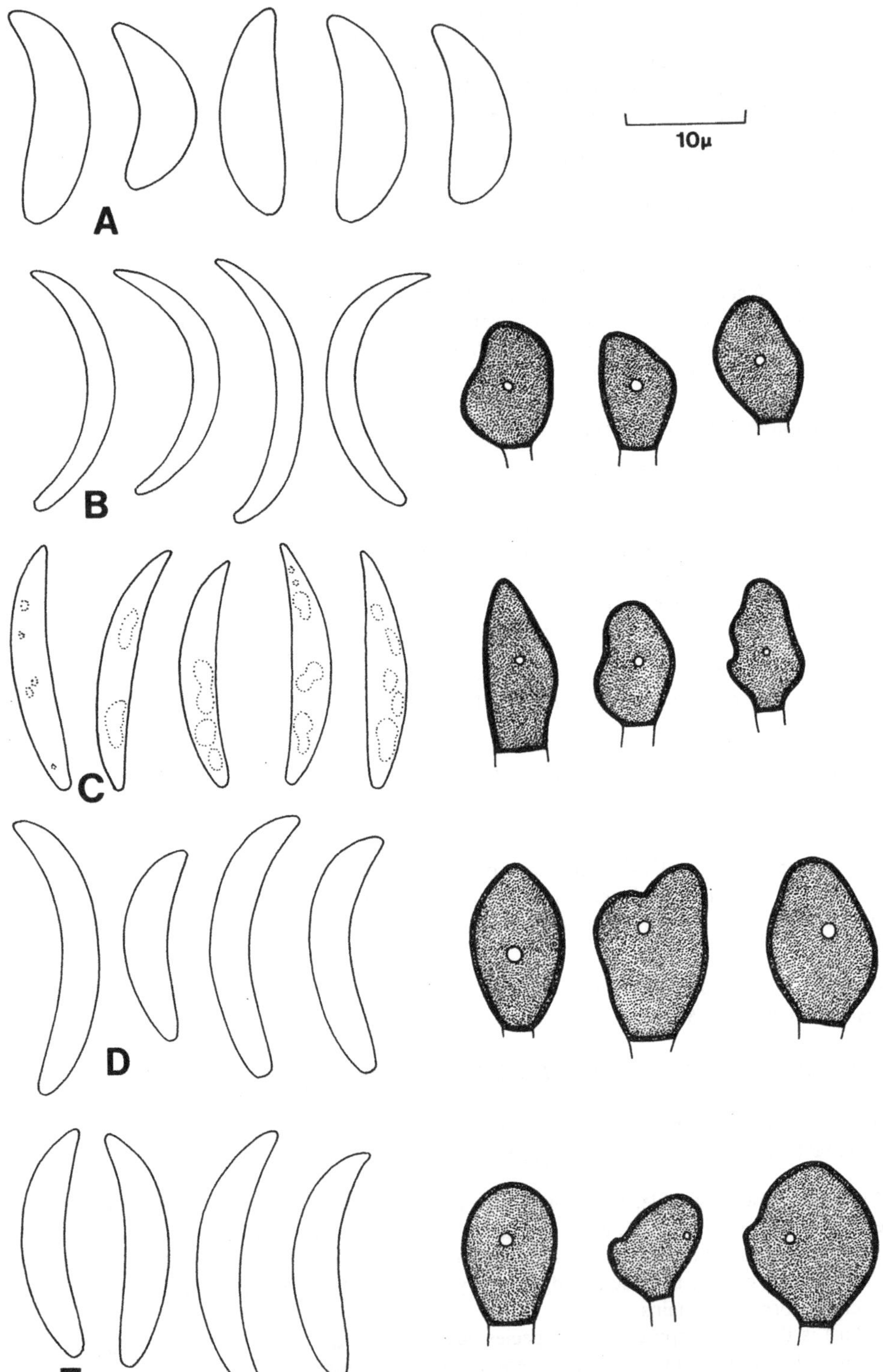

Fig. 320. A, *Colletotrichum phyllachoroides*, conidia; B, *C. dematium*, conidia and appressoria; C, *C. truncatum*, conidia and appressoria; D, *C. capsici*, conidia and appressoria; E, *C. circinans*, conidia and appressoria.

It is known from a wide variety of hosts in different families and is found extensively in temperate regions, less so in tropical to subtropical areas. In herb. IMI it is represented by 216 collections from 37 countries on 118 different host genera.

C. CIRCINANS (Berk.) Vogl., *Ann. R. Accad. Agr. Torino* **49**: 175 (1907).
Vermicularia circinans Berk., *Gardn. Chron.* **38**: 595 (1851).
Colletotrichum dematium (Pers. ex Fr.) Grove forma *circinans* (Berk.) Arx, *Phytopath. Z.* **29**: 461 (1957).

Colonies dark brown, aerial mycelium fawn to vinaceous buff through which globose to pulvinate black sclerotia sometimes protrude, conidial masses ochreous to umber. Sclerotia abundant. Setae abundant. Conidia falcate, fusiform apices, acute, guttulate, 19-21 × 3.5 µ. Appressoria abundant, clavate or circular, sometimes irregular, 10-14.5 × 6-6.5 µ, often becoming complex.

Sagasta Azpeitia (1969) gives an account of symptoms and development of *C. circinans*.

On and from *Allium cepa*, *A. tuberosum*, *A. fistulosum*, Brunei, N. Borneo, Netherlands, New Guinea, Papua New Guinea, Hong Kong, Nigeria, Zambia, Uganda, Kenya, Ethiopia, India, Pakistan, Mauritius, Cyprus, U.K. (IMI 214121, 191002, 190174, 88363, 100188, 99798, 91382b, 75030, 73882, 81542, 20987, 20988, 20986, 117041b, 86955a, 58544, 137171, 156856).

C. TRUNCATUM (Schw.) Andrus & Moore, *Phytopath.* **25**: 121 (1935). (Fig. 320C)
Vermicularia truncata Schw., *Syn. Amer. Bor.* 1865 (1832).
An additional 4 synonyms were included by Arx (1957b) under the name *C. dematium* (Pers. ex Fr.) Grove forma *truncatum* (Schw.) Arx (as '*truncata*').

Colonies dark brown to black, with a felt of greyish mycelium through which the tufted to irregular, often confluent black sclerotia protrude, occasionally with scattered tufts of pale grey to white mycelium, conidial masses honey. Sclerotia abundant. Setae abundant. Conidia falcate, apices obtuse, guttulate, 15.5-24 × 3.5-4 µ. Appressoria abundant, clavate or circular, sometimes irregular, 11-16 × 8-9.5 µ, often becoming complex.

Merlo (1969) provided an account of the morphological and cultural characters of *C. truncatum* from *Acacia longifolia*. There is also the previous treatment by Tiffany & Gilman (1954).

On and from *Alysicarpus zeyheri*, *A.* sp., *Arachis hypogaea*, *Amaranthus viridis*, *Cajanus cajan*, *C. indicus*, *Canavalia ensiformis*, *Centrosema pubescens*, *Cocculus pendulus*, *Glycine max*, *G. soja*, *Lotus corniculatus*, *Medicago sativa*, *Mimosa invisa*, *Pachyrhizus erosus*, *Phaseolus acutifolius*, *Phaseolus acutifolius* var. *latifolius*, *P. atropurpureus*, *P. aureus*, *P. lunatus*, *P. mungo*, *P. quadrangulatus*, *P. vulgaris*, *P.* sp., *Pisum sativum*, *Onobrychis* sp., *Trifolium alexandrinum*, *Vigna sinensis*, *V. unguiculata*, from Tanzania, Malawi, S. Rhodesia, Kenya, Nigeria, Zambia, Uganda, Sierra Leone, Malawi, S. Africa, Ghana, S. Arabia, Pakistan, India, New Britain, Guinea, Sabah, Hong Kong, Selangor, N. Borneo, Malaysia, Thailand, Honduras, Trinidad, Cuba, Venezuela, Brazil, U.S.A., represented in herb. IMI by 60 collections.

C. CAPSICI (Syd.) Butl. & Bisby, *Imp. Counc. agr. Res. India Sci. Monogr.* 1: 152 (1931). (Fig. 320D)
Colletotrichum boerhaviae Pavgi & U.P. Singh, *Mycopath. Mycol. appl.* **27**: 82 (1965).
Colletotrichum bryoniae (Ferr.) Maire, *Bull. Soc. H.N. de l'Afr. Nord* 183 (1917).
Colletotrichum oligochaetum Cav. var. *bryoniae* Ferr., *Annls mycol.* **10**: 293 (1912).
Colletotrichum corylifoliae Pavgi & U.P. Singh, *Mycopath. Mycol. appl.* **24**: 356 (1964).

Colonies of dense whitish to dark grey aerial mycelium, reverse dark brown, conidial masses pale buff to salmon. Sclerotia absent. Setae abundant. Conidia falcate, fusiform, apices acute, 18-23 × 3.5-4 µ. Appressoria abundant, medium brown, clavate to circular, edge usually entire, 9-14 × 6.5-11.5 µ, often becoming complex and forming long closely branched chains.

This species was thought by Arx (1957b) to be the same as *C. dematium* (Pers. ex Fr.) Grove, but it can be differentiated by the wider conidia. It has been fully described and illustrated by Mordue (1971) who considered *C. dematium* to be a saprophytic taxon and *C. capsici* pathogenic.

It is known from a wide range of hosts in different families and is found extensively in the tropics and subtropics of Africa, Asia, America and Australasia. In herb. IMI it is represented by 473 collections from 176 different host genera.

52. SPHACELOMA De Bary, *Ann. Oenologie* **4**: 165 (1874).
Manginia Viala & Pacottet, *C.R. Acad. Sci. Paris* **139**: 16 (1904).
Melanobasidium Maubl., *Bull. trimest. Soc. mycol. Fr.* **22**: 64 (1906).
Melanobasis Clements & Shear nom. nov., *Genera of Fungi*: 370 (1931).
? *Melanodochium* Syd., *Annls mycol.* **36**: 310 (1938).
Kurosawaia Hara, *A List of Japanese fungi*: 172 (1954).
Melanophora von Arx, *Verh. K. Akad. Wet. Amst.* **51**: 71 (1957).
Sp. typ.: *S. ampelinum* de Bary
Illustrations: Sutton & Pollack (1973).

Mycelium immersed, branched, septate, hyaline to pale brown. *Conidiomata* acervular, foliicolous or caulicolous, initially separate but often coalescing, composed of hyaline to pale brown textura angularis. *Conidiophores* infrequently formed, 1-2 septate, cylindrical, unbranched, pale brown or hyaline, formed from the upper cells of the conidiomata. *Conidiogenous cells* enteroblastic, polyphialidic, with 1-3 loci, integrated or discrete, indeterminate, pale brown or hyaline, ampulliform to doliiform. *Conidia* hyaline, aseptate, ellipsoid, smooth, eguttulate.

The number of taxa described in *Sphaceloma* exceeds 150. Species are mainly differentiated according to host symptoms, there being few significant morphological differences between them. *Elsinoë* Racib. functions as the teleomorphosis but not all *Sphaceloma* species have been correlated with their *Elsinoë* states. Knowledge of the genus largely rests upon the many works of Jenkins and Bitancourt separately and in cooperation with one another, and the exsiccati distributed as *Myriangiales selecti exsiccati* (Jenkins & Bitancourt, 1956, 1968). For published references to the genus see Stevenson (1971). Sutton & Pollack (1973) expanded the generic synonymy and provided the first accurate data on conidiogenesis in the group.

S. AMPELINUM de Bary, *ibid.*: 165 (1874).

Teleomorphosis: *Elsinoë ampelina* Shear, *Phytopath.* **19**: 677 (1929).

Conidiomata up to 60 µ diam. Conidiogenous cells 3.5-6 × 3-5.5 µ. Conidia 4-7.5 × 2-3.5 µ.

On leaves, stems and fruit of *Vitis vinifera*, *V.* sp., U.S.A., Chile, Brazil, Cuba, Australia, S. Rhodesia, Papua New Guinea, India, Zambia (IMI 20582-20590, 29525, 6773, 6774, 18928, 52512, 52513, 74086, 119655, 123534, 129679, 178549).

53. GLOEOSPORIDIELLA Petrak, *Hedwigia* **62**: 318 (1921).
Sp. typ.: *G. ribis* (Lib.) Petrak (syn. *Leptothyrium ribis* Lib.).

Mycelium immersed, branched, septate, hyaline. *Conidiomata* acervular, epidermal to subepidermal, separate or confluent, ± circular or irregular, composed of thin-walled, hyaline to pale brown textura angularis. *Ostiole* absent, dehiscence by irregular rupture of overlying host tissue, conidial masses yellowish. *Conidiophores* hyaline, branched at the base and above, septate, smooth, cylindrical or irregular, formed from the upper pseudoparenchyma. *Conidiogenous cells* enteroblastic, phialidic, determinate, integrated, cylindrical, to botuliform, hyaline, smooth, channel often wide, collarette occasionally flared. *Macroconidia* hyaline, aseptate, thin-walled, eguttulate, falcate, fusiform, occasionally straight, apex obtuse, base truncate. *Microconidia* hyaline, aseptate, thin-walled, eguttulate, straight, cylindrical, apex obtuse, base truncate. Both macro- and microconidia are phialidic.

Arx (1970) recognised 6 species, of which three were on *Ribes* spp. Rimpau (1962) confirmed corellation of *G. ribis* with *Drepanopeziza ribis* (Kleb.) Höhn. and *G. variabilis* with *D. variabilis* Müller, Hütter & Schüepp. He distinguished within *D. ribis* five formae speciales based on differential ability to infect various *Ribes* spp. Several more taxa have been referred to the genus.

G. RIBIS (Lib.) Petrak, *ibid.*: 318 (1921). (Fig. 321A, B, E)
Leptothyrium ribis Lib., *Pl. Crypt. Ard.* exs. 258 (1834).
Gloeosporium ribis (Lib.) Mont. & Desm. in Kickx, *Fl. Crypt. Flandr.* **2**: 95 (1867).
Gloeosporium ribis Mont., *Ann. Sci. nat.*, ser. 3, **12**: 296 (1849).
Cryptosporium ribis Fckl, *Jb. nassau. Ver. Naturk.* **23-24**: 111 (1870).
Gloeosporium curvatum Oud., *Ned. Kruidk. Arch.* **2**: 171 (1867).

Gloeosporidiella ribis (Lib.) Petrak f. sp. *rubri* Kleb. *Haupt-u. Nebenfruchtf. Ascom.* 380 (1918).
Gloeosporidiella ribis (Lib.) Mont. & Desm. forma *rubri* Wallner, Thümen, *Fungi austr.* exs. 676 (1872).
Gloeosporidiella ribis (Lib.) Petrak f. sp. *nigri* Kleb., *Haupt-u. Nebenfruchtf. Ascom.* 380 (1918).
Gloeosporidiella ribis (Lib.) Petrak f. sp. *grossulariae* Kleb., *Haupt-u. Nebenfruchtf. Ascom.* 380 (1918).
Gloeosporium ribis (Lib.) Mont. & Desm. var. *macrosporum* Savul. & Sandu, *Herb. myc. rom.* fasc. 4, exs. 271 (1929).
Gloeosporidiella ribis (Lib.) Petrak f. sp. *petraei* Rimpau, *Phytopath. Z.* **43**: 279 (1962).

Teleomorphosis: *Drepanopeziza ribis* (Kleb.) Höhn., *Annls mycol.* **15**: 332 (1917).

Conidiomata 200-300 µ diam. Conidiophores up to 33 µ long × 3-5 µ wide. Macroconidia curved, 17.5-24 × 5.5-7 µ; microconidia 4-6.5 × 2-3 µ.

Causing leaf lesions on *Ribes cereum, R. grossularia, R. hirtellum, R. inerme, R. nigrum, R. prostratum, R. rubrum, R.* sp., U.K., Italy, Romania, Germany, Hungary, Austria, Sweden, Eire, U.S.S.R., Sask., Ont. and Que., Canada, Wyo. and Calif., U.S.A., New Zealand (IMI 21787, 21778, 21779, 21785, 33490, 33492, 57015, 63027, 98011, 97284, 182792, 11983, 11527, 11686, 21781, 113527, 154391, 125970, 25905, 27041, 6921, 80563, 80564, 21780, 50361, 56601, 60857, 86842, 69088, 29743, 21784, 21786, 50774, 21782, 21783, 16576, 107237, 33490).

G. VARIABILIS (Laub.) Nannf., *Nov. Act. Reg. Soc. Sci. Upsal.* ser. 4, **8**: 171 (1932). (Fig. 321C)
Gloeosporium variabile Laub., *Nat. Z. Land-u. Forstw.* **2**: 56 (1904).
Gloeosporidiella variabilis (Laub.) Arx, *Verh. K. Ned. Akad. Wet. Amst.* **51**: 146 (1957).

Teleomorphosis: *Drepanopeziza variabilis* Müller, Hütter & Schüepp, *Sydowia* **12**: 412 (1958).

Conidiomata up to 500 µ diam. Conidiophores 10-25 × 2.5-5 µ. Conidia curved, 18-24 × 4.5-5 µ. Microconidia absent.

Causing leaf lesions on *Ribes alpinum*, Sweden (IMI 33491).

G. BARTHOLOMAEI (Dearn.) Arx, *Verh. K. Ned. Akad. Wet. Amst.* **51**: 62 (1957). (Fig. 321D)
Gloeosporium bartholomaei Dearn., *Mycologia* **9**: 356 (1917).

Conidiomata up to 350 µ diam. Conidiophores 16-28 × 3.5-5 µ. Conidia straight, occasionally curved, 22.5-27 × 5.5-7 µ. Microconidia absent.

Causing leaf lesions on *Ribes bracteosum* (IMI 21687, isotype).

G. turgida (Berk. & Br.) comb. nov.
Cryptosporium turgidum Berk. & Br., *Ann. Mag. Nat. Hist.* **7**: 129 (1881).
Stagonospora turgida (Berk. & Br.) Sacc., *Syll. fung.* **3**: 447 (1884).
Cryptosporiopsis turgida (Berk. & Br.) Pirozynski & Morgan-Jones, *Trans. Br. mycol. Soc.* **51**: 187 (1968).
Cryptosporium fraxini Rostr., *K. danske Vidensk. Selsk. Sbr.* **4**: 38 (1904).

Conidiomata 222-500 µ diam. Conidiophores 30-35 × 3-4 µ. Conidia falcate, 20-28 × 4-5 µ.

On bark of *Fraxinus* sp., U.K. (IMI 105567, 119422).

G. nobilis (Sacc.) comb. nov.
Gloeosporium nobile Sacc., *Michelia* **2**: 153 (1880).
Cryptocline nobile (Sacc.) Arx, *Verh. K. Akad. Wet. Amst.* ser. 2, **51**: 109 (1957).
Cryptosporiopsis nobilis (Sacc.) Pirozynski & Morgan-Jones, *Trans. Br. mycol. Soc.* **51**: 189 (1968).

Conidiomata up to 300 µ diam. Conidiophores up to 40 µ long × 2.5-3 µ. Conidia falcate, 17-19.5 × 4-4.5 µ.

On leaves of *Laurus nobilis*, Italy, Portugal (IMI 21742, 16573).

54. PLEUROPLACONEMA Petrak, *Annls mycol.* **21**: 300 (1923).
Sp. typ.: *P. sambuci* Petrak

Mycelium immersed, branched, septate, hyaline. *Conidiomata* eustromatic, peridermal, separate, irregular in shape,

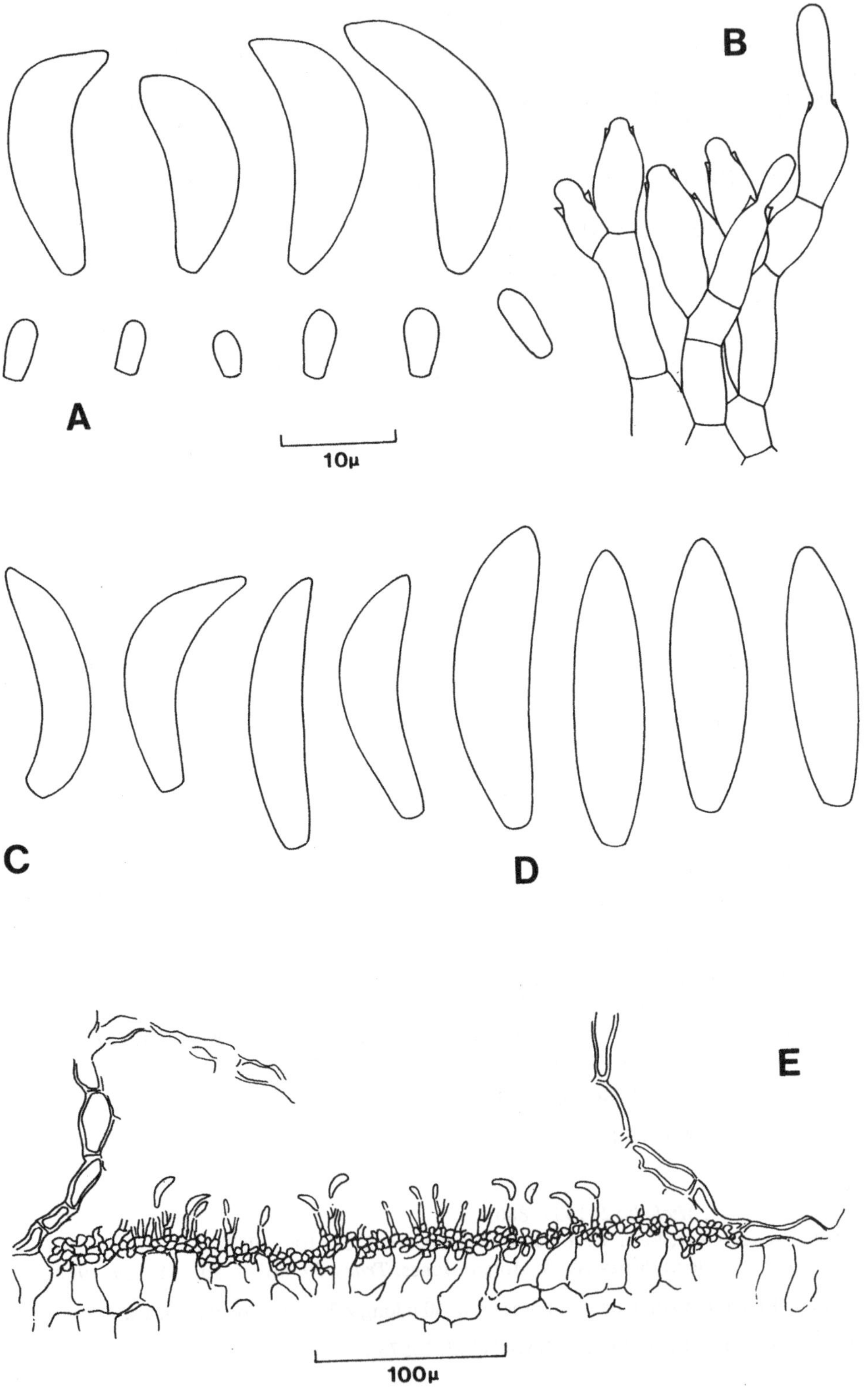

Fig. 321. A, B, E, *Gloeosporidiella ribis*, macroconidia and microconidia, conidiophores, vertical section of a conidioma; C, *G. variabilis*, conidia; D, *G. bartholomaei*, conidia.

irregularly convoluted, hyaline, internal shape dictated by peridermal cells; walls of hyaline, thin-walled textura angularis. Dehiscence irregular, by breakdown of periderm. *Conidiophores* septate, hyaline, branched irregularly and penicillately, twisted and contorted with one another, formed from the inner cells of the locular walls. *Conidiogenous cells* enteroblastic, phialidic, integrated, determinate, hyaline, formed as short or long branches laterally or from the apices of the conidiophores, apertures and collarettes minute. *Conidia* acropleurogenous, hyaline, aseptate, thin-walled, straight or slightly curved, often 1-2 guttulate.

Apart from the type, only one other species has been described in the genus, *P. punicae* Chaudhuri & Singh (1934). Since this was reported with distinct pycnidia it seems most unlikely that it is congeneric with *P. sambuci*, but an alternative genus cannot yet be suggested in the absence of the type.

P. SAMBUCI Petrak, *ibid.*: 300 (1923). (Fig. 322)

Conidiomata up to 220 μ diam. Conidiogenous cells 4-10 × 2 μ, tapered to the apices. Conidia 3-3.5 × 1-1.5 μ.

On dead branches of *Sambucus racemosa*, Germany (IMI 24791).

55. CYTOGLOEUM Petrak, *Annls mycol.* **23**: 77 (1925).
Sp. typ.: *C. tiliae* Petrak

Mycelium immersed, branched, septate, pale brown to hyaline. *Conidiomata* acervular, separate, slightly pulvinate in the centre, subperidermal, composed of small-celled, thin-walled textura intricata. Dehiscence by irregular rupture of the bark. *Conidiophores* hyaline, branched at the base and frequently along their length, septate, smooth, cylindrical, straight

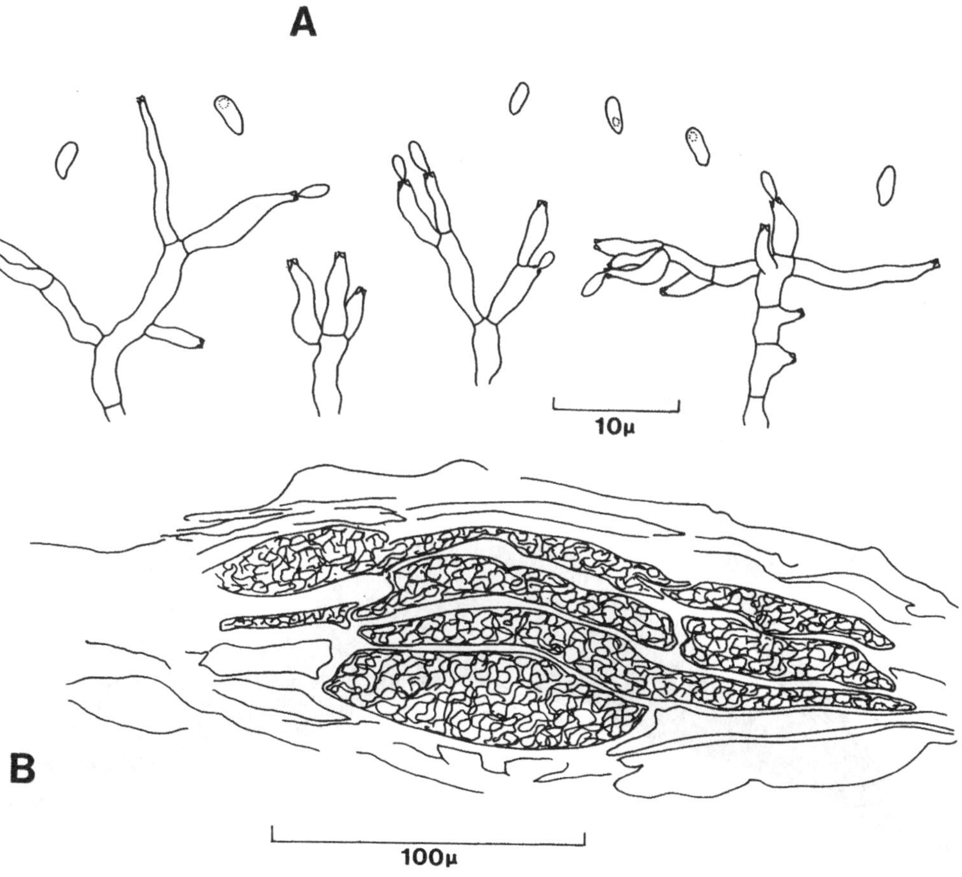

Fig. 322. *Pleuroplaconema sambuci*, A, conidia and conidiophores; B, vertical section of conidiomata.

or flexuous, with conidia formed at the apices of main and lateral branches, formed from the upper acervular cells. *Conidiogenous cells* enteroblastic, phialidic, determinate, integrated, tapered slightly towards the apices, hyaline, smooth, collarette and channel minute, apical periclinal wall thickened. *Conidia* hyaline, aseptate, thin-walled, eguttulate, smooth, cylindrical, straight or occasionally slightly curved, apex obtuse, base acute.

The genus is monotypic.

C. TILIAE Petrak, *ibid.*: 77 (1925). (Fig. 323)

Conidiomata up to 1050 µ diam., pseudoparenchyma deeper in the middle than at the periphery. Conidiophores up to 55 µ long × 2-3 µ wide. Conidia 6-8 × 1.5-2 µ.

On twigs of *Tilia* sp., Germany (IMI 24169, holotype).

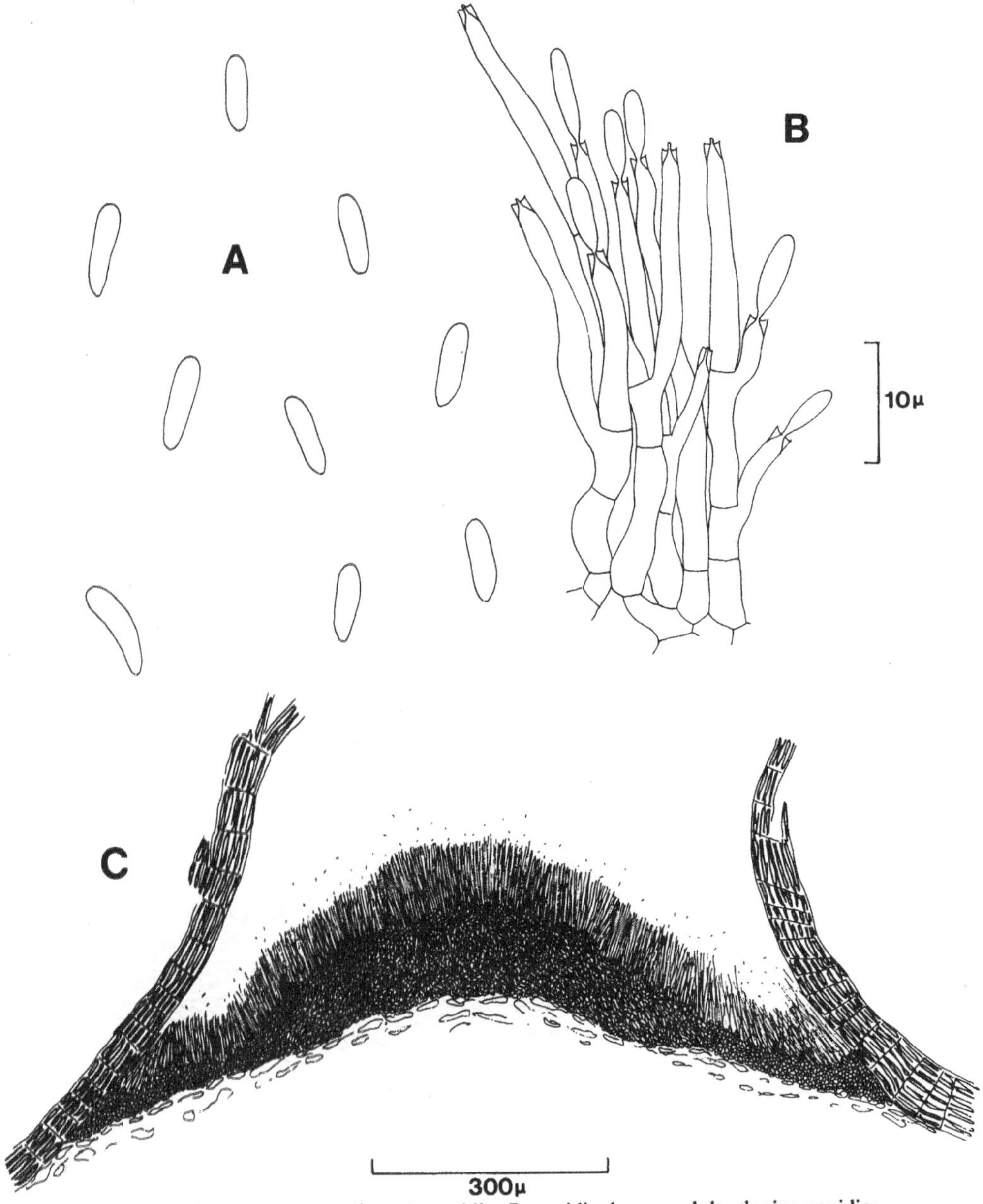

Fig. 323. *Cytogloeum tiliae*, A, conidia; B, conidiophores and developing conidia; C, vertical section of a conidioma.

56. ASCHERSONIA Mont., *Ann. Sci. nat.*, ser. 3, **10**: 121 (1848).
Bizozzeriella Speg., *An. Soc. Cient. Arg.* **26**: 73 (1888).
Underwoodina O. Kuntze, *Rev. Gen. Pl.* **3**: 538 (1893).
Lectotype sp.: *A. taitensis* Mont., fide Petch (1921).
Illustrations: Petch (1921).

Mycelium superficial, pruinose, fibrillose, membranous, hyaline, branched, septate, when present forming a hypothallus at the base of the stroma. *Conidiomata* stromatic, superficial, entomogenous on Lecaniidae and Aleyrodidae (scale insects), separate or aggregated, extremely variable in shape and pigmentation, subglobose, hemispherical or pulvinate, composed of very thick-walled, hyaline, branched, septate textura intricata which only rarely becomes more deeply pigmented at the periphery, multilocular, the locules tubular, flask-shaped, oval or globose, simple or convoluted, lacking any really distinct wall differentiating them from the stromatal tissue. *Ostioles* absent, dehiscence by wide, irregular openings in the upper stroma. *Conidiophores* septate, branched irregularly, cylindrical or tapered towards the apices, hyaline, smooth, formed from the inner tissue of the locular walls. *Conidiogenous cells* enteroblastic, phialidic, integrated, determinate, hyaline, cylindrical, collarette and channel minute, periclinal wall thickened. *Paraphyses* often present, filiform, hyaline, aseptate, thin-walled, fusiform or narrowly oval, multiguttulate.

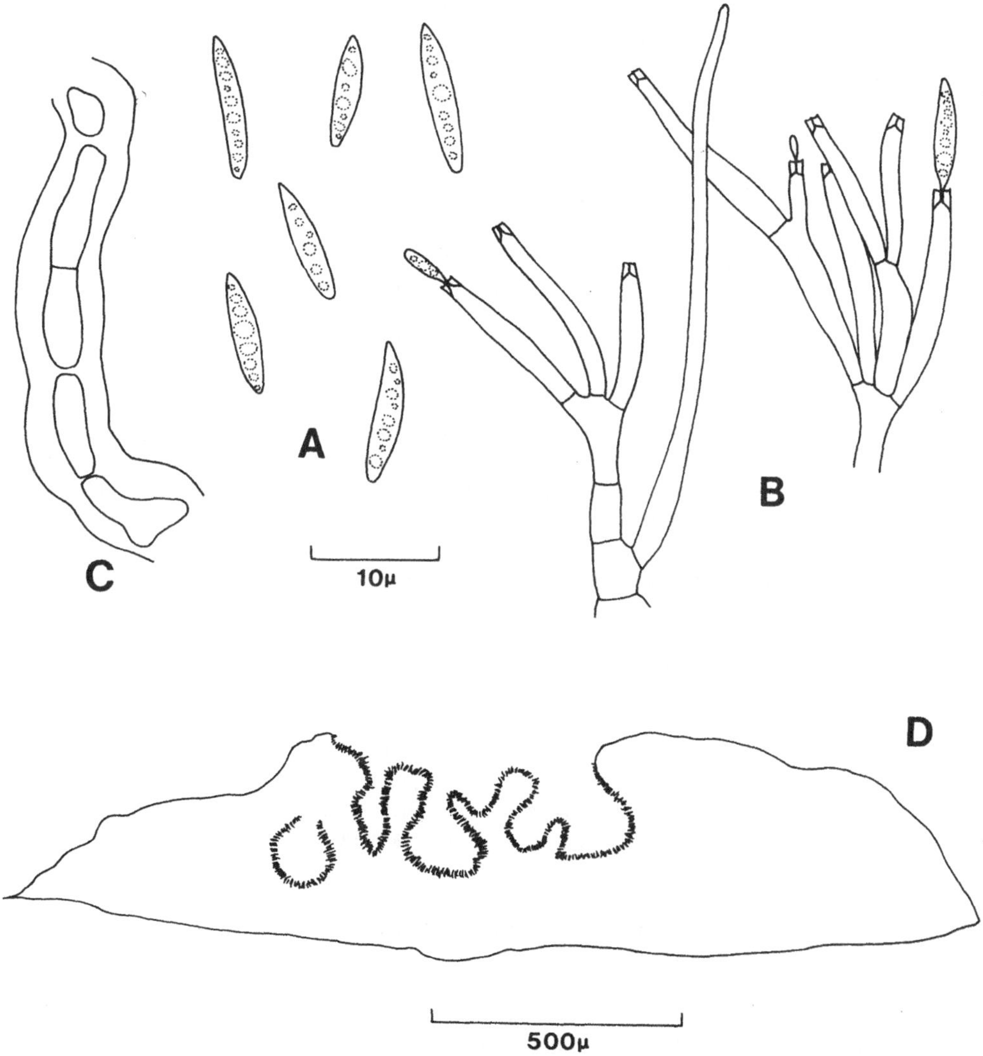

Fig. 324. *Aschersonia aleyrodis*, A, conidia; B, conidiophores and paraphysis; C, hyphae comprising textura intricata; D, vertical section of a conidioma.

Petch (1921) has provided the most comprehensive account of the genus and the species of *Hypocrella* Sacc. with which members are correlated. He recognised 25 named *Aschersonia* species, including *A. aleyrodis* Webber which is briefly described and illustrated here. The genus at present contains about 50 described taxa.

A. ALEYRODIS Webber, *Bull. U.S. Dept. Ag.* 13: 20 (1897). (Fig. 324)

Stromata flattened, pulvinate, minutely tomentose, creamy white, 2 mm diam., 2 mm thick; hypothallus thin, membranous. Locules flask-shaped to irregular or convoluted, opening widely. Conidiophores up to 40 µ long × 2-3 µ wide. Paraphyses up to 60 µ long and 1 µ wide. Conidia 9-12 × 2 µ.

On *Coccids* on *Citrus aurantium*, *C. paradisi*, *C.* sp., *Aleyrodiscus minimus* on *Psidium*, indet. on *Isoglossa lactea*, *Aleyrodis* sp. on indet., indet. on *Psidium guajava*, Jamaica, British Solomon Isl., India, Cuba, Kenya, Costa Rica, San Domingo and U.S.A. (IMI 27490, 30106, 148715, 27489, 169490, 157601, 78217, 146957).

57. PYCNIDIELLA Höhn., *Sber. Akad. Wiss. Wien* **124**: 91 (1915).
Sp. typ.: *P. resinae* (Ehrenb. ex Fr.) Höhn. (syn. *Sphaeria resinae* Ehrenb. ex Fr.).

Mycelium immersed and superficial, branched, septate, hyaline. *Conidiomata* eustromatic, superficial, separate or aggregated, yellow, globose to conic, unilocular, convoluted or imperfectly divided; walls thick, of hyaline to pale brown, thin-walled textura intricata. *Ostiole* present but later opening widely. *Conidiophores* septate, branched at the base and above, irregular, hyaline, smooth, with acropleurogenous conidia, formed from the inner cells of the stromatal wall. *Conidiogenous cells* enteroblastic, phialidic, determinate, discrete, cylindrical or tapered towards the apices, hyaline, smooth, channel and collarette minute but apical periclinal wall thickened. *Conidia* globose, smooth, thin-walled, hyaline, aseptate.

P. RESINAE (Ehrenb. ex Fr.) Höhn., *ibid.*: 91 (1915). (Fig. 325)
Cytispora resinae Ehrenb., *Sylv. Berol.*: 15 (1818).
Sphaeria resinae Ehrenb. ex Fr., *Syst. Myc.* **2**: 453 (1823).
Tubercularia resinae (Ehrenb. ex Fr.) Thüm., *Fungi Austriaci* no. 79 (1871).
Zythia resinae (Ehrenb. ex Fr.) Karst., *Medd. Soc. Fauna flora fenn.* **14**: 104 (1887).

Teleomorphosis: *Tromera resinae* (Fr.) Korb., *Parerg. lich.*: 453 (1865).

Conidiomata up to 450 µ diam. Conidiogenous cells up to 28 × 3 µ. Conidia 2 µ diam.

On resin of *Pinus sylvestris*, *P. contorta*, *P. banksiana*, *P.* sp., *Cupressus macrocarpa*, *Abies excelsa*, *Pseudotsuga menziesii*, *Picea mariana*, *P. sitchensis*, *P.* sp., U.K., Austria, Italy, Kenya, Ont., Canada, U.S.A. (IMI 68124b, 68126, 68125a, 70248c, 89856, 194563, 69235a, 69238a, 104894, 11973, 57710, 74200b, 74201b, 73338, 68738, 70854a, 73276, 122897, 53590, 91003).

58. ZYTHIOSTROMA Höhn., *Mitt. Bot. Inst. tech. Hochsch. Wien* **8**: 89 (1931).
Sp. typ.: *Z. mougeotti* (Fr.) Höhn. (syn. *Sphaeria mougeotii* Fr.).

Mycelium immersed, branched, septate, pale brown to hyaline. *Conidiomata* eustromatic, subcortical, finally superficial, separate or aggregated, globose to collabent, often collapsing irregularly, red, irregularly multilocular, locules convoluted; walls all composed of pale brown textura angularis, somewhat thicker-walled and slightly darker in the outer region. *Ostiole* single, papillate, circular. *Conidiophores* septate, branched irregularly, cylindrical, tapered to the apices, hyaline, smooth, formed from the inner wall of the locules. *Conidiogenous cells* enteroblastic, phialidic, integrated, determinate, hyaline, cylindrical, apical or on small or long lateral branches immediately below septa, channel and collarette minute, periclinal wall thick. *Conidia* acropleurogenous, hyaline, aseptate, straight or slightly curved, thin-walled, eguttulate.

Petch (1943) gave a brief account of the genus, later amplified by Booth (1959).

Z. MOUGEOTII (Fr.) Höhn., *ibid.*: 88 (1931). (Fig. 326A, B)
Sphaeria mougeotii Fr., *Elench. fung.* **2**: 100 (1828).
Sphaeronaemella mougeotii (Fr.) Sacc., *Syll. fung.* **3**: 617 (1884).
Zythia mougeotii (Fr.) Jaczewski, *Nouv. Mem. Soc. Nat. Mosc.* **15**: 367 (1898).
Sphaeronaema hederae Fckl, *Fung. rhen.* exs. VIII No. 775 (1863).

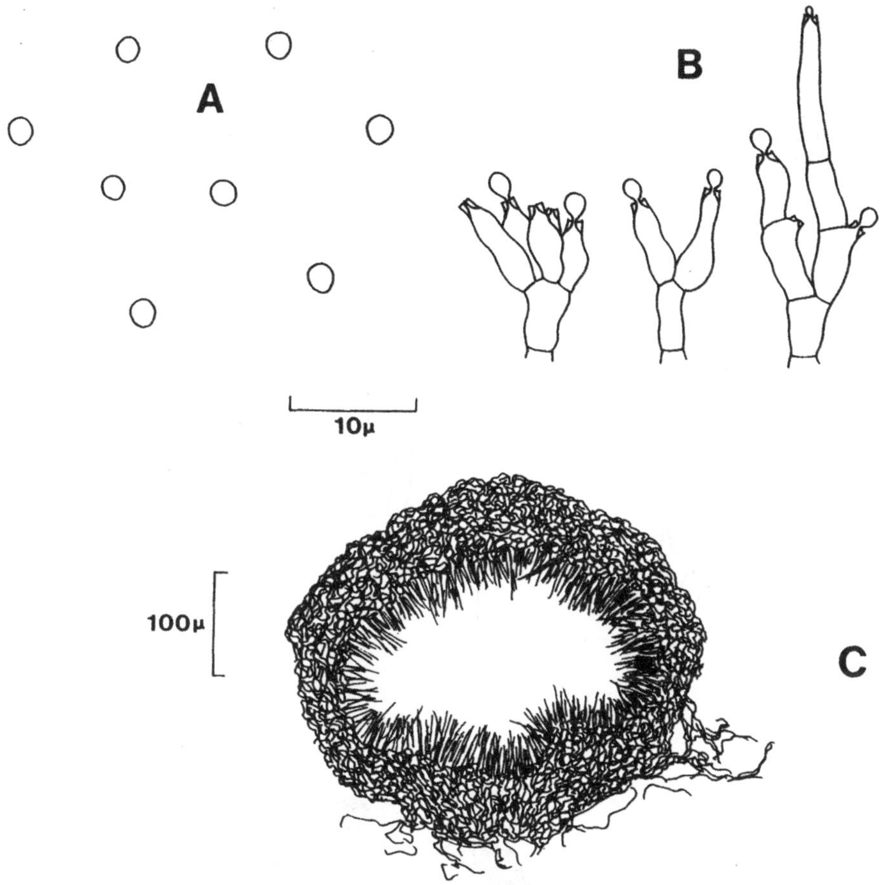

Fig. 325. *Pycnidiella resinae*, A, conidia; B, conidiophores and developing conidia; C, vertical section of a closed conidioma.

Teleomorphosis: *Nectria sinopica* Fr., *Summa Veg. Scand.*: 388 (1849).

Conidiomata 350-500 μ diam., 450-500 μ high, red; ostiole 100-150 μ diam. Conidia 2-3 × 1 μ.

On dead branches of *Hedera helix*, U.K. (IMI 217872, 217870, 217874, 217875, 217784, 217873, 217871, 9766, 62651, 66666, 7475, 19096, 91526, 71245, 71244, 76789a, 105427, 62652a, 53560).

Z. PINASTRI (Karst.) Höhn., *Mitt. bot. Inst. tech. Hochsch. Wien* **8**: 90 (1931). (Fig. 326C)
Zythia pinastri Karst., *Rev. mycol.* **7**: 106 (1885).
Diplozythia scolecospora Bub. pro parte, *Annls mycol.* **2**: 399 (1904).

Teleomorphosis: *Scoleconectria cucurbitula* (Tode ex Fr.) Booth, *Mycol. Pap.* 73: 15 (1959).

Conidiomata 250-400 μ diam., dark reddish brown. Conidia 4-6 × 1 μ.

On branches of *Pinus sylvestris*, *P. radiata*, *P. rigida*, *P. excelsa*, *P. massoniana*, *P. pinea*, *P. strobus*, *P. nigra*, *P.* sp.. *Abies balsamea*, *A. grandis*, *Cedrus deodara*, and *Taxus* sp., Austria, Netherlands, U.K., Pakistan, Hong Kong, Canada, N.J., Md., Calif., U.S.A., represented in herb. IMI by 54 collections.

59. ENDOTHIELLA Sacc., *Annls mycol.* **4**: 273 (1906).
Allantozythiella Danilova, *Bot. Mater.* **7**: 140 (1951).
Calopactis H. & P. Syd., *Annls mycol.* **10**: 82 (1912).
Sp. typ.: *E. gyrosa* Sacc.

Mycelium immersed, branched, septate, hyaline to pale brown. *Conidiomata* pseudostromatic, immersed but strongly erumpent, separate or aggregated, reddish to orange brown, often with a pruinose appearance, globose, pulvinate or rostrate, unilocular or complex and multilocular or convoluted; walls thick, at the base, sides and between locules of pale brown to hyaline textura angularis, in the upper region of progressively darker brown more thick-walled textura porrecta. *Ostioles* one or more to each locule, circular, papillate or apical on rostrate extensions of the stromata. *Conidiophores* irregularly septate and branched at the base and above, hyaline, smooth, formed from the inner cells of the locular walls. *Conidiogenous cells* enteroblastic, phialidic, determinate, integrated, apical or more frequently lateral on small branches

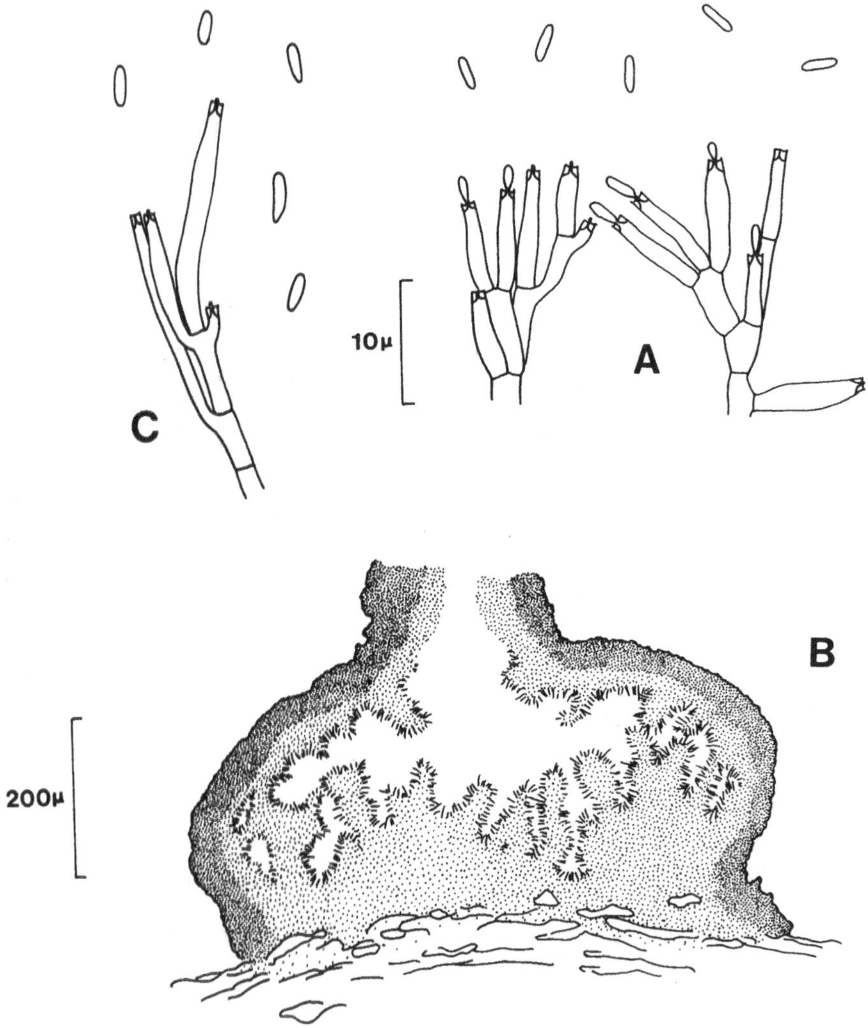

Fig. 326. A, B, *Zythiostroma mougeotii*, conidia, conidiophores, vertical section of a conidioma; C, *Z. pinastri*, conidia and a conidiophore.

immediately below septa, hyaline, smooth, markedly tapered to the apices, channel, collarette and periclinal thickening barely discernible. *Conidia* hyaline, aseptate, smooth, thin-walled, eguttulate or guttulate, ellipsoid.

Sutton (1977a) has discussed the generic synonymy. The genus was introduced to take the anamorphoses of *Endothia* species and so far it contains 4 described taxa, although all the imperfect states of published *Endothia* species have not received formal names in *Endothiella*. The conidiomata are quite variable in shape and size, from simple pycnidial to complex stromatic to rostrate pycnidial, but all have the typical red-orange brown pigmentation. This feature and the minute conidiogenous apparatus distinguish it from *Kaskaskia* Born & Crane which has brown conidiomata and easily seen, well-defined phialides.

ENDOTHIELLA state of ENDOTHIA EUGENIAE (Nutman & Roberts) J. Reid & Booth, *Can. J. Bot.* **47**: 1059 (1969). (Fig. 327)

Conidiomata comparatively simple, but rostrate, 600 µ high × 400 µ diam. Conidiophores up to 30 µ long × 2 µ wide at the base. Conidia 3.5-4.5 × 1.5-2 µ.

responsible for the sudden death disease (acute dieback) of cloves (Booth & Gibson, 1973).

On branches of *Eugenia caryophyllata*, Tanzania, Malaysia (IMI 156210, 44945, 45452, 45450, 45449, 45448a, 45453, 58569, 58568, 58388, 58567, 56425a, 49266, 45440, 45443, 45445, 45446, 44952, 44953, 44951, 44949, 44950, 44954, holotype).

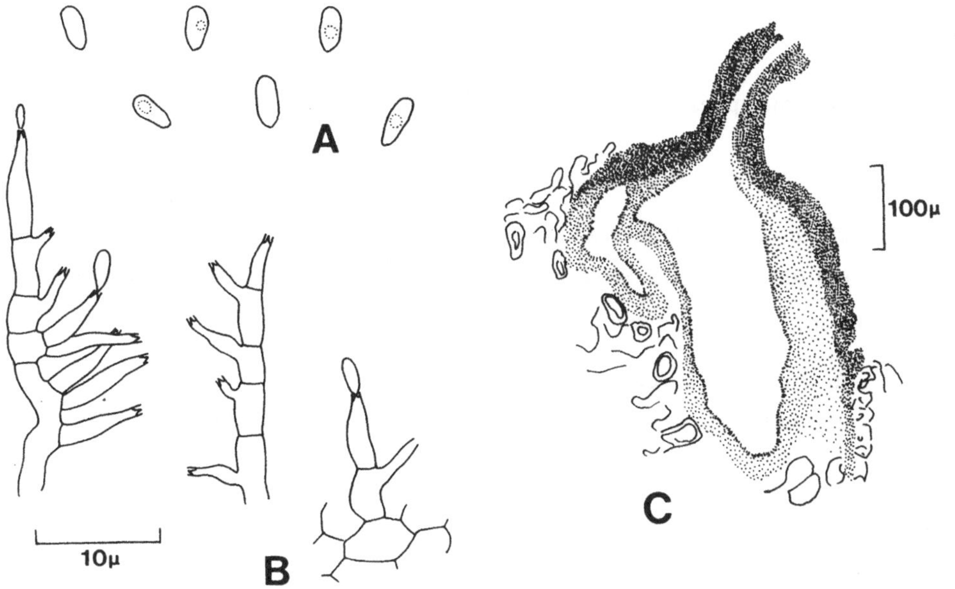

Fig. 327. *Endothiella* state of *Endothia eugeniae*, A, conidia; B, conidiophores, C, vertical section of a conidioma.

60. CHAETOMELLA Fckl, *Jb. nassau. Ver. naturk.* **23-24**: 402 (1870).
Volutellospora Mathur & Thirum., *Sydowia* **18**: 38, *1964* (1965).
Harikrishnaella Singh & Sarbhoy, *Sydowia* **25**: 66, *1971* (1972).
Original spp.: *C. oblonga* Fckl, *C. atra* Fckl
Lectotype sp.: *C. oblonga* Fckl, fide Höhnel (1915e).

Mycelium superficial or immersed, hyaline to pale brown, branched, septate. *Conidiomata* pycnidial, separate, globose but opening widely, sessile or very shortly stipitate, dark brown to black but pale when young, superficial, unilocular, thick-walled; wall of distinct layers of textura angularis, an outer region 1 cell thick of light brown pseudoparenchyma, a median region 4-6 cells thick of dark brown, thick-walled pseudoparenchyma arranged radially, and an inner layer several cells thick of thin-walled, hyaline pseudoparenchyma. *Ostiole* (raphe) single, longitudinal. *Setae* brown, smooth, thick-walled, septate, unbranched, apices coiled, acuminate or obtuse. *Conidiophores* hyaline, branched verticillately or irregularly, filiform, septate, smooth, with acropleurogenous conidia and formed from the inner cells of the pycnidial wall. *Conidiogenous cells* enteroblastic, phialidic, determinate, integrated, filiform, hyaline, smooth, channel and collarette minute. *Conidia* hyaline, aseptate, cymbiform to allantoid, ± guttulate.

The genus has been revised by Stolk (1963) and Sutton & Sarbhoy (1976) and the latter authors provided a key to the species.

C. ACUTISETA Sutton & Sarbhoy, *Trans. Br. mycol. Soc.* **66**: 299 (1976). (Fig. 329D)

Conidiomata 600-850 × 500 µ. Setae 4-6 septate, 500-750 × 12-15 µ, acute, pale brown to hyaline. Conidia 7.5-10 × 1.5-2 µ.

On leaves of *Acer* sp., twigs of *Quercus* sp. and from soil, Pakistan, U.K., Canada (IMI 81967 (holotype), 188506, 145683).

C. ACUTISETA var. NIGRA Sutton & Sarbhoy, *Trans. Br. mycol. Soc.* **66**: 299 (1976). (Fig. 329B, C)

Conidiomata 450-720 × 400 µ. Setae of two types, 1) acute, blackish brown, acuminate, 250-400 × 10.5-12 µ and 2) obtuse, dark brown, flexuous, 450-850 × 6-7 µ. Conidia 7.5-9.5 × 1.5-2 µ.

From soil, New Caledonia (IMI 104157, holotype).

C. CIRCINOSETA Stolk, *Trans. Br. mycol. Soc.* **46**: 413 (1963). (Fig. 329A, F)

Conidiomata 300-625 × 250-400 µ. Setae of two types, 1) dark reddish brown, 12 septate, 700-1000 × 10-16 µ, coiled 1-4 times at the apices, and 2) 4-6 septate, brown, 60-180 × 5-10 µ, pale brown, clavate. Conidia 8.5-10.5 × 1.5-2 µ.

From soil and on leaf litter of *Eucalyptus* spp., Brazil (IMI 192514n, 186982a, 190855k, 101650 (holotype)).

C. OBLONGA Fckl, *Jb. nassau. Ver. Naturk.* **23-24**: 402 (1870).

Conidiomata 160-500 × 120-250 µ. Setae 200-320 × 8-10 µ, clavate, dark brown. Conidia 8-11 × 1.5-2 µ.

On leaves of *Rosa* sp., bark of *Quercus* and *Chaemaecyparis lawsoniana*.

C. RAPHIGERA Swift, *Mycologia* **22**: 165 (1930). (Figs. 328, 329E)

Conidiomata 150-450 × 100-200 µ. Setae 40-170 × 2-6 µ, pale brown, clavate, often hooked at the apices. Conidia 4-7 × 1.5-2.5 µ.

Recorded from a wide variety of substrates throughout the world (Sutton & Sarbhoy, 1976) but most commonly from soil. Represented in herb. IMI by 30 collections.

61. CORNUCOPIELLA Höhn., *Sber. Akad. Wiss. Wien* **124**: 118 (1915).
Sp. typ.: *C. mirabilis* Höhn.

Mycelium immersed and superficial, branched, septate, dark brown. *Conidiomata* dark brown, superficial, cylindrical, sessile, thick-walled; lateral wall consisting of brown, thick-walled, smooth textura porrecta and a short basal foot of similar tissue. *Ostiole* absent, conidiomata open from early stages in development. *Conidiophores* hyaline, repeatedly branched at the base and above, often in association with septa, long, filiform, cylindrical, formed from the base and sides of the inner layer of the conidioma. *Conidiogenous cells* enteroblastic, phialidic, integrated, determinate, apertures apical on main or lateral branches, hyaline, smooth, cylindrical, collarettes and apertures minute. *Conidia* hyaline, aseptate, thin-walled, smooth, eguttulate, ellipsoid.

The genus is monotypic.

C. MIRABILIS Höhn., *ibid.*: 118 (1915). (Fig. 330)

Conidiomata up to 150 µ high × 60 µ wide. Conidiophores up to 40 µ long × 1.5 µ wide. Conidia 2.5-3.5 × 1.5 µ.

On bare wood of *Fagus*, Austria (IMI 202938 ex FH).

62. ASCOCHYTOPSIS P. Henn., Engler's *Bot. Jahrb.* B38: 117 (1905).
Scleroparodia Petrak, *Annls mycol.* **32**: 341 (1934).
Cryptogene Syd., *Annls mycol.* **37**: 240 (1939).
Cryptogenella Syd., *Annls mycol.* **37**: 241 (1939).
Sp. typ.: *A. vignae* P. Henn.

Mycelium immersed, branched, septate, hyaline to pale brown. *Conidiomata* superficial, sessile or short-stipitate,

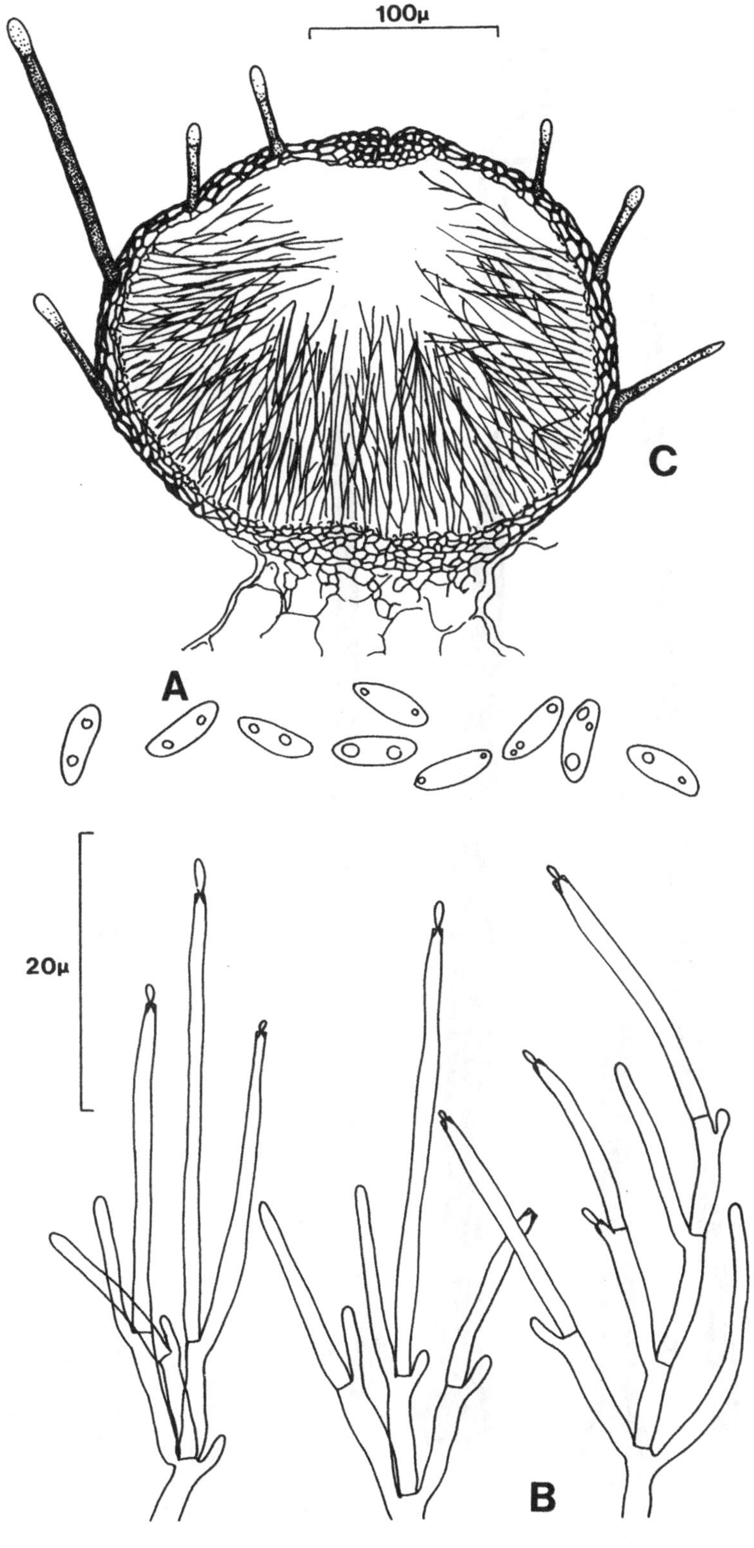

Fig. 328. *Chaetomella raphigera*, A, conidia; B, conidiophores; C, vertical section of a closed conidioma.

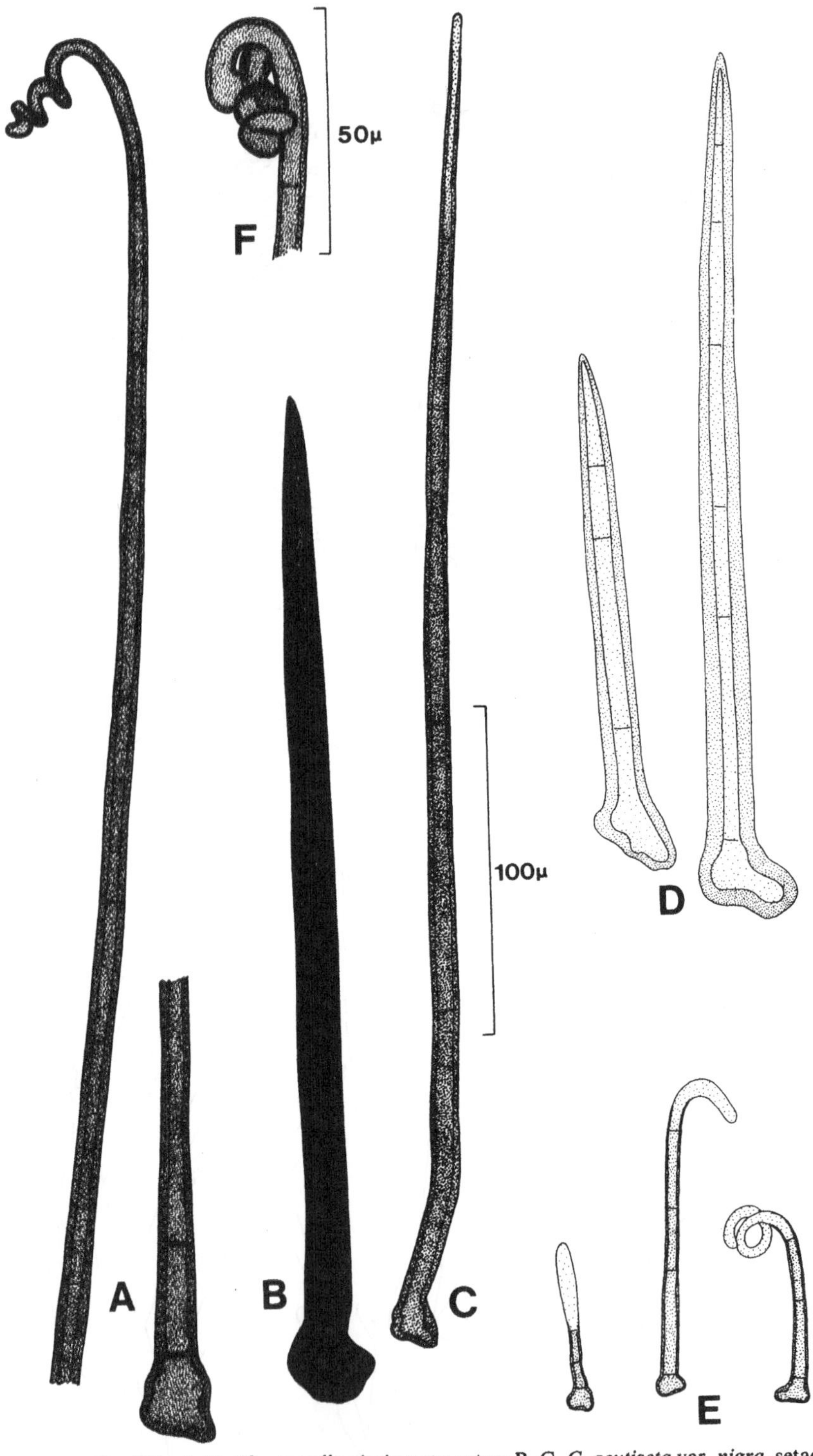

Fig. 329. A, F, *Chaetomella circinoseta*, setae; B, C, *C. acutiseta* var. *nigra*, setae; D, *C. acutiseta*, setae; E, *C. raphigera*, setae.

separate, globose, dark brown to black, unilocular, occasionally confluent, the basal region of small, dark brown textura angularis; walls thick, composed of an outer layer of dark brown, thick-walled textura globulosa, the cells often protruding, and an inner layer of paler, thin-walled isodiametric textura angularis. *Ostiole* absent, dehiscence by irregular breakdown of upper wall, finally cupulate. *Conidiophores* when present, septate, hyaline, cylindrical, branched only at the base, formed from the inner cells of the wall. *Conidiogenous cells* enteroblastic, phialidic, determinate, integrated or discrete, hyaline,

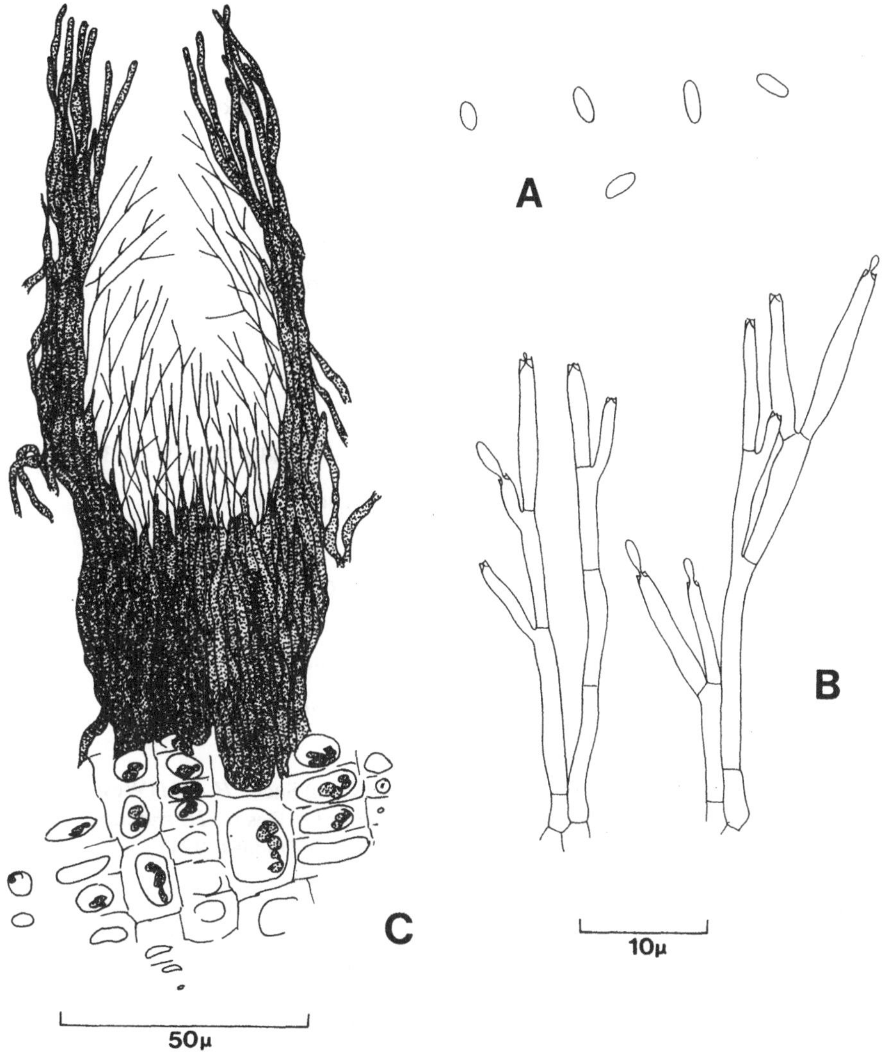

Fig. 330. *Cornucopiella mirabilis*, A, conidia; B, conidiophores; C, vertical section of a conidioma.

smooth, tapered gradually to the apices, collarette and aperture minute. *Conidia* acrogenous, hyaline, aseptate, straight or falcate, thin-walled, eguttulate, base obtuse, apex tapered.

Höhnel (1910e, 1911c, 1916f) repeatedly reduced *Ascochytopsis* to synonymy with *Oncospora* Kalchbr. but whereas the latter was described from *Capparis citrifolia*, all the taxa here included in *Ascochytopsis* are from Leguminosae. This feature combined with the characteristic conidiomata, conidial ontogeny and morphology are sufficient to keep *Ascochytopsis* separate. It is quite possible they represent conidial states of *Parodiella* species.

KEY

1. Conidia falcate 2
1. Conidia ± straight 3
   2. Conidia 28-40 × 2.5-3.5 µ .... *A. parodiellae*
   2. Conidia 15-19 × 2 µ .... *A. vignae*
3. Conidia 3-4 × 1-1.5 µ .... *A. leguminosarum*
3. Conidia 7.5-13 × 1 µ .... *A. sydowii*

A. VIGNAE P. Henn., *ibid.*: 117 (1905). (Fig. 332B, C, D)

Conidiomata up to 200 µ diam. Conidiophores 7-13 × 3 µ. Conidia 15-19 × 2 µ.

On living leaves of *Vigna* cf. *vexillata* (IMI 196496, holotype).

A. leguminosarum (Petrak) comb. nov.
Scleroparodia leguminosarum Petrak, *Annls mycol.* **32**: 341 (1934).

Conidiomata 100-180 µ diam. Conidiogenous cells 5-8 µ diam. Conidia 3-4 × 1-1.5 µ.

On living leaves of *Phaseolus lunatus*.

A. parodiellae (Syd.) comb. nov. (Fig. 331)
Cryptogenella parodiellae Syd., *Annls mycol.* **37**: 241 (1939).

Conidiomata 100-200 µ diam. Conidiogenous cells 9-16 × 3-4 µ. Conidia 28-40 × 2.5-3.5 µ.

On living leaves of *Dolichos lupiniflorus*, S. Rhodesia (IMI 6364, holotype).

A. sydowii nom. nov. (Fig. 332A)
Cryptogene parodiellae Syd., *Annls mycol.* **37**: 240 (1939), non *Cryptogenella parodiellae* Syd. (= *Ascochytopsis parodiellae* (Syd.) Sutton).

Conidiomata 180-300 µ diam. Conidiogenous cells 5-20 × 2-3.5 µ. Conidia 7.5-13 × 1 µ.

On leaves of *Dolichos lupiniflorus*, S. Rhodesia (IMI 6363, holotype).

63. HAINESIA Ell. & Sacc. apud Sacc., *Syll. fung.* **3**: 698 (1884).
Original spp.: *H. rhoina* (Sacc.) Ell. & Sacc., *H. tremellina* (Sacc.) Sacc., *H. rubi* (Westd.) Sacc.
Lectotype sp.: *H. lythri* (Desm.) Höhn. (syn. *Dacryomyces lythri* Desm., *H. rhoina* (Sacc.) Ell. & Sacc.), fide Höhnel (1918a), Clements & Shear (1931), Petch (1943), Shear & Dodge (1921).

Mycelium immersed, branched, septate, pale brown to hyaline. *Conidiomata* semi-immersed, finally superficial, separate, very pale brown, initially globose, later opening widely to become cupulate; basal wall of large, pale brown thin-walled textura angularis, lateral walls of pale, thin-walled, textura intricata to porrecta which become progressively more hyaline towards the apex of the conidioma, finally separating and splayed out as free hyphae. *Conidiophores* hyaline, branched at the base and above, septate, filiform, long, with acropleurogenous conidia, formed all round the inner wall of the conidioma. *Conidiogenous cells* enteroblastic, phialidic, determinate, integrated, filiform, hyaline, smooth, channel and collarette minute. *Conidia* hyaline, aseptate, cymbiform to allantoid, ± guttulate, acute at each end.

The most detailed account of the genus is by Shear & Dodge (1921) following the work of Höhnel (1918a). There is general agreement on *H. rhoina* as the lectotype species and this name was replaced by *H. lythri* (Desm.) Höhn. by Shear & Dodge (1921).

About 30 species have been described in the genus, and since the type is cosmopolitan it seems quite likely that many are the same species.

H. LYTHRI (Desm.) Höhn., *Sber. Akad. Wiss. Wien* **115**: 687 (1906). (Fig. 333)
Dacromyces lythri Desm., *Pl. Crypt. Fr.* no. 1545 (1846).
Sphaeronaema corneum Cke & Ell., *Grevillea* **6**: 84 (1878).
Gloeosporium tremellinum Sacc., *Michelia* **2**: 168 (1880).

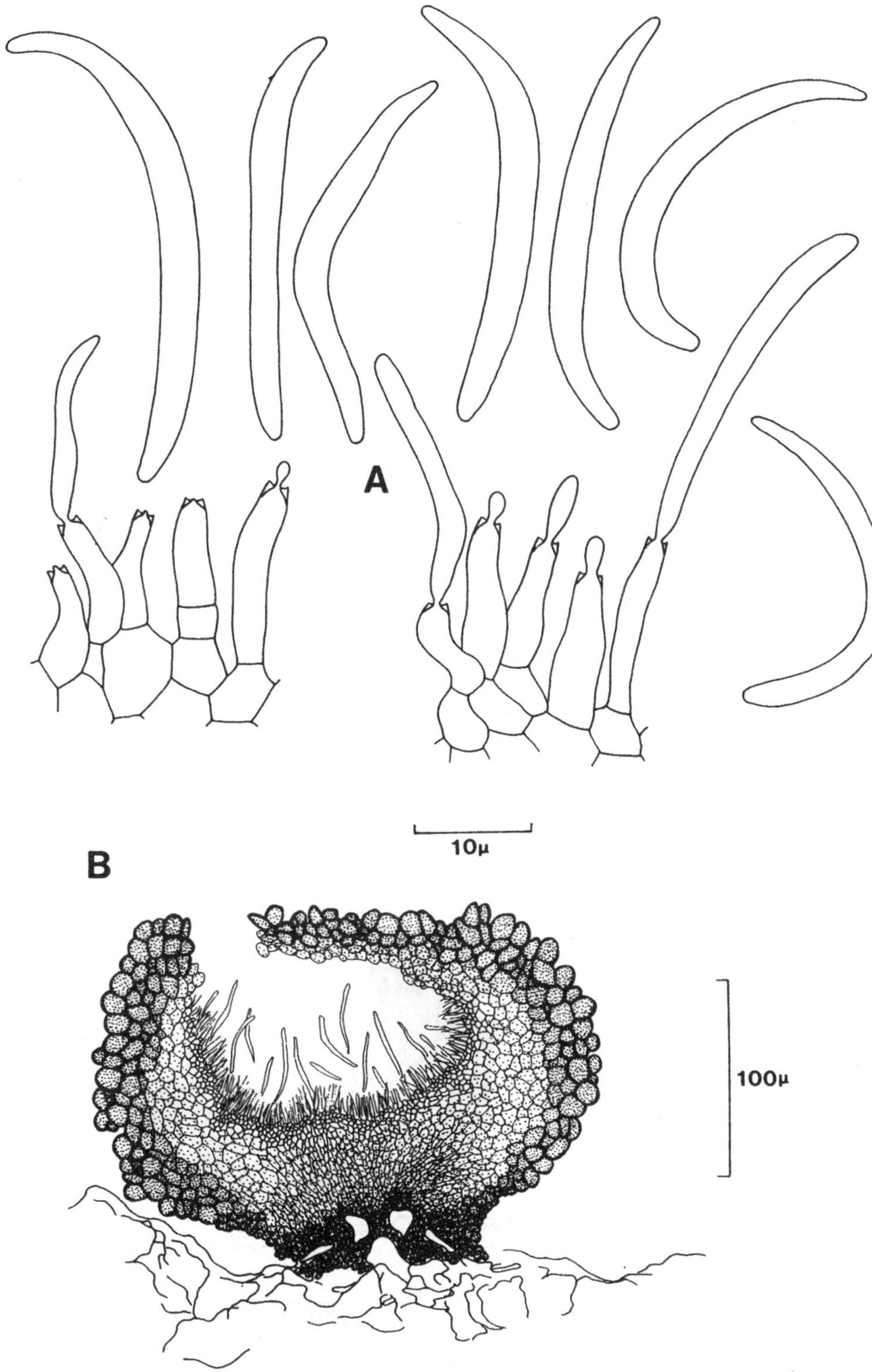

Fig. 331. *Ascochytopsis parodiellae*, A, conidia and conidiophores with developing conidia; B, vertical section of a conidioma.

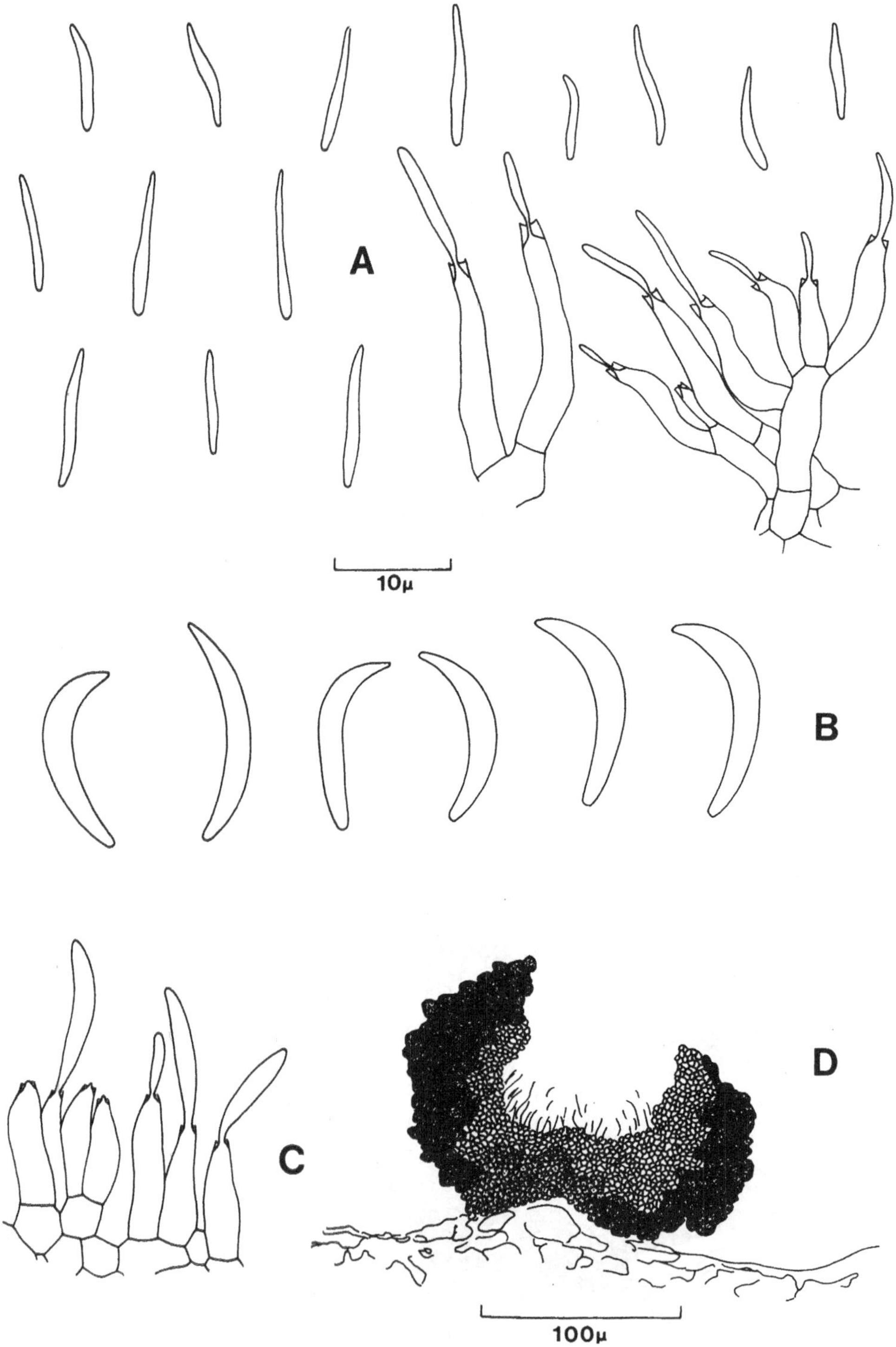

Fig. 332. A, *Ascochytopsis sydowii*, conidia and conidiophores with developing conidia; B, C, D, *A. vignae*; conidia, conidiogenous cells and developing conidia; vertical section of a conidioma.

Hainesia tremellina (Sacc.) Sacc., *Syll. fung.* **3**: 699 (1884).
Gloeosporium rhoinum Sacc., *Fungi italici pl.* 1035 (1881).
Hainesia rhoina (Sacc.) Ell. & Sacc., *Syll. fung.* **3**: 699 (1884).
Hymenula rhoina (Sacc.) Bub. & Kabát apud Kabát & Bub., *Fungi Imp. Exs.* no. 749 (1910).
Tubercularia rhois Halsted, *Bull. Torrey bot. Cl.* **20**: 251 (1893).
Hainesia epilobii Eliasson, *Bih. K. Sv. Vet. Akad. Handl.* 3, **22**: 16 (1896).
Hainesia castaneae Oud., *Ned. Kruid. Arch. Ver. Med. Ned. Bot. Ver.* 3, **2**: 755 (1902).
Hainesia rostrupii Oud., *ibid.*: 756 (1902).
Tubercularia zythioides C. Massal., *Madonna Verona* **2**: 39 (1908).
Patellina fragariae Stev. & Peterson, *Phytopath.* **6**: 264 (1916).
Hainesia lunata Kranz, *Sydowia* **22**: 364 (1969).
Hainesia jabalpurensis Sahni, *Mycopath. Mycol. appl.* **36**: 283 (1968).

Teleomorphosis: *Pezizella oenotherae* (Cke & Ell.) Sacc., *Syll. fung.* **8**: 278 (1889).

Conidiomata up to 200 μ diam. Conidiophores up to 50 μ long × 1-2 μ wide. Conidia 5-7.5 × 1.5-2 μ.
The species and diseases with which it is associated are described by Sutton & Gibson (1977).

On leaves of a wide variety of hosts, including *Acer pseudoplatanus, Calluna* sp., *Dissotis paucistellata, Epilobium angustifolium, Eucalyptus saligna, E.* sp., *Fragaria vesca, F.* sp., *Paeonia suffructicosa, Rosa sentifolia, R.* sp., *Rubus thyrsoides, R.* sp., *Woodfordia fruticosa* and from water, G.B., Germany, Australia, New Zealand, Guinea, Zambia, Malawi, Korea, Malaysia, India, Singapore (IMI 47075, 190108, 98567, holotype *H. lunata*, 107968, 167048, 160397, 158404, 159932, 168429, 175854, 17681, 47512, 172556, 191770b, 80261, 132983, 16436, 59322, 110566, holotype *H. jabalpurensis*, 112818, 188330, 194263, 200045, 196779a).

64. AGYRIELLOPSIS Höhn., *Annls mycol.* **1**: 404 (1903).
Sp. typ.: *A. caeruleo-atra* Höhn.

Mycelium partly immersed, partly superficial, branched, septate, pale brown to hyaline. *Conidiomata* eustromatic, superficial, separate, pulvinate to verruciform, dark brown to black, becoming excipuliform; outer tissue of brown, thin-

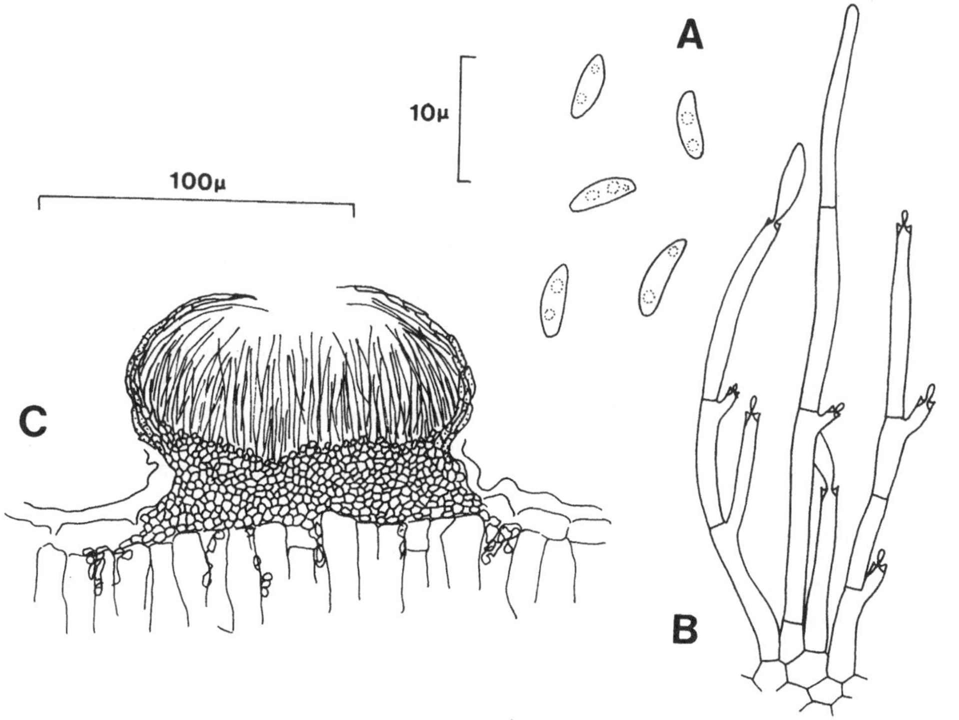

Fig. 333. *Hainesia lythri*, A, conidia; B, conidiophores; C, vertical section of a conidioma.

walled textura angularis, darker at the periphery, basal tissue of thin-walled hyaline pseudoparenchyma; multilocular, each locule separated by vertical columns of loosely arranged pseudoparenchyma and hyphae. Dehiscence by irregular breakdown of the overlying tissue. *Conidiophores* hyaline, septate, branched, irregular, smooth, cylindrical, straight or irregular, formed from the base and the locular walls. *Conidiogenous cells* enteroblastic, phialidic, determinate, integrated on distinct conidiophores, or long irregular hyphae, or discrete, lageniform or cylindrical, hyaline, with acropleurogenous conidia, channel and collarette minute. *Conidia* hyaline, ellipsoid, aseptate, smooth, thin-walled, eguttulate.

The holotype was insufficient to allow an accurate description of the conidiomata. This has therefore been fabricated from the original description and fragments of tissue on von Höhnel's slide. The genus is monotypic.

A. CAERULEO-ATRA Höhn., *ibid.*: 404 (1903). (Fig. 334)

Conidiomata 400-500 µ wide × 250 µ high. Conidiophores variable in length. Conidiogenous cells up to 7 µ long. Conidia 2.5-3 × 1 µ.

On bare wood of *Abies pectinata*, Czechoslovakia (holotype, FH).

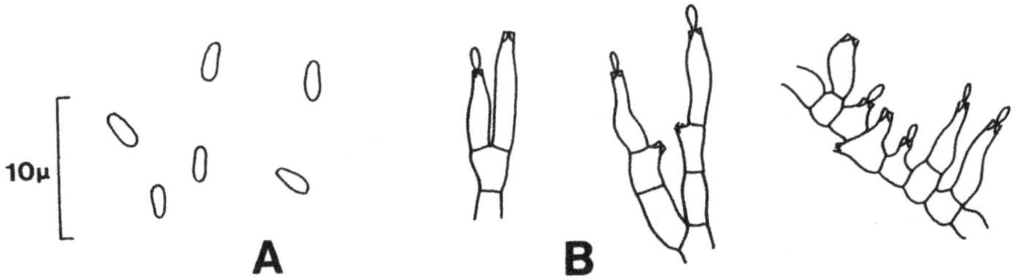

Fig. 334. *Agryiellopsis caeruleo-atra* **A, conidia, B, conidiophores.**

65. WAYDORA Sutton nom. nov. apud Sutton, V.G. Rao & Mhaskar, *Trans. Br. mycol. Soc.* **67**: 248 (1976).
Pulvinaria Rodway, *Pap. Proc. R. Soc. Tasmania 1917*: 110 (1918), non Bonorden, 1851.
Sp. typ.: *W. typica* (Rodway) Sutton (syn. *Pulvinaria typica* Rodway).
Illustration: Sutton, Rao & Mhaskar (1976).

Mycelium immersed, branched, septate, brown. *Conidiomata* eustromatic, superficial, sessile, separate, pulvinate to globose, carbonous, blackish brown, consisting of 5-8 vertically divided segments (as in an orange) surrounding a central column; base immersed, composed of dark brown, thick-walled textura angularis; central column of thin-walled, pale brown textura intricata; each stromatal locule with a distinct wall consisting of 4 separate outer layers and 3 separate inner layers where it joins the central column; outer layer of dark blackish brown, thick-walled textura angularis, next layer of thick-walled, pale brown textura oblita, inner layer of dark brown, thin-walled textura angularis becoming smaller in the ostiolar region, finally there is an inner layer of small-celled, hyaline textura angularis from which the conidiogenous apparatus arises. *Ostiole* single to each locule, distinct, circular, nonpapillate. *Conidiophores* septate, branched at the base and along their lengths irregularly, ± cylindrical to irregular, hyaline, smooth, formed from the inner wall of the locules. *Conidiogenous cells* enteroblastic, phialidic, integrated, determinate, hyaline, cylindrical or tapered to the apices, collarette and channel minute. *Conidia* aseptate, hyaline, smooth, fusiform, straight to allantoid, eguttulate.

The genus is monotypic.

W. TYPICA (Rodway) Sutton, *ibid.*: 248 (1976).
Pulvinaria typica Rodway, *Pap. Proc. R. Soc. Tasmania 1917*: 110 (1918).

Conidiomata 600-800 µ diam.; cavities up to 400 µ wide. Conidiophores up to 30 µ long. Conidia 6-7 × 1 µ.

On capsules of *Eucalyptus globulus*, *E. grandis*, *E. robusta*, *E. saligna*, *E.* sp., Tasmania, Australia, Brazil, Bolivia, Fla., U.S.A. (IMI 188270, holotype ex K, 188855c, 192518k, 194557b, 202855b, 194558g, 202852a, 194555b, 194556b, 194559h.

66. STILBOPHOMA Petrak, *Bot. Arch.* **43**: 93 (1941).
Sp. typ.: *S. microspora* Petrak

Mycelium immersed, branched, septate, pale brown to hyaline. *Conidiomata* eustromatic, epidermal to subepidermal, erumpent, hemispherical to pulvinate, separate, unilocular, carbonous, blackish brown, pushing up a flap of epidermis; consisting of a thick dome-shaped upper wall of small-celled, dark brown textura angularis, darkest at the periphery, becoming medium brown towards the cavity but lined finally with a layer of very dark brown pseudoparenchyma, basal wall or hypostroma composed of thin-walled, pale brown textura angularis. *Ostiole* not observed but the centre of the stroma often appears papillate. *Conidiophores* long, filiform, cylindrical or irregular, probably branched at the base, hyaline, with many septa delimiting small cells, aggregated into synnemata within the locule, formed from the basal hypostroma. *Conidiogenous cells* enteroblastic, phialidic, determinate, integrated, hyaline, smooth, with lateral apertures immediately below transverse septa, apparatus extremely small, the possibility of polyphialides being present could not be verified due to minute dimensions. *Conidia* hyaline, aseptate, cylindrical or slightly curved, thin-walled, smooth, ± guttulate.

The genus had remained monotypic until the addition of *S. inaequalis*.

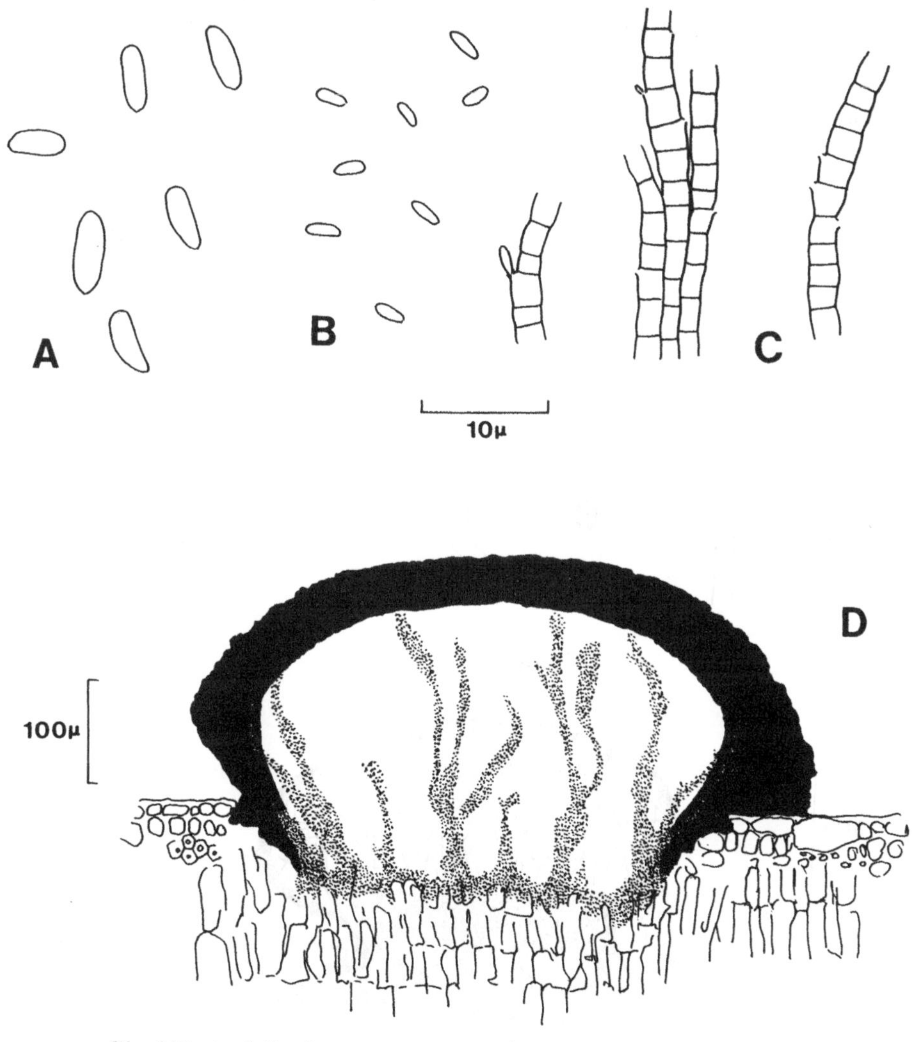

Fig. 335. A, *Stilbophoma inaequalis*, conidia; B, C, D, *S. microspora*, conidia, conidiophores, vertical section of a conidioma.

S. MICROSPORA Petrak, *ibid.*: 93 (1941). (Fig. 335B, C, D)

Conidiomata up to 600 µ diam. Conidia 2-3 × 1 µ.

On fronds of *Borassus* sp., *Palmae*, India (IMI 33195, holotype, 204597).

S. inaequalis (Sacc. & Trott.) comb. nov. (Fig. 335A)
Plenodomus inaequalis Sacc. & Trott., *Bull. Soc. bot. Belg.* **38**: 167 (1899).

Conidiomata up to 450 µ diam. Conidia 4.5-7 × 2-2.5 µ.

On fronds of *Pandanus* sp., Congo (IMI 187914, holotype).

67. PRAGMOPYCNIS Sutton & Funk, *Can. J. Bot.* **53**: 522 (1975).
Sp. typ.: *P. pithya* Sutton & Funk

Mycelium immersed, branched, septate, hyaline to pale brown. *Conidiomata* eustromatic, separate, sparse, superficial, shortly stipitate, subglobose, black, unilocular, thick-walled; walls of dark brown, thick-walled textura angularis on the outside, becoming thinner walled and paler towards the inside. *Ostiole* absent, dehiscence by irregular breakdown of the upper wall. *Conidiophores* septate, hyaline, branched irregularly and ramifying throughout the locule, mainly originating from the base of the conidioma. *Conidiogenous cells* enteroblastic, polyphialidic, integrated, indeterminate, terminal or interclary, hyaline, apertures and collarettes minute. *Conidia* hyaline, aseptate, guttulate, fusiform, allantoid, smooth.

The genus contains a single species.

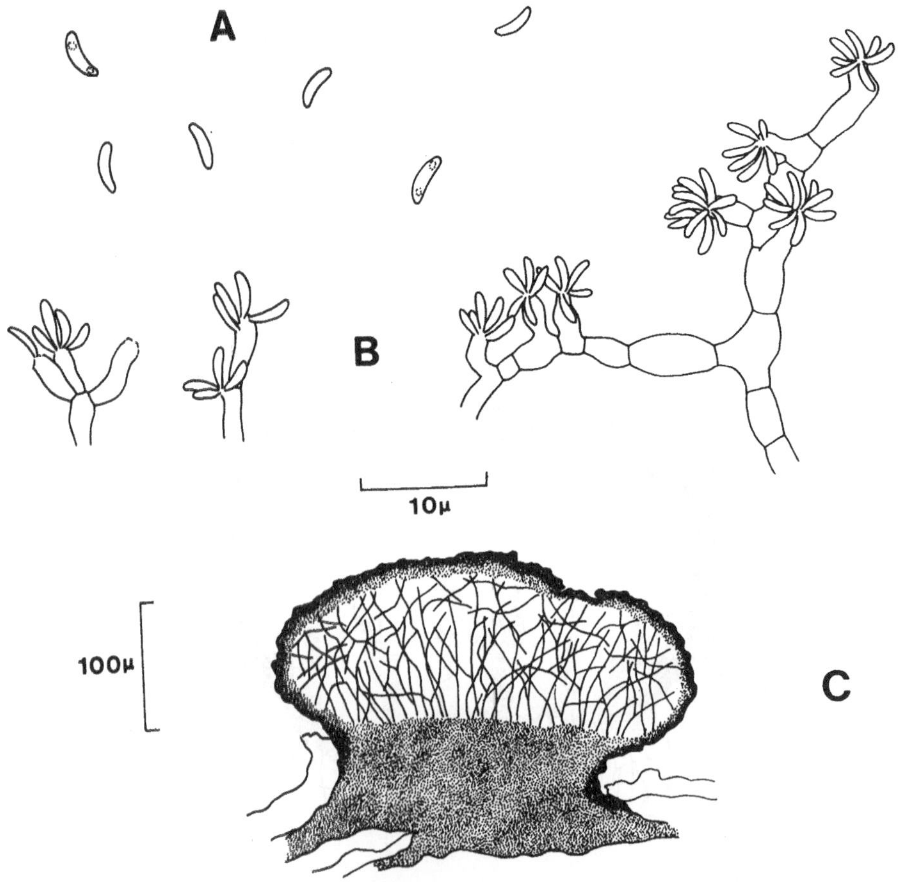

Fig. 336. *Pragmopycnis pithya*, A, conidia; B, conidiophores with developing conidia; C, vertical section of a closed conidioma.

P. PITHYA Sutton & Funk, *ibid.*: 522 (1975). (Fig. 336)

Teleomorphosis: *Pragmopora pithya* (Fr.) Groves, *Can. J. Bot.* **45**: 176 (1967).

Conidiomata 200-500 μ diam. Conidiophores up to 2 μ wide. Conidia 3-4 × 1 μ.

On twigs of *Pseudotsuga menziesii*, B.C., Canada, N.Y., U.S.A. (IMI 164477, isotype, DAVFP 20123, 20640, 20639, DAOM 113865).

68. SIRODOTHIS Clements, *Genera of Fungi*: 176 (1909).
Pleurophomella Höhn., *Sber. Akad. Wiss. Wien* **123**: 123 (1914).
Sp. typ.: *S. populnea* (Thüm.) Sutton & Funk (syn. *Dothiorella populnea* Thüm., *S. populi* Clements).
Illustrations: Groves (1952); Sutton & Funk (1975).

Mycelium immersed, branched, septate, pale brown. *Conidiomata* eustromatic, caulicolous, initially immersed then strongly erumpent, consisting of one to several, caespitose, unilocular, globose, pycnidium-like structures originating from a common basal stroma, dark brown to black, glabrous or pruinose, thick-walled; walls composed of brown, thick-walled textura angularis which is progressivley thicker, darker and more sclerotioid towards the periphery; basal stroma of uniformly very dark brown textura angularis. *Ostioles* absent dehiscence by irregular breakdown of the overlying tissue. *Conidiophores* filiform, septate, hyaline, branched mostly at the base formed from the inner cells of the locular walls. *Conidiogenous cells* enteroblastic, phialidic, integrated, determinate, apical or intercalary, apertures arising immediately below transverse septa, collarette, channel and periclinal thickening minute. *Conidia* hyaline, aseptate, eguttulate, cylindrical.

The generic synonymy has been explained by Sutton & Funk (1975), and although only four species have been accepted by formal combination of epithets in *Sirodothis*, there are several more species which function as the unnamed *Sirodothis* states of *Tympanis* species (as *Pleurophomella* states in Groves, 1952). The distinctions between these anamorphoses of *Tympanis* species cannot be used to separate taxa so without the teleomorphosis it is often difficult or impossible to determine which taxon one is dealing with.

S. POPULNEA (Thüm.) Sutton & Funk, *Can. J. Bot.* **53**: 524 (1975).
Dothiorella populnea Thüm., *Bull. Soc. Imper. Nat. Moscou* **55**: 226 (1880).
Pleurophomella populnea (Thüm.) Petrak, *Sydowia* **15**: 212 (1962), not validly published.
Pleurophomella spermatiospora Höhn., *Sber. Akad. Wiss. Wien* **125**: 44 (1916).
Sirodothis populi Clements, *Genera of Fungi*: 176 (1903).

Teleomorphosis: *Tympanis spermatiospora* (Nyl.) Nyl., *Notis. Sällsk. Faun. Fl. fenn.* **10**: 70 (1868).

Conidiomatal locules in groups of up to 15, each 200-400 μ diam. Conidiophores 13-32 × 2-2.5 μ. Conidia 4-4.5 × 1 μ.

On branches of *Populus tremula*, *P. tremuloides*, *P.* sp., Colo., U.S.A., Ont., Canada, Australia (IMI 184815, holotype ex BPI, 206793, 73228).

Additional taxa accepted in *Sirodothis* but not separable by morphological characteristics from *S. populnea* are *S. saligna* (Höhn.) Sutton & Funk, *S. inversa* (Fr.) Sutton & Funk and *S. columnaris* (Wallr.) Sutton & Funk.

69. KASKASKIA Born & Crane, *Phytopathology* **62**: 926 (1972).
Sp. typ.: *K. gleditsiae* Born & Crane

Mycelium immersed, branched, septate, hyaline to pale brown. *Conidiomata* eustromatic, immersed, becoming erumpent, ± sessile, separate or confluent, pulvinate to irregular, carbonous, blackish brown, tissue of pale brown textura angularis, especially in the well-developed basal region, darker, thick-walled and sclerotioid at the periphery and surrounding the individual locules; locules single, pycnidial like, sometimes convoluted, lacking ostioles. Dehiscence by irregular breakdown of the locular walls. *Conidiophores* hyaline, septate, smooth, consisting of a basal cell with radiating branches from the apex, branches also septate and branched irregularly, filiform, with acrogenous conidia, formed from the inner cells of the locular walls. *Conidiogenous cells* enteroblastic, phialidic, integrated, determinate, cylindrical, hyaline, smooth, channel minute, collarette minimal, periclinal wall thickened. *Conidia* hyaline, aseptate, thin-walled, smooth, eggutulate, oblong to elliptical.

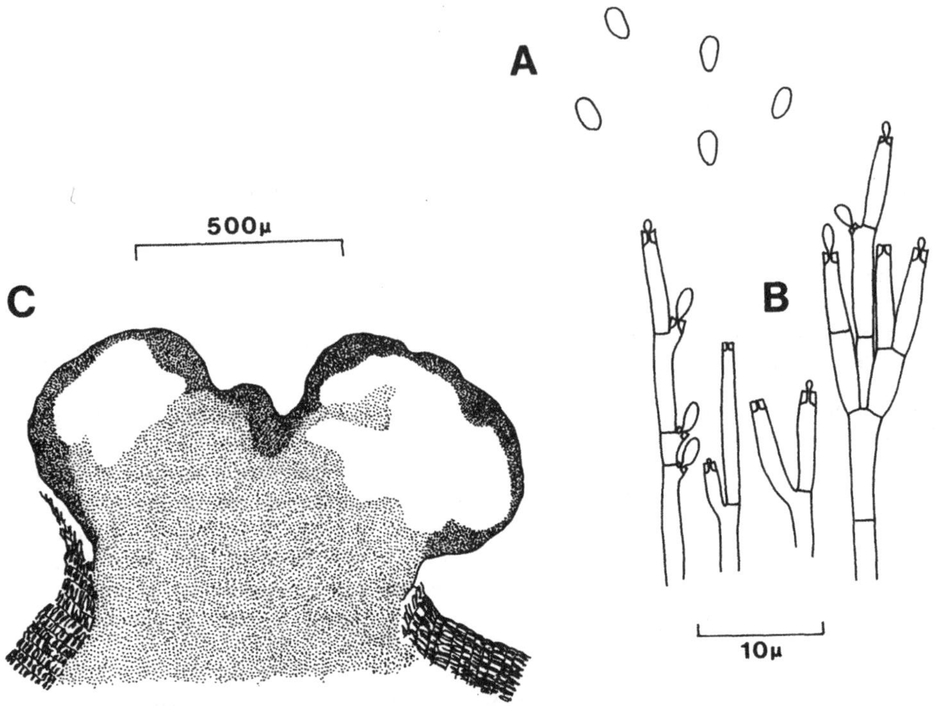

Fig. 337. *Kaskaskia gleditseae*, A, conidia; B, conidiophores; C, vertical section of a conidioma.

The diversity in conidiophore morphology in *K. gleditsiae* invites comparison with several genera of stromatic coelomycetes, notably *Endothiella*. It can be distinguished from the latter by the brown-black rather than reddish brown conidiomata, the larger phialides with prominent periclinal thickenings, and the erect conidiophores arising from textura angularis. In *Endothiella* the conidiophores are almost like textura intricata themselves and frequently remain horizontal, thus producing small vertical conidiogenous branches.

K. GLEDITSIAE Born & Crane, *ibid.*: 927 (1972). (Fig. 337)

Conidiomata 250-860 µ diam.; locules 530-940 × 300-860 µ. Conidiophores 28-98 × 1.5 µ. Conidia 2-4 × .5-1 µ.

Causing lesions on bark of *Gleditsia triacanthos* var. *inermis*, Ill., U.S.A. (IMI 151725, isotype).

70. CAMAROPYCNIS Cash, *Mycologia* **37**: 314 (1945).
Sp. typ.: *C. libocedri* Cash

Mycelium superficial, branched, septate, dark brown, forming a felty mass between needles. *Conidiomata* eustromatic, superficial with a thick hypostroma, ± shortly stipitate, separate, pulvinate to conical or subglobose, carbonous, dark brown, smooth, becoming wrinkled when dry; tissue of pale brown, thin-walled textura angularis, darker and thicker-walled at the periphery and in the hypostroma, multilocular and convoluted, each locule separated by thin-walled, hyaline pseudoparenchyma. *Ostiole* single, central, circular. *Conidiophores* hyaline, branched irregularly, septate, smooth, with acropleurogenous conidia, formed from the inner cells of the locular walls. *Conidiogenous* cells enteroblastic, phialidic, determinate, integrated cylindrical, tapered at the apices, hyaline, smooth, channel and collarette minute. *Conidia* hyaline, cylindrical, aseptate, smooth, thin-walled, apex obtuse, base ± truncate, eguttulate.

The genus is monotypic.

C. LIBOCEDRI Cash, *ibid.*: 314 (1945). (Fig. 338)

Conidiomata 700-1000 µ diam. Locules 50-100 µ diam. Conidiophores up to 30 µ long. Conidia 15.5-19 × 2.5-4 µ.

On branches of *Libocedrus decurrens* and *Pinus monophylla*, Calif., U.S.A. (IMI 51912*a*, 142176, 142177, 142279*b*, 142178).

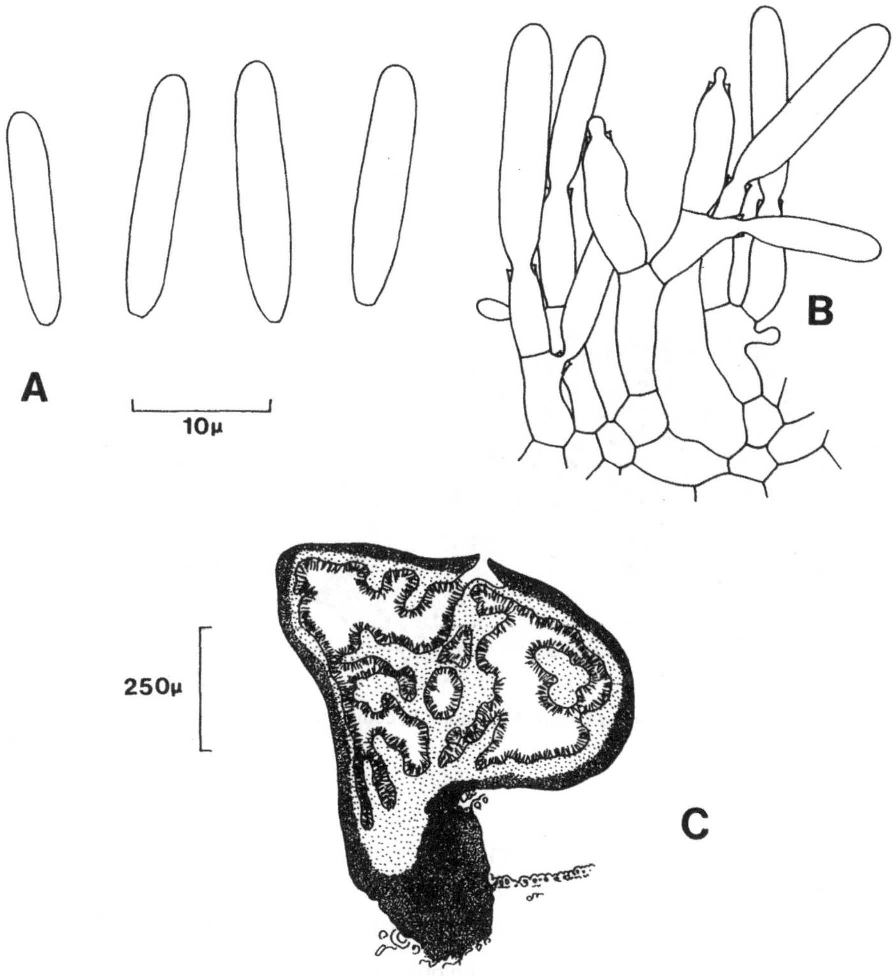

Fig. 338. *Camaropycnis libocedri*, A, conidia; B, conidiophores and developing conidia; C, vertical section of a conidioma.

71. CYTOSPORELLA Sacc., *Syll. fung.* **3**: 251 (1884).
'*Cytispora (Cytisporella)*' Sacc., *Michelia* **2**: 100 (1880).
Original spp.: *C. sycina* (Sacc.) Sacc., *C. scheidweileri* (Westd.) Sacc., *C. aesculi* (Westd.) Sacc., *C. pisiformis* (Fr.) Sacc., *C. sphaerosperma* (Westd.) Sacc., *C. mendax* Sacc. & Roum.
Lectotype sp.: *C. sycina* (Sacc.) Sacc. (syn. *Cytospora sycina* Sacc.), fide Clements & Shear (1931).

Mycelium immersed, septate, pale brown to hyaline, branched. *Conidiomata* eustromatic, separate, ± pulvinate, dark brown, peridermal to subperidermal, multilocular, locules convoluted and at varying levels but concentrated in the centre of the stroma; base broad, composed of host tissue and thin-walled pale brown textura angularis; lateral and upper walls thick, of dark brown thick-walled cells becoming darker at the periphery; locules separated by dark brown textura intricata from which conidiophores arise. *Ostiole* absent, dehiscence by irregular breakdown of upper wall. *Conidiophores* hyaline, septate, branched at the base and above, smooth, with acropleurogenous conidia, formed from the inner cells of the locular walls. *Conidiogenous cells* enteroblastic, phialidic, indeterminate, integrated, ± straight, hyaline, smooth, often formed as small lateral branches immediately below transverse septa, collarette and channel minute. *Conidia* hyaline, aseptate, thin-walled, eguttulate, smooth, ellipsoidal.

This genus, which contains more than 60 described species, has been used for *Cytospora*-like taxa with ovoid rather than allantoid conidia. However the stromata are of quite different structure and the conidiophores are not branched in the same manner. There is no modern account of the genus, even though taxa are regularly added to it.

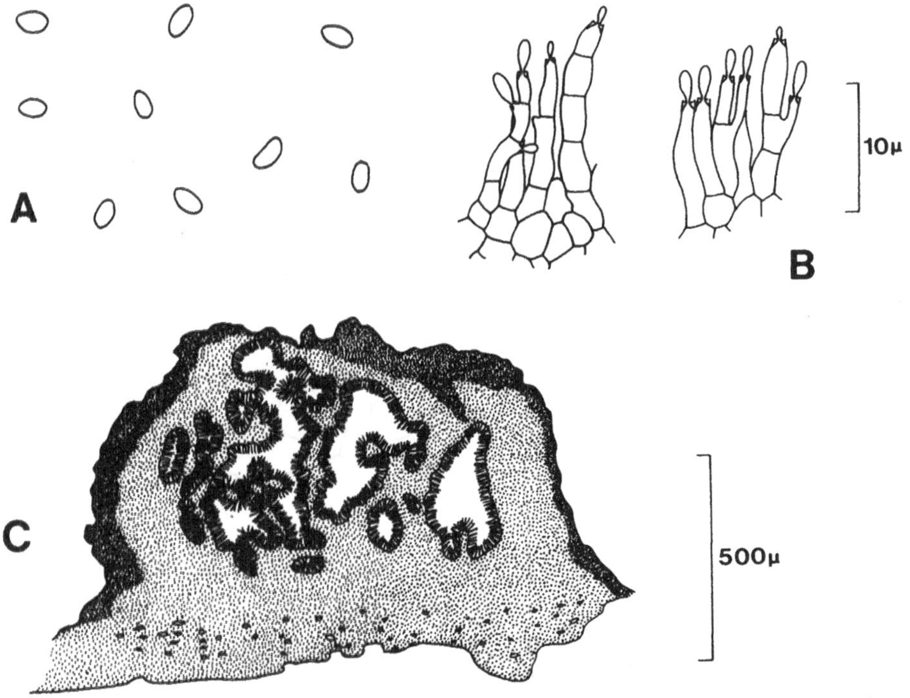

Fig. 339. *Cytosporella sycina*, A, conidia; B, conidiophores; C, vertical section of a closed conidioma.

C. SYCINA (Sacc.) Sacc., *ibid.*: 251 (1884). (Fig. 339)
Cytospora sycina Sacc., *Michelia* **2**: 100 (1880).

Stromata 1500 μ wide × 1100 μ high. Conidiophores 10-13 × 1-2 μ. Conidia 2-3 × 1-2 μ.

On dead twigs of *Ficus* sp., France (IMI 153412, holotype ex PC).

72. DOTHIORINA Höhn., *Sber. Akad. Wiss. Wien* **120**: 464 (1911).
Sp. typ.: *D. tulasnei* (Sacc.) Höhn. (syn. *Dothiorella tulasnei* Sacc.).
Illustrations: Dixon (1957); Riedl (1977).

Mycelium immersed, branched, septate, associated with green stain of wood. *Conidiomata* eustromatic, superficial, sessile, separate or aggregated and confluent, subspherical, pulvinate to moriform, fleshy and aeruginous when fresh, drying to blackish aeruginous or black, irregularly multilocular, locules at different levels; tissue of gelatinous anastomosing greenish textura oblita, somewhat parallel between locules and dark brown to black at the periphery. *Ostioles* numerous, almost papillate, circular, one to each locule. *Conidiophores* hyaline to light green, repeatedly branched, septate, with granular ornamentation towards the base, with acrogenous conidia, formed from the inner cells of the locular walls. *Conidiogenous cells* enteroblastic, phialidic, determinate, integrated, tapered at the apices, hyaline, smooth, channel and collarette minute. *Conidia* hyaline, with light green contents, aseptate, thin-walled, biguttulate, straight, cylindrical.

Dixon (1975) has given an expanded account of *D. tulasnei*. The species is figured not only with the normal phialides and conidia described above but also with phialides possessing a relatively long collarette within which 3-4 conidia are formed, as in typical *Chalara* species. If this is indeed a correct interpretation of conidiogenesis then it is one of the few examples of this type in pycnidial, acervular or stromatic genera that I know. Berthet (1964) described normal phialides in *D. tulasnei* and also gave a critical account of conidial states of other discomycetes such as *Belonidium albidium* Grelet & Crozals and *Calycella sulfurina* (Quèl.) Boud. where the *Chalara* type of phialide was found, on superficial hyphae or in pycnidia. Dixon's figures therefore require confirmation. Riedl (1977) redescribed *D. tulasnei* and *D. discoidea* (Berk. & Br.) Höhn., and added a third species, *D. subcarnea*.

D. TULASNEI (Sacc.) Höhn., *ibid.*: 464 (1911).
Dothiorella tulasnei Sacc., *Syll. fung.* **3**: 39 (1884).

Teleomorphosis: *Chlorociboria aeruginascens* (Nyl.) Kanouse ex Ram., Korf & Bat., *Mycologia* **49**: 858 (1958), and subsp. *brasiliensis* (Berk. & Cke) Dixon, *Mycotaxon* **1**: 213 (1975).

Conidiomata up to 2 mm diam. Conidiophores up to 40 µ long. Conidia 3-5 × 1-1.5 µ.

On dead green stained wood.

73. MELASMIA Lév., *Ann. Sci. nat.*, ser. 3, **5**: 276 (1846).
Sp. typ.: *M. acerina* Lév.

Mycelium immersed, branched, septate, hyaline to pale brown. *Conidiomata* eustromatic, immersed, epidermal, circular to irregular, at first discrete, later confluent forming a composite conidioma which is more or less circular, black, applanate, unilocular or multilocular, consisting of an upper wall several cells thick, of dark brown, thick-walled textura angularis situated on the inner face of the upper wall of the epidermis, and a thicker lower wall of pale brown, thin-walled, larger celled textura angularis amongst which are scattered dark brown sclerotioid cells; the conidioma is periodically divided by thick vertical columns of dark brown textura intricata. *Ostiole* absent, dehiscence by irregular rupture of the upper wall. *Conidiophores* hyaline, branched and septate at the base and irregularly above, smooth, ± straight, tapered to the apices, restricted to the lower wall. *Conidiogenous cells* enteroblastic, phialidic, integrated, occasionally discrete, determinate, hyaline, ± straight, tapered to the apices, collarette minute, channel comparatively long, periclinal wall thickened. *Conidia* hyaline, aseptate, thin-walled, eguttulate, straight or slightly curved, cylindrical or slightly tapered to the base.

More than 60 species have been described in the genus but there is no modern revision.

M. ACERINA Lév., *ibid.*: 276 (1846). (Fig. 340)

Teleomorphosis: *Rhytisma acerinum* (Pers. ex St. Am.) Fr., *Syst. myc.* **2**: 569 (1823).

Conidiomata finally up to 10 mm diam. Conidiophores 13-21 × 1.5-2 µ near the apex. Conidia 8-10 × 1 µ.
This species has been consistently reported in the literature (Allescher, 1901; Diedicke, 1914a; Nannfeldt, 1932; Terrier, 1942; Grove, 1937; Dennis, 1968) but this account represents the first accurate information on conidiogenesis. No later synonyms are known.

Causing tar spot on leaves of *Acer campestre, A. opulifolium, A. platanoides, A. pseudoplatanus, A. saccharinum, A. monspelianus, A. italicum, A. villosum, A. rubrum, A. caudatum, A. caesium, A. picitans, A.* spp., G.B., Finland, Austria, Hungary, Italy, Romania, Czechoslovakia, Turkey, India, Ont., Canada, Md., Mass., Ga., W. Va., Wisc., U.S.A. (IMI 23174, 29263, 23178, 23179, 23175, 23180, 28962, 57010, 57084, 140484, 23172, 23173, 23171, 120530, 11123, 11125, 11124, 152444, 175906, 201542, 197958, 164394, 164343, 167011, 168484, 167022, 177433, 60392, 69536, 42633, 17557, 12709, 6936, 32459, 32460, 32513, 89463, 23177, 156304, 23176, 27070, 27069, 27068, 27071).

74. PILIDIUM Kunze apud Kunze & Schmidt, *Mykol. Hefte* **2**: 92 (1823).
Sclerotiopsis Speg., *An. Soc. Cient. Argent.* **13**: 14 (1882).
Sp. typ.: *P. acerinum* Kunze

Mycelium immersed, hyaline to pale brown, branched, septate. *Conidiomata* eustromatic, separate, epidermal to subepidermal, depressed, flattened, wrinkled, dark brown, unilocular, thick-walled, composed of reddish-brown, thick-walled textura angularis, the basal wall thinner than the upper wall, hyaline towards the conidiophores. *Ostiole* absent, dehiscence by irregularly stellate rupture. *Conidiophores* 1-2 septate, cylindrical, markedly tapered at the apices, branched at the base and above, hyaline, formed only from the basal wall. *Conidiogenous cells* enteroblastic, phialidic, integrated, less often discrete, determinate, hyaline, tapered to apices, aperture and collarette minute. *Conidia* falcate, aseptate, hyaline, apex acute, base obtuse, thin-walled, eguttulate.

Sclerotiopsis Speg. was maintained as a separate genus by Shear & Dodge (1921) who considered *S. concava* (Desm.) Shear & Dodge as the pycnidial state of *Pezizella lythri* (Desm.) Shear & Dodge. Petrak (1951a) regarded *S. australasica* as distinct from *S. concava* (with which Shear & Dodge (1921) had placed it in synonymy) and transferred

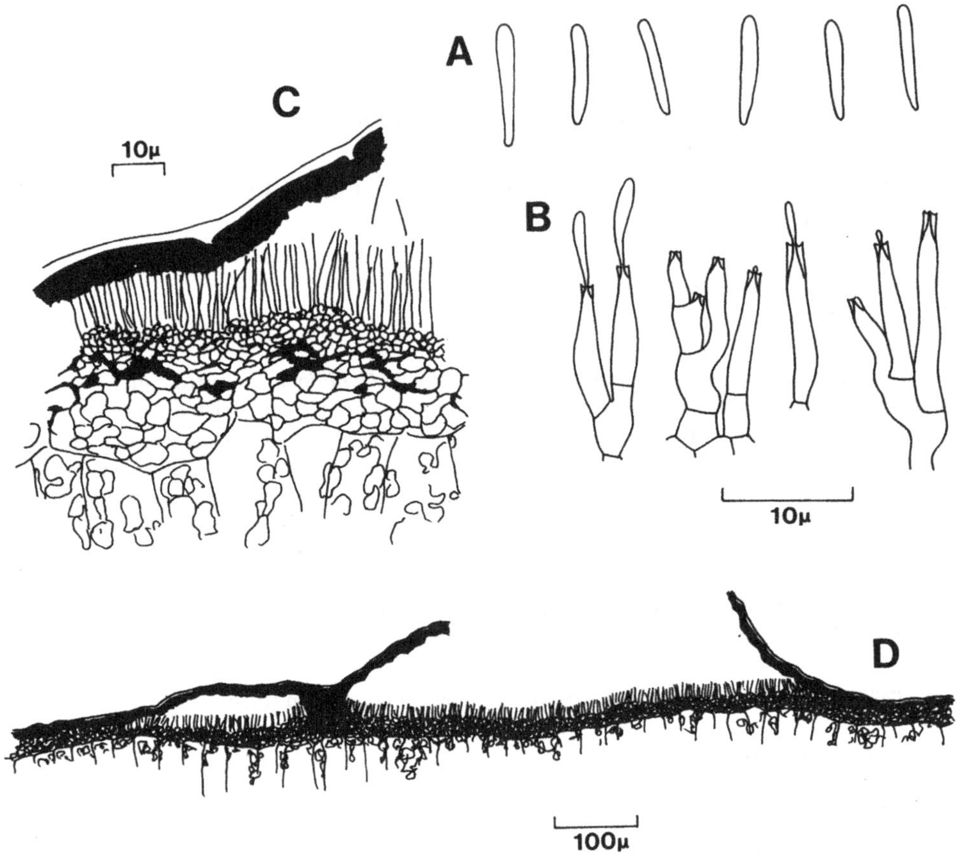

Fig. 340. *Melasmia acerina*, A, conidia; B, conidiophores with developing conidia; C, detail of conidiomatal wall; D, vertical section of a conidioma.

the name to *Pilidium*. There are however only two distinct species, *P. acerinum* and *P. concavum* (Desm.) Höhn. which are separated by conidial size.

P. ACERINUM Kunze apud Kunze & Schmidt, *ibid.*: 92 (1823). (Fig. 341B, C)
Leptothyrium acerinum (Kunze) Cda, *Icones fungorum* **2**: 25 (1838).
Fusicoccum coronatum Karst., *Hedwigia* **23**: 21 (1884).
Leptothyrium medium Cke, *Grevillea* **13**: 98 (1885).
Leptothyrium medium Cke var. *castaneicola* Cke, *Grevillea* **13**: 98 (1885).

Conidiomata up to 200 μ diam. Conidiophores 12-19 × 2 μ. Conidia 14-16 × 2 μ.

Leaves of *Betula* sp., *Eucalyptus gobulus, E. saligna, E.* sp., *Quercus* sp., *Acer opulifolium*, Italy, G.B., Calif., U.S.A., Brazil, Spain, New Zealand (IMI 22057, 35336, 207753a, 188576, 142248, 194555j, 188328b, 197538e, 188824, 188510a).

P. CONCAVUM (Desm.) Höhn., *Sber. Akad. Wiss. Wien* **124**: 148 (1915). (Fig. 341A)
Ceuthospora concava Desm., *Ann. Sci. nat.*, ser. 3, **8**: 17 (1847).
Sclerotiopsis concava (Desm.) Shear & Dodge, *Mycologia* **13**: 155 (1921).
Sclerotiopsis australasica Speg., *Ann. Soc. cient. Argent.* **13**: 14 (1882).
Pilidium australasicum (Speg.) Petrak, *Sydowia* **5**: 341 (1951).
Leptothyrium macrothecium Fckl, *Jb. nassau. Ver. Naturk.* **23-24**: 383 (1870).
Leptothyrium protuberans Sacc., *Michelia* **2**: 351 (1881).
Sporonema dubium C. Massal., *Nuovo G. bot. ital.* **21**: 166 (1889).
Sporonema quercicola C. Massal., *Nuovo G. bot. ital.* **21**: 166 (1889).

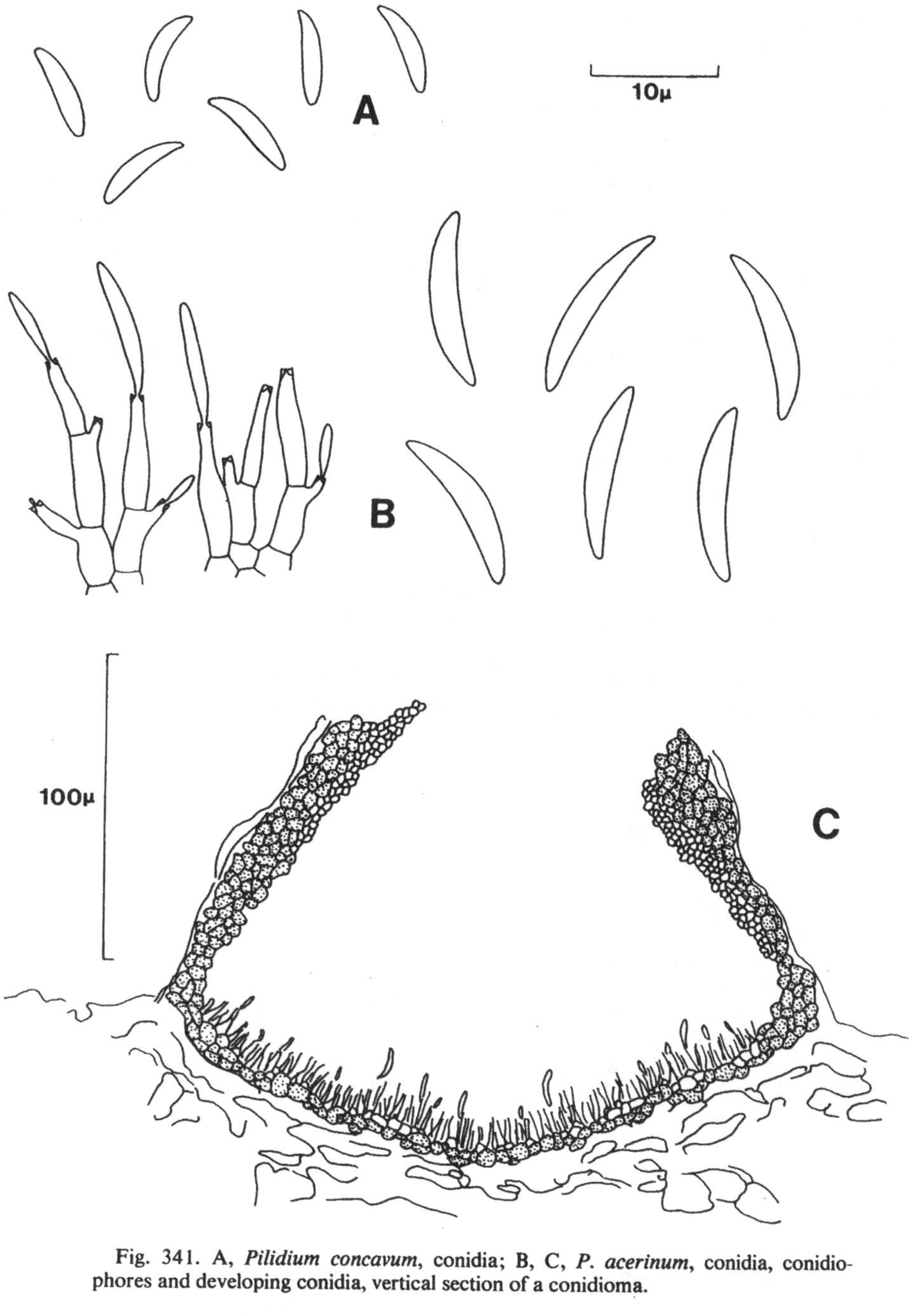

Fig. 341. A, *Pilidium concavum*, conidia; B, C, *P. acerinum*, conidia, conidiophores and developing conidia, vertical section of a conidioma.

Leptothyrium borzianum Tassi, *Rev. mycol.* **18**: 171 (1896).
Sclerotiopsis potentillae Oud., *Ned. Kruid. Arch. 3, Ver. Med. Ned. Bot. Ver.* **2**: 248 (1900).
Sclerotiopsis pelargonii Scalia, *Mycetes Siculi Novi* **2**: 2 (1903).
Sclerotiopsis rubi C. Massal., *Malpighia* **20**: 166 (1906).
Sporonema pulvinatum Shear, *Bull. Torrey bot. Cl.* **34**: 308 (1907).

Conidiomata 80-150 μ diam. Conidiophores 10-20 × 1.5 μ. Conidia 4-9 × 1.5-2 μ.

Substrata, see Shear & Dodge (1921). In herb. IMI on *Epilobium montanum*, *Quercus* sp., *Rhus cotinus*, Eire, Austria (IMI 202708b, 203153, 24785).

75. SIROPLACODIUM Petrak, *Ann. naturh. Mus. Wien* **1**: 509 (1940).
Sp. typ.: *S. atrum* Petrak

Mycelium immersed, abundant, branched, septate, hyaline. *Conidiomata* separate, rarely confluent, dark brown to black, applanate, lenticular, unilocular, consisting of an upper wall one cell thick, of dark brown, thick-walled textura angularis, and a lower wall of paler brown, thin-walled pseudoparenchyma from which the conidiophores arise. *Ostiole* absent, dehiscence by a longitudinal slit. *Conidiophores* branched at the base and above, hyaline, septate, cylindrical, parallel, smooth, formed from the lower wall. *Conidiogenous cells* enteroblastic, phialidic, determinate, integrated or discrete, hyaline, collarette long, flared, channel minute, periclinal wall thickened. *Conidia* hyaline, aseptate, straight, cylindrical, apex obtuse, base ± truncate, occasionally with a minute guttule, thin-walled, smooth.

Petrak (1958a) added two more species, *S. umbelliferarum* with conidia 9-16 × 2-2.5 µ, and *S. longisporum* with conidia 16-26 × 2-3 µ, and Petrak (1958b) transferred *Placosphaeria shastensis* Sprague & Cooke to *Siroplacodium*, but none of these species has been re-examined.

S. ATRUM Petrak, *ibid.*: 509 (1940). (Fig. 342)

Conidiomata up to 400 µ long × 160 µ wide. Conidiophores up to 24 µ long × 2-3 µ wide. Conidia 7.5-10 × 2 µ.

On dead stems of Gramineae (IMI 201889, holotype ex ZT).

76. CYTONAEMA Höhn., *Sber. Akad. Wiss. Wien* **123**: 131 (1914).
Sp. typ.: *C. spinellum* (Kalchbr.) Höhn. (syn. *Sphaeronaema spinellum* Kalchbr.).

Mycelium immersed, pale brown to hyaline, branched, septate. *Conidiomata* eustromatic, peridermal, separate, globose, brown, unilocular, thick-walled, with a rostrate apex extending above the substrate, glabrous; wall consisting of an

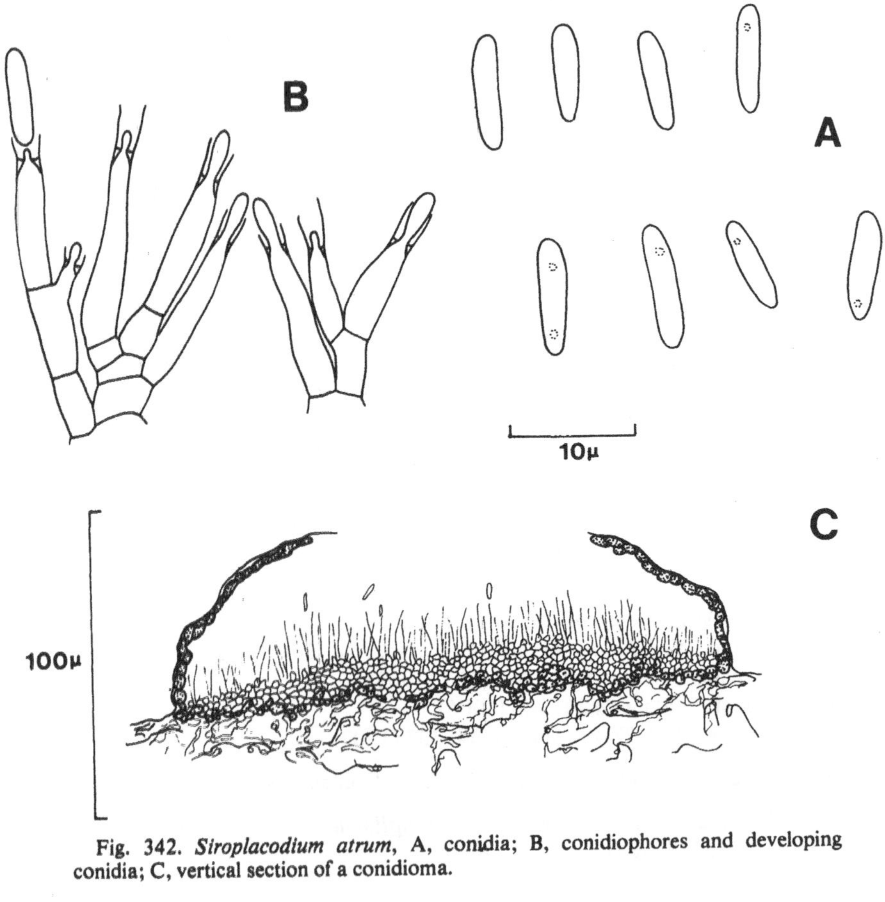

Fig. 342. *Siroplacodium atrum*, A, conidia; B, conidiophores and developing conidia; C, vertical section of a conidioma.

outer layer of thick-walled, dark brown textura intricata, a middle layer of paler, thick-walled cells and an inner region of thin-walled, hyaline textura angularis. *Ostiole* central, circular, furfuraceous, situated at the apex of a long rostrum the walls of which are of large-celled dark brown, elongated textura intricata which is sclerotic on the outside. *Conidiophores* completely line the locule and ostiolar channel, hyaline, septate, branched at the base and frequently above, cylindrical, tapered towards the apices. *Conidiogenous cells* enteroblastic, phialidic, integrated, determinate, hyaline, formed as short or long lateral branches from the conidiophores immediately below septa, aperture and collarette minute. *Conidia* acropleurogenous, hyaline, aseptate, thin-walled, straight or allantoid, eguttulate.

Höhnel (1914b) originally thought *Cytonaema* was close to *Cytospora* and Défago (1944) placed it in synonymy with the latter genus. It differs from *Cytospora* however in the simple unilocular conidiomata, hyaline conidia and rostrate ostiole. *Cytospora* is multilocular and stromatic, with very pale brown conidia in mass, and ostioles are never situated on rostrate extensions of the stromata.

C. SPINELLUM (Kalchbr.) Höhn., *ibid.*: 132 (1914). (Fig. 343)
Sphaeronaema spinellum Kalchbr., *Bot. Z.* **20**: 199 (1862).

Conidiomata up to 550 μ diam; rostrum up to 600 μ long. Conidiophores up to 30 μ long × 2-3 μ wide. Conidia 2.5-3.5 × 1 μ.

On branches of *Salix alba*, *S.* sp., Poland, Austria (IMI 31052, 11410).

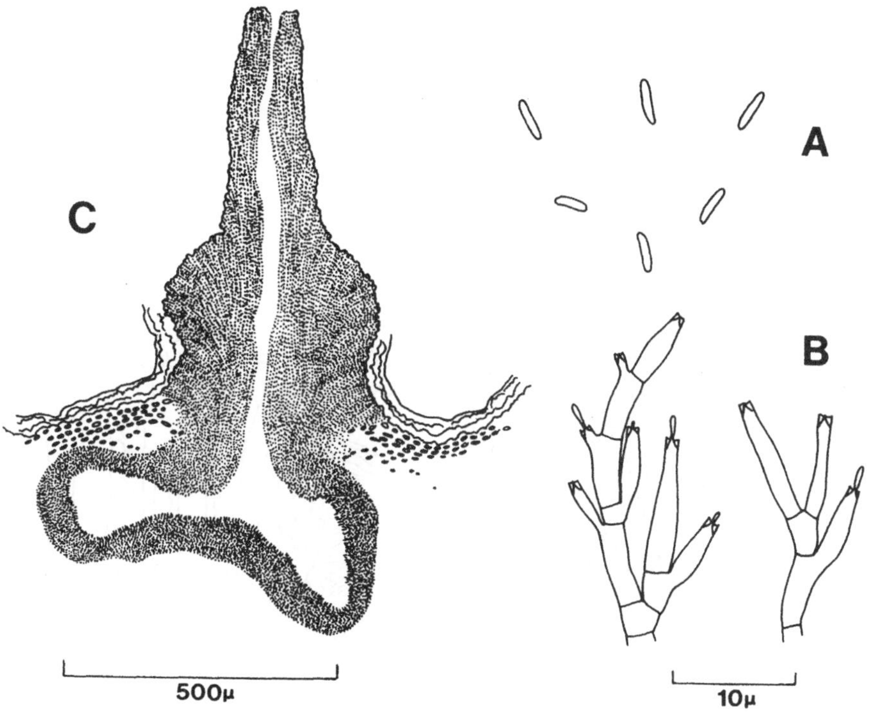

Fig. 343. *Cytonaema spinellum*, A, conidia; B, conidiophores; C, vertical section of a conidioma.

77. RHODOSTICTA Woronichin apud Tranzschel & Serebrianikow, *Mycotheca Rossica*, Fasc. III & IV, Index (Schedae): 6 (1911).
Sp. typ.: *R. caraganae* Woronichin

Mycelium immersed, branched, septate, hyaline, thick-walled. *Conidiomata* pseudostromatic, immersed, yellowish brown, irregular in shape but often occupying the whole leaf including upper and lower epidermises and palisade tissue, composed of textura intricata, multilocular, locules simple, globose to slightly irregular, walls relatively thin, of large thin-walled cells. *Ostioles* single to each locule, circular to irregular, with slightly pale brown ostiolar cells. *Conidiophores* hyaline, branched at the base, septate, smooth with acropleurogenous conidia, formed from the inner cells of the locular walls. *Conidiogenous cells* enteroblastic, phialidic, determinate, integrated or discrete, cylindrical to lageniform, hyaline, smooth, apertures apical on main axis, also on small lateral branches immediately below septa, channel minute, collarette distinct, periclinal wall considerably thickened. *Conidia* hyaline, aseptate, smooth, eguttulate, cuneiform to obovoid.

The single species was described as the conidial state of *Physalosporina caraganae* Woronichin. Höhnel (1917b) has reduced *Physalosporina* to synonymy with *Polystigma* DC. ex Fr. and referred *P. caraganae* to this genus. The macroconidial state of *Polystigma* is *Polystigimina* Sacc., therefore *Rhodosticta* probably represents the microconidial state.

R. CARAGANAE Woronichin, *ibid.*: 6 (1911). (Fig. 344)

Lesions yellowish brown, occupying part or whole of leaf. Locules up to 150 µ diam. Conidiophores 10-17 × 3.5-6 µ. Conidia 4-6 × 2 µ.

Causing leaf lesions on *Caragana frutescens*, U.S.S.R. (IMI 199145, isotype ex LE).

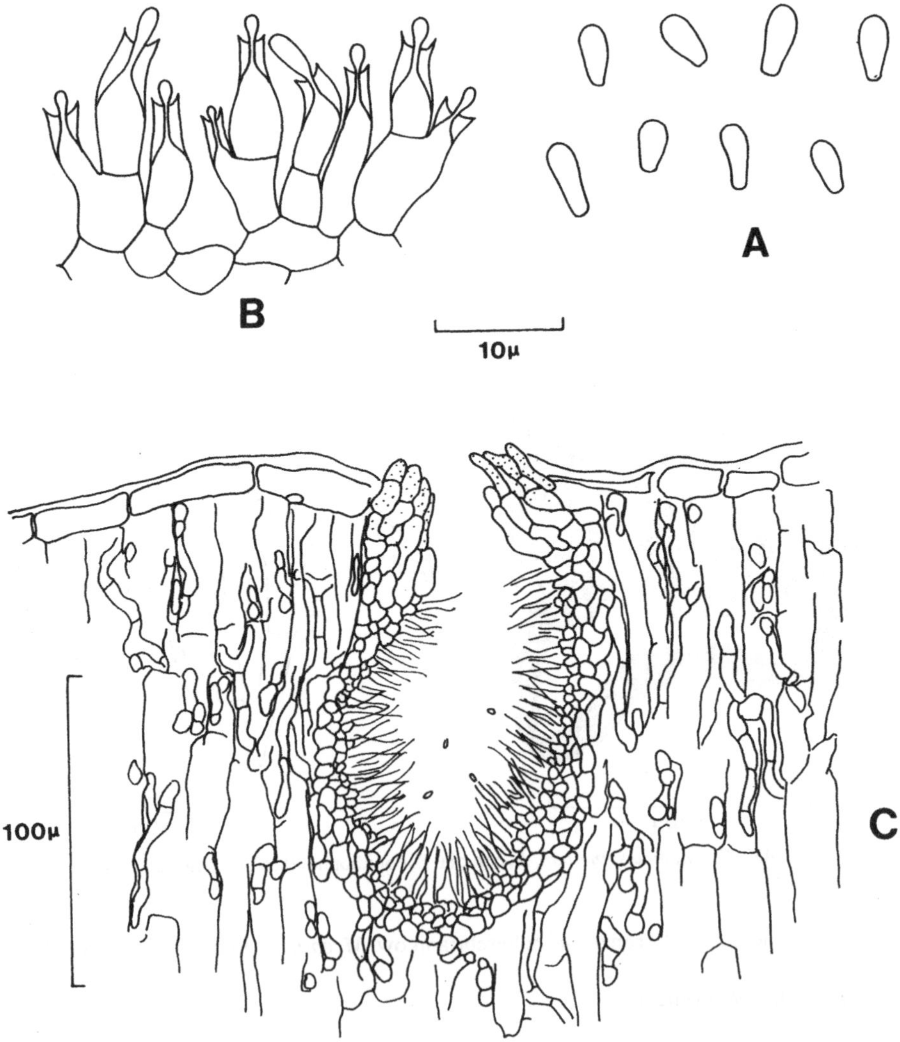

Fig. 344. *Rhodosticta caraganae*, A, conidia; B, conidiophores; C, vertical section of a conidioma.

78. KENDRICKOMYCES Sutton, Rao & Mhaskar, *Trans. Br. mycol. Soc.* **67**: 243 (1976).
Sp. typ.: *K. indicus* Sutton, Rao & Mhaskar

Mycelium immersed, branched, septate, hyaline to pale brown. *Conidiomata* eustromatic, immersed, globose to collabent, at maturity furfuraceous, multilocular, thick-walled; walls of complex structure: outer peripheral wall of dark brown, thick-walled textura angularis, middle layer of textura oblita, cells of which become progressively looser in the upper region of the conidiomata: locules initiated in the upper layer, aggregated into 1-2 separate groups within the stroma, vertically elongated, regularly arranged, each with an ostiolar channel merging into a common ostiole. *Ostioles* 1-2 to each stroma, papillate, circular. *Conidiophores* septate, irregularly branched, hyaline, smooth, formed from the inner cells of the locular walls. *Conidiogenous cells* enteroblastic, phialidic, integrated, determinate, hyaline, tapered towards the apices, collarette, channel and periclinal thickening minute. *Conidia* hyaline, aseptate, smooth, guttulate, allantoid.

The genus is monotypic.

K. INDICUS Sutton, Rao & Mhaskar, *ibid.*: 244 (1976). (Fig. 345)

Conidiomata .1-.15 mm diam. Locules 320-530 µ diam. × 60-80 µ high. Conidiophores up to 18 µ long × 1.5-2 µ wide. Conidia 4.5-6.5 × 1.5 µ.

On dead branches of *Terminalia belerica*, India (IMI 190465, holotype).

79. PHOMOPSIS (Sacc.) Sacc., *Annls mycol.* **3**: 166 (1905).
Phoma Westd. subgen. *Phomopsis* Sacc., *Syll. fung.* **3**: 66 (1884).
Myxolibertella Höhn., *Annls mycol.* **1**: 526 (1903).
? *Petasodes* Clements, *Genera of Fungi*: 133 (1909).
Chromocytospora Speg., *An. Mus. nac. Buenos Aires* **20**: 392 (1911).
Lasiostroma Griff. & Maubl., *Bull. trimest. Soc. mycol. Fr.* **27**: 472 (1911).
? *Vestergrenia* (Sacc. & P. Syd.) ? Died., *Annls mycol.* **11**: 528 (1913).
Leptostromella Sacc. subgen. *Vestergrenia* Sacc. & P. Syd., *Syll. fung.* **14**: 998 (1899).
Endogloea Höhn., *Z. f. Garungsph.* **5**: 207 (1915).
Malacostroma Höhn., *Mitt. bot. Inst. tech. Hochsch. Wien* **7**: 56 (1930).
Leucophomopsis Höhn., *Ber. dtsch. Bot. Ges.* **35**: 255 (1917).
Phomopsella Höhn., *Hedwigia* **62**: 85 (1920).
Cyclophomopsis Höhn., *Hedwigia* **62**: 86 (1920).
Placophomopsis Grove, *J. Bot., Lond.* **59**: 315 (1921).
? *Pseudophomopsis* Höhn., *Mitt. bot. Inst. tech. Hochsch. Wien* **3**: 28 (1926).
Cyphellopycnis Tehon & Stout, *Mycologia* **21**: 189 (1929).
Fusicytospora Gutner, *Acta Inst. Bot. Acad. Sci. URSS*, ser. 2, **2**: 473 (1934).
Phomopsioides Costa & Camara, *Portug. Acta biol.* **4**: 335 (1954).
Subramanella Srivastava apud Srivastava, Zakia & Govindarajan, *Mycologia* **54**: 5 (1962).
Original spp.: *P. lamii* Sacc. & D. Sacc., *P. pritchardiae* (Cke & Harkn.) Sacc.
Lectotype sp.: None yet designated.

Mycelium immersed, branched, septate, hyaline to pale brown. *Conidiomata* eustromatic, immersed, brown to dark brown, separate or aggregated and confluent, globose, ampulliform or applanate, unilocular, multilocular or convoluted, thick-walled; walls of brown, thin- or thick-walled textura angularis, often somewhat darker in the upper region, lined by a layer of smaller-celled tissue. *Ostiole* single, or several in complex conidiomata, circular, often papillate. *Conidiophores* branched and septate at the base and above, occasionally short and only 1-2 septate, more frequently multiseptate and filiform, hyaline, formed from the inner cells of the locular walls. *Conidiogenous cells* enteroblastic, phialidic, determinate, integrated, rarely discrete, hyaline, cylindrical, apertures apical on long or short lateral and main branches of the conidiophores, collarette, channel and periclinal thickening minute. *Conidia* of two basic types, but in some species with intermediates between the two: α-conidia hyaline, fusiform, straight, usually biguttulate (one guttule at each end) but sometimes with more guttules, aseptate; β-conidia hyaline, filiform, straight or more often hamate, eguttulate, aseptate.

Although Sutton (1977a) has discussed the extensive generic synonymy under the individual generic names, this must be viewed as tentative because of the absence of a lectotype species for *Phomopsis* itself. So far the generic name has not been correctly typified, the decision resting between *P. lamii* Sacc. & D. Sacc. and *P. pritchardiae* (Cke & Harkn.) Sacc.

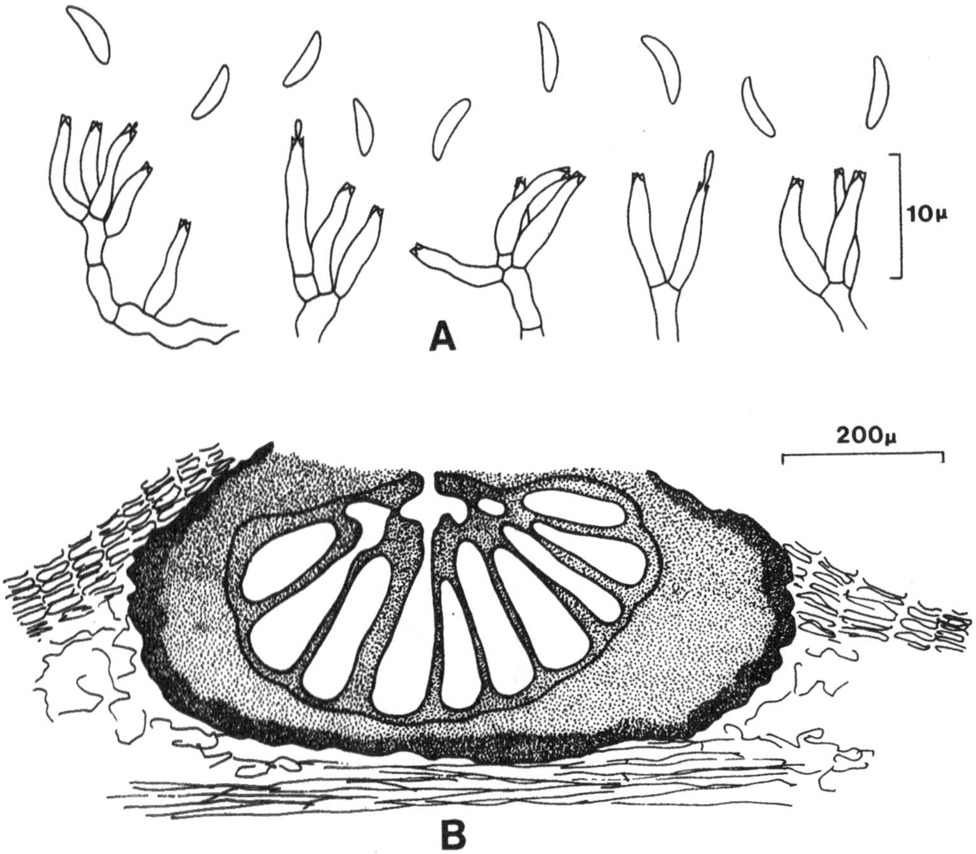

Fig. 345. *Kendrickomyces indicus*, A, conidia and conidiophores; B, vertical section of a conidioma.

Neither of these two original species appears to have been re-examined since their original descriptions, so the arbitrary selection of either could lead to unfortunate consequences should it eventually be proved that the species is not congeneric with the many taxa now accepted in the genus. The situation is further complicated by the fact that *Myxolibertella* clearly antedates what we now know as *Phomopsis* and to preserve the use of *Phomopsis* in its present sense it will need to be formally conserved against *Myxolibertella*, provided the problem of lectotypification can be solved. The widespread occurrence of *Phomopsis* species as plant pathogens and the extensive literature dealing with this aspect of their biology are factors in favour of conservation of the generic name.

So far more than 400 taxa have been described in *Phomopsis* of which herb. IMI holdings amount to 222. There has been no modern revisionary treatment as a whole. Hahn (1928, 1930, 1933, 1943) dealt with several taxa on coniferous substrates, and Wehmeyer (1933) in his revision of *Diaporthe* species, which are the teleomorphoses of *Phomopsis*, illustrated and described many species that had been correlated with their ascomatal states. Reliable documentation of other individual species is scattered throughout the literature (Archer, 1973; Arnold & Carter, 1974; Early & Punithalingam, 1972; Funk, 1975; Hawksworth & Punithalingam, 1973b, 1975; Punithalingam, 1973, 1974a, b, 1975; Punithalingam & Gibson, 1978; Sutton, 1964, 1967c, 1968d; van Kesteren, 1967; and many other papers). These articles are all fully illustrated and provide a guide to the diversity found amongst species in the genus.

Clearly the genus is in need of revisionary studies on several counts. The problem of typification must be solved and type studies should be linked with an analysis of behaviour in artificial culture. Futhermore, the correlation of *Diaporthe* teleomorphoses and *Phomopsis* anamorphoses is essential to any future delimitation of taxa in the genus.

A selection of species is presented below. Most have been described and illustrated previously but they reflect the type of variation found in the genus.

P. ABDITA (Sacc.) Trav., *F. It. Crypt.* (Pyrenomycetes) **2**(1): 237 (1906). (Fig. 346A)
Phoma abdita Sacc., *Michelia* **2**: 96 (1880).

Conidiomata up to 100 µ, globose to linear, unilocular or convoluted. Conidiophores filiform to subulate, branched and septate mainly at the base, less so above, up to 30 µ long × 2-3 µ wide at the base. Conidia: α-conidia fusiform to ellipsoid, sometimes attenuated towards the base, 2-4 guttulate, 8-10 × 2 µ.

On twigs and branches of *Melia azaderach*, Cuba, India, France (IMI 96542 ex PAD, holotype *P. abdita*, 92667, 127450).

P. archeri nom. nov. (Fig. 346B)
Phomopsis pittospori S.A. Archer, *Trans. Br. mycol. Soc.* **61**: 221 (1973), non *P. pittospori* (Cke & Harkn.) Grove, *Bull. misc. Inf. R. bot. Gdn Kew 1919*: 181 (1919).

Conidiomata up to 1 mm diam., globose to subglobose. Conidiophores sparingly septate and branched, filiform, up to 15 µ long. Conidia: α-conidia ellipsoid, less often fusiform, each end obtuse, 0-2 guttulate, 5.5-9 × 2-2.5 µ; β-conidia straight, curved or hamate, 15-19 µ long.

These dimensions are somewhat lower than those reported in the original account where α-conidia were described as 7-10 × 2.5 µ and β-conidia as 20-30 µ long.

In *P. pittospori* (Cke & Harkn.) Grove, which was originally described from stems of *Pittosporum* in California, the α-conidia are much narrower (6-8 × 1.5 µ), although the β-conidia and conidiophores are of similar size.

From leaves of *Pittosporum tenuifolium*, U.K. (IMI 164584a, b, 152586, holotype).

P. arnoldiae nom. nov.
Phomopsis elaeagni (Carter & Sacamano) Arnold & Carter, *Mycologia* **66**: 193 (1974), non *P. elaeagni* Sacc., *Nuovo G. bot. Ital.* N.S., **22**: 50 (1915), nec *P. elaeagni* (Sacc.) Petrak, *Annls mycol.* **19**: 48 (1921), nec *P. elaeagni* Sandu-Ville, *Stud. Cercet. ştiinţ. Iaşi, Biol. Ştiinţe agric.* **13**: 260 (1962).
Fusicoccum elaeagni Carter & Sacamano, *Mycologia* **59**: 535 (1967).

Conidiomata up to 900 µ wide × 500 µ high, multi-locular. Conidiophores 6-16 × 1-2 µ, simple or occasionally branched and septate at the base. Conidia: α-conidia ellipsoid to fusiform often biguttulate, 5.5-11 × 1.5-2 µ; β-conidia hamate, 15-20 µ long.

Causing canker and dieback of *Elaeagnus angustifolia*, Mo., Ill., U.S.A., Ont., Canada.

P. DIACHENII Sacc., *Annls mycol.* **13**: 118 (1915). (Fig. 346D)
Cyphellopycnis pastinacae Tehon & Stout, *Mycologia* **21**: 189 (1929).

Conidiomata up to 400 µ diam., multilocular, multiostiolate, linear. Conidiophores filiform, sparingly branched, up to 10 µ long. Conidia: α-conidia fusiform, 2-3 guttulate, 7.5-16.5 × 2.5-4.5 µ; β-conidia filiform to hamate, 15-20 µ long.

On stems, dried seed and fruit of *Pastinaca sativa*, N.Y., Ill., Mich., U.S.A., U.K. (IMI 91184 ex PAD, holotype *P. diachenii*, 91014 ex ILLS, holotype *C. pastinacae*, 22564, 102590).

P. lantanae (Costa & Camara) comb. nov. (Fig. 346C)
Phomopsioides lantanae Costa & Camara, *Agron. lusit.* **17**: 157 (1955).

Conidiomata 100-150 µ diam., simple, unilocular. Conidiophores sparingly branched and septate at the base, up to 15 µ long × 3 µ wide at the base. Conidia fusiform, biguttulate, 6.5-9 × 2.5 µ.

On leaves and stems of *Lantana camara*, India (IMI 95103, 211613).

P. OBSCURANS (Ell. & Ev.) Sutton, *Trans. Br. mycol. Soc.* **48**: 615 (1965). (Fig. 346E)
Phoma obscurans Ell. & Ev., *Proc. Acad. nat. Sci. Philad.* 357 (1894).
Sphaeropsis obscurans (Ell. & Ev.) O. Kze, *Rev. Gen. Pl.* **3**(2): 529 (1898).
Phyllosticta obscurans (Ell. & Ev.) Tassi, *Bull. Ort. bot. Siena* **5**: 13 (1906).
Dendrophoma obscurans (Ell. & Ev.) Anderson, *Bull. Ill. agric. Exp. St.* 229: 135 (1920).

Conidiomata simple, unilocular, 140-210 µ diam. *Conidiophores* filiform, branched extensively, up to 35 µ long. Conidia: α-conidia biguttulate 5.5-7.5 × 1.5-2 µ, β-conidia absent.

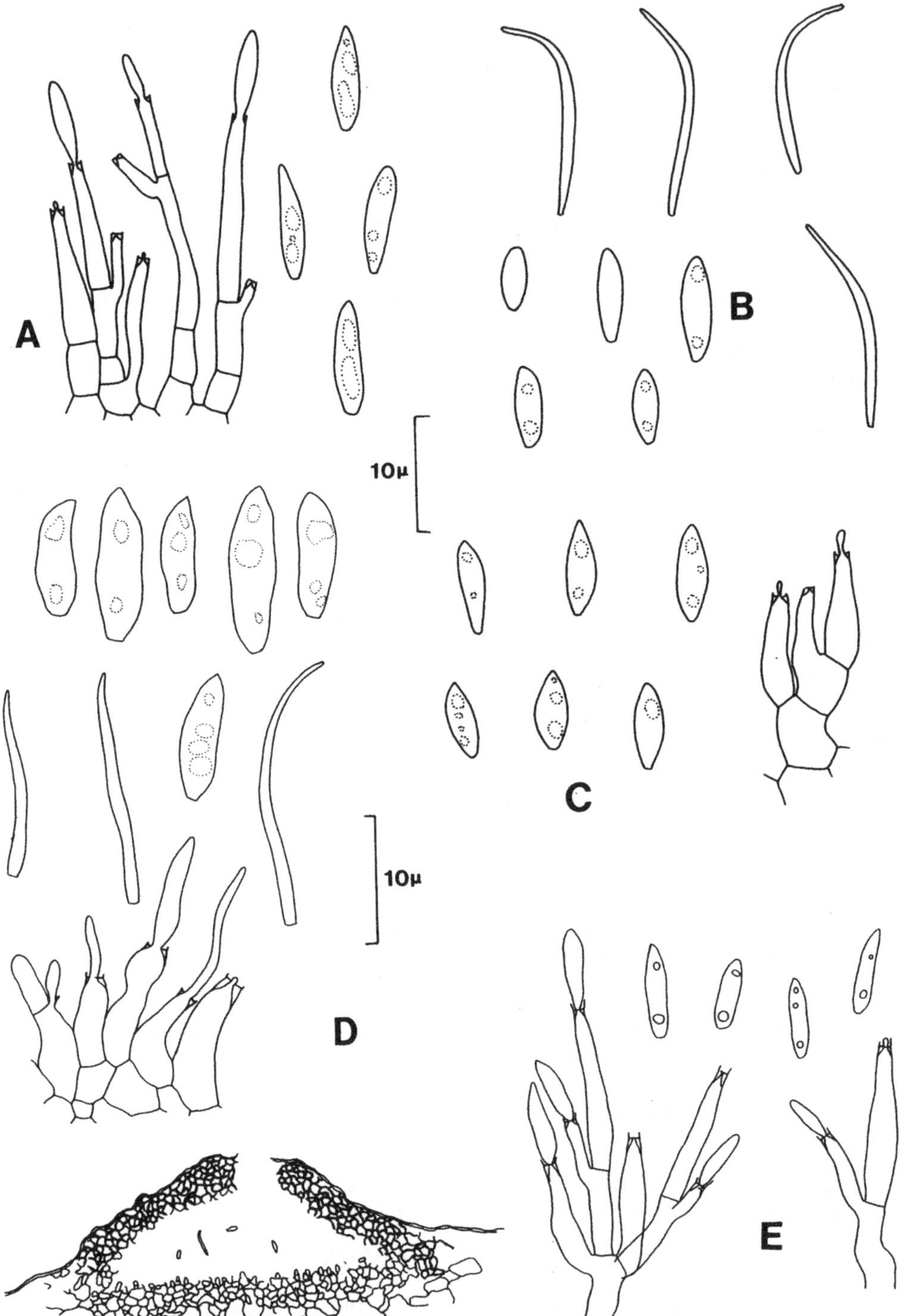

Fig. 346. A, *Phomopsis abdita*, α-conidia and conidiophores; B, *P. archeri*, α- and β-conidia; C, *P. lantanae*, α-conidia and a conidiophore; D, *P. diachenii*, α-, β-conidia, conidiophores, vertical section of a conidioma; E, *P. obscurans*, α-conidia and conidiophores.

Causing lesions on *Fragaria vesca*, *F. x ananassa*, *F.* sp., S. Africa, Australia, U.S.A., New Zealand, Brunei, Egypt, Togo, Ethiopia, Zambia, Pakistan, Tanzania, S. Rhodesia, Nepal, Tonga, India, Malawi, Mauritius, represented in herb. IMI by 35 collections.

P. STIPATA (Lib.) Sutton, *Trans. Br. mycol. Soc.* **52**: 509 (1969). (Fig. 347C)
Ascochyta stipata Lib., *Pl. Crypt. Ard.* fasc. IV, no. 354 (1837).
Septoria stipata (Lib.) Sacc., *Syll. fung.* **10**: 352 (1902).
Libertina stipata (Lib.) Höhn., *Annls mycol.* **22**: 197 (1924).
Ascochyta effusa Lib., *Pl. Crypt. Ard.* Fasc. IV, no. 355 (1837).
Septoria effusa (Lib.) Desm., *Ann. Sci. nat.* **8**: 23 (1847).
Libertina effusa (Lib.) Höhn., *Ber. dtsch. Bot. Ges.* **38**: 107 (1920).
Phomopsis effusa (Lib.) Sutton, *Trans. Br. mycol. Soc.* **50**: 356 (1967), non *P. effusa* (Rob.) Höhn., 1918.
Phomopsis libertiae Morelet, nom. nov., *Bull. Soc. Sci. nat. Archéol. Toulon Var* 178: 6 (1968).

Conidiomata 120-185 µ diam., unilocular, simple. Conidiophores filiform to subulate, branched at the base and above, 15-40 × 6 µ tapering to 1 µ wide. Conidia: α-conidia filiform, sigmoid to hamate or simply curved, 16.5-24 × 1.5-2 µ; β-conidia absent.

On leaves of *Prunus padus*, *P. cerasus*, France, Austria, Siberia, U.S.S.R. (IMI 101344, isotype *A. stipata* ex K, 20368, 13029, 30559, 24749, 24748).

P. terminaliae (P. Henn.) comb. nov. (Fig. 347A)
Phoma terminaliae P. Henn., *Hedwigia* **48**: 14 (1908), non *Phomopsis terminaliae* Sawada, *Spec. Publ. Coll. Agric., Nat. Taiwan Univ.* 8: 132 (1959), non rite publ.

Conidiomata up to 250 µ diam., globose to subglobose. Conidiophores mostly branched at the base, filiform, subulate, up to 20 µ long. Conidia: α-conidia fusiform to ellipsoid, often obtuse at each end, 2-4 guttulate, 9-10.5 × 2.5 µ; β-conidia absent.

P. terminaliae Sawada was not validly published. It was described with multiguttulate conidia measuring 10-15 × 3-4 µ, which is quite a bit larger than in *P. terminaliae* (P. Henn.) Sutton.

On leaves and decaying fruits of *Terminalia stuhlmannii*, *T. mollis*, Zambia (IMI 100563, 119823).

P. tersa (Sacc.) comb. nov. (Fig. 347B)
Phoma tersa Sacc., *Bol. Soc. Biol. Coimbra* **11**: 16 (1893).

Conidiomata up to 650 µ diam., flat, effuse, multilocular. Conidiophores filiform, subulate, branched and septate only at the base, up to 15 µ long. Conidia: α-conidia straight, cylindrical, obtuse at each end, sometimes slightly constricted in the middle, prominently biguttulate, 6.5-7.5 × 2.5 µ; β-conidia absent.

On and from leaves, fruit and stems of *Passiflora edulis*, *P.* sp., Portugal, Malta, Sarawak, Sri Lanka, Fiji (IMI 91833, holotype ex PAD, 69193, 190710, 19582, 215118).

80. DISCOSPORIUM Höhn., *Z. f. Garungsph.* **5**: 196 (1915).
Chondroplea Kleb., *Phytopath. Z.* **6**: 291 (1933).
Sp. typ.: *D. populeum* (Sacc.) Sutton (syn. *Phoma populea* Sacc., *D. hyalinum* (Ell.) Höhn.).

Mycelium immersed, branched, septate, pale brown to hyaline. *Conidiomata* eustromatic, immersed, separate or aggregated, subperidermal, flattened globose, medium brown, unilocular or multilocular, thick-walled, composed of host and textura angularis and intricata fungal tissue, convoluted. Dehiscence by irregular breakdown of upper tissues. *Conidiophores* hyaline, branched mostly from the base, septate, smooth, tapered gradually towards the apices, formed from the inner cells of the conidiomatal wall. *Conidiogenous cells* enteroblastic, phialidic, indeterminate, integrated or discrete, cylindrical, hyaline, smooth, with 1-3 percurrent proliferations, apertures wide, collarettes minute. *Conidia* obovoid to napiform, aseptate, base truncate, thick-walled, individually hyaline but greyish brown in mass, smooth.

The type species is a pathogen that has popularly been known as *Chondroplea populea* and a recent account has been given by Booth, Gibson & Sutton (1973) and a bibliography by Petrak (1957). Petrak (1957) also noted that the type species of *Discosporium* was identical with *C. populea* and since it has precedence over *Chondroplea* it must be adopted

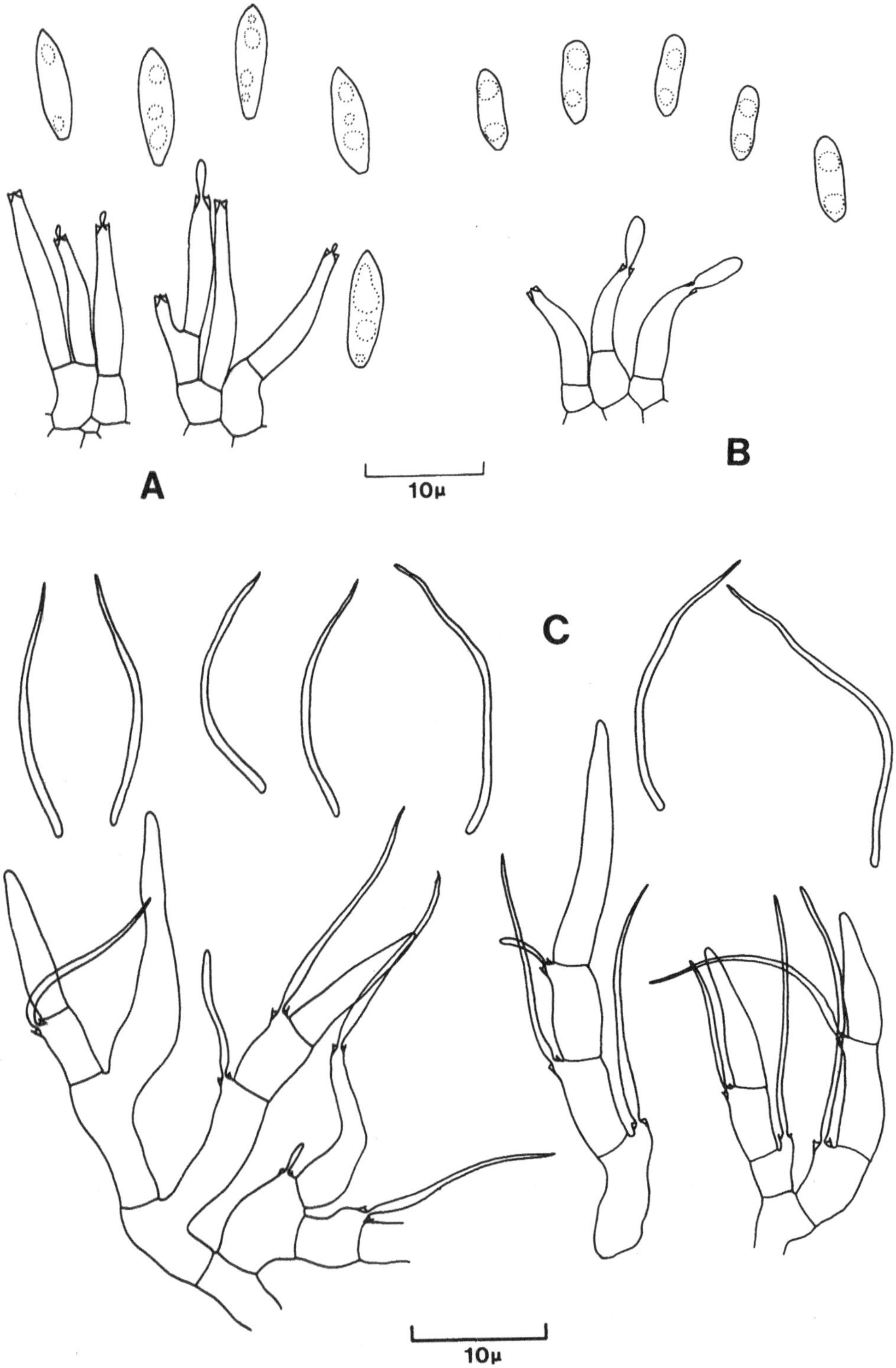

Fig. 347. A, *Phomopsis terminaliae*, α-conidia and conidiophores; B, *P. tersa*, α-conidia and conidiophores; C, *P. stipata*, β-conidia and conidiophores.

with the change in nomenclature suggested by Sutton (1977a). Arx (1957a, 1970) erroneously included *Discosporium* as a synonym of *Cryptosporiopsis* Bub. & Kabát.

D. POPULEUM (Sacc.) Sutton, *Mycol. Pap.* 141: 75 (1977). (Fig. 348).
Phoma populea Sacc., *Michelia* **1**: 358 (1878).
Dothiorella populea (Sacc.) Sacc., *Syll. fung.* **3**: 237 (1884).
Discula populea (Sacc.) Höhn., *Z. f. Garungsphy.* **5**: 20 (1915).
Chondroplea populea (Sacc.) Kleb., *Phytopath. Z.* **6**: 291 (1933).
Melanconium hyalinum Ell., *Bull. Torrey bot. Cl.* **9**: 133 (1882).
Myxosporium hyalinum (Ell.) Wint., *Fungi Europaei*: 3749 (1886).
Gloeosporium hyalinum (Ell.) Dearn., *Bull. Can. Dept. Agr.* n. ser., **71**: 65 (1926).
Discosporium hyalinum (Ell.) Höhn., *Z. f. Garungsph.* **5**: 196 (1915).
Myxosporium ellisii Sacc., *Syll. fung.* **3**: 724 (1884).
Dothichiza populea Sacc. & Briard apud Sacc., *Syll. fung.* **3**: 672 (1884).
Cytosporella populi Oud., *Ned. Kruidk. Arch.*, ser. 2, **5**: 494 (1889).

Teleomorphosis: *Cryptodiaporthe populea* (Sacc.) Butin, *Sydowia* **11**: 31 (1958).

Conidiomata 700-800 µ diam. Conidiophores 15-20 × 3-4 µ. Conidia 9-11 × 7-8 µ.

On branches of *Populus* spp., *Salix* spp., Germany, Hungary, Austria, Romania, U.K., Canada, U.S.A. (IMI 21648, 11683, 13142, 154355, 21645, 33151, 73122, 16659a).

D. tremuloides (Ell. & Ev.) comb. nov. (Fig. 349)
Gloeosporium tremuloides Ell. & Ev., *Proc. Acad. nat. Sci. Phil.* 370 (1894).
Leptomelanconium tremuloides (Ell. & Ev.) Arx, *Verh. K. Akad. Wet.* **51**: 142 (1957).
Cryptocline tremuloides (Ell. & Ev.) Sutton, *Mycol. Pap.* 123: 42 (1971).

Conidiomata up to 100 µ diam. Conidiophores 6-30 × 3-4 µ. Conidia very pale brown, 10-13 × 5-6 µ.
Doubt concerning the generic relationships of this taxon have already been expressed by Sutton & Chao (1970) and Sutton (1971b). Although the species is foliicolous and does not show the range in conidiomatal morphology as is seen in *D. populeum* (Sacc.) Sutton, it does agree in conidial morphology and conidiogenesis, so I believe the species is more satisfactorily placed in *Discosporium* than in *Cryptocline* where I had referred it previously.

On leaves of *Populus tremuloides*, Wisc., U.S.A. (IMI 144356, holotype *G. tremuloides* ex NY).

D. eugeniae (Allesch.) comb. nov. (Fig. 350)
Gloeosporium eugeniae Allesch. in Rabenh., *Krypt. Fl. Dtsch.* **7**: 473 (1902).

Conidiomata up to 250 µ diam. Conidiogenous cells 14-20 × 2.5-4.5 µ. Conidia hyaline, pyriform to obpyriform, 11-13.5 × 5-6.5 µ.

Precisely the same arguments can be advanced for the inclusion of this species in *Discosporium* rather than *Cryptocline* or *Monostichella* as were suggested for *D. tremuloides*.

Causing lesions on leaves of *Eugenia jambolana*, Kenya (IMI 110641, 110640b).

81. PHACIDIOPYCNIS Potebnia, *Z. Pflanzenkr.* **22**: 143 (1912).
Discosporiopsis Petrak, *Annls mycol.* **19**: 217 (1921).
? *Chondrostroma* Syd., *Annls mycol.* **38**: 470 (1940).
Leptoteichon Kleb., *Phytopath. Z.* **6**: 297 (1933).
Operculella Kheswalla, *Ind. J. agr. Sci.* **11**: 317 (1914).
Sp. typ.: *P. piri* (Fckl) Weindlmayr (syn. *Cytospora piri* Fckl, *P. malorum* Potebnia).

Mycelium immersed, branched, septate, hyaline to pale brown. *Conidiomata* eustromatic, immersed to superficial, dark brown, circular, unilocular to multilocular; walls thick, of pale brown, thin-walled textura angularis, somewhat darker above. *Ostiole* absent, dehiscence by circumscissile or irregular rupture. *Conidiophores* hyaline, branched sparingly, 1-2 septate, smooth, tapered to the apices, formed all round the stromatal cavity. *Conidiogenous cells* enteroblastic, phialidic,

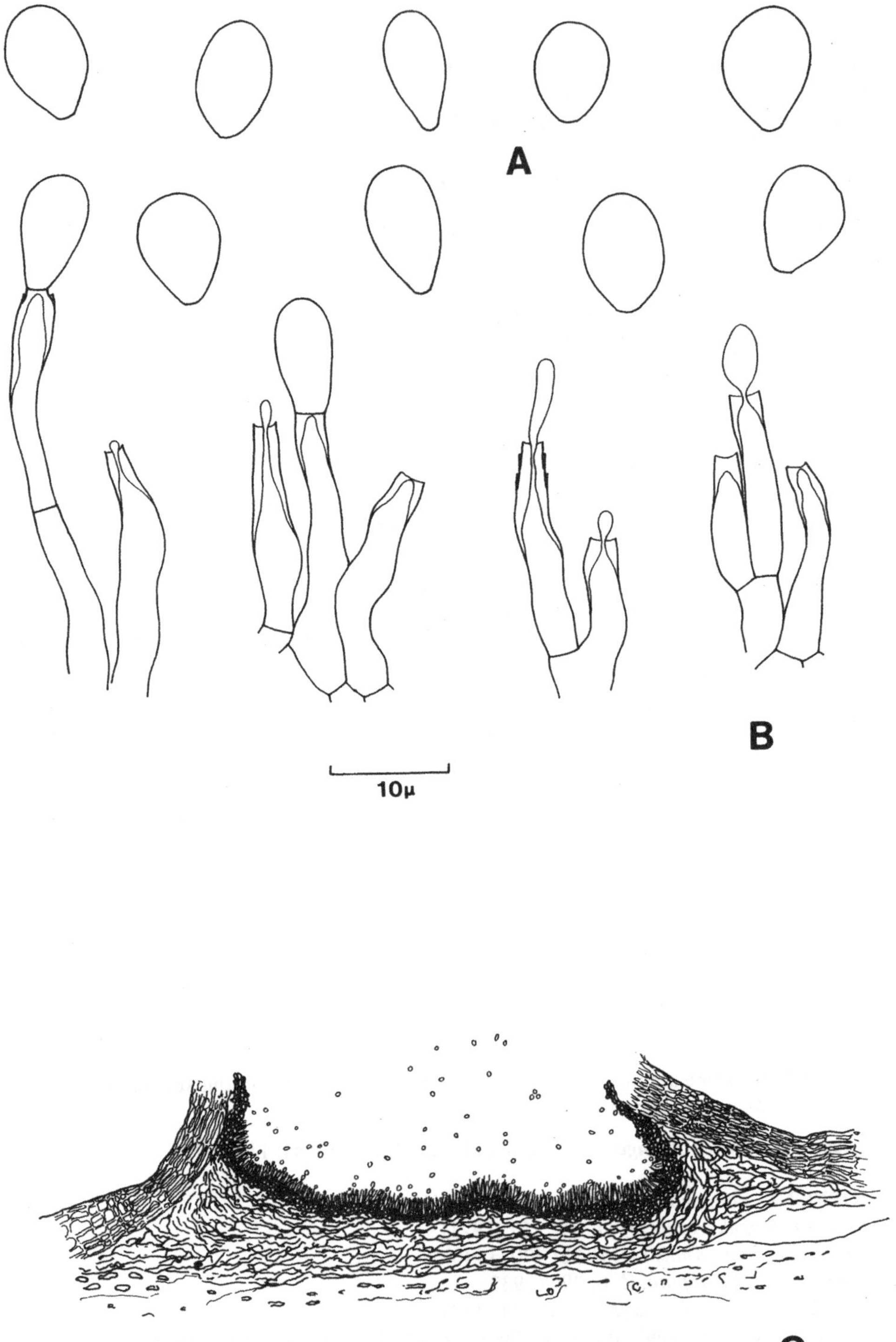

Fig. 348. *Discosporium populeum*, A, conidia; B, conidiophores and developing conidia; C, vertical section of a conidioma.

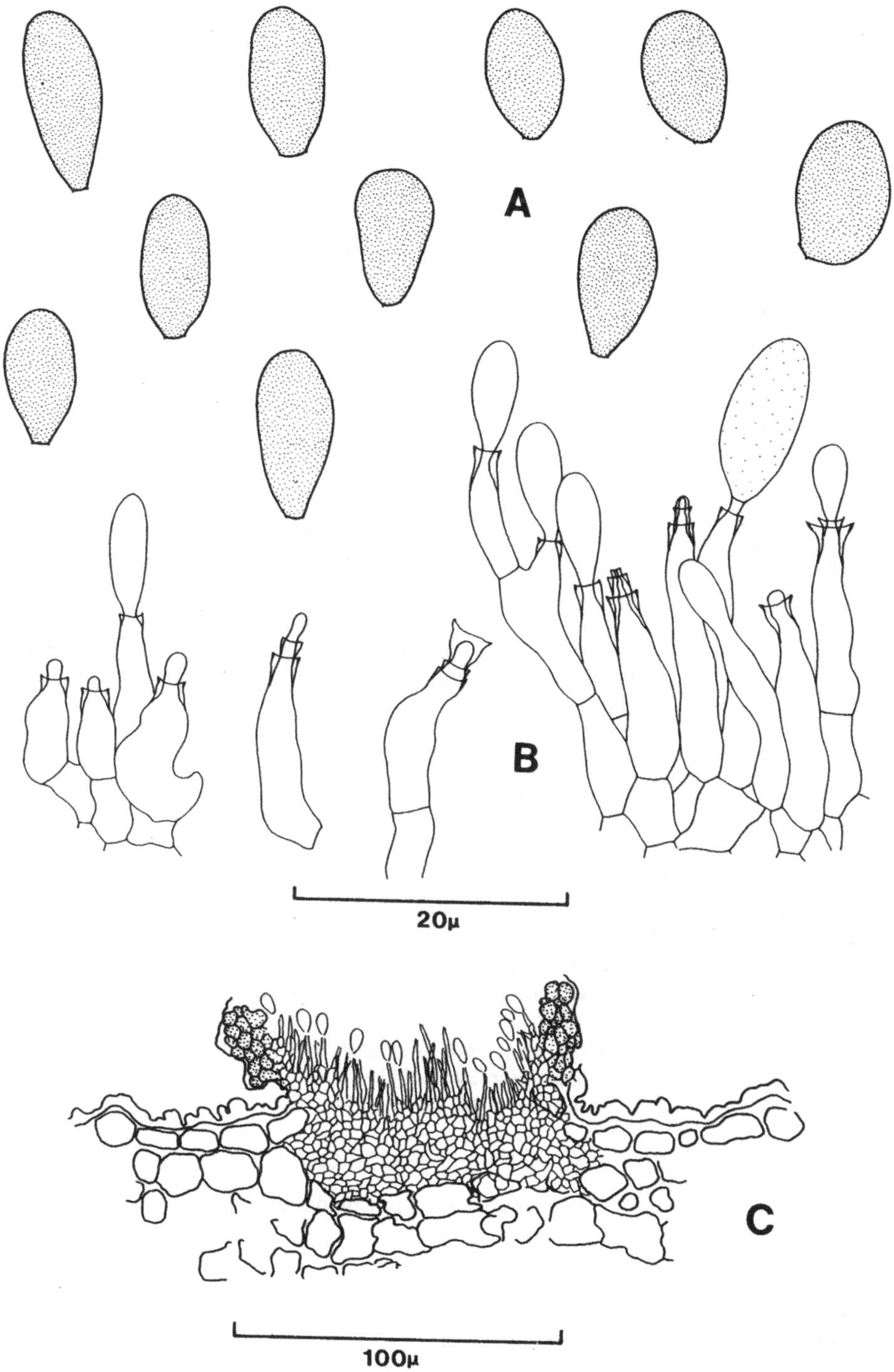

Fig. 349. *Discosporium tremuloides*, A, conidia, B, conidiogenous cells and developing conidia; C, vertical section of a conidioma.

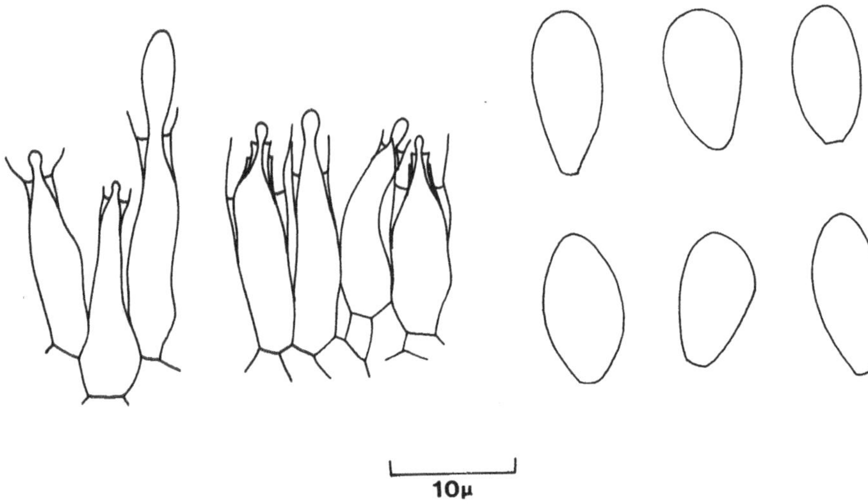

Fig. 350. *Discosporium eugeniae*, conidia and conidiophores.

determinate, discrete, less often integrated, lageniform or cylindrical, hyaline, smooth, channel and collarette minute, periclinal wall markedly thickened. *Conidia* hyaline, aseptate, thin-walled, irregularly guttulate, straight or curved, often lobed, very variable.

The generic synonymy has been fully discussed by Sutton (1977a), and the type species has been repeatedly renamed, as the synonymy shows. Additional taxa already placed in *Phacidiopycnis* include *P. pseudotsugae* (Wils.) Hahn, *P. boycei* (Hahn) Funk and *P. balsamicola* Funk (1970).

P. PIRI (Fckl) Weindlmayr, *Sydowia* **18**: 27 (1965).
Cytospora piri Fckl, *Jb. nassau. Ver. Naturk.* **15**: 52 (1860).
Myxosporium piri (Fckl) Fckl, *Jb. nassau. Ver. Naturk.* **23-24**: 399 (1870).
Discula piri (Fckl) Höhn., *Mitt. Bot. Inst. tech. Hochsch. Wien.* **3**: 53 (1926).
Gloeosporium melanconioides Pk, *53rd Rep. N.Y. St. Mus. Bot.*: 849 (1899).
Cryptosporiopsis piri (Fckl) Petrak, *Annls mycol.* **23**: 125 (1925).
Discosporiopsis piri (Fckl) Petrak, *Annls mycol.* **19**: 217 (1921).
Discosporium piri (Fckl) Hohn., *Z.f. Garunsph.* **5**: 194 (1915).
Phacidiopycnis melanconioides (Pk) Arx, *Verh. K. Ned. Akad. Wet. Amst.* **51**: 103 (1957).
Pyrenochaeta furfuracea Rostr., *Plantepatol.*: 571 (1902).
Phacidiopycnis furfuracea (Rostr.) Jørstad, *Mld. Stat. Plantepat. Imt.* **1**: 22 (1945).
Phacidiopycnis malorum Potebnia, *Z. Pflanzenkr.* **22**: 129 (1912).
Fuckelia conspicua Marchal, *Bull. Soc. Bot. belg.* **54**: 109 (1921).

Teleomorphosis: *Potebniamyces discolor* (Mont. & Sacc.) Smerlis, *Can. J. Bot.* **40**: 352 (1962).

Conidiomata 400-800 µ diam. Conidiophores 5-8 × 2-3 µ. Conidia ellipsoid, 10-14 × 6-9 µ.

On *Pyrus communis*, G.B. (IMI 57608).

P. padwickii (Kheswalla) comb. nov. (Fig. 351A)
Operculella padwickii Kheswalla, *Ind. J. agr. Sci.* **11**: 317 (1941).

Conidiomata 300-750 µ diam. Conidiogenous cells 8-15 × 2.5-5.5 µ. Conidia rather irregular, 10.5-14.5 × 4.5-12.5 µ.

From *Cicer arietinum*, India (IMI 167505, 22427, 192210, 192211).

P. tuberivora (Güssow & Foster) comb. nov. (Fig. 351B)
Phomopsis tuberivora Güssow & Foster, *Can. J. Res.* **4**: 253 (1932).

Conidiomata up to 800 µ diam. Conidiophores 15-30 × 2-3 µ. Conidia fusiform, 9.5-12 × 4.5-5.5 µ.

From *Solanum tuberosum, Medicago sativa, Hoya* and *Olearia traversii*, U.K., N.Z. (IMI 38386, 62681, 62682, 79162, 79161, 159972, 160106, 182311, 170746).

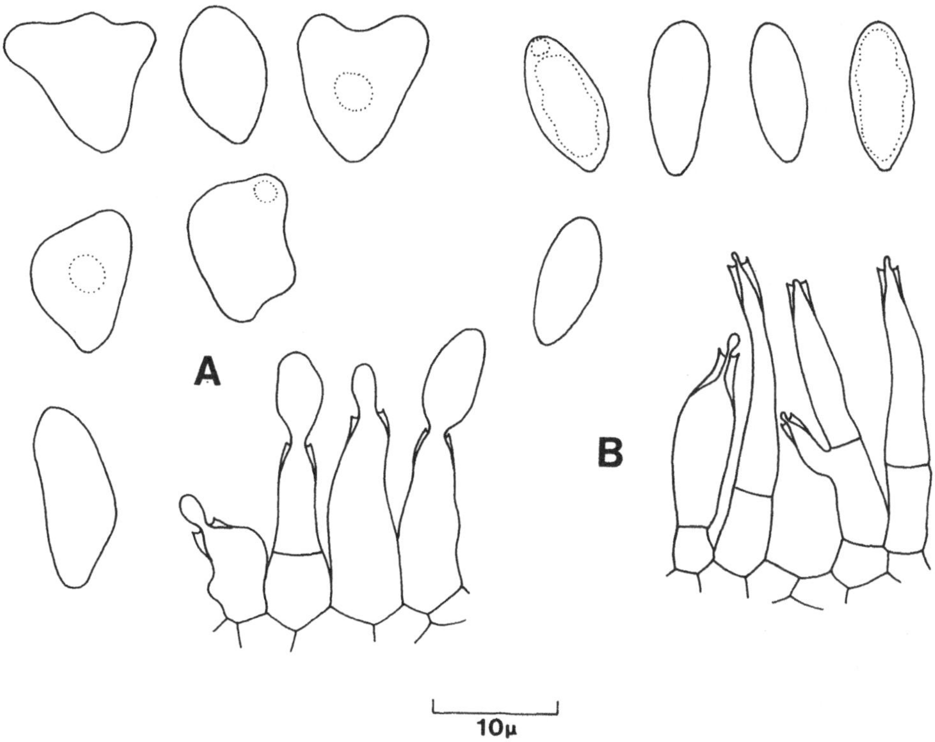

Fig. 351. A, *Phacidiopycnis padwickii*, conidia and conidiophores; B, *P. tuberivora*, conidia and conidiophores.

82. CYTOSPORA Ehrenb. ex Fr. (as '*Cytispora*'), *Syst. Myc.* **2** (2): 540 (1823).
Cytospora Ehrenb., *Sylv. Mycol. Berol.*: 28 (1818).
Bostrychia Fr. ex Fr., *Syst. Myc.* **1**: LII (1821).
Bostrychia Fr., *K. Vet. Akad. Handl.* **39**: 117 (1818).
Psecadia Fr., *Summa Veg. Scand.*: 414 (1849).
Lamyella Fr., *Summa Veg. Scand.*: 410 (1849).
Torsellia Fr., *Summa Veg. Scand.*: 412 (1849).
Hypoplasta Preuss, *Linnaea* **26**: 71 (1855).
Cytophoma Höhn., *Sber. Akad. Wiss. Wien* **123**: 133 (1914).
Monopycnis Naumov, *Bull. Soc. Oural. d'Amis Sci. nat.* **35**: 36 (1915).
Pleuronaema Höhn., *Hedwigia* **59**: 257 (1917).
Cytosporopsis Höhn., *Annls mycol.* **16**: 123 (1918).
? *Allantophomopsis* Petrak, *Annls mycol.* **23**: 103 (1925).
Leucocytospora (Höhn.) Höhn., *Mitt. bot. Inst. tech. Hochsch. Wien* **4**: 73 (1927).
Cytospora Ehrenb. ex Fr. subgen. *Leucocytospora* Höhn., *Annls mycol.* **16**: 130 (1918).
Cyclocytospora Höhn., *Mitt. bot. Inst. tech. Hochsch. Wien* **5**: 17 (1928).
Original spp.: 18 species listed
Lectotype sp.: *C. chrysosperma* (Pers.) Fr. (syn. *Naemaspora chrysosperma* Pers.), fide Donk (1964).

Mycelium immersed, septate, pale brown to hyaline, branched. *Conidiomata* eustromatic, separate, subperidermal, conical, erumpent, and dark brown, multi-locular and convoluted, the locules radiating and enlarging from the centre, separated by walls of pale to dark brown textura angularis or intricata, but uniting in the ostiolar region; basal, lateral and upper walls of dark brown, thick-walled textura intricata, becoming very small-celled textura angularis in the conidiogenous region. *Ostiole* single, circular, prominent, often surrounded by a furfuraceous ring of textura globulosa. *Conidiophores* hyaline, septate, branched irregularly at the base and above, smooth, mostly with apical conidia but sometimes acropleurogenous, formed from the inner cells of the locular walls. *Conidiogenous cells* enteroblastic, phialidic, determinate, integrated, straight, hyaline, smooth, occasionally formed as very small lateral branches immediately below transverse septa but more often as long distinct branches, collarette and channel minute. *Conidia* formed in distinct, variously coloured masses (as globose droplets or tendrils), hyaline, aseptate, thin-walled, eguttulate, smooth, allantoid.

The generic synonymy has been fully discussed by Sutton (1977a) under each of the synonyms listed above.

Cytospora species are the anamorphoses of *Valsa* Fr. and sometimes occur with their ascomatal states, although they are more frequently found alone. There have been few critical studies of either *Valsa* species, their *Cytospora* states, or both combined. A limited amount of work on *Valsa* and its *Cytospora* species occurring in the USSR was produced by Gutner (1934). Apart from these and the treatment of British species by Grove (1935), the only modern accounts have been by Gvritishvili (1965, 1967, 1968, 1969, 1973a, b, c). The need is evident therefore for correlative studies involving cultural work linked to type studies to determine the range of variation amongst species. So far more than 400 species have been described in the genus, and many more have been placed incorrectly in other genera, such as *Dendrophoma* Sacc.

C. CHRYSOSPERMA Pers. ex Fr., *ibid.*: 542 (1823). (Fig. 352)
Naemospora chrysosperma Pers., *Syn. fung.* 108 (1801).

Teleomorphosis: *Valsa sordida* Nitschke, *Pyren. Germ.*: 203 (1870).

Conidiomata 1-1.5 mm diam. Conidiophores up to 20 µ long × 1.5-2 µ. Conidia 4.5-6 × 1 µ.

On twigs and branches of *Populus nigra*, *P. tremuloides*, *P. sargentii*, *P. trichocarpa*, *P.* sp., Ont., Canada, Wyo., Kans., Calif., U.S.A., U.K., Italy, Austria (IMI 25897, 35638, 35640, 40071, 154353, 142188, 12938, 97167, 98033, 30425).

83. PHLYCTEMA Desm., *Ann. Sci. nat.*, ser. 3, **8**: 16 (1847).
Allantozythia Höhn., *Mykol. Unters.* **3**: 322 (1923).
Sp. typ.: *P. vagabunda* Desm.

Mycelium immersed, branched, septate, hyaline. *Conidiomata* eustromatic, immersed, then erumpent and sporodochial, separate, yellowish brown, pulvinate, circular, unilocular but convoluted, thick-walled; wall of small-celled textura angularis, at the base darker brown and thicker-walled than at the sides. *Ostiole* absent, dehiscence by rupture of the overlying tissues. *Conidiophores* hyaline, septate, branched irregularly, cylindrical to filiform, formed from the walls lining the conidiomata. *Conidiogenous cells* enteroblastic, phialidic, integrated or discrete, determinate, hyaline, with a wide channel and minute collarette and periclinal wall thickening. *Conidia* hyaline, aseptate, fusiform, eguttulate, straight, slightly curved or irregular.

Höhnel (1920b), Petrak (1924) and von Arx (1957a; 1970) have reviewed the genus, the latter author accepting three species distinguished by conidial sizes of 14-30 × 2-4 µ (*P. vagabunda*), 20-38 × 2-4 µ (*P. caulium*), and 28-47 × 5-7 µ (*P. ptychospermatis*). More than 60 taxa however have been described in the genus and Höhnel (1920b) only eliminated 18 species which he referred to *Phomopsis*, *Septoria*, *Rhabdospora* etc., so the genus is still in need of some revision.

P. VAGABUNDA Desm., *ibid.*: 16 (1847). (Fig. 353)
synonyms in von Arx (1957; 1970).

Teleomorphosis: *Pezicula alba* Guthrie, *Trans. Br. mycol. Soc.* **42**: 504 (1959).

Conidiomata up to 600 µ diam. Conidiophores up to 30 µ long × 3.5-4 µ wide, often intermixed with sterile hyphae or paraphyses. Conidia 15-18 × 2.5-3 µ.

On fruit of *Malus pumila*, *M. sylvestris*, *M.* sp., *Aconitium napellus*, *Erigeron* sp., Tasmania, N.S.W., W.A., Australia, New Zealand, Canada, France, Germany, Eire, U.K. (IMI 68994, 111897a, 49935, 21683, 98476, 98129, 21683, 31108, 90605 ex PC, holotype *P. vagabunda*, 205438).

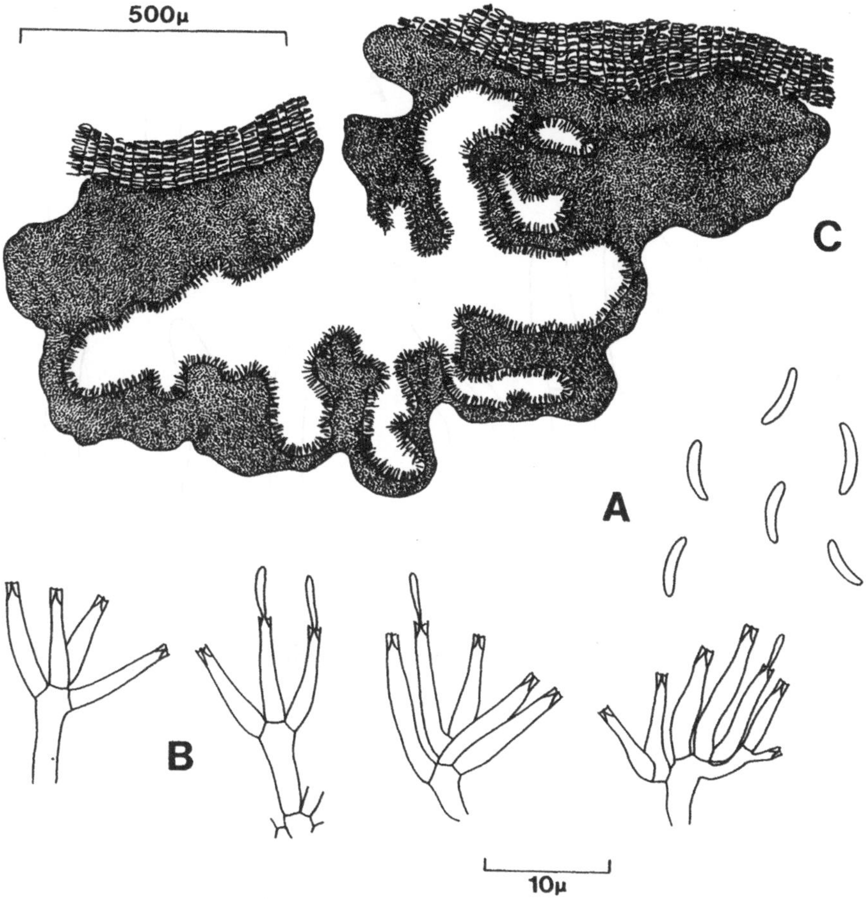

Fig. 352. *Cytospora chrysosperma*, A, conidia; B, conidiophores; C, vertical section of a conidioma.

84. BLENNORIA Moug. & Fr. in Fr., *Syst. Orb. Veg.*: 366 (1825).
Sp. typ.: *B. buxi* Fr.
Illustration: Sutton (1972b).

Mycelium immersed, branched, septate, hyaline to pale brown. *Conidiomata* eustromatic, amphigenous, epidermal to subepidermal, separate, pulvinate, oval, circular or irregular, carbonous, dark brown; tissue of pale brown, thin-walled textura angularis, somewhat darker and thicker-walled at the periphery; multilocular, each locule separated by vertically arranged cells. Dehiscence not observed, but papillate swellings over the surface indicate ostiolar development. *Conidiophores* hyaline, branched irregularly, septate, smooth, with acropleurogenous conidia, formed from the inner cells of the locular walls. *Conidiogenous cells* enteroblastic, phialidic, determinate, integrated or discrete, cylindrical, hyaline, smooth, channel and collarette minute. *Conidia* hyaline, cylindrical, aseptate, smooth, thin-walled, base abruptly tapered, apex obtuse, eguttulate.

This genus has often been regarded as synonymous with *Ceuthospora* Grev. but Sutton (1972b) and Morgan-Jones, Nag Raj & Kendrick (1972g) independently demonstrated their differences. Apart from quite different conidiomata, eustromatic in *Blennoria* and pseudostromatic in *Ceuthospora*, the conidia of *Blennoria* lack any appendage whilst those of *Ceuthospora* develop a minute apical mucilaginous fan-shaped appendage. So far 14 taxa have been assigned to the genus.

B. BUXI Fr., *Syst. myc.* **3**: 472 (1832). (Fig. 354)

Conidiomata up to 1200 μ diam. Conidiophores 25 × 2.5 μ. Conidia 12-13 × 2.5-3 μ.

On dead leaves of *Buxus sempervirens*, U.K., Switzerland (IMI 49362, 49361, 105572, 127975).

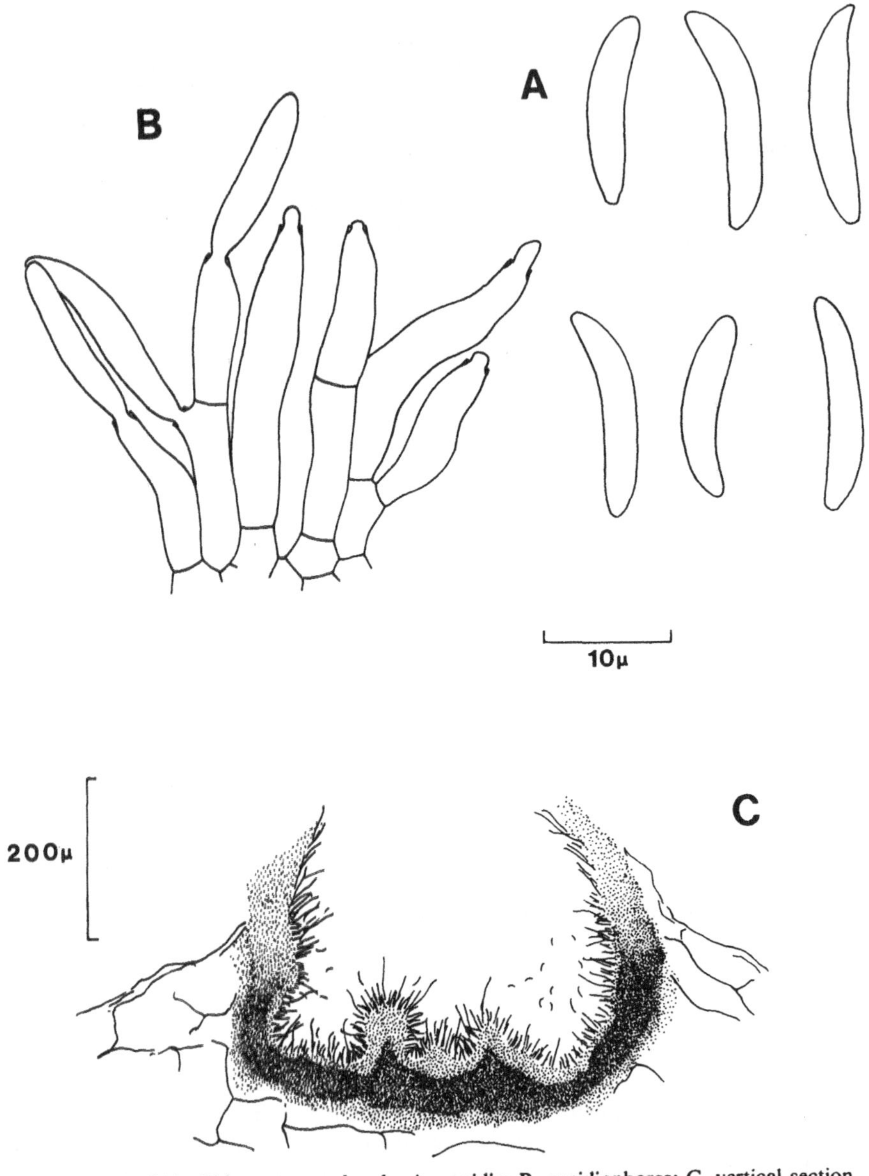

Fig. 353. *Phlyctema vagabunda*, A, conidia; B, conidiophores; C, vertical section of a conidioma.

85. PLECTOPHOMELLA Moesz, *Mag. Bot. Lap.* **21**: 13 (1922).
Apocytospora Höhn., *Mitt. bot. Inst. tech. Hochsch. Wien* **1**: 43 (1924).
Sp. typ.: *P. visci* Moesz

Mycelium immersed, septate, branched, brown. *Conidiomata* eustromatic, subepidermal, separate or aggregated, globose, dark brown, in culture divided and multilocular; walls thick, composed of dark brown, thick-walled textura angularis, interlocular tissue hyaline, thin-walled. *Ostiole* central, circular, papillate. *Conidiophores* hyaline, irregularly branched and septate. *Conidiogenous cells* enteroblastic, phialidic, determinate, discrete, doliiform to ampulliform, hyaline, smooth, collarette and channel minute, formed from the inner cells of the stromatal wall. *Paraphyses* filiform, flexuous, septate, branched. *Conidia* hyaline, aseptate, straight or slightly curved, smooth, guttulate or eguttulate.

The genus is monotypic.

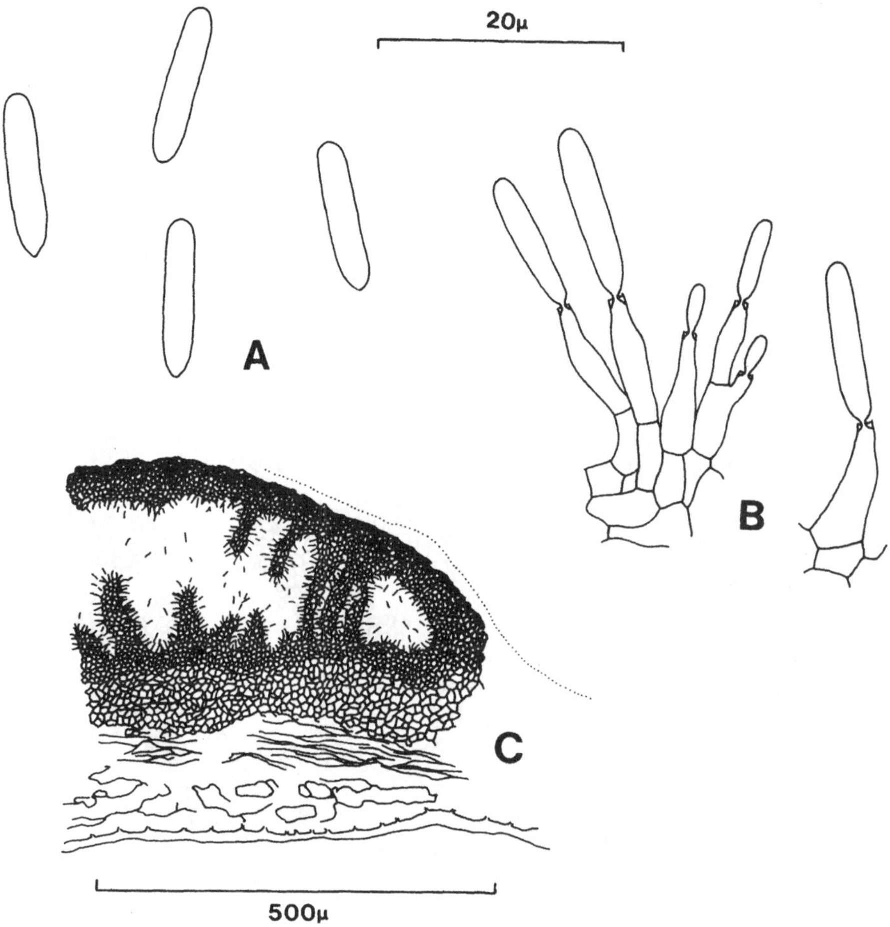

Fig. 354. *Blennoria buxi*, A, conidia; B, conidiophores and developing conidia; C, vertical section of a closed conidioma.

P. VISCI Moesz, *ibid.*: 13 (1922). (Fig. 355)
Apocytospora visci Höhn., *Mitt. bot. Inst. techn. Hochsch. Wien* **1**: 43 (1924).
Phyllosticta visci Sacc., *Michelia* **1**: 125 (1878).

Conidiomata 250-320 µ diam. Paraphyses 6-12 µ long. Conidia 4-5 × 1.5-2 µ.

From leaves of *Viscum album*, France (IMI 199474).

86. PLEUROCYTOSPORA Petrak, *Annls mycol.* **21**: 256 (1923).
Original spp.: *P. vestita* Petrak, *P. lycii* Petrak
Sp. typ.: *P. vestita* Petrak, fide Clements & Shear (1931).

Mycelium immersed, branched, septate, hyaline to brown. *Conidiomata* eustromatic, immersed, peridermal, becoming erumpent, separate, irregular in shape, dark brown, irregularly multilocular and convoluted; outer stromatal tissue composed of dark brown, loose, thin-walled textura globulosa, walls of locules of similar tissue but cells much smaller and compact, becoming hyaline and thinner-walled towards the conidiogenous region. Dehiscence irregular by breakdown of the upper wall, usually at one point. *Conidiophores* septate, hyaline, branched mainly at the base, less frequently above, cylindrical, tapered towards the apices, formed from the inner wall of the locules. *Conidiogenous cells* enteroblastic, phialidic, integrated or discrete, determinate, hyaline, formed as short or long lateral branches from the conidiophores immediately below septa, aperture and collarette minute. *Conidia* acropleurogenous, hyaline, aseptate, thin-walled, straight or allantoid, eguttulate.

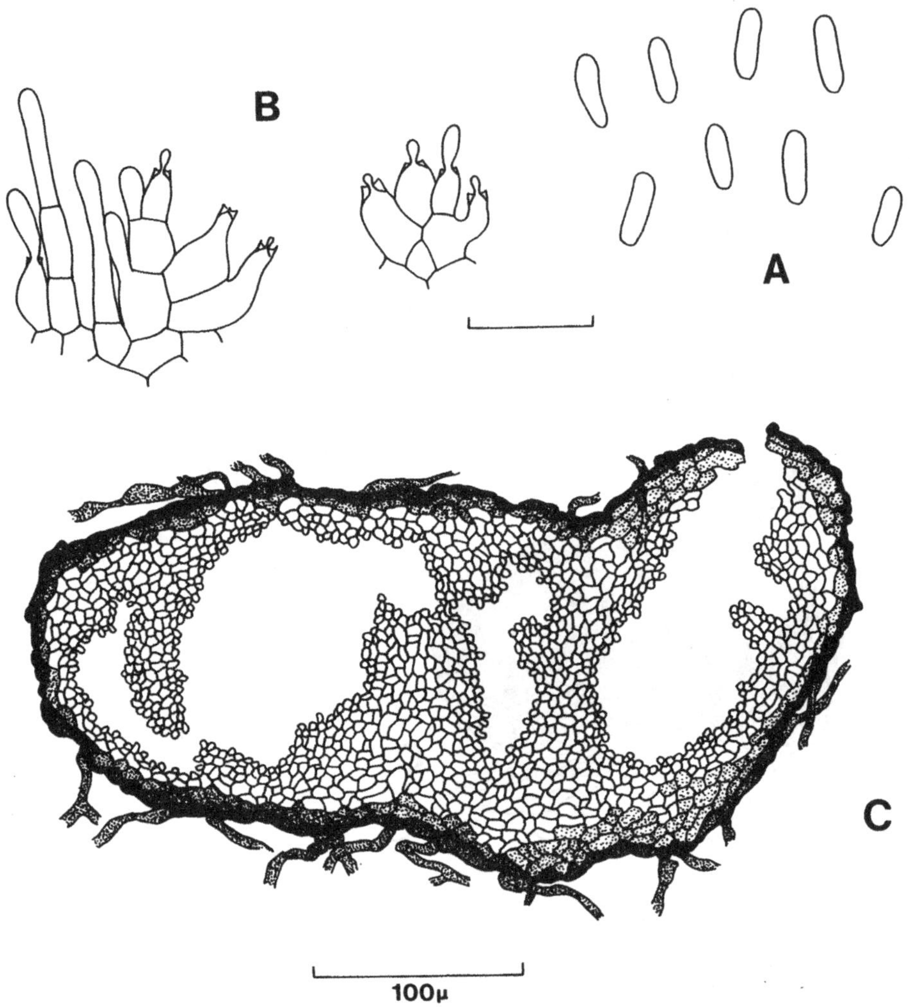

Fig. 355. *Plectophomella visci*, A, conidia; B, conidiophores; C, vertical section of a conidioma.

P. VESTITA Petrak, *ibid.*: 257 (1923). (Fig. 356A)

Conidiophores 25-40 µ long × 1.5-2.5 µ wide. Conidia 3.5-4 × 1-1.5 µ.

On branches of *Ribes rubrum*, Czechoslovakia (IMI 33611).

P. LYCII Petrak, *ibid.*: 260 (1923). (Fig. 356B, C)

Conidiophores up to 70 µ long × 2-3 µ wide. Conidia 2.5-3.5 × 1 µ.

On branches of *Lycium halimifolium*, *L.* sp., Austria, Germany (IMI 24793, 30378).

87. SPORONEMA Desm., *Ann. Sci. nat.*, ser. 3, **8**: 182 (1847).
Sp. typ.: *S. phacidioides* Desm.

Mycelium immersed, branched, septate, hyaline. *Conidiomata* eustromatic, foliicolous, separate or confluent, immersed, unilocular, convoluted, rarely multilocular, brown, thick-walled; walls at the base and sides of pale brown to hyaline, thin-walled textura angularis, in the upper region of thick-walled, darker brown tissue. *Ostiole* absent, dehiscence by breakdown of the central region of the upper wall. *Conidiophores* hyaline, septate, branched irregularly from the base

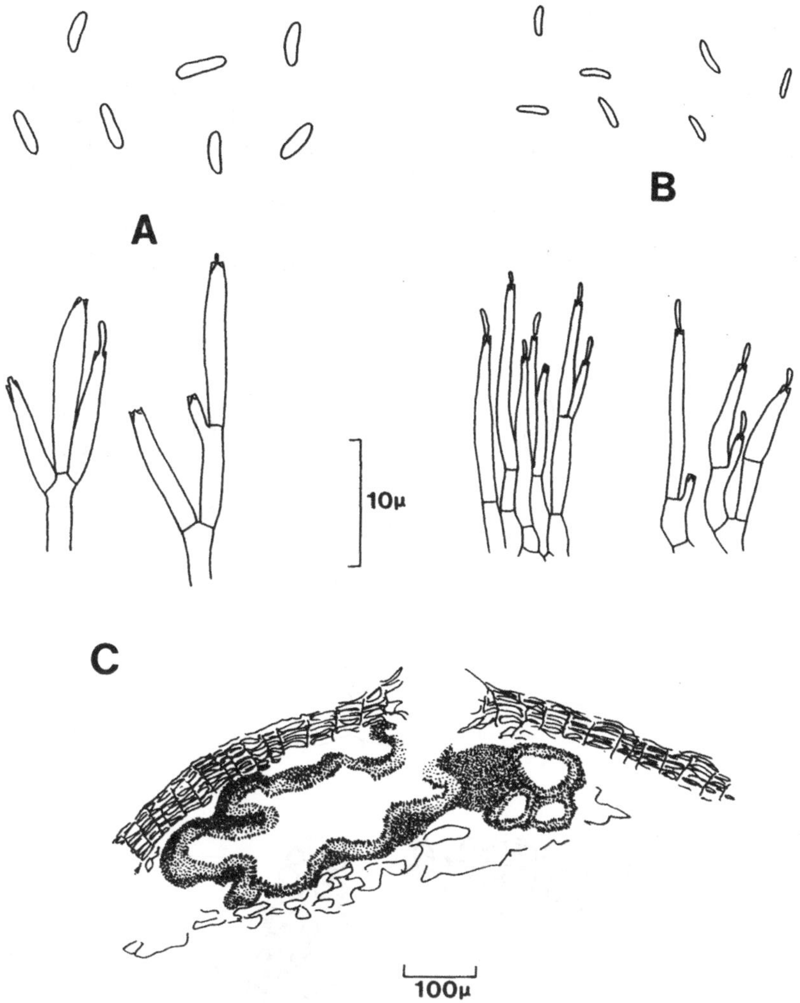

Fig. 356. A, *Pleurocytospora vestita*, conidia and conidiophores; B, C, *P. lycii*, conidia, conidiophores, vertical section of a conidioma.

and above, cylindrical or irregularly curved, formed from the inner cells of the locular wall. *Conidiogenous cells* enteroblastic, phialidic, integrated, rarely discrete, determinate, hyaline, tapered towards the apices, loci terminal on the main conidiophore and short lateral branches, collarette minimal, channel and periclinal thickening present. *Conidia* hyaline, aseptate, eguttulate, smooth, fusiform to cylindrical or irregular.

More than 40 taxa have been described in *Sporonema* and so far there have been two attempts to revise some of these species. Höhnel (1910d) redescribed *S. phacidioides* and removed four species to other genera, and more recently Limber (1955) studied and compared 18 species. Three were maintained in *Sporonema, S. camelliae* Earle and *S. oxycocci* Shear which were also kept in *Sporonema* are clearly *Coleophoma* species, three were placed in *Placosphaeria*, three in *Sclerotiopsis* (now known as *Pilidium*), and two in *Phomopsis* and one each in *Sirexcipula* and *Siropatella*. There is still justification for a further revision in the light of modern concepts of conidiogenesis.

S. PHACIDIOIDES Desm., *ibid.*: 182 (1847). (Fig. 357)
Ascochyta medicaginis Fckl, *Jb. nassau. Ver. Naturk.* **23-24**: 388 (1870).
Phyllosticta medicaginis (Fckl) Sacc., *Syll. fung.* **3**: 42 (1884).
Gloeosporium morianum Sacc., *Syll. fung.* **10**: 458 (1892).

Teleomorphosis: *Leptotrochila medicaginis* (Fckl) Schüepp, *Phytopath. Z.* **36**: 253 (1959).

Conidiomata up to 150 µ diam × 100 µ deep. Conidiophores up to 35 µ long × 4 µ wide. Conidiogenous cells 7-15 × 2-3 µ narrowing to 1.5 µ. Conidia 4-5.5 × 2-2.5 µ.

On leaves of *Medicago sativa, M. lupulina, M. polymorpha* var. *vulgaris, M. polymorpha, M. varia, M. falcata, M. arborea, M. denticulata*, Italy, Czechoslovakia, Austria, U.K., Cyprus, U.S.S.R., Latvia, Ariz., Wisc., N.Y., Colo., Mont., U.S.A., Ont., Sask., Man., Canada, Dominica, Ethiopia, Iraq, Malawi, Tanzania, Zambia, S. Rhodesia, Israel, Kenya, India, N.S.W., Australia, New Zealand (IMI 29738, 29737, 182772, 29741, 29742, 59347, 12458, 29740, 63013, 96887, 148603, 205416, 214599, 178284, 167002, 223331, 25986, 12811, 135419, 132106, 126851, 166994, 110772, 113528, 114351, 116612, 29739, 69200, 50878, 34686, 68173, 94049, 94048, 61161, 82435, 120756, 8850, 88500, 74528, 76319, 76320, 121759, 121758, 100701, 100702, 78922, 27042, 27040).

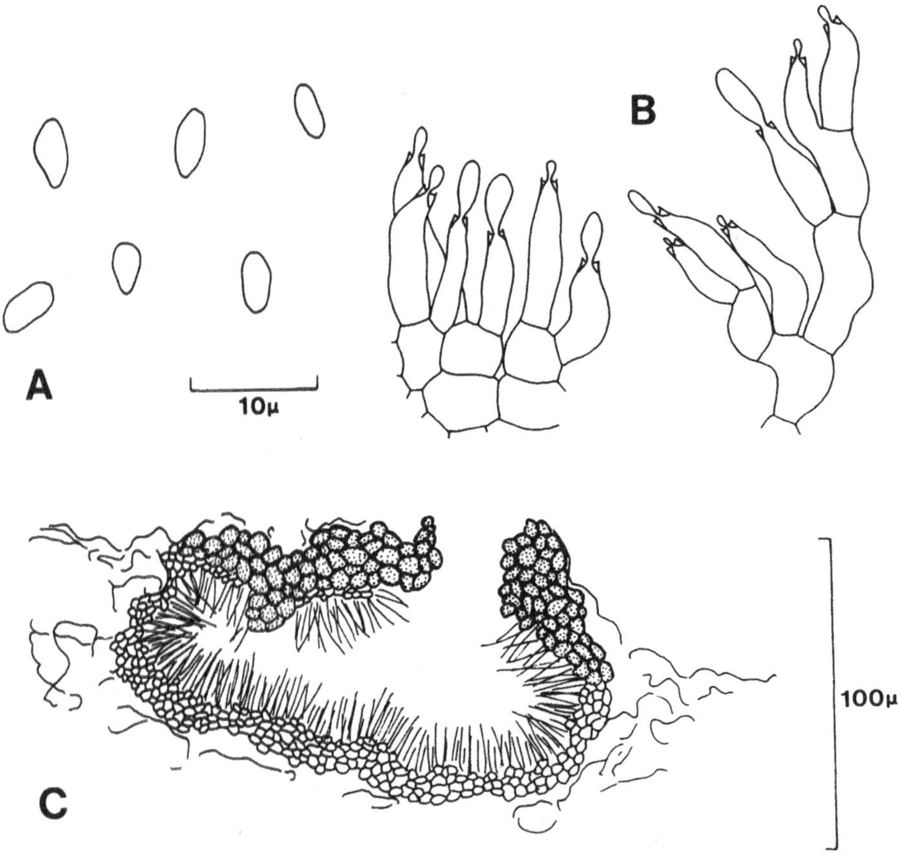

Fig. 357. *Sporonema phacidioides*, A, conidia; B, conidiophores and developing conidia; C, vertical section of a conidioma.

88. MICRODISCULA Höhn., *Sber. Akad. Wiss. Wien* **124**: 142 (1915).
Sp. typ.: *M. rubicola* (Bres.) Höhn. (syn. *Sphaeronaema rubicola* Bres.).

Mycelium immersed, branched, septate, hyaline. *Conidiomata* eustromatic, globose, separate, dark brown, subepidermal, multilocular and convoluted, thick-walled; wall composed of hyaline, thick-walled textura intricata, becoming brown in the upper peripheral region with a dark brown crust. *Ostiole* single, circular, ± central. *Conidiophores* hyaline, branched at the base and above, septate, smooth, cylindrical, branches straight, formed from the locular walls. *Conidiogenous cells* enteroblastic, phialidic, integrated, determinate, hyaline, terminal or lateral, cylindrical, collarette and aperture minute. *Conidia* hyaline, aseptate, thin-walled, eguttulate, straight, fusiform to cylindrical.

A second species, *M. phragmitis* (Westd.) Höhn., is known, but so far has remained unexamined by me.

M. RUBICOLA (Bres.) Höhn., *ibid.*: 142 (1915). (Fig. 358)
Sphaeronaema rubicola Bres., *Hedwigia* **39**: 325 (1900).

Conidiomata up to 625 µ diam. Conidiophores up to 35 µ long × 1-1.5 µ wide. Conidia 3-3.5 × 1 µ.

On dead stems of *Rubus fruticosus*, Germany (IMI 201968, holotype ex K).

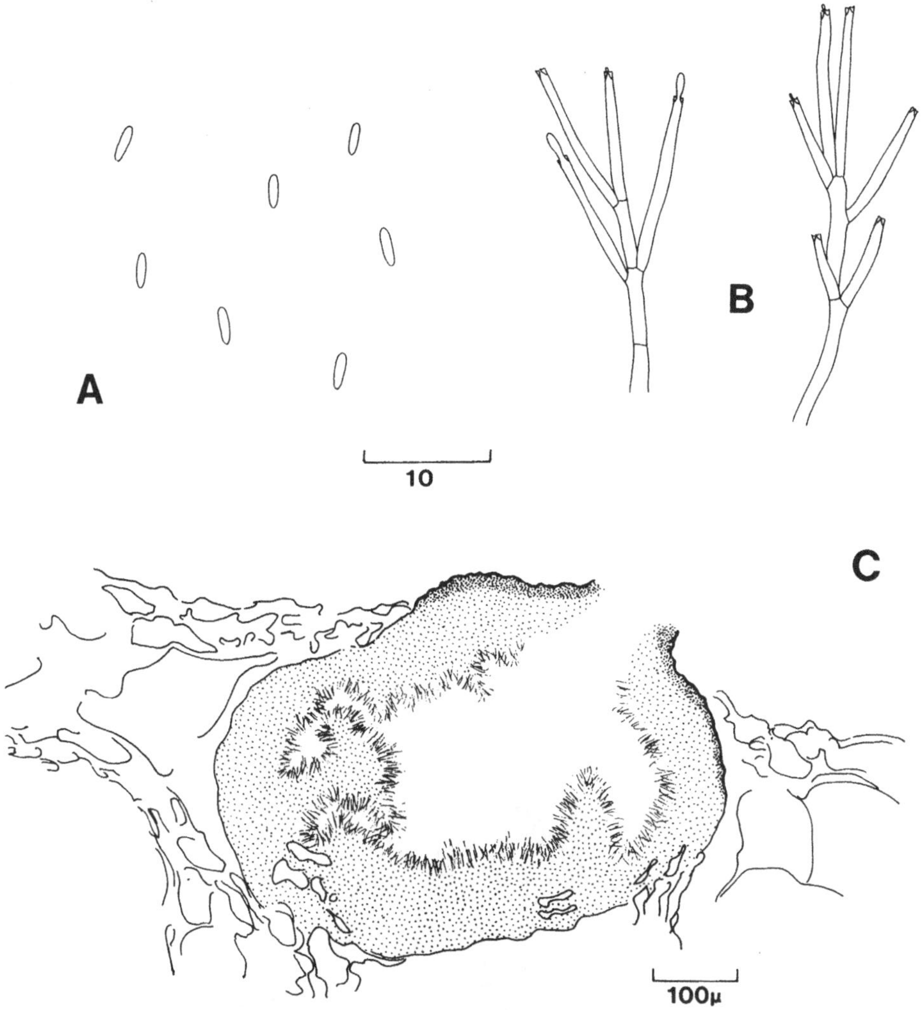

Fig. 358. *Microdiscula rubicola*, A, conidia, B, conidiophores; C, vertical section of a conidioma.

89. AMPHICYTOSTROMA Petrak, *Annls mycol.* **19**: 63 (1921).
Sp. typ.: *A. tiliae* (Sacc.) Petrak (syn. *Cytospora tiliae* Sacc.).

Mycelium immersed, branched, septate, pale brown to hyaline. *Conidiomata* stromatic, produced in association with the perfect state, the locules of which are basal with long ostiolar channels. The stromata are periclinal, completely surrounding the ostiolar channels but not extending as deeply as the locules. Stromata subperidermal, multilocular but not convoluted, basal tissue of dense, pale brown textura intricata, the overlying tissue very diffuse, of little structural organisation, mainly consisting of sterile flexuous hyphae formed from very pale brown loose pseudoparenchyma. *Ostioles* absent, dehiscence by rupture of the periderm near the apex of the ascocarp ostioles. *Conidiophores* restricted to the basal region, hyaline, septate, smooth, irregularly branched at the base and above, filiform, with acropleurogenous conidia.

Conidiogenous cells enteroblastic, phialidic, integrated, determinate, cylindrical, hyaline, smooth, channel and collarette minute, periclinal wall thickened. *Conidia* hyaline, aseptate, thin-walled, smooth, eguttulate, cylindrical, straight or slightly curved.

The genus is monotypic.

A. TILIAE (Sacc.) Petrak, *Annls mycol.* **19**: 63 (1921). (Fig. 359)
Cytospora tiliae Sacc., *Michelia* **1**: 519 (1879).

Teleomorphosis: *Amphiporthe hranicensis* (Petrak) Petrak, *Sydowia* **24**: 257 (1971).

Stromata up to 700 µ wide. Conidiophores up to 20 µ long × 1.5-2 µ wide. Conidia 4.5-6.5 × 1.5 µ.

On branches of *Tilia* spp., U.K., Germany (IMI 13126, 13137, 13517, 13518, 15231, 30401, 31007, 33249, 40957).

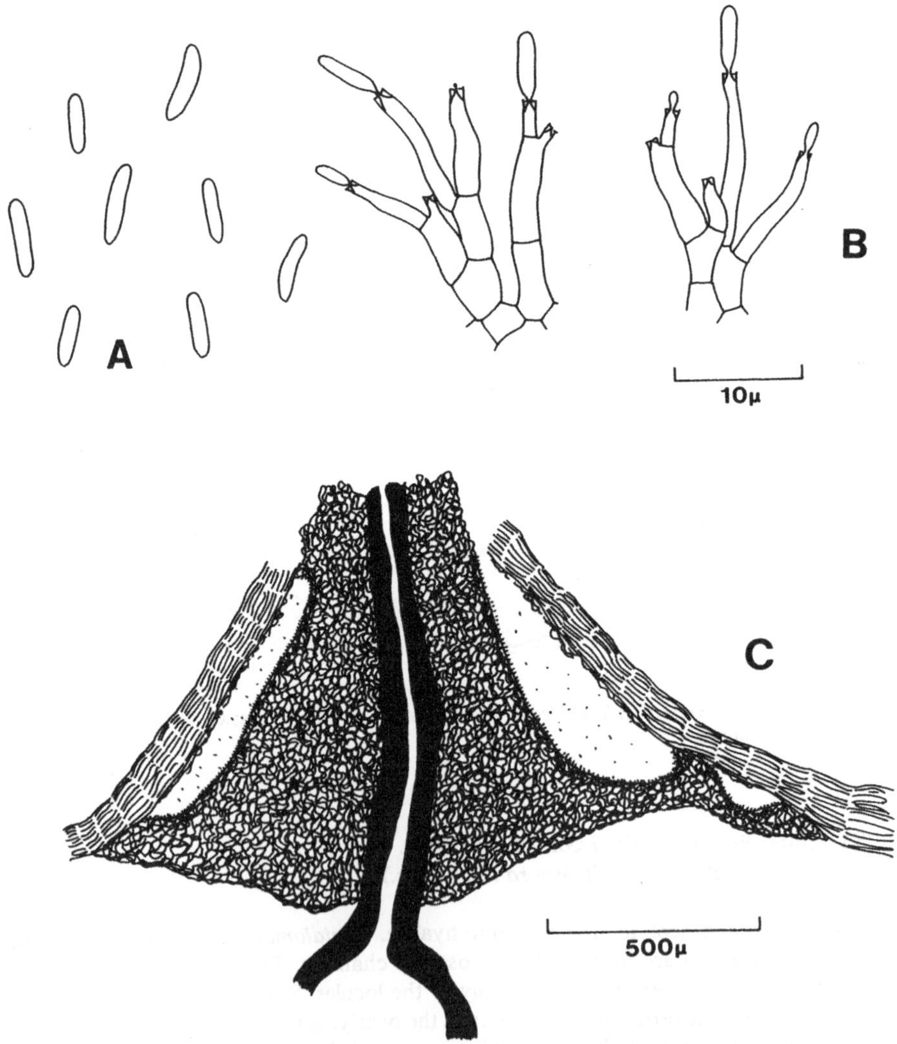

Fig. 359. *Amphicytostroma tiliae*, A, conidia; B, conidiophores and developing conidia; C, vertical section of a conidioma.

90. DWAYALOMELLA Brisson, Pirozynski & Pauzé, *Can. J. Bot.* **53**: 2867 (1975).
Sp. typ.: *D. vaccinii* Brisson, Pirozynski & Pauzé
Illustration: Brisson, Pirozynski & Pauzé (1975).

Mycelium hyaline, septate, branched. *Conidiomata* acervular, black, circular to elliptical, becoming cupulate, separate, composed of pale brown textura intricata. *Setae* marginal, numerous, simple, brown, unbranched, rigid, straight, thick-walled, smooth, aseptate, acute at the apex. *Conidiophores* absent. *Conidiogenous cells* enteroblastic, phialidic, determinate, discrete, hyaline, smooth, cylindrical, channel and collarette minute, periclinal wall thickened, formed from the base and lateral walls. *Conidia* cylindrical, medianly 1 euseptate, smooth, thin-walled, hyaline, apex ± obtuse with 2-3 divergent subapical, unbranched, filiform, cellular appendages, base obtuse, slightly protuberant with 2-3 divergent, unbranched, filiform, cellular appendages.

The genus is monotypic.

D. VACCINII Brisson, Pirozynski & Pauzé, *ibid.*: 2867 (1975).

Conidiomata 100-1000 µ. Setae up to 150 µ long × 3-4 µ wide. Conidiogenous cells up to 40 × 3-4 µ. Conidia 26-35 × 3.5-5 µ; appendages 20-35 µ long.

From stems of *Vaccinium angustifolium*, Que., Canada.

91. **NANOSCHEMA** gen. nov.
(etym. nano-(G) dwarf, et schema (G)-form, shape).

Mycelium immersum. Conidiomata tubularia, separata, superficialia, substrato per hypostromate parvo affixa; paries basalis ex textura angulari compositus, paries lateralis ex textura porrecta compositus sed apicem conidiomatum versus fimbriatus. Ostiolum absens. Conidiophora hyalina, septata, cylindrica, irregulariter ramosa, ex base conidiomatum formata. Cellulae conidiogenae enteroblasticae, phialidicae, integratae, determinatae. Conidia hyalina, medio 1 septata, eguttulata, ellipsoidea, recta vel leniter curvata.

Sp. typ.: *N. elaeocarpi* Sutton

Mycelium immersed. *Conidiomata* tubular, separate, superficial, attached to the substrate by a small hypostroma; basal wall of textura angularis, lateral wall of textura porrecta, apex fimbriate. *Ostiole* absent. *Conidiophores* hyaline, septate, cylindrical, irregularly branched, formed from the base of the conidiomata. *Conidiogenous cells* enteroblastic, phialidic, integrated, determinate. *Conidia* hyaline, medianly 1 septate, eguttulate, ellipsoid, straight or slightly curved.

The genera of coelomycetes with tubular to infundibular conidiomata have been fully illustrated and described by Sutton (1973a, 1975b). These are *Satchmopsis* Sutton & Hodges, *Shawiella* Hansf., *Stevensonula* Petrak and *Ojibwaya* Sutton. However they do not form a homogeneous group either on the basis of conidiomatal structure, conidiogenesis or on conidial morphology. The tissue types and gross morphology of the conidiomata are quite varied. *Stevensonula* has a fimbriate margin similar to that in *Nanoschema* but in all other respects the genera differ. In conidiogenesis only *Satchmopsis* is phialidic but this genus is quite distinct from *Nanoschema* for the conidiogenous cells are ampulliform, the conidia narrowly cylindrical and aseptate, and the conidiomata infundibular with a distinct margin. None of the genera resemble *Nanoschema* in conidial morphology.

N. elaeocarpi sp. nov. (Fig. 360)

Mycelium immersum, ramosum, septatum hyalinum. Conidiomata tubularia, separata, superficialia, pallide brunnea, substrato per hypostromata parvo affixa, usque ad 120 µ alta × 60 µ crassa; paries basalis ex textura angulari brunnea, parietibus tenuibus, regionem conidiogenam versus hyalinus; paries lateralis ex textura porrecta pallide brunnea, parietibus tenuibus sed apicem conidiomatum versus fimbriatus. Ostiolum absens. Conidiophora hyalina, aseptata, irregulariter ramosa, cylindrica vel apicem versus deminuta, usque ad 35 µ longa × 2-3 µ crassa, ex base conidiomatum formata. Cellulae conidiogenae enteroblasticae, phialidicae, integratae, determinatae, cylindricae, 15-20 × 2-3 µ. Conidia hyalina, medio 1 septata, laevia, eguttulata, recta vel leniter curvata, ellipsoidea, 8.5-12 × 3-4 µ.

In foliis *Elaeocarpi sphaerici*, Dala, British Solomon Islands Protectorate, 24 June 1974, G. Jackson PP 353, IMI 187436, holotypus.

Associated with and probably causing large irregular-shaped lesions up to 2 cm diam., each with dark brown to purplish brown margin and cream necrotic tissue. *Mycelium* immersed, branched, septate, hyaline.

Conidiomata tubular, separate, superficial, pale brown, attached to the substrate by a small hypostroma up to 120 µ high × 60 µ wide, basal wall consisting of brown, thin-walled textura angularis becoming more hyaline towards the conidiogenous region, lateral wall of thin-walled, pale brown textura porrecta the elements of which become progressively more separate towards the apex of the conidioma which is shortly fimbriate. Ostiole absent. Conidiophores hyaline, septate, irregularly branched, cylindrical or gradually tapered to the apices, up to 35 µ long × 2-3 µ wide, formed from the base and the lower regions of the lateral wall. Conidiogenous cells enteroblastic, phialidic, integrated, determinate, cylindrical, periclinal thickening very prominent, channel and collarette minute, 15-20 × 2-3 µ. Conidia hyaline, medianly 1 euseptate, smooth, eguttulate, straight or slightly curved, ellipsoid, 8.5-12 × 3-4 µ.

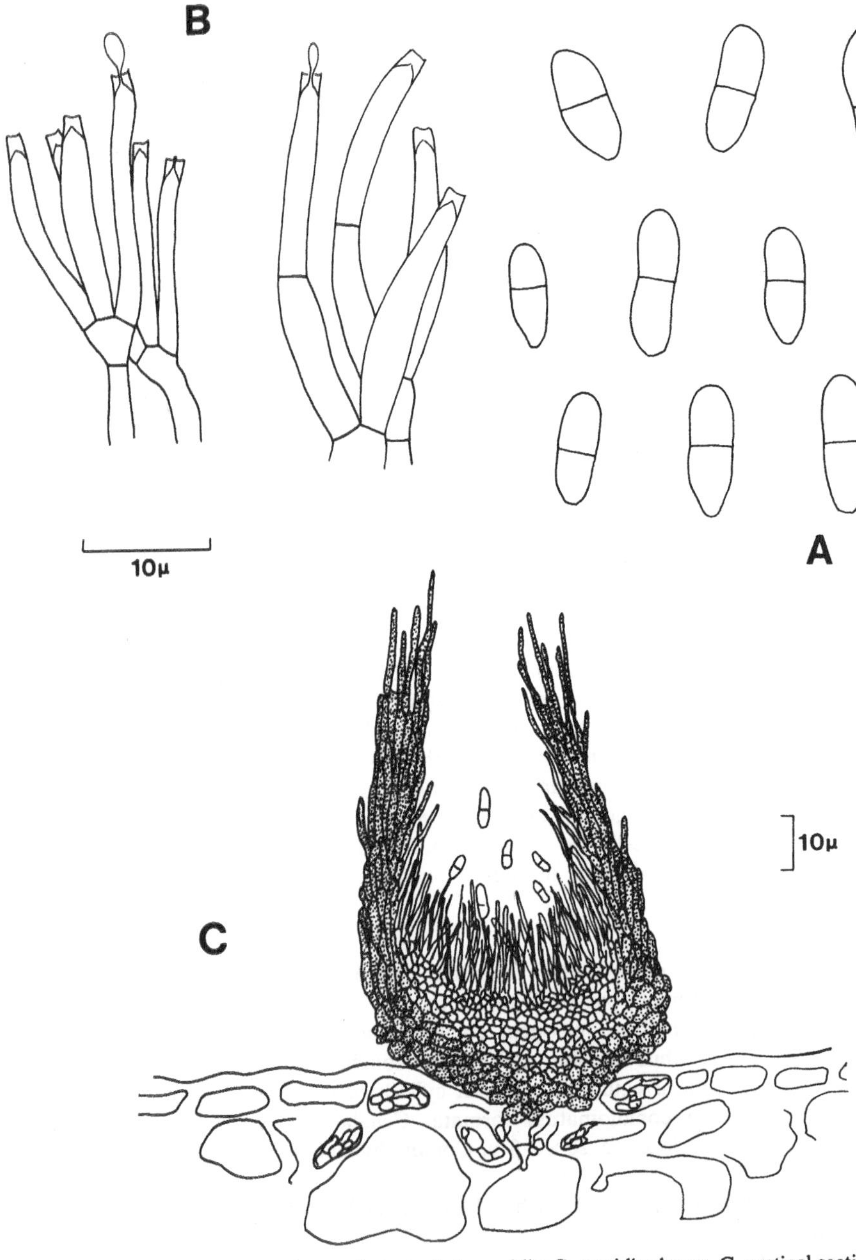

Fig. 360. *Nanoschema elaeocarpi*, A, conidia; B, conidiophores, C, vertical section of a conidioma.

92. FUCKELIA Bonorden, *Abhandlungen aus dem Gebeite der Mykologie*: 135 (1864).
Sp. typ.: *F. ribesia* (Lk) Sutton (syn. *Sphaeria ribesia* Lk, *Fuckelia ribis* Bonorden).

Mycelium immersed, branched, septate, pale brown. *Conidiomata* eustromatic, initially peridermal to subperidermal, separate or caespitose, grey to grey brown, consisting of a basal, ± central foot (at the base of pseudostromatic, but above consisting of fungal pseudoparenchyma), surmounted at the apex and subapical region by ± globose punctate to wrinkled conidiogenous structures; the latter are multilocular, convoluted and composed of dark brown thick-walled textura angularis on the outside with interlocular walls of pale brown thin-walled textura angularis or intricata. *Ostioles* absent, dehiscence by breakdown of upper walls of locules. *Conidiophores* hyaline, septate, branched irregularly above and below, irregular in shape but generally tapered, formed from the walls lining the locules. *Conidiogenous cells* enteroblastic, phialidic, determinate, integrated, less often discrete, apical or lateral on small branches immediately below septa, hyaline, smooth, tapered to an often recurved apex, channel and collarette minute, periclinal wall thickened. *Conidia* hyaline, smooth, thin-walled, guttulate, medianly 1 euseptate, ellipsoid.

The recent account of *Godronia* by Groves (1965) places the status of *Fuckelia* and other conidial states in perspective.

F. RIBESIA (Lk) Sutton, *Mycol. Pap.* 141: 87 (1977). (Fig. 361)
Sphaeria ribesia Lk, *Handb. z. Erk. der Gew.* **3**: 376 (1833).
Fuckelia ribis Bonorden, *Abhandlungen aus dem Gebeite der Mykologie*: 135 (1864).

Teleomorphosis: *Godronia ribesia* (Fr.) Seaver, *Mycologia* **37**: 339 (1945).

Conidiomata .5-1.5 mm diam. × .2-1.0 mm high. Conidiogenous cells 10-13 × 3-3.5 µ wide at the base. Conidia 7.5-8.5 × 3-3.5 µ.

On stems of *Ribes* sp., *R. rubrum*, U.K., Austria (IMI 80749, 30451).

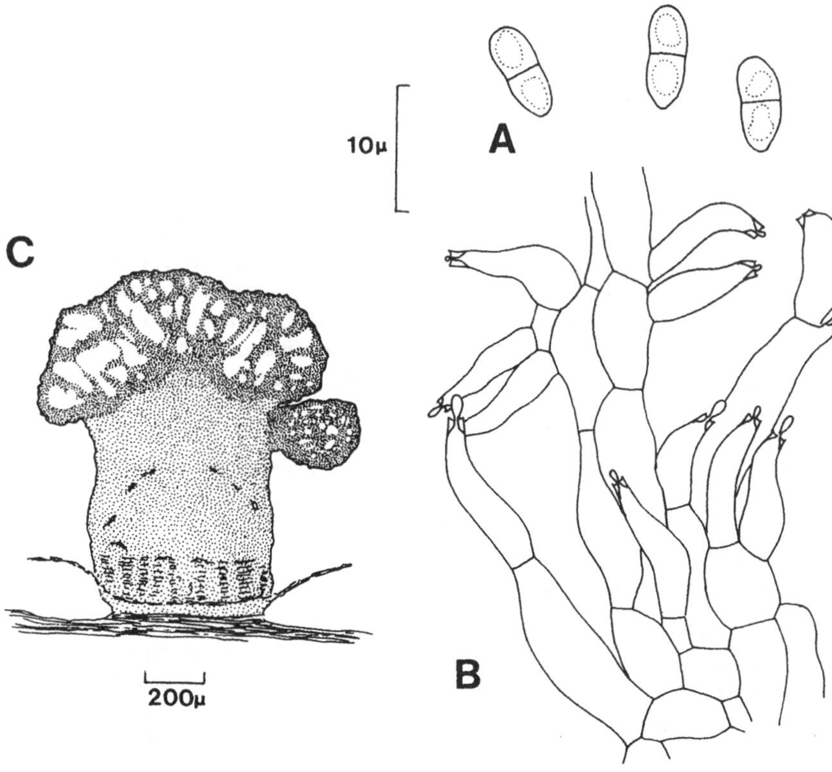

Fig. 361. *Fuckelia ribesia*, A, conidia; B, conidiophores, C, vertical section of a closed conidioma.

93. CORNICULARIELLA Karst., *Hedwigia* **23**: 57 (1884).
Cornularia Sacc., *Syll. fung.* **3**: 598 (1884).
Chondropodium Höhn., *Sber. Akad. Wiss. Wien* **125**: 45 (1916).
Sp. typ.: *C. abietis* Karst.

Mycelium immersed, subperidermal, branched, septate, hyaline to pale brown. *Conidiomata* eustromatic, base immersed, subperidermal, separate, single or in groups of 2-4, conical, the upper part erumpent, unilocular; walls thick, composed of a thin outer layer of dark brown but thin-walled textura angularis and inner tissue of hyaline to pale brown textura oblita. *Ostiole* single, circular. *Conidiophores* hyaline, repeatedly branched irregularly, septate, smooth, with fertile and infertile branches, formed from the locular walls and the channel towards the ostiole. *Conidiogenous cells* enteroblastic, phialidic, determinate, integrated, subulate, hyaline, smooth, channel wide, collarette minute. *Conidia* hyaline, medianly 1 euseptate, smooth, thin-walled, eguttulate, straight or markedly curved and falcate, tapered to apex and base.

Di Cosmo (1978) accepted seven species in *Cornicolariella* and provided a key.

C. SPINA (Berk. & Rav.) di Cosmo, *Can. J. Bot.* **56**: 1676 (1978). (Fig. 362)
Chondropodium spinum (Berk. & Rav.) Höhn., *ibid.*: 46 (1916).
Sphaeronaema spinum Berk. & Rav., *Grevillea* **2**: 177 (1874).
Cornularia spina Sacc. & Syd., *Syll. fung.* **16**: 982 (1902).
Sphaeronaema fraxini Pk, *Ann. Rept. N.Y. St. Mus.* **29**: 71 (1878).
Sphaerographium fraxini (Pk) Sacc., *Syll. fung.* **3**: 598 (1883).

Teleomorphosis: *Durandiella fraxini* (Schw. ex Fr.) Seaver, *N. Am. Cup Fungi, Inoperculates*: 367 (1951).

Conidiomata up to 2 mm high × 650 µ wide, tapered to 150 µ wide at the apex. Conidiogenous cells 12-26 × 2-4 µ wide at the base. Conidia 1 septate (3 septate in Groves (1954)), 45-75 × 3.5-4 µ.

On dead branches of *Fraxinus americana*, *F.* sp., N.Y., U.S.A., Ont., Que., Canada (IMI 26895, 29765, 29695, 29696, 32276, 25981, 25980, 29697).

94. GELATINOSPORIUM Pk, *25th Rep. St. Mus. N.Y.* 84 (1871).
Gelatinosporis Clements & Shear nom. nov., *Genera of Fungi*: 367 (1931).
Original spp.: *G. abietinum* Pk, *G. betulinum* Pk
Lectotype sp.: *G. betulinum* Pk, fide Höhnel (1916g).

Mycelium immersed, branched, septate, pale brown to hyaline. *Conidiomata* eustromatic, separate or gregarious, black, pulvinate, erumpent, unilocular or multilocular and convoluted with a distinct upper wall, thick-walled; wall of pale brown textura oblita, more hyaline towards conidiogenous region. *Ostiole* absent, dehiscence by breakdown and splitting of upper wall. *Conidiophores* branched and septate at the base and above, hyaline, filiform, formed from the inner cells of the locular wall. *Conidiogenous cells* enteroblastic, phialidic, determinate, discrete or integrated, apertures apical, collarette, channel and periclinal thickening minute. *Conidia* hyaline, 1 septate, smooth, thin-walled, curved or sigmoid, tapered at the apex, eguttulate.

Di Cosmo (1978) maintained *Gelatinosporium* distinct from *Foveostroma* mainly on the basis of differences in stromatal structure.

G. BETULINUM Pk, *ibid.*: 84 (1871). (Fig. 363)
Gelatinosporis betulina (Pk) Clements & Shear, *Genera of Fungi*: 367 (1931).

Conidiomata pulvinate, comparatively simple, locule convoluted, up to 1500 µ high × 800 µ diam. Conidiophores 14-32 × 2-3 µ. Conidia 30-40 × 2-3 µ, falcate or irregularly sigmoid, medianly 1 septate, eguttulate.

This species was selected as lectotype of *Gelatinosporium* Pk by Höhnel (1916g) and its generic synonymy with *Micropera* briefly discussed by Sutton (1977b). *G. betulinum* is distinct from *M. betulina* which was described with smaller conidia (18-20 × 3 µ), and from *G. fulvum* Pk (anamorphosis of *Dermea molliuscula* (Schw.) Cash) which Groves (1946) described with conidia measuring 50-75 × 2.5-3.5 µ.

On branches of *Betula lutea*, N.Y., U.S.A. (IMI 197179 holotype ex NY).

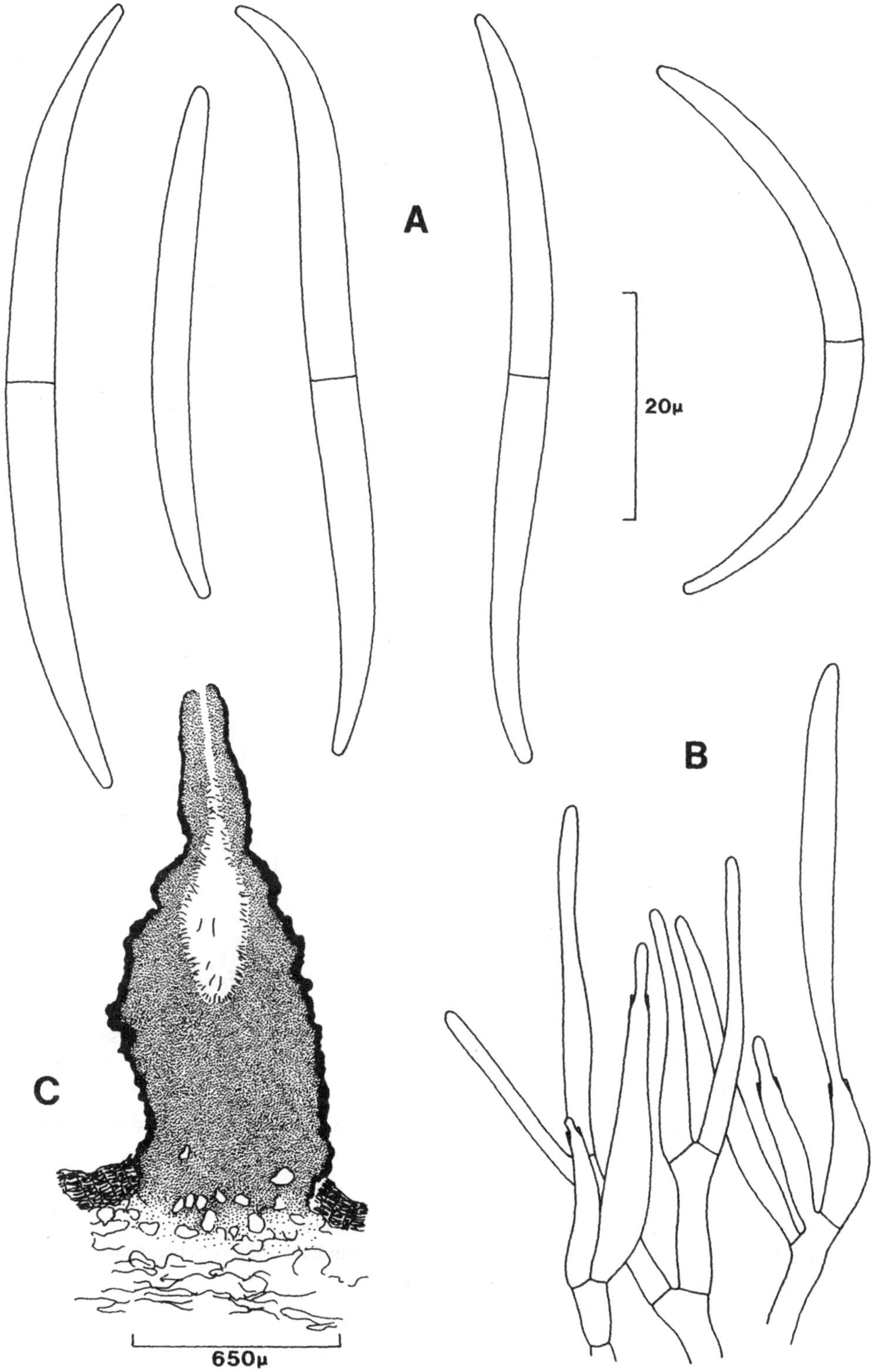

Fig. 362. *Cornicullariella spina*, A, conidia; B, conidiophores and developing conidia; C, vertical section of a conidioma.

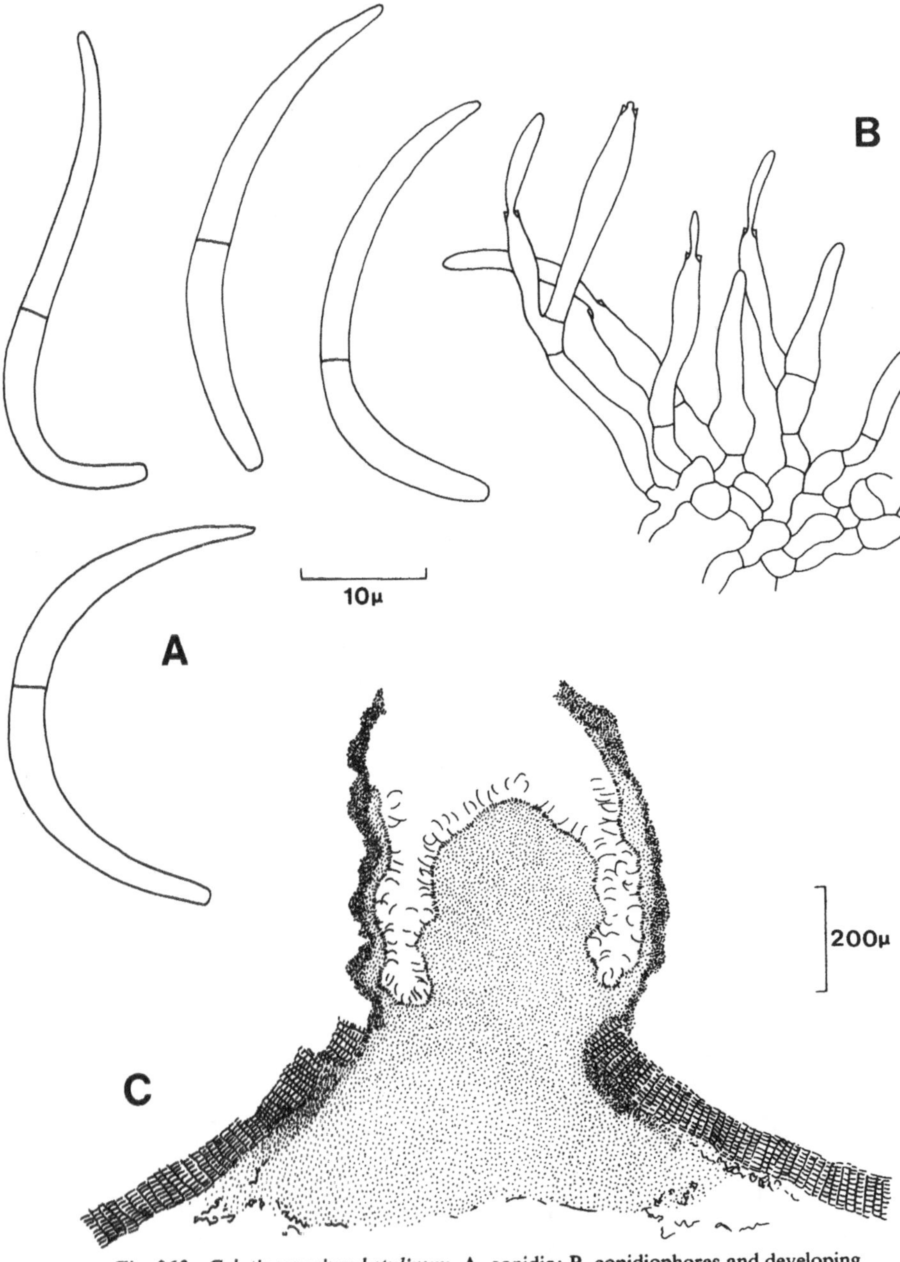

Fig. 363. *Gelatinosporium betulinum*, A, conidia; B, conidiophores and developing conidia; C, vertical section of a conidioma.

95. **FOVEOSTROMA** di Cosmo nom. nov., *Can. J. Bot.* **56**: 1682 (1978).
Micropera Lév., *Ann. sci. nat.*, ser. 3, **5**: 283 (1846), non Lindley, 1832 (Orchidaceae).
? *Micula* Duby, *Hedwigia* **1**: 8 (1858).
Type sp.: *F. drupacearum* (Lév.) di Cosmo (syn. *Micropera drupacearum* Lév.).

Mycelium immersed, branched, septate, hyaline to pale brown. *Conidiomata* eustromatic, separate, occasionally confluent, pulvinate, often furfuraceous or pruinose, finally opening widely, light brown, unilocular, multilocular or convoluted, locules situated in the upper part of the conidioma; tissue at the base of pale brown textura intricata, becoming more textura angularis in the upper region, darker and more thick-walled at the periphery. *Ostiole* absent, dehiscence by irregular breakdown of the overlying wall. *Conidiophores* branched and septate at the base and above, hyaline, filiform, formed from the inner cells of the locular wall. *Conidiogenous cells* enteroblastic, phialidic, determinate, discrete or more often integrated, apertures apical on comparatively long branches of the conidiophore, hyaline, smooth, collarette, channel and periclinal thickening minute. *Conidia* hyaline, 1 septate, smooth, thin-walled, curved or sigmoid, tapered towards the apex and the base, guttulate or eguttulate.

Di Cosmo (1978) has revised *Corniculariella* Karst., and in doing so proposed replacing *Micropera* with *Foveostroma* nom. nov. He specifically rejected *Gelatinosporium* as a synonym of *Micropera* which had previously been suggested by Groves (1954) and Sutton (1977a). There is also an earlier name in contention, that being *Micula* Duby. This was tentatively placed as a synonym of *Micropera* by Sutton (1977a). Until further work is done on establishing the generic limits and generic synonymy, the name *Foveostroma* is taken up, but, if subsequently it is proved that either *Micula* or *Gelatinosporium* are the same as *Foveostroma* then obviously the latter name will have to be superceded by the oldest available name.

So far about 40 taxa have been assigned to *Micropera*. Although correlations with *Dermea* have been frequently reported (Groves, 1968b; 1937, 1946; Funk, 1976) the genus remains largely unrevised apart from the description of six species by Petch (1943) and the treatment of coniferous species by Funk (1976). Whereas the conidiomata are extremely variable, ranging from simple acervuli to multilocular stromata, and cannot be usefully employed to distinguish species, it seems that conidia of the different species are highly distinctive (Funk, 1976).

F. DRUPACEARUM (Lév.) di Cosmo, *Can. J. Bot.* **56**: 1682 (1978). (Fig. 364)
Micropera drupacearum Lév., *Ann. Sci. nat.*, ser. 3, **5**: 283 (1846).
Micropera cerasi Sacc., *Mycologiae Venetae*: 160 (1873).
Micropera cerasi Bonord., *Abh. Nat. Ges. Hall.* **8**: 133 (1864).
Micropera roseola Lév., *Ann. sci. nat.*, ser. 3, **5**: 283 (1846).

Teleomorphosis: *Dermea cerasi* (Pers. ex Fr.) Fr., *Syst. Orb. Veg.* 115 (1825).

Conidiomata irregular, pruinose, yellowish brown, .5-2.0 mm diam. × .2-1.0 mm high. Conidiophores up to 25 µ long × 1.5-2.5 µ. Conidia falcate or sigmoid, fusiform, medianly 1 septate, often guttulate, 43-65 × 3 µ.

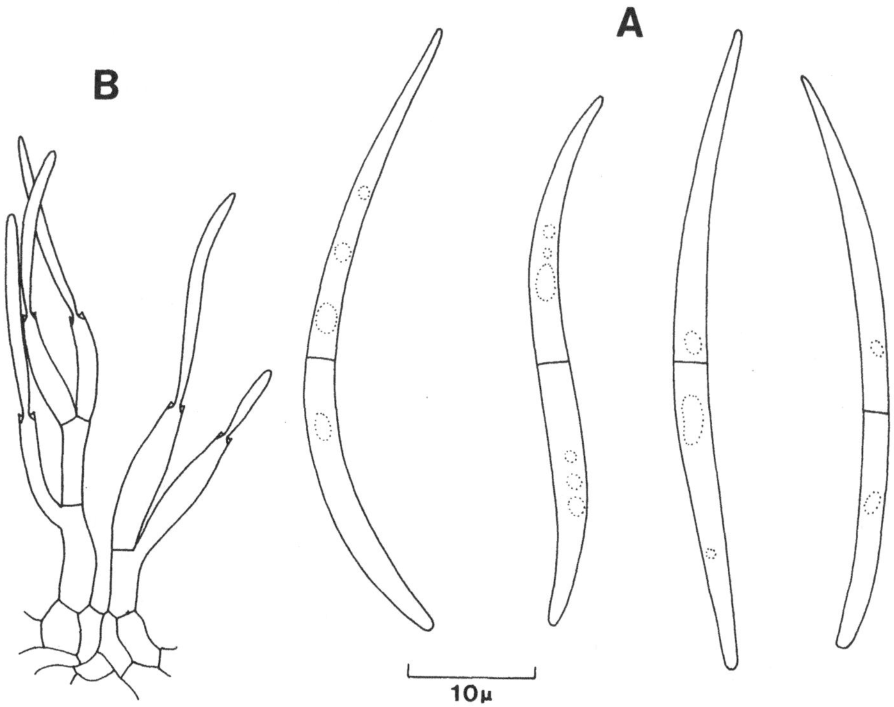

Fig. 364. *Foveostroma drupacearum*, A, conidia; B, conidiophores and developing conidia.

On twigs of *Prunus emarginata, P. avium, P. cerasus, P. serotina, P. pennsylvanica, P.* sp., U.K., Ont., B.C., N.S., Canada, Wash., U.S.A., Germany, France, Austria (IMI 29271, 22376, 57006, 11496, 12552, 13032, 30441, 22378, 22377, 26763, 26762, 73262, 51849, 25979, 76273, 29687, 29686, 29688, 361).

96. CYLINDROGLOEUM Petrak, *Annls mycol.* **39**: 276 (1941).
Sp. typ.: *C. arcticum* Petrak

Mycelium immersed, branched, septate, hyaline. *Conidiomata* acervular, subcuticular to epidermal, separate or confluent, formed of pale brown to hyaline, thin-walled textura angularis; cuticle not noticeably discoloured. Dehiscence irregular. *Conidiophores* 1-2 septate, branched mainly at the base, hyaline, rather irregular, formed from the upper cells of the pseudoparenchyma. *Conidiogenous cells* enteroblastic, phialidic, discrete or integrated, determinate, lageniform to cylindrical, hyaline, smooth, collarette minute, channel wide. *Conidia* hyaline, thin-walled, eguttulate, smooth, straight or slightly curved, cylindrical or fusiform, 0-1 septate.

This description is based on *C. trillii* (Ell. & Ev.) Arx which is not the type species. Morgan-Jones (1971b) described conidiogenesis in *C. trillii* as phialidic. Until *C. arcticum* is examined the genus cannot be assigned confidently to synonymy with *Cylindrosporella* Höhn., from which *C. trillii* certainly appears to be no different, apart from the occasional development of a single median septum in the conidia.

C. TRILLII (Ell. & Ev.) Arx, *Verh. K. Ned. Akad. Wet. Afd. Natuurk.* **51**: 143 (1957). (Fig. 365).
Gloeosporium trillii Ell. & Ev., *Proc. Acad. nat. Sci. Phil.* 371 (1894).

Conidiomata 80-200 µ diam. Conidiogenous cells 5-8 × 2-3 µ. Conidia 11-20 × 2-3 µ.

Causing leaf lesions on *Trillium sessile*, U.S.A. (IMI 107245, holotype).

97. DIPLOZYTHIELLA Died., *Annls mycol.* **14**: 215 (1916).
Sp. typ.: *D. bambusina* Died.

Mycelium immersed, branched, septate, hyaline. *Conidiomata* eustromatic, immersed, separate or aggregated, hypodermal, ± globose to irregular, very pale brown, unilocular; wall composed of very pale brown to hyaline textura angularis, with tissue thicker in the ostiolar region. *Ostiole* very indistinct, probably no more than apical rupturing of upper stromatal cells. *Conidiophores* absent. *Conidiogenous cells* completely lining the locular wall, enteroblastic, phialidic, determinate, discrete, cylindrical to lageniform, hyaline, smooth, collarette minute but channel prominent and periclinal wall thickened. *Conidia* hyaline, thin-walled, medianly 1 euseptate, ± constricted, apex and base obtuse, fusiform, straight, often guttulate.

The genus is monotypic.

D. BAMBUSINA Died., *ibid.*: 215 (1916). (Fig. 366)

Conidiomata up to 165 µ diam. Conidiogenous cells 5-11.5 × 3-4 µ. Conidia 14.5-21 × 3 µ.

Causing lesions on leaves of *Bambusa*, India (IMI 95771).

98. CEUTHODIPLOSPORA Died., *Annls mycol.* **10**: 149 (1912).
Sp. typ.: *C. robiniae* (Bub.) Died. (syn. *Cytodiplospora robiniae* Bub.).

Mycelium immersed, branched, septate, dark brown. *Conidiomata* stromatic, separate, globose, dark brown to black, immersed, unilocular or occasionally with 2-3 locules, thick-walled; walls of dark brown, thick-walled textura angularis at the periphery, becoming smaller-celled, thin-walled and paler towards the hyaline conidiogenous region; tissue darker near the ostiole and paler at the base. *Ostiole* single, circular, central, papillate. *Conidiophores* absent. *Conidiogenous cells* enteroblastic, phialidic, hyaline, doliiform, smooth, channel and collarette minute, periclinal wall occasionally thickened, formed from the inner cells of the stromatic wall. *Conidia* hyaline, smooth, thin-walled, eguttulate, aseptate with 1 median euseptum, then slightly constricted, cylindrical to ellipsoid.

Clements & Shear (1931) placed *Ceuthodiplospora* in synonymy with *Cytodiplospora* Oud. Arx (1963) regarded the latter generic name as a synonym of *Discella* Berk. & Br. but *Diplodina* Westd. appears to be the earliest available name

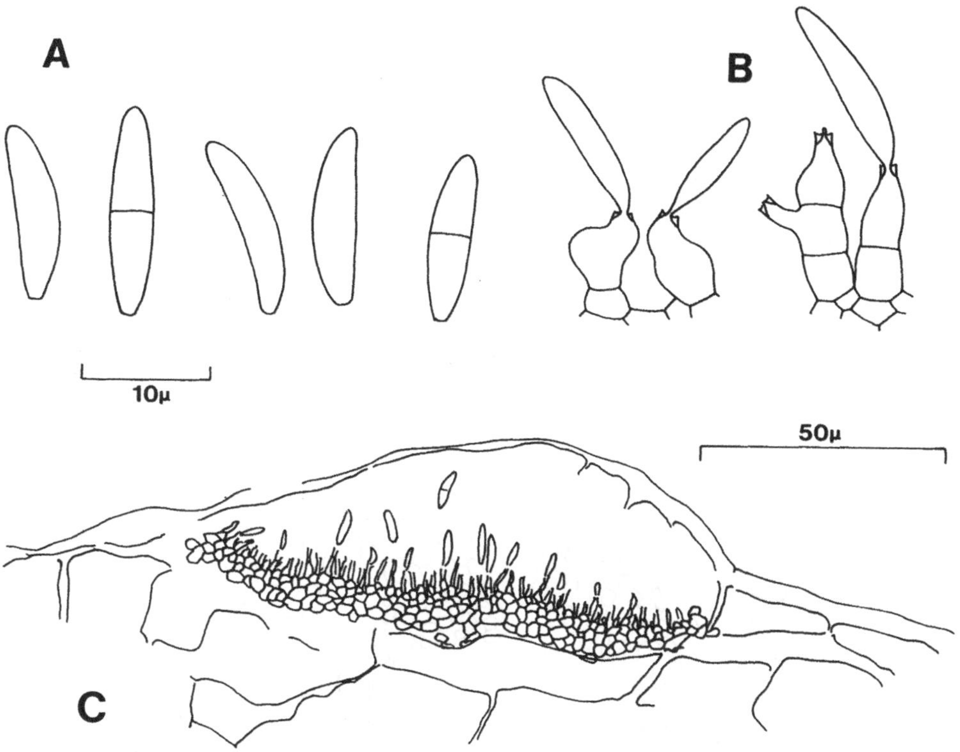

Fig. 365. *Cylindrogloeum trillii*, A, conidia; B, conidiophores and developing conidia; C, vertical section of a conidioma.

for these fungi. *C. robiniae* however differs in being stromatic with relatively small doliiform phialides and as such is maintained as distinct. The genus is monotypic.

C. ROBINIAE (Bub.) Died., *ibid.*: 149 (1912). (Fig. 367)
Cytodiplospora robiniae Bub., *Annls mycol.* **4**: 116 (1906).

Teleomorphosis: *Pleomassaria robiniae* Bub., *Annls mycol.* **4**: 111 (1906).

Conidiomata up to 400 μ diam. Conidiogenous cells 5-7.5 × 4.5-6 μ. Conidia 5.5-9.5 × 3-3.5 μ.

On twigs of *Robinia pseudacacia*, Czechoslovakia (IMI 201347, holotype ex BPI).

99. PLACONEMINA Petrak, *Annls mycol.* **19**: 197 (1921).
Sp. typ.: *P. dothideoides* (Mont.) Petrak (syn. *Ascospora dothideoides* Mont.).

Mycelium immersed, branched, septate, pale to dark brown. *Conidiomata* pseudostromatic, immersed, separate or aggregated, hypodermal, ± globose to irregular or linear, dark brown, unilocular but convoluted; wall composed of pale brown thin-walled textura angularis at the base, becoming dark brown and thick-walled in the upper region to form an epidermal and subepidermal clypeus. *Ostiole* single, circular, central, ± papillate. *Conidiophores* absent. *Conidiogenous cells* completely lining the locular wall, enteroblastic, phialidic, determinate, discrete, ampulliform, hyaline, smooth, collarette and channel minute, periclinal wall thickened. *Conidia* hyaline, thin-walled, medianly 1 euseptate, apex obtuse, base flattened, ± straight, cylindrical to fusiform.

The genus is monotypic. It only appears to differ from *Diplozythiella* in the thick-walled, dark brown nature of the conidiomata.

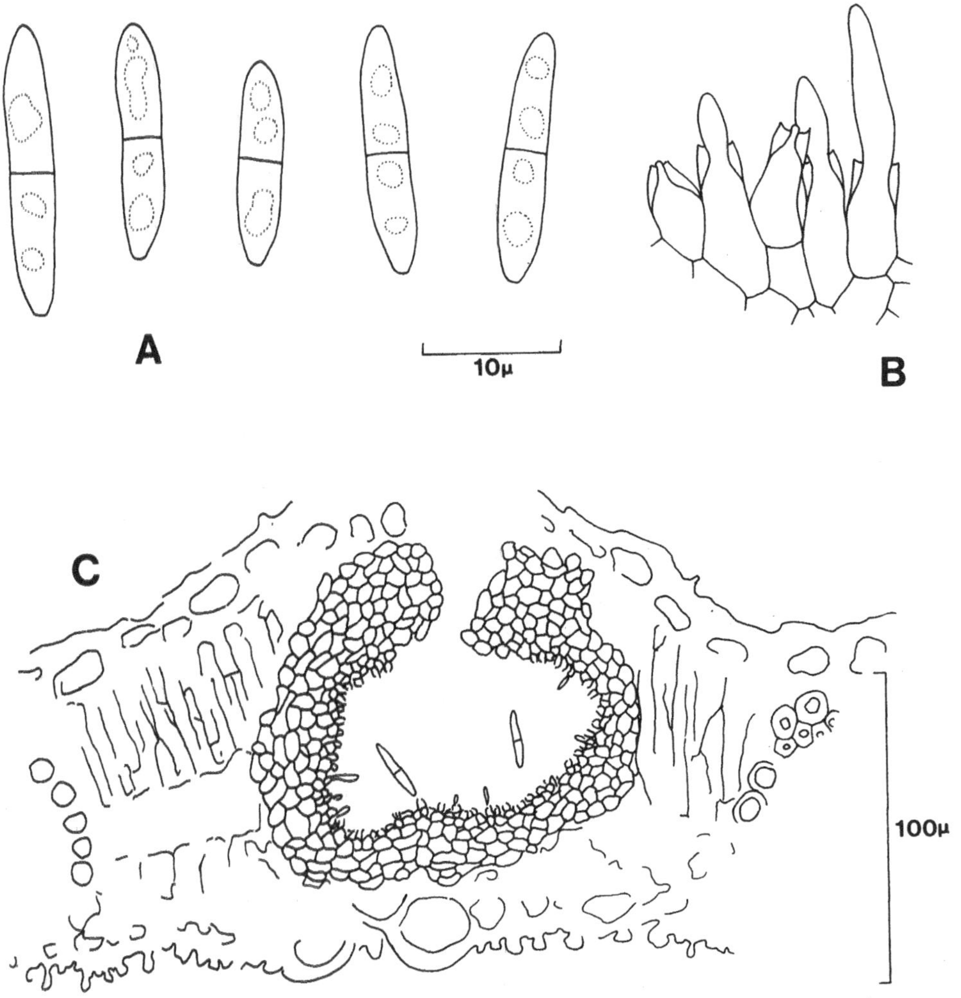

Fig. 366. *Diplozythiella bambusina*, A, conidia; B, conidiogenous cells and developing conidia; C, vertical section of a conidioma.

P. DOTHIDEOIDES (Mont.) Petrak, *ibid.*: 197 (1921). (Fig. 368)
Ascospora dothideoides Mont., *Sylloge gen. spec. Crypt.* 275 (1856).
Placosphaeria dothideoides (Mont.) Sacc., *Syll. fung.* **3**: 246 (1884).

Conidiomata 1250 µ long × 250 µ wide. Conidiogenous cells 5.5-7 × 4.5-6 µ. Conidia 7.5-13.5 × 2.5-3 µ.

On culms of *Phragmites communis*, Czechoslovakia (IMI 24225).

100. DAVISIELLA Petrak, *Annls mycol.* **22**: 134 (1924).
Sp. typ.: *D. elymina* (Davis) Petrak (syn. *Cytodiplospora elymina* Davis).

Mycelium immersed, branched, septate, hyaline. *Conidiomata* either enclosed by perithecia or immersed in clypear tissue of *Phyllachora*, globose to pyriform, hyaline, separate, rarely aggregated, thin-walled, composed of hyaline, small-celled textura angularis. *Ostiole* single, circular, central. *Conidiophores* hyaline, septate, branched irregularly either as vertically or horizontally arranged hyphae, formed from the inner cells of the conidiomatal wall. *Conidiogenous cells* enteroblastic, phialidic, discrete or more frequently integrated, cylindrical to doliiform, determinate, apertures apical, or lateral and then often immediately below septa, channel comparatively wide, collarette minute. *Conidia* hyaline, medianly 1 euseptate or aseptate, thin-walled, smooth, minutely guttulate, irregular in shape, straight or curved, apex obtuse, base truncate, fusiform.

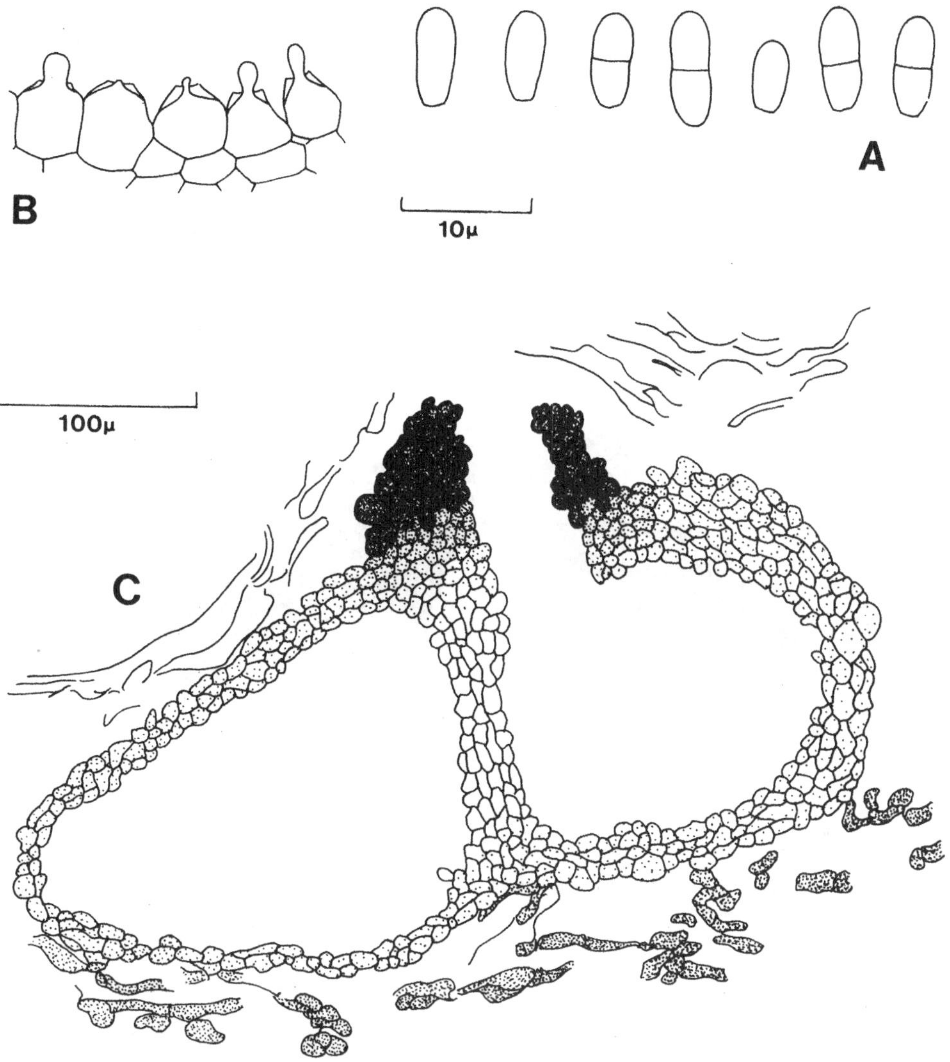

Fig. 367. *Ceuthodiplospora robiniae*, A, conidia; B, conidiogenous cells; C, vertical section of a conidioma.

The conidiomatal wall is simple when the fungus is immersed in clypeus tissue, but when found in perithecia there is a layer of very thin-walled, large-celled, hyaline tissue between the perithecial wall and the hyperparasite. Whether this belongs to the host or the hyperparasite is difficult to establish.

Apart from the type, two other species have been described in the genus. *D. domingensis* was recorded from several *Phyllachora* species on different grasses in Dominica by Petrak & Ciferri (1932). Although the hyperparasitic habit is correct and the conidial size of 13-20 × 2-2.5 µ seems reasonable for the genus, the conidiophores (8-12 × 2-2.5 µ) are somewhat longer than in *D. elymina*. This is the only reservation I have at including the species in *Davisiella*. The other species, *D. botryodiplodiae* was described from conidiomata of *Botryodiplodia theobromae* Pat. by Ahmad (1961), but from the brief description and in the absence of the type, its relationships cannot be determined.

D. ELYMINA (Davis) Petrak, *ibid.*: 134 (1924). (Fig. 369)

Cytodiplospora elymina Davis, *Trans. Wisc. Acad. Sci. Arts Lett.* **19**: 701 (1919).

Conidiomata ⌓ 100 µ diam. Conidiophores up to 30 µ long. Conidiogenous cells 5-7.5 × 2.5-4 µ. Conidia 12-15.5 × 2.5-3 µ.

In *Phyllachora* on *Muhlenbergia racemosa*, Wisc., U.S.A. (IMI 202984, ex WIS).

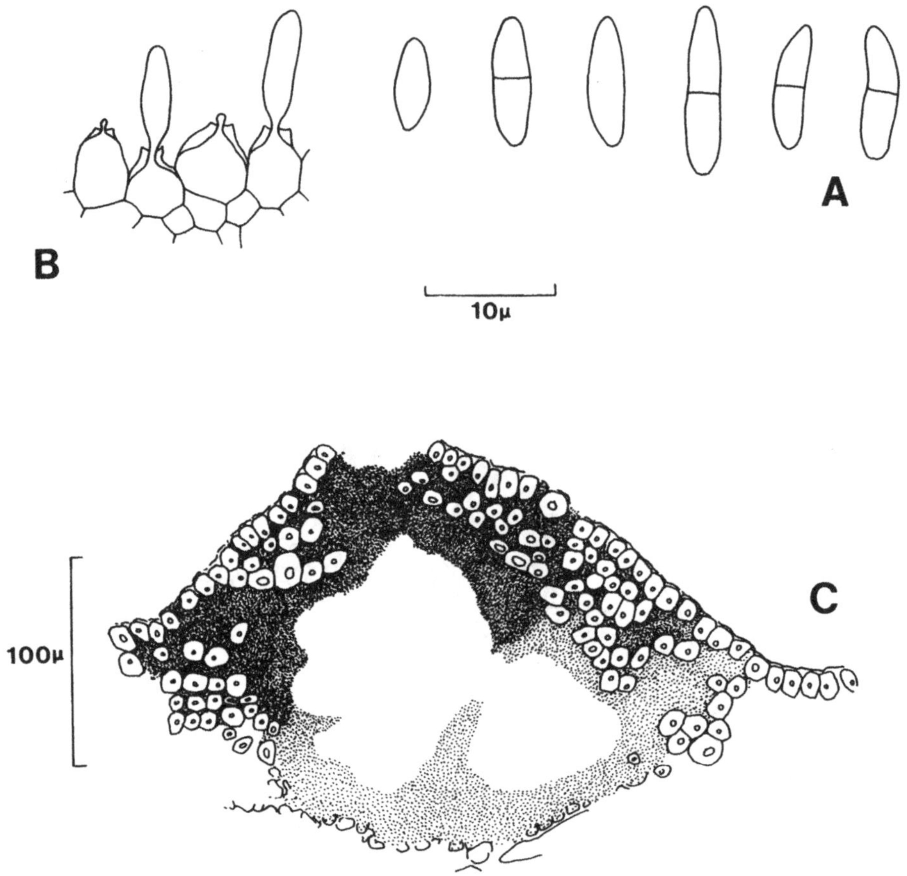

Fig. 368. *Placonemina dothideoides*, A, conidia; B, conidiogenous cells; C, vertical section of a closed conidioma.

101. **HELHONIA** Sutton, nom. nov.
(etym. anagr. Höhnel)
Ceuthosporella Höhn., *Mitt. bot. Inst. techn. Hochsch. Wien* **6**: 117 (1929), non *Ceuthosporella* Petrak & H. Sydow (1923).
Sp. typ.: *H. rhamnigena* (Fautrey) Sutton (syn. *Phoma rhamnigena* Fautrey).

Mycelium immersed, branched, septate, pale brown to hyaline. *Conidiomata* eustromatic, separate or aggregated, subglobose, pale brown, subepidermal to epidermal, unilocular, thick-walled above, of pale brown, thin-walled textura angularis, thin-walled below, of hyaline, small-celled textura angularis. Dehiscence irregular, ostiole absent. *Conidiophores* hyaline, branched only at the base, septate, smooth, with acrogenous conidia, ± cylindrical, formed from the inner cells of the stroma wall. *Conidiogenous cells* enteroblastic, phialidic, determinate, integrated, less often discrete, hyaline, smooth, collarette and channel minute. *Conidia* hyaline, 1 euseptate, thin-walled, guttulate, smooth, cylindrical, straight, tapered at one or both ends.

The generic name *Ceuthosporella* was first published by Höhnel (1916a), but as a nomen nudum. The formal description appeared much later (Höhnel, 1929). However, between these two dates, Petrak & H. Sydow (1923) also introduced a *Ceuthosporella*, but based on quite a different type species. Thinking that their name was a later homonym of Höhnel's nomen nudum, they proposed the nomen novum, *Microcyta* Petrak & H. Sydow (1927). This is superfluous because the first valid publication of *Ceuthosporella* is Petrak & H. Sydow (1927) not Höhnel (1929).

Höhnel (1929) studied an isotype of *Phoma rhamnigena* Fautrey and found the substrate to be *Sambucus nigra* not *Staphylea pinnata*. This is insufficient to justify changing the epithet from '*rhamnigena*' to '*sambuci*'. The genus is monotypic.

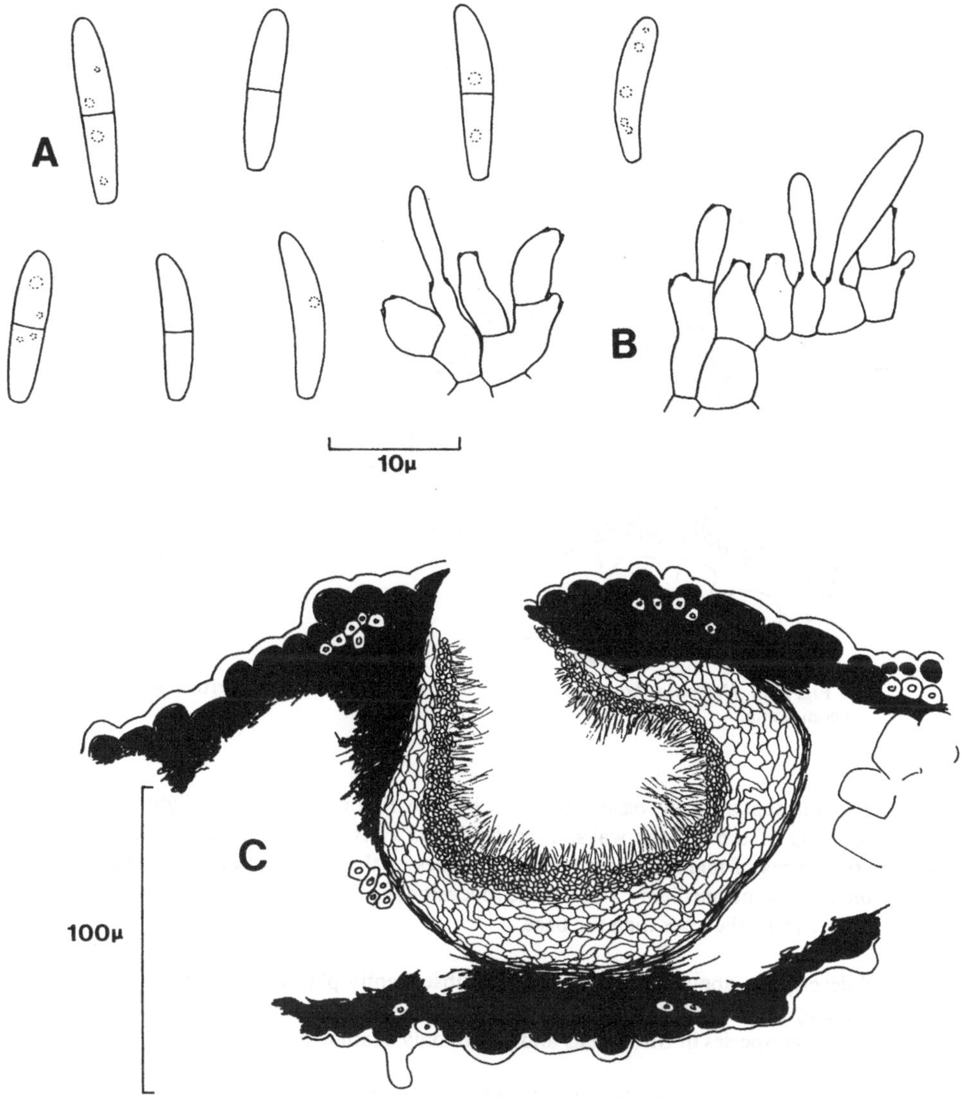

Fig. 369. *Daviesiella elymina*, A, conidia; B, conidiophores, C, vertical section of a conidioma.

H. rhamnigena (Fautrey) comb. nov. (Fig. 370)
Phoma rhamnigena Fautrey, *Rev. mycol.* **12**: 166 (1890).
Ceuthosporella sambuci Höhn., *Mitt. bot. Inst. tech. Hochsch. Wien* **6**: 118 (1929).

Conidiomata up to 200 µ diam. Conidiophores 5-10 × 2.5 µ. Conidia 12.5-16.5 × 1.5 µ.

On stems of *Sambucus nigra*, France (isotype, FH).

102. SIROCOCCUS Preuss, *Linnaea* **26**: 716 (1855).
Sp. typ.: *S. strobilinus* Preuss

Mycelium immersed, branched, septate, hyaline to pale brown. *Conidiomata* eustromatic, initially immersed then strongly erumpent, separate, occasionally confluent, dark brown, globose, flattened globose or pulvinate, unilocular, multilocular or convoluted, thick-walled; wall composed of pale brown, thin-walled textura angularis except towards the periphery where cells are darker brown and thicker-walled. *Ostiole* absent, dehiscence by breakdown of the upper wall.

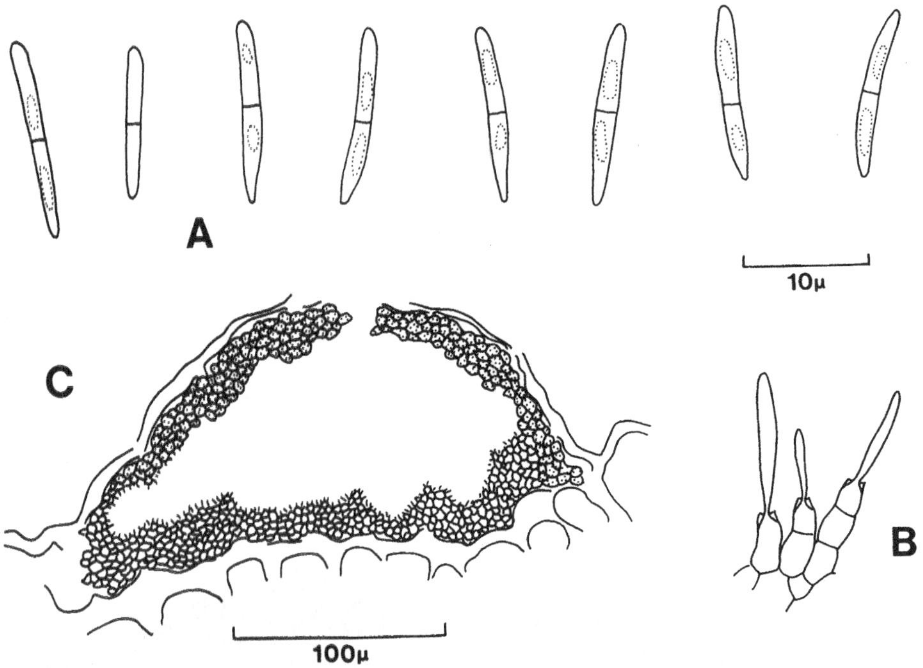

Fig. 370. *Helhonia rhamnigena*, A, conidia; B, conidiophores; C, vertical section of a conidioma.

Conidiophores branched and septate at the base and above, hyaline, filiform, straight or slightly bent, tapered to the apices, formed from the inner cells of the locular walls. *Conidiogenous cells* enteroblastic, phialidic, determinate, integrated, hyaline, apertures at the apices of main or short lateral branches of the conidiophores, collarette, channel and periclinal thickening minute. *Conidia* hyaline, medianly 1-euseptate (occasionally 2-3 septate in culture), fusiform, straight or curved, eguttulate, tapered more towards the apex than the base.

Since its original description about 30 taxa have been subsequently placed in *Sirococcus*. Several of these have been removed to other genera by Höhnel (1914a) although the true relationships of the majority still need to be worked out. Petrak (1947c) accepted three species in *Sirococcus* and I have only examined two of them.

S. SPIRAEAE (Lebedeva) Petrak, *Sydowia* **1**: 155 (1947). (Fig. 371C)
Sirodiplospora spiraeae Lebedeva, *Not. Sist. Inst. Crypt. Hort. Bot. Petrop.* **1**: 62 (1922).
Sirexcipulina moravica Petrak, *Annls mycol.* **21**: 278 (1923).

Conidiomata up to 350 μ diam. Conidiophores 30-40 μ long × 2-2.5 μ wide, tapering to 1.5 μ. Conidia 11.5-13.5 × 2.5-3 μ.

On twigs of *Spiraea sorbifolia*, *S.* sp., Latvia, Leningrad, U.S.S.R. (IMI 94182, isotype, 31124, 33641).

S. STROBILINUS Preuss, *Linnaea* **26**: 716 (1853). (Fig. 371A, B, D)
Sporonema strobilinum Desm., *Ann. Sci. nat.*, ser. 3, **18**: 368 (1825).
Sirococcus strobilinus (Desm.) Petrak, *Sydowia* **1**: 155 (1947).
Plenodomus strobilinus (Desm.) Höhn., *Sber. Akad. Wiss. Wien* **119**: 647 (1910).
Discella strobilina (Desm.) Died., *Krypt. Fl. Mk Brandenb.* **9**: 752 (1915).
Ascochyta strobilina (Cda) Wollenw., *Annls mycol.* **15**: 31 (1917).
Fusarium strobilinum Cda, *Icon. fung.* **1**: 4 (1837).
Phoma conigena Karst., *Medd. Soc. Fauna Fl. fenn.* **16**: 29 (1888).
Phoma conophila Sacc., *Syll. fung.* **10**: 163 (1892).
Ascochyta conorum P. Henn., *Hedwigia* **43**: 73 (1904).
Ascochyta piniperda Lindau in Engler & Prantl, *Nat. Pl.* **1** (1): 368 (1900).

Conidiomata up to 600 μ diam. × 250-300 μ deep. Conidiophores up to 50 μ long × 2.5 μ wide, tapering to 1.5 μ. Conidia 12-16 × 3 μ.

The distinctions between this species and *S. spiraeae* are relatively minor (slightly larger conidia, smaller conidiophores).

On cones and needles of *Picea excelsa, P. sitchensis, Pinus contorta, Tsuga heterophylla, Pseudotsuga douglasii*, U.K., France, Switzerland, B.C., Canada, Latvia, U.S.S.R. (IMI 85097, isotype ex PC, 12971, 210867, 184528, 211250, 163501, 214839, 211525, 211238, 159497).

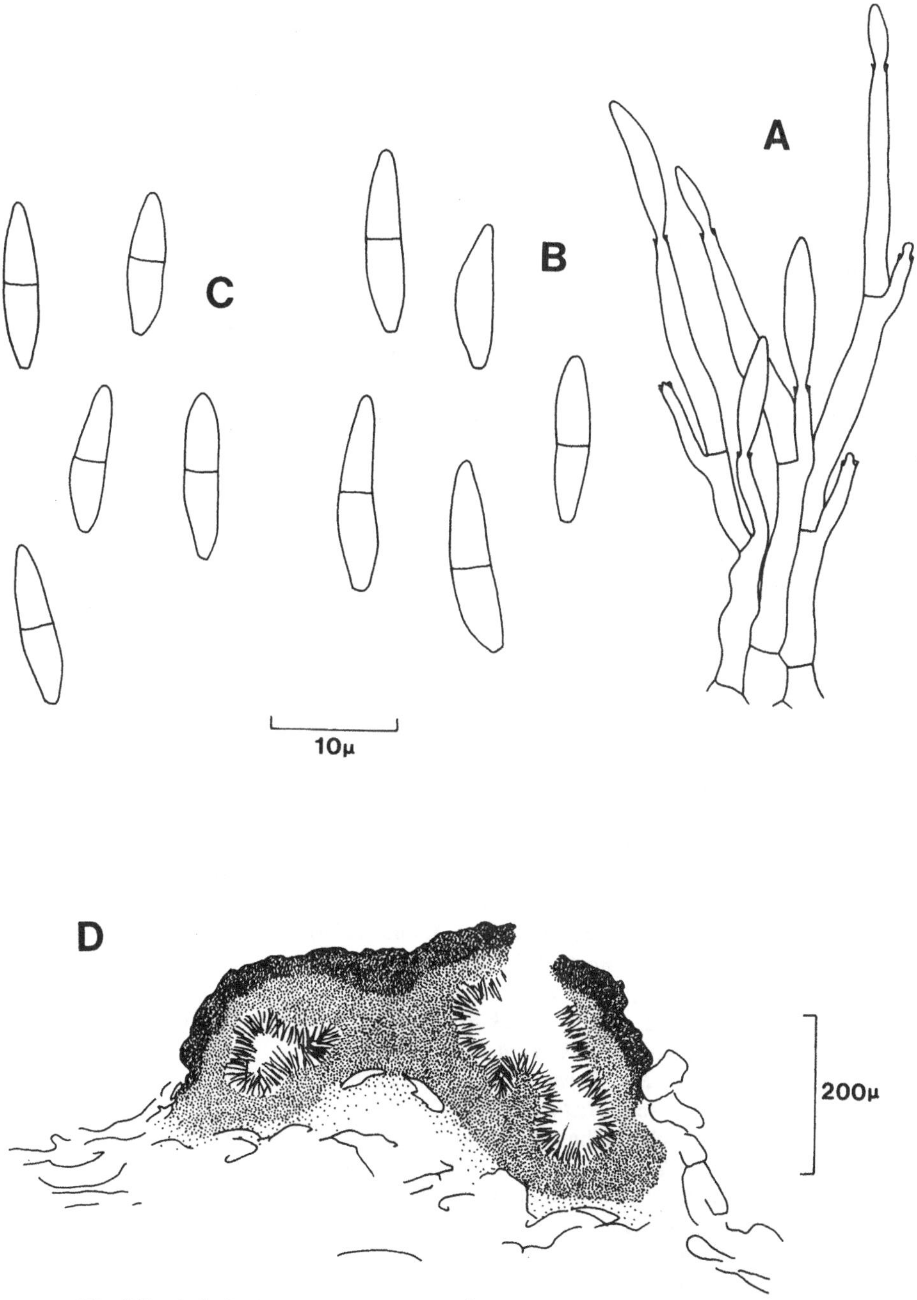

Fig. 371. A, B, D, *Sirococcus strobilinus*, conidiophores, conidia, vertical section of a conidioma; C, *S. spiraeae*, conidia.

103. DIPLODINA Westd., *Bull. Acad. r. Belg. Cl. Sci. nat.*, II, **2**: 562 (1857).
Septomyxa Sacc., *Syll. fung.* **3**: 766 (1884).
Cytodiplospora Oud., *Ned. Kruidk. Arch.*, ser. 2, **6**: 292 (1894).
? *Diplodinula* Tassi, *Bull. Lab. Ort. Bot. Siena* **5**: 41 (1902).
? *Macrodendrophoma* T. Johnston, *Sci. Proc. R. Soc. Dublin* **10**: 158 (1904).
Fioriella Sacc. & D. Sacc., *Syll. fung.* **18**: 432 (1906).
? *Diplosclerophoma* Petrak, *Annls mycol.* **21**: 293 (1923).
? *Diploplenodomopsis* Petrak, *Annls mycol.* **21**: 208 (1923).
Sp. typ.: *D. microsperma* (Johnston) Sutton (syn. *Stilbospora microsperma* Johnston, *D. salicis* Westd.).

Mycelium immersed, branched, septate, hyaline to pale brown. *Conidiomata* eustromatic, immersed, subperidermal, separate, occasionally confluent, applanate, dark brown, multilocular or convoluted; wall composed of an outer layer of hyaline, thin-walled textura angularis and an inner layer of pale brown slightly thicker walled textura angularis which gives rise to a layer of hyaline textura intricata which bears the conidiophores. Dehiscence irregular, by rupture of the overlying bark. *Conidiophores* hyaline, smooth, variably septate and branched, either doliiform or cylindrical and tapered, formed from the tissue lining the locular walls. *Conidiogenous cells* enteroblastic, phialidic, hyaline, smooth, integrated, determinate, with a distinct apical aperture on main or side branches of the conidiophore, collarette and channel minute, periclinal wall thickened. *Conidia* hyaline, smooth, thin-walled, fusiform, 0-2 septate, usually medianly 1 septate, eguttulate, straight or sometimes slightly curved.

The generic synonymy has been discussed by Sutton (1977a) under the individual names. The resulting interpretation of *Diplodina* differs from that previously accepted, and now encompasses common species which used to be known in *Discella* (*D. carbonacea*) and *Septomyxa* (*S. tulasnei*). Apart from a fundamental similarity in conidiomata, conidiogenesis and conidial morphology, such species have similar teleomorphoses in *Cryptodiaporthe* Petrak.

KEY

1. Conidiophores often long, filiform, conidia 17-19 × 3-3.5 μ; on *Aesculus* .... *D. aesculi*
1. Conidiophores very irregularly branched 2
2. Conidia 13.5-15.5 × 3-3.5 μ; on *Acer* .... *D. acerina*
2. Conidia 16-20.5 × 4-4.5 μ, often 0-2 septate; on *Salix* .... *D. microsperma*

D. acerina (Pass.) comb. nov. (Fig. 372C)
Gloeosporium acerinum Pass. in Thüm., *Mycotheca Univ.* exs. 93 (1875).
Septogloeum acerinum (Pass.) Sacc., *Michelia* **2**: 541 (1882).
Discella acerina (Pass.) von Arx, *Proc. Acad. Sci. Amst.*, ser. C, **66**: 176 (1963).
Gloeosporium acerinum Westd., *Herb. crypt. belg.* exs. 979 (1845).
Marssonia acerina (Westd.) Bres. in Allesch. & Schnabl, *Fungi bav.* exs. 286 (1892).
Marssonina acerina (Westd.) Magnus, *Hedwigia* **45**: 90 (1906).
Septomyxa acerina (Westd.) Höhn., *Sber. Akad. Wiss. Wien* **125**: 105 (1916).
Hendersonia collapsa Cke & Ell., *Grevillea* **5**: 30 (1876).
Stagonospora collapsa (Cke & Ell.) Sacc., *Syll. fung.* **3**: 448 (1884).
Myxosporium tulasnei Sacc., *Syll. fung.* **3**: 723 (1884).
Septomyxa tulasnei (Sacc.) Höhn., *Annls mycol.* **1**: 527 (1903).
Marssonina tulasnei (Sacc.) Died., *Krypt. Fl. Mk Brandenb.* **9**, Pilze VII: 822 (1915).
Discella pseudoplatani Oud., *Ned. Kruidk. Arch.* 2, **5**: 507 (1889).
Septomyxa negundinis Allesch., *Ber. Bayer. bot. Ges.* **5**: 22 (1897).
Myxosporium spaethianum Allesch., *Hedwigia* **36**: 163 (1897).
Septomyxa spaethianum (Allesch.) Höhn., *Sber. Akad. Wiss. Wien* **125**: 89 (1916).
Septomyxa negundinis Oud., *Ned. Kruidk. Arch.* 3, **1**: 512 (1898).
Cytodiplospora aceris Oud., *Contr. Fl. Myc. Pays Bas* **17**: 266 (1901).
Myxolibertella aceris Höhn., *Annls mycol.* **1**: 526 (1903).
Marssonia decolorans Kab. & Bub., *Ost. bot. Z.* 10 (1904).
Fusicoccum aceris Grove, *J. Bot., Lond.* 189 (1916).
Fusicoccum obtusulum Grove, *British Stem- and Leaf-Fungi* **1**: 249 (1935).
Gloeosporium aceris-tataricae Savul. & Sandu, *Hedwigia* **75**: 209 (1936).

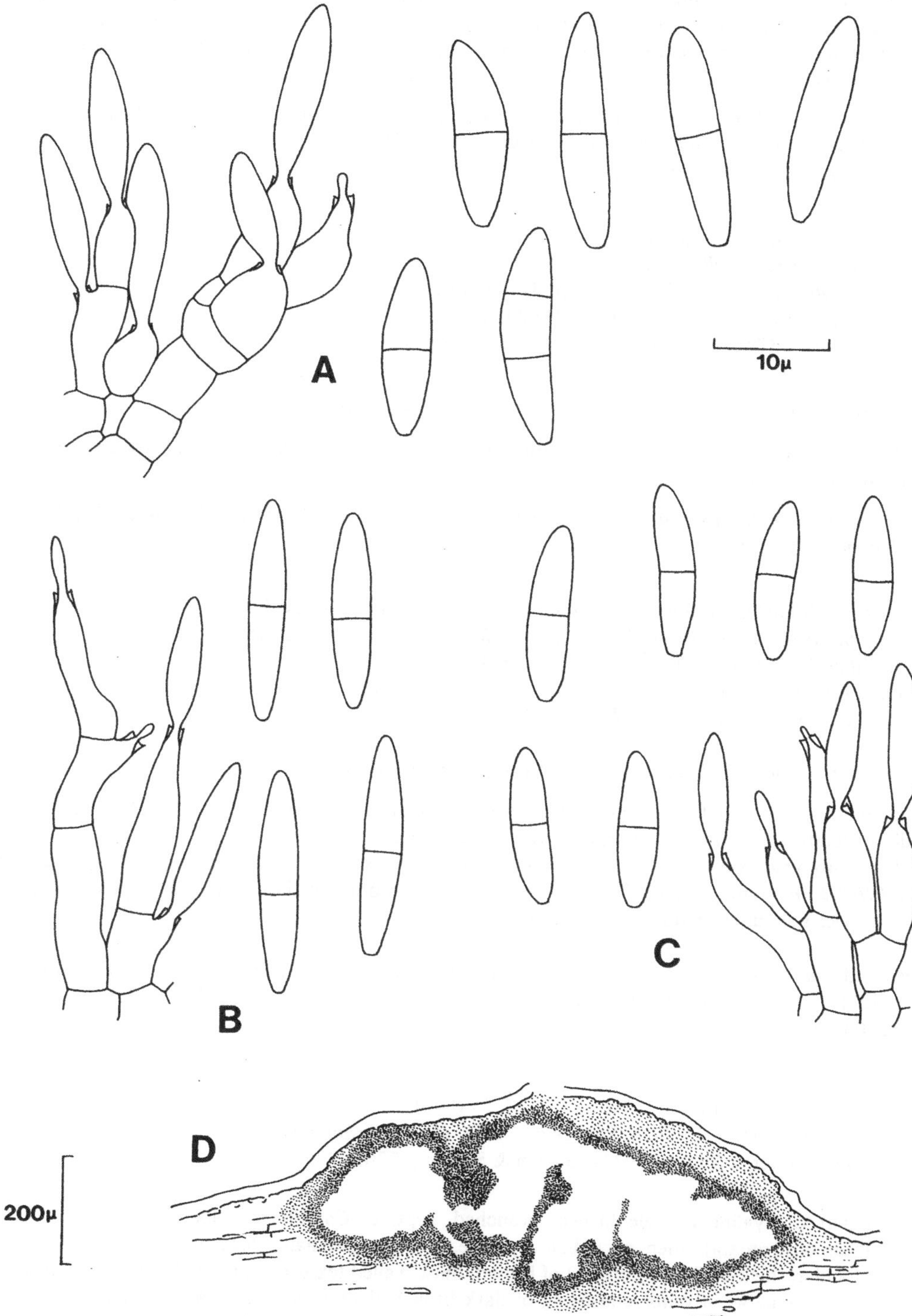

Fig. 372. A, D, *Diplodina microsperma*, conidia, conidiophores, vertical section of a conidioma; B, *D. aesculi*, conidia and conidiophores; C, *D. acerina*, conidia and conidiophores.

Teleomorphosis: *Cryptodiaporthe hystrix* (Tode) Petrak, *Annls mycol.* **19**: 119 (1921).

Conidiomata up to 1 mm diam. Conidiophores up to 25 µ long × 2.5-4 µ. Conidia 1 septate, 13.5-15.5 × 3-3.5 µ.

This species has usually been known under the name *Septomyxa tulasnei* (Sacc.) Höhn., but von Arx (1963) effectively reduced *Septomyxa* to synonymy with *Discella*, and Sutton (1977a) has advanced arguments showing that *Diplodina* is the correct generic name to adopt for this species.

On twigs of *Acer pseudoplatanus, A. negundo, A. ginnala, A.* sp., Germany, Austria, U.K., Latvia (IMI 188562, 188527, 207298, 27473b, 22475, 80076, 80240a, 13822b, 193008, 130957, 31152, 31102, 214361, 193266, 12737, 12399, 22722, 22723, 22152, 22724, 133253).

D. aesculi (Sacc.) comb. nov. (Fig. 372B)
Septomyxa aesculi Sacc., *Syll. fung.* **3**: 766 (1884).
Phomopsis aesculi (Sacc.) Lind, *Danish Fungi*: 422 (1913).
Fusicoccum aesculanum Sacc., *Annls mycol.* **12**: 293 (1914).
Phomopsis aesculana (Sacc.) Petrak, *Annls mycol.* **19**: 208 (1921).

Teleomorphosis: *Cryptodiaporthe aesculi* (Fckl) Petrak, *Annls mycol.* **19**: 119 (1921).

Conidiomata up to 1.5 mm diam. Conidiophores up to 35 µ long × 4 µ, tapering to 2.5 µ wide. Conidia 1 septate, long fusiform, 17-19.5 × 3-3.5 µ.

On bark of *Aesculus hippocastanum, A.* sp., Austria, Luxemburg (IMI 91335 ex P, 33624, 13034, 24210, holotype *F. aesculanum*, 31101, 31100).

D. MICROSPERMA (Johnston) Sutton, *Mycol. Pap.* 141: 69 (1977). (Fig. 372A, D)
Stilbospora microsperma Johnston, *A Flora of Berwick upon Tweed* **2**: 192 (1831).
Diplodina salicis Westd., *Bull. Acad. r. Belg. Cl. Sci.* II **2**: 562 (1857).
Discella salicis (Westd.) Boerema, *Neth. J. Pl. Path.* **76**: 165 (1970).
Septomyxa salicis Grove, *J. Bot., Lond.* 147 (1922).

Teleomorphosis: *Cryptodiaporthe salicella* (Fr.) Petrak, *Annls mycol.* **19**: 180 (1921).

Conidiomata up to 1 mm diam. Conidiophores up to 20 µ long × 3-6 µ wide or simply doliiform, 8 × 6 µ. Conidia usually 1 septate, sometimes 0 or 2 septate, 16-20.5 × 4-4.5 µ.

For many years known as *Discella carbonacea* (Fr.) Berk. & Br., it was subsequently changed to *Discella salicis* (Westd.) Boerema (1970a). The basionym for this combination, however, is the type of *Diplodina*, so it is clear the species should be referred to the latter genus.

On twigs and branches of *Salix babylonica, S. capraea, S. cordata, S.* spp., S. Dak., U.S.A., Austria, Sweden, U.K., U.S.S.R. (IMI 35326, 21635, 16514, 21633, 31078, 21634, 96705, 79807, 21638, 80754, 80762, 102637b, 105648, 213596, 21636, 63056).

104. DILOPHOSPORA Desm., *Ann. Sci. nat.*, ser. 2, **14**: 5 (1840).

Sp. typ.: *D. alopecuri* (Fr.) Fr. (syn. *Sphaeria alopecuri* Fr., *D. graminis* Desm.).
Illustrations: Walker & Sutton (1974); Gibson & Sutton (1976).

Mycelium immersed, hyaline to pale brown, branched, septate. *Conidiomata* like solitary pycnidia or loosely aggregated into linear eustromata; when pycnidial, separate, globose, papillate, dark brown, immersed, subepidermal, unilocular or multilocular, thick-walled, composed of brown, thick-walled textura angularis becoming thinner and paler to the inside; stromata linear between veins, consisting of dark brown, thick-walled textura angularis in the epidermis and hypodermis and textura intricata connecting the pycnidia. *Ostioles* single to each locule, central, circular, papillate. *Conidiophores* absent. *Conidiogenous cells* enteroblastic, phialidic, determinate, discrete, ampulliform to doliiform, hyaline, smooth, channel and collarette minute, periclinal wall thick, formed from the inner cells of the locular walls. *Conidia* hyaline, 0-3 euseptate, continuous, apex and base obtuse, thin-walled, eguttulate, smooth, cylindrical, at each end with several simple, dichotomously, or irregularly branched extracellular setulae.

Walker & Sutton (1974) described the type species and discussed the affinities of other taxa placed in *Dilophospora*, together with the status of suspected relationships with *Lidophia* Walker & Sutton (syn. *Dilophia* Sacc., 1883, non T. Thomson, 1853, Cruciferae). Gibson & Sutton (1976) provided a fuller illustration and a more detailed account of the disease this species causes.

D. ALOPECURI (Fr.) Fr., *Summa Veg. Scand.* 419 (1849).

Conidiomata 160-220 μ diam. Conidiogenous cells 4-8 × 2.5-5 μ. Conidia up to 13 μ long × 1.5-2 μ wide; appendages 6 μ long.

On leaves and culms, causing twist of *Agrostis* spp., *Holcus lanatus, H. mollis, Hordeum* sp., *Triticum vulgare* var. *erythrosperma*, Gramineae indet., G.B., New Zealand, Australia, Canada, Roumania (IMI 93656, 93039, 136633, 63572, 83650, 61515, 74018, 87359, 120475, 21613, 16845).

105. TITAEOSPORA Bub., *Annls mycol.* **14**: 345 (1916).
Sp. typ.: *T. equiseti* (Desm.) Vassiljevsky (syn. *Libertella equiseti* Desm., *T. detospora* (Sacc.) Bub.).

Mycelium immersed, branched, septate, hyaline. *Conidiomata* acervular, separate or confluent, circular to elongated, subepidermal, yellowish brown, composed of pale brown to hyaline, thin-walled textura angularis. *Conidiophores* irregularly branched at the base and above, hyaline, septate, smooth, formed from the upper cells of the pseudoparenchyma. *Conidiogenous cells* enteroblastic, phialidic, determinate, integrated or discrete, cylindrical, hyaline, channel wide, collarette occasionally prominent, periclinal thickening not marked. *Conidia* hyaline, smooth, thin-walled, eguttulate, medianly 1-euseptate, fusiform, almost falcate, apex obtuse, base truncate, often anastomosing in the acervulus by germination tubes produced just above the base. *Paraphyses* long, filiform, hyaline, often branched, attenuated above, septate, formed from the acervular tissue and also conidiogenous cells.

Vassiljevsky & Karakulin (1950) redescribed the genus and provided the basic synonymy which was subsequently accepted by Arx (1970). He related *Titaeospora* to *Digitosporium* Gremmen but the latter differs by the holoblastic, digitate, pale brown conidia formed in stromata. Conidia in *Titaeospora* are basically falcate but become interconnected by basal germination tubes that anastomose.

Apart from the type species another three taxa have been described in *Titaeospora*. Neither *T. daphniphylli* Ramakr. & Sund. nor *T. pterolobii* Ramakr. & Sund. appear to be correctly placed in this genus since conidial anastomoses were not reported.

T. EQUISETI (Desm.) Vassiljevsky apud Vassiljevsky & Karakulin, *Fungi Imperfecti Parasitici.* Pars II. Melanconiales: 594 (1950). (Fig. 373)
Libertella equiseti Desm., *Ann. Sci. nat.*, ser. 3, **8**: 179 (1847).
Septoria equiseti Desm. in Moug., *Stirp. Voges.* 1264 (1848).
Rhabdospora equiseti (Desm.) Allesch. in Rabenh., *Krypt. Fl.* **6**: 901 (1901).
Phloeospora equiseti (Desm.) Höhn., *Mitt. bot. Inst. tech. Hochsch. Wien* **5**: 3 (1928).
Gloeosporium equiseti Ell. & Ev., *J. Mycol.* **4**: 52 (1888).
Septogloeum equiseti (Ell. & Ev.) Died., *Krypt. Fl. Mk. Brandenb.* **9**: 835 (1915).
Septoria detospora Sacc., *Michelia* **2**: 529 (1882).
Rhabdospora detospora (Sacc.) Allesch. in Rabenh., *Krypt. Fl.* **6**: 901 (1901).
Titaeospora detospora (Sacc.) Bub., *Annls mycol.* **14**: 345 (1916).
Septogloeum equiseti Tehon, *Mycologia* **29**: 445 (1934).

Conidiomata up to 200 μ diam. Conidiophores 12-20 × 3-4 μ. Conidia 27-38 × 3-3.5 μ.

On stems and leaves of *Equisetum arvense, E. limosum, E. limosum* var. *minus, E. laevigatum, E. hiemale, E. ramosissimum, E.* spp., France, Austria, Germany, U.K., India, Calif., Wyo., N. Dak., U.S.A. (IMI 5342-5346 ex K, 11620, 16575, 33643, 97072, 5341, 5317, 31162).

106. PYCNOFUSARIUM Punithalingam apud Hawksworth & Punithalingam, *Trans. Br. mycol. Soc.* **61**: 63 (1973).
Sp. typ.: *P. rusci* Hawksworth & Punithalingam

Mycelium immersed, branched, septate, hyaline. *Conidiomata* acervular, subepidermal, deep-seated, separate, formed of a relatively small basal aggregation of hyaline, thin-walled textura intricata. *Conidiophores* hyaline, branched, septate,

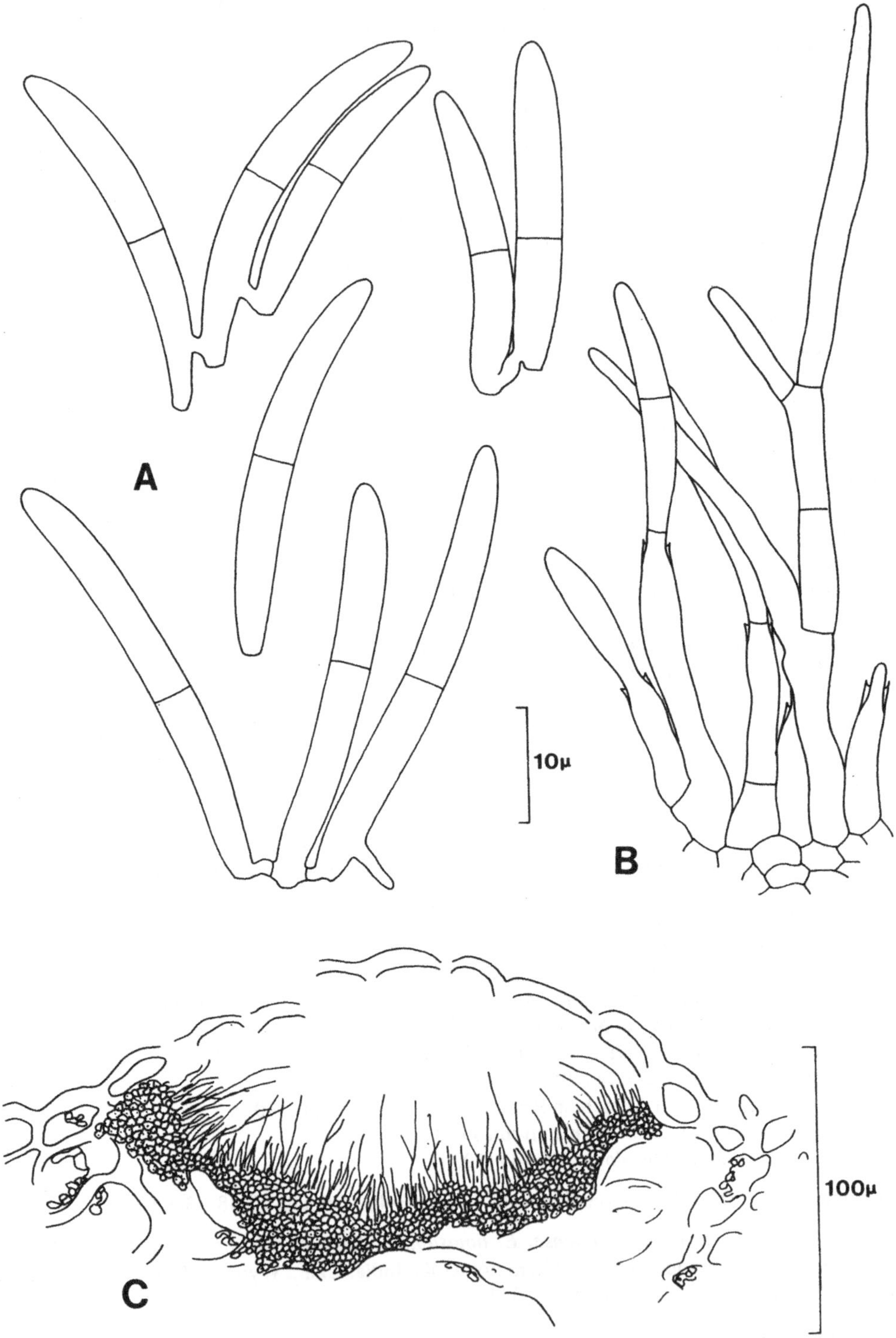

Fig. 373. *Titaeospora equiseti*, A, conidia; B, conidiophores and developing conidia; C, vertical section of a conidioma.

smooth, formed from the basal pseudoparenchyma and appearing as a globose mass of hyphae. *Conidiogenous cells* enteroblastic, phialidic, determinate, integrated, cylindrical, hyaline. *Conidia* 1-2 euseptate, continuous, base obtuse with a lateral, cellular, exogenous short appendage (as in the foot cell of *Fusarium*), apex obtuse to subacute, thin-walled, smooth, ± guttulate, curved, fusiform.

The genus differs little from *Fusarium* and can only really be differentiated by the deep-seated origin of the stroma. Its disposition must await studies on growth in culture where its relationships to known species of *Fusarium* may be more readily apparent.

P. RUSCI Hawksworth & Punithalingam, *ibid.*: 63 (1973). (Fig. 374)

Conidiomata mostly epiphyllous, scarlet to reddish-yellow, 100-151 µ diam. Conidiophores 25-50 × 3-4 µ. Conidia 30-35 × 2.5-4 µ.

On dead cladodes of *Ruscus aculeatus*, U.K. (IMI 166113, holotype).

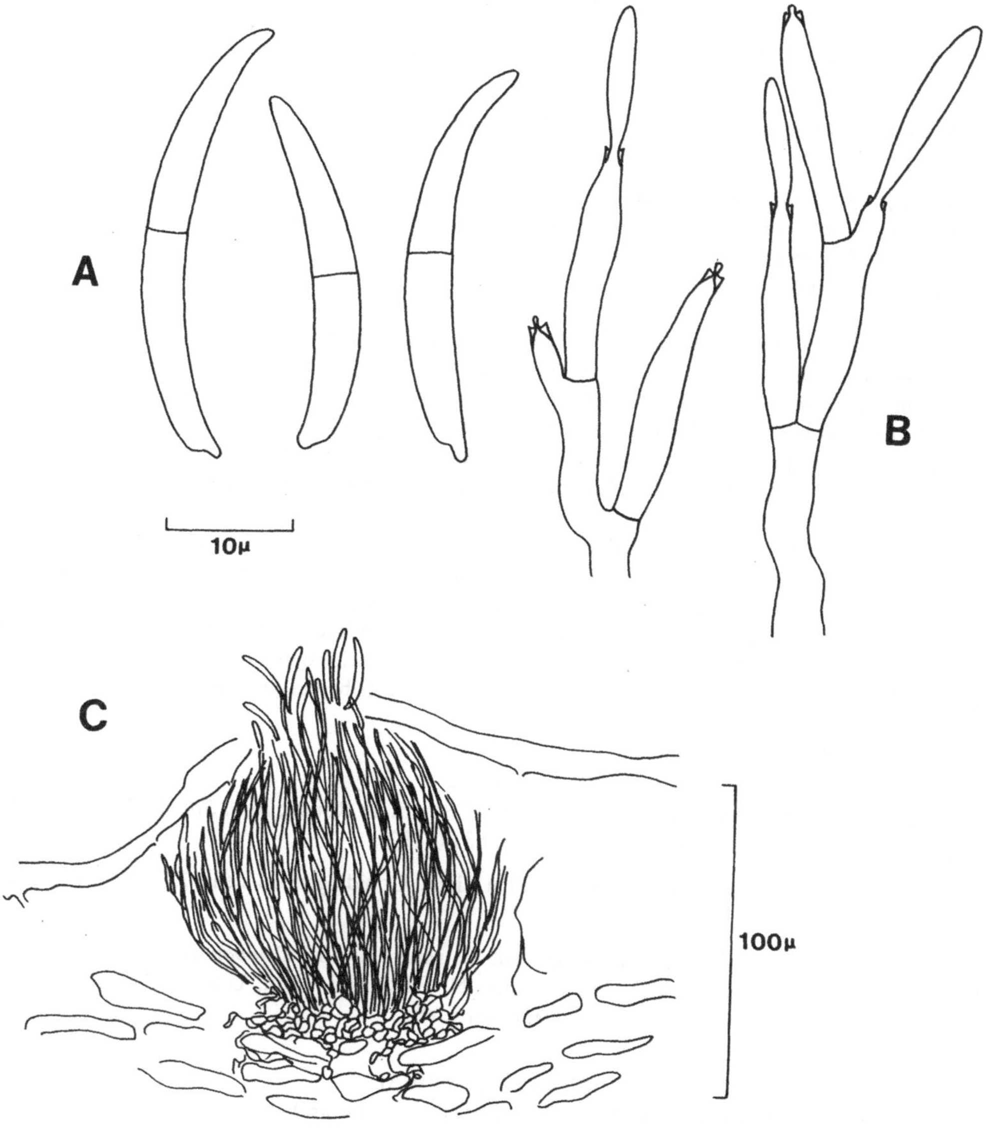

Fig. 374. *Pycnofusarium rusci*, A, conidia; B, conidiophores and developing conidia; C, vertical section of a conidioma.

U

107. BOTRYOCREA Petrak, *Sydowia* **3**: 140 (1949).
Sp. typ.: *B. sclerotioides* (Höhn.) Petrak (syn. *Stagonopsis sclerotioides* Höhn.).

Mycelium immersed or superficial, hyaline, branched, septate. *Conidiomata* separate or sometimes 2-3 aggregated, reddish brown with a grey-white tomentum, globose to irregular, superficial, hypocreoid; wall consists of an outer layer of hyaline, loose, branched hyphae, a middle layer 1-2 cells thick of golden brown, thick-walled textura prismatica, and an inner layer 1-3 cells thick of hyaline, thin-walled textura angularis. *Ostiole* absent, dehiscence by breakdown of conidiomatal wall. *Conidiophores* hyaline, branched above and below, septate, smooth, filiform, usually straight, formed from the inner cells of the conidiomatal wall. *Conidiogenous cells* enteroblastic, phialidic, integrated, more rarely discrete, determinate, hyaline, smooth, tapered to the apices, collarette and channel minute, periclinal wall thickened. *Conidia* fusarioid, hyaline, 3-euseptate, thin-walled, smooth, eguttulate, with a basal foot cell, apex abruptly tapered and curved.

This distinctive genus with characteristic *Fusarium* conidia is so far monotypic.

B. SCLEROTIOIDES (Höhn.) Petrak, *ibid.*: 140 (1949). (Fig. 375)
Stagonopsis sclerotioides Höhn., *Ann. Nat. Hofm. Wien* **20**: 368 (1905).

Conidiomata up to 1000 µ diam. × 400-500 µ high. Conidiophores up to 65 µ long × 2.5-3.5 µ. Conidia 18-21 × 3.5-4 µ.

On rotting stems of *Astragalus*, Turkey (IMI 202939, holotype ex FH).

108. CHEILARIA Lib., *Pl. Crypt. Ard.* cent. III, no. 63 (1834), *Ann. Sci. nat.*, ser. 2, **7**: 125 (1837).
Original spp.: *C. aceris* Lib., *C. agrostis* Lib., *C. heraclei* Lib., *C. urticae* Lib.
Lectotype sp.: *C. agrostis* Lib., fide Höhnel (1916c).
Illustration: Gibson & Sutton (1976).

Mycelium immersed, branched, septate, pale to dark brown, aggregated in the epidermal cells to form an immersed stroma. *Conidiomata* acervular, epidermal, separate or confluent, formed of pale to dark brown, thin-walled textura angularis and textura epidermoidea, small in relation to total area occupied by the stroma. *Conidiophores* absent. *Conidiogenous cells* enteroblastic, phialidic, discrete, determinate, cylindrical to lageniform or doliiform, hyaline, smooth, collarette and channel minute, periclinal wall markedly thickened, formed from the upper cells of the pseudoparenchyma. *Conidia* hyaline, thin-walled, eguttulate, smooth, falcate, fusiform, 0-3 (usually 2) euseptate.

Four species were originally placed in *Cheilaria* by Libert (1834). Subsequently *C. aceris* was made the type species of *Didymosporina* Höhn., *C. urticae* the type species of *Apomelasmia* Grove and *Cheilariopsis* Petrak, and *C. heraclei* the type species of *Anaphysmene* Bub. This left *C. agrostis* as the only original species in *Cheilaria* and Höhnel (1916c) accepted it as the lectotype. Petrak (1959b) concurred. *C. agrostis* is perhaps better known as *Septogloeum oxysporum* Bomm., Rouss. & Sacc. and detailed accounts have been provided by Sprague (1950) and Vassiljevsky & Karakulin (1950). It causes a blotch and char spot of various Gramineae and is described in detail by Gibson & Sutton (1976).

C. AGROSTIS Lib., *ibid.*: (1837). (Fig. 376)
Labrella agrostis (Lib.) Sacc., *Syll. fung.* **10**: 421 (1892).
Septogloeum oxysporum Bomm., Rouss. & Sacc. apud Bomm. & Rouss., *Bull. Soc. R. Bot. Belg.* **29**: 294 (1891).
Placosphaeria graminis Sacc. & Roum., *Michelia* **2**: 631 (1882).

Lesions 2-5 cm long. Conidiomata up to 100 µ diam. Conidiogenous cells 7.5-12.5 × 4.5-6 µ. Conidia 22-25 × 3.5-4 µ.

Causing leaf lesions on *Calamagrostis inexpensa*, *C. varia*, *C.* sp., *Cymbopogon giganteus*, *Milium effusum*, *Triticium* sp., U.K., Germany, Switzerland, Sudan, N. Dak., and Wyo., U.S.A. (IMI 98044, 92505, 108562, 62380, 131759, 56801, 152241 (isotype *Cheilaria agrostis*)).

109. MYRIELLINA Höhn., *Sber. Akad. Wiss., Wien* **124**: 100 (1915).
Sp. typ.: *M. cydoniae* (Desm.) Höhn. (syn. *Cheilaria cydoniae* Desm.).

Mycelium immersed, branched, septate, pale brown. *Conidiomata* acervular, subepidermal, separate, formed of pale

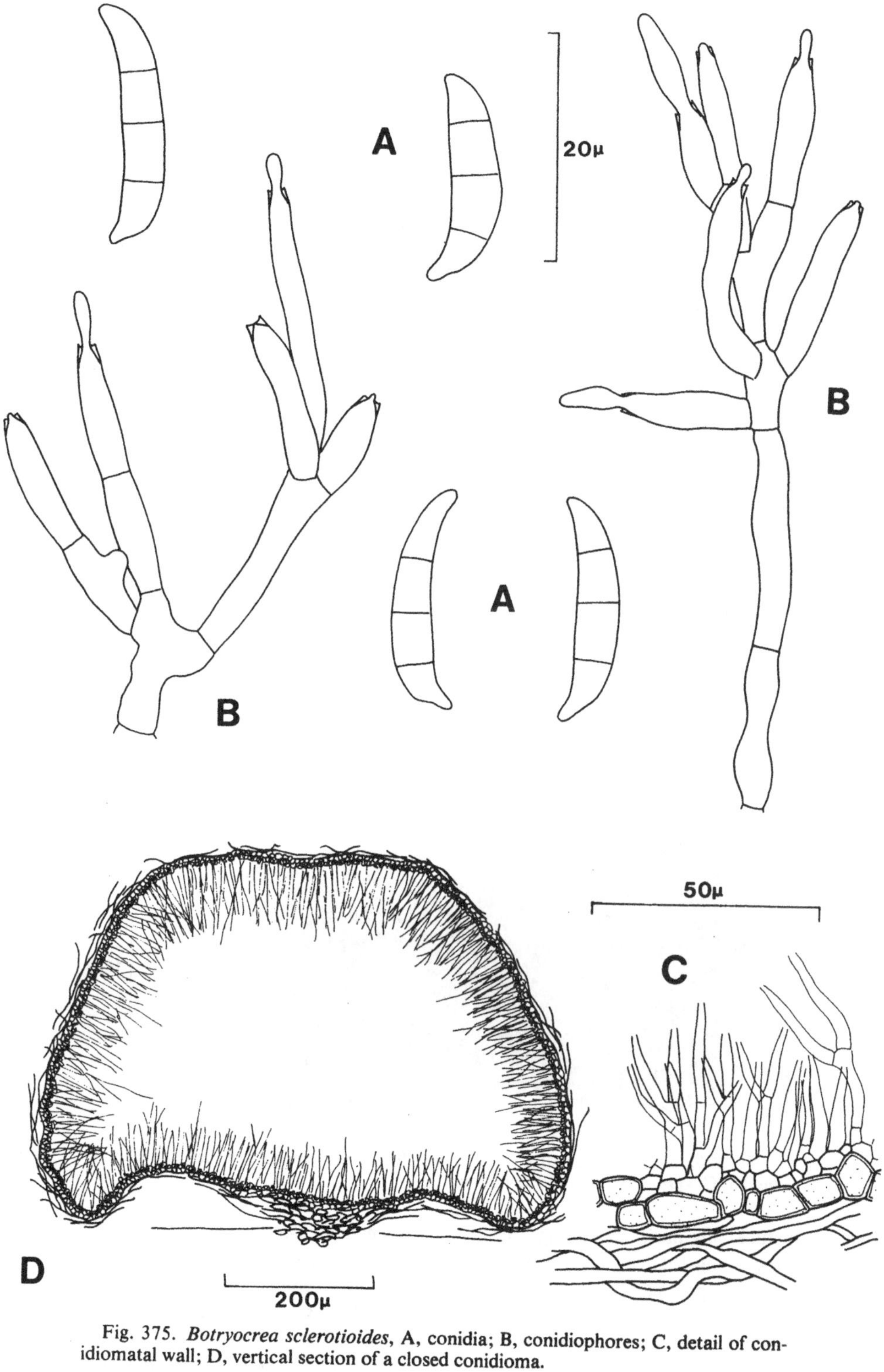

Fig. 375. *Botryocrea sclerotioides*, A, conidia; B, conidiophores; C, detail of conidiomatal wall; D, vertical section of a closed conidioma.

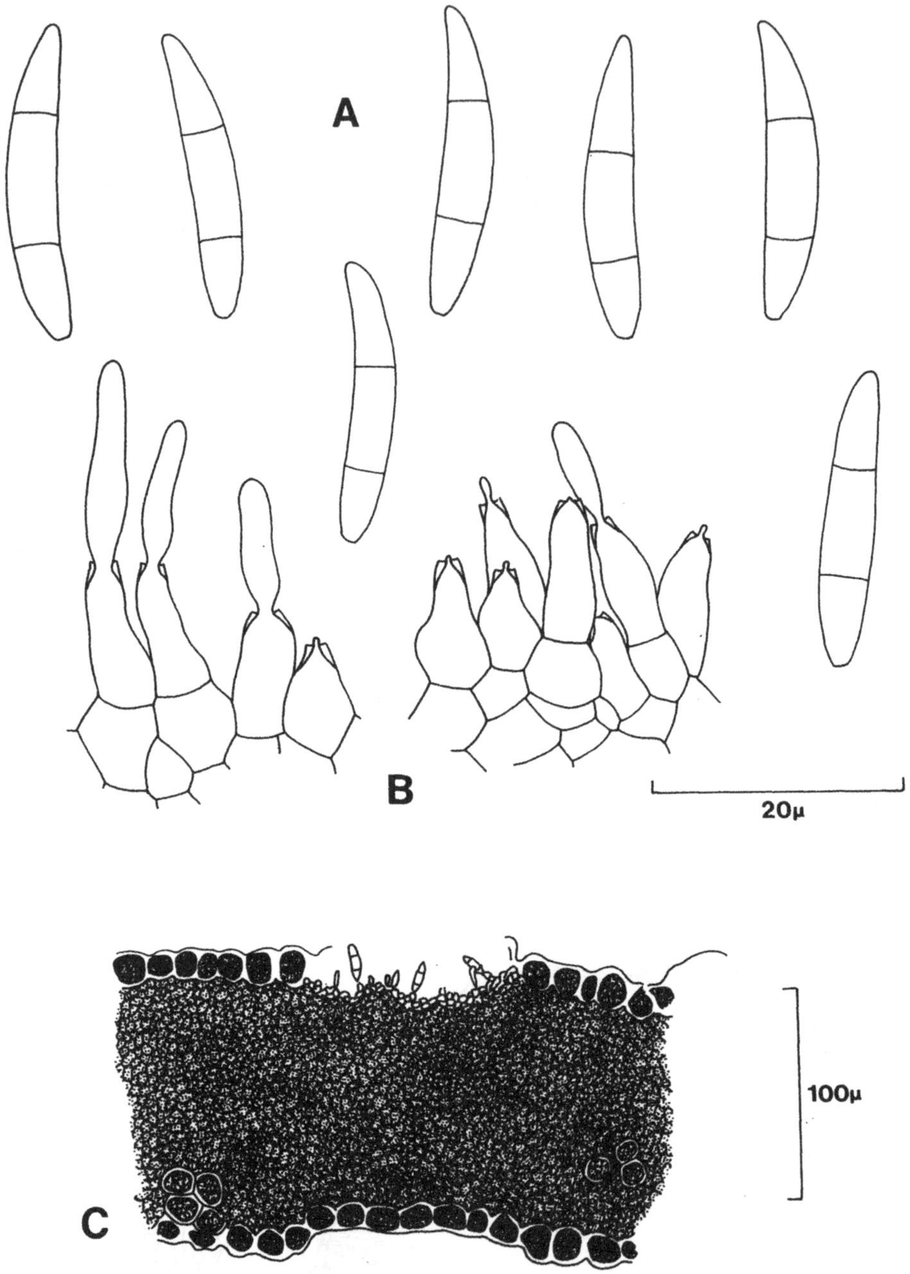

Fig. 376. *Cheilaria agrostis*, A, conidia; B, conidiogenous cells and developing conidia; C, vertical section of a conidioma.

brown, thin-walled textura angularis extending into the hypodermis. Dehiscence irregular. *Conidiophores* hyaline, branched at the base and along the length immediately below transverse septa, smooth, septate, cylindrical, straight or irregular, formed from the upper acervular cells. *Conidiogenous cells* enteroblastic, phialidic, integrated, determinate, forming conidia from apical apertures or lateral apertures immediately below transverse septa, cylindrical, hyaline, smooth, collarette conspicuous or minute, channel minute. *Conidia* hyaline, 1-3 euseptate, straight, curved or irregular, fusiform to cylindrical, apex obtuse, base ± truncate, occasionally branched below a median septum, smooth, thin-walled, eguttulate.

The genus is monotypic.

M. CYDONIAE (Desm.) Höhn., *ibid.*: 100 (1915). (Fig. 377)
Cylindrosporium cydoniae (Mont.) Schoschiaschvili apud Vassiljevsky & Karakulin, *Fungi Imperfecti Parasitici* II. Melanconiales: 561 (1950).
Cheilaria cydoniae Desm., *Ann. Sci. nat.*, ser. 3, **8**: 26 (1847).
Phyllosticta cydoniae (Desm.) Sacc., *Syll. fung.* **3**: 5 (1884).
Gloeosporium cydoniae Mont. in Castagne, *Catalogue des Plantes des environs de Marseille. suppl.* 59 (1851).
Septogloeum cydoniae (Mont.) Peglion, *Riv. Pat. veg. dei sign. A. Berlese* **2**: 335 (1894).

Conidiomata up to 120 µ diam. Conidiophores 21-30 × 3-5 µ. Conidia 9-17 × 2.5-3 µ.

Causing lesions on *Cydonia vulgaris*, Italy (IMI 21708).

110. SEPTOGLOEUM Sacc., *Michelia* **2**: 11 (1880).
Sp. typ.: *S. carthusianum* (Sacc.) Sacc. (syn. *Gloeosporium carthusianum* Sacc.).
Illustrations: Sutton & Pollack (1974).

Mycelium immersed, branched, septate, hyaline. *Conidiomata* acervular, epidermal to subepidermal, separate or confluent, formed of pale brown thin-walled pseudoparenchyma. Dehiscence irregular. *Conidiophores* short, stout, 1-2 septate, hyaline, smooth, branched at the base, formed from the upper pseudoparenchyma. *Conidiogenous cells* enteroblastic, phialidic, discrete or integrated, determinate, cylindrical, doliiform to obpyriform, hyaline, smooth, channel wide, collarette minimal, periclinal wall thickened. *Conidia* hyaline, 1-3 euseptate, thin-walled, smooth, eguttulate, base truncate, apex obtuse, straight or curved, constricted, obovoid.

More than 120 species have been described in *Septogloeum* but the genus was redescribed by Sutton & Pollack (1974) and restricted to two species. There are still however many taxa in *Septogloeum* that should be referred elsewhere, but until type species of *Cylindrosporium* auct., *Phloeospora* Wallr. and related genera are undertaken, there is little hope of rectifying this anomaly.

KEY

Conidia consistently 1-septate .... *S. thomasianum*
Conidia curved, 0-3 septate .... *S. carthusianum*

S. CARTHUSIANUM (Sacc.) Sacc., *ibid.*: 11 (1880).
Gloeosporium carthusianum Sacc., *Michelia* **1**: 93 (1877).
Ramularia carthusiana (Sacc.) Höhn., *Sber. Akad. Wiss. Wien* **118**: 80 (1909).

Conidiomata up to 100 µ diam. Conidiophores 13-21 × 7-9 µ. Conidia 26-34.5 × 10-12 µ, curved, 0-3 septate.

On leaves of *Euonymus europaeus*, Germany, Italy, Romania (IMI 101150 (isotype), 101151, 50931, 186536).

S. THOMASIANUM (Sacc.) Höhn. apud Lind, *Danish Fungi*: 448 (1913).
Gloeosporium thomasianum Sacc., *Michelia* **1**: 129 (1879).
Marssonia thomasiana (Sacc.) Sacc., *Syll. fung.* **3**: 768 (1884).
Marssonina thomasiana (Sacc.) Magn., *Hedwigia* **45**: 88 (1906).

Conidiomata up to 100 µ diam. Conidiophores 10-15 µ × 5-9 µ. Conidia 22-25 × 8-11 µ, straight, obovoid, 1-septate.

On leaves of *Celastrus scandens*, *Euonymus europaeus*, *E. latifolius*, Austria, Canada, Romania, Sweden (IMI 99445, 61601, 24760, 33623, 11621).

111. PHLYCTAENIELLA Petrak, *Annls mycol.* **20**: 323 (1922).
Sp. typ.: *P. polonica* Petrak

Mycelium immersed, branched, septate, hyaline. *Conidiomata* eustromatic, separate, immersed, pale brown, globose, unilocular, scarcely erumpent; wall several cells thick, at the base and sides of hyaline, thin-walled, small-celled textura angularis, above of larger pale brown tissue. *Ostiole* indistinct, dehiscence by rupture of the upper wall. *Conidiophores*

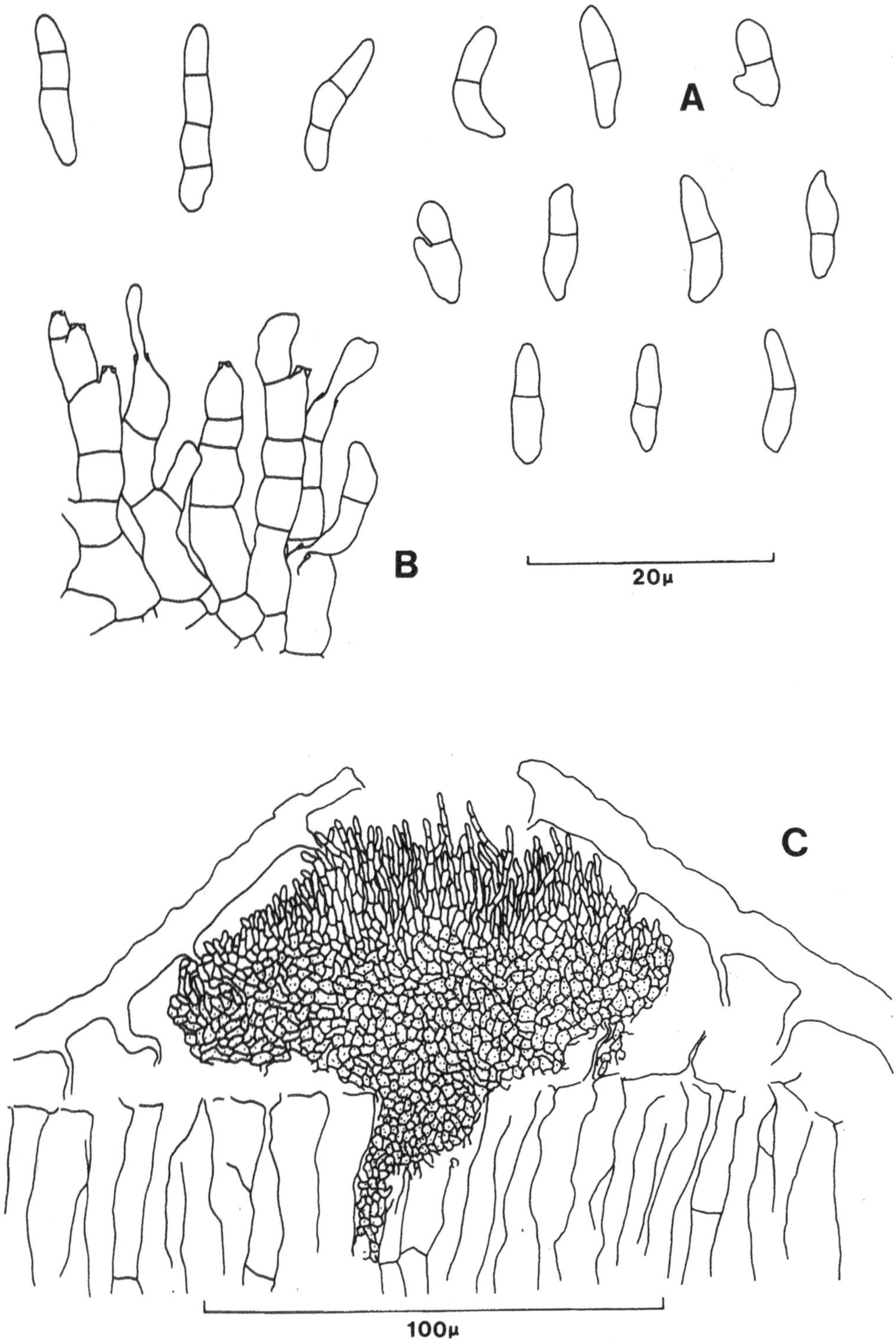

Fig. 377. *Myriellina cydoniae*, A, conidia; B, conidiophores and developing conidia; C, vertical section of a conidioma.

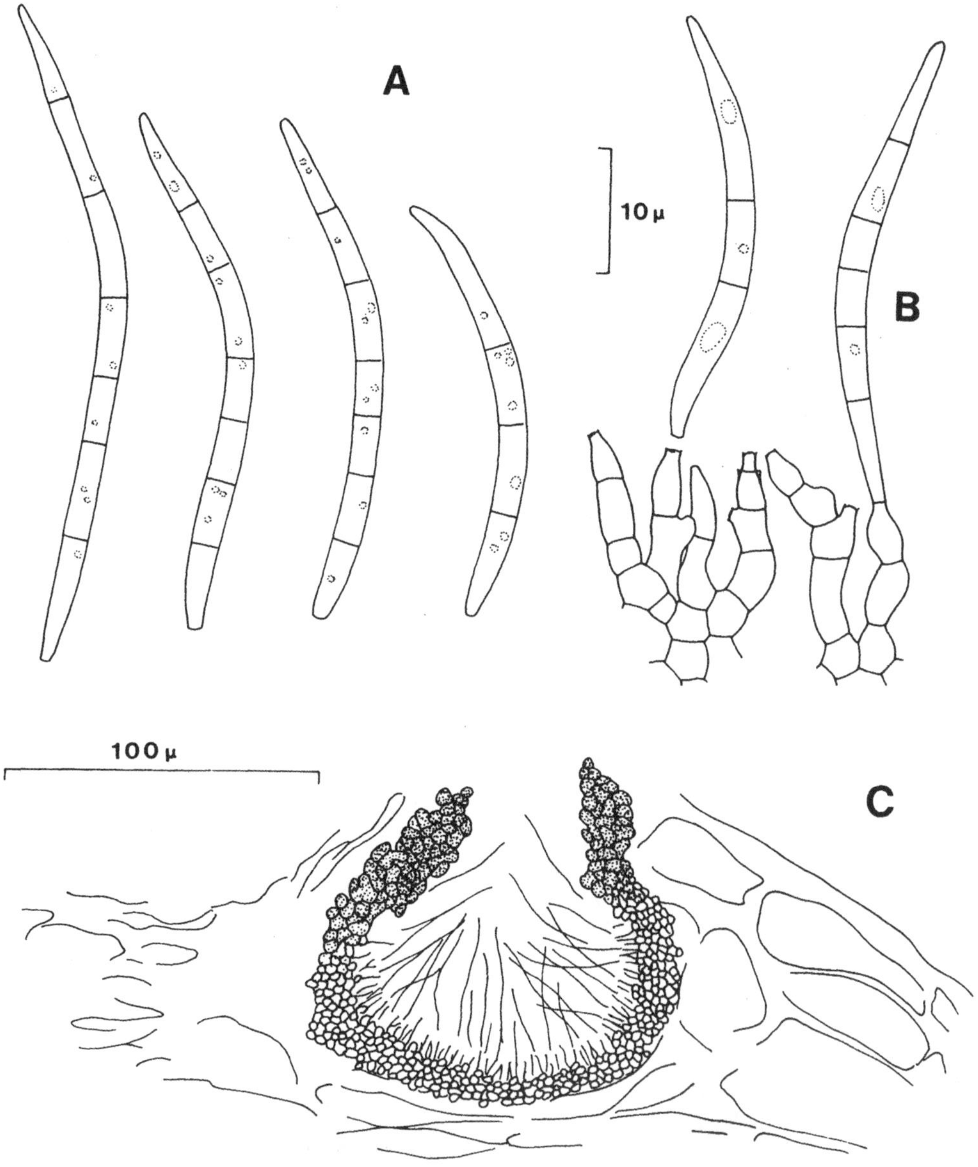

Fig. 378. *Phlyctaeniella humuli*, A, conidia; B, conidiophores and developing conidia; C, vertical section of a conidioma.

hyaline, smooth, septate, irregularly branched, especially at the base, formed from the inner cells of the stroma wall. *Conidiogenous cells* enteroblastic, phialidic, integrated or discrete, determinate, only rarely becoming percurrent, hyaline, markedly tapered at the apices, smooth, with apical or lateral apertures, collarette minute, channels comparatively wide. *Conidia* hyaline, smooth, thin-walled, irregularly guttulate, filiform, straight, curved or irregular, multiseptate.

The type species has not been examined. The above generic description is based on *P. humuli* Petrak, a species added later. If *P. humuli* and *P. polonica* are not congeneric then this generic concept will need to be changed.

P. HUMULI Petrak, *Annls mycol.* **27**: 397 (1929). (Fig. 378)

Conidiomata up to 130 µ diam. Conidiophores up to 20 µ long × 2.5-3 µ wide. Conidia 2-6 euseptate, 34-55 × 2.5 µ.

On stems of *Humulus lupulus*, Austria (IMI 202260 ex ZT).

112. PERIPERIDIUM Darker, *Can. J. Bot.* **41**: 1392 (1963).
Sp. typ.: *P. acicola* Darker

Mycelium immersed, hyaline to pale brown, branched, septate. *Conidiomata* eustromatic, separate, epidermal to subepidermal, ± depressed, flattened, dark brown, unilocular, thick-walled; wall composed of greenish brown, thick-walled textura angularis above, basal wall of thinner hyaline tissue. *Ostiole* absent, dehiscence by irregular rupture. *Conidiophores* 1-2 septate, cylindrical, tapered towards the apices, branched at the base, hyaline, formed only from the basal wall. *Conidiogenous cells* enteroblastic, phialidic, integrated or discrete, determinate, hyaline, aperture and collarette minute. *Conidia* falcate, 0-2 euseptate, hyaline, apex and base obtuse, thin-walled, eguttulate.

The genus is monotypic.

P. ACICOLA Darker, *ibid.*: 1392 (1963). (Fig. 379)

Teleomorphosis: *Micraspis acicola* Darker, *Can. J. Bot.* **41**: 1390 (1963).

Conidiomata up to 190 μ diam. Conidiophores 8-14 × 2-2.5 μ. Conidia 20-28.5 × 2 μ.

On needles of *Picea mariana*, Ont., Canada (IMI 193728, holotype ex DAOM, 193721).

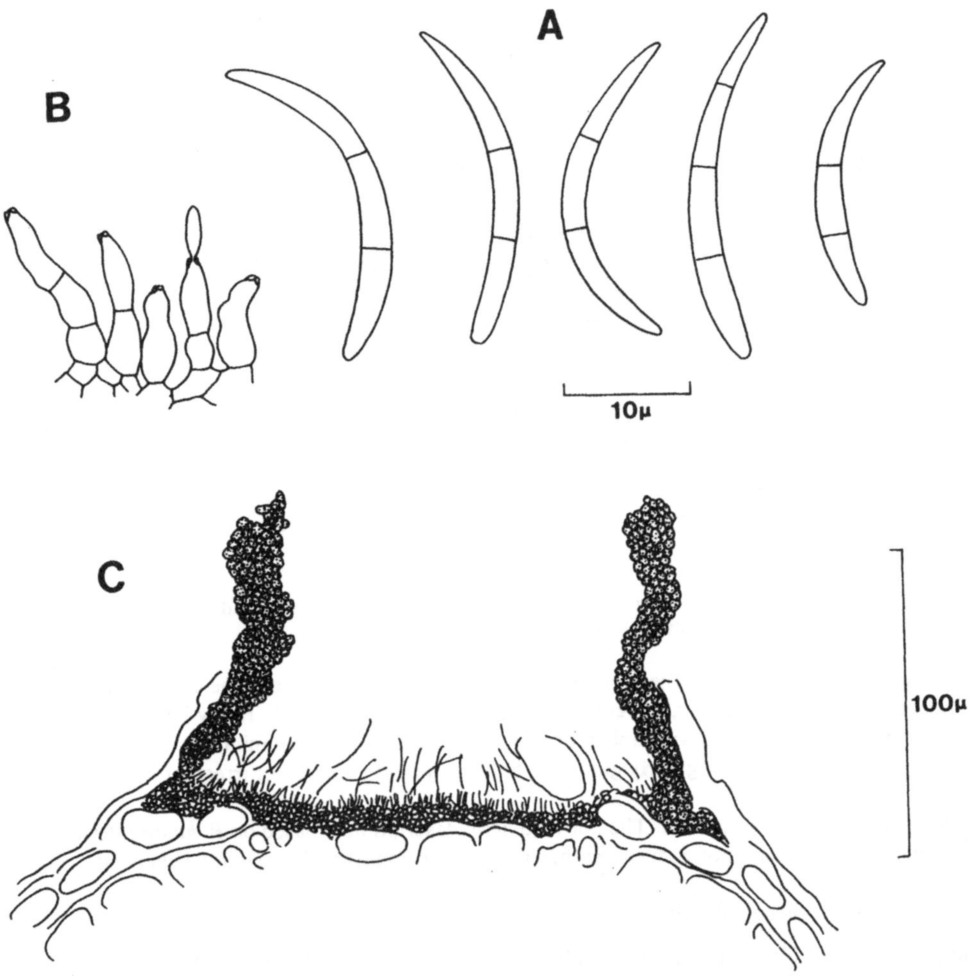

Fig. 379. *Periperdium acicola*, A, conidia; B, conidiophores, C, vertical section of a conidioma.

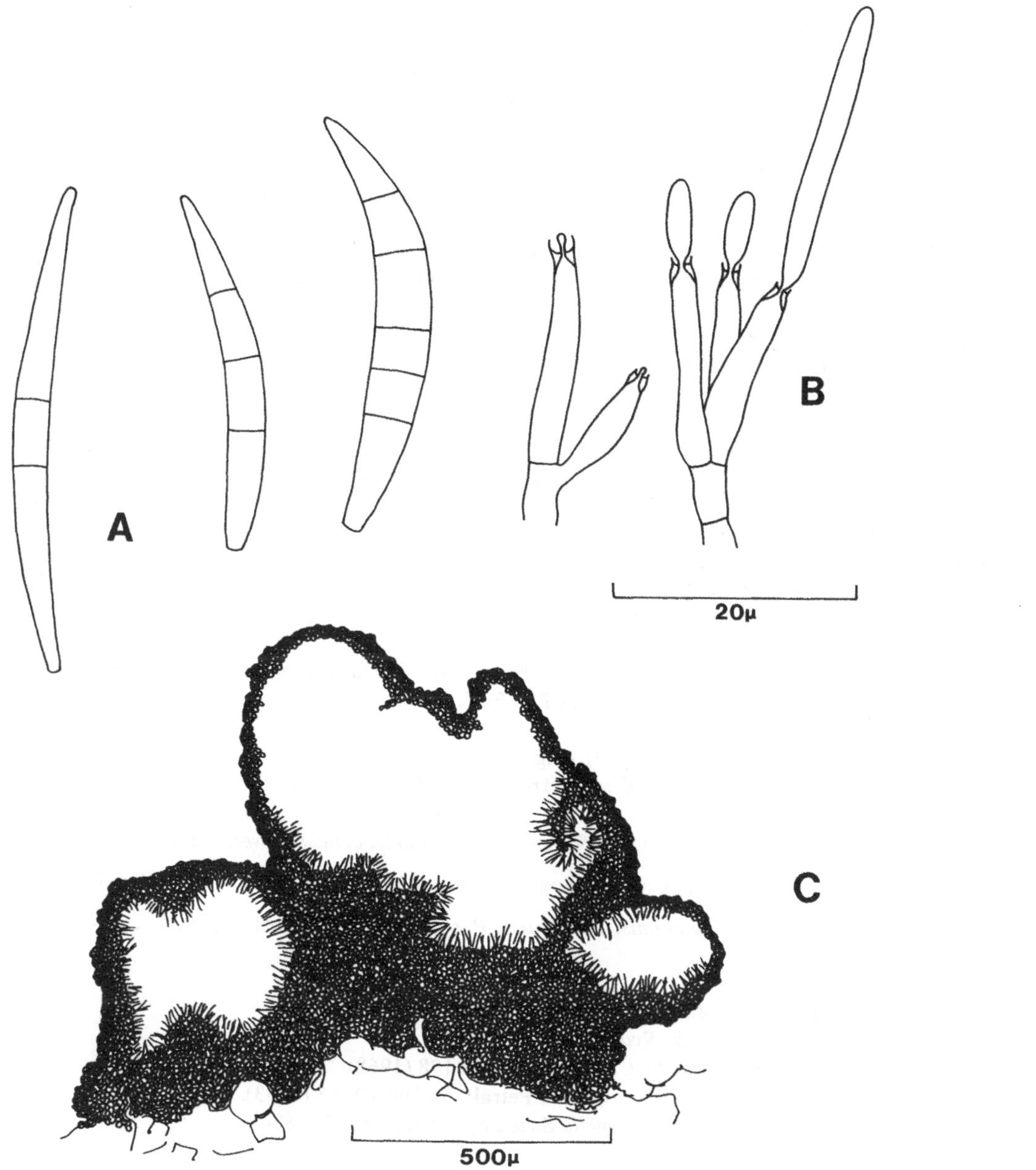

Fig. 380. *Brunchorstia pinea*, A, conidia; B, conidiophores and developing conidia; C, vertical section of a conidioma.

113. BRUNCHORSTIA Eriksson, *Bot. Centralbl.* **12**: 298 (1891).
Sp. typ.: *B. pinea* (Karst.) Höhn. (syn. *Septoria pinea* Karst., *B. destruens* Eriksson).

Mycelium immersed, branched, septate, pale brown. *Conidiomata* stromatic, immersed, becoming erumpent, separate or aggregated and confluent, pulvinate, globose or irregular, dark brown, irregularly multilocular; walls and base thick, composed of outer thick-walled dark brown textura angularis and an inner region of thinner walled paler pseudoparenchyma. Dehiscence irregular, by breakdown of upper wall. *Conidiophores* septate, irregularly branched, cylindrical or tapered towards the apex, hyaline, smooth, formed from the inner wall of the locules. *Conidiogenous cells* enteroblastic, integrated or discrete, determinate, hyaline, cylindrical, with a minute channel, distinct periclinal thickening and a small collarette. *Conidia* acrogenous, hyaline, 2-5 euseptate, continuous, curved, fusiform, smooth, thin-walled, eguttulate.

Punithalingam & Gibson (1973) have provided an account of the species, its synonymy and the dieback disease it causes.

B. PINEA (Karst.) Höhn., *Sber. Akad. Wiss. Wien* **124**: 143 (1915). (Fig. 380)

Teleomorphosis: *Gremmeniella abietina* (Lagerb.) Morelet, *Bull. Soc. Sci. nat. Archéol. Toulon Var* **183**: 9 (1969).

Conidiomata up to 600 µ wide × 1400 µ deep. Conidiophores 10-15 × 2-3 µ. Conidia 25-40 × 3-3.5 µ.

On branches of *Picea, Pinus* and *Pseudotsuga* spp., U.K., Roumania (IMI 172705, 56681, 56680, 106914, 50069, 50070). Synonymy and host range in Punithalingam & Gibson (1973).

114. CYTOPLACOSPHAERIA Petrak, *Annls mycol.* **17**: 79 (1919).
Sp. typ.: *C. rimosa* (Oud.) Petrak (syn. *Placosphaeria rimosa* Oud.).

Mycelium immersed, branched, septate, hyaline. *Conidiomata* eustromatic, like solitary pycnidia or loosely aggregated into linear stromata with 2-5 locules; externally lenticular, separate, dark brown to black, immersed, subepidermal, unilocular or multilocular, locules single or sparingly divided and convoluted, thick-walled; walls of dark brown, thick-walled textura angularis, becoming thinner and paler towards the conidiogenous region, overlying wall thicker and darker. *Ostiole* single to each locule, circular, papillate. *Conidiophores* absent. *Conidiogenous cells* enteroblastic, phialidic, determinate, discrete, ampulliform to doliiform, hyaline, smooth, channel and collarette minute, periclinal wall thickened, formed from the inner cells of the locular walls. *Conidia* hyaline, 1-3 septate, continuous, apex and base obtuse, thin-walled, eguttulate, smooth, straight or bent irregularly.

Petrak (1924) regarded *Cytoplacosphaeria* as merging with *Cheilaria* Lib. since it was tentatively considered to be the conidial state of *Scirrhia rimosa* (Alb. & Schw. ex Fr.) Fckl.

C. RIMOSA (Oud.) Petrak, *ibid.*: 79 (1919). (Fig. 381)
Placosphaeria rimosa Oud., *Ned. Kruidk. Arch.*, 2 ser., **5**: 493 (1889).

? Teleomorphosis: *Scirrhia rimosa* (Alb. & Schw. ex Fr.) Fckl, *Jb. nassau. ver. Naturk.* **23-24**: 220 (1870).

Conidiomata up to .15 cm long. Conidiogenous cells 4.5-9.5 × 3-4 µ. Conidia 13-20 × 3 µ.

On culms of *Phragmites communis, P. vulgaris, P.* sp., Latvia, Czechoslovakia, U.K. (IMI 21573, 33476, 94369, 95185e, 95175).

115. AMEROSPORIUM Speg., *An. Soc. Cient. Arg.* **13**: 20 (1882).
Amerosporina (Petrak) Petrak, *Sydowia* **18**: 377 (1965).
? *Thyriochaetum* Frolov, *Nov. Sist. niz. Rast. 1968*: 198 (1968).
Amerosporium Speg. subgen. *Amerosporina* Petrak, *Sydowia* **7**: 68 (1953).
Chaetomella Fckl subgen. *Euchaetomella* Sacc., *Syll. fung.* **3**: 321 (1884).
Sp. typ.: *A. polynematoides* Speg.

Mycelium immersed, branched, septate, hyaline to pale brown. *Conidiomata* initially spherical, later collapsed irregularly, usually superficial, separate, dark brown to black or greenish black, sessile, unilocular, initially closed, later dehiscing longitudinally in the upper wall; wall pseudoparenchymatous, at the base of textura angularis, hyaline to pale brown, at the sides of elongated cells giving rise on the inner-side to paraphysis, but becoming thick-walled and dark brown on the outside; the overlying wall similar to the lateral walls but with an extra layer 3-5 cells thick of paler brown isodiametric pseudoparenchyma. *Ostiole* absent. *Setae* sparse, irregularly inserted, straight, or slightly curved, unbranched, dark brown, thick-walled, verrucose, septate, ± tapered to the obtuse apices. *Conidiophores* absent. *Conidiogenous cells* enteroblastic, phialidic, determinate, discrete hyaline, smooth, cylindrical, straight or slightly curved, occasionally producing conidia in a mucilaginous sheath, channel minute, periclinal wall thickened, collarette minimal, restricted to the basal wall. *Conidia* pale brown to olivaceous, fusiform, aseptate, smooth, apex conical, base often flattened. *Paraphyses* restricted to the lateral wall, incurved, tapered markedly towards the subacute apices, septate, branched below the septa, hyaline.

Amerosporium has been the subject of considerable confusion and although more than 50 taxa have been described in

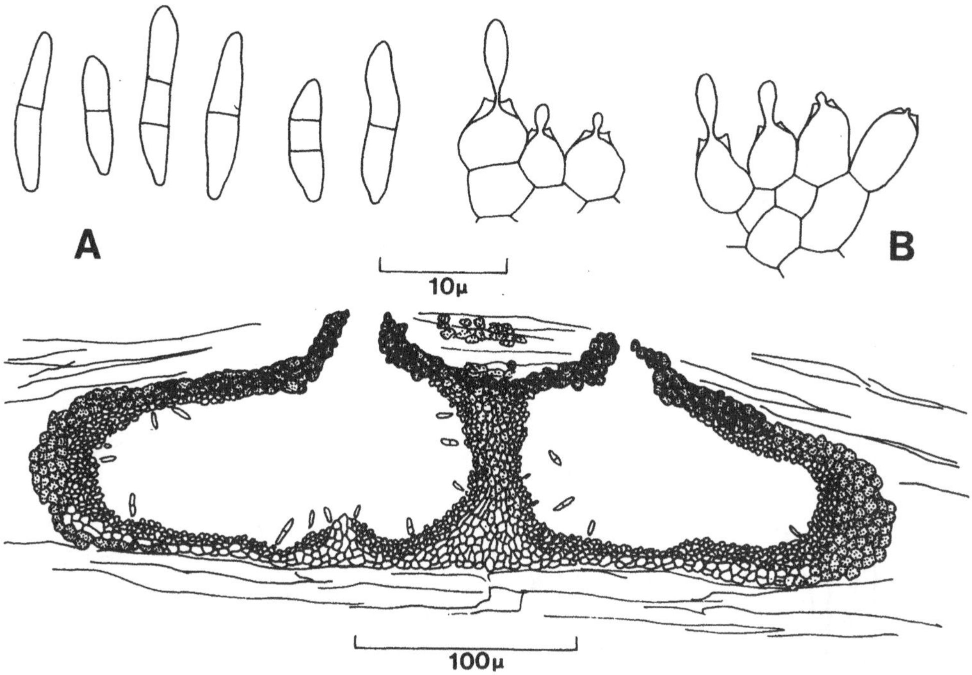

Fig. 381. *Cytoplacosphaeria rimosa*, A, conidia; B, conidiophores; C, vertical section of a conidioma.

the genus, only two species can at the moment be retained in it with confidence. Petrak & Sydow (1935) divided it into subgenus *Euamerosporium* and subgenus *Phaeopolynema* (Speg.) Petrak & Syd. (type *A. argentinense* (Speg.) Petrak & Syd.). Later, Petrak (1953a) introduced the new subgenus *Amerosporina* Petrak (type *A. concinnum* Petrak) and subsequently a fourth subgenus *Amerosporula* Petrak (1965). However, Sutton (1977a) placed *Phaeopolynema* as a synonym of *Hymenopsis* Sacc., and Petrak (1965) elevated *Amerosporina* to generic rank with *A. concinna* (Petrak) Petrak as the type species. The type of *A. concinnum* has been examined and conforms well with the generic characteristics of *Amerosporium* even though Petrak thought the conidia were catenate. Whereas *Amerosporina* was considered of unknown affinity by Sutton (1977a) it is now formally reduced to synonymy with *Amerosporium*.

KEY

Conidia 9-11 × 2 µ .... *A. polynematoides*
Conidia 13-15 × 2.5-3 µ .... *A. concinnum*

A. POLYNEMATOIDES Speg., *ibid.*: 20 (1882). (Fig. 382A, C, D, E)
Chaetomella cycadina Ramachandra Rao & Baheker, *Mycopath. Mycol. appl.* **23**: 266 (1964).

Conidiomata 400-600 µ diam. Conidiogenous cells 13.5-21 × 1.5-2 µ. Conidia 9-11 × 2 µ.

On *Conium maculatum*, Argentina (IMI 103757, holotype of *A. polynematoides* ex LPS); on *Cycas revoluta*, India (IMI 110662, holotype of *Chaetomella cycadina* ex MUA), also on *Bambusa, Boerhaevia diffusa, Bougainvillea glabra, Cassia, Clerodendrum enormi, Cynodon dactylon, Ficus, Indigofera, Lasiurus hirsutus, Thunbergia, Typha elephantinus*, and from soil, from India, Pakistan, Iraq, Zambia, Fla., U.S.A. (IMI 199265b, 133438, 133439, 112457, 112456, 160056a, 198374, 198379a, 177100, 100735, 106614, 217563, 209286, 166000, 203022).

A. CONCINNUM Petrak, *Sydowia* **7**: 68 (1953). (Fig. 382B)
Amerosporina concinna (Petrak) Petrak, *Sydowia* **18**: 377 (1968).

Conidiomata 300-800 µ diam. Setae up to 150 × 6.5 µ. Conidia 13-15 × 2.5-3 µ.

On *Echinops*, Iran (IMI 204124, isotype of *A. concinnum* ex BPI).

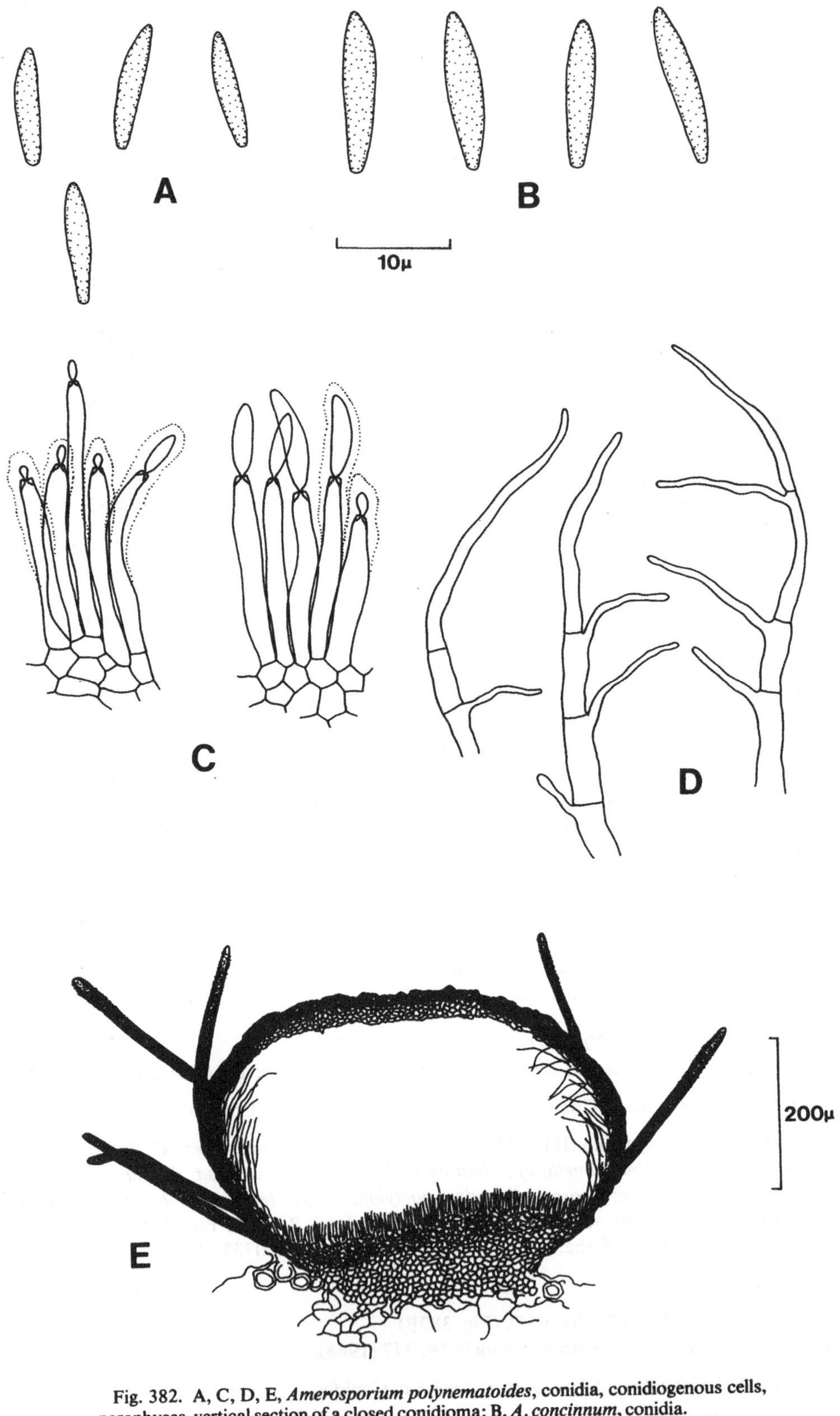

Fig. 382. A, C, D, E, *Amerosporium polynematoides*, conidia, conidiogenous cells, paraphyses, vertical section of a closed conidioma; B, *A, concinnum*, conidia.

116. VANDERYSTIELLA P. Henn., *Ann. Mus. Congo Belg.* **2**: 229 (1908).
Deightonia Petrak, *Sydowia* **1**: 114 (1947).
Sp. typ.: *V. leopoldvilleana* P. Henn.

Mycelium immersed, branched, septate, pale brown to hyaline. *Conidiomata* acervular, epidermal to subepidermal, separate or confluent, hypophyllous, formed of pale brown, thin-walled textura angularis. Dehiscence irregular. *Conidiophores* absent. *Conidiogenous cells* enteroblastic, phialidic, determinate, discrete, cylindrical, hyaline, smooth, channel wide, collarette prominent, periclinal wall thickened, formed from the upper cells of the acervulus. *Conidia* pale brown, aseptate, base truncate, smooth, eguttulate, biconic, rhomboid.

The genus is monotypic.

V. LEOPOLDVILLEANA P. Henn., *ibid.*: 229 (1908). (Fig. 383)
Deightonia rhombispora Petrak, *Sydowia* **1**: 114 (1947).

Conidiomata up to 70 µ diam. Conidiogenous cells 11-19 × 3-3.5 µ. Conidia 12-14 × 8.5-9 µ.

Causing lesions on leaves of *Leptoderris brachyptera*, Sierra Leone (IMI 17758, holotype *D. rhombispora*).

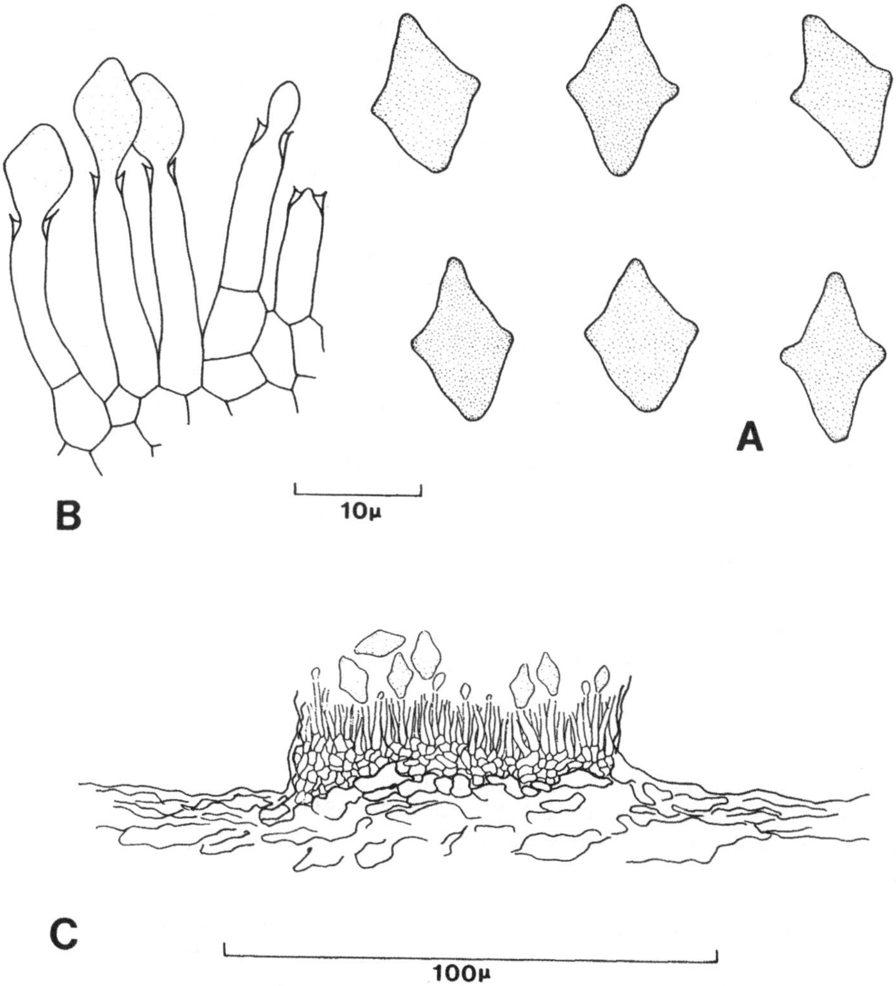

Fig. 383. *Vanderystiella leopoldvilleana*, A, conidia; B, conidiophores and developing conidia; C, vertical section of a conidioma.

117. GREENERIA Scribner & Viala, *C.R. Acad. Sci. Paris* **105**: 473 (1887).
Frankiella Speschnew, *Trav. Jard. bot. Tiflis* **5**: 7 (1900).
Sp. typ.: *G. uvicola* (Berk. & Curt.) Punithalingam (syn. *Phoma uvicola* Berk. & Curt., *Greeneria fuliginea* Scribner & Viala).
Illustration: Sutton & Gibson (1977).

Mycelium immersed, branched, septate, pale brown to hyaline. *Conidiomata* acervular, subepidermal, separate or confluent, formed of pale brown, thin-walled textura angularis. Dehiscence irregular. *Conidiophores* branched irregularly from the base and above, septate, hyaline, cylindrical, tapered to the apices, smooth, formed from the upper pseudoparenchyma. *Conidiogenous cells* enteroblastic, phialidic, discrete or more frequently integrated, determinate, cylindrical and tapered to the apices, hyaline, smooth, collarette and channel minute. *Conidia* very pale brown, smooth, thin-walled, base truncate, apex obtuse, cylindrical or fusiform or oval.

This species has been fully described and illustrated by Sutton & Gibson (1977). The genus is monotypic.

G. UVICOLA (Berk. & Curt.) Punithalingam, *Mycol. Pap.* 136: 6 (1974).
Phoma uvicola Berk. & Curt., *Grevillea* **2**: 82 (1873).
Myrothecium convexum Berk. & Curt., *Grevillea* **3**: 99 (1875).
Greeneria fuliginea Scribner & Viala, *C.R. Acad. Sci., Paris* **105**: 473 (1887).
Melanconium fuligineum (Scribner & Viala) Cavara, *Atti Ist. Bot. Univ. Pavia* **1**: 4 (1888).
Frankiella viticola Speschnew, *Trav. Jard. bot. Tiflis* **5**: 7 (1900).
Phyllosticta frankiana Sacc. & Syd., *Syll. fung.* **16**: 826 (1902).

Conidiomata up to 170 µ diam. Conidiophores up to 30 × 3 µ. Conidia 7.5-10 × 3-4 µ.

Causing bitter fruit rot and on leaves of *Vitis vinifera*, U.S.A., India, Hong Kong, South Africa (IMI 185913 ex NY, 185914 ex NY, 185915 ex NY, 184804 ex BPI, 183277 ex K, 183294 ex K, 168796, 129724, 35888, 190731).

118. FAIRMANIELLA Petrak & Syd., *Repert. spec. nov. regni veg. Beih.* **42**: 481 (1927).
Sp. typ.: *F. leprosa* (Fairm.) Petrak & Syd. (syn. *Coniothyrium leprosum* Fairm.).

Mycelium immersed, branched, septate, dark brown. *Conidiomata* acervular, subepidermal, separate, formed of medium brown, thick-walled textura epidermoidea, in the upper region of more or less vertically orientated, paler, thin-walled textura prismatica. Dehiscence irregular. *Conidiophores* absent. *Conidiogenous cells* enteroblastic, phialidic, determinate, discrete, ampulliform, hyaline, smooth, channel and collarette minute. *Conidia* pale brown, aseptate, obtuse to truncate at the base, apex obtuse, thick-walled, elliptical to ovate, punctulate.

The genus has been redescribed by Sutton (1971b) and remains monotypic.

F. LEPROSA (Fairm.) Petrak & Syd., *ibid.*: 481 (1927). (Fig. 384)

Conidiomata 100-200 µ diam. × 80 µ deep. Conidiogenous cells 3-5 × 2.5-3.5 µ. Conidia 5-7.5 × 3-4 µ.

Usually associated with lesions on *Eucalyptus robusta, E. citriodora, E. fasciculata, E. globulus, E.* sp., Hawaii, U.S.A., Chile, Australia, New Zealand, Zambia (IMI 148757b ex BISH, 147102 ex K, 62069, 57994, holotype, 21659, 129781, 94677).

119. CYTOPLEA Bizz & Sacc. in Bizz., *Flor. Venet. Critt.* **1**: 401 (1885).
Neopycnodothis T. Kobayashi, *Ann. Phytopath. Soc. Japan* **30**: 154 (1965).
Sp. typ.: *C. arundinacea* (Sacc.) Petrak & Syd. (syn. *Coniothyrium arundinaceum* Sacc., *Cytoplea arundinicola* Bizz. & Sacc.).

Mycelium immersed, branched, septate, pale brown to hyaline. *Conidiomata* eustromatic, immersed but strongly erumpent, black, circular to lenticular, pulvinate, epidermal to subepidermal, multilocular, locules separated by vertical columns of hyaline to pale brown pseudoparenchyma; upper wall of dark brown, sclerotioid textura angularis extending over the whole area of the conidioma, extending at the sides to become almost clypeate; walls of locules and dividing walls of hyaline to pale brown, thin-walled pseudoparenchyma. *Ostiole* indistinct, dehiscence by rupture of the central region in

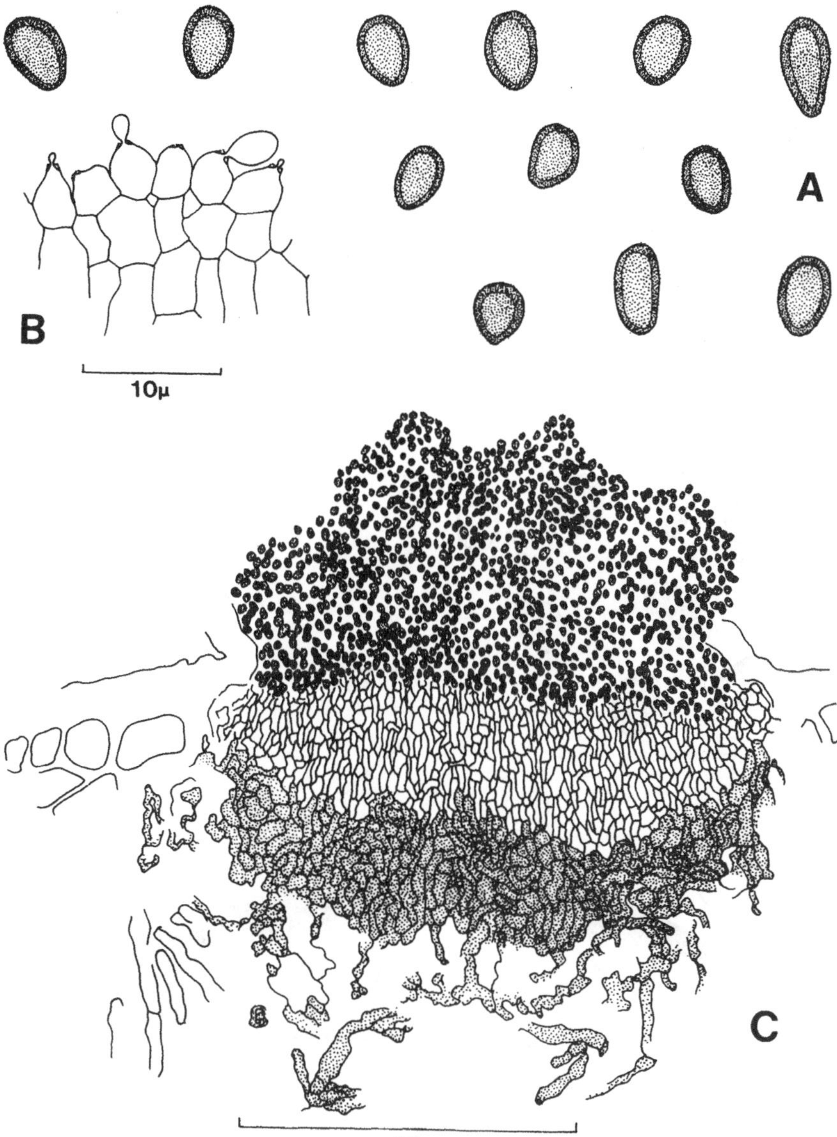

Fig. 384. *Fairmaniella leprosa*, A, conidia; B, conidiogenous cells; C, vertical section of a conidioma.

the upper wall. *Conidiophores* absent. *Conidiogenous cells* enteroblastic, phialidic, separate, determinate, discrete, hyaline, smooth, ampulliform or short cylindrical, channel and collarette minute, periclinal wall thickened. *Conidia* oval to ellipsoid, brown, thin or thick-walled, often biguttulate, verrucose to warty, aseptate.

The only full treatment of *Cytoplea* has been by Petrak & Sydow (1927). The genus was divided into two subgenera – *Cytoplea* subgen. *Eucytoplea* and *Cytoplea* subgen. *Cyclothyrium*. However, *C. juglandis*, presumably type of subgen. *Cyclothyrium*, is not congeneric with *C. arundinacea*, the type of subgen. *Eucytoplea*. Consequently *Cyclothyrium* is taken up here for *C. juglandis* and related species. Unfortunately species have been added to *Cytoplea* in the mistaken impression that *C. juglandis* is typical of the genus (Grove, 1937, Chesters, 1938). A thorough revision is required.

C. ARUNDINACEA (Sacc.) Petrak & Syd., *Rep. spec. nov. regni veg.* **42**: 439 (1927). (Fig. 385B, C, E)
Coniothyrium arundinaceum Sacc., *Michelia* **1**: 203 (1878).
Cytoplea arundinicola Bizz. & Sacc. in Bizz., *Flor. Venet. Critt.* **1**: 401 (1885).

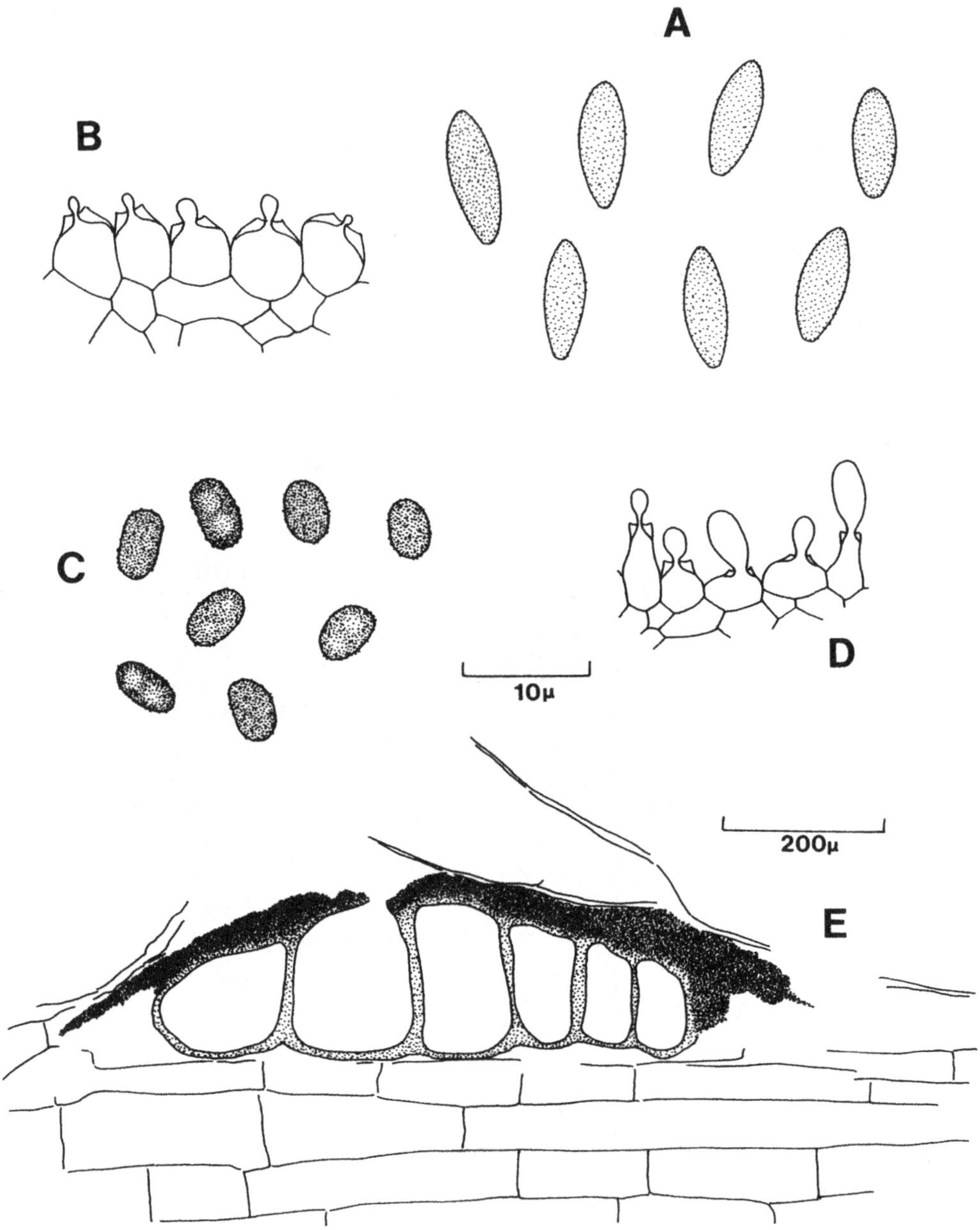

Fig. 385. A, D, *Cytoplea orientalis*, conidia, conidiogenous cells; B, C, E, *C. arundinacea*, conidiogenous cells, conidia; vertical section of a conidioma.

Stromata up to 800 µ diam. × 250 µ high. Conidiogenous cells 3.5-7 × 3-5 µ. Conidia 4.5-6 × 3-3.5 µ.

On dead stems of *Arundo donax*, Italy (IMI 199459, holotype of *C. arundinaceum* ex K).

C. phyllostachydis (Kobayashi) comb. nov.
Neopycnodothis phyllostachydis Kobayashi, *Ann. Phytopath. Soc. Japan* **30**: 154 (1965).

Stromata 2-5 mm diam. Conidia 14.5-19.5 × 8-10 µ.

The type has been unavailable, but Kobayashi (1965) records the species on *Phyllostachys edulis* from Japan.

C. orientalis sp. nov. (Fig. 385A, D)

Mycelium immersum, ramosum, septatum, brunneum, pallide brunneum vel hyalinum. *Conidiomata* usque ad 750 µ diam., eustromatica, initio immersa, demum erumpentia, intervenia, amphigena, sed raro hypophylla atro brunnea, lenticularia, pulvinata, epidermalia, ruptilia, multilocularia, loculis 50 µ diam.; parietes ex cellulis atro brunneis, sclerotioideis formatis (textura angularis). *Ostiolum* et conidiophora absentia. *Cellulae conidiogenae* enteroblasticae, phialidicae, separatae, determinatae, discretae, hyalinae, laeves, ampulliformes vel breve cylindricae, 5.5-7.5 × 4.5-6 µ. *Conidia* fusiformia, pallide brunnea, parietibus tenuibus, verruculosa, aseptata, 9-11 × 3-4 µ.

In foliis vivis *Phragmitis karkae*, Swamp nr. Aropa Pln., Kieta, Papua New Guinea, D. Shaw et G. Jackson 9356, 5 May 1974, IMI 199713, holotypus, isotypus in PM.

Lesions greyish brown with a paler brown centre, lenticular to ellipsoid, margin indistinct, separate, rarely confluent, up to .7 cm long × .3 cm wide. Conidiomata up to 750 µ diam., eustromatic, at first immersed, then strongly erumpent and appearing superficial, interveinal, amphigenous, more rarely hypophyllous, dark brown, lenticular, pulvinate, epidermal, multilocular, the locules at one level and about 50 µ diam.; walls composed of dark brown, thin-walled textura angularis. Ostioles absent, dehiscence by irregular breakdown of the upper wall. Conidiophores absent. Conidiogenous cells enteroblastic, phialidic, separate, determinate, discrete, hyaline, smooth, ampulliform to short cylindrical, 5.5-7.5 × 4.5-6 µ. Conidia fusiform, pale brown, thin-walled, verruculose, aseptate, 9-11 × 3-4 µ.

This species is readily separated from *C. arundinacea* and *C. phyllostachydis* by conidial size and shape. It is associated with distinct leaf lesions on *Phragmites kaka* whereas the other two species are culm-inhabiting.

On leaves of *Phragmites kaka*, Papua New Guinea, Burma (IMI 199713, holotype, 164838).

120. EPITHYRIUM (Sacc.) Sacc. & Trotter, *Syll. fung.* **25**: 249 (1931).

Coniothyrium Cda subgen. *Epithyrium* Sacc., *Syll. fung.* **10**: 268 (1892).
Biatoridina Schczedrova, *Bot. Zh. SSR* **49**: 1315 (1964).

Original spp.: *E. rugulosum* Sacc., *E. gleditschiae* (Casali) Sacc. & Trotter, *E. gregarium* (Died.) Sacc., *E. fusco-atrum* (Died.) Sacc., *E. obscurum* (Pass.) Sacc., *E. populi* (Oud.) Sacc., *E. innatum* (Karst.) Sacc., *E. resinae* (Sacc. & Berl.) Sacc., *E. domesticum* (P. Henn.) Sacc.
Lectotype sp.: *E. resinae* (Sacc. & Berl.) Sacc. (syn. *Coniothyrium resinae* Sacc. & Berl.).

Mycelium immersed and superficial, branched, septate, brown. *Conidiomata* eustromatic, superficial or semi-immersed, separate or aggregated, dark brown to black, globose, unilocular, convoluted or multilocular; walls thick, of brown, thick-walled textura intricata becoming angularis towards the inside. *Ostiole* absent, dehiscence by irregular breakdown of the upper tissues, often opening widely. *Conidiophores* hyaline, branched and septate at the base, smooth, formed from the inner cells of the locular wall. *Conidiogenous cells* enteroblastic, phialidic, determinate, discrete or integrated, lageniform or cylindrical, hyaline, smooth, channel and collarette minute, periclinal thickening marked. *Conidia* globose, smooth, thick-walled, pale brown, aseptate.

Only four species were originally placed in *Coniothyrium* subgen. *Epithyrium* and these were also included when it was raised to generic rank. They were *C. obscurum* Pass., *C. populi* Oud., *C. innatum* Karst. and *C. resinae* Sacc. & Berl. Petrak & Sydow (1924) placed *C. obscurum* in *Microdiplodia* Allesch. and *C. innatum* as a synonym of *Anthostoma melanotes* (Berk. & Br.) Sacc. Later, Petrak & Sydow (1927) returned *E. populi* to *Coniothyrium* Cda and combined *C. resinae* into *Lichenoconium* Petrak & Syd. Thus selection of *E. innatum* would place *Epithyrium* as a later synonym of the ascomycete genus *Anthostoma* while adopting *E. obscurum* would place *Epithyrium* as a synonym of *Microdiplodia*, a genus with 1-septate conidia. Neither conforms with the intention of Saccardo (1892) or Saccardo apud Trotter (1931). *Coniothyrium* and its segregates are imperfectly known as generic circumscription is concerned so selection of *E. populi* would not in any way determine the true affinities of the genus at this time. By selecting *E. resinae*, *Epithyrium* could become a later synonym for *Lichenoconium*. However, this genus was based on *L. lichenicola* (Karst.) Petrak & Syd. (Clements & Shear, 1931) and since accepted species are lichenicolous it seems most improbable that *E. resinae* is congeneric because it is the anamorphosis of a lichen. On the basis of *E. resinae*, *Epithyrium* could be recognised as a distinct genus. The species was separated from conidial *Tromera resinae* (Fr.) Körb. by the dark brown conidiomata and it is clear from the description that this is the fungus later described as the *Biatoridina pinastri* Schczedrova, conidial state of *Tromera difformis* (Fr.) Rehm.

E. RESINAE (Sacc. & Berl.) Sacc., apud Sacc. & Trotter, *Syll. fung.* **25**: 249 (1931). (Fig. 386)
Coniothyrium resinae Sacc. & Berl., *Atti R. Ist. Venet. Sci. Lett. Art.* **6**, ser. 3, 739 (1885).
Lichenoconium resinae (Sacc. & Berl.) Petrak & Syd., *Beih. Rep. spec. nov. regni veg.* **42**: 436 (1927).
Biatoridina pinastri Schczedrova, *Bot. Zh. SSR* **49**: 1315 (1964).

Teleomorphosis: *Tromera difformis* (Fr.) Rehm in Rabenh., *Krypt. Fl.* **3**(1): 307 (1896).

Conidiomata 180-200 μ diam. Conidiogenous cells 6-8 × 2.5-3 μ. Conidia 2.5-3 μ diam.

On resinous exudate of *Picea mariana, P. glauca, P.* sp., *Pinus sylvestris, P.* sp., U.K., Alta, Canada (IMI 68387, 70854b, 68125b, 69232, 73337a, 69012, 74201c).

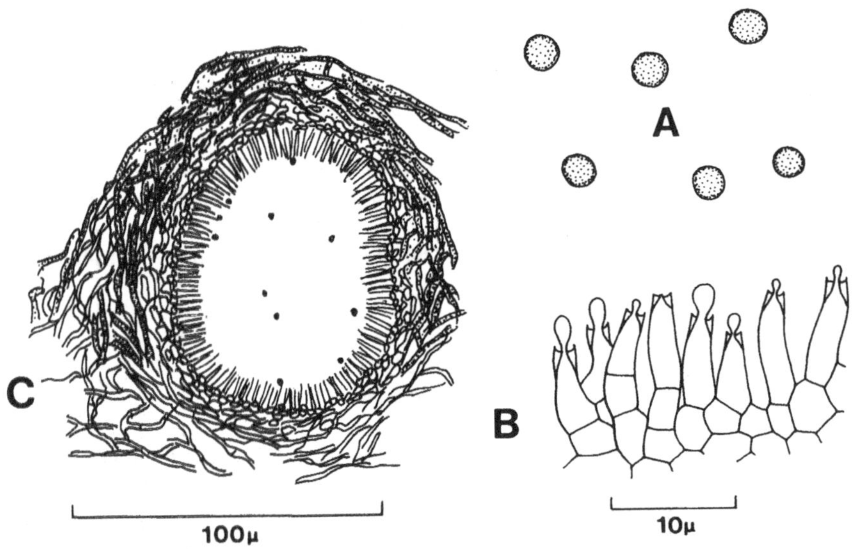

Fig. 386. *Epithyrium resinae*, A, conidia; B, conidiophores and developing conidia; C, vertical section of a closed conidioma.

121. NEOMELANCONIUM Petrak, *Annls mycol.* **38**: 208 (1940).
Sp. typ.: *N. gelatosporum* (Zimm.) Petrak (syn. *Melanconium gelatosporum* Zimm.).

Mycelium immersed, branched, septate, hyaline to pale brown. *Conidiomata* eustromatic, immersed, subperidermal, separate, irregularly multilocular; walls composed of small-celled, hyaline to pale brown, thin-walled textura angularis. Dehiscence irregular. *Conidiophores* absent. *Conidiogenous cells* enteroblastic, phialidic, indeterminate, discrete, cylindrical, hyaline, smooth, thick-walled, formed from the walls of the locules, with 1-3 percurrent proliferations. *Conidia* aseptate, globose, dark brown, smooth, thick-walled, base truncate.

This generic description is based on *N. deightonii* Petrak since *N. gelatosporum* has not been examined. Petrak (1954), in describing *N. deightonii*, enlarged and amended his concept of the genus and proposed the subgeneric name *Neomelanconiopsis* Petrak.

N. gelatosporum was described with conidia formed in a mucilaginous sheath whereas *N. deightonii* conidia were not. If this is indicative of a fundamental difference between the two species, confirmed by a study of *N. gelatosporum*, then it will be necessary to raise *Neomelanconium* subgen. *Neomelanconiopsis* to generic rank for *N. deightonii*.

N. DEIGHTONII Petrak, *Sydowia* **8**: 51 (1954). (Fig. 387)

Conidiomata 1.5-3 mm diam. Conidiogenous cells 15-21 × 6-9 μ. Conidia 17-21 × 14.5-17 μ.

On branches of *Spondias mombin*, Sierra Leone (IMI 51713, holotype).

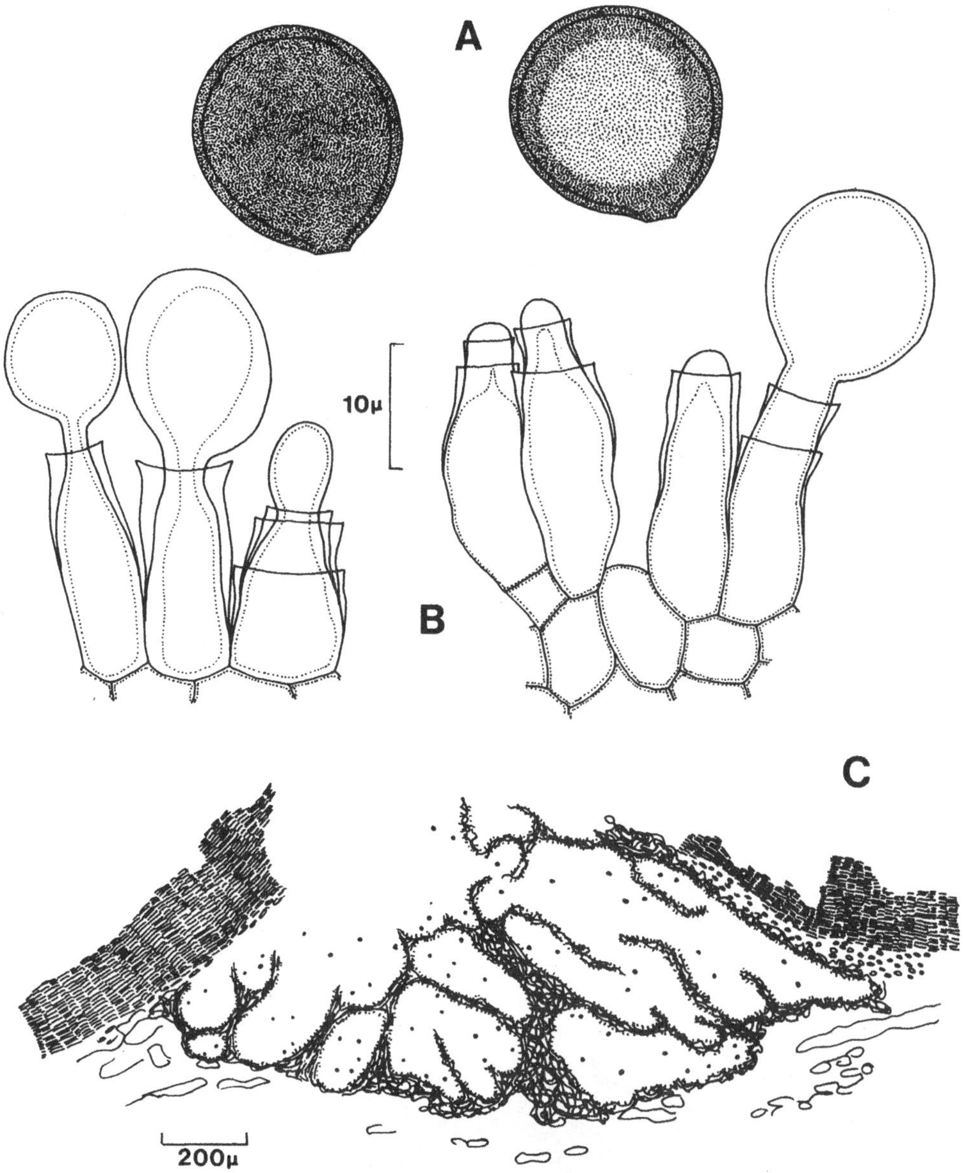

Fig. 387. *Neomelanconium deightonii*, A, conidia; B, conidiogenous cells and developing conidia; C, vertical section of a conidioma.

122. CYCLOTHYRIUM Petrak, *Annls mycol.* **21**: 5 (1923).

Cytoplea Bizz. & Sacc. subgen. *Cyclothyrium* (Petrak) Petrak & Syd., *Beih. rep. spec. nov. regn. veg.* **42**: 449 (1927).

Coniothyriopsiella Bender nom. nov., *Mycologia* **24**: 410 (1932).

Coniothyriopsis Petrak, *Annls mycol.* **21**: 5 (1923), non Speg., 1910.

Original spp.: *C. ulmigenum* (Tul.) Petrak, *C. incrustans* (Sacc.) Petrak

Lectotype sp.: *C. juglandis* (Schum. ex Rabenh.) Sutton (syn. *Cytospora juglandis* Schum. ex Rabenh.), fide Clements & Shear (1931).

Mycelium immersed, branched, septate, pale brown. *Conidiomata* eustromatic, immersed, subperidermal, separate, dark brown to black, multilocular and convoluted, walls of locules composed of an outer layer of thin-walled, brown, isodiametric textura angularis and an inner layer of similar hyaline tissue. *Ostiole* indistinct, dehiscence by rupture of the overlying host tissues. *Conidiophores* hyaline, branched at the base, less so above, septate, smooth, hyaline, formed from the entire inner walls of the locules. *Conidiogenous cells* enteroblastic, phialidic, indeterminate, frequently growing percurrently to produce more than one aperture at a higher level, integrated or more often discrete, straight, cylindrical, hyaline, smooth, channel and collarette small, periclinal thickening very marked. *Conidia* pale brown, smooth, aseptate, thick-walled, often medianly guttulate, short cylindrical, apex obtuse, base ± truncate.

Petrak & Sydow (1927) reduced *Cyclothyrium* to subgeneric rank within *Cytoplea* Bizz. & Sacc. and placed the type species name into synonymy with *Cytoplea juglandis* (Schum.) Petrak. However, *C. juglandis* is not congeneric with *C. arundinacea* (Sacc.) Petrak & Syd., the type species of *Cytoplea*. Consequently *Cyclothyrium* is recognised as a genus in its own right. *Coniothyriopsiella insitiva* (Sacc.) Petrak was also combined into *Cytoplea* by Petrak & Sydow (1927) but again, since it is congeneric with *C. juglandis* it must be referred to *Cyclothyrium*. Petrak & Sydow (1927) and Petrak (1925) regarded *C. insitiva* as conidial *Pseudothyridaria insitiva* Petrak.

About 30 species have been described in *Cytoplea*, which is the name formerly used for this group, but it is evident that a revision of the genus is required in the light of the generic synonymy listed by Sutton (1977a).

C. JUGLANDIS (Schum. ex Rabenh.) Sutton, *Mycol. Pap.* 141: 56 (1977). (Fig. 388)
Naemospora juglandis Schum., *Enum. Pl. Saell.* **2**: 178 (1803).
Cytospora juglandis Schum. ex Rabenh., *Krypt. Fl.* 148 (1844).
synonymy as cited in Petrak & Sydow (1927).

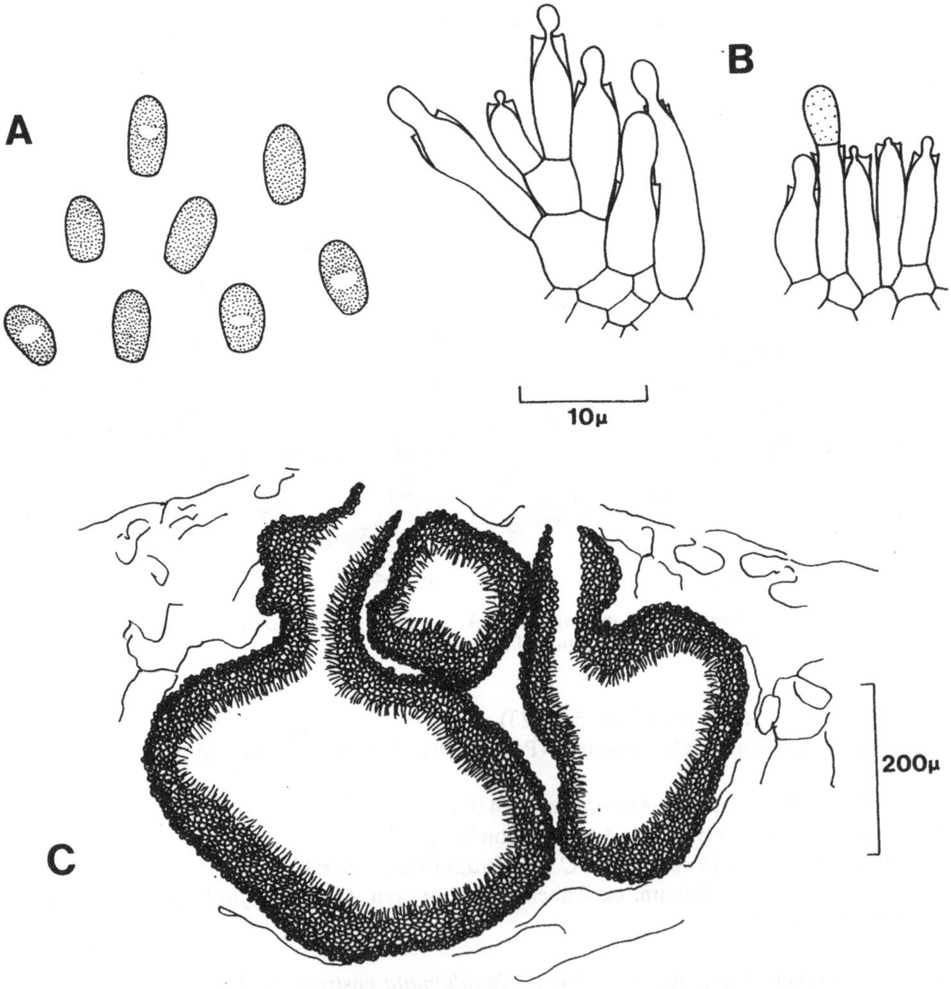

Fig. 388. *Cyclothyrium juglandis*, A, conidia; B, conidiophores and developing conidia; C, vertical section of a conidioma.

Teleomorphosis: *Thyridaria rubro-notata* (Berk. & Br.) Sacc., *Syll. fung.* **2**: 141 (1883).

Stromata 600-700 µ diam. Conidiophores and conidiogenous cells 12-16 × 2.5-4 µ. Conidia 5-6 × 3-3.5 µ.

On branches of *Juglans regia*, *Hedera* sp., *Acer pseudoplatanus*, *Ulmus* sp., *Gymnosporia* sp., *Rosa* sp., *Salix* sp., U.K., India (IMI 13953, 187465, 180470, 141949, 90633, 133441, 107957, 13952, 13951, 33648, 21337, 192685, 192660, 192705, 192673, 192681, 56938, 41562).

123. PHAEOCYTOSTROMA Petrak, *Annls mycol.* **19**: 45 (1921).
Pleocyta Petrak & H. Syd., *Beih. rep. spec. nov. regni veg.* **42**: 454 (1927).
Phaeocytosporella Stout, *Mycologia* **22**: 280 (1930).
Sp. typ.: *P. ambiguum* (Mont.) Petrak (syn. *Sphaeropsis ambigua* Mont., *P. istrica* Petrak).
Illustrations: Sutton (1964a).

Mycelium immersed, branched, septate, light or dark brown. *Conidiomata* eustromatic, separate, rarely confluent, immersed, unilocular, multilocular or convoluted, dark brown, thick-walled; wall of dark brown, thick-walled textura angularis. *Ostiole* single, circular, central, slightly papillate, or dehiscence by breakdown of upper wall of the conidioma. *Conidiophores* hyaline, septate and branched at the base, less so above, cylindrical to filiform, formed from the inner cells of the locular walls. *Conidiogenous cells* enteroblastic, phialidic, determinate, discrete or integrated, cylindrical, hyaline, with apical periclinal thickening, little or no collarette, channel narrow or wide. *Conidia* aseptate, brown, eguttulate, smooth, thin or thick-walled, base truncate, apex obtuse, cylindrical or ellipsoid. *Paraphyses* hyaline, septate, branched, filiform, obtuse at the apices.

The genus was treated in detail by Sutton (1964a) who accepted three species and explained the generic synonymy. It is comprised almost entirely of graminicolous species.

KEY

1. Conidia less than 5 µ wide 2
1. Conidia more than 5 µ wide 4
   2. Conidia less than 11 µ long; on *Calamus* .... *P. sacchari* var. *calami*
   2. Conidia more than 11 µ long 3
3. Conidia 11-14.5 × 3.5-4 µ; mainly on *Saccharum* .... *P. sacchari* var. *sacchari*
3. Conidia 15.5-19 × 4-5 µ; on *Pennisetum* .... *P. sacchari* var. *penniseti*
   4. Conidia 28-34.5 × 14-17 µ .... *P. megalosporum*
   4. Conidia 14-19 × 7-10.5 µ .... *P. plurivorum*
   4. Conidia 12-15 × 5.5-6.5 µ .... *P. ambiguum*

P. AMBIGUUM (Mont.) Petrak apud Petrak & H. Syd., *Beih. rep. spec. nov. regni veg.* **42**: 457 (1927). (Fig. 389E)
Sphaeropsis ambigua Mont., *Ann. Sci. nat.*, 3 ser., **12**: 308 (1849).
Phaeocytostroma istrica Petrak, *Annls mycol.* **19**: 45 (1921).
Phaeocytosporella zeae Stout, *Mycologia* **22**: 280 (1930).

Conidiomata up to 550 µ long × 300 µ wide. Conidiophores 15-35 µ long × 2-3.5 µ wide. Conidia ellipsoid to pyriform, medium brown, 12-15 × 5.5-6.5 µ. Paraphyses 40-65 × 2.5-3.5 µ.

On stems of *Zea mays* and isolated from *Sorghum bicolor*, France, Australia (IMI 100722, holotype *S. ambigua* ex PC, 206294).

P. MEGALOSPORUM Sutton, *Mycol. Pap.* 97: 28 (1964). (Fig. 389D)

Conidiomata up to 2 mm diam. Conidiophores 15-35 µ long × 2.5-3.5 µ wide. Conidia fusiform to pyriform, thick-walled, dark brown, 28-34.5 × 14-17 µ. Paraphyses 25-60 × 2.5-3.5 µ.
From paddyfield soil, India (IMI 92336, holotype).

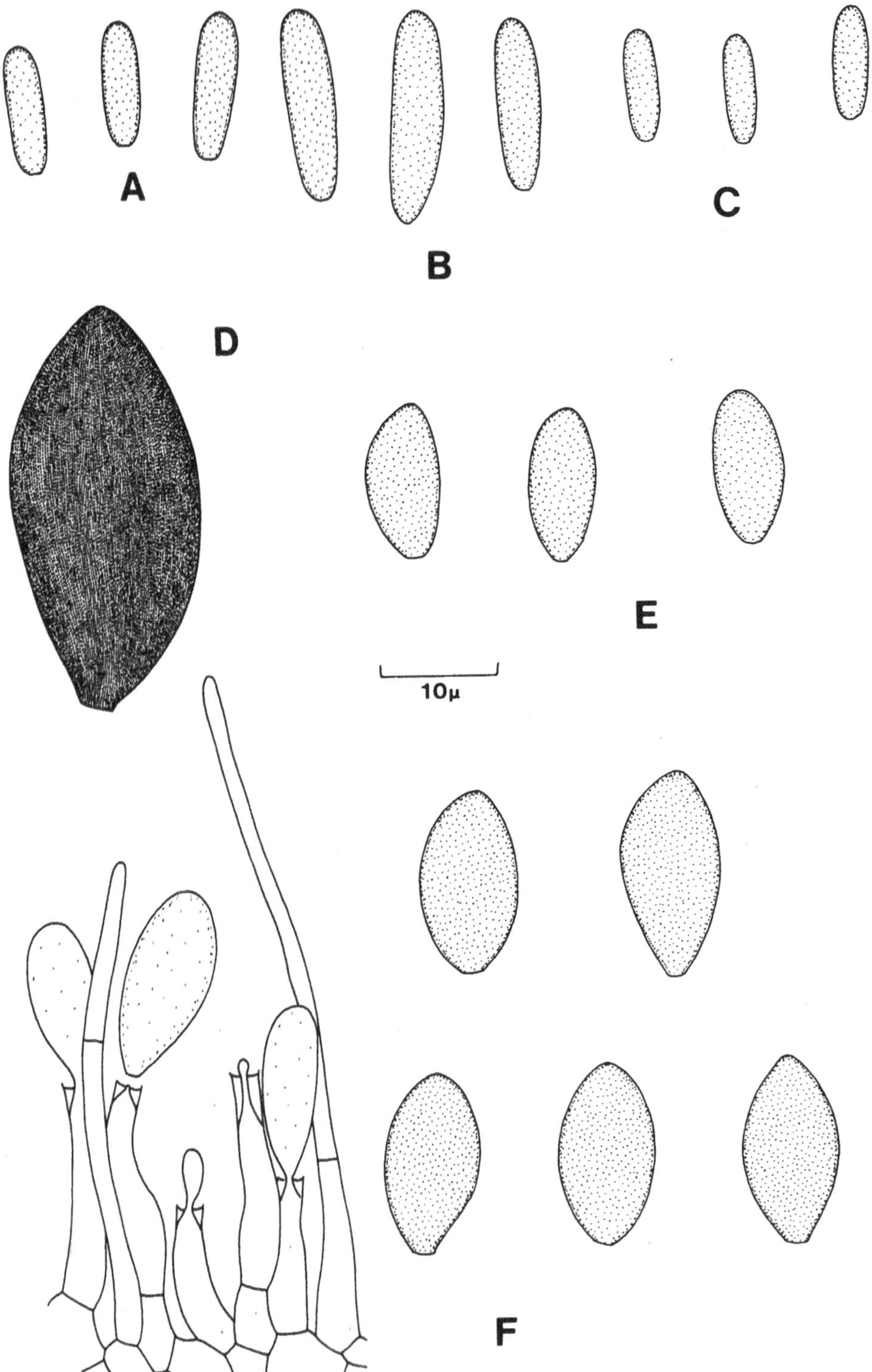

Fig. 389. A, *Phaeocytostroma sacchari*, conidia; B, *P. sacchari* var. *penniseti*, conidia; C, *P. sacchari* var. *calami*, conidia; D, *P. megalosporum*, conidia; E, *P. ambiguum*, conidia; F, *P. plurivorum*, conidiophores, paraphyses and conidia.

P. plurivorum sp. nov. (Fig. 389F)

Conidiomata immersa, nigra, globosa vel lenticularia, multiloculata, usque ad 700 µ longa × 300-400 µ crassa; parietibus fusco-brunneis ex textura angulari compositis. Massae conidiorum fuliginosae copiosae. Cellulae conidiogenae enteroblasticae, phialidicae, determinatae, discretae vel in conidiophoris incorporatae, hyalinae, cylindricae vel filiformes, 13.5-20 × 3-4 µ. Conidia pyriformia vel fusiformia, laevia, brunnea, apicem fere papillata, basim truncata, 14-19 × 7-10.5 µ. Paraphyses hyalina, sparse ramosa, septata, flexuosa, 30-60 µ longa × 1.5-2 µ crassa.

In culmis emortuis *Zeae maydis*, Mgorogo, Tanzania, G.B. Wallace 1448, 23 Jan. 1932, IMI 22520, holotypus.

Conidiomata eustromatic, separate, rarely confluent, often lenticular, multilocular, dark brown, up to 700 µ long × 300-400 µ wide; walls thick, dark brown, of textura angularis except for textura porrecta dividing the locules. Conidiogenous cells enteroblastic, phialidic, determinate, discrete or integrated, hyaline, cylindrical or filiform, with a narrow channel, minute collarette and distinct periclinal thickening, 13.5-20 × 3-4 µ. Conidia pyriform to fusiform, apex obtuse or almost papillate, base truncate, smooth, brown, aseptate, 14-19 × 7-10.5 µ. Paraphyses hyaline, rarely branched at the base, septate, flexuose, 30-60 × 1.5-2 µ.

The concept of *P. ambiguum* advanced by Sutton (1964a) included both *P. ambiguum* sensu stricto and *P. plurivorum*. However, the consistent difference in conidial shape and size necessitate the separation of these two taxa.

On dead culms of *Zea mays*, from *Ipomoea batatas*, sugarcane field soil, grassland soil, Tanzania, Egypt, Mauritius, Sask., Canada (IMI 22520, holotype, 150326, 205591, 205592, 205593, 151140, 84557).

P. SACCHARI (Ell. & Ev.) Sutton, *Mycol. Pap.* 97: 26 (1964). (Fig. 389A)
Trullula sacchari Ell. & Ev. apud Cockerell, *J. Inst. Jamaica* **1**: 159 (1892).
Pleocyta sacchari (Massee) Petrak & H. Syd., *Beih. rep. spec. nov. regni veg.* **42**: 455 (1927).
Melanconium sacchari Massee apud Speg., *Rev. Fac. Agron., La Plata* **2**: 242 (1896).
Coniothyrium sacchari (Massee) Prill. & Delacr., *Bull. trimest. Soc. mycol. Fr.* **13**: 113 (1897).
Microsphaeropsis bakeri Syd., *Annls mycol.* **14**: 369 (1916).
Haplosporella melanconioides Sacc., *Atti Accad. Ven.-Trent.-Istr.* **10**: 73 (1917).

Conidiomata up to 650 µ long × 350 µ wide. Conidiophores 5-20 × 1.5-2 µ. Conidia cylindrical, pale brown, 9-14.5 × 3.5-5 µ.

Causing rind disease of sugarcane (Sutton & Waterston, 1966).

On culms of *Saccharum officinarum, S. munja, S.* sp., *Chasmopodium caudatum, Sorghum vulgare, Apluda mutica, Zea mays*, indet. Gramineae, Peru, Venezuela, Jamaica, Cuba, Trinidad, Sierra Leone, Nigeria, Tanzania, Kenya, Uganda, Malawi, S. Africa, India, Pakistan, Sri Lanka, Papua New Guinea, Qd., Australia, Noumea (92918 ex NY, holotype of *T. sacchari*, 103346b, 103369, 210015, 185762, 165931, 98402, 185720, 131842, 126180, 100190, 210044, 151299, 148860, 147113, 144461, 131823, 131819, 138177, 141794, 156346, 136540, 126164, 120394, 113476, 119346, 34459a, 102804, 22688, 63708, 22682, 89241, 34472, 22685, 22686, 22687, 22683, 104271, 22689, 22684, 82464, 81146).

P. sacchari (Ell. & Ev.) Sutton var. **calami** (Panwar & Bohra) comb. nov. (Fig. 389C)
Phaeocytostroma calami Panwar & Bohra, *Curr. Sci.* **41**: 787 (1972).

Conidiomata up to 1000 µ diam., often with black dendritic hyphae radiating from the ostiole. Conidiophores up to 21 µ long × 2-3 µ wide. Conidia pale brown, cylindrical, 8-9 × 2.5-3 µ. Paraphyses up to 40 µ long × 2-2.5 µ wide, branched.

On dead culm of *Calamus rotang*, India (IMI 162201, holotype).

P. SACCHARI (Ell. & Ev.) Sutton var. PENNISETI Sutton, *Mycol. Pap.* 97: 28 (1964). (Fig. 389B)

Conidiomata up to 450 µ long × 300 µ wide. Conidiophores 10-25 × 2.5-4 µ. Conidia pale brown, cylindrical, 15.5-19 × 4-5 µ. Paraphyses 40-70 × 2 µ.

On culms of *Pennisetum purpureum*, Zambia (IMI 93900, holotype).

124. HENDERSONINA Butler apud Butler & Khan, *Mem. Dept. Agric. Ind.* **6**: 198 (1913).
Sp. typ.: *H. sacchari* Butler

Mycelium immersed, branched, septate, hyaline to pale brown. *Conidiomata* eustromatic, immersed to semi-erumpent, separate or aggregated, unilocular or irregularly multilocular and convoluted, dark brown to black, papillate, ± globose or conical; walls composed of brown, thin-walled textura angularis, dark brown on the outside, becoming paler towards the conidiogenous region and the interlocular walls. *Ostiole* indistinct, dehiscence irregular. *Conidiophores* septate, branched at the base and above, formed from the inner cells of the locular walls. *Conidiogenous cells* (not observed in authentic material, description only from unnamed congeneric material on *Erianthus*) enteroblastic, phialidic, determinate, integrated or discrete, cylindrical to lageniform, collarette and channel minute, periclinal wall thickened. *Conidia* smooth, 0-1 euseptate, evenly brown, fusiform to ± cylindrical, with a dark scar at each end, eguttulate.

Material identified as *H. sacchari* by Dr. E.J. Butler (IMI 60370) although plentiful as dried cultures, does not show either stages in conidiogenesis or any secondary conidia. Butler & Khan (1913) reported hyaline, filiform, straight or curved, aseptate secondary conidia. Also the conidial size in IMI 60370 of 21-28 × 5.5-9.5 µ is in excess of that quoted in the original account. Other collections from *Saccharum* and *Erianthus* conform more closely to Butler's account but still do not have secondary conidia. There are therefore grounds for reticence over the extent to which *H. sacchari* varies both in morphology and the hosts it is capable of colonizing.

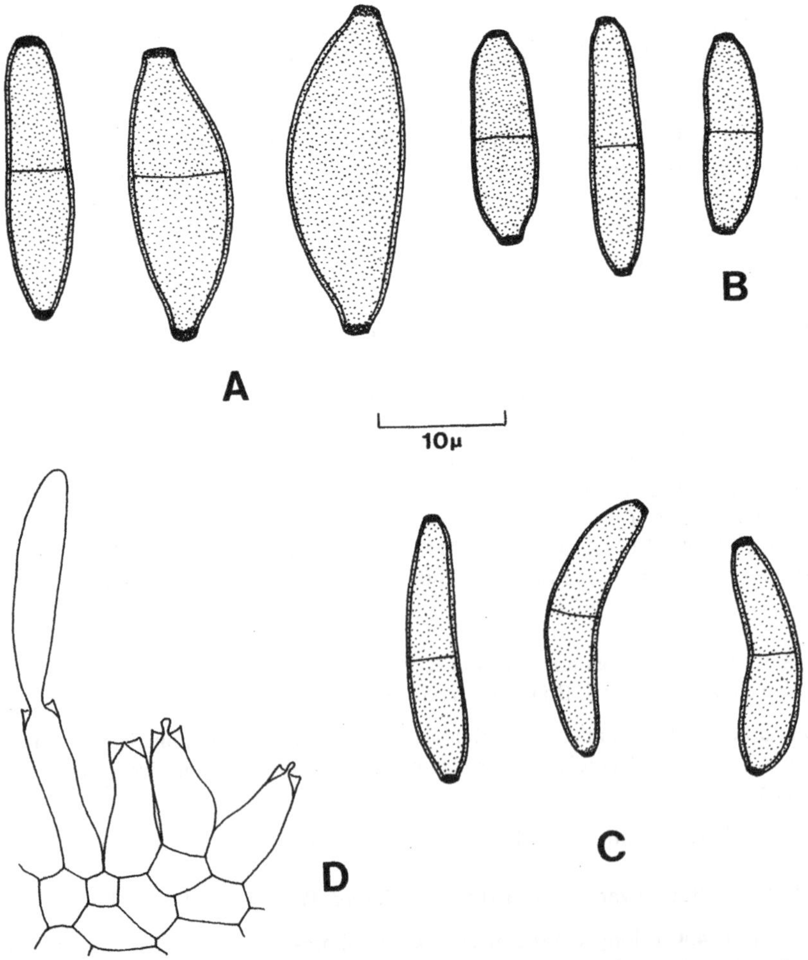

Fig. 390. *Hendersonina sacchari*, A, conidia (IMI 60370); B, conidia (IMI 62555); C, conidia (IMI 82322); D, conidiogenous cells and developing conidia.

H. SACCHARI Butler, *ibid.*: 198 (1913). (Fig. 390)

Conidiomata 1-2 mm diam. Conidia 21-28 × 5.5-9.5 (IMI 60370), 19-29 × 4-5 µ (IMI 82322), 17-24 × 4-5 µ (IMI 62555) (15-24 × 3.75-5 µ in original account).

From *Saccharum officinarum*, *Erianthus ravennae*, India, Bangladesh, Pakistan (IMI 60370, 62555, 82322).

125. PLACODIPLODIA Bub., *Ber. dtsch. Bot. Ges.* **34**: 305 (1916).
Sp. typ.: *P. copelandii* Bub.

Mycelium immersed, pale brown to hyaline, branched, septate. *Conidiomata* eustromatic, immersed, erumpent, separate or aggregated and linearly confluent, dark brown to black, navicular, unilocular; upper wall of dark brown, thick-walled textura angularis, lower wall of similar but hyaline, thin-walled tissue. *Ostiole* absent, dehiscence by a longitudinal fissure. *Conidiophores* absent. *Conidiogenous cells* enteroblastic, phialidic, determinate, discrete, ampulliform to doliiform, hyaline, smooth, collarette and channel minute. *Conidia* pale brown, medianly 1 euseptate, cylindrical to ellipsoid, both ends obtuse, smooth, eguttulate.

The holotype of *P. copelandii* was too old for full details of conidiogenesis to be elucidated. One phialide was found. The genus is monotypic.

P. COPELANDII Bub., *ibid.*: 305 (1916). (Fig. 391)

Conidiomata up to 450 µ long × 150 µ wide. Conidiogenous cell 5 × 5 µ. Conidia 10-13 × 3.5-4 µ.

On stems of *Cyathea* sp., Philippines (IMI 194253, holotype ex BPI).

126. ENDOCORYNEUM Petrak, *Annls mycol.* **20**: 334 (1922).
Sp. typ.: *E. loculosum* (Sacc.) Petrak (syn. *Coryneum loculosum* Sacc.).

Mycelium immersed, hyaline to pale brown, branched, septate. *Conidiomata* eustromatic, immersed, subperidermal, erumpent, separate, oval to elongated, dark brown, multilocular, locules at same level; walls of loose, branched, septate, smooth, thin-walled, hyaline textura intricata except at the periphery where tissue is textura angularis, dark brown and

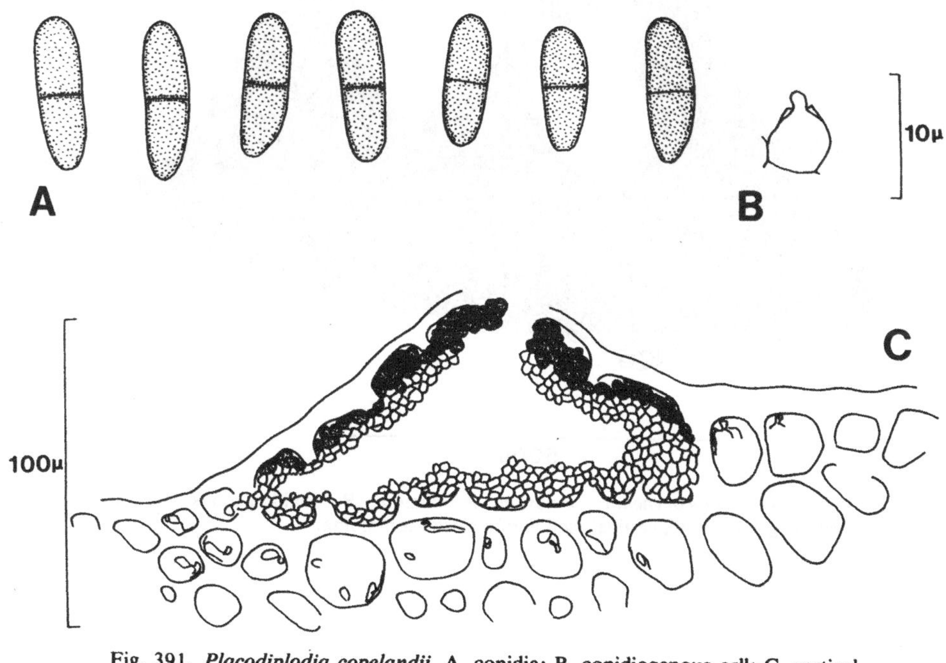

Fig. 391. *Placodiplodia copelandii*, A, conidia; B, conidiogenous cell; C, vertical section of a conidioma.

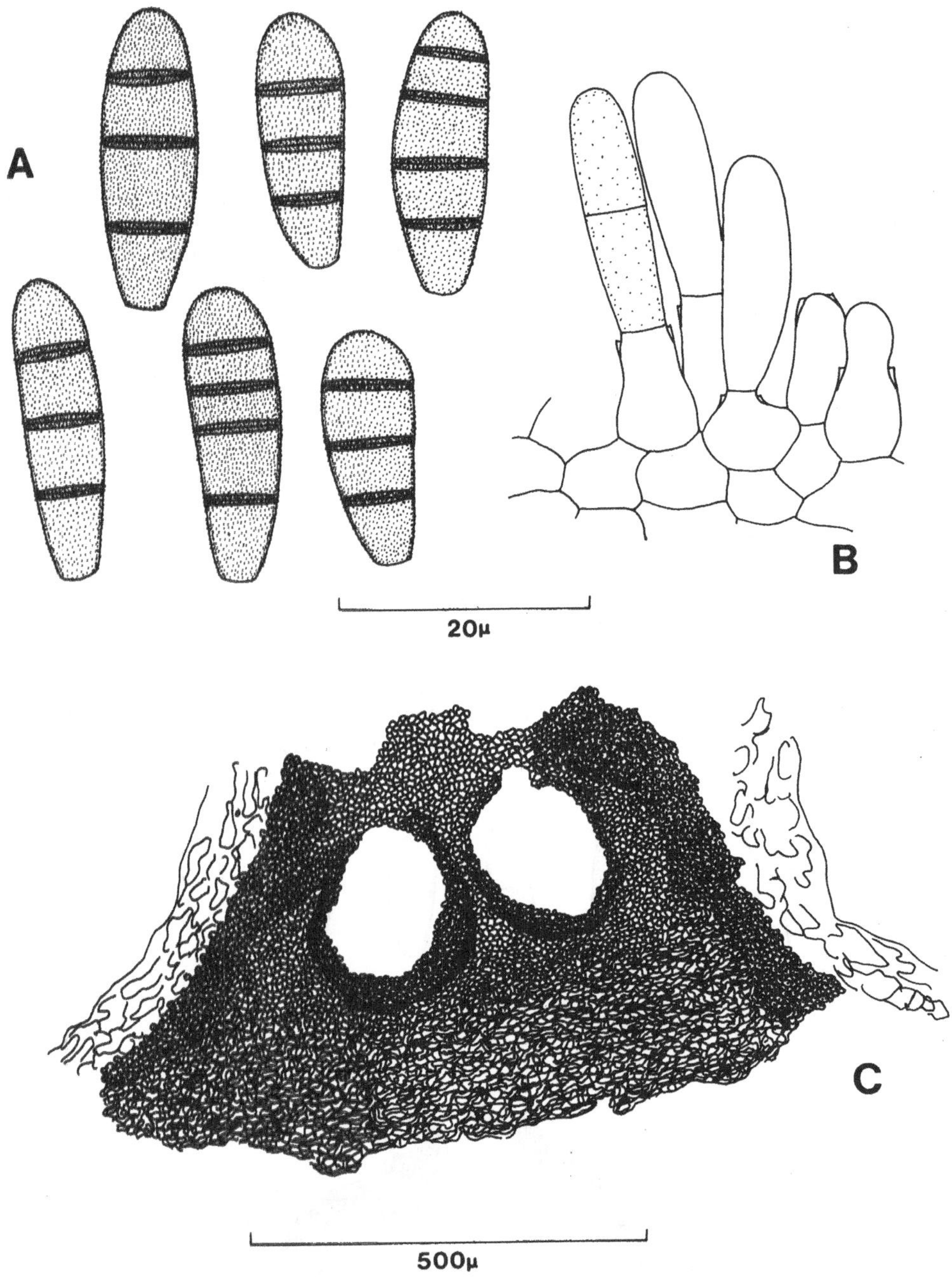

Fig. 392. *Endocoryneum loculosum*, A, conidia; B, conidiogenous cells and developing conidia; C, vertical section of a closed conidioma.

compact; locules globose, each with an ostiole. *Conidiophores* absent. *Conidiogenous cells* enteroblastic, phialidic, discrete, determinate, doliiform to cylindrical, hyaline, collarette minimal, channel wide, formed from cells lining the locules. *Conidia* straight, cylindrical to clavate, 3(-4) euseptate, continuous, obtuse at the apex, base truncate, verruculose, pale brown, septa darker.

The only species was illustrated by Sutton (1975a).

E. LOCULOSUM (Sacc.) Petrak, *ibid.*: 334 (1922). (Fig. 392)

Conidiomata 1700 µ long × 900 µ wide × 700 µ deep. Locules 210 µ diam. Conidiogenous cells 6.5-10.5 × 4-8 µ. Conidia 19-25 × 6.5-8 µ.

On branches of *Fraxinus excelsior*, Austria (IMI 179081, holotype ex PAD).

127. HENDERSONIOPSIS Höhn., *Annls mycol.* **16**: 123 (1918).
Sp. typ.: *H. thelebola* (Sacc.) Höhn. (syn. *Stilbospora thelebola* Sacc.).

Mycelium immersed, branched, septate, pale brown to hyaline. *Conidiomata* eustromatic, separate, ± pulvinate, dark brown, subperidermal, multilocular and convoluted, often mixed with ectostroma of the perfect state; locular wall tissue of irregular, pale to medium brown, thin-walled textura angularis with thin-walled, hyaline conidiogenous tissue towards the inside, often surrounded by pseudostromatic tissue. Dehiscence by a central ostiole surrounded by cream white tissue, conidial mass dark olivaceous green. *Macroconidiophores* hyaline to pale brown, smooth, branched and septate only at the base, cylindrical, formed from the inner cells of the locular walls. *Macroconidiogenous cells* enteroblastic, occasionally percurrent, indeterminate, integrated or discrete, smooth, pale brown to hyaline, with a prominent flared collarette and percurrently proliferating to give 1-3 additional collarettes, distinct channel and periclinal thickening. *Macroconidia* pale brown, smooth, fusiform, base truncate, apex obtuse to conic, often slightly curved, 2-7 euseptate. *Microconidiophores* hyaline, smooth, branched and septate mainly at the base, tapered to the apices, formed from the inner locular wall cells. *Microconidiogenous cells* enteroblastic, phialidic, integrated or discrete, channel and collarette minute. *Microconidia* hyaline, aseptate, smooth, falcate, fusiform, eguttulate.

The correlation between stages suggested by Tulasne & Tulasne (1863) was later proved culturally by Wehmeyer (1938). A full account of the species was provided by Wehmeyer (1941). The genus contains a single species.

H. THELEBOLA (Sacc.) Höhn., *ibid.*: 124 (1918). (Fig. 393)
Stilbospora thelebola Sacc., *Michelia* **2**: 542 (1882).

Teleomorphosis: *Melanconis thelebola* (Fr.) Sacc., *Syll. fung.* **1**: 605 (1882).

Conidiomata up to 1200 µ wide × 950 µ deep. Macroconidiophores up to 40 µ long × 4.5-6 µ. Macroconidia 31-53 × 10-12 µ. Microconidiophores up to 21 µ long × 1.5-2 µ wide. Microconidia 6.5-8 × 1.5-2 µ.

On twigs and branches of *Alnus incana*, *A. glutinosa*, *A. cordata*, *A.* spp., G.B., Austria, Latvia, Ont., Canada, U.S.S.R. (IMI 200663, 16712, 58629, 19092a, 58396, 58627, 58628, 16704, 94618, 26965, 24106, 11664, 57159, 30352, 105287b).

128. MASSARIOTHEA Syd., *Annls mycol.* **37**: 249 (1939).
Apocoryneum Sutton, *Mycol. Pap.* 138: 148 (1975).
Sp. typ.: *M. themedae* Syd.

Mycelium superficial or immersed, branched, septate, brown. *Conidiomata* eustromatic, separate or aggregated, globose to rostrate, dark brown to black, immersed or superficial, unilocular or multilocular, thick-walled; wall of dark brown thick-walled textura angularis to intricata on the outside and thin-walled, pale brown to hyaline textura angularis on the inside. *Ostiole* central, circular, sometimes papillate to rostrate. *Conidiogenous cells* enteroblastic, phialidic, determinate, discrete, doliiform to lageniform, hyaline, smooth, collarette minute, channel ± wide, formed from the inner cells of the locular and pycnidial walls. *Conidia* brown, 3-7 distoseptate with markedly reduced lumina, continuous, straight or curved, base truncate, apex obtuse, smooth, cylindrical, both ends with a thickened periclinal wall. *Paraphyses* branched, septate, hyaline, filiform.

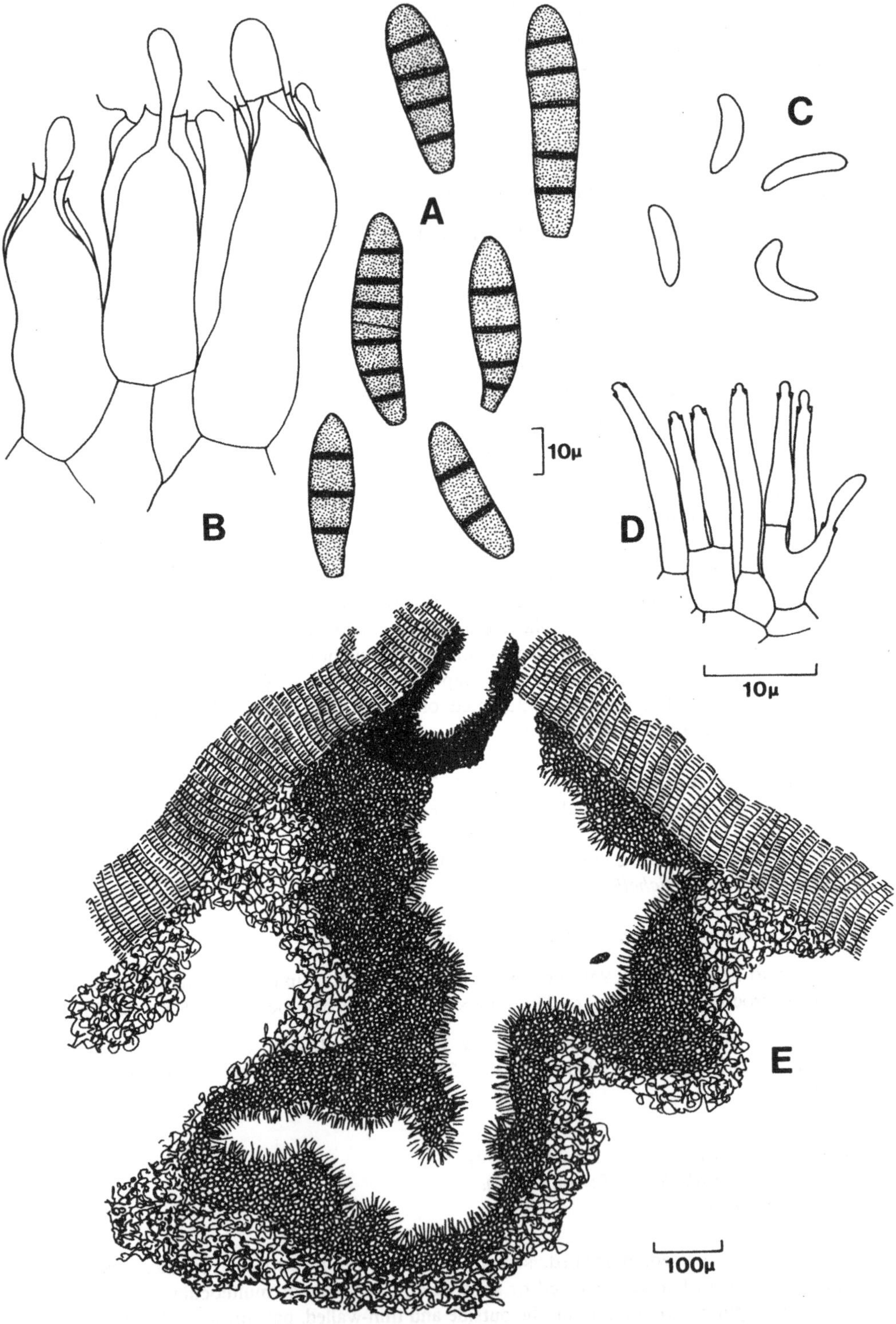

Fig. 393. *Hendersoniopsis thelebola*, A, macroconidia; B, macroconidiogenous cells; C, microconidia; D, microconidiogenous cells; E, vertical section of a conidioma.

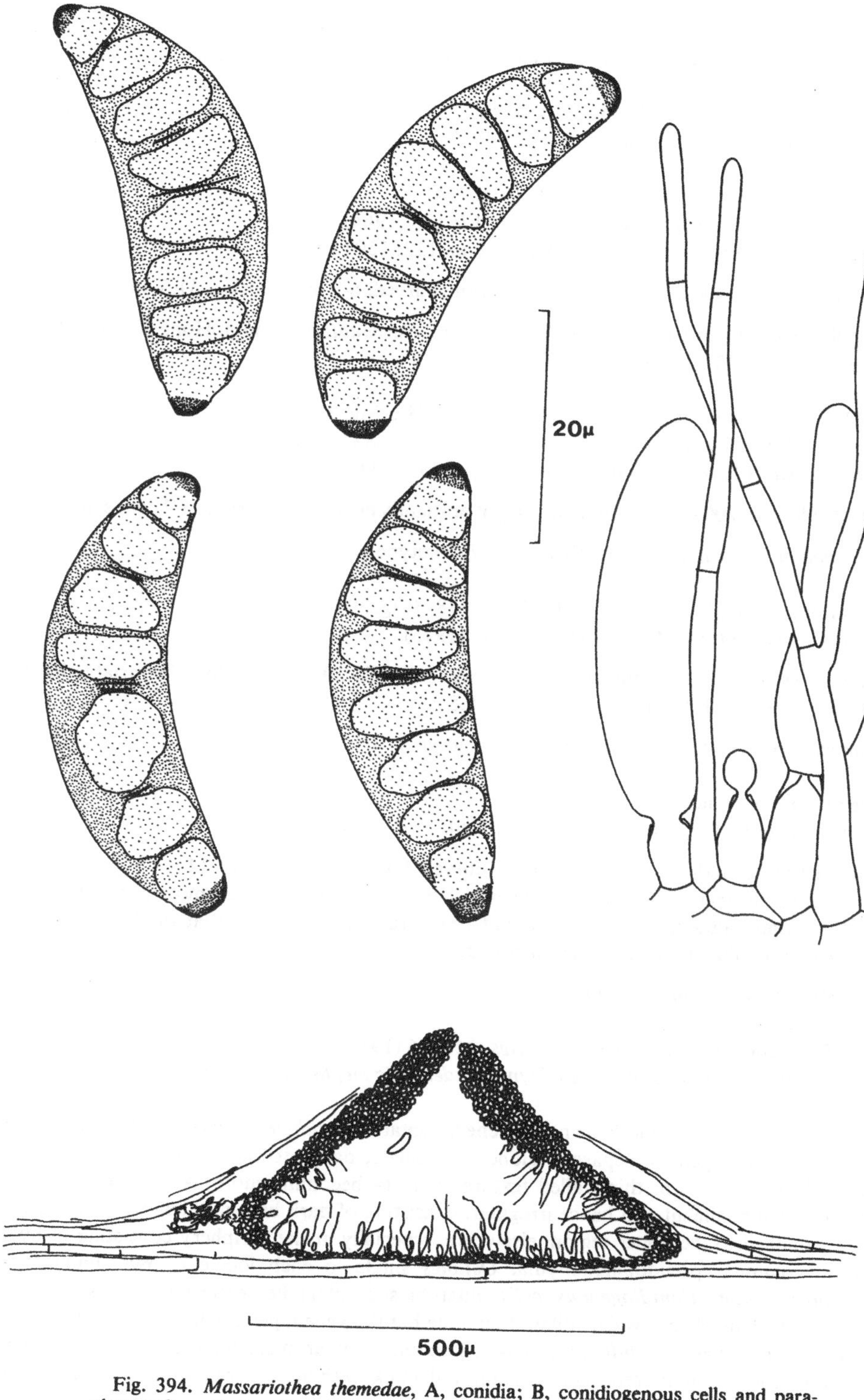

Fig. 394. *Massariothea themedae*, A, conidia; B, conidiogenous cells and paraphyses; C, vertical section of a conidioma.

KEY

1. Conidiomata multilocular, rostrate; conidia variable in septation .... *M. shawiae*
1. Conidiomata unilocular, ostiole rarely papillate; conidia 5-7 distoseptate 2
   2. Conidiomata 200 µ diam. Conidia 48-69 × 12.5 µ .... *M. paspali*
   2. Conidiomata 600 µ diam. Conidia 31.5-44 × 10.5-12 µ .... *M. themedae*
   2. Conidiomata 300-450 µ diam. Conidia 40-52 × 12-14 µ .... *M. botulispora*

M. THEMEDAE Syd., *Annls mycol.* **37**: 249 (1939). (Fig. 394)
Stilbospora graminicola Shukla, *Indian Phytopath.* **17**: 178 (1964).
Apocoryneum graminicola (Shukla) Sutton, *Mycol. Pap.* 138: 150 (1975).

Conidiomata flattened, 600 µ diam. Conidia fusiform, curved, 6-7 distoseptate, 31.5-44 × 10.5-12 µ.

On leaves of *Sorghum vulgare*, Gramineae, *Themeda* sp., *Neyraudia* sp., India, Hong Kong (IMI 74895, holotype *S. graminicola*, 100043, 204509, 230260).

M. PASPALI (Ell. & Ev.) Sutton, *Nova Hedwigia* **29**: 785 (1978). (Fig. 395)
Coryneum paspali Ell. & Ev., *Proc. Acad. nat. Sci. Phil. 1891*: 82 (1891).
Apocoryneum paspali (Ell. & Ev.) Sutton, *Mycol. Pap.* 138: 148 (1975).

Conidiomata globose, 200 µ diam. Conidia straight or slightly curved, cylindrical, 6-7 distoseptate, 48-69 × 12.5 µ.

On culms of *Paspalum platycaule*, U.S.A. (IMI 178767, holotype).

M. SHAWIAE (Sutton) Sutton, *Nova Hedwigia* **29**: 785 (1978). (Fig. 396)
Apocoryneum shawiae Sutton, *Mycol. Pap.* 138: 153 (1975).

Conidiomata 600 µ diam.; multilocular. Conidiogenous cells 50 × 2-3 µ. Conidia variable in shape, gradually tapered to the apex, 3-7 distoseptate, 31-52 × 7.5-8.5 µ.

From *Sorghum vulgare*, New Guinea (IMI 167141e, holotype).

M. botulispora (Teng) comb. nov.
Hendersonia botulispora Teng, *Sinensia* **7**: 769 (1936).

Conidiomata 300-450 µ diam. Conidia strongly curved, fusiform, 40-52 × 12-14 µ.

The original illustrations show this species to have slightly wider, much more strongly curved conidia than existing species, and although the type has not been examined it is clear that the fungus should be placed in *Massariothea* rather than retained in *Hendersonia*.

On leaves of *Phragmites communis*, China.

129. CAMAROGRAPHIUM Bub., *Ber. dtsch. Bot. Ges.* **34**: 306 (1916).
Sp. typ.: *C. stephensii* (Berk. & Br.) Bub. (syn. *Hendersonia stephensii* Berk. & Br.).

Mycelium immersed, hyaline to pale brown, branched, septate. *Conidiomata* like solitary pycnidia or aggregated loosely into linear eustromata; pycnidia separate, globose, papillate, dark brown, immersed, subepidermal, unilocular, thick-walled, composed of pale brown, thin-walled textura intricata becoming more like textura angularis towards the inside; stromata linear, often confluent and hence irregular, consisting of dark brown, thick-walled textura angularis connecting the pycnidia, confined to the upper region, elsewhere, especially at the periphery consisting of hyaline, irregular, loosely aggregated to columnar pseudoparenchyma. *Ostioles* single to each locule or pycnidium, central, circular, papillate. *Conidiophores* absent. *Conidiogenous cells* enteroblastic, phialidic, determinate, discrete, ampulliform to doliiform, hyaline, smooth, channel very wide, collarette minute but thickened, periclinal wall thick, formed from the inner cells of the pycnidial or locular walls. *Conidia* pale brown, smooth, irregular in shape, ellipsoid, globose, obpyriform, base ± obtuse, guttulate, distoseptate, lumina reduced, periclinal wall thick, with up to 5 transverse septa and many longitudinal or oblique septa.

Grove (1937) accepted 3 species in *Camarographium*; apart from the type, they are *C. metableticum* (Trail) Grove and *C. abietis* (Wils. & Anders.) Grove. On the basis of conidiomatal structure alone they are not congeneric with *C.*

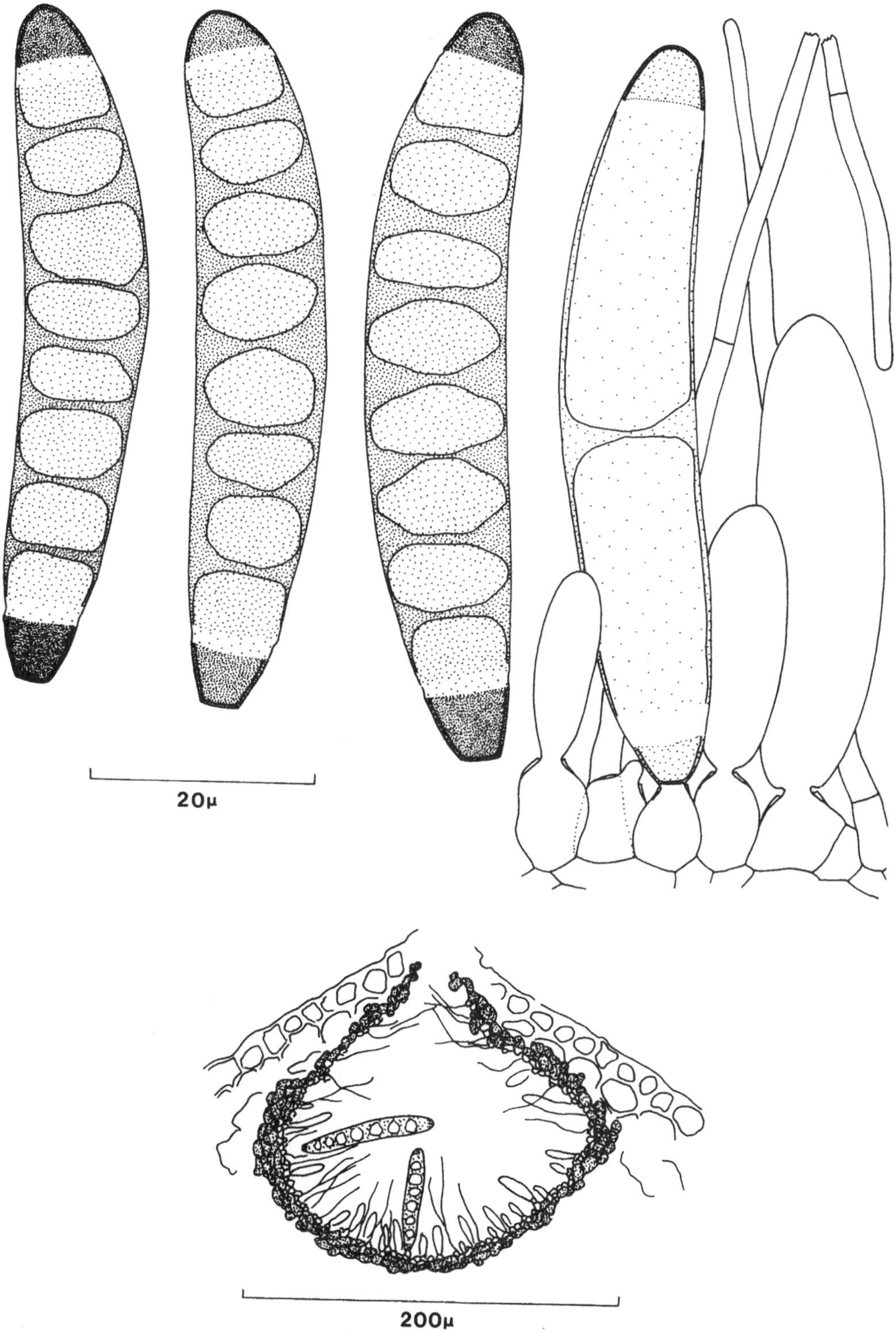

Fig. 395. *Massariothea paspali*, A, conidia; B, conidiogenous cells and paraphyses; C, vertical section of a conidioma.

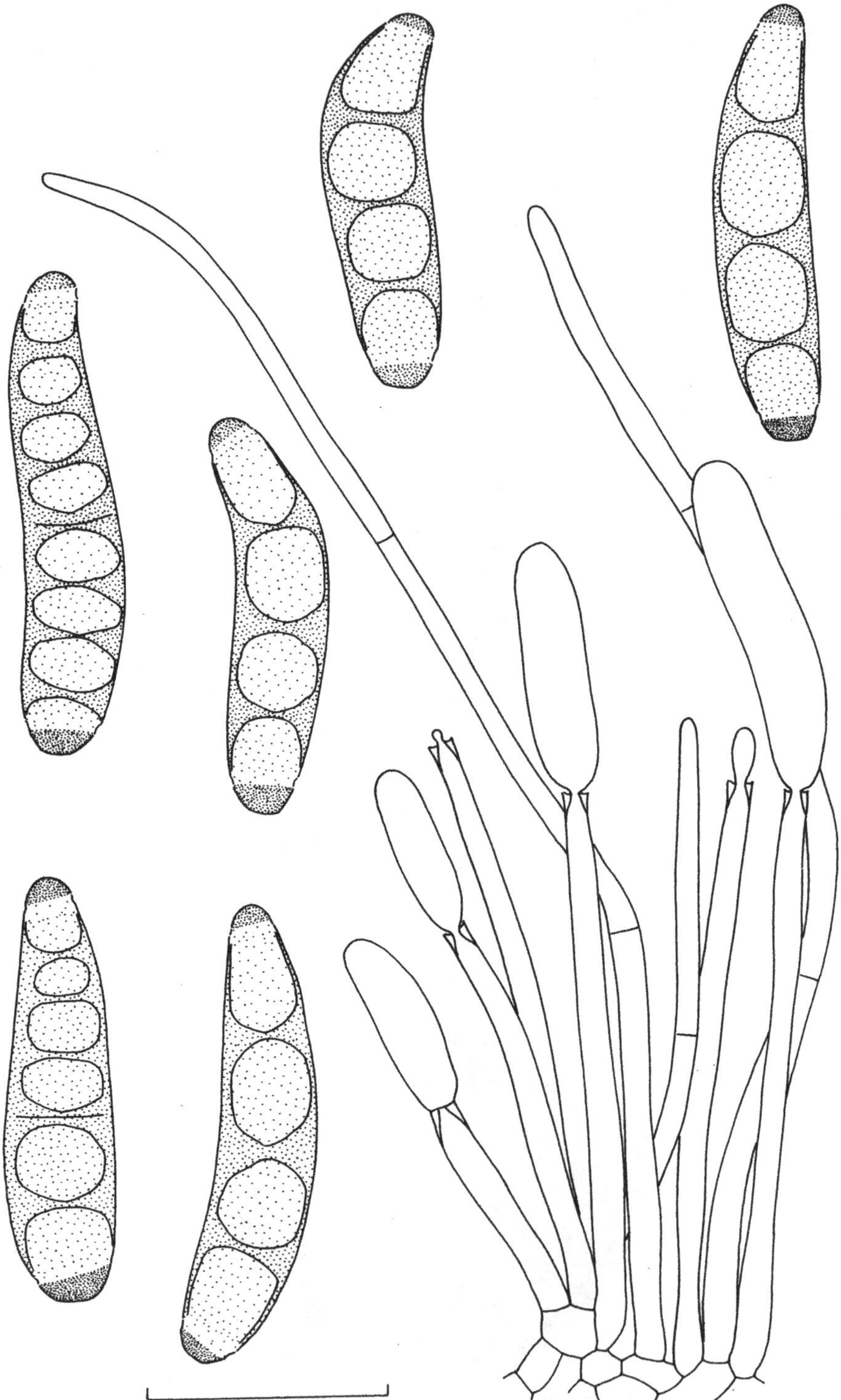

Fig. 396. *Massariothea shawiae*, A, conidia; B, conidiogenous cells and paraphyses.

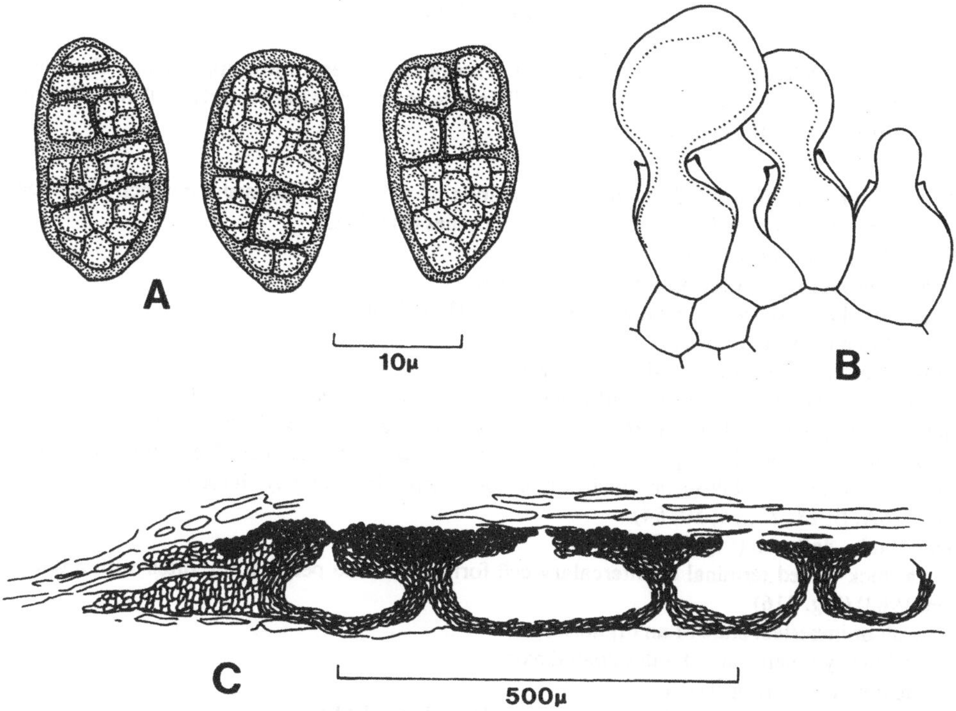

Fig. 397. *Camarographium stephensii*, A, conidia; B, conidiogenous cells and developing conidia; C, vertical section of a conidiomata.

stephensii, but it is premature to either place them in existing genera or propose new generic names for them until many more generic names based on brown muriform conidia have been revised.

C. STEPHENSII (Berk. & Br.) Bub., *ibid.*: 306 (1916). (Fig. 397)
Hendersonia stephensii Berk. & Br., *Ann. Hist. nat.* **7**: 95 (1851).
Camarosporium stephensii (Berk. & Br.) Sacc., *Syll. fung.* **3**: 469 (1884).

Conidiomata and stromatic locules up to 300 μ diam. Conidiogenous cells 10-12 × 8-11 μ. Conidia 40-52 × 22-28 μ.

On petioles of *Pteridium aquilinum*, U.K., Yugoslavia (IMI 194180 det. Bubák, 15472, 27476a, 49063, 49760, 181543, 107249).

GLOSSARY

The glossary includes definitions of terms used in the descriptions of genera and species, and also the characters used in both dichotomous and synoptic keys. Examples which illustrate some terms particularly well are indicated by citation of relevant figure number following the definitions.

acervulus (adj. acervular) – an immersed conidioma consisting of a flat layer of pseudoparenchyma upon which conidia are initiated and produced while still covered by the host tissues (Fig. 72C)
acrogenous (conidia) – formed at the apex (Fig. 29B)
acropetal (conidia) – produced in succession towards the apex
acropleurogenous (conidia) – formed at the apex and the sides (Fig. 328B)
acuminate – tapered to a point (Fig. 13G)
allantoid (conidia) – sausage-shaped, curved (Fig. 352A)
ampulliform (conidiogenous cell) – globose to subglobose with a very short papilla (Fig. 10B)
anamorph (anamorphosis) – asexual or imperfect state of a fungus (Hennebert & Weresub, 1977)
annellide (adj. annellidic) – a holoblastic percurrently-proliferating conidiogenous cell (Kendrick, 1971) (Fig. 144B)
appendage (conidia) – cellular, setulose or mucilaginous ornamentation, usually filiform but sometimes infundibular or capitate (Figs. 16A, B; 18; 22; 24; 50; 54)
applanate (conidioma) – flattened (Fig. 71D)
appressorium – a thick-walled terminal or intercalary cell formed prior to penetration, often with a circular germination pore (Sutton, 1963) (Fig. 316)
asexual – anamorph or imperfect state of an organism
asymmetrical – lacking symmetry about a designated axis
attenuated – gradually tapered (Fig. 136A)
bacillar (conidia) – cylindrical, minute, extremely narrow (<1 μ) (Fig. 218A)
basipetal (conidia) – produced in succession towards the base (Fig. 5B)
biconic (conidia) – conical at each end (Fig. 383A)
black-banded (conidia) – septa which are dark brown to black and much thicker than normal septa (Fig. 76A)
blastic (conidiogenesis) – blown out, with enlargement of the conidial initial before delimitation by a septum (Kendrick, 1971)
capitate (appendage) – mucilaginous cap-like appendage (Fig. 76A)
catenate (conidia) – developing in chains (Fig. 4A, B, C)
cellular (appendage) – an appendage, at least when immature, with cellular contents and a cell wall, later frequently lacking contents; originates as an outgrowth of the cell and maintains cellular continuity, although sometimes delimited by a septum (Figs. 24A; 25)
channel (phialide) – zone at the phialide apex in which the area occupied by the protoplast is usually of considerably reduced diameter and surrounded by periclinal thickening: the conidial primordium emerges through the channel to produce a new conidium (Fig. 213B)
cheiroid (conidia) – approximately the shape of a hand
clavate – club-shaped, wider towards the apex than the base (Fig. 141A)
clypeus (adj. clypeate) – shield-shaped pseudoparenchymatic tissue around an ostiole, often of larger diameter than the conidioma (Fig. 258C)
coelomycetes – collective term for anamorphic fungi with pycnidial, acervular, stromatic or pycnothyrial conidiomata.
collabent (conidioma) – collapsed in the middle
collarette (phialide) – the remains of the outer conidiogenous cell wall left on the conidiogenous cell after secession of the first conidium (Fig. 249C)
concolourous (conidia) – cells of even pigmentation (Figs. 32A; 40A)
confluent (conidiomata) – becoming joined by hyphae, cells or stromatic tissue (Fig. 397C)
conidiogenesis – the process by which conidia are formed from conidiogenous cells
conidiogenous cell – the actual cell from which a conidium is formed, rather than the cells which support it which then comprise the conidiophore (Figs. 10B; 37C; 70B; 234B (integrated c. cells))
conidioma – any hyphal structure which bears conidia, e.g. separate conidiophore, synnema, acervulus, pycnidium, sporodochium etc.
conidiophore – a hypha or cell which bears a conidiogenous cell (Fig. 29B)
conidium – an asexual spore which is liberated from a conidiogenous cell
constricted (conidia) – periclinal wall indented to varying degrees where it meets transverse septa (Fig. 8A)

continuous (conidia) – not constricted at the transverse septa (Fig. 12A)
convoluted – a conidioma in which one or more cavities are of irregular shape and imperfectly separated (Fig. 95C)
corniform (appendage) – horn-shaped, when branched with several branches on one side (Fig. 27A)
cuneiform (conidia) – wedge-shaped (Fig. 199A, B)
cupulate (conidioma) – cup-shaped (Figs. 132C; 271C)
cylindrical – round in cross section and of equal width throughout its length (Fig. 4B)
cymbiform (conidia) – broadly reniform (Fig. 96A)
deltoid (conidia) – triangular in shape (Fig. 255A)
dendroid – with many lateral irregular branches (Fig. 48C)
denticle – a cylindrical projection on which a conidium is formed, a term usually used in sympodial conidiogenesis (Fig. 139B, D)
depressed – sunken (Fig. 15C)
determinate (conidiogenous cell) – ceasing to grow in length on production of a terminal or apical conidium (Fig. 10B)
deuteromycete – an anamorphic (asexual, imperfect) fungus
dichotomous – branching, often successively into two approximately equal arms (Fig. 148A-C)
dichotomous key – natural key with two or more choices to consider at any point leading to the next choice, and eventually, the name of the taxon; supposedly reflecting a degree of phylogenetic relationship between taxa
digitate (conidia) – comprised of finger-like processes (Fig. 67A)
discoid (conidioma) – flattened cupulate (Fig. 85C)
discrete (conidiogenous cell) – not subtended by a conidiophore (Fig. 10B)
disjunctor cell – a thin-walled cell produced between developing conidia, and by fragmentation of which conidia secede (Fig. 9B)
distoseptate – of a conidium in which eusepta are lacking, lumina are reduced and surrounded by sac-like walls distinct from the outer wall (Luttrell, 1963); Ellis (1971, 1976) has used pseudoseptate (Fig. 208)
doliiform (conidiogenous cell) – barrel-shaped to broadly subglobose (Figs. 16C; 17E; 28B)
effuse – spread out
eguttulate (conidia) – lacking guttules (Fig. 3A)
endogenous (basal appendage) – originating inside the conidiogenous cell and therefore truly basal (Fig. 149B)
enteroblastic (conidiogenesis) – blastic with only the inner wall of the conidiogenous cell contributing towards formation of the conidial wall (Fig. 213B)
enterothallic (conidiogenesis) – thallic with only the inner wall of the conidiogenous cell contributing towards formation of the conidial wall
euseptate – of a conidium having true multilayered septa, lumina not reduced (Luttrell, 1963) (Figs. 5A; 7A-D)
eustroma – a mass of fungal cells or closely interwoven hyphae comprising a distinctly structured tissue, but *not* applied to either simple pycnidia or acervuli in the context of this account
eustroma, simple – unilocular stroma, only very rarely slightly convoluted (Fig. 94C)
eustroma, complex – multilocular stroma, either convoluted or with distinctly separate regularly or irregularly arranged locules (Fig. 99C)
eustroma, complex, effuse – multilocular effuse stroma with distinctly separate locules (Fig. 95C)
eustroma, complex, pulvinate – superficial ± cushion-shaped multilocular stroma, either convoluted or with distinctly separate regularly or irregularly arranged locules (Fig. 80D)
excentric – displaced to one side (Fig. 144A)
exogenous (basal appendage) – originating outside the conidiogenous cell and therefore suprabasally inserted (Fig. 157B, C, F)
extracellular (appendages) – formed outside the conidial wall and therefore discontinuous with it (Fig. 18)
falcate (conidia) – curved and tapered to each end (Fig. 42A-D)
filiform (conidia) – slender, thread-like (Fig. 41A)
fimbriate (conidioma) – with free hyphae at the apex (Fig. 360C)
furfuraceous – powdery, scurfy (Fig. 94C)
fusiform (conidia) – tapered to each end (Fig. 48A)
glabrous – smooth
guttulate (conidia) – one or more spherical oil droplets within a cell (Fig. 2A)
hamate (conidia) – hooked at the apex (Fig. 120A)
holoblastic (conidiogenesis) – blastic with both outer and inner walls of the conidiogenous cell contributing towards formation of the conidial wall (Kendrick, 1971)
holomorph (holomorphosis) – a whole fungus in all its states (Hennebert & Weresub, 1977)

holothallic (conidiogenesis) – thallic with both outer and inner walls of the conidiogenous cell contributing towards formation of the conidial wall (Fig. 4A-C)
hyphopodium – a short, lateral appressorium which adheres to the surface of the substrate
hypostroma – used for the pseudoparenchyma at the base of superficial conidiomata by which they are attached to the substrate, and, for the fungal tissue immersed in the substrate in pycnothyrial genera (Fig. 188D)
immersed (mycelium, conidioma) – enclosed or formed in either fungal or host tissue (Fig. 13G)
imperfect – anamorph or asexual state of a fungus
infundibuliform (conidioma, mucilaginous appendage) – funnel-shaped (Fig. 18B, E)
integrated (conidiogenous cell) – subtended by a conidiophore consisting of a single cell or several cells (Fig. 5B)
labiate – lobed, or having lips, hence bilatiate (2 lobes), trilabiate (3 lobes), etc.
lageniform (conidiogenous cell) – broad at the base, ± abruptly tapered towards the apex (flask-shaped) (Fig. 301B)
linear (conidioma) – longer than broad
locule (conidioma) – cavity enclosed by fungal, host or fungal/host tissue
locus (conidiogenous) – the specific area on a conidiogenous cell at which conidial formation occurs
longitudinal (conidioma) – longer than broad
lumen (cell) – the protoplasmic membrane and enclosed protoplast (Fig. 208)
mammillate (conidium) – breast-shaped, apex conical with a papillate tip
median – in the middle (Fig. 8A)
monoblastic (conidiogenous cell) – holoblastic with one conidiogenous locus (Fig. 12B)
monophialidic (conidiogenous cell) – phialide with a single conidiogenous locus (Fig. 215B)
multilocular – a conidioma in which there are several separate cavities (Fig. 74C)
multiseptate (conidia) – with several septa (Fig. 12A)
muriform (conidia) – with transverse and longitudinal septa (Fig. 7C, D)
mycelium – a group of discrete hyphae comprising the vegetative state of a fungus
obpyriform (conidia) – shape of a pear upside down, with the broad end near the base
operculate (dehiscence) – conidiomata dehiscing by a circular flap of tissue produced by circumscissile rupture of overlying tissues
ostiole – a circular pore in a conidioma through which conidia are dispersed after secession from conidiogenous cells (Figs. 11C; 215C)
papillate (ostiole) – nipple-like (Fig. 14C)
paraphysis – sterile hypha which is free at the apex; often produced amongst fertile conidiophores or conidiogenous cells (Fig. 22C)
percurrent (conidiogenous cell) – growing straight on through the open end left by the secession of the preceding conidium (Fig. 42E)
perfect – teleomorph or sexual state of a fungus
periclinal thickening (of phialide) – unstainable region surrounding the phialidic channel, referred to as the annulus in *Alternaria* by Campbell (1970), and Knox-Davis (1979) (Fig. 213B)
peripheral (locules) – situated in the outer regions of the stroma and often subtended or supported by a considerable amount of sterile stroma below (Fig. 80D)
persistent (conidiogenous cell) – carried away at the base of the conidium at secession (Fig. 90)
phialide – an enteroblastic conidiogenous cell in which conidia are developed in a new wall which is not derived from any existing layers of the conidiogenous cell wall (Kendrick, 1971) (Fig. 213B)
pleurogenous (conidia) – formed along the sides (Fig. 230B)
polyblastic (conidiogenous cell) – holoblastic with several conidiogenous loci (Fig. 41B)
polyphialidic (conidiogenous cell) – phialide with several conidiogenous loci (Fig. 222E)
protuberate (conidia) – with short projections not qualifying as appendages (Fig. 70A)
pseudostroma – a mass of fungal cells or closely interwoven hyphae combined with host cells to produce a distinct tissue (Fig. 100C)
pycnidium (adj. pycnidial) – a globose to lageniform, ostiolate, brown, thin-walled conidioma composed of textura angularis; usually 2-3 cells thick but sometimes much thicker (Figs. 12C; 14C)
pycnothyrium – superficial, shield-shaped, hemispherical, brown conidioma, sometimes comprised of only an upper wall, sometimes also with a basal wall
pyriform (conidia) – pear-shaped with the broad end at the apex (Fig. 254A)
raphe – specialized longitudinal dehiscence mechanism, only known in *Chaetomella* (Fig. 328C)
reniform (conidia) – kidney-shaped (Fig. 92C)
repent (mycelium) – creeping

reticulate (mycelium) – net-like, anastomosing (Fig. 48C)
reticulate (ornamentation) – marked by lines or ridges which finally appear to be net-like (Fig. 99A)
rostrate (conidia, conidiomata) – beaked and attenuated towards the apex (Figs. 230C; 35A)
rupture, irregular (conidiomata) – dehiscence by irregular breakdown of upper wall, or disorganised fracture of overlying tissues
scar (conidial) – that part of the septum involved in secession which forms the base of the conidium (Fig. 16A)
scyphoid (conidia) – cup-like
semi-immersed (conidioma) – initially immersed, finally the greater part of development appearing to be superficial, or, half superficial – half immersed (Fig. 101C)
seta – a modified hypha, usually thick-walled, free at the apex, produced as ornament to conidiomata or mycelium (Fig. 13G)
setula – a very thin, hair-like conidial appendage which is extracellular in origin (Fig. 268A)
sexual – teleomorph or perfect state of a fungus
sheath – a complete mucilaginous covering of conidia, conidiogenous cells, conidiophores or paraphyses; often splitting or becoming everted to leave appendage/s on the conidia (Fig. 256A, C)
smooth – lacking any ornamentation either on the wall or as appendages
spathulate (appendages) – inflated at the apex, like a spoon
sporodochium – a pulvinate stroma, with conidiophores or conidiogenous cells covering its upper surface which are not covered by the substrate during development
stellate (conidia, dehiscence) – star-shaped (Fig. 65A)
stipitate (conidiomata) – with a stalk (Fig. 230C, D)
subapical – inserted just below the apex (Fig. 144A)
subcuticular (conidiomata) – situated between the cuticle and the upper epidermal wall (Fig. 111C)
subicle – mycelial felt or network of hyphae at the base of conidiomata
superficial (mycelium, conidioma) – occurring on the surface of the substrate not enclosed by host or fungal tissue (Fig. 27D)
suprabasal – inserted just above the basal scar (Fig. 157)
symmetrical – of equal morphology about a designated axis
sympodial (conidiogenous cell) – proliferating, the main axis elongating by growth of a succession of apices, each of which develops behind and to one side of the previous apex where growth had ceased with the production of a conidium (Fig. 120B)
synchronous (conidiogenous cell) – a polyblastic conidiogenous cell in which all the loci develop conidia at the same time, often difficult to distinguish from sympodial development
synnema – a compacted group of erect and fused hyphae, the apices and occasionally intercalary cells of which function conidiogenously
synoptic key – artificial key in which each taxon is given a number; all characters of the taxa are listed, with genus numbers entered for those that are positive; by eliminating successive characters, at random if desired, eventually the single number left gives the name of the taxon
teleomorph (teleomorphosis) – sexual or perfect state of a fungus (Hennebert & Weresub, 1977)
tentacular – several filiform mucilaginous appendages originating from more or less the same point (Fig. 18A)
textura – tissue structure of conidiomata (Korf, 1973)
thallic (conidiogenesis) – conidiogenesis in which conidium development occurs without enlargement of the conidial initial, or when enlargement does occur it takes place after delimitation by a septum or septa (Fig. 14A-C)
tubular (conidioma) – cylindrical, not tapered to the apex (Fig. 360C)
tunicate (conidia) – covered in a rigid layer, exterior to the cell wall, as opposed to a mucilaginous sheath (Fig. 261A, B)
turbinate (conidia) – top-shaped, flattened at the apex, abruptly tapered at the base
unilocular (conidioma) – conidioma in which there is a single, simple, undivided cavity (Fig. 70C)
ventricose (conidia) – very broadly fusiform (Fig. 94A)
verruculose – finely ornamented (Fig. 8A)
versicoloured – septate conidia in which one or more cells is of different pigmentation (and usually of different wall thickness) than the rest (Fig. 149A)

REFERENCES

AGNIHOTHRUDU, V. (1962). Notes on fungi from North-East India X. *Chaetospermum camelliae* sp. nov. on tea (*Camellia sinensis* (L.) O. Kuntze). *Mycopath. Mycol. Appl.* **16**: 113-116.

AHMAD, S. (1961). Further contributions to the Fungi of West Pakistan. I. *Biologia, Lahore* **6**: 117-136.

AHMAD, S. (1967). Contributions to the Fungi of West Pakistan. VI. *Biologia, Lahore* **13**: 15-42.

AHMAD, S. (1971). Contributions to the Fungi of West Pakistan. X. *Biologia, Lahore* **17**: 1-26.

AINSWORTH, G.C. (1966). A general purpose classification of fungi. *Bibliography of Systematic Mycology* **1**: 1-4.

AINSWORTH, G.C. (1971). *Dictionary of the Fungi.* Ed. 6. Commonwealth Mycological Institute, Kew.

ALLESCHER, A. (1900). Rabenhorst's *Kryptogamen-Flora* **1**(6): 714-882.

ALLESCHER, A. (1901). Rabenhorst's *Kryptogamen-Flora* **1**(7): 370-373.

ALLESCHER, A. (1902). Rabenhorst's *Kryptogamen-Flora* **1**(7): 661-662.

ALLESCHER, A. (1903). Rabenhorst's *Kryptogamen-Flora* **1**(7): 711.

ST.-AMANS, M. DE. (1821). *Flore Agenaise.*

ANON. (1977). Nomenclature of fungi pathogenic to man and animals. *Med. Res. Counc. Mem.* 23(4th ed.): 1-26.

ARCHER, S.A. (1973). *Phomopsis pittospori* sp. nov., associated with leaf-spotting of *Pittosporum tenuifolium*. *Trans. Br. mycol. Soc.* **61**: 221-225.

ARNOLD, R.H. & CARTER, J.C. (1974). *Fusicoccum elaeagni*, the cause of a canker and dieback of russian olive, redescribed and redisposed to the genus *Phomopsis*. *Mycologia* **66**: 191-197.

ARNAUD, G. (1918). Les Astérinées. *Annls Ecole Agr. Montpell.*, N.S. **16**: 1-288.

ARNAUD, G. (1954). Mycologie concrète: Genera II (Suite et fin). *Bull. trimest. Soc. mycol. Fr.* **69**: 265-306.

ARX, J.A. VON (1949). Beiträge zur Kenntnis der Gattung *Mycosphaerella*. *Sydowia* **3**: 28-38.

ARX, J.A. VON (1951). Ueber die Gattung *Laestadia* und die Gnomoniaceen. *Antonie van Leeuwenhoek* **17**: 259-272.

ARX, J.A. VON (1957a). Revision der zu *Gloeosporium* Gestellten Pilze. *Verh. K. Ned. Akad. Wet. Amst.* **51**: 1-153.

ARX, J.A. VON (1957b). Die Arten der Gattung *Colletotrichum* Corda. *Phytopath. Z.* 29: 413-468.

ARX, J.A. VON (1963). Revision der zu *Gloeosporium* gestellten Pilze. Nachträge und Berichtigungen. *Verh. K. Ned. Akad. Wet. Amst.* **66**: 172-182.

ARX, J.A. VON (1970). A revision of the fungi classified as *Gloeosporium*. *Bibl. Myc.* **24**: 1-203.

ARX, J.A. VON & MÜLLER, E. (1954). Die Gattungen der amerosporen Pyrenomyceten. *Beitr. Krypfl. Schw.* **11**: 1-434.

ARX, J.A. VON & VAN DER VELDEN, F.J.J.A. (1961). Das *Colletotrichum* der Gurkengewächse. *Phytopath. Z.* **41**: 228-235.

AVERNA-SACCA, R. (1926). As manifestações pathologicans que acompanham o/desenuolvimento da broca *Stephanoderes hampei* Ferr. (= *St. Coffeae* Hag.) nos fructos ou nos sementes do cafeeiro. *Commissão de Estudo e Debellacao do Praga Cafeeira* 15: 46-49.

BACHE-WIIG, S. (1940). Contribution to the life history of a systemic fungus parasite, *Cryptomycina pteridis*. *Mycologia* **32**: 214-250.

BARRON, G.L. (1968). *The Genera of Hyphomycetes from Soil.* Williams & Wilkins Co.

BATISTA, A.C. & CIFERRI, R. (1963). The sooty mold of the family Asbolisiaceae. *Publ. Inst. Mic. Univ. Recife* 163: 1-229.

BATISTA, A.C. & PERES, G.E.P. (1961). *Belaina* n. gen. e outros Sphaeropsidaceae. *Mem. Soc. Broteriana* **14**: 49-65.

BATISTA, A.C., PERES, G.E.P. & BEZERRA, J.L. (1962). Novos fungos de Fumagina, da família Phaeosaccardinulaceae. *Broteria* N.S. **31**: 93-118.

BATISTA, A.C., PERES, G.E.P. & MAIA, H. DA S. (1959). Novas espécies de *Gelatosphaera, Asteromidium* e *Septoria* a antigas espécies de Sphaeropsidaceae. *Publ. Inst. Mic. Univ. Recife* 214: 1-29.

BATISTA, A.C., PERES, G.E.P. & MAIA, H. (1960). Revisao de alguns fungos do genero *Asteromella*. *Saccardoa* **1**: 17-29.

BATISTA, A.C. & VITAL, A.F. (1957). Contribuicao ao estudo dos fungus Sphaeropsidales. *An. Soc. Biol. Pern.* **15**: 413-427.

BAUSA ALCALDE, M. (1948). Hongos microscopicos de los alvededores de La Alberca (Salamanca). *An. Jard. bot. Madrid* **7**: 457-500.

BAUSA ALCALDE, M. (1952). Algunos micromicetos recolectados por el Prof. Caballero Segares en Valencia. *An. Inst. bot. A.J. Cavanilles* **10**: 227-255.

BAYLISS-ELLIOT, J.S. (1915). Studies in Discomycetes. I. *Trans. Br. mycol. Soc.* **5**: 417-421.

BERKELEY, M.J. (1860). *Outlines of British Fungology.* Lorell Reeve, London.

BERKELEY, M.J. & BROOME, C.E. (1850). XL. – Notices of British fungi. *Ann. Mag. Nat. Hist.*, 2 ser., **5**: 455-466.

BERTHET, P. (1964). Formes conidiennes de divers Discomycètes. *Bull. trimest. Soc. mycol. Fr.* **80**: 125-149.

BEZERRA, J.L. & POROCA, D.J.M. (1969). Alguns Deuteromycotina do Brasil. *Publ. Inst. Mic. Univ. Recife* 639: 1-10.

BIGA, M.L.B., CIFFERRI, R. & BESTAGNO, G. (1959). Ordinamento artificiale delle specie del genere *Coniothyrium* Corda. *Sydowia* **12**: 258-320.

BOEREMA, G.H. (1963). *Phoma herbarum* Westend., the type-species of the form-genus *Phoma* Sacc. *Persoonia* **3**: 9-15.

BOEREMA, G.H. (1967a). The *Phoma* organisms causing gangrene of potatoes. *Neth. J. Pl. Path.* **73**: 190-192.

BOEREMA, G.H. (1967b). Über die Identität der *Phoma radicis* Arten. *Nova Hedwigia* **14**: 57-60.

BOEREMA, G.H. (1970a). Nomenclatural remarks on three plant parasitic fungi. *Neth. J. Pl. Path.* **76**: 165-170.

BOEREMA, G.H. (1970b). Additional notes on *Phoma herbarum*. *Persoonia* **6**: 15-48.

BOEREMA, G.H. (1976). The *Phoma* species studied in culture by Dr. R.W.G. Dennis. *Trans. Br. mycol. Soc.* **67**: 289-319.

BOEREMA, G.H. & BOLLEN, G.J. (1975). Conidiogenesis and conidial septation as differentiating criteria between *Phoma* and *Ascochyta*. *Persoonia* **8**: 111-144.

BOEREMA, G.H. & DE JONG, C.B. (1968). Über die samenbürtigen *Phoma* – Arten von *Valerianella*. *Phytopath. Z.* **61**: 362-371.

BOEREMA, G.H. & DORENBOSCH, M.M.J. (1968). Some *Phoma* species recently described from marine soils in India. *Trans. Br. mycol. Soc.* **56**: 145-146.

BOEREMA, G.H. & DORENBOSCH, M.M.J. (1970). On *Phoma macrostomum* Mont., a ubiquitous species on woody plants. *Persoonia* **6**: 49-58.

BOEREMA, G.H. & DORENBOSCH, M.M.J. (1973). The *Phoma* and *Ascochyta* species described by Wollenweber and Hochapfel in their study of fruit-rotting. *Stud. Mycol., Baarn* 3: 1-50.

BOEREMA, G.H., DORENBOSCH, M.M.J. & KESTEREN, H.A. VAN (1965). Remarks on species of *Phoma* referred to *Peyronellaea*. *Persoonia* **4**: 47-68.

BOEREMA, G.H., DORENBOSCH, M.M.J. & KESTEREN, H.A. VAN (1968). Remarks on species of *Phoma* referred to *Peyronellaea*. *Persoonia* **5**: 201-205.

BOEREMA, G.H., DORENBOSCH, M.M.J. & KESTEREN, H.A. VAN (1971). Remarks on species of *Phoma* referred to *Peyronellaea* – III. *Persoonia* **6**: 171-177.

BOEREMA, G.H., DORENBOSCH, M.M.J. & KESTEREN, H.A. VAN (1973). Remarks on species of *Phoma* referred to *Peyronellaea* – IV. *Persoonia* **7**: 131-139.

BOEREMA, G.H., DORENBOSCH, M.M.J. & KESTEREN, H.A. VAN (1977). Remarks on species of *Phoma* referred to *Peyronellaea* – V. *Kew. Bull.* **31**: 533-544.

BOEREMA, G.H., DORENBOSCH, M.M.J. & LEFFRING, L. (1965). A comparative study of the black stem fungi on lucerne and red clover and the footrot fungus on pea. *Neth. J. Pl. Path.* **71**: 79-89.

BOEREMA, G.H. & GRIFFIN, J.J. (1974). *Phoma* species from *Viburnum*. *Trans. Br. mycol. Soc.* **63**: 109-114.

BOEREMA, G.H. & HÖWELER, L.H. (1967). *Phoma exigua* Desm. and its varieties. *Persoonia* **5**: 15-28.

BOEREMA, G.H. & KESTEREN, H.A. VAN (1963). The nomenclature of two fungi parasitizing *Brassica*. *Persoonia* **3**: 17-28.

BOEREMA, G.H. & VALCKX, A.G.M. (1970). Enkele bijzondare schimmelaantestingen III. (Mycologische Waarnemingen no. 15). *Gewashescherming* **1**: 65-68.

BOEREMA, G.H. & VERHOEVEN, A.A. (1972). Check-list for scientific names of common parasitic fungi. Series 1a: Fungi on trees and shrubs. *Neth. J. Pl. Path.* **78** (suppl. 1): 1-63.

BOEREMA, G.H. & VERHOEVEN, A.A. (1973). Check-list for scientific names of common parasitic fungi. Series 1b: Fungi on trees and shrubs; additions. *Neth. J. Pl. Path.* **79**: 165-179.

BOOTH. C. (1957). Studies of Pyrenomycetes: I. Four species of *Chaetosphaeria*, two with *Catenularia* conidia. II. *Melanopsamma pomiformis* and its *Stachybotrys* conidia. *Mycol. Pap.* 68: 1-27.

BOOTH, C. (1959). Studies of Pyrenomycetes: IV. *Nectria* (Part I). *Mycol. Pap.* 73: 1-115.

BOOTH, C. (1967). CMI Descriptions of Pathogenic Fungi and Bacteria No. 140.

BOOTH, C. (1971). *The genus Fusarium*. Commonwealth Mycological Institute, Kew.

BOOTH, C. & CORNFORD, C.E. (1959). British records, 41. *Pleospora betae*. *Trans. Br. mycol. Soc.* **42**: 119-120.

BOOTH, C. & GIBSON, I.A.S. (1973). CMI Descriptions of Pathogenic Fungi and Bacteria no. 363.

BOOTH, C., GIBSON, I.A.S. & SUTTON, B.C. (1973). CMI Descriptions of Pathogenic Fungi and Bacteria No. 364.

BORELLI, D. (1976). *Pyrenochaeta mackinnonii* nova species agente de micetoma. *Castellania* **4**: 227-234.

BRETON, A. & FAUREL, L. (1970). Étude comparative des *Monodia elegans* nov. gen., nov. sp. et *Pullospora tetrachaeta* Faur. et Schott., Sphaeropsidales coprophiles. *Revue Mycol.* **35**: 22-40.

BREWER, J.G. & BOEREMA, G.H. (1965). Electron microscope observations on the development of pycnidiospores in *Phoma* and *Ascochyta* spp. *Verh. K. Ned. Akad. wet. Amst.* **68**: 86-97.

BRISSON, J.D., PIROZYNSKI, K.A. & PAUZÉ, J.F. (1975). A new Coelomycete (*Dwayalomella vaccinii* gen. and sp. nov.) from *Vaccinium angustifolium* in Canada. *Can. J. Bot.* **53**: 2866-2871.

BROOKES, F.T. & SEARLE, G.O. (1921). An investigation of some tomato diseases. *Trans. Br. mycol. Soc.* **7**: 173-197.

BUBÁK, F. (1906a). Neue oder kritische Pilze. *Annls mycol.* **4**: 105-124.

BUBÁK, F. (1906b). Einige neue Pilze aus nord America. *J. Mycol.* **12**: 52-56.

BUBÁK, F. (1911). Ein neuer Pilz mit sympodialer Konidienbildung. *Ber. Dtsch. Bot. Ges.* **29**: 381-385.

BUBÁK, F. (1916). Systematische Untersuchungen einiger, Farne bewohneden Pilze. *Ber. Dtsch. Bot. Ges.* **35**: 295-332.

BUBÁK, F. & KABÁT, J.E. (1907). Mykologische Beiträge IV. *Hedwigia* **46**: 288-298.

BUTIN, H. (1964). Über zwei Nebenfruchtformen von *Sydowia polyspora* (Bref. et v. Tav.) Müller. *Sydowia* **17**: 114-118.

BUTIN, H. (1968). A new species of *Ceratocystis* causing blue-stain in *Araucaria araucana*. *Can. J. Bot.* **46**: 61-63.

BUTIN, H. (1977). Taxonomy and morphology of *Ascodichaena rugosa* gen. et sp. nov. *Trans. Br. mycol. Soc.* **69**: 249-254.

BUTIN, H. & SCHNEIDER, R. (1976). *Kabatina populi* nov. spec. *Phytopath. Z.* **85**: 39-42.

BUTLER, E.J. & KHAN, A.H. (1931). Some new sugarcane diseases. *Mem. Dept. Agric. Ind.* **6**: 181-208.

BYZOVA, Z.M., VASYAGINA, M.P., DEEVA, N.G., KALYMBETOV, B.K., PISAREVA, N.F. & SHVARTSMAN, S.R. (1970). Fungi imperfecti (Deuteromycetes) 2. Sphaeropsidales. *Flora sporovykh rastenii Kazakhstana* **5**(2): 1-384.

CAMARA, E. DE S. DA (1950). Catalogus Fungorum Juresi (Serra do Gerês) Ad Micofloram Lusitanicam. *Agron. lusit.* **12**: 89-122.

CAMPBELL, R. (1970). An electron microscope study of exogenously dormant spores, spore germination, hyphae and conidiophores of *Alternaria brassicicola*. *Arch. Mikrobiol.* **69**: 60-68.

CARMICHAEL, J.W. (1955). Lacto-fuchsin: A new medium for mounting fungi. *Mycologia* **47**: 611.

CARROLL, F.E., MÜLLER, E. & SUTTON, B.C. (1977). Preliminary studies on the incidence of needle endophytes in some European conifers. *Sydowia* **29**: 87-103.

CASH, E.K. (1952). A record of the fungi named by J.B. Ellis. *Spec. Pub. Div. Mycol. Dis. Surv., U.S.D.A.* 1-518.

CASTELLANI, E. & GERMANO, G. (1977). Le Stagonosporae graminicole. *Annali Fac. Sci. agr. Univ. Torino* **10**: 1-135 (of reprint).

CHAUDHURI, H. & SINGH, J. (1934). Une nouvelle maladie du Grenadier (*Punica granatum* Linn.). *Bull. trimest. Soc. mycol. Fr.* **50**: 153-161.

CHESTERS, C.G.C. (1938). Studies on British Pyrenomycetes. II. A comparative study of *Melanomma pulvis-pyrius* (Pers.) Fuckel, *Melanomma fuscidulum* Sacc. and *Thyridaria rubro-notata* (B. & Br.) Sacc. *Trans. Br. mycol. Soc.* **22**: 116-150.

CHOUDHURY, M. & AHMED, Q.A. (1969). Physiologic specialization of *Colletotrichum corchori* Ikata and Yoshida, the causal organism of an anthracnose of Jute (*Corchorus capsularis* L.). *Mycopath. Mycol. appl.* **38**: 161-168.

CHRISTENSEN, M. & RAPER, K.B. (1978). Synoptic key to *Aspergillus nidulans* group species and related *Emericella* species. *Trans. Br. mycol. Soc.* **71**: 177-191.

CICCARONE, A. (1947). Morfologia di *Neobarclaya natalensis* Syd. *Nuova G. bot. ital.*, N.S. **54**: 1-15.

CICCARONE, A. (1963). Observazioni sistemiche in *Diachorella onobrychidis* (DC.) v. Höhnel. *Phytopath. medit.* **2**: 239-250.

CICCARONE, A. (1966). Observazioni su *Placosphaeria bambusacearum* Sacc. et Syd., *Placonema bambusacearum* (Sacc. et Syd.) Petr. e *Shanoria bambusacearum* (Shanor) Subram. et Ramakr. e proposta della nuova combinazione: *Shanoria bambusacearum* (Sacc. et Syd.) Ciccarone. *Sydowia* **19**: 176-180.

CIFERRI, R. (1962). Schedae mycologicae 35-98 (Con otto travole fuori testo.). *Atti Ist. Bot. Univ. Pavia* **19**: 85-139.

CLEMENTS, F.E. & SHEAR, C.L. (1931). *The Genera of Fungi*. Hafner Press, New York.

CONNERS, I.L. (1959). Species of *Leptothyrium* and *Kabatia* on *Lonicera*. *Can. J. Bot.* **37**: 419-429.

COOK, A.A. (1975). *Diseases of Tropical and Subtropical Fruits and Nuts*. Hafner Press, New York.

COOKE, M.C. (1871). *Handbook of British Fungi*. Macmillan: London.

COOKE, M.C. (1886). British Sphaeropsideae. Provisional list of species hitherto found in the British Islands. *Grevillea* **14**: 132-138.

COOKE, W.B. & HAWKSWORTH, D.L. (1970). A preliminary list of the Families proposed for Fungi (including the lichens). *Mycol. Pap.* 121: 1-86.

CORDA, A.C.I. (1829). Sturm's *Deutschlands Flora* **3**: 105-136.

CORDA, A.C.I. (1840). Icones Fungorum hucusque cognitorum **4**: 1-63.

CRANE, J.L. (1971). Illinois fungi I. *Chaetosticta*. *Can. J. Bot.* **49**: 31-34.

CROXALL, H.E. (1950). Studies on British Pyrenomycetes. III. The British species of the genus *Diatrypella* Cesati and de Notaris. *Trans. Br. mycol. Soc.* **33**: 45-72.

CUNNELL, G.J. (1956). Some pycnidial fungi on *Carex*. *Trans. Br. mycol. Soc.* **39**: 21-47.

CUNNELL, G.J. (1957a). On *Neottiospora caricina* (Desm.) Höhnel. *Trans. Br. mycol. Soc.* **40**: 433-442.

CUNNELL, G.J. (1957b). *Stagonospora* spp. on *Phragmites communis* Trin. *Trans. Br. mycol. Soc.* **40**: 443-455.

CUNNELL, G.J. (1958). On *Robillarda phragmitis* sp. nov. *Trans. Br. mycol. Soc.* **41**: 405-412.

CUNNELL, G.J. (1961). *Stagonospora macropycnidia* sp. nov. *Trans. Br. mycol. Soc.* **44**: 87-90.

CURZI, M. & BARBIANI, M. (1927). Fungi aternensis. *Atti Ist. Bot. Univ. Pavia* **3**: 147-202.

DARKER, G.D. (1967). A revision of the Genera of the Hypodermataceae. *Can. J. Bot.* **45**: 1399-1444.

DEFAGO, G. (1944). Seconde contribution à la connaissance des Valsées von Höhnel. *Phytopath. Z.* **14**: 103-147.

DENNIS, R.W.G. (1943). A black rot of apples new to Britain. *Gardeners' Chron.* **114**: 221.

DENNIS, R.W.G. (1946). Notes on some British fungi ascribed to *Phoma* and related genera. *Trans. Br. mycol. Soc.* **29**: 11-42.

DENNIS, R.W.G. (1968). ***British Ascomycetes***. 2nd Ed. Cramer, Vaduz.

DENNIS, R.W.G. (1974). New or interesting British Microfungi, II. *Kew Bull.* **29**: 157-179.

DIAS, M.R. DE SOUSA & CAMARA, E. DE SOUSA DA (1955). Fungi lusitaniae XI. *Agron. lusit.* **17**: 101-113.

DIAS, M.R. DE SOUSA & TEIXEIRA, R.S. DA SILVA (1963). Fungi lusitaniae XVIII. *Agron. lusit.* **24**: 33-38.

DI COSMO, F. (1978). A revision of *Corniculariella*. *Can. J. Bot.* **56**: 1665-1690.

DIEDICKE, H. (1912). Die Abteiling Hyalodidymae der Sphaerioideen. *Annls mycol.* **10**: 135-152.

DIEDICKE, H. (1914a). *Kryptogamenflora Mark Brandenburg* **9**, Pilze VII: 417-640.

DIEDICKE, H. (1914b). *Kryptogamenflora Mark Brandenburg* **9**: Pilze VII: 641-800.

DINGLEY, J.M. & GILMOUR, J.W. (1972). *Colletotrichum acutatum* Simmds. f. sp. *pinea* associated with "terminal crook" disease of *Pinus* spp. *N.Z. Jl For. Sci.* **2**: 192-201.

DIXON, J.R. (1975). *Chlorosplenium* and its segregates. II. The Genera *Chlorociboria* and *Chlorencoelia*. *Mycotaxon* **1**: 191-237.

DODGE, B.O. (1936). *Schizoparme straminea* and *Nectriella versoniana* identical. *Mycologia* **28**: 95-98.

DODGE, B.O. (1938). A further study of the dry-rot disease of *Opuntia*. *Mycologia* **33**: 82-96.

DOMSCH, K.H. & GAMS, W. (1970). *Pilze aus Agraboden*. G. Fischer, Stuttgart.

DONG, XUAN B. (1972). Contributions à l'étude taxonomique des Hyphomycetes (Deuteromycetes). I Esquisse d'un nouvelle classification. *Ces. Mykol.* **26**: 155-166.

DONK, M.A. (1964). Nomina conservanda proposita I. Proposals in Fungi, Deuteromycetes. *Regn. veg.* **34**: 7-15.

DONK, M.A. (1966). *Cicinnobolus* Bary (Fungi, Sphaeropsidales). *Taxon* **15**: 149-151.

DONK, M.A. (1968). Report of the committee for Fungi and Lichens 1964-1968. *Taxon* **17**: 578-581.

DORENBOSCH, M.M.J. (1970). Key to nine ubiquitous soil-borne *Phoma*-like fungi. *Persoonia* **6**: 1-14.

DORENBOSCH, M.M.J. & BOEREMA, G.H. (1973). About *Phoma liliana* Chandra & Tandon II. *Mycopath. Mycol. appl.* **50**: 255-256.

DORENBOSCH, M.M.J. & HÖWELER, L.H. (1968). About *Phoma liliana* Chandra & Tandon. *Mycopath. Mycol. appl.* **35**: 264-267.

DRING, D.M. (1961). Studies on *Mycosphaerella brassicicola* (Duby) Oudem. *Trans. Br. mycol. Soc.* **44**: 253-264.

DYKO, B.J. & SUTTON, B.C. (1979a). Two unusual Deuteromycetes. *Trans. Br. mycol. Soc.* **72**: 411-417.

DYKO, B.J. & SUTTON, B.C. (1979b). A revision of *Linodochium, Pseudocenangium, Septopatella* and *Siroscyphella*. *Can. J. Bot.* **57**: 370-385.

EARLY, M.P. & PUNITHALINGAM, E. (1972). *Phomopsis anacardii* sp. nov. on *Anacardium occidentale*. *Trans. Br. mycol. Soc.* **59**: 345-347.

ELLIS, M.B. (1950). A modification of the 'Necol' Technique for mounting microfungi. *Trans. Br. mycol. Soc.* **33**: 22.

ELLIS, M.B. (1971). *Dematiaceous Hyphomycetes*. Commonwealth Mycological Institute, Kew.

ELLIS, M.B. (1976). *More Dematiaceous Hyphomycetes*. Commonwealth Mycological Institute, Kew.

ERIKSSON, B. (1974). On Deuteromycetes on Diapensiales and Ericales in Fennoscandia. *Svensk Bot. Tidskr.* **68**: 235-253.

ERIKSSON, O. (1966). On *Eudarluca caricis* (Fr.) O. Eriks., comb. nov., a cosmopolitan Uredinicolous Pyrenomycete. *Bot. Notiser* **119**: 33-69.

FARR, M.L. (1968). The nomenclatural status and synonymy of *Phyllachora bambusina, Lateropeltis bambusarum* and their conidial states. *Mycologia* **60**: 924-931.

FARR, M.L. (1973). An annotated list of Spegazzini's fungus taxa. *Bibl. Mycol.* **35**: 1-1661.

FARR, M.L. & HORNER, H.T. (1968). Fungi on *Selaginella*. *Nova Hedwigia* **15**: 239-283.

FAUREL, L. & SCHOTTER, G. (1965). Notes mycologiques. IV. Champignons coprophiles du Sahara central et notamment de la Tefedest. *Rev. Mycol.* **30**: 141-165.

FIRMAN, I.D. & WALLER, J.M. (1977). Coffee Berry Disease and other *Colletotrichum* diseases of coffee. *Phytopath. Pap.* 20: 1-53.

FONSEKA, R.N. (1960). The morphology of *Chaetospermum chaetosporum*. *Trans. Br. mycol. Soc.* **43**: 631-636.

FRAGOSO, R.G. (1915). Hongos parasitos de la florula hispanense, nuevos o poco conocidos. *Bol. Real Soc. Esp.* **15**: 121-133.

FRIES, E.M. (1923). *Systema mycologicum* **2**: 276-620.

FRIES, E.M. (1828). *Elenchus fungorum*: 1-154.

FRIES, E.M. (1832). *Systema mycologicum* **3**: 261-524.

FRIES, E.M. (1849). *Summa veg. Scand.*: 1-572.

FUCKEL, L. (1870). Symbolae mycologicae. *Jb. nassau. Ver. Naturk.* **23-24**: 1-459.

FUNK, A. (1970). Taxonomy of *Phomopsis boycei* and its relationship to *Potebniamyces balsamicola*. *Can. J. Bot.* **48**: 1023-1025.

FUNK, A. (1975). New microfungi on coastal Douglas fir. *Can. J. Bot.* **53**: 2297-2302.
FUNK, A. (1976). The genus *Dermea* and related conidial states on Douglas fir. *Can. J. Bot.* **54**: 2852-2856.
FUNK, A. & SUTTON, B.C. (1972). A disease of conifer buds in Western Canada associated with *Dichomera gemmicola* n. sp. *Can. J. Bot.* **50**: 1513-1518.
GARRETT, S.D. (1970). *Pathogenic Root-infecting Fungi.* Cambridge University Press, Cambridge.
GIBBS, J.N. (1969). Inoculum sources for coffee berry disease. *Ann. Appl. Biol.* **64**: 515-522.
GIBSON, I.A.S. & SUTTON, B.C. (1976). CMI Descriptions of Pathogenic Fungi and Bacteria Nos. 488, 490.
GILMAN, J.C. (1956). *A Manual of Soil Fungi* (2nd ed.). Iowa St. Coll. Press.
GOLATKAR, V.V. (1976). *Harknessia*: a new generic record in India. *Curr. Sci.* **45**: 269-270.
GREGORY, P.H. (1973). *Microbiology of the Atmosphere* (2nd ed.) 1-377. Leonard Hill; Bucks, England.
GREMMEN, J. (1975). Further notes on Discomycetous Fungi on coniferous hosts. *Sydowia* Beih. 1: 179-182.
GRIFFITHS, D.A. & CAMPBELL, W.P. (1972). Fine structure of conidial development in *Colletotrichum atramentarium.* *Trans. Br. mycol. Soc.* **59**: 483-489.
GRIFFITHS, D.A. & SWART, H.J. (1974a). Conidial structure in *Pestalotia pezizoides. Trans. Br. mycol. Soc.* **63**: 169-173.
GRIFFITHS, D.A. & SWART, H.J. (1974b). Conidial structure in two species of *Pestalotiopsis. Trans. Br. mycol. Soc.* **62**: 295-304.
GROVE, W.B. (1919). Mycological notes IV. *J. Bot. Lond.* **57**: 206-210.
GROVE, W.B. (1935). *British Stem- and Leaf-Fungi* **1**: 1-448. C.U.P.
GROVE, W.B. (1937). *British Stem- and Leaf-Fungi* **2**: 1-406. C.U.P.
GROVES, J.W. (1937). Three Dermateaceae occurring on *Nemopanthus. Mycologia* **29**: 66-80.
GROVES, J.W. (1938). *Dermatea acerina* and *Pezicula acericola. Mycologia* **30**: 416-430.
GROVES, J.W. (1939). Some *Pezicula* species and their conidial stages. *Can. J. Bot.* C, **17**: 125-143.
GROVES, J.W. (1940). Three *Pezicula* species occurring on *Alnus. Mycologia* **32**: 112-123.
GROVES, J.W. (1941). *Pezicula carnea* and *Pezicula subcarnea. Mycologia* **33**: 510-522.
GROVES, J.W. (1946). North American species of *Dermea. Mycologia* **38**: 351-431.
GROVES, J.W. (1952). The genus *Tympanis. Can. J. Bot.* **30**: 571-651.
GROVES, J.W. (1954). The genus *Durandiella. Can. J. Bot.* **32**: 116-144.
GROVES, J.W. (1965). The genus *Godronia. Can. J. Bot.* **43**: 1195-1276.
GROVES, J.W. (1968a). Two new species of *Ascocalyx. Can. J. Bot.* **46**: 1273-1278.
GROVES, J.W. (1968b). *Micropera boycei*, the conidial state of *Dermea pseudotsugae. Mycologia* **60**: 718-719.
GUBA, E.F. (1961). *Monograph of Monochaetia and Pestalotia.* Harvard Univ. Press, Cambridge, Mass.
GUTNER, L.S. (1934). Materials for a monograph of the genus *Cytospora. Acta Inst. Bot. Acad. Sci. URSS*, ser. 2, Fasc. 2: 411-484.
GVRITISHVILI, M.N. (1965). Novye dlya Gruzii vidy gribov iz roda *Cytospora.* Ehr. Soobshch. 2. (Spp. of fungi of the genus *Cytospora* new to Georgia). *Trudy Inst. Zeshch. Rast. Gruz SSR* **17**: 191-196.
GVRITISHVILI, M.N. (1967). Nekotorye taksonomicheskie i nomenklaturmye zametki po rodu *Cytospora* Fr. (Some taxonomic and nomenclatural notes on the genus *Cytospora*). *Nov. Sist. Niz. Rast.* 272-274.
GVRITISHVILI, M.N. (1968). *Cytospora schulzeri* Sacc. et Syd. v. SSR. *Mikol. i Fitopatol.* **2**: 87-92.
GVRITISHVILI, M.N. (1969). *Cytospora sacculus* (Schw.) Gvrit. v. SSR. *Mikol. i Fitopatol.* **3**: 207-213.
GVRITISHVILI, M.N. (1973a). Novye vidy grubov *Cytospora* Fr. soob. 1. *Mikol. i Fitopatol.* **7**: 135-136.
GVRITISHVILI, M.N. (1973b). Novye dannye po rodu *Cytospora* Fr. (New data on the genus *Cytospora* Fr.). *Trudy Inst. Zeshch. Rast. Gruz SSR* **24**: 204-208.
GVRITISHVILI, M.N. (1973c). Novye vidy gribov rod a *Cytospora* Fr. II. (New species of the genus *Cytospora.* II.). *Mikol. i Fitopatol.* **7**: 544-549.
HAHN, G.G. (1928). *Phomopsis conorum* (Sacc.) Died. – an old fungus of the Douglas fir and other conifers. *Trans. Br. mycol. Soc.* **13**: 278-286.
HAHN, G.G. (1930). Life-history studies of the species of *Phomopsis* occurring on conifers. Part. I. *Trans. Br. mycol. Soc.* **15**: 32-93.
HAHN, G.G. (1933). An undescribed *Phomopsis* from Douglas fir on the Pacific coast. *Mycologia* **25**: 369-375.
HAHN, G.G. (1943). Taxonomy, distribution and pathology of *Phomopsis occulta* and *P. juniperivora. Mycologia* **35**: 112-129.
HANSFORD, C.G. (1945). Contributions towards the fungus flora of Uganda. – VIII. New records and revisions (continued). *Proc. Linn. Soc. Lond.* **157** (1944-5): 20-41.
HANSFORD, C.G. (1946). The foliicolus ascomycetes, their parasites and associated fungi. *Mycol. Pap.* 15: 1-240.
HARA, K. (1954). *A list of Japanese Fungi hitherto known* Ed. 4: 1-447.
HAWKSWORTH, D.L. (1974). *Mycologist's Handbook.* Commonwealth Mycological Institute, Kew.

HAWKSWORTH, D.L. (1976). New and interesting microfungi from Slapton, South Devonshire: Deuteromycotina III. *Trans. Br. mycol. Soc.* **67**: 51-59.

HAWKSWORTH, D.L. (1977). Taxonomic and biological observations on the genus *Lichenoconium* (Sphaeropsidales). *Persoonia* **9**: 159-198.

HAWKSWORTH, D.L. & PUNITHALINGAM, E. (1973a). Typification and nomenclature of *Dichaena* Fr., *Heterographa* Fée, *Polymorphum* Chev., *Psilospora* Rabenh., and *Psilosporina* Died. *Trans. Br. mycol. Soc.* **60**: 501-509.

HAWKSWORTH, D.L. & PUNITHALINGAM, E. (1973b). New and interesting microfungi from Slapton, South Devonshire: Deuteromycotina. *Trans. Br. mycol. Soc.* **61**: 57-69.

HAWKSWORTH, D.L. & PUNITHALINGAM, E. (1975). New and interesting microfungi from Slapton, South Devonshire: Deuteromycotina II. *Trans. Br. mycol. Soc.* **64**: 89-99.

HENNEBERT, G.L. & WERESUB, L.K. (1977). Terms for states and forms of fungi, their names and types. *Mycotaxon* **6**: 207-211.

HERMANIDES-NIJHOFF, E.J. (1977). *Aureobasidium* and allied genera. *Stud. Mycol., Baarn* 15: 141-177.

HESS, H. & MÜLLER, E. (1951). Zur Entwicklungsgeschichte von *Dothidella insculpta* (Wallr.) Theiss. et Syd. *Ber. Schw. Bot. Ges.* **61**: 5-34.

HIGGINS, B.B. (1920). Morphology and life history of some ascomycetes with special reference to the presence and function of spermatia. *Amer. J. Bot.* **7**: 435-444.

HIGGINS, B.B. (1929). Morphology and life history of some ascomycetes. II. *Amer. J. Bot.* **16**: 287-296.

HIGGINS, B.B. (1936). Morphology and life history of some ascomycetes III. *Amer. J. Bot.* **23**: 598-602.

HINO, I. & KATUMOTO, K. (1961). *Icones Fungorum Bambusicolorum Japonicorum.* The Fuji Botanic Garden.

HÖHNEL, F. VON (1909). Fragmente zur Mykologie no. 464. Über *Endobotrya elegans* B. et Br. *Sber. Akad. Wiss. Wien* **118**: 1534-1536.

HÖHNEL, F. VON (1910a). Fragmente zur Mykologie no. 538. *Cystotricha striola* B. et Br. *Sber. Akad. Wiss. Wien* **119**: 632-633.

HÖHNEL, F. VON (1910b). Fragmente zur Mykologie no. 540. Über *Lasmenia* Speg. *Sber. Akad. Wiss. Wien* **119**: 634-635.

HÖHNEL, F. VON (1910c). Fragmente zur Mykologie no. 544. *Oncospora bullata* Kalchbr. et Cooke. *Sber. Akad. Wis. Wien* **119**: 639-640.

HÖHNEL, F. VON (1910d). Fragmente zur Mykologie no. 547. *Sporonema* Desm. *Sber. Akad. Wiss. Wien* **119**: 642-648.

HÖHNEL, F. VON (1910e). Fragmente zur Mykologie no. 542. Über die Gattung *Melophia* Sacc. *Sber. Akad. Wiss. Wien* **119**: 636-638.

HÖHNEL, F. VON (1911a). Zur Systematik der Sphaeropsideen und Melanconieen. *Annls mycol.* **9**: 258-265.

HÖHNEL, F. VON (1911b). Fragmente zur Mykologie no. 698. *Cicinnobella parodiellicola* P. Henn. *Sber. Akad. Wiss. Wien* **120**: 440.

HÖHNEL, F. VON (1911c). Fragmente zur Mykologie no. 665. *Ascochytopsis vignae* P. Henn. *Sber. Akad. Wiss. Wien* **120**: 395-396.

HÖHNEL, F. VON (1914a). Fragmente zur Mykologie no. 858. Über die Gattung *Sirococcus* Preuss. *Sber. Akad. Wiss. Wien* **123**: 119-126.

HÖHNEL, F. VON (1914b). Fragmente zur Mykologie no. 862. Über *Sphaeronema spinella* Kalchbrenner. *Sber. Akad. Wiss. Wien* **123**: 130-132.

HÖHNEL, F. VON (1915a). Fragmente zur Mykologie no. 910. Über die Gattung *Trullula* Cesati. *Sber. Akad. Wiss. Wien* **124**: 95-99.

HÖHNEL, F. VON (1915b). Fragmente zur Mykologie no. 928. Über die Gattung *Leptostroma* Fries. *Sber. Akad. Wiss. Wien* **124**: 126-129.

HÖHNEL, F. VON (1915c). Fragmente zur Mykologie no. 912. Über *Sphaerographium* Saccardo. *Sber. Akad. Wiss. Wien* **124**: 100-103.

HÖHNEL, F. VON (1915d). Beitrage zur Mykologie: 9. Über die Gattung *Myxosporium* Link. *Z. f. Garunsph.* **5**: 191-215.

HÖHNEL, F. VON (1915e). Fragmente zur Mykologie no. 896. Über die Gattung *Chaetomella* Fuck. *Sber. Akad. Wiss. Wien* **124**: 77-78.

HÖHNEL, F. VON (1916a). Mykologisches. XXIV. *Öst. bot. Z.* **66**: 94-112.

HÖHNEL, F. VON (1916b). Fragmente zur Mykologie no. 981. Über *Gloeosporium* und *Marsonia*. *Sber. Akad. Wiss. Wien* **125**: 94-96.

HÖHNEL, F. VON (1916c). Fragmente zur Mykologie no. 964. Über die Gattung *Cheilaria* Libert. *Sber. Akad. Wiss. Wien* **125**: 59-62.

HÖHNEL, F. VON (1916d). Fragmente zur Mykologie no. 976. Über *Cheilaria aceris* Libert. *Sber. Akad. Wiss. Wien* **125**: 81-84.

HÖHNEL, F. VON (1916e). Fragmente zur Mykologie no. 961. Über *Asteroma padi* DC. *Sber. Akad. Wiss. Wien* **125**: 50-53.

HÖHNEL, F. VON (1916f). Fragmente zur Mykologie no. 965. Über *Micropera pinastri* Sacc. und *Oncospora abietina* Oud. et Fautrey. *Sber. Akad. Wiss. Wien* **125**: 39-42.

HÖHNEL, F. VON (1916g). Fragmente zur Mykologie no. 957. Über die Gattung *Gelatinosporium* Peck. *Sber. Akad. Wiss. Wien* **125**: 42-44.

HÖHNEL, F. VON (1917a). Fragmente zur Mykologie no. 1017. Über *Pyrenopeziza agrostemmatis* Fuckel. *Sber. Akad. Wiss. Wien* **126**: 333-335.

HÖHNEL, F. VON (1917b). Mycologische Fragmente. *Annls mycol.* **15**: 293-383.

HÖHNEL, F. VON (1918a). Fungi imperfecti. Beitrage zur Kenntnis derselben. *Hedwigia* **60**: 129-209.

HÖHNEL, F. VON (1918b). Fragmente zur Mykologie no. 1100. Über *Tryblidiopycnis pinastri* (P.) und deren Nebenfrucht. *Sber. Akad. Wiss. Wien* **127**: 560-563.

HÖHNEL, F. VON (1920a). Fungi imperfecti. Beitrage zur Kenntnis derselben. *Hedwigia* **62**: 56-89.

HÖHNEL, F. VON (1920b). Über die Gattung *Phlyctaena* Desmazières. *Ber. Dtsch. Bot. Ges.* **38**: 102-110.

HÖHNEL, F. VON (1923a). System der Fungi imperfecti Fuckel. *Mykol. Unters.* **1**: 301-369.

HÖHNEL, F. VON (1923b). Fragmente zur Mykologie no. 961. Über *Asterma padi* DC. *Sber. Akad. Wiss. Wien* **125**: 50-53.

HÖHNEL, F. VON (1924). Beitrage zur Kenntnis der Gattung *Cylindrosporium* Grev. *Annls mycol.* **22**: 191-203.

HÖHNEL, F. VON (1927). Über die Nebenfruchtform von *Pezicula rubi* (Libert) Niessl. *Mitt. bot. Inst. tech. Hochsch. Wien* **4**: 105-106.

HÖHNEL, F. VON (1929). Über *Phoma rhamnigena* Fautrey. *Mitt. bot. Inst. tech. Hochsch. Wien* **6**: 116-118.

HOLLIDAY, P. & PUNITHALINGAM, E. (1970). CMI Descriptions of Pathogenic Fungi and Bacteria Nos. 271, 272, 275, 276, 278.

HUDSON, H.J. (1968). The ecology of fungi on plant remains above the soil. *New Phytol.* **67**: 837-874.

HUDSON, H.J. & SUTTON, B.C. (1964). *Trisulcosporium* and *Tetranacrium*, two new genera of fungi imperfecti. *Trans. Br. mycol. Soc.* **47**: 197-203.

HUGHES, G.C. (1975). Studies of fungi in oceans and estuaries since 1961. I. Lignicolous, caulicolous and foliicolous species. *Oceanogr. Mar. Biol. Ann. Rev.* **13**: 69-180.

HUGHES, S.J. (1952). Studies on microfungi XIV. *Stigmella, Camptomeris, Polythrincium* and *Fusicladiella*. *Mycol. Pap.* 49: 1-25.

HUGHES, S.J. (1953a). Conidiophores, conidia, and classification. *Can. J.Bot.* **31**: 577-659.

HUGHES, S.J. (1953b). Fungi from the Gold Coast. II. *Mycol. Pap.* 50: 1-104.

HUGHES, S.J. (1958). Revisiones hyphomycetum aliquot cum appendice de nominibus rejiciendis. *Can. J. Bot.* **36**: 727-838.

IVORY, M.H. (1967). A new variety of *Dothistroma pini* in Kenya. *Trans. Br. mycol. Soc.* **50**: 289-297.

JEFFREY, C. (1977). *Biological Nomenclature.* (2nd ed.) Edward Arnold: London.

JENKINS, A.E. & BITANCOURT, A.A. (1956). Myriangiales selecti exsiccati, parts 1-10, numbers 1-500 (1940-1954), indexes and other lists pertaining to this collection. *Rev. Soc. bras. Agron.* **12**: 41-68.

JENKINS, A.E. & BITANCOURT, A.A. (1968). "Myriangiales selecti exsiccati" fascicle (nos. 501-550). *Mycologia* **60**: 719-720.

JENKINS, S.F., WINSTEAD, N.N. & McCOMBS, C.L. (1962). New physiological races of *Glomerella cingulata* var. *orbiculare* (syn. *C. lagenarium, C. orbiculare*) based on cucurbit differentials. *Phytopathology* **52**: 15.

JØRSTAD, I. (1945). Parasittopene på Kultur-og nyttevekster i Norge I. Sekksporesopper (Ascomycetes) og konidiesopper (Fungi Imperfecti). *Meld. Statens Plantepatol. Inst.* **1**: 1-142.

JØRSTAD, I. (1965). *Septoria* and septorioid fungi on dicotyledons in Norway. *Skr. Nor. Videnskaps-Akad. Oslo, I: Math.-Naturv. Kl.* **22**: 1-110.

JØRSTAD, I. (1967). *Septoria* and septorioid fungi on Gramineae in Norway. *Skr. Nor. Videnskaps-Akad. Oslo, I. Math.-Naturv. Kl.* **24**: 1-63.

KALANI, I.K. (1963). Some new records of *Ciliochorella* spp. from India. *Mycopath. Mycol. appl.* **19**: 238-240.

KALANI, I.K. (1964). *Stauronema indicum* sp. nov. from India. *Curr. Sci.* **33**: 622-623.

KEISSLER, K. (1930). Rabenhorst's *Kryptogamen-Flora* **8**: Die Flechtenparasiten: 1-712.

KENDRICK, W.B. (ed.) (1971). *Taxonomy of Fungi Imperfecti.* Univ. Toronto Press.

KENDRICK, W.B. (ed.) (1979). *The Whole Fungus.* Nat. Mus. Canada, and Kananaskis Foundation.

KHAN, S.R. & KIMBROUGH, J.W. (1974). Taxonomic position of *Termitaria* and *Mattirolella* (entomogenous Deuteromycetes). *Amer. J. Bot.* **61**: 395-399.

KHAN, S.R. & STRANGE, R.N. (1975). Evidence for the role of a fungal stimulant as a determinant of differential susceptibility of jute cultivars to *Colletotrichum corchori*. *Physiol. Pl. Path.* **5**: 157-164.

KNIGHT-FISCHER, D.E. (1960). Studies on *Didymella lycopersici* Kleb., the causal fungus of stem rot disease of tomatoes. *Trans. Br. mycol. Soc.* **43**: 519-522.

KNOX-DAVIES, P.S. (1970). The nuclei of *Alternaria brassicicola*. *Trans. Br. mycol. Soc.* **72**: 81-90.

KOBAYASHI, T. (1964). A new species of *Phellostroma* associated with stem and twig blight of *Tsuga sieboldii* Carr. *Trans. mycol. Soc. Jap.* **5**: 6-8.

KOBAYASHI, T. (1965). *Neopycnodothis* (Sphaerioidaceae, Fungi Imperfecti), a new genus of bamboo inhabitant. *Ann. Phytopath. Soc. Japan* **30**: 153-155.

KOBAYASHI, T. (1967). Critical revision on the genera *Rhizosphaera* Mangin et Hariot and *Rhizophoma* Petrak & Sydow, a little known fungous group associated with needle disease of conifers. *Bull. Govt. For. St.* 204: 91-112.

KORF, R.P. (1952). A monograph of the Arachnopezizeae. *Lloydia* **14**: 129-180.

KORF, R.P. (1972). Synoptic key to the genera of the Pezizales. *Mycologia* **64**: 937-994.

KORF, R.P. (1973). Chapter 9. Discomycetes and Tuberales, pp. 249-319. In *The Fungi* **IVA** (eds. G.C. Ainsworth, F.K. Sparrow & A.S. Sussman). Academic Press.

KOWALSKI, J. (1968). Investigations on some biological properties of fungi attacking leaves of *D. purpurea* and *D. lanata. Acta mycol., Warszawa* **4**: 23-38.

KRUG, J.C. (1977). *Kabatia valpellinensis* reported from Switzerland. *Sydowia* **29**: 275-277.

KRUSCHEWA, R. (1958). Parasitic fungi new for Bulgaria. *Trav. Sci. Fac. Agron. Sofia* **7**: 487-494.

LAUNDON, G.F. (1971). Records of fungal plant diseases in New Zealand – 2. *N.Z. Jl. Bot.* **9**: 610-624.

LEACH, C.M. (1962). Sporulation of diverse species of fungi under near-ultraviolet radiation. *Can. J. Bot.* **40**: 151-161.

LEAKEY, C.L.A. (1964). *Dactuliophora*, a new genus of Mycelia sterilia from tropical Africa. *Trans. Br. mycol. Soc.* **47**: 341-350.

LIBERT, M.A. (1834). Plantae cryptogamicae, quas in Arduenna collegit. Leodii. Cent. III.

LIMBER, D.P. (1955). Studies in the genus *Sporonema. Mycologia* **47**: 389-402.

LINDER, D.H. (1944). I. Classification of the Marine fungi. *Farlowia* **1**: 401-467.

LUC, M. (1953). Sur trois Champignons du Palmier à huile en Côte d'Ivoire. *Rev. Mycol.* **18**, Suppl. colon. 2: 94-102.

LUTTRELL, E.S. (1963). Taxonomic criteria in *Helminthosporium. Mycologia* **55**: 643-674.

MAAS, P.W.T. (1965). The identity of the footrot fungus of flax. *Neth. J. Pl. Path.* **71**: 113-121.

MADELIN, M.F. (1968). Chapter 9. Fungal Parasites of Invertebrates I. Entomogenous fungi, pp. 226-238. In *The Fungi* **III** (eds. G.C. Ainsworth & A.S. Sussman). Academic Press.

MAGNUS, P. (1906). Notwendige Umänderung des Namens der Pilzgattung *Marssonia* Fisch. *Hedwigia* **45**: 88-91.

MÄKELÄ, K. (1977). *Septoria* and *Selenophoma* species on Gramineae in Finland. *Annls Agr. fenn.* **16**: 256-276.

MARASAS, W.F.O., PAUER, G.D. & BOEREMA, G.H. (1974). A serious leaf blotch disease of groundnuts (*Arachis hypogaea* L.) in southern Africa caused by *Phoma arachidicola* sp. nov. *Phytophylactica* **6**: 195-202.

MASON, E.W. (1937). Annotated account of fungi received at the Imperial Mycological Institute, List II. (Fascicle 3-General Part). *Mycol. Pap.* 4: 69-99.

MELNIK, V.A. (1970). Novye i interesnye vidy sferopsidnykh gribov. (New and interesting spp. of Sphaeropsidales). *Nov. Sist. niz. Rast.* **7**: 238-242.

MELNIK, V.A. (1971). Taxonomy of the genus *Ascochyta* Lib. *Mikol. i Fitopatol.* **5**: 15-22.

MELNIK, V.A. (1976). Rod *Apiocarpella* Syd. *Nov. Sist. Niz. Rast.* **13**: 93-98.

MELNIK, V.A. (1977). *Opredelitel' gribov roda Ascochyta Lib.* (Key to fungi of the genus *Ascochyta* Lib.). Leningrad: U.S.S.R.

MÉRAT, F.V. (1821). *Nouvelle Flore des Environs de Paris* (ed. 10). Paris.

MERLO, P.A. (1969). Antracnosis de la "*Acacia longifolia*" ocasionada por "*Colletotrichum dematium* f. *truncata*" in San Clemente del Tuya, prov. de Buenos Aires (Argentina). *Revta Fac. Agron. Univ. Nac. La Plata* **45**: 53-59.

MESSAIEN, C.M., LAFON, R., & MALOT, P. (1969). Nécroses de racines, pourritures de tiges et verse parasitaire du Mais. *Ann. epiphyt.* ser. C, **10**: 441-474.

MONTAGNE, J.F.C. (1834). Notice sur les plantes cryptogames récemment découvertes en France contenant aussi l'indication précise des localités de quelques espèces les plus rares de la flore française. *Ann. Sci. nat.*, ser. 2, **1**: 295-307.

MORDUE, J.E.M. (1967). CMI Descriptions of Pathogenic Fungi and Bacteria nos. 131, 132, 133.

MORDUE, J.E.M. (1971). CMI Descriptions of Pathogenic Fungi and Bacteria nos. 315, 316, 317.

MORELET, M. (1968a). De Aliquibus in Mycologia Novatatibus (2ᵉ note). *Bull. Soc. Sci. nat. Archéol. Youlon Var.* 176: 6-7.

MORELET, M. (1968b). De Aliquibus in Mycologia Novitatibus. *Bull. Soc. Sci. nat. Archéol. Toulon Var* 175: 5-6.

MORELET, M. (1973a). De aliquibus in mycologia novitatibus (8ᵉ note). *Bull. Soc. Sci. nat. Archéol. Toulon Var* 203: 12.

MORELET, M. (1973b). De Aliquibus in Mycologia Novitatibus (9³ note). *Bull. Soc. Sci. nat. Archéol. Toulon Var* 204: 8.

MORGAN-JONES, G. (1964). Taxonomic and biological studies in the Coelomycetes. Ph.D. Thesis, University of Nottingham.

MORGAN-JONES, G. (1967). CMI Descriptions of Pathogenic Fungi and Bacteria nos. 134, 135.

MORGAN-JONES, G. (1971a). Conidium ontogeny in Coelomycetes, I. Some amerosporous species which possess annellides. *Can. J. Bot.* **49**: 1921-1929.

MORGAN-JONES, G. (1971b). Conidium ontogeny in Coelomycetes. II. Some Melanconiales which possess phialides. *Can. J. Bot.* **49**: 1931-1937.
MORGAN-JONES, G. (1971c). Conidium ontogeny in Coelomycetes. III. Meristem thalloconidia. *Can. J. Bot.* **49**: 1939-1940.
MORGAN-JONES, G. (1971d). An addition to the genus *Leptomelanconium* Petrak. *Can. J. Bot* **49**: 1011-1013.
MORGAN-JONES, G. (1971e). A new species of *Dinemasporium* from Ontario. *Can. J. Bot.* **49**: 1363-1365.
MORGAN-JONES, G. (1971f). An addition to the genus *Discogloeum* Petrak. *Can. J. Bot.* **49**: 1461-1462.
MORGAN-JONES, G. (1973). Genera coelomycetarum. VII. *Crytocline* Petrak. *Can. J. Bot.* **51**: 309-325.
MORGAN-JONES, G. (1975a). Concerning some species of *Microsphaeropsis. Can. J. Bot.* **52**: 2575-2579.
MORGAN-JONES, G. (1975b). Notes on Coelomycetes. I. *Mycotaxon* **2**: 167-183.
MORGAN-JONES, G. & KENDRICK, B. (1972). Icones generum coelomycetum III. *Univ. Waterloo Biol. Ser.* 5: 1-42.
MORGAN-JONES, G., NAG RAJ, T.R. & KENDRICK, W.B. (1972a). Conidium ontogeny in Coelomycetes. IV. Percurrently proliferating phialides. *Can. J. Bot.* **50**: 2009-2014.
MORGAN-JONES, G., NAG RAJ, T.R. & KENDRICK, W.B. (1972b). Genera coelomycetarum V. *Alpakesa* and *Bartalinia. Can. J. Bot.* **50**: 877-882.
MORGAN-JONES, G., NAG RAJ, T.R. & KENDRICK, B. (1972c). Genera coelomycetarum. VI. *Kellermania. Can. J. Bot.* **50**: 1641-1648.
MORGAN-JONES, G., NAG RAJ, T.R. & KENDRICK, W.B. (1972d). Icones genera coelomycetarum II. *Univ. Waterloo Biol. Ser.* 4: 1-40.
MORGAN-JONES, G., NAG RAJ, T.R. & KENDRICK, W.B. (1972e). Icones generum coelomycetum V. *Univ. Waterloo Biol. Ser.* 7: 1-52.
MORGAN-JONES, G., NAG RAJ, T.R. & KENDRICK, W.B. (1972f). Icones generum coelomycetum IV. *Univ. Waterloo Biol. Ser.* 6: 1-42.
MORGAN-JONES, G., NAG RAJ, T.R. & KENDRICK, W.B. (1972g). Icones genera coelomycetarum I. *Univ. Waterloo Biol. Ser.* 3: 1-42.
MÜLLER, E. & ARX, J.A. VON (1962). Die Gattungen der didymosporen Pyrenomyceten. *Beitr. Kryptogamenfl. Schweiz* **11**: 1-922.
MÜLLER, E. & SHOEMAKER, R.A. (1964). *Nummospora hexamera* and *Cheiromyces acutus*, new Fungi imperfecti. *Nova Hedwigia* **7**: 1-4.
MUNJAL, R.L. (1966). The genus *Ciliochhorella* in India. *Indian Phytopath.* **19**: 198-200.
NAG RAJ, T.R. (1973a). Genera coelomycetum. X. *Ellisiella, Samukuta* and *Sakireeta. Can. J. Bot.* **51**: 2463-2472.
NAG RAJ, T.R. (1973b). Genera coelomycetum. IX. *Brycekendrickia* gen. nov. and *Vasudevella. Can. J. Bot.* **51**: 1337-1341.
NAG RAJ, T.R. (1974). Icones generum coelomycetum VI. *Univ. Waterloo Biol. Ser.* 13: 1-41.
NAG RAJ, T.R. (1975). Genera coelomycetum XI. *Hyalotia, Hyalotiella* and *Hyalotiopsis. Can. J. Bot.* **53**: 1615-1624.
NAG RAJ, T.R. (1977a). Miscellaneous microfungi II. *Can. J. Bot.* **55**: 757-765.
NAG RAJ, T.R. (1977b). *Ypsilonia, Acanthotheciella*, and *Kazulia* gen. nov. *Can. J. Bot.* **55**: 1599-1622.
NAG RAJ, T.R. (1978a). Genera coelomycetum XVI. *Fibulocoela* form-gen. nov., a coelomycete with basidiomycetous affinities. *Can. J. Bot.* **56**: 1485-1491.
NAG RAJ, T.R. (1978b). Genera coelomycetum XIV. *Allelochaeta, Basilocula, Ceuthosira, Microgloeum, Neobarclaya, Polynema, Pycnidiochaeta*, and *Xenodomus. Can. J. Bot.* **56**: 686-707.
NAG RAJ, T.R. & KENDRICK, W.B. (1971). Genera coelomycetarum. I. *Urohendersonia. Can. J. Bot.* **49**: 1853-1862.
NAG RAJ, T.R. & KENDRICK, W.B. (1972a). Genera coelomycetarum. III. *Pestalozziella. Can. J. Bot.* **50**: 607-617.
NAG RAJ, T.R. & KENDRICK, W.B. (1972b). Genera coelomycetarum. II. *Doliomyces. Can. J. Bot.* **50**: 45-48.
NAG RAJ, T.R. & MORGAN-JONES, G. (1973). Genera coelomycetum. VIII. *Rhabdogloeopsis* Petrak and *Rhabdogloeum* Sydow. *Can. J. Bot.* **51**: 565-569.
NAG RAJ, T.R., MORGAN-JONES, G. & KENDRICK, W.B. (1972). Genera coelomycetarum. IV. *Pseudorobillarda* gen. nov., a generic segregate of *Robillarda* Sacc. *Can. J. Bot.* **50**: 861-867.
NAG RAJ, T.R., MORGAN-JONES, G. & KENDRICK, W.B. (1973). *Pseudorobillarda* Nag Raj et al., a later homonym of *Pseudorobillarda* Morelet. *Can. J. Bot.* **51**: 688-689.
NANNFELDT, J.A. (1932). Studien über die Morphologie und Systematik der Nichtlichenisierten inoperculaten Discomyceten. *Nov. Acta Reg. Soc. Scient. Upsal.*, ser. 4, **8**(2): 1-368.
NEERGAARD, P. (1978). Seed Pathology I and II. Macmillan Press Ltd: London.
NEES, T.F.L. & HENRY, A. (1837). *Das system der Pilze* **1**: 1-74. Bonn.
OUDEMANS, C.A.J.A. (1922). *Enumeratio Systematica Fungorum* **4**: 1-1231. The Hague.
OVERHOLTS, L.O. (1926). Mycological notes for 1924. *Mycologia* **18**: 31-38.

PAPENDORF, M.C. & DU TOIT, J.W. (1967). *Melanophoma*: a new genus of the Sphaeropsidales. *Trans. Br. mycol. Soc.* **50**: 503-506.

PEACE, T.R. (1962). *Pathology of Trees and Shrubs, with Special Reference to Britain*. Clarendon Press: Oxford.

PETCH, T. (1921). Studies in entomogenous fungi: II. The genera *Hypocrella* and *Aschersonia*. *Ann. R. bot. Gdn., Peradeniya* **7**: 167-278.

PETCH, T. (1943). British Nectrioideae and allied genera. *Trans. Br. mycol. Soc.* **26**: 53-70.

PETRAK, F. (1921). Mykologische Notizen. II. *Annls mycol.* **19**: 17-128.

PETRAK, F. (1922). Beiträge zur Kenntnis der Pilzflora der südlichen Alpenländer und Norditaliens. *Annls mycol.* **20**: 126-159.

PETRAK, F. (1923). Mykologische Notizen. VI. *Annls. mycol.* **21**: 182-335.

PETRAK, F. (1924). Mykologische Notizen VII. *Annls mycol.* **22**: 1-182.

PETRAK, F. (1925). Mykologische Notizen VIII. *Annls mycol.* **23**: 1-143.

PETRAK, F. (1944). Über die Gattungen *Chaetopyrena* Pass., *Sclerochaeta* v. Höhn., *Sclerochaetella* v. Höhn., *Vermiculariella* Oud., *Chaetosphaeronema* Moesz und *Pseudophoma* v. Höhn. *Annls mycol.* **42**: 58-71.

PETRAK, F. (1947a). Kritische Bemerkungen über einige, in letzter Zeit als neu beschriebene Askomyzeten und Fungi imperfecti. *Sydowia* **1**: 61-79.

PETRAK, F. (1947b). Über die Gattungen *Eriosporopsis* n. gen. und *Eriospora* Berk. et Br. *Sydowia* **1**: 94-97.

PETRAK, F. (1947c). Über die Gattungen *Sirococcus* Preuss, *Peckia* Clint. und *Sirodomus* n. gen. *Sydowia* **1**: 151-158.

PETRAK, F. (1949). Beiträge zur Pilzflora Irans. *Sydowia* **3**: 268-332.

PETRAK, F. (1950). Über *Loranthomyces* v. Höhn. und einige andere Gattungen der Trichothyriaceen. *Sydowia* **4**: 163-174.

PETRAK, F. (1951a). Ergebnisse einer Revision der Grundtypen verschiedener Gattungen der Askomyzeten und Fungi imperfecti II. *Sydowia* **5**: 328-356.

PETRAK, F. (1951b). Über die Gattungen *Hendersonula* Speg. und *Hendersonulina* n. gen. *Sydowia* **5**: 418-422.

PETRAK, F. (1952a). *Stevensonula* n. gen., eine neue Gattung der Parasphaeropsideen. *Sydowia* **6**: 1-4.

PETRAK, F. (1952b). Ergebnisse einer Revision der Grundtypen verschiedener Gattungen der Askomyzeten und Fungi imperfecti III. *Sydowia* **6**: 336-343.

PETRAK, F. (1953a). Beitrage zur Kenntnis der Pilzflora Irans. *Sydowia* **7**: 50-78.

PETRAK, F. (1953b). Ergebnisse einer Revision der Grundtypen verschiedener Gattungen der Askomyzeten und Fungi imperfecti. IV. *Sydowia* **7**: 295-308.

PETRAK, F. (1954). Über eine westafrikanische Art der Gattung *Neomelanconium* Petr. *Sydowia* **8**: 51-53.

PETRAK, F. (1957). Über die Gattungen *Dothichiza* Lib. und *Chondroplea* Kleb. *Sydowia* **10**: 201-235.

PETRAK, F. (1958a). Mykologische Bemerkungen. *Sydowia* **11**: 337-353.

PETRAK, F. (1958b). Zwei neue Arten der Gattung *Siroplacodium*, aus den kurdischan Hochgebirgen. *Sydowia* **11**: 317-319.

PETRAK, F. (1959a). Ergebnisse einer Revision der Grundtypen verscheidener Gattungen der Askomyzeten und Fungi imperfecti. *Sydowia* **13**: 1-9.

PETRAK, F. (1959b). Über die Gattung *Cheilaria* Lib. *Sydowia* **13**: 57-63.

PETRAK, F. (1961). *Macrophoma sapinea* (Fr.) Petr. in den Föhrenwäldern des niederösterreichischen Waldviertels. *Sydowia* **15**: 309-316.

PETRAK, F. (1962). Mykologische Bemerkungen. *Sydowia* **15**: 204-217.

PETRAK, F. (1963a). Ergebnisse einer Revision der Grundtypen verschiedener Gattungen der Askomyzeten und Fungi imperfecti. *Sydowia* **16**: 353-361.

PETRAK, F. (1963b). Über die gattung *Megaloseptoria* Naumov. *Sydowia* **16**: 373-376.

PETRAK, F. (1965). Über die Gattungen *Chaetomella* Fuck., *Amerosporium* Speg. und *Volutellospora* Mathur et Thirumalachar. *Sydowia* **18**: 373-379.

PETRAK, F. (1968a). Ergebnisse einer Revision der Grundtypen verschiedener Gattungen der Askomyzeten und Fungi imperfecti. *Sydowia* **21**: 240-248.

PETRAK, F. (1968b). Über die Gattung *Sphaerellopsis* Cooke. *Sydowia* **20**: 200-202.

PETRAK, F. (1972). Über Irrtümer bei der Beschreibung oder einreihung neuer arten von Askomyzeten und Fungi imperfecti, deren ursochen und verhütug. *Sydowia* **25**: 222-224.

PETRAK, F. & CIFERRI, R. (1932). Fungi dominicani II. *Annls mycol.* **30**: 149-353.

PETRAK, F. & SYDOW, H. (1923). Kritisch-systematisch Originaluntersuchungen über Pyrenomyzeten, Sphaeropsideen und Melanconieen. *Annls mycol.* **21**: 349-384.

PETRAK, F. & SYDOW, H. (1924). Kritisch-systematische Originaluntersuchungen über Pyrenomyzeten, Sphaeropsideen und Melanconieen. *Annls mycol.* **22**: 318-386.

PETRAK, F. & SYDOW, H. (1925). Kritisch-systematische Originaluntersuchungen über Pyrenomyzeten, Sphaeropsideen und Melanconieen. *Annls mycol.* **23**: 209-294.

PETRAK, F. & SYDOW, H. (1927). Die Gattungen der Pyrenomyzeten, Sphaeropsideen und Melanconieen. *Beih. Rep. spec. nov. regni veg.* **42**: 1-551.

PETRAK, F. & SYDOW, H. (1935). Kritisch-systematische Originaluntersuchungen über Pyrenomyzeten, Sphaeropsideen und Melanconieen VI. *Annls mycol.* **33**: 157-193.

PETRAK, F. & SYDOW, H. (1936). Kritisch-systematische Originaluntersuchungen über Pyrenomyzeten, Sphaeropsideen und Melanconieen. VII. *Annls mycol.* **34**: 11-52.

PEYRONEL, B. (1922). Champignons nouveaux des Vallées Vaudoises du Piémont. *Bull. trimest. Soc. mycol. Fr.* **38**: 140-143.

PIROZYNSKI, K.A. (1971). Note on *Geastrumia polystigmatis*. *Mycologia* **63**: 897-901.

PIROZYNSKI, K.A. & MORGAN-JONES, G. (1968). Notes on Microfungi III. *Trans. Br. mycol. Soc.* **51**: 185-206.

PIROZYNSKI, K.A. & SHOEMAKER, R.A. (1970). *Seimatosporium* leaf spot of *Ledum* and *Rhododendron*. *Can. J. Bot.* **48**: 2199-2203.

PIROZYNSKI, K.A. & SHOEMAKER, R.A. (1971). Some Coelomycetes with appendaged conidia. *Can. J. Bot.* **49**: 529-541.

PIROZYNSKI, K.A. & SHOEMAKER, R.A. (1972). *Vestigium*, a new genus of Coelomycetes. *Can. J. Bot.* **50**: 1163-1164.

PLATA, P., MARTIN, R., PRENDES, C. & BLESA, C. (1974). Contribution to the knowledge of the principal banana diseases in the Canary Islands. 2 – Fungi causing transport disorders. *Vieraea* **3**: 62-71.

POLLACK, F.G. & SUTTON, B.C. (1972). *Colletogloeum fouquieriae*, a new fungus on Ocotillo. *Mycologia* **64**: 702-707.

POTEBNIA, A. (1910). Beiträge zur Micromycetenflora Mittel-Russlands. *Annls mycol.* **8**: 42-93.

PUNITHALINGAM, E. (1967). CMI Descriptions of pathogenic fungi and bacteria nos. 136, 137, 138, 139, 140.

PUNITHALINGAM, E. (1970). Studies on Sphaeropsidales in culture. *Mycol. Pap.* 119: 1-24.

PUNITHALINGAM, E. (1973). Two new species of *Phomopsis*. *Trans. Br. mycol. Soc.* **60**: 157-160.

PUNITHALINGAM, E. (1974a). Studies on Sphaeropsidales in culture II. *Mycol. Pap.* 136: 1-63.

PUNITHALINGAM, E. (1974b). New species of *Phomopsis*. *Trans. Br. mycol. Soc.* **63**: 229-236.

PUNITHALINGAM, E. (1975). Some new species and combinations in *Phomopsis*. *Trans. Br. mycol. Soc.* **64**: 427-435.

PUNITHALINGAM, E. (1976a). *Phoma oculo-hominis* sp. nov. from corneal ulcer. *Trans. Br. mycol. Soc.* **67**: 142-143.

PUNITHALINGAM, E. (1976b). CMI Descriptions of pathogenic fungi and bacteria no. 519.

PUNITHALINGAM, E. (1979). Graminicolous *Ascochyta* species. *Mycol. Pap.* 142: 1-214.

PUNITHALINGAM, E. & ENGLISH, M.P. (1975). *Pyrenochaeta unguis-hominis* sp. nov. on human toenails. *Trans. Br. mycol. Soc.* **64**: 539-541.

PUNITHALINGAM, E. & GIBSON, I.A.S. (1973). CMI Descriptions of pathogenic fungi and bacteria nos. 367, 368, 369.

PUNITHALINGAM, E. & GIBSON, I.A.S. (1978). *Phomopsis* species from *Aquilaria agallocha*. *Nova Hedwigia* **29**: 251-255.

PUNITHALINGAM, E. & HOLLIDAY, P. (1972). CMI Descriptions of pathogenic fungi and bacteria nos. 333, 335, 339.

PUNITHALINGAM, E. & WALLER, J.M. (1969). CMI Descriptions of pathogenic fungi and bacteria no. 400.

PUNITHALINGAM, E. & WATERSTON, J.M. (1970). CMI Descriptions of pathogenic fungi and bacteria nos. 273, 277.

PURKAYASTHA, R.P. & RAY, C. (1976). The detection of phytoalexins in jute leaves after infection by *Colletotrichum corchorum* and their possible role in lesion development. *Physiol. Pl. Path.* **6**: 265-273.

RACOVITZA, A. (1959). Étude systèmatique et biologique des champignons bryophiles. *Mem. Mus. Hist. nat., Paris* **10**: 1-288.

RADULESCU, E., NEGRU, A. & DOCEA, E. (1973). *Septoriozele diu Romania*. Bucharest, Edit. Acad. Rep. Soc. Rom. Lei, 325pp.

RAM, A., RAM, C. & ROCHA, H.M. (1973). *Colletotrichum crassipes* associated with cocoa wilting in Bahia, Brazil. *Revista Theobroma* **3**: 33-40.

RAMAKRISHNAN, T.S., SRINIVASAN, K.V. & SUNDARAM, N.V. (1953). Additions to fungi of Madras XIV. *Proc. Indian Acad. Sci.* Sect. B, **37**: 83-95.

RAO, D. & RAO, P.R. (1964). *Barnettella* – a new member of the Melanconiaceae. *Mycopath. Mycol. appl.* **22**: 55-58.

RAO, D. & RAO, V. (1973). Studies on *Barnettella*. *Ind. Phytopath.* **26**: 233-236.

RAWLINSON, C.J., SUTTON, B.C. & MUTHYALU, G. (1978). Taxonomy and biology of *Pyrenopeziza brassicae* sp. nov. (*Cylindrosporium concentricum*), a pathogen of winter oilseed rape (*Brassica napus* ssp. *oleifera*). *Trans. Br. mycol. Soc.* **71**: 425-439.

REID, D.A. (1966). Two new fungi from New Guinea. *Aust. J. Bot.* **14**: 31-34.

REUSSER, F.A. (1964). Über einige Arten der Gattung *Guignardia* Viala et Ravaz. *Phytopath. Z.* **51**: 205-240.

RIBALDI, M. (1953). Necrosi corticale e aeterazione cromatica del ligno prodatte dul Pino austriaco da *Phomatosphaeropsis pinicola* n. gen. et n. sp. *Ann. Sper. agr.* N.S. **7**: 837-851.

RIEDL, H. (1974). *Dothioropsis corni*, n. gen., n. spec., ein Bindeglied zwischen Melanconiales und Sphaeropsidales. *Phyton (Austria)* **16**: 221-230.

RIEDL, H. (1977). Die Gattung *Dothiorina* v. Hoehnel. *Sydowia* **29**: 146-154.
RIMPAU, R.H. (1962). Untersuchungen uber die Gattung *Drepanopeziza* (Kleb.) v. Höhn. *Phytopath. Z.* **43**: 257-306.
ROGERS, D.P. (1949). Nomina conservanda proposita and nomina confusa-Fungi. *Farlowia* **3**: 425-493.
ROGERS, D.P. (1959). On *Cicinobolus*. *Mycologia* **51**: 96-98.
ROY, R.Y., DWIVEDI, R.S. & KHANNA, P.K. (1965). *Vinculum*, a new genus of Deuteromycetes. *Trans. Br. mycol. Soc.* **48**: 113-115.
RUPPRECHT, H. (1958). Beiträge zur Kenntnis der Fungi Imperfecti – II. *Sydowia* **11**: 424-434.
RUPPRECHT, H. (1959). Beiträge zur Kenntnis der Fungi Imperfecti III. *Sydowia* **13**: 10-22.
SACCARDO, P.A. (1880a). Conspectus generum fungorum italiae inferiorum nempe ad Sphaeropsideas, Melanconieas et Hyphomyceteas pertinentium, systemate sporologico dispositoru. *Michelia* **2**: 1-38.
SACCARDO, P.A. (1880b). Fungi gallici. ser. II. *Michelia* **2**: 38-135.
SACCARDO, P.A. (1884). *Syll. fung.* **3**: 1-860.
SACCARDO, P.A. (1886). Fungi Italici Sphaeropsideae no. 1459.
SACCARDO, P.A. (1892). *Syll. fung.* **10**: 1-964.
SACCARDO, P.A. (1895). *Syll. fung.* **11**: 1-753.
SACCARDO, P.A. & SYDOW, P. (1899). *Syll. fung.* **14**: 1-1316.
SACCARDO, P.A. & SYDOW, P. (1902). *Syll. fung.* **16**: 1-1291.
SACCARDO, P.A. & TROTTER, A. (1913). *Syll. fung.* **22**: 1-1612.
SAGASTA AZPEITIA, E.M. (1969). Notas sobre el hongo *Colletotrichum circinans* (Berk.) Vogl., productor de "anthracnosis" en Cebollas. *Boln Patol. veg. Ent. agric.* **30**: 147-153.
SAVILE, D.B.O. (1962). Collection and care of botanical specimens. *Publ. Can. Dept. Agr., Res. Br.* 1113.
SCHEFFER, R.P. (1950). Anthracnose leafspot of crucifers. *Tech. Bull. N.C. agric. Exp. Sta.* 92: 1-26.
SCHEIBER, E., ZENTMYER, G.A., MITCHELL, D.J. & ROHEIM, J. (1970). Coffee berry necrosis in Guatemala and Costa Rica. *Phytopathology* **60**: 1542.
SCHNEIDER, R. (1976). Taxonomie der Pyknidienpilzgattung *Pyrenochaeta*. *Ber. Dtsch. bot. Ges.* **89**: 507-514.
SCHNEIDER, R. (1979). Die Gattung *Pyrenochaeta* De Notaris. *Mitt. Biol. Bundes. Land. -u. Forstw.* 189: 1-73.
SCHÜEPP, H. (1959). Untersuchungen über Pseudopezizoideae sensu Nannfeldt. *Phytopath. Z.* **36**: 213-269.
SEAVER, F.J. (1951). *The North American Cup-Fungi (Inoperculates)*. 428 pp. New York.
SEELER, E.V. Jr. (1943). Several fungicolous fungi. *Farlowia* **1**: 119-133.
SESHADRI, V.S. & SHAW, C.G. (1978). Validity of *Muelleromyces variisporus* (Died.) Ullasa and *Kamatella longipedicellata* (T.S. & K. Ramakr.) Ullasa. *Mycotaxon* **7**: 453-456.
SHEAR, C.L. & DODGE, B.O. (1921). The life history and identity of "*Patellina fragariae*", "*Leptothyrium macrothecium*", and "*Peziza oenotherae*". *Mycologia* **13**: 135-170.
SHEAR, C.L. & STEVENS, N.E. (1935). *Sphaeria zeae* (*Diplodia zeae*) and confused species. *Mycologia* **27**: 467-477.
SHOEMAKER, R.A. (1963). Generic correlations and concepts: *Griphosphaerioma* and *Labridella*. *Can. J. Bot.* **41**: 1419-1423.
SHOEMAKER, R.A. (1964a). *Seimatosporium* (= *Cryptostictis*) parasites of *Rosa*, *Vitis*, and *Cornus*. *Can. J. Bot.* **42**: 411-417.
SHOEMAKER, R.A. (1964b). Conidial states of some *Botryosphaeria* species on *Vitis* and *Quercus*. *Can. J. Bot.* **42**: 1297-1301.
SHOEMAKER, R.A. (1964c). Staining asci and annellophores. *Stain Technol.* **39**: 120-121.
SHOEMAKER, R.A. (1967). *Cucurbitaria piceae* and associated Sphaeropsidales on spruce buds. *Can. J. Bot.* **45**: 1243-1248.
SHOEMAKER, R.A. & MÜLLER, E. (1964). Generic correlations and concepts: *Clathridium* (= *Griphosphaeria*) and *Seimatosporium* (= *Sporocadus*). *Can. J. Bot.* **42**: 403-407.
SHOEMAKER, R.A., MÜLLER, E. & MORGAN-JONES, G. (1966). Fuckel's *Massaria marginata* and *Seiridium marginatum* Nees ex Steudel. *Can. J. Bot.* **44**: 247-254.
SIMMONDS, J.H. (1941). Latent infection in tropical fruits discussed in relation to the part played by species of *Gloeosporium* and *Colletotrichum*. *Proc. R. Soc. Qd.* **52**: 99-121.
SIMMONDS, J.H. (1965). A study of the species of *Colletotrichum* causing ripe fruit rots in Queensland. *Qd. J. agric. Anim. Sci.* **22**: 437-459.
SINGH, S.M. (1977). A new species of *Colletogloeum*. *Ind. Phytopath.* **29**: 70-72.
SIVANESAN, A. & GIBSON, I.A.S. (1976). CMI Descriptions of pathogenic fungi and bacteria no. 481.
SPEER, E.O. (1972). Untersuchungen an einigen parasitischen Pilzen und ihren Wirtspflanzen. *Sydowia* **25**: 1-34.
SPEGAZZINI, C. (1880). Fungi Argentini. Pugillus tertius. *An. Soc. Cient. Arg.* **10**: 122-168.
SPEGAZZINI, C. (1887). Fungi fuegiani. *Bol. Acad. Nac. Cienc. Cordoba* **11**: 3-176.

SPEGAZZINI, C. (1888). Fungi Guaranitici. Pugillus II. *An. Soc. Cient. Arg.* **26**: 1-76.

SPRAGUE, R. (1950). *Diseases of Cereals and Grasses in North America*. 1-538. Ronald Press.

SPRAGUE, R. & JOHNSON, A.G. (1950). Species of *Selenophoma* on North American grasses. *Oreg. St. Monogr. Stud. Bot.* 10.

STARBÄCK, K. (1895). Discomyceten-Studien. *Kgl. Sv. Vetenskaps – Akad. Handl. Bih.* **21**: 1-42.

STEVENS, N.E. (1933). Two apple black rot fungi in the United States. *Mycologia* **25**: 536-548.

STEVENS, F.L. & DALBEY, N.E. (1919). New or noteworthy Porto Rican fungi. *Mycologia* **11**: 4-9.

STEVENSON, J.A. (1971). An account of fungus exsiccati. *Beih. Nova Hedwigia* 36: 1-563.

STEYAERT, R.L. (1949). Contribution à l'étude monographique de *Pestalotia* de Not. et *Monochaetia* Sacc. (*Truncatella* gen. nov. et *Pestalotiopsis* gen. nov.). *Bull. Jard. Bot. Brux.* **19**: 285-354.

STEYAERT, R.L. (1953a). New and old species of *Pestalotiopsis*. *Trans. Br. mycol. Soc.* **36**: 81-89.

STEYAERT, R.L. (1953b). *Pestalotiopsis* from the Gold Coast and Togoland. *Trans. Br. mycol. Soc.* **36**: 235-242.

STEYAERT, R.L. (1954). Concerning some South African *Pestalotiopsis* Steyaert. *Bothalia* **6**: 379-383.

STEYAERT, R.L. (1955). *Pestalotia, Pestalotiopsis* et *Truncatella*. *Bull. Jard. bot. Brux.* **25**: 191-199.

STEYAERT, R.L. (1961). Type specimens of Spegazzini's collections in the *Pestalotiopsis* and related genera. (Fungi Imperfecti: Melanconiales). *Darwiniana, B. Aires* **12**: 157-175.

STOLK, A.C. (1963). The genus *Chaetomella* Fuckel. *Trans. Br. mycol. Soc.* **46**: 409-425.

STOWELL, E.A. & BACKUS, M.P. (1966). Morphology and cytology of *Diplocarpon maculatum* in *Crataegus*. I. The *Entomosporium* stage. *Mycologia* **58**: 949-960.

SUBHEDAR, A. & RAO, V.G. (1976). Some new Sphaeropsidaceous fungi. *J. Univ. Poona* **48**: 63-69.

SUBRAMANIAN, C.V. (1962). A classification of the Hyphomycetes. *Curr. Sci.* **31**: 409-411.

SUBRAMANIAN, C.V. & RAMAKRISHNAN, K. (1953). *Plagionema*, a new genus of the Sphaeropsidales. *J. Ind. bot. Soc.* **32**: 131-136.

SUBRAMANIAN, C.V. & RAMAKRISHNAN, K. (1956). *Ciliochorella* Sydow, *Plagionema* Subram. & Ramakr., and *Shanoria* gen. nov. *Trans. Br. mycol. Soc.* **39**: 314-318.

SUBRAMANIAN, C.V. & REDDY, R.K.C. (1974). The genus *Discosia* I. Taxonomy. *Kavaka* **2**: 57-89.

SUTTON, B.C. (1962). *Colletotrichum dematium* (Pers. ex Fr.) Grove and *C. trichellum* (Fr. ex Fr.) Duke. *Trans. Br. mycol. Soc.* **45**: 222-232.

SUTTON, B.C. (1963). Coelomycetes. II. *Neobarclaya, Mycohypallage, Bleptosporium* and *Cryptostictis*. *Mycol. Pap.* 88: 1-50.

SUTTON, B.C. (1964a). Coelomycetes III. *Annellolacinia* gen. nov., *Aristastoma, Phaeocytostroma, Seimatosporium* etc. *Mycol. Pap.* 97: 1-42.

SUTTON, B.C. (1964b). *Phoma* and related genera. *Trans. Br. mycol. Soc.* **47**: 497-509.

SUTTON, B.C. (1964c). *Melanconium* Link ex Fries. *Persoonia* **3**: 193-198.

SUTTON, B.C. (1965a). Studies on the taxonomy of *Colletotrichum* Cda with especial reference to *C. graminicola* (Ces.) Wilson, Ph.D. Thesis, University of London.

SUTTON, B.C. (1965b). Typification of *Dendrophoma* and a reassessment of *D. obscurans*. *Trans. Br. mycol. Soc.* **48**: 611-616.

SUTTON, B.C. (1966). Development of fructifications in *Colletotrichum graminicola* (Ces.) Wils. and related species. *Can. J. Bot.* **44**: 887-897.

SUTTON, B.C. (1967a). Redescription of *Ajrekarella* Kamat & Kalani. *Mycopath. Mycol. appl.* **33**: 76-80.

SUTTON, B.C. (1967b). Two new genera of the Sphaeropsidales and their relationships with *Diachorella, Strasseria* and *Plagiorhadbus*. *Can. J. Bot.* **45**: 1249-1263.

SUTTON, B.C. (1967c). *Libertina*, a synonym of *Phomopsis*. *Trans. Br. mycol. Soc.* **50**: 355-358.

SUTTON, B.C. (1968a). The appressoria of *Colletotrichum graminicola* and *C. falcatum*. *Can. J. Bot.* **46**: 873-876.

SUTTON, B.C. (1968b). *Kellermania* and its generic segregates. *Can. J. Bot.* **46**: 181-196.

SUTTON, B.C. (1968c). *Polynema*, an earlier name for *Neobarclaya*. *Mycologia* **60**: 201-203.

SUTTON, B.C. (1968d). *Phomopsis schini* (Carranza) comb. nov. *Trans. Br. mycol. Soc.* **51**: 616-618.

SUTTON, B.C. (1969a). Forest microfungi. III. The heterogeneity of *Pestalotia* de Not. section sexloculatae Klebahn sensu Guba. *Can. J. Bot.* **48**: 2083-2094.

SUTTON, B.C. (1969b). Type studies of *Coniella, Anthasthoopa* and *Cyclodomella* *Can. J. Bot.* **47**: 603-608.

SUTTON, B.C. (1969c). *Minimidochium setosum* n. gen. n. sp. and *Dinemasporium aberrans* n. sp. from West Africa. *Can. J. Bot.* **47**: 2095-2100.

SUTTON, B.C. (1971a). Conidium ontogeny in pycnidial and acervular fungi. In *Taxonomy of Fungi Imperfecti* (ed. W.B. Kendrick) pp. 263-278.

SUTTON, B.C. (1971b). Coelomycetes. IV. The genus *Harknessia* and similar fungi on *Eucalyptus*. *Mycol. Pap.* 123: 1-46.

SUTTON, B.C. (1972a). *Wakefieldia punctata* Arnaud and *Blennoria buxi* Fr. sensu Arnaud. *Trans. Br. mycol. Soc.* **59**: 285-294.

SUTTON, B.C. (1972b). Nomenclature of *Ceuthospora, Pyrenophora* and *Blennoria* (Fungi). *Taxon* **21**: 319-326.

SUTTON, B.C. (1973a). Ch. 11. Coelomycetes, pp. 513-582 in *The Fungi* **IVA** (Eds. G.C. Ainsworth, F.K. Sparrow & A.S. Sussman). Academic Press.

SUTTON, B.C. (1973b). Hyphomycetes from Manitoba and Saskatchewan, Canada. *Mycol. Pap.* 132: 1-143.

SUTTON, B.C. (1974). Miscellaneous coelomycetes on *Eucalyptus. Nova Hedwigia* **25**: 161-172.

SUTTON, B.C. (1975a). Coelomycetes V. *Coryneum. Mycol. Pap.* 138: 1-224.

SUTTON, B.C. (1975b). *Eucalyptus microfungi: Satchmopsis* gen. nov. and new species of *Coniella, Coniothyrium* and *Harknessia. Nova Hedwigia* **26**: 1-16.

SUTTON, B.C. (1975c). *Diploceras*, another generic synonym of *Seimatosporium. Trans. Br. mycol. Soc.* **64**: 483-487.

SUTTON, B.C. (1975d). *Wojnowicia* and *Angiopomopsis. Ces. Myk.* **29**: 97-104.

SUTTON, B.C. (1977a). Coelomycetes VI. Nomenclature of generic names proposed for Coelomycetes. *Mycol. Pap.* 141: 1-253.

SUTTON, B.C. (1977b). Nomenclature and taxonomy of *Shanoria, Placonema* and *Neoplaconema* (Sphaeropsidales). *Kew Bull.* **31**: 461-464.

SUTTON, B.C. (1978a). Two undescribed deuteromycete microfungi from the Azores. *Kew Bull.* **33**: 343-346.

SUTTON, B.C. (1978b). New and interesting hyphomycetes from Tampa, Florida. *Mycologia* **70**: 784-801.

SUTTON, B.C. (in press). Suprageneric classification of the Deuteromycotina. *Proc. Int. Symp. Fungal Taxonomy, Madras, 1973.*

SUTTON, B.C. & ALCORN, J. (1974). *Neottiosporina. Austr. J. Bot.* **22**: 517-530.

SUTTON, B.C. & CHAO, R.L.C. (1970). *Leptomelanconium. Trans. Br. mycol. Soc.* **55**: 37-44.

SUTTON, B.C., GHAFFAR, A. & ABBAS, S.Q. (1972). *Seimatosporiopsis* gen. nov. (Sphaeropsidales). *Trans. Br. mycol. Soc.* **59**: 295-300.

SUTTON, B.C. & FUNK, A. (1975). Conidial states of some *Pragmopora* and *Tympanis* species. *Can. J. Bot.* **53**: 521-526.

SUTTON, B.C. & GIBSON, I.A.S. (1977). CMI Descriptions of pathogenic fungi and bacteria nos. 535, 536, 258.

SUTTON, B.C. & HODGES, C. (1977). *Harknessia araucariae* from Brazil. *Mycologia* **69**: 829-831.

SUTTON, B.C. & MARASAS, W.F.O. (1976). Observations on *Neottiosporina* and *Tiarosporella. Trans. Br. mycol. Soc.* **67**: 69-76.

SUTTON, B.C. & PIROZYNSKI, K.A. (1963). Notes on British Microfungi I. *Trans. Br. mycol. Soc.* **46**: 505-522.

SUTTON, B.C. & PIROZYNSKI, K.A. (1965). Notes on microfungi. II. *Trans. Br. mycol. Soc.* **48**: 349-366.

SUTTON, B.C. & POLLACK, F.G. (1973). *Gloeosporium cercocarpi* and *Sphaceloma cercocarpi. Mycologia* **65**: 1125-1134.

SUTTON, B.C. & POLLACK, F.G. (1974). Microfungi on *Cercocarpus. Mycopath. Mycol. appl.*. **52**: 331-351.

SUTTON, B.C., RAO, V.G. & MHASKAR, D.N. (1976). *Kendrickomyces* gen. nov. and *Waydora* nom. nov., two unusual stromatic Coelomycetes. *Trans. Br. mycol. Soc.* **67**: 243-249.

SUTTON, B.C. & SANDHU, D. (1969a). Electron microscopy of conidium development and secession in *Cryptosporiopsis* sp., *Phoma fumosa, Melanconium bicolor*, and *M. apiocarpum. Can. J. Bot.* **47**: 745-749.

SUTTON, B.C. & SANDHU, D. (1969b). *Phragmotrichum pini* (W.B. Cooke) comb. nov. *Trans. Br. mycol. Soc.* **52**: 67-71.

SUTTON, B.C. & SARBHOY, A.K. (1976). Revision of *Chaetomella* and comments upon *Vermiculariopsis* and *Thyriochaetum. Trans. Br. mycol. Soc.* **66**: 297-303.

SUTTON, B.C. & WATERSTON, J.M. (1966). CMI Descriptions of pathogenic fungi and bacteria nos. 88, 89, 90.

SUTTON, B.C. & WATERSTON, J.M. (1970). CMI Descriptions of pathogenic fungi and bacteria no. 228.

SWART, H.J. (1972). Australian leaf-inhabiting fungi. III. Observations on *Harknessia. Trans. Br. mycol. Soc.* **59**: 309-311.

SWART, H.J. & GRIFFITHS, D.A. (1974). Australian leaf-inhabiting fungi. V. Two species of *Seimatosporium* on *Eucalyptus. Trans. Br. mycol. Soc.* **62**: 359-366.

SYDOW, H. (1942). Sydow, Mycotheca germanica Fasc. LXIX-LXXII (nos. 3401-3600). *Annls mycol.* **40**: 193-218.

SYDOW, H. & MITTER, J.H. (1935). Fungi indici – II. *Annls mycol.* **33**: 46-71.

SYDOW, H. & PETRAK, F. (1922). Ein Beitrag zur Kenntnis der Pilzflora Nordamerikas, insbesondere der nordwestlichen Staaten. *Annls mycol.* **20**: 178-218.

SYDOW, H. & SYDOW, P. (1900). Beiträge zur Kenntnis der Pilzflora der Mark Brandenburg. III. *Hedwigia Beibl.* **39**: (1)-(6).

SYDOW, H., SYDOW, P. & BUTLER, E.J. (1916). Fungi Indiae orientalis. Pars V. *Annls mycol.* **14**: 177-220.

TALDE, U.K. (1975). A new species of *Furcaspora* from India. *Ind. Phytopath.* **27**: 630-631.

TASSI, F. (1904). Origine e sviluppo delle Leptostromaceae. *Bull. Lab. Ort. Bot., Siena* **6**: 3-124.

TERRIER, C.-A. (1942). Essai sur la systématique des Phacidiaceae (Fr.) sensu Nannfeldt (1932). 1-99. Thesis, Ecole polytech. fed. Zurich.

TETEREVNIKOVA-BABAYAN, D.N. (1974). Review of the *Septoria* species on the Rosaceae in the USSR. *Sovr. uspek. mikol. i. likhenol. Sov. Pribalt.* 172-175.

TETEREVNIKOVA-BABAYAN, D.N. (1976). A critical survey of *Septoria* species parasitizing Salicaceae. *Septoria* species inhabiting willow. *Biol. Zh. Arm.* **29**: 3-11.

THAUNG, M.M. (1975). Miscellaneous fungi from Burma. *Trans. Br. mycol. Soc.* **64**: 307-312.

THOMSON, J.R. (1936). *Cylindrosporium concentricum* Grev. *Trans. Br. mycol. Soc.* **20**: 123-132.

THYR, B.D. & SHAW, C.G. (1964). Identity of the fungus causing red band disease on *Pines*. *Mycologia* **56**: 103-109.

TIFFANY, L.H. & GILMAN, J.C. (1954). Species of *Colletotrichum* from legumes. *Mycologia* **46**: 52-75.

TROTTER, A. (1931). Saccardo's *Syll. fung.* **25**: 1-1093.

TUBAKI, K. (1958). Studies on the Japanese Hyphomycetes. V. Leaf and stem group with a discussion of the classification of hyphomycetes and their perfect stages. *J. Hattori Bot. Lab.* **20**: 142-244.

TULASNE, L.R. & TULASNE, C. (1863). *Selecta Fungorum Carpologia* **2**: 1-202. Paris.

TULLOCH, M. (1972). The genus *Myrothecium* Tode ex Fr. *Mycol. Pap.* 130: 1-42.

TWYMAN, E.S. (1947). Notes on the dieback of oak caused by *Colpoma quercinum* (Fr.) Wallr. *Trans. Br. mycol. Soc.* **29**: 234-241.

ULLASA, B.A. & RAO, V.G. (1973). *Deshpandiella*, a new genus of the Diaporthaceae (Ascomycetes). *Bull. Torrey Bot. Cl.* **100**: 41-43.

URRIES, M.J. (1955). Novedades micologicas de la Flora Camaria. *An. Inst. bot. A.J. Cavanilles* **14**: 153-170.

VAN DER AA, H.A. (1972). Centraalbureau voor Schimmelcultures, Baarn and Delft, Progress Report 1971. *Verh. K. ned. Akad. Wet. Amst.* **61**: 50-69.

VAN DER AA, H.A. (1973). Studies in *Phyllosticta* I. *Stud. Mycol. Baarn* 5: 1-110.

VAN DER AA, H.A. (1975). The perfect state of *Sarcophoma miribelii*. *Persoonia* **8**: 283-289.

VAN DER AA, H.A. (1977). Centraalbureau voor Schimmelcultures, Baarn and Delft. Progress Report 1976. *Verh. K. ned. Akad. Wet. Amst.* **68**: 1-24.

VAN KESTEREN, H.A. (1967). 'Black root rot' in Cucurbitaceae caused by *Phomopsis sclerotioides* nov. spec. *Neth. J. Pl. Path.* **73**: 112-116.

VAN KESTEREN, H.A. (1971). Aandacht gevraagd voor de "paarse vlekhenziekte" bij *Sedum*. *Gewasheschermina* **2**: 74.

VAN KESTEREN, H.A. (1972). The causal organism of purple blotch disease of *Sedum*. *Neth. J. Pl. Path.* **78**: 116-118.

VAN VLOTEN, H. & GREMMEN, J. (1953). Studies in the discomycete genera *Crumenula* De Not. and *Cenangium* Fr. *Acta bot. neerl.* **2**: 226-241.

VASANT RAO & SUTTON, B.C. (1976). Synnematous fungi I. *Kavaka* **3**: 21-28.

VASSILJEVSKY, N.I. & KARAKULIN, B.P. (1950). Fungi Imperfecti Parasitici. Pars. II. Melanconiales: 1-679. Moscow, Leningrad, USSR.

VEGH, I., BOURGEOIS, M., BOUSCEUET, J.F. & VELASTEGUI, J. (1974). Contribution à l'étude du *Phoma exigua* associé au dépérissement de la pervenche mineure (*Vinca minor* L.) médicinale. *Bull. trimest. Soc. mycol. Fr.* **90**: 121-133.

VIEGAS, A.P. (1944). Alguns fungos do Brasil II. Ascomycetos. *Bragantia* **4**: 5-392.

VIEGAS, A.P. (1945). Alguns fungos do Brasil XI. Fungi imperfecti. *Bragantia* **5**: 717-779.

VISWANATHAN, T.S. (1959). A new species of *Ciliochorella* on *Eucalyptus globulus* L. *Curr. Sci.* **28**: 171.

WAGER, V.A. (1970). *Flower Garden Diseases and Pests*. Purnell, Cape Town.

WAKEFIELD, E.M. (1940). Nomina generica conservanda. Contributions from the nomenclature committee of the British Mycological Society III. *Trans. Br. mycol. Soc.* **24**: 282-293.

WALKER, J. & SUTTON, B.C. (1974). *Dilophia* Sacc. and *Dilophospora* Desm. *Trans. Br. mycol. Soc.* **62**: 231-241.

WALLROTH, F.G. (1833). Flora Cryptogamica Germaniae **2**: 1-923.

WEBSTER, J. (1955). Graminicolous Pyrenomycetes V. Conidial states of *Leptosphaeria michotii*, *L. microscopica*, *Pleospora vagans* and the perfect state of *Dinemasporium graminium*. *Trans. Br. mycol. Soc.* **38**: 347-356.

WEHMEYER, L.E. (1933). The genus *Diaporthe* Nitschke and its segregates. *Univ. Mich. Stud.* **9**: 1-161.

WEHMEYER, L.E. (1938). Cultural life histories of *Melanconis* and *Pseudovalsa* III. *Pap. Mich. Acad. Sci. Arts Lett.* **23**: 233-242.

WEHMEYER, L.E. (1941). A revision of *Melanconis, Pseudovalsa, Prosthecium*, and *Titania*. *Univ. Mich. Stud. Sci. Ser.* **14**: 1-161.

WEINDLMAYR, J. (1965). Beiträge zu einer Revision der Gattung *Myxosporium* Link. 2 Mitteilung. *Sydowia* **18**: 26-32.

WHETZEL, H.H. & WHITE, W.L. (1940). *Mollisia tetrica, Peziza sejournei* and the genera *Phaeociboria* and *Pycnopeziza*. *Mycologia* **32**: 609-620.

WHITE, W.L. (1936). A new species of *Chondropodium* on *Pseudotsuga taxifolia*. *Mycologia* **28**: 433-438.

WHITE, W.L. & WHETZEL, H.H. (1938). Pleomorphic life cycles in a new genus of the Helotiaceae. *Mycologia* **30**: 187-203.

WHITNEY, H.S., REID, J. & PIROZYNSKI, K.A. (1975). Some new fungi associated with needle blight of conifers. *Can. J. Bot.* **53**: 3051-3063.

WILSON, G.W. (1914). The identity of the anthracnose of grasses. *Phytopath.* **4**: 106-112.

WILSON, M. (1922). Notes on new or rare British Fungi. *Trans. Br. mycol. Soc.* **7**: 79-85.

WOLF, F.A. (1912). The perfect stage of *Actinonema rosae*. *Bot. Gaz.* **54**: 218-234.

WOLF, F.A. (1924). Strawberry leaf scorch. *J. Elisha Mitchell Sci. Soc.* **39**: 141-163.

ZAMBETTAKIS, E.C. (1954). Recherches sur la systématique des "Sphaeropsidales-Phaeodidymae". *Bull. trimest. Soc. mycol. Fr.* **70**: 219-350.

ZARZYCKA, H. (1976). Physiologic races of *Colletotrichum lini* (Westend.) Tochinai under Polish conditions. *Rozprawy Naukowe* **68**: 1-83.

HOST INDEX

Geranium
Coniella australiensis
Pestalozziella subsessilis
Ginkgo
Ceuthospora melaleuca
Gleditsia
Kaskaskia gleditsiae
Glochidion
Colletogloeum glochidiicola
Glyceria
Chaetospermum chaetosporum
Glycine
Colletotrichum lindemuthianum
Colletotrichum truncatum
Phoma medicaginis
Glycyrrhiza
Selenophoma juncea
Gnaphalium
Heteropatella umbilicata
Goebelia
Schwarzmannia goebeliae
Gossypium
Phoma medicaginis var. *pinodella*
Phoma pomorum
Gramineae
Barnettella indica
Massariothea themedae
Phaeocytostroma sacchari
Pseudoseptoria stomaticola
Stroplacodium atrum
Stauronema cruciferum
Wojnowicia hirta
Grevillea
Seimatosporium grevilleae
Seimatosporium hakeae
Shawiella grevilleae
Grewia
Aplosporella beaumontiana
Guajacum
Colletotrichum crassipes
Gunnera
Harknessia gunnerae
Gymnosporia
Cyclothyrium juglandis

Hakea
Seimatosporium hakeae
Haloguignardia
Gloesporidina cecidii
Hamamelis
Seimatosporium consocium
Hardenbergia
Seimatosporium kennedyae
Harungana
Idiocercus pirozynskii
Hedera
Acarosporium sympodiale
Ceuthospora hederae
Cheirospora botryospora
Coleophoma cylindrospora
Colletotrichum trichellum
Cryptocline paradoxa
Cyclothyrium juglandis
Phoma cava
Phoma hedericola
Zythiostroma mougeotii
Hedysarum
Diachorella onobrychidis
Phloeosporella hedysari
Helianthus
Phoma macdonaldii
Heliconia
Phoma tropica
Heracleum
Anaphysmene heraclei
Phoma complanata
Sclerostagonospora heraclei
Hesperis
Phoma lingam
Heteromeles
Entomosporium mespili
Hevea
Chaetospermum carneum
Colletotrichum crassipes
Coniella granati
Libertella heveae
Phoma leveillei
Hibiscus
Coniella musaiaensis var. *hibisci*
Phoma fimeti
Hicoria
Asteroma caryae
Holcus
Dilophospora alopecuri
Homo
Pyrenochaeta unguis-hominis
Hordeum
Apiocarpella macrospora
Dilophospora alopecuri
Pseudoseptoria donacis
Pseudoseptroa everhartii
Hoya
Pseudoseptoria everhartii
Humulus
Phlyctaeniella humuli
Hymenaea
Erythrogloeum hymenaeae
Hymenocardia
Coniella petrakii
Geastrumia polystigmatis
Hyparrhenia
Monochaetiella hyparrheniae
Hypericum
Seimatosporium hypericinum
Hyphaene
Microsphaeropsis olivacea

Ilex
Ceuthospora phacidioides
Coleophoma cylindrospora
Pyrenochaeta ilicis
Imperata
Colletotrichum caudatum
Indigofera
Amerosporium polynematoides
Inga
Aplosporella beaumontiana
Ipomoea
Phaeocytostroma plurivorum
Iris
Neottiospora caricina
Isoglossa
Aschersonia aleyrodis
Jasminum
Ajrekarella polychaetriae
Jatropha
Pestalotiopsis guepinii
Juglans
Cyclothyrium juglandis
Juncus
Crandallia juncicola
Psammina bommerae
Septoriella junci
Juniperus
Coleophoma cylindrospora
Kabatina thujae var. *juniperi*
Microsphaeropsis olivacea
Phoma leveillei
Sclerophoma pythiophila
Seimatosporium foliicola
Seiridium juniperi
Seiridium unicorne

Kennedya
Seimatosporium kennedyae
Klainedoxa
Harknessia deightonii

Lablab
Colletotrichum lindemuthianum
Laburnum
Microsphaeropsis olivacea
Lagenaria
Colletotrichum orbiculare
Lamium
Phoma nebulosa
Landolphia
Dialaceniopsis landolphiae
Lantana
Phomopsis lantanae
Larix
Cryptosporiopsis abietna
Sphaeropsis sapinea
Lasiurus
Amerosporium polynematoides
Lathyrus
Ascochyta pisi
Colletotrichum lindemuthianum
Diachorella lathyri
Phoma medicaginis
Pseudorobillarda agrostis
Lauraceae
Phaeodomus erumpens
Laurus
Coleophoma cylindrospora
Gloeosporidiella nobilis
Pestalotiopsis guepinii
Lecanora
Vouauxiella lichenicola
Vouauxiella verrucosa
Ledum
Seimatosporium arbuti
Thoracella ledi
Lens
Hymenopsis trochiloides
Lepidium
Giulia tenuis
Leptoderris
Vanderystiella leopoldvilleana

FUNGUS INDEX

(Names in italics are synonyms, in bold face are new taxa or new combinations, and in capitals are accepted genera with full descriptions.)

Printed and bound by CPI Group (UK) Ltd, Croydon, CR0 4YY
16/09/2026
14963433-0008